# Probability Theory

## *An Analytic View, Second Edition*

This second edition of Daniel W. Stroock's text is suitable for first-year graduate students with a good grasp of introductory undergraduate probability. It provides a reasonably thorough introduction to modern probability theory with an emphasis on the mutually beneficial relationship between probability theory and analysis. It includes more than 750 exercises and offers new material on Levy processes, large deviations theory, Gaussian measures on a Banach space, and the relationship between a Wiener measure and partial differential equations.

The first part of the book deals with independent random variables, Central Limit phenomena, the general theory of weak convergence and several of its applications, as well as elements of both the Gaussian and Markovian theories of measures on function space. The introduction of conditional expectation values is postponed until the second part of the book, where it is applied to the study of martingales. This part also explores the connection between martingales and various aspects of classical analysis and the connections between a Wiener measure and classical potential theory.

Dr. Daniel W. Stroock is the Simons Professor of Mathematics Emeritus at the Massachusetts Institute of Technology. He has published many articles and is the author of six books, most recently *Partial Differential Equations for Probabilists* (2008).

# Probability Theory

## *An Analytic View*

### *Second Edition*

Daniel W. Stroock

*Massachusetts Institute of Technology*

CAMBRIDGE
UNIVERSITY PRESS

CAMBRIDGE UNIVERSITY PRESS
Cambridge, New York, Melbourne, Madrid, Cape Town,
Singapore, São Paulo, Delhi, Mexico City

Cambridge University Press
32 Avenue of the Americas, New York, NY 10013-2473, USA

www.cambridge.org
Information on this title: www.cambridge.org/9780521132503



First edition published 1994
First paperback edition 2000
Second edition published 2011
Reprinted 2011, 2012

*A catalog record for this publication is available from the British Library.*

*Library of Congress Cataloging in Publication Data*

Stroock, Daniel W.
Probability theory : an analytic view/ Daniel W. Stroock. – 2nd ed.
p. cm.
Includes bibliographical references and index.
ISBN 978-0-521-76158-1 (hardback) – ISBN 978-0-521-13250-3 (pbk.)
1. Probabilities. I. Title.
QA 273.S763 2010
519.2–dc22 2010027652

ISBN 978-0-521-76158-1 Hardback
ISBN 978-0-521-13250-3 Paperback

**This book is dedicated to my teachers:**

*M. Kac, H.P. McKean, Jr., and S.R.S. Varadhan*

# Contents

# Preface

## From the Preface to the First Edition

When writing a graduate level mathematics book during the last decade of the twentieth century, one probably ought not inquire too closely into one's motivation. In fact, if ones own pleasure from the exercise is not sufficient to justify the effort, then one should seriously consider dropping the project. Thus, to those who (either before or shortly after opening it) ask *for whom was this book written*, my pale answer is *me*; and, for this reason, I thought that I should preface this preface with an explanation of who I am and what were the peculiar educational circumstances that eventually gave rise to this somewhat peculiar book.

My own introduction to probability theory began with a private lecture from H.P. McKean, Jr. At the time, I was a (more accurately, *the*) graduate student of mathematics at what was then called The Rockefeller Institute for Biological Sciences. My official mentor there was M. Kac, whom I had cajoled into becoming my adviser after a year during which I had failed to insert even one micro-electrode into the optic nerves of innumerable limuli. However, as I soon came to realize, Kac had accepted his role on the condition that it would not become a burden. In particular, he had no intention of wasting much of his own time on a reject from the neurophysiology department. On the other hand, he was most generous with the time of his younger associates, and that is how I wound up in McKean's office. Never one to bore his listeners with a lot of dull preliminaries, McKean launched right into a wonderfully lucid explanation of P. Lévy's interpretation of the infinitely divisible laws. I have to admit that my appreciation of the lucidity of his lecture arrived nearly a decade after its delivery, and I can only hope that my reader will reserve judgment of my own presentation for an equal length of time.

In spite of my perplexed state at the end of McKean's lecture, I was sufficiently intrigued to delve into the readings that he suggested at its conclusion. Knowing that the only formal mathematics courses that I would be taking during my graduate studies would be given at N.Y.U. and guessing that those courses would be oriented toward partial differential equations, McKean directed me to material which would help me understand the connections between partial differential equations and probability theory. In particular, he suggested that I start with the, then recently translated, two articles by E.B. Dynkin which had appeared originally in the famous 1956 volume of *Teoriya Veroyatnostei i ee Primeneniya*. Dynkin's articles turned out to be a godsend. They were beautifully crafted to tell the reader enough so that he could understand the ideas and not so much that he would become bored by them. In addition, they gave me an introduction to a host of ideas and techniques (e.g., stopping times and the strong Markov property), all of which Kac himself consigned to the category of overelaborated measure theory. In fact, it would be reasonable to say that my thesis was simply the application of techniques which I picked up from Dynkin to a problem that I picked up by reading some notes by Kac. Of course, along the way I profited immeasurably from continued contact with McKean, a large number of courses at N.Y.U. (particularly ones taught by M. Donsker, F. John, and L. Nirenberg), and my increasingly animated conversations with S.R.S. Varadhan.

As I trust the preceding description makes clear, my graduate education was anything but deprived; I had ready access to some of the very best analysts of the day. On the other hand, I never had a *proper* introduction to my field, probability theory. The first time that I ever summed independent random variables was when I was summing them in front of a class at N.Y.U. Thus, although I now admire the magnificent body of mathematics created by A.N. Kolmogorov, P. Lévy, and the other twentieth-century heroes of the field, I am not a *dyed-in-the-wool* probabilist (i.e., what Donsker would have called a true *coin-tosser*). In particular, I have never been able to develop sufficient sensitivity to the distinction between a *proof* and a *probabilistic proof*. To me, a proof is clearly *probabilistic* only if its punch-line comes down to an argument like $P(A) \leq P(B)$ because $A \subseteq B$; and there are breathtaking examples of such arguments. However, to base an entire book on these examples would require a level of genius that I do not possess. In fact, I myself enjoy probability theory best when it is inextricably interwoven with other branches of mathematics and not when it is presented as an entity unto itself. For this reason, the reader should not be surprised to discover that he finds some of the material presented in this book *does not belong here*; but I hope that he will make an effort to figure out why I disagree with him.

## Preface to the Second Edition

My favorite "preface to a second edition" is the one that G.N. Watson wrote for the second edition of his famous treatise on Bessel functions. The first edition appeared in 1922, the second came out in 1941, and Watson had originally intended to stay abreast of developments and report on them in the second edition. However, in his preface to the second edition Watson admits that his interest in the topic had "waned" during the intervening years and apologizes that, as a consequence, the new edition contains less new material than he had thought it would.

My excuse for not incorporating more new material into this second edition is related to but somewhat different from Watson's. In my case, what has waned is not my interest in probability theory but instead my ability to assimilate the transformations that the subject has undergone. When I was a student, probabilists were still working out the ramifications of Kolmogorov's profound insights into the connections between probability and analysis, and I have spent my career investigating and exploiting those connections. However, about the time when the first edition of this book was published, probability theory began a return to its origins in combinatorics, a topic in which my abilities are woefully deficient. Thus, although I suspect that, for at least a decade, the most exciting developments in the field will have a strong combinatorial component, I have not attempted to prepare my readers for those developments. I repeat that my decision not to incorporate more combinatorics into this new edition in no way reflects my assessment of the direction in which probability is likely to go but instead reflects my assessment of my own inability to do justice to the beautiful combinatorial ideas that have been introduced in the recent past.

In spite of the preceding admission, I believe that the material in this book remains valuable and that, no matter how probability theory evolves, the ideas and techniques presented here will play an important role. Furthermore, I have made some substantive changes. In particular, I have given more space to infinitely divisible laws and their associated Lévy processes, both of which are now developed in $\mathbb{R}^N$ rather than just in $\mathbb{R}$. In addition, I have added an entire chapter devoted to Gaussian measures in infinite dimensions from the perspective of the Segal–Gross school. Not only have recent developments in Malliavin calculus and conformal field theory sparked renewed interest in this topic, but it seems to me that most modern texts pay either no or too little attention to this beautiful material. Missing from the new edition is the treatment of singular integrals. I included it in the first edition in the hope that it would elucidate the similarity between cancellations that underlie martingale theory, especially Burkholder's Inequality, and Calderon–Zygmund theory. I still believe that these similarities are worth thinking about, but I have decided that my explanation of them led me too far astray and was more of a distraction than a pedagogically valuable addition.

Besides those mentioned above, minor changes have been made throughout. For one thing, I have spent a lot of time correcting old errors and, undoubtedly, inserting new ones. Secondly, I have made several organizational changes as well as others that are remedial. A summary of the contents follows.

## Summary

**1**: Chapter 1 contains a sampling of the standard, pointwise convergence theorems dealing with partial sums of independent random variables. These include the Weak and Strong Laws of Large Numbers as well as Hartman–Wintner's Law of the Iterated Logarithm. In preparation for the Law of the Iterated Logarithm, Cramér's theory of large deviations from the Law of Large Numbers is developed in § 1.4. Everything here is very standard, although I feel that my passage from the bounded to the general case of the Law of the Iterated Logarithm has been considerably smoothed by the ideas that I learned during a conversation with M. Ledoux.

**2**: The whole of Chapter 2 is devoted to the classical Central Limit Theorem. After an initial (and slightly flawed) derivation of the basic result via moment considerations, Lindeberg's general version is derived in § 2.1. Although Lindeberg's result has become a *sine qua non* in the writing of probability texts, the Berry–Esseen estimate has not. Indeed, until recently, the Berry–Esseen estimate required a good many somewhat tedious calculations with characteristic functions (i.e., Fourier transforms), and most recent authors seem to have decided that the rewards did not justify the effort. I was inclined to agree with them until P. Diaconis brought to my attention E. Bolthausen's adaptation of C. Stein's techniques (the so-called *Stein's method*) to give a proof that is not only brief but also, to me, aesthetically pleasing. In any case, no use of Fourier methods is made in the derivation given in § 2.2. On the other hand, Fourier techniques are introduced in § 2.3, where it is shown that even elementary Fourier analytic tools lead to important extensions of the basic Central Limit Theorem to more than one dimension. Finally, in § 2.4, the Central Limit Theorem is applied to the study of Hermite multipliers and (following Wm. Beckner) is used to derive both E. Nelson's hypercontraction estimate for the Mehler kernel as well as Beckner's own estimate for the Fourier transform. I am afraid that, with this flagrant example of *the sort of thing that does not belong here*, I may be trying the patience of my purist colleagues. However, I hope that their indignation will be somewhat assuaged by the fact that the rest of the book is essentially independent of the material in § 2.4.

**3**: This chapter is devoted to the study of infinitely divisible laws. It begins in § 3.1 with a few refinements (especially The Lévy Continuity Theorem) of the Fourier techniques introduced in § 2.3. These play a role in § 3.2, where the Lévy–Khinchine formula is first derived and then applied to the analysis of stable laws.

**4**: In Chapter 4 I construct the Lévy processes (a.k.a. independent increment processes) corresponding to infinitely divisible laws. Secton 4.1 provides the requisite information about the pathspace $D(\mathbb{R}^N)$ of right-continuous paths with left limits, and § 4.2 gives the construction of Lévy processes with discontinuous paths, the ones corresponding to infinitely divisible laws having no Gaussian part. Finally, in § 4.3 I construct Brownian motion, the Lévy process with continuous paths, following the prescription given by Lévy.

**5**: Because they are not needed earlier, conditional expectations do not appear until Chapter 5. The advantage gained by this postponement is that, by the time I introduce them, I have an ample supply of examples to which conditioning can be applied; the disadvantage is that, with considerable justice, many probabilists feel that one is not doing *probability theory* until one is conditioning. Be that as it may, Kolmogorov's definition is given in § 5.1 and is shown to extend naturally both to $\sigma$-finite measure spaces as well as to random variables with values in a Banach space. Section 5.2 presents Doob's basic theory of real-valued, discrete parameter martingales: Doob's Inequality, his Stopping Time Theorem, and his Martingale Convergence Theorem. In the last part of § 5.2, I introduce reversed martingales and apply them to DeFinetti's theory of exchangeable random variables.

**6**: Chapter 6 opens with extensions of martingale theory in two directions: to $\sigma$-finite measures and to random variables with values in a Banach space. The results in § 6.1 are used in § 6.2 to derive Birkhoff's Individual Ergodic Theorem and a couple of its applications. Finally, in § 6.3 I prove Burkholder's Inequality for martingales with values in a Hilbert space. The derivation that I give is essentially the same as Burkholder's second proof, the one that gives optimal constants.

**7**: Section 7.1 provides a brief introduction to the theory of martingales with a continuous parameter. As anyone at all familiar with the topic knows, anything approaching a full account of this theory requires much more space than a book like this can give it. Thus, I deal with only its most rudimentary aspects, which, fortunately, are sufficient for the applications to Brownian motion that I have in mind. Namely, in § 7.2 I first discuss the intimate relationship between continuous martingales and Brownian motion (Lévy's martingale characterization of Brownian motion), then derive the simplest (and perhaps most widely applied) case of the Doob–Meyer Decomposition Theory, and finally show what Burkholder's Inequality looks like for continuous martingales. In the concluding section, § 7.3, the results in §§ 7.1–7.2 are applied to derive the Reflection Principle for Brownian motion.

**8**: In § 8.1 I formulate the description of Brownian motion in terms of its Gaussian, as opposed to its independent increment, properties. More precisely, following Segal and Gross, I attempt to convince the reader that Wiener measure (i.e., the distribution of Brownian motion) would like to be the standard Gauss measure on the Hilbert space $H^1(\mathbb{R}^N)$ of absolutely continuous paths with a square integrable derivative, but, for technical reasons, cannot live there and has to settle for a Banach space in which $H^1(\mathbb{R}^N)$ is densely embedded. Using Wiener measure as the model, in § 8.2 I show that, at an abstract level, any non-degenerate, centered Gaussian measure on an infinite dimensional, separable Banach space shares the same structure as Wiener measure in the sense that there is always a densely embedded Hilbert space, known as the Cameron–Martin space, for which it would like to be the standard Gaussian measure but on which it does not fit. In order to carry out this program, I need and prove Fernique's Theorem for Gaussian measures on a Banach space. In § 8.3 I begin by going in the opposite direction, showing how to pass from a Hilbert space $H$ to a Gaussian measure on a Banach space $E$ for which $H$ is the Cameron–Martin space. The rest of § 8.3 gives two applications: one to "pinned Brownian" motion and the second to a very general statement of orthogonal invariance for Gaussian measures. The main goal of § 8.4 is to prove a large deviations result, known as Schilder's Theorem, for abstract Wiener spaces; and once I have Schilder's Theorem, I apply it to derive a version of Strassen's Law of the Iterated Logarithm. Starting with the Ornstein–Uhlenbeck process, I construct in § 8.5 a family of Gaussian measures known in the mathematical physics literature as Euclidean free fields. In the final section, § 8.6, I first show how to construct Banach space–valued Brownian motion and then derive the original form of Strassen's Law of the Iterated Logarithm in that context.

**9**: The central topic here is the abstract theory of weak convergence of probability measures on a Polish space. The basic theory is developed in § 9.1. In § 9.2 I apply the theory to prove the existence of regular conditional probability distributions, and in § 9.3 I use it to derive Donsker's Invariance Principle (i.e., the pathspace statement of the Central Limit Theorem).

**10**: Chapter 10 is an introduction to the connections between probability theory and partial differential equations. At the beginning of § 10.1 I show that martingale theory provides a link between probability theory and partial differential equations. More precisely, I show how to represent in terms of Wiener integrals solutions to parabolic and elliptic partial differential equations in which the Laplacian is the principal part. In the second part of § 10.1, I use this link to calculate various Wiener integrals. In § 10.2 I introduce the Markov property of Wiener measure and show how it not only allows one to evaluate other Wiener integrals in terms of solutions to elliptic partial differential equations but also enables one to prove interesting facts about solutions to such equations as a consequence of their representation in terms of Wiener integrals. Continuing in the same spirit, I show in § 10.2 how to represent solutions to the Dirichlet problem in terms of Wiener integrals, and in § 10.3 I use Wiener measure to construct and discuss heat kernels related to the Laplacian.

**11**: The final chapter is an extended example of the way in which probability theory meshes with other branches of analysis, and the example that I have chosen is the marriage between Brownian motion and classical potential theory. Like an ideal marriage, this one is simultaneously intimate and mutually beneficial to both partners. Indeed, the more one knows about it, the more convinced one becomes that the properties of Brownian paths are a perfect reflection of properties of harmonic functions, and vice versa. In any case, in § 11.1 I sharpen the results in § 10.2.3 and show that, in complete generality, the solution to the Dirichlet problem is given by the Wiener integral of the boundary data evaluated at the place where Brownian paths exit from the region. Next, in § 11.2, I discuss the Green function for a region and explain how its existence reflects the recurrence and transience properties of Brownian paths. In preparation for § 11.4, § 11.3 is devoted to the Riesz Decomposition Theorem for excessive functions. Finally, in § 11.4, I discuss the capacity of regions, derive Chung's representation of the capacitory measure in terms of the last place where a Brownian path visits a region, apply the probabilistic interpretation of capacity to give a derivation of Wiener's test for regularity, and conclude with two asymptotic calculations in which capacity plays a crucial role.

## Suggestions about the Use of This Book

In spite of the realistic assessment contained in the first paragraph of its preface, when I wrote the first edition of this book I harbored the naïve hope that it might become *the standard* graduate text in probability theory. By the time that I started preparing the second edition, I was significantly older and far less naïve about its prospects. Although the first edition has its admirers, it has done little to dent the sales record of its competitors. In particular, the first edition has seldom been adopted as the text for courses in probability, and I doubt that the second will be either. Nonetheless, I close this preface with a few suggestions for anyone who does choose to base a course on it.

I am well aware that, except for those who find their way into the poorly stocked library of some prison camp, few copies of this book will be read from cover to cover. For this reason, I have attempted to organize it in such a way that, with the help of the table of dependence that follows, a reader can select a path which does not require his reading all the sections preceding the information he is seeking. For example, the contents of §§ 1.1–1.2, § 1.4, § 2.1, § 2.3, and § 5.1–5.2 constitute the backbone of a one semester, graduate level introduction to probability theory. What one attaches to this backbone depends on the speed with which these sections are covered and the content of the courses for which the course is the introduction. If the goal is to prepare the students for a career as a "quant" in what is left of the financial industry, an obvious choice is § 4.3 and as much of Chapter 7 as time permits, thereby giving one's students a reasonably solid introduction to Brownian motion. On the other hand, if one wants the students to appreciate that white noise is not the only noise that they may encounter in life, one might defer the discussion of Brownian motion and replace it with the material in Chapter 3 and §§ 4.1–4.2.

Alternatively, one might use this book in a more advanced course. An introduction to stochastic processes with an emphasis on their relationship to partial differential equations can be constructed out of Chapters 6, 7, 10, and 11, and § 4.3 combined with Chapter 8 could be used to provide background for a course on Gaussian processes.

Whatever route one takes through this book, it will be a great help to your students for you to suggest that they consult other texts. Indeed, it is a familiar fact that the third book one reads on a subject is always the most lucid, and so one should suggest at least two other books. Among the many excellent choices available, I mention Wm. Feller's *An Introduction to Probability Theory and Its Applications, Vol. II,* and M. Loéve's classic *Probability Theory.* In addition, for background, precision (including accuracy of attribution), and supplementary material, R. Dudley's *Real Analysis and Probability* is superb. Finally, an ever growing list of errata can be found at

www.mit-math.edu/~dws/prob2errata.pdf

# Table of Dependence

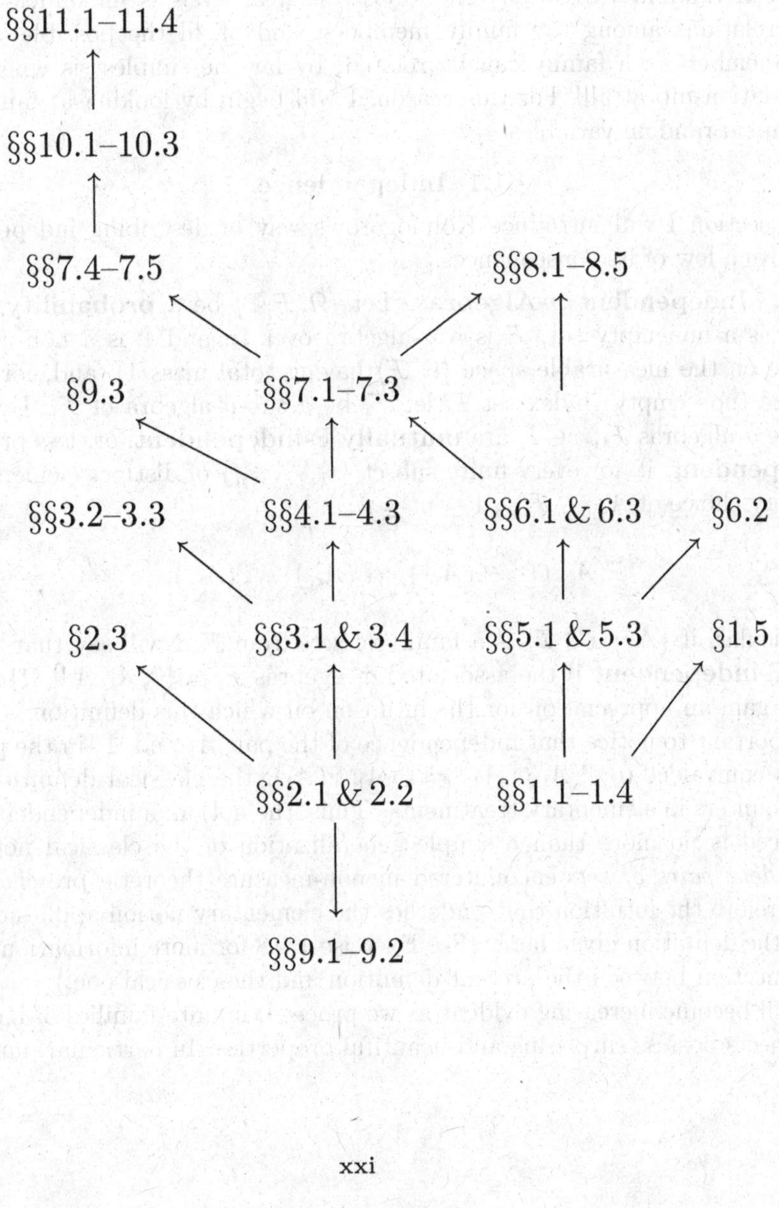

# Chapter 1

## Sums of Independent Random Variables

In one way or another, most probabilistic analysis entails the study of large families of random variables. The key to such analysis is an understanding of the relations among the family members; and of all the possible ways in which members of a family can be related, by far the simplest is when there is no relationship at all! For this reason, I will begin by looking at families of *independent* random variables.

### § 1.1 Independence

In this section I will introduce Kolmogorov's way of describing independence and prove a few of its consequences.

**§ 1.1.1. Independent $\sigma$-Algebras.** Let $(\Omega, \mathcal{F}, \mathbb{P})$ be a **probability space** (i.e., $\Omega$ is a nonempty set, $\mathcal{F}$ is a $\sigma$-algebra over $\Omega$, and $\mathbb{P}$ is a non-negative measure on the measurable space $(\Omega, \mathcal{F})$ having total mass 1), and, for each $i$ from the (non-empty) index set $\mathcal{I}$, let $\mathcal{F}_i$ be a sub-$\sigma$-algebra of $\mathcal{F}$. I will say that the $\sigma$-algebras $\mathcal{F}_i$, $i \in \mathcal{I}$, are **mutually $\mathbb{P}$-independent**, or, less precisely, **$\mathbb{P}$-independent**, if, for every finite subset $\{i_1, \dots, i_n\}$ of distinct elements of $\mathcal{I}$ and every choice of $A_{i_m} \in \mathcal{F}_{i_m}$, $1 \le m \le n$,

$$\mathbb{P}(A_{i_1} \cap \cdots \cap A_{i_n}) = \mathbb{P}(A_{i_1}) \cdots \mathbb{P}(A_{i_n}). \tag{1.1.1}$$

In particular, if $\{A_i : i \in \mathcal{I}\}$ is a family of sets from $\mathcal{F}$, I will say that $A_i$, $i \in \mathcal{I}$, are **$\mathbb{P}$-independent** if the associated $\sigma$-algebras $\mathcal{F}_i = \{\emptyset, A_i, A_i\complement, \Omega\}$, $i \in \mathcal{I}$, are. To gain an appreciation for the intuition on which this definition is based, it is important to notice that independence of the pair $A_1$ and $A_2$ in the present sense is equivalent to $\mathbb{P}(A_1 \cap A_2) = \mathbb{P}(A_1)\mathbb{P}(A_2)$, the classical definition that one encounters in elementary treatments. Thus, the notion of independence just introduced is no more than a simple generalization of the classical notion of *independent pairs of sets* encountered in non-measure theoretic presentations, and therefore the intuition that underlies the elementary notion applies equally well to the definition given here. (See Exercise 1.1.8 for more information about the connection between the present definition and the classical one.)

As will become increasing evident as we proceed, infinite families of independent objects possess surprising and beautiful properties. In particular, mutually independent $\sigma$-algebras tend to *fill up space* in a sense made precise by the following beautiful thought experiment designed by A.N. Kolmogorov. Let $\mathcal{I}$ be any index set, take $\mathcal{F}_\emptyset = \{\emptyset, \Omega\}$, and, for each non-empty subset $\Lambda \subseteq \mathcal{I}$, let

$$\mathcal{F}_\Lambda = \bigvee_{i\in\Lambda} \mathcal{F}_i \equiv \sigma\left(\bigcup_{i\in\mathcal{I}} \mathcal{F}_i\right)$$

be the $\sigma$-algebra generated by $\bigcup_{i\in\Lambda}\mathcal{F}_i$ (i.e., $\mathcal{F}_\Lambda$ is the smallest $\sigma$-algebra containing $\bigcup_{i\in\Lambda}\mathcal{F}_i$). Next, define the **tail $\sigma$-algebra** $\mathcal{T}$ to be the intersection over all finite $\Lambda \subseteq \mathcal{I}$ of the $\sigma$-algebras $\mathcal{F}_{\Lambda\complement}$. When $\mathcal{I}$ itself is finite, $\mathcal{T} = \{\emptyset, \Omega\}$ and is therefore $\mathbb{P}$-**trivial** in the sense that $\mathbb{P}(A) \in \{0,1\}$ for every $A \in \mathcal{T}$. The interesting remark made by Kolmogorov is that even when $\mathcal{I}$ is infinite, $\mathcal{T}$ is $\mathbb{P}$-trivial whenever the original $\mathcal{F}_i$'s are $\mathbb{P}$-independent. To see this, for a given non-empty $\Lambda \subseteq \mathcal{I}$, let $\mathcal{C}_\Lambda$ denote the collection of sets of the form $A_{i_1} \cap \cdots \cap A_{i_n}$ where $\{i_1, \dots, i_n\}$ are distinct elements of $\Lambda$ and $A_{i_m} \in \mathcal{F}_{i_m}$ for each $1 \le m \le n$. Clearly $\mathcal{C}_\Lambda$ is closed under intersection and $\mathcal{F}_\Lambda = \sigma(\mathcal{C}_\Lambda)$. In addition, by assumption, $\mathbb{P}(A\cap B) = \mathbb{P}(A)\mathbb{P}(B)$ for all $A \in \mathcal{C}_\Lambda$ and $B \in \mathcal{C}_{\Lambda\complement}$. Hence, by Exercise 1.1.12, $\mathcal{F}_\Lambda$ is independent of $\mathcal{F}_{\Lambda\complement}$. But this means that $\mathcal{T}$ is independent of $\mathcal{F}_F$ for every finite $F \subseteq \mathcal{I}$, and therefore, again by Exercise 1.1.12, $\mathcal{T}$ is independent of

$$\mathcal{F}_\mathcal{I} = \sigma\left(\bigcup\{\mathcal{F}_F : F \text{ a finite subset of } \Lambda\}\right).$$

Since $\mathcal{T} \subseteq \mathcal{F}_\mathcal{I}$, this implies that $\mathcal{T}$ is *independent of itself*; that is, $\mathbb{P}(A\cap B) = \mathbb{P}(A)\mathbb{P}(B)$ for all $A, B \in \mathcal{T}$. Hence, for every $A \in \mathcal{T}$, $\mathbb{P}(A) = \mathbb{P}(A)^2$, or, equivalently, $\mathbb{P}(A) \in \{0,1\}$, and so I have now proved the following famous result.

THEOREM 1.1.2 (**Kolmogorov's 0–1 Law**). *Let $\{\mathcal{F}_i : i \in \mathcal{I}\}$ be a family of $\mathbb{P}$-independent sub-$\sigma$-algebras of $(\Omega, \mathcal{F}, \mathbb{P})$, and define the tail $\sigma$-algebra $\mathcal{T}$ accordingly, as above. Then, for every $A \in \mathcal{T}$, $\mathbb{P}(A)$ is either 0 or 1.*

To develop a feeling for the kind of conclusions that can be drawn from Kolmogorov's 0–1 Law (cf. Exercises 1.1.18 and 1.1.19 as well), let $\{A_n : n \ge 1\}$ be a sequence of subsets of $\Omega$, and recall the notation

$$\overline{\lim_{n\to\infty}} A_n \equiv \bigcap_{m=1}^{\infty} \bigcup_{n\ge m} A_n = \{\omega : \omega \in A_n \text{ for infinitely many } n \in \mathbb{Z}^+\}.$$

Obviously, $\overline{\lim}_{n\to\infty} A_n$ is measurable with respect to the tail field determined by the sequence of $\sigma$-algebras $\{\emptyset, A_n, A_n\complement, \Omega\}$, $n \in \mathbb{Z}^+$; and therefore, if the $A_n$'s are $\mathbb{P}$-independent elements of $\mathcal{F}$, then

$$\mathbb{P}\left(\overline{\lim_{n\to\infty}} A_n\right) \in \{0,1\}.$$

In words, this conclusion can be summarized as follows: *for any sequence of $\mathbb{P}$-independent events $A_n$, $n \in \mathbb{Z}^+$, either $\mathbb{P}$-almost every $\omega \in \Omega$ is in infinitely many $A_n$'s or $\mathbb{P}$-almost every $\omega \in \Omega$ is in at most finitely many $A_n$'s.* A more quantitative statement of this same fact is contained in the second part of the following useful result.

LEMMA 1.1.3 (**Borel–Cantelli Lemma**). *Let $\{A_n : n \in \mathbb{Z}^+\} \subseteq \mathcal{F}$ be given. Then*

$$\sum_{n=1}^{\infty} \mathbb{P}(A_n) < \infty \implies \mathbb{P}\left(\overline{\lim_{n\to\infty}} A_n\right) = 0. \tag{1.1.4}$$

*In fact, if the $A_n$'s are $\mathbb{P}$-independent sets, then*

$$\sum_{n=1}^{\infty} \mathbb{P}(A_n) = \infty \iff \mathbb{P}\left(\overline{\lim_{n\to\infty}} A_n\right) = 1. \tag{1.1.5}$$

*(See part* (**iii**) *of Exercise 5.2.40 and Lemma 11.4.14 for generalizations.)*

PROOF: The first assertion, which is due to E. Borel, is an easy application of countable additivity. Namely, by countable additivity,

$$\mathbb{P}\left(\overline{\lim_{n\to\infty}} A_n\right) = \lim_{m\to\infty} \mathbb{P}\left(\bigcup_{n\geq m} A_n\right) \leq \lim_{m\to\infty} \sum_{n\geq m} \mathbb{P}(A_n) = 0$$

if $\sum_{n=1}^{\infty} \mathbb{P}(A_n) < \infty$.

To complete the proof of (1.1.5) when the $A_n$'s are independent, note that, by countable additivity, $\mathbb{P}\left(\overline{\lim}_{n\to\infty} A_n\right) = 1$ if and only if

$$\lim_{m\to\infty} \mathbb{P}\left(\bigcap_{n\geq m} A_n\complement\right) = \mathbb{P}\left(\bigcup_{m=1}^{\infty} \bigcap_{n\geq m} A_n\complement\right) = \mathbb{P}\left(\left(\overline{\lim_{n\to\infty}} A_n\right)\complement\right) = 0.$$

But, by independence and another application of countable additivity, for any given $m \geq 1$ we have that

$$\mathbb{P}\left(\bigcap_{n=m}^{\infty} A_n\complement\right) = \lim_{N\to\infty} \prod_{n=m}^{N} (1 - \mathbb{P}(A_n)) \leq \lim_{N\to\infty} \exp\left[-\sum_{n=m}^{N} \mathbb{P}(A_n)\right] = 0$$

if $\sum_{n=1}^{\infty} \mathbb{P}(A_n) = \infty$. (In the preceding, I have used the trivial inequality $1 - t \leq e^{-t}$, $t \in [0, \infty)$.) $\square$

A second, and perhaps more transparent, way of dealing with the contents of the preceding is to introduce the non-negative random variable $N(\omega) \in$

$\mathbb{Z}^+ \cup \{\infty\}$, that counts the number of $n \in \mathbb{Z}^+$ such that $\omega \in A_n$. Then, by Tonelli's Theorem,[1] $\mathbb{E}^{\mathbb{P}}[N] = \sum_{n=1}^{\infty} \mathbb{P}(A_n)$, and so Borel's contribution is equivalent to the $\mathbb{E}^{\mathbb{P}}[N] < \infty \implies \mathbb{P}(N < \infty) = 1$, which is obvious, whereas, when combined with Kolmogorov's 0–1 Law, Cantelli's contribution is that, for mutually independent $A_n$'s, $\mathbb{P}(N < \infty) > 0 \implies \mathbb{E}^{\mathbb{P}}[N] < \infty$, which is not obvious.

**§ 1.1.2. Independent Functions.** Having described what it means for the $\sigma$-algebras to be $\mathbb{P}$-independent, I will now transfer the notion to random variables on $(\Omega, \mathcal{F}, \mathbb{P})$. Namely, for each $i \in \mathcal{I}$, let $X_i$ be a **random variable** (i.e., a measurable function on $(\Omega, \mathcal{F})$) with values in the measurable space $(E_i, \mathcal{B}_i)$). I will say that the random variables $X_i$, $i \in \mathcal{I}$, are **(mutually) $\mathbb{P}$-independent** if the $\sigma$-algebras

$$\sigma(X_i) = X_i^{-1}(\mathcal{B}_i) \equiv \{X_i^{-1}(B_i) : B_i \in \mathcal{B}_i\}, \; i \in \mathcal{I},$$

are $\mathbb{P}$-independent. If $B(E;\mathbb{R}) = B\big((E,\mathcal{B});\mathbb{R}\big)$ denotes the space of bounded measurable $\mathbb{R}$-valued functions on the measurable space $(E,\mathcal{B})$, then it should be clear that $\mathbb{P}$-independence of $\{X_i : i \in \mathcal{I}\}$ is equivalent to the statement that

$$\mathbb{E}^{\mathbb{P}}\big[f_{i_1} \circ X_{i_1} \cdots f_{i_n} \circ X_{i_n}\big] = \mathbb{E}^{\mathbb{P}}\big[f_{i_1} \circ X_{i_1}\big] \cdots \mathbb{E}^{\mathbb{P}}\big[f_{i_n} \circ X_{i_n}\big]$$

for all finite subsets $\{i_1, \dots, i_n\}$ of distinct elements of $\mathcal{I}$ and all choices of $f_{i_1} \in B\big(E_{i_1};\mathbb{R}\big), \dots, f_{i_n} \in B\big(E_{i_n};\mathbb{R}\big)$. Finally, if $\mathbf{1}_A$ given by

$$\mathbf{1}_A(\omega) \equiv \begin{cases} 1 & \text{if } \omega \in A \\ 0 & \text{if } \omega \notin A \end{cases}$$

denotes the **indicator function** of the set $A \subseteq \Omega$, notice that the family of sets $\{A_i : i \in \mathcal{I}\} \subseteq \mathcal{F}$ is $\mathbb{P}$-independent if and only if the random variables $\mathbf{1}_{A_i}$, $i \in \mathcal{I}$, are $\mathbb{P}$-independent.

Thus far I have discussed only the abstract notion of independence and have yet to show that the concept is not vacuous. In the modern literature, the standard way to construct lots of independent quantities is to take products of probability spaces. Namely, if $(E_i, \mathcal{B}_i, \mu_i)$ is a probability space for each $i \in \mathcal{I}$, one sets $\Omega = \prod_{i \in \mathcal{I}} E_i$; defines $\pi_i : \Omega \longrightarrow E_i$ to be the natural projection map for each $i \in \mathcal{I}$; takes $\mathcal{F}_i = \pi_i^{-1}(\mathcal{B}_i)$, $i \in \mathcal{I}$, and $\mathcal{F} = \bigvee_{i \in \mathcal{I}} \mathcal{F}_i$; and shows that there is a unique probability measure $\mathbb{P}$ on $(\Omega, \mathcal{F})$ with the properties that

$$\mathbb{P}\big(\pi_i^{-1}\Gamma_i\big) = \mu_i(\Gamma_i) \quad \text{for all} \quad i \in \mathcal{I} \text{ and } \Gamma_i \in \mathcal{B}_i$$

[1] Throughout this book, I use $\mathbb{E}^{\mathbb{P}}[X, A]$ to denote the expected value under $\mathbb{P}$ of $X$ over the set $A$. That is, $\mathbb{E}^{\mathbb{P}}[X, A] = \int_A X \, d\mathbb{P}$. Finally, when $A = \Omega$, I will write $\mathbb{E}^{\mathbb{P}}[X]$. Tonelli's Theorem is the version of Fubini's Theorem for non-negative functions. Its virtue is that it applies whether or not the integrand is integrable.

and the $\sigma$-algebras $\mathcal{F}_i$, $i \in \mathcal{I}$, are $\mathbb{P}$-independent. Although this procedure is extremely powerful, it is rather mechanical. For this reason, I have chosen to defer the details of the product construction to Exercises 1.1.14 and 1.1.16 and to, instead, spend the rest of this section developing a more hands-on approach to constructing independent sequences of real-valued random variables. Indeed, although the product method is more ubiquitous and has become the construction of choice, the one that I am about to present has the advantage that it shows independent random variables can arise "naturally" and even in a familiar places.

**§ 1.1.3. The Rademacher Functions.** Until further notice, take $(\Omega, \mathcal{F}) = \big([0,1), \mathcal{B}_{[0,1)}\big)$ (when $E$ is a metric space, I use $\mathcal{B}_E$ to denote the Borel field over $E$) and $\mathbb{P}$ to be the restriction $\lambda_{[0,1)}$ of Lebesgue measure $\lambda_{\mathbb{R}}$ to $[0,1)$. Next define the **Rademacher functions** $R_n$, $n \in \mathbb{Z}^+$, on $\Omega$ as follows. Take the **integer part** $\lfloor t \rfloor$ of $t \in \mathbb{R}$ to be the largest integer dominated by $t$, and consider the function $R : \mathbb{R} \longrightarrow \{-1, 1\}$ given by

$$R(t) = \begin{cases} -1 & \text{if } \ t - \lfloor t \rfloor \in \left[0, \frac{1}{2}\right) \\ \ \ 1 & \text{if } \ t - \lfloor t \rfloor \in \left[\frac{1}{2}, 1\right). \end{cases}$$

The function $R_n$ is then defined on $[0,1)$ by

$$R_n(\omega) = R\big(2^{n-1}\omega\big), \qquad n \in \mathbb{Z}^+ \text{ and } \omega \in [0,1).$$

I will now show that the Rademacher functions are $\mathbb{P}$-independent. To this end, first note that every real-valued function $f$ on $\{-1,1\}$ is of the form $\alpha + \beta x$, $x \in \{-1,1\}$, for some pair of real numbers $\alpha$ and $\beta$. Thus, all that I have to show is that

$$\mathbb{E}^{\mathbb{P}}\big[(\alpha_1 + \beta_1 R_1) \cdots (\alpha_n + \beta_n R_n)\big] = \alpha_1 \cdots \alpha_n$$

for any $n \in \mathbb{Z}^+$ and $(\alpha_1, \beta_1), \dots, (\alpha_n, \beta_n) \in \mathbb{R}^2$. Since this is obvious when $n = 1$, I will assume that it holds for $n$ and need only check that it must also hold for $n+1$, and clearly this comes down to checking that

$$\mathbb{E}^{\mathbb{P}}\big[F(R_1, \dots, R_n)\, R_{n+1}\big] = 0$$

for any $F : \{-1,1\}^n \longrightarrow \mathbb{R}$. But $(R_1, \dots, R_n)$ is constant on each interval

$$I_{m,n} \equiv \left[\frac{m}{2^n}, \frac{m+1}{2^n}\right), \quad 0 \le m < 2^n,$$

whereas $R_{n+1}$ integrates to 0 on each $I_{m,n}$. Hence, by writing the integral over $\Omega$ as the sum of integrals over the $I_{m,n}$'s, we get the desired result.

At this point I have produced a countably infinite sequence of independent **Bernoulli random variables** (i.e., two-valued random variables whose range is usually either $\{-1,1\}$ or $\{0,1\}$) with mean value 0. In order to get more general random variables, I will combine our Bernoulli random variables together in a clever way.

Recall that a random variable $U$ is said to be **uniformly distributed** on the finite interval $[a,b]$ if

$$\mathbb{P}(U \le t) = \frac{t-a}{b-a} \quad \text{for } t \in [a,b].$$

Lemma 1.1.6. *Let $\{X_\ell : \ell \in \mathbb{Z}^+\}$ be a sequence of $\mathbb{P}$-independent $\{0,1\}$-valued Bernoulli random variables with mean value $\frac{1}{2}$ on some probability space $(\Omega, \mathcal{F}, \mathbb{P})$, and set*

$$U = \sum_{\ell=1}^{\infty} \frac{X_\ell}{2^\ell}.$$

*Then $U$ is uniformly distributed on $[0,1]$.*

Proof: Because the assertion only involves properties of distributions, it will be proved in general as soon as I prove it for a particular realization of independent, mean value $\frac{1}{2}$, $\{0,1\}$-valued Bernoulli random variables. In particular, by the preceding discussion, I need only consider the random variables

$$\epsilon_n(\omega) \equiv \frac{1 + R_n(\omega)}{2}, \quad n \in \mathbb{Z}^+ \text{ and } \omega \in [0,1),$$

on $\big([0,1), \mathcal{B}_{[0,1)}, \lambda_{[0,1)}\big)$. But, as is easily checked (cf. part (**i**) of Exercise 1.1.11), for each $\omega \in [0,1)$, $\omega = \sum_{n=1}^{\infty} 2^{-n}\epsilon_n(\omega)$. Hence, the desired conclusion is trivial in this case. $\square$

Now let $(k,\ell) \in \mathbb{Z}^+ \times \mathbb{Z}^+ \longmapsto n(k,\ell) \in \mathbb{Z}^+$ be any one-to-one mapping of $\mathbb{Z}^+ \times \mathbb{Z}^+$ onto $\mathbb{Z}^+$, and set

$$Y_{k,\ell} = \frac{1 + R_{n(k,\ell)}}{2}, \qquad (k,\ell) \in \left(\mathbb{Z}^+\right)^2.$$

Clearly, each $Y_{k,\ell}$ is a $\{0,1\}$-valued, Bernoulli random variable with mean value $\frac{1}{2}$, and the family $\left\{Y_{k,\ell} : (k,\ell) \in \left(\mathbb{Z}^+\right)^2\right\}$ is $\mathbb{P}$-independent. Hence, by Lemma 1.1.6, each of the random variables

$$U_k \equiv \sum_{\ell=1}^{\infty} \frac{Y_{k,\ell}}{2^\ell}, \qquad k \in \mathbb{Z}^+,$$

is uniformly distributed on $[0,1)$. In addition, the $U_k$'s are obviously mutually independent. Hence, I have now produced a sequence of mutually independent random variables, each of which is uniformly distributed on $[0,1)$. To complete our program, I use the time-honored transformation that takes a uniform random variable into an arbitrary one. Namely, given a **distribution function** $F$ on $\mathbb{R}$ (i.e., $F$ is a right-continuous, non-decreasing function that tends to 0 at $-\infty$ and 1 at $+\infty$), define $F^{-1}$ on $[0,1]$ to be the left-continuous inverse of $F$. That is,

$$F^{-1}(t) = \inf\{s \in \mathbb{R} : F(s) \geq t\}, \qquad t \in [0,1].$$

(Throughout, the infemum over the empty set is taken to be $+\infty$.) It is then an easy matter to check that when $U$ is uniformly distributed on $[0,1)$ the random variable $X = F^{-1} \circ U$ has distribution function $F$:

$$\mathbb{P}(X \leq t) = F(t), \qquad t \in \mathbb{R}.$$

Hence, after combining this with what we already know, I have now completed the proof of the following theorem.

THEOREM 1.1.7. *Let $\Omega = [0,1)$, $\mathcal{F} = \mathcal{B}_{[0,1)}$, and $\mathbb{P} = \lambda_{[0,1)}$. Then, for any sequence $\{F_k : k \in \mathbb{Z}^+\}$ of distribution functions on $\mathbb{R}$, there exists a sequence $\{X_k : k \in \mathbb{Z}^+\}$ of $\mathbb{P}$-independent random variables on $(\Omega, \mathcal{F}, \mathbb{P})$ with the property that $\mathbb{P}(X_k \leq t) = F_k(t)$, $t \in \mathbb{R}$, for each $k \in \mathbb{Z}^+$.*

## Exercises for § 1.1

EXERCISE 1.1.8. As I pointed out, $\mathbb{P}(A_1 \cap A_2) = \mathbb{P}(A_1)\mathbb{P}(A_2)$ if and only if the $\sigma$-algebra generated by $A_1$ is $\mathbb{P}$-independent of the one generated by $A_2$. Construct an example to show that the analogous statement is false when dealing with three, instead of two, sets. That is, just because $\mathbb{P}(A_1 \cap A_2 \cap A_3) = \mathbb{P}(A_1)\mathbb{P}(A_2)\mathbb{P}(A_3)$, show that it is not necessarily true that the three $\sigma$-algebras generated by $A_1$, $A_2$, and $A_3$ are $\mathbb{P}$-independent.

EXERCISE 1.1.9. This exercise deals with three elementary, but important, properties of independent random variables. Throughout, $(\Omega, \mathcal{F}, \mathbb{P})$ is a given probability space.

**(i)** Let $X_1$ and $X_2$ be a pair of $\mathbb{P}$-independent random variables with values in the measurable spaces $(E_1, \mathcal{B}_1)$ and $(E_2, \mathcal{B}_2)$, respectively. Given a $\mathcal{B}_1 \times \mathcal{B}_2$-measurable function $F : E_1 \times E_2 \longrightarrow \mathbb{R}$ that is bounded below, use Tonelli's or Fubini's Theorem to show that

$$x_2 \in E_2 \longmapsto f(x_2) \equiv \mathbb{E}^{\mathbb{P}}\big[F(X_1, x_2)\big] \in \mathbb{R}$$

is $\mathcal{B}_2$-measurable and that

$$\mathbb{E}^{\mathbb{P}}\big[F(X_1, X_2)\big] = \mathbb{E}^{\mathbb{P}}\big[f(X_2)\big].$$

**(ii)** Suppose that $X_1, \dots, X_n$ are $\mathbb{P}$-independent, real-valued random variables. If each of the $X_m$'s is $\mathbb{P}$-integrable, show that $X_1 \cdots X_n$ is also $\mathbb{P}$-integrable and that

$$\mathbb{E}^{\mathbb{P}}\big[X_1 \cdots X_n\big] = \mathbb{E}^{\mathbb{P}}\big[X_1\big] \cdots \mathbb{E}^{\mathbb{P}}\big[X_n\big].$$

**(iii)** Let $\{X_n : n \in \mathbb{Z}^+\}$ be a sequence of independent random variables taking values in some separable metric space $E$. If $\mathbb{P}(X_n = x) = 0$ for all $x \in E$ and $n \in \mathbb{Z}^+$, show that $\mathbb{P}(X_m = X_n \text{ for some } m \neq n) = 0$.

EXERCISE 1.1.10. As an application of Lemma 1.1.6 and part (**ii**) of Exercise 1.1.9, prove the identity

$$\sin z = z\prod_{n=1}^{\infty}\cos(2^{-n}z) \quad \text{for all } z\in\mathbb{C}.$$

EXERCISE 1.1.11. Define $\{\epsilon_n(\omega) : n\geq 1\}$ for $\omega\in[0,1)$ as in the proof of Lemma 1.1.6.

(**i**) Show that $\{\epsilon_n(\omega) : n\geq 1\}$ is the unique sequence $\{\alpha_n : n\geq 1\}\subseteq\{0,1\}^{\mathbb{Z}^+}$ such that $\omega-\sum_{m=1}^{n}2^{-m}\alpha_m < 2^{-n}$, and conclude that $\epsilon_1(\omega)=\lfloor 2\omega\rfloor$ and $\epsilon_{n+1}(\omega)=\lfloor 2^{n+1}\omega\rfloor-2\lfloor 2^n\omega\rfloor$ for $n\geq 1$.

(**ii**) Define $F:[0,1)\longrightarrow[0,1)^2$ by

$$F(\omega)=\left(\sum_{n=1}^{\infty}2^{-n}\epsilon_{2n-1}(\omega),\sum_{n=1}^{\infty}2^{-n}\epsilon_{2n}(\omega)\right),$$

and show that $\lambda_{[0,1)^2}=F_*\lambda_{[0,1)}$. That is, $\lambda_{[0,1)}\big(\{\omega : F(\omega)\in\Gamma\}\big)=\lambda^2_{[0,1)}(\Gamma)$ for all $\Gamma\in\mathcal{B}_{[0,1)^2}$.

(**iii**) Define $G:[0,\infty)^2\longrightarrow[0,1)$ by

$$G\big((\omega_1,\omega_2)\big)=\sum_{n=1}^{\infty}\frac{2\epsilon_n(\omega_1)+\epsilon_n(\omega_2)}{4^n},$$

and show that $\lambda_{[0,1)}=G_*\lambda_{[0,1)^2}$.

Parts (**ii**) and (**iii**) are special cases of a general principle that says, under very general circumstances, measures can be transformed into one another.

EXERCISE 1.1.12. Given a non-empty set $\Omega$, recall[2] that a collection $\mathcal{C}$ of subsets of $\Omega$ is called a **$\pi$-system** if $\mathcal{C}$ is closed under finite intersections. At the same time, recall that a collection $\mathcal{L}$ is called a **$\lambda$-system** if $\Omega\in\mathcal{L}$, $A\cup B\in\mathcal{L}$ whenever $A$ and $B$ are disjoint members of $\mathcal{L}$, $B\setminus A\in\mathcal{L}$ whenever $A$ and $B$ are members of $\mathcal{L}$ with $A\subseteq B$, and $\bigcup_1^{\infty}A_n\in\mathcal{L}$ whenever $\{A_n : n\geq 1\}$ is a non-decreasing sequence of members of $\mathcal{L}$. Finally, recall (cf. Lemma 2.1.12 in *op. cit.* that if $\mathcal{C}$ is a $\pi$-system, then the $\sigma$-algebra $\sigma(\mathcal{C})$ generated by $\mathcal{C}$ is the smallest $\mathcal{L}$-system $\mathcal{L}\supseteq\mathcal{C}$.

Show that if $\mathcal{C}$ is a $\pi$-system and $\mathcal{F}=\sigma(\mathcal{C})$, then two probability measures $\mathbb{P}$ and $\mathbb{Q}$ are equal on $\mathcal{F}$ if they are equal on $\mathcal{C}$. Next use this to see that if $\{\mathcal{C}_i : i\in\mathcal{I}\}$ is a family of $\pi$-systems contained in $\mathcal{F}$ and if (1.1.1) holds when the $A_i$'s are from the $\mathcal{C}_i$'s, then the family of $\sigma$-algebras $\{\sigma(\mathcal{C}_i) : i\in\mathcal{I}\}$ is independent.

[2] See, for example, § 3.1 in the author's *Essentials of Integration Theory for Analysis*, # 262 in the Springer-Verlag G.T.M. series (2011).

EXERCISE 1.1.13. In this exercise I discuss two criteria for determining when random variables on the probability space $(\Omega, \mathcal{F}, \mathbb{P})$ are independent.

**(i)** Let $X_1, \dots, X_n$ be bounded, real-valued random variables. Using Weierstrass's Approximation Theorem, show that the $X_m$'s are $\mathbb{P}$-independent if and only if

$$\mathbb{E}^{\mathbb{P}}\left[X_1^{m_1} \cdots X_n^{m_n}\right] = \mathbb{E}^{\mathbb{P}}\left[X_1^{m_1}\right] \cdots \mathbb{E}^{\mathbb{P}}\left[X_n^{m_n}\right]$$

for all $m_1, \dots, m_n \in \mathbb{N}$.

**(ii)** Let $\mathbf{X} : \Omega \longrightarrow \mathbb{R}^m$ and $\mathbf{Y} : \Omega \longrightarrow \mathbb{R}^n$ be random variables. Show that $\mathbf{X}$ and $\mathbf{Y}$ are $\mathbb{P}$-independent if and only if

$$\mathbb{E}^{\mathbb{P}}\left[\exp\left[\sqrt{-1}\left(\left(\boldsymbol{\alpha}, \mathbf{X}\right)_{\mathbb{R}^m} + \left(\boldsymbol{\beta}, \mathbf{Y}\right)_{\mathbb{R}^n}\right)\right]\right]$$
$$= \mathbb{E}^{\mathbb{P}}\left[\exp\left[\sqrt{-1}\left(\boldsymbol{\alpha}, \mathbf{X}\right)_{\mathbb{R}^m}\right]\right] \mathbb{E}^{\mathbb{P}}\left[\exp\left[\sqrt{-1}\left(\boldsymbol{\beta}, \mathbf{Y}\right)_{\mathbb{R}^n}\right]\right]$$

for all $\boldsymbol{\alpha} \in \mathbb{R}^m$ and $\boldsymbol{\beta} \in \mathbb{R}^n$.

**Hint**: The *only if* assertion is obvious. To prove the *if* assertion, first check that $\mathbf{X}$ and $\mathbf{Y}$ are independent if

$$\mathbb{E}^{\mathbb{P}}\left[f(\mathbf{X})\, g(\mathbf{Y})\right] = \mathbb{E}^{\mathbb{P}}\left[f(\mathbf{X})\right] \mathbb{E}^{\mathbb{P}}\left[g(\mathbf{Y})\right]$$

for all $f \in C_c^\infty(\mathbb{R}^m; \mathbb{C})$ and $g \in C_c^\infty(\mathbb{R}^n; \mathbb{C})$. Second, given such $f$ and $g$, apply elementary Fourier analysis to write

$$f(\mathbf{x}) = \int_{\mathbb{R}^m} e^{\sqrt{-1}\,(\boldsymbol{\alpha}, \mathbf{x})_{\mathbb{R}^m}}\, \varphi(\boldsymbol{\alpha})\, d\boldsymbol{\alpha} \quad \text{and} \quad g(\mathbf{y}) = \int_{\mathbb{R}^n} e^{\sqrt{-1}\,(\boldsymbol{\beta}, \mathbf{y})_{\mathbb{R}^n}}\, \psi(\boldsymbol{\beta})\, d\boldsymbol{\beta},$$

where $\varphi$ and $\psi$ are smooth functions with **rapidly decreasing** (i.e., tending to 0 as $|\mathbf{x}| \to \infty$ faster than any power of $(1 + |\mathbf{x}|)^{-1}$) derivatives of all orders. Finally, apply Fubini's Theorem.

EXERCISE 1.1.14. Given a pair of measurable spaces $(E_1, \mathcal{B}_1)$ and $(E_2, \mathcal{B}_2)$, recall that their product is the measurable space $\left(E_1 \times E_2, \mathcal{B}_1 \times \mathcal{B}_2\right)$, where $\mathcal{B}_1 \times \mathcal{B}_2$ is the $\sigma$-algebra over the Cartesian product space $E_1 \times E_2$ generated by the sets $\Gamma_1 \times \Gamma_2$, $\Gamma_i \in \mathcal{B}_i$. Further, recall that, for any probability measures $\mu_i$ on $(E_i, \mathcal{B}_i)$, there is a unique probability measure $\mu_1 \times \mu_2$ on $\left(E_1 \times E_2, \mathcal{B}_1 \times \mathcal{B}_2\right)$ such that

$$(\mu_1 \times \mu_2)\left(\Gamma_1 \times \Gamma_2\right) = \mu_1(\Gamma_1)\mu_2(\Gamma_2) \quad \text{for } \Gamma_i \in \mathcal{B}_i.$$

More generally, for any $n \geq 2$ and measurable spaces $\{(E_i, \mathcal{B}_i) : 1 \leq i \leq n\}$, one takes $\prod_1^n \mathcal{B}_i$ to be the $\sigma$-algebra over $\prod_1^n E_i$ generated by the sets $\prod_1^n \Gamma_i$, $\Gamma_i \in \mathcal{B}_i$. Since $\prod_1^{n+1} E_i$ and $\prod_1^{n+1} \mathcal{B}_i$ can be identified with $(\prod_1^n E_i) \times E_{n+1}$ and

$(\prod_1^n \mathcal{B}_i) \times \mathcal{B}_{n+1}$, respectively, one can use induction to show that, for every choice of probability measures $\mu_i$ on $(E_i, \mathcal{B}_i)$, there is a unique probability measure $\prod_1^n \mu_i$ on $(\prod_1^n E_i, \prod_1^n \mathcal{B}_i)$ such that

$$\left(\prod_1^n \mu_i\right)\left(\prod_1^n \Gamma_i\right) = \prod_1^n \mu_i(\Gamma_i), \quad \Gamma_i \in \mathcal{B}_i.$$

The purpose of this exercise is to generalize the preceding construction to infinite collections. Thus, let $\mathcal{I}$ be an infinite index set, and, for each $i \in \mathcal{I}$, let $(E_i, \mathcal{B}_i)$ be a measurable space. Given $\emptyset \neq \Lambda \subseteq \mathcal{I}$, use $\mathbf{E}_\Lambda$ to denote the Cartesian product space $\prod_{i \in \Lambda} E_i$ and $\pi_\Lambda$ to denote the natural projection map taking $\mathbf{E}_\mathcal{I}$ onto $\mathbf{E}_\Lambda$. Further, let $\mathcal{B}_\mathcal{I} = \prod_{i \in \mathcal{I}} \mathcal{B}_i$ stand for the $\sigma$-algebra over $\mathbf{E}_\mathcal{I}$ generated by the collection $\mathcal{C}$ of subsets

$$\pi_F^{-1}\left(\prod_{i \in F} \Gamma_i\right), \quad \Gamma_i \in \mathcal{B}_i,$$

as $F$ varies over non-empty, finite subsets of $\mathcal{I}$ (abbreviated by $\emptyset \neq F \subset\subset \mathcal{I}$). In the following steps, I outline a proof that, for every choice of probability measures $\mu_i$ on the $(E_i, \mathcal{B}_i)$'s, there is a unique probability measure $\prod_{i \in \mathcal{I}} \mu_i$ on $(\mathbf{E}_\mathcal{I}, \mathcal{B}_\mathcal{I})$ with the property that

$$\left(\prod_{i \in \mathcal{I}} \mu_i\right)\left(\pi_F^{-1}\left(\prod_{i \in F} \Gamma_i\right)\right) = \prod_{i \in F} \mu_i(\Gamma_i), \quad \Gamma_i \in \mathcal{B}_i, \tag{1.1.15}$$

for every $\emptyset \neq F \subset\subset \mathcal{I}$. Not surprisingly, the probability space

$$\left(\prod_{i \in \mathcal{I}} E_i, \prod_{i \in \mathcal{I}} \mathcal{B}_i, \prod_{i \in \mathcal{I}} \mu_i\right)$$

is called the **product** over $\mathcal{I}$ of the spaces $(E_i, \mathcal{B}_i, \mu_i)$; and when all the factors are the same space $(E, \mathcal{B}, \mu)$, it is customary to denote it by $(E^\mathcal{I}, \mathcal{B}^\mathcal{I}, \mu^\mathcal{I})$, and if, in addition, $\mathcal{I} = \{1, \dots, N\}$, one uses $(E^N, \mathcal{B}^N, \mu^N)$.

**(i)** After noting (cf. Exercise 1.1.12) that two probability measures that agree on a $\pi$-system agree on the $\sigma$-algebra generated by that $\pi$-system, show that there is at most one probability measure on $(\mathbf{E}_\mathcal{I}, \mathcal{B}_\mathcal{I})$ that satisfies the condition in (1.1.15). Hence, the problem is purely one of existence.

**(ii)** Let $\mathcal{A}$ be the *algebra* over $\mathbf{E}_\mathcal{I}$ generated by $\mathcal{C}$, and show that there is a *finitely* additive $\mu : \mathcal{A} \longrightarrow [0, 1]$ with the property that

$$\mu\left(\pi_F^{-1}(\Gamma_F)\right) = \left(\prod_{i \in F} \mu_i\right)(\Gamma_F), \quad \Gamma_F \in \mathcal{B}_F,$$

for all $\emptyset \neq F \subset\subset \mathcal{I}$. Hence, all that one has to do is check that $\mu$ admits a $\sigma$-additive extension to $\mathcal{B}_{\mathcal{I}}$, and, by a standard extension theorem, this comes down to checking that $\mu(A_n) \searrow 0$ whenever $\{A_n : n \geq 1\} \subseteq \mathcal{A}$ and $A_n \searrow \emptyset$. Thus, let $\{A_n : n \geq 1\}$ be a non-increasing sequence from $\mathcal{A}$, and assume that $\mu(A_n) \geq \epsilon$ for some $\epsilon > 0$ and all $n \in \mathbb{Z}^+$. One must show that $\bigcap_1^\infty A_n \neq \emptyset$.

**(iii)** Referring to the last part of **(ii)**, show that there is no loss in generality to assume that $A_n = \pi_{F_n}^{-1}(\Gamma_{F_n})$, where, for each $n \in \mathbb{Z}^+$, $\emptyset \neq F_n \subset\subset \mathcal{I}$ and $\Gamma_{F_n} \in \mathcal{B}_{F_n}$. In addition, show that one may assume that $F_1 = \{i_1\}$ and that $F_n = F_{n-1} \cup \{i_n\}$, $n \geq 2$, where $\{i_n : n \geq 1\}$ is a sequence of distinct elements of $\mathcal{I}$. Now, make these assumptions, and show that it suffices to find $a_\ell \in E_{i_\ell}$, $\ell \in \mathbb{Z}^+$, with the property that, for each $m \in \mathbb{Z}^+$, $(a_1, \dots, a_m) \in \Gamma_{F_m}$.

**(iv)** Continuing **(iii)**, for each $m, n \in \mathbb{Z}^+$, define $g_{m,n} : \mathbf{E}_{F_m} \longrightarrow [0,1]$ so that

$$g_{m,n}(\mathbf{x}_{F_m}) = \mathbf{1}_{\Gamma_{F_n}}(x_{i_1}, \dots, x_{i_n}) \quad \text{if } n \leq m$$

and

$$g_{m,n}(\mathbf{x}_{F_m}) = \int_{\mathbf{E}_{F_n \setminus F_m}} \mathbf{1}_{\Gamma_{F_n}}(\mathbf{x}_{F_m}, \mathbf{y}_{F_n \setminus F_m}) \left( \prod_{\ell=m+1}^{n} \mu_{i_\ell} \right) (d\mathbf{y}_{F_n \setminus F_m}) \quad \text{if } n > m.$$

After noting that, for each $m$ and $n$, $g_{m,n+1} \leq g_{m,n}$ and

$$g_{m,n}(\mathbf{x}_{F_m}) = \int_{E_{i_{m+1}}} g_{m+1,n}(\mathbf{x}_{F_m}, y_{i_{m+1}}) \, \mu_{i_{m+1}}(dy_{i_{m+1}}),$$

set $g_m = \lim_{n\to\infty} g_{m,n}$ and conclude that

$$g_m(\mathbf{x}_{F_m}) = \int_{E_{i_{m+1}}} g_{m+1}(\mathbf{x}_{F_m}, y_{i_{m+1}}) \, \mu_{i_{m+1}}(dy_{i_{m+1}}).$$

In addition, note that

$$\int_{E_{i_1}} g_1(x_{i_1}) \, \mu_{i_1}(dx_{i_1}) = \lim_{n\to\infty} \int_{E_{i_1}} g_{1,n}(x_{i_1}) \, \mu_{i_1}(dx_{i_1}) = \lim_{n\to\infty} \mu(A_n) \geq \epsilon,$$

and proceed by induction to produce $a_\ell \in E_{i_\ell}$, $\ell \in \mathbb{Z}^+$, so that

$$g_m\big((a_1, \dots, a_m)\big) \geq \epsilon \quad \text{for all } m \in \mathbb{Z}^+.$$

Finally, check that $\{a_m : m \geq 1\}$ is a sequence of the sort for which we were looking at the end of part **(iii)**.

EXERCISE 1.1.16. Recall that if $\Phi$ is a measurable map from one measurable space $(E, \mathcal{B})$ into a second one $(E', \mathcal{B}')$, then the **distribution** of $\Phi$ under a measure $\mu$ on $(E, \mathcal{B})$ is the **pushforward** measure $\Phi_*\mu$ (also denoted by $\mu \circ \Phi^{-1}$) defined on $(E', \mathcal{B}')$ by

$$\Phi_*\mu(\Gamma) = \mu\big(\Phi^{-1}(\Gamma)\big) \quad \text{for} \quad \Gamma \in \mathcal{B}'.$$

Given a non-empty index set $\mathcal{I}$ and, for each $i \in \mathcal{I}$, a measurable space $(E_i, \mathcal{B}_i)$ and an $E_i$-valued random variable $X_i$ on the probability space $(\Omega, \mathcal{F}, \mathbb{P})$, define $\mathbf{X} : \Omega \longrightarrow \prod_{i \in \mathcal{I}} E_i$ so that $\mathbf{X}(\omega)_i = X_i(\omega)$ for each $i \in \mathcal{I}$ and $\omega \in \Omega$. Show that $\{X_i : i \in \mathcal{I}\}$ is a family of $\mathbb{P}$-independent random variables if and only if $\mathbf{X}_*\mathbb{P} = \prod_{i \in \mathcal{I}} (X_i)_*\mathbb{P}$. In particular, given probability measures $\mu_i$ on $(E_i, \mathcal{B}_i)$, set

$$\Omega = \prod_{i \in \mathcal{I}} E_i, \quad \mathcal{F} = \prod_{i \in \mathcal{I}} \mathcal{B}_i, \quad \mathbb{P} = \prod_{i \in \mathcal{I}} \mu_i,$$

let $X_i : \Omega \longrightarrow E_i$ be the natural projection map from $\Omega$ onto $E_i$, and show that $\{X_i : i \in \mathcal{I}\}$ is a family of mutually $\mathbb{P}$-independent random variables such that, for each $i \in \mathcal{I}$, $X_i$ has distribution $\mu_i$.

EXERCISE 1.1.17. Although it does not entail infinite product spaces, an interesting example of the way in which the preceding type of construction can be effectively applied is provided by the following elementary version of a *coupling* argument.

(**i**) Let $(\Omega, \mathcal{B}, \mathbb{P})$ be a probability space and $X$ and $Y$ a pair of $\mathbb{P}$-square integrable $\mathbb{R}$-valued random variables with the property that

$$\big(X(\omega) - X(\omega')\big)\big(Y(\omega) - Y(\omega')\big) \ge 0 \quad \text{for all } (\omega, \omega') \in \Omega^2.$$

Show that

$$\mathbb{E}^{\mathbb{P}}[XY] \ge \mathbb{E}^{\mathbb{P}}[X]\,\mathbb{E}^{\mathbb{P}}[Y].$$

**Hint**: Define $X_i$ and $Y_i$ on $\Omega^2$ for $i \in \{1, 2\}$ so that $X_i(\boldsymbol{\omega}) = X(\omega_i)$ and $Y_i(\boldsymbol{\omega}) = Y(\omega_i)$ when $\boldsymbol{\omega} = (\omega_1, \omega_2)$, and integrate the inequality

$$0 \le \big(X(\omega_1) - X(\omega_2)\big)\big(Y(\omega_1) - Y(\omega_2)\big) = \big(X_1(\boldsymbol{\omega}) - X_2(\boldsymbol{\omega})\big)\big(Y_1(\boldsymbol{\omega}) - Y_2(\boldsymbol{\omega})\big)$$

with respect to $\mathbb{P}^2$.

(**ii**) Suppose that $n \in \mathbb{Z}^+$ and that $f$ and $g$ are $\mathbb{R}$-valued, Borel measurable functions on $\mathbb{R}^n$ that are non-decreasing with respect to each coordinate (separately). Show that if $\mathbf{X} = (X_1, \dots, X_n)$ is an $\mathbb{R}^n$-valued random variable on a probability space $(\Omega, \mathcal{B}, \mathbb{P})$ whose coordinates are mutually $\mathbb{P}$-independent, then

$$\mathbb{E}^{\mathbb{P}}\big[f(\mathbf{X})\,g(\mathbf{X})\big] \ge \mathbb{E}^{\mathbb{P}}\big[f(\mathbf{X})\big]\,\mathbb{E}^{\mathbb{P}}\big[g(\mathbf{X})\big]$$

so long as $f(\mathbf{X})$ and $g(\mathbf{X})$ are both $\mathbb{P}$-square integrable.

**Hint**: First check that the case when $n = 1$ reduces to an application of (**i**). Next, describe the general case in terms of a multiple integral, apply Fubini's Theorem, and make repeated use of the case when $n = 1$.

EXERCISE 1.1.18. A $\sigma$-algebra is said to be **countably generated** if it contains a countable collection of sets that generate it. The purpose of this exercise is to show that just because a $\sigma$-algebra is itself countably generated does not mean that all its sub-$\sigma$-algebras are.

Let $(\Omega, \mathcal{F}, \mathbb{P})$ be a probability space and $\{A_n : n \in \mathbb{Z}^+\} \subseteq \mathcal{F}$ a sequence of $\mathbb{P}$-independent sub-subsets of $\mathcal{F}$ with the property that $\alpha \le \mathbb{P}(A_n) \le 1 - \alpha$ for some $\alpha \in (0,1)$. Let $\mathcal{F}_n$ be the sub-$\sigma$-algebra generated by $A_n$. Show that the tail $\sigma$-algebra $\mathcal{T}$ determined by $\{\mathcal{F}_n : n \in \mathbb{Z}^+\}$ cannot be countably generated.

**Hint**: Show that $C \in \mathcal{T}$ is an **atom** in $\mathcal{T}$ (i.e., $B = C$ whenever $B \in \mathcal{T} \setminus \{\emptyset\}$ is contained in $C$) only if one can write

$$C = \varliminf_{n\to\infty} C_n \equiv \bigcup_{m=1}^{\infty} \bigcap_{n \ge m} C_n,$$

where, for each $n \in \mathbb{Z}^+$, $C_n$ equals either $A_n$ or $A_n\complement$. Conclude that every atom in $\mathcal{T}$ must have $\mathbb{P}$-measure 0. Now suppose that $\mathcal{T}$ were generated by $\{B_\ell : \ell \in \mathbb{N}\}$. By Kolmogorov's 0–1 Law, $\mathbb{P}(B_\ell) \in \{0,1\}$ for every $\ell \in \mathbb{N}$. Take

$$\hat{B}_\ell = \begin{cases} B_\ell & \text{if } \ P(B_\ell) = 1 \\ B_\ell\complement & \text{if } \ P(B_\ell) = 0 \end{cases} \quad \text{and set} \quad C = \bigcap_{\ell \in \mathbb{N}} \hat{B}_\ell.$$

Note that, on the one hand, $\mathbb{P}(C) = 1$, while, on the other hand, $C$ is an atom in $\mathcal{T}$ and therefore has probability 0.

EXERCISE 1.1.19. Here is an interesting application of Kolmogorov's 0–1 Law to a property of the real numbers.

(**i**) Referring to the discussion preceding Lemma 1.1.6 and part (**i**) of Exercise 1.1.11, define the transformations $T_n : [0,1) \longrightarrow [0,1)$ for $n \in \mathbb{Z}^+$ so that

$$T_n(\omega) = \omega - \frac{R_n(\omega)}{2^n}, \quad \omega \in [0,1),$$

and notice (cf. the proof of Lemma 1.1.6) that $T_n(\omega)$ simply *flips* the $n$th coefficient in the binary expansion $\omega$. Next, let $\Gamma \in \mathcal{B}_{[0,1)}$, and show that $\Gamma$ is measurable with respect to the $\sigma$-algebra $\sigma(\{R_n : n > m\})$ generated by $\{R_n : n > m\}$ if and only if $T_n(\Gamma) = \Gamma$ for each $1 \le n \le m$. In particular, conclude that $\lambda_{[0,1)}(\Gamma) \in \{0,1\}$ if $T_n\Gamma = \Gamma$ for every $n \in \mathbb{Z}^+$.

(**ii**) Let $\mathfrak{F}$ denote the set of all finite subsets of $\mathbb{Z}^+$, and for each $F \in \mathfrak{F}$, define $T^F : [0,1) \longrightarrow [0,1)$ so that $T^\emptyset$ is the identity mapping and

$$T^{F\cup\{m\}} = T^F \circ T_m \quad \text{for each } F \in \mathfrak{F} \text{ and } m \in \mathbb{Z}^+ \setminus F.$$

As an application of (**i**), show that for every $\Gamma \in \mathcal{B}_{[0,1)}$ with $\lambda_{[0,1)}(\Gamma) > 0$,

$$\lambda_{[0,1)}\left(\bigcup_{F\in\mathfrak{F}} T^F(\Gamma)\right) = 1.$$

In particular, this means that if $\Gamma$ has positive measure, then almost every $\omega \in [0,1)$ can be moved to $\Gamma$ by *flipping* a finite number of the coefficients in the binary expansion of $\omega$.

## § 1.2 The Weak Law of Large Numbers

Starting with this section, and for the rest of this chapter, I will be studying what happens when one averages independent, real-valued random variables. The remarkable fact, which will be confirmed repeatedly, is that the limiting behavior of such averages depends hardly at all on the variables involved. Intuitively, one can explain this phenomenon by pretending that the random variables are building blocks that, in the averaging process, first get homothetically shrunk and then reassembled according to a regular pattern. Hence, by the time that one passes to the limit, the peculiarities of the original blocks get lost.

Throughout the discussion, $(\Omega, \mathcal{F}, \mathbb{P})$ will be a probability space on which there is a sequence $\{X_n : n \geq 1\}$ of real-valued random variables. Given $n \in \mathbb{Z}^+$, use $S_n$ to denote the partial sum $X_1 + \cdots + X_n$ and $\overline{S}_n$ to denote the average:

$$\frac{S_n}{n} = \frac{1}{n}\sum_{\ell=1}^{n} X_\ell.$$

§ **1.2.1. Orthogonal Random Variables.** My first result is a very general one; in fact, it even applies to random variables that are not necessarily independent and do not necessarily have mean 0.

LEMMA 1.2.1. *Assume that*

$$\mathbb{E}^{\mathbb{P}}[X_n^2] < \infty \text{ for } n \in \mathbb{Z}^+ \quad \text{and} \quad \mathbb{E}^{\mathbb{P}}[X_k X_\ell] = 0 \text{ if } k \neq \ell.$$

*Then, for each* $\epsilon > 0$,

$$\epsilon^2\, \mathbb{P}\left(|\overline{S}_n| \geq \epsilon\right) \leq \mathbb{E}^{\mathbb{P}}[\overline{S}_n^2] = \frac{1}{n^2}\sum_{\ell=1}^{n} \mathbb{E}^{\mathbb{P}}[X_\ell^2] \quad \text{for } n \in \mathbb{Z}^+. \tag{1.2.2}$$

*In particular, if*

$$M \equiv \sup_{n\in\mathbb{Z}^+} \mathbb{E}^{\mathbb{P}}[X_n^2] < \infty,$$

*then*

$$\epsilon^2\, \mathbb{P}\left(|\overline{S}_n| \geq \epsilon\right) \leq \mathbb{E}^{\mathbb{P}}[\overline{S}_n^2] \leq \frac{M}{n}, \qquad n \in \mathbb{Z}^+ \text{ and } \epsilon > 0; \tag{1.2.3}$$

*and so* $\overline{S}_n \longrightarrow 0$ *in* $L^2(\mathbb{P};\mathbb{R})$ *and therefore also in* $\mathbb{P}$*-probability.*

PROOF: To prove the equality in (1.2.2), note that, by orthogonality,

$$\mathbb{E}^{\mathbb{P}}[S_n^2] = \sum_{\ell=1}^{n} \mathbb{E}^{\mathbb{P}}[X_\ell^2].$$

The rest is just an application of **Chebyshev's inequality**, the estimate that results after integrating the inequality

$$\epsilon^2 \mathbf{1}_{[\epsilon,\infty)}(|Y|) \le Y^2 \mathbf{1}_{[\epsilon,\infty)}(|Y|) \le Y^2$$

for any random variable $Y$. □

§ **1.2.2. Independent Random Variables.** Although Lemma 1.2.1 does not use independence, independent random variables provide a ready source of orthogonal functions. To wit, recall that for any $\mathbb{P}$-square integrable random variable $X$, its **variance** $\mathrm{Var}(X)$ satisfies

$$\mathrm{Var}(X) \equiv \mathbb{E}^{\mathbb{P}}\Big[\big(X - \mathbb{E}^{\mathbb{P}}[X]\big)^2\Big] = \mathbb{E}^{\mathbb{P}}[X^2] - \big(\mathbb{E}^{\mathbb{P}}[X]\big)^2 \le \mathbb{E}^{\mathbb{P}}[X^2].$$

In particular, if the random variables $X_n$, $n \in \mathbb{Z}^+$, are $\mathbb{P}$-square integrable and $\mathbb{P}$-independent, then the random variables

$$\hat{X}_n \equiv X_n - \mathbb{E}^{\mathbb{P}}[X_n], \quad n \in \mathbb{Z}^+,$$

are still $\mathbb{P}$-square integrable, have mean value 0, and therefore are orthogonal. Hence, the following statement is an immediate consequence of Lemma 1.2.1.

THEOREM 1.2.4. *Let $\{X_n : n \in \mathbb{Z}^+\}$ be a sequence of $\mathbb{P}$-independent, $\mathbb{P}$-square integrable random variables with mean value $m$ and variance dominated by $\sigma^2$. Then, for every $n \in \mathbb{Z}^+$ and $\epsilon > 0$,*

$$\epsilon^2\, \mathbb{P}\Big(\big|\overline{S}_n - m\big| \ge \epsilon\Big) \le \mathbb{E}^{\mathbb{P}}\Big[\big(\overline{S}_n - m\big)^2\Big] \le \frac{\sigma^2}{n}. \tag{1.2.5}$$

*In particular, $\overline{S}_n \longrightarrow m$ in $L^2(\mathbb{P};\mathbb{R})$ and therefore in $\mathbb{P}$-probability.*

As yet I have made only minimal use of independence: all that I have done is subtract off the mean of independent random variables and thereby made them orthogonal. In order to bring the full force of independence into play, one has to exploit the fact that one can compose independent random variables with any (measurable) functions without destroying their independence; in particular, *truncating* independent random variables does not destroy independence. To see how such a property can be brought to bear, I will now consider the problem of extending the last part of Theorem 1.2.4 to $X_n$'s that are less than $\mathbb{P}$-square integrable. In order to understand the statement, recall that a family of random variables $\{X_i : i \in \mathcal{I}\}$ is said to be **uniformly $\mathbb{P}$-integrable** if

$$\lim_{R \nearrow \infty} \sup_{i \in \mathcal{I}} \mathbb{E}^{\mathbb{P}}\Big[|X_i|,\ |X_i| \ge R\Big] = 0.$$

As the proof of the following theorem illustrates, the importance of this condition is that it allows one to simultaneously approximate the random variables $X_i$, $i \in \mathcal{I}$, by bounded random variables.

THEOREM 1.2.6 (**The Weak Law of Large Numbers**). *Let $\{X_n : n \in \mathbb{Z}^+\}$ be a uniformly $\mathbb{P}$-integrable sequence of $\mathbb{P}$-independent random variables. Then*

$$\frac{1}{n}\sum_1^n \big(X_m - \mathbb{E}^{\mathbb{P}}[X_m]\big) \longrightarrow 0 \text{ in } L^1(\mathbb{P};\mathbb{R})$$

*and therefore also in $\mathbb{P}$-probability. In particular, if $\{X_n : n \in \mathbb{Z}^+\}$ is a sequence of $\mathbb{P}$-independent, $\mathbb{P}$-integrable random variables that are identically distributed, then $\overline{S}_n \longrightarrow \mathbb{E}^{\mathbb{P}}[X_1]$ in $L^1(\mathbb{P};\mathbb{R})$ and $\mathbb{P}$-probability. (Cf. Exercise 1.2.11.)*

PROOF: Without loss in generality, I will assume that $\mathbb{E}^{\mathbb{P}}[X_n] = 0$ for every $n \in \mathbb{Z}^+$.

For each $R \in (0,\infty)$, define $f_R(t) = t\,\mathbf{1}_{[-R,R]}(t)$, $t \in \mathbb{R}$,

$$m_n^{(R)} = \mathbb{E}^{\mathbb{P}}\big[f_R \circ X_n\big], \quad X_n^{(R)} = f_R \circ X_n - m_n^{(R)}, \quad \text{and} \quad Y_n^{(R)} = X_n - X_n^{(R)},$$

and set

$$\overline{S}_n^{(R)} = \frac{1}{n}\sum_{\ell=1}^n X_\ell^{(R)} \quad \text{and} \quad \overline{T}_n^{(R)} = \frac{1}{n}\sum_{\ell=1}^n Y_\ell^{(R)}.$$

Since $\mathbb{E}[X_n] = 0 \implies m_n^{(R)} = -\mathbb{E}\big[X_n,\, |X_n| > R\big]$,

$$\begin{aligned}
\mathbb{E}^{\mathbb{P}}\big[|\overline{S}_n|\big] &\le \mathbb{E}^{\mathbb{P}}\big[|\overline{S}_n^{(R)}|\big] + \mathbb{E}^{\mathbb{P}}\big[|\overline{T}_n^{(R)}|\big] \\
&\le \mathbb{E}^{\mathbb{P}}\big[|\overline{S}_n^{(R)}|^2\big]^{\frac{1}{2}} + 2 \max_{1\le\ell\le n} \mathbb{E}^{\mathbb{P}}\big[|X_\ell|,\, |X_\ell| \ge R\big] \\
&\le \frac{R}{\sqrt{n}} + 2 \max_{\ell\in\mathbb{Z}^+} \mathbb{E}^{\mathbb{P}}\big[|X_\ell|,\, |X_\ell| \ge R\big];
\end{aligned}$$

and therefore, for each $R > 0$,

$$\varlimsup_{n\to\infty} \mathbb{E}^{\mathbb{P}}\big[|\overline{S}_n|\big] \le 2 \sup_{\ell\in\mathbb{Z}^+} \mathbb{E}^{\mathbb{P}}\big[|X_\ell|,\, |X_\ell| \ge R\big].$$

Hence, because the $X_\ell$'s are uniformly $\mathbb{P}$-integrable, we get the desired convergence in $L^1(\mathbb{P};\mathbb{R})$ by letting $R \nearrow \infty$. □

§ **1.2.3. Approximate Identities.** The name of Theorem 1.2.6 comes from a somewhat invidious comparison with the result in Theorem 1.4.9. The reason why the appellation *weak* is not entirely fair is that, although The Weak Law is indeed less *refined* than the result in Theorem 1.4.9, it is every bit as useful as the one in Theorem 1.4.9 and maybe even more important when it comes to applications. What The Weak Law provides is a ubiquitous technique for constructing an **approximate identity** (i.e., a sequence of measures that approximate a point mass) and measuring how fast the approximation is taking place. To illustrate how clever selections of the random variables entering The Weak Law can lead to interesting applications, I will spend the rest of this section discussing S. Bernstein's approach to Weierstrass's Approximation Theorem.

For a given $p \in [0,1]$, let $\{X_n : n \in \mathbb{Z}^+\}$ be a sequence of $\mathbb{P}$-independent $\{0,1\}$-valued Bernoulli random variables with mean value $p$. Then

$$\mathbb{P}(S_n = \ell) = \binom{n}{\ell} p^\ell (1-p)^{n-\ell} \quad \text{for} \quad 0 \le \ell \le n.$$

Hence, for any $f \in C([0,1];\mathbb{R})$, the $n$th **Bernstein polynomial**

$$B_n(p;f) \equiv \sum_{\ell=0}^{n} \binom{n}{\ell} f\left(\frac{\ell}{n}\right) p^\ell (1-p)^{n-\ell} \tag{1.2.7}$$

of $f$ at $p$ is equal to

$$\mathbb{E}^{\mathbb{P}}\left[f \circ \overline{S}_n\right].$$

In particular,

$$\begin{aligned}\left|f(p) - B_n(p;f)\right| &= \left|\mathbb{E}^{\mathbb{P}}\left[f(p) - f \circ \overline{S}_n\right]\right| \le \mathbb{E}^{\mathbb{P}}\left[\left|f(p) - f \circ \overline{S}_n\right|\right] \\ &\le 2\|f\|_{\mathrm{u}} P\left(\left|\overline{S}_n - p\right| \ge \epsilon\right) + \rho(\epsilon; f),\end{aligned}$$

where $\|f\|_{\mathrm{u}}$ is the **uniform norm** of $f$ (i.e., the supremum of $|f|$ over the domain of $f$) and

$$\rho(\epsilon; f) \equiv \sup\left\{|f(t) - f(s)| : 0 \le s < t \le 1 \text{ with } t - s \le \epsilon\right\}$$

is the modulus of continuity of $f$. Noting that $\mathrm{Var}(X_n) = p(1-p) \le \frac{1}{4}$ and applying (1.2.5), we conclude that, for every $\epsilon > 0$,

$$\left\|f(p) - B_n(p;f)\right\|_{\mathrm{u}} \le \frac{\|f\|_{\mathrm{u}}}{2n\epsilon^2} + \rho(\epsilon; f).$$

In other words, for all $n \in \mathbb{Z}^+$,

$$\left\|f - B_n(\,\cdot\,;f)\right\|_{\mathrm{u}} \le \beta(n;f) \equiv \inf\left\{\frac{\|f\|_{\mathrm{u}}}{2n\epsilon^2} + \rho(\epsilon;f) : \epsilon > 0\right\}. \tag{1.2.8}$$

Obviously, (1.2.8) not only shows that, as $n \to \infty$, $B_n(\,\cdot\,;f) \longrightarrow f$ uniformly on $[0,1]$, it even provides a rate of convergence in terms of the modulus of continuity of $f$. Thus, we have done more than simply prove Weierstrass's theorem; we have produced a rather explicit and tractable sequence of approximating polynomials, the sequence $\{B_n(\,\cdot\,;f) : n \in \mathbb{Z}^+\}$. Although this sequence is, by no means, the most efficient one,[1] as we are about to see, the Bernstein polynomials have a lot to recommend them. In particular, they have the feature that they provide non-negative polynomial approximations to non-negative functions. In fact, the following discussion reveals much deeper non-negativity preservation properties possessed by the Bernstein approximation scheme.

In order to bring out the virtues of the Bernstein polynomials, it is important to replace (1.2.7) with an expression in which the coefficients of $B_n(\,\cdot\,;f)$ (as polynomials) are clearly displayed. To this end, introduce the **difference operator** $\Delta_h$ for $h>0$ given by

$$[\Delta_h f](t) = \frac{f(t+h)-f(t)}{h}.$$

A straightforward inductive argument (using Pascal's Identity for the binomial coefficients) shows that

$$(-h)^m\,[\Delta_h^m f](t) = \sum_{\ell=0}^{m}(-1)^\ell\binom{m}{\ell}f(t+\ell h) \quad \text{for} \quad m\in\mathbb{Z}^+,$$

where $\Delta_h^{(m)}$ denotes the $m$th iterate of the operator $\Delta_h$. Taking $h=\frac{1}{n}$, we now see that

$$\begin{aligned}
B_n(p;f) &= \sum_{\ell=0}^{n}\sum_{k=0}^{n-\ell}\binom{n}{\ell}\binom{n-\ell}{k}(-1)^k f(\ell h)p^{\ell+k}\\
&= \sum_{r=0}^{n}p^r\sum_{\ell=0}^{r}\binom{n}{\ell}\binom{n-\ell}{r-\ell}(-1)^{r-\ell}f(\ell h)\\
&= \sum_{r=0}^{n}(-p)^r\binom{n}{r}\sum_{\ell=0}^{r}\binom{r}{\ell}(-1)^\ell f(\ell h)\\
&= \sum_{r=0}^{n}\binom{n}{r}(ph)^r\,[\Delta_h^r f](0),
\end{aligned}$$

where $\Delta_h^0 f \equiv f$. Hence, we have proved that

$$B_n(p;f) = \sum_{\ell=0}^{n} n^{-\ell}\binom{n}{\ell}[\Delta_{\frac{1}{n}}^\ell f](0)p^\ell \quad \text{for} \quad p\in[0,1]. \tag{1.2.9}$$

The marked resemblance between the expression on the right-hand side of (1.2.9) and a Taylor polynomial is more than coincidental. To demonstrate how

[1] See G.G. Lorentz's *Bernstein Polynomials*, Chelsea Publ. Co. (1986) for a lot more information.

one can exploit the relationship between the Bernstein and Taylor polynomials, say that a function $\varphi \in C^\infty\big((a,b);\mathbb{R}\big)$ is **absolutely monotone** if its $m$th derivative $D^m\varphi$ is non-negative for every $m \in \mathbb{N}$. Also, say that $\varphi \in C^\infty([0,1];[0,1])$ is a **probability generating function** if there exists a $\{u_n : n \in \mathbb{N}\} \subseteq [0,1]$ such that

$$\sum_{n=0}^{\infty} u_n = 1 \quad \text{and} \quad \varphi(t) = \sum_{n=0}^{\infty} u_n t^n \quad \text{for} \quad t \in [0,1].$$

Obviously, every probability generating function is absolutely monotone on $(0,1)$. The somewhat surprising (remember that most infinitely differentiable functions do not admit power series expansions) fact which I am about to prove is that, apart from a multiplicative constant, the converse is also true. In fact, one does not need to know, a priori, that the function is smooth so long as it satisfies a discrete version of absolute monotonicity.

THEOREM 1.2.10. *Let $\varphi \in C\big([0,1];\mathbb{R}\big)$ with $\varphi(1) = 1$ be given. Then the following are equivalent:*
**(i)** *$\varphi$ is a probability generating function,*
**(ii)** *the restriction of $\varphi$ to $(0,1)$ is absolutely monotone;*
**(iii)** *$\big[\Delta^m_{\frac{1}{n}}\varphi\big](0) \geq 0$ for every $n \in \mathbb{N}$ and $0 \leq m \leq n$.*

PROOF: The implication **(i)** $\Longrightarrow$ **(ii)** is trivial. To see that **(ii)** implies **(iii)**, first observe that if $\psi$ is absolutely monotone on $(a,b)$ and $h \in (0, b-a)$, then $\Delta_h\psi$ is absolutely monotone on $(a, b-h)$. Indeed, because $D \circ \Delta_h\psi = \Delta_h \circ D\psi$ on $(a, b-h)$, we have that

$$h\big[D^m \circ \Delta_h\psi\big](t) = \int_t^{t+h} D^{m+1}\psi(s)\,ds \geq 0, \quad t \in (a, b-h),$$

for any $m \in \mathbb{N}$. Returning to the function $\varphi$, we now know that $\Delta^m_h\varphi$ is absolutely monotone on $(0, 1-mh)$ for all $m \in \mathbb{N}$ and $h > 0$ with $mh < 1$. In particular,

$$[\Delta^m_h\varphi](0) = \lim_{t \searrow 0}[\Delta^m_h\varphi](t) \geq 0 \quad \text{if} \quad mh < 1,$$

and so $\big[\Delta^m_h\varphi\big](0) \geq 0$ when $h = \frac{1}{n}$ and $0 \leq m < n$. Moreover, since

$$[\Delta^n_{\frac{1}{n}}\varphi](0) = \lim_{h \nearrow \frac{1}{n}} [\Delta^n_h\varphi](0),$$

we also know that $\big[\Delta^n_h\varphi\big](0) \geq 0$ when $h = \frac{1}{n}$, and this completes the proof that **(ii)** implies **(iii)**.

Finally, assume that **(iii)** holds and set $\varphi_n = B_n(\,\cdot\,;\varphi)$. Then, from (1.2.9) and the equality $\varphi_n(1) = \varphi(1) = 1$, we see that each $\varphi_n$ is a probability generating function. Thus, in order to complete the proof that **(iii)** implies **(i)**, all that one has to do is check that a uniform limit of probability generating functions is itself a probability generating function. To this end, write

$$\varphi_n(t) = \sum_{\ell=0}^{\infty} u_{n,\ell} t^\ell, \quad t \in [0,1] \text{ for each } n \in \mathbb{Z}^+.$$

Because the $u_{n,\ell}$'s are all elements of $[0,1]$, one can use a diagonalization procedure to choose $\{n_k : k \in \mathbb{Z}^+\}$ so that

$$\lim_{k\to\infty} u_{n_k,\ell} = u_\ell \in [0,1]$$

exists for each $\ell \in \mathbb{N}$. But, by Lebesgue's Dominated Convergence Theorem, this means that

$$\varphi(t) = \lim_{k\to\infty} \varphi_{n_k}(t) = \sum_{\ell=0}^{\infty} u_\ell t^\ell \quad \text{for every} \quad t \in [0,1).$$

Finally, by the Monotone Convergence Theorem, the preceding extends immediately to $t = 1$, and so $\varphi$ is a probability generating function. (Notice that the argument just given does not even use the assumed uniform convergence and shows that the pointwise limit of probability generating functions is again a probability generating function.) □

The preceding is only one of many examples in which The Weak Law leads to useful ways of forming an approximate identity. A second example is given in Exercises 1.2.12 and 1.2.13. My treatment of these is based on that of Wm. Feller.[2]

### Exercises for § 1.2

EXERCISE 1.2.11. Although, for historical reasons, The Weak Law is usually thought of as a theorem about convergence in $\mathbb{P}$-probability, the forms in which I have presented it are clearly results about convergence in either $\mathbb{P}$-mean or even $\mathbb{P}$-square mean. Thus, it is interesting to discover that one can replace the uniform integrability assumption made in Theorem 1.2.6 with a *weak uniform integrability* assumption if one is willing to settle for convergence in $\mathbb{P}$-probability. Namely, let $X_1, \dots, X_n, \dots$ be mutually $\mathbb{P}$-independent random variables, assume that

$$F(R) \equiv \sup_{n\in\mathbb{Z}^+} R\mathbb{P}\Big(|X_n| \ge R\Big) \longrightarrow 0 \quad \text{as } R \nearrow \infty,$$

[2] Wm. Feller, *An Introduction to Probability Theory and Its Applications, Vol. II*, Wiley, Series in Probability and Math. Stat. (1968). Feller provides several other similar applications of The Weak Law, including the ones in the following exercises.

and set

$$m_n = \frac{1}{n}\sum_{\ell=1}^{n} \mathbb{E}^{\mathbb{P}}\Big[X_\ell,\ |X_\ell| \le n\Big], \quad n \in \mathbb{Z}^+.$$

Show that, for each $\epsilon > 0$,

$$\begin{aligned}\mathbb{P}\Big(\big|\overline{S}_n - m_n\big| \ge \epsilon\Big) &\le \frac{1}{(n\epsilon)^2}\sum_{\ell=1}^{n} \mathbb{E}^{\mathbb{P}}\Big[X_\ell^2,\ |X_\ell| \le n\Big] + \mathbb{P}\Big(\max_{1\le\ell\le n} |X_\ell| > n\Big)\\ &\le \frac{2}{n\epsilon^2}\int_0^n F(t)\,dt + F(n),\end{aligned}$$

and conclude that $\big|\overline{S}_n - m_n\big| \longrightarrow 0$ in $\mathbb{P}$-probability. (See part **(ii)** of Exercises 1.4.26 and 1.4.27 for a partial converse to this statement.)

**Hint**: Use the formula

$$\operatorname{Var}(Y) \le \mathbb{E}^{\mathbb{P}}\big[Y^2\big] = 2\int_{[0,\infty)} t\,\mathbb{P}(|Y| > t)\,dt.$$

EXERCISE 1.2.12. Show that, for each $T \in [0,\infty)$ and $t \in (0,\infty)$,

$$\lim_{n\to\infty} e^{-nt}\sum_{0\le k\le nT} \frac{(nt)^k}{k!} = \begin{cases} 1 & \text{if } T > t \\ 0 & \text{if } T < t.\end{cases}$$

**Hint**: Let $X_1, \dots, X_n, \dots$ be $\mathbb{P}$-independent, $\mathbb{N}$-valued **Poisson random variables** with mean value $t$. That is, the $X_n$'s are $\mathbb{P}$-independent and

$$\mathbb{P}\big(X_n = k\big) = e^{-t}\frac{t^k}{k!} \quad \text{for } k \in \mathbb{N}.$$

Show that $S_n$ is an $\mathbb{N}$-valued Poisson random variable with mean value $nt$, and conclude that, for each $T \in [0,\infty)$ and $t \in (0,\infty)$,

$$e^{-nt}\sum_{0\le k\le nT} \frac{(nt)^k}{k!} = P\big(\overline{S}_n \le T\big).$$

EXERCISE 1.2.13. Given a right-continuous function $F : [0,\infty) \longrightarrow \mathbb{R}$ of bounded variation with $F(0) = 0$, define its **Laplace transform** $\varphi(\lambda)$, $\lambda \in [0,\infty)$, by the Riemann–Stieltjes integral:

$$\varphi(\lambda) = \int_{[0,\infty)} e^{-\lambda t}\,dF(t).$$

Using Exercise 1.2.12, show that

$$\sum_{k\le nT} \frac{(-n)^k}{k!}\big[D^k\varphi\big](n) \longrightarrow F(T) \quad \text{as } n \to \infty$$

for each $T \in [0,\infty)$ at which $F$ is continuous. Conclude, in particular, that $F$ can be recovered from its Laplace transform. Although this is not the most practical recovery method, it is distinguished by the fact that it does not involve complex analysis.

## § 1.3 Cramér's Theory of Large Deviations

From Theorem 1.2.4, we know that if $\{X_n : n \in \mathbb{Z}^+\}$ is a sequence of $\mathbb{P}$-independent, $\mathbb{P}$-square integrable random variables with mean value 0, and if the averages $\overline{S}_n$, $n \in \mathbb{Z}^+$, are defined accordingly, then, for every $\epsilon > 0$,

$$\mathbb{P}\left(|\overline{S}_n| \geq \epsilon\right) \leq \frac{\max_{1\leq m\leq n} \mathrm{Var}(X_m)}{n\epsilon^2}, \qquad n \in \mathbb{Z}^+.$$

Thus, so long as

$$\frac{\mathrm{Var}(X_n)}{n} \longrightarrow 0 \text{ as } n \to \infty,$$

the $\overline{S}_n$'s are becoming more and more concentrated near 0, and the rate at which this concentration is occurring can be estimated in terms of the variances $\mathrm{Var}(X_n)$. In this section, we will see that, by placing more stringent integrability requirements on the $X_n$'s, one can gain more information about the rate at which the $\overline{S}_n$'s are concentrating at 0.

In all of this analysis, the trick is to see how independence can be combined with 0 mean value to produce unexpected cancellations; and, as a preliminary warm-up exercise, I begin with the following.

Theorem 1.3.1. *Let $\{X_n : n \in \mathbb{Z}^+\}$ be a sequence of $\mathbb{P}$-independent, $\mathbb{P}$-integrable random variables with mean value 0, and assume that*

$$M_4 \equiv \sup_{n\in\mathbb{Z}^+} \mathbb{E}^{\mathbb{P}}\left[X_n^4\right] < \infty.$$

*Then, for each $\epsilon > 0$,*

$$\epsilon^4 \mathbb{P}\left(|\overline{S}_n| \geq \epsilon\right) \leq \mathbb{E}^{\mathbb{P}}\left[\overline{S}_n^{\,4}\right] \leq \frac{3M_4}{n^2}, \quad n \in \mathbb{Z}^+. \tag{1.3.2}$$

*In particular, $\overline{S}_n \longrightarrow 0$ $\mathbb{P}$-almost surely.*

Proof: Obviously, in order to prove (1.3.2), it suffices to check the second inequality, which is equivalent to $\mathbb{E}^{\mathbb{P}}\left[S_n^4\right] \leq 3M_4 n^2$. But

$$\mathbb{E}^{\mathbb{P}}\left[S_n^4\right] = \sum_{m_1,\dots,m_4=1}^{n} \mathbb{E}^{\mathbb{P}}\left[X_{m_1}\cdots X_{m_4}\right],$$

and, by Schwarz's Inequality, each of these terms is dominated by $M_4$. In addition, of these terms, the only ones that do not vanish have either all their factors the same or two pairs of equal factors. Thus, the number of non-vanishing terms is $n + 3n(n-1) = 3n^2 - 2n$.

Given (1.3.2), the proof of the last part becomes an easy application of the Borel–Cantelli Lemma. Indeed, for any $\epsilon > 0$, we know from (1.3.2) that

$$\sum_{n=1}^{\infty} \mathbb{P}\left(|\overline{S}_n| \geq \epsilon\right) < \infty,$$

and therefore, by (1.1.4), that $\mathbb{P}\left(\overline{\lim}_{n\to\infty} |\overline{S}_n| \geq \epsilon\right) = 0$. □

REMARK 1.3.3. The final assertion in Theorem 1.3.1 is a primitive version of The Strong Law of Large Numbers. Although The Strong Law will be taken up again, and considerably refined, in Section 1.4, the principle on which its proof here was based is an important one: namely, *control more moments and you will get better estimates; get better estimates and you will reach more refined conclusions.*

With the preceding adage in mind, I will devote the rest of this section to examining what one can say when one has all moments at one's disposal. In fact, from now on, I will be assuming that $X_1, \dots, X_n, \dots$ are independent random variables with common distribution $\mu$ having the property that the **moment generating function**

$$M_\mu(\xi) \equiv \int_{\mathbb{R}} e^{\xi\, x}\, \mu(dx) < \infty \quad \text{for all } \xi \in \mathbb{R}. \tag{1.3.4}$$

Obviously, (1.3.4) is more than sufficient to guarantee that the $X_n$'s have moments of all orders. In fact, as an application of Lebesgue's Dominated Convergence Theorem, one sees that $\xi \in \mathbb{R} \longmapsto M(\xi) \in (0,\infty)$ is infinitely differentiable and that

$$\mathbb{E}^{\mathbb{P}}\left[X_1^n\right] = \int_{\mathbb{R}} x^n\, \mu(dx) = \frac{d^n M}{d\xi^n}(0) \quad \text{for all } n \in \mathbb{N}.$$

In the discussion that follows, I will use $m$ and $\sigma^2$ to denote, respectively, the common mean value and variance of the $X_n$'s.

In order to develop some intuition for the considerations that follow, I will first consider an example, which, for many purposes, is *the canonical example* in probability theory. Namely, let $g : \mathbb{R} \longrightarrow (0,\infty)$ be the **Gauss kernel**

$$g(y) \equiv \frac{1}{\sqrt{2\pi}} \exp\left[-\frac{|y|^2}{2}\right], \quad y \in \mathbb{R}, \tag{1.3.5}$$

and recall that a random variable $X$ is **standard normal** if

$$\mathbb{P}(X \in \Gamma) = \int_\Gamma g(y)\, dy, \quad \Gamma \in \mathcal{B}_{\mathbb{R}}.$$

In spite of their somewhat insultingly bland moniker, standard normal random variables are the building blocks for the most honored family in all of probability theory. Indeed, given $m \in \mathbb{R}$ and $\sigma \in [0,\infty)$, the random variable $Y$ is said to be **normal** (or **Gaussian**) **with mean value** $m$ **and variance** $\sigma^2$ (often this is abbreviated by saying that $X$ is an $N(m,\sigma^2)$**-random variable** and written as $X \in N(m,\sigma^2)$) if and only if the distribution of $Y$ is $\gamma_{m,\sigma^2}$, where $\gamma_{m,\sigma^2}$ is the distribution of the variable $\sigma X + m$ when $X$ is standard normal. That is, $Y$ is an $N(m,\sigma^2)$ random variable if, when $\sigma = 0$, $\mathbb{P}(Y = m) = 1$ and, when $\sigma > 0$,

$$\mathbb{P}(Y \in \Gamma) = \int_\Gamma \frac{1}{\sigma} g\left(\frac{y-m}{\sigma}\right) dy \quad \text{for } \Gamma \in \mathcal{B}_{\mathbb{R}}.$$

There are two obvious reasons for the honored position held by Gaussian random variables. In the first place, they certainly have finite moment generating functions. In fact, since

$$\int_{\mathbb{R}} e^{\xi y}\, g(y)\, dy = \exp\left(\frac{\xi^2}{2}\right), \quad \xi \in \mathbb{R},$$

it is clear that

$$M_{\gamma_{m,\sigma^2}}(\xi) = \exp\left[\xi m + \frac{\sigma^2\xi^2}{2}\right]. \tag{1.3.6}$$

Secondly, *they add nicely*. To be precise, it is a familiar fact from elementary probability theory that *if $X$ is an $N(m,\sigma^2)$-random variable and $\hat{X}$ is an $N(\hat{m},\hat{\sigma}^2)$-random variable that is independent of $X$, then $X+\hat{X}$ is an $N\big(m+\hat{m},\sigma^2+\hat{\sigma}^2\big)$-random variable.* In particular, if $X_1,\dots,X_n$ are mutually independent, standard normal random variables, then $\overline{S}_n$ is an $N\left(0,\frac{1}{n}\right)$-random variable. That is,

$$\mathbb{P}\big(\overline{S}_n \in \Gamma\big) = \sqrt{\frac{n}{2\pi}} \int_\Gamma \exp\left[-\frac{n|y|^2}{2}\right] dy.$$

Thus (cf. Exercise 1.3.16), for any $\Gamma$ we see that

$$\lim_{n\to\infty} \frac{1}{n} \log\Big[P\big(\overline{S}_n \in \Gamma\big)\Big] = -\text{ess inf}\left\{\frac{|y|^2}{2} : y \in \Gamma\right\}, \tag{1.3.7}$$

where the "ess" in (1.3.7) stands for *essential* and means that what follows is taken *modulo a set of measure* 0. (Hence, apart from a minus sign, the right-hand side of (1.3.7) is the greatest number dominated by $\frac{|y|^2}{2}$ for Lebesgue-almost every $y \in \Gamma$.) In fact, because

$$\int_x^\infty g(y)\, dy \le x^{-1} g(x) \quad \text{for all } x \in (0,\infty),$$

we have the rather precise upper bound

$$\mathbb{P}\big(|\overline{S}_n| \ge \epsilon\big) \le \sqrt{\frac{2}{n\pi\epsilon^2}} \exp\left[-\frac{n\epsilon^2}{2}\right] \quad \text{for } \epsilon > 0.$$

At the same time, it is clear that, for $0 < \epsilon < |a|$,

$$P\big(|\overline{S}_n - a| < \epsilon\big) \ge \sqrt{\frac{2\epsilon^2 n}{\pi}} \exp\left[-\frac{n(|a|+\epsilon)^2}{2}\right].$$

More generally, if the $X_n$'s are mutually independent $N(m,\sigma^2)$-random variables, then one finds that

$$\mathbb{P}\big(|\overline{S}_n - m| \geq \sigma\epsilon\big) \leq \sqrt{\frac{2}{n\pi\epsilon^2}}\,\exp\left[-\frac{n\epsilon^2}{2}\right] \quad \text{for } \epsilon > 0;$$

and, for $0 < \epsilon < |a|$ and sufficiently large $n$'s,

$$P\big(|\overline{S}_n - (m+a)| < \sigma\epsilon\big) \geq \sqrt{\frac{2\epsilon^2 n}{\pi}}\,\exp\left[-\frac{n(|a|+\epsilon)^2}{2}\right].$$

Of course, in general one cannot hope to know such explicit expressions for the distribution of $\overline{S}_n$. Nonetheless, on the basis of the preceding, one can start to see what is going on. Namely, when the distribution $\mu$ falls off rapidly outside of compacts, averaging $n$ independent random variables with distribution $\mu$ has the effect of *building an exponentially deep well in which the mean value m lies at the bottom.* More precisely, if one believes that the Gaussian random variables *are normal* in the sense that they are typical, then one should conjecture that, even when the random variables are not normal, the behavior of $\mathbb{P}\big(|\overline{S}_n - m| \geq \epsilon\big)$ for large $n$'s should resemble that of Gaussians with the same variance; and it is in the verification of this conjecture that the moment generating function $M_\mu$ plays a central role. Namely, although an expression in terms of $\mu$ for the distribution of $S_n$ is seldom readily available, the moment generating function for $S_n$ is easily expressed in terms of $M_\mu$. To wit, as a trivial application of independence, we have

$$\mathbb{E}^{\mathbb{P}}\left[e^{\xi S_n}\right] = M_\mu(\xi)^n, \quad \xi \in \mathbb{R}.$$

Hence, by Markov's Inequality applied to $e^{\xi S_n}$, we see that, for any $a \in \mathbb{R}$,

$$\mathbb{P}\big(\overline{S}_n \geq a\big) \leq e^{-n\xi a}\,M_\mu(\xi)^n = \exp\big[-n\big(\xi a - \Lambda_\mu(\xi)\big)\big], \quad \xi \in [0,\infty),$$

where

(1.3.8) $$\Lambda_\mu(\xi) \equiv \log\big(M_\mu(\xi)\big)$$

is the **logarithmic moment generating function of** $\mu$. The preceding relation is one of those lovely situations in which a single quantity is dominated by a whole family of quantities, which means that one should optimize by minimizing over the dominating quantities. Thus, we now have

(1.3.9) $$\mathbb{P}\big(\overline{S}_n \geq a\big) \leq \exp\left[-n \sup_{\xi\in[0,\infty)} \big(\xi a - \Lambda_\mu(\xi)\big)\right].$$

Notice that (1.3.9) is really very good. For instance, when the $X_n$'s are $N(m,\sigma^2)$-random variables and $\sigma > 0$, then (cf. (1.3.6)) the preceding leads quickly to the estimate

$$\mathbb{P}\big(\overline{S}_n - m \geq \epsilon\big) \leq \exp\left(-\frac{n\epsilon^2}{2\sigma^2}\right),$$

which is essentially the upper bound at which we arrived before.

Taking a hint from the preceding, I now introduce the **Legendre transform**

$$I_\mu(x) \equiv \sup\{\xi x - \Lambda_\mu(\xi) : \xi \in \mathbb{R}\}, \qquad x \in \mathbb{R}, \tag{1.3.10}$$

of $\Lambda_\mu$ and, before proceeding further, make some elementary observations about the structure of the functions $\Lambda_\mu$ and $I_\mu$.

Lemma 1.3.11. *The function $\Lambda_\mu$ is infinitely differentiable. In addition, for each $\xi \in \mathbb{R}$, the probability measure $\nu_\xi$ on $\mathbb{R}$ given by*

$$\nu_\xi(\Gamma) = \frac{1}{M_\mu(\xi)} \int_\Gamma e^{\xi x}\,\mu(dx) \quad \text{for } \Gamma \in \mathcal{B}_{\mathbb{R}}$$

*has moments of all orders,*

$$\int_{\mathbb{R}} x\,\nu_\xi(dx) = \Lambda'_\mu(\xi), \quad \text{and} \quad \int_{\mathbb{R}} x^2\,\nu_\xi(dx) - \left(\int_{\mathbb{R}} x\,\nu_\xi(dx)\right)^2 = \Lambda''_\mu(\xi).$$

*Next, the function $I_\mu$ is a $[0,\infty]$-valued, lower semicontinuous, convex function that vanishes at $m$. Moreover,*

$$I_\mu(x) = \sup\{\xi x - \Lambda_\mu(\xi) : \xi \geq 0\} \quad \text{for } x \in [m,\infty)$$

*and*

$$I_\mu(x) = \sup\{\xi x - \Lambda_\mu(\xi) : \xi \leq 0\} \quad \text{for } x \in (-\infty,m].$$

*Finally, if*

$$\alpha = \inf\{x \in \mathbb{R} : \mu\big((-\infty,x]\big) > 0\} \text{ and } \beta = \sup\{x \in \mathbb{R} : \mu\big([x,\infty)\big) > 0\},$$

*then $I_\mu$ is smooth on $(\alpha,\beta)$ and identically $+\infty$ off of $[\alpha,\beta]$. In fact, either $\mu(\{m\}) = 1$ and $\alpha = m = \beta$ or $m \in (\alpha,\beta)$, in which case $\Lambda'_\mu$ is a smooth, strictly increasing mapping from $\mathbb{R}$ onto $(\alpha,\beta)$,*

$$I_\mu(x) = \Xi_\mu(x)\,x - \Lambda_\mu\big(\Xi_\mu(x)\big),\ x \in (\alpha,\beta), \quad \text{where} \quad \Xi_\mu = \big(\Lambda'_\mu\big)^{-1}$$

*is the inverse of $\Lambda'_\mu$, $\mu(\{\alpha\}) = e^{-I_\mu(\alpha)}$ if $\alpha > -\infty$, and $\mu(\{\beta\}) = e^{-I_\mu(\beta)}$ if $\beta < \infty$.*

PROOF: For notational convenience, I will drop the subscript "$\mu$" during the proof. Further, note that the smoothness of $\Lambda$ follows immediately from the positivity and smoothness of $M$, and the identification of $\Lambda'(\xi)$ and $\Lambda''(\xi)$ with the mean and variance of $\nu_\xi$ is elementary calculus combined with the remark following (1.3.4). Thus, I will concentrate on the properties of the function $I$.

As the pointwise supremum of functions that are linear, $I$ is certainly lower semicontinuous and convex. Also, because $\Lambda(0) = 0$, it is obvious that $I \geq 0$. Next, by Jensen's Inequality,

$$\Lambda(\xi) \geq \xi \int_{\mathbb{R}} x\,\mu(dx) = \xi\, m,$$

and, therefore, $\xi x - \Lambda(\xi) \leq 0$ if $x \leq m$ and $\xi \geq 0$ or if $x \geq m$ and $\xi \leq 0$. Hence, because $I$ is non-negative, this proves the one-sided extremal characterizations of $I_\mu(x)$ depending on whether $x \geq m$ or $x \leq m$.

Turning to the final part, note first that there is nothing more to do in the case when $\mu(\{m\}) = 1$. Thus, assume that $\mu(\{m\}) < 1$, in which case it is clear that $m \in (\alpha, \beta)$ and that none of the measures $\nu_\xi$ is degenerate (i.e., concentrate at one point). In particular, because $\Lambda''(\xi)$ is the variance of the $\nu_\xi$, we know that $\Lambda'' > 0$ everywhere. Hence, $\Lambda'$ is strictly increasing and therefore admits a smooth inverse $\Xi$ on its image. Furthermore, because $\Lambda'(\xi)$ is the mean of $\nu_\xi$, it is clear that the image of $\Lambda'$ is contained in $(\alpha, \beta)$. At the same time, given an $x \in (\alpha, \beta)$, note that

$$e^{-\xi x} \int_{\mathbb{R}} e^{\xi y}\,\mu(dy) \longrightarrow \infty \quad \text{as } |\xi| \to \infty,$$

and therefore $\xi \rightsquigarrow \xi x - \Lambda(\xi)$ achieves a maximum at some point $\xi_x \in \mathbb{R}$. In addition, by the first derivative test, $\Lambda'(\xi_x) = x$, and so $\xi_x = \Xi^{-1}(x)$. Finally, suppose that $\beta < \infty$. Then

$$e^{-\xi\beta} \int_{\mathbb{R}} e^{\xi y}\,\mu(dy) = \int_{(-\infty,\beta]} e^{-\xi(\beta - y)}\,\mu(dy) \searrow \mu(\{\beta\}) \quad \text{as } \xi \to \infty,$$

and therefore $e^{-I(\beta)} = \inf_{\xi \geq 0} e^{-\xi\beta} M(\xi) = \mu(\{\beta\})$. Since the same reasoning applies when $\alpha > -\infty$, we are done. □

THEOREM 1.3.12 (**Cramér's Theorem**). *Let $\{X_n : n \geq 1\}$ be a sequence of $\mathbb{P}$-independent random variables with common distribution $\mu$, assume that the associated moment generating function $M_\mu$ satisfies (1.3.4), set $m = \int_{\mathbb{R}} x\,\mu(dx)$, and define $I_\mu$ accordingly, as in (1.3.10). Then,*

$$\mathbb{P}\big(\overline{S}_n \geq a\big) \leq e^{-nI_\mu(a)} \quad \text{for all } a \in [m, \infty),$$
$$\mathbb{P}\big(\overline{S}_n \leq a\big) \leq e^{-nI_\mu(a)} \quad \text{for all } a \in (-\infty, m].$$

*Moreover, for $a \in (\alpha, \beta)$ (cf. Lemma 1.3.11), $\epsilon > 0$, and $n \in \mathbb{Z}^+$,*

$$\mathbb{P}\Big(\big|\overline{S}_n - a\big| < \epsilon\Big) \geq \left(1 - \frac{\Lambda''_\mu(\Xi_\mu(a))}{n\epsilon^2}\right) \exp\Big[-n\Big(I_\mu(a) + \epsilon|\Xi_\mu(a)|\Big)\Big],$$

*where $\Lambda_\mu$ is the function given in (1.3.8) and $\Xi_\mu \equiv (\Lambda_\mu{}')^{-1}$.*

PROOF: To prove the first part, suppose that $a \in [m, \infty)$, and apply the second part of Lemma 1.3.11 to see that the exponent in (1.3.9) equals $-nI_\mu(a)$, and, after replacing $\{X_n : n \geq 1\}$ by $\{-X_n : n \geq 1\}$, one also gets the desired estimate when $a \leq m$.

To prove the lower bound, let $a \in [m, \beta)$ be given, and set $\xi = \Xi_\mu(a) \in [0, \infty)$. Next, recall the probability measure $\nu_\xi$ described in Lemma 1.3.11, and remember that $\nu_\xi$ has mean value $a = \Lambda'_\mu(\xi)$ and variance $\Lambda''_\mu(\xi)$. Further, if $\{Y_n : n \in \mathbb{Z}^+\}$ is a sequence of independent, identically distributed random variables with common distribution $\nu_\xi$, then it is an easy matter to check that, for any $n \in \mathbb{Z}^+$ and every $\mathcal{B}_{\mathbb{R}^n}$-measurable $F : \mathbb{R}^n \longrightarrow [0, \infty)$,

$$\mathbb{E}^{\mathbb{P}}\Big[F(Y_1, \dots, Y_n)\Big] = \frac{1}{M_\mu(\xi)^n}\mathbb{E}^{\mathbb{P}}\Big[e^{\xi S_n} F(X_1, \dots, X_n)\Big].$$

In particular, if

$$T_n = \sum_{\ell=1}^n Y_\ell \quad \text{and} \quad \overline{T}_n = \frac{T_n}{n},$$

then, because $I_\mu(a) = \xi a - \Lambda_\mu(\xi)$,

$$\begin{aligned}\mathbb{P}\Big(\big|\overline{S}_n - a\big| < \epsilon\Big) &= M(\xi)^n \mathbb{E}^{\mathbb{P}}\Big[e^{-\xi T_n}, \ \big|\overline{T}_n - a\big| < \epsilon\Big] \\ &\geq e^{-n\xi(a+\epsilon)} M(\xi)^n \, \mathbb{P}\Big(\big|\overline{T}_n - a\big| < \epsilon\Big) \\ &= \exp\Big[-n\Big(I_\mu(a) + \xi\epsilon\Big)\Big]\mathbb{P}\Big(\big|\overline{T}_n - a\big| < \epsilon\Big).\end{aligned}$$

But, because the mean value and variance of the $Y_n$'s are, respectively, $a$ and $\Lambda''_\mu(\xi)$, (1.2.5) leads to

$$\mathbb{P}\Big(\big|\overline{T}_n - a\big| \geq \epsilon\Big) \leq \frac{\Lambda''_\mu(\xi)}{n\epsilon^2}.$$

The case when $a \in (\alpha, m]$ is handled in the same way. □

Results like the ones obtained in Theorem 1.3.12 are examples of a class of results known as **large deviations estimates.** They are *large deviations* because the probability of their occurrence is exponentially small. Although large deviation estimates are available in a variety of circumstances,[1] in general one has to settle for the cruder sort of information contained in the following.

[1] In fact, some people have written entire books on the subject. See, for example, J.-D. Deuschel and D. Stroock, *Large Deviations*, now available from the A.M.S. in the Chelsea Series.

COROLLARY 1.3.13. *For any* $\Gamma \in \mathcal{B}_{\mathbb{R}}$,

$$-\inf_{x\in\Gamma^\circ} I_\mu(x) \le \varliminf_{n\to\infty} \frac{1}{n}\log\Big[P\big(\overline{S}_n \in \Gamma\big)\Big]$$
$$\le \varlimsup_{n\to\infty} \frac{1}{n}\log\Big[P\big(\overline{S}_n \in \Gamma\big)\Big] \le -\inf_{x\in\overline{\Gamma}} I_\mu(x).$$

*(I use* $\Gamma^\circ$ *and* $\overline{\Gamma}$ *to denote the interior and closure of a set* $\Gamma$*. Also, recall that I take the infemum over the empty set to be* $+\infty$*.)*

PROOF: To prove the upper bound, let $\Gamma$ be a closed set, and define $\Gamma_+ = \Gamma \cap [m,\infty)$ and $\Gamma_- = \Gamma \cap (-\infty, m]$. Clearly,

$$\mathbb{P}\big(\overline{S}_n \in \Gamma\big) \le 2\mathbb{P}\big(\overline{S}_n \in \Gamma_+\big) \vee \mathbb{P}\big(\overline{S}_n \in \Gamma_-\big).$$

Moreover, if $\Gamma_+ \ne \emptyset$ and $a_+ = \min\{x : x \in \Gamma_+\}$, then, by Lemma 1.3.11 and Theorem 1.3.12,

$$I_\mu(a_+) = \inf\big\{I_\mu(x) : x \in \Gamma_+\big\} \quad \text{and} \quad P\big(\overline{S}_n \in \Gamma_+\big) \le e^{-nI_\mu(a_+)}.$$

Similarly, if $\Gamma_- \ne \emptyset$ and $a_- = \max\{x : x \in \Gamma_-\}$, then

$$I_\mu(a_-) = \inf\big\{I_\mu(x) : x \in \Gamma_-\big\} \quad \text{and} \quad P\big(\overline{S}_n \in \Gamma_-\big) \le e^{-nI_\mu(a_-)}.$$

Hence, either $\Gamma = \emptyset$, and there is nothing to do anyhow, or

$$\mathbb{P}\big(\overline{S}_n \in \Gamma\big) \le 2\exp\big[-n\inf\big\{I_\mu(x) : x \in \Gamma\big\}\big], \quad n \in \mathbb{Z}^+,$$

which certainly implies the asserted upper bound.

To prove the lower bound, assume that $\Gamma$ is a non-empty open set. What I have to show is that

$$\varliminf_{n\to\infty} \frac{1}{n}\log\Big[P\big(\overline{S}_n \in \Gamma\big)\Big] \ge -I_\mu(a)$$

for every $a \in \Gamma$. If $a \in \Gamma \cap (\alpha,\beta)$, choose $\delta > 0$ so that $(a-\delta, a+\delta) \subseteq \Gamma$ and use the second part of Theorem 1.3.12 to see that

$$\varliminf_{n\to\infty} \frac{1}{n}\log\Big[P\big(\overline{S}_n \in \Gamma\big)\Big] \ge -I_\mu(a) - \epsilon\big|\Xi_\mu(a)\big|$$

for every $\epsilon \in (0,\delta)$. If $a \notin [\alpha,\beta]$, then $I_\mu(a) = \infty$, and so there is nothing to do. Finally, if $a \in \{\alpha,\beta\}$, then $\mu(\{a\}) = e^{-I_\mu(a)}$ and therefore

$$\mathbb{P}\big(\overline{S}_n \in \Gamma\big) \ge P\big(\overline{S}_n = a\big) \ge e^{-nI_\mu(a)}. \quad \square$$

REMARK 1.3.14. The upper bound in Theorem 1.3.12 is often called **Chernoff's Inequality**. The idea underlying its derivation is rather mundane by comparison to the subtle idea underlying the proof of the lower bound. Indeed, it may not be immediately obvious what that idea was! Thus, consider once again the second part of the proof of Theorem 1.3.12. What I had to do is estimate the probability that $\overline{S}_n$ lies in a neighborhood of $a$. When $a$ is the mean value $m$, such an estimate is provided by the Weak Law. On the other hand, when $a \neq m$, the Weak Law for the $X_n$'s has very little to contribute. Thus, what I did is replace the original $X_n$'s by random variables $Y_n$, $n \in \mathbb{Z}^+$, whose mean value is $a$. Furthermore, the transformation from the $X_n$'s to the $Y_n$'s was sufficiently simple that it was easy to estimate $X_n$-probabilities in terms of $Y_n$-probabilities. Finally, the Weak Law applied to the $Y_n$'s gave strong information about the rate of approach of $\frac{1}{n}\sum_{\ell=1}^n Y_\ell$ to $a$.

I close this section by verifying the conjecture (cf. the discussion preceding Lemma 1.3.11) that the Gaussian case is *normal*. In particular, I want to check that the *well around* $m$ in which the distribution of $\overline{S}_n$ becomes concentrated looks Gaussian, and, in view of Theorem 1.3.12, this comes down to the following.

THEOREM 1.3.15. *Let everything be as in Lemma 1.3.11, and assume that the variance $\sigma^2 > 0$. There exists a $\delta \in (0,1]$ and a $K \in (0,\infty)$ such that $[m-\delta, m+\delta] \subseteq (\alpha,\beta)$ (cf. Lemma 1.3.11), $\left|\Lambda''_\mu(\Xi(x))\right| \le K$,*

$$\left|\Xi_\mu(x)\right| \le K|x-m|, \quad \text{and} \quad \left|I_\mu(x) - \frac{(x-m)^2}{2\sigma^2}\right| \le K|x-m|^3$$

*for all $x \in [m-\delta, m+\delta]$. In particular, if $0 < \epsilon < \delta$, then*

$$\mathbb{P}\left(|\overline{S}_n - m| \ge \epsilon\right) \le 2\exp\left[-n\left(\frac{\epsilon^2}{2\sigma^2} - K\epsilon^3\right)\right],$$

*and if $|a-m| < \delta$ and $\epsilon > 0$, then*

$$\mathbb{P}\left(|\overline{S}_n - a| < \epsilon\right) \ge \left(1 - \frac{K}{n\epsilon^2}\right)\exp\left[-n\left(\frac{|a-m|^2}{2\sigma^2} + K|a-m|\left(\epsilon + |a-m|^2\right)\right)\right].$$

PROOF: Without loss in generality (cf. Exercise 1.3.17), I will assume that $m = 0$ and $\sigma^2 = 1$. Since, in this case, $\Lambda_\mu(0) = \Lambda'_\mu(0) = 0$ and $\Lambda''_\mu(0) = 1$, it follows that $\Xi_\mu(0) = 0$ and $\Xi'_\mu(0) = 1$. Hence, we can find an $M \in (0,\infty)$ and a $\delta \in (0,1]$ with $\alpha < -\delta < \delta < \beta$ for which $\left|\Xi_\mu(x) - x\right| \le M|x|^2$ and $\left|\Lambda_\mu(\xi) - \frac{\xi^2}{2}\right| \le M|\xi|^3$ whenever $|x| \le \delta$ and $|\xi| \le (M+1)\delta$, respectively. In particular, this leads immediately to $\left|\Xi_\mu(x)\right| \le (M+1)|x|$ for $|x| \le \delta$, and the estimate for $I_\mu$ comes easily from the preceding combined with equation $I_\mu(x) = \Xi(x)x - \Lambda_\mu(\Xi_\mu(x))$. □

## Exercises for § 1.3

EXERCISE 1.3.16. Let $(E, \mathcal{F}, \mu)$ be a measure space and $f$ a non-negative, $\mathcal{F}$-measurable function. If either $\mu(E) < \infty$ or $f$ is $\mu$-integrable, show that

$$\|f\|_{L^p(\mu;\mathbb{R})} \longrightarrow \|f\|_{L^\infty(\mu;\mathbb{R})} \quad \text{as } p \to \infty.$$

**Hint**: Handle the case $\mu(E) < \infty$ first, and treat the case when $f \in L^1(\mu;\mathbb{R})$ by considering the measure $\nu(dx) = f(x)\,\mu(dx)$.

EXERCISE 1.3.17. Referring to the notation used in this section, assume that $\mu$ is a non-degenerate (i.e., it is not concentrated at a single point) probability measure on $\mathbb{R}$ for which (1.3.4) holds. Next, let $m$ and $\sigma^2$ be the mean and variance of $\mu$, use $\nu$ to denote the distribution of

$$x \in \mathbb{R} \longmapsto \frac{x-m}{\sigma} \in \mathbb{R} \quad \text{under } \mu,$$

and define $\Lambda_\nu$, $I_\nu$, and $\Xi_\nu$ accordingly. Show that

$$\begin{aligned} \Lambda_\mu(\xi) &= \xi m + \Lambda_\nu(\sigma\xi), \qquad \xi \in \mathbb{R}, \\ I_\mu(x) &= I_\nu\left(\frac{x-m}{\sigma}\right), \qquad x \in \mathbb{R}, \\ \text{Image}(\Lambda'_\mu) &= m + \sigma\,\text{Image}(\Lambda'_\nu), \\ \Xi_\mu(x) &= \frac{1}{\sigma}\Xi_\nu\left(\frac{x-m}{\sigma}\right), \qquad x \in \text{Image}(\Lambda'_\mu). \end{aligned}$$

EXERCISE 1.3.18. Continue with the same notation as in the preceding.

**(i)** Show that $I_\nu \le I_\mu$ if $M_\mu \le M_\nu$.

**(ii)** Show that

$$I_\mu(x) = \frac{(x-m)^2}{2\sigma^2}, \qquad x \in \mathbb{R},$$

when $\mu$ is the $N(m,\sigma^2)$ distribution with $\sigma > 0$, and show that

$$I_\mu(x) = \frac{x-a}{b-a}\log\frac{x-a}{(1-p)(b-a)} + \frac{b-x}{b-a}\log\frac{b-x}{p(b-a)}, \qquad x \in (a,b),$$

when $a < b$, $p \in (0,1)$, and $\mu(\{a\}) = 1 - \mu(\{b\}) = p$.

**(iii)** When $\mu$ is the centered Bernoulli distribution given by $\mu(\{\pm 1\}) = \frac{1}{2}$, show that $M_\mu(\xi) \le \exp\left[\frac{\xi^2}{2}\right]$, $\xi \in \mathbb{R}$, and conclude that $I_\mu(x) \ge \frac{x^2}{2}$, $x \in \mathbb{R}$. More generally, given $n \in \mathbb{Z}^+$, $\{\sigma_k : 1 \le k \le n\} \subseteq \mathbb{R}$, and independent random variables $X_1, \dots, X_n$ with this $\mu$ as their common distribution, let $\nu$ denote the distribution of $S \equiv \sum_1^n \sigma_k X_k$ and show that $I_\nu(x) \geq \frac{x^2}{2\Sigma^2}$, where $\Sigma^2 \equiv \sum_1^n \sigma_k^2$. In particular, conclude that

$$\mathbb{P}(|S| \geq a) \leq 2\exp\left[-\frac{a^2}{2\Sigma^2}\right], \quad a \in [0,\infty).$$

EXERCISE 1.3.19. Although it is not exactly the direction in which I have been going, it seems appropriate to include here a derivation of **Stirling's formula**. Namely, recall **Euler's Gamma function**:

$$\Gamma(t) \equiv \int_{[0,\infty)} x^{t-1} e^{-x}\, dx, \qquad t \in (-1,\infty). \tag{1.3.20}$$

The goal of this exercise is to prove that

$$\Gamma(t+1) \sim \sqrt{2\pi t}\left(\frac{t}{e}\right)^t \quad \text{as } t \nearrow \infty, \tag{1.3.21}$$

where the *tilde* "$\sim$" means that the two sides are **asymptotic** to one another in the sense that their ratio tends to 1. (See Exercise 2.1.16 for another approach.)

The first step is to make the problem look like one to which Exercise 1.3.16 is applicable. Thus, make the substitution $x = ty$, and apply Exercise 1.3.16 to see that

$$\left(\frac{\Gamma(t+1)}{t^{t+1}}\right)^{\frac{1}{t}} = \left(\int_{[0,\infty)} y^t\, e^{-ty}\, dy\right)^{\frac{1}{t}} \longrightarrow e^{-1}.$$

This is, of course, far less than we want to know. Nonetheless, it does show that all the *action* is going to take place near $y = 1$ and that the principal factor in the asymptotics of $\frac{\Gamma(t+1)}{t^{t+1}}$ is $e^{-t}$. In order to highlight these observations, make the substitution $y = z + 1$ and obtain

$$\frac{\Gamma(t+1)}{t^{t+1}e^{-t}} = \int_{(-1,\infty)} (1+z)^t\, e^{-tz}\, dz.$$

Before taking the next step, introduce the function $R(z) = \log(1+z) - z + \frac{z^2}{2}$ for $z \in (-1,1)$, and check that $R(z) \leq 0$ if $z \in (-1,0]$ and that $|R(z)| \leq \frac{|z|^3}{3(1-|z|)}$ everywhere in $(-1,1)$. Now let $\delta \in (0,1)$ be given, and show that

$$\int_{-1}^{-\delta} (1+z)^t\, e^{-tz}\, dz \leq \exp\left[-\frac{t\delta^2}{2}\right]$$

and

$$\begin{aligned}\int_\delta^\infty (1+z)^t\, e^{-tz}\, dz &\leq \left[(1+\delta)e^{-\delta}\right]^{t-1} \int_\delta^\infty (1+z)e^{-z}\, dz \\ &\leq 2\exp\left[1 - \frac{t\delta^2}{2} + \frac{\delta^3}{3(1-\delta)}\right].\end{aligned}$$

Next, write $(1+z)^t e^{-tz} = e^{-\frac{tz^2}{2}} e^{tR(z)}$. Then

$$\int_{|z|\le\delta} (1+z)^t e^{-tz}\,dz = \int_{|z|\le\delta} e^{-\frac{tz^2}{2}}\,dz + E(t,\delta),$$

where

$$E(t,\delta) \equiv \int_{|z|\le\delta} e^{-\frac{tz^2}{2}}\left(e^{tR(z)}-1\right)dz = \int_0^\delta e^{-\frac{tz^2}{2}}\left(e^{tR(z)}+e^{-R(z)}-2\right)dz.$$

Check that

$$\left|\int_{|z|\le\delta} e^{-\frac{tz^2}{2}}\,dz - \sqrt{\frac{2\pi}{t}}\right| = t^{-\frac{1}{2}}\int_{|z|\ge t^{\frac{1}{2}}\delta} e^{-\frac{z^2}{2}}\,dz \le \frac{2}{t\delta}e^{-\frac{t\delta^2}{2}}.$$

At the same time, by Taylor's Theorem, $e^{tR(z)}+e^{tR(-z)}-2 = t\big(R(z)+R(-z)\big)+F(t,z)$, where $|F(t,z)| \le \frac{t^2}{2}\big(R(z)^2) + R(-z)^2\big)e^{tR(-|z|)}$. Since $|R(z)+R(-z)| = \sum_{m=2}^\infty \frac{z^{2m}}{m} \le \frac{z^4}{2(1-z^2)}$, one can dominate $|E(t,\delta)|$ uniformly for $0<\delta\le\frac{1}{2}$ by a constant times

$$t\int_0^\delta z^4 e^{-\frac{tz^2}{2}}\,dz + t^2\int_0^\delta z^6 e^{-\frac{tz^2}{2}}\,dz,$$

and so there is a $C<\infty$ such that $|E(t,\delta)| \le Ct^{-\frac{3}{2}}$ for all $t\ge 1$ and $0<\delta\le\frac{1}{2}$. Finally, take $\delta = \sqrt{3t^{-1}\log t}$ to arrive at

$$\left|\frac{\Gamma(t+1)}{t^{t+1}e^{-t}} - \sqrt{\frac{2\pi}{t}}\right| \le Ct^{-\frac{3}{2}}$$

for $t\ge 9$, and from this get (1.3.21).

EXERCISE 1.3.22. Inspired by T.H. Carne,[2] here is a rather different sort of application of large deviation estimates. Namely, the goal is to show that for each $n\ge 2$ and $1\le m<n$ there exists an $(m-1)$st order polynomial $p_{m,n}$ with the property that

$$\left|x^n - p_{m,n}(x)\right| \le 2\exp\left[-\frac{m^2}{2n}\right] \quad \text{for } x\in[-1,1].$$

**(i)** Given a $\mathbb{C}$-valued $f$ on $\mathbb{Z}$, define $\mathcal{A}f:\mathbb{Z}\longrightarrow\mathbb{C}$ by

$$\mathcal{A}f(n) = \frac{f(n+1)+f(n-1)}{2}, \quad n\in\mathbb{Z},$$

and show that, for any $n\ge 1$, $\mathcal{A}^n f = \mathbb{E}^{\mathbb{P}}\big[f(S_n)\big]$, where $S_n$ is the sum of $n$ $\mathbb{P}$-independent, $\{-1,1\}$-valued Bernoulli random variables with mean value 0.

[2] T.H. Carne, "A transformation formula for Markov chains," *Bull. Sc. Math.*, **109**, pp. 399–405 (1985). As Carne points out, what he is doing is the discrete analog of Hadamard's representation, via the Weierstrass transform, of solutions to heat equations in terms of solutions to the wave equations.

**(ii)** Show that, for each $z \in \mathbb{C}$, there is a unique sequence $\{Q(m,z) : m \in \mathbb{Z}\} \subseteq \mathbb{C}$ satisfying $Q(0,z) = 1$,

$$Q(-m,z) = Q(m,z), \quad \text{and } \big[\mathcal{A}Q(\,\cdot\,,z)\big](m) = zQ(m,z) \text{ for all } m \in \mathbb{Z}.$$

In fact, show that, for each $m \in \mathbb{Z}^+$: $Q(m,\,\cdot\,)$ is a polynomial of degree $m$ and

$$Q(m,\cos\theta) = \cos(m\theta), \quad \theta \in \mathbb{C}.$$

In particular, this means that $|Q(n,x)| \le 1$ for all $x \in [-1,1]$. (It also means that $Q(n,\,\cdot\,)$ is the $n$th **Chebychev polynomial**.)

**(iii)** Using induction on $n \in \mathbb{Z}^+$, show that

$$\big[\mathcal{A}^n Q(\,\cdot\,,z)\big](m) = z^n Q(m,z), \quad m \in \mathbb{Z} \text{ and } z \in \mathbb{C},$$

and conclude that

$$z^n = \mathbb{E}\Big[Q\big(S_n,z\big)\Big], \quad n \in \mathbb{Z}^+ \quad \text{and} \quad z \in \mathbb{C}.$$

In particular, if

$$p_{m,n}(z) \equiv \mathbb{E}\Big[Q\big(S_n,z\big),\, \big|S_n\big| < m\Big] = 2^{-n} \sum_{|2\ell-n|<m} \binom{n}{\ell} Q(2\ell-n,z),$$

conclude that (cf. Exercise 1.3.18)

$$\sup_{x\in[-1,1]} |x^n - p_{m,n}(x)| \le P\big(|S_n| \ge m\big) \le 2\exp\left[-\frac{m^2}{2n}\right] \quad \text{for all } 1 \le m \le n.$$

**(iv)** Suppose that $A$ is a self-adjoint contraction on the real or complex Hilbert space $H$ (i.e., $(f,Ag)_H = \overline{(g,Af)_H}$ and $\|Af\|_H \le \|f\|_H$ for all $f,\, g \in H$). Next, assume that $\big(f, A^\ell g\big)_H = 0$ for some $f,\, g \in H$ and each $0 \le \ell < m$. Show that

$$\big|\big(f, A^n g\big)_H\big| \le 2\|f\|_H \|g\|_H \exp\left[-\frac{m^2}{2n}\right] \quad \text{for } n \ge m.$$

(See Exercise 2.3.30 for an application.)

**Hint**: Note that $\big(f, p_{m,n}(A)g\big)_H = 0$, and use the Spectral Theorem to see that, for any polynomial $p$,

$$\|p(A)f\|_H \le \sup_{x\in[-1,1]} |p(x)|\, \|f\|_H, \quad f \in H.$$

## § 1.4 The Strong Law of Large Numbers

In this section I will discuss a few almost sure convergence properties of averages of independent random variables. Thus, once again, $\{X_n : n \geq 1\}$ will be a sequence of independent random variables on a probability space $(\Omega, \mathcal{F}, P)$, and $S_n$ and $\overline{S}_n$ will be, respectively, the sum and average of $X_1, \ldots, X_n$. Throughout this section, the reader should notice how much more immediately important a role independence (as opposed to orthogonality) plays than it did in Section 1.2.

To get started, I point out that, for both $\{S_n : n \geq 1\}$ and $\{\overline{S}_n : n \geq 1\}$, the set on which convergence occurs has $\mathbb{P}$-measure either 0 or 1. In fact, we have the following simple application of Kolmogorov's 0–1 Law (Theorem 1.1.2).

LEMMA 1.4.1. *For any sequence $\{a_n : n \in \mathbb{Z}^+\} \subseteq \mathbb{R}$ and any sequence $\{b_n : n \in \mathbb{Z}^+\} \subseteq (0, \infty)$ that converges to an element of $(0, \infty]$, the set on which*

$$\lim_{n\to\infty} \frac{S_n - a_n}{b_n} \quad \text{exists in } \mathbb{R}$$

*has $\mathbb{P}$-measure either 0 or 1. In fact, if $b_n \longrightarrow \infty$ as $n \to \infty$, then both*

$$\overline{\lim_{n\to\infty}} \frac{S_n - a_n}{b_n} \quad \text{and} \quad \underline{\lim_{n\to\infty}} \frac{S_n - a_n}{b_n}$$

*are $\mathbb{P}$-almost surely constant.*

PROOF: Simply observe that all of the events and functions involved can be expressed in terms of $\{S_{m+n} - S_m : n \geq 1\}$ for each $m \in \mathbb{Z}^+$ and are therefore tail-measurable. □

The following beautiful statement, which was proved originally by Kolmogorov, is the driving force behind many of the almost sure convergence results about both $\{S_n : n \geq 1\}$ and $\{\overline{S}_n : n \geq 1\}$.

THEOREM 1.4.2. *If the $X_n$'s are independent, $\mathbb{P}$-square integrable random variables, and if*

$$\sum_{n=1}^{\infty} \operatorname{Var}(X_n) < \infty, \tag{1.4.3}$$

*then*

$$\sum_{n=1}^{\infty} \big(X_n - \mathbb{E}^{\mathbb{P}}[X_n]\big) \quad \text{converges } \mathbb{P}\text{-almost surely.}$$

Note that, since

$$\sup_{n\geq N} \mathbb{P}\left(\left|\sum_{\ell=N}^{n} \big(X_\ell - \mathbb{E}^{\mathbb{P}}[X_\ell]\big)\right| \geq \epsilon\right) \leq \frac{1}{\epsilon^2} \sum_{\ell=N}^{\infty} \operatorname{Var}(X_\ell), \tag{1.4.4}$$

(1.4.3) certainly implies that the series $\sum_{n=1}^{\infty}\left(X_n - \mathbb{E}^{\mathbb{P}}[X_n]\right)$ converges in $\mathbb{P}$-measure. Thus, all that I am attempting to do here is replace a convergence in measure statement with an almost sure one. Obviously, this replacement would be trivial if the "$\sup_{n\geq N}$" in (1.4.3) appeared on the other side of $\mathbb{P}$. The remarkable fact which we are about to prove is that, in the present situation, the "$\sup_{n\geq N}$" can be brought inside!

THEOREM 1.4.5 (**Kolmogorov's Inequality).** *If the $X_n$'s are independent and $\mathbb{P}$-square integrable, then*

$$\mathbb{P}\left(\sup_{n\geq 1}\left|\sum_{\ell=1}^{n}\left(X_\ell - \mathbb{E}^{\mathbb{P}}[X_\ell]\right)\right| \geq \epsilon\right) \leq \frac{1}{\epsilon^2}\sum_{n=1}^{\infty}\mathrm{Var}(X_n) \tag{1.4.6}$$

*for each $\epsilon > 0$. (See Exercise 1.4.21 for more information.)*

PROOF: Without loss in generality, assume that each $X_n$ has mean value 0.
Given $1 \leq n < N$, note that

$$S_N^2 - S_n^2 = \left(S_N - S_n\right)^2 + 2\left(S_N - S_n\right)S_n \geq 2\left(S_N - S_n\right)S_n;$$

and therefore, since $S_N - S_n$ has mean value 0 and is independent of the $\sigma$-algebra $\sigma(\{X_1,\dots,X_n\})$,

$$(*) \qquad \mathbb{E}^{\mathbb{P}}\left[S_N^2,\, A_n\right] \geq \mathbb{E}^{\mathbb{P}}\left[S_n^2,\, A_n\right] \quad \text{for any } A_n \in \sigma\left(\{X_1,\dots,X_n\}\right).$$

In particular, if $A_1 = \{|S_1| > \epsilon\}$ and

$$A_{n+1} = \left\{\left|S_{n+1}\right| > \epsilon \text{ and } \max_{1\leq\ell\leq n}\left|S_\ell\right| \leq \epsilon\right\}, \qquad n \in \mathbb{Z}^+,$$

then, the $A_n$'s are mutually disjoint,

$$B_N \equiv \left\{\max_{1\leq n\leq N}\left|S_n\right| > \epsilon\right\} = \bigcup_{n=1}^{N} A_n,$$

and so (*) implies that

$$\begin{aligned}\mathbb{E}^{\mathbb{P}}\left[S_N^2,\, B_N\right] &= \sum_{n=1}^{N}\mathbb{E}^{\mathbb{P}}\left[S_N^2,\, A_n\right] \geq \sum_{n=1}^{N}\mathbb{E}^{\mathbb{P}}\left[S_n^2,\, A_n\right] \\ &\geq \epsilon^2\sum_{n=1}^{N}\mathbb{P}(A_n) = \epsilon^2\mathbb{P}(B_N).\end{aligned}$$

Thus,

$$\epsilon^2 \mathbb{P}\left(\sup_{n\ge1}|S_n| > \epsilon\right) = \lim_{N\to\infty} \epsilon^2 \mathbb{P}(B_N) \le \lim_{N\to\infty} \mathbb{E}^{\mathbb{P}}[S_N^2] \le \sum_{n=1}^{\infty} \mathbb{E}^{\mathbb{P}}[X_n^2],$$

and so the result follows after one takes left limits with respect to $\epsilon > 0$. □

PROOF OF THEOREM 1.4.2: Again assume that the $X_n$'s have mean value 0. By (1.4.6) applied to $\{X_{N+n} : n \in \mathbb{Z}^+\}$, we see that (1.4.3) implies

$$\mathbb{P}\left(\sup_{n>N}|S_n - S_N| \ge \epsilon\right) \le \frac{1}{\epsilon^2} \sum_{n=N+1}^{\infty} \mathbb{E}^{\mathbb{P}}[X_n^2] \longrightarrow 0 \quad \text{as} \quad N \to \infty$$

for every $\epsilon > 0$, and this is equivalent to the $\mathbb{P}$-almost sure Cauchy convergence of $\{S_n : n \ge 1\}$. □

In order to convert the conclusion in Theorem 1.4.2 into a statement about $\{\overline{S}_n : n \ge 1\}$, I will need the following elementary *summability* fact about sequences of real numbers.

LEMMA 1.4.7 (**Kronecker**). *Let $\{b_n : n \in \mathbb{Z}^+\}$ be a non-decreasing sequence of positive numbers that tend to $\infty$, and set $\beta_n = b_n - b_{n-1}$, where $b_0 \equiv 0$. If $\{s_n : n \ge 1\} \subseteq \mathbb{R}$ is a sequence that converges to $s \in \mathbb{R}$, then*

$$\frac{1}{b_n}\sum_{\ell=1}^{n} \beta_\ell s_\ell \longrightarrow s.$$

*In particular, if $\{x_n : n \ge 1\} \subseteq \mathbb{R}$, then*

$$\sum_{n=1}^{\infty} \frac{x_n}{b_n} \text{ converges in } \mathbb{R} \implies \frac{1}{b_n}\sum_{\ell=1}^{n} x_\ell \longrightarrow 0 \text{ as } n \to \infty.$$

PROOF: To prove the first part, assume that $s = 0$, and for given $\epsilon > 0$ choose $N \in \mathbb{Z}^+$ so that $|s_\ell| < \epsilon$ for $\ell \ge N$. Then, with $M = \sup_{n\ge1}|s_n|$,

$$\left|\frac{1}{b_n}\sum_{\ell=1}^{n}\beta_\ell s_\ell\right| \le \frac{Mb_N}{b_n} + \epsilon \longrightarrow \epsilon \quad \text{as } n \to \infty.$$

Turning to the second part, set $y_\ell = \frac{x_\ell}{b_\ell}$, $s_0 = 0$, and $s_n = \sum_{\ell=1}^n y_\ell$. After summation by parts,

$$\frac{1}{b_n}\sum_{\ell=1}^{n} x_\ell = s_n - \frac{1}{b_n}\sum_{\ell=1}^{n}\beta_\ell s_{\ell-1};$$

and so, since $s_n \longrightarrow s \in \mathbb{R}$ as $n \to \infty$, the first part gives the desired conclusion. □

After combining Theorem 1.4.2 with Lemma 1.4.7, we arrive at the following interesting statement.

COROLLARY 1.4.8. *Assume that $\{b_n : n \geq 1\} \subseteq (0, \infty)$ increases to infinity as $n \to \infty$, and suppose that $\{X_n : n \geq 1\}$ is a sequence of independent, $\mathbb{P}$-square integrable random variables. If*

$$\sum_{n=1}^{\infty} \frac{\operatorname{Var}(X_n)}{b_n^2} < \infty,$$

*then*

$$\frac{1}{b_n} \sum_{\ell=1}^{n} \Big(X_\ell - \mathbb{E}^{\mathbb{P}}[X_\ell]\Big) \longrightarrow 0 \quad \mathbb{P}\text{-almost surely.}$$

As an immediate consequence of the preceding, we see that $\overline{S}_n \longrightarrow m$ $\mathbb{P}$-almost surely if the $X_n$'s are identically distributed and $\mathbb{P}$-square integrable. In fact, without very much additional effort, we can also prove the following much more significant refinement of the last part of Theorem 1.3.1.

THEOREM 1.4.9 (**Kolmogorov's Strong Law**). *Let $\{X_n : n \in \mathbb{Z}^+\}$ be a sequence of $\mathbb{P}$-independent, identically distributed random variables. If $X_1$ is $\mathbb{P}$-integrable and has mean value $m$, then, as $n \to \infty$, $\overline{S}_n \longrightarrow m$ $\mathbb{P}$-almost surely and in $L^1(\mathbb{P}; \mathbb{R})$. Conversely, if $\overline{S}_n$ converges (in $\mathbb{R}$) on a set of positive $\mathbb{P}$-measure, then $X_1$ is $\mathbb{P}$-integrable.*

PROOF: Assume that $X_1$ is $\mathbb{P}$-integrable and that $\mathbb{E}^{\mathbb{P}}[X_1] = 0$. Next, set $Y_n = X_n \mathbf{1}_{[0,n]}(|X_n|)$, and note that

$$\sum_{n=1}^{\infty} \mathbb{P}(Y_n \neq X_n) = \sum_{n=1}^{\infty} \mathbb{P}(|X_n| > n)$$
$$\leq \sum_{n=1}^{\infty} \int_{n-1}^{n} \mathbb{P}(|X_1| > t)\, dt = \mathbb{E}^{\mathbb{P}}\big[|X_1|\big] < \infty.$$

Thus, by the first part of the Borel–Cantelli Lemma,

$$\mathbb{P}\big(\exists n \in \mathbb{Z}^+ \ \forall N \geq n \ Y_N = X_N\big) = 1.$$

In particular, if $\overline{T}_n = \frac{1}{n}\sum_{\ell=1}^{n} Y_\ell$ for $n \in \mathbb{Z}^+$, then, for $\mathbb{P}$-almost every $\omega \in \Omega$, $\overline{T}_n(\omega) \longrightarrow 0$ if and only if $\overline{S}_n(\omega) \longrightarrow 0$. Finally, to see that $\overline{T}_n \longrightarrow 0$ $\mathbb{P}$-almost surely, first observe that, because $\mathbb{E}^{\mathbb{P}}[X_1] = 0$, by the first part of Lemma 1.4.7,

$$\lim_{n\to\infty} \frac{1}{n} \sum_{\ell=1}^{n} \mathbb{E}^{\mathbb{P}}[Y_\ell] = \lim_{n\to\infty} \mathbb{E}^{\mathbb{P}}\big[X_1, |X_1| \leq n\big] = 0,$$

and therefore, by Corollary 1.4.8, it suffices for us to check that

$$\sum_{n=1}^{\infty} \frac{\mathbb{E}^{\mathbb{P}}[Y_n^2]}{n^2} < \infty.$$

To this end, set

$$C = \sup_{\ell \in \mathbb{Z}^+} \ell \sum_{n=\ell}^{\infty} \frac{1}{n^2},$$

and note that

$$\begin{aligned}\sum_{n=1}^{\infty} \frac{\mathbb{E}^{\mathbb{P}}[Y_n^2]}{n^2} &= \sum_{n=1}^{\infty} \frac{1}{n^2} \sum_{\ell=1}^{n} \mathbb{E}^{\mathbb{P}}\big[X_1^2,\, \ell - 1 < |X_1| \le \ell\big] \\ &= \sum_{\ell=1}^{\infty} \mathbb{E}^{\mathbb{P}}\big[X_1^2,\, \ell - 1 < |X_1| \le \ell\big] \sum_{n=\ell}^{\infty} \frac{1}{n^2} \\ &\le C \sum_{\ell=1}^{\infty} \frac{1}{\ell} \mathbb{E}^{\mathbb{P}}\big[X_1^2,\, \ell - 1 < |X_1| \le \ell\big] \le C\, \mathbb{E}^{\mathbb{P}}\big[|X_1|\big] < \infty.\end{aligned}$$

Thus, the $\mathbb{P}$-almost sure convergence is now established, and the $L^1(\mathbb{P};\mathbb{R})$-convergence result was proved already in Theorem 1.2.6.

Turning to the converse assertion, first note that (by Lemma 1.4.1) if $\overline{S}_n$ converges in $\mathbb{R}$ on a set of positive $\mathbb{P}$-measure, then it converges $\mathbb{P}$-almost surely to some $m \in \mathbb{R}$. In particular,

$$\overline{\lim_{n\to\infty}} \frac{|X_n|}{n} = \overline{\lim_{n\to\infty}} \big|\overline{S}_n - \overline{S}_{n-1}\big| = 0 \quad \mathbb{P}\text{-almost surely;}$$

and so, if $A_n \equiv \big\{|X_n| > n\big\}$, then $\mathbb{P}\big(\overline{\lim}_{n\to\infty} A_n\big) = 0$. But the $A_n$'s are mutually independent, and therefore, by the second part of the Borel–Cantelli Lemma, we now know that $\sum_{n=1}^{\infty} \mathbb{P}(A_n) < \infty$. Hence,

$$\mathbb{E}^{\mathbb{P}}\big[|X_1|\big] = \int_0^{\infty} \mathbb{P}(|X_1| > t)\, dt \le 1 + \sum_{n=1}^{\infty} \mathbb{P}\big(|X_n| > n\big) < \infty. \quad \square$$

REMARK 1.4.10. A reason for being interested in the converse part of Theorem 1.4.9 is that it provides a reconciliation between the measure theory vs. frequency schools of probability theory.

Although Theorem 1.4.9 is the centerpiece of this section, I want to give another approach to the study of the almost sure convergence properties of $\{S_n : n \ge 1\}$. In fact, following P. Lévy, I am going to show that $\{S_n : n \ge 1\}$ *converges* $\mathbb{P}$*-almost surely if it converges in* $\mathbb{P}$-measure. Hence, for example, Theorem 1.4.2 can be proved as a direct consequence of (1.4.4), without appeal to Kolmogorov's Inequality.

The key to Lévy's analysis lies in a version of the *reflection principle*, whose statement requires the introduction of a new concept. Given an $\mathbb{R}$-valued random variable $Y$, say that $\alpha \in \mathbb{R}$ is a **median** of $Y$ and write $\alpha \in \operatorname{med}(Y)$, if

$$\mathbb{P}(Y \le \alpha) \wedge \mathbb{P}(Y \ge \alpha) \ge \tfrac{1}{2}. \tag{1.4.11}$$

Notice that (as distinguished from a *mean value*) every $Y$ admits a median; for example, it is easy to check that

$$\alpha \equiv \inf\{t \in \mathbb{R} : \mathbb{P}(Y \le t) \ge \tfrac{1}{2}\}$$

is a median of $Y$. In addition, it is clear that

$$\mathrm{med}(-Y) = -\mathrm{med}(Y) \quad \text{and} \quad \mathrm{med}\,(\beta + Y) = \beta + \mathrm{med}\,(Y) \text{ for all } \beta \in \mathbb{R}.$$

On the other hand, the notion of median is flawed by the fact that, in general, a random variable will admit an entire non-degenerate interval of medians. In addition, it is neither easy to compute the medians of a sum in terms of the medians of the summands nor to relate the *medians* of an integrable random variable to its *mean value*. Nonetheless, at least if $Y \in L^p(\mathbb{P};\mathbb{R})$ for some $p \in [1,\infty)$, the following estimate provides some information. Namely, since, for $\alpha \in \mathrm{med}(Y)$ and $\beta \in \mathbb{R}$,

$$\frac{|\alpha-\beta|^p}{2} \le |\alpha-\beta|^p\Big(\mathbb{P}(Y \ge \alpha) \wedge \mathbb{P}(Y \le \alpha)\Big) \le \mathbb{E}^{\mathbb{P}}\big[|Y-\beta|^p\big],$$

we see that, for any $p \in [1,\infty)$ and $Y \in L^p(\mathbb{P};\mathbb{R})$,

$$|\alpha-\beta| \le \Big(2\mathbb{E}^{\mathbb{P}}\big[|Y-\beta|^p\big]\Big)^{\frac{1}{p}} \text{ for all } \beta \in \mathbb{R} \text{ and } \alpha \in \mathrm{med}\,(Y).$$

In particular, if $Y \in L^2(\mathbb{P};\mathbb{R})$ and $m$ is the mean value of $Y$, then

$$|\alpha - m| \le \sqrt{2\mathrm{Var}(Y)} \quad \text{for all } \alpha \in \mathrm{med}(Y). \tag{1.4.12}$$

Theorem 1.4.13 (**Lévy's Reflection Principle**). *Let $\{X_n : n \in \mathbb{Z}^+\}$ be a sequence of $\mathbb{P}$-independent random variables, and, for $k \le \ell$, choose $\alpha_{\ell,k} \in \mathrm{med}\,(S_\ell - S_k)$. Then, for any $N \in \mathbb{Z}^+$ and $\epsilon > 0$,*

$$\mathbb{P}\left(\max_{1\le n\le N}(S_n + \alpha_{N,n}) \ge \epsilon\right) \le 2\mathbb{P}(S_N \ge \epsilon), \tag{1.4.14}$$

*and therefore*

$$\mathbb{P}\left(\max_{1\le n\le N}|S_n + \alpha_{N,n}| \ge \epsilon\right) \le 2\mathbb{P}(|S_N| \ge \epsilon). \tag{1.4.15}$$

Proof: Clearly (1.4.15) follows by applying (1.4.14) to both the sequences $\{X_n : n \ge 1\}$ and $\{-X_n : n \ge 1\}$ and then adding the two results.

To prove (1.4.14), set $A_1 = \{S_1 + \alpha_{N,1} \geq \epsilon\}$ and

$$A_{n+1} = \left\{ \max_{1\leq \ell \leq n} (S_\ell + \alpha_{N,\ell}) < \epsilon \text{ and } S_{n+1} + \alpha_{N,n+1} \geq \epsilon \right\}$$

for $1 \leq n < N$. Obviously, the $A_n$'s are mutually disjoint and

$$\bigcup_{n=1}^{N} A_n = \left\{ \max_{1\leq n\leq N} (S_n + \alpha_{N,n}) \geq \epsilon \right\}.$$

In addition,

$$\{S_N \geq \epsilon\} \supseteq A_n \cap \{S_N - S_n \geq \alpha_{N,n}\} \quad \text{for each } 1 \leq n \leq N.$$

Hence,

$$\begin{aligned}\mathbb{P}(S_N \geq \epsilon) &\geq \sum_{n=1}^{N} \mathbb{P}\Big(A_n \cap \{S_N - S_n \geq \alpha_{N,n}\}\Big) \\ &\geq \frac{1}{2}\sum_{n=1}^{N} \mathbb{P}(A_n) = \frac{1}{2}\mathbb{P}\left( \max_{1\leq n\leq N} (S_n + \alpha_{N,n}) \geq \epsilon \right),\end{aligned}$$

where, in the passage to the last line, I have used the independence of the sets $A_n$ and $\{S_N - S_n \geq \alpha_{N,n}\}$. □

COROLLARY 1.4.16. *Let $\{X_n : n \in \mathbb{Z}^+\}$ be a sequence of independent random variables, and assume that $\{S_n : n \in \mathbb{Z}^+\}$ converges in $\mathbb{P}$-measure to an $\mathbb{R}$-valued random variable $S$. Then $S_n \longrightarrow S$ $\mathbb{P}$-almost surely. (Cf. Exercise 1.4.25 as well.)*

PROOF: What I must show is that, for each $\epsilon > 0$, there is an $M \in \mathbb{Z}^+$ such that

$$\sup_{N\geq 1} \mathbb{P}\Big( \max_{1\leq n\leq N} |S_{n+M} - S_M| \geq \epsilon \Big) < \epsilon.$$

To this end, let $0 < \epsilon < 1$ be given, and choose $M \in \mathbb{Z}^+$ so that

$$\mathbb{P}\left( |S_{n+M} - S_{k+M}| \geq \frac{\epsilon}{2} \right) < \frac{\epsilon}{2} \quad \text{for all } 1 \leq k < n.$$

Next, for a given $N \in \mathbb{Z}^+$, choose $\alpha_{N,n} \in \operatorname{med}(S_{M+N} - S_{M+n})$ for $0 \leq n \leq N$. Then $|\alpha_{N,n}| \leq \frac{\epsilon}{2}$, and so, by (1.4.15) applied to $\{X_{M+n} : n \geq 1\}$,

$$\begin{aligned}P\left( \max_{1\leq n\leq N} |S_{M+n} - S_M| \geq \epsilon \right) &\leq P\left( \max_{1\leq n\leq N} |S_{M+n} - S_M + \alpha_{N,n}| \geq \frac{\epsilon}{2} \right) \\ &\leq 2P\Big( |S_{M+N} - S_M| \geq \frac{\epsilon}{2} \Big) < \epsilon. \quad □\end{aligned}$$

REMARK 1.4.17. The most beautiful and startling feature of Lévy's line of reasoning is that it requires *no* integrability assumptions. Of course, in many applications of Corollary 1.4.16, integrability considerations enter into the proof that $\{S_n : n \geq 1\}$ converges in $\mathbb{P}$-measure. Finally, a word of caution may be in order. Namely, the result in Corollary 1.4.16 applies to the quantities $S_n$ themselves; it does *not* apply to associated quantities like $\overline{S}_n$. Indeed, suppose that $\{X_n : n \geq 1\}$ is a sequence of independent, identically distributed random variables that satisfy

$$\mathbb{P}\big(X_n \leq -t\big) = \mathbb{P}\big(X_n \geq t\big) = \Big(\big(1+t^2\big)\log\big(e^4+t^2\big)\Big)^{-\frac{1}{2}} \quad \text{for all } t \geq 0.$$

On the one hand, by Exercise 1.2.11, we know that the associated averages $\overline{S}_n$ tend to 0 in probability. On the other hand, by the second part of Theorem 1.4.9, we know that the sequence $\{\overline{S}_n : n \geq 1\}$ diverges almost surely.

## Exercises for § 1.4

EXERCISE 1.4.18. Let $X$ and $Y$ be non-negative random variables, and suppose that

$$\mathbb{P}\big(X \geq t\big) \leq \frac{1}{t}\mathbb{E}^{\mathbb{P}}\Big[Y, X \geq t\Big], \quad t \in (0,\infty). \tag{1.4.19}$$

Show that

$$\Big(\mathbb{E}^{\mathbb{P}}\big[X^p\big]\Big)^{\frac{1}{p}} \leq \frac{p}{p-1}\Big(\mathbb{E}^{\mathbb{P}}\big[Y^p\big]\Big)^{\frac{1}{p}}, \quad p \in (1,\infty). \tag{1.4.20}$$

**Hint**: First, reduce to the case when $X$ is bounded. Next, recall that, for any measure space $(E, \mathcal{F}, \mu)$, any non-negative, measurable $f$ on $(E, \mathcal{F})$, and any $\alpha \in (0,\infty)$,

$$\int_E f(x)^\alpha\,\mu(dx) = \alpha \int_{(0,\infty)} t^{\alpha-1}\mu(f > t)\,dt = \alpha \int_{(0,\infty)} t^{\alpha-1}\mu(f \geq t)\,dt.$$

Use this together with (1.4.19) to justify the relation

$$\begin{aligned}\mathbb{E}^{\mathbb{P}}\big[X^p\big] &\leq p\int_{(0,\infty)} t^{p-2}\,\mathbb{E}^{\mathbb{P}}\big[Y, X \geq t\big]\,dt \\ &= p\mathbb{E}^{\mathbb{P}}\left[Y\int_0^X t^{p-2}\,dt\right] = \frac{p}{p-1}\,\mathbb{E}^{\mathbb{P}}\big[X^{p-1}Y\big],\end{aligned}$$

and arrive at (1.4.20) after an application of Hölder's Inequality.

EXERCISE 1.4.21. Let $\{X_n : n \geq 1\}$ be a sequence of mutually independent, $\mathbb{P}$-square integrable random variables with mean value 0, and assume that $\sum_1^\infty \mathbb{E}[X_n^2] < \infty$. Let $S$ denote the random variable (guaranteed by Theorem 1.4.2) to which $\{S_n : n \geq 1\}$ converges $\mathbb{P}$-almost surely, and, using elementary orthogonality considerations, check that $S_n \longrightarrow S$ in $L^2(\mathbb{P};\mathbb{R})$ as well. Next, after examining the proof of Kolmogorov's Inequality (cf. (1.4.6)), show that

$$\mathbb{P}\left(\sup_{n\in\mathbb{Z}^+} |S_n|^2 \geq t\right) \leq \frac{1}{t}\mathbb{E}^{\mathbb{P}}\left[S^2,\ \sup_{n\in\mathbb{Z}^+} |S_n|^2 \geq t\right], \quad t > 0.$$

Finally, by applying (1.4.20), show that

$$\mathbb{E}^{\mathbb{P}}\left[\sup_{n\in\mathbb{Z}^+} |S_n|^{2p}\right] \leq \left(\tfrac{p}{p-1}\right)^p \mathbb{E}^{\mathbb{P}}\left[|S|^{2p}\right], \quad p \in (1,\infty), \tag{1.4.22}$$

and conclude from this that, for each $p \in (2,\infty)$, $\{S_n : n \geq 1\}$ converges to $S$ in $L^p(P)$ if and only if $S \in L^p(P)$.

EXERCISE 1.4.23. If $X \in L^2(\mathbb{P};\mathbb{R})$, then it is easy to characterize its mean $m$ as the $c \in \mathbb{R}$ that minimizes $\mathbb{E}^{\mathbb{P}}\left[(X-c)^2\right]$. Assuming that $X \in L^1(\mathbb{P};\mathbb{R})$, show that $\alpha \in \mathrm{med}(X)$ if and only if

$$\mathbb{E}^{\mathbb{P}}\left[|X-\alpha|\right] = \min_{c\in\mathbb{R}} \mathbb{E}^{\mathbb{P}}\left[|X-c|\right].$$

**Hint**: Show that, for any $a, b \in \mathbb{R}$,

$$\mathbb{E}^{\mathbb{P}}\left[|X-b|\right] - \mathbb{E}^{\mathbb{P}}\left[|X-a|\right] = \int_a^b \left[\mathbb{P}(X \leq t) - \mathbb{P}(X \geq t)\right] dt.$$

EXERCISE 1.4.24. Let $\{X_n : n \geq 1\}$ be a sequence of $\mathbb{P}$-square integrable random variables that converges in probability to a random variable $X$, and assume that $\sup_{n\geq 1} \mathrm{Var}(X_n) < \infty$. Show that $X$ is square integrable and that $\mathbb{E}^{\mathbb{P}}\left[|X_n - X|\right] \longrightarrow 0$. In particular, if, in addition, $\mathrm{Var}(X_n) \longrightarrow \mathrm{Var}(X)$, show that $\mathbb{E}^{\mathbb{P}}\left[|X_n - X|^2\right] \longrightarrow 0$.

**Hint**: Let $\alpha_n \in \mathrm{med}(X_n)$, and show that $\alpha_+ = \overline{\lim}_{n\to\infty} \alpha_n$ and $\alpha_- = \underline{\lim}_{n\to\infty} \alpha_n$ are both elements of $\mathrm{med}(X)$. Combine this with (1.4.12) to conclude that $\sup_{n\geq 1}\left|\mathbb{E}^{\mathbb{P}}[X_n]\right| < \infty$ and therefore that $\sup_{n\geq 1} \mathbb{E}^{\mathbb{P}}[X_n^2] < \infty$.

EXERCISE 1.4.25. The following variant of Theorem 1.4.13 is sometimes useful and has the advantage that it avoids the introduction of medians. Namely, show that, for any $t \in (0,\infty)$ and $n \in \mathbb{Z}^+$,

$$\mathbb{P}\left(\max_{1\leq m\leq n} |S_n| \geq 2t\right) \leq \frac{\mathbb{P}(|S_n| > t)}{1 - \max_{1\leq m\leq n} \mathbb{P}(|S_n - S_m| > t)}.$$

Note that this can be used in place of (1.4.15) when proving results like the one in Corollary 1.4.16.

EXERCISE 1.4.26. A random variable $X$ is said to be **symmetric** if $-X$ has the same distribution as $X$ itself. Obviously, the most natural choice of median for a symmetric random variable is 0; and thus, because sums of independent, symmetric random variables are again symmetric, (1.4.14) and (1.4.15) are particularly useful when the $X_n$'s are symmetric, since the $\alpha_{\ell,k}$'s can then be taken to be 0. In this connection, consider the following interesting variation on the theme of Theorem 1.4.13.

**(i)** Let $X_1, \dots, X_n, \dots$ be independent, symmetric random variables, set $M_n(\omega) = \max_{1\le \ell\le n} |X_\ell(\omega)|$, let $\tau_n(\omega)$ be the smallest $1 \le \ell \le n$ with the property that $|X_\ell(\omega)| = M_n(\omega)$, and define

$$Y_n(\omega) = X_{\tau_n(\omega)}(\omega) \quad \text{and} \quad \hat{S}_n = S_n - Y_n.$$

Show that

$$\omega \in \Omega \longmapsto \big(\hat{S}_n(\omega), Y_n(\omega)\big) \in \mathbb{R}^2 \quad \text{and} \quad \omega \in \Omega \longmapsto \big(-\hat{S}_n(\omega), Y_n(\omega)\big) \in \mathbb{R}^2$$

have the same distribution, and conclude first that

$$\begin{aligned}\mathbb{P}(Y_n \ge t) &\le \mathbb{P}(Y_n \ge t \;\&\; \hat{S}_n \ge 0) + \mathbb{P}(Y_n \ge t \;\&\; \hat{S}_n \le 0)\\ &= 2\mathbb{P}(Y_n \ge t \;\&\; \hat{S}_n \ge 0) \le 2\mathbb{P}(S_n \ge t),\end{aligned}$$

for all $t \in \mathbb{R}$, and then that

$$\mathbb{P}\Big(\max_{1\le \ell\le n} |X_\ell| \ge t\Big) \le 2\mathbb{P}\big(|S_n| \ge t\big), \quad t \in [0,\infty).$$

**(ii)** Continuing in the same setting, add the assumption that the $X_n$'s are identically distributed, and use part **(i)** to show that

$$\begin{aligned}&\lim_{n\to\infty} \mathbb{P}\big(|\overline{S}_n| \le C\big) = 1 \quad \text{for some } C \in (0,\infty)\\ &\qquad \implies \lim_{n\to\infty} n\mathbb{P}\big(|X_1| \ge n\big) = 0.\end{aligned}$$

**Hint**: Note that

$$\mathbb{P}\Big(\max_{1\le \ell\le n} |X_\ell| > t\Big) = 1 - \mathbb{P}(|X_1| \le t)^n$$

and that $\frac{1-(1-x)^n}{x} \longrightarrow n$ as $x \searrow 0$.

In conjunction with Exercise 1.2.11, this proves that *if* $\{X_n : n \ge 1\}$ *is a sequence of independent, identically distributed symmetric random variables, then* $\overline{S}_n \longrightarrow 0$ *in* $\mathbb{P}$*-probability if and only if* $\lim_{n\to\infty} n\mathbb{P}\big(|X_1| \ge n\big) = 0$.

EXERCISE 1.4.27. Let $X$ and $X'$ be a pair of independent random variables that have the same distribution, let $\alpha$ be a median of $X$, and set $Y = X - X'$.

**(i)** Show that $Y$ is symmetric and that

$$\mathbb{P}\big(|X-\alpha| \geq t\big) \leq 2\mathbb{P}\big(|Y| \geq t\big) \quad \text{for all} \quad t \in [0,\infty),$$

and conclude that, for any $p \in (0,\infty)$,

$$2^{-\frac{1}{p}\vee 1}\mathbb{E}^{\mathbb{P}}\big[|X|^p\big]^{\frac{1}{p}} \leq \Big(\big(2\mathbb{E}^{\mathbb{P}}\big[|Y|^p\big]\big)^{\frac{1}{p}} + |\alpha|\Big).$$

In particular, $|X|^p$ is integrable if and only if $|Y|^p$ is.

**(ii)** The result in **(i)** leads to my final refinement of The Weak Law of Large Numbers. Namely, let $\{X_n : n \geq 1\}$ be a sequence of independent, identically distributed random variables. By combining Exercise 1.2.11, part **(ii)** in Exercise 1.4.26, and part **(i)** above, show that[1]

$$\begin{aligned} &\lim_{n\to\infty} \mathbb{P}\Big(\big|\overline{S}_n\big| \leq C\Big) = 1 \quad \text{for some } C \in (0,\infty) \\ &\qquad \Longrightarrow \lim_{n\to\infty} n\mathbb{P}\big(|X_1| \geq n\big) = 0 \\ &\qquad \Longrightarrow \overline{S}_n - \mathbb{E}^{\mathbb{P}}\big[X_1,\, |X_1| \leq n\big] \longrightarrow 0 \text{ in } \mathbb{P}\text{-probability.} \end{aligned}$$

EXERCISE 1.4.28. Let $\{X_n : n \geq 1\}$ be a sequence of mutually independent, identically distributed, $\mathbb{P}$-integrable random variables with mean value $m$. As we already know, when $m > 0$, the partial sums $S_n$ tend, $\mathbb{P}$-almost surely, to $+\infty$ at an asymptotic linear rate $m$; and, of course, when $m < 0$, the situation is similar at $-\infty$. On the other hand, when $m = 0$, we know that, if $|S_n|$ tends to $\infty$ at all, then, $\mathbb{P}$-almost surely, it does so at a strictly sublinear rate. In this exercise, you are to sharpen this statement by proving that

$$m = 0 \Longrightarrow \varliminf_{n\to\infty} |S_n| < \infty \quad \mathbb{P}\text{-almost surely.}$$

The beautiful argument given below is due to Y. Guivarc'h, but its full power cannot be appreciated in the present context (cf. Exercise 6.2.19). Furthermore, a classic result (cf. Exercise 5.2.43) due to K.L. Chung and W.H. Fuchs gives a much sharper conclusion for the independent random variables. Their result says that $\varliminf_{n\to\infty} |S_n| = 0$ $\mathbb{P}$-almost surely.

In order to prove the assertion here, assume that $\lim_{n\to\infty} |S_n| = \infty$ with positive $\mathbb{P}$-probability, use Kolmogorov's 0–1 Law to see that $|S_n| \longrightarrow \infty$ $\mathbb{P}$-almost surely, and proceed as follows.

---

[1] These ideas are taken from the book by Wm. Feller cited at the end of § 1.2. They become even more elegant when combined with a theorem due to E.J.G. Pitman, which is given in Feller's book.

**(i)** Show that there must exist an $\epsilon > 0$ with the property that

$$\mathbb{P}\big(\forall \ell > k \; |S_\ell - S_k| \geq \epsilon\big) \geq \epsilon$$

for some $k \in \mathbb{Z}^+$ and therefore that

$$\mathbb{P}(A) \geq \epsilon, \quad \text{where } A \equiv \big\{\omega : \forall \ell \in \mathbb{Z}^+ \; |S_\ell(\omega)| \geq \epsilon\big\}.$$

**(ii)** For each $\omega \in \Omega$ and $n \in \mathbb{Z}^+$, set

$$\Gamma_n(\omega) = \Big\{t \in \mathbb{R} : \exists 1 \leq \ell \leq n \; |t - S_\ell(\omega)| < \tfrac{\epsilon}{2}\Big\}$$

and

$$\Gamma_n'(\omega) = \Big\{t \in \mathbb{R} : \exists 1 \leq \ell \leq n \; |t - S_\ell'(\omega)| < \tfrac{\epsilon}{2}\Big\},$$

where $S_n' \equiv \sum_{\ell=1}^n X_{\ell+1}$. Next, let $R_n(\omega)$ and $R_n'(\omega)$ denote the Lebesgue measure of $\Gamma_n(\omega)$ and $\Gamma_n'(\omega)$, respectively; and, using the translation invariance of Lebesgue measure, show that

$$R_{n+1}(\omega) - R_n'(\omega) \geq \epsilon \mathbf{1}_{A'}(\omega), \\ \text{where } A' \equiv \big\{\omega : \forall \ell \geq 2 \; |S_\ell(\omega) - S_1(\omega)| \geq \epsilon\big\}.$$

On the other hand, show that

$$\mathbb{E}^{\mathbb{P}}\big[R_n'\big] = \mathbb{E}^{\mathbb{P}}\big[R_n\big] \quad \text{and} \quad \mathbb{P}(A') = \mathbb{P}(A),$$

and conclude first that

$$\epsilon\mathbb{P}(A) \leq \mathbb{E}^{\mathbb{P}}\big[R_{n+1} - R_n\big], \quad n \in \mathbb{Z}^+,$$

and then that

$$\epsilon\mathbb{P}(A) \leq \varliminf_{n\to\infty} \frac{1}{n}\mathbb{E}^{\mathbb{P}}\big[R_n\big].$$

**(iii)** In view of parts **(i)** and **(ii)**, what remains to be done is show that

$$m = 0 \implies \varliminf_{n\to\infty} \frac{1}{n}\mathbb{E}^{\mathbb{P}}\big[R_n\big] = 0.$$

But, clearly, $0 \leq R_n(\omega) \leq n\epsilon$. Thus, it is enough to show that, when $m = 0$, $\frac{R_n}{n} \longrightarrow 0$ $\mathbb{P}$-almost surely; and, to this end, first check that

$$\frac{S_n(\omega)}{n} \longrightarrow 0 \implies \frac{R_n(\omega)}{n} \longrightarrow 0,$$

and, finally, apply The Strong Law of Large Numbers.

EXERCISE 1.4.29. As I have already said, for many applications The Weak Law of Large Numbers is just as good as and even preferable to the Strong Law. Nonetheless, here is an application in which the full strength of the Strong Law plays an essential role. Namely, I want to use the Strong Law to produce examples of continuous, strictly increasing functions $F$ on $[0,1]$ with the property that their derivative

$$F'(x) \equiv \lim_{y\to x} \frac{F(y)-F(x)}{y-x} = 0 \quad \text{at Lebesgue-almost every } x \in (0,1).$$

By familiar facts about functions of a real variable, one knows that such functions $F$ are in one-to-one correspondence with non-atomic, Borel probability measures $\mu$ on $[0,1]$ which charge every non-empty open subset but are singular to Lebesgue's measure. Namely, $F$ is the distribution function determined by $\mu$: $F(x) = \mu\big((-\infty, x]\big)$.

**(i)** Set $\Omega = \{0,1\}^{\mathbb{Z}^+}$, and, for each $p \in (0,1)$, take $M_p = (\beta_p)^{\mathbb{Z}^+}$, where $\beta_p$ on $\{0,1\}$ is the Bernoulli measure with $\beta_p(\{1\}) = p = 1 - \beta_p(\{0\})$. Next, define

$$\omega \in \Omega \longmapsto Y(\omega) \equiv \sum_{n=1}^{\infty} 2^{-n}\omega_n \in [0,1],$$

and let $\mu_p$ denote the $M_p$-distribution of $Y$. Given $n \in \mathbb{Z}^+$ and $0 \le m < 2^n$, show that

$$\mu_p\big([m2^{-n}, (m+1)2^{-n}]\big) = p^{\ell_{m,n}}(1-p)^{n-\ell_{m,n}},$$

where $\ell_{m,n} = \sum_{k=1}^{n} \omega_k$ and $(\omega_1, \dots, \omega_n) \in \{0,1\}^n$ is determined by $m2^{-n} = \sum_{k=1}^{n} 2^{-k}\omega_k$. Conclude, in particular, that $\mu_p$ is non-atomic and charges every non-empty open subset of $[0,1]$.

**(iii)** Given $x \in [0,1)$ and $n \in \mathbb{Z}^+$, define

$$\epsilon_n(x) = \begin{cases} 1 & \text{if } 2^{n-1}x - \lfloor 2^{n-1}x \rfloor \ge \frac{1}{2} \\ 0 & \text{if } 2^{n-1}x - \lfloor 2^{n-1}x \rfloor < \frac{1}{2}, \end{cases}$$

where $\lfloor s \rfloor$ denotes the integer part of $s$. If $\{\epsilon_n : n \ge 1\} \subseteq \{0,1\}$ satisfies $x = \sum_1^{\infty} 2^{-m}\epsilon_m$, show that $\epsilon_m = \epsilon_m(x)$ for all $m \ge 1$ if and only if $\epsilon_m = 0$ for infinitely many $m \ge 1$. In particular, conclude first that $\omega_n = \epsilon_n\big(Y(\omega)\big)$, $n \in \mathbb{Z}^+$, for $M_p$-almost every $\omega \in \Omega$ and, second, by the Strong Law, that

$$\frac{1}{n}\sum_{m=1}^{n} \epsilon_n(x) \longrightarrow p \quad \text{for } \mu_p\text{-almost every } x \in [0,1].$$

Thus, $\mu_{p_1} \perp \mu_{p_2}$ whenever $p_1 \ne p_2$.

**(iv)** By Lemma 1.1.6, we know that $\mu_{\frac{1}{2}}$ is Lebesgue measure $\lambda_{[0,1]}$ on $[0,1]$. Hence, we now know that $\mu_p \perp \lambda_{[0,1]}$ when $p \neq \frac{1}{2}$. In view of the introductory remarks, this completes the proof that, for each $p \in (0,1) \setminus \{\frac{1}{2}\}$, the function $F_p(x) = \mu_p\big((-\infty, x]\big)$ is a strictly increasing, continuous function on $[0,1]$ whose derivative vanishes at Lebesgue-almost every point. Here, one can do better. Namely, referring to part **(iii)**, let $\Delta_p$ denote the set of $x \in [0,1)$ such that

$$\lim_{n\to\infty} \frac{1}{n}\Sigma_n(x) = p, \quad \text{where } \Sigma_n(x) \equiv \sum_{m=1}^{n} \epsilon_m(x).$$

We know that $\Delta_{\frac{1}{2}}$ has Lebesgue measure 1. Show that, for each $x \in \Delta_{\frac{1}{2}}$ and $p \in (0,1) \setminus \{\frac{1}{2}\}$, $F_p$ is differentiable with derivative 0 at $x$.

**Hint**: Given $x \in [0,1)$, define

$$L_n(x) = \sum_{m=1}^{n} 2^{-m}\epsilon_m(x) \quad \text{and} \quad R_n(x) = L_n(x) + 2^{-n}.$$

Show that

$$F_p\big(R_n(x)\big) - F_p\big(L_n(x)\big) = M_p\left(\sum_{m=1}^{n} 2^{-m}\omega_m = L_n(x)\right) = p^{\Sigma_n(x)}(1-p)^{n-\Sigma_n(x)}.$$

When $p \in (0,1) \setminus \{\frac{1}{2}\}$ and $x \in \Delta_{\frac{1}{2}}$, use this together with $4p(1-p) < 1$ to show that

$$\lim_{n\to\infty} n \log\left(\frac{F_p\big(R_n(x)\big) - F_p\big(L_n(x)\big)}{R_n(x) - L_n(x)}\right) < 0.$$

To complete the proof, for given $x \in \Delta_{\frac{1}{2}}$ and $n \geq 2$ such that $\Sigma_n(x) \geq 2$, let $m_n(x)$ denote the largest $m < n$ such that $\epsilon_m(x) = 1$, and show that $\frac{m_n(x)}{n} \longrightarrow 1$ as $n \to \infty$. Hence, since $2^{-n-1} < h \leq 2^{-n}$ implies that

$$\frac{F_p(x) - F_p(x-h)}{h} \leq 2^{n-m_n(x)+1}\frac{F_p\big(R_n(x)\big) - F_p\big(L_n(x)\big)}{R_n(x) - L_n(x)},$$

one concludes that $F_p$ is left-differentiable at $x$ and has left derivative equal to 0 there. To get the same conclusion about right derivatives, simply note that $F_p(x) = 1 - F_{1-p}(1-x)$.

**(v)** Again let $p \in (0,1) \setminus \{\frac{1}{2}\}$ be given, but this time choose $x \in \Delta_p$. Show that

$$\lim_{h\searrow 0} \frac{F_p(x+h) - F_p(x)}{h} = +\infty.$$

The argument is similar to the one used to handle part **(iv)**. However, this time the role played by the inequality $4pq < 1$ is played here by $(2p)^p(2q)^q > 1$ when $q = 1 - p$.

## § 1.5 Law of the Iterated Logarithm

Let $X_1, \dots, X_n, \dots$ be a sequence of independent, identically distributed random variables with mean value 0 and variance 1. In this section, I will investigate exactly how large $\{S_n : n \in \mathbb{Z}^+\}$ can become as $n \to \infty$. To get a feeling for what one should be expecting, first note that, by Corollary 1.4.8, for any non-decreasing $\{b_n : n \geq 1\} \subseteq (0, \infty)$,

$$\frac{S_n}{b_n} \longrightarrow 0 \quad \mathbb{P}\text{-almost surely if} \quad \sum_{n=1}^{\infty} \frac{1}{b_n^2} < \infty.$$

Thus, for example, $S_n$ grows more slowly than $n^{\frac{1}{2}} \log n$. On the other hand, if the $X_n$'s are $N(0,1)$-random variables, then so are the random variables $\frac{S_n}{\sqrt{n}}$; and therefore, for every $R \in (0, \infty)$,

$$\mathbb{P}\left( \overline{\lim_{n\to\infty}} \frac{S_n}{\sqrt{n}} \geq R \right) = \lim_{N\to\infty} \mathbb{P}\left( \bigcup_{n\geq N} \left\{ \frac{S_n}{\sqrt{n}} \geq R \right\} \right) \geq \lim_{N\to\infty} \mathbb{P}\left( \frac{S_N}{\sqrt{N}} \geq R \right) > 0.$$

Hence, at least for normal random variables, one can use Lemma 1.4.1 to see that

$$\overline{\lim_{n\to\infty}} \frac{S_n}{\sqrt{n}} = \infty \quad \mathbb{P}\text{-almost surely;}$$

and so $S_n$ grows faster than $n^{\frac{1}{2}}$.

If, as we did in Section 1.3, we proceed on the assumption that Gaussian random variables are typical, we should expect the growth rate of the $S_n$'s to be something between $n^{\frac{1}{2}}$ and $n^{\frac{1}{2}} \log n$. What, in fact, turns out to be the precise growth rate is

$$\Lambda_n \equiv \sqrt{2n \log_{(2)}(n \vee 3)}, \tag{1.5.1}$$

where $\log_{(2)} x \equiv \log\bigl(\log x\bigr)$ (*not* the logarithm with base 2) for $x \in [e, \infty)$. That is, one has **The Law of the Iterated Logarithm**:

$$\overline{\lim_{n\to\infty}} \frac{S_n}{\Lambda_n} = 1 \quad \mathbb{P}\text{-almost surely.} \tag{1.5.2}$$

This remarkable fact was discovered first for Bernoulli random variables by Khinchine, was extended by Kolmogorov to random variables possessing $2+\epsilon$ moments, and eventually achieved its final form in the work of Hartman and Wintner. The approach that I will adopt here is based on ideas (taught to me by M. Ledoux) introduced originally to handle generalizations of (1.5.2) to random variables with values in a Banach space.[1] This approach consists of two steps. The first establishes a preliminary version of (1.5.2) that, although it is far cruder than (1.5.2) itself, will allow me to justify a reduction of the general case to the case of bounded random variables. In the second step, I deal with bounded random variables and more or less follow Khinchine's strategy for deriving (1.5.2) once one has estimates like the ones provided by Theorem 1.3.15.

In what follows, I will use the notation

$$\Lambda_\beta = \Lambda_{\lfloor\beta\rfloor} \quad \text{and} \quad \tilde{S}_\beta = \frac{S_{\lfloor\beta\rfloor}}{\Lambda_\beta} \quad \text{for } \beta \in [3,\infty),$$

where $\lfloor\beta\rfloor$ is the integer part of $\beta$.

LEMMA 1.5.3. *Let $\{X_n : n \geq 1\}$ be a sequence of independent, identically distributed random variables with mean value* 0 *and variance* 1. *Then, for any $a \in (0,\infty)$ and $\beta \in (1,\infty)$*,[2]

$$\varlimsup_{n\to\infty} |\tilde{S}_n| \leq a \quad (\text{a.s.}, \mathbb{P}) \quad \textit{if} \quad \sum_{m=1}^{\infty} \mathbb{P}\Big(\big|\tilde{S}_{\beta^m}\big| \geq a\,\beta^{-\frac{1}{2}}\Big) < \infty.$$

PROOF: Let $\beta \in (1,\infty)$ be given and, for each $m \in \mathbb{N}$ and $1 \leq n \leq \beta^m$, let $\alpha_{m,n}$ be a median (cf. (1.4.11)) of $S_{\lfloor\beta^m\rfloor} - S_n$. Noting that, by (1.4.12), $|\alpha_{m,n}| \leq \sqrt{2\beta^m}$, we know that

$$\begin{aligned}\varlimsup_{n\to\infty} |\tilde{S}_n| &= \varlimsup_{m\to\infty} \max_{\beta^{m-1}\leq n\leq\beta^m} |\tilde{S}_n| \leq \beta^{\frac{1}{2}} \varlimsup_{m\to\infty} \max_{\beta^{m-1}\leq n\leq\beta^m} \frac{|S_n|}{\Lambda_{\beta^m}} \\ &\leq \beta^{\frac{1}{2}} \varlimsup_{m\to\infty} \max_{n\leq\beta^m} \frac{|S_n+\alpha_{m,n}|}{\Lambda_{\beta^m}},\end{aligned}$$

and therefore

$$\mathbb{P}\Big(\varlimsup_{n\to\infty} |\tilde{S}_n| \geq a\Big) \leq \mathbb{P}\left(\varlimsup_{m\to\infty} \max_{n\leq\beta^m} \frac{|S_n+\alpha_{m,n}|}{\Lambda_{\beta^m}} \geq a\,\beta^{-\frac{1}{2}}\right).$$

But, by Theorem 1.4.13,

$$\mathbb{P}\left(\max_{n\leq\beta^m} \frac{|S_n+\alpha_{m,n}|}{\Lambda_{\beta^m}} \geq a\beta^{-\frac{1}{2}}\right) \leq 2\mathbb{P}\left(\big|\tilde{S}_{\beta^m}\big| \geq a\beta^{-\frac{1}{2}}\right),$$

and so the desired result follows from the Borel–Cantelli Lemma. □

---

[1] See §§ 8.4.2 and 8.6.3 and, for much more information, M. Ledoux and M. Talagrand, *Probability in Banach Spaces*, Springer-Verlag, Ergebnisse Series 3.Folge·Band 23 (1991).

[2] Here and elsewhere, I use (a.s.,$\mathbb{P}$) to abbreviate "$\mathbb{P}$-almost surely."

LEMMA 1.5.4. *For any sequence $\{X_n : n \geq 1\}$ of independent, identically distributed random variables with mean value 0 and variance $\sigma^2$,*

$$\overline{\lim_{n\to\infty}} \left|\tilde{S}_n\right| \leq 8\sigma \quad (\text{a.s.}, \mathbb{P}). \tag{1.5.5}$$

PROOF: Without loss in generality, I assume throughout that $\sigma = 1$; and, for the moment, I will also assume that the $X_n$'s are symmetric (cf. Exercise 1.4.26). By Lemma 1.5.3, we will know that (1.5.5) holds with 8 replaced by 4 once I show that

$$\sum_{m=0}^{\infty} \mathbb{P}\Big(\left|\tilde{S}_{2^m}\right| \geq 2^{\frac{3}{2}}\Big) < \infty. \tag{*}$$

In order to take maximal advantage of symmetry, let $(\Omega, \mathcal{F}, \mathbb{P})$ be the probability space on which the $X_n$'s are defined, use $\{R_n : n \geq 1\}$ to denote the sequence of Rademacher functions on $[0,1)$ introduced in Section 1.1, and set $\mathbb{Q} = \lambda_{[0,1)} \times \mathbb{P}$ on $\big([0,1) \times \Omega, \mathcal{B}_{[0,1)} \times \mathcal{F}\big)$. It is then an easy matter to check that symmetry of the $X_n$'s is equivalent to the statement that

$$\omega \in \Omega \longrightarrow \big(X_1(\omega), \dots, X_n(\omega), \dots\big) \in \mathbb{R}^{\mathbb{Z}^+}$$

has the same distribution under $\mathbb{P}$ as

$$(t,\omega) \in [0,1) \times \Omega \longmapsto \big(R_1(t)X_1(\omega), \dots, R_n(t)X_n(\omega), \dots\big) \in \mathbb{R}^{\mathbb{Z}^+}$$

has under $\mathbb{Q}$. Next, using the last part of (**iii**) in Exercise 1.3.18 with $\sigma_k = X_k(\omega)$, note that

$$\begin{aligned}\lambda_{[0,1)}&\left(\left\{t \in [0,1) : \left|\sum_{n=1}^{2^m} R_n(t)X_n(\omega)\right| \geq a\right\}\right) \\ &\leq 2\exp\left[-\frac{a^2}{2\sum_{n=1}^{2^m} X_n(\omega)^2}\right], \quad a \in [0,\infty) \text{ and } \omega \in \Omega.\end{aligned}$$

Hence, if

$$A_m \equiv \left\{\omega \in \Omega : \frac{1}{2^m}\sum_{n=1}^{2^m} X_m(\omega)^2 \geq 2\right\}$$

and

$$F_m(\omega) \equiv \lambda_{[0,1)}\left(\left\{t \in [0,1) : \left|\sum_{n=1}^{2^m} R_n(t)X_n(\omega)\right| \geq 2^{\frac{3}{2}}\Lambda_{2^m}\right\}\right),$$

then, by Tonelli's Theorem,

$$\mathbb{P}\left(\left\{\omega \in \Omega : \left|S_{2^m}(\omega)\right| \geq 2^{\frac{3}{2}}\Lambda_{2^m}\right\}\right) = \int_\Omega F_m(\omega)\,\mathbb{P}(d\omega)$$

$$\leq 2\int_\Omega \exp\left[-\frac{8\Lambda_{2^m}^2}{2\sum_{n=1}^{2^m} X_n(\omega)^2}\right]\mathbb{P}(d\omega) \leq 2\exp\left[-4\log_{(2)} 2^m\right] + 2\mathbb{P}(A_m).$$

Thus, (*) comes down to proving that $\sum_{m=0}^\infty \mathbb{P}(A_m) < \infty$; and, in order to check this, I argue in much the same way as I did when I proved the converse statement in Kolmogorov's Strong Law. Namely, set

$$T_m = \sum_{n=1}^{2^m} X_n^2, \quad B_m = \left\{\frac{T_{m+1} - T_m}{2^m} \geq 2\right\}, \quad \text{and} \quad \overline{T}_m = \frac{T_m}{2^m}$$

for $m \in \mathbb{N}$. Clearly, $\mathbb{P}(A_m) = P(B_m)$. Moreover, the sets $B_m$, $m \in \mathbb{N}$, are mutually independent; and therefore, by the Borel–Cantelli Lemma, I need only check that

$$\mathbb{P}\left(\varlimsup_{m\to\infty} B_m\right) = P\left(\varlimsup_{m\to\infty} \frac{T_{m+1} - T_m}{2^m} \geq 2\right) = 0.$$

But, by the Strong Law, we know that $\overline{T}_m \longrightarrow 1$ (a.s., $\mathbb{P}$), and therefore it is clear that

$$\frac{T_{m+1} - T_m}{2^m} \longrightarrow 1 \quad \text{(a.s., } \mathbb{P}\text{)}.$$

I have now proved (1.5.5) with 4 replacing 8 for symmetric random variables. To eliminate the symmetry assumption, again let $(\Omega, \mathcal{F}, \mathbb{P})$ be the probability space on which the $X_n$'s are defined, let $(\Omega', \mathcal{F}', \mathbb{P}')$ be a second copy of the same space, and consider the random variables

$$(\omega, \omega') \in \Omega \times \Omega' \longmapsto Y_n(\omega, \omega') \equiv X_n(\omega) - X_n(\omega')$$

under the measure $\mathbb{Q} \equiv \mathbb{P} \times \mathbb{P}'$. Since the $Y_n$'s are obviously (cf. part (**i**) of Exercise 1.4.21) symmetric, the result which I have already proved says that

$$\varlimsup_{n\to\infty} \frac{|S_n(\omega) - S_n(\omega')|}{\Lambda_n} \leq 2^{\frac{5}{2}} \leq 8 \quad \text{for } \mathbb{Q}\text{-almost every } (\omega, \omega') \in \Omega \times \Omega'.$$

Now suppose that $\varlimsup_{n\to\infty} \frac{|S_n|}{\Lambda_n} > 8$ on a set of positive $\mathbb{P}$-measure. Then, by Kolmogorov's 0–1 Law, there would exist an $\epsilon > 0$ such that

$$\varlimsup_{n\to\infty} \frac{|S_n(\omega)|}{\Lambda_n} \geq 8 + \epsilon \quad \text{for } \mathbb{P}\text{-almost every } \omega \in \Omega;$$

and so, by Fubini's Theorem,[3] we would have that, for $\mathbb{Q}$-almost every $(\omega, \omega') \in \Omega \times \Omega'$, there is a $\{n_m(\omega) : m \in \mathbb{Z}^+\} \subseteq \mathbb{Z}^+$ such that $n_m(\omega) \nearrow \infty$ and

$$\varlimsup_{m\to\infty} \frac{\left|S_{n_m(\omega)}(\omega')\right|}{\Lambda_{n_m(\omega)}} \geq \varlimsup_{m\to\infty} \frac{\left|S_{n_m(\omega)}(\omega)\right|}{\Lambda_{n_m(\omega)}} - \varlimsup_{m\to\infty} \frac{\left|S_{n_m(\omega)}(\omega) - S_{n_m(\omega)}(\omega')\right|}{\Lambda_{n_m(\omega)}} \geq \epsilon.$$

But, again by Fubini's Theorem, this would mean that there exists a $\{n_m : m \in \mathbb{Z}^+\} \subseteq \mathbb{Z}^+$ such that $n_m \nearrow \infty$ and $\varlimsup_{m\to\infty} \frac{\left|S_{n_m}(\omega')\right|}{\Lambda_{n_m}} \geq \epsilon$ for $\mathbb{P}'$-almost every $\omega' \in \Omega'$, and obviously this contradicts

$$\mathbb{E}^{\mathbb{P}'}\left[\left(\frac{S_n}{\Lambda_n}\right)^2\right] = \frac{1}{2\log_{(2)} n} \longrightarrow 0. \quad \square$$

We have now got the *crude statement* alluded to above. In order to get the more precise statement contained in (1.5.2), I will need the following application of the results in § 1.3.

LEMMA 1.5.6. *Let $\{X_n : n \geq 1\}$ be a sequence of independent random variables with mean value 0, variance 1, and common distribution $\mu$. Further, assume that (1.3.4) holds. Then, for each $R \in (0, \infty)$ there is an $N(R) \in \mathbb{Z}^+$ such that*

$$\mathbb{P}\left(\left|\tilde{S}_n\right| \geq R\right) \leq 2\exp\left[-\left(1 - K\sqrt{\frac{8R\log_{(2)} n}{n}}\right) R^2 \log_{(2)} n\right] \tag{1.5.7}$$

*for $n \geq N(R)$. In addition, for each $\epsilon \in (0, 1]$, there is an $N(\epsilon) \in \mathbb{Z}^+$ such that, for all $n \geq N(\epsilon)$ and $|a| \leq \frac{1}{\epsilon}$,*

$$\mathbb{P}\left(\left|\tilde{S}_n - a\right| < \epsilon\right) \geq \frac{1}{2}\exp\left[-\left(a^2 + 4K|a|\epsilon\right) \log_{(2)} n\right]. \tag{1.5.8}$$

*In both (1.5.7) and (1.5.8), the constant $K \in (0, \infty)$ is the one in Theorem 1.3.15.*

PROOF: Set

$$\lambda_n = \frac{\Lambda_n}{n} = \left(\frac{2\log_{(2)}(n \vee 3)}{n}\right)^{\frac{1}{2}}.$$

To prove (1.5.7), simply apply the upper bound in the last part of Theorem 1.3.15 to see that, for sufficiently large $n \in \mathbb{Z}^+$,

$$\mathbb{P}\left(\left|\tilde{S}_n\right| \geq R\right) = \mathbb{P}\left(\left|\overline{S}_n\right| \geq R\lambda_n\right) \leq 2\exp\left[-n\left(\frac{(R\lambda_n)^2}{2} - K(R\lambda_n)^3\right)\right].$$

[3] This is Fubini at his best and subtlest. Namely, I am using Fubini to switch between *horizontal* and *vertical* sets of measure 0.

To prove (1.5.8), first note that

$$\mathbb{P}\left(\left|\tilde{S}_n - a\right| < \epsilon\right) = \mathbb{P}\left(\left|\overline{S}_n - a_n\right| < \epsilon_n\right),$$

where $a_n = a\lambda_n$ and $\epsilon_n = \epsilon\lambda_n$. Thus, by the lower bound in the last part of Theorem 1.3.15,

$$\begin{aligned} P\left(\left|\tilde{S}_n - a\right| < \epsilon\right) &\geq \left(1 - \frac{K}{n\epsilon_n^2}\right) \exp\left[-n\left(\frac{a_n^2}{2} + K|a_n|(\epsilon_n + a_n^2)\right)\right] \\ &\geq \left(1 - \frac{K}{2\epsilon^2 \log_{(2)} n}\right) \exp\left[-\left(a^2 + 2K|a|(\epsilon + a^2\lambda_n)\right)\log_{(2)} n\right] \end{aligned}$$

for sufficiently large $n$'s. □

THEOREM 1.5.9 (**Law of Iterated Logarithm**). *The equation (1.5.2) holds for any sequence $\{X_n : n \geq 1\}$ of independent, identically distributed random variables with mean value 0 and variance 1. In fact, $\mathbb{P}$-almost surely, the set of limit points of $\left\{\frac{S_n}{\Lambda_n} : n \geq 1\right\}$ coincides with the entire interval $[-1,1]$. Equivalently, for any $f \in C(\mathbb{R};\mathbb{R})$,*

$$\overline{\lim_{n\to\infty}} f\left(\frac{S_n}{\Lambda_n}\right) = \sup_{t\in[-1,1]} f(t) \quad (\text{a.s.}, \mathbb{P}). \tag{1.5.10}$$

*(Cf. Exercise 1.5.12 for a converse statement and §§ 8.4.2 and 8.6.3 for related results.)*

PROOF: I begin with the observation that, because of (1.5.5), I may restrict my attention to the case when the $X_n$'s are bounded random variables. Indeed, for any $X_n$'s and any $\epsilon > 0$, an easy truncation procedure allows us to find an $\psi \in C_b(\mathbb{R};\mathbb{R})$ such that $Y_n \equiv \psi \circ X_n$ again has mean value 0 and variance 1 while $Z_n \equiv X_n - Y_n$ has variance less than $\epsilon^2$. Hence, if the result is known when the random variables are bounded, then, by (1.5.5) applied to the $Z_n$'s,

$$\overline{\lim_{n\to\infty}} \left|\tilde{S}_n(\omega)\right| \leq 1 + \overline{\lim_{n\to\infty}} \left|\frac{\sum_{m=1}^n Z_m(\omega)}{\Lambda_n}\right| \leq 1 + 8\epsilon,$$

and, for $a \in [-1,1]$,

$$\underline{\lim_{n\to\infty}} \left|\tilde{S}_n(\omega) - a\right| \leq \overline{\lim_{n\to\infty}} \left|\frac{\sum_{m=1}^n Z_m(\omega)}{\Lambda_n}\right| \leq 8\epsilon$$

for $\mathbb{P}$-almost every $\omega \in \Omega$.

In view of the preceding, from now on I may and will assume that the $X_n$'s are bounded. To prove that $\overline{\lim}_{n\to\infty} \tilde{S}_n \le 1$ (a.s., $\mathbb{P}$), let $\beta \in (1,\infty)$ be given, and use (1.5.7) to see that

$$\mathbb{P}\Big(\big|\tilde{S}_{\beta^m}\big| \ge \beta^{\frac{1}{2}}\Big) \le 2\exp\Big[-\beta^{\frac{1}{2}}\log_{(2)}\lfloor\beta^m\rfloor\Big]$$

for all sufficiently large $m \in \mathbb{Z}^+$. Hence, by Lemma 1.5.3 with $a = \beta$, we see that $\overline{\lim}_{n\to\infty} \big|\tilde{S}_n\big| \le \beta$ (a.s., $\mathbb{P}$) for every $\beta \in (1,\infty)$. To complete the proof, I must still show that, for every $a \in (-1,1)$ and $\epsilon > 0$,

$$\mathbb{P}\Big(\underline{\lim}_{n\to\infty} \big|\tilde{S}_n - a\big| < \epsilon\Big) = 1.$$

Because I want to get this conclusion as an application of the second part of the Borel–Cantelli Lemma, it is important that we be dealing with independent events, and for this purpose I use the result just proved to see that, for every integer $k \ge 2$,

$$\begin{aligned}\underline{\lim}_{n\to\infty} \big|\tilde{S}_n - a\big| &\le \inf_{k\to\infty} \underline{\lim}_{m\to\infty} \big|\tilde{S}_{k^m} - a\big| \\ &= \inf_{k\to\infty} \underline{\lim}_{m\to\infty} \left|\frac{S_{k^m} - S_{k^{m-1}}}{\Lambda_{k^m}} - a\right| \quad \mathbb{P}\text{-almost surely.}\end{aligned}$$

Thus, because the events

$$A_{k,m} \equiv \left\{\left|\frac{S_{k^m} - S_{k^{m-1}}}{\Lambda_{k^m}} - a\right| < \epsilon\right\}, \quad m \in \mathbb{Z}^+,$$

are independent for each $k \ge 2$, all that I need to do is check that

$$\sum_{m=1}^{\infty} \mathbb{P}(A_{k,m}) = \infty \quad \text{for sufficiently large } k \ge 2.$$

But

$$\mathbb{P}(A_{k,m}) = \mathbb{P}\left(\left|\tilde{S}_{k^m - k^{m-1}} - \frac{\Lambda_{k^m}\, a}{\Lambda_{k^m - k^{m-1}}}\right| < \frac{\Lambda_{k^m}\,\epsilon}{\Lambda_{k^m - k^{m-1}}}\right),$$

and, because

$$\lim_{k\to\infty} \max_{m\in\mathbb{Z}^+} \left|\frac{\Lambda_{k^m}}{\Lambda_{k^m - k^{m-1}}} - 1\right| = 0,$$

everything reduces to showing that

$$(*) \qquad \sum_{m=1}^{\infty} \mathbb{P}\Big(\big|\tilde{S}_{k^m - k^{m-1}} - a\big| < \epsilon\Big) = \infty$$

for each $k \ge 2$, $a \in (-1,1)$, and $\epsilon > 0$. Finally, referring to (1.5.8), choose $\epsilon_0 > 0$ so small that $\rho \equiv a^2 + 4K\epsilon_0|a| < 1$, and conclude that, when $0 < \epsilon < \epsilon_0$,

$$\mathbb{P}\Big(\big|\tilde{S}_n - \big| < \epsilon\Big) \ge \frac{1}{2}\exp\Big[-\rho\log_{(2)} n\Big]$$

for all sufficiently large $n$'s, from which (*) is easy. □

REMARK 1.5.11. The reader should notice that the Law of the Iterated Logarithm provides a *naturally occurring* sequence of functions that converge in measure but not almost everywhere. Indeed, it is obvious that $\tilde{S}_n \longrightarrow 0$ in $L^2(\mathbb{P};\mathbb{R})$, but the Law of the Iterated Logarithm says that $\{\tilde{S}_n : n \geq 1\}$ is wildly divergent when looked at in terms of $\mathbb{P}$-almost sure convergence.

## Exercises for § 1.5

EXERCISE 1.5.12. Let $\{X_n : n \geq 1\}$ be a sequence of mutually independent, identically distributed random variables for which

$$\mathbb{P}\left(\varlimsup_{n\to\infty} \frac{|S_n|}{\Lambda_n} < \infty\right) > 0. \tag{1.5.13}$$

In this exercise I[4] will outline a proof that $X_1$ is $\mathbb{P}$-square integrable, $\mathbb{E}^{\mathbb{P}}[X_1] = 0$, and

$$\varlimsup_{n\to\infty} \frac{S_n}{\Lambda_n} = -\varliminf_{n\to\infty} \frac{S_n}{\Lambda_n} = \mathbb{E}^{\mathbb{P}}[X_1^2]^{\frac{1}{2}} \quad \text{(a.s., }\mathbb{P}). \tag{1.5.14}$$

**(i)** Using Lemma 1.4.1, show that there is a $\sigma \in [0, \infty)$ such that

$$\varlimsup_{n\to\infty} \frac{|S_n|}{\Lambda_n} = \sigma \quad \text{(a.s., }\mathbb{P}). \tag{1.5.15}$$

Next, assuming that $X_1$ is $\mathbb{P}$-square integrable, use The Strong Law of Large Numbers together with Theorem 1.5.9 to show that $\mathbb{E}^{\mathbb{P}}[X_1] = 0$ and

$$\sigma = \mathbb{E}^{\mathbb{P}}[X_1^2]^{\frac{1}{2}} = \varlimsup_{n\to\infty} \frac{S_n}{\Lambda_n} = -\varliminf_{n\to\infty} \frac{S_n}{\Lambda_n} \quad \text{(a.s., }\mathbb{P}).$$

In other words, everything comes down to proving that (1.5.13) implies that $X_1$ is $\mathbb{P}$-square integrable.

**(ii)** Assume that the $X_n$'s are symmetric. For $t \in (0, \infty)$, set

$$\check{X}_n^t = X_n \mathbf{1}_{[0,t]}(|X_n|) - X_n \mathbf{1}_{(t,\infty)}(|X_n|),$$

and show that

$$(\check{X}_1^t, \ldots, \check{X}_n^t, \ldots) \quad \text{and} \quad (X_1, \ldots, X_n, \ldots)$$

[4] I follow Wm. Feller "An extension of the law of the iterated logarithm to variables without variance," *J. Math. Mech.*, **18** #4, pp. 345–355 (1968), although V. Strassen was the first to prove the result.

have the same distribution. Conclude first that, for all $t \in [0,1)$,

$$\varlimsup_{n\to\infty} \frac{\left|\sum_{m=1}^n X_n \mathbf{1}_{[0,t]}\big(|X_n|\big)\right|}{\Lambda_n} \le \sigma \quad (\text{a.s.}, \mathbb{P}),$$

where $\sigma$ is the number in (1.5.15), and second that

$$\mathbb{E}^{\mathbb{P}}\big[X_1^2\big] = \lim_{t\nearrow\infty} \mathbb{E}^{\mathbb{P}}\Big[X_1^2,\, |X_1| \le t\Big] \le \sigma^2.$$

**Hint**: Use the equation

$$X_n \mathbf{1}_{[0,t]}\big(|X_n|\big) = \frac{X_n + \check{X}_n^t}{2},$$

and apply part (**i**).

(**iii**) For general $\{X_n : n \ge 1\}$, produce an independent copy $\{X_n' : n \ge 1\}$ (as in the proof of Lemma 1.5.4), and set $Y_n = X_n - X_n'$. After checking that

$$\varlimsup_{n\to\infty} \frac{\left|\sum_{m=1}^n Y_m\right|}{\Lambda_n} \le 2\sigma \quad (\text{a.s.}, \mathbb{P}),$$

conclude first that $\mathbb{E}^{\mathbb{P}}\big[Y_1^2\big] \le 4\sigma^2$ and then (cf. part (**i**) of Exercise1.4.27) that $\mathbb{E}^{\mathbb{P}}\big[X_1^2\big] < \infty$. Finally, apply (**i**) to arrive at $\mathbb{E}^{\mathbb{P}}\big[X_1\big] = 0$ and (1.5.14).

EXERCISE 1.5.16. Let $\{\tilde{s}_n : n \ge 1\}$ be a sequence of real numbers which possess the properties that

$$\varlimsup_{n\to\infty} \tilde{s}_n = 1, \quad \varliminf_{n\to\infty} \tilde{s}_n = -1, \quad \text{and} \quad \lim_{n\to\infty} \big|\tilde{s}_{n+1} - \tilde{s}_n\big| = 0.$$

Show that the set of subsequential limit points of $\{\tilde{s}_n : n \ge 1\}$ coincides with $[-1,1]$. Apply this observation to show that, in order to get the final statement in Theorem 1.5.9, I need only have proved (1.5.10) for the function $f(x) = x$, $x \in \mathbb{R}$.

**Hint**: In proving the last part, use the square integrability of $X_1$ to see that

$$\sum_{n=1}^{\infty} \mathbb{P}\left(\frac{X_n^2}{n} \ge 1\right) < \infty,$$

and apply the Borel–Cantelli Lemma to conclude that $\tilde{S}_n - \tilde{S}_{n-1} \longrightarrow 0$ (a.s., $\mathbb{P}$).

EXERCISE 1.5.17. Let $\{\mathbf{X}_n : n \ge 1\}$ be a sequence of $\mathbb{R}^N$-valued, identically distributed random variables on $(\Omega, \mathcal{F}, \mathbb{P})$ with the property that, for each $\mathbf{e} \in \mathbb{S}^{N-1} = \{\mathbf{x} \in \mathbb{R}^N : |\mathbf{x}| = 1\}$, $\big(\mathbf{e}, \mathbf{X}_1\big)_{\mathbb{R}^N}$ has mean value 0 and variance 1. Set $\mathbf{S}_n = \sum_{m=1}^n \mathbf{X}_m$ and $\tilde{\mathbf{S}}_n = \frac{\mathbf{S}_n}{\Lambda_n}$, and show that $\varlimsup_{n\to\infty} |\tilde{\mathbf{S}}_n| = 1$ $\mathbb{P}$-almost surely. Here are some steps that you might want to follow.

**(i)** Let $\{\mathbf{e}_k : k \geq 1\}$ be a countable, dense subset of $\mathbb{S}^{N-1}$ for which $\{\mathbf{e}_1, \dots, \mathbf{e}_N\}$ is orthonormal, and suppose that the sequence $\{\tilde{\mathbf{s}}_n : n \geq 1\} \subseteq \mathbb{R}^N$ has the property that $\overline{\lim}_{n\to\infty}\big|(\mathbf{e}_k, \tilde{\mathbf{s}}_n)_{\mathbb{R}^N}\big| = 1$ for each $k \geq 1$. Note that $|\tilde{\mathbf{s}}_n| \leq N^{\frac{1}{2}} \max_{1\leq k\leq N}\big|(\mathbf{e}_k, \tilde{s}_n)_{\mathbb{R}^N}\big|$, and conclude that $C \equiv \sup_{n\geq 1} |\tilde{\mathbf{s}}_n| \in [1, \infty)$.

**(ii)** Continuing **(i)**, for a given $\epsilon > 0$, choose $\ell \geq 1$ so that $\mathbb{S}^{N-1} \subseteq \bigcup_{k=1}^{\ell} B\big(\mathbf{e}_k, \frac{\epsilon}{C}\big)$. Show that

$$|\tilde{\mathbf{s}}_n| \leq \max_{1\leq k\leq \ell}\big|(\mathbf{e}, \tilde{\mathbf{s}}_n)_{\mathbb{R}^N}\big| + \epsilon,$$

and conclude first that $\overline{\lim}_{n\to\infty} |\tilde{\mathbf{s}}_n| \leq 1 + \epsilon$ and then that $\overline{\lim}_{n\to\infty} |\tilde{\mathbf{s}}_n| \leq 1$. At the same time, since $|\tilde{\mathbf{s}}_n| \geq \big|(\mathbf{e}_1, \tilde{\mathbf{s}}_n)_{\mathbb{R}^N}\big|$, show that $\overline{\lim}_{n\to\infty} |\tilde{\mathbf{s}}_n| \geq 1$. Thus $\overline{\lim}_{n\to\infty} |\tilde{\mathbf{s}}_n| = 1$.

**(iii)** Let $\{\mathbf{e}_k : k \geq 1\}$ be as in **(i)**, and apply Theorem 1.5.9 to show that, for $\mathbb{P}$-almost all $\omega \in \Omega$, the sequence $\{\tilde{\mathbf{S}}_n(\omega) : n \geq 1\}$ satisfies the condition in **(i)**. Thus, by **(ii)**, $\overline{\lim}_{n\to\infty} |\tilde{\mathbf{S}}_n(\omega)| = 1$ for $\mathbb{P}$-almost every $\omega \in \Omega$.

# Chapter 2
## The Central Limit Theorem

In the preceding chapter I dealt with averages of random variables and showed that, in great generality, those averages converge almost surely or in probability to a constant. At least when all the random variables have the same distribution and moments of all orders, one way of rationalizing this phenomenon is to recognize that the mean value is conserved whereas all higher centered moments are driven to 0 when one averages. Of course, the reason why it is easy to conserve the first moment is that the mean of the sum is the sum of the means. Thus, if one is going to attempt to find a simple normalization procedure that conserves a quantity involving more than the mean value, one should seek a quantity that shares this additivity property.

With this in mind, one is led to ask what happens if one normalizes in a way that conserves the variance. For this purpose, suppose that $\{X_n : n \in \mathbb{Z}^+\}$ is a sequence of mutually independent, identically distributed random variables with mean value 0 and variance 1, and set $S_n = \sum_1^n X_k$. Then $\breve{S}_n \equiv n^{-\frac{1}{2}} S_n$ again has mean value 0 and variance 1. On the other hand, because of Theorem 1.5.9, we know that, with probability 1, $\overline{\lim}_{n\to\infty} \breve{S}_n = \infty = -\underline{\lim}_{n\to\infty} \breve{S}_n$. Hence, from the point of view of either almost sure convergence or even convergence in probability, there is no hope that the $\breve{S}_n$'s will converge.

Nonetheless, the random variables $\{\breve{S}_n : n \geq 1\}$ possess remarkable stability when viewed from a distributional perspective. Indeed, if the $X_n$'s standard normal, then so are the $\breve{S}_n$'s, and therefore $\breve{S}_n \in N(0,1)$ for all $n \geq 1$. More generally, even if the $X_n$'s are not Gaussian, fixing their mean value and variance in this way forces *all* their moments to stabilize. To be precise, assume that $X_1$ has finite moments of all orders, that its mean is 0, and that its variance is 1. Trivially, $L_1 \equiv \lim_{n\to\infty} \mathbb{E}^{\mathbb{P}}[\breve{S}_n] = 0$ and $L_2 \equiv \lim_{n\to\infty} \mathbb{E}^{\mathbb{P}}[\breve{S}_n^2] = 1$. Next, assume that $L_\ell \equiv \lim_{n\to\infty} \mathbb{E}^{\mathbb{P}}[\breve{S}_n^\ell]$ exists for $1 \leq \ell \leq m$, where $m \geq 2$. I will show now that $L_{m+1} \equiv \lim_{n\to\infty} \mathbb{E}^{\mathbb{P}}[\breve{S}_n^{m+1}]$ exists and is equal to $mL_{m-1}$. To this end, first note that, since $\mathbb{E}^{\mathbb{P}}[X_n] = 0$ and the $X_n$'s are independent and identically distributed,

$$\mathbb{E}^{\mathbb{P}}\big[S_n^{m+1}\big] = n\mathbb{E}^{\mathbb{P}}\big[X_n\big(X_n + S_{n-1}\big)^m\big] = n\sum_{j=0}^{m}\binom{m}{j}\mathbb{E}^{\mathbb{P}}\big[X_n^{j+1}\big]\mathbb{E}^{\mathbb{P}}\big[S_{n-1}^{m-j}\big]$$

$$= nm\mathbb{E}^{\mathbb{P}}\big[S_{n-1}^{m-1}\big] + n\sum_{j=2}^{m}\binom{m}{j}\mathbb{E}^{\mathbb{P}}\big[X_n^{j+1}\big]\mathbb{E}^{\mathbb{P}}\big[S_{n-1}^{m-j}\big].$$

Thus, after dividing through by $n^{\frac{m+1}{2}}$, one gets the desired conclusion when $n \to \infty$. Starting from $L_1 = 0$ and $L_2 = 1$, one now can use induction to check that $L_{2m-1} = 0$ and $L_{2m} = \prod_{\ell=1}^{m}(2\ell - 1) = \frac{(2m)!}{2^m m!}$ for all $m \in \mathbb{Z}^+$. That is,

$$\lim_{n\to\infty} \mathbb{E}^{\mathbb{P}}\big[\breve{S}_n^{2m-1}\big] = 0 \quad \text{and} \quad \lim_{n\to\infty} \mathbb{E}^{\mathbb{P}}\big[\breve{S}_n^{2m}\big] = \prod_{\ell=1}^{m}(2\ell - 1) = \frac{(2m)!}{2^m m!},$$

for all $m \in \mathbb{Z}^+$. In other words, at least when the $X_n$'s have moments of all orders, $\lim_{n\to\infty} \mathbb{E}^{\mathbb{P}}\big[\breve{S}_n^m\big]$ exists and is independent of the particular choice of random variables. In particular, since for the Gaussian case, $\mathbb{E}^{\mathbb{P}}[\breve{S}_n^m] = \mathbb{E}^{\mathbb{P}}[X_1^m]$, we conclude that all moments of the $\breve{S}_n$'s converge to the corresponding moments of a standard normal random variable.

In this chapter we will see that the preceding stabilization result is just one manifestation of a general principle known as the *Central Limit phenomenon.*

## § 2.1 The Basic Central Limit Theorem

In this section I will derive the basic Central Limit Theorem using a beautiful argument which was introduced by J. Lindeberg. Throughout, $\langle \varphi, \mu\rangle$ denotes the integral of a function $\varphi$ against a measure $\mu$.

**§ 2.1.1. Lindeberg's Theorem.** Let $\{X_n : n \geq 1\}$ be a sequence of independent, square integrable random variables with mean value 0, and set $\breve{S}_n = n^{-\frac{1}{2}}\sum_{m=1}^{n} X_m$. At least when the $X_n$'s are identically distributed and have moments of all orders and variance 1, we just saw that (recall that $\gamma_{m,\sigma^2}$ is the distribution of an $N(m, \sigma^2)$-random variable)

$$\lim_{n\to\infty} \mathbb{E}^{\mathbb{P}}\big[\varphi(\breve{S}_n)\big] = \langle \varphi, \gamma_{0,1}\rangle \tag{2.1.1}$$

for any polynomial $\varphi : \mathbb{R} \longrightarrow \mathbb{C}$. In this subsection, I will prove a result that shows that, under much more general conditions, (2.1.1) holds for all $\varphi \in C^3(\mathbb{R}; \mathbb{C})$ with bounded second and third order derivatives.

In the following statement,

$$\sigma_m = \sqrt{\operatorname{Var}(X_m)} > 0, \quad \Sigma_n = \sqrt{\operatorname{Var}(S_n)} = \sqrt{\sum_{m=1}^{n} \sigma_m^2}, \quad \text{and } \breve{S}_n \equiv \frac{S_n}{\Sigma_n}. \tag{2.1.2}$$

Notice that when the $X_k$'s are identically distributed and have variance 1, the $\breve{S}_n$ in (2.1.2) is consistent with the notation used above. Finally, set

$$r_n = \max_{1\leq m\leq n} \frac{\sigma_m}{\Sigma_n} \quad \text{and} \quad g_n(\epsilon) = \frac{1}{\Sigma_n^2}\sum_{m=1}^{n} \mathbb{E}^{\mathbb{P}}\Big[X_m^2,\, |X_m| \geq \epsilon\Sigma_n\Big] \tag{2.1.3}$$

for $\epsilon > 0$. Clearly, in the identically distributed case, $r_n = n^{-\frac{1}{2}}$ and

$$g_n(\epsilon) = \sigma_1^{-2}\mathbb{E}^{\mathbb{P}}\Big[X_1^2,\, |X_1| \geq n^{\frac{1}{2}}\sigma_1\epsilon\Big] \longrightarrow 0 \quad \text{as } n \to \infty \text{ for each } \epsilon > 0.$$

THEOREM 2.1.4 (**Lindeberg**). *Refer to the preceding, and let $\varphi$ be an element of $C^3(\mathbb{R};\mathbb{R})$ with bounded second and third order derivatives. Then, for each $\epsilon > 0$,*

$$\text{(2.1.5)} \qquad \Big|\mathbb{E}^{\mathbb{P}}\big[\varphi(\breve{S}_n)\big] - \langle \varphi, \gamma_{0,1}\rangle\Big| \leq \left(\frac{\epsilon}{6} + \frac{r_n}{2}\right)\|\varphi'''\|_{\mathrm{u}} + g_n(\epsilon)\|\varphi''\|_{\mathrm{u}}.$$

*In particular, because*

$$\text{(2.1.6)} \qquad r_n^2 \leq \epsilon^2 + g_n(\epsilon), \quad \epsilon > 0,$$

*(2.1.1) holds if $g_n(\epsilon) \longrightarrow 0$ as $n \to \infty$ for each $\epsilon > 0$.*

PROOF: Choose $N(0,1)$-random variables $Y_1, \dots, Y_n$ which are both mutually independent and independent of the $X_m$'s. (After augmenting the probability space, if necessary, this can be done as an application of either Theorem 1.1.7 or Exercise 1.1.14.) Next, set

$$\breve{Y}_k = \frac{\sigma_k Y_k}{\Sigma_n} \quad \text{and} \quad \breve{T}_n = \sum_1^n \breve{Y}_k,$$

and observe that $\breve{T}_n$ is again an $N(0,1)$-random variable and therefore that

$$\Delta \equiv \Big|\mathbb{E}^{\mathbb{P}}\big[\varphi(\breve{S}_n)\big] - \langle \varphi, \gamma_{0,1}\rangle\Big| = \Big|\mathbb{E}^{\mathbb{P}}\big[\varphi(\breve{S}_n)\big] - \mathbb{E}^{\mathbb{P}}\big[\varphi(\breve{T}_n)\big]\Big|.$$

Further, set $\breve{X}_k = \frac{X_k}{\Sigma_n}$, and define

$$U_m = \sum_{1\leq k\leq m-1} \breve{Y}_k + \sum_{m+1\leq k\leq n} \breve{X}_k \quad \text{for } 1 \leq m \leq n,$$

where a sum over the empty set is taken to be 0. It is then clear that

$$\Delta \leq \sum_1^n \Delta_m \quad \text{where } \Delta_m \equiv \Big|\mathbb{E}^{\mathbb{P}}\big[\varphi(U_m + \breve{X}_m)\big] - \mathbb{E}^{\mathbb{P}}\big[\varphi(U_m + \breve{Y}_m)\big]\Big|.$$

Moreover, if

$$R_m(\xi) \equiv \varphi(U_m + \xi) - \varphi(U_m) - \xi\varphi'(U_m) - \tfrac{\xi^2}{2}\varphi''(U_m), \quad \xi \in \mathbb{R},$$

then (because both $\breve{X}_m$ and $\breve{Y}_m$ are independent of $U_m$ and have the same first two moments)

$$\Delta_m = \left|\mathbb{E}^{\mathbb{P}}\left[R_m(\breve{X}_m)\right] - \mathbb{E}^{\mathbb{P}}\left[R_m(\breve{Y}_m)\right]\right| \le \left|\mathbb{E}^{\mathbb{P}}\left[R_m(\breve{X}_m)\right]\right| + \left|\mathbb{E}^{\mathbb{P}}\left[R_m(\breve{Y}_m)\right]\right|.$$

In order to complete the derivation of (2.1.5), note that, by Taylor's Theorem,

$$|R_m(\xi)| \le \left(\|\varphi'''\|_{\mathrm{u}}\tfrac{|\xi|^3}{6}\right) \wedge \left(\|\varphi''\|_{\mathrm{u}}|\xi|^2\right);$$

and therefore, for each $\epsilon > 0$,

$$\begin{aligned}
&\sum_1^n \mathbb{E}^{\mathbb{P}}\left[|R_m(\breve{X}_m)|\right] \\
&\le \frac{\|\varphi'''\|_{\mathrm{u}}}{6}\sum_1^n \mathbb{E}^{\mathbb{P}}\left[|\breve{X}_m|^3,\, |X_m| \le \epsilon\Sigma_n\right] + \|\varphi''\|_{\mathrm{u}}\sum_1^n \mathbb{E}^{\mathbb{P}}\left[\breve{X}_m^2,\, |X_m| \ge \epsilon\Sigma_n\right] \\
&\le \frac{\epsilon\|\varphi'''\|_{\mathrm{u}}}{6}\sum_1^n \frac{\sigma_m^2}{\Sigma_n^2} + \|\varphi''\|_{\mathrm{u}} g_n(\epsilon) = \frac{\epsilon\|\varphi'''\|_{\mathrm{u}}}{6} + \|\varphi''\|_{\mathrm{u}} g_n(\epsilon),
\end{aligned}$$

while

$$\sum_1^n \mathbb{E}^{\mathbb{P}}\left[|R_m(\breve{Y}_n)|\right] \le \frac{\|\varphi'''\|_{\mathrm{u}}}{6}\mathbb{E}^{\mathbb{P}}\left[|Y_1|^3\right]\sum_1^n \frac{\sigma_m^3}{\Sigma_n^3} \le \frac{3^{\frac{3}{4}} r_n \|\varphi'''\|_{\mathrm{u}}}{6}.$$

Hence, (2.1.5) is now proved.

Given (2.1.5), all that remains is to prove (2.1.6). However, for any $1 \le m \le n$ and $\epsilon > 0$,

$$\sigma_m^2 = \mathbb{E}^{\mathbb{P}}\left[X_m^2,\, |X_m| < \epsilon\Sigma_n\right] + \mathbb{E}^{\mathbb{P}}\left[X_m^2,\, |X_m| \ge \epsilon\Sigma_n\right] \le \Sigma_n^2\left(\epsilon^2 + g_n(\epsilon)\right). \quad \square$$

The condition that $g_n(\epsilon) \longrightarrow 0$ for each $\epsilon > 0$ is often called *Lindeberg's condition* because it was introduced by J. Lindeberg and it was he who proved that it is a sufficient condition for (2.1.1) to hold for all (cf. Theorem 2.1.8) $\varphi \in C_{\mathrm{b}}(\mathbb{R}^N;\mathbb{C})$. Later, Feller proved that (2.1.1) for all $\varphi \in C_{\mathrm{b}}(\mathbb{R}^N;\mathbb{R})$ plus $r_n \to 0$ imply that Lindeberg's condition holds. Together, these two results are known as the **Lindeberg–Feller Theorem**. See Exercise 2.3.20 for a proof of Feller's part.

**§ 2.1.2. The Central Limit Theorem.** If one is not concerned about rates of convergence, then the differentiability requirement on $\varphi$ can be dropped from the last part of Theorem 2.1.4. In order to understand the reason for this, it is helpful to couch the statement of Theorem 2.1.4 entirely in terms of measures. Thus, let $\mu_n$ denote the distribution of $\breve{S}_n$. Then, under Lindeberg's condition, Theorem 2.1.4 allows one to say that $\langle\varphi, \mu_n\rangle \longrightarrow \langle\varphi, \gamma_{0,1}\rangle$ for all $\varphi \in C^3(\mathbb{R}^N;\mathbb{C})$ with bounded second and third order derivatives. Because we are dealing with statements about integration and integration is a very forgiving operation, this sort of result self-improves. To be precise, I prove the following lemma.

LEMMA 2.1.7. *Suppose that* $\{\mu_n : n \geq 1\}$ *is a sequence of (non-negative) locally finite*[1] *Borel measures on* $\mathbb{R}^N$ *and that* $\mu$ *is a locally finite Borel measure on* $\mathbb{R}^N$ *with the property that* $\langle \varphi, \mu_n \rangle \longrightarrow \langle \varphi, \mu \rangle$ *for all* $\varphi \in C_c^\infty(\mathbb{R}^N;\mathbb{R})$. *Then, for any* $\psi \in C(\mathbb{R}^N;[0,\infty))$, $\langle \psi, \mu \rangle \leq \underline{\lim}_{n\to\infty} \langle \psi, \mu_n \rangle$. *Moreover, if* $\psi \in C(\mathbb{R}^N;[0,\infty))$ *is* $\mu_n$*-integrable for each* $n \in \mathbb{Z}^+$ *and if* $\langle \psi, \mu_n \rangle \longrightarrow \langle \psi, \mu \rangle \in [0,\infty)$, *then for any sequence* $\{\varphi_n : n \geq 1\} \subseteq C(\mathbb{R}^N;\mathbb{C})$ *that converges uniformly on compacts to a* $\varphi \in C(\mathbb{R}^N;\mathbb{C})$ *and satisfies* $|\varphi_n| \leq C\psi$ *for some* $C < \infty$ *and all* $n \geq 1$, $\langle \varphi_n, \mu_n \rangle \longrightarrow \langle \varphi, \mu \rangle$.

PROOF: Choose $\rho \in C_c^\infty(B(\mathbf{0},1);[0,\infty))$ with total integral 1, and set $\rho_\epsilon(\mathbf{x}) = \epsilon^{-N}\rho(\epsilon^{-1}\mathbf{x})$ for $\epsilon > 0$. Also, choose $\eta \in C_c^\infty(B(\mathbf{0},2);[0,1])$ so that $\eta = 1$ on $\overline{B(\mathbf{0},1)}$, and set $\eta_R(\mathbf{x}) = \eta(R^{-1}\mathbf{x})$ for $R > 0$.

Begin by noting that $\langle \varphi, \mu_n \rangle \longrightarrow \langle \varphi, \mu \rangle$ for all $\varphi \in C_c^\infty(\mathbb{R}^N;\mathbb{C})$. Next, suppose that $\varphi \in C_c(\mathbb{R}^N;\mathbb{C})$, and, for $\epsilon > 0$, set $\varphi_\epsilon = \rho_\epsilon \star \varphi$, the **convolution**

$$\int_{\mathbb{R}^N} \rho_\epsilon(\mathbf{x}-\mathbf{y})\varphi(\mathbf{y})\,d\mathbf{y}$$

of $\rho_\epsilon$ with $\varphi$. Then, for each $\epsilon > 0$, $\varphi_\epsilon \in C_c^\infty(\mathbb{R}^N;\mathbb{C})$ and therefore $\langle \varphi_\epsilon, \mu_n \rangle \longrightarrow \langle \varphi_\epsilon, \mu \rangle$. In addition, there is an $R > 0$ such that $\operatorname{supp}(\varphi_\epsilon) \subseteq B(0,R)$ for all $\epsilon \in (0,1]$. Hence,

$$\overline{\lim_{n\to\infty}} \big|\langle \varphi, \mu_n \rangle - \langle \varphi, \mu \rangle\big| \leq 2\langle \eta_R, \mu \rangle \|\varphi_\epsilon - \varphi\|_{\mathrm{u}}.$$

Since $\lim_{\epsilon\searrow 0} \|\varphi_\epsilon - \varphi\|_{\mathrm{u}} = 0$, we have now shown that $\langle \varphi, \mu_n \rangle \longrightarrow \langle \varphi, \mu \rangle$ for all $\varphi \in C_c(\mathbb{R}^N;\mathbb{C})$.

Now suppose that $\psi \in C(\mathbb{R}^N;[0,\infty))$, and set $\psi_R = \eta_R\psi$, where $\eta_R$ is as above. Then, for each $R > 0$, $\langle \psi_R, \mu \rangle = \lim_{n\to\infty} \langle \psi_R, \mu_n \rangle \leq \underline{\lim}_{n\to\infty} \langle \psi, \mu_n \rangle$. Hence, by Fatou's Lemma, $\langle \psi, \mu \rangle \leq \underline{\lim}_{R\to\infty} \langle \psi_R, \mu \rangle \leq \underline{\lim}_{n\to\infty} \langle \psi, \mu_n \rangle$.

Finally, suppose that $\psi \in C(\mathbb{R}^N;[0,\infty))$ is $\mu_n$-integrable for each $n \in \mathbb{Z}^+$ and that $\langle \psi, \mu_n \rangle \longrightarrow \langle \psi, \mu \rangle \in [0,\infty)$. Given $\{\varphi_n : n \geq 1\} \subseteq C(\mathbb{R}^N;\mathbb{C})$ satisfying $|\varphi_n| \leq C\psi$ and converging uniformly on compacts to $\varphi$, one has

$$\big|\langle \varphi_n, \mu_n \rangle - \langle \varphi, \mu \rangle\big| \leq \big|\langle \varphi_n - \varphi, \mu_n \rangle\big| + \big|\langle \varphi, \mu_n \rangle - \langle \varphi, \mu \rangle\big|.$$

Moreover, for each $R > 0$,

$$\begin{aligned}
&\overline{\lim_{n\to\infty}} \big|\langle \varphi_n - \varphi, \mu_n \rangle\big| \\
&\leq \overline{\lim_{n\to\infty}} \sup_{\mathbf{x}\in B(\mathbf{0},2R)} |\varphi_n(\mathbf{x}) - \varphi(\mathbf{x})| \langle \eta_R, \mu_n \rangle + \overline{\lim_{n\to\infty}} \big|\langle (1-\eta_R)(\varphi_n - \varphi), \mu_n \rangle\big| \\
&\leq 2C \overline{\lim_{n\to\infty}} \langle (1-\eta_R)\psi, \mu_n \rangle = \lim_{n\to\infty} 2C\big(\langle \psi, \mu_n \rangle - \langle \eta_R\psi, \mu_n \rangle\big) = 2C\langle (1-\eta_R)\psi, \mu \rangle,
\end{aligned}$$

[1] A Borel measure on a topological space is locally finite if it gives finite measure to compacts.

and similarly

$$\begin{aligned}&\varlimsup_{n\to\infty}\bigl|\langle\varphi,\mu_n\rangle-\langle\varphi,\mu\rangle\bigr|\\&\le\varlimsup_{n\to\infty}\bigl|\langle\eta_R\varphi,\mu_n\rangle-\langle\eta_R\varphi,\mu\rangle\bigr|+C\varlimsup_{n\to\infty}\langle(1-\eta_R)\psi,\mu_n\rangle+C\langle(1-\eta_R)\psi,\mu\rangle\\&=2C\langle(1-\eta_R)\psi,\mu\rangle.\end{aligned}$$

Finally, because $\psi$ is $\mu$-integrable, $\langle(1-\eta_R)\psi,\mu\rangle\longrightarrow 0$ as $R\to\infty$ by Lebesgue's Dominated Convergence Theorem, and so we are done. □

By combining Theorem 2.1.4 with the preceding, we have the following version of the famous **Central Limit Theorem.**

THEOREM 2.1.8 (**Central Limit Theorem**). *With the setting the same as it was in Theorem 2.1.4, assume that $g_n(\epsilon)\longrightarrow 0$ as $n\to\infty$ for each $\epsilon>0$. Then*

$$\lim_{n\to\infty}\mathbb{E}^{\mathbb{P}}\bigl[\varphi_n(\check{S}_n)\bigr]=\langle\varphi,\gamma_{0,1}\rangle$$

*whenever $\{\varphi_n : n\ge 1\}\subseteq C(\mathbb{R};\mathbb{C})$ satisfies*

$$\sup_{n\ge1}\sup_{y\in\mathbb{R}}\frac{|\varphi_n(y)|}{1+|y|^2}<\infty$$

*and tends to $\varphi$ uniformly on compacts. Moreover, if $-\infty\le a<b\le\infty$, then*

$$\lim_{n\to\infty}\mathbb{P}\bigl(a\le\check{S}_n\le b\bigr)=\gamma_{0,1}\bigl((a,b]\bigr)=\frac{1}{\sqrt{2\pi}}\int_a^b\exp\left[-\frac{y^2}{2}\right]dy.\tag{2.1.9}$$

*(See Exercise 2.1.10 for more information about the identically distributed case.)*

PROOF: Take $\mu_n$ to be the distribution of $\check{S}_n$. By Theorem 2.1.4, we know that $\langle\varphi,\mu_n\rangle\longrightarrow\langle\varphi,\gamma_{0,1}\rangle$ for all $\varphi\in C_c^\infty(\mathbb{R}^N;\mathbb{R})$. In addition, we know that $\langle\psi,\mu_n\rangle=2=\langle\psi,\gamma_{0,1}\rangle$ when $\psi(y)=1+y^2$. Hence, the first assertion is an application of Lemma 2.1.7.

Turning to the second assertion, let $a<b$ be given. To prove (2.1.9), choose $\{\varphi_k : k\ge1\}\subseteq C_b(\mathbb{R};\mathbb{R})$ and $\{\psi_k : k\ge1\}\subseteq C_b(\mathbb{R};\mathbb{R})$ so that $0\le\varphi_k\nearrow\mathbf{1}_{(a,b)}$ and $1\ge\psi_k\searrow\mathbf{1}_{[a,b]}$ as $k\to\infty$. Then,

$$\varliminf_{n\to\infty}\mathbb{P}\bigl(a<\check{S}_n<b\bigr)\ge\lim_{n\to\infty}\mathbb{E}^{\mathbb{P}}\bigl[\varphi_k(\check{S}_n)\bigr]=\int_{\mathbb{R}}\varphi_k(y)\,\gamma_{0,1}(dy)\longrightarrow\gamma_{0,1}\bigl((a,b)\bigr)$$

as $k\to\infty$, and, similarly,

$$\varlimsup_{n\to\infty}\mathbb{P}\bigl(a\le\check{S}_n\le b\bigr)\le\lim_{n\to\infty}\mathbb{E}^{\mathbb{P}}\bigl[\psi_k(\check{S}_n)\bigr]=\int_{\mathbb{R}}\psi_k(y)\,\gamma_{0,1}(dy)\longrightarrow\gamma_{0,1}\bigl([a,b]\bigr).$$

Finally, note that $\gamma_{0,1}\bigl((a,b)\bigr)=\gamma_{0,1}\bigl([a,b]\bigr)$. □

## Exercises for § 2.1

EXERCISE 2.1.10. Let $\{X_n : n \geq 1\}$ be a sequence of independent, identically distributed random variables, set $\breve{S}_n = n^{-\frac{1}{2}}\sum_{m=1}^n X_m$, and assume that

$$\overline{\lim_{n\to\infty}}\ \mathbb{E}^{\mathbb{P}}\big[\breve{S}_n^2 \wedge R^2\big] \leq 1 \quad \text{for every } R \in [0,\infty).$$

In particular, by Lemma 2.1.7, this will certainly be the case whenever (2.1.1) holds for every $\varphi \in C_c(\mathbb{R};\mathbb{R})$. The purpose of this exercise is to show that the $X_n$'s are $\mathbb{P}$-square integrable, have mean value 0, and variance no more than 1; and the method which I will use is based on the same line of reasoning as was given in Exercise 1.5.12.

**(i)** Assuming that $X_1 \in L^2(\mathbb{P};\mathbb{R})$, show that $\mathbb{E}^{\mathbb{P}}[X_1] = 0$ and $\mathbb{E}^{\mathbb{P}}[X_1^2] \leq 1$. In particular, use this together with the result in part **(i)** of Exercise 1.4.27 to see that it suffices to handle the case when the $X_n$'s are symmetric.

**(ii)** In this and the succeeding parts of this exercise, we will be assuming that the $X_n$'s are symmetric. Following the same route as was suggested in **(ii)** of Exercise 1.5.12, set

$$\check{X}_n^t = X_n\mathbf{1}_{[0,t]}\big(|X_n|\big) - X_n\mathbf{1}_{(t,\infty)}\big(|X_n|\big), \quad n \in \mathbb{Z}^+,$$

and recall that $\big(\check{X}_1^t,\dots,\check{X}_n^t,\dots\big)$ and $\big(X_1,\dots,X_n,\dots\big)$ have the same distribution for each $t \in (0,\infty)$. Use this together with our basic assumption to see that $\lim_{R\to\infty}\sup_{\substack{n\in\mathbb{Z}^+\\ t\in(0,\infty)}} \mathbb{P}\big(A_n(t,R)\big) = 0$, where

$$A_n(t,R) \equiv \left\{\left|\sum_1^n X_k\right| \vee \left|\sum_1^n \check{X}_k^t\right| \geq n^{\frac{1}{2}}R\right\}.$$

**(iii)** Continuing in the setting of part **(ii)**, set

$$\breve{S}_n^t = \frac{1}{n^{\frac{1}{2}}}\sum_1^n X_k\mathbf{1}_{[0,t]}\big(|X_k|\big).$$

After noting that the $X_n\mathbf{1}_{[0,t]}\big(|X_n|\big)$'s are symmetric, check (cf. the proof of Theorem 1.3.1) that $\mathbb{E}^{\mathbb{P}}\big[|\breve{S}_n^t|^4\big] \leq 3t^4$. In particular, conclude that, for each $t \in (0,\infty)$, there is an $R(t) \in (0,\infty)$ such that

$$\mathbb{E}^{\mathbb{P}}\big[|\breve{S}_n^t|^2,\, A_n\big(t,R(t)\big)\big] \leq 3^{\frac{1}{2}}t^2\,\mathbb{P}\big(A_n\big(t,R(t)\big)\big)^{\frac{1}{2}} \leq 1 \quad \text{for all } n \in \mathbb{Z}^+.$$

**(iv)** Given $t \in (0,\infty)$, choose $R(t) \in (0,\infty)$ as in the preceding. Taking into account the identity

$$\breve{S}_n^t = \frac{\sum_1^n X_k + \sum_1^n \breve{X}_k^t}{2n^{\frac{1}{2}}},$$

show that

$$\begin{aligned}\mathbb{E}^{\mathbb{P}}\big[X_1^2,\, |X_1| \le t\big] = \mathbb{E}^{\mathbb{P}}\big[|\breve{S}_n^t|^2\big] &\le \mathbb{E}^{\mathbb{P}}\big[|\breve{S}_n^t|^2,\, A_n(t,R(t))\complement\big] + 1 \\ &\le \mathbb{E}^{\mathbb{P}}\Big[\breve{S}_n^2 \wedge R(t)^2\Big] + 1\end{aligned}$$

for all $n \in \mathbb{Z}^+$ and $t \in (0,\infty)$. In particular, use this and our basic hypothesis to conclude first that $\mathbb{E}^{\mathbb{P}}\big[X_1^2,\, |X_1| \le t\big] \le 2$ for all $t \in (0,\infty)$ and then that $X_1$ is square $\mathbb{P}$-integrable.

**(v)** Show that (2.1.1) holds for all $\varphi \in C_c^\infty(\mathbb{R};\mathbb{R})$ if and only if $X_1$ has mean value 0 and variance 1.

EXERCISE 2.1.11. An interesting way in which to interpret The Central Limit Theorem is as the solution to a certain *fixed point problem.* Namely, let $\mathcal{P}$ denote the set of probability measures $\mu$ on $(\mathbb{R};\mathcal{B}_{\mathbb{R}})$ with the properties that

$$\int_{\mathbb{R}} x^2\,\mu(dx) = 1 \quad \text{and} \quad \int_{\mathbb{R}} x\,\mu(dx) = 0.$$

Next, define $T_2\mu$ for $\mu \in \mathcal{P}$ to be the probability measure on $(\mathbb{R},\mathcal{B}_{\mathbb{R}})$ given by

$$T_2\mu(\Gamma) = \iint_{\mathbb{R}^2} \mathbf{1}_\Gamma\left(\frac{x+y}{\sqrt{2}}\right)\mu(dx)\mu(dy) \quad \text{for} \quad \Gamma \in \mathcal{B}_{\mathbb{R}}.$$

After checking that $T_2$ maps $\mathcal{P}$ into itself, use The Central Limit Theorem to show that, for every $\mu \in \mathcal{P}$,

$$\lim_{n\to\infty} \int_{\mathbb{R}} \varphi\, dT_2^n\mu = \int_{\mathbb{R}} \varphi\, d\gamma_{0,1}, \quad \varphi \in C_b(\mathbb{R};\mathbb{C}).$$

Conclude, in particular, that $\gamma_{0,1}$ is the one and only element $\mu$ of $\mathcal{P}$ with the property that $T_2\mu = \mu$ and that this fixed point is *attracting.* (See Exercise 2.3.21 for more information.)

EXERCISE 2.1.12. Here is another indication of the remarkable stability of normal random variables. Namely, I will outline here a derivation[2] of the **Lévy–Cramér Theorem** which says that if $X$ and $Y$ are independent random variables whose sum is normal (with some mean and variance), then both $X$ and $Y$ are normal.

[2] This derivation is based on a note by Z. Sasvári, who himself borrowed some of the ideas from A. Rényi. I know of no derivation that does not rely on complex analysis and would be very interested in learning one.

**(i)** Assume that $X + Y \in N(a, \sigma^2)$, and, by subtracting $a$ from $X$, reduce to the case in which $X + Y \in N(0, \sigma^2)$. Next, show that there is nothing more to do when $\sigma = 0$ and that one can always reduce to the case $\sigma = 1$ when $\sigma > 0$. Thus, from now on, assume that $X + Y \in N(0, 1)$.

**(ii)** Choose $r \in (0, \infty)$ so that $\mathbb{P}(|X| \vee |Y| \geq r) \leq \frac{1}{2}$, and conclude that

$$\mathbb{P}(|X| \geq r + R) \vee \mathbb{P}(|Y| \geq r + R) \leq 4 \exp\left[-\frac{R^2}{2}\right], \quad R \in (0, \infty).$$

In particular, show that the moment generating functions $z \in \mathbb{C} \longmapsto M(z) = \mathbb{E}^{\mathbb{P}}\left[e^{zX}\right] \in \mathbb{C}$ and $z \in \mathbb{C} \longmapsto N(z) = \mathbb{E}^{\mathbb{P}}\left[e^{zY}\right] \in \mathbb{C}$ exist and are entire functions. Further, note that $M(z)N(z) = \exp\left[\frac{z^2}{2}\right]$, and conclude that $M$ and $N$ never vanish. Finally, from the fact that $X + Y$ has mean 0, show that one can reduce to the case in which both $X$ and $Y$ have mean 0. Thus, from now on, we assume that $M'(0) = 0 = N'(0)$.

**(iii)** Because $M$ never vanishes and $M(0) = 1$, elementary complex analysis (cf. Lemma 3.2.3) guarantees that there is a unique entire function $\theta : \mathbb{C} \longrightarrow \mathbb{C}$ such that $\theta(0) = 0$ and $M(z) = e^{\theta(z)}$ for all $z \in \mathbb{C}$. Further, from $M'(0) = 0$, note that $\theta'(0) = 0$. Thus,

$$\theta(z) = \sum_{n=2}^{\infty} c_n z^n \quad \text{where } n!c_n = \frac{d^n}{dx^n} \log\left(\mathbb{E}^{\mathbb{P}}\left[e^{xX}\right]\right)\bigg|_{x=0} \in \mathbb{R}.$$

Finally, note that $N(z) = \exp\left[\frac{z^2}{2} - \theta(z)\right]$.

**(iv)** As an application of Hölder's Inequality, observe that $x \in \mathbb{R} \longmapsto \theta(x) \in \mathbb{R}$ and $x \in \mathbb{R} \longmapsto \frac{x^2}{2} - \theta(x) \in \mathbb{R}$ are both convex. Thus, since $\theta'(0) = 0$, both these functions are non-increasing on $(-\infty, 0]$ and non-decreasing on $[0, \infty)$. Use this observation to check that

$$\theta(x) \geq 0 \leq \frac{x^2}{2} - \theta(x) \quad \text{for all } x \in \mathbb{R}.$$

Next, use the preceding in conjunction with the trivial remarks

$$\exp\left[\mathfrak{Re}\left(\theta(z)\right)\right] = \left|\mathbb{E}^{\mathbb{P}}\left[e^{zX}\right]\right| \leq e^{\theta(x)}$$

and

$$\exp\left[\mathfrak{Re}\left(\tfrac{z^2}{2} - \theta(z)\right)\right] = \left|\mathbb{E}^{\mathbb{P}}\left[e^{zY}\right]\right| \leq \exp\left[\tfrac{x^2}{2} - \theta(x)\right]$$

to arrive at

$$-y^2 \leq 2\mathfrak{Re}\left(\theta(z)\right) \leq x^2 \quad \text{for } z = x + \sqrt{-1}\,y \in \mathbb{C}.$$

In particular, this means that

$$\left|\mathfrak{Re}\big(\theta(z)\big)\right| \le \frac{|z|^2}{2}, \quad z \in \mathbb{C}.$$

(**v**) To complete the program, use Cauchy's integral formula to show that, for each $n \in \mathbb{Z}^+$ and $r > 0$, on the one hand,

$$c_n r^n = \frac{1}{2\pi}\int_0^{2\pi} \theta\left(re^{\sqrt{-1}\,\theta}\right) e^{-\sqrt{-1}\,n\theta}\, d\theta, \quad r > 0,$$

while, on the other hand (since $\overline{\theta(z)} = \theta(\bar{z})$ and therefore $\partial_z \overline{\theta(z)} = 0$),

$$0 = \int_0^{2\pi} \overline{\theta\left(re^{\sqrt{-1}\,\theta}\right)}\, e^{-\sqrt{-1}\,n\theta}\, d\theta.$$

Hence,

$$c_n r^n = \frac{1}{\pi}\int_0^{2\pi} \mathfrak{Re}\left(\theta\left(re^{\sqrt{-1}\,\theta}\right)\right) e^{-\sqrt{-1}\,n\theta}\, d\theta, \quad n \in \mathbb{Z}^+ \text{ and } r > 0.$$

Finally, in combination with the estimate obtained in (**iv**) and the fact that $c_0 = c_1 = 0$, this leads to the conclusion that $c_n = 0$ for $n \ne 2$ and therefore that $\theta(z) = c_2 z^2$ with $0 \le c_2 \le \frac{1}{2}$.

EXERCISE 2.1.13. An important result that is closely related to The Central Limit Theorem is the following observation, which occupies a central position in the development of classical statistical mechanics.[3]

(**i**) For each $n \in \mathbb{Z}^+$, let $\lambda_n$ denote the normalized surface measure on the $(n-1)$-dimensional sphere

$$\mathbb{S}^{n-1}\left(\sqrt{n}\right) = \left\{\mathbf{x} \in \mathbb{R}^n : |\mathbf{x}| = n^{\frac{1}{2}}\right\},$$

and denote by $\lambda_n^{(1)}$ the distribution of the coordinate $x_1$ under $\lambda_n$. Check that, when $n \ge 2$, $\lambda_n^{(1)}(dt) = f_n(t)\, dt$, where

$$f_n(t) = \frac{\omega_{n-2}}{n^{\frac{1}{2}}\omega_{n-1}}\left(1 - \frac{t^2}{n}\right)^{\frac{n-3}{2}} \mathbf{1}_{(-1,1)}\left(n^{-\frac{1}{2}}t\right),$$

[3] Although E. Borel seems to have thought he was the first to discover this result and rhapsodizes about it a good deal in "Sur les principes de la cinétique des gaz," *Ann. l'École Norm. sup.*, 3e **t. 23**, it appears already in the 1866 article "Über die Entwicklungen einer Funktion von beliebig vielen Variabeln nach Laplaceshen Funktionen höherer Ordnung," *J. Reine u. Angewandte Math.*, by F. Mehler and is only a small part of what Mehler discovered there. Be that as it may, Borel deserves credit for recognizing the significance of this result for statistical mechanics.

and $\omega_{k-1}$ denotes the surface area of the $(k-1)$-dimensional unit sphere in $\mathbb{R}^k$. Using polar coordinates to compute the right-hand side of

$$(2\pi)^{\frac{k}{2}} = \int_{\mathbb{R}^k} e^{-\frac{|\mathbf{x}|^2}{2}}\, dx,$$

first check that

$$\omega_{k-1} = \frac{2\pi^{\frac{k}{2}}}{\Gamma\left(\frac{k}{2}\right)},$$

where $\Gamma(t)$ is Euler's $\Gamma$-function (cf. (1.3.20)), and then apply Stirling's formula (cf. (1.3.21)) to see that

$$\frac{\omega_{n-2}}{n^{\frac{1}{2}}\omega_{n-1}} \longrightarrow \frac{1}{\sqrt{2\pi}} \quad \text{as} \quad n \to \infty.$$

Now, using $g$ to denote the density for the standard Gauss distribution (i.e., the Gauss kernel in (1.3.5)), apply these computations to show that

$$\sup_{n\geq 3}\sup_{t\in\mathbb{R}} \frac{f_n(t)}{g(t)} < \infty \quad \text{and that} \quad \frac{f_n(t)}{g(t)} \longrightarrow 1 \text{ uniformly on compacts.}$$

In particular, conclude that, for any $\varphi \in L^1(\gamma_{0,1};\mathbb{R})$,

$$\int_{\mathbb{R}} \varphi\, d\lambda_n^{(1)} \longrightarrow \int_{\mathbb{R}} \varphi\, d\gamma_{0,1}. \tag{2.1.14}$$

**(ii)** A less computational approach to the same calculation is the following. Let $\{X_n : n \geq 1\}$ be a sequence of independent $N(0,1)$ random variables, and set $R_n = \sqrt{X_1^2 + \cdots + X_n^2}$. First note that $\mathbb{P}(R_n = 0) = 0$ and then that the distribution of

$$\boldsymbol{\theta}_n \equiv \frac{n^{\frac{1}{2}}(X_1,\ldots,X_n)}{R_n}$$

is $\lambda_n$. Next, use The Strong Law of Large Numbers to see that $\frac{R_n^2}{n} \longrightarrow 1$ (a.s., $\mathbb{P}$) and conclude that, for any $N \in \mathbb{Z}^+$,

$$\lim_{n\to\infty} \mathbb{E}^{\mathbb{P}}\left[\varphi\left(\boldsymbol{\theta}_n^{(N)}\right)\right] = \mathbb{E}^{\mathbb{P}}\left[\varphi(X_1,\ldots,X_N)\right], \quad \varphi \in C_c(\mathbb{R}^N;\mathbb{R}),$$

where, for $n \geq N$, $\boldsymbol{\theta}_n^{(N)} \in \mathbb{R}^N$ denotes the projection of $\boldsymbol{\theta}_n \in \mathbb{R}^n$ onto its first $N$ coordinates. Conclude that if $\lambda_n^{(N)}$ on $\left(\mathbb{R}^N, \mathcal{B}_{\mathbb{R}^N}\right)$ denotes the distribution of $\mathbf{x} = (x_1,\ldots,x_n) \in \mathbb{R}^n \longmapsto \mathbf{x}^{(N)} \equiv (x_1,\ldots,x_N) \in \mathbb{R}^N$ under $\lambda_n$, then

$$\lim_{n\to\infty} \int_{\mathbb{R}^N} \varphi\, d\lambda_n^{(N)} = \int_{\mathbb{R}^N} \varphi\, d\gamma_{0,1}^N \quad \text{for all} \quad \varphi \in C_b(\mathbb{R}^N;\mathbb{C}).$$

**(iii)** By considering the case when $N = 2$, show that, for any $\varphi \in C_b(\mathbb{R};\mathbb{R})$,

$$\lim_{n\to\infty} \int_{\mathbf{S}^{n-1}(\sqrt{n})} \left( \frac{1}{n}\sum_{k=1}^{n} \varphi(x_k) - \int_{\mathbb{R}} \varphi\, d\gamma_{0,1} \right)^2 \lambda_n(d\mathbf{x}) = 0. \tag{2.1.15}$$

Notice that the non-computational argument has the advantage that it immediately generalizes the earlier result to cover $\lambda_n^{(N)}$ for all $N \in \mathbb{Z}^+$, not just $N = 1$ (cf. Exercise 2.3.24). On the other hand, the conclusion is weaker in the sense that convergence of the densities has been replaced by convergence of integrals with bounded continuous integrands and that no estimate on the rate of convergence is provided. More work is required to restore the stronger statements when $N \geq 2$.

When couched in terms of statistical mechanics, this result can be interpreted as a derivation of the Maxwell distribution of velocities for an **ideal gas** of free particles of mass 2 and having average energy 1.

EXERCISE 2.1.16. The most frequently encountered applications of Stirling's formula (cf. (1.3.21)) are to cases when $t \in \mathbb{Z}^+$. That is, one is usually interested in the formula

$$n! \sim \sqrt{2\pi n}\left(\frac{n}{e}\right)^n. \tag{2.1.17}$$

Here is a derivation of (2.1.17) as an application of The Central Limit Theorem. Namely, take $\{X_n : n \geq 1\}$ to be a sequence of independent, random variables with $\mathbb{P}(X_n > x) = \exp(-(x+1)^+)$, $x \in \mathbb{R}$ for all $n \in \mathbb{Z}^+$. For $n \geq 1$, note that

$$\begin{aligned}\mathbb{P}\left(\breve{S}_{n+1} \in [0, \tfrac{1}{4}]\right) &= \frac{1}{n!}\int_{1+n}^{1+4^{-1}\sqrt{n}+n} x^n e^{-x}\, dx \\ &= \frac{n^{n+\frac{1}{2}}e^{-n}}{n!}\int_{n^{-\frac{1}{2}}}^{n^{-\frac{1}{2}}+\frac{1}{4}\sqrt{1+n^{-1}}} \left(1+n^{-\frac{1}{2}}y\right)^n e^{-\sqrt{n}\,y}\, dy.\end{aligned}$$

By The Central Limit Theorem,

$$\mathbb{P}\left(\breve{S}_n \in [0, \tfrac{1}{4}]\right) \longrightarrow \frac{1}{\sqrt{2\pi}}\int_0^{\frac{1}{4}} e^{-\frac{x^2}{2}}\, dx.$$

At the same time, an elementary computation shows that

$$\int_{n^{-\frac{1}{2}}}^{n^{-\frac{1}{2}}+\frac{1}{4}\sqrt{1+n^{-1}}} \left(1+n^{-\frac{1}{2}}y\right)^n e^{-\sqrt{n}\,y}\, dy \longrightarrow \int_0^{\frac{1}{4}} e^{-\frac{x^2}{2}}\, dx,$$

and clearly (2.1.17) follows from these. In fact, if one applies the Berry–Esseen estimate proved in the next section, one finds that

$$\frac{\sqrt{2\pi n}\,\left(\frac{n}{e}\right)^n}{n!} = 1 + \mathcal{O}\big(n^{-\frac{1}{2}}\big).$$

However, this last observation is not very interesting since we saw in Exercise 1.3.19 that the true correction term is of order $n^{-1}$.[4]

## § 2.2 The Berry–Esseen Theorem via Stein's Method

As we will see in the next section, the principles underlying the passage from Theorem 2.1.4 to Theorem 2.1.8 are very general. In fact, as we will see in Chapter 9, some of these principles can be formulated in such a way that they extend to a very abstract setting. However, rather than delve into such extensions here, I will devote this section to a closer examination of the situation at hand. Specifically, in this section we are going to see how to make the final part of Theorem 2.1.8 quantitative.

From (2.1.5), we get a rate of convergence in terms of the second and third derivatives of $\varphi$. In fact, if we assume that

$$\tau_k \equiv \big(\mathbb{E}^{\mathbb{P}}\big[|X_k|^3\big]\big)^{\frac{1}{3}} < \infty, \quad 1 \le k \le n, \tag{2.2.1}$$

then (cf. the proof of Theorem 2.1.4), by using the estimates

$$\big|R_m(\xi)\big| \le \frac{\|\varphi'''\|_{\mathrm{u}}|\xi|^3}{6} \quad \text{and} \quad \sigma_k \le \tau_k,$$

one sees that (2.1.5) can be replaced by

$$\left|\mathbb{E}^{\mathbb{P}}\big[\varphi(\check{S}_n)\big] - \int_{\mathbb{R}} \varphi\, d\gamma_{0,1}\right| \le \frac{2\|\varphi'''\|_{\mathrm{u}}}{3} \frac{\sum_1^n \tau_k^3}{\Sigma_n^3} \tag{2.2.2}$$

when the $X_k$'s have third moments.

Although both (2.1.5) and (2.2.2) are interesting, neither one of them can be used to get very much information about the rate at which the distribution functions

$$x \in \mathbb{R} \longmapsto F_n(x) \equiv P\big(\check{S}_n \le x\big) \in [0,1] \tag{2.2.3}$$

[4] As this exercise demonstrates, Stirling's formula is intimately connected to The Central Limit Theorem. In fact, apart from the constant $\sqrt{2\pi}$, what we now call Stirling's formula was discovered first by DeMoivre while he was proving The Central Limit Theorem for Bernoulli random variables. For more information, see, for example, Wm. Feller's discussion of Stirling's formula in his *Introduction to Probability Theory and Its Applications, Vol. I*, Wiley, Series in Probability and Math. Stat. (1968).

are tending to the **error function**

$$G(x) \equiv \gamma_{0,1}\big((-\infty, x]\big) = \frac{1}{\sqrt{2\pi}} \int_{-\infty}^{x} e^{-\frac{t^2}{2}}\, dt. \tag{2.2.4}$$

To see how (2.1.5) and (2.2.2) must be modified in order to gain such information, first observe that

$$\begin{aligned} &\int_{\mathbb{R}} \varphi'(x)\big(F_n(x) - G(x)\big)\, dx \\ &\qquad = \mathbb{E}^{\mathbb{P}}\big[\varphi(\check{S}_n)\big] - \int_{\mathbb{R}} \varphi(y)\, \gamma_{0,1}(dy), \quad \varphi \in C^1_{\mathrm{b}}(\mathbb{R};\mathbb{R}). \end{aligned} \tag{2.2.5}$$

(To prove (2.2.5), reduce to the case in which $\varphi \in C^1_{\mathrm{c}}(\mathbb{R};\mathbb{R})$ and $\varphi(0) = 0$; and for this case apply either Fubini's Theorem or integration by parts over the intervals $(-\infty, 0]$ and $[0, \infty)$ separately.) Hence, in order to get information about the distance between $F_n$ and $G$, we will have to learn how to replace the right-hand sides of (2.1.5) and (2.2.2) with expressions that depend only on the first derivative of $\varphi$. For example, if the dependence is on $\|\varphi'\|_{\mathrm{u}}$, then we get information about the $L^1(\mathbb{R};\mathbb{R})$ distance between $F_n$ and $G$, whereas if the dependence is on $\|\varphi'\|_{L^1(\mathbb{R};\mathbb{R})}$, then the information will be about the uniform distance between $F_n$ and $G$.

§ **2.2.1. $L^1$-Berry–Esseen.** The basic idea that I will use to get estimates in terms of $\varphi'$ was introduced by C. Stein and is an example of a procedure known as **Stein's method.**[1] In the case at hand, his method stems from the trivial observation that if $\mu$ is a Borel probability measure on $\mathbb{R}$ and $g$ is the Gauss kernel in (1.3.5), then $\mu = \gamma_{0,1}$ if and only if $\partial\left(\frac{\mu}{g}\right) = 0$ in the sense of Schwartz distribution theory. Equivalently, if $A_+$ is the raising operator (cf. § 2.4.1) given by $A_+\varphi(x) = x\varphi(x) - \partial\varphi(x)$, then, because $\langle A_+\varphi, \mu\rangle = \left\langle \varphi g, \partial\left(\frac{\mu}{g}\right)\right\rangle$, $\mu = \gamma_{0,1}$ if and only if $\langle A_+\varphi, \mu\rangle = 0$ for sufficiently many test functions $\varphi$. In fact, as will be shown in what follows, $\mu$ will be close to $\gamma_{0,1}$ if, in an appropriate sense, $\langle A_+\varphi, \mu\rangle$ is small.

To make mathematics out of the preceding, I will need the following.

LEMMA 2.2.6. *Let $\varphi \in C^1(\mathbb{R};\mathbb{R})$, assume that $\|\varphi'\|_{\mathrm{u}} < \infty$, set $\tilde{\varphi} = \varphi - \langle \varphi, \gamma_{0,1}\rangle$, and define*

$$x \in \mathbb{R} \longmapsto f(x) \equiv e^{\frac{x^2}{2}} \int_{-\infty}^{x} \tilde{\varphi}(t) e^{-\frac{t^2}{2}}\, dt \in \mathbb{R}. \tag{2.2.7}$$

*Then $f \in C^2_{\mathrm{b}}(\mathbb{R};\mathbb{R})$,*

$$\|f\|_{\mathrm{u}} \le 2\|\varphi'\|_{\mathrm{u}}, \quad \|f'\|_{\mathrm{u}} \le 3\sqrt{\tfrac{\pi}{2}}\|\varphi'\|_{\mathrm{u}}, \quad \|f''\|_{\mathrm{u}} \le 6\|\varphi'\|_{\mathrm{u}}, \tag{2.2.8}$$

[1] Stein provided an introduction, by way of examples, to his own method in *Approximate Computation of Expectations*, I.M.S., Lec. Notes & Monograph Series # **7** (1986).

and

$$f'(x) - xf(x) = \tilde{\varphi}(x), \quad x \in \mathbb{R}. \tag{2.2.9}$$

PROOF: The facts that $f \in C^1(\mathbb{R};\mathbb{R})$ and that (2.2.9) holds are elementary applications of The Fundamental Theorem of Calculus. Moreover, knowing that $f \in C^1(\mathbb{R};\mathbb{R})$ and using (2.2.9), we see that $f \in C^2(\mathbb{R};\mathbb{R})$ and, in fact, that

$$f''(x) - xf'(x) = f(x) + \varphi'(x), \quad x \in \mathbb{R}. \tag{2.2.10}$$

To prove the estimates in (2.2.8), first note that, because $\tilde{\varphi}$ and therefore $f$ are unchanged when $\varphi$ is replaced by $\varphi - \varphi(0)$, I may and will assume that $\varphi(0) = 0$ and therefore that $|\varphi(t)| \le \|\varphi'\|_{\mathrm{u}}|t|$. In particular, this means that

$$\left|\int_{\mathbb{R}} \varphi\, d\gamma_{0,1}\right| \le \|\varphi'\|_{\mathrm{u}} \int_{\mathbb{R}} |t|\,\gamma_{0,1}(dt) = \|\varphi'\|_{\mathrm{u}}\sqrt{\tfrac{2}{\pi}}.$$

Next, observe that, because $\int_{\mathbb{R}} \tilde{\varphi}(t)e^{-\frac{t^2}{2}}\,dt = 0$, an alternative expression for $f$ is

$$f(x) = -e^{\frac{x^2}{2}} \int_x^\infty \tilde{\varphi}(t)e^{-\frac{t^2}{2}}\,dt, \quad x \in \mathbb{R}.$$

Thus, by using the original expression for $f(x)$ when $x \in (-\infty, 0)$ and the alternative one when $x \in [0,\infty)$, we see first that

$$|f(x)| \le e^{\frac{x^2}{2}} \int_{|x|}^\infty \left|\tilde{\varphi}(-t\operatorname{sgn}(x))\right| e^{-\frac{t^2}{2}}\,dt, \quad x \in \mathbb{R},$$

and then that

$$|f(x)| \le \|\varphi'\|_{\mathrm{u}} e^{\frac{x^2}{2}} \int_{|x|}^\infty \left(t + \sqrt{\tfrac{2}{\pi}}\right) e^{-\frac{t^2}{2}}\,dt.$$

But, since

$$\frac{d}{dx}\left(e^{\frac{x^2}{2}} \int_x^\infty e^{-\frac{t^2}{2}}\,dt\right) \le e^{\frac{x^2}{2}} \int_x^\infty t\,e^{-\frac{t^2}{2}}\,dt - 1 = 0 \quad \text{for } x \in [0,\infty),$$

we have that, for $x \in \mathbb{R}$,

$$|x|e^{\frac{x^2}{2}} \int_{|x|}^\infty e^{-\frac{t^2}{2}}\,dt \le e^{\frac{x^2}{2}} \int_{|x|}^\infty te^{-\frac{t^2}{2}}\,dt = 1 \text{ and } e^{\frac{x^2}{2}} \int_{|x|}^\infty e^{-\frac{t^2}{2}}\,dt \le \sqrt{\tfrac{\pi}{2}}; \tag{2.2.11}$$

which means that I have now proved the first estimate in (2.2.8). To prove the other two estimates there, derive from (2.2.10)

$$\frac{d}{dx}\left(e^{-\frac{x^2}{2}} f'(x)\right) = e^{-\frac{x^2}{2}}\left(f(x) + \varphi'(x)\right)$$

and therefore that

$$f'(x) = e^{\frac{x^2}{2}} \int_{-\infty}^{x} \big(f(t)+\varphi'(t)\big)e^{-\frac{t^2}{2}}\,dt = -e^{\frac{x^2}{2}} \int_{x}^{\infty} \big(f(t)+\varphi'(t)\big)e^{-\frac{t^2}{2}}\,dt, \quad x \in \mathbb{R}.$$

Thus, reasoning as I did above and using the first estimate in (2.2.8) and the relations in (2.2.9), (2.2.10), and (2.2.11), one arrives at the second and third estimates in (2.2.8). □

I now have the ingredients needed to apply Stein's method to the following example of a Berry–Esseen type of estimate.

THEOREM 2.2.12 (**$\mathbf{L^1}$-Berry–Esseen Estimate**). *Continuing in the setting of Theorem 2.1.4, one has that for all $\epsilon > 0$ (cf. (2.1.3), (2.2.3), and (2.2.4))*

$$\|F_n - G\|_{L^1(\mathbb{R};\mathbb{R})} \le 6(r_n + \epsilon) + 3\sqrt{2\pi}\, g_n(2\epsilon). \tag{2.2.13}$$

*Moreover, if (cf. (2.2.1)) $\tau_m < \infty$ for each $1 \le m \le n$, then*

$$\|F_n - G\|_{L^1(\mathbb{R};\mathbb{R})} \le \left(6r_n + \frac{3\sum_{m=1}^n \tau_m^3}{\Sigma_n^3}\right) \wedge \left(\frac{9\sum_{m=1}^n \tau_m^3}{\Sigma_n^3}\right). \tag{2.2.14}$$

*In particular, if $\sigma_m^2 = 1$ and $\tau_m \le \tau < \infty$ for each $1 \le m \le n$, then*

$$\|F_n - G\|_{L^1(\mathbb{R};\mathbb{R})} \le \frac{6 + 2\tau^3}{\sqrt{n}} \le \frac{8\tau^3}{\sqrt{n}}.$$

PROOF: Let $\varphi \in C^1(\mathbb{R};\mathbb{R})$ having bounded first derivative be given, and define $f$ accordingly, as in (2.2.7). Everything turns on the equality in (2.2.9). Indeed, because of that equality, we know that the right-hand side of (2.2.5) is equal to

$$\mathbb{E}^{\mathbb{P}}\big[f'(\check{S}_n)\big] - \mathbb{E}^{\mathbb{P}}\big[\check{S}_n f(\check{S}_n)\big] = \sum_{m=1}^n \Big(\check{\sigma}_m^2 \mathbb{E}^{\mathbb{P}}\big[f'(\check{S}_n)\big] - \mathbb{E}^{\mathbb{P}}\big[\check{X}_m f(\check{S}_n)\big]\Big),$$

where I have set $\check{\sigma}_m = \frac{\sigma_m}{\Sigma_n}$ and $\check{X}_m = \frac{X_m}{\Sigma_n}$. Next, define

$$\check{T}_{n,m}(t) = \check{S}_n + (t-1)\check{X}_m \quad \text{for } t \in [0,1],$$

note that $\check{T}_{n,m}(0)$ is independent of $\check{X}_m$, and conclude that

$$\begin{aligned}
\mathbb{E}^{\mathbb{P}}\big[\check{X}_m f(\check{S}_n)\big] &= \int_0^1 \mathbb{E}^{\mathbb{P}}\big[\check{X}_m^2 f'(\check{T}_{n,m}(t))\big]\,dt \\
&= \check{\sigma}_m^2 \mathbb{E}^{\mathbb{P}}\big[f'(\check{T}_{n,m}(0))\big] + \int_0^1 \mathbb{E}^{\mathbb{P}}\big[\check{X}_m^2\big(f'(\check{T}_{n,m}(t)) - f'(\check{T}_{n,m}(0)\big)\big]\,dt
\end{aligned}$$

for each $1 \le m \le n$. Hence, we now see that

$$\mathbb{E}^{\mathbb{P}}\left[\varphi(\check{S}_n)\right] - \int_{\mathbb{R}} \varphi\, d\gamma_{0,1} = \sum_{m=1}^{n} \check{\sigma}_m^2 A_m - \sum_{m=1}^{n} \int_0^1 B_m(t)\, dt, \tag{2.2.15}$$

where

$$A_m \equiv \mathbb{E}^{\mathbb{P}}\Big[f'(\check{S}_n) - f'\big(\check{T}_{n,m}(0)\big)\Big]$$

and

$$B_m(t) \equiv \mathbb{E}^{\mathbb{P}}\Big[\check{X}_m^2\Big(f'\big(\check{T}_{n,m}(t)\big) - f'\big(\check{T}_{n,m}(0)\big)\Big)\Big].$$

Obviously, by Taylor's Theorem and Hölder's Inequality, for each $1 \le m \le n$,

$$|A_m| \le \check{\sigma}_m \|f''\|_{\mathrm{u}} \le \left(r_n \wedge \frac{\tau_m}{\Sigma_n}\right) \|f''\|_{\mathrm{u}} \tag{*}$$

while, for each $t \in [0,1]$ and $\epsilon > 0$,

$$\big|B_m(t)\big| \le 2\epsilon t \check{\sigma}_m^2 \|f''\|_{\mathrm{u}} + 2\frac{\|f'\|_{\mathrm{u}}}{\Sigma_n^2}\mathbb{E}^{\mathbb{P}}\Big[X_m^2,\, |X_m| \ge 2\epsilon\Sigma_n\Big].$$

Thus, after summing over $1 \le m \le n$, integrating with respect to $t \in [0,1]$, and using (2.2.5), (2.2.15), and (*), we arrive at

$$\left|\int_{\mathbb{R}} \varphi'(x)\big(F_n(x) - G(x)\big)\, dx\right| \le (r_n + \epsilon)\|f''\|_{\mathrm{u}} + 2g_n(2\epsilon)\|f'\|_{\mathrm{u}},$$

which, in conjunction with the estimates in (2.2.8), leads immediately to the estimate in (2.2.13). In order to get (2.2.14), simply note that

$$\big|B_m(t)\big| \le t\int_0^1 \mathbb{E}^{\mathbb{P}}\Big[|\check{X}_m|^3 \big|f''\big(\check{T}_{n,m}(st)\big)\big|\Big]\, ds \le t\|f''\|_{\mathrm{u}}\frac{\tau_m^3}{\Sigma_n^3},$$

and again use (2.2.15), (2.2.8), and (*). □

§ **2.2.2. The Classical Berry–Esseen Theorem.** The result in Theorem 2.2.12 is already significant. However, it is not the classical Berry–Esseen Theorem, which is the analogous statement about $\|F_n - G\|_{\mathrm{u}}$.

In order to prove the classical result via Stein's method, we must learn how to replace the $\|\varphi'''\|_{\mathrm{u}}$ in Lindeberg's Theorem by $\|\varphi'\|_{L^1(\mathbb{R};\mathbb{R})}$. It turns out that this replacement is far more challenging than replacing $\|\varphi'''\|_{\mathrm{u}}$ by $\|\varphi'\|_{\mathrm{u}}$, which was the replacement needed to prove Theorem 2.2.12. The argument that I will use is a clever induction procedure that was introduced into this context by E. Bolthausen.[2] But, before I can apply Bolthausen's argument, I will need the following variation on Lemma 2.2.6.

[2] The Berry–Esseen Theorem appears as a warm-up exercise in Bolthausen's "An estimate of the remainder term in a combinatorial central limit theorem," *Z. Wahr. Gebiete* **66**, pp. 379–386 (1984).

LEMMA 2.2.16. *Let $\varphi \in C^1(\mathbb{R};\mathbb{R})$, and define $f$ accordingly, as in (2.2.7). Then $\|f\|_u \le \sqrt{\frac{\pi}{8}}\|\varphi'\|_{L^1(\mathbb{R};\mathbb{R})}$ and $\|f'\|_u \le \|\varphi'\|_{L^1(\mathbb{R};\mathbb{R})}$.*

PROOF: I will assume, throughout, that $\|\varphi'\|_{L^1(\mathbb{R};\mathbb{R})} = 1$. Observe that, by the Fundamental Theorem of Calculus, (cf. the notation in Lemma 2.2.6)

$$\tilde{\varphi}(x) = -\int_{\mathbb{R}} \tilde{\varphi}_y(x)\,\varphi'(y)\,dy, \quad \text{where } \varphi_y = \mathbf{1}_{(-\infty,y]},$$

and so (cf. (2.2.4))

$$f(x) = -\int_{\mathbb{R}} \psi_y(x)\,\varphi'(y)\,dy, \quad \text{where } \psi_y(x) = \sqrt{2\pi}e^{\frac{x^2}{2}}\big(G(x\wedge y) - G(x)G(y)\big) \ge 0.$$

At the same time, these, together with (2.2.9), give

$$f'(x) = -\int_{\mathbb{R}} \Big(x\psi_y(x) + \tilde{\varphi}_y(x)\Big)\,\varphi'(y)\,dy.$$

Hence, the desired estimates come down to checking that

$$e^{\frac{x^2}{2}}\big(G(x\wedge y) - G(x)G(y)\big) \le \tfrac{1}{4},$$

and

$$\Big|\sqrt{2\pi}xe^{\frac{x^2}{2}}\Big(G(x\wedge y) - G(x)G(y)\Big) + \mathbf{1}_{(-\infty,y]}(x) - G(y)\Big| \le 1$$

for all $(x,y) \in \mathbb{R}\times\mathbb{R}$. But, if $x \le y$,

$$G(x\wedge y) - G(x)G(y) \le G(x) - G(x)^2 = \frac{1}{4}\Big(1 - 4\big(G(x) - \tfrac{1}{2}\big)^2\Big)$$

and

$$\begin{aligned}\big(G(x) - \tfrac{1}{2}\big)^2 &= \frac{1}{2\pi}\left(\int_0^{|x|} e^{-\frac{\xi^2}{2}}\,d\xi\right)^2 \\ &\ge \frac{1}{8\pi}\iint_{\xi^2+\eta^2\le x^2} e^{-\frac{\xi^2+\eta^2}{2}}\,d\xi d\eta = \frac{1}{4}\Big(1 - e^{-\frac{x^2}{2}}\Big),\end{aligned}$$

which proves the first inequality. To get the second one, it suffices to consider each of the four cases $0 \le x \le y$, $x \ge 0$ & $y < x$, $y < x < 0$, and $x < 0$ & $y \ge x$ separately and take into account that, from the first part of (2.2.11),

$$x \ge 0 \implies \sqrt{2\pi}xe^{\frac{x^2}{2}}\big(1 - G(x)\big) \le 1 \quad \text{and} \quad x < 0 \implies \sqrt{2\pi}|x|e^{\frac{x^2}{2}}G(x) \le 1. \quad \square$$

As distinguished from Lemma 2.2.6, Lemma 2.2.16 contains no estimate on $\|f''\|_u$. Indeed, there is no such estimate in terms of $\|\varphi'\|_{L^1(\mathbb{R};\mathbb{R})}$. As a consequence, the proof of the following is much more involved than that of Theorem 2.2.12

THEOREM 2.2.17 (**Classical Berry–Esseen Estimate**). *Let everything be as in Theorem 2.1.4, and assume that (cf. (2.2.1)) $\tau_m < \infty$ for each $1 \le m \le n$. Then (cf. (2.2.3) and (2.2.4))*

$$\|F_n - G\|_{\mathrm{u}} \le 10 \frac{\sum_1^n \tau_m^3}{\Sigma_n^3}. \tag{2.2.18}$$

*In particular, if $\sigma_m = 1$ for all $1 \le m \le n$, then (2.2.18) can be replaced by*

$$\|F_n - G\|_{\mathrm{u}} \le 10 \frac{\sum_1^n \tau_m^3}{n^{\frac{3}{2}}} \le 10 \frac{\max_{1\le m\le n} \tau_m^3}{\sqrt{n}}. \tag{2.2.19}$$

PROOF: For each $n \in \mathbb{Z}^+$, let $\beta_n$ denote the smallest number $\beta$ with the property that

$$\|F_n - G\|_{\mathrm{u}} \le \beta \frac{\sum_1^n \tau_m^3}{\Sigma_n^3}$$

for all choices of random variables satisfying the hypotheses under which (2.2.18) is to be proved. My strategy is to give an inductive proof that $\beta_n \le 10$ for all $n \in \mathbb{Z}^+$; and, because $\Sigma_1 \le \tau_1$ and therefore $\beta_1 \le 1$, I need only be concerned with $n \ge 2$.

Given $n \ge 2$ and $X_1, \dots, X_n$, define $\breve{X}_m$, $\breve{\sigma}_m$, and $\breve{T}_{n,m}(t)$ for $1 \le m \le n$ and $t \in [0,1]$ as in the proof of Theorem 2.2.12. Next, for each $1 \le m \le n$, set

$$\Sigma_{n,m} = \sqrt{\Sigma_n^2 - \sigma_m^2}, \quad \breve{\tau}_m = \frac{\tau_m}{\Sigma_n}, \quad \rho_n = \sum_1^n \breve{\tau}_m^3, \quad \text{and } \rho_{n,m} = \sum_{\substack{1\le \ell\le n\\ \ell\ne m}} \left(\frac{\tau_\ell}{\Sigma_{n,m}}\right)^3.$$

Finally, set

$$S_{n,m} = \sum_{\substack{1\le \ell\le n\\ \ell\ne m}} X_\ell \quad \text{and} \quad \breve{S}_{n,m} = \frac{S_{n,m}}{\Sigma_{n,m}},$$

and let $x \in \mathbb{R} \longmapsto F_{n,m}(x) \equiv P(\breve{S}_{n,m} \le x) \in [0,1]$ denote the distribution function for $\breve{S}_{n,m}$. Notice that, by definition, $\|F_{n,m} - G\|_{\mathrm{u}} \le \beta_{n-1}\rho_{n,m}$ for each $1 \le m \le n$. Furthermore, because (cf. (2.1.3))

$$\frac{\Sigma_{n,m}^2}{\Sigma_n^2} = 1 - \breve{\sigma}_m^2 \ge 1 - r_n^2 \quad \text{and} \quad \rho_{n,m} \le \left(\frac{\Sigma_n}{\Sigma_{n,m}}\right)^3 \rho_n,$$

we know first that

$$\rho_{n,m} \le \frac{\rho_n}{(1 - r_n^2)^{\frac{3}{2}}}, \quad 1 \le m \le n,$$

and therefore that

$$\text{(2.2.20)} \qquad \max_{1\le m\le n} \|F_{n,m}-G\|_{\mathrm{u}} \le \frac{\rho_n\beta_{n-1}}{(1-r_n^2)^{\frac{3}{2}}}.$$

Now let $\varphi \in C_{\mathrm{b}}^2(\mathbb{R};\mathbb{R})$ with $\|\varphi''\|_{L^1(\mathbb{R})}<\infty$ be given, define $f$ accordingly, as in (2.2.7), and let

$$\{A_m : 1\le m\le n\} \quad\text{and}\quad \{B_m(t): 1\le m\le n \;\&\; t\in[0,1]\}$$

be the associated quantities appearing in (2.2.15). By (2.2.9), we have that

$$\begin{aligned}
|A_m| &\le \Big|\mathbb{E}^{\mathbb{P}}\Big[\check{X}_m f(\check{S}_n)\Big]\Big| + \Big|\mathbb{E}^{\mathbb{P}}\Big[\check{T}_{n,m}(0)\Big(f(\check{S}_n)-f(\check{T}_{n,m}(0))\Big)\Big]\Big| \\
&\qquad + \Big|\mathbb{E}^{\mathbb{P}}\Big[\varphi(\check{S}_n)-\varphi(\check{T}_{n,m}(0))\Big]\Big| \\
&\le \mathbb{E}^{\mathbb{P}}\big[|\check{X}_m|\big]\|f\|_{\mathrm{u}} + \mathbb{E}^{\mathbb{P}}\big[|\check{X}_m\,\check{T}_{n,m}(0)|\big]\|f'\|_{\mathrm{u}} \\
&\qquad + \int_0^1 \big|\mathbb{E}^{\mathbb{P}}\big[\check{X}_m\varphi'(\check{T}_{n,m}(\xi))\big]\big|\,d\xi \\
&\le \check{\sigma}_m\left(\|f\|_{\mathrm{u}} + \frac{\Sigma_{n,m}}{\Sigma_n}\|f'\|_{\mathrm{u}}\right) + \max_{\xi\in[0,1]}\big|\mathbb{E}^{\mathbb{P}}\big[\check{X}_m\varphi'(\check{T}_{n,m}(\xi))\big]\big| \\
&\le \check{\sigma}_m\big(\|f\|_{\mathrm{u}}+\|f'\|_{\mathrm{u}}\big) + \max_{\xi\in[0,1]}\big|\mathbb{E}^{\mathbb{P}}\big[\check{X}_m\varphi'(\check{T}_{n,m}(\xi))\big]\big|.
\end{aligned}$$

Similarly, from (2.2.9)) and the independence of $\check{X}_m$ from $\check{T}_{m,n}(0)$, one sees that $|B_m(t)|$ is dominated by

$$\begin{aligned}
&t\Big|\mathbb{E}^{\mathbb{P}}\Big[\check{X}_m^3 f(\check{T}_{n,m}(t))\Big]\Big| + \Big|\mathbb{E}^{\mathbb{P}}\Big[\check{X}_m^2\check{T}_{n,m}(0)\Big(f(\check{T}_{n,m}(t))-f(\check{T}_{n,m}(0))\Big)\Big]\Big| \\
&\qquad + \Big|\mathbb{E}^{\mathbb{P}}\Big[\check{X}_m^2\Big(\varphi(\check{T}_{n,m}(t))-\varphi(\check{T}_{n,m}(0))\Big)\Big]\Big| \\
&\le t\mathbb{E}^{\mathbb{P}}\big[|\check{X}_m|^3\big]\|f\|_{\mathrm{u}} + t\mathbb{E}^{\mathbb{P}}\big[|\check{X}_m|^3\big]\mathbb{E}^{\mathbb{P}}\big[|\check{T}_{n,m}(0)|\big]\|f'\|_{\mathrm{u}} \\
&\qquad + t\int_0^1 \big|\mathbb{E}^{\mathbb{P}}\big[\check{X}_m^3\varphi'(\check{T}_{n,m}(t\xi))\big]\big|\,d\xi \\
&\le t\check{\tau}_m^3\big(\|f\|_{\mathrm{u}}+\|f'\|_{\mathrm{u}}\big) + t\max_{\xi\in[0,1]}\big|\mathbb{E}^{\mathbb{P}}\big[\check{X}_m^3\varphi'(\check{T}_{n,m}(\xi))\big]\big|.
\end{aligned}$$

In order to handle the second term in the last line of each of these calculations, introduce the function

$$(\xi,\omega,y)\in[0,1]\times\Omega\times\mathbb{R}\longmapsto \psi(\xi,\omega,y)\equiv\varphi'\left(\xi\check{X}_m(\omega)+\frac{\Sigma_{n,m}}{\Sigma_n}y\right)\in\mathbb{R}.$$

Then, because $\breve{X}_m$ is independent of $\breve{T}_{n,m}(0)$,

$$\left|\mathbb{E}^{\mathbb{P}}\left[\breve{X}_m^k\varphi'(\breve{T}_{n,m}(\xi))\right] - \int_\Omega \breve{X}_m(\omega)^k\left(\int_{\mathbb{R}}\psi(\xi,\omega,y)\,\gamma_{0,1}(dy)\right)\mathbb{P}(d\omega)\right|$$
$$\leq \int_\Omega \left|\breve{X}_m(\omega)\right|^k\left|\int_{\mathbb{R}}\psi(\xi,\omega,y)\,dF_{n,m}(y) - \int_{\mathbb{R}}\psi(\xi,\omega,y)\,dG(y)\right|\mathbb{P}(d\omega)$$
$$= \int_\Omega \left|\breve{X}_m(\omega)\right|^k\left|\int_{\mathbb{R}}\psi'(t,\omega,y)\big(G(y)-F_{n,m}(y)\big)\,dy\right|\mathbb{P}(d\omega)$$
$$\leq \frac{\beta_{n-1}\rho_n}{(1-r_n^2)^{\frac{3}{2}}}\mathbb{E}^{\mathbb{P}}\left[\left|\breve{X}_m\right|^k\right]\|\varphi''\|_{L^1(\mathbb{R};\mathbb{R})} \leq \frac{\breve{\tau}_m^k\beta_{n-1}\|\varphi''\|_{L^1(\mathbb{R};\mathbb{R})}\rho_n}{(1-r_n^2)^{\frac{3}{2}}}, \quad k\in\{1,3\},$$

where I have used $\psi'(t,\omega,y)$ to denote the first derivative of $y\in\mathbb{R}\longmapsto\psi(\xi,\omega,y)$, applied (2.2.5) and (2.2.20), and noted that

$$\|\psi'(\xi,\omega,\cdot)\|_{L^1(\mathbb{R};\mathbb{R})} = \|\varphi''\|_{L^1(\mathbb{R};\mathbb{R})} \quad \text{for all } (\xi,\omega)\in[0,1]\times\Omega.$$

At the same time, because

$$\|\psi(\xi,\omega,\cdot)\|_{L^1(\mathbb{R};\mathbb{R})} = \frac{\Sigma_n}{\Sigma_{n,m}}\|\varphi'\|_{L^1(\mathbb{R};\mathbb{R})} \quad \text{for all } (\xi,\omega)\in[0,1]\times\Omega,$$

we have that, for each $\xi\in[0,1]$,

$$\left|\int_\Omega \breve{X}_m(\omega)^k\left(\int_{\mathbb{R}}\psi(\xi,\omega,y)\,\gamma_{0,1}(dy)\right)\mathbb{P}(d\omega)\right| \leq \frac{\|\varphi'\|_{L^1(\mathbb{R};\mathbb{R})}\,\breve{\tau}_m^k}{\big(2\pi(1-r_n^2)\big)^{\frac{1}{2}}}.$$

Hence, by combining these estimates, we arrive at

$$|A_m| \leq \breve{\tau}_m\left(\|f\|_{\mathrm{u}} + \|f'\|_{\mathrm{u}} + \frac{\|\varphi'\|_{L^1(\mathbb{R};\mathbb{R})}}{\big(2\pi(1-r_n^2)\big)^{\frac{1}{2}}} + \frac{\beta_{n-1}\rho_n}{(1-r_n^2)^{\frac{3}{2}}}\|\varphi''\|_{L^1(\mathbb{R};\mathbb{R})}\right)$$

and

$$|B_m(t)| \leq t\breve{\tau}_m^3\left(\|f\|_{\mathrm{u}} + \|f'\|_{\mathrm{u}} + \frac{\|\varphi'\|_{L^1(\mathbb{R};\mathbb{R})}}{\big(2\pi(1-r_n^2)\big)^{\frac{1}{2}}} + \frac{\beta_{n-1}\rho_n}{(1-r_n^2)^{\frac{3}{2}}}\|\varphi''\|_{L^1(\mathbb{R};\mathbb{R})}\right)$$

for all $1\leq m\leq n$ and $t\in[0,1]$, and, after putting these together with (2.2.5) and (2.2.15), we conclude that

$$\left|\int_{\mathbb{R}}\varphi'(y)\big(G(y)-F_n(y)\big)\,dy\right|$$
$$\leq \frac{3}{2}\Bigg(\|f\|_{\mathrm{u}} + \|f'\|_{\mathrm{u}} \tag{2.2.21}$$
$$+ \frac{\|\varphi'\|_{L^1(\mathbb{R};\mathbb{R})}}{\big(2\pi(1-r_n^2)\big)^{\frac{1}{2}}} + \frac{\beta_{n-1}\|\varphi''\|_{L^1(\mathbb{R};\mathbb{R})}\rho_n}{(1-r_n^2)^{\frac{3}{2}}}\Bigg)\rho_n.$$

I next apply (2.2.21) to a special class of $\varphi$'s. Namely, set

$$h(x) = \begin{cases} 1 & \text{if } x < 0 \\ 1-x & \text{if } x \in [0,1] \\ 0 & \text{if } x > 1, \end{cases}$$

and define

$$h_\epsilon(x) = \epsilon^{-1} \int_{\mathbb{R}} \eta(\epsilon^{-1}y) h(x-y)\, dy \quad \text{for } \epsilon > 0 \text{ and } x \in \mathbb{R},$$

where $\eta \in C_c^\infty(\mathbb{R};[0,\infty))$ satisfies $\int_{\mathbb{R}} \eta(y)\, dy = 1$. Finally, let $a \in \mathbb{R}$ be given, and set

$$\varphi_{\epsilon,L}(x) = h_\epsilon\left(\tfrac{x-a}{L\rho_n}\right), \quad x \in \mathbb{R} \text{ and } \epsilon,\ L > 0.$$

It is then an easy matter to check that $\|\varphi'_{\epsilon,L}\|_{L^1(\mathbb{R};\mathbb{R})} = 1$ while $\|\varphi''_{\epsilon,L}\|_{L^1(\mathbb{R};\mathbb{R})} \le \frac{2}{L\rho_n}$. Hence, by plugging the estimates from Lemma 2.2.16 into (2.2.21) and then letting $\epsilon \searrow 0$, we find that, for each $L > 0$,

$$\begin{aligned} &\sup_{a\in\mathbb{R}} \left| \frac{1}{L\rho_n} \int_a^{a+L\rho_n} (G(y) - F_n(y))\, dy \right| \\ &\qquad \le \frac{3}{2}\left(1 + \sqrt{\frac{\pi}{8}} + \frac{1}{(2\pi(1-r_n^2))^{\frac{1}{2}}} + \frac{2\beta_{n-1}}{(1-r_n^2)^{\frac{3}{2}} L}\right)\rho_n. \end{aligned} \tag{2.2.22}$$

But

$$\frac{1}{L\rho_n}\int_{a-L\rho_n}^a F_n(y)\, dy \le F_n(a) \le \frac{1}{L\rho_n}\int_a^{a+L\rho_n} F_n(y)\, dy,$$

while

$$0 \le \frac{1}{L\rho_n}\int_a^{a+L\rho_n} G(y)\, dy - G(a) = \frac{1}{L\rho_n}\int_a^{a+L\rho_n} (a + L\rho_n - y)\, \gamma_{0,1}(dy) \le \frac{L\rho_n}{\sqrt{8\pi}},$$

and, similarly,

$$0 \le G(a) - \frac{1}{L\rho_n}\int_{a-L\rho_n}^a G(y)\, dy \le \frac{L\rho_n}{\sqrt{8\pi}}.$$

Thus, from (2.2.22), we first obtain, for each $L \in (0,\infty)$,

$$\|F_n - G\|_{\mathrm{u}} \le \left(\frac{3}{2} + \sqrt{\frac{9\pi}{32}} + \frac{3}{(8\pi(1-r_n^2))^{\frac{1}{2}}} + \frac{3\beta_{n-1}}{(1-r_n^2)^{\frac{3}{2}} L} + \frac{L}{(8\pi)^{\frac{1}{2}}}\right)\rho_n,$$

and then, after minimizing with respect to $L \in (0,\infty)$,

$$
\begin{aligned}
\|F_n - G\|_{\mathrm{u}} \le \Bigg(\frac{3}{2} + \sqrt{\frac{9\pi}{32}} + \sqrt{\frac{9}{8\pi}}\,\big(1-r_n^2\big)^{-\frac12} \\
+ \sqrt[4]{\frac{18}{\pi}}\,\beta_{n-1}^{\frac12}\big(1-r_n^2\big)^{-\frac34}\Bigg)\rho_n.
\end{aligned}
\tag{2.2.23}
$$

In order to complete the proof starting from (2.2.23), we have to consider the two cases determined by whether $\rho_n \ge \frac{1}{10}$ or $\rho_n < \frac{1}{10}$. Because $\|F_n - G\|_{\mathrm{u}} \le 1$, it is obvious that we can take $\beta_n \le 10$ in the first case. On the other hand, if $\rho_n \le \frac{1}{10}$ and we assume that $\beta_{n-1} \le 10$, then, because

$$
\rho_n = \frac{1}{\Sigma_n^3}\sum_1^n \mathbb{E}^{\mathbb{P}}\big[|X_m|^3\big] \ge \frac{1}{\Sigma_n^3}\sum_1^n \mathbb{E}^{\mathbb{P}}\big[X_m^2\big]^{\frac32} = \sum_1^n \check{\sigma}_m^3 \ge r_n^3,
$$

(2.2.23) says that $\|F_n - G\|_{\mathrm{u}} \le 10\rho_n$. Hence, in either case, $\beta_{n-1} \le 10 \implies \beta_n \le 10$. □

It is clear from the preceding derivation (in particular, the final step) that the constant 10 appearing in (2.2.18) and (2.2.19) can be replaced by the smallest $\beta > 1$ that satisfies the equation

$$
\beta = \frac{3}{2} + \sqrt{\frac{9\pi}{32}} + \sqrt{\frac{9}{8\pi}}\,\big(1-\beta^{-\frac23}\big)^{-\frac12} + \sqrt[4]{\frac{18}{\pi}}\,\beta^{\frac12}\big(1-\beta^{-\frac23}\big)^{-\frac34}.
$$

Numerical experimentation indicates that 10 is quite a good approximation to the actual solution of this minimization problem. However, it should be recognized that, with sufficient diligence and entirely different techniques, one can show that the 10 in (2.2.18) can be replaced by a number that is less than 1. Thus, I do not claim that Stein's method gives the best result, only that it gives whatever it gives with relatively little pain.

## Exercises for § 2.2

EXERCISE 2.2.24. It is important to know that, at least qualitatively, one cannot do better than Berry–Esseen. To see this, consider independent, symmetric, $\{-1,1\}$-valued Bernoulli random variables, and define $F_n$ accordingly. Next, observe that when $t_n = -(2n+1)^{-\frac12}$,

$$
F_{2n+1}(t_n) - G(t_n) = \frac{1}{\sqrt{2\pi}}\int_{t_n}^0 e^{-\frac{x^2}{2}}\,dx
$$

and therefore that $\overline{\lim}_{n\to\infty} n^{\frac12}\|F_n - G\|_{\mathrm{u}} \ge \frac{1}{\sqrt{2\pi}}$. In particular, since $\tau_m = 1$ for these Bernoulli random variables, we conclude that the constant in the Berry–Esseen estimate cannot be smaller than $(2\pi)^{-\frac12}$.

EXERCISE 2.2.25. Because the derivation of Theorem 2.2.12 is so elegant and simple, one wonders whether (2.2.14) can be used as the starting point for a proof of (2.2.19). Unfortunately, the following naïve idea falls considerably short of the mark.

Let $X_1, \dots, X_n$ satisfy the hypotheses of Theorem 2.2.17. Starting from (2.2.14) and proceeding as I did in the passage from (2.2.22) to (2.2.23), show that for every $L > 0$

$$\|F_n - G\|_{\mathrm{u}} \le \frac{6\sum_1^n \tau_m^3}{L\Sigma_n^3} + \frac{L}{\sqrt{8\pi}},$$

and conclude that

$$\|F_n - G\|_{\mathrm{u}} \le \left(\frac{72}{\pi}\right)^{\frac{1}{4}} \left(\frac{\sum_1^n \tau_m^3}{\Sigma_n^3}\right)^{\frac{1}{2}}.$$

Obviously, this is unacceptably poor when $\Sigma_n^{-3}\sum_1^n \tau_m^3$ is small.

## § 2.3 Some Extensions of The Central Limit Theorem

In most modern treatments of The Central Limit Theorem, Fourier analysis plays a central role. Indeed, the Fourier transform makes the argument so simple that it can mask what is really happening. However, now that we know Lindeberg's argument, it is time to introduce Fourier techniques and begin to see how they facilitate reasoning involving independent random variables.

**§ 2.3.1. The Fourier Transform.** The *Fourier transform* of finite, $\mathbb{C}$-valued, Borel measure $\mu$ on $\mathbb{R}^N$ is the function $\hat{\mu} : \mathbb{R}^N \longrightarrow \mathbb{C}$ given by

$$\hat{\mu}(\boldsymbol{\xi}) = \int_{\mathbb{R}^N} \exp\left[\sqrt{-1}\,(\boldsymbol{\xi}, \mathbf{x})_{\mathbb{R}^N}\right] \mu(d\mathbf{x}) \quad \text{for} \quad \mathbf{x} \in \mathbb{R}^N. \tag{2.3.1}$$

When $\mu$ is a probability measure which is the distribution of an $\mathbb{R}^N$-valued random variable $\mathbf{X}$, probabilists usual call its Fourier transform the **characteristic function** of $\mathbf{X}$, and when $\mu$ admits a density $\varphi$ with respect to Lebesgue measure $\lambda_{\mathbb{R}^N}$, one uses

$$\hat{\varphi}(\boldsymbol{\xi}) = \int_{\mathbb{R}^N} \exp\left[\sqrt{-1}\,(\boldsymbol{\xi}, \mathbf{x})_{\mathbb{R}^N}\right] \varphi(\mathbf{x})\, d\mathbf{x} \quad \text{for} \quad \boldsymbol{\xi} \in \mathbb{R}^N \tag{2.3.2}$$

in place of $\hat{\mu}$ to denote its Fourier transform.

Obviously, $\hat{\mu}$ is a continuous function that is bounded by the total variation $\|\mu\|_{\mathrm{var}}$ of $\mu$; and only slightly less obvious[1] is the fact that, for $\varphi \in C_{\mathrm{c}}^{\infty}(\mathbb{R}^N;\mathbb{C})$, $\hat{\varphi} \in C^{\infty}(\mathbb{R}^N;\mathbb{C})$ and that $\hat{\varphi}$ as well as all its derivatives are **rapidly decreasing** (i.e., they tend to 0 at infinity faster than $\left(1 + |\boldsymbol{\xi}|^2\right)^{-1}$ to any power).

---

[1] One uses integration by parts to check that $\widehat{\partial^{\alpha}\varphi}(\boldsymbol{\xi}) = (-\sqrt{-1}\boldsymbol{\xi})^{\alpha}\hat{\varphi}(\boldsymbol{\xi})$ and concludes that $|\boldsymbol{\xi}|^n|\hat{\varphi}(\boldsymbol{\xi})|$ is bounded by $\sum_{\|\alpha\|=n} \|\partial^{\alpha}\varphi\|_{L^1(\mathbb{R}^N)}$.

LEMMA 2.3.3. *Let $\mu$ be a finite, $\mathbb{C}$-valued Borel measure on $\mathbb{R}^N$. Then, for every $\varphi \in C_{\mathrm{b}}(\mathbb{R}^N;\mathbb{C}) \cap L^1(\mathbb{R}^N;\mathbb{C})$ with $\hat{\varphi} \in L^1(\mathbb{R}^N;\mathbb{C})$,*

$$\langle \varphi, \mu \rangle = \int_{\mathbb{R}^N} \varphi \, d\mu = \frac{1}{(2\pi)^N} \int_{\mathbb{R}^N} \hat{\varphi}(\boldsymbol{\xi}) \overline{\hat{\mu}(\boldsymbol{\xi})} \, d\boldsymbol{\xi}. \tag{2.3.4}$$

*Moreover, given a sequence $\{\mu_n : n \in \mathbb{Z}^+\}$ of Borel probability measures and a Borel probability measure $\mu$ on $\mathbb{R}^N$, $\widehat{\mu_n} \longrightarrow \hat{\mu}$ uniformly on compacts if $\langle \varphi, \mu_n \rangle \longrightarrow \langle \varphi, \mu \rangle$ for every $\varphi \in C_{\mathrm{c}}(\mathbb{R}^N, \mathbb{R})$. Conversely, if $\widehat{\mu_n}(\boldsymbol{\xi}) \longrightarrow \hat{\mu}(\boldsymbol{\xi})$ pointwise, then $\langle \varphi_n, \mu_n \rangle \longrightarrow \langle \varphi, \mu \rangle$ whenever $\{\varphi_n : n \geq 1\}$ is a uniformly bounded sequence in $C_{\mathrm{b}}(\mathbb{R}^N;\mathbb{C})$ that tends to $\varphi$ uniformly on compacts. (Cf. Theorem 3.1.8 for more refined information on this subject.)*

PROOF: Choose $\rho \in C_{\mathrm{c}}^{\infty}(\mathbb{R}^N;[0,\infty))$ to be an even function with $\int_{\mathbb{R}^N} \rho \, d\mathbf{x} = 1$, and set $\rho_\epsilon(\mathbf{x}) = \epsilon^{-N}\rho(\epsilon^{-1}\mathbf{x})$ for $\epsilon \in (0,\infty)$. Next, define $\psi_\epsilon$ for $\epsilon \in (0,\infty)$ to be the **convolution** $\rho_\epsilon \star \mu$ of $\rho_\epsilon$ with $\mu$. That is,

$$\psi_\epsilon(\mathbf{x}) = \int_{\mathbb{R}^N} \rho_\epsilon(\mathbf{x} - \mathbf{y}) \, \mu(d\mathbf{y}) \quad \text{for} \quad \mathbf{x} \in \mathbb{R}^N.$$

It is then easy to check that $\psi_\epsilon \in C_{\mathrm{b}}(\mathbb{R}^N;\mathbb{C})$ and $\|\psi_\epsilon\|_{L^1(\mathbb{R}^N;\mathbb{R})} \leq \|\mu\|_{\mathrm{var}}$ for every $\epsilon \in (0,\infty)$. In addition, one sees (by Fubini's Theorem) that $\hat{\psi}_\epsilon(\boldsymbol{\xi}) = \hat{\rho}(\epsilon\,\boldsymbol{\xi})\hat{\mu}(\boldsymbol{\xi})$. Thus, for any $\varphi \in C_{\mathrm{b}}(\mathbb{R}^N;\mathbb{C}) \cap L^1(\mathbb{R}^N;\mathbb{C})$, Fubini's Theorem followed by the classical Parseval Identity (cf. Exercise 2.4.37) yields

$$\langle \varphi_\epsilon, \mu \rangle = \int_{\mathbb{R}^N} \varphi(\mathbf{x}) \, \psi_\epsilon(\mathbf{x}) \, d\mathbf{x} = \frac{1}{(2\pi)^N} \int_{\mathbb{R}^N} \hat{\rho}(\epsilon\,\boldsymbol{\xi}) \, \hat{\varphi}(\boldsymbol{\xi}) \, \hat{\mu}(-\boldsymbol{\xi}) \, d\boldsymbol{\xi},$$

where $\varphi_\epsilon \equiv \rho_\epsilon \star \varphi$ is the convolution of $\rho_\epsilon$ with $\varphi$. Since, as $\epsilon \searrow 0$, $\varphi_\epsilon \longrightarrow \varphi$ while $\hat{\rho}(\epsilon\,\boldsymbol{\xi}) \longrightarrow 1$ boundedly and pointwise, (2.3.4) now follows from Lebesgue's Dominated Convergence Theorem.

Turning to the second part of the theorem, first suppose that $\langle \varphi, \mu_n \rangle \longrightarrow \langle \varphi, \mu \rangle$ for every $\varphi \in C_{\mathrm{c}}(\mathbb{R}^N;\mathbb{R})$, and let $\boldsymbol{\xi}_n \longrightarrow \boldsymbol{\xi}$ in $\mathbb{C}$. Then, by the last part of Lemma 2.1.7 applied to $\varphi_n(\mathbf{x}) = e^{\sqrt{-1}\,(\boldsymbol{\xi}_n,\mathbf{x})_{\mathbb{R}^N}}$ and $\varphi(\mathbf{x}) = e^{\sqrt{-1}\,(\boldsymbol{\xi},\mathbf{x})_{\mathbb{R}^N}}$, $\widehat{\mu_n}(\boldsymbol{\xi}_n) \longrightarrow \hat{\mu}(\boldsymbol{\xi})$. Hence, $\widehat{\mu_n} \longrightarrow \hat{\mu}$ uniformly on compacts. Conversely, suppose that $\widehat{\mu_n} \longrightarrow \hat{\mu}$ pointwise. Again by Lemma 2.1.7, we need only check that $\langle \varphi, \mu_n \rangle \longrightarrow \langle \varphi, \mu \rangle$ when $\varphi \in C_{\mathrm{c}}^{\infty}(\mathbb{R}^N;\mathbb{C})$. But, for such a $\varphi$, $\hat{\varphi}$ is smooth and rapidly decreasing, and therefore the result follows immediately from the first part of the present lemma together with Lebesgue's Dominated Convergence Theorem. □

REMARK 2.3.5. Although it may seem too obvious to mention, an important, and rather amazing, consequence of Lemma 2.3.3 is that *a finite Borel measure on $\mathbb{R}^N$ is completely determined by its 1-dimensional marginals.* To understand this remark, recall that for a linear subspace $L$ of $\mathbb{R}^N$, the **marginal distribution** of $\mu$ on $L$ is the measure $(\Pi_L)_*\mu$, where $\Pi_L$ denotes orthogonal projection onto $L$. In particular, if $\mathbf{e} \in \mathbb{S}^{N-1}$ and $\mu_{\mathbf{e}}$ is the marginal distribution of $\mu$ on the 1-dimensional subspace spanned by $\mathbf{e}$, then $\hat{\mu}(\xi\mathbf{e}) = \widehat{\mu_{\mathbf{e}}}(\xi)$. Hence, the Fourier transform of $\mu$ is determined by the Fourier transforms of $\{\mu_{\mathbf{e}} : \mathbf{e} \in \mathbb{S}^{N-1}\}$, and therefore, by Lemma 2.3.3, $\mu$ can be recovered from its 1-dimensional marginals. Of course, one should be careful when applying this observation. For instance, when applied to an $\mathbb{R}^N$-valued random variable $\mathbf{X} = (X_1, \dots, X_N)$, it says that the distribution of $\mathbf{X}$ can be recovered from a knowledge of the distributions of $(\mathbf{e}, \mathbf{X})_{\mathbb{R}^N}$ for all $\mathbf{e} \in \mathbb{S}^{N-1}$, but it does *not* say that the distributions of the coordinates $X_i$, $1 \le i \le N$, determine the distribution of $\mathbf{X}$.

§ **2.3.2. Multidimensional Central Limit Theorem.** The great virtue of the Fourier transform is that it behaves so well under operations built out of translation. In applications to probability theory, this virtue is of particular importance when adding independent random variables. Specifically, if $\mathbf{X}$ and $\mathbf{Y}$ are independent, then the characteristic function of $\mathbf{X} + \mathbf{Y}$ is the product of the characteristic functions of $\mathbf{X}$ and $\mathbf{Y}$. This observation, combined with Lemma 2.3.3 leads to the following easy proof of The Central Limit Theorem for independent, identically distributed, $\mathbb{R}$-valued random variables $\{X_n : n \ge 1\}$ with mean value 0 and variance 1. Namely, if $\mu_n$ is the distribution of $\breve{S}_n$, then

$$\hat{\mu}_n(\xi) = \left(\hat{\mu}\left(\tfrac{\xi}{\sqrt{n}}\right)\right)^n = \left(1 - \frac{\xi^2}{2n} + o\left(\tfrac{1}{n}\right)\right)^n \longrightarrow e^{-\frac{\xi^2}{2}} = \widehat{\gamma_{0,1}}(\xi)$$

for every $\xi \in \mathbb{R}$.

Actually, as we are about to see, a slight variation on the preceding will allow us to lift the results that we already know for $\mathbb{R}$-valued random variables to random variables with values in $\mathbb{R}^N$. However, before I can state this result, I must introduce the analogs of the mean value and variance for vector-valued random variables. Thus, given an $\mathbb{R}^N$-valued random variable $\mathbf{X}$ on the probability space $(\Omega, \mathcal{F}, \mathbb{P})$ with $|\mathbf{X}| \in L^1(\mathbb{P}; \mathbb{R})$, the **mean value** $\mathbb{E}^{\mathbb{P}}[\mathbf{X}]$ of $\mathbf{X}$ is the $\mathbf{m} \in \mathbb{R}^N$ that is determined by the property that

$$(\boldsymbol{\xi}, \mathbf{m})_{\mathbb{R}^N} = \mathbb{E}^{\mathbb{P}}\left[(\boldsymbol{\xi}, \mathbf{X})_{\mathbb{R}^N}\right] \quad \text{for all} \quad \boldsymbol{\xi} \in \mathbb{R}^N.$$

Similarly, if $|\mathbf{X}|$ is $\mathbb{P}$-square integrable, then the **covariance** $\mathbf{cov}(\mathbf{X})$ of $\mathbf{X}$ is the symmetric linear transformation $\mathbf{C}$ on $\mathbb{R}^N$ determined by

$$(\boldsymbol{\xi}, \mathbf{C}\,\boldsymbol{\eta})_{\mathbb{R}^N} = \mathbb{E}^{\mathbb{P}}\left[(\boldsymbol{\xi}, \mathbf{X} - \mathbb{E}^{\mathbb{P}}[\mathbf{X}])_{\mathbb{R}^N}\,(\boldsymbol{\eta}, \mathbf{X} - \mathbb{E}^{\mathbb{P}}[\mathbf{X}])_{\mathbb{R}^N}\right] \quad \text{for } \boldsymbol{\xi},\, \boldsymbol{\eta} \in \mathbb{R}^N.$$

Notice that $\mathbf{cov}(\mathbf{X})$ is not only symmetric but is also non-negative definite, since for each $\boldsymbol{\xi} \in \mathbb{R}^N$, $(\boldsymbol{\xi}, \mathbf{cov}(\mathbf{X})\,\boldsymbol{\xi})_{\mathbb{R}^N}$ is nothing but the variance of $(\boldsymbol{\xi}, \mathbf{X})_{\mathbb{R}^N}$. Finally, given $\mathbf{m} \in \mathbb{R}^N$ and a symmetric, non-negative $\mathbf{C} \in \mathbb{R}^N \otimes \mathbb{R}^N$, I will use $\gamma_{\mathbf{m},\mathbf{C}}$ to denote the Borel probability measure on $\mathbb{R}^N$ determined by the property that

$$\int_{\mathbb{R}^N} \varphi\, d\gamma_{\mathbf{m},\mathbf{C}} = \int_{\mathbb{R}^N} \varphi(\mathbf{m} + \mathbf{C}^{\frac{1}{2}}\mathbf{y})\, \gamma_{0,1}^N(d\mathbf{y}), \quad \varphi \in C_{\mathrm{b}}(\mathbb{R}^N; \mathbb{R}), \tag{2.3.6}$$

where $\mathbf{C}^{\frac{1}{2}}$ is the non-negative definite, symmetric square root of $\mathbf{C}$.

Clearly, an $\mathbb{R}^N$-valued random variable $\mathbf{Y}$ has distribution $\gamma_{\mathbf{m},\mathbf{C}}$ if and only if, for each $\boldsymbol{\xi} \in \mathbb{R}^N$, $(\boldsymbol{\xi}, \mathbf{Y})_{\mathbb{R}^N}$ is a normal random variable with mean value $(\boldsymbol{\xi}, \mathbf{m})_{\mathbb{R}^N}$ and variance $(\boldsymbol{\xi}, \mathbf{C}\,\boldsymbol{\xi})_{\mathbb{R}^N}$. For this reason, $\gamma_{\mathbf{m},\mathbf{C}}$ is called the **normal or Gaussian distribution** with mean value $\mathbf{m}$ and covariance $\mathbf{C}$. For the same reason, a random variable with $\gamma_{\mathbf{m},\mathbf{C}}$ as its distribution is called a **normal or Gaussian random variable** with mean value $\mathbf{m}$ and covariance $\mathbf{C}$, or, more briefly, an $N(\mathbf{m}, \mathbf{C})$-random variable. Finally, one can use this characterization to see that

$$\widehat{\gamma_{\mathbf{m},\mathbf{C}}}(\boldsymbol{\xi}) = \exp\left[\sqrt{-1}(\boldsymbol{\xi}, \mathbf{m}) - \tfrac{1}{2}(\boldsymbol{\xi}, \mathbf{C}\boldsymbol{\xi})_{\mathbb{R}^N}\right]. \tag{2.3.7}$$

In the following statements, I will be assuming that $\{\mathbf{X}_n : n \in \mathbb{Z}^+\}$ is a sequence of mutually independent, square $\mathbb{P}$-integrable, $\mathbb{R}^N$-valued random variables on the probability space $(\Omega, \mathcal{F}, P)$. Further, I will assume that, for each $n \in \mathbb{Z}^+$, $\mathbf{X}_n$ has mean value $\mathbf{0}$ and strictly positive definite covariance $\mathbf{cov}(\mathbf{X}_n)$. Finally, for $n \in \mathbb{Z}^+$, set

$$\mathbf{S}_n = \sum_{m=1}^{n} \mathbf{X}_m, \quad \mathbf{C}_n \equiv \mathbf{cov}(\mathbf{S}_n) = \sum_{m=1}^{n} \mathbf{cov}(\mathbf{X}_m),$$
$$\Sigma_n = \big(\det(\mathbf{C}_n)\big)^{\frac{1}{2N}} \text{ and } \breve{\mathbf{S}}_n = \frac{\mathbf{S}_n}{\Sigma_n}.$$

Notice that when $N = 1$, the above use of the notation $\Sigma_n$ and $\breve{\mathbf{S}}_n$ is consistent with that in § 2.1.1.

With these preparations, I am ready to prove the following multidimensional generalization of Theorem 2.1.8.

THEOREM 2.3.8. *Referring to the preceding, assume that the limit*

$$\mathbf{A} \equiv \lim_{n\to\infty} \frac{\mathbf{C}_n}{\Sigma_n^2} \tag{2.3.9}$$

*exists and that*

$$\lim_{n\to\infty} \frac{1}{\Sigma_n^2} \sum_{m=1}^{n} \mathbb{E}^{\mathbb{P}}\left[|\mathbf{X}_m|^2, \, |\mathbf{X}_m| \geq \epsilon\Sigma_n\right] = 0 \quad \textit{for each } \epsilon > 0. \tag{2.3.10}$$

*Then, for every sequence* $\{\varphi_n : n \geq 1\} \subseteq C(\mathbb{R}^N; \mathbb{C})$ *that satisfies*

$$\sup_{n\geq 1} \sup_{\mathbf{y}\in\mathbb{R}^N} \frac{|\varphi_n(\mathbf{y})|}{1 + |\mathbf{y}|^2} < \infty \tag{2.3.11}$$

*and converges uniformly on compacts to* $\varphi$,

$$(2.3.12) \qquad \lim_{n\to\infty} \mathbb{E}^{\mathbb{P}}\big[\varphi_n(\check{\mathbf{S}}_n)\big] = \langle \varphi, \gamma_{\mathbf{0},\mathbf{A}}\rangle.$$

*In particular, when the* $\mathbf{X}_n$*'s are uniformly square* $\mathbb{P}$*-integrable random variables with mean value* $\mathbf{0}$ *and common covariance* $\mathbf{C}$,

$$\lim_{n\to\infty} \mathbb{E}^{\mathbb{P}}\left[\varphi_n\left(\frac{\mathbf{S}_n}{\sqrt{n}}\right)\right] = \langle \varphi, \gamma_{\mathbf{0},\mathbf{C}}\rangle$$

*whenever* $\{\varphi_n : n \geq 1\} \subseteq C(\mathbb{R}^N;\mathbb{C})$ *satisfies (2.3.11) and converges to* $\varphi$ *uniformly on compacts.*

PROOF: Given $\mathbf{e} \in \mathbb{S}^{N-1}$, set

$$\Sigma_n(\mathbf{e}) = \sqrt{(\mathbf{e}, \mathbf{C}_n\mathbf{e})_{\mathbb{R}^N}} \quad \text{and} \quad \rho_n(\mathbf{e}) = \frac{\Sigma_n(\mathbf{e})}{\Sigma_n}.$$

Then, $\rho(\mathbf{e}) \equiv \inf_{n\geq 1}\rho_n(\mathbf{e}) \in (0,1]$ and $\rho_n(\mathbf{e}) \longrightarrow \sqrt{(\mathbf{e},\mathbf{A}\mathbf{e})_{\mathbb{R}^N}}$ as $n \to \infty$. In particular, if $(\mathbf{e}_1, \dots, \mathbf{e}_N)$ is an orthonormal basis in $\mathbb{R}^N$, then

$$\begin{aligned}\mathbb{E}^{\mathbb{P}}\big[|\check{\mathbf{S}}_n|^2\big] &= \sum_{i=1}^N \mathbb{E}^{\mathbb{P}}\big[(\mathbf{e}_i, \check{\mathbf{S}}_n)_{\mathbb{R}^N}^2\big] = \sum_{i=1}^N \rho_n(\mathbf{e}_i)^2 \\ &\longrightarrow \sum_{i=1}^N (\mathbf{e}_i, \mathbf{A}\mathbf{e}_i)_{\mathbb{R}^N} = \int_{\mathbb{R}^N} |\mathbf{y}|^2\,\gamma_{\mathbf{0},\mathbf{A}}(dy).\end{aligned}$$

Hence, by Lemmas 2.1.7 and 2.3.3 plus (2.3.7), all that we have to do is check that

$$(*) \qquad f_n(\boldsymbol{\xi}) \equiv \mathbb{E}^{\mathbb{P}}\left[e^{\sqrt{-1}\,(\boldsymbol{\xi},\check{\mathbf{S}}_n)_{\mathbb{R}^N}}\right] \longrightarrow e^{-\frac{1}{2}(\boldsymbol{\xi},\mathbf{A}\boldsymbol{\xi})_{\mathbb{R}^N}}$$

for each $\boldsymbol{\xi} \in \mathbb{R}^N$.

When $\boldsymbol{\xi} = \mathbf{0}$, (*) is trivial. Thus, assume that $\boldsymbol{\xi} \neq \mathbf{0}$, set $\mathbf{e} = \frac{\boldsymbol{\xi}}{|\boldsymbol{\xi}|}$, and take $\check{S}_n(\mathbf{e}) = \frac{(\mathbf{e},\mathbf{S}_n)_{\mathbb{R}^N}}{\Sigma_n(\mathbf{e})}$. Because

$$\begin{aligned}&\frac{1}{\Sigma_n(\mathbf{e})^2}\sum_{m=1}^n \mathbb{E}^{\mathbb{P}}\big[(\mathbf{e},\mathbf{X}_m)_{\mathbb{R}^N}^2,\ |(\mathbf{e},\mathbf{X}_m)_{\mathbb{R}^N}| \geq \epsilon\Sigma_n(\mathbf{e})\big] \\ &\quad\leq \frac{1}{\rho(\mathbf{e})^2\Sigma_n^2}\sum_{m=1}^n \mathbb{E}^{\mathbb{P}}\big[(\mathbf{e},\mathbf{X}_m)_{\mathbb{R}^N}^2,\ |(\mathbf{e},\mathbf{X}_m)_{\mathbb{R}^N}| \geq \epsilon\rho(\mathbf{e})\Sigma_n(\mathbf{e})\big]\end{aligned}$$

tends to 0 for each $\epsilon > 0$, Theorem 2.1.8 combined with Lemma 2.3.3 guarantees that, for any $\eta \in \mathbb{R}$,

$$\mathbb{E}^{\mathbb{P}}\left[e^{\sqrt{-1}\,\eta_n \breve{S}_n(\mathbf{e})}\right] \longrightarrow e^{-\frac{1}{2}|\eta|^2}$$

for any $\{\eta_n : n \geq 1\} \subseteq \mathbb{R}$ that tends to $\eta$. In particular, if $\eta = \sqrt{(\boldsymbol{\xi}, \mathbf{A}\boldsymbol{\xi})_{\mathbb{R}^N}}$ and $\eta_n = \rho_n(\mathbf{e})|\boldsymbol{\xi}|$, we find that

$$f_n(\boldsymbol{\xi}) = \mathbb{E}^{\mathbb{P}}\left[e^{\sqrt{-1}\,\eta_n \breve{S}_n(\mathbf{e})}\right] \longrightarrow e^{-\frac{1}{2}(\boldsymbol{\xi}, \mathbf{A}\boldsymbol{\xi})_{\mathbb{R}^N}}.$$

When $\mathbf{C}$ is non-degenerate, the final part is a trivial application of the initial part. When $\mathbf{C}$ is degenerate but not zero, one can reduce to the non-degenerate case by projecting onto the span of its eigenvectors with strictly positive eigenvalues, and when $\mathbf{C} = 0$, there is nothing to do. □

**§ 2.3.3. Higher Moments.** In this subsection I will show that when the $X_n$'s possess higher moments, then (2.1.1) remains true for $\varphi$'s that can grow faster than $1 + |y|^2$. As an initial step in this direction, I give the following simple example.

LEMMA 2.3.13. *Suppose that $\{X_n : n \geq 1\}$ is a sequence of independent, identically distributed random variables with mean value 0 and variance 1. If $\mathbb{E}^{\mathbb{P}}[X_1^{2\ell}] < \infty$ for some $\ell \in \mathbb{Z}^+$, then (2.1.1) holds for any $\varphi \in C(\mathbb{R}^N;\mathbb{C})$ that satisfies*

$$\sup_{y \in \mathbb{R}} \frac{|\varphi(y)|}{1 + |y|^{2\ell}} < \infty. \tag{2.3.14}$$

PROOF: Refer to the discussion in the introduction to this chapter, and observe that the argument there shows that

$$\lim_{n \to \infty} \mathbb{E}^{\mathbb{P}}\left[\breve{S}_n^{2\ell}\right] = \frac{(2\ell)!}{2^\ell \ell!} = \int_{\mathbb{R}} y^{2\ell}\, \gamma_{0,1}(dy)$$

whenever the $2\ell$th moment of $X_1$ is finite. Hence the desired conclusion is an application of the last part of Lemma 2.1.7 with $\psi(y) = 1 + |y|^{2\ell}$. □

In most situations one cannot carry out the computations needed to give a direct proof that the last part of Lemma 2.1.7 applies, and for this reason the following lemma is often useful.

LEMMA 2.3.15. *Suppose that $\{\mu_n : n \geq 1\}$ is a sequence of finite (non-negative) Borel measures on $\mathbb{R}^N$, and assume $\mu$ is a finite Borel measure with the property that $\langle \varphi, \mu_n \rangle \longrightarrow \langle \varphi, \mu \rangle$ for all $\varphi \in C_{\mathrm{b}}^{\infty}(\mathbb{R}^N;\mathbb{R})$. If, for some $\psi \in C(\mathbb{R}^N;[0,\infty))$ and $p \in (1,\infty)$,*

$$\sup_{n \geq 1} \langle \psi^p, \mu_n \rangle < \infty, \tag{2.3.16}$$

*then $\langle \varphi_n, \mu_n \rangle \longrightarrow \langle \varphi, \mu \rangle$ whenever $\{\varphi_n : n \geq 1\} \subseteq C(\mathbb{R}^N;\mathbb{C})$ is a sequence that satisfies $|\varphi_n| \leq \psi$ for all $n \in \mathbb{Z}^+$ and converges to $\varphi$ uniformly on compacts.*

PROOF: By Lemma 2.1.7, all that we have to prove is that $\langle\psi,\mu_n\rangle \longrightarrow \langle\psi,\mu\rangle$. For this purpose, note that, under our present hypotheses, Lemma 2.1.7 shows that $\langle\psi,\mu\rangle \le \underline{\lim}_{n\to\infty}\langle\psi,\mu_n\rangle < \infty$ and that $\langle\psi\wedge R,\mu_n\rangle \longrightarrow \langle\psi\wedge R,\mu\rangle \le \langle\psi,\mu\rangle$ for each $R > 0$. Thus, it suffices to observe that

$$\sup_{n\ge1}\langle(\psi-\psi\wedge R),\mu_n\rangle = \sup_{n\ge1}\int_{\{\psi>R\}}\psi\,d\mu_n \le R^{1-p}\sup_{n\ge1}\langle\psi^p,\mu_n\rangle \longrightarrow 0$$

as $R\to\infty$. $\square$

Knowing Lemma 2.3.15, one's problem is to find conditions under which one can show that $\sup_{n\ge1}\mathbb{E}^{\mathbb{P}}[\psi(\check{\mathbf{S}}_n)] < \infty$ for an interesting class of non-negative $\psi$'s. One such class is provided by the notion of a sub-Gaussian random variable. Given $\beta\in[0,\infty)$, an $\mathbb{R}^N$-valued random variable $\mathbf{X}$ is said to be **$\beta$-sub-Gaussian** if

$$\mathbb{E}^{\mathbb{P}}\left[e^{(\boldsymbol{\xi},\mathbf{X})_{\mathbb{R}^N}}\right] \le e^{\frac{\beta^2|\boldsymbol{\xi}|^2}{2}}, \quad \boldsymbol{\xi}\in\mathbb{R}^N. \tag{2.3.17}$$

The origin of this terminology should be clear: if $X\in N(0,\sigma^2)$, then equality holds in (2.3.17) with $\beta=\sigma$.

LEMMA 2.3.18. *Let $\mathbf{X}$ be an $\mathbb{R}^N$-valued random variable. If $\mathbf{X}$ is a $\beta$-sub-Gaussian, then $\mathbb{E}^{\mathbb{P}}[X]=0$, $\mathrm{Cov}(\mathbf{X})\le\beta^2 I$,*

$$\mathbb{P}(|\mathbf{X}|\ge R) \le 2Ne^{-\frac{R^2}{2N\beta^2}}, \quad R>0,$$

*and, for each $\alpha\in[0,\beta^{-1})$,*

$$\mathbb{E}^{\mathbb{P}}\left[e^{\frac{\alpha^2|\mathbf{X}|^2}{2}}\right] \le \left(1-(\alpha\beta)^2\right)^{-\frac{N}{2}}.$$

*Conversely, if $A\equiv\mathbb{E}^{\mathbb{P}}\left[e^{\frac{\alpha^2|\mathbf{X}|^2}{2}}\right]<\infty$ for some $\alpha\in(0,\infty)$ and $\mathbb{E}^{\mathbb{P}}[\mathbf{X}]=\mathbf{0}$, then $\mathbf{X}$ is $\beta$-sub-Gaussian when $\beta=\frac{\sqrt{2(1+A)}}{\alpha}$. In particular, if $\mathbf{X}$ is a bounded random variable with mean value $\mathbf{0}$, then $\mathbf{X}$ is $\beta$-sub-Gaussian with $\beta\le2\|\mathbf{X}\|_{L^\infty(\mathbb{P};\mathbb{R}^N)}$. Finally, if $\mathbf{X}_1,\dots,\mathbf{X}_n$ are independent random variables, and, for each $1\le m\le n$, $\mathbf{X}_m$ is $\beta_m$-sub-Gaussian, then for any $a_1,\dots,a_n\in\mathbb{R}$, $\sum_{m=1}^n a_m\mathbf{X}_m$ is $\beta$-sub-Gaussian when $\beta=\sqrt{\sum_{m=1}^n(a_m\beta_m)^2}$.*

PROOF: Since the moment generating function of the sum of independent random variables is the product of the moment generating functions of the summands, the final assertion is essentially trivial.

To prove the first assertion, use Lebesgue's Dominated Convergence Theorem to justify

$$\mathbb{E}^{\mathbb{P}}\left[(\mathbf{e},\mathbf{X})_{\mathbb{R}^N}\right] = \lim_{t\searrow0}t^{-1}\left(\mathbb{E}^{\mathbb{P}}\left[e^{t(\mathbf{e},\mathbf{X})_{\mathbb{R}^N}}\right]-1\right) \le \lim_{t\searrow0}\frac{e^{\frac{\beta^2t^2}{2}}-1}{t}=0$$

and

$$\mathbb{E}^{\mathbb{P}}\big[(\mathbf{e},\mathbf{X})_{\mathbb{R}^N}^2\big] = \lim_{t\searrow 0} \frac{\mathbb{E}^{\mathbb{P}}\big[e^{t(\mathbf{e},\mathbf{X})_{\mathbb{R}^N}}\big] + \mathbb{E}^{\mathbb{P}}\big[e^{-t(\mathbf{e},\mathbf{X})_{\mathbb{R}^N}}\big] - 2}{t^2} \le 2\lim_{t\searrow 0} \frac{e^{\frac{\beta^2 t^2}{2}} - 1}{t^2} = \beta^2$$

for any $\mathbf{e} \in \mathbb{S}^{N-1}$. Next, from

$$\mathbb{P}\big((\mathbf{e},\mathbf{X})_{\mathbb{R}^N} \ge R\big) \le e^{-tR}\mathbb{E}^{\mathbb{P}}\big[e^{t(\mathbf{e},\mathbf{X})_{\mathbb{R}^N}}\big] \le \exp\left(-tR + \frac{\beta^2 t^2}{2}\right)$$

for any $\ge 0$ and $\mathbf{e} \in \mathbb{S}^{N-1}$, one gets $\mathbb{P}\big((\mathbf{e},\mathbf{X})_{\mathbb{R}^N} \ge R\big) \le e^{-\frac{R^2}{2\beta^2}}$ by minimizing over $t \ge 0$. Since

$$\mathbb{P}\big(|\mathbf{X}| \ge R\big) \le 2N \max_{\mathbf{e}\in\mathbb{S}^{N-1}} \mathbb{P}\big((\mathbf{e},\mathbf{X})_{\mathbb{R}^N} \ge N^{-\frac{1}{2}}R\big),$$

the estimate for $\mathbb{P}(|\mathbf{X}| \ge R)$ follows. To get the estimate on $\mathbb{E}^{\mathbb{P}}\big[e^{\frac{\alpha^2|\mathbf{X}|^2}{2}}\big]$, use Tonelli's Theorem to see that

$$\mathbb{E}^{\mathbb{P}}\big[e^{\frac{\alpha^2|\mathbf{X}|^2}{2}}\big] = \int_{\mathbb{R}} \mathbb{E}^{\mathbb{P}}\big[e^{(\boldsymbol{\xi},\mathbf{X})_{\mathbb{R}^N}}\big]\,\gamma_{0,\alpha^2 I}(d\boldsymbol{\xi}) \le \int_{\mathbb{R}} e^{\frac{\beta^2|\boldsymbol{\xi}|^2}{2}}\,\gamma_{0,\alpha^2 I}(d\boldsymbol{\xi}) = \big(1-(\alpha\beta)^2\big)^{-\frac{N}{2}}.$$

Now assume that $A = \mathbb{E}^{\mathbb{P}}\big[e^{\frac{\alpha^2|\mathbf{X}|^2}{2}}\big] < \infty$ for some $\alpha \in (0,\infty)$ and that $\mathbb{E}^{\mathbb{P}}[\mathbf{X}] = \mathbf{0}$. Then

$$\begin{aligned}\mathbb{E}^{\mathbb{P}}\big[e^{(\boldsymbol{\xi},\mathbf{X})_{\mathbb{R}^N}}\big] &= 1 + \int_0^1 (1-t)\mathbb{E}^{\mathbb{P}}\big[(\boldsymbol{\xi},\mathbf{X})_{\mathbb{R}^N}^2 e^{t(\boldsymbol{\xi},\mathbf{X})_{\mathbb{R}^N}}\big]\,dt \le 1 + \frac{|\boldsymbol{\xi}|^2}{2}\mathbb{E}^{\mathbb{P}}\big[|\mathbf{X}|^2 e^{|\boldsymbol{\xi}||\mathbf{X}|}\big] \\ &\le 1 + \frac{|\boldsymbol{\xi}|^2}{2} e^{\frac{|\boldsymbol{\xi}|^2}{\alpha^2}}\mathbb{E}^{\mathbb{P}}\big[|\mathbf{X}|^2 e^{\frac{\alpha^2|\mathbf{X}|^2}{4}}\big] \le 1 + A\frac{|\boldsymbol{\xi}|^2}{\alpha^2} e^{\frac{|\boldsymbol{\xi}|^2}{\alpha^2}} \le \left(1 + \frac{A|\boldsymbol{\xi}|^2}{\alpha^2}\right) e^{\frac{|\boldsymbol{\xi}|^2}{\alpha^2}},\end{aligned}$$

from which it is clear that $\mathbf{X}$ is $\beta$-sub-Gaussian for the prescribed $\beta$. In particular, if $K = \|\mathbf{X}\|_{L^\infty(\mathbb{P};\mathbb{R}^N)} \in (0,\infty)$, then $\mathbb{E}^{\mathbb{P}}\big[e^{\frac{\alpha^2|\mathbf{X}|^2}{2}}\big] \le e^{\frac{\alpha^2 K^2}{2}}$ for all $\alpha \ge 0$, and so, if, in addition, $\mathbf{X}$ has mean value $\mathbf{0}$, then $\mathbf{X}$ is $\beta$-sub-Gaussian for $\beta = \sqrt{t^{-1}(1+e^{tK^2})}$ for all $t > 0$. Taking $t = K^{-2}$, we see that $\beta = K\sqrt{1+e} \le 2K$. When $\|\mathbf{X}\|_{L^\infty(\mathbb{P};\mathbb{R}^N)} = 0$, there is nothing to do. □

By combining Lemmas 2.3.15 and 2.3.18 with Theorem 2.3.8, we get the following.

THEOREM 2.3.19. *Working in the setting and with the notation in Theorem 2.3.8, assume that, for each $n \in \mathbb{Z}^+$,*

$$\mathbb{E}^{\mathbb{P}}\big[e^{(\boldsymbol{\xi},\mathbf{X}_n)_{\mathbb{R}^N}}\big] \le e^{\beta_n|\boldsymbol{\xi}|^2}, \quad \boldsymbol{\xi} \in \mathbb{R}^N,$$

*where $\beta_n \in (0,\infty)$. If*

$$\beta \equiv \sup_{n\geq 1} \frac{\sqrt{\sum_{m=1}^{n}\beta_m^2}}{\Sigma_n} < \infty,$$

*then (2.3.12) holds for any $\varphi \in C(\mathbb{R}^N;\mathbb{C})$ satisfying*

$$|\varphi(\mathbf{y})| \leq Ce^{\frac{\alpha^2|\mathbf{y}|^2}{2}}, \quad \mathbf{y}\in\mathbb{R}^N,$$

*for some $C<\infty$ and $\alpha \in \left(0,\frac{1}{\beta}\right)$. In particular, if the $\mathbf{X}_n$'s are identically distributed with covariance $\mathbf{C}$ and if $\mathbb{E}^{\mathbb{P}}\left[e^{\alpha^2|\mathbf{X}_1|^2}\right] < \infty$ for some $\alpha\in(0,\infty)$, then, for any $\varphi\in C(\mathbb{R}^N;\mathbb{C})$,*

$$\varlimsup_{|\mathbf{y}|\to\infty} |\mathbf{y}|^{-2}\log\bigl(1+|\varphi(\mathbf{y})|\bigr) = 0 \implies \lim_{n\to\infty}\mathbb{E}^{\mathbb{P}}\bigl[\varphi(n^{-\frac{1}{2}}\mathbf{S}_n)\bigr] = \langle\varphi,\gamma_{\mathbf{0},\mathbf{C}}\rangle.$$

## Exercises for § 2.3

EXERCISE 2.3.20. Here is a proof of Feller's part of the Lindeberg–Feller Theorem. Referring to Theorem 2.1.4 and the discussion proceeding it, assume that, as $n\to\infty$,

$$r_n \longrightarrow 0 \quad \text{and} \quad \mathbb{E}^{\mathbb{P}}\left[e^{\sqrt{-1}\xi\check{S}_n}\right] \longrightarrow e^{-\frac{\xi^2}{2}} \text{ for all } \xi\in\mathbb{R}.$$

**(i)** Show that

$$\left|1-\mathbb{E}^{\mathbb{P}}\left[e^{\sqrt{-1}\frac{\xi X_m}{\Sigma_n}}\right]\right| \leq \frac{\xi^2\sigma_m^2}{2\Sigma_{m,n}^2},$$

and conclude that, for each $R>0$, there is an $N_R$ such that

$$\max_{1\leq m\leq n}\left|1-\mathbb{E}^{\mathbb{P}}\left[e^{\sqrt{-1}\frac{\xi X_m}{\Sigma_n}}\right]\right| \leq \frac{1}{2} \quad \text{for } n\geq N_R \text{ and } |\xi|\leq R.$$

**(ii)** Take the branch of the logarithm given by $\log\zeta = -\sum_{k=1}^{\infty}\frac{(1-\zeta)^k}{k}$ for $\zeta\in\mathbb{C}$ with $|1-\zeta|<1$, and check that $\bigl|(1-\zeta)+\log\zeta\bigr| \leq |1-\zeta|^2$ for $|1-\zeta|\leq\frac{1}{2}$. Conclude first that

$$\left|\sum_{m=1}^{n}\mathbb{E}^{\mathbb{P}}\left[1-e^{\sqrt{-1}\frac{\xi X_m}{\Sigma_n}}\right] + \sum_{m=1}^{n}\log\mathbb{E}^{\mathbb{P}}\left[e^{\sqrt{-1}\frac{\xi X_m}{\Sigma_n}}\right]\right| \leq \frac{R^4 r_n^2}{4}$$

for $n\geq N_R$ and $|\xi|\leq R$, and then that

$$\Delta_n(\xi) \equiv \frac{\xi^2}{2} - \sum_{m=1}^{n}\mathbb{E}^{\mathbb{P}}\left[1-\cos\frac{\xi X_m}{\Sigma_n}\right] \longrightarrow 0$$

uniformly for $\xi$'s in compacts.

**(iii)** Given $\epsilon > 0$, show that

$$\sum_{m=1}^{n} \mathbb{E}^{\mathbb{P}}\left[1 - \cos\frac{\xi X_m}{\Sigma_n}, |X_m| < \epsilon\Sigma_n\right] \leq \frac{\xi^2}{2\Sigma_n^2}\sum_{m=1}^{n}\mathbb{E}^{\mathbb{P}}\left[X_m^2, |X_m| < \epsilon\Sigma_n\right]$$
$$\leq \frac{\xi^2}{2} - \frac{\xi^2}{2}g_n(\epsilon)$$

and that

$$\sum_{m=1}^{n} \mathbb{E}^{\mathbb{P}}\left[1 - \cos\frac{\xi X_m}{\Sigma_n}, |X_m| \geq \epsilon\Sigma_n\right] \leq \frac{1}{2\epsilon^2}.$$

Finally, combine these and apply **(ii)** to get $\overline{\lim}_{n\to\infty}\xi^2 g_n(\epsilon) \leq 2\epsilon^{-2}$ for all $\xi \in \mathbb{R}$.

EXERCISE 2.3.21. It is of some interest to know that the second moment assumption can be removed from the hypotheses in Exercise 2.1.11 and that the result there extends to Borel probability measures on $\mathbb{R}^N$. To explain what I have in mind, first use that exercise to see that if $\sigma^2 = \int_{\mathbb{R}} x^2\,\mu(dx) < \infty$, then $\mu = T_2\mu \implies \mu \in N(0,\sigma^2)$. What I want to do now is remove the a priori assumption that $\int_{\mathbb{R}} x^2\,\mu(dx) < \infty$. That is, I want to show that, for *any* probability measure $\mu$ on $\mathbb{R}$, $\mu = T_2\mu \iff \mu \in N(0,\sigma^2)$ for some $\sigma \in [0,\infty)$. Since the "$\Longleftarrow$" direction is obvious, and, by the discussion above, the "$\implies$" direction is already covered when $\int_{\mathbb{R}} x^2\,\mu(dx) < \infty$, all that remains is to show that

$$\mu = T_2\mu \implies \int_{\mathbb{R}} x^2\,\mu(dx) < \infty. \tag{2.3.22}$$

See Exercise 2.4.33 for an interesting application of this result.

**(i)** Check (2.3.22) first under the condition that $\mu$ is symmetric (i.e., $\mu(-\Gamma) = \mu(\Gamma)$ for all $\Gamma \in \mathcal{B}_{\mathbb{R}}$). Indeed, if $\mu$ is symmetric, show that

$$\hat{\mu}(\xi) = \int_{\mathbb{R}} \cos(\xi x)\,\mu(dx), \quad \xi \in \mathbb{R}.$$

At the same time, show that

$$\mu = T_2\mu \implies \hat{\mu}(2^{-\frac{1}{2}}\xi) = \hat{\mu}(\xi)^{\frac{1}{2}}, \quad \xi \in \mathbb{R}.$$

Conclude from these two that $\hat{\mu} > 0$ everywhere and that

$$\int_{\mathbb{R}} \cos(2^{-\frac{n}{2}}\xi x)\,\mu(dx) = \hat{\mu}(\xi)^{2^{-n}}, \quad n \in \mathbb{N} \text{ and } \xi \in \mathbb{R}.$$

Finally, note that $1 - x \leq -\log x$ for $x \in (0,1]$, apply this to the preceding to get

$$2^n\int_{\mathbb{R}}\left(1 - \cos(2^{-\frac{n}{2}}x)\right)\mu(dx) \leq -\log(\hat{\mu}(1)) < \infty, \quad n \in \mathbb{N},$$

and arrive at

$$\int_{\mathbb{R}} x^2 \, \mu(dx) \leq -2 \log(\hat{\mu}(1))$$

after an application of Fatou's Lemma.
**(ii)** To complete the program, let $\mu$ be any solution to $\mu = T_2\mu$, and define $\nu$ by

$$\nu(\Gamma) = \iint\limits_{\mathbb{R}^2} \mathbf{1}_\Gamma(x - y) \, \mu(dx)\mu(dy).$$

Check that $\nu$ is symmetric and that $\nu = T_2\nu$. Hence, by **(i)**, $\int_{\mathbb{R}} x^2 \, \nu(dx) < \infty$ (in fact, $\nu$ is centered normal). Finally, use this and part **(i)** of Exercise 1.4.27 to deduce that $\int_{\mathbb{R}} x^2 \, \mu(dx) < \infty$.
**(iii)** Make the obvious extension of $T_2$ to Borel probability measures $\mu$ on $\mathbb{R}^N$. That is,

$$T_2\mu(\Gamma) = \iint\limits_{\mathbb{R}^N \times \mathbb{R}^N} \mathbf{1}_\Gamma \left( \frac{\mathbf{x} + \mathbf{y}}{2^{\frac{1}{2}}} \right) \mu(d\mathbf{x})\mu(d\mathbf{y}) \quad \text{for } \Gamma \in \mathcal{B}_{\mathbb{R}^N}.$$

Using the result just proved when $N = 1$, show that $\mu = T_2\mu$ if and only if $\mu = \gamma_{\mathbf{0},\mathbf{C}}$ for some non-negative definite, symmetric $\mathbf{C}$.

EXERCISE 2.3.23. In connection with the preceding exercise, define $T_\alpha\mu$ for $\alpha \in (0, \infty)$ and Borel probability measures $\mu$ on $\mathbb{R}^N$, so that

$$T_\alpha\mu(\Gamma) = \iint\limits_{\mathbb{R}^N \times \mathbb{R}^N} \mathbf{1}_\Gamma(2^{-\frac{1}{\alpha}}(\mathbf{x} + \mathbf{y})) \, \mu(d\mathbf{x})\mu(d\mathbf{y}), \quad \Gamma \in \mathcal{B}_{\mathbb{R}^N}.$$

The problem under consideration here is that of determining for which $\alpha$'s there exist nontrivial (i.e., $\mu \neq \delta_{\mathbf{0}}$) solutions to the fixed point equation $\mu = T_\alpha\mu$. Begin by reducing the problem to the case when $N = 1$. Next, repeat the initial argument given in part **(ii)** of Exercise 2.3.21 to see that there is some solution if and only if there is one that is symmetric. Assuming that $\mu$ is a non-trivial, symmetric solution, use the reasoning in part **(i)** there to see that

$$\int_{\mathbb{R}} x^2 \, \mu(dx) = \begin{cases} \infty & \text{if } \alpha \in (0, 2) \\ 0 & \text{if } \alpha \in (2, \infty). \end{cases}$$

In particular, when $\alpha \in (2, \infty)$, there are no non-trivial solutions to $\mu = T_\alpha\mu$. (See § 3.2.3 for more on this topic.)

EXERCISE 2.3.24. Return to the setting of Exercise 2.1.13. After noting that, so long as $\mathbf{e} \in \mathbf{S}^{n-1}$, the distribution of

$$\mathbf{x} \in \mathbf{S}^{n-1}(\sqrt{n}) \longmapsto (\mathbf{e}, \mathbf{x})_{\mathbb{R}^n} \in \mathbb{R}$$

is independent of $\mathbf{e}$, use Lemma 2.3.3 to prove that the assertion in (2.1.15) follows as a consequence of the one in (2.1.14).

EXERCISE 2.3.25. Begin by checking the identity (cf. (1.3.20))

$$\int_0^\infty t^s e^{-\frac{t^2}{2\beta^2}}\,dt = 2^{\frac{s-1}{2}}\beta^{s+1}\Gamma\left(\frac{s+1}{2}\right)$$

for all $\beta \in (0,\infty)$ and $s \in (-1,\infty)$. Use the preceding to see that, for each $p \in (0,\infty)$,

$$\mathbb{E}^{\mathbb{P}}\left[|X|^p\right] = \sqrt{\frac{2^p}{\pi}}\Gamma\left(\frac{p+1}{2}\right)\sigma^p \quad \text{if } X \in N(0,\sigma^2). \tag{2.3.26}$$

The goal of the exercise is to show that the moments of sub-Gaussian random variable display similar behavior.

**(i)** Suppose that $X$ is $\beta$-sub-Gaussian, and show that, for each $p \in (0,\infty)$,

$$\mathbb{E}^{\mathbb{P}}\left[|X|^p\right] \le K_p\beta^p \quad \text{where } K_p \equiv p2^{\frac{p}{2}}\Gamma\left(\frac{p}{2}\right) = 2^{\frac{p}{2}+1}\Gamma\left(\frac{p}{2}+1\right).$$

**(ii)** Again suppose that $X$ is $\beta$-sub-Gaussian, and let $\sigma^2$ be its variance. Show that

$$\mathbb{E}^{\mathbb{P}}\left[|X|^p\right] \ge K_4^{-(1-\frac{p}{2})^+}\left(\frac{\sigma}{\beta}\right)^{2+|p-2|}\beta^p$$

for each $p \in (0,\infty)$.

**Hint**: When $p \ge 2$, the inequality is trivial. To prove it when $p < 2$, show that, for any $q \in (1,\infty)$,

$$\sigma^2 \le \mathbb{E}^{\mathbb{P}}\left[|X|^p\right]^{\frac{1}{q}}\mathbb{E}^{\mathbb{P}}\left[|X|^{\frac{2q-p}{q-1}}\right]^{\frac{1}{q'}},$$

where $q' = \frac{q}{q-1}$ is the Hölder conjugate of $q$.

**(iii)** Suppose that $X_1,\dots,X_n$ are independent and that, for each $1 \le m \le n$, $X_m$ is $\beta_m$-sub-Gaussian and has variance $\sigma_m^2$. Given $\{a_1,\dots,a_n\} \subseteq \mathbb{R}$, set

$$S = \sum_{m=1}^n a_m X_m, \quad \Sigma = \sqrt{\sum_{m=1}^n (a_m\sigma_m)^2}, \quad \text{and} \quad B = \sqrt{\sum_{m=1}^n (a_m\beta_m)^2},$$

and show that, for each $p \in (0,\infty)$,

$$K_4^{-(1-\frac{p}{2})^+}\left(\frac{\Sigma}{B}\right)^{2+|p-2|}B^p \le \mathbb{E}^{\mathbb{P}}\left[|S|^p\right] \le K_pB^p.$$

In particular, if $\beta_m = \beta$ and $\sigma_m = \sigma$ for all $1 \le m \le n$, then

$$K_4^{-(1-\frac{p}{2})^+}\left(\frac{\sigma}{\beta}\right)^{2+|p-2|}(\beta A)^p \le \mathbb{E}^{\mathbb{P}}\left[|S|^p\right] \le K_p(\beta A)^p, \quad \text{where } A = \sqrt{\sum_{m=1}^n a_m^2}.$$

**(iv)** The most famous case of the situation discussed in **(iii)** is when the $X_m$'s are symmetric Bernoulli (i.e., $\mathbb{P}(X_m = \pm 1) = \frac{1}{2}$). First use **(iii)** in Exercise 1.3.17 or direct computation to check that $X_m$ is 1-sub-Gaussian, and then conclude that

$$(2.3.27) \qquad K_4^{-(1-\frac{p}{2})^+}\left(\sum_{m=1}^n a_m^2\right)^{\frac{p}{2}} \le \mathbb{E}^P\left[\left|\sum_{m=1}^n a_m X_m\right|^p\right] \le K_p\left(\sum_{m=1}^n a_m^2\right)^{\frac{p}{2}}$$

for all $\{a_1, \dots, a_n\} \subseteq \mathbb{R}$. This fact is known as **Khinchine's Inequality**.

EXERCISE 2.3.28. Let $X_1, \dots, X_n$ be independent, symmetric (Exercise 1.4.26) random variables, and set $S = \sum_1^n X_m$. Show that, for each $p \in (0, \infty)$ (cf. part **(ii)** in Exercise 2.3.25),

$$K_4^{-(1-\frac{p}{2})^+}\mathbb{E}^{\mathbb{P}}\left[\left(\sum_1^n X_m^2\right)^{\frac{p}{2}}\right] \le \mathbb{E}^{\mathbb{P}}\left[|S|^p\right] \le K_p \mathbb{E}^{\mathbb{P}}\left[\left(\sum_1^n X_m^2\right)^{\frac{p}{2}}\right].$$

**Hint**: Refer to the beginning of the proof of Lemma 1.1.6, and let $R_1, \dots, R_n$ be the Rademacher functions on $[0,1)$, set $\mathbb{Q} = \lambda_{[0,1)} \times \mathbb{P}$ on $([0,1) \times \Omega, \mathcal{B}_{[0,1)} \times \mathcal{F})$, and observe that

$$\omega \in \Omega \longmapsto S(\omega) \equiv \sum_1^n X_m(\omega)$$

has the same distribution under $\mathbb{P}$ as

$$(t, \omega) \in [0,1) \times \Omega \longmapsto T(t, \omega) \equiv \sum_1^n R_m(t) X_m(\omega)$$

does under $\mathbb{Q}$. Next, apply Khinchine's inequality to see that, for each $\omega \in \Omega$,

$$K_4^{-(1-\frac{p}{2})^+}\left(\sum_1^n X_m(\omega)^2\right)^{\frac{p}{2}} \le \int_{[0,1)} |T(t,\omega)|^p\, dt \le K_p\left(\sum_1^n X_m(\omega)^2\right)^{\frac{p}{2}},$$

and complete the proof by taking the $\mathbb{P}$-integral of this with respect to $\omega$.

At least when $p \in (1, \infty)$, I will show later that this sort of inequality holds in much greater generality. Specifically, see Burkholder's Inequality in Theorem 6.3.6.

EXERCISE 2.3.29. Suppose that $\mathbf{X}$ is an $\mathbb{R}^N$-valued Gaussian random variable with mean value $\mathbf{0}$ and covariance $\mathbf{C}$.

**(i)** Show that if $\mathbf{A} : \mathbb{R}^N \longrightarrow \mathbb{R}^N$ is a linear transformation, then $\mathbf{AX}$ is an $N(\mathbf{0}, \mathbf{ACA}^\top)$ random variable, where $\mathbf{A}^\top$ is the adjoint transformation.

**(ii)** Given a linear subspace $L$ of $\mathbb{R}^N$, let $\mathcal{F}_L$ be the $\sigma$-algebra generated by $\{(\boldsymbol{\xi}, \mathbf{X})_{\mathbb{R}^N} : \boldsymbol{\xi} \in L\}$, and take $L^{\perp_{\mathbf{C}}}$ to be the subspace of $\boldsymbol{\eta}$ such that $(\boldsymbol{\eta}, \mathbf{C}\boldsymbol{\xi})_{\mathbb{R}^N} = 0$ for all $\boldsymbol{\xi} \in L$. Show that $\mathcal{F}_L$ is independent of $\mathcal{F}_{L^{\perp_{\mathbf{C}}}}$.

**Hint**: Show that, because of linearity, it suffices to check that

$$\mathbb{E}^{\mathbb{P}}\left[e^{\sqrt{-1}(\boldsymbol{\xi},\mathbf{X})_{\mathbb{R}^N}} e^{\sqrt{-1}(\boldsymbol{\eta},\mathbf{X})_{\mathbb{R}^N}}\right] = \mathbb{E}^{\mathbb{P}}\left[e^{\sqrt{-1}(\boldsymbol{\xi},\mathbf{X})_{\mathbb{R}^N}}\right]\mathbb{E}^{\mathbb{P}}\left[e^{\sqrt{-1}(\boldsymbol{\eta},\mathbf{X})_{\mathbb{R}^N}}\right]$$

for all $\boldsymbol{\xi} \in L$ and $\boldsymbol{\eta} \in L^{\perp_{\mathbf{C}}}$.

**(iii)** Suppose that $N = N_1 + N_2$, where $N_i \in \mathbb{Z}^+$ for $i \in \{1, 2\}$, write $\mathbb{R}^N \ni \mathbf{x} = \begin{pmatrix} \mathbf{x}_{(1)} \\ \mathbf{x}_{(2)} \end{pmatrix} \in \mathbb{R}^{N_1} \times \mathbb{R}^{N_2}$, and take $L = \{\mathbf{x} : \mathbf{x}_{(1)} = \mathbf{0}_{(1)}\}$. Show that if $\Pi$ is a linear transformation taking $\mathbb{R}^N$ onto $L$ that satisfies $(\boldsymbol{\xi} - \Pi\boldsymbol{\xi}, \mathbf{C}\boldsymbol{\eta})_{\mathbb{R}^N} = 0$ for all $\boldsymbol{\xi} \in \mathbb{R}^N$ and $\boldsymbol{\eta} \in L$, then $\Pi^\top \mathbf{X}$ is independent of $(\mathbf{I} - \Pi^\top)\mathbf{X}$.

**(iv)** Write

$$\mathbf{C} = \begin{pmatrix} \mathbf{C}_{(11)} & \mathbf{C}_{(12)} \\ \mathbf{C}_{(21)} & \mathbf{C}_{(22)} \end{pmatrix},$$

where the block structure corresponds to $\mathbb{R}^N = \mathbb{R}^{N_1} \times \mathbb{R}^{N_2}$, and assume that $\mathbf{C}_{(22)}$ is non-degenerate. Show that the one and only transformation $\Pi$ of the sort in part **(iii)** is given by

$$\Pi = \begin{pmatrix} \mathbf{0}_{(11)} & \mathbf{0}_{(12)} \\ \mathbf{C}_{(22)}^{-1}\mathbf{C}_{(21)} & \mathbf{I}_{(22)} \end{pmatrix},$$

and therefore that

$$\Pi^\top = \begin{pmatrix} \mathbf{0}_{(11)} & \mathbf{C}_{(12)}\mathbf{C}_{(22)}^{-1} \\ \mathbf{0}_{(21)} & \mathbf{I}_{(22)} \end{pmatrix}.$$

**Hint**: Note that $(\Pi\boldsymbol{\xi})_{(2)} = \mathbf{C}_{(22)}^{-1}\mathbf{C}_{(21)}\boldsymbol{\xi}_{(1)}$ if $\boldsymbol{\xi}_{(2)} = \mathbf{0}_{(2)}$ and $\Pi\boldsymbol{\xi} = \boldsymbol{\xi}$ if $\boldsymbol{\xi}_{(2)} = \mathbf{0}_{(2)}$.

**(v)** Continuing with the assumption that $\mathbf{C}_{(22)}$ is non-degenerate, show that

$$\mathbf{X} = \begin{pmatrix} \mathbf{C}_{(12)}\mathbf{C}_{(22)}^{-1}\mathbf{Y} \\ \mathbf{Y} \end{pmatrix} + \begin{pmatrix} \mathbf{Z} \\ \mathbf{0} \end{pmatrix},$$

where $\mathbf{Y}$ is an $\mathbb{R}^{N_2}$-valued, $N(\mathbf{0}, \mathbf{C}_{(22)})$-random variable, $\mathbf{Z}$ is an $\mathbb{R}^{N_1}$-valued $N(\mathbf{0}, \mathbf{B})$ random variable with $\mathbf{B} = \mathbf{C}_{(11)} - \mathbf{C}_{(12)}\mathbf{C}_{(22)}^{-1}\mathbf{C}_{(21)}$, and $\mathbf{Y}$ is independent of $\mathbf{Z}$. Conclude that, for any measurable $F : \mathbb{R}^{N_1} \times \mathbb{R}^{N_2} \longrightarrow \mathbb{R}$ that is bounded below, $\mathbb{E}^{\mathbb{P}}\left[F(\mathbf{X}_{(1)}, \mathbf{X}_{(2)})\right]$ equals

$$\int_{\mathbb{R}^{N_2}} \left( \int_{\mathbb{R}^{N_1}} F(\mathbf{x}_{(1)}, \mathbf{x}_{(2)})\, \gamma_{\mathbf{C}_{(12)}\mathbf{C}_{(22)}^{-1}\mathbf{x}_{(2)}, \mathbf{B}}(d\mathbf{x}_{(1)}) \right) \gamma_{\mathbf{0}, C_{(22)}}(d\mathbf{x}_{(2)}).$$

EXERCISE 2.3.30. Given $h \in L^2(\mathbb{R}^N;\mathbb{C})$, recall that the $(n+2)$-fold convolution $h^{\star(n+2)}$ is a bounded continuous function for each $n \in \mathbb{N}$. Next, assume that $h(-\mathbf{x}) = \overline{h(\mathbf{x})}$ for almost every $\mathbf{x} \in \mathbb{R}^N$ and that $h \equiv 0$ off of $B_{\mathbb{R}^N}(\mathbf{0},1)$. As an application of part **(iii)** in Exercise 1.3.22, show that

$$\left|h^{\star(n+2)}(\mathbf{x})\right| \le 2\|h\|^2_{L^2(\mathbb{R}^N;\mathbb{C})}\|h\|^n_{L^1(\mathbb{R}^N;\mathbb{C})}\exp\left[-\frac{\left((|\mathbf{x}|-2)^+\right)^2}{2n}\right].$$

**Hint**: Note that $h \in L^1(\mathbb{R}^N;\mathbb{C})$, assume that $M \equiv \|h\|_{L^1(\mathbb{R}^N;\mathbb{C})} > 0$, and define $Af = M^{-1}h \star f$ for $f \in L^2(\mathbb{R}^N;\mathbb{C})$. Show that $A$ is a self-adjoint contraction on $L^2(\mathbb{R}^N;\mathbb{C})$, check that

$$h^{\star(n+2)}(\mathbf{x}) = M^n\left(\mathbf{T}_{\mathbf{x}}h, A^n h\right)_{L^2(\mathbb{R}^N;\mathbb{C})},$$

where $\mathbf{T}_{\mathbf{x}}h \equiv h(\,\cdot + \mathbf{x})$, and note that

$$\left(\mathbf{T}_{\mathbf{x}}h, A^\ell h\right)_{L^2(\mathbb{R}^N;\mathbb{C})} = 0 \quad \text{if } \ell \le |\mathbf{x}| - 2.$$

## § 2.4 An Application to Hermite Multipliers

This section does not really belong here and should probably be skipped by those readers who want to restrict their attention to purely probabilistic matters. On the other hand, for those who want to see how probability theory interacts with other branches of mathematical analysis, the present section may come as something of a revelation.

**§ 2.4.1. Hermite Multipliers.** The topic of this section will be a class of linear operators called Hermite multipliers, and what will be discussed are certain boundedness properties of these operators. The setting is as follows. For $n \in \mathbb{N}$, define

(2.4.1) $$H_n(x) = (-1)^n e^{\frac{x^2}{2}}\frac{d^n}{dx^n}\left(e^{-\frac{x^2}{2}}\right), \quad x \in \mathbb{R}.$$

Clearly, $H_n$ is an $n$th order, real, monic (i.e., 1 is the coefficient of the highest order term) polynomial. Moreover, if we define the **raising operator** $A_+$ on $C^1(\mathbb{R};\mathbb{C})$ by

$$[A_+\varphi](x) = -e^{\frac{x^2}{2}}\frac{d}{dx}\left(e^{-\frac{x^2}{2}}\varphi(x)\right) = -\frac{d\varphi}{dx}(x) + x\varphi(x), \quad x \in \mathbb{R},$$

then

(2.4.2) $$H_{n+1} = A_+ H_n \quad \text{for all } n \in \mathbb{N}.$$

At the same time, if $\varphi$ and $\psi$ are continuously differentiable functions whose first derivatives are **tempered** (i.e., have at most polynomial growth at infinity), then

$$(\varphi, A_+\psi)_{L^2(\gamma_{0,1};\mathbb{C})} = (A_-\varphi, \psi)_{L^2(\gamma_{0,1};\mathbb{C})}, \tag{2.4.3}$$

where $A_-$ is the **lowering operator** given by $A_-\varphi = \frac{d\varphi}{dx}$. After combining (2.4.2) with (2.4.3), we see that, for all $0 \le m \le n$,

$$(H_m, H_n)_{L^2(\gamma_{0,1};\mathbb{C})} = (H_m, A_+^n H_0)_{L^2(\gamma_{0,1};\mathbb{C})} = (A_-^n H_m, H_0)_{L^2(\gamma_{0,1};\mathbb{C})} = m!\,\delta_{m,n},$$

where, at the last step, I have used the fact that $H_m$ is a monic $m$th order polynomial. Hence, the (normalized) **Hermite polynomials**

$$\overline{H}_n(x) = \frac{H_n(x)}{\sqrt{n!}} = \frac{(-1)^n}{\sqrt{n!}} e^{\frac{x^2}{2}} \frac{d^n}{dx^n}\left(e^{-\frac{x^2}{2}}\right), \quad x \in \mathbb{R},$$

form an orthonormal set in $L^2(\gamma_{0,1};\mathbb{C})$. (Indeed, they are one choice of the orthogonal polynomials relative to the Gauss weight.)

LEMMA 2.4.4. *For each $\lambda \in \mathbb{C}$, set*

$$H(x;\lambda) = \exp\left[\lambda x - \frac{\lambda^2}{2}\right], \quad x \in \mathbb{R}.$$

*Then*

$$H(x;\lambda) = \sum_{n=0}^{\infty} \frac{\lambda^n}{n!} H_n(x), \quad x \in \mathbb{R}, \tag{2.4.5}$$

*where the convergence is both uniform on compact subsets of $\mathbb{R}\times\mathbb{C}$ and, for $\lambda$'s in compact subsets of $\mathbb{C}$, uniform in $L^2(\gamma_{0,1};\mathbb{C})$. In particular, $\{\overline{H}_n : n \in \mathbb{N}\}$ is an orthonormal basis in $L^2(\gamma_{0,1};\mathbb{C})$.*

PROOF: By (2.4.1) and Taylor's expansion for the function $e^{-\frac{x^2}{2}}$, it is clear that (2.4.5) holds for each $(x,\lambda)$ and that the convergence is uniform on compact subsets of $\mathbb{R}\times\mathbb{C}$. Furthermore, because the $H_n$'s are orthogonal, the asserted uniform convergence in $L^2(\gamma_{0,1};\mathbb{C})$ comes down to checking that

$$\lim_{m\to\infty} \sup_{|\lambda|\le R} \sum_{n=m}^{\infty} \left|\frac{\lambda^n}{n!}\right|^2 \|H_n\|^2_{L^2(\gamma_{0,1};\mathbb{C})} = 0$$

for every $R \in (0,\infty)$, and obviously this follows from our earlier calculation that $\|H_n\|^2_{L^2(\gamma_{0,1};\mathbb{C})} = n!$.

To prove the assertion that $\{\overline{H}_n : n \in \mathbb{N}\}$ forms an orthonormal basis in $L^2(\gamma_{0,1};\mathbb{C})$, it suffices to check that any $\varphi \in L^2(\gamma_{0,1};\mathbb{C})$ that is orthogonal to all of the $H_n$'s must be 0. But, because of the $L^2(\gamma_{0,1};\mathbb{C})$ convergence in (2.4.5), we would have that

$$\int_{\mathbb{R}} \varphi(x)\, e^{\lambda x}\, \gamma_{0,1}(dx) = 0, \quad \lambda \in \mathbb{C},$$

for such a $\varphi$. Hence, if

$$\psi(x) = \frac{e^{-\frac{x^2}{2}}\varphi(x)}{\sqrt{2\pi}}, \quad x \in \mathbb{R},$$

then $\|\psi\|_{L^1(\mathbb{R};\mathbb{C})} = \|\varphi\|_{L^1(\gamma_{0,1};\mathbb{C})} \le \|\varphi\|_{L^2(\gamma_{0,1};\mathbb{C})} < \infty$ and (cf. (2.3.2)) $\hat{\psi} \equiv 0$, which, by the $L^1(\mathbb{R};\mathbb{C})$ Fourier inversion formula

$$\frac{1}{2\pi}\int_{\mathbb{R}} e^{-\alpha|\xi|}\, e^{-\sqrt{-1}x\xi}\hat{\psi}(\xi)\, d\xi \xrightarrow{\alpha\searrow 0} \psi \quad \text{in } L^1(\mathbb{R};\mathbb{C}),$$

means that $\psi$ and therefore $\varphi$ vanish Lebesgue-almost everywhere. □

Now that we know $\{\overline{H}_n : n \in \mathbb{N}\}$ is an orthonormal basis, I can uniquely determine a normal operator $\mathcal{H}_\theta$ for each $\theta \in \mathbb{C}$ by specifying that

$$\mathcal{H}_\theta H_n = \theta^n\, H_n \quad \text{for each } n \in \mathbb{N}.$$

The operator $\mathcal{H}_\theta$ is called the **Hermite multiplier** with parameter $\theta$, and clearly

$$\mathrm{Dom}(\mathcal{H}_\theta) = \left\{\varphi \in L^2(\gamma_{0,1};\mathbb{C}) : \sum_{n=1}^{\infty} |\theta|^{2n}\left|\left(\varphi, \overline{H}_n\right)_{L^2(\gamma_{0,1};\mathbb{C})}\right|^2 < \infty\right\}$$

$$\mathcal{H}_\theta\varphi = \sum_{n=0}^{\infty} \theta^n\left(\varphi, \overline{H}_N\right)_{L^2(\gamma_{0,1};\mathbb{C})}\overline{H}_n, \quad \varphi \in \mathrm{Dom}(\mathcal{H}_\theta).$$

In particular, $\mathcal{H}_\theta$ is a contraction if and only if $\theta$ is an element of the closed unit disk $\mathbb{D}$ in $\mathbb{C}$, and it is unitary precisely when $\theta \in \mathbb{S}^1 \equiv \partial\mathbb{D}$. Also, the adjoint of $\mathcal{H}_\theta$ is $\mathcal{H}_{\overline{\theta}}$, and so it is self-adjoint if and only if $\theta \in \mathbb{R}$.

As we are about to see, there are special choices of $\theta$ for which the corresponding Hermite multiplier has interesting alternative interpretations and unexpected additional properties. For example, consider the **Mehler kernel**[1]

$$M(x,y;\theta) = \frac{1}{\sqrt{1-\theta^2}}\exp\left[-\frac{(\theta x)^2 - 2\theta xy + (\theta y)^2}{2(1-\theta^2)}\right]$$

[1] This kernel appears in the 1866 article by Mehler referred to in the footnote in Exercise 2.1.13. It arises there as the generating function for spherical harmonics on the sphere $\mathbb{S}^\infty\left(\sqrt{\infty}\right)$.

for $\theta \in (0,1)$ and $x, y \in \mathbb{R}$. By a straightforward Gaussian computation (i.e., "complete the square" in the exponential) one can easily check that

$$\int_{\mathbb{R}} H(y;\lambda)\, M(x,y;\theta)\, \gamma_{0,1}(dy) = H(x;\theta\lambda)$$

for all $\theta \in (0,1)$ and $(x,\lambda) \in \mathbb{R}\times\mathbb{C}$. In conjunction with (2.4.5), this means that

$$\mathcal{H}_\theta\varphi = \int_{\mathbb{R}} M(\,\cdot\,,y;\theta)\,\varphi(y)\,\gamma_{0,1}(dy), \quad \theta \in (0,1) \text{ and } \varphi \in L^2(\gamma_{0,1};\mathbb{C}), \tag{2.4.6}$$

and from here it is not very difficult to prove the following properties of $\mathcal{H}_\theta$ for $\theta \in (0,1)$.

Lemma 2.4.7. *For each $\varphi \in L^2(\gamma_{0,1};\mathbb{C})$, $(\theta,x) \in (0,1)\times\mathbb{R} \longmapsto \mathcal{H}_\theta\varphi(x) \in \mathbb{C}$ may be chosen to be a continuous function that is non-negative if $\varphi \geq 0$ Lebesgue-almost everywhere. In addition, for each $\theta \in (0,1)$ and every $p \in [1,\infty]$,*

$$\|\mathcal{H}_\theta\varphi\|_{L^p(\gamma_{0,1};\mathbb{C})} \leq \|\varphi\|_{L^p(\gamma_{0,1};\mathbb{C})}. \tag{2.4.8}$$

Proof: The first assertions are immediate consequences of the representation in (2.4.6). To prove the second assertion, observe that $\mathcal{H}_\theta\mathbf{1} = \mathbf{1}$ and therefore, as a special case of (2.4.6),

$$\int_{\mathbb{R}} M(x,y;\theta)\,\gamma_{0,1}(dy) = 1 \quad \text{for all} \quad \theta \in (0,1) \text{ and } x \in \mathbb{R}.$$

Hence, by (2.4.6) and Jensen's Inequality, for any $p \in [1,\infty)$,

$$\big|[\mathcal{H}_\theta\varphi](x)\big|^p \leq \int_{\mathbb{R}} M(x,y;\theta)\,|\varphi(y)|^p\,\gamma_{0,1}(dy).$$

At the same time, by symmetry, $\int_{\mathbb{R}} M(x,y;\theta)\,\gamma_{0,1}(dx) = 1$ for all $(\theta,y) \in (0,1)\times\mathbb{R}$, and therefore

$$\int_{\mathbb{R}} \big|[\mathcal{H}_\theta\varphi](x)\big|^p\,\gamma_{0,1}(dx) \leq \iint_{\mathbb{R}\times\mathbb{R}} M(x,y;\theta)\,|\varphi(y)|^p\,\gamma_{0,1}(dx)\gamma_{0,1}(dy) = \int_{\mathbb{R}} |\varphi|^p\,d\gamma_{0,1}.$$

Hence, (2.4.8) is now proved for $p \in [1,\infty)$. The case when $p = \infty$ is even easier and is left to the reader. □

The conclusions drawn in Lemma 2.4.7 from the Mehler representation in (2.4.6) are interesting but not very deep (cf. Exercise 2.4.36). A deeper fact is the relationship between Hermite multipliers and the Fourier transform. For the purposes of this analysis, it is best to define the **Fourier operator** $\mathcal{F}$ by

$$[\mathcal{F}f](\xi) = \int_{\mathbb{R}} e^{\sqrt{-1}\,2\pi\xi x} f(x)\,dx, \quad \xi \in \mathbb{R}, \tag{2.4.9}$$

for $f \in L^1(\mathbb{R};\mathbb{C})$. The advantage of this choice is that, without the introduction of any further factors of $\sqrt{2\pi}$, the Parseval Identity (cf. Exercise 2.4.37) becomes the statement that $\mathcal{F}$ determines a unitary operator on $L^2(\mathbb{R};\mathbb{C})$. In order to relate $\mathcal{F}$ to Hermite multipliers, observe that, after analytically continuing the result of another simple Gaussian computation,

$$\int_{\mathbb{R}} e^{\zeta x} e^{-\pi x^2}\,dx = e^{\frac{\zeta^2}{4\pi}} \quad \text{for all } \zeta \in \mathbb{C},$$

we see from (2.4.5) that

$$\sum_{n=0}^{\infty} \frac{\lambda^n}{n!} \int_{\mathbb{R}} e^{\sqrt{-1}\,2\pi\xi x} H_n\big(\sqrt{2\pi p}\,x\big) e^{-\pi x^2}\,dx$$
$$= e^{-\pi\xi^2} \exp\left[\frac{(p-1)\lambda^2}{2} + \sqrt{-1}\,\lambda\sqrt{2\pi p}\,\xi\right] = e^{-\pi\xi^2} \sum_{n=0}^{\infty} \frac{\lambda^n}{n!} \theta_p^n H_n\big(\sqrt{2\pi p'}\,\xi\big),$$

where $p' = \frac{p}{p-1}$ is the **Hölder conjugate** of $p$ and $\theta_p \equiv \sqrt{-1}\,(p-1)^{\frac{1}{2}}$. Thus, we have now proved that, for each $p \in (1,\infty)$ and $n \in \mathbb{N}$,

$$\int_{\mathbb{R}} e^{\sqrt{-1}\,2\pi\xi x} H_n\big(\sqrt{2\pi p}\,x\big) e^{-\pi x^2}\,dx = \theta_p^n\, H_n\big(\sqrt{2\pi p'}\,x\big)\, e^{-\pi\xi^2}. \tag{2.4.10}$$

In particular, when $p = 2$, (2.4.10) says that

$$\mathcal{F}h_n = \big(\sqrt{-1}\big)^n h_n, \quad n \in \mathbb{N}, \tag{2.4.11}$$

where $h_n$ is the $n$th (un-normalized) **Hermite function** given by

$$h_n(x) = H_n\big((4\pi)^{\frac{1}{2}} x\big) e^{-\pi x^2}, \quad n \in \mathbb{N} \text{ and } x \in \mathbb{R}. \tag{2.4.12}$$

More generally, (2.4.10) leads to the following relationship between $\mathcal{F}$ and Hermite multipliers. Namely, for each $p \in (1,\infty)$, define $\mathcal{U}_p$ on $L^p(\gamma_{0,1};\mathbb{C})$ by

$$[\mathcal{U}_p\varphi](x) = p^{\frac{1}{2p}} \varphi\big((2\pi p)^{\frac{1}{2}} x\big) e^{-\pi x^2}, \quad x \in \mathbb{R}.$$

It is then an easy matter to check that $\mathcal{U}_p$ is an isometric surjection from $L^p(\gamma_{0,1};\mathbb{C})$ onto $L^p(\mathbb{R};\mathbb{C})$. In addition, (2.4.10) can now be interpreted as the statement that, for every $p \in (1,\infty)$ and every polynomial $\varphi$,

$$\mathcal{F} \circ \mathcal{U}_p\varphi = A_p \mathcal{U}_{p'} \circ \mathcal{H}_{\theta_p}\varphi \quad \text{where} \quad A_p \equiv \left(\frac{p^{\frac{1}{p}}}{(p')^{\frac{1}{p'}}}\right)^{\frac{1}{2}}. \tag{2.4.13}$$

See Exercise 2.4.35, where it is shown that $A_p < 1$ for $p \in (0,1)$.

**§ 2.4.2. Beckner's Theorem.** Having completed this brief introduction to Hermite multipliers, I will now address a problem to which The Central Limit Theorem has something to contribute. The problem is that of determining the set of $(\theta, p, q) \in \mathbb{D} \times (1, \infty) \times (0, \infty)$ with $p \le q$ for which $\mathcal{H}_\theta$ determines a contraction from $L^p(\gamma_{0,1}; \mathbb{C})$ into $L^q(\gamma_{0,1}; \mathbb{C})$. In view of the preceding discussion, when $\theta \in (0, 1)$, a solution to this problem has implications for the Mehler transform; and, when $q = p'$, the solution tells us about the Fourier operator. The role that The Central Limit Theorem plays in this analysis is hidden in the following beautiful criterion, which was first discovered by Wm. Beckner.[2]

THEOREM 2.4.14 (**Beckner**). *Let $\theta \in \mathbb{D}$ and $1 \le p \le q < \infty$ be given. Then*

$$\|\mathcal{H}_\theta \varphi\|_{L^q(\gamma_{0,1};\mathbb{C})} \le \|\varphi\|_{L^p(\gamma_{0,1};\mathbb{C})} \quad \text{for all} \quad \varphi \in L^2(\gamma_{0,1}; \mathbb{C}) \tag{2.4.15}$$

*if*

$$\left( \frac{|1 - \theta\zeta|^q + |1 + \theta\zeta|^q}{2} \right)^{\frac{1}{q}} \le \left( \frac{|1 - \zeta|^p + |1 + \zeta|^p}{2} \right)^{\frac{1}{p}} \tag{2.4.16}$$

*for every $\zeta \in \mathbb{C}$.*

That (2.4.16) implies (2.4.15) is trivial is quite remarkable. Indeed, it takes a problem in infinite dimensional analysis and reduces it to a calculus question about functions on the complex plane. Even though, as we will see later, this reduction leads to highly non-trivial problems in calculus, Theorem 2.4.14 has to be considered a major step toward understanding the contraction properties of Hermite multipliers.[3]

The first step in the proof of Theorem 2.4.14 is to interpret (2.4.16) in operator theoretic language. For this purpose, let $\beta$ denote the standard Bernoulli probability measure on $(\mathbb{R}, \mathcal{B}_\mathbb{R})$. That is, $\beta(\{\pm 1\}) = \frac{1}{2}$. Next, use $\chi_\emptyset$ to denote the function on $\mathbb{R}$ that is constantly equal to 1 and $\chi_{\{1\}}$ to stand for the identity function on $\mathbb{R}$ (i.e., $\chi_{\{1\}}(x) = x$, $x \in \mathbb{R}$). It is then clear that $\chi_\emptyset$ and $\chi_{\{1\}}$ constitute an orthonormal basis in $L^2(\beta; \mathbb{C})$; in fact, they are the orthogonal polynomials there. Hence, for each $\theta \in \mathbb{C}$, we can define the **Bernoulli multiplier** $\mathcal{K}_\theta$ as the unique normal operator on $L^2(\beta; \mathbb{C})$ prescribed by

$$\mathcal{K}_\theta \chi_F = \begin{cases} \chi_\emptyset & \text{if} \quad F = \emptyset \\ \theta \chi_{\{1\}} & \text{if} \quad F = \{1\}. \end{cases}$$

---

[2] See Beckner's "Inequalities in Fourier analysis," *Ann. Math.*, # **102** #1, pp. 159–182 (1975).

[3] Later, in his article "Gaussian kernels have only Gaussian maximizers," *Invent. Math.* **12**, pp. 179–208 (1990), E. Lieb essentially killed this line of research. His argument, which is entirely different from the one discussed here, handles not only the Hermite multipliers but essentially every operator whose kernel can be represented as the exponential of a second order polynomial.

Furthermore, (2.4.16) is equivalent to the statement that

$$\|\mathcal{K}_\theta\varphi\|_{L^q(\beta;\mathbb{C})} \le \|\varphi\|_{L^p(\beta,\mathbb{C})} \quad \text{for all} \quad \varphi \in L^2(\beta;\mathbb{C}). \tag{2.4.17}$$

Indeed, it is obvious that (2.4.16) is equivalent to (2.4.17) restricted to $\varphi$'s of the form $x \in \mathbb{R} \longmapsto 1 + \zeta x$ as $\zeta$ runs over $\mathbb{C}$; and from this, together with the observation that every element of $L^2(\beta;\mathbb{C})$ can be represented in the form $a\chi_\emptyset + b\chi_{\{1\}}$ as $(a,b)$ runs over $\mathbb{C}^2$, one quickly concludes that (2.4.16) implies (2.4.17) for general $\varphi \in L^2(\beta;\mathbb{C})$.

I next want to show that (2.4.17) can be parlayed into a seemingly more general statement. To this end, define the $n$-fold tensor product operator $\mathcal{K}_\theta^{\otimes n}$ on $L^2(\beta^n;\mathbb{C})$ as follows. For $F \subseteq \{1,\dots,n\}$ set $\chi_F \equiv 1$ if $F = \emptyset$ and define

$$\chi_F(\mathbf{x}) = \prod_{j\in F} \chi_{\{1\}}(x_j) \quad \text{for} \quad \mathbf{x} = (x_1,\dots,x_n) \in \mathbb{R}^n$$

if $F \neq \emptyset$. Note that $\{\chi_F : F \subseteq \{1,\dots,n\}\}$ is an orthonormal basis for $L^2(\beta^n;\mathbb{C})$, and define $\mathcal{K}_\theta^{\otimes n}$ to be the unique normal operator on $L^2(\beta^n;\mathbb{C})$ for which

$$\mathcal{K}_\theta^{\otimes n}\chi_F = \theta^{|F|}\chi_F, \quad F \subseteq \{1,\dots,n\}, \tag{2.4.18}$$

where $|F|$ is used here to denote the number of elements in the set $F$. Alternatively, one can describe $\mathcal{K}_\theta^{\otimes n}$ inductively on $n \in \mathbb{Z}^+$ by saying that $\mathcal{K}_\theta^{\otimes 1} = \mathcal{K}_\theta$ and that, for $\Phi \in C(\mathbb{R}^{n+1};\mathbb{C})$ and $(\mathbf{x},y) \in \mathbb{R}^n \times \mathbb{R}$,

$$\big[\mathcal{K}_\theta^{\otimes(n+1)}\Phi\big](\mathbf{x},y) = \big[\mathcal{K}_\theta\Psi(\mathbf{x},\,\cdot\,)\big](y) \quad \text{where} \quad \Psi(\mathbf{x},y) = \big[\mathcal{K}_\theta^{\otimes n}\Phi(\,\cdot\,,y)\big](\mathbf{x}).$$

It is this alternative description that makes it easiest to see the extension of (2.4.17) alluded to above. Namely, what I will now show is that, for every $n \in \mathbb{Z}^+$,

$$(2.4.17) \implies \big\|\mathcal{K}_\theta^{\otimes n}\Phi\big\|_{L^q(\beta^n;\mathbb{C})} \le \|\Phi\|_{L^p(\beta^n;\mathbb{C})}, \quad \Phi \in L^2(\beta^n;\mathbb{C}). \tag{2.4.19}$$

Obviously, there is nothing to do when $n = 1$. Next, assume (2.4.19) for $n$, let $\Phi \in C(\mathbb{R}^{n+1};\mathbb{C})$ be given, and define $\Psi$ as in the second description of $\mathcal{K}_\theta^{\otimes(n+1)}\Phi$. Then, by (2.4.17) applied to $\Psi(\mathbf{x},\,\cdot\,)$ for each $\mathbf{x} \in \mathbb{R}^n$ and by the induction hypothesis applied to $\Phi(\,\cdot\,,y)$ for each $y \in \mathbb{R}$, we have that

$$\begin{aligned}
\|\mathcal{K}_\theta^{\otimes(n+1)}\Phi\|^q_{L^q(\beta^{n+1};\mathbb{C})} &= \int_{\mathbb{R}^n}\left(\int_{\mathbb{R}}\Big|\big[\mathcal{K}_\theta\Psi(\mathbf{x},\,\cdot\,)\big](y)\Big|^q\,\beta(dy)\right)\beta^n(d\mathbf{x})\\
&\le \int_{\mathbb{R}^n}\left(\int_{\mathbb{R}}|\Psi(\mathbf{x},y)|^p\,\beta(dy)\right)^{\frac{q}{p}}\beta^n(d\mathbf{x}) = \left\|\int_{\mathbb{R}^n}|\Psi(\,\cdot\,,y)|^p\,\beta(dy)\right\|^{\frac{q}{p}}_{L^{\frac{q}{p}}(\beta^n;\mathbb{C})}\\
&\le \left(\int_{\mathbb{R}^n}\big\|\,|\Psi(\,\cdot\,,y)|^p\,\big\|_{L^{\frac{q}{p}}(\beta^n;\mathbb{C})}\beta(dy)\right)^{\frac{q}{p}} = \left(\int_{\mathbb{R}}\|\Psi(\,\cdot\,,y)\|^p_{L^q(\beta^n;\mathbb{C})}\beta(dy)\right)^{\frac{q}{p}}\\
&\le \left(\int_{\mathbb{R}}\|\Phi(\,\cdot\,,y)\|^p_{L^p(\beta^n;\mathbb{C})}\beta(dy)\right)^{\frac{q}{p}} = \|\Phi\|^q_{L^p(\beta^{n+1};\mathbb{C})},
\end{aligned}$$

where, in the passage to the third line, I have used the continuous form of Minkowski's Inequality (it is at this point that the only essential use of the hypothesis $p \le q$ is made).

I am now ready to take the main step in the proof of Theorem 2.4.14.

LEMMA 2.4.20. *Define* $\mathcal{A}_n : L^2(\beta;\mathbb{C}) \longrightarrow L^2(\beta^n;\mathbb{C})$ *by*

$$[\mathcal{A}_n\varphi](\mathbf{x}) = \varphi\left(\frac{\sum_{\ell=1}^n x_\ell}{\sqrt{n}}\right) \quad \text{for} \quad \mathbf{x} \in \mathbb{R}^n.$$

*Then, for every pair of tempered* $\varphi$ *and* $\psi$ *from* $C(\mathbb{R};\mathbb{C})$,

$$\|\varphi\|_{L^p(\gamma_{0,1};\mathbb{C})} = \lim_{n\to\infty} \|\mathcal{A}_n\varphi\|_{L^p(\beta^n;\mathbb{C})} \quad \text{for every} \quad p \in [1,\infty) \tag{2.4.21}$$

*and*

$$\Big(\mathcal{H}_\theta\varphi,\psi\Big)_{L^2(\gamma_{0,1};\mathbb{C})} = \lim_{n\to\infty}\Big(\mathcal{K}_\theta^{\otimes n}\circ\mathcal{A}_n\varphi,\mathcal{A}_n\psi\Big)_{L^2(\beta^n;\mathbb{C})} \tag{2.4.22}$$

*for every* $\theta \in (0,1)$. *Moreover, if, in addition, either* $\varphi$ *or* $\psi$ *is a polynomial, then* (2.4.22) *continues to hold for all* $\theta \in \mathbb{C}$.

PROOF: Let $\varphi$ and $\psi$ be tempered elements of $C(\mathbb{R};\mathbb{C})$, and define

$$f_n(\theta) = \big(\mathcal{K}_\theta^{\otimes n}\circ\mathcal{A}_n\varphi,\mathcal{A}_n\psi\big)_{L^2(\beta^n;\mathbb{C})} \quad \text{and} \quad f(\theta) = \big(\mathcal{H}_\theta\varphi,\psi\big)_{L^2(\gamma_{0,1};\mathbb{C})}$$

for $n \in \mathbb{Z}^+$ and $\theta \in \mathbb{C}$. I begin by showing that

$$\lim_{n\to\infty} f_n(\theta) = f(\theta), \quad \theta \in (0,1). \tag{2.4.23}$$

Notice that (2.4.23) is (2.4.22) for $\theta \in (0,1)$ and that (2.4.21) follows from (2.4.22) with $\varphi = \mathbf{1}$, $\psi = |\varphi|^p$, and any $\theta \in (0,1)$.

In order to prove (2.4.23), I will need to introduce other expressions for $f(\theta)$ and the $f_n(\theta)$'s. To this end, set

$$\mathbf{C}_\theta = \begin{pmatrix} 1 & \theta \\ \theta & 1 \end{pmatrix},$$

and, using (2.4.6), observe (cf. (2.3.6)) that

$$f(\theta) = \int_{\mathbb{R}^2} \varphi(x)\,\overline{\psi(y)}\,\gamma_{\mathbf{0},\mathbf{C}_\theta}(dx\times dy).$$

Next, let, for each $x \in \mathbb{R}\setminus\{0\}$, define $k_\theta(x,\cdot\,)$ to be the probability measure on $\mathbb{R}$ such that $k_\theta(x,\{\pm\mathrm{sgn}x\}) = \frac{1\pm\theta}{2}$, and set $k_\theta(0,\cdot\,) = \beta$. Then it is easy to check that $\int_{\mathbb{R}} \chi_{\{0\}}(y)\, k_\theta(\pm 1, dy) = \chi_{\{0\}}(\pm 1)$ and $\int_{\mathbb{R}} \chi_{\{1\}}(y)\, k_\theta(\pm 1, dy) = \theta\chi_{\{1\}}(\pm 1)$ and therefore $\mathcal{K}_\theta\varphi(\pm 1) = \int_{\mathbb{R}} \varphi(y)\, k_\theta(\pm 1, dy)$ for all $\varphi$. Hence, if $\boldsymbol{\beta}_\theta$ is the probability measure on $\mathbb{R}^2$ determined by $\boldsymbol{\beta}_\theta(dx \times dy) = k_\theta(x, dy)\, \beta(dx)$ or, equivalently,

$$\boldsymbol{\beta}_\theta(\{(\pm 1, \pm 1)\}) = \tfrac{1+\theta}{4} \quad \text{and} \quad \boldsymbol{\beta}_\theta(\{(\pm 1, \mp 1)\}) = \tfrac{1-\theta}{4},$$

then

$$\left(\mathcal{K}_\theta\varphi, \psi\right)_{L^2(\beta;\mathbb{C})} = \int_{\mathbb{R}^2} \varphi(x)\,\overline{\psi(y)}\, \boldsymbol{\beta}_\theta(dx \times dy).$$

Proceeding by induction, it follows that

$$\left(\mathcal{K}_\theta^{\otimes n}\Phi, \Psi\right)_{L^2(\beta;\mathbb{C})} = \int_{\mathbb{R}^2} \cdots \int_{\mathbb{R}^2} \Phi(\mathbf{x})\,\overline{\Psi(\mathbf{y})}\, \boldsymbol{\beta}_\theta(dx_1 \times dy_1) \cdots \boldsymbol{\beta}_\theta(dx_n \times dy_n)$$

for all $\Phi, \Psi \in C(\mathbb{R}^n;\mathbb{C})$. Hence, if (cf. Exercise 1.1.14) $\Omega = \left(\mathbb{R}^2\right)^{\mathbb{Z}^+}$, $\mathcal{F} = \mathcal{B}_\Omega$, and $\mathbb{P}_\theta = \left(\boldsymbol{\beta}_\theta\right)^{\mathbb{Z}^+}$, then

$$f_n(\theta) = \mathbb{E}^{\mathbb{P}_\theta}\left[F\left(\frac{\sum_1^n \mathbf{Z}_m}{\sqrt{n}}\right)\right],$$

where $F(\mathbf{z}) \equiv \varphi(x)\,\overline{\psi(y)}$ for $\mathbf{z} = (x, y) \in \mathbb{R}^2$ and $\mathbf{Z}_n(\omega) = \mathbf{z}_n$, $n \in \mathbb{Z}^+$, when $\omega = (\mathbf{z}_1, \dots, \mathbf{z}_n, \dots) \in \Omega$. Further, under $\mathbb{P}_\theta$, the $\mathbf{Z}_n$'s are mutually independent, identically distributed $\mathbb{R}^2$-valued random variables with mean value $\mathbf{0}$ and covariance $\mathbf{C}_\theta$. In addition, $\mathbf{Z}_1$ is bounded, and therefore the last part of Theorem 2.3.19 applies and guarantees that (2.4.23) holds.

To complete the proof, suppose that $\varphi$ is a polynomial of degree $k$. It is then an easy matter to check that

$$\left(\mathcal{A}_n\varphi, \chi_F\right)_{L^2(\beta^n;\mathbb{C})} = 0 \quad \text{if } |F| > k,$$

and therefore (cf. (2.4.18)) $\theta \in \mathbb{C} \longmapsto f_n(\theta) \in \mathbb{C}$ is also a polynomial of degree no more than $k$. Moreover, because

$$|f_n(\theta)| = \left|\sum_F \theta^{|F|}\left(\mathcal{A}_n\varphi, \chi_F\right)_{L^2(\beta^n;\mathbb{C})}\left(\chi_F, \mathcal{A}_n\psi\right)_{L^2(\beta^n;\mathbb{C})}\right|,$$

we also know that

$$\left|f_n(\theta)\right| \leq \left(|\theta| \vee 1\right)^k \left\|\mathcal{A}_n\varphi\right\|_{L^2(\beta^n;\mathbb{C})} \left\|\mathcal{A}_n\psi\right\|_{L^2(\beta^n;\mathbb{C})}, \quad n \in \mathbb{Z}^+ \text{ and } \theta \in \mathbb{C}.$$

Hence, because of (2.4.21) with $p = 2$, $\{f_n : n \in \mathbb{Z}^+\}$ is a family of entire functions on $\mathbb{C}$ that are uniformly bounded on compact subsets. At the same time, because $(\varphi, H_m)_{L^2(\gamma_{0,1};\mathbb{C})} = 0$ for $m > k$, $f$ is also a polynomial of degree at most $k$, and therefore (2.4.23) already implies that the convergence extends to the whole of $\mathbb{C}$ and is uniform on compacts. Finally, in the case when $\psi$, instead of $\varphi$, is a polynomial, simply note that

$$\left(\mathcal{K}_\theta^{\otimes n}\circ\mathcal{A}_n\varphi,\mathcal{A}_n\psi\right)_{L^2(\beta^n;\mathbb{C})}=\overline{\left(\mathcal{K}_{\bar\theta}^{\otimes n}\circ\mathcal{A}_n\psi,\mathcal{A}_n\varphi\right)_{L^2(\beta^n;\mathbb{C})}}$$

and $\left(\mathcal{H}_\theta\varphi,\psi\right)_{L^2(\gamma_{0,1};\mathbb{C})}=\overline{\left(\mathcal{H}_{\bar\theta}\psi,\varphi\right)_{L^2(\gamma_{0,1};\mathbb{C})}}$, and apply the preceding. □

Proof of Theorem 2.4.14: Assume that (2.4.16) holds for a given pair $1<p\le q<\infty$ and $\theta\in\mathbb{D}$. We then know that (2.4.19) holds for every $n\in\mathbb{Z}^+$. Hence, by Lemma 2.4.20, if $\varphi$ and $\psi$ are tempered elements of $C(\mathbb{R};\mathbb{C})$ and at least one of them is a polynomial, then

$$\begin{aligned}\left|\left(\mathcal{H}_\theta\varphi,\psi\right)_{L^2(\gamma_{0,1};\mathbb{C})}\right|&=\lim_{n\to\infty}\left|\left(\mathcal{K}_\theta^{\otimes n}\circ\mathcal{A}_n\varphi,\mathcal{A}_n\psi\right)_{L^2(\beta^n;\mathbb{C})}\right|\\&\le\varlimsup_{n\to\infty}\left\|\mathcal{A}_n\varphi\right\|_{L^p(\beta^n;\mathbb{C})}\left\|\mathcal{A}_n\psi\right\|_{L^{q'}(\beta^n;\mathbb{C})}=\|\varphi\|_{L^p(\gamma_{0,1};\mathbb{C})}\|\psi\|_{L^{q'}(\gamma_{0,1};\mathbb{C})}.\end{aligned}$$

In other words, we now know that, for all tempered $\varphi$ and $\psi$ from $C(\mathbb{R};\mathbb{C})$,

$$\left|\left(\mathcal{H}_\theta\varphi,\psi\right)_{L^2(\gamma_{0,1};\mathbb{C})}\right|\le\|\varphi\|_{L^p(\gamma_{0,1};\mathbb{C})}\|\psi\|_{L^{q'}(\gamma_{0,1};\mathbb{C})}\tag{2.4.24}$$

so long as one or the other is a polynomial.

To complete the proof when $p\in(1,2]$, note that, for any fixed polynomial $\varphi$, (2.4.24) for every tempered $\psi\in C(\mathbb{R};\mathbb{C})$ guarantees that the inequality in (2.4.15) holds for that $\varphi$. At the same time, because $p\in(1,2]$ and the polynomials are dense in $L^2(\gamma_{0,1};\mathbb{C})$, (2.4.15) follows immediately from its own restriction to polynomials.

Finally, assume that $p\in[2,\infty)$ and therefore that $q'\in(1,2]$. Then, again because the polynomials are dense in $L^2(\gamma_{0,1};\mathbb{C})$, (2.4.24) for a fixed tempered $\varphi\in C(\mathbb{R};\mathbb{C})$ and all polynomials $\psi$ implies (2.4.15) first for all tempered continuous $\varphi$'s and thence for all $\varphi\in L^2(\gamma_{0,1};\mathbb{C})$. □

### § 2.4.3. Applications of Beckner's Theorem.

I will now apply Theorem 2.4.14 to two important examples. The first example involves the case when $\theta\in(0,1)$ and shows that the contraction property proved in Lemma 2.4.7 can be improved to say that, for each $p\in(1,\infty)$ and $\theta\in(0,1)$, there is a $q=q(p,\theta)\in(p,\infty)$ such that $\mathcal{H}_\theta$ is a contraction on $L^p(\gamma_{0,1};\mathbb{C})$ into $L^q(\gamma_{0,1};\mathbb{C})$. Such an operator is said to be **hypercontractive**, and the fact that $\mathcal{H}_\theta$ is hypercontractive was first proved by E. Nelson in connection with his renowned construction of a non-trivial, two-dimensional quantum field.[4] The proof that I will give is entirely different from Nelson's and is much closer to the ideas introduced by L. Gross[5] as they were developed by Beckner.

[4] Nelson's own proof appeared in his "The free Markov field," *J. Fnal. Anal.* **12**, pp. 12–21 (1974).

[5] See Gross's "Logarithmic Sobolev inequalities," *Amer. J. Math.* **97** #4, pp. 1061–1083 (1975). In this paper, Gross introduced the idea of proving estimates on $\mathcal{H}_\theta$ from the corresponding estimates for $\mathcal{K}_\theta$. In this connection, have a look at Exercises 2.4.39 and 2.4.41.

THEOREM 2.4.25 (**Nelson**). *Let $\theta \in (0,1)$ and $p \in (1,\infty)$ be given, and set*

$$q(p,\theta) = 1 + \frac{p-1}{\theta^2}.$$

*Then*

$$\|\mathcal{H}_\theta\varphi\|_{L^q(\gamma_{0,1};\mathbb{C})} \le \|\varphi\|_{L^p(\gamma_{0,1};\mathbb{C})}, \quad \varphi \in L^2(\gamma_{0,1};\mathbb{C}), \tag{2.4.26}$$

*for every $1 \le q \le q(p,\theta)$. Moreover, if $q > q(p,\theta)$, then*

$$\sup\left\{\|\mathcal{H}_\theta\varphi\|_{L^q(\gamma_{0,1};\mathbb{C})} : \|\varphi\|_{L^p(\gamma_{0,1};\mathbb{C})} = 1\right\} = \infty. \tag{2.4.27}$$

PROOF: I will leave the proof of (2.4.27) as an exercise. (Try taking $\varphi$'s of the form $e^{\lambda x^2}$.) Also, because $\gamma_{0,1}$ is a probability measure and therefore the left-hand side of (2.4.26) is non-decreasing as a function of $q$, I will restrict my attention to the proof of (2.4.26) for $q = q(p,\theta)$. Hence, by Theorem 2.4.14, what I have to do is prove (2.4.16) for every $1 < p < q < \infty$ and $\theta \in (0,1)$ that are related by

$$\theta = \left(\frac{p-1}{q-1}\right)^{\frac{1}{2}}. \tag{2.4.28}$$

I begin with the case when $1 < p < q \le 2$, and I will first consider $\zeta \in [0,1)$. Introducing the generalized binomial coefficients $\binom{r}{0} = 1$ and

$$\binom{r}{\ell} \equiv \frac{r(r-1)\cdots(r-\ell+1)}{\ell!} \quad \text{for} \quad r \in \mathbb{R} \text{ and } \ell \in \mathbb{Z}^+,$$

one can write

$$\frac{|1-\theta\zeta|^q + |1+\theta\zeta|^q}{2} = 1 + \sum_{k=1}^{\infty}\binom{q}{2k}(\theta\zeta)^{2k}$$

and

$$\frac{|1-\zeta|^p + |1+\zeta|^p}{2} = 1 + \sum_{k=1}^{\infty}\binom{p}{2k}\zeta^{2k}.$$

Noting that, because $q \le 2$, $\binom{q}{2k} \ge 0$ for every $k \in \mathbb{Z}^+$, and using the fact that, because $\frac{p}{q} \in (0,1)$, $(1+x)^{\frac{p}{q}} \le 1 + \frac{p}{q}x$ for all $x \ge 0$, we see that

$$\left(\frac{|1-\theta\zeta|^q + |1+\theta\zeta|^q}{2}\right)^{\frac{p}{q}} \le 1 + \frac{p}{q}\sum_{k=1}^{\infty}\binom{q}{2k}(\theta\zeta)^{2k}.$$

Hence, I will have completed the case under consideration once I check that

$$\frac{p}{q}\sum_{k=1}^{\infty}\binom{q}{2k}(\theta\zeta)^{2k} \le \sum_{k=1}^{\infty}\binom{p}{2k}\zeta^{2k},$$

and clearly this will follow if I show that

$$\frac{p}{q}\binom{q}{2k}\theta^{2k} \le \binom{p}{2k} \quad \text{for each} \quad k \in \mathbb{Z}^+.$$

But the choice of $\theta$ in (2.4.28) makes the preceding an equality when $k = 1$, and, when $k \ge 2$,

$$\frac{\frac{p}{q}\binom{q}{2k}\theta^{2k}}{\binom{p}{2k}} \le \prod_{j=2}^{2k-1}\frac{j-q}{j-p} \le 1,$$

since $1 < p < q \le 2$.

At this point, I have proved (2.4.15) for $1 < p < q \le 2$ and $\theta$ given by (2.4.28) when $\zeta \in (0,1)$. Continuing with this choice of $p$, $q$, and $\theta$, note that (2.4.15) extends immediately to $\zeta \in [-1,1]$ by continuity and symmetry. Finally, for general $\zeta \in \mathbb{C}$, set

$$a = \frac{|1-\zeta|+|1+\zeta|}{2}, \quad b = \frac{|1-\zeta|-|1+\zeta|}{2}, \quad \text{and } c = \frac{b}{a} \in [-1,1].$$

Then

$$|1 \pm \theta\zeta| = \left|\tfrac{1+\theta}{2}(1\pm\zeta) + \tfrac{1-\theta}{2}(1\mp\zeta)\right| \le a \mp \theta b,$$

and, therefore, by the preceding applied to $c$, we have that

$$\left(\frac{|1-\theta\zeta|^q+|1+\theta\zeta|^q}{2}\right)^{\frac{1}{q}} \le a\left(\frac{|1-\theta c|^q+|1+\theta c|^q}{2}\right)^{\frac{1}{q}}$$

$$\le a\left(\frac{|1-c|^p+|1+c|^p}{2}\right)^{\frac{1}{p}} = \left(\frac{|a-b|^p+|a+b|^p}{2}\right)^{\frac{1}{p}} = \left(\frac{|1-\zeta|^p+|1+\zeta|^p}{2}\right)^{\frac{1}{p}}.$$

Hence, I have now completed the case when $1 < p < q \le 2$ and $\theta$ is given by (2.4.28).

To handle the other cases, I will use the equivalence of (2.4.16) and (2.4.17). Thus, what we already know is that (2.4.17) holds for $1 < p < q \le 2$ and the $\theta$ in (2.4.28). Next, suppose that $2 \le p < q < \infty$. Then, since $1 < q' < p' \le 2$ and

$$\frac{p-1}{q-1} = \frac{q'-1}{p'-1},$$

an application to $q'$ and $p'$ of the result that we already have yields

$$\begin{aligned}\|\mathcal{K}_\theta\varphi\|_{L^q(\beta;\mathbb{C})} &= \sup\left\{\left(\mathcal{K}_\theta\varphi,\psi\right)_{L^2(\beta;\mathbb{C})} : \psi\in L^2(\beta;\mathbb{C}) \text{ with } \|\psi\|_{L^{q'}(\beta)}=1\right\}\\ &= \sup\left\{\left(\varphi,\mathcal{K}_\theta\psi\right)_{L^2(\beta;\mathbb{C})} : \psi\in L^2(\beta;\mathbb{C}) \text{ with } \|\psi\|_{L^{q'}(\beta)}=1\right\}\\ &\le \|\varphi\|_{L^p(\beta;\mathbb{C})},\end{aligned}$$

where the $\theta$ is the one given in (2.4.28). Thus, the only case that remains is the one when $1<p\le 2\le q<\infty$. But, in this case, set $\xi=(p-1)^{\frac12}$, $\eta=(q-1)^{-\frac12}$, and observe that, because the associated $\theta$ in (2.4.28) is the product of $\xi$ with $\eta$, $\mathcal{K}_\theta=\mathcal{K}_\eta\circ\mathcal{K}_\xi$ and therefore

$$\|\mathcal{K}_\theta\varphi\|_{L^q(\beta;\mathbb{C})}\le\|\mathcal{K}_\xi\varphi\|_{L^2(\beta;\mathbb{C})}\le\|\varphi\|_{L^p(\beta;\mathbb{C})}. \quad \square$$

As my second, and final, application of Theorem 2.4.14, I present the theorem of Beckner for which he concocted Theorem 2.4.14 in the first place. The result was conjectured originally by H. Weyl, who guessed, on the basis of $\mathcal{F}h_0=(\sqrt{-1})^n h_0$, that the norm $\|\mathcal{F}\|_{p\to p'}$ of $\mathcal{F}$ as an operator on $L^p(\mathbb{R};\mathbb{C})$ to $L^{p'}(\mathbb{R};\mathbb{C})$ should be achieved by $h_0$. Weyl's conjecture was partially verified by I. Babenko, who proved it when $p'$ is an even integer. In particular, when combined with the Riesz–Thorin Interpolation Theorem, Babenko's result already shows (cf. Exercise 2.4.35) that $\|\mathcal{F}\|_{p\to p'}<1$ for $p\in(0,1)$.

THEOREM 2.4.29 (**Beckner**). *For each $p\in[1,2]$,*

$$\|\mathcal{F}f\|_{L^{p'}(\mathbb{R};\mathbb{C})}\le A_p\|f\|_{L^p(\mathbb{R};\mathbb{C})},\quad f\in L^p(\mathbb{R};\mathbb{C})\cap L^2(\mathbb{R};\mathbb{C}), \tag{2.4.30}$$

*where $\mathcal{F}$ is the Fourier operator in (2.4.9), $A_1=1=A_2$, and $A_p$ is the constant in (2.4.13) when $p\in(0,1)$. Moreover, if $f$ is the Gauss kernel $e^{-\pi x^2}$, then (2.4.30) is an equality.*

PROOF: Because of (2.4.11), the second part is a straightforward computation that I leave to the reader. Also, I will only consider (2.4.30) when $p\in(1,2)$, the other cases being well known (cf. Exercise 2.4.37).

Because of (2.4.13), the proof of (2.4.30) comes down to showing that

$$\|\mathcal{H}_{\theta_p}\varphi\|_{L^{p'}(\gamma_{0,1};\mathbb{C})}\le\|\varphi\|_{L^p(\gamma_{0,1};\mathbb{C})},\quad \varphi\in L^p(\gamma_{0,1};\mathbb{C}), \tag{2.4.31}$$

where $\theta_p=\sqrt{-1}\,(p-1)^{\frac12}$. Indeed, by (2.4.13), (2.4.31) implies that

$$\|\mathcal{F}\mathcal{U}_p\varphi\|_{L^{p'}(\mathbb{R};\mathbb{C})}\le A_p\|\varphi\|_{L^p(\gamma_{0,1};\mathbb{C})} \tag{2.4.32}$$

for all polynomials $\varphi$. Next, if $\varphi\in L^2(\gamma_{0,1};\mathbb{C})$ and $\{\varphi_n : n\ge 1\}$ is a sequence of polynomials which tend to $\varphi$ in $L^2(\gamma_{0,1};\mathbb{C})$, then, because $p\in(1,2)$, it is easy to check that $\varphi_n \longrightarrow \varphi$ in $L^p(\gamma_{0,1};\mathbb{C})$ and $\mathcal{U}_p\varphi_n \longrightarrow \mathcal{U}_p\varphi$ in $L^2(\mathbb{R};\mathbb{C})$; and therefore, since $\mathcal{F}$ is a bounded on $L^2(\mathbb{R};\mathbb{C})$, Fatou's Lemma shows that (2.4.32) continues to hold for all $\varphi \in L^2(\gamma_{0,1};\mathbb{C})$. Now let $f \in C_c(\mathbb{R};\mathbb{C})$, and set $\varphi = \mathcal{U}_p^{-1}f$. Then, (2.4.32) implies that (2.4.30) holds for $f$. Finally, if $f \in L^2(\mathbb{R};\mathbb{C}) \cap L^p(\mathbb{R};\mathbb{C})$, choose $\{f_n : n \geq 1\} \subseteq C_c(\mathbb{R};\mathbb{C})$ so that $f_n \longrightarrow f$ in both $L^2(\mathbb{R};\mathbb{C})$ and $L^p(\mathbb{R};\mathbb{C})$, and conclude that (2.4.30) continues to hold.

By Theorem 2.4.14, (2.4.31) will follow as soon as I prove (2.4.16) for $\theta_p$. For this purpose, write

$$\zeta = \xi + \sqrt{-1}\,(p-1)^{-\frac{1}{2}}\eta, \quad \text{where} \quad \xi,\, \eta \in \mathbb{R}.$$

Then, because $p'-1 = (p-1)^{-1}$, proving (2.4.16) for $\theta_p$ becomes the problem of checking that

$$\begin{aligned}&\left(\frac{\left[(1-\eta)^2 + (p-1)\xi^2\right]^{\frac{p'}{2}} + \left[(1+\eta)^2 + (p-1)\xi^2\right]^{\frac{p'}{2}}}{2}\right)^{\frac{1}{p'}} \\ (*)\qquad &\leq \left(\frac{\left[(1-\xi)^2 + (p'-1)\eta^2\right]^{\frac{p}{2}} + \left[(1+\xi)^2 + (p'-1)\eta^2\right]^{\frac{p}{2}}}{2}\right)^{\frac{1}{p}}\end{aligned}$$

for all $\xi,\, \eta \in \mathbb{R}$.

To prove (*), consider, for each $\alpha \in (0,\infty)$, the function $g_\alpha : [0,\infty)^2 \longrightarrow [0,\infty)$ defined by $g_\alpha(x,y) = \left[x^{\frac{1}{\alpha}} + y^{\frac{1}{\alpha}}\right]^\alpha$. It is an easy matter to check that $g_\alpha$ is concave or convex depending on whether $\alpha \in [1,\infty)$ or $\alpha \in (0,1)$. In particular, since $\frac{p'}{2} \in (1,\infty)$, when we set $\alpha = \frac{p'}{2}$, $x_\pm = |1 \pm \eta|^{p'}$, and $y = (p-1)^{\frac{p'}{2}}|\xi|^{p'}$, we get

$$\begin{aligned}&\frac{\left[(1-\eta)^2 + (p-1)\xi^2\right]^{\frac{p'}{2}} + \left[(1+\eta)^2 + (p-1)\xi^2\right]^{\frac{p'}{2}}}{2} \\ &= \frac{g_\alpha(x_-,y) + g_\alpha(x_+,y)}{2} \leq g_\alpha\left(\frac{x_- + x_+}{2}, y\right) \\ &= \left[\left(\frac{|1-\eta|^{p'} + |1+\eta|^{p'}}{2}\right)^{\frac{2}{p'}} + (p-1)\xi^2\right]^{\frac{p'}{2}};\end{aligned}$$

and similarly, because $\frac{p}{2} \in (0,1)$,

$$\frac{\left[(1-\xi)^2 + (p'-1)\eta^2\right]^{\frac{p}{2}} + \left[(1+\xi)^2 + (p'-1)\eta^2\right]^{\frac{p}{2}}}{2}$$
$$\geq \left[\left(\frac{|1-\xi|^p + |1+\xi|^p}{2}\right)^{\frac{2}{p}} + (p'-1)\eta^2\right]^{\frac{p}{2}}.$$

Thus, (*) will be proved once I show that

$$(**)\quad \left(\frac{|1-\eta|^{p'} + |1+\eta|^{p'}}{2}\right)^{\frac{2}{p'}} + (p-1)\xi^2 \leq \left(\frac{|1-\xi|^p + |1+\xi|^p}{2}\right)^{\frac{2}{p}} + (p'-1)\eta^2.$$

But because (cf. Theorems 2.4.14 and 2.4.25) we know that (2.4.16) holds with $p$ replaced by 2, $q = p'$, and $\theta = (p-1)^{\frac{1}{2}}$, the left side of (**) is dominated by

$$(p-1)\xi^2 + \frac{\left(1-(p'-1)^{\frac{1}{2}}\eta\right)^2 + \left(1+(p'-1)^{\frac{1}{2}}\eta\right)^2}{2} = 1 + (p-1)\xi^2 + (p'-1)\eta^2.$$

At the same time, again by (2.4.16), only this time with $p$, 2, and the same choice of $\theta$, we see that the right-hand side of (**) dominates

$$(p'-1)\eta^2 + \frac{\left(1-(p-1)^{\frac{1}{2}}\xi\right)^2 + \left(1+(p-1)^{\frac{1}{2}}\xi\right)^2}{2} = 1 + (p-1)\xi^2 + (p'-1)\eta^2. \quad \square$$

## Exercises for § 2.4

EXERCISE 2.4.33. Define $S : \mathbb{R}^2 \longrightarrow \mathbb{R}$ so that $S(x_1, x_2) = \frac{x_1 + x_2}{\sqrt{2}}$, let $\pi_1$ and $\pi_2$ be the natural projection maps given by $\pi_i(x_1, x_2) = x_i$ for $i \in \{1, 2\}$, and let $\lambda_{\mathbb{R}}$ denote Lebesgue measure on $\mathbb{R}$. The goal of this exercise is to prove that if $f : \mathbb{R} \longrightarrow \mathbb{R}$ is a Borel measurable function with the property that

$$(2.4.34)\quad f \circ S = \frac{f \circ \pi_1 + f \circ \pi_2}{\sqrt{2}} \quad \lambda_{\mathbb{R}}^2\text{-almost everywhere,}$$

then there is an $\alpha \in \mathbb{R}$ such that $f(x) = \alpha x$ for $\lambda_{\mathbb{R}^1}$-almost every $x \in \mathbb{R}$. Here are steps which one can take to prove this result.

**(i)** After noticing that (2.4.34) holds when $\lambda_{\mathbb{R}}$ is replaced by $\gamma_{0,1}$, apply Exercise 2.3.21 to see that the $\gamma_{0,1}$-distribution of $x \rightsquigarrow f(x)$ is $\gamma_{0,\alpha}$ for some $\alpha \in [0, \infty)$. Conclude, in particular, that $f \in L^2(\gamma_{0,1}; \mathbb{R})$.

**(ii)** For each $n \geq 0$, let $Z^{(n)}$ denote $\text{span}\big(\{H_n \circ \pi_1 H_{n-m} \circ \pi_2 : 0 \leq m \leq n\}\big)$. Show that $Z^{(m)} \perp Z^{(n)}$ in $L^2(\gamma_{0,1}^2;\mathbb{R})$ when $m \neq n$ and the span of $\bigcup_{n=0}^{\infty} Z^{(n)}$ is dense in $L^2(\gamma_{0,1}^2;\mathbb{R})$. Conclude from these that if $F \in L^2(\gamma_{0,1}^2;\mathbb{R})$, then $F = \sum_{n=0}^{\infty} \Pi_n F$, where $\Pi_n$ denotes orthogonal projection onto $Z^{(n)}$ and the series converges in $L^2(\gamma_{0,1}^2;\mathbb{R})$.

**(iii)** Using the generating (2.4.5), show that

$$H_n \circ S = 2^{-\frac{n}{2}} \sum_{m=0}^{n} \binom{n}{m} H_m \circ \pi_1 H_{n-m} \circ \pi_2,$$

and use this to conclude that for any $\varphi \in L^2(\gamma_{0,1};\mathbb{R})$,

$$\Pi_n(\varphi \circ S) = \frac{\big(\varphi, H_n\big)_{L^2(\gamma_{0,1};\mathbb{R})}}{n!} H_n \circ S.$$

**(iv)** Show that if $\varphi \in L^2(\gamma_{0,1};\mathbb{R})$, then

$$\Pi_n\left(\frac{\varphi \circ \pi_1 + \varphi \circ \pi_2}{\sqrt{2}}\right) = \frac{\big(\varphi, H_n\big)_{L^2(\gamma_{0,1};\mathbb{R})}}{2^{\frac{1}{2}} n!}\big(H_n \circ \pi_1 + H_n \circ \pi_2\big).$$

**(v)** By combining **(iii)** and **(iv)**, show that

$$(*) \qquad \frac{\big(f, H_n\big)_{L^2(\gamma_{0,1};\mathbb{R})}}{n!} H_n \circ S = \frac{\big(f, H_n\big)_{L^2(\gamma_{0,1};\mathbb{R})}}{2^{\frac{1}{2}} n!}\big(H_n \circ \pi_1 + H_n \circ \pi_2\big).$$

From this, show that $\big(f, H_n\big)_{L^2(\gamma_{0,1};\mathbb{R})} = 0$ unless $n = 1$. When $n = 0$, this is obvious. When $n \geq 2$, one can argue that, if $\big(f, H_n\big)_{L^2(\gamma_{0,1};\mathbb{R})} \neq 0$, then (*) implies that $H_n' \circ \pi_1 = H_n' \circ \pi_2$, which is possible only if $H_n'$ is constant. Finally, since $f = \sum_{n=0}^{\infty} \frac{1}{n!}\big(f, H_n\big)_{L^2(\gamma_{0,1};\mathbb{R})} H_n$, it follows that

$$f(x) = \big(f, H_1\big)_{L^2(\gamma_{0,1};\mathbb{R})} H_1(x) = \left(\int \xi f(\xi)\, \gamma_{0,1}(d\xi)\right) x$$

for $\gamma_{0,1}$-almost every $x \in \mathbb{R}$.

EXERCISE 2.4.35. Because the Fourier operator $\mathcal{F}$ (cf. (2.4.9)) is a contraction from $L^1(\mathbb{R};\mathbb{C})$ to $L^\infty(\mathbb{R};\mathbb{C})$ as well as from $L^2(\mathbb{R};\mathbb{C})$ into $L^2(\mathbb{R};\mathbb{C})$, the Riesz–Thorin Interpolation Theorem guarantees that it is a contraction from $L^p(\mathbb{R};\mathbb{C})$ into $L^{p'}(\mathbb{R};\mathbb{C})$ for each $p \in (0,1)$. However, this is a case in which Riesz–Thorin gives a less than optimal result. Indeed, show that

$$t \in \left(\tfrac{1}{2}, 1\right) \longmapsto \log A_{\frac{1}{t}} \in \mathbb{R}$$

is a strictly convex function that tends to 0 at both end points and is therefore strictly negative. Hence, $A_p < 1$ for $p \in (1,2)$.

EXERCISE 2.4.36. The inequality in (2.4.8) is an example of a general principle. Namely, if $(E,\mathcal{B})$ is any measurable space, then a map $(x,\Gamma) \in E\times\mathcal{B} \longmapsto \Pi(x,\Gamma)\in[0,1]$ is called a **transition probability** whenever $x\in E\longmapsto \Pi(x,\Gamma)$ is $\mathcal{B}$-measurable for each $\Gamma\in\mathcal{B}$ and $\Gamma\in\mathcal{B}\longmapsto\Pi(x,\Gamma)$ is a probability measure on $(E,\mathcal{B})$ for each $x\in E$. Given a transition probability $\Pi(x,\,\cdot\,)$, define the linear operator $\Pi$ on $B(E;\mathbb{C})$ (the space of bounded, $\mathcal{B}$-measurable $\varphi: E\longrightarrow\mathbb{C}$) by

$$[\Pi\varphi](x) = \int_E \varphi(y)\,\Pi(x,dy), \quad x\in E, \quad \text{for } \varphi\in B(E;\mathbb{C}).$$

Check that $\Pi$ takes $B(E;\mathbb{C})$ into itself and that $\|\Pi\varphi\|_{\mathrm{u}}\le\|\varphi\|_{\mathrm{u}}$. Next, given a $\sigma$-finite measure $\mu$ on $(E,\mathcal{B})$, say that $\mu$ is **$\Pi$-invariant** if

$$\mu(\Gamma) = \int_E \Pi(x,\Gamma)\,\mu(dx) \quad \text{for all } \Gamma\in\mathcal{B}.$$

Using Jensen's Inequality, first show that, for each $p\in[1,\infty)$,

$$\big|[\Pi\varphi](x)\big|^p \le \big[\Pi|\varphi|^p\big](x), \quad x\in E,$$

and then that, for any $\Pi$-invariant $\mu$,

$$\|\Pi\varphi\|_{L^p(\mu;\mathbb{C})}\le\|\varphi\|_{L^p(\mu;\mathbb{C})}, \quad \varphi\in B(E;\mathbb{C}).$$

Finally, show that $\mu$ is $\Pi$-invariant if it is **$\Pi$-reversing** in the sense that

$$\int_{\Gamma_1}\Pi(x,\Gamma_2)\,\mu(dx) = \int_{\Gamma_2}\Pi(y,\Gamma_1)\,\mu(dy) \quad \text{for all} \quad \Gamma_1,\Gamma_2\in\mathcal{B}.$$

EXERCISE 2.4.37. Recall the Hermite functions $h_n$, $n\in\mathbb{N}$, in (2.4.12) and define the **normalized Hermite functions** $\overline{h}_n$, $n\in\mathbb{N}$ by

$$\overline{h}_n = \frac{2^{\frac14}}{(n!)^{\frac12}}h_n, \quad n\in\mathbb{N}.$$

By noting that (cf. the discussion following (2.4.12)) $\overline{h}_n = \mathcal{U}_2\overline{H}_n$, show that $\{\overline{h}_n : n\in\mathbb{N}\}$ constitutes an orthonormal basis in $L^2(\mathbb{R};\mathbb{C})$, and from this together with (2.4.11), arrive at **Parseval's Identity**:

$$\|\mathcal{F}f\|_{L^2(\mathbb{R};\mathbb{C})} = \|f\|_{L^2(\mathbb{R};\mathbb{C})}, \quad f\in L^1(\mathbb{R};\mathbb{C})\cap L^2(\mathbb{R};\mathbb{C}),$$

and conclude that $\mathcal{F}$ determines a unique unitary operator $\overline{\mathcal{F}}$ on $L^2(\mathbb{R};\mathbb{C})$ such that $\overline{\mathcal{F}}f = \mathcal{F}f$ for $f\in L^1(\mathbb{R};\mathbb{C})\cap L^2(\mathbb{R};\mathbb{C})$. Finally, use this to verify the $L^2$-Fourier inversion formula $\overline{\mathcal{F}}^{-1} = \tilde{\mathcal{F}}$, where $[\tilde{\mathcal{F}}f](x)\equiv[\mathcal{F}f](-x)$, $x\in\mathbb{R}$, for $f\in L^1(\mathbb{R};\mathbb{C})\cap L^2(\mathbb{R};\mathbb{C})$.

EXERCISE 2.4.38. By the same reasoning as I used to prove Theorem 2.4.29, show that, for any pair $1 < p \le 2 \le q < \infty$ and any complex number $\theta = \xi + \sqrt{-1}\,\eta$, (2.4.16) and therefore (2.4.15) hold if both $(q-1)\eta^2 + \xi^2 \le 1$ and

$$(q-2)(\xi\eta)^2 \le \left[1 - \xi^2 - (q-1)\eta^2\right]\left[(p-1) - (q-1)\alpha^2 - \beta^2\right].$$

EXERCISE 2.4.39. L. Gross had a somewhat different approach to the proof of (2.4.26). As in the proof that I have given, he reduced everything to checking (2.4.17). However, he did this in a different way. Namely, given $b \in (0,1)$, he set $f(x) = 1 + bx$ and introduced the functions

$$f_t(x) \equiv \left[\mathcal{K}_{e^{-t}} f\right](x) = \tfrac{1+e^{-t}}{2} f(x) + \tfrac{1-e^{-t}}{2} f(-x), \quad (t,x) \in [0,\infty) \times \mathbb{R},$$

and $q(t) = 1 + (p-1)e^{2t}$, $t \in [0,\infty)$, and proved that

$$(*) \qquad \frac{d}{dt}\|f_t\|_{L^{q(t)}(\beta;\mathbb{C})} \le 0.$$

Following the steps below, see if you can reproduce Gross's calculation.

(**i**) Set

$$F(t) = \|f_t\|_{L^{q(t)}(\beta;\mathbb{C})},$$

and, by somewhat tedious but completely elementary differential calculus, show that

$$\begin{aligned}\tfrac{dF}{dt}(t) = \frac{F(t)^{1-q(t)}}{q(t)^2}\Bigg[&-\dot q(t)\int_{\mathbb{R}} f^{q(t)} \log\left(\tfrac{f_t}{F(t)}\right)^{q(t)} d\beta \\ &+ \tfrac{q(t)^2}{2}\int_{\mathbb{R}} f_t(x)^{q(t)-1}\big(f_t(-x) - f_t(x)\big)\,\beta(dx)\Bigg].\end{aligned}$$

Next, check that

$$\begin{aligned}&\int_{\mathbb{R}} f_t(x)^{q(t)-1}\big(f_t(-x) - f_t(x)\big)\,\beta(dx) \\ &\qquad = -\tfrac{1}{2}\int_{\mathbb{R}} \big(f_t(x)^{q(t)-1} - f_t(-x)^{q(t)-1}\big)\big(f_t(x) - f_t(-x)\big)\,\beta(dx),\end{aligned}$$

and, after verifying that

$$(\xi^{q-1} - \eta^{q-1})(\xi - \eta) \ge \frac{4(q-1)\big(\xi^{\frac{q}{2}} - \eta^{\frac{q}{2}}\big)^2}{q^2}, \qquad \xi, \eta \in (0,\infty) \text{ and } q \in (1,\infty),$$

conclude that

$$(**) \qquad \begin{aligned}\tfrac{dF}{dt}(t) \le \frac{F(t)^{1-q(t)}}{q(t)^2}\Bigg[&-\dot q(t)\int_{\mathbb{R}} f^{q(t)} \log\left(\tfrac{f_t}{F(t)}\right)^{q(t)} d\beta \\ &+ (q(t)-1)\int_{\mathbb{R}} \Big(f_t(x)^{\frac{q(t)}{2}} - f_t^{\frac{q(t)}{2}}(-x)\Big)^2 \beta(dx)\Bigg].\end{aligned}$$

(**ii**) Prove the **Logarithmic Sobolev Inequality**

$$(2.4.40) \qquad \int_{\mathbb{R}} \varphi^2 \log\left(\frac{\varphi}{\|\varphi\|_{L^2(\beta;\mathbb{C})}}\right)^2 d\beta \le 2\int_{\mathbb{R}} \big(\varphi(x) - \varphi(-x)\big)^2\,\beta(dx)$$

for strictly positive $\varphi$'s on $\mathbb{R}$.

**Hint**: Reduce to the case when $\varphi(x) = 1 + bx$ for some $b \in (0,1)$, and, in this case, check that (2.4.40) is the elementary calculus inequality

$$(1+b)^2\log(1+b) + (1-b)^2\log(1-b) - (1+b^2)\log(1+b^2) \le 2b^2, \quad b \in (0,1).$$

**(iii)** By plugging (2.4.40) into (**), arrive at (*), and conclude that (2.4.17) holds for $\theta \in (0,1)$ and $q = 1 + \frac{p-1}{\theta^2}$.

EXERCISE 2.4.41. The major difference between Gross's and Beckner's approaches to proving Nelson's Theorem 2.4.25 is that Gross based his proof on the equivalence of contraction results like (2.4.17) and (2.4.15) to Logarithmic Sobolev Inequalities like (2.4.40). In Exercise 2.4.38, I outlined how one passes from a Logarithmic Sobolev Inequality to a contraction result. The object of this exercise is to go in the opposite direction. Specifically, starting from (2.4.26), show that

$$\int_{\mathbb{R}} \varphi^2 \log\left(\frac{\varphi}{\|\varphi\|_{L^2(\gamma_{0,1};\mathbb{C})}}\right)^2 d\gamma_{0,1} \le 2\int_{\mathbb{R}} |\varphi'|^2\,\gamma_{0,1}(dx) \tag{2.4.42}$$

for non-negative, continuously differentiable $\varphi \in L^2(\gamma_{0,1};\mathbb{C}) \setminus \{0\}$ with $\varphi' \in L^2(\gamma_{0,1};\mathbb{C})$. See Exercise 8.4.8 for another derivation.

EXERCISE 2.4.43. As an application of Theorem 2.4.25, show that

$$\|H_n\|_{L^p(\gamma_{0,1};\mathbb{C})} \le \sqrt{n!(p-1)} \quad \text{for } n \in \mathbb{N} \text{ and } p \in [2,\infty).$$

To see that this estimate is quite good, show that $\|H_1\|^p_{L^p(\gamma_{0,1};\mathbb{C})} = \frac{2^{\frac{p}{2}}}{\pi^{\frac{1}{2}}}\Gamma\left(\frac{p+1}{2}\right)^{\frac{1}{p}}$, and apply Stirling's formula (1.3.21) to conclude that $\|H_1\|_{L^p(\gamma_{0,1};\mathbb{C})} \sim \left(\frac{p-1}{e}\right)^{\frac{1}{2}}$ as $p \to \infty$.

# Chapter 3
## Infinitely Divisible Laws

The results in this chapter are an attempt to answer the following question. Given an $\mathbb{R}^N$-valued random variable $\mathbf{Y}$ with the property that, for each $n \in \mathbb{Z}^+$, $\mathbf{Y} = \sum_{m=1}^n \mathbf{X}_m$, where $\mathbf{X}_1, \dots, \mathbf{X}_n$ are independent and identically distributed, what can one say about the distribution of $\mathbf{Y}$?

Recall that the **convolution** $\nu_1 \star \nu_2$ of two finite Borel measures $\nu_1$ and $\nu_2$ on $\mathbb{R}^N$ is given by

$$\nu_1 \star \nu_2(\Gamma) = \iint_{\mathbb{R}^N \times \mathbb{R}^N} \mathbf{1}_\Gamma(\mathbf{x}+\mathbf{y})\, \nu_1(d\mathbf{x})\nu_2(d\mathbf{y}), \quad \Gamma \in \mathcal{B}_{\mathbb{R}^N},$$

and that the distribution of the sum of two independent random variables is the convolution of their distributions. Thus, the analytic statement of our problem is that of describing those probability measures $\mu$ that, for each $n \geq 1$, can be written as the $n$-fold convolution power $\mu_{\frac{1}{n}}^{\star n}$ of some probability measure $\mu_{\frac{1}{n}}$. I will say that such a $\mu$ is **infinitely divisible** and will use $\mathcal{I}(\mathbb{R}^N)$ to denote the class of infinitely divisible measures on $\mathbb{R}^N$. Since the Fourier transform takes convolution into ordinary multiplication, the Fourier formulation of this problem is that of describing those Borel probability measures on $\mathbb{R}^N$ whose Fourier transform $\hat{\mu}$ has, for each $n \in \mathbb{Z}^+$, an $n$th root which is again the Fourier transform of a Borel probability measure on $\mathbb{R}^N$.

Not surprisingly, the Fourier formulation of the problem is, in many ways, the most amenable to analysis, and it is the formulation in terms of which I will solve it in this chapter. On the other hand, this formulation has the disadvantage that, although it yields a quite satisfactory description of $\hat{\mu}$, it leaves open the problem of extracting information about $\mu$ from properties of $\hat{\mu}$. For this reason, the following chapter will be devoted to developing a probabilistic understanding of the analytic answer obtained in this chapter.

### § 3.1 Convergence of Measures on $\mathbb{R}^N$

In order to carry out our program, I will need two important facts about the convergence of probability measures on $\mathbb{R}^N$. The first of these is a minor modification of the classical Helly–Bray Theorem, and the second is an improvement, due to Lévy, of Lemma 2.3.3.

Say that the sequence $\{\mu_n : n \geq 1\} \subseteq \mathbf{M}_1(\mathbb{R}^N)$ **converges weakly** to $\mu \in \mathbf{M}_1(\mathbb{R}^N)$ and write $\mu_n \Longrightarrow \mu$ when $\langle \varphi, \mu_n \rangle \longrightarrow \langle \varphi, \mu \rangle$ for all $\varphi \in C_{\mathrm{b}}(\mathbb{R}^N; \mathbb{C})$, and apply Lemma 2.3.3 to check that $\mu_n \Longrightarrow \mu$ if and only if $\widehat{\mu_n}(\boldsymbol{\xi}) \longrightarrow \hat{\mu}(\boldsymbol{\xi})$ for every $\boldsymbol{\xi} \in \mathbb{R}^N$.

**§ 3.1.1. Sequential Compactness in $\mathbf{M}_1(\mathbb{R}^N)$.** Given a subset $S$ of $\mathbf{M}_1(\mathbb{R}^N)$, I will say that $S$ is **sequentially relatively compact** if, for every sequence $\{\mu_n : n \geq 1\} \subseteq S$, there a subsequence $\{\mu_{n_m} : m \geq 1\}$ and a $\mu \in \mathbf{M}_1(\mathbb{R}^N)$ such that $\mu_{n_m} \Longrightarrow \mu$.

THEOREM 3.1.1. *A subset $S$ of $\mathbf{M}_1(\mathbb{R}^N)$ is sequentially relatively compact if and only if*

$$\lim_{R\to\infty} \sup_{\mu \in S} \mu\big(B(0,R)\complement\big) = 0. \tag{3.1.2}$$

PROOF: I begin by pointing out that there is a countable set $\{\varphi_k : k \in \mathbb{Z}^+\} \subseteq C_{\mathrm{c}}(\mathbb{R}^N; \mathbb{R})$ of linear independent functions whose span is dense, with respect to uniform convergence, in $C_{\mathrm{c}}(\mathbb{R}^N; \mathbb{R})$. To see this, choose $\eta \in C_{\mathrm{c}}\big(\mathbb{R}^N; [0,1]\big)$ so that $\eta = 1$ on $\overline{B(\mathbf{0},1)}$ and 0 off $B(0,2)$, and set $\eta_R(\mathbf{y}) = \eta(R^{-1}\mathbf{y})$ for $R > 0$. Next, for each $\ell \in \mathbb{Z}^+$, apply the Stone–Weierstrass Theorem to choose a countable dense subset $\{\psi_{j,\ell} : j \in \mathbb{Z}^+\}$ of $C\big(\overline{B(\mathbf{0},2\ell)}; \mathbb{R}\big)$, and set $\varphi_{j,\ell} = \eta_\ell \psi_{j,\ell}$. Clearly $\{\varphi_{j,\ell} : (j,\ell) \in (\mathbb{Z}^+)^2\}$ is dense in $C_{\mathrm{c}}(\mathbb{R}^N; \mathbb{R})$. Finally, using lexicographic ordering of $(\mathbb{Z}^+)^2$, extract a linearly independent subset $\{\varphi_k : k \in \mathbb{Z}^+\}$ by taking $\varphi_k = \varphi_{j_k,\ell_k}$, where $(j_1,\ell_1) = (1,1)$ and $(j_{k+1},\ell_{k+1})$ is the first $(j,\ell)$ such that $\varphi_{j,\ell}$ is linearly independent of $\{\varphi_1, \dots, \varphi_k\}$.

Given a sequence $\{\mu_n : n \geq 1\} \subseteq S$, we can use a diagonalization procedure to find a subsequence $\{\mu_{n_m} : m \geq 1\}$ such that $a_k = \lim_{m\to\infty} \langle \varphi_k, \mu_{n_m} \rangle$ exists for every $k \in \mathbb{Z}^+$. Next, define the linear functional $\Lambda$ on the span of $\{\varphi_k : k \in \mathbb{Z}^+\}$ so that $\Lambda(\varphi_k) = a_k$. Notice that if $\varphi = \sum_{k=1}^K \alpha_k \varphi_k$, then

$$|\Lambda(\varphi)| = \lim_{m\to\infty} \left| \sum_{k=1}^K \alpha_k \langle \varphi_k, \mu_{n_m} \rangle \right| = \lim_{m\to\infty} \big|\langle \varphi, \mu_{n_m} \rangle\big| \leq \|\varphi\|_{\mathrm{u}},$$

and similarly that $\Lambda(\varphi) = \lim_{m\to\infty} \langle \varphi, \mu_{n_m} \rangle \geq 0$ if $\varphi \geq 0$. Hence, $\Lambda$ admits a unique extension as a non-negativity preserving linear functional on $C_{\mathrm{c}}(\mathbb{R}^N; \mathbb{R})$ that satisfies $|\Lambda(\varphi)| \leq \|\varphi\|_{\mathrm{u}}$ for all $\varphi \in C_{\mathrm{c}}(\mathbb{R}^N; \mathbb{R})$.

Now assume that (3.1.2) holds. For each $\ell \in \mathbb{Z}^+$, apply the Riesz Representation Theorem to produce a non-negative Borel measure $\nu_\ell$ supported on $\overline{B(\mathbf{0},2\ell)}$ so that $\langle \varphi, \nu_\ell \rangle = \Lambda(\eta_\ell \varphi)$ for $\varphi \in C_{\mathrm{c}}(\mathbb{R}^N; \mathbb{R})$. Since $\langle \varphi, \nu_{\ell+1} \rangle = \Lambda(\varphi) = \langle \varphi, \nu_\ell \rangle$ whenever $\varphi$ vanishes off of $B(\mathbf{0},\ell)$, it is clear that

$$\nu_{\ell+1}\big(\Gamma \cap B(\mathbf{0},\ell+1)\big) \geq \nu_{\ell+1}\big(\Gamma \cap B(\mathbf{0},\ell)\big) = \nu_\ell\big(\Gamma \cap B(\mathbf{0},\ell)\big) \quad \text{for all } \Gamma \in \mathcal{B}_{\mathbb{R}^N}.$$

Hence, if

$$\mu(\Gamma) \equiv \lim_{\ell\to\infty} \mu_\ell\big(\Gamma \cap B(\mathbf{0},\ell)\big) = \sum_{\ell=1}^{\infty} \mu_\ell\Big(\Gamma \cap \big(B(\mathbf{0},\ell) \setminus B(\mathbf{0},\ell-1)\big)\Big),$$

then $\mu$ is a non-negative Borel measure on $\mathbb{R}^N$ whose restriction to $B(\mathbf{0},\ell)$ is $\nu_\ell$ for each $\ell \in \mathbb{Z}^+$. In particular, $\mu(\mathbb{R}^N) \le 1$ and $\langle \varphi, \mu\rangle = \lim_{m\to\infty}\langle \varphi, \mu_{n_m}\rangle$ for every $\varphi \in C_c(\mathbb{R}^N;\mathbb{R})$. Thus, by Lemma 2.1.7, all that remains is to check that $\mu(\mathbb{R}^N) = 1$. But

$$\begin{aligned}\mu(\mathbb{R}^N) \ge \langle \eta_\ell, \mu\rangle = \lim_{m\to\infty} \langle \eta_\ell, \mu_{n_m}\rangle &\ge \overline{\lim_{m\to\infty}}\, \mu_{n_m}\big(\overline{B(\mathbf{0},\ell)}\big) \\ &= 1 - \varliminf_{m\to\infty} \mu_{n_m}\big(B(\mathbf{0},\ell)\complement\big),\end{aligned}$$

and, by (3.1.2), the final term tends to 0 as $\ell \to \infty$.

To prove the converse assertion, suppose that $S$ is sequentially relatively compact. If (3.1.2) failed, then we could find an $\theta \in (0,1)$ and, for each $n \in \mathbb{Z}^+$, a $\mu_n \in S$ such that $\mu_n\big(B(\mathbf{0},n)\big) \le \theta$. By sequential relative compactness, this would mean that there is a subsequence $\{\mu_{n_m} : m \ge 1\} \subseteq S$ and a $\mu \in \mathbf{M}_1(\mathbb{R}^N)$ such that $\mu_{n_m} \Longrightarrow \mu$ and $\mu_{n_m}\big(B(\mathbf{0},n_m)\big) \le \theta$. On the other hand, for any $R > 0$,

$$\mu\big(B(\mathbf{0},R)\big) \le \langle \eta_R, \mu\rangle \le \overline{\lim_{m\to\infty}}\, \mu_{n_m}\big(B(\mathbf{0},n_m)\big) \le \theta,$$

and so we would arrive at the contradiction $1 = \lim_{R\to\infty} \mu\big(B(\mathbf{0},R)\big) \le \theta$. □

§ **3.1.2. Lévy's Continuity Theorem.** My next goal is to find a test in terms of the Fourier transform to determine when (3.1.2) holds.

LEMMA 3.1.3. *Define*

$$s(r) = \inf_{\theta \ge r}\left(1 - \frac{\sin\theta}{\theta}\right) \quad \text{for } r \in (0,\infty).$$

*Then $s$ is a strictly positive, non-decreasing, continuous function that tends to 0 as $r \searrow 0$. Moreover, if $\mu \in \mathbf{M}_1(\mathbb{R}^N)$, then, for all $(r,R) \in (0,\infty)^2$,*

$$\big|1 - \hat\mu(r\mathbf{e})\big| \le rR + 2\mu\big(\{\mathbf{y} : |(\mathbf{e},\mathbf{y})_{\mathbb{R}^N}| \ge R\}\big) \quad \text{for all } \mathbf{e} \in \mathbb{S}^{N-1}, \tag{3.1.4}$$

*and*

$$\begin{aligned}\mu\big(B(\mathbf{0},N^{\frac12}R)\complement\big) &\le N \sup_{\mathbf{e}\in\mathbb{S}^{N-1}} \mu\big(\{\mathbf{y} : |(\mathbf{e},\mathbf{y})_{\mathbb{R}^N}| \ge R\}\big) \\ &\le \frac{N}{s(rR)} \max\big\{\big|1 - \hat\mu(\boldsymbol{\xi})\big| : |\boldsymbol{\xi}| \le r\big\}.\end{aligned} \tag{3.1.5}$$

*In particular, for any $S \subseteq \mathbf{M}_1(\mathbb{R}^N)$, (3.1.2) holds if and only if*

$$\lim_{|\boldsymbol{\xi}|\searrow 0} \sup_{\mu\in S} \left|1 - \hat{\mu}(\boldsymbol{\xi})\right| = 0. \tag{3.1.6}$$

PROOF: Given (3.1.4) and (3.1.5), the final assertion is obvious. To prove (3.1.4), simply observe that $\left|1 - e^{\sqrt{-1}(r\mathbf{e},\mathbf{y})_{\mathbb{R}^N}}\right| \leq 2 \wedge \left(r|(\mathbf{e},\mathbf{y})_{\mathbb{R}^N}|\right)$.

Turning to (3.1.5), note that

$$\left|1 - \hat{\mu}(\boldsymbol{\xi})\right| \geq \int_{\mathbb{R}^N} \left(1 - \cos(\boldsymbol{\xi},\mathbf{y})_{\mathbb{R}^N}\right) \mu(d\mathbf{y}).$$

Thus, for each $\mathbf{e} \in \mathbb{S}^{N-1}$,

$$\begin{aligned}\frac{1}{r}\int_0^r \left|1 - \hat{\mu}(t\mathbf{e})\right| dt &\geq \int_{\mathbb{R}^N\setminus\{\mathbf{0}\}} \left(1 - \frac{\sin\left(r(\mathbf{e},\mathbf{y})_{\mathbb{R}^N}\right)}{r(\mathbf{e},\mathbf{y})_{\mathbb{R}^N}}\right) \mu(d\mathbf{y}) \\ &\geq s(rR)\mu\left(\{\mathbf{y} : |(\mathbf{e},\mathbf{y})_{\mathbb{R}^N}| \geq R\}\right),\end{aligned}$$

and therefore

$$\sup_{\boldsymbol{\xi}\in B(\mathbf{0},r)} \left|1 - \hat{\mu}(\boldsymbol{\xi})\right| \geq s(rR)\mu\left(\{\mathbf{y} : |(\mathbf{e},\mathbf{y})_{\mathbb{R}^N}| \geq R\}\right). \tag{3.1.7}$$

Since the first inequality in (3.1.5) is obvious, there is nothing more to be done. □

I am now ready to prove Lévy's crucial improvement to Lemma 2.3.3.

THEOREM 3.1.8 (**Lévy's Continuity Theorem**). *Let $\{\mu_n : n \geq 1\} \subseteq \mathbf{M}_1(\mathbb{R}^N)$, and assume that $f(\boldsymbol{\xi}) = \lim_{n\to\infty} \hat{\mu}_n(\boldsymbol{\xi})$ exists for each $\boldsymbol{\xi} \in \mathbb{R}^N$. Then there is a $\mu \in \mathbf{M}_1(\mathbb{R}^N)$ such that $f = \hat{\mu}$ if and only if there is a $\delta > 0$ for which $\lim_{n\to\infty} \sup_{|\boldsymbol{\xi}|\leq\delta} \left|\hat{\mu}_n(\boldsymbol{\xi}) - f(\boldsymbol{\xi})\right| = 0$, in which case $\mu_n \Longrightarrow \mu$. (See part* (**iv**) *of Exercise 3.1.9 for another version.)*

PROOF: The only assertion not already covered by Lemmas 2.1.7 and 2.3.3 is the "if" part of the equivalence. But, if $\hat{\mu}_n \longrightarrow f$ uniformly in a neighborhood of $\mathbf{0}$, then it is easy to check that $\sup_{n\geq 1} |1 - \hat{\mu}_n(\boldsymbol{\xi})|$ must tend to zero as $|\boldsymbol{\xi}| \to 0$. Hence, by the last part of Lemma 3.1.3 and Theorem 3.1.1, we know that there exists a $\mu$ and a subsequence $\{\mu_{n_m} : m \geq 1\}$ such that $\mu_{n_m} \Longrightarrow \mu$. Since $\hat{\mu}$ must equal $f$, Lemma 2.3.3 says that $\mu_n \Longrightarrow \mu$. □

## Exercises for § 3.1

EXERCISE 3.1.9. One might think that to address the sort of problem posed at the beginning of this chapter it would be helpful to know which functions $f : \mathbb{R}^N \longrightarrow \mathbb{C}$ are the Fourier transforms of a probability measure. Such a characterization is the content of **Bochner's Theorem**, whose proof will be outlined in this exercise. Unfortunately, his characterization looks more useful than it is in practice. For instance, I will not use it to solve our problem, and it is difficult to see how its use would simplify matters.

In order to state Bochner's Theorem, say that a function $f : \mathbb{R}^N \longrightarrow \mathbb{C}$ is **non-negative definite** if, for each $n \geq 1$ and $\boldsymbol{\xi}_1, \dots, \boldsymbol{\xi}_n \in \mathbb{R}^N$, the matrix $\big(\big(f(\boldsymbol{\xi}_i - \boldsymbol{\xi}_j)\big)\big)_{1\leq i,j\leq n}$ is Hermitian and non-negative definite. Equivalently,[1]

$$\sum_{i,j=1}^{n} f(\boldsymbol{\xi}_i - \boldsymbol{\xi}_j)\zeta_i\bar{\zeta}_j \geq 0 \quad \text{for all } \zeta_1, \dots, \zeta_n \in \mathbb{C}.$$

Then Bochner's Theorem is the statement that *$f = \hat{\mu}$ for some $\mu \in \mathbf{M}_1(\mathbb{R}^N)$ if and only if $f(\mathbf{0}) = 1$ and $f$ is a continuous, non-negative definite function.*

**(i)** It is ironic that the necessity assertion is the more useful even though it is nearly trivial. Indeed, if $f = \hat{\mu}$, then it is obvious that $f(\mathbf{0}) = 1$ and that $f$ is continuous. To see that it is also non-negative definite, write

$$\sum_{i,j=1}^{n} e^{\sqrt{-1}(\boldsymbol{\xi}_i - \boldsymbol{\xi}_j, \mathbf{x})_{\mathbb{R}^N}}\zeta_i\bar{\zeta}_j = \left|\sum_{i=1}^{n} e^{\sqrt{-1}(\boldsymbol{\xi}_i, \mathbf{x})_{\mathbb{R}^N}}\zeta_i\right|^2,$$

and integrate in $\mathbf{x}$ with respect to $\mu$.

**(ii)** The first step in proving the sufficiency is to use the non-negative definiteness assumption to show that $f(-\mathbf{x}) = \overline{f(\mathbf{x})}$ and $|f(\mathbf{x})| \leq f(\mathbf{0})$ for all $\mathbf{x} \in \mathbb{R}^N$. Obviously, this proves that $\|f\|_{\mathrm{u}} \leq 1$. Second, using a standard Riemann approximation procedure and the continuity of $f$, check that, for any rapidly decreasing, continuous $\hat{\psi} : \mathbb{R}^N \longrightarrow \mathbb{C}$,

$$\iint_{\mathbb{R}^N \times \mathbb{R}^N} f(\mathbf{x} - \boldsymbol{\eta})\hat{\psi}(\mathbf{x})\overline{\hat{\psi}(\boldsymbol{\eta})}\, d\mathbf{x}\, d\boldsymbol{\eta} \geq 0.$$

In particular, when $f \in L^1(\mathbb{R}^N; \mathbb{C})$, set

$$m(\mathbf{x}) = (2\pi)^{-N} \int_{\mathbb{R}^N} e^{-\sqrt{-1}(\mathbf{x}, \boldsymbol{\xi})_{\mathbb{R}^N}} f(\boldsymbol{\xi})\, d\boldsymbol{\xi},$$

[1] Recall that a non-negative definite operator on a complex Hilbert space is always Hermitian.

and use Parseval's Identity and Fubini's Theorem, together with elementary manipulations, to arrive at

$$(2\pi)^N \int_{\mathbb{R}^N} m(\mathbf{x})\,\psi(\mathbf{x})^2\,d\mathbf{x} = \iint_{\mathbb{R}^N\times\mathbb{R}^N} f(\boldsymbol{\xi}-\boldsymbol{\eta})\hat{\psi}(\boldsymbol{\xi})\overline{\hat{\psi}(\boldsymbol{\eta})}\,d\boldsymbol{\xi}\,d\boldsymbol{\eta} \geq 0$$

for all $\psi \in L^1(\mathbb{R}^N;\mathbb{R}) \cap C_b(\mathbb{R}^N;\mathbb{R})$ with $\hat{\psi} \in L^1(\mathbb{R}^N;\mathbb{R})$. Conclude that $m$ is non-negative, and use this to complete the proof in the case when $f \in L^1(\mathbb{R}^N;\mathbb{C})$.

**(iii)** It remains only to pass from the case when $f \in L^1(\mathbb{R}^N;\mathbb{C})$ to the general case. For each $t \in (0,\infty)$, set $f_t(\mathbf{x}) = e^{-t\frac{|\mathbf{x}|^2}{2}} f(\mathbf{x})$. Clearly, $f_t(\mathbf{0}) = 1$ and $f_t \in C_b(\mathbb{R}^N;\mathbb{C}) \cap L^1(\mathbb{R}^N;\mathbb{C})$. In addition, show that

$$\sum_{i,j=1}^n f_t(\boldsymbol{\xi}_i - \boldsymbol{\xi}_j)\zeta_i\bar{\zeta}_j = \int_{\mathbb{R}^N} \left(\sum_{i,j=1}^n f(\boldsymbol{\xi}_i-\boldsymbol{\xi}_j)\zeta_i(\mathbf{x})\bar{\zeta}_j(\mathbf{x})\right)\gamma_{\mathbf{0},t\mathbf{I}}(d\mathbf{x}) \geq 0,$$

where $\zeta_i(\mathbf{x}) \equiv \zeta_i e^{\sqrt{-1}(\boldsymbol{\xi}_i,\mathbf{x})_{\mathbb{R}^N}}$. Hence, $f_t$ is also non-negative definite, and so, by part **(ii)**, we know that $f_t = \widehat{\mu_t}$ for some $\mu_t \in \mathbf{M}_1(\mathbb{R}^N)$. Finally, apply Lévy's Continuity Theorem to see that $\mu_t \Longrightarrow \mu$, where $\mu \in \mathbf{M}_1(\mathbb{R}^N)$ satisfies $f = \hat{\mu}$.

**(iv)** Let $\{\mu_n : n \geq 1\}$ and $f$ be as in Theorem 3.1.8. Combining Bochner's Theorem with Lemma 2.1.7, show that there exists a $\mu \in \mathbf{M}_1(\mathbb{R}^N)$ such that $f = \hat{\mu}$ and $\mu_n \Longrightarrow \mu$ if and only if $f$ is continuous.

EXERCISE 3.1.10. Suppose that $f$ is a non-negative definite function with $f(\mathbf{0}) = 1$. As we have just seen, if $f$ is continuous, then $f = \hat{\mu}$ for some $\mu \in \mathbf{M}_1(\mathbb{R}^N)$.

**(i)** Assuming that $f = \hat{\mu}$, show that

$$(*) \qquad \|f\|_{\mathrm{u}} \leq 1 \quad \text{and} \quad |f(\boldsymbol{\eta}) - f(\boldsymbol{\xi})|^2 \leq 2\big[1 - \mathfrak{Re}\big(f(\boldsymbol{\eta}-\boldsymbol{\xi})\big)\big], \quad \boldsymbol{\xi},\boldsymbol{\eta} \in \mathbb{R}^N.$$

Next, show that (*) follows directly from non-negative definiteness, whether or not $f$ is continuous. Thus, a non-negative definite function is uniformly continuous everywhere if it is continuous at the origin.

**Hint**: Both parts of (*) follow from the fact that

$$A = \begin{pmatrix} 1 & \overline{f(\boldsymbol{\xi})} & \overline{f(\boldsymbol{\eta})} \\ f(\boldsymbol{\xi}) & 1 & f(\boldsymbol{\xi}-\boldsymbol{\eta}) \\ f(\boldsymbol{\eta}) & \overline{f(\boldsymbol{\xi}-\boldsymbol{\eta})} & 1 \end{pmatrix}$$

is non-negative definite. To get the second part, consider the quadratic form $(\mathbf{v}, A\mathbf{v})_{\mathbb{C}^3}$ with $\mathbf{v} = (v_1, 1, -1)$.[2]

[2] This choice of $\mathbf{v}$ was suggested to me by Linan Chen.

**(ii)** To understand how essential a role continuity plays in Bochner's criterion, show that $f = \mathbf{1}_{\{\mathbf{0}\}}$ is non-negative definite. Even though this $f$ cannot be the Fourier transform of any $\mu \in \mathbf{M}_1(\mathbb{R}^N)$, it is nonetheless the "Fourier transform" of a non-negativity preserving linear functional, one for which there is no Riesz representation. To be more precise, consider the linear functional $\Lambda$ on the space of functions $\varphi \in C_{\mathrm{b}}(\mathbb{R}^N;\mathbb{C})$ for which

$$\Lambda\varphi \equiv \lim_{R\to\infty} \frac{1}{|B(\mathbf{0},R)|}\int_{B(\mathbf{0},R)} \varphi(\mathbf{x})\,d\mathbf{x} \quad \text{exists,}$$

and show that $f(\boldsymbol{\xi}) = \Lambda(e_{\boldsymbol{\xi}})$, where $e_{\boldsymbol{\xi}}(\mathbf{x}) = e^{\sqrt{-1}(\boldsymbol{\xi},\mathbf{x})_{\mathbb{R}^N}}$.

EXERCISE 3.1.11. It is important to recognize the extent to which Lévy's Continuity Theorem and, as a by-product, Bochner's Theorem, are strictly finite dimensional results. For example, let $H$ be an infinite dimensional, separable, real Hilbert space, and define $f(h) = e^{-\frac{1}{2}\|h\|_H^2}$. Obviously, $f$ is a continuous and $f(0) = 1$. Show that it is also non-negative definite in the sense that $\big(\big(f(h_i - h_j)\big)\big)_{1\le i,j\le n}$ is a non-negative definite, Hermitian matrix for each $n \in \mathbb{Z}^+$ and $h_1,\dots,h_n \in H$. Now suppose that there were a Borel probability measure $\mu$ on $H$ such that

$$\hat{\mu}(h) \equiv \int_H e^{\sqrt{-1}(h,x)_H}\,\mu(dx) = f(h), \quad h \in H.$$

Show that, for any orthonormal basis $\{e_i : i \in \mathbb{Z}^+\}$ in $H$, the functions $X_i(h) = (e_i,h)_H$, $i \in \mathbb{Z}^+$, would be, under $\mu$, a sequence of independent, $N(0,1)$-random variables, and conclude from this that

$$\int_H e^{-\|h\|_H^2}\,\mu(dh) = \prod_{i\in\mathbb{Z}^+} \mathbb{E}^{\mu}\big[e^{-X_i^2}\big] = 0.$$

Hence, no such $\mu$ can exist. See Chapter 8 for a much more thorough account of this topic.

**Hint**: The non-negative definiteness of $f$ can be seen as a consequence of the analogous result for $\mathbb{R}^n$.

EXERCISE 3.1.12. The **Riemann–Lebesgue Lemma** says that $\hat{f}(\boldsymbol{\xi}) \longrightarrow 0$ as $|\boldsymbol{\xi}| \to \infty$ if $f \in L^1(\mathbb{R}^N;\mathbb{C})$. Thus $\hat{\mu}(\xi) \longrightarrow 0$ as $|\xi| \to \infty$ if $\mu \in \mathbf{M}_1(\mathbb{R})$ is absolutely continuous. In this exercise we will examine situations in which $\mu \in \mathbf{M}_1(\mathbb{R})$ but $\hat{\mu}(\xi) \not\longrightarrow 0$ as $|\xi| \to \infty$.

**(i)** Given a symmetric $\mu \in \mathbf{M}_1(\mathbb{R})$, show that $\hat{\mu}$ is real valued, and use Bochner's Theorem to show that $\hat{\mu}(\xi)$ cannot tend to a strictly negative number as $|\xi| \to \infty$.
**Hint**: Let $\alpha > 0$, and suppose that $\hat{\mu}(\xi) \longrightarrow -2\alpha$ as $|\xi| \to \infty$. Choose $R > 0$ so that $\hat{\mu}(\xi) \le -\alpha$ for $|\xi| \ge R$ and $n \in \mathbb{Z}^+$ so that $(n-1)\alpha > 1$. Set $A = \big(\big(\hat{\mu}(\ell R - kR)\big)\big)_{1\le k,\ell\le n}$, and show that $A$ cannot be non-negative definite.

**(ii)** Show that $\hat{\mu}(\xi) \not\to 0$ if $\mu$ has an atom (i.e., $\mu(\{x\}) > 0$ for some $x \in \mathbb{R}$).

**Hint**: Reduce to the case in which $\mu$ is symmetric, $\mu(\{0\}) > 0$, and therefore that $\mu = p\delta_0 + q\nu$, where $p \in (0,1]$, $q = 1-p$, and $\nu \in \mathbf{M}_1(\mathbb{R})$ is symmetric. If $p = 1$, $\hat{\mu}(\xi) = 1$ for all $\xi$. If $p \in (0,1)$, then $\hat{\mu}(\xi) \longrightarrow 0$ as $|\xi| \to \infty$ implies $\hat{\nu}(\xi) \longrightarrow -\frac{p}{q} < 0$.

**(iii)** To produce an example that is non-atomic, refer to Exercise 1.4.29, take $p \in (0,1) \setminus \{\frac{1}{2}\}$, and let $\mu = \mu_p$, where $\mu_p$ is the measure described in that exercise. Show that $\mu$ is a non-atomic element of $\mathbf{M}_1(\mathbb{R})$ for which $\hat{\mu} \not\to 0$ as $|\xi| \to \infty$.

**Hint**: Show that $\hat{\mu}$ never vanishes and that $\hat{\mu}(2^m\pi)$ is independent of $m \in \mathbb{Z}^+$.

## § 3.2 The Lévy–Khinchine Formula

Throughout, $\mathcal{I}(\mathbb{R}^N)$ will be the set of $\mu \in \mathbf{M}_1(\mathbb{R}^N)$ that are infinitely divisible. My strategy for characterizing $\mathcal{I}(\mathbb{R}^N)$ will be to start from an easily understood subset of $\mathcal{I}(\mathbb{R}^N)$ and to get the rest by taking weak limits.

The elements of $\mathcal{I}(\mathbb{R}^N)$ that first come to mind are the Gaussian measures (cf. (2.3.6)) $\gamma_{\mathbf{m},\mathbf{C}}$. Indeed, if $\mathbf{m} \in \mathbb{R}^N$ and $\mathbf{C}$ is a symmetric, non-negative definite transformation on $\mathbb{R}^N$, then it is clear from (2.3.7) that $\gamma_{\mathbf{m},\mathbf{C}} = \gamma^{\star n}_{\frac{\mathbf{m}}{n},\frac{\mathbf{C}}{n}}$. Unfortunately, this is not a good starting place because it is too rigid: limits of Gaussians are again Gaussian. Indeed, suppose that $\gamma_{\mathbf{m}_n,\mathbf{C}_n} \Longrightarrow \mu$. Then

$$e^{\sqrt{-1}\,(\boldsymbol{\xi},\mathbf{m}_n)_{\mathbb{R}^N} - \frac{1}{2}(\boldsymbol{\xi},\mathbf{C}_n\boldsymbol{\xi})_{\mathbb{R}^N}} \longrightarrow \hat{\mu}(\boldsymbol{\xi}) \quad \text{for all } \boldsymbol{\xi} \in \mathbb{R}^N,$$

and so $\mu = \gamma_{\mathbf{m},\mathbf{C}}$, where $\mathbf{m} = \lim_{n\to\infty} \mathbf{m}_n$ and $\mathbf{C} = \lim_{n\to\infty} \mathbf{C}_n$. In other words, one cannot use weak convergence to escape the class of Gaussian measures.

A more fruitful choice is to start with the Poisson measures. Recall that if $\nu$ is a probability measure on $\mathbb{R}^N$ and $\alpha \in [0,\infty)$, then the **Poisson measure** with jump distribution $\nu$ and jumping rate $\alpha$ (see § 4.2 for an explanation of this terminology) is the measure

$$\pi_{\alpha,\nu} = e^{-\alpha} \sum_{n=0}^{\infty} \frac{\alpha^n}{n!} \nu^{\star n}.$$

To see that $\pi_{\alpha,\nu}$ is infinitely divisible, note that

$$\widehat{\pi_{\alpha,\nu}}(\boldsymbol{\xi}) = \exp\left(\alpha \int \left(e^{\sqrt{-1}\,(\boldsymbol{\xi},\mathbf{y})_{\mathbb{R}^N}} - 1\right) \nu(d\mathbf{y})\right),$$

and therefore that $\pi_{\alpha,\nu} = \pi^{\star n}_{\frac{\alpha}{n},\nu}$. To see why the Poisson measures provide a more hopeful choice of starting point, let $\mathbf{m} \in \mathbb{R}^N$ and a non-negative definite, symmetric $\mathbf{C}$ be given, and choose $(\mathbf{e}_1, \dots, \mathbf{e}_N)$ to be an orthonormal basis of eigenvectors for $\mathbf{C}$. Next, set $m_i = (\mathbf{m}, \mathbf{e}_i)_{\mathbb{R}^N}$ and $\sigma_i = \sqrt{(\mathbf{e}_i, \mathbf{C}\mathbf{e}_i)_{\mathbb{R}^N}}$, and take

$$\nu_n = \frac{1}{2N}\left(\sum_{i=1}^N \delta_{\frac{m_i \mathbf{e}_i}{n}} + \frac{1}{2}\sum_{i=1}^N \left(\delta_{\frac{\sigma_i \mathbf{e}_i}{\sqrt{n}}} + \delta_{-\frac{\sigma_i \mathbf{e}_i}{\sqrt{n}}}\right)\right).$$

Then the Fourier transform of $\pi_{2Nn,\nu_n}$ is

$$\exp\left(\sum_{i=1}^N n\left(e^{\frac{\sqrt{-1}m_i(\boldsymbol{\xi},\mathbf{e}_i)_{\mathbb{R}^N}}{n}} - 1\right) + \sum_{i=1}^N n\left(\cos\frac{\sigma_i(\boldsymbol{\xi},\mathbf{e}_i)_{\mathbb{R}^N}}{n^{\frac{1}{2}}} - 1\right)\right),$$

which tends to $\widehat{\gamma_{\mathbf{m},\mathbf{C}}}(\boldsymbol{\xi})$ as $n \to \infty$, and so $\pi_{2Nn,\nu_n} \Longrightarrow \gamma_{\mathbf{m},\mathbf{C}}$ as $n \to \infty$. Thus, one can use weak convergence to break out to the class of Poisson measures.

As I will show in the next subsection, the preceding is a special case of a result (cf. Theorem 3.2.7) that says that every infinitely divisible measure is the weak limit of Poisson measures. However, before proving that result, it will be convenient to alter our description of Poisson measures. For one thing, it should be clear that, without loss in generality, I may always assume that the jump distribution $\nu$ assigns no mass to $\mathbf{0}$. Indeed, if $\nu(\{0\}) = 1$, then $\pi_{\alpha,\nu} = \delta_{\mathbf{0}} = \pi_{0,\nu'}$ no matter how $\alpha$ and $\nu'$ are chosen. If $\beta = \nu(\{0\}) \in (0,1)$, then $\pi_{\alpha,\nu} = \pi_{\alpha',\nu'}$, where $\alpha' = \alpha(1-\beta)$ and $\nu' = (1-\beta)^{-1}(\nu - \beta\delta_{\mathbf{0}})$. In addition, although the segregation of the rate and jumping distribution provides probabilistic insight, there is no essential reason for doing so. Thus, nothing is lost if one replaces $\pi_{\alpha,\nu}$ by $\pi_M$, where $M$ is the finite measure $\alpha\mu$, in which case

$$\widehat{\pi_M}(\boldsymbol{\xi}) = \exp\left(\int \left(e^{\sqrt{-1}(\boldsymbol{\xi},\mathbf{y})_{\mathbb{R}^N}} - 1\right)M(d\mathbf{y})\right).$$

With these considerations in mind, let $\mathfrak{M}_0(\mathbb{R}^N)$ be the space of non-negative, finite Borel measures $M$ on $\mathbb{R}^N$ with $M(\{\mathbf{0}\}) = 0$, and set $\mathcal{P}(\mathbb{R}^N) = \{\pi_M : M \in \mathfrak{M}_0(\mathbb{R}^N)\}$, the space of Poisson measures on $\mathbb{R}^N$.

**§ 3.2.1. $\mathcal{I}(\mathbb{R}^N)$ Is the Closure of $\mathcal{P}(\mathbb{R}^N)$.** Let $\overline{\mathcal{P}(\mathbb{R}^N)}$ be the closure of $\mathcal{P}(\mathbb{R}^N)$ under weak convergence. That is, $\mu \in \overline{\mathcal{P}(\mathbb{R}^N)}$ if and only if there exists a sequence $\{M_n : n \geq 1\} \subseteq \mathfrak{M}_0(\mathbb{R}^N)$ such that $\pi_{M_n} \Longrightarrow \mu$. My goal here is to prove that

$$\mathcal{I}(\mathbb{R}^N) = \overline{\mathcal{P}(\mathbb{R}^N)}. \tag{3.2.1}$$

Before turning to the proof of (3.2.1), I need the following simple lemma about non-vanishing, $\mathbb{C}$-valued functions. In its statement, and elsewhere,

$$\log\zeta = -\sum_{m=1}^{\infty}\frac{(1-\zeta)^m}{m} \quad \text{for } \zeta \in \mathbb{C} \text{ with } |1-\zeta| < 1 \tag{3.2.2}$$

is the principle branch of logarithm function on the open unit disk around 1 in the complex plane.

LEMMA 3.2.3. *Let $R \in (0,\infty)$ be given. If $f \in C\big(\overline{B(\mathbf{0},R)};\mathbb{C}\setminus\{0\}\big)$ with $f(\mathbf{0}) = 1$, then there is a unique $\ell_f \in C\big(\overline{B(\mathbf{0};R)};\mathbb{C}\big)$ such that $\ell_f(\mathbf{0}) = 0$ and $f = e^{\ell_f}$. Moreover, if $\boldsymbol{\xi} \in \overline{B(\mathbf{0};R)}$, $r \in (0,\infty)$, and $\left|1 - \frac{f(\boldsymbol{\eta})}{f(\boldsymbol{\xi})}\right| < 1$ for all $\boldsymbol{\eta} \in \overline{B(\boldsymbol{\xi},r)\cap B(\mathbf{0},R)}$, then, for each $\boldsymbol{\eta} \in \overline{B(\boldsymbol{\xi},r)\cap B(\mathbf{0},R)}$,*

$$\ell_f(\boldsymbol{\eta}) - \ell_f(\boldsymbol{\xi}) = \log\frac{f(\boldsymbol{\eta})}{f(\boldsymbol{\xi})},$$

*and therefore*

$$|\ell_f(\boldsymbol{\eta}) - \ell_f(\boldsymbol{\xi})| \le 2\left|1 - \frac{f(\boldsymbol{\eta})}{f(\boldsymbol{\xi})}\right| \quad \text{if } \left|1 - \frac{f(\boldsymbol{\eta})}{f(\boldsymbol{\xi})}\right| \le \frac{1}{2}.$$

*Finally, if $\tilde f$ is a second element of $C\big(\overline{B(\mathbf{0};R)};\mathbb{C}\setminus\{0\}\big)$ with $\tilde f(\mathbf{0}) = 1$ and if $\left|1 - \frac{\tilde f(\boldsymbol{\xi})}{f(\boldsymbol{\xi})}\right| \le \frac{1}{2}$ for all $\boldsymbol{\xi} \in \overline{B(\mathbf{0},R)}$, then*

$$|\ell_{\tilde f}(\boldsymbol{\xi}) - \ell_f(\boldsymbol{\xi})| \le 2\left|1 - \frac{\tilde f(\boldsymbol{\xi})}{f(\boldsymbol{\xi})}\right| \quad \text{for } \boldsymbol{\xi} \in \overline{B(\mathbf{0},R)}.$$

*In particular, if $\{f_n : n \ge 1\} \subseteq C\big(\overline{B(\mathbf{0},R)};\mathbb{C}\setminus\{0\}\big)$ with $f_n(\mathbf{0}) = 1$ for all $n \ge 1$, and if $f_n \longrightarrow f \in C\big(\overline{B(\mathbf{0};R)};\mathbb{C}\setminus\{0\}\big)$ uniformly on $\overline{B(\mathbf{0},R)}$, then $f(0) = 1$ and $\ell_{f_n} \longrightarrow \ell_f$ uniformly on $\overline{B(\mathbf{0};R)}$.*

PROOF: To prove the existence and uniqueness of $\ell_f$, begin by observing that there exists an $M \in \mathbb{Z}^+$ and $0 = r_0 < r_1 < \cdots < r_M = R$ such that

$$\left|1 - \frac{f(\boldsymbol{\xi})}{f\left(\frac{r_{m-1}\boldsymbol{\xi}}{|\boldsymbol{\xi}|}\right)}\right| \le \frac{1}{2} \quad \text{for } 1 \le m \le M \text{ and } \boldsymbol{\xi} \in \overline{B(\mathbf{0},r_m)}\setminus\overline{B(\mathbf{0},r_{m-1})}.$$

Thus we can define a function $\ell_f$ on $\overline{B(\mathbf{0},R)}$ so that $\ell_f(\mathbf{0}) = 0$ and

$$\ell_f(\boldsymbol{\xi}) = \ell_f\left(\frac{r_{m-1}\boldsymbol{\xi}}{|\boldsymbol{\xi}|}\right) + \log\frac{f(\boldsymbol{\xi})}{f\left(\frac{r_{m-1}\boldsymbol{\xi}}{|\boldsymbol{\xi}|}\right)}$$
$$\text{if } 1 \le m \le M \text{ and } \boldsymbol{\xi} \in \overline{B(\mathbf{0},r_m)}\setminus\overline{B(\mathbf{0},r_{m-1})}.$$

Furthermore, working by induction on $1 \le m \le M$, one sees that this $\ell_f$ is continuous and satisfies $f = e^{\ell_f}$. Finally, for any $\ell \in C\big(\overline{B(\mathbf{0},R)};\mathbb{C}\big)$ satisfying $\ell(\mathbf{0}) = 0$ and $f = e^{\ell}$, $(\sqrt{-1}2\pi)^{-1}(\ell - \ell_f)$ is a continuous, $\mathbb{Z}$-valued function that vanishes at $\mathbf{0}$, and therefore $\ell = \ell_f$.

Next suppose that $\boldsymbol{\xi} \in B(\mathbf{0}, R)$ and that

$$\left|1 - \frac{f(\boldsymbol{\eta})}{f(\boldsymbol{\xi})}\right| < 1 \quad \text{for all } \boldsymbol{\eta} \in \overline{B(\boldsymbol{\xi}, r) \cap B(\mathbf{0}, R)}.$$

Set

$$\ell(\boldsymbol{\eta}) = \ell_f(\boldsymbol{\xi}) + \log \frac{f(\boldsymbol{\eta})}{f(\boldsymbol{\xi})} \quad \text{for } \boldsymbol{\eta} \in B(\boldsymbol{\xi}, r) \cap B(\mathbf{0}, R),$$

and check that $\boldsymbol{\eta} \rightsquigarrow (\sqrt{-1}2\pi)^{-1}\big(\ell(\boldsymbol{\eta}) - \ell_f(\boldsymbol{\eta})\big)$ is a continuous, $\mathbb{Z}$-valued function that vanishes at $\boldsymbol{\xi}$. Hence, $\ell = \ell_f$ on $B(\mathbf{0}, R) \cap B(\boldsymbol{\xi}, r)$, and therefore on $\overline{B(\mathbf{0}, R) \cap B(\boldsymbol{\xi}, r)}$. Since $|\log(1-\zeta)| \le 2|\zeta|$ if $|\zeta| \le \frac{1}{2}$, this completes the proof of the asserted properties of $\ell_f$.

Turning to the comparison between $\ell_f$ and $\ell_{\tilde{f}}$ when $\left|1 - \frac{\tilde{f}(\boldsymbol{\xi})}{f(\boldsymbol{\xi})}\right| \le \frac{1}{2}$ for all $\boldsymbol{\xi} \in \overline{B(\mathbf{0}, R)}$, set $\ell(\boldsymbol{\xi}) = \ell_f(\boldsymbol{\xi}) + \log \frac{\tilde{f}(\boldsymbol{\xi})}{f(\boldsymbol{\xi})}$, check that $\ell(\mathbf{0}) = 0$ and $\tilde{f} = e^{\ell}$, and conclude that $\ell_{\tilde{f}} - \ell_f = \log \frac{\tilde{f}}{f}$. From this, the asserted estimate for $|\ell_{\tilde{f}} - \ell_f|$ is immediate. □

Lemma 3.2.4. *Define $r \rightsquigarrow s(r)$ as in Lemma 3.1.3, and let $\mu \in \mathbf{M}_1(\mathbb{R}^N)$ and $0 < r < R$ be given. If $|1 - \hat{\mu}(\boldsymbol{\xi})| \le \frac{1}{2}$ for all $\boldsymbol{\xi} \in \overline{B(\mathbf{0}, r)}$ and there is an $\nu \in \mathbf{M}_1(\mathbb{R}^N)$ such that $\mu = \nu^{\star n}$ for some*

$$n \ge \frac{16}{s\left(\frac{r}{4R}\right)}, \tag{3.2.5}$$

*then $|\hat{\mu}(\boldsymbol{\xi})| \ge 2^{-n}$ for all $\boldsymbol{\xi} \in \overline{B(\mathbf{0}, R)}$.*

Proof: First apply Lemma 3.2.3 to see that, because $\hat{\mu}(\boldsymbol{\xi}) = \hat{\nu}(\boldsymbol{\xi})^n$, neither $\hat{\mu}$ nor $\hat{\nu}$ vanishes anywhere on $\overline{B(\mathbf{0}, r)}$ and therefore that there are unique $\ell, \tilde{\ell} \in C(\overline{B(\mathbf{0}, r)}; \mathbb{C})$ such that $\ell(\mathbf{0}) = 0 = \tilde{\ell}(\mathbf{0})$, $\hat{\mu} = e^{\ell}$, and $\hat{\nu} = e^{\tilde{\ell}}$ on $\overline{B(\mathbf{0}, r)}$. Further, since $\hat{\mu} = e^{n\tilde{\ell}}$, uniqueness requires that $\tilde{\ell} = \frac{1}{n}\ell$. Next, observe that, because $\ell = \log \hat{\mu}$ and $|1 - \hat{\mu}| \le \frac{1}{2}$ on $\overline{B(\mathbf{0}, r)}$, $|\ell| \le 2$ there. Hence, because $\mathfrak{Re}\,\ell \le 0$, $|1 - \hat{\nu}| = \left|1 - e^{\frac{1}{n}\ell}\right| \le \frac{2}{n}$ on $\overline{B(\mathbf{0}, r)}$. Using this in (3.1.7), we have, for any $\rho > 0$ and $\mathbf{e} \in \mathbb{S}^{N-1}$, that

$$\nu\big(\{\mathbf{y} : |(\mathbf{e}, \mathbf{y})_{\mathbb{R}^N}| \ge \rho\}\big) \le \frac{1}{s(r\rho)} \max_{\boldsymbol{\xi} \in \overline{B(\mathbf{0}, r)}} \big|1 - \hat{\nu}(\boldsymbol{\xi})\big| \le \frac{2}{ns(r\rho)}, \tag{3.2.6}$$

which, by (3.1.4), leads to $\big|1 - \hat{\nu}(\boldsymbol{\xi})\big| \le \rho R + \frac{4}{ns(r\rho)}$ for $\boldsymbol{\xi} \in \overline{B(\mathbf{0}, R)}$. Finally take $\rho = \frac{1}{4R}$, and use (3.2.5) and $\hat{\mu}(\boldsymbol{\xi}) = \hat{\nu}(\boldsymbol{\xi})^n$ to check that this gives the desired conclusion. □

I now have everything that I need to prove the equality (3.2.1).

THEOREM 3.2.7. *For each $\mu \in \mathcal{I}(\mathbb{R}^N)$ there is a unique $\ell_\mu \in C(\mathbb{R}^N;\mathbb{C})$ satisfying $\ell_\mu(\mathbf{0}) = 0$ and $\hat{\mu} = e^{\ell_\mu}$. Moreover, for each $n \in \mathbb{Z}^+$, $e^{\frac{1}{n}\ell_\mu}$ is the Fourier transform of the unique $\mu_{\frac{1}{n}} \in \mathbf{M}_1(\mathbb{R}^N)$ such that $\mu = \mu_{\frac{1}{n}}^{\star n}$. In addition, if $M_n \in \mathfrak{M}_0(\mathbb{R}^N)$ is defined by*

$$M_n(\Gamma) \equiv n\mu_{\frac{1}{n}}\big(\Gamma \cap (\mathbb{R}^N \setminus \{\mathbf{0}\})\big) \quad \text{for } \Gamma \in \mathcal{B}_{\mathbb{R}^N}, \tag{3.2.8}$$

*then $\pi_{M_n} \Longrightarrow \mu$. Finally, $\mathcal{I}(\mathbb{R}^N)$ is closed in the sense that $\mu \in \mathcal{I}(\mathbb{R}^N)$ if there exists a sequence $\{\mu_k : k \geq 1\} \subseteq \mathcal{I}(\mathbb{R}^N)$ such that $\mu_k \Longrightarrow \mu$. In particular, $\mu_{\frac{1}{n}}$ is uniquely determined and (3.2.1) holds.*

PROOF: Let $\mu \in \mathcal{I}(\mathbb{R}^N)$ be given. Since there is an $r > 0$ such that $|1-\hat{\mu}(\boldsymbol{\xi})| \leq \frac{1}{2}$ for all $\boldsymbol{\xi} \in \overline{B(\mathbf{0},r)}$ and, for all $n \in \mathbb{Z}^+$, $\mu = \mu_{\frac{1}{n}}^{\star n}$ for some $\mu_{\frac{1}{n}} \in \mathbf{M}_1(\mathbb{R}^N)$, Lemma 3.2.4 guarantees that $\hat{\mu}$ never vanishes. Hence, by Lemma 3.2.3, both the existence and uniqueness of $\ell_\mu$ follow. Moreover, if $\mu = \mu_{\frac{1}{n}}^{\star n}$, then, from $\hat{\mu}(\boldsymbol{\xi}) = \widehat{\mu_{\frac{1}{n}}}(\boldsymbol{\xi})^n$, we know first that $\widehat{\mu_{\frac{1}{n}}}$ never vanishes and then that $\ell_\mu = n\ell$, where $\ell$ is the unique element of $C(\mathbb{R}^N;\mathbb{C})$ satisfying $\ell(\mathbf{0}) = 0$ and $\widehat{\mu_{\frac{1}{n}}} = e^\ell$. In particular, this proves that $\mu_{\frac{1}{n}} = e^{\frac{1}{n}\ell}$ for any $\mu_{\frac{1}{n}}$ with $\mu = \mu_{\frac{1}{n}}^{\star n}$, and so there is at most one such $\mu_{\frac{1}{n}}$.

Now define $M_n$ as in the statement, and observe that

$$\widehat{\pi_{M_n}}(\boldsymbol{\xi}) = \exp\Big(n\big(\hat{\mu}_{\frac{1}{n}}(\boldsymbol{\xi}) - 1\big)\Big) = \exp\Big(n\big(e^{\frac{1}{n}\ell_\mu(\boldsymbol{\xi})} - 1\big)\Big) \longrightarrow e^{\ell_\mu(\boldsymbol{\xi})} = \hat{\mu}(\boldsymbol{\xi})$$

as $n \to \infty$. Hence, $\pi_{M_n} \Longrightarrow \mu$. In particular, this proves that $\mathcal{I}(\mathbb{R}^N) \subseteq \overline{\mathcal{P}(\mathbb{R}^N)}$, and therefore, since we already know that $\mathcal{P}(\mathbb{R}^N) \subseteq \mathcal{I}(\mathbb{R}^N)$, the final statement will follow once we check that $\mathcal{I}(\mathbb{R}^N)$ is closed.

To prove that $\mathcal{I}(\mathbb{R}^N)$ is closed, suppose that $\{\mu_k : k \geq 1\} \subseteq \mathcal{I}(\mathbb{R}^N)$ and that $\mu_k \Longrightarrow \mu$. The first step in checking that $\mu \in \mathcal{I}(\mathbb{R}^N)$ is to show that $\hat{\mu}$ never vanishes. To this end, use the fact that $\hat{\mu}_k \longrightarrow \hat{\mu}$ uniformly on compacts to see that there must exist an $r > 0$ such that $|1 - \hat{\mu}_k(\boldsymbol{\xi})| \leq \frac{1}{2}$ for all $k \in \mathbb{Z}^+$ and $\boldsymbol{\xi} \in \overline{B(\mathbf{0},r)}$. Hence, because each of the $\mu_k$'s is infinitely divisible, one can use Lemma 3.2.4 to see that, for each $R \in (0,\infty)$,

$$\inf\{|\hat{\mu}_k(\boldsymbol{\xi})| : k \in \mathbb{Z}^+ \text{ and } \boldsymbol{\xi} \in \overline{B(\mathbf{0},R)}\} > 0,$$

and clearly this is more than enough to show that $\hat{\mu}$ never vanishes. Thus we can choose a unique $\ell \in C(\mathbb{R}^N;\mathbb{C})$ so that $\ell(\mathbf{0}) = 0$ and $\hat{\mu} = e^\ell$. Moreover, if $\ell_k = \ell_{\mu_k}$, then, by Lemma 3.2.3, $\ell_k \longrightarrow \ell$ uniformly on compacts. Now let $n \in \mathbb{Z}^+$ be given, and choose $\{\mu_{k,\frac{1}{n}} : k \geq 1\} \subseteq \mathbf{M}_1(\mathbb{R}^N)$ so that $\mu_k = \mu_{k,\frac{1}{n}}^{\star n}$. Then we know that $\widehat{\mu_{k,\frac{1}{n}}} = e^{\frac{1}{n}\ell_k}$, and so, as $k \to \infty$, $\hat{\mu}_{k,\frac{1}{n}} \longrightarrow e^{\frac{1}{n}\ell}$ uniformly on compacts. Hence, by Lévy's Continuity Theorem, $e^{\frac{1}{n}\ell} = \hat{\mu}_{\frac{1}{n}}$ for some $\mu_{\frac{1}{n}} \in \mathbf{M}_1(\mathbb{R}^N)$. Since this means that $\mu = \mu_{\frac{1}{n}}^{\star n}$, we have shown that $\mu \in \mathcal{I}(\mathbb{R}^N)$. $\square$

**§ 3.2.2. The Formula.** Theorem 3.2.7 provides interesting information, but it fails to provide a concrete characterization of the infinitely divisible laws. In this subsection I will give an explicit formula for $\hat{\mu}$ when $\mu \in \mathcal{I}(\mathbb{R}^N)$, which, in view of the first part of Theorem 3.2.7, is equivalent to characterizing the functions in $\{\ell_\mu : \mu \in \mathcal{I}(\mathbb{R}^N)\}$.

In order to understand what follows, it may be helpful to first guess what the characterization might be. We already know two families of measures which are contained in $\mathcal{I}(\mathbb{R}^N)$: the Gaussian measures $\gamma_{\mathbf{m},\mathbf{C}}$ for $\mathbf{m} \in \mathbb{R}^N$ and symmetric, non-negative definite $\mathbf{C} \in \mathrm{Hom}(\mathbb{R}^N;\mathbb{R}^N)$, and the Poisson measures $\pi_M$ for $M \in \mathfrak{M}_0(\mathbb{R}^N)$. Further, it is obvious that $\mu, \nu \in \mathcal{I}(\mathbb{R}^N) \implies \mu \star \nu \in \mathcal{I}(\mathbb{R}^N)$, and we know that $\mu \in \mathcal{I}(\mathbb{R}^N)$ if $\mu_n \Longrightarrow \nu$ for some $\{\mu_n : n \geq 1\} \subseteq \mathcal{I}(\mathbb{R}^N)$. Finally, Theorem 3.2.7 tells us that every element of $\mathcal{I}(\mathbb{R}^N)$ is the limit of Poisson measures. Thus, by Lévy's Continuity Theorem, we should be asking what sort of functions can arise as the locally uniform limit of functions of the form

$$(*) \qquad \boldsymbol{\xi} \rightsquigarrow \ell = \sqrt{-1}(\boldsymbol{\xi}, \mathbf{m})_{\mathbb{R}^N} - \tfrac{1}{2}(\boldsymbol{\xi}, \mathbf{C}\boldsymbol{\xi})_{\mathbb{R}^N} + \int_{\mathbb{R}^N} \left[e^{\sqrt{-1}(\boldsymbol{\xi},\mathbf{y})_{\mathbb{R}^N}} - 1\right] M(d\mathbf{y}),$$

and, as I already noted, only the Poisson component $M$ offers much flexibility.

With this in mind, I introduce for each $\alpha \in [0,\infty)$ the class $\mathfrak{M}_\alpha(\mathbb{R}^N)$ of Borel measures $M$ on $\mathbb{R}^N$ such that

$$M(\{\mathbf{0}\}) = 0 \quad \text{and} \quad \int_{\mathbb{R}^N} \frac{|\mathbf{y}|^\alpha}{1+|\mathbf{y}|^\alpha} M(dy) < \infty.$$

When $M \in \mathfrak{M}_0(\mathbb{R}^N)$, the function $\ell$ in (*) equals $\ell_\mu$ for $\mu = \gamma_{\mathbf{m},\mathbf{C}} \star \pi_M$. More generally, even if $M \in \mathfrak{M}_\alpha(\mathbb{R}^N) \setminus \mathfrak{M}_0(\mathbb{R}^N)$, for each $r > 0$, $M_r$ given by $M(d\mathbf{y}) = \mathbf{1}_{[r,\infty)}(|\mathbf{y}|)M(d\mathbf{y})$ is an element of $\mathfrak{M}_0(\mathbb{R}^N)$. Furthermore, if $M \in \mathfrak{M}_1(\mathbb{R}^N)$, then it is clear that, as $r \searrow 0$,

$$\int_{\mathbb{R}^N} \left[e^{\sqrt{-1}(\boldsymbol{\xi},\mathbf{y})_{\mathbb{R}^N}} - 1\right] M_r(d\mathbf{y}) \longrightarrow \int_{\mathbb{R}^N} \left[e^{\sqrt{-1}(\boldsymbol{\xi},\mathbf{y})_{\mathbb{R}^N}} - 1\right] M(d\mathbf{y})$$

uniformly on compacts. Thus, by Lévy's Continuity Theorem, when $M \in \mathfrak{M}_1(\mathbb{R}^N)$, the function $\ell$ in (*) is $\ell_\mu$ for a $\mu \in \mathcal{I}(\mathbb{R}^N)$. In order to handle $M \in \mathfrak{M}_\alpha(\mathbb{R}^N)$ for $\alpha > 1$, we must make the integrand $M$-integrable at $\mathbf{0}$ by subtracting off the next term in the Taylor expansion of $e^{\sqrt{-1}(\boldsymbol{\xi},\mathbf{y})_{\mathbb{R}^N}}$. Thus, choose a Borel measurable function $\eta : \mathbb{R}^N \longrightarrow [0,1]$ that equals 1 in a neighborhood of $\mathbf{0}$, and set $\ell_r(\boldsymbol{\xi})$ equal to

$$\sqrt{-1}(\boldsymbol{\xi}, \mathbf{m})_{\mathbb{R}^N} - \tfrac{1}{2}(\boldsymbol{\xi}, \mathbf{C}\boldsymbol{\xi})_{\mathbb{R}^N} + \int_{\mathbb{R}^N} \left[e^{\sqrt{-1}(\boldsymbol{\xi},\mathbf{y})_{\mathbb{R}^N}} - 1 - \sqrt{-1}\eta(\mathbf{y})(\boldsymbol{\xi},\mathbf{y})_{\mathbb{R}^N}\right] M_r(d\mathbf{y}).$$

Because

$$\ell_r(\boldsymbol{\xi}) = \sqrt{-1}(\boldsymbol{\xi}, \mathbf{m}_r)_{\mathbb{R}^N} - \tfrac{1}{2}(\boldsymbol{\xi}, \mathbf{C}\boldsymbol{\xi})_{\mathbb{R}^N} + \int_{\mathbb{R}^N} \left[e^{\sqrt{-1}(\boldsymbol{\xi},\mathbf{y})_{\mathbb{R}^N}} - 1\right] M_r(d\mathbf{y}),$$
$$\text{where } \mathbf{m}_r = \mathbf{m} - \int_{\mathbb{R}^N} \eta(\mathbf{y})\mathbf{y}\, M_r(d\mathbf{y}),$$

we know that $\ell_r = \ell_{\mu_r}$ for $\mu_r = \gamma_{\mathbf{m}_r,\mathbf{C}} \star \pi_{M_r}$. In addition, if $M \in \mathfrak{M}_2(\mathbb{R}^N)$ and $\ell(\xi)$ equals

$$\sqrt{-1}(\boldsymbol{\xi}, \mathbf{m})_{\mathbb{R}^N} - \tfrac{1}{2}(\boldsymbol{\xi}, \mathbf{C}\boldsymbol{\xi})_{\mathbb{R}^N} + \int_{\mathbb{R}^N} \left[e^{\sqrt{-1}(\boldsymbol{\xi},\mathbf{y})_{\mathbb{R}^N}} - 1 - \sqrt{-1}\eta(\mathbf{y})(\boldsymbol{\xi},\mathbf{y})_{\mathbb{R}^N}\right] M(d\mathbf{y}),$$

then $\ell_r \longrightarrow \ell$ uniformly on compacts. Hence, again by Lévy's Continuity Theorem, we know that, for each $M \in \mathfrak{M}_2(\mathbb{R}^N)$, the function

$$(**)\qquad \begin{aligned} \boldsymbol{\xi} \rightsquigarrow \ell(\boldsymbol{\xi}) \equiv \sqrt{-1}(\boldsymbol{\xi},\mathbf{m})_{\mathbb{R}^N} &- \tfrac{1}{2}(\boldsymbol{\xi}, \mathbf{C}\boldsymbol{\xi})_{\mathbb{R}^N} \\ &+ \int_{\mathbb{R}^N} \left[e^{\sqrt{-1}(\boldsymbol{\xi},\mathbf{y})_{\mathbb{R}^N}} - 1 - \sqrt{-1}\eta(\mathbf{y})(\boldsymbol{\xi},\mathbf{y})_{\mathbb{R}^N}\right] M(d\mathbf{y}) \end{aligned}$$

equals $\ell_\mu$ for some $\mu \in \mathcal{I}(\mathbb{R}^N)$.

One might think that repeated application of the same procedure would show that one need not stop at $\mathfrak{M}_2(\mathbb{R}^N)$ and that more singular $M$'s can occur in the representation of $\ell$. More precisely, one might try accommodating $M$'s from $\mathfrak{M}_3(\mathbb{R}^N)$ by subtracting off the next term in the Taylor expansion. That is, one would replace

$$\int_{\mathbb{R}^N} \left[e^{\sqrt{-1}(\boldsymbol{\xi},\mathbf{y})_{\mathbb{R}^N}} - 1 - \sqrt{-1}\eta(\mathbf{y})(\boldsymbol{\xi},\mathbf{y})_{\mathbb{R}^N}\right] M_r(d\mathbf{y})$$

by

$$\int_{\mathbb{R}^N} \left[e^{\sqrt{-1}(\boldsymbol{\xi},\mathbf{y})_{\mathbb{R}^N}} - 1 - \sqrt{-1}\eta(\mathbf{y})(\boldsymbol{\xi},\mathbf{y})_{\mathbb{R}^N} + \tfrac{1}{2}\eta(\mathbf{y})(\boldsymbol{\xi},\mathbf{y})^2_{\mathbb{R}^N}\right] M_r(d\mathbf{y})$$

in the expression for $\ell_r$. However, to re-write this $\ell_r$ in the form given in (*), one would have to replace $\mathbf{C}$ by

$$\mathbf{C} - \int_{\mathbb{R}^N} \eta(\mathbf{y})\mathbf{y} \otimes \mathbf{y}\, M_r(d\mathbf{y}),$$

which would destroy non-negative definiteness as $r \searrow 0$.

The preceding discussion is evidence for the conjecture that the functions $\ell$ of the form in (**) coincide with $\{\ell_\mu : \mu \in \mathcal{I}(\mathbb{R}^N)\}$, and the rest of this subsection is devoted to the verification of this conjecture. Because of their role here, elements of $\mathfrak{M}_2(\mathbb{R}^N)$ are called **Lévy measures**.

The strategy that I will adopt derives from the observation that $\ell_\mu(\boldsymbol{\xi}) = \lim_{n\to\infty} n\big(\hat{\mu}_{\frac{1}{n}}(\boldsymbol{\xi}) - 1\big)$. Thus, if we can understand the operation

$$A_\mu\varphi = \lim_{n\to\infty} n\big(\langle\varphi, \mu_{\frac{1}{n}}\rangle - \varphi(0)\big)$$

for a sufficiently rich class of functions $\varphi$, then we can understand $\ell_\mu(\boldsymbol{\xi})$ by applying $A_\mu$ to $\mathbf{x} \rightsquigarrow e^{\sqrt{-1}(\boldsymbol{\xi},\mathbf{x})_{\mathbb{R}^N}}$. Even though $\mathbf{x} \rightsquigarrow e^{\sqrt{-1}(\boldsymbol{\xi},\mathbf{x})_{\mathbb{R}^N}}$ is not an element, for technical reasons, it turns out that the class of $\varphi$'s on which it is easiest to understand $A_\mu$ is the Schwartz test function space $\mathscr{S}(\mathbb{R}^N;\mathbb{C})$ (the space of smooth $\mathbb{C}$-valued functions that, together with all of their derivatives, are rapidly decreasing). The basic reason why $\mathscr{S}(\mathbb{R}^N;\mathbb{C})$ is well suited to our analysis is that the Fourier transform maps $\mathscr{S}(\mathbb{R}^N;\mathbb{C})$ onto itself. Further, once we understand how $A_\mu$ acts on $\mathscr{S}(\mathbb{R}^N;\mathbb{C})$, it is a relatively simple matter to use that understanding to compute $\ell_\mu(\boldsymbol{\xi})$.

LEMMA 3.2.9. *Let $\mu \in \mathcal{I}(\mathbb{R}^N)$ be given. For each $r \in (0,\infty)$ there exists a $C(r) < \infty$ such that $|\ell_\mu(\boldsymbol{\xi})| \le C(r)(1+|\boldsymbol{\xi}|^2)$ for all $\boldsymbol{\xi} \in \mathbb{R}^N$ whenever $\mu \in \mathcal{I}(\mathbb{R}^N)$ satisfies $|1-\hat{\mu}(\boldsymbol{\xi})| \le \frac{1}{2}$ for $\boldsymbol{\xi} \in \overline{B(\mathbf{0},r)}$. Moreover,*

$$\begin{aligned}A_\mu(c\mathbf{1}+\varphi) &\equiv \lim_{n\to\infty} n\Big(\langle c\mathbf{1}+\varphi, \mu_{\frac{1}{n}}\rangle - \big(c+\varphi(\mathbf{0})\big)\Big)\\ &= \frac{1}{(2\pi)^N}\int_{\mathbb{R}^N} \overline{\ell_\mu(\boldsymbol{\xi})}\hat{\varphi}(\boldsymbol{\xi})\,d\boldsymbol{\xi}\end{aligned} \tag{3.2.10}$$

*for each $c \in \mathbb{C}$ and $\varphi \in \mathscr{S}(\mathbb{R}^N;\mathbb{C})$.*

PROOF: Suppose that $\mu \in \mathcal{I}(\mathbb{R}^N)$ satisfies $|1-\hat{\mu}(\boldsymbol{\xi})| \le \frac{1}{2}$ for $\boldsymbol{\xi} \in \overline{B(\mathbf{0},r)}$. Applying (3.1.4) and the second inequality in (3.2.6) with $\nu = \mu_{\frac{1}{n}}$, we know that, for any $(\rho, R) \in (0,\infty)^2$,

$$\sup_{|\boldsymbol{\xi}|\le R} |1-\widehat{\mu_{\frac{1}{n}}}(\boldsymbol{\xi})| \le \rho R + \frac{4}{ns(r\rho)}.$$

Hence, if $R \ge r$, then, by taking $\rho = \frac{1}{4R}$, we obtain $\sup_{|\boldsymbol{\xi}|\le R}|1-\widehat{\mu_{\frac{1}{n}}}(\boldsymbol{\xi})| \le \frac{1}{2}$ and therefore $\sup_{|\boldsymbol{\xi}|\le R}|\frac{1}{n}\ell_\mu(\boldsymbol{\xi})| \le 2$ if $n$ satisfies (3.2.5). Finally, observe that there is an $\epsilon > 0$ such that $s(t) \ge \epsilon t^2$ for $t \in (0,1]$, and therefore that $|\ell_\mu(\boldsymbol{\xi})| \le 2\left(1+\frac{64R^2}{\epsilon r^2}\right)$ for $|\boldsymbol{\xi}| \le R$, which completes the proof of the first assertion.

Clearly it suffices to prove (3.2.10) when $c = 0$. Thus, let $\varphi \in \mathscr{S}(\mathbb{R}^N;\mathbb{C})$ be given. Then, by (2.3.4),

$$\begin{aligned}(2\pi)^N n\big(\langle\varphi,\mu_{\frac{1}{n}}\rangle - \varphi(\mathbf{0})\big) &= \int_{\mathbb{R}^N} n\big(e^{\frac{1}{n}\overline{\ell_\mu(\boldsymbol{\xi})}} - 1\big)\hat{\varphi}(\boldsymbol{\xi})\,d\boldsymbol{\xi}\\ &= \int_0^1\left(\int_{\mathbb{R}^N} e^{\frac{t}{n}\overline{\ell_\mu(\boldsymbol{\xi})}}\,\overline{\ell_\mu(\boldsymbol{\xi})}\hat{\varphi}(\boldsymbol{\xi})\,d\boldsymbol{\xi}\right)dt \longrightarrow \int_{\mathbb{R}^N}\overline{\ell_\mu(\boldsymbol{\xi})}\hat{\varphi}(\boldsymbol{\xi})\,d\boldsymbol{\xi},\end{aligned}$$

where (keeping in mind that $|e^{\frac{1}{n}\ell_\mu}| = |\hat{\mu}_{\frac{1}{n}}(\boldsymbol{\xi})| \le 1$, $\ell_\mu(\boldsymbol{\xi})$ has a most quadratic growth, and $\hat{\varphi}(\boldsymbol{\xi})$ is rapidly decreasing) the passage to the second line is justified by Fubini's Theorem and the limit is an application of Lebesgue's Dominated Convergence Theorem. □

Lemma 3.2.9, especially (3.2.10), provides us with two critical pieces of information about $A_\mu$. Namely, it tells us that $A_\mu$ satisfies the **minimum principle** and that it is **quasi-local**. To be precise, set $\mathbf{D} = \mathbb{R}\oplus\mathscr{S}(\mathbb{R}^N;\mathbb{R})$. That is, $\varphi \in \mathbf{D}$ if and only if there is a $\varphi(\infty) \in \mathbb{R}$ such that $\varphi - \varphi(\infty)\mathbf{1} \in \mathscr{S}(\mathbb{R}^N;\mathbb{R})$. I will say that a real-valued linear functional $A$ on $\mathbf{D}$ satisfies the minimum principle if

$$A\varphi \ge 0 \text{ if } \varphi \in \mathbf{D} \text{ and } \varphi(\mathbf{0}) = \min_{\mathbf{x}\in\mathbb{R}^N} \varphi(\mathbf{x}) \tag{3.2.11}$$

and that $A$ is quasi-local if

$$\lim_{R\to\infty} A\varphi_R = 0 \quad \text{for all } \varphi \in \mathbf{D}, \tag{3.2.12}$$

where $\varphi_R(\mathbf{x}) = \varphi\left(\frac{\mathbf{x}}{R}\right)$ for $R > 0$. Notice that, by applying the minimum principle to both $\mathbf{1}$ and $-\mathbf{1}$, one knows that $A\mathbf{1} = 0$.

To see that $A_\mu$ satisfies both these conditions, first observe that if $\varphi(\mathbf{0}) = \min_{\mathbf{x}\in\mathbb{R}^N} \varphi(\mathbf{x})$, then $\langle \varphi, \mu_{\frac{1}{n}}\rangle - \varphi(\mathbf{0}) \ge 0$ for all $n \in \mathbb{Z}^+$, and therefore that $A_\mu\varphi \ge 0$. Secondly, to check that $A_\mu$ is quasi-local, note that it suffices to treat $\varphi \in \mathscr{S}(\mathbb{R}^N;\mathbb{R})$ and that for such a $\varphi$, $\widehat{\varphi_R}(\boldsymbol{\xi}) = R^N\hat{\varphi}(R\boldsymbol{\xi})$. Thus,

$$(2\pi)^N A_\mu\varphi_R = \int_{\mathbb{R}^N} \overline{\ell_\mu(R^{-1}\boldsymbol{\xi})}\hat{\varphi}(\boldsymbol{\xi})\, d\boldsymbol{\xi} \longrightarrow 0,$$

since $\ell_\mu(\mathbf{0}) = 0$ and $\boldsymbol{\xi} \rightsquigarrow \sup_{R\ge1} |\ell_\mu(R^{-1}\boldsymbol{\xi})\hat{\varphi}(\boldsymbol{\xi})|$ is rapidly decreasing.

As I am about to show, these two properties allow us to say a great deal about $A_\mu$. Before explaining this, first observe that if $M \in \mathfrak{M}_\alpha(\mathbb{R}^N)$, then, for every Borel measurable $\varphi : \mathbb{R}^N \longrightarrow \mathbb{C}$,

$$\sup_{\mathbf{y}\in\mathbb{R}^N\setminus\{\mathbf{0}\}} \frac{|\varphi(\mathbf{y})|}{1\wedge|\mathbf{y}|^\alpha} < \infty \implies \varphi \in L^1(M;\mathbb{C}). \tag{3.2.13}$$

Using (3.2.13), one can easily check that if $\varphi \in C^2_{\mathrm{b}}(\mathbb{R}^N;\mathbb{C})$ and $\eta \in \mathscr{S}(\mathbb{R}^N;\mathbb{R})$ equals 1 in a neighborhood of $\mathbf{0}$, then

$$\mathbf{y} \rightsquigarrow \varphi(y) - \varphi(0) - \eta(\mathbf{y})\big(\mathbf{y}, \nabla\varphi(\mathbf{0})\big)_{\mathbb{R}^N}$$

is $M$-integrable for every $M \in \mathfrak{M}_2(\mathbb{R}^N)$.

Second, in preparation for the proof of the next lemma, I have to introduce the following partition of unity for $\mathbb{R}^N \setminus \{\mathbf{0}\}$. Choose $\psi \in C^\infty(\mathbb{R}^N;[0,1])$ so that

$\psi$ has compact support in $B(\mathbf{0},2)\setminus\overline{B(\mathbf{0},\frac{1}{4})}$ and $\psi(\mathbf{y})=1$ when $\frac{1}{2}\le|\mathbf{y}|\le 1$, and set $\psi_m(\mathbf{y})=\psi(2^m\mathbf{y})$ for $m\in\mathbb{Z}$. Then, if $\mathbf{y}\in\mathbb{R}^N$ and $2^{-m-1}\le|\mathbf{y}|\le 2^{-m}$, $\psi_m(\mathbf{y})=1$ and $\psi_n(\mathbf{y})=0$ unless $-m-2\le n\le -m+1$. Hence, if $\Psi(\mathbf{y})=\sum_{m\in\mathbb{Z}}\psi_m(\mathbf{y})$ for $\mathbf{y}\in\mathbb{R}^N\setminus\{\mathbf{0}\}$, then $\Psi$ is a smooth function with values in $[1,4]$; and therefore, for each $m\in\mathbb{Z}$, the function $\chi_m$ given by $\chi_m(\mathbf{0})=0$ and $\chi_m(\mathbf{y})=\frac{\psi_m(\mathbf{y})}{\Psi(\mathbf{y})}$ for $\mathbf{y}\in\mathbb{R}^N\setminus\{\mathbf{0}\}$ is a smooth, $[0,1]$-valued function that vanishes off of $B(\mathbf{0},2^{-m+1})\setminus\overline{B(\mathbf{0},2^{-m-2})}$. In addition, for each $\mathbf{y}\in\mathbb{R}^N\setminus\{0\}$, $\sum_{m\in\mathbb{Z}}\chi_m(\mathbf{y})=1$ and $\chi_m(\mathbf{y})=0$ unless $2^{-m-2}\le|\mathbf{y}|\le 2^{-m+1}$.

Finally, given $n\in\mathbb{Z}^+$ and $\varphi\in C^n(\mathbb{R}^N;\mathbb{C})$, define $\nabla^n\varphi(\mathbf{x})$ to be the multilinear map on $(\mathbb{R}^N)^n$ into $\mathbb{C}$ by

$$\big[\nabla^n\varphi(x)\big](\boldsymbol{\xi}_1,\dots,\boldsymbol{\xi}_n)=\frac{\partial^n}{\partial_{t_1}\cdots\partial t_n}\varphi\left(\mathbf{x}+\sum_{m=1}^n t_m\boldsymbol{\xi}_m\right)\bigg|_{t_1=\cdots=t_n=0}.$$

Obviously, $\nabla\varphi$ and $\nabla^2\varphi$ can be identified as the gradient of $\varphi$ and Hessian of $\varphi$.

LEMMA 3.2.14. *Let $\mathbf{D}$ be the space of functions described above. If $A:\mathbf{D}\longrightarrow\mathbb{R}$ is a linear functional on $\mathbf{D}$ that satisfies (3.2.11) and (3.2.12), then there is a unique $M\in\mathfrak{M}_2(\mathbb{R}^N)$ such that $A\varphi=\int_{\mathbb{R}^N}\varphi(\mathbf{y})\,M(d\mathbf{y})$ whenever $\varphi$ is an element of $\mathscr{S}(\mathbb{R}^N;\mathbb{R})$ for which $\varphi(\mathbf{0})=0$, $\nabla\varphi(\mathbf{0})=\mathbf{0}$, and $\nabla^2\varphi(\mathbf{0})=\mathbf{0}$. Next, given $\eta\in C_c^\infty(\mathbb{R}^N;[0,1])$ satisfying $\eta=1$ in a neighborhood of $\mathbf{0}$, set $\eta_{\boldsymbol{\xi}}(\mathbf{y})=\eta(\mathbf{y})(\boldsymbol{\xi},\mathbf{y})_{\mathbb{R}^N}$ for $\boldsymbol{\xi}\in\mathbb{R}^N$, and define $\mathbf{m}^\eta\in\mathbb{R}^N$ and $\mathbf{C}\in\mathrm{Hom}(\mathbb{R}^N;\mathbb{R}^N)$ by*

$$(3.2.15)\quad (\boldsymbol{\xi},\mathbf{m}^\eta)=A\eta_{\boldsymbol{\xi}}\quad\text{and}\quad \big(\boldsymbol{\xi},\mathbf{C}\boldsymbol{\xi}'\big)_{\mathbb{R}^N}=A(\eta_{\boldsymbol{\xi}}\eta_{\boldsymbol{\xi}'})-\int_{\mathbb{R}^N}(\eta_{\boldsymbol{\xi}}\eta_{\boldsymbol{\xi}'})(\mathbf{y})\,M(d\mathbf{y}).$$

*Then $\mathbf{C}$ is symmetric, non-negative definite, and independent of the choice of $\eta$. Finally, for any $\varphi\in\mathbf{D}$,*

$$(3.2.16)\quad \begin{aligned}A\varphi&=\tfrac{1}{2}\mathrm{Trace}\big(\mathbf{C}\nabla^2\varphi(\mathbf{0})\big)+\big(\mathbf{m}^\eta,\nabla\varphi(\mathbf{0})\big)_{\mathbb{R}^N}\\&\quad+\int_{\mathbb{R}^N}\Big(\varphi(\mathbf{y})-\varphi(\mathbf{0})-\eta(\mathbf{y})\big(\mathbf{y},\nabla\varphi(\mathbf{0})\big)_{\mathbb{R}^N}\Big)\,M(d\mathbf{y}).\end{aligned}$$

PROOF: Referring to the partition of unity described above, define $\Lambda_m\varphi=A(\chi_m\varphi)$ for $\varphi\in C^\infty\big(\overline{B(\mathbf{0},2^{-m+1})}\setminus B(\mathbf{0},2^{-m-2});\mathbb{R}\big)$, where

$$\chi_m\varphi(\mathbf{y})=\begin{cases}\chi_m(\mathbf{y})\varphi(\mathbf{y}) & \text{if } 2^{-m-2}\le|\mathbf{y}|\le 2^{-m+1}\\ 0 & \text{otherwise.}\end{cases}$$

Clearly $\Lambda_m$ is linear. In addition, if $\varphi\ge 0$, then $\chi_m\varphi\ge 0=\chi_m\varphi(\mathbf{0})$, and so, by (3.2.11), $\Lambda_m\varphi\ge 0$. Similarly, for any $\varphi\in C^\infty\big(\overline{B(\mathbf{0},2^{-m+1})}\setminus B(\mathbf{0},2^{-m-2});\mathbb{R}\big)$, $\|\varphi\|_{\mathrm{u}}\chi_m\pm\chi_m\varphi\ge 0=\big(\|\varphi\|_{\mathrm{u}}\chi_m\pm\chi_m\varphi\big)(\mathbf{0})$, and therefore $|\Lambda_m\varphi|\le K_m\|\varphi\|_{\mathrm{u}}$, where $K_m = A\chi_m$. Hence, $\Lambda_m$ admits a unique extension as a continuous linear functional on $C\big(\overline{B(\mathbf{0},2^{-m+1})}\setminus B(\mathbf{0},2^{-m-2});\mathbb{R}\big)$ that is non-negativity preserving and has norm $K_m$; and so, by the Riesz Representation Theorem, we now know that there is a unique non-negative Borel measure $M_m$ on $\mathbb{R}^N$ such that $M_m$ is supported on $\overline{B(\mathbf{0},2^{-m+1})}\setminus B(\mathbf{0},2^{-m-2})$, $K_m = M_m(\mathbb{R}^N)$, and $A(\chi_m\varphi) = \int_{\mathbb{R}^N}\varphi(\mathbf{y})\,M_m(d\mathbf{y})$ for all $\varphi\in\mathscr{S}(\mathbb{R}^N;\mathbb{R})$.

Now define the non-negative Borel measure $M$ on $\mathbb{R}^N$ by $M = \sum_{m\in\mathbb{Z}} M_m$. Clearly, $M(\{\mathbf{0}\}) = 0$. In addition, if $\varphi\in C_c^\infty(\mathbb{R}^N\setminus\{\mathbf{0}\};\mathbb{R})$, then there is an $n\in\mathbb{Z}^+$ such that $\chi_m\varphi\equiv 0$ unless $|m|\le n$. Thus,

$$
\begin{aligned}
A\varphi &= \sum_{m=-n}^{n} A(\chi_m\varphi) = \sum_{m=-n}^{n}\int_{\mathbb{R}^N}\varphi(\mathbf{y})\,M_m(d\mathbf{y})\\
&= \int_{\mathbb{R}^N}\left(\sum_{m=-n}^{n}\chi_m(\mathbf{y})\varphi(\mathbf{y})\right)M(d\mathbf{y}) = \int_{\mathbb{R}^N}\varphi(\mathbf{y})\,M(d\mathbf{y}),
\end{aligned}
$$

and therefore

$$
A\varphi = \int_{\mathbb{R}^N}\varphi(\mathbf{y})\,M(d\mathbf{y}) \tag{3.2.17}
$$

for $\varphi\in C_c^\infty(\mathbb{R}^N\setminus\{\mathbf{0}\};\mathbb{R})$.

Before taking the next step, observe that, as an application of (3.2.11), if $\varphi_1,\varphi_2\in\mathbf{D}$, then

$$
\varphi_1\le\varphi_2 \text{ and } \varphi_1(\mathbf{0}) = \varphi_2(\mathbf{0}) \implies A\varphi_1\le A\varphi_2. \tag{*}
$$

Indeed, by linearity, this reduces to the observation that, by (3.2.11), if $\varphi\in\mathbf{D}$ is non-negative and $\varphi(\mathbf{0}) = 0$, then $A\varphi\ge 0$.

With these preparations, I can show that, for any $\varphi\in\mathbf{D}$,

$$
\varphi\ge 0 = \varphi(\mathbf{0}) \implies \int_{\mathbb{R}^N}\varphi(\mathbf{y})\,M(d\mathbf{y})\le A\varphi. \tag{**}
$$

To check this, apply (*) to $\varphi_n = \sum_{m=-n}^{n}\chi_m\varphi$ and $\varphi$, and use (3.2.17) together with the Monotone Convergence Theorem to conclude that

$$
\int_{\mathbb{R}^N}\varphi(\mathbf{y})\,M(d\mathbf{y}) = \lim_{n\to\infty}\int_{\mathbb{R}^N}\varphi_n(\mathbf{y})\,M(d\mathbf{y}) = \lim_{n\to\infty}A\varphi_n\le A\varphi.
$$

Now let $\eta$ be as in the statement of the lemma, and set $\eta_R(\mathbf{y}) = \eta(R^{-1}\mathbf{y})$ for $R>0$. By (**) with $\varphi(\mathbf{y}) = |\mathbf{y}|^2\eta(\mathbf{y})$ we know that

$$
\int_{\mathbb{R}^N}|\mathbf{y}|^2\eta(\mathbf{y})\,M(d\mathbf{y})\le A\varphi<\infty.
$$

At the same time, by (3.2.17) and (*),

$$\int_{\mathbb{R}^N} (1-\eta(\mathbf{y}))\eta_R(\mathbf{y})\, M(d\mathbf{y}) \le A(\mathbf{1}-\eta)$$

for all $R > 0$, and therefore, by Fatou's Lemma,

$$\int_{\mathbb{R}^N} (1-\eta(\mathbf{y}))\, M(d\mathbf{y}) \le A(\mathbf{1}-\eta) < \infty.$$

Hence, I have proved that $M \in \mathfrak{M}_2(\mathbb{R}^N)$.

I am now in a position to show that (3.2.17) continues to hold for any $\varphi \in \mathscr{S}(\mathbb{R}^N;\mathbb{R})$ that vanishes along with its first and second order derivatives at $\mathbf{0}$. To this end, first suppose that $\varphi$ vanishes in a neighborhood of $\mathbf{0}$. Then, for each $R > 0$, (3.2.17) applies to $\eta_R\varphi$, and so

$$\int_{\mathbb{R}^N} \eta_R(\mathbf{y})\varphi(\mathbf{y})\, M(d\mathbf{y}) = A(\eta_R\varphi) = A\varphi + A\big((\mathbf{1}-\eta_R)\varphi\big).$$

By (*) applied to $\pm(\mathbf{1}-\eta_R)\varphi$ and $(\mathbf{1}-\eta_R)\|\varphi\|_{\mathrm{u}}$,

$$\big|A\big((\mathbf{1}-\eta_R)\varphi\big)\big| \le \|\varphi\|_{\mathrm{u}} A(\mathbf{1}-\eta_R) = -\|\varphi\|_{\mathrm{u}} A\eta_R \longrightarrow 0 \quad \text{as } R\to\infty,$$

where I used (3.2.12) to get the limit assertion. Thus,

$$A\varphi = \lim_{R\to\infty} \int_{\mathbb{R}^N} \eta_R(\mathbf{y})\varphi(\mathbf{y})\, M(d\mathbf{y}) = \int_{\mathbb{R}^N} \varphi(\mathbf{y})\, M(d\mathbf{y}),$$

because, since $M$ is finite on the support of $\varphi$ and therefore $\varphi$ is $M$-integrable, Lebesgue's Dominated Convergence Theorem applies. I still have to replace the assumption that $\varphi$ vanishes in a neighborhood of $\mathbf{0}$ by the assumption that it vanishes to second order there. For this purpose, first note that, by (3.2.13), $\varphi$ is certainly $M$-integrable, and therefore

$$\int_{\mathbb{R}^N} \varphi(\mathbf{y})\, M(d\mathbf{y}) = \lim_{R\searrow 0} A\big((\mathbf{1}-\eta_R)\varphi\big) = A\varphi - \lim_{R\searrow 0} A(\eta_R\varphi).$$

By our assumptions about $\varphi$ at $\mathbf{0}$, we can find a $C < \infty$ such that $|\eta_R\varphi(\mathbf{y})| \le CR|\mathbf{y}|^2\eta(\mathbf{y})$ for all $R \in (0,1]$. Hence, by (*) and the $M$-integrability of $|\mathbf{y}|^2\eta(\mathbf{y})$, there is a $C' < \infty$ such that $|A(\eta_R\varphi)| \le C'R$ for small $R > 0$, and therefore $A(\eta_R\varphi) \longrightarrow 0$ as $R \searrow 0$.

To complete the proof from here, let $\varphi \in \mathscr{S}(\mathbb{R}^N;\mathbb{R})$ be given, and set

$$\tilde{\varphi}(\mathbf{x}) = \varphi(\mathbf{x}) - \varphi(\mathbf{0}) - \eta(\mathbf{x})\big(\mathbf{x}, \nabla\varphi(\mathbf{0})\big)_{\mathbb{R}^N} - \tfrac{1}{2}\eta(\mathbf{x})^2\big(\mathbf{x}, \nabla^2\varphi(\mathbf{0})\mathbf{x}\big)_{\mathbb{R}^N}.$$

Then, by the preceding, (3.2.17) holds for $\tilde{\varphi}$ and, after one re-arranges terms, says that (3.2.16) holds. Thus, the properties of $\mathbf{C}$ are all that remain to be proved. That $\mathbf{C}$ is symmetric requires no comment. In addition, from (*), it is clearly non-negative definite. Finally, to see that it is independent of the $\eta$ chosen, let $\eta'$ be a second choice, note that $\eta'_{\boldsymbol{\xi}} = \eta_{\boldsymbol{\xi}}$ in a neighborhood of $\mathbf{0}$, and apply (3.2.17). □

REMARK 3.2.18. A careful examination of the proof of Lemma 3.2.14 reveals a lot. Specifically, it shows why the operation performed by the linear functional $A$ cannot be of order greater than 2. The point is, that, because of the minimum principle, $A$ acts as a bounded, non-negative linear functional on the difference between $\varphi$ and its second order Taylor polynomial, and, because of quasi-locality, this action can be represented by integration against a non-negative measure. The reason why the second order Taylor polynomial suffices is that second order polynomials are, apart from constants, the lowest order polynomials that can have a definite sign.

In order to complete the program, I need to introduce the notion of a **Lévy system**, which is a triple $(\mathbf{m}, \mathbf{C}, M)$ consisting of an $\mathbf{m} \in \mathbb{R}^N$, a symmetric, non-negative definite transformation $\mathbf{C}$ on $\mathbb{R}^N$, and a Lévy measure $M \in \mathfrak{M}_2(\mathbb{R}^N)$. Given a Lévy system $(\mathbf{m}, \mathbf{C}, M)$ and a Borel measurable $\eta : \mathbb{R}^N \longrightarrow [0,1]$ satisfying

$$\left( \sup_{\mathbf{y} \in B(\mathbf{0},1)\setminus\{\mathbf{0}\}} |\mathbf{y}|^{-1}(1-\eta(\mathbf{y})) \right) \vee \left( \sup_{\mathbf{y} \notin B(\mathbf{0},1)} \eta(\mathbf{y})|\mathbf{y}| \right) < \infty, \tag{3.2.19}$$

we will need to know that

$$\begin{gathered} \frac{1}{1+|\boldsymbol{\xi}|^2} \int_{\mathbb{R}^N} \left| e^{\sqrt{-1}(\boldsymbol{\xi},\mathbf{y})_{\mathbb{R}^N}} - 1 - \sqrt{-1}\eta(\mathbf{y})(\boldsymbol{\xi},\mathbf{y})_{\mathbb{R}^N} \right| M(d\mathbf{y}) \\ \text{is bounded and tends to 0 as } |\boldsymbol{\xi}| \to \infty. \end{gathered} \tag{3.2.20}$$

To see this, note that, for each $r \in (0,1]$,

$$\begin{aligned} &\int_{\mathbb{R}^N} \left| e^{\sqrt{-1}(\boldsymbol{\xi},\mathbf{y})_{\mathbb{R}^N}} - 1 - \sqrt{-1}\eta(\mathbf{y})(\boldsymbol{\xi},\mathbf{y})_{\mathbb{R}^N} \right| M(d\mathbf{y}) \\ &\le \int_{B(\mathbf{0},r)} \left| e^{\sqrt{-1}(\boldsymbol{\xi},\mathbf{y})_{\mathbb{R}^N}} - 1 - \sqrt{-1}(\boldsymbol{\xi},\mathbf{y})_{\mathbb{R}^N} \right| M(d\mathbf{y}) \\ &\quad + |\boldsymbol{\xi}| \int_{B(\mathbf{0},r)} (1-\eta(\mathbf{y}))|\mathbf{y}|\, M(d\mathbf{y}) + \int_{B(\mathbf{0},r)\complement} (2 + |\boldsymbol{\xi}|\eta(\mathbf{y})|\mathbf{y}|)\, M(\mathbf{y}) \\ &\le \frac{|\boldsymbol{\xi}|^2}{2} \int_{B(\mathbf{0},r)} |\mathbf{y}|^2 \, M(d\mathbf{y}) + |\boldsymbol{\xi}| \int_{B(\mathbf{0},r)} \frac{1-\eta(\mathbf{y})}{\mathbf{y}|} |\mathbf{y}|^2 \, M(d\mathbf{y}) \\ &\quad + |\boldsymbol{\xi}| \int_{B(\mathbf{0},r)\complement} \eta(\mathbf{y})|\mathbf{y}|\, M(d\mathbf{y}) + 2M(B(\mathbf{0},r)\complement). \end{aligned}$$

Obviously, this proves the boundedness in (3.2.20). In addition, it shows that, for each $r \in (0,1]$, the limit there as $|\boldsymbol{\xi}| \to \infty$ is dominated by $\frac{1}{2}\int_{B(\mathbf{0},r)} |\mathbf{y}|^2 \, M(d\mathbf{y})$, which tends to 0 as $r \searrow 0$.

Knowing (3.2.20), we can define

$$\begin{aligned}\ell^\eta_{(\mathbf{m},\mathbf{C},M)}(\boldsymbol{\xi}) =&\sqrt{-1}(\mathbf{m},\boldsymbol{\xi})_{\mathbb{R}^N} - \tfrac{1}{2}(\boldsymbol{\xi},\mathbf{C}\boldsymbol{\xi})_{\mathbb{R}^N}\\ &+\int_{\mathbb{R}^N}\Big(e^{\sqrt{-1}(\boldsymbol{\xi},\mathbf{y})_{\mathbb{R}^N}} - 1 - \sqrt{-1}\eta(\mathbf{y})(\boldsymbol{\xi},\mathbf{y})_{\mathbb{R}^N}\Big)\,M(d\mathbf{y})\end{aligned} \tag{3.2.21}$$

for any Lévy system $(\mathbf{m},\mathbf{C},M)$ and any Borel measurable $\eta:\mathbb{R}^N\longrightarrow[0,1]$ that satisfies (3.2.19). Furthermore, because $\ell^\eta_{(\mathbf{m},\mathbf{C},M_r)}\longrightarrow\ell^\eta_{(\mathbf{m},\mathbf{C},M)}$ uniformly on compacts when $M_r(d\mathbf{y}) = \mathbf{1}_{[r,\infty)}(|\mathbf{y}|)\,M(d\mathbf{y})$, it is clear that $\ell^\eta_{(\mathbf{m},\mathbf{C},M)}$ is continuous.

Theorem 3.2.22 (**Lévy–Khinchine**). *For each $\mu\in\mathcal{I}(\mathbb{R}^N)$, there is a unique $\ell_\mu\in C(\mathbb{R}^N;\mathbb{C})$ such that $\ell_\mu(\mathbf{0})=0$ and $\hat\mu = e^{\ell_\mu}$, and, for each $n\in\mathbb{Z}^+$, $e^{\frac{1}{n}\ell_\mu}$ is the Fourier transform of the unique $\mu_{\frac{1}{n}}\in\mathbf{M}_1(\mathbb{R}^N)$ satisfying $\mu=\mu_{\frac{1}{n}}^{\star n}$. Next, let $\eta:\mathbb{R}^N\longrightarrow[0,1]$ be a Borel measurable function that satisfies (3.2.19). Then, for each $\mu\in\mathcal{I}(\mathbb{R}^N)$, there is a unique Lévy system $(\mathbf{m}^\eta_\mu,\mathbf{C}_\mu,M_\mu)$ such that $\ell_\mu = \ell^\eta_{(\mathbf{m}^\eta_\mu,\mathbf{C}_\mu,M_\mu)}$, and, for each Lévy system $(\mathbf{m},\mathbf{C},M)$, there is a unique $\mu\in\mathcal{I}(\mathbb{R}^N)$ such that $\ell_\mu=\ell^\eta_{(\mathbf{m},\mathbf{C},M)}$. In fact, if $\mu\in\mathcal{I}(\mathbb{R}^N)$, then*

$$\int_{\mathbb{R}^N}\varphi(\mathbf{y})\,M_\mu(d\mathbf{y}) = \lim_{n\to\infty} n\langle\varphi,\mu_{\frac{1}{n}}\rangle$$

*for all $\varphi\in\mathscr{S}(\mathbb{R}^N;\mathbb{C})$ that satisfy $\lim_{|\mathbf{y}|\searrow 0}|\mathbf{y}|^{-2}|\varphi(\mathbf{y})|=0$,*

$$\mathbf{C}_\mu = \lim_{n\to\infty} n\int_{\mathbb{R}^N}\eta_0(\mathbf{y})^2\mathbf{y}\otimes\mathbf{y}\,\mu_{\frac{1}{n}}(d\mathbf{y}) - \int_{\mathbb{R}^N}\eta_0(\mathbf{y})^2\mathbf{y}\otimes\mathbf{y}\,M_\mu(d\mathbf{y}),$$

*and*

$$\mathbf{m}^{\eta_0}_\mu = \lim_{n\to\infty} n\int_{\mathbb{R}^N}\eta_0(\mathbf{y})\mathbf{y}\,\mu_{\frac{1}{n}}(d\mathbf{y})$$

*for any if $\eta_0\in C^\infty_c\big(\mathbb{R}^N;[0,1]\big)$ satisfying $\eta_0=1$ in a neighborhood of $\mathbf{0}$. Finally, for any Borel measurable $\eta:\mathbb{R}^N\longrightarrow[0,1]$ satisfying (3.2.19),*

$$\mathbf{m}^\eta_\mu = \mathbf{m}^{\eta_0}_\mu + \int_{\mathbb{R}^N}\big(\eta(\mathbf{y})-\eta_0(\mathbf{y})\big)\,M_\mu(d\mathbf{y}).$$

Proof: The initial assertion is covered by Theorem 3.2.7.

To prove the second assertion, let $\eta\in C^\infty_c\big(\mathbb{R}^N;[0,1]\big)$ with $\eta=1$ in $B(\mathbf{0},1)$ be given. For $\mu\in\mathcal{I}(\mathbb{R}^N)$, I will show that $\ell_\mu=\ell^\eta_{(\mathbf{m}^\eta,\mathbf{C},M)}$, where $\mathbf{m}^\eta$, $\mathbf{C}$, and $M$ are determined from (cf. (3.2.10)) $A_\mu$ as in Lemma 3.2.14. To this end, define $e_{\boldsymbol{\xi}}$ for $\boldsymbol{\xi}\in\mathbb{R}^N$ by $e_{\boldsymbol{\xi}}(\mathbf{x}) = e^{\sqrt{-1}(\boldsymbol{\xi},\mathbf{x})_{\mathbb{R}^N}}$, and set $\eta_R(\mathbf{x})=\eta(R^{-1}\mathbf{x})$ for $R>0$. The idea is to show that, as $R \to \infty$, $A_\mu(\eta_R e_{\boldsymbol{\xi}})$ tends to both $\ell_\mu(\boldsymbol{\xi})$ and to $\ell^\eta_{(\mathbf{m}^\eta,\mathbf{C},M)}(\boldsymbol{\xi})$. To check the first of these, use (3.2.10) to see that

$$(2\pi)^N A_\mu(\eta_R e_{\boldsymbol{\xi}}) = \int_{\mathbb{R}^N} \overline{\ell_\mu(\boldsymbol{\xi}')\widehat{\eta_R}(\boldsymbol{\xi}'+\boldsymbol{\xi})}\, d\boldsymbol{\xi}' = \int_{\mathbb{R}^N} \overline{\ell_\mu(R^{-1}\boldsymbol{\xi}'-\boldsymbol{\xi})\hat{\eta}(\boldsymbol{\xi}')}\, d\boldsymbol{\xi}'.$$

Hence, since $\ell_\mu$ is continuous and, by Lemma 3.2.9, $\sup_{R\ge 1}|\ell_\mu(R^{-1}\boldsymbol{\xi})\hat{\eta}(\boldsymbol{\xi})|$ is rapidly decreasing, Lebesgue's Dominated Convergence Theorem says that

$$\lim_{R\to\infty} A_\mu(\eta_R e_{\boldsymbol{\xi}}) = \overline{\ell_\mu(-\boldsymbol{\xi})}(2\pi)^{-N} \lim_{R\to\infty}\int_{\mathbb{R}^N} \hat{\eta}_R(\boldsymbol{\xi}')\, d\boldsymbol{\xi}' = \ell_\mu(\boldsymbol{\xi}).$$

To prove that $A_\mu(\eta_R e_{\boldsymbol{\xi}})$ also tends to $\ell^\eta_{(\mathbf{m}^\eta,\mathbf{C},M)}(\boldsymbol{\xi})$, use (3.2.16) to write

$$A_\mu(\eta_R e_{\boldsymbol{\xi}}) = \ell^\eta_{(\mathbf{m}^\eta,\mathbf{C},M)}(\boldsymbol{\xi}) - \sqrt{-1}\int_{\mathbb{R}^N} \big(1-\eta_R(\mathbf{y})\big) e_{\boldsymbol{\xi}}(\mathbf{y})\, M(d\mathbf{y}),$$

and observe that the last term is dominated by $M\big(B(\mathbf{0},R)\complement\big) \longrightarrow 0$.

So far we know that, for each $\mu \in \mathcal{I}(\mathbb{R}^N)$, there is a Lévy system $(\mathbf{m}^\eta,\mathbf{C},M)$ such that $\ell_\mu(\boldsymbol{\xi}) = \ell^\eta_{(\mathbf{m}^\eta,\mathbf{C},M)}$. Moreover, in the preliminary discussion at the beginning of this subsection, it was shown that, for each Lévy system $(\mathbf{m},\mathbf{C},M)$, there exists a $\mu \in \mathcal{I}(\mathbb{R}^N)$ for which $\ell^\eta_{(\mathbf{m},\mathbf{C},M)} = \ell_\mu$.

Finally, let $\eta_0$ be as in the statement of this theorem. Given $\mu \in \mathcal{I}(\mathbb{R}^N)$, let $\mathbf{m}^{\eta_0}_\mu \in \mathbb{R}^N$, $\mathbf{C}_\mu \in \mathrm{Hom}(\mathbb{R}^N;\mathbb{R}^N)$, and $M_\mu \in \mathfrak{M}_2(\mathbb{R}^N)$ be associated with $A_\mu$ as in (3.2.16) of Lemma 3.2.14 when $\eta = \eta_0$. As we have just seen, $\ell_\mu = \ell^{\eta_0}_{(\mathbf{m}^{\eta_0}_\mu,\mathbf{C}_\mu,M_\mu)}$. Further, by that lemma and (3.2.10), we know that

$$\int_{\mathbb{R}^N} \varphi(\mathbf{y})\, M_\mu(d\mathbf{y}) = A_\mu\varphi = \lim_{n\to\infty} n\langle \varphi, \mu_{\frac{1}{n}}\rangle$$

for any $\varphi \in \mathscr{S}(\mathbb{R}^N;\mathbb{R})$ that vanishes to second order at $\mathbf{0}$. In addition, by that same lemma and (3.2.10), we know that

$$\mathbf{C}_\mu = \lim_{n\to\infty} n\int_{\mathbb{R}^N} \eta_0(\mathbf{y})^2 \mathbf{y}\otimes\mathbf{y}\, \mu_{\frac{1}{n}}(d\mathbf{y}) - \int_{\mathbb{R}^N} \eta_0(\mathbf{y})^2\mathbf{y}\otimes\mathbf{y}\, M_\mu(d\mathbf{y})$$

and that

$$\mathbf{m}^{\eta_0}_\mu = \lim_{n\to\infty} n\int_{\mathbb{R}^N} \eta_0(\mathbf{y})\mathbf{y}\, \mu_{\frac{1}{n}}(d\mathbf{y}).$$

In particular, $\mathbf{m}^{\eta_0}$, $\mathbf{C}_\mu$, and $M_\mu$ are all uniquely determined by $\mu$ and $\eta_0$. In addition, if $\eta : \mathbb{R}^N \longrightarrow [0,1]$ is any other Borel measurable function satisfying (3.2.19), then the preceding combined with

$$\ell^\eta_{(\mathbf{m},\mathbf{C},M)}(\boldsymbol{\xi}) = \ell^{\eta_0}_{(\mathbf{m},\mathbf{C},M)}(\boldsymbol{\xi}) + \sqrt{-1}\int_{\mathbb{R}^N} \big(\eta_0(\mathbf{y})-\eta(\mathbf{y})\big)(\boldsymbol{\xi},\mathbf{y})_{\mathbb{R}^N}\, M(d\mathbf{y})$$

shows that $\ell^\eta_{(\mathbf{m},\mathbf{C},M)} = \ell_\mu$ if and only if $\mathbf{m} = \mathbf{m}^{\eta_0}_\mu + \int_{\mathbb{R}^N}\big(\eta(\mathbf{y})-\eta_0(\mathbf{y})\big)\, M(d\mathbf{y})$, $\mathbf{C} = \mathbf{C}_\mu$, and $M = M_\mu$. □

The expression in (3.2.21) for $\ell_\mu$ in terms of a Lévy system is known as the **Lévy–Khinchine formula**.

## Exercises for § 3.2

EXERCISE 3.2.23. Referring to (3.2.21), suppose that $\mu \in \mathcal{I}(\mathbb{R}^N)$ with $\ell_\mu = \ell^\eta_{(\mathbf{m},\mathbf{C},M)}$ for some Lévy system $(\mathbf{m}, \mathbf{C}, M)$ whose Lévy measure $M$ satisfies $\int_{|\mathbf{y}|\geq 1} e^{\lambda|\mathbf{y}|}\, M(d\mathbf{y}) < \infty$ for all $\lambda \in (0,\infty)$. Show that $\ell_\mu$ admits a unique extension as an analytic function on $\mathbb{C}^N$ and that $\ell_\mu(\boldsymbol{\xi})$ continues to be given by (3.2.21) when the $\mathbb{R}^N$-inner product of $(\xi_1,\dots,\xi_N) \in \mathbb{C}^N$ with $(\xi'_1,\dots,\xi'_N) \in \mathbb{C}^N$ is $\sum_{i=1}^N \xi_i\xi'_i$. Further, show that

$$\int_{\mathbb{R}^N} e^{(\boldsymbol{\xi},\mathbf{y})_{\mathbb{R}^N}}\, \mu(d\mathbf{y}) = e^{\ell_\mu(-\sqrt{-1}\boldsymbol{\xi})} \quad \text{for all } \boldsymbol{\xi} \in \mathbb{C}^N.$$

**Hint**: The first part is completely elementary complex analysis. To handle the second part, begin by arguing that it is enough to treat the cases when either $M = 0$ or $\mathbf{C} = 0$. The case $M = 0$ is trivial, and the case when $\mathbf{C} = \mathbf{0}$ can be further reduced to the one in which $\mu = \pi_M$ for an $M \in \mathfrak{M}_0(\mathbb{R}^N)$ with compact support in $\mathbb{R}^N \setminus \{\mathbf{0}\}$. Finally, use the representation $\pi_M = e^{-\alpha}\sum_{m=0}^\infty \frac{\alpha^m}{m!}\nu^{\star m}$ to complete the computation in this case.

EXERCISE 3.2.24. Given $\mu \in \mathcal{I}(\mathbb{R}^N)$ and knowing (3.2.20), show that

$$\big(\boldsymbol{\xi}, \mathbf{C}_\mu\boldsymbol{\xi}\big)_{\mathbb{R}^N} \equiv -2 \lim_{t\to\infty} t^{-2}\ell_\mu(t\boldsymbol{\xi}) \quad \text{for all } \mu \in \mathcal{I}(\mathbb{R}^N) \text{ and } \boldsymbol{\xi} \in \mathbb{R}^N.$$

Similarly, when $M_\mu \in \mathfrak{M}_1(\mathbb{R}^N)$, show that

$$\mathbf{m}_\mu \equiv \mathbf{m}^\eta_\mu - \int_{\mathbb{R}^N} \eta(\mathbf{y})\mathbf{y}\, M_\mu(d\mathbf{y})$$

is independent of the choice of $\eta$ satisfying (3.2.19) and, for each $\boldsymbol{\xi} \in \mathbb{R}^N$,

$$\big(\boldsymbol{\xi}, \mathbf{m}_\mu\big) = -\sqrt{-1} \lim_{t\to\infty} t^{-1}\Big(\ell_\mu(t\boldsymbol{\xi}) + \tfrac{t^2}{2}\big(\boldsymbol{\xi}, \mathbf{C}_\mu\boldsymbol{\xi}\big)_{\mathbb{R}^N}\Big) \quad \text{and}$$

$$\ell_\mu(\boldsymbol{\xi}) = -\tfrac{1}{2}\big(\boldsymbol{\xi}, \mathbf{C}_\mu\boldsymbol{\xi}\big)_{\mathbb{R}^N} + \sqrt{-1}\big(\boldsymbol{\xi}, \mathbf{m}_\mu\big)_{\mathbb{R}^N} + \int_{\mathbb{R}^N}\Big(e^{\sqrt{-1}(\boldsymbol{\xi},\mathbf{y})_{\mathbb{R}^N}} - 1\Big)\, M_\mu(d\mathbf{y}).$$

Finally, if $\mu \in \mathcal{I}(\mathbb{R}^N)$ is symmetric, show that $M_\mu$ is also symmetric and that

$$\ell_\mu(\boldsymbol{\xi}) = -\frac{1}{2}(\boldsymbol{\xi}, \mathbf{C}_\mu\boldsymbol{\xi}) + \int_{\mathbb{R}^N}\Big(\cos(\boldsymbol{\xi},\mathbf{y})_{\mathbb{R}^N} - 1\Big)\, M_\mu(d\mathbf{y}).$$

EXERCISE 3.2.25. Given $\mu \in \mathcal{I}(\mathbb{R})$, show that $\mu\big((-\infty,0)\big) = 0$ if and only if $C_\mu = 0$, $M_\mu \in \mathfrak{M}_1(\mathbb{R})$, $M_\mu\big((-\infty,0)\big) = 0$, and (cf. the preceding exercise) $m_\mu \geq 0$. The following are steps that you might follow.

**(i)** To prove the "if" assertion, set $M^r(dy) = \mathbf{1}_{[r,\infty)}(y)\, M_\mu(dy)$ for $r > 0$, and show that $\delta_{m_\mu} \star \pi_{M^r}\big((-\infty,0)\big) = 0$ for all $r > 0$ and $\delta_{m_\mu} \star \pi_{M^r} \Longrightarrow \mu$ as $r \searrow 0$. Conclude from these that $\mu\big((-\infty,0)\big) = 0$.

**(ii)** Now assume that $\mu\big((-\infty,0)\big) = 0$. To see that $C_\mu = 0$, show that if $\sigma > 0$, then $\gamma_{0,\sigma^2} \star \nu\big((-\infty,0)\big) > 0$ for any $\nu \in \mathbf{M}_1(\mathbb{R})$.

**(iii)** Continuing **(ii)**, show that $\mu\big((-\infty,0)\big) \ge \mu_{\frac{1}{n}}\big((-\infty,0)\big)^n$, and conclude first that $\mu_{\frac{1}{n}}\big((-\infty,0)\big) = 0$ for all $n \in \mathbb{Z}^+$ and then that

$$M_\mu\big((-\infty,0)\big) = 0 \text{ and } m_\mu^\eta \ge \int_{\mathbb{R}} \eta_0(y) y\, M_\mu(dy).$$

Finally, deduce from these that $M_\mu \in \mathfrak{M}_1(\mathbb{R})$ and that $m_\mu \ge 0$.

**(iv)** Suppose that $X \in N(0,1)$, and show that the distribution of $|X|$ cannot be infinitely divisible.

EXERCISE 3.2.26. The **Gamma distributions** is an interesting source of infinitely divisible laws. Namely, consider the family $\{\mu_t : t \in (0,\infty)\} \subseteq \mathbf{M}_1(\mathbb{R})$ given by

$$\mu_t(dx) = \mathbf{1}_{(0,\infty)}(x) \frac{x^{t-1}e^{-x}}{\Gamma(t)}\, dx.$$

**(i)** Show by direct computation that

$$\mu_s \star \mu_t(dx) = \frac{B(s,t)}{\Gamma(s)\Gamma(t)} \mathbf{1}_{(0,\infty)}(x) x^{s+t-1} e^{-x}\, dx,$$

where

$$B(s,t) \equiv \int_{(0,1)} \xi^{s-1}(1-\xi)^{t-1}\, d\xi$$

is Euler's **Beta function**, and conclude that $\mu_{s+t} = \mu_s \star \mu_t$. In particular, one gets, as a dividend, the famous identity $B(s,t) = \frac{\Gamma(s)\Gamma(t)}{\Gamma(s+t)}$.

**(ii)** As a consequence of **(i)**, we know that the $\mu_t$'s are infinitely divisible. Show that their Lévy–Khinchine representation is

$$\widehat{\mu_t}(\xi) = \exp\left[ t \int_{(0,\infty)} \left( e^{\sqrt{-1}\,\xi y} - 1 \right) e^{-y} \frac{dy}{y} \right].$$

EXERCISE 3.2.27. Given a $\mu \in \mathbf{M}_1(\mathbb{R}^N)$ for which there exists a strictly increasing sequence $\{n_m : m \ge 1\} \subseteq \mathbb{Z}^+$ and a sequence $\{\mu_{\frac{1}{n_m}} : m \ge 1\} \subseteq \mathbf{M}_1(\mathbb{R}^N)$ such that $\mu = \mu_{\frac{1}{n_m}}^{\star n_m}$ for all $m \ge 1$, show that $\mu \in \mathcal{I}(\mathbb{R}^N)$.

**Hint**: First use Lemma 3.2.4 to show that $\hat\mu$ never vanishes and therefore that there is a unique $\ell_\mu \in C(\mathbb{R}^N;\mathbb{C})$ such that $\ell_\mu(\mathbf{0}) = 0$ and $\hat\mu = e^{\ell_\mu}$. Next, proceed as in the proof of Theorem 3.2.7 to show that $\mu \in \overline{\mathcal{P}(\mathbb{R}^N)}$, and apply that theorem to conclude that $\mu \in \mathcal{I}(\mathbb{R}^N)$.

## § 3.3 Stable Laws

Recall from Exercise 2.3.23 the maps $T_\alpha : \mathbf{M}_1(\mathbb{R}^N) \longrightarrow \mathbf{M}_1(\mathbb{R}^N)$ given by the prescription

$$T_\alpha\mu(\Gamma) = \iint_{\mathbb{R}^N\times\mathbb{R}^N} \mathbf{1}_\Gamma\left(\frac{\mathbf{x}+\mathbf{y}}{2^{\frac{1}{\alpha}}}\right)\mu(d\mathbf{x})\mu(d\mathbf{y}),$$

and let $F_\alpha(\mathbb{R}^N)$ denote the set of non-trivial fixed points of $T_\alpha$. That is, $F_\alpha(\mathbb{R}^N) = \{\mu \in \mathbf{M}_1(\mathbb{R}^N) \setminus \{\delta_\mathbf{0}\} : \mu = T_\alpha\mu\}$. If $\mu \in F_\alpha(\mathbb{R}^N)$ and $\mu_{2^{-n}}$ denotes the distribution of $\mathbf{x} \rightsquigarrow 2^{-\frac{n}{\alpha}}\mathbf{x}$ under $\mu$, then $\mu = \mu_{2^{-n}}^{\star 2^n}$ for all $n$. Hence, by the result in Exercises 3.2.27, $\mu \in \mathcal{I}(\mathbb{R}^N)$, and so $F_\alpha(\mathbb{R}^N) \subseteq \mathcal{I}(\mathbb{R}^N)$ for all $\alpha \in (0,\infty)$. In this section, I will study the Lévy systems associated with elements of $F_\alpha(\mathbb{R}^N)$.

**§ 3.3.1. General Results.** Knowing that $F_\alpha(\mathbb{R}^N) \subseteq \mathcal{I}(\mathbb{R}^N)$, we can phrase the condition $\mu = T_\alpha\mu$ in terms of the associated Lévy systems. Namely, $\mu \in F_\alpha(\mathbb{R}^N)$ if and only if $\mu \in \mathcal{I}(\mathbb{R}^N) \setminus \{\delta_\mathbf{0}\}$ and $\ell_\mu(\boldsymbol{\xi}) = 2\ell_\mu(2^{-\frac{1}{\alpha}}\boldsymbol{\xi})$ for all $\boldsymbol{\xi} \in \mathbb{R}^N$. Next, using this and Exercise 3.2.24, we see that, for $\mu \in F_\alpha(\mathbb{R}^N)$,

$$\ell_\mu(\boldsymbol{\xi}) = 2^{-n}\ell_\mu(2^{\frac{n}{\alpha}}\boldsymbol{\xi}) = 2^{n(\frac{2}{\alpha}-1)}2^{-\frac{2n}{\alpha}}\ell_\mu(2^{\frac{n}{\alpha}}\boldsymbol{\xi}) \longrightarrow \begin{cases} 0 & \text{if } \alpha > 2 \\ -\frac{1}{2}(\boldsymbol{\xi}, \mathbf{C}_\mu\boldsymbol{\xi})_{\mathbb{R}^N} & \text{if } \alpha = 2 \end{cases}$$

as $n \to \infty$. Thus, we have already recovered the results in Exercises 2.3.21 and 2.3.23.

I next will examine $F_\alpha(\mathbb{R}^N)$ for $\alpha \in (0,2)$ in greater detail. For this purpose, define $\hat{T}_\alpha M$ for $M \in \mathfrak{M}_2(\mathbb{R}^N)$ to be the Borel measure determined by

$$\int_{\mathbb{R}^N} \varphi(\mathbf{y})\,\hat{T}_\alpha M(d\mathbf{y}) = 2\int_{\mathbb{R}^N} \varphi(2^{-\frac{1}{\alpha}}\mathbf{y})\,M(d\mathbf{y}) \tag{3.3.1}$$

for Borel measurable $\varphi : \mathbb{R}^N \longrightarrow [0,\infty)$. It is easy to check that $\hat{T}_\alpha$ maps $\mathfrak{M}_2(\mathbb{R}^N)$ into itself.

LEMMA 3.3.2. *For any $\alpha \in (0,2)$,*

$$\mu \in F_\alpha(\mathbb{R}^N) \cup \{\delta_\mathbf{0}\} \iff \begin{cases} \mathbf{C}_\mu = \mathbf{0}, \quad M_\mu = \hat{T}_\alpha M_\mu, \text{ and} \\ (1 - 2^{1-\frac{1}{\alpha}})\mathbf{m}_\mu^\eta = \int_{\mathbb{R}^N} \big(\eta(\mathbf{y}) - \eta(2^{\frac{1}{\alpha}}\mathbf{y})\big)\mathbf{y}\,M_\mu(d\mathbf{y}). \end{cases}$$

*In addition, if $M \in \mathfrak{M}_2(\mathbb{R}^N) \setminus \{0\}$ satisfies $M = \hat{T}_\alpha M$ for some $\alpha \in (0,2)$, then $M \in \mathfrak{M}_\beta(\mathbb{R}^N)$ for all $\beta > \alpha$ but $M \notin \mathfrak{M}_\alpha(\mathbb{R}^N)$.*

PROOF: From the uniqueness of the Lévy system associated with an element of $\mathcal{I}(\mathbb{R}^N)$, it is clear that, for any $\mu \in \mathcal{I}(\mathbb{R}^N)$, $M_{T_\alpha\mu} = \hat{T}_\alpha M_\mu$, $\mathbf{C}_{T_\alpha\mu} = 2^{1-\frac{2}{\alpha}}\mathbf{C}_\mu$, and

$$\mathbf{m}_{T_\alpha\mu}^\eta = 2^{1-\frac{1}{\alpha}}\mathbf{m}_\mu^\eta + \int_{\mathbb{R}^N} \big(\eta(\mathbf{y}) - \eta(2^{\frac{1}{\alpha}}\mathbf{y})\big)\mathbf{y}\,\hat{T}_\alpha M_\mu(d\mathbf{y}).$$

Hence, $\mu \in F_\alpha(\mathbb{R}^N) \cup \{\delta_{\mathbf{0}}\}$ if and only if $M_\mu = \hat{T}_\alpha M_\mu$, $\mathbf{C}_\mu = 2^{1-\frac{2}{\alpha}}\mathbf{C}_\mu$, and, for any $\eta$ satisfying (3.2.19),

$$(1 - 2^{1-\frac{1}{\alpha}})\mathbf{m}_\mu^\eta = \int_{\mathbb{R}^N} (\eta(\mathbf{y}) - \eta(2^{\frac{1}{\alpha}}\mathbf{y}))\mathbf{y}\, M_\mu(d\mathbf{y}).$$

In particular, when $\alpha \in (0,2)$, $\mathbf{C}_\mu = \mathbf{0}$, and so the first assertion follows.

The second assertion turns on the fact that, for all $n \in \mathbb{Z}^+$,

$$M = \hat{T}_\alpha M \implies M\big(\overline{B(\mathbf{0}, 2^{-\frac{n}{\alpha}})} \setminus B(\mathbf{0}, 2^{-\frac{n+1}{\alpha}})\big) = 2^n M\big(\overline{B(\mathbf{0},1)} \setminus B(\mathbf{0}, 2^{-\frac{1}{\alpha}})\big).$$

From this we see that $\kappa \equiv M\big(\overline{B(\mathbf{0},1)} \setminus B(\mathbf{0}, 2^{-\frac{1}{\alpha}})\big) > 0$ unless $M = 0$ and that the $M$-integral of $|\mathbf{y}|^\beta$ over $B(\mathbf{0},1)$ is bounded below by $2^{-1}\kappa \sum_{n=0}^\infty 2^{n(1-\frac{\beta}{\alpha})}$ and above by $\kappa \sum_{n=0}^\infty 2^{n(1-\frac{\beta}{\alpha})}$. $\square$

THEOREM 3.3.3. *$\mu \in F_2(\mathbb{R}^N)$ if and only if $\mu = \gamma_{\mathbf{0},\mathbf{C}}$ for some non-negative definite, symmetric $\mathbf{C} \in \mathrm{Hom}(\mathbb{R}^N;\mathbb{R}^N) \setminus \{0\}$. If $\alpha \in (1,2)$, then $\mu \in F_\alpha(\mathbb{R}^N)$ if and only if $\mu \in \mathcal{I}(\mathbb{R}^N)$ and $\ell_\mu(\boldsymbol{\xi})$ equals*

$$\frac{\sqrt{-1}}{1-2^{1-\frac{1}{\alpha}}} \int_{2^{-\frac{1}{\alpha}} < |\mathbf{y}| \le 1} (\boldsymbol{\xi}, \mathbf{y})_{\mathbb{R}^N}\, M(d\mathbf{y}) + \int_{\mathbb{R}^N} \Big(e^{\sqrt{-1}(\boldsymbol{\xi},\mathbf{y})_{\mathbb{R}^N}} - 1 - \sqrt{-1}\,\mathbf{1}_{[0,1]}(|\mathbf{y}|)(\boldsymbol{\xi},\mathbf{y})_{\mathbb{R}^N}\Big) M(d\mathbf{y})$$

*for some $M \in \Big(\bigcap_{\beta>\alpha} \mathfrak{M}_\beta(\mathbb{R}^N)\Big) \setminus \mathfrak{M}_\alpha(\mathbb{R}^N)$ satisfying $M = \hat{T}_\alpha M$. If $\alpha \in (0,1)$, then $\mu \in F_\alpha(\mathbb{R}^N)$ if and only if $\mu \in \mathcal{I}(\mathbb{R}^N)$ and $\ell_\mu(\boldsymbol{\xi})$ equals*

$$\int_{\mathbb{R}^N} \Big(e^{\sqrt{-1}(\boldsymbol{\xi},\mathbf{y})_{\mathbb{R}^N}} - 1\Big) M(d\mathbf{y})$$

*for some $M \in \Big(\bigcap_{\beta>\alpha} \mathfrak{M}_\beta(\mathbb{R}^N)\Big) \setminus \mathfrak{M}_\alpha(\mathbb{R}^N)$ satisfying $M = \hat{T}_\alpha M$. Finally, $\mu \in F_1(\mathbb{R}^N)$ if and only if $\mu \in \mathcal{I}(\mathbb{R}^N)$ and either $\mu = \delta_{\mathbf{m}}$ for some $\mathbf{m} \in \mathbb{R}^N \setminus \{\mathbf{0}\}$ or*

$$\ell_\mu(\boldsymbol{\xi}) = \sqrt{-1}(\mathbf{m}, \boldsymbol{\xi})_{\mathbb{R}^N} + \int_{\mathbb{R}^N} \Big(e^{\sqrt{-1}(\boldsymbol{\xi},\mathbf{y})_{\mathbb{R}^N}} - 1 - \sqrt{-1}\,\mathbf{1}_{[0,1]}(|\mathbf{y}|)(\boldsymbol{\xi},\mathbf{y})_{\mathbb{R}^N}\Big) M(d\mathbf{y})$$

*for some $\mathbf{m} \in \mathbb{R}^N$ and $M \in \Big(\bigcap_{\beta\in(1,2]} \mathfrak{M}_\beta(\mathbb{R}^N)\Big) \setminus \mathfrak{M}_1(\mathbb{R}^N)$ satisfying $M = \hat{T}_1 M$ and*

$$\int_{\frac{1}{2} < |\mathbf{y}| \le 1} \mathbf{y}\, M(d\mathbf{y}) = \mathbf{0}.$$

PROOF: The first assertion requires no comment. When $\alpha \in (0,2)$, the "if" assertions can be proved by checking that, in each case, $\ell_\mu(\boldsymbol{\xi}) = 2\ell_\mu(2^{-\frac{1}{\alpha}}\boldsymbol{\xi})$. When $\alpha \in [1,2)$, the "only if" assertion follows immediately from Lemma 3.3.2 with $\eta = \mathbf{1}_{\overline{B(\mathbf{0},1)}}$, and when $\alpha \in (0,1)$, it follows from that lemma combined with the observation that

$$M = \hat{T}_\alpha M \implies (1 - 2^{1-\frac{1}{\alpha}}) \int_{\overline{B(\mathbf{0},1)}} \mathbf{y}\, M(dy) = \int_{\{2^{-\frac{1}{\alpha}} < |\mathbf{y}| \le 1\}} \mathbf{y}\, M(d\mathbf{y}). \quad \square$$

§ **3.3.2. $\alpha$-Stable Laws.** The most studied elements of $F_\alpha(\mathbb{R}^N)$ are the **$\alpha$-stable laws**: those $\mu \in \mathcal{I}(\mathbb{R}^N)\backslash\{\delta_{\mathbf{0}}\}$ such that $\ell_\mu(t\boldsymbol{\xi}) = t^\alpha \ell_\mu(\boldsymbol{\xi})$ for all $t \in (0,\infty)$, not just for $t = 2^{\frac{1}{\alpha}}$. Equivalently, if $\mu \in \mathbf{M}_1(\mathbb{R}^N)$ is $\alpha$-stable if and only if $\mu \in \mathcal{I}(\mathbb{R}^N) \setminus \{\delta_{\mathbf{0}}\}$ and, for all non-negative, Borel measurable functions $\varphi$,

$$\int_{\mathbb{R}^N} \varphi(\mathbf{y})\, \mu_t(d\mathbf{y}) = \int_{\mathbb{R}^N} \varphi(t^{\frac{1}{\alpha}}\mathbf{y})\, \mu(d\mathbf{y}), \quad t \in (0,\infty),$$

where $\widehat{\mu_t}(\boldsymbol{\xi}) = e^{t\ell_\mu(\boldsymbol{\xi})}$. Thus, there are no $\alpha$-stable laws if $\alpha > 2$, and $\mu$ is 2-stable if and only if $\mu = \gamma_{\mathbf{0},\mathbf{C}}$ for some $\mathbf{C} \neq \mathbf{0}$. To examine the $\alpha$-stable laws when $\alpha \in (0,2)$, I will need the computations contained in the following lemmas.

LEMMA 3.3.4. *Assume that $M \in \mathfrak{M}_2(\mathbb{R}^N)$ and that $\alpha \in (0,2)$, and define the finite Borel measure $\nu$ on $\mathbb{S}^{N-1}$ by*

$$\langle \varphi, \nu \rangle = \frac{1}{\Gamma(2-\alpha)} \int_{\mathbb{R}^N \setminus \{\mathbf{0}\}} \varphi\left(\tfrac{\mathbf{y}}{|\mathbf{y}|}\right) |\mathbf{y}|^2 e^{-|\mathbf{y}|}\, M(d\mathbf{y})$$

*for bounded, Borel measurable $\varphi : \mathbb{S}^{N-1} \longrightarrow \mathbb{C}$. Then, $M$ satisfies*

$$\int_{\mathbb{R}^N} \varphi(t\mathbf{y})\, M(d\mathbf{y}) = t^\alpha \int_{\mathbb{R}^N} \varphi(\mathbf{y})\, M(d\mathbf{y}), \quad t \in (0,\infty) \tag{3.3.5}$$

*for all $\varphi \in C_c\big(\mathbb{R}^N \setminus \{\mathbf{0}\}; \mathbb{R}\big)$ if and only if*

$$\int_{\mathbb{R}^N} \varphi(\mathbf{y})\, M(d\mathbf{y}) = \int_{\mathbb{S}^{N-1}} \left( \int_{(0,\infty)} \varphi(r\boldsymbol{\omega}) \frac{dr}{r^{1+\alpha}} \right) \nu(d\boldsymbol{\omega})$$

*for all $\varphi \in C_c\big(\mathbb{R}^N \setminus \{\mathbf{0}\}; \mathbb{R}\big)$.*

PROOF: The "if" assertion is obvious. In addition, the "only if" assertion will follow once I prove it for $\varphi$'s such that $\varphi(\mathbf{y}) = \varphi_1\left(\frac{\mathbf{y}}{|\mathbf{y}|}\right)\varphi_2(|\mathbf{y}|)$, where $\varphi_1 \in C\big(\mathbb{S}^{N-1}; [0,\infty)\big)$ and $\varphi_2 \in C_c\big((0,\infty); \mathbb{R}\big)$. Given $\varphi_1 \in C\big(\mathbb{S}^{N-1}; [0,\infty)\big)$, determine the Borel measure $\rho$ on $(0,\infty)$ by

$$\langle \varphi_2, \rho \rangle = \int_{\mathbb{R}^N \setminus \{\mathbf{0}\}} \varphi_1\left(\tfrac{\mathbf{y}}{|\mathbf{y}|}\right) \varphi_2(|\mathbf{y}|) |\mathbf{y}|^2\, M(d\mathbf{y})$$

for $\varphi_2 \in C_c\big((0,\infty);\mathbb{R}\big)$. Then (3.3.5) implies that

$$\int_{(0,\infty)} e^{-tr}\,\rho(dr) = t^{\alpha-2}\int_{(0,\infty)} e^{-r}\,\rho(dr) = t^{\alpha-2}\Gamma(2-\alpha)\langle\varphi,\nu\rangle$$

for $t \in (0,\infty)$. Hence, since

$$\int_{(0,\infty)} r^{1-\alpha}e^{-tr}\,dr = \Gamma(2-\alpha)t^{\alpha-2}, \quad t \in (0,\infty),$$

uniqueness of the Laplace transform (cf. Exercise 1.2.12) implies that $\rho(dr) = \langle\varphi_1,\nu\rangle r^{1-\alpha}\,dr$, and therefore that

$$\int_{\mathbb{R}^N\setminus\{\mathbf{0}\}} \varphi_1\big(\tfrac{\mathbf{y}}{|\mathbf{y}|}\big)\varphi_2(|\mathbf{y}|)\,M(d\mathbf{y}) = \int_{(0,\infty)} \frac{\varphi_2(r)}{r^2}\,\rho(dr) = \langle\varphi_1,\nu\rangle\int_{(0,\infty)} \varphi_1(r)\,\frac{dr}{r^{1+\alpha}}. \quad \square$$

Lemma 3.3.6. *Let $\mu \in \mathcal{I}(\mathbb{R}^N)$. Then $\mu$ is 2-stable if and only if $\mu = \gamma_{\mathbf{0},\mathbf{C}}$ for some symmetric, non-negative definite $\mathbf{C} \neq \mathbf{0}$; $\mu$ is $\alpha$-stable for some $\alpha \in (0,1)$ if and only if there is a finite, non-negative Borel measure $\nu \neq 0$ on $\mathbb{S}^{N-1}$ such that*

$$\ell_\mu(\boldsymbol{\xi}) = \int_{\mathbb{S}^{N-1}}\left(\int_{(0,\infty)} \big[e^{\sqrt{-1}(\boldsymbol{\xi},r\boldsymbol{\omega})_{\mathbb{R}^N}} - 1\big]\,\frac{dr}{r^{1+\alpha}}\right)\nu(d\boldsymbol{\omega});$$

*$\mu$ is 1-stable if and only if there exists a finite, non-negative, Borel measure $\nu$ on $\mathbb{S}^{N-1}$ and an $\mathbf{m} \in \mathbb{R}^N$ satisfying*

$$|\mathbf{m}| + \nu(\mathbb{S}^{N-1}) > 0 \text{ and } \int_{\mathbb{S}^{N-1}} \boldsymbol{\omega}\,\nu(d\boldsymbol{\omega}) = \mathbf{0}$$

*such that $\ell_\mu(\boldsymbol{\xi})$ equals*

$$\sqrt{-1}(\boldsymbol{\xi},\mathbf{m})_{\mathbb{R}^N} + \int_{\mathbb{S}^{N-1}}\left(\int_{(0,\infty)} \big[e^{\sqrt{-1}(\boldsymbol{\xi},r\boldsymbol{\omega})_{\mathbb{R}^N}} - 1 - \sqrt{-1}\mathbf{1}_{[0,1]}(r)(\boldsymbol{\xi},r\boldsymbol{\omega})_{\mathbb{R}^N}\big]\,\frac{dr}{r^2}\right)\nu(d\boldsymbol{\omega});$$

*and $\mu$ is $\alpha$-stable for some $\alpha \in (1,2)$ if and only if there is a finite, non-negative, Borel measure $\nu \neq 0$ on $\mathbb{S}^{N-1}$ such that $\ell_\mu(\boldsymbol{\xi})$ equals*

$$\frac{\sqrt{-1}}{1-\alpha}\int_{\mathbb{S}^{N-1}} (\boldsymbol{\xi},\boldsymbol{\omega})_{\mathbb{R}^N}\,\nu(d\boldsymbol{\omega}) + \int_{\mathbb{S}^{N-1}}\left(\int_{(0,\infty)} \big[e^{\sqrt{-1}(\boldsymbol{\xi},r\boldsymbol{\omega})_{\mathbb{R}^N}} - 1 - \sqrt{-1}\mathbf{1}_{[0,1]}(r)(\boldsymbol{\xi},r\boldsymbol{\omega})_{\mathbb{R}^N}\big]\,\frac{dr}{r^{1+\alpha}}\right)\nu(d\boldsymbol{\omega}).$$

PROOF: The sufficiency part of each case is easy to check directly or as a consequence of Theorem 3.3.3. To prove the necessity, first check that if $\mu$ is $\alpha$-stable and therefore $\ell_\mu(t\boldsymbol{\xi}) = t^\alpha \ell_\mu(\boldsymbol{\xi})$, then $M$ must have the scaling property in (3.3.5) and therefore have the form described in Lemma 3.3.4. Second, when $M$ has this form, simply check that in each case the result in Theorem 3.3.3 translates into the result here. □

In the following, $\mathbb{C}_+$ denotes the open upper half-space $\{\zeta \in \mathbb{C} : \mathfrak{Im}(\zeta) > 0\}$ in $\mathbb{C}$, and $\overline{\mathbb{C}_+}$ denotes its closure $\{\zeta \in \mathbb{C} : \mathfrak{Im}(\zeta) \geq 0\}$. In addition, given $\zeta \in \mathbb{C}$ and $\alpha \in (0,2)$, we take $\zeta^\alpha \equiv |\zeta|^\alpha e^{\sqrt{-1}\alpha \arg\zeta}$, where $\arg\zeta$ is 0 if $\zeta = 0$ and is the unique $\theta \in (-\pi, \pi]$ such that $\zeta = |\zeta| e^{\sqrt{-1}\theta}$ if $\zeta \neq 0$.

LEMMA 3.3.7. *If $\alpha \in (0,1)$, then*

$$\int_{(0,\infty)} \frac{e^{\sqrt{-1}\zeta r} - 1}{r^{1+\alpha}}\, dr = -\frac{\Gamma(1-\alpha)}{\alpha} \left( \frac{\zeta}{\sqrt{-1}} \right)^\alpha \quad \text{for } \zeta \in \overline{\mathbb{C}_+}.$$

*In particular,*

$$a_\alpha \equiv \int_{(0,\infty)} \frac{\cos r - 1}{r^{1+\alpha}}\, dr = \begin{cases} \frac{\Gamma(2-\alpha)}{\alpha(\alpha-1)} \cos \frac{\alpha\pi}{2} & \text{if } \alpha \in (1,2) \\ -\frac{\Gamma(1-\alpha)}{\alpha} \cos \frac{\alpha\pi}{2} & \text{if } \alpha \in (0,1) \\ -\frac{\pi}{2} & \text{if } \alpha = 1 \end{cases}$$

*and*

$$b_\alpha \equiv \int_{(0,\infty)} \frac{\sin r}{r^{1+\alpha}}\, dr = \frac{\Gamma(1-\alpha)}{\alpha} \sin \frac{\alpha\pi}{2} \quad \text{if } \alpha \in (0,1).$$

PROOF: Let $f_\alpha(\zeta)$ denote the integral on the left-hand side of the first equation. Clearly $f_\alpha$ is continuous on $\overline{\mathbb{C}_+}$ and analytic on $\mathbb{C}_+$. In addition, for $\xi > 0$,

$$f_\alpha(\sqrt{-1}\xi) = \xi^\alpha \int_{(0,\infty)} \frac{e^{-r} - 1}{r^{1+\alpha}}\, dr = -\frac{\xi^\alpha}{\alpha} \int_{(0,\infty)} r^{-\alpha} e^{-r}\, dr = -\frac{\Gamma(1-\alpha)}{\alpha} \xi^\alpha.$$

Hence, since $-\frac{\Gamma(1-\alpha)}{\alpha}\left(\frac{\zeta}{\sqrt{-1}}\right)^\alpha$ is also analytic in $\mathbb{C}_+$, $f_\alpha(\zeta) = -\frac{\Gamma(1-\alpha)}{\alpha}\left(\frac{\zeta}{\sqrt{-1}}\right)^\alpha$ for $\zeta \in \mathbb{C}_+$, and, by continuity, this equality extends to $\overline{\mathbb{C}_+}$. Hence, $c = \frac{\Gamma(1-\alpha)}{\alpha}$ and $\theta = -\frac{\alpha\pi}{2}$.

When $\alpha \in (0,1)$, the values of $a_\alpha$ and $b_\alpha$ follow immediately from the evaluation of $f_\alpha(1)$. When $\alpha \in (1,2)$, one can find the value of $a_\alpha$ by first observing that

$$\int_{(0,\infty)} \frac{\cos(\xi r) - 1}{r^{1+\alpha}}\, dr = \xi^\alpha \int_{(0,\infty)} \frac{\cos r - 1}{r^{1+\alpha}}\, dr \quad \text{for } \xi \in (0,\infty),$$

and then differentiating this with respect to $\xi$ to get

$$\alpha \int_{(0,\infty)} \frac{\cos r - 1}{r^{1+\alpha}}\, dr = -\int_{(0,\infty)} \frac{\sin r}{r^\alpha}\, dr = -b_{\alpha-1}.$$

To evaluate $a_1$, simply note that

$$a_1 = \lim_{\alpha\searrow 1} a_\alpha = -\lim_{\alpha\searrow 1} \frac{\Gamma(2-\alpha)}{\alpha}\frac{\cos\frac{\alpha\pi}{2}}{1-\alpha} = -\frac{\pi}{2}. \quad \square$$

THEOREM 3.3.8. *Let $\mu \in \mathcal{I}(\mathbb{R}^N)$. If $\alpha \in (0,2)\setminus\{1\}$, then $\mu$ is $\alpha$-stable if and only if there exists a finite, non-negative, Borel measure $\nu \neq 0$ on $\mathbb{S}^{N-1}$ such that*

$$\ell_\mu(\boldsymbol{\xi}) = (-1)^{\mathbf{1}_{(0,1)}(\alpha)} \int_{\mathbb{S}^{N-1}} \left(\frac{(\boldsymbol{\xi},\boldsymbol{\omega})_{\mathbb{R}^N}}{\sqrt{-1}}\right)^\alpha \nu(d\boldsymbol{\omega}).$$

*On the other hand, $\mu$ is 1-stable if and only if there exist an $\mathbf{m} \in \mathbb{R}^N$ and a finite, non-negative, Borel measure $\nu$ on $\mathbb{S}^{N-1}$ such that $|\mathbf{m}| + \nu(\mathbb{S}^{N-1}) > 0$,*

$$\int_{\mathbb{S}^{N-1}} \boldsymbol{\omega}\,\nu(d\boldsymbol{\omega}) = \mathbf{0},$$

*and*

$$\ell_\mu(\boldsymbol{\xi}) = \sqrt{-1}(\boldsymbol{\xi},\mathbf{m})_{\mathbb{R}^N} - \sqrt{-1}\int_{\mathbb{S}^{N-1}} (\boldsymbol{\xi},\boldsymbol{\omega})_{\mathbb{R}^N} \log(\boldsymbol{\xi},\boldsymbol{\omega})_{\mathbb{R}^N}\,\nu(d\boldsymbol{\omega}),$$

*where $\zeta\log\zeta = \zeta\log|\zeta| + \sqrt{-1}\zeta\arg\zeta$ for $\zeta \in \mathbb{C}$.*

PROOF: When $\alpha \in (0,1)$, the conclusion is a simple application of the corresponding results in Lemmas 3.3.6 and 3.3.7. When $\alpha \in (1,2)$, one has to massage the corresponding expression in Lemma 3.3.6. Specifically, begin with the observation that

$$\begin{aligned}&\frac{\sqrt{-1}\xi}{1-\alpha} + \int_{(0,\infty)} \left[e^{\sqrt{-1}\xi r} - 1 - \sqrt{-1}\xi\mathbf{1}_{[0,1]}(r)r\right]\frac{dr}{r^{1+\alpha}}\\ &= |\xi|^\alpha\left(\int_{(0,\infty)}\left[e^{\sqrt{-1}\mathrm{sgn}(\xi)r} - 1 - \sqrt{-1}\mathrm{sgn}(\xi)\mathbf{1}_{[0,1]}(r)r\right]\frac{dr}{r^{1+\alpha}} + \frac{\sqrt{-1}\mathrm{sgn}(\xi)}{1-\alpha}\right)\end{aligned}$$

for $\xi \in \mathbb{R}$. Thus, we can write the expression for $\ell_\mu(\boldsymbol{\xi})$ as

$$\int_{\mathbb{S}^{N-1}} |(\boldsymbol{\xi},\boldsymbol{\omega})_{\mathbb{R}^N}|^\alpha g_\alpha(\mathrm{sgn}(\boldsymbol{\xi},\boldsymbol{\omega})_{\mathbb{R}^N})\,\nu(d\boldsymbol{\omega}),$$

where (cf. Lemma 3.3.7)

$$\begin{aligned}g_\alpha(\pm 1) &= \int_{(0,\infty)}\left[e^{\pm\sqrt{-1}r} - 1 \mp \sqrt{-1}\mathbf{1}_{[0,1]}(r)r\right]\frac{dr}{r^{1+\alpha}} \pm \frac{\sqrt{-1}}{1-\alpha}\\ &= a_\alpha \pm \sqrt{-1}\int_{(0,\infty)}\left(\sin r - \mathbf{1}_{[0,1]}(r)r\right)\frac{dr}{r^{1+\alpha}} \pm \frac{\sqrt{-1}}{1-\alpha}.\end{aligned}$$

Next use integration by parts over the intervals $(0,1]$ and $[1,\infty)$ to check that

$$\int_{(0,\infty)} \left(\sin r - \mathbf{1}_{[0,1]}(r)r\right)\frac{dr}{r^{1+\alpha}} = \frac{1}{\alpha-1} + \frac{1}{\alpha}\int_{(0,\infty)} \frac{\cos r - 1}{r^\alpha}\,dr = \frac{1}{\alpha-1} + \frac{a_{\alpha-1}}{\alpha}.$$

Hence, since $\frac{a_{\alpha-1}}{\alpha} = -\frac{\Gamma(2-\alpha)}{\alpha(\alpha-1)}\sin\frac{\alpha\pi}{2}$,

$$g_\alpha(\pm 1) = \frac{\Gamma(2-\alpha)}{\alpha(\alpha-1)}e^{\mp\frac{\alpha\pi}{2}},$$

and therefore

$$g_\alpha\left(\operatorname{sgn}(\mathbf{x},\boldsymbol{\omega})_{\mathbb{R}^N}\right)\left|(\boldsymbol{\xi},\boldsymbol{\omega})_{\mathbb{R}^N}\right|^\alpha = \frac{\Gamma(2-\alpha)}{\alpha(\alpha-1)}\left(\frac{(\boldsymbol{\xi},\boldsymbol{\omega})_{\mathbb{R}^N}}{\sqrt{-1}}\right)^\alpha.$$

Thus, all that we need to do is replace the $\nu$ in Theorem 3.3.8 by $\frac{1-\alpha}{\Gamma(1-\alpha)}\nu$.

Turning to the case $\alpha = 1$, note that, because of the mean zero condition on $\nu$,

$$\begin{aligned}
&\int_{\mathbb{S}^{N-1}}\left(\int_{(0,\infty)}\left[e^{\sqrt{-1}(\boldsymbol{\xi},\boldsymbol{\omega})_{\mathbb{R}^N} r} - 1 - \sqrt{-1}\mathbf{1}_{[0,1]}(r)r(\boldsymbol{\xi},\boldsymbol{\omega})_{\mathbb{R}^N}\right]\frac{dr}{r^2}\right)\nu(d\boldsymbol{\omega})\\
&\quad= \lim_{\alpha\nearrow 1}\int_{\mathbb{S}^{N-1}}\left(\int_{(0,\infty)}\left[e^{\sqrt{-1}(\boldsymbol{\xi},\boldsymbol{\omega})_{\mathbb{R}^N} r} - 1\right]\frac{dr}{r^{1+\alpha}}\right)\nu(d\boldsymbol{\omega})\\
&\quad= -\lim_{\alpha\nearrow 1}\frac{\Gamma(1-\alpha)}{\alpha}\int_{\mathbb{S}^{N-1}}\left(\frac{(\boldsymbol{\xi},\boldsymbol{\omega})_{\mathbb{R}^N}}{\sqrt{-1}}\right)^\alpha\nu(d\boldsymbol{\omega})\\
&\quad= \sqrt{-1}\lim_{\alpha\nearrow 1}\frac{1}{1-\alpha}\int_{\mathbb{S}^{N-1}}\left[(\boldsymbol{\xi},\boldsymbol{\omega})_{\mathbb{R}^N}^\alpha - (\boldsymbol{\xi},\boldsymbol{\omega})_{\mathbb{R}^N}\right]\nu(d\boldsymbol{\omega})\\
&\quad= -\sqrt{-1}\int_{\mathbb{S}^{N-1}}(\boldsymbol{\xi},\boldsymbol{\omega})_{\mathbb{R}^N}\log(\boldsymbol{\xi},\boldsymbol{\omega})_{\mathbb{R}^N}\,\nu(d\boldsymbol{\omega}),
\end{aligned}$$

where I have used $(1-\alpha)\Gamma(1-\alpha) = \Gamma(2-\alpha) \longrightarrow 1$. □

I close this section with a discussion of the most commonly encoutered stable laws.

COROLLARY 3.3.9. *For any $\alpha \in (0,2]$, $\mu$ is a symmetric and $\alpha$-stable law if and only if there is a finite, non-negative, symmetric, Borel measure $\nu \neq 0$ on $\mathbb{S}^{N-1}$ such that*

$$\ell_\mu(\boldsymbol{\xi}) = -\int_{\mathbb{S}^{N-1}}\left|(\boldsymbol{\xi},\boldsymbol{\omega})_{\mathbb{R}^N}\right|^\alpha\nu(d\boldsymbol{\omega}).$$

*Moreover, $\mu$ is a rotationally invariant, symmetric, $\alpha$-stable law if and only if $\ell_\mu(\boldsymbol{\xi}) = -t|\boldsymbol{\xi}|^\alpha$ for some $t \in (0,\infty)$.*

PROOF: If $\mu$ is 2-stable, then $\mu = \gamma_{\mathbf{0},\mathbf{C}}$ for some $\mathbf{C} \neq 0$ and is therefore symmetric. In addition, by defining $\nu$ on $\mathbb{S}^{N-1}$ so that

$$\langle \varphi, \nu \rangle = \frac{1}{2} \int_{\mathbb{R}^N} |\mathbf{y}|^2 \varphi\left(\frac{\mathbf{y}}{|\mathbf{y}|}\right) \gamma_{\mathbf{0},\mathbf{C}}(d\boldsymbol{\omega}),$$

we see that

$$\ell_\mu(\boldsymbol{\xi}) = -\frac{1}{2}(\boldsymbol{\xi}, \mathbf{C}\boldsymbol{\xi})_{\mathbb{R}^N} = \int_{\mathbb{S}^{N-1}} \left|(\boldsymbol{\xi}, \boldsymbol{\omega})_{\mathbb{R}^N}\right|^2 \nu(d\boldsymbol{\omega}).$$

If $\alpha \in (0,2) \setminus \{1\}$, then, for every non-zero, symmetric $\nu$ on $\mathbb{S}^{N-1}$,

$$-\int_{\mathbb{S}^{N-1}} \left|(\boldsymbol{\xi}, \boldsymbol{\omega})_{\mathbb{R}^N}\right|^\alpha \nu(d\boldsymbol{\omega}) = (-1)^{\mathbf{1}_{(0,1)}(\alpha)} \left|\csc \frac{\alpha\pi}{2}\right| \int_{\mathbb{S}^{N-1}} \left(\frac{(\boldsymbol{\xi}, \boldsymbol{\omega})_{\mathbb{R}^N}}{\sqrt{-1}}\right)^\alpha \nu(d\boldsymbol{\omega})$$

is $\ell_\mu(\boldsymbol{\xi})$ for a symmetric, $\alpha$-stable $\mu$. Conversely, if $\mu$ is symmetric and $\alpha$-stable for some $\alpha \in (0,1)$, then, because $\ell_\mu(\boldsymbol{\xi}) = \ell_\mu(-\boldsymbol{\xi})$, the associated $\nu$ in Theorem 3.3.8 can be chosen to be symmetric, in which case $\ell_\mu(\boldsymbol{\xi})$ equals

$$(-1)^{\mathbf{1}_{(0,1)}(\alpha)} \int_{\mathbb{S}^{N-1}} \left(\frac{(\boldsymbol{\xi}, \boldsymbol{\omega})_{\mathbb{R}^N}}{\sqrt{-1}}\right)^\alpha \nu(\boldsymbol{\omega}) = -\left|\cos \frac{\alpha\pi}{2}\right| \int_{\mathbb{S}^{N-1}} \left|(\boldsymbol{\xi}, \boldsymbol{\omega})_{\mathbb{R}^N}\right|^\alpha \nu(d\boldsymbol{\omega}).$$

To handle the case when $\alpha = 1$, first suppose that $\nu \neq 0$ on $\mathbb{S}^{N-1}$ is symmetric. Then

$$-\int_{\mathbb{S}^{N-1}} \left|(\boldsymbol{\xi}, \boldsymbol{\omega})_{\mathbb{R}^N}\right| \nu(d\boldsymbol{\omega}) = 2 \int_{\{\boldsymbol{\omega}:(\boldsymbol{\xi},\boldsymbol{\omega})_{\mathbb{R}^N}<0\}} (\boldsymbol{\xi}, \boldsymbol{\omega})_{\mathbb{R}^N} \nu(d\boldsymbol{\omega})$$

$$= \frac{2}{\pi\sqrt{-1}} \int_{\{\boldsymbol{\omega}:(\boldsymbol{\xi},\boldsymbol{\omega})_{\mathbb{R}^N}<0\}} \left[(\boldsymbol{\xi}, \boldsymbol{\omega})_{\mathbb{R}^N} \log(\boldsymbol{\xi}, \boldsymbol{\omega})_{\mathbb{R}^N} + (\boldsymbol{\xi}, -\boldsymbol{\omega})_{\mathbb{R}^N} \log(\boldsymbol{\xi}, -\boldsymbol{\omega})_{\mathbb{R}^N}\right] \nu(d\boldsymbol{\omega})$$

$$= -\sqrt{-1}\frac{1}{\pi} \int_{\mathbb{S}^{N-1}} (\boldsymbol{\xi}, \boldsymbol{\omega})_{\mathbb{R}^N} \log(\boldsymbol{\xi}, \boldsymbol{\omega})_{\mathbb{R}^N} \nu(d\boldsymbol{\omega}),$$

which is $\ell_\mu(\boldsymbol{\xi})$ for a symmetric, 1-stable $\mu$. Conversely, if $\mu$ is symmetric and 1-stable, one can use $\ell_\mu(\boldsymbol{\xi}) = \ell_\mu(-\boldsymbol{\xi})$ to see that $\mathbf{m} = 0$ and $\nu$ in the expression for $\ell_\mu(\boldsymbol{\xi})$ in Theorem 3.3.8 can be taken to be symmetric. Hence, by the preceding calculation, $\ell_\mu(\boldsymbol{\xi})$ has the desired form.

Finally, if $\mu$ is a rotationally invariant, symmetric, $\alpha$-stable law, then $\ell_\mu(\boldsymbol{\xi})$ is a rotationally invariant function of $\boldsymbol{\xi}$ and therefore the preceding leads to

$$\ell_\mu(\boldsymbol{\xi}) = -\int_{\mathbb{S}^{N-1}} \left(\int_{\mathbb{S}^{N-1}} |\boldsymbol{\xi}|^\alpha \left|(\boldsymbol{\omega}, \boldsymbol{\omega}')\right|^\alpha \nu(d\boldsymbol{\omega}')\right) \overline{\lambda_{\mathbb{S}^{N-1}}}(d\boldsymbol{\omega}) = -t|\boldsymbol{\xi}|^\alpha,$$

where $\overline{\lambda_{\mathbb{S}^{N-1}}}$ is normalized surface measure on $\mathbb{S}^{N-1}$ and

$$t = \nu(\mathbb{S}^{N-1}) \int_{\mathbb{S}^{N-1}} \left|(\mathbf{e}, \boldsymbol{\omega})_{\mathbb{R}^N}\right|^\alpha \overline{\lambda_{\mathbb{S}^{N-1}}}(d\boldsymbol{\omega})$$

for any $\mathbf{e} \in \mathbb{S}^{N-1}$. Conversely, by taking $\nu$ to be an appropriate multiple of $\overline{\lambda_{\mathbb{S}^{N-1}}}$, one sees that, for any $t \in (0, \infty)$, $-t|\boldsymbol{\xi}|^\alpha$ is $\ell_\mu(\boldsymbol{\xi})$ for a symmetric, $\alpha$-stable $\mu$. $\square$

## Exercises for § 3.3

EXERCISE 3.3.10. Given $\alpha \in (0,2)$, define $S_\alpha \nu$ for finite, non-negative, Borel measures $\nu$ on $\overline{B(\mathbf{0},1)} \setminus B(\mathbf{0},2^{-\frac{1}{\alpha}})$ by

$$S_\alpha \nu(\Gamma) = \sum_{m\in\mathbb{Z}} 2^{-m} \int_{\mathbb{R}^N} \mathbf{1}_\Gamma(2^{\frac{m}{\alpha}}\mathbf{y})\,\nu(d\mathbf{y}),$$

and show that this map is one-to-one and onto the set of $\mathbf{M} \in \mathfrak{M}_2(\mathbb{R}^N)$ satisfying (cf. (3.3.1)) $M = \hat{T}_\alpha M$. Conclude that, for each $\alpha \in (0,2)$, $F_\alpha(\mathbb{R}^N)$ contains lots of elements!

EXERCISE 3.3.11. Here are a few further properties of elements of $F_\alpha(\mathbb{R}^N)$.

**(i)** Show that there is $\mu \in F_\alpha(\mathbb{R}^N)$ such that $\mu(\{\mathbf{y} : (\mathbf{e},\mathbf{y})_{\mathbb{R}^N} < 0\}) = 0$ for some $\mathbf{e} \in \mathbb{S}^{N-1}$ if and only if $\alpha \in (0,1)$.

**Hint**: Reduce to the case when $N = 1$, and look at Exercise 3.2.24.

**(ii)** If $\mu \in F_1(\mathbb{R}^N)$, show that, for every $\mathbf{e} \in \mathbb{S}^{N-1}$, $\mu(\{\mathbf{y} : (\mathbf{e},\mathbf{y})_{\mathbb{R}^N} < 0\}) > 0 \iff \mu(\{\mathbf{y} : (\mathbf{e},\mathbf{y})_{\mathbb{R}^N} > 0\}) > 0$.

**(iii)** If $\alpha \in (1,2)$, show that for each $\epsilon > 0$ there is a $\mu \in F_\alpha(\mathbb{R})$ such that $\mu((-\infty,-\epsilon]) = 0$.

EXERCISE 3.3.12. Take $N = 1$. This exercise is about an important class of stable laws known as **one-sided stable laws**: stable laws that are supported on $[0,\infty)$.

**(i)** Show that there exists a one-sided $\alpha$-stable law only if $\alpha \in (0,1)$.

**(ii)** If $\alpha \in (0,1)$, show that $\mu$ is a one-sided $\alpha$-stable law if and only if $\ell_\mu(\xi) = -t\left(\frac{\xi}{\sqrt{-1}}\right)^\alpha$ for some $t \in (0,\infty)$.

**(iii)** Let $\alpha \in (0,1)$, and use $\nu_t^\alpha$ to denote the one-sided $\alpha$-stable law with $\ell_{\nu_t^\alpha}(\xi) = -t\left(\frac{\xi}{\sqrt{-1}}\right)^\alpha$. Show that

$$\int_{[0,\infty)} e^{\sqrt{-1}\zeta y}\,\nu_t^\alpha(dy) = \exp\left[-t\left(\frac{\zeta}{\sqrt{-1}}\right)^\alpha\right] \quad \text{for } \zeta \in \mathbb{C} \text{ with } \mathfrak{Im}(\zeta) \ge 0.$$

In particular, use Exercise 1.2.12 to conclude that $\nu_t^\alpha$ is characterized by the facts that it is supported on $[0,\infty)$ and its Laplace transform is given by

$$\int_{[0,\infty)} e^{-\lambda y}\nu_t^\alpha(dy) = e^{-t\lambda^\alpha}, \quad \lambda \ge 0.$$

EXERCISE 3.3.13. Given $\alpha \in (0,2]$, let $\mu_t^\alpha$ denote the symmetric $\alpha$-stable law, described in Corollary 3.3.9, with $\ell_{\mu_t^\alpha}(\boldsymbol{\xi}) = -t|\boldsymbol{\xi}|^\alpha$. Clearly $\mu_t^2 = \gamma_{\mathbf{0},2tI}$. When $\alpha \in (0,2)$, show that

$$\mu_t^\alpha = \int_{[0,\infty)} \gamma_{\mathbf{0},2\tau\mathbf{I}}\, \nu_t^{\frac{\alpha}{2}}(d\tau),$$

where $\nu_t^{\frac{\alpha}{2}}$ is the one-sided $\frac{\alpha}{2}$-stable law in part **(iii)** of the preceding exercise. This representation is an example of **subordination**, and, as we will see in Exercise 3.3.17, can be used to good effect.

EXERCISE 3.3.14. Because their Fourier transforms are rapidly decreasing, we know that each of the measures $\nu_t^\alpha$ in part **(iii)** of Exercise 3.3.11 admits a smooth density with respect to Lebesgue measure $\lambda_{\mathbb{R}}$ on $\mathbb{R}$. In this exercise, we examine these densities.

**(i)** For $\alpha \in (0,1)$, set

$$h_t^\alpha = \frac{d\nu_t^\alpha}{d\lambda_{\mathbb{R}}} \quad \text{for } t \in (0,\infty), \tag{3.3.15}$$

and show that

$$\int_0^\infty e^{-\lambda\tau} h_t^\alpha(\tau)\, d\tau = e^{-t\lambda^\alpha}, \quad \lambda \in [0,\infty),$$

and that $h_t^\alpha(\tau) \equiv t^{-\frac{1}{\alpha}} h_1^\alpha(t^{-\frac{1}{\alpha}}\tau)$.

**(ii)** Only when $\alpha = \frac{1}{2}$ is an explicit expression for $h_1^\alpha$ readily available. To find this expression, first note that, by the uniqueness of the Laplace transform (cf. Exercise 1.2.12) and **(i)**, $h_1^{\frac{1}{2}}$ is uniquely determined by

$$\int_0^\infty e^{-\lambda^2\tau} h_1^{\frac{1}{2}}(\tau)\, d\tau = e^{-\lambda}, \quad \lambda \in [0,\infty).$$

Next, show that

$$\int_0^\infty \tau^{-\frac{1}{2}} e^{-(\frac{a^2}{\tau}+b^2\tau)}\, d\tau = \frac{\pi^{\frac{1}{2}} e^{-2ab}}{b} \quad \text{and} \quad \int_0^\infty \tau^{-\frac{3}{2}} e^{-(\frac{a^2}{\tau}+b^2\tau)}\, d\tau = \frac{\pi^{\frac{1}{2}} e^{-2ab}}{a}$$

for all $(a,b) \in (0,\infty)^2$, and conclude from the second of these that

$$h_1^{\frac{1}{2}}(\tau) = \frac{\mathbf{1}_{(0,\infty)}(\tau) e^{-\frac{1}{4\tau}}}{\sqrt{4\pi}\tau^{\frac{3}{2}}}. \tag{3.3.16}$$

**Hint**: To prove the first identity, try the change of variables $x = a\tau^{-\frac{1}{2}} - b\tau^{\frac{1}{2}}$, and get the second by differentiating the first with respect to $a$.

EXERCISE 3.3.17. In this exercise we will discuss the densities of the symmetric stable laws $\mu_t^\alpha$ for $\alpha \in (0,2)$ (cf. Exercise 3.3.13). Once again, we know that each $\mu_t^\alpha$ admits a smooth density with respect to Lebesgue measure $\lambda_{\mathbb{R}^N}$ on $\mathbb{R}^N$. Further, it is clear that this density is symmetric and that

$$\frac{d\mu_t^\alpha}{d\lambda_{\mathbb{R}^N}}(\mathbf{x}) = t^{-\frac{1}{\alpha}} \frac{d\mu_1^\alpha}{d\lambda_{\mathbb{R}^N}}(t^{-\frac{1}{\alpha}}\mathbf{x}) \quad \text{for } t \in (0,\infty).$$

**(i)** Referring to Exercise 3.3.14 and using Exercise 3.3.12, show that

$$\text{(3.3.18)} \qquad \frac{d\mu_1^\alpha}{d\lambda_{\mathbb{R}^N}}(\mathbf{x}) = \frac{1}{(4\pi)^{\frac{N}{2}}} \int_0^\infty \tau^{-\frac{N}{2}} e^{-\frac{|\mathbf{x}|^2}{4\tau}} h_1^{\frac{\alpha}{2}}(\tau)\, d\tau.$$

**(ii)** Because we have an explicit expression for $h_1^{\frac{1}{2}}$, we can use (3.3.18) to get an explicit expression for $\frac{d\mu_1^1}{d\lambda_{\mathbb{R}^N}}$. In fact, show that

$$\text{(3.3.19)} \qquad \frac{d\mu_t^1}{d\lambda_{\mathbb{R}^N}}(\mathbf{x}) = \pi_t^{\mathbb{R}^N}(\mathbf{x}) \equiv \frac{2t^N}{\omega_N (t^2 + |\mathbf{x}|^2)^{\frac{N+1}{2}}}, \quad (t,\mathbf{x}) \in (0,\infty)\times\mathbb{R}^N,$$

where $\omega_N = 2\pi^{\frac{N+1}{2}}\Gamma\left(\frac{N+1}{2}\right)^{-1}$ is the surface area of $\mathbb{S}^N$ in $\mathbb{R}^{N+1}$. The function $\pi_1^{\mathbb{R}}$ is the density for what probabilists call the **Cauchy distribution**. For general $N$'s, $(t,\mathbf{x}) \in (0,\infty)\times\mathbb{R}^N \longmapsto \pi_t^{\mathbb{R}^N}(\mathbf{x})$ is what analysts call the **Poisson kernel** for the right half-space in $\mathbb{R}^{N+1}$. That is (cf. Exercise 10.2.22), if $f \in C_b(\mathbb{R}^N;\mathbb{R})$, then

$$(t,\mathbf{x}) \rightsquigarrow u_f(t,\mathbf{x}) = \int_{\mathbb{R}^N} f(\mathbf{x}-\mathbf{y})\, \pi_t^{\mathbb{R}^N}(\mathbf{y})\, d\mathbf{y}$$

is the unique, bounded harmonic extension of $f$ to the right half-space.

**(iii)** Given $\alpha \in (0,2)$, show that

$$\|f\|_\alpha^2 \equiv \iint_{\mathbb{R}^N\times\mathbb{R}^N} e^{-|\mathbf{y}-\mathbf{x}|^\alpha} f(\mathbf{x})\overline{f(\mathbf{y})}\, d\mathbf{x}d\mathbf{y} = \int_{\mathbb{R}^N} |\hat{f}(\boldsymbol{\xi})|^2\, \mu_1^\alpha(d\boldsymbol{\xi})$$

for $f \in L^1(\mathbb{R}^N;\mathbb{C})$. This can be used to prove that $\|\cdot\|_\alpha$ determines a Hilbert norm on $C_c(\mathbb{R}^N;\mathbb{C})$.

# Chapter 4
## Lévy Processes

Although analysis was the engine that drove the proofs in Chapter 3, probability theory can do a lot to explain the meaning of the conclusions drawn there. Specifically, in this chapter I will develop an intuitively appealing way of thinking about a random variable $\mathbf{X}$ whose distribution is infinitely divisible, an $\mathbf{X}$ for which $\mathbb{E}^{\mathbb{P}}\big[e^{\sqrt{-1}\,(\boldsymbol{\xi},\mathbf{X})_{\mathbb{R}^N}}\big]$ equals

$$\exp\Bigg(\sqrt{-1}(\boldsymbol{\xi},\mathbf{m}) - \tfrac{1}{2}(\boldsymbol{\xi},\mathbf{C}\boldsymbol{\xi})_{\mathbb{R}^N} + \int_{\mathbb{R}^N}\Big[e^{\sqrt{-1}\,(\boldsymbol{\xi},\mathbf{y})_{\mathbb{R}^N}} - 1 - \sqrt{-1}\,\mathbf{1}_{[0,1]}(|\mathbf{y}|)(\boldsymbol{\xi},\mathbf{y})_{\mathbb{R}^N}\Big]\,M(d\mathbf{y})\Bigg)$$

for some $\mathbf{m} \in \mathbb{R}^N$, some symmetric, non-negative definite $\mathbf{C} \in \mathrm{Hom}(\mathbb{R}^N;\mathbb{R}^N)$, and Lévy measure $M \in \mathfrak{M}_2(\mathbb{R}^N)$. In most of this chapter I will deal with the case when there is no Gaussian component. That is, I will be assuming that $\mathbf{C} = 0$. Because it is distinctly different, I will treat the Gaussian component separately in the final section. However, I begin with some general comments that apply to the considerations in the whole chapter.

The key idea, which seems to have been Lévy's, is to develop a dynamic picture of $\mathbf{X}$. To understand the origin of his idea, denote by $\mu \in \mathcal{I}(\mathbb{R}^N)$ the distribution of $\mathbf{X}$, and define $\ell_\mu$ accordingly, as in Theorem 3.2.7. Then, for each $t \in [0,\infty)$, there is a unique $\mu_t \in \mathcal{I}(\mathbb{R}^N)$ for which $\widehat{\mu_t} = e^{t\ell_\mu}$, and so $\mu_{s+t} = \mu_s \star \mu_t$ for all $s,t \in [0,\infty)$. Lévy's idea was to associate with $\{\mu_t : t \geq 0\}$ a family of random variables $\{\mathbf{Z}(t) : t \geq 0\}$ that would reflect the structure of $\{\mu_t : t \geq 0\}$. Thus, $\mathbf{Z}(0) = \mathbf{0}$ and, for each $(s,t) \in [0,\infty)$, $\mathbf{Z}(s+t) - \mathbf{Z}(s)$ should be independent of $\{\mathbf{Z}(\tau) : \tau \in [0,s]\}$ and have distribution $\mu_t$. In other words, $\{\mathbf{Z}(t) : t \geq 0\}$ should be the continuous parameter analog of the sums of independent, identically distributed random variables. Indeed, given any $\tau > 0$, let $\{\mathbf{X}_m : m \geq 0\}$ be a sequence of independent random variables with distribution $\mu_\tau$. Then $\{\mathbf{Z}(n\tau) : n \geq 0\}$ should have the same distribution as $\{\mathbf{S}_n : n \geq 0\}$, where $\mathbf{S}_n = \sum_{1\leq m\leq n}\mathbf{X}_m$. This observation suggests that one should think about $t \rightsquigarrow \mathbf{Z}(t)$ as a evolution that, when one understands its dynamics, will reveal information about $\mathbf{Z}(1)$ and therefore $\mu$.

For reasons that should be obvious now, an evolution $\{\mathbf{Z}(t) : t \in [0,\infty)\}$ of the sort described above used to be called a **process with independent, homogeneous increments**, the term "process" being the common one for continuous families of random variables and the adjective "homogeneous" referring to the fact that the distribution of the increment $\mathbf{Z}(t) - \mathbf{Z}(s)$ for $0 \le s < t$ depends only on the length $t - s$ of the time interval over which it is taken. In more recent times, a process with independent, homogeneous increments is said to be a **Lévy process**, and so I will adopt this more modern terminology.

Assuming that the family $\{\mathbf{Z}(t) : t \in [0,\infty)\}$ exists, notice that we already know what the joint distribution of $\{\mathbf{Z}(t_k) : k \in \mathbb{N}\}$ must be for any choice of $0 = t_0 < \cdots < t_k < \cdots$. Indeed, $\mathbf{Z}(0) = \mathbf{0}$ and

$$\mathbb{P}\big(\mathbf{Z}(t_k) - \mathbf{Z}(t_{k-1}) \in \Gamma_k,\ 1 \le k \le K\big) = \prod_{k+1}^{K} \mu_{t_k - t_{k-1}}(\Gamma_k)$$

for any $K \in \mathbb{Z}^+$ and $\Gamma_1, \dots, \Gamma_K \in \mathcal{B}_{\mathbb{R}^N}$. Equivalently, $\mathbb{P}\big(\mathbf{Z}(t_k) \in \Gamma_k,\ 0 \le k \le K\big)$ equals

$$\mathbf{1}_{\Gamma_0}(\mathbf{0}) \int \cdots \int_{(\mathbb{R}^N)^K} \prod_{k=1}^{K} \mathbf{1}_{\Gamma_k}\Big(\sum_{j=1}^{k} \mathbf{y}_j\Big) \prod_{k=1}^{K} \mu_{t_k - t_{k-1}}(d\mathbf{y}_k)$$

for any $K \in \mathbb{Z}^+$ and $\Gamma_0, \dots, \Gamma_K \in \mathcal{B}_{\mathbb{R}^N}$. My goal is this chapter is to show that each $\mu \in \mathcal{I}(\mathbb{R}^N)$ admits a Lévy process $\{\mathbf{Z}_\mu(t) : t \ge 0\}$ and that the construction of the associated Lévy process improves our understanding of $\mu$.

Unfortunately, before I can carry out this program, I need to deal with a few technical, bookkeeping matters.

## § 4.1 Stochastic Processes, Some Generalities

Given an index $\mathcal{A}$ with some nice structure and a family $\{X(\alpha) : \alpha \in \mathcal{A}\}$ of random variables on a probability space $(\Omega, \mathcal{F}, \mathbb{P})$ taking values in some measurable space $(E, \mathcal{B})$, it is often helpful to think about $\{X(\alpha) : \alpha \in A\}$ in terms of the map $\omega \in \Omega \longmapsto \mathbf{X}(\,\cdot\,,\omega) \in E^{\mathcal{A}}$. For instance, if $\mathcal{A}$ is linearly ordered, then $\omega \rightsquigarrow X(\,\cdot\,,\omega)$ can be thought of as a random evolution. More generally, when probabilists want to indicate that they are thinking about $\{X(\alpha) : \alpha \in \mathcal{A}\}$ as the map $\omega \rightsquigarrow X(\,\cdot\,,\omega)$, they call $\{X(\alpha) : \alpha \in \mathcal{A}\}$ a **stochastic process** on $\mathcal{A}$ with **state space** $(E, \mathcal{B})$.

The **distribution of a stochastic process** is the probability measure $X_*\mathbb{P}$ on[1] $(E^{\mathcal{A}}, \mathcal{B}^{\mathcal{A}})$ obtained by pushing $\mathbb{P}$ forward under the map $\omega \rightsquigarrow X(\,\cdot\,,\omega)$. Hence two stochastic processes $\{X(\alpha) : \alpha \in \mathcal{A}\}$ and $\{Y(\alpha) : \alpha \in \mathcal{A}\}$ on $(E, \mathcal{B})$ have the same distribution if and only if

$$\mathbb{P}\big(X(\alpha_k) \in \Gamma_k,\ 0 \le k \le K\big) = \mathbb{P}\big(Y(\alpha_k) \in \Gamma_k,\ 0 \le k \le K\big)$$

[1] Recall that $\mathcal{B}^{\mathcal{A}}$ is the $\sigma$-algebra over $E^{\mathcal{A}}$ that is generated by all the maps $\psi \in E^{\mathcal{A}} \longmapsto \psi(\alpha) \in E$ as $\alpha$ runs over $\mathcal{A}$.

for all $K \in \mathbb{Z}^+$, $\{\alpha_0, \dots, \alpha_K\} \subseteq \mathcal{A}$, and $\Gamma_0, \dots, \Gamma_K \in \mathcal{B}$.

As long as $\mathcal{A}$ is countable, there are no problems because $E^{\mathcal{A}}$ is a reasonably tame object and $\mathcal{B}^{\mathcal{A}}$ contains lots of its subsets. However, when $\mathcal{A}$ is uncountable, $E^{\mathcal{A}}$ is a ridiculously large space and $\mathcal{B}^{\mathcal{A}}$ will be too meager to contain many of the subsets in which one is interested. The point is that for $B$ to be in $\mathcal{B}^{\mathcal{A}}$ there must (cf. Exercise 4.1.11) be a countable subset $\{\alpha_k : k \in \mathbb{N}\}$ of $\mathcal{A}$ such that one can determine whether or not $\psi \in B$ by knowing $\{\psi(\alpha_k) : k \in \mathbb{N}\}$. Thus (cf. Exercise 4.1.11), for instance, $C([0,\infty);\mathbb{R}) \notin \mathcal{B}_{\mathbb{R}}^{[0,\infty)}$.

Probabilists expended a great deal of effort to overcome the problem raised in the preceding paragraph. For instance, using a remarkable piece of measure theoretic reasoning, J.L. Doob[2] proved that in the important case when $\mathcal{A} = [0,\infty)$ and $E = \mathbb{R}$, one can always make a modification, what he called the "separable modification," so that sets like $C([0,\infty);\mathbb{R})$ become measurable. However, in recent times, probabilists have tried to simplify their lives by constructing their processes in such a way that these unpleasant measurability questions never arise. That is, if they suspect that the process should have some property that is not measurable with respect to $\mathcal{B}^{\mathcal{A}}$, they avoid constructions based on general principles, like Kolmogorov's Extension Theorem (cf. part **(iii)** of Exercise 9.1.17), and instead adopt a construction procedure that produces the process with the desired properties already present.

The rest of this chapter contains important examples of this approach, and the rest of this section contains a few technical preparations.

§ **4.1.1. The Space $D(\mathbb{R}^N)$.** Unless its Lévy measure $M$ is zero, a Lévy process for $\mu \in \mathcal{I}(\mathbb{R}^N)$ cannot be constructed so that it has continuous paths. In fact, if $M \neq 0$, then $t \rightsquigarrow \mathbf{Z}_\mu(t)$ will be almost never continuous. Nonetheless, $\{\mathbf{Z}_\mu(t) : t \geq 0\}$ can be constructed so that its paths are reasonably nice. Specifically, its paths can be made to be right-continuous everywhere and have no oscillatory discontinuities. For this reason, I introduce the space $D(\mathbb{R}^N)$ of paths $\boldsymbol{\psi} : [0,\infty) \longrightarrow \mathbb{R}^N$ such that $\boldsymbol{\psi}(t) = \boldsymbol{\psi}(t+) \equiv \lim_{\tau \searrow t} \boldsymbol{\psi}(\tau)$ for each $t \in [0,\infty)$ and $\boldsymbol{\psi}(t-) \equiv \lim_{\tau \nearrow t} \boldsymbol{\psi}(\tau)$ exists in $\mathbb{R}^N$ for each $t \in (0,\infty)$. Equivalently, $\boldsymbol{\psi}(0) = \boldsymbol{\psi}(0+)$, and, for each $t \in (0,\infty)$ and $\epsilon > 0$, there is a $\delta \in (0,t)$ such that $\sup\{|\boldsymbol{\psi}(t) - \boldsymbol{\psi}(\tau)| : \tau \in (t, t+\delta)\} < \epsilon$ and $\sup\{|\boldsymbol{\psi}(t-) - \boldsymbol{\psi}(\tau)| : \tau \in (t-\delta, t)\} < \epsilon$.

The following lemma presents a few basic properties possessed by elements of $D(\mathbb{R}^N)$. In its statement, for $n \in \mathbb{N}$ and $\tau \in (0,\infty)$, $\lfloor \tau \rfloor_n^+ = \min\{m2^{-n} : m \in \mathbb{Z}^+ \text{ and } m \geq 2^n\tau\}$ and $\lfloor \tau \rfloor_n^- = \lfloor \tau \rfloor_n^+ - 2^{-n} = \max\{m2^{-n} : m \in \mathbb{N} \text{ and } m < 2^n\tau\}$. In addition, for $0 \leq a < b$,

$$\|\boldsymbol{\psi}\|_{[a,b]} \equiv \sup_{t \in [a,b]} |\boldsymbol{\psi}(t)| \tag{4.1.1}$$

---

[2] See Chapter II of Doob's *Stochastic Processes*, Wiley (1953).

is the uniform norm of $\psi \restriction [a,b]$, and

$$\operatorname{var}_{[a,b]}(\psi) = \sup\Big\{\sum_{k=1}^{K} |\psi(t_k) - \psi(t_{k-1})| : K \in \mathbb{Z}^+ \text{ and } a = t_0 < t_1 < \cdots < t_K = b\Big\} \tag{4.1.2}$$

is the total variation of $\psi \restriction [a,b]$.

LEMMA 4.1.3. *If $\psi \in D(\mathbb{R}^N)$, then, for each $t > 0$, $\|\psi\|_{[0,t]} < \infty$, and for each $r > 0$, the set*

$$J(t,r,\psi) \equiv \{\tau \in (0,t] : |\psi(\tau) - \psi(\tau-)| \geq r\}$$

*is finite subset of $(0,t]$. In addition, there exists an $n(t,r,\psi) \in \mathbb{N}$ such that, for every $n \geq n(t,r,\psi)$ and $m \in \mathbb{Z}^+ \cap (0, 2^n]$,*

$$\big|\psi(m2^{-n}t) - \psi((m-1)2^{-n}t)\big| \geq r \implies m2^{-n} = \big\lfloor \tfrac{\tau}{t} \big\rfloor_n^+ \text{ for some } \tau \in J(t,r,\psi).$$

*Finally,*

$$\|\psi\|_{[0,t]} = \lim_{n\to\infty} \max\{|\psi(m2^{-n}t)| : m \in \mathbb{N} \cap [0,2^n]\}$$

*and*

$$\operatorname{var}_{[0,t]}(\psi) = \lim_{n\to\infty} \sum_{m \in \mathbb{Z}^+ \cap [0,2^n]} \big|\psi(m2^{-n}t) - \psi((m-1)2^{-n}t)\big|.$$

PROOF: Begin by noting that it suffices to treat the case when $t = 1$, since one can always reduce to this case by replacing $\psi$ with $\tau \rightsquigarrow \psi(t\tau)$.

If $\|\psi\|_{[0,1]}$ were infinite, then we could find a sequence $\{\tau_n : n \geq 1\} \subseteq [0,1]$ such that $|\psi(\tau_n)| \longrightarrow \infty$, and clearly, without loss in generality, we could choose this sequence so that $\tau_n \longrightarrow \tau \in [0,1]$ and $\{\tau_n : n \geq 1\}$ is either strictly decreasing or strictly increasing. But, in the first case this would contradict right-continuity, and in the second it would contradict the existence of left limits. Thus, $\|\psi\|_{[0,1]}$ must be finite.

Essentially the same reasoning shows that $J(1,r,\psi)$ is finite. If it were not, then we could find a sequence $\{\tau_n : n \geq 0\}$ of distinct points in $(0,1]$ such that $|\psi(\tau_n) - \psi(\tau_n-)| \geq r$, and again we could choose them so that they were either strictly increasing or strictly decreasing. If they were strictly increasing, then $\tau_n \nearrow \tau$ for some $\tau \in (0,1]$ and, for each $n \in \mathbb{Z}^+$, there would exist a $\tau_n' \in (\tau_{n-1}, \tau_n)$ such that $|\psi(\tau_n) - \psi(\tau_n')| \geq \frac{r}{2}$, which would contradict the existence of a left limit at $\tau$. Similarly, right-continuity would be violated if the $\tau_n$'s were decreasing.

Although it has the same flavor, the proof of the existence of $n(1,r,\psi)$ is a bit trickier. Let $0 < \tau_1 < \cdots \tau_K \leq 1$ be the elements of $J(1,r,\psi)$. If $n(1,r,\psi)$

failed to exist, then we could choose a subsequence $\{(m_j,n_j) : j \geq 1\}$ from $\mathbb{Z}^+ \times \mathbb{N}$ so that $\{n_j : j \geq 1\}$ is strictly increasing and $t_j \equiv m_j 2^{-n_j} \in (0,1]$ satisfies $|\psi(t_j) - \psi(t_j - 2^{-n_j})| \geq r$ for all $j \in \mathbb{Z}^+$, but $t_j \neq \lfloor \tau_k \rfloor^+_{n_j}$ for any $j \in \mathbb{Z}^+$ and $1 \leq k \leq K$. If $t_j = t$ infinitely often for some $t$, then we would have the contradiction that $t \notin J(1,r,\psi)$ and yet $|\psi(t) - \psi(t-)| \geq r$. Hence, I will assume that the $t_j$'s are distinct. Further, without loss in generality, I assume that $\{t_j : j \geq 1\}$ is a subset of one of the intervals $(0,\tau_1)$, $(\tau_{k-1},\tau_k)$ for some $2 \leq k \leq K$, or of $(\tau_K, 1]$. Finally, I may and will assume that either $t_j \nearrow t \in (0,1]$ or that $t_j \searrow t \in [0,1)$. But, since $|\psi(t_j) - \psi(t_j - 2^{-n_j})| \geq r$, $t_j \nearrow t$ contradicts the existence of $\psi(t-)$. Similarly, if $t_j \searrow t$ and $t_j - 2^{-n_j} \geq t$ for infinitely many $j's$, then we get a contradiction with right-continuity at $t$. Thus, the only remaining case is when $t_j \searrow t$ and $t_j - 2^{-n_j} < t \leq t_j$ for all but a finite number of $j$'s, in which case we get the contradiction that $t \notin J(1,r,\psi)$ and yet

$$|\psi(t) - \psi(t-)| = \lim_{j\to\infty} |\psi(t_j) - \psi(t_j - 2^{-n_j})| \geq r.$$

To prove the assertion about $\|\psi\|_{[0,1]}$, simply observe that, by monotonicity, the limit exists and that, by right-continuity, for any $t \in [0,1]$,

$$|\psi(t)| = \lim_{n\to\infty} |\psi(\lfloor t \rfloor^+_n)| \leq \lim_{n\to\infty} \max_{0\leq m\leq 2^n} |\psi(m2^{-n})| \leq \|\psi\|_{[0,1]}.$$

The assertion about $\mathrm{var}_{[0,1]}(\psi)$ is proved in essentially the same manner, although now the monotonicity comes from the triangle inequality and the first equality in the preceding must be replaced by $|\psi(t) - \psi(t-)| = \lim_{n\to\infty} |\psi(\lfloor t \rfloor^+_n) - \psi(\lfloor t \rfloor^-_n)|$. □

I next give $D(\mathbb{R}^N)$ the topological structure corresponding to uniform convergence on compacts, or, equivalently, the topological structure for which

$$\rho(\psi,\psi') \equiv \sum_{n=1}^{\infty} 2^{-n} \frac{\|\psi - \psi'\|_{[0,n]}}{1 + \|\psi - \psi'\|_{[0,n]}}$$

is a metric. Because it is not separable (cf. Exercise 4.1.10), this topological structure is less than ideal. Nonetheless, the metric $\rho$ is complete. To see that it is, first observe that $|\psi(\tau-)| \leq \|\psi\|_{[0,t]}$ for all $0 < \tau \leq t$. Thus, if $\sup_{\ell>k} \rho(\psi_\ell,\psi_k) \longrightarrow 0$ as $k \to \infty$, then there exist paths $\psi : [0,\infty) \longrightarrow \mathbb{R}^N$ and $\tilde{\psi} : (0,\infty) \longrightarrow \mathbb{R}^N$ such that

$$\sup_{\tau\in[0,t]} |\psi_k(\tau) - \psi(\tau)| \longrightarrow 0 \quad \text{and} \quad \sup_{\tau\in(0,t]} |\psi_k(\tau-) - \tilde{\psi}(\tau)| \longrightarrow 0$$

for each $t > 0$. Therefore, if $t \geq \tau_n \searrow \tau$, then

$$\overline{\lim_{n\to\infty}}\, |\psi(\tau) - \psi(\tau_n)| \leq 2\|\psi - \psi_k\|_{[0,t]} + \overline{\lim_{n\to\infty}}\, |\psi_k(\tau) - \psi_k(\tau_n)| \leq 2\|\psi - \psi_k\|_{[0,t]}$$

for all $k \in \mathbb{Z}^+$, and so $\psi$ is right-continuous. Essentially the same argument shows that $\psi(\tau-) = \tilde{\psi}(\tau)$ for $\tau > 0$, which means, of course, that $\psi \in D(\mathbb{R}^N)$ and that $\sup_{\tau\in(0,t]} |\psi_k(\tau-) - \psi(\tau-)| \longrightarrow 0$ for each $t > 0$.

One might think that I would take the measurable structure on $D(\mathbb{R}^N)$ to be the one given by the Borel field $\mathcal{B}_{D(\mathbb{R}^N)}$ determined by uniform convergence on compacts. However, this is not the choice I will make. Instead, the measurable structure I choose for $D(\mathbb{R}^N)$ is the one that $D(\mathbb{R}^N)$ inherits as a subset of $(\mathbb{R}^N)^{[0,\infty)}$. That is, I take for $D(\mathbb{R}^N)$ the measurable structure given by the $\sigma$-algebra $\mathcal{F}_{D(\mathbb{R}^N)} = \sigma(\{\psi(t) : t \in [0,\infty)\})$, the $\sigma$-algebra generated by the maps $\psi \in D(\mathbb{R}^N) \longmapsto \psi(t) \in \mathbb{R}^N$ as $t$ runs over $[0,\infty)$. The reason for my insisting on this choice is that I want two $D(\mathbb{R}^N)$-valued stochastic processes $\{\mathbf{X}(t) : t \geq 0\}$ and $\{\mathbf{Y}(t) : t \geq 0\}$ to induce the same measure on $D(\mathbb{R}^N)$ if they have the same distribution. Seeing as (cf. Exercise 4.1.11) $\mathcal{F}_{D(\mathbb{R}^N)} \subsetneqq \mathcal{B}_{D(\mathbb{R}^N)}$, this would not be true were I to choose the Borel structure.

Because $\mathcal{F}_{D(\mathbb{R}^N)} \neq \mathcal{B}_{D(\mathbb{R}^N)}$, $\mathcal{F}_{D(\mathbb{R}^N)}$-measurability does not follow from topological properties like continuity. Nonetheless, many functions related to the topology of $D(\mathbb{R}^N)$ are $\mathcal{F}_{D(\mathbb{R}^N)}$-measurable. For example, the last part of Lemma 4.1.3 proves that both $\psi \rightsquigarrow \|\psi\|_{[0,t]}$, which is continuous, and $\psi \rightsquigarrow \operatorname{var}_{[0,t]}(\psi)$, which is lower semicontinuous, are both $\mathcal{F}_{D(\mathbb{R}^N)}$-measurable for all $t \in [0,\infty)$. In the next subsection, I will examine other important functions on $D(\mathbb{R}^N)$ and will show that they, too, are $\mathcal{F}_{D(\mathbb{R}^N)}$-measurable.

§ **4.1.2. Jump Functions.** Let $\mathfrak{M}_\infty(\mathbb{R}^N)$ be the space of non-negative Borel measures $M$ on $\mathbb{R}^N$ with the properties that $M(\{\mathbf{0}\}) = 0$ and $M(B(\mathbf{0},r)\complement) < \infty$ for all $r > 0$. A **jump function** is a map $t \in [0,\infty) \longmapsto j(t,\cdot) \in \mathfrak{M}_\infty(\mathbb{R}^N)$ with the property that, for each $\Delta \in \mathcal{B}_{\mathbb{R}^N}$ with $\mathbf{0} \notin \bar{\Delta}$, $j(0,\Delta) = 0$, $t \rightsquigarrow j(t,\Delta)$ is a non-decreasing, piecewise constant element of $D(\mathbb{R}^N)$ such that $j(t,\Delta) - j(t-,\Delta) \in \{0,1\}$ for each $t > 0$.

LEMMA 4.1.4. *A map $t \rightsquigarrow j(t,\cdot)$ is a non-zero jump function if and only if there exists a set $\emptyset \neq J \subset (0,\infty)$ that is finite or countable and a set $\{\mathbf{y}_\tau : \tau \in J\} \subseteq \mathbb{R}^N \setminus \{\mathbf{0}\}$ such that $\{\tau \in J \cap (0,t] : |\mathbf{y}_\tau| \geq r\}$ is finite for each $(t,r) \in (0,\infty)^2$ and*

$$j(t,\cdot) = \sum_{\tau\in J} \mathbf{1}_{[\tau,\infty)}(t)\delta_{\mathbf{y}_\tau}. \tag{4.1.5}$$

*In particular, if $t \rightsquigarrow j(t,\cdot)$ is a jump function and $t > 0$, then, either $j(t,\cdot) = j(t-,\cdot)$ or $j(t,\cdot) - j(t-,\cdot) = \delta_{\mathbf{y}}$ for some $\mathbf{y} \in \mathbb{R}^N \setminus \{\mathbf{0}\}$.*

PROOF: It should be obvious that if $J$ and $\{\mathbf{y}_\tau : \tau \in J\}$ satisfy the stated conditions, then the $t \rightsquigarrow j(t,\cdot)$ given by (4.1.5) is a jump function. To go in the other direction, suppose that $t \rightsquigarrow j(t,\cdot)$ is a jump function, and, for each $r > 0$, set $f_r(t) = j(t, \mathbb{R}^N \setminus B(\mathbf{0},r))$. Because $t \rightsquigarrow f_r(t)$ is a non-decreasing, piecewise constant, right-continuous function satisfying $f_r(0) = 0$ and $f_r(t) - f_r(t-) \in$

$\{0,1\}$ for each $t > 0$, it has at most a countable number of discontinuities, and at most $f_r(t)$ of them can occur in any interval $(0,t]$. Furthermore, if $f_r$ has a discontinuity at $\tau$, then $j(\tau, B(\mathbf{0},r)) - j(\tau-, B(\mathbf{0},r)) = 0$, and so the measure $\nu_\tau = j(\tau,\,\cdot\,) - j(\tau-,\,\cdot\,)$ is a $\{0,1\}$-valued probability measure on $\mathbb{R}^N$ that assigns mass 0 to $B(\mathbf{0},r)$. Hence (cf. Exercise 4.1.15) $f_r(\tau) \neq f_r(\tau-) \implies \nu_\tau = \delta_{\mathbf{y}}$ for some $\mathbf{y}_\tau \in \mathbb{R}^N \setminus B(\mathbf{0},r)$. From these considerations, it follows easily that if $J(r) = \{\tau \in (0,\infty) : f_r(\tau) \neq f_r(\tau-)\}$ and if, for each $\tau \in J(r)$, $\mathbf{y}_\tau \in \mathbb{R}^N \setminus B(\mathbf{0},r)$ is chosen so that $j(\tau,\,\cdot\,) - j(\tau-,\,\cdot\,) = \delta_{\mathbf{y}_\tau}$, then $J(r) \cap (0,t]$ is finite for all $t > 0$ and

$$j(t,\,\cdot\,) \upharpoonright B(\mathbf{0},r)\complement = \sum_{\tau \in J(r)} \mathbf{1}_{[\tau,\infty)}(t)\delta_{\mathbf{y}_\tau}.$$

Thus, if $J = \bigcup_{r>0} J(r)$, then $J$ is at most countable, $\{(\tau, \mathbf{y}_\tau) : \tau \in J\}$ has the required finiteness property, and (4.1.5) holds. $\square$

The reason for my introducing jump functions is that every element $\psi \in D(\mathbb{R}^N)$ determines a jump function $t \rightsquigarrow j(t,\,\cdot\,,\psi)$ by the prescription

$$\begin{aligned} j(t,\Gamma,\psi) &= \sum_{\tau \in J(t,\psi)} \mathbf{1}_\Gamma\big(\psi(\tau) - \psi(\tau-)\big), \\ &\text{where } J(t,\psi) \equiv \{\tau \in (0,t] : \psi(\tau) \neq \psi(\tau-)\}, \end{aligned} \tag{4.1.6}$$

for $\Gamma \subseteq \mathbb{R}^N \setminus \{\mathbf{0}\}$. To check that $j(t,\,\cdot\,,\psi)$ is well defined and is a jump function, take $J(\psi) = \bigcup_{t>0} J(t,\psi)$ and $\mathbf{y}_\tau = \psi(\tau) - \psi(\tau-)$ when $\tau \in J(\psi)$, note that, by Lemma 4.1.3, $J(\psi)$ is at most countable and that $\{(\tau, \mathbf{y}_\tau) : \tau \in J(\psi)\}$ has the finiteness required in Lemma 4.1.4, and observe that (4.1.5) holds when $j(t,\,\cdot\,) = j(t,\,\cdot\,,\psi)$ and $J = J(\psi)$.

Because it will be important for us to know that the distribution of a $D(\mathbb{R}^N)$-valued stochastic process determines the distribution of the jump functions for its paths, we will make frequent use of the following lemma.

LEMMA 4.1.7. *If $\varphi : \mathbb{R}^N \longrightarrow \mathbb{R}$ is a $\mathcal{B}_{\mathbb{R}^N}$-measurable function that vanishes in a neighborhood of $\mathbf{0}$, then $\varphi$ is $j(t,\,\cdot\,,\psi)$-integrable for all $(t,\psi) \in [0,\infty) \times D(\mathbb{R}^N)$, and*

$$(t,\psi) \in [0,\infty) \times D(\mathbb{R}^N) \longmapsto \int_{\mathbb{R}^N} \varphi(\mathbf{y})\, j(t, d\mathbf{y}, \psi) \in \mathbb{R}$$

*is a $\mathcal{B}_{[0,\infty)} \times \mathcal{F}_{D(\mathbb{R}^N)}$-measurable function that, for each $\psi$, is right-continuous and piecewise constant as a function of $t$. Finally, for all Borel measurable $\varphi : \mathbb{R}^N \longrightarrow [0,\infty)$, $(t,\psi) \in [0,\infty) \times D(\mathbb{R}^N) \longmapsto \int_{\mathbb{R}^N} \varphi(\mathbf{y}) j(t, d\mathbf{y}, \psi) \in [0,\infty]$ is $\mathcal{B}_{[0,\infty)} \times \mathcal{F}_{D(\mathbb{R}^N)}$-measurable.*

PROOF: The final assertion is an immediate consequence of the earlier one plus the Monotone Convergence Theorem.

Let $r > 0$ be given. If $\varphi$ is a Borel measurable function that vanishes on $B(\mathbf{0},r)$, then it is immediate from the first part of Lemma 4.1.3 that $\varphi$ is

$j(t,\cdot,\psi)$-integrable for all $(t,\psi)\in[0,\infty)\times D(\mathbb{R}^N)$ and, for each $\psi\in D(\mathbb{R}^N)$, $t\rightsquigarrow\int_{\mathbb{R}^N}\varphi(\mathbf{y})\,j(t,d\mathbf{y},\psi)$ is right-continuous and piecewise constant. Thus, it suffices to show that, for each $t\in(0,\infty)$,

$$(*)\qquad \psi\rightsquigarrow\int_{\mathbb{R}^N}\varphi(\mathbf{y})\,j(t,d\mathbf{y},\psi)\quad\text{is } \mathcal{F}_{D(\mathbb{R}^N)}\text{-measurable.}$$

Moreover, it suffices to do this when $t=1$ and $\varphi$ is continuous, since rescaling time allows one to replace $t$ by 1 and the set of $\varphi$'s for which (*) is true is closed under pointwise convergence. But, by the second part of Lemma 4.1.3, we know that

$$\sum_{m=1}^{2^n}\varphi\Big(\psi(m2^{-n})-\psi((m-1)2^{-n})\Big)=\sum_{\tau\in J(1,r,\psi)}\varphi\Big(\psi(\lfloor\tau\rfloor_n^+)-\psi(\lfloor\tau\rfloor_n^-)\Big)$$

for $n\geq n(1,r,\psi)$, and therefore

$$\int_{\mathbb{R}^N}\varphi(\mathbf{y})\,j(1,d\mathbf{y},\psi)=\lim_{n\to\infty}\sum_{m=1}^{2^n}\varphi\Big(\psi(m2^{-n})-\psi((m-1)2^{-n})\Big).\quad\square$$

Here are some properties of a path $\psi\in D(\mathbb{R}^N)$ which are determined by its relationship to its jump function. First, it should be obvious that $\psi\in C(\mathbb{R}^N)\equiv C\big([0,\infty);\mathbb{R}^N\big)$ if and only if $j(t,\cdot,\psi)=0$ for all $t>0$. At the opposite extreme, say that a $\psi$ is an **absolutely pure jump path** if and only if (cf. § 3.2.2) $j(t,\cdot,\psi)\in\mathfrak{M}_1(\mathbb{R}^N)$ and $\psi(t)=\int\mathbf{y}\,j(t,d\mathbf{y},\psi)$ for all $t>0$. Among the absolutely pure jump paths are those that are the piecewise constant paths: those absolutely pure jump $\psi$'s for which $j(t,\cdot,\psi)\in\mathfrak{M}_0(\mathbb{R}^N)$, $t>0$. Because of Lemma 4.1.7, each of these properties is $\mathcal{F}_{D(\mathbb{R}^N)}$-measurable. In particular, if $\{\mathbf{Z}(t):t\geq0\}$ is a $D(\mathbb{R}^N)$-valued stochastic process whose paths almost surely have any one of these properties, then the paths of every $D(\mathbb{R}^N)$-valued stochastic process with the same distribution as $\{\mathbf{Z}(t):t\geq0\}$ will almost surely possess that property.

Finally, I need to address the question of when a jump function is the jump function for some $\psi\in D(\mathbb{R}^N)$.

THEOREM 4.1.8. *Let $t\rightsquigarrow j(t,\cdot)$ be a non-zero jump function, and set $j^\Gamma(t,d\mathbf{y})=\mathbf{1}_\Gamma(\mathbf{y})j(t,d\mathbf{y})$ for $\Gamma\in\mathcal{B}_{\mathbb{R}^N}$. If $\Delta\in\mathcal{B}_{\mathbb{R}^N}$ with $\mathbf{0}\notin\bar{\Delta}$ and if $\psi^\Delta(t)=\int_\Delta\mathbf{y}\,j(t,d\mathbf{y})$, then $\psi^\Delta$ is a piecewise constant element of $D(\mathbb{R}^N)$, $j(t,\cdot,\psi^\Delta)=j^\Delta(t,\cdot)$, and $j(t,\cdot,\psi-\psi^\Delta)=j^{\mathbb{R}^N\setminus\Delta}(t,\cdot)=j(t,\cdot)-j^\Delta(t,\cdot)$ for any $\psi\in D(\mathbb{R}^N)$ whose jump function is $t\rightsquigarrow j(t,\cdot)$. Finally, suppose that $\{\psi_m:m\geq0\}\subseteq D(\mathbb{R}^N)$ and a non-decreasing sequence $\{\Delta_m:m\geq0\}\subseteq\mathcal{B}_{\mathbb{R}^N}$ satisfy the conditions that $\mathbb{R}^N\setminus\{\mathbf{0}\}=\bigcup_{m=0}^\infty\Delta_m$ and, for each $m\in\mathbb{N}$, $\mathbf{0}\notin\overline{\Delta_m}$ and $j(t,\cdot,\psi_m)=j^{\Delta_m}(t,\cdot)$, $t\geq0$. If $\psi_m\longrightarrow\psi$ uniformly on compacts, then $j(t,\cdot,\psi)=j(t,\cdot)$, $t\geq0$.*

PROOF: Throughout the proof I will use the notation introduced in Lemma 4.1.4.

Assuming that $\mathbf{0} \notin \bar{\Delta}$, we know that

$$j^{\Delta}(t, \cdot\,) = \sum_{\tau \in J} \mathbf{1}_{[\tau,\infty)}(t)\mathbf{1}_{\Delta}(\mathbf{y}_\tau)\delta_{\mathbf{y}_\tau},$$

where, for each $t > 0$, there are only finitely many non-vanishing terms. At the same time,

$$\psi^{\Delta}(t) = \sum_{\tau \in J} \mathbf{1}_{[\tau,\infty)}(t)\mathbf{1}_{\Delta}(\mathbf{y}_\tau)\mathbf{y}_\tau \text{ and } j(t, \cdot\,, \psi - \psi^{\Delta}) = \sum_{\tau \in J} \mathbf{1}_{[\tau,\infty)}(t)\mathbf{1}_{\mathbb{R}^N \setminus \Delta}(\mathbf{y}_\tau)\delta_{\mathbf{y}_\tau}$$

if $j(t, \cdot\,, \psi) = j(t, \cdot\,)$. Thus, all that remains is to prove the final assertion. To this end, suppose that $j(t, \cdot\,, \psi) \neq j(t-, \cdot\,, \psi)$. Since $\|\psi - \psi_m\|_{[0,t]} \longrightarrow 0$, there exists an $m$ such that $\psi_m(t) \neq \psi_m(t-)$ and therefore that $j(t, \cdot\,) - j(t-, \cdot\,) = \delta_{\mathbf{y}}$ for some $y \in \Delta_m$. Since this means that $\psi_n(t) - \psi_n(t-) = \mathbf{y}$ for all $n \geq m$, it follows that $\psi(t) - \psi(t-) = \mathbf{y}$ and therefore that $j(t, \cdot\,, \psi) - j(t-, \cdot\,, \psi) = \delta_{\mathbf{y}} = j(t, \cdot\,) - j(t-, \cdot\,)$. Conversely, suppose that $j(t, \cdot\,) \neq j(t-, \cdot\,)$ and choose $m$ so that $j(t, \cdot\,) - j(t-, \cdot\,) = \delta_{\mathbf{y}}$ for some $\mathbf{y} \in \Delta_m$. Then $\psi_n(t) - \psi_n(t-) = \mathbf{y}$ for all $n \geq m$. Thus, since this means that $\psi(t) - \psi(t-) = \mathbf{y}$, we again have that $j(t, \cdot\,, \psi) - j(t-, \cdot\,, \psi) = \delta_{\mathbf{y}} = j(t, \cdot\,) - j(t-, \cdot\,)$. After combining these, we see that $j(t, \cdot\,, \psi) - j(t-, \cdot\,, \psi) = j(t, \cdot\,) - j(t-, \cdot\,)$ for all $t > 0$, from which it is an easy step to $j(t, \cdot\,) = j(t, \cdot\,, \psi)$ for all $t \geq 0$. □

## Exercises for § 4.1

EXERCISE 4.1.9. When dealing with uncountable collections of random variables, it is important to understand what functions are measurable with respect to them. To be precise, suppose that $\{X_i : i \in \mathcal{I}\}$ is a non-empty collection of functions on some space $\Omega$ with values in some measurable space $(E, \mathcal{B})$, and let $\mathcal{F} = \sigma(\{X_i : i \in \mathcal{I}\})$ be the $\sigma$-algebra over $\Omega$ which they generate. Show that $A \in \mathcal{F}$ if and only if there is a sequence $\{i_m : m \in \mathbb{Z}^+\} \subseteq \mathcal{I}$ and an $\Gamma \in \mathcal{B}^{\mathbb{Z}^+}$ such that

$$A = \{\omega \in \Omega : (X_{i_1}(\omega), \dots, X_{i_m}(\omega), \dots) \in \Gamma\}.$$

More generally, if $f : \Omega \longrightarrow \mathbb{R}$, show that $f$ is $\mathcal{F}$-measurable if and only if there is a sequence $\{i_m : m \in \mathbb{Z}^+\} \subseteq \mathcal{I}$ and a $\mathcal{F}^{\mathbb{Z}^+}$-measurable $F : E^{\mathbb{Z}^+} \longrightarrow \mathbb{R}$ such that

$$f(\omega) = F(X_{i_1}(\omega), \dots, X_{i_m}(\omega), \dots).$$

**Hint**: Make use of Exercise 1.1.12.

EXERCISE 4.1.10. Let $\mathbf{e} \in \mathbb{S}^{N-1}$, set $\psi_t(\tau) = \mathbf{1}_{[t,\infty)}(\tau)\mathbf{e}$ for $t \in [0,1]$, and show that $\|\psi_t - \psi_s\|_{[0,1]} = 1$ for all $s \neq t$ from $[0,1]$. Conclude from this that $D(\mathbb{R}^N)$ is not separable in the topology of uniform convergence on compacts.

EXERCISE 4.1.11. Using Exercise 4.1.9, show that a function $\varphi : D(\mathbb{R}^N) \longrightarrow \mathbb{R}$ is $\mathcal{F}_{D(\mathbb{R}^N)}$-measurable if and only if there exists an $(\mathbb{R}^N)^{\mathbb{N}}$-measurable function $\Phi : (\mathbb{R}^N)^{\mathbb{N}} \longrightarrow \mathbb{R}$ and a sequence $\{t_k : k \in \mathbb{N}\} \subseteq [0,\infty)$ such that

$$\varphi(\psi) = \Phi(\psi(t_0), \dots, \psi(t_k), \dots), \quad \psi \in D(\mathbb{R}^N).$$

Next, define $\psi_t$ as in Exercise 4.1.10, and use that exercise together with the preceding to show that the open set $\{\psi \in D(\mathbb{R}^N) : \exists\, t \in [0,1]\ \|\psi - \psi_t\|_{[0,1]} < 1\}$ is not $\mathcal{F}_{D(\mathbb{R}^N)}$-measurable. Conclude that $\mathcal{B}_{D(\mathbb{R}^N)} \supsetneqq \mathcal{F}_{D(\mathbb{R}^N)}$. Similarly, conclude that neither $D(\mathbb{R}^N)$ nor $C(\mathbb{R}^N)$ is a measurable subset of $(\mathbb{R}^N)^{[0,\infty)}$. On the other hand, as we have seen, $C(\mathbb{R}^N) \in \mathcal{F}_{D(\mathbb{R}^N)}$.

EXERCISE 4.1.12. Show that

$$\operatorname{var}_{[0,t]}(\psi) \geq \int_{\mathbb{R}^N} |\mathbf{y}|\, j(t, d\mathbf{y}, \psi), \quad (t,\psi) \in [0,\infty) \times D(\mathbb{R}^N). \tag{4.1.13}$$

**Hint**: This is most easily seen from the representation of $j(t,\,\cdot\,,\psi)$ in terms of point masses at the discontinuities of $\psi$. One can use this representation to show that, for each $r > 0$,

$$\operatorname{var}_{[0,t]}(\psi) \geq \sum_{\tau \in J(t,r,\psi)} |\psi(\tau) - \psi(\tau-)| = \int_{|\mathbf{y}|\geq r} |\mathbf{y}|\, j(t, d\mathbf{y}, \psi), \quad (t,\psi) \in [0,\infty).$$

EXERCISE 4.1.14. If $\psi$ is an absolutely pure jump path, show that $\operatorname{var}_{[0,t]}(\psi) = \int |\mathbf{y}|\, j(t, d\mathbf{y}, \psi)$ and therefore that $\psi$ has locally bounded variation. Conversely, if $\psi \in D(\mathbb{R}^N)$ has locally bounded variation, show that $\psi$ is an absolutely pure jump path if and only if $\operatorname{var}_{[0,t]}(\psi) = \int |\mathbf{y}|\, j(t, d\mathbf{y}, \psi)$. Finally, if $\psi \in D(\mathbb{R}^N)$ and $j(t,\,\cdot\,,\psi) \in \mathfrak{M}_1(\mathbb{R}^N)$ for all $t \geq 0$, set $\psi_c(t) \equiv \psi(t) - \int \mathbf{y}\, j(t, dy, \psi)$ and show that $\psi_c \in C(\mathbb{R}^N)$ and

$$\operatorname{var}_{[0,t]}(\psi) = \operatorname{var}_{[0,t]}(\psi_c) + \int |\mathbf{y}|\, j(t, d\mathbf{y}, \psi).$$

EXERCISE 4.1.15. If $\nu \in \mathbf{M}_1(\mathbb{R}^N)$, show that $\nu(\Gamma) \in \{0,1\}$ for all $\Gamma \in \mathcal{B}_{\mathbb{R}^N}$ if and only if $\nu = \delta_{\mathbf{y}}$ for some $\mathbf{y} \in \mathbb{R}^N$.

**Hint**: Begin by showing that it suffices to handle the case when $N = 1$. Next, assuming that $N = 1$, show that $\nu$ is compactly supported, let $m$ be its mean value, and show that $\nu = \delta_m$.

## § 4.2 Discontinuous Lévy Processes

In this section I will construct the Lévy processes corresponding to those $\mu \in \mathcal{I}(\mathbb{R}^N)$ with no Gaussian component. That is,

$$\begin{aligned} \hat{\mu}(\boldsymbol{\xi}) = \exp\Big( & \sqrt{-1}(\boldsymbol{\xi}, \mathbf{m}_\mu)_{\mathbb{R}^N} \\ & + \int_{\mathbb{R}^N} \Big[ e^{\sqrt{-1}(\boldsymbol{\xi},\mathbf{y})} - 1 - \sqrt{-1}\,\mathbf{1}_{[0,1]}(|\mathbf{y}|)(\boldsymbol{\xi},\mathbf{y})_{\mathbb{R}^N} \Big] M_\mu(d\mathbf{y}) \Big). \end{aligned} \tag{4.2.1}$$

Because they are the building blocks out of which all such processes are made, I will treat separately the case when $\mu$ is a Poisson measure $\pi_M$ for some $M \in \mathfrak{M}_0(\mathbb{R}^N)$ and will call the corresponding Lévy process the **Poisson process** associated with $M$.

§ **4.2.1. The Simple Poisson Process.** I begin with the case when $N = 1$ and $M = \delta_1$, for which $\pi_M$ is the **simple Poisson measure** $e^{-1}\sum_{m=0}^{\infty}\frac{1}{m!}\delta_m$ whose Fourier transform is $\exp\left(e^{\sqrt{-1}\xi} - 1\right)$.

To construct the Poisson process associated with $\delta_1$, start with a sequence $\{\tau_m : m \geq 1\}$ of independent, **unit exponential** random variables on a probability space $(\Omega, \mathcal{F}, \mathbb{P})$. That is,

$$\mathbb{P}(\{\omega : \tau_1(\omega) > t_1, \dots, \tau_n(\omega) > t_n\}) = \exp\left(-\sum_{m=1}^{n} t_m^+\right)$$

for all $n \in \mathbb{Z}^+$ and $(t_1, \dots, t_n) \in \mathbb{R}^n$. Without loss in generality, I may and will assume that $\tau_m(\omega) > 0$ for all $m \in \mathbb{Z}^+$ and $\omega \in \Omega$. In addition, by The Strong Law of Large Numbers, I may and will assume that $\sum_{m=1}^{\infty}\tau_m(\omega) = \infty$ for all $\omega \in \Omega$. Next, set $T_0(\omega) = 0$ and $T_n(\omega) = \sum_{m=1}^{n}\tau_m(\omega)$, and define

$$N(t,\omega) = \max\{n \in \mathbb{N} : T_n(\omega) \leq t\} = \sum_{n=1}^{\infty}\mathbf{1}_{[T_n(\omega),\infty)}(t) \quad \text{for } t \in [0,\infty). \tag{4.2.2}$$

Clearly $t \rightsquigarrow N(t,\omega)$ is a non-decreasing, right-continuous, piecewise constant, $\mathbb{N}$-valued path that starts at 0 and, whenever it jumps, jumps by $+1$. In particular, $N(\cdot,\omega) \in D(\mathbb{R}^N)$, $N(t,\omega) - N(t-,\omega) \in \{0,1\}$ for all $t \in (0,\infty)$, and (cf. (4.1.6)) $j(t, \cdot, N(\cdot,\omega)) = N(t,\omega)\delta_1$.

Because $\mathbb{P}(N(t) = n) = \mathbb{P}(T_n \leq t < T_{n+1})$, $\mathbb{P}(N(t) = 0) = \mathbb{P}(\tau_1 > t) = e^{-t}$, and, when $n \geq 1$ (below $|\Gamma|$ denotes the Lebesgue measure of $\Gamma \in \mathcal{B}_{\mathbb{R}^n}$)

$$\mathbb{P}(N(t) = n), = \int\!\!\cdots\!\!\int_A e^{-\sum_{m=1}^{n+1}\tau_m}\, d\tau_1 \cdots d\tau_{n+1} = e^{-t}|B|,$$

where $A = \left\{(\tau_1, \dots, \tau_{n+1}) \in (0,\infty)^{n+1} : \sum_{m=1}^{n}\tau_m \leq t < \sum_{m=1}^{n+1}\tau_m\right\}$ and $B = \left\{(\tau_1, \dots, \tau_n) \in (0,\infty)^n : \sum_{m=1}^{n}\tau_m \leq t\right\}$. By making the change of variables $s_m = \sum_{j=1}^{m}\tau_j$ and remarking that the associated Jacobian is 1, one sees that $|B| = |C|$, where $C = \left\{(s_1, \dots, s_n) \in \mathbb{R}^n : 0 < s_1 < \cdots < s_n \leq t\right\}$. Since $|C| = \frac{t^n}{n!}$, we have shown that the $\mathbb{P}$-distribution of $N(t)$ is the Poisson measure $\pi_{t\delta_1}$. In particular, $\pi_{\delta_1}$ is the $\mathbb{P}$-distribution of $N(1)$.

I now want to use the same sort of calculation to show that $\{N(t) : t \in [0,\infty)\}$ is a **simple Poisson process**, that is, a Lévy process for $\pi_{\delta_1}$. (See Exercise 4.2.18 for another, perhaps preferable, approach.)

LEMMA 4.2.3. *For any $(s,t) \in [0,\infty)$, the $\mathbb{P}$-distribution of the increment $N(s+t)-N(s)$ is $\pi_{t\delta_1}$. In addition, for any $K \in \mathbb{Z}^+$ and $0 = t_0 < t_1 < \cdots < t_K$, the increments $\{N(t_k) - N(t_{k-1}) : 1 \le k \le K\}$ are independent.*

PROOF: What I have to show is that, for all $K \in \mathbb{Z}^+$, $0 = n_0 \le \cdots \le n_K$, and $0 = t_0 < t_1 < \cdots < t_K$,

$$\begin{aligned}\mathbb{P}\big(N(t_k) - N(t_{k-1}) &= n_k - n_{k-1},\, 1 \le k \le K\big)\\ &= \prod_{k=1}^{K} \frac{e^{-(t_k-t_{k-1})}(t_k - t_{k-1})^{n_k-n_{k-1}}}{(n_k - n_{k-1})!},\end{aligned}$$

which is equivalent to checking that

$$\mathbb{P}\big(N(t_k) = n_k,\, 1 \le k \le K\big) = e^{-t_K} \prod_{k=1}^{K} \frac{(t_k - t_{k-1})^{n_k-n_{k-1}}}{(n_k - n_{k-1})!}.$$

We will work by induction on $K$, and the case $K = 1$ is already covered. Thus, assume that $K \ge 2$ and the the result has been proved for values smaller that $K$. There is nothing to do when $n_K = 0$, and so we will assume that $n_K \ge 1$. If $n_1 = 0$, set $K_0 = \min\{k : n_k \ge 1\}$. Then $2 \le K_0 \le K$, and so, by the induction hypothesis,

$$\begin{aligned}&\mathbb{P}\big(N(t_k) = n_k,\, 1 \le k \le K\big) = \mathbb{P}\big(\tau_1 > t_{K_0-1}\ \&\ T_{n_k} \le t_k < T_{n_k+1},\, K_0 \le k \le K\big)\\ &\quad = \mathbb{P}\big(\tau_1 \in (t_{K_0-1}, t_{K_0}]\ \&\ T_{n_k} - \tau_1 \le t_k - \tau_1 < T_{n_k+1} - \tau_1,\, K_0 \le k \le K\big)\\ &\quad = \int_{t_{K_0-1}}^{t_{K_0}} e^{-t}\mathbb{P}\big(T_{n_k-1} \le t_k - t < T_{n_k},\, K_0 \le k \le K\big)\, dt\\ &\quad = e^{-t_K} \prod_{K_0<k\le K} \frac{(t_k - t_{k-1})^{n_k-n_{k-1}}}{(n_k - n_{k-1})!} \int_{t_{K_0-1}}^{t_{K_0}} \frac{(t_{K_0} - t)^{n_{K_0}-1}}{(n_{K_0} - 1)!}\, dt\\ &\quad = e^{-t_K} \prod_{1\le k\le K} \frac{(t_k - t_{k-1})^{n_k-n_{k-1}}}{(n_k - n_{k-1})!}.\end{aligned}$$

If $n_1 \ge 1$, then, by the either by the induction hypothesis or the preceding,

$$\begin{aligned}&\mathbb{P}\big(N(t_k) = n_k,\, 1 \le k \le K\big)\\ &= \mathbb{P}\big(T_{n_1} \le t_1\ \&\ T_{n_1+1} - T_{n_1} > t_1 - T_{n_1}\\ &\qquad\qquad \&\ T_{n_k} - T_{n_1} \le t_k - T_{n_1} < T_{n_k+1} - T_{n_1}\ 2 \le k \le K\big)\\ &= \frac{1}{(n_1-1)!}\int_0^{t_1} t^{n_1-1}e^{-t}\mathbb{P}\big(N(t_1 - t) = 0\ \&\ N(t_k - t) = n_k - n_1,\, 2 \le k \le K\big)\, dt\\ &= \frac{e^{-t_K}}{(n_1-1)!}\int_0^{t_1} t^{n_1-1} \prod_{k=2}^{K} \frac{(t_k - t_{k-1})^{n_k-n_{k-1}}}{(n_k - n_{k-1})!}\, dt = e^{-t_K} \prod_{k=1}^{K} \frac{(t_k - t_{k-1})^{n_k-n_{k-1}}}{(n_k - n_{k-1})!},\end{aligned}$$

which completes the proof. □

The simple Poisson process $\{N(t) : t \geq 0\}$ is aptly named. It starts at 0, waits a unit exponential holding time before jumping to 1, sits at 1 for another, independent, unit exponential holding time before jumping to 2, etc. Thus, since $\pi_{\delta_1}$ is the distribution of this process at time 1, we now have an appealing picture of the way in which simple Poisson random variables arise.

Given $\alpha \in [0, \infty)$, I will say that a $D(\mathbb{R})$-valued process whose distribution is the same as that of $\{N(\alpha t) : t \geq 0\}$ is a **simple Poisson process run at rate** $\alpha$.

§ **4.2.2. Compound Poisson Processes.** I next want to build a Poisson process associated with a general $M \in \mathfrak{M}_0(\mathbb{R}^N)$. If $M = 0$, there is nothing to do, since the corresponding process will simply sit at $\mathbf{0}$ for all time. If $M \neq 0$, I write it as $\alpha\nu$, where $\alpha = M(\mathbb{R}^N)$ and $\nu = \frac{M}{\alpha}$. After augmenting the probability space if necessary, I introduce a sequence $\{\mathbf{X}_n : n \geq 1\}$ of mutually independent, $\nu$-distributed, random variables that are independent of the unit exponential random variables $\{\tau_m : m \geq 1\}$ out of which I built the simple Poisson process $\{N(t) : t \geq 0\}$ in the preceding subsection. Further, since $M(\{\mathbf{0}\}) = 0$, I may and will assume that none of the $\mathbf{X}_n$'s is ever $\mathbf{0}$. Finally, set

$$\mathbf{Z}_M(t, \omega) = \sum_{1 \leq n \leq N(\alpha t, \omega)} \mathbf{X}_n(\omega), \tag{4.2.4}$$

with the understanding that a sum over the empty set is $\mathbf{0}$.

Clearly, the process $\{\mathbf{Z}_M(t) : t \geq 0\}$ is nearly as easily understood as is the simple Poisson process. Like the simple Poisson process, its paths are right-continuous, start at $\mathbf{0}$, and are piecewise constant. Further, its holding times and jumps are all independent of one another. The difference is that its holding times are now $\alpha$-exponential random variables (i.e., exponential with mean value $\frac{1}{\alpha}$) and its jumps are random variables with distribution $\nu$. In particular,

$$j(t, \cdot\,, \mathbf{Z}_M(\cdot\,, \omega)) = \sum_{1 \leq n \leq N(\alpha t, \omega)} \delta_{\mathbf{X}_n(\omega)} = \sum_{n=1}^{\infty} \mathbf{1}_{[T_n(\omega), \infty)}(t)\delta_{\mathbf{X}_n(\omega)}. \tag{4.2.5}$$

I now want to check that $\{\mathbf{Z}_M(t) : t \geq 0\}$ is a Lévy process for $\pi_M$ and, as such, deserves to be called a Poisson process associated with $M$: the one with **rate** $M(\mathbb{R}^N)$ and **jump distribution** $\frac{M}{M(\mathbb{R}^N)}$. That is, I want to show that, for each $0 = t_0 < t_1 < \cdots t_K$, the random variables $\mathbf{Z}_M(t_k) - \mathbf{Z}_M(t_{k-1})$, $1 \leq k \leq K$, are mutually independent and that the $k$th one has distribution $\pi_{(t_k - t_{k-1})M}$. Equivalently, I need to check that, for any $\boldsymbol{\xi}_1, \ldots, \boldsymbol{\xi}_K \in \mathbb{R}^N$,

$$\mathbb{E}^{\mathbb{P}}\left[\exp\left(\sqrt{-1}\sum_{k=1}^{K}\left(\boldsymbol{\xi}_k, \mathbf{Z}_M(t_k) - \mathbf{Z}_M(t_{k-1})\right)_{\mathbb{R}^N}\right)\right] = \prod_{k=1}^{K} \widehat{\pi_{\tau_k M}}(\boldsymbol{\xi}_k),$$

where $\tau_k = t_k - t_{k-1}$. But, because of our independence assumptions, the above expectation is equal to

$$\sum_{n_K \geq \cdots \geq n_1 \geq 0} \mathbb{P}\big(N(\alpha t_k) - N(\alpha t_{k-1}) = n_k - n_{k-1},\ 1 \leq k \leq K\big)$$

$$\times\, \mathbb{E}^{\mathbb{P}}\left[\exp\left(\sqrt{-1}\sum_{k=1}^{K} \sum_{n_{k-1}+1 < m \leq n_k} (\boldsymbol{\xi}_k, \mathbf{X}_m)_{\mathbb{R}^N}\right)\right]$$

$$= \sum_{n_K \geq \cdots \geq n_1 \geq 0} \prod_{k=1}^{K} \frac{e^{-\alpha\tau_k} \tau_k^{n_k - n_{k-1}}}{(n_k - n_{k-1})!} \hat{\nu}(\boldsymbol{\xi}_k)^{n_k - n_{k-1}} = \prod_{k=1}^{K} \widehat{\pi_{\tau_k M}}(\boldsymbol{\xi}_k).$$

Any stochastic process $\{\mathbf{Z}(t) : t \geq 0\}$ with right-continuous, piecewise constant paths and the same distribution as the process $\{\mathbf{Z}_M(t) : t \geq 0\}$ just constructed is called a **Poisson process** associated with $M$.

Here is a beautiful and important procedure for transforming one Poisson process into another.

LEMMA 4.2.6. *Suppose that $F : \mathbb{R}^N \longrightarrow \mathbb{R}^{N'}$ is a Borel measurable function that takes the origin in $\mathbb{R}^N$ into the origin in $\mathbb{R}^{N'}$, and, for $M \in \mathfrak{M}_0(\mathbb{R}^N)$, define $M^F \in \mathfrak{M}_0(\mathbb{R}^{N'})$ by*

$$M^F(\Gamma) = M\Big(F^{-1}(\Gamma \setminus \{\mathbf{0}\})\Big) \quad \text{for } \Gamma \in \mathcal{B}_{\mathbb{R}^{N'}}.$$

*If $\{\mathbf{Z}(t) : t \geq 0\}$ is a Poisson process associated with $\pi_M$ and*

$$\mathbf{Z}^F(t,\omega) = \int_{\mathbb{R}^N} F(\mathbf{y})\, \jmath\big(t, d\mathbf{y}, \mathbf{Z}(\,\cdot\,,\omega)\big) \quad \text{for } (t,\omega) \in [0,\infty) \times \Omega, \tag{4.2.7}$$

*then $\{\mathbf{Z}^F(t) : t \geq 0\}$ is a Poisson process associated with $\pi_{M^F}$. Moreover, if, for each $i$ in an index set $\mathcal{I}$, $F_i : \mathbb{R}^N \longrightarrow \mathbb{R}^{N_i}$ is a Borel measurable satisfying $F_i(\mathbf{0}) = \mathbf{0}$ and, for each $\mathbf{y} \in \mathbb{R}^N$, there is at most one $i \in \mathcal{I}$ for which $F_i(\mathbf{y}) \neq \mathbf{0}$, then the processes $\big\{\{\mathbf{Z}^{F_i}(t) : t \geq 0\} : i \in \mathcal{I}\big\}$ are mutually independent.*

PROOF: In proving the first part, I will, without loss in generality, assume that (cf. (4.2.4)) $\mathbf{Z} = \mathbf{Z}_M$. But then, by (4.2.5),

$$\mathbf{Z}^F(t,\omega) = \sum_{1 \leq n \leq N(\alpha t,\omega)} F\big(\mathbf{X}_n(\omega)\big),$$

from which the first assertion follows immediately from the same computation with which I just showed that $\{\mathbf{Z}_M(t) : t \geq 0\}$ is a Poisson process associated with $M$.

To prove the second assertion, I begin by observing that it suffices to treat the case when $\mathcal{I} = \{1,2\}$. To see this, suppose that we knew the result in that case, and let $n > 2$ and a set $\{i_1, \dots, i_n\}$ of distinct elements from $\mathcal{I}$ be given. By taking $F_1 = (F_{i_1}, \dots, F_{i_{n-1}})$, $F_2 = F_{i_n}$, and applying the assumed result, we would have that $\{\mathbf{Z}^{F_{i_n}}(t) : t \geq 0\}$ is independent of $\big\{\big(\mathbf{Z}^{F_{i_1}}(t), \dots, \mathbf{Z}^{F_{i_{n-1}}}(t)\big) : t \geq 0\big\}$. Hence, proceeding by induction, we would be able to show that the processes $\big\{\{\mathbf{Z}^{F_{i_m}}(t) : t \geq 0\} : 1 \leq m \leq n\big\}$ are independent.

Now assume that $\mathcal{I} = \{1,2\}$. What I have to check is that, for any $K \in \mathbb{Z}^+$, $0 = t_0 < t_1 < \cdots < t_K$, and $\{(\boldsymbol{\xi}_k^1, \boldsymbol{\xi}_k^2) : 1 \leq k \leq K\} \subseteq \mathbb{R}^{N_1} \times \mathbb{R}^{N_2}$,

$$\begin{aligned}
\mathbb{E}^{\mathbb{P}}\Bigg[\exp\Bigg(\sqrt{-1}\sum_{k=1}^K \Big[\big(\boldsymbol{\xi}_k^1, \mathbf{Z}^{F_1}(t_k) - \mathbf{Z}^{F_1}(t_{k-1})\big)_{\mathbb{R}^{N_1}} \\
+ \big(\boldsymbol{\xi}_k^2, \mathbf{Z}^{F_2}(t_k) - \mathbf{Z}^{F_2}(t_{k-1})\big)_{\mathbb{R}^{N_2}}\Big]\Bigg)\Bigg] \\
= \mathbb{E}^{\mathbb{P}}\left[\exp\left(\sqrt{-1}\sum_{k=1}^K \big(\boldsymbol{\xi}_k^1, \mathbf{Z}^{F_1}(t_k) - \mathbf{Z}^{F_1}(t_{k-1})\big)_{\mathbb{R}^{N_1}}\right)\right] \\
\times \mathbb{E}^{\mathbb{P}}\left[\exp\left(\sqrt{-1}\sum_{k=1}^K \big(\boldsymbol{\xi}_k^2, \mathbf{Z}^{F_2}(t_k) - \mathbf{Z}^{F_2}(t_{k-1})\big)_{\mathbb{R}^{N_2}}\right)\right].
\end{aligned}$$

For this purpose, take $F : \mathbb{R}^N \longrightarrow \mathbb{R}^{N_1+N_2}$ to be given by $F(\mathbf{y}) = \big(F_1(\mathbf{y}), F_2(\mathbf{y})\big)$, and set $\boldsymbol{\xi}_k = (\boldsymbol{\xi}_k^1, \boldsymbol{\xi}_k^2)$. Then the first expression in the preceding equals

$$\begin{aligned}
\mathbb{E}^{\mathbb{P}}\left[\exp\left(\sqrt{-1}\sum_{k=1}^K \big(\boldsymbol{\xi}_k, \mathbf{Z}^F(t_k) - \mathbf{Z}^F(t_{k-1})\big)_{\mathbb{R}^{N_1+N_2}}\right)\right] \\
= \prod_{k=1}^K \mathbb{E}^{\mathbb{P}}\Big[\exp\Big(\sqrt{-1}\big(\boldsymbol{\xi}_k, \mathbf{Z}^F(t_k - t_{k-1})\big)_{\mathbb{R}^{N_1+N_2}}\Big)\Big],
\end{aligned}$$

since $\{\mathbf{Z}^F(t) : t \geq 0\}$ has independent, homogeneous increments. Hence, it suffices to observe that, for any $t > 0$ and $\boldsymbol{\xi} = (\boldsymbol{\xi}^1, \boldsymbol{\xi}^2)$,

$$\begin{aligned}
\mathbb{E}^P\Big[\exp\Big(\big(\boldsymbol{\xi}, \mathbf{Z}^F(t)\big)_{\mathbb{R}^{N_1+N_2}}\Big)\Big] = \exp\left(t\int_{\mathbb{R}^N}\Big(e^{\sqrt{-1}(\boldsymbol{\xi}, F(\mathbf{y}))_{\mathbb{R}^{N_1+N_2}}} - 1\Big)\,M(d\mathbf{y})\right) \\
= \exp\left(t\int_{\mathbb{R}^N}\Big(e^{\sqrt{-1}(\boldsymbol{\xi}^1, F_1(\mathbf{y}))_{\mathbb{R}^{N_1}}} - 1\Big)\,M(d\mathbf{y})\right) \\
\times \exp\left(t\int_{\mathbb{R}^N}\Big(e^{\sqrt{-1}(\boldsymbol{\xi}^2, F_2(\mathbf{y}))_{\mathbb{R}^{N_2}}} - 1\Big)\,M(d\mathbf{y})\right) \\
= \mathbb{E}^P\Big[\exp\Big(\big(\boldsymbol{\xi}^1, \mathbf{Z}^{F_1}(t)\big)_{\mathbb{R}^{N_1}}\Big)\Big]\mathbb{E}^P\Big[\exp\Big(\big(\boldsymbol{\xi}^2, \mathbf{Z}^{F_2}(t)\big)_{\mathbb{R}^{N_2}}\Big)\Big]. \quad \square
\end{aligned}$$

As an essentially immediate consequence of Lemma 4.2.6 and Theorem 4.1.8, we have the following important conclusion.

THEOREM 4.2.8. *If $\{\mathbf{Z}(t) : t \geq 0\}$ is a Poisson process associated with $M \in \mathfrak{M}_0(\mathbb{R}^N)$, then, for each $\Delta \in \mathcal{B}_{\mathbb{R}^N \setminus \{\mathbf{0}\}}$, $\{j(t, \Delta, \mathbf{Z}(\,\cdot\,)) : t \geq 0\}$ is a simple Poisson process run at rate $M(\Delta)$. Moreover, if*

$$\mathbf{Z}^\Delta(t) = \int_\Delta \mathbf{y}\, j(t, d\mathbf{y}, \mathbf{Z}) \text{ and } M^\Delta(\Gamma) = M(\Delta \cap \Gamma) \text{ for } \Gamma \in \mathcal{B}_{\mathbb{R}^N},$$

*then $\{\mathbf{Z}^\Delta(t) : t \geq 0\}$ is the Poisson process associated with $M^\Delta$ and $j(t, \Gamma, \mathbf{Z}^\Delta) = j(t, \Gamma \cap \Delta, \mathbf{Z})$ for all $(t, \Gamma) \in [0, \infty) \times \mathcal{B}_{\mathbb{R}^N}$. Finally, if $\{\Delta_i : i \in \mathcal{I}\}$ is a family of mutually disjoint Borel subsets of $\mathbb{R}^N \setminus \{\mathbf{0}\}$, then both the Poisson processes $\{\{\mathbf{Z}^{\Delta_i}(t) : t \geq 0\} : i \in \mathcal{I}\}$ as well as the jump processes $\{\{j(t, \Delta_i, \mathbf{Z}) : t \geq 0\} : i \in \mathcal{I}\}$ are mutually independent.*

The result in Theorem 4.2.8 says that the jumps of a Poisson process can be decomposed into a family of mutually independent, simple Poisson processes run at rates determined by the $M$-measure of the jump sizes. The next result can be thought of as a re-assembly procedure that complements this decomposition result.

THEOREM 4.2.9. *If $\{\{\mathbf{Z}_k(t) : t \geq 0\} : 1 \leq k \leq K\}$ are mutually independent Poisson processes associated with $\{M_k : 1 \leq k \leq K\} \subseteq \mathfrak{M}_0(\mathbb{R}^N)$, then*

$$\left\{\mathbf{Z}(t) \equiv \sum_{k=1}^K \mathbf{Z}_k(t) : t \geq 0\right\} \text{ is a Poisson process associated with } M \equiv \sum_{k=1}^K M_k.$$

*Next, suppose that the $M_k$'s are mutually singular in the sense that, for each $k$, there exists a $\Delta_k \in \mathcal{B}_{\mathbb{R}^N \setminus \{\mathbf{0}\}}$ with the properties that $\Delta_k \cap \Delta_\ell = \emptyset$ and $M_k(\Delta_k\complement) = 0 = M_\ell(\Delta_k)$ for $\ell \neq k$. Then, for $\mathbb{P}$-almost every $\omega \in \Omega$,*

$$j(t, \cdot\,, \mathbf{Z}(\,\cdot\,, \omega)) = \sum_{k=1}^K j(t, \cdot\,, \mathbf{Z}_k(\,\cdot\,, \omega)), \quad t \in [0, \infty).$$

*Equivalently, for $\mathbb{P}$-almost every $\omega \in \Omega$ and all $t \geq 0$, there is at most one $k$ such that $\mathbf{Z}_k(t, \omega) \neq \mathbf{Z}_k(t-, \omega)$.*

PROOF: Clearly, $\{\mathbf{Z}(t) : t \geq 0\}$ starts at $\mathbf{0}$ and has independent increments. In addition, for any $s, t \in [0, \infty)$ and $\boldsymbol{\xi} \in \mathbb{R}^N$,

$$\begin{aligned}
\mathbb{E}^{\mathbb{P}}\left[e^{\sqrt{-1}(\boldsymbol{\xi}, \mathbf{Z}(s+t) - \mathbf{Z}(s))_{\mathbb{R}^N}}\right] &= \prod_{k=1}^K \mathbb{E}^{\mathbb{P}}\left[e^{\sqrt{-1}(\boldsymbol{\xi}, \mathbf{Z}_k(s+t) - \mathbf{Z}_k(s))_{\mathbb{R}^N}}\right] \\
&= \prod_{k=1}^K \exp\left(t \int_{\mathbb{R}^N} \left(e^{\sqrt{-1}(\boldsymbol{\xi}, \mathbf{y})_{\mathbb{R}^N}} - 1\right) M_k(d\mathbf{y})\right) \\
&= \exp\left(t \int_{\mathbb{R}^N} \left(e^{\sqrt{-1}(\boldsymbol{\xi}, \mathbf{y})_{\mathbb{R}^N}} - 1\right) M(d\mathbf{y})\right).
\end{aligned}$$

Now assume that the $M_k$'s are as in the final part of the statement, and choose $\Delta_k$'s accordingly. Without loss in generality, I will assume that $\mathbb{R}^N \setminus \{\mathbf{0}\} = \bigcup_{k=1}^K \Delta_k$. Also, because the assertion depends only on the joint distribution of the processes involved, I may and will assume that

$$\mathbf{Z}_k(t) = \int_{\Delta_k} \mathbf{y}\, \jmath(t, d\mathbf{y}, \mathbf{Z}) \quad \text{for } 1 \le k \le K,$$

since then $\mathbf{Z}(t) = \sum_{k=1}^K \mathbf{Z}_k(t)$, and, by Theorem 4.2.8, the $\mathbf{Z}_k$'s are independent and the $k$th one is a Poisson process associated with $M_k$. But with this choice, another application of Theorem 4.2.8 shows that $\jmath(t, \Gamma, \mathbf{Z}_k) = \jmath(t, \Gamma \cap \Delta_k, \mathbf{Z})$, and therefore

$$\jmath(t, \Gamma, \mathbf{Z}) = \sum_{k=1}^K \jmath(t, \Gamma, \mathbf{Z}_k), \quad t \in [0, \infty). \quad \square$$

Because the paths of a Poisson process are piecewise constant, they certainly have finite variation on each compact time interval. The first part of the next lemma provides an estimate of that variation. The estimate in the second part will be used in § 4.2.5.

LEMMA 4.2.10. *If $\{\mathbf{Z}(t) : t \ge 0\}$ is a Poisson process associated with $M \in \mathfrak{M}_0(\mathbb{R}^N)$, then*

$$\mathbb{E}^{\mathbb{P}}\big[\mathrm{var}_{[0,t]}(\mathbf{Z})\big] = t \int_{\mathbb{R}^N} |\mathbf{y}|\, M(dy).$$

*In addition, if $\int_{\mathbb{R}^N} |\mathbf{y}|\, M(d\mathbf{y}) < \infty$ and $\bar{\mathbf{Z}}(t) = \mathbf{Z}(t) - \int_{\mathbb{R}^N} \mathbf{y}\, M(d\mathbf{y})$, then*

$$\mathbb{P}(\|\bar{\mathbf{Z}}\|_{[0,t]} \ge R) \le \frac{N^2 t}{R^2} \mathbb{E}^{\mathbb{P}}\big[|\bar{\mathbf{Z}}(t)|^2\big] = \frac{N^2 t}{R^2} \int_{\mathbb{R}^N} |\mathbf{y}|^2\, M(d\mathbf{y}).$$

PROOF: Again I will assume that (cf. (4.2.4)) $\mathbf{Z} = \mathbf{Z}_M$, in which case

$$\mathrm{var}_{[0,t]}(\mathbf{Z}) = \sum_{1 \le m \le N(\alpha t)} |\mathbf{X}_m|.$$

Hence (cf. the notation used in § 4.1.1)

$$\mathbb{E}^{\mathbb{P}}\big[\mathrm{var}_{[0,t]}(\mathbf{Z})\big] = \mathbb{E}^{\mathbb{P}}\big[N(\alpha t)\big]\mathbb{E}^{\mathbb{P}}\big[|\mathbf{X}_1|\big] = \alpha t \int_{\mathbb{R}^N} |\mathbf{y}|\, \nu(d\mathbf{y}) = t \int_{\mathbb{R}^N} |\mathbf{y}|\, M(d\mathbf{y}).$$

Turning to the second part, begin by observing that

$$\begin{aligned}\mathbb{P}(\|\bar{\mathbf{Z}}\|_{[0,t]} > R) &= \lim_{n\to\infty} \mathbb{P}\left( \max_{1 \le m \le 2^n} \big|\bar{\mathbf{Z}}(m2^{-n}t)\big| > R\right) \\ &\le N \lim_{n\to\infty} \sup_{\mathbf{e} \in \mathbb{S}^{N-1}} \mathbb{P}\left( \max_{1 \le m \le 2^n} \big|\big(\mathbf{e}, \bar{\mathbf{Z}}(m2^{-n}t)\big)_{\mathbb{R}^N}\big| > N^{-\frac{1}{2}} R\right).\end{aligned}$$

Next, given $\mathbf{e} \in \mathbb{S}^{N-1}$ and $n \geq 1$, write

$$\left(\mathbf{e}, \bar{\mathbf{Z}}(m2^{-n}t)\right)_{\mathbb{R}^N} = \sum_{1 \leq \ell \leq m} \left(\mathbf{e}, \bar{\mathbf{Z}}(\ell 2^{-n}t) - \bar{\mathbf{Z}}((\ell-1)2^{-n}t)\right)_{\mathbb{R}^N},$$

and apply Kolmogorov's Inequality to conclude that

$$\mathbb{P}\left(\max_{1 \leq m \leq 2^n} \left|\left(\mathbf{e}, \bar{\mathbf{Z}}(m2^{-n}t)\right)_{\mathbb{R}^N}\right| > N^{-\frac{1}{2}}R\right) \leq NR^{-2}\mathbb{E}^{\mathbb{P}}\left[\left(\mathbf{e}, \bar{\mathbf{Z}}(t)\right)^2_{\mathbb{R}^N}\right].$$

Thus, we will be done once I check that $\mathbb{E}^{\mathbb{P}}\left[|\bar{\mathbf{Z}}_M(t)|^2\right] = t\int_{\mathbb{R}^N} |\mathbf{y}|^2\, M(d\mathbf{y})$. To this end, first note that $\mathbb{E}^{\mathbb{P}}\left[|\bar{\mathbf{Z}}(t)|^2\right] = \mathbb{E}^{\mathbb{P}}\left[|\mathbf{Z}(t)|^2\right] - \alpha^2 t^2 |\mathbf{m}|^2$, where $\mathbf{m} = \int_{\mathbb{R}^N} \mathbf{y}\,\nu(d\mathbf{y})$. At the same time, if $\bar{\mathbf{X}}_m = \mathbf{X}_m - \mathbf{m}$, then $\mathbb{E}^{\mathbb{P}}\left[|\mathbf{Z}(t)|^2\right]$ equals

$$\mathbb{E}^{\mathbb{P}}\left[\left|\sum_{1 \leq m \leq N(\alpha t)} \mathbf{X}_m\right|^2\right] = \mathbb{E}^{\mathbb{P}}\left[\left|\sum_{1 \leq m \leq N(\alpha t)} \bar{\mathbf{X}}_m\right|^2\right] + |\mathbf{m}|^2 \mathbb{E}^{\mathbb{P}}\left[N(\alpha t)^2\right]$$

$$= \alpha t \mathbb{E}^{\mathbb{P}}\left[|\bar{\mathbf{X}}_1|^2\right] + |\mathbf{m}|^2\left(\alpha^2 t^2 + \alpha t\right) = \alpha t \mathbb{E}^{\mathbb{P}}\left[|\mathbf{X}_1|^2\right] + \alpha^2 t^2 |\mathbf{m}|^2.$$

Thus, since $\alpha \mathbb{E}^{\mathbb{P}}\left[|\mathbf{X}_1|^2\right] = \int_{\mathbb{R}^N} |\mathbf{y}|^2\, M(d\mathbf{y})$, the desired equality follows. □

§ **4.2.3. Poisson Jump Processes.** Rather than attempting to construct more general Lévy processes directly, I will first construct their jump processes and then construct them out of their jumps. With this idea in mind, given a probability space $(\Omega, \mathcal{F}, \mathbb{P})$, I will say that $(t, \omega) \rightsquigarrow j(t, \cdot, \omega)$ is a **Poisson jump process** associated with $M \in \mathfrak{M}_\infty(\mathbb{R}^N)$ if, for each $\omega \in \Omega$, $t \rightsquigarrow j(t, \cdot, \omega)$ is a jump function, and for each $n \in \mathbb{Z}^+$ and collection $\{\Delta_1, \dots, \Delta_n\}$ of mutually disjoint Borel subsets sets of $\mathbb{R}^N$ satisfying $\mathbf{0} \notin \bigcup_{i=1}^n \overline{\Delta_i}$, $\{\{j(t, \Delta_i) : t \geq 0\} : 1 \leq i \leq n\}$ are mutually independent, simple Poisson processes, the $i$th of which is run at rate $M(\Delta_i)$ for each $1 \leq i \leq n$. By starting with simple functions and passing to limits, one can easily check that

$$(t, \omega) \in [0, \infty) \times \Omega \longmapsto \int \varphi(\mathbf{y})\, j(t, d\mathbf{y}, \omega) \in [0, \infty]$$

is measurable for every Borel measurable function $\varphi : \mathbb{R}^N \longrightarrow [0, \infty]$. Therefore, if $F : \mathbb{R}^N \longrightarrow \mathbb{R}^{N'}$ is a Borel measurable function, and, for $T > 0$,

$$\Omega(T) \equiv \left\{\omega : \int |F(\mathbf{y})|\, j(T, d\mathbf{y}, \omega) < \infty\right\},$$

then both the set $\Omega(T)$ and the function

$$(t, \omega) \in [0, T] \times \Omega(T) \rightsquigarrow \int F(\mathbf{y})\, j(t, \mathbf{y}, \omega) \in \mathbb{R}^{N'}$$

are measurable. Note that if $|F(\mathbf{y})|$ vanishes for $\mathbf{y}$'s in a neighborhood of $\mathbf{0}$, then $\Omega(T) = \Omega$ for all $T > 0$.

My goal in this subsection is to prove the following existence result.

THEOREM 4.2.11. *For each $M \in \mathfrak{M}_\infty(\mathbb{R}^N)$ there exists an associated Poisson jump process. (See § 9.2.2 for another approach.)*

PROOF: Set $A_0 = \mathbb{R}^N\setminus\overline{B(\mathbf{0},1)}$ and $A_k = \overline{B(\mathbf{0},2^{-k+1})}\setminus\overline{B(\mathbf{0},2^{-k})}$ for $k \in \mathbb{Z}^+$, and define $M_k(d\mathbf{y}) = \mathbf{1}_{A_k}(\mathbf{y})\,M(d\mathbf{y})$. Next, choose mutually independent Poisson processes $\{\{\mathbf{Z}_k(t) : t \geq 0\} : k \in \mathbb{N}\}$ so that the $k$th one is associated with $M_k$, and set $j_k(t,\cdot,\omega) = j(t,\cdot,\mathbf{Z}_k(\cdot,\omega))$. Without loss in generality, I may and will assume that $j_k(t, A_k\complement, \omega) = 0$ for all $(t,\omega) \in [0,\infty)\times\Omega$ and $k \in \mathbb{N}$. In addition, by Theorem 4.2.9, if $\mathbf{Z}^{(m)}(t) = \sum_{k=0}^m \mathbf{Z}_k(t)$, then we know that, for $\mathbb{P}$-almost every $\omega \in \Omega$,

$$j^{(m)}(t,\cdot,\omega) \equiv j(t,\cdot,\mathbf{Z}^{(m)}(\cdot,\omega)) = \sum_{k=0}^m j_k(t,\cdot,\omega), \quad t \geq 0.$$

Hence, I may and will assume that

$$t \rightsquigarrow j(t,\cdot,\omega) \equiv \sum_{k=1}^\infty j_k(t,\cdot,\omega)$$

is a jump function for all $\omega \in \Omega$. Finally, suppose that $\{\Delta_i : 1 \leq i \leq n\} \subseteq \mathcal{B}_{\mathbb{R}^N}$ are disjoint and that $\mathbf{0} \notin \bigcup_{i=1}^n \overline{\Delta_i}$. Choose $m \in \mathbb{N}$ so that $(\bigcup_1^n \Delta_i)\cap B(\mathbf{0},2^{-m}) = \emptyset$, and note that, $\mathbb{P}$-almost surely, $j(t,\Delta_i,\omega) = j^{(m)}(t,\Delta_i,\omega)$ for all $t \geq 0$ and $1 \leq i \leq n$. Hence, the required independence property is a consequence of the last part of Theorem 4.2.8. □

In preparation for the next section, I prove the following.

LEMMA 4.2.12. *Let $F : \mathbb{R}^N \longrightarrow \mathbb{R}^{N'}$ be a Borel measurable function satisfying $F(\mathbf{0}) = \mathbf{0}$ and $\mathbf{0} \notin \overline{F^{-1}(\mathbb{R}^{N'}\setminus B(\mathbf{0},r))}$ for any $r > 0$. For any $M \in \mathfrak{M}_\infty(\mathbb{R}^N)$, $M^F \in \mathfrak{M}_\infty(\mathbb{R}^{N'})$. Moreover, if $\{j(t,\cdot) : t \geq 0\}$ is a Poisson jump process associated with $M$ and $j^F(t,\Gamma,\omega) \equiv j(t,F^{-1}(\Gamma\setminus\{\mathbf{0}\}),\omega)$, then $\{j^F(t,\cdot) : t \geq 0\}$ is a Poisson jump process associated with (cf. Lemma 4.2.6) $M^F$. Finally, if $\mathbf{0} \notin \overline{F^{-1}(\mathbb{R}^{N'}\setminus\{\mathbf{0}\})}$ and*

$$\mathbf{Z}^F(t,\omega) \equiv \int \mathbf{y}\, j^F(t,d\mathbf{y},\omega) = \int F(\mathbf{y})\, j(t,d\mathbf{y},\omega),$$

*then $M^F \in \mathfrak{M}_0(\mathbb{R}^{N'})$, $\{\mathbf{Z}^F(t) : t \geq 0\}$ is a Poisson process associated with $M^F$, and $j(t,\cdot,\mathbf{Z}^F(\cdot,\omega)) = j^F(t,\cdot,\omega)$.*

PROOF: Clearly $M^F \in \mathfrak{M}_0(\mathbb{R}^N)$. To complete proof of the first assertion, suppose that $\{\Delta_1,\dots,\Delta_n\}$ are disjoint Borel subsets of $\mathbb{R}^{N'}$ and that $\mathbf{0} \notin \bigcup_{i=1}^n \overline{\Delta_i}$. Then $\{F^{-1}(\Delta_1),\dots,F^{-1}(\Delta_n)\}$ satisfy the same conditions as subsets of $\mathbb{R}^N$, and therefore, since $j^F(t,\Delta_i,\omega) = j(t, F^{-1}(\Delta_i),\omega)$, $\{\{j^F(t,\Delta_i) : t \geq 0\} : 1 \leq i \leq n\}$ has the required properties.

Turning to the second assertion, first note that $M^F \in \mathfrak{M}_0(\mathbb{R}^{N'})$ is an immediate consequence of $\mathbf{0} \notin \overline{F^{-1}(\mathbb{R}^{N'} \setminus \{\mathbf{0}\})}$ and that the equality $j\big(t, \cdot\,, \mathbf{Z}^F(\cdot\,,\omega)\big) = j^F(t, \cdot\,,\omega)$ is a trivial application of the final part of Theorem 4.1.8. To prove that $\{\mathbf{Z}^F(t) : t \geq 0\}$ is a Poisson process associated with $M^F$, use Theorem 4.2.8 to see that $\{j^F(t,\cdot\,) : t \geq 0\}$ has the same distribution as the jump process for a Poisson process $\{\mathbf{Z}(t) : t \geq 0\}$ associated with $M^F$. Hence, since $\mathbf{Z}(t) = \int \mathbf{y}\, j(t, d\mathbf{y}, \mathbf{Z})$, $\{\mathbf{Z}^F(t) : t \geq 0\}$ has the same distribution as $\{\mathbf{Z}(t) : t \geq 0\}$. □

**§ 4.2.4. Lévy Processes with Bounded Variation.** Although the contents of the previous section provide the machinery with which to construct a Lévy process for any $\mu$ with Fourier transform given by (4.2.1), for reasons made clear in the next lemma, I will treat the special case when $M \in \mathfrak{M}_1(\mathbb{R}^N)$ here and will deal with $M \in \mathfrak{M}_2(\mathbb{R}^N) \setminus \mathfrak{M}_1(\mathbb{R}^N)$ in the following subsection.

LEMMA 4.2.13. *Let $\{j(t,\cdot\,) : t \geq 0\}$ be a Poisson jump process associated with $M \in \mathfrak{M}_2(\mathbb{R}^N)$, and set $V(t,\omega) = \int |\mathbf{y}|\, j(t, d\mathbf{y},\omega)$. Then $V(t) < \infty$ almost surely or $V(t) = \infty$ almost surely for all $t > 0$, depending on whether $M$ is or is not in $\mathfrak{M}_1(\mathbb{R}^N)$. (See Exercise 4.3.21 to see that the same conclusion holds for any $M \in \mathfrak{M}_\infty(\mathbb{R}^N)$.)*

PROOF: Since $\int_{|\mathbf{y}|>1} |\mathbf{y}|\, j(t, d\mathbf{y},\omega) < \infty$ for all $(t,\omega) \in [0,\infty) \times \Omega$, the question is entirely about the finiteness of $V_0(t,\omega) \equiv \int_{\overline{B(\mathbf{0},1)}} |\mathbf{y}|\, j(t, d\mathbf{y},\omega)$. To study this question, set $A_k = \overline{B(\mathbf{0}, 2^{-k+1})} \setminus \overline{B(\mathbf{0}, 2^{-k})}$, $F_k(\mathbf{y}) = |\mathbf{y}|\mathbf{1}_{A_k}(\mathbf{y})$, and $V_k(t,\omega) = \int_{A_k} |\mathbf{y}|\, j(t, d\mathbf{y},\omega)$ for $k \geq 1$. Clearly, the processes $\{\{V_k(t) : t \geq 0\} : k \in \mathbb{Z}^+\}$ are mutually independent. In addition, for each $k$, $t \rightsquigarrow V_k(t)$ is non-decreasing and, by the second part of Lemma 4.2.12, $\{V_k(t) : t \geq 0\}$ is a Poisson process associated with $M^{F_k}$. Thus, by Lemma 4.2.10,

$$a_k \equiv \mathbb{E}^{\mathbb{P}}[V_k(t)] = t \int_{A_k} |\mathbf{y}|\, M(d\mathbf{y}) \text{ and } b_k \equiv \operatorname{Var}\big(V_k(t)\big) = t \int_{A_k} |\mathbf{y}|^2\, M(d\mathbf{y}).$$

From the first of these, it follows that

$$\mathbb{E}^{\mathbb{P}}\left[\int_{\overline{B(\mathbf{0},1)}} |\mathbf{y}|\, j(t, d\mathbf{y})\right] = \sum_{k=1}^{\infty} \mathbb{E}^{\mathbb{P}}[V_k(t)] = \int_{\overline{B(\mathbf{0},1)}} |\mathbf{y}|\, M(dy),$$

which finishes the case when $M \in \mathfrak{M}_1(\mathbb{R}^N)$. When $M \in \mathfrak{M}_2(\mathbb{R}^N) \setminus \mathfrak{M}_1(\mathbb{R}^N)$, set $\bar{V}_k(t) = V_k(t) - ta_k$. Then, for each $t > 0$, $\{\bar{V}_k(t) : k \in \mathbb{Z}^+\}$ is a sequence of mutually independent random variables with mean value 0. Furthermore,

$$\sum_{k=1}^{\infty} \operatorname{Var}\big(\bar{V}_k(t)\big) = t \sum_{k=1}^{\infty} b_k = t \int_{\overline{B(\mathbf{0},1)}} |\mathbf{y}|^2\, M(d\mathbf{y}) < \infty.$$

Hence, by Theorem 1.4.2, $\sum_{k=1}^{\infty} \bar{V}_k(t)$ converges $\mathbb{P}$-almost surely. But, because $M \notin \mathfrak{M}_1(\mathbb{R}^N)$, $\sum_{k=1}^{\infty} a_k = \infty$, and so, for each $t > 0$, $\sum_{k=1}^{\infty} V_k(t)$ must diverge $\mathbb{P}$-almost surely. □

Before stating the main result of the subsection, I want to introduce the notion of a **generalized Poisson measure**. Namely, if $M \in \mathfrak{M}_1(\mathbb{R}^N) \setminus \mathfrak{M}_0(\mathbb{R}^N)$ and $\pi_M$ is the element of $\mathcal{I}(\mathbb{R}^N)$ whose Fourier transform is given by

$$\exp\left(\int \left(e^{\sqrt{-1}(\boldsymbol{\xi},\mathbf{y})_{\mathbb{R}^N}} - 1\right) M(d\mathbf{y})\right),$$

or, equivalently, $\widehat{\pi_M}$ is given by (4.2.1) with $\mathbf{m} = \int_{B(\mathbf{0},1)} \mathbf{y}\, M(d\mathbf{y})$, then I will call $\pi_M$ the generalized Poisson measure for $M$. Similarly, if $\{\mathbf{Z}(t) : t \geq 0\}$ is a Lévy process for a generalized Poisson measure $\pi_M$, I will say that it is a **generalized Poisson process** associated with $M$.

THEOREM 4.2.14. *Suppose that $M \in \mathfrak{M}_1(\mathbb{R}^N)$ and that $\{j(t, \cdot) : t \geq 0\}$ is a Poisson jump process associated with $M$. Set $\mathcal{N} = \{\omega : \exists t > 0\ j(t, \cdot, \omega) \notin \mathfrak{M}_1(\mathbb{R}^N)\}$, and define $(t, \omega) \rightsquigarrow \mathbf{Z}_M(t, \omega)$ so that*

$$\mathbf{Z}_M(t, \omega) = \begin{cases} \int \mathbf{y}\, j(t, d\mathbf{y}, \omega) & \text{if } \omega \notin \mathcal{N} \\ \mathbf{0} & \text{if } \omega \in \mathcal{N}. \end{cases}$$

*Then $\mathbb{P}(\mathcal{N}) = 0$ and $\{\mathbf{Z}_M(t) : t \geq 0\}$ is a (possibly generalized) Poisson process associated with $M$. In particular, $t \rightsquigarrow \mathbf{Z}_M(t, \omega)$ is absolutely pure jump for all $\omega \in \Omega$, and $\{j(t, \cdot, \mathbf{Z}_M) : t \geq 0\}$ is a Poisson jump process associated with $M$. Finally, if $\mu \in \mathcal{I}(\mathbb{R}^N)$ has Fourier transform given by (4.2.1), then*

$$\left\{ t\left(\mathbf{m} - \int_{B(\mathbf{0},1)} \mathbf{y}\, M(d\mathbf{y})\right) + \mathbf{Z}_M(t) : t \geq 0 \right\}$$

*is a Lévy process for $\mu$.*

PROOF: That $\mathbb{P}(\mathcal{N}) = 0$ follows from Lemma 4.2.13. To prove that $\{\mathbf{Z}_M(t) : t \geq 0\}$ is a Lévy process for $\pi_M$, set

$$\mathbf{Z}^{(r)}(t, \omega) = \int_{|\mathbf{y}|>r} \mathbf{y}\, j(t, d\mathbf{y}, \omega)$$

for $r > 0$. By Lemma 4.2.12, $\{\mathbf{Z}^{(r)}(t) : t \geq 0\}$ is a Poisson process associated with $M^{(r)}(d\mathbf{y}) \equiv \mathbf{1}_{(r,\infty)}(\mathbf{y})\, M(d\mathbf{y})$. In addition, if $\omega \notin \mathcal{N}$, then $\mathbf{Z}^{(r)}(\cdot, \omega) \longrightarrow \mathbf{Z}_M(\cdot, \omega)$ uniformly on compacts, from which it is easy to check that $\{\mathbf{Z}_M(t) : t \geq 0\}$ is a Poisson process associated with $M$ and that the process in the last assertion is a Lévy process for the $\mu$ whose Fourier transform is given by (4.2.1) with this $M$. Finally, by the last part of Theorem 4.1.8, $j(t, \cdot, \mathbf{Z}_M(\cdot, \omega)) = j(t, \cdot, \omega)$ when $\omega \notin \mathcal{N}$, from which it is clear that $\{j(t, \cdot, \mathbf{Z}_M) : t \geq 0\}$ is a Poisson jump process associated with $M$. □

§ **4.2.5. General, Non-Gaussian Lévy Processes.** In this subsection I will complete the construction of Lévy processes with no Gaussian component.

THEOREM 4.2.15. *For each* $\mathbf{m} \in \mathbb{R}^N$ *and* $M \in \mathfrak{M}_2(\mathbb{R}^N)$ *there is a Lévy process for the* $\mu \in \mathcal{I}(\mathbb{R}^N)$ *whose Fourier transform is given by (4.2.1). Moreover, if* $\{\mathbf{Z}(t) : t \geq 0\}$ *is such a process, then* $\{j(t, \cdot, \mathbf{Z}) : t \geq 0\}$ *is a Poisson jump process associated with* $M$*. Finally, if, for* $r \in (0, 1]$,

$$\mathbf{Z}^{(r)}(t) = \int_{|\mathbf{y}|>r} \mathbf{y}\, j(t, d\mathbf{y}, \mathbf{Z}) - t \int_{r<|\mathbf{y}|\leq 1} \mathbf{y}\, M(d\mathbf{y}),$$

*then*

$$\mathbb{P}\left( \sup_{\tau\in[0,t]} \left|\mathbf{Z}(\tau) - \tau\mathbf{m} - \mathbf{Z}^{(r)}(\tau)\right| \geq \epsilon \right) \leq \frac{N^2 t}{\epsilon^2} \int_{\overline{B(\mathbf{0},r)}} |\mathbf{y}|^2\, M(d\mathbf{y}).$$

PROOF: Without loss in generality, I will assume that $\mathbf{m} = \mathbf{0}$.

By Theorem 4.2.11, we know that there is a Poisson jump process $\{j(t, \cdot) : t \geq 0\}$ associated with $M$. Take

$$\bar{j}(t, d\mathbf{y}, \omega) = j(t, d\mathbf{y}, \omega) - t\mathbf{1}_{\overline{B(\mathbf{0},1)}}(\mathbf{y}) M(d\mathbf{y}),$$

and define

$$\mathbf{Z}^{(r)}(t, \omega) = \int_{|\mathbf{y}|>r} \mathbf{y}\, \bar{j}(t, d\mathbf{y}, \omega), \quad (t, \omega) \in [0, \infty) \times \Omega,$$

for $r \in (0, 1]$. By Theorem 4.2.14, we know that $\{\mathbf{Z}^{(r)}(t) : t \geq 0\}$ is a Lévy process for $\mu^{(r)}$, where

$$\widehat{\mu^{(r)}}(\boldsymbol{\xi}) = \exp\left( \int_{|\mathbf{y}|>r} \left[ e^{\sqrt{-1}(\boldsymbol{\xi},\mathbf{y})_{\mathbb{R}^N}} - 1 - \sqrt{-1}\,\mathbf{1}_{[0,1]}(\mathbf{y}) (\boldsymbol{\xi}, \mathbf{y})_{\mathbb{R}^N} \right] M(d\mathbf{y}) \right).$$

Furthermore, by the second part of Lemma 4.2.10, we know that, for $0 < r < r' \leq 1$,

$$(*) \qquad \mathbb{P}\left(\|\mathbf{Z}^{(r')} - \mathbf{Z}^{(r)}\|_{[0,t]} \geq \epsilon\right) \leq \frac{N^2 t}{\epsilon^2} \int_{r<|\mathbf{y}|\leq r'} |\mathbf{y}|^2\, M(d\mathbf{y}).$$

Hence, if $1 \geq r_m \searrow 0$ is chosen so that

$$\int_{\overline{B(\mathbf{0},r_m)}} |\mathbf{y}|^2\, M(d\mathbf{y}) \leq 2^{-m},$$

then

$$\begin{aligned}\mathbb{P}\left( \sup_{n>m} \|\mathbf{Z}^{(r_n)} - \mathbf{Z}^{(r_m)}\|_{[0,t]} \geq \tfrac{1}{m} \right) &\leq \sum_{n\geq m} \mathbb{P}\left(\|\mathbf{Z}^{(r_{n+1})} - \mathbf{Z}^{(r_n)}\|_{[0,t]} \geq (m+1)^{-2}\right) \\ &\leq N^2 t \sum_{n=m}^{\infty} (n+1)^4 2^{-n},\end{aligned}$$

and therefore, by the first part of the Borel–Cantelli Lemma,

$$\mathbb{P}\left(\exists m\ \forall n \geq m\ \|\mathbf{Z}^{(r_n)} - \mathbf{Z}^{(r_m)}\|_{[0,t]} \leq \tfrac{1}{m+1}\right) = 1.$$

We now know that there is a $\mathbb{P}$-null set $\mathcal{N}$ such that, for any $\omega \notin \mathcal{N}$, there exists a $\mathbf{Z}(\,\cdot\,,\omega) \in D(\mathbb{R}^N)$ to which $\{\mathbf{Z}^{(r_m)}(\,\cdot\,,\omega) : n \geq 0\}$ converges uniformly on compacts. Thus, if we take $\mathbf{Z}(t,\omega) = \mathbf{0}$ for $(t,\omega) \in [0,\infty) \times \mathcal{N}$, then it is an easy matter to check that $\{\mathbf{Z}(t) : t \geq 0\}$ is a Lévy process for the $\mu \in \mathcal{I}(\mathbb{R}^N)$ whose Fourier transform is given by (4.2.1) with $\mathbf{m} = \mathbf{0}$. In addition, since, by Theorem 4.1.8, we know that $t \rightsquigarrow j(t,\,\cdot\,,\omega)$ is the jump function for $t \rightsquigarrow \mathbf{Z}(t,\omega)$ when $\omega \notin \mathcal{N}$, it is clear that $\{j(t,\,\cdot\,,\mathbf{Z}) : t \geq 0\}$ is a Poisson jump process associated with $M$. Finally, to prove the estimate in the concluding assertion, observe that, for $\omega \notin \mathcal{N}$, the path $t \rightsquigarrow \mathbf{Z}^{(r)}(t,\omega)$ used in our construction coincides with the path described in the statement. Thus, the desired estimate is an easy consequence of the one in (*) above. □

COROLLARY 4.2.16. *Let $\mu \in \mathcal{I}(\mathbb{R}^N)$ with Fourier transform given by (4.2.1), and suppose that $\{\mathbf{Z}(t) : t \geq 0\}$ is a Lévy process for $\mu$. Then, depending on whether or not $M \in \mathfrak{M}_1(\mathbb{R}^N)$, either $\mathbb{P}$-almost all or $\mathbb{P}$-almost none of the paths $t \rightsquigarrow \mathbf{Z}(t)$ has locally bounded variation. Moreover, if $M \in \mathfrak{M}_1(\mathbb{R}^N)$, then, $\mathbb{P}$-almost surely,*

$$t \rightsquigarrow \mathbf{Z}(t) - t\left(\mathbf{m} - \int_{B(\mathbf{0},1)} \mathbf{y}\, M(d\mathbf{y})\right) \quad \text{is an absolutely pure jump path.}$$

PROOF: From Theorem 4.2.14, we already know that $t \rightsquigarrow \mathbf{Z}(t) - t\mathbf{m}$ is almost surely an absolutely pure jump path if $M \in \mathfrak{M}_1(\mathbb{R}^N)$, and so $t \rightsquigarrow \mathbf{Z}(t)$ is almost surely of locally bounded variation. Conversely, if $t \rightsquigarrow \mathbf{Z}(t)$ has locally bounded variation with positive probability, then, by (4.1.13), $j(t,\,\cdot\,,\mathbf{Z}) \in \mathfrak{M}_1(\mathbb{R}^N)$ with positive probability. But then, since $\{j(t,\,\cdot\,,\mathbf{Z}) : t \geq 0\}$ is a Poisson jump process associated with $M$, it follows from Lemma 4.2.13 that $M \in \mathfrak{M}_1(\mathbb{R}^N)$. □

COROLLARY 4.2.17. *Let $\mu$ and $\{\mathbf{Z}(t) : t \geq 0\}$ be as in Corollary 4.2.16. Given $\Delta \in \mathcal{B}_{\mathbb{R}^N}$ with $\mathbf{0} \notin \bar{\Delta}$, set*

$$\mathbf{Z}^{\Delta}(t) = \int_{\Delta} \mathbf{y}\, j(t, d\mathbf{y}, \mathbf{Z}),\ M^{\Delta}(d\mathbf{y}) = \mathbf{1}_{\Delta}(\mathbf{y}) M(d\mathbf{y}),\ \text{and } \mathbf{m}^{\Delta} = \int_{B(\mathbf{0},1)} \mathbf{y}\, M^{\Delta}(d\mathbf{y}).$$

*Then $\{\mathbf{Z}^{\Delta}(t) : t \geq 0\}$ is a Poisson process associated with $M^{\Delta}$, $\{\mathbf{Z}(t) - \mathbf{Z}^{\Delta}(t) : t \geq 0\}$ is a Lévy process for the element of $\mathcal{I}(\mathbb{R}^N)$ whose Fourier transform is*

$$\exp\Bigg(\sqrt{-1}(\boldsymbol{\xi}, \mathbf{m} - \mathbf{m}^{\Delta})_{\mathbb{R}^N} + \int_{\mathbb{R}^N \setminus \Delta} \left[e^{\sqrt{-1}(\boldsymbol{\xi},\mathbf{y})_{\mathbb{R}^N}} - 1 - \sqrt{-1}\mathbf{1}_{[0,1]}(|\mathbf{y}|)(\boldsymbol{\xi}, \mathbf{y})_{\mathbb{R}^N}\right] M(d\mathbf{y})\Bigg),$$

*and $\{\mathbf{Z}(t) - \mathbf{Z}^{\Delta}(t) : t \geq 0\}$ is independent of $\{j(t,\,\cdot\,,\mathbf{Z}^{\Delta}) : t \geq 0\}$, and therefore of $\{\mathbf{Z}^{\Delta}(t) : t \geq 0\}$ as well.*

PROOF: That $\{\mathbf{Z}^\Delta(t) : t \geq 0\}$ is a Poisson process associated with $M^\Delta$ is an immediate consequence of Lemma 4.2.12. Next, define $\mathbf{Z}^{(r)}(t)$ as in Theorem 4.2.15. Then, for all $r \in (0,1]$,

$$\mathbf{Z}^{(r)}(t) - \mathbf{Z}^\Delta(t) = \int_{|\mathbf{y}|>r} \mathbf{1}_{\mathbb{R}^N\setminus\Delta}(\mathbf{y})\mathbf{y}\, j(t,d\mathbf{y}) - t\int_{r<|\mathbf{y}|\leq 1} \mathbf{y}\, M(d\mathbf{y}).$$

In particular, this means that $\{\mathbf{Z}^{(r)}(t)-\mathbf{Z}^\Delta(t) : t \geq 0\}$ has independent, homogeneous increments and (cf. Theorem 4.1.8) is independent of $\{j(t,\cdot\,,\mathbf{Z}^\Delta) : t \geq 0\}$. Thus, since, as $r \searrow 0$, $\mathbf{Z}^{(r)}(t) \longrightarrow \mathbf{Z}(t) - t\mathbf{m}$ in probability, it follows that $\{\mathbf{Z}(t) - \mathbf{Z}^\Delta(t) : t \geq 0\}$ is independent of $\{j(t,\cdot\,,\mathbf{Z}^\Delta) : t \geq 0\}$. In addition,

$$e^{-\sqrt{-1}t(\boldsymbol{\xi},\mathbf{m}-\mathbf{m}^\Delta)_{\mathbb{R}^N}}\mathbb{E}^{\mathbb{P}}\big[e^{\sqrt{-1}(\boldsymbol{\xi},\mathbf{Z}(t)-\mathbf{Z}^\Delta(t))_{\mathbb{R}^N}}\big] = \lim_{r\searrow 0}\mathbb{E}^{\mathbb{P}}\big[e^{\sqrt{-1}(\boldsymbol{\xi},\mathbf{Z}^{(r)}(t)-\mathbf{Z}^\Delta(t)+t\mathbf{m}^\Delta)_{\mathbb{R}^N}}\big]$$

$$= \lim_{r\searrow 0}\exp\left(\int_{(\Delta\cup\overline{B(\mathbf{0},r)})\complement}\Big[e^{\sqrt{-1}(\boldsymbol{\xi},\mathbf{y})_{\mathbb{R}^N}} - 1 - \sqrt{-1}\mathbf{1}_{[0,1]}\big(|\mathbf{y}|\big)\big(\boldsymbol{\xi},\mathbf{y}\big)_{\mathbb{R}^N}\Big]\,M(d\mathbf{y})\right)$$

$$= \exp\left(\int_{\mathbb{R}^N\setminus\Delta}\Big[e^{\sqrt{-1}(\boldsymbol{\xi},\mathbf{y})_{\mathbb{R}^N}} - 1 - \sqrt{-1}\mathbf{1}_{[0,1]}\big(|\mathbf{y}|\big)\big(\boldsymbol{\xi},\mathbf{y}\big)_{\mathbb{R}^N}\Big]\,M(d\mathbf{y})\right).$$

Hence, it follows that $\{\mathbf{Z}(t) - \mathbf{Z}^\Delta(t) : t \geq 0\}$ is a Lévy process for the specified element of $\mathcal{I}(\mathbb{R}^N)$. □

## Exercises for § 4.2

EXERCISE 4.2.18. Here is another proof that the process $\{N(t) : t \geq 0\}$ in § 4.2.1 has independent, homogeneous increments. Refer to the notation used there.

**(i)** Given $n \in \mathbb{Z}^+$ and measurable functions $f : [0,\infty)^{n+1} \longmapsto [0,\infty)$ and $g : [0,\infty)^n \longrightarrow \mathbb{R}$, show that

$$\mathbb{E}^{\mathbb{P}}\big[f(\tau_1,\dots,\tau_{n+1}),\ \tau_{n+1} > g(\tau_1,\dots,\tau_n)\big]$$
$$= \mathbb{E}^{\mathbb{P}}\big[e^{-g(\tau_1,\dots,\tau_n)^+} f\big(\tau_1,\dots,\tau_n,\tau_{n+1} + g(\tau_1,\dots,\tau_n)^+\big)\big].$$

**(ii)** Let $K \in \mathbb{Z}^+$, $0 = n_0 \leq n_1 \leq \cdots \leq n_K$, and $0 = t_0 \leq t_1 < \cdots < t_K = s$ be given, and set $A = \{N(t_k) = n_k,\ 1 \leq k \leq K\}$. Show that $A = B \cap \{\tau_{n_K+1} > s - T_{n_K}\}$, where $B \in \sigma(\{\tau_1,\dots,\tau_{n_K}\})$, and apply **(i)** to see that $\mathbb{P}(A) = \mathbb{E}^{\mathbb{P}}\big[e^{(s-T_{n_K})},\ B\big]$.

**(iii)** Let $n \in \mathbb{Z}^+$ and $t > 0$ be given, and set $h(\xi) = \mathbb{P}(T_{n-1} > \xi)$. Referring to **(ii)** and again using **(i)**, show that

$$\mathbb{P}\big(A \cap \{N(s+t) - N(s) < n\}\big) = \mathbb{E}^{\mathbb{P}}\big[h(t+s-T_{n_K+1}),\ B \cap \{\tau_{n_K+1} > s - T_{n_K}\}\big]$$
$$= \mathbb{E}^{\mathbb{P}}\big[e^{-(s-T_{n_K})}h(t-\tau_{n_K+1}),\ B\big] = \mathbb{E}^{\mathbb{P}}\big[h(t-\tau_{n_K+1})\big]\mathbb{E}^{\mathbb{P}}\big[e^{-(s-T_{n_K})},\ B\big]$$
$$= \mathbb{P}\big(N(t) < n\big)\mathbb{P}(A).$$

EXERCISE 4.2.19. Let $\{N(t) : t \ge 0\}$ be a simple Poisson process, and show that $\lim_{t\to\infty} \frac{N(t)}{t} = 1$ $\mathbb{P}$-almost surely.

**Hint**: First use The Strong Law of Large Numbers to show that $\lim_{n\to\infty} \frac{N(n)}{n} = 1$ $\mathbb{P}$-almost surely. Second, use

$$\mathbb{P}\left(\sup_{n\le t\le n+1} \frac{N(t)-N(n)}{t} \ge \epsilon\right) \le \mathbb{P}\big(N(1) \ge n\epsilon\big) \le \frac{2}{\epsilon^2 n^2}$$

to see that

$$\lim_{t\to\infty} \left|\frac{N(t)}{t} - \frac{N(\lfloor t\rfloor)}{\lfloor t\rfloor}\right| = 0 \quad \mathbb{P}\text{-almost surely.}$$

EXERCISE 4.2.20. Assume that $\mu \in \mathcal{I}(\mathbb{R})$ has its Fourier transform given by (4.2.1), and let $\{Z(t) : t \ge 0\}$ be a Lévy process for $\mu$. Using Exercise 3.2.25, show that $t \rightsquigarrow Z(t)$ is non-decreasing if and only if $M \in \mathfrak{M}_1(\mathbb{R})$, $M\big((-\infty, 0)\big) = 0$, and $m \ge \int_{[-1,1]} y\, M(dy)$.

EXERCISE 4.2.21. Let $\{j(t, \cdot) : t \ge 0\}$ be a Poisson jump process associated with some $M \in \mathfrak{M}_\infty(\mathbb{R}^N)$, and suppose that $F : \mathbb{R}^N \longrightarrow \mathbb{R}$ is a Borel measurable, $M$-integrable function that vanishes at $\mathbf{0}$.

**(i)** Let $\mathcal{N}$ be the set of $\omega \in \Omega$ for which there is a $t > 0$ such that $F$ is not $j(t, \cdot, \omega)$-integrable, and show that $\mathbb{P}(\mathcal{N}) = 0$.

**(ii)** Show that (cf. Lemma 4.2.6) $M^F \in \mathfrak{M}_1(\mathbb{R})$ and that, in fact,

$$\int |y|\, M^F(dy) = \int |F(\mathbf{y})|\, M(d\mathbf{y}) < \infty.$$

Next, define

$$Z^F(t, \omega) = \begin{cases} \int F(\mathbf{y})\, j(t, d\mathbf{y}, \omega) & \text{if } \omega \notin \mathcal{N} \\ 0 & \text{if } \omega \in \mathcal{N}, \end{cases}$$

and show that $\{Z^F(t) : t \ge 0\}$ is a (possibly generalized) Poisson process associated with $M^F$.

**(iii)** Show that

$$\lim_{t\to\infty} \frac{Z^F(t)}{t} = \int F(\mathbf{y})\, M(d\mathbf{y}) \quad \mathbb{P}\text{-almost surely.}$$

**Hint**: Begin by using Lemma 4.2.10 to show that it suffices to handle $F$'s that vanish in a neighborhood of $\mathbf{0}$. When $F$ vanishes in a neighborhood of $\mathbf{0}$, use Lemma 4.2.12 to see that $\{Z^F(t) : t \ge 0\}$ is a Poisson process associated with $M^F$. Finally, use the representation of a Poisson process in terms of a simple Poisson process and independent random variables, and apply The Strong Law of Large Numbers together with the result in Exercise 4.2.19.

EXERCISE 4.2.22. Let $\{\mathbf{Z}(t) : t \geq 0\}$ be a Lévy process for the $\mu \in \mathcal{I}(\mathbb{R}^N)$ with Fourier transform given by (4.2.1), and set $\bar{\mathbf{Z}}(t) = \mathbf{Z}(t) - t\mathbf{m}$. Show that for all $R \in [1, \infty)$ and $t \in (0, \infty)$, $\mathbb{P}(\|\bar{\mathbf{Z}}\|_{[0,t]} \geq R)$ is dominated by $t$ times

$$\frac{4N}{R^2} \int_{B(\mathbf{0},1)} |\mathbf{y}|^2 \, M(d\mathbf{y}) + \frac{2}{R} \int_{1<|\mathbf{y}|\leq\sqrt{R}} |\mathbf{y}| \, M(d\mathbf{y}) + M\big(\overline{B(\mathbf{0}, \sqrt{R})}\complement\big).$$

**Hint**: Write $\bar{\mathbf{Z}}(t) = \mathbf{Z}_1(t) + \mathbf{Z}_2(t) + \mathbf{Z}_3(t)$, where

$$\mathbf{Z}_2(t) = \int_{1<|\mathbf{y}|\leq\sqrt{R}} \mathbf{y} \, j(t, d\mathbf{y}, \mathbf{Z}) \quad \text{and} \quad \mathbf{Z}_3(t) = \int_{|\mathbf{y}|>\sqrt{R}} \mathbf{y} \, j(t, d\mathbf{y}, \mathbf{Z}).$$

Then,

$$\mathbb{P}(\|\mathbf{Z}\|_{[0,t]} \geq R) \leq \mathbb{P}(\|\mathbf{Z}_1\|_{[0,t]} \geq \tfrac{R}{2}) + \mathbb{P}(\|\mathbf{Z}_2\|_{[0,t]} \geq \tfrac{R}{2}) + \mathbb{P}(\|\mathbf{Z}_3\|_{[0,t]} \neq 0).$$

Apply the estimates in Lemma 4.2.10 to control the first two terms on the right, and use

$$\mathbb{P}\Big(j(t, \mathbb{R}^N \setminus \overline{B(\mathbf{0}, \sqrt{R})}, \mathbf{Z}) \neq 0\Big) = 1 - e^{-tM(\mathbb{R}^N \setminus \overline{B(\mathbf{0},\sqrt{R})})}$$

to control the third.

EXERCISE 4.2.23. Let $\nu$ be a locally finite Borel measure on $\mathbb{R}^N$. A **Poisson point process** with intensity measure $\nu$ is a random, locally finite, purely atomic measure-valued random variable $\omega \rightsquigarrow P(\,\cdot\,, \omega)$ with the properties that, for any bounded $\Gamma \in \mathcal{B}_{\mathbb{R}^N}$, $P(\Gamma)$ is a Poisson random variable with mean value $\nu(\Gamma)$ and, for any $n \geq 2$ and family $\{\Gamma_1, \dots, \Gamma_n\}$ of mutually disjoint, bounded, Borel subsets of $\mathbb{R}^N$, $\{P(\Gamma_1), \dots, P(\Gamma_n)\}$ are mutually independent. The purpose of this exercise is to show how one always construct such a Poisson point process for any $\nu$.

(**i**) Define $F : \mathbb{R}^N \longrightarrow \mathbb{R}^N$ so that $F(\mathbf{0}) = \mathbf{0}$ and $F(\mathbf{y}) = \frac{\mathbf{y}}{|\mathbf{y}|^2}$ for $\mathbf{y} \neq \mathbf{0}$. Clearly, $F$ is one-to-one and onto, and both $F$ and $F^{-1}$ are Borel measurable. Assuming that $\nu(\{\mathbf{0}\}) = 0$, show that $M \equiv F_*\nu \in \mathfrak{M}_\infty(\mathbb{R}^N)$ and that $\nu = (F^{-1})_*M$.

(**ii**) Continue to assume that $\nu(\{\mathbf{0}\}) = 0$, let $\{j(t, \cdot\,) : t \geq 0\}$ be a Poisson jump process associated with the $M$ in (**i**), and set $P(\,\cdot\,, \omega) = (F^{-1})_* j(1, \cdot\,, \omega)$. Show that $\omega \rightsquigarrow P(\,\cdot\,, \omega)$ is a Poisson point process with intensity $\nu$.

(**vi**) In order to handle $\nu$'s that charge $\mathbf{0}$, suppose $\nu(\{\mathbf{0}\}) > 0$. Choose a point $\mathbf{x} \in \mathbb{R}^N$ for which $\nu(\{\mathbf{x}\}) = 0$, define $\nu'(\Gamma) = \nu(\mathbf{x} + \Gamma)$, note that $\nu'(\{\mathbf{0}\}) = 0$, and construct a Poisson point process $\omega \rightsquigarrow P'(\,\cdot\,, \omega)$ with intensity measure $\nu'$. Finally, define $P(\Gamma, \omega) = P'(\Gamma - \mathbf{x}, \omega)$, and check that $\omega \rightsquigarrow P(\,\cdot\,, \omega)$ is a Poisson point process with intensity measure $\nu$.

EXERCISE 4.2.24. Let $M \in \mathfrak{M}_2(\mathbb{R}^N)$ be given, and assume that there exists a decreasing sequence $\{r_n : n \geq 0\} \subseteq (0,1]$ with $r_n \searrow 0$ such that

$$\mathbf{m} = \lim_{n\to\infty} \int_{r_n < |\mathbf{y}| \leq 1} \mathbf{y}\, M(d\mathbf{y})$$

exists. Let $\mu \in \mathcal{I}(\mathbb{R}^N)$ have Fourier transform given by (4.2.1) with this $\mathbf{m}$ and $M$. If $\{\mathbf{Z}(t) : t \geq 0\}$ is a Lévy process for $\mu$, set

$$\mathbf{Z}_n(t,\omega) = \int_{|\mathbf{y}|>r_n} \mathbf{y}\, j\big(t, d\mathbf{y}, \mathbf{Z}(\,\cdot\,,\omega)\big),$$

and show that $\lim_{n\to\infty} \mathbb{P}\big(\|\mathbf{Z} - \mathbf{Z}_n\|_{[0,t]} \geq \epsilon\big) = 0$ for all $t \geq 0$ and $\epsilon > 0$. Thus, after passing to a subsequence $\{n_m : m \geq 0\}$ if necessary, one sees that, $\mathbb{P}$-almost surely,

$$\mathbf{Z}(t,\omega) = \lim_{m\to\infty} \int_{|\mathbf{y}|>r_{n_m}} \mathbf{y}\, j\big(t, d\mathbf{y}, \mathbf{Z}(\,\cdot\,,\omega)\big),$$

where the convergence is uniform on finite time intervals. In particular, one can say that $\mathbb{P}$-almost all the paths $t \rightsquigarrow \mathbf{Z}(t,\omega)$ are "conditionally pure jump."

## § 4.3 Brownian Motion, the Gaussian Lévy Process

What remains of the program in this chapter is the construction of a Lévy process for the standard, normal distribution $\gamma_{\mathbf{0},\mathbf{I}}$, the infinitely divisible law whose Fourier transform is $e^{-\frac{|\boldsymbol{\xi}|^2}{2}}$. Indeed, if $\{\mathbf{Z}_{\gamma_{0,I}}(t) : t \geq 0\}$ is such a process and $\{\mathbf{Z}_\mu(t) : t \geq 0\}$ is a Lévy process for the $\mu \in \mathcal{I}(\mathbb{R}^N)$ whose Fourier transform is given by (4.2.1), and if $\{\mathbf{Z}_{\gamma_{0,I}}(t) : t \geq 0\}$ is independent of $\{\mathbf{Z}_\mu(t) : t \geq 0\}$, then it is an easy matter to check that $\mathbf{C}^{\frac{1}{2}}\mathbf{Z}_{\gamma_{0,I}}(t) + \mathbf{Z}_\mu(t)$ will be a Lévy process for $\gamma_{\mathbf{0},\mathbf{C}} \star \mu$, whose Fourier transform is

$$\exp\Bigg(\sqrt{-1}(\boldsymbol{\xi},\mathbf{m})_{\mathbb{R}^N} - \tfrac{1}{2}(\boldsymbol{\xi},\mathbf{C}\boldsymbol{\xi})_{\mathbb{R}^N} + \int_{\mathbb{R}^N} \Big[e^{\sqrt{-1}(\boldsymbol{\xi},\mathbf{y})_{\mathbb{R}^N}} - 1 - \sqrt{-1}\,\mathbf{1}_{[0,1]}(|\mathbf{y}|)(\boldsymbol{\xi},\mathbf{y})_{\mathbb{R}^N}\Big]\, M(d\mathbf{y})\Bigg).$$

Because one of its earliest applications was as a mathematical model for the motion of "Brownian particles," [1] such a Lévy process for $\gamma_{0,1}$ is called a **Brownian motion**. In recognition of its provenance, I will adopt this terminology and will use the notation $\{\mathbf{B}(t) : t \geq 0\}$ instead of $\{\mathbf{Z}_{\gamma_{0,I}}(t) : t \geq 0\}$.

[1] R. Brown, an eighteenth century English botanist, observed the motion of pollen particles in a dilute gas. His observations were interpreted by A. Einstein as evidence for the kinetic theory of gases. In his famous 1905 paper, Einstein took the first steps in a program, eventually completed by N. Wiener in 1923, to give a mathematical model of what Brown had seen.

Before getting into the details, it may be helpful to think a little about what sorts of properties we should expect the paths $t \rightsquigarrow \mathbf{B}(t)$ will possess. For this purpose, set $M_n = n\big(\delta_{n^{-\frac{1}{2}}} + \delta_{-n^{-\frac{1}{2}}}\big)^N$, and recall that we have seen already that $\pi_{M_n} \Longrightarrow \gamma_{\mathbf{0},\mathbf{I}}$. Since a Poisson process associated with $M_n$ has nothing but jumps of size $n^{-\frac{1}{2}}$, if one believes that the Lévy process for $\gamma_{\mathbf{0},\mathbf{I}}$ should be, in some sense, the limit of such Poisson processes, then it is reasonable to guess that its paths will have jumps of size 0. That is, they will be continuous.

Although the prediction that the paths of $\{\mathbf{B}(t) : t \geq 0\}$ will be continuous is correct, it turns out that, because it is based on the Central Limit Theorem, the heuristic reasoning just given does not lead to the easiest construction. The problem is that The Central Limit Theorem gives convergence of distributions, not random variables, and therefore one should not expect the paths, as opposed to their distributions, of the approximating Poisson processes to converge. For this reason, it is easier to avoid The Central Limit Theorem and work with Gaussian random variables from the start, and that is what I will do here. The Central Limit approach is the content of § 9.3.

**§ 4.3.1. Deconstructing Brownian Motion.** My construction of Brownian motion is based on an idea of Lévy's; and in order to explain Lévy's idea, I will begin with the following line of reasoning.

Assume that $\{\mathbf{B}(t) : t \geq 0\}$ is a Brownian motion in $\mathbb{R}^N$. That is, $\{\mathbf{B}(t) : t \geq 0\}$ starts at $\mathbf{0}$, has independent increments, any increment $\mathbf{B}(s+t) - \mathbf{B}(s)$ has distribution $\gamma_{\mathbf{0},t\mathbf{I}}$, and the paths $t \rightsquigarrow \mathbf{B}(t)$ are continuous. Next, given $n \in \mathbb{N}$, let $t \rightsquigarrow \mathbf{B}_n(t)$ be the polygonal path obtained from $t \rightsquigarrow \mathbf{B}(t)$ by linear interpolation during each time interval $[m2^{-n}, (m+1)2^{-n}]$. Thus,

$$\mathbf{B}_n(t) = \mathbf{B}(m2^{-n}) + 2^n\left(t - m2^{-n}\right)\Big(\mathbf{B}\big((m+1)2^{-n}\big) - \mathbf{B}(m2^{-n})\Big)$$

for $m2^{-n} \leq t \leq (m+1)2^{-n}$. The distribution of $\{\mathbf{B}_0(t) : t \geq 0\}$ is very easy to understand. Namely, if $\mathbf{X}_{m,0} = \mathbf{B}(m) - \mathbf{B}(m-1)$ for $m \geq 1$, then the $\mathbf{X}_{m,0}$'s are independent, standard normal $\mathbb{R}^N$-valued random variables, $\mathbf{B}_0(m) = \sum_{1 \leq m \leq n} \mathbf{X}_{m,0}$, and $\mathbf{B}_0(t) = (m-t)\mathbf{B}_0(m-1) + (t-m+1)\mathbf{B}_0(m)$ for $m-1 \leq t \leq m$. To understand the relationship between successive $\mathbf{B}_n$'s, observe that $\mathbf{B}_{n+1}(m2^{-n}) = \mathbf{B}_n(m2^{-n})$ for all $m \in \mathbb{N}$ and that

$$\begin{aligned}
\mathbf{X}_{m,n+1} &\equiv 2^{\frac{n}{2}+1}\Big(\mathbf{B}_{n+1}\big((2m-1)2^{-n-1}\big) - \mathbf{B}_n\big((2m-1)2^{-n-1}\big)\Big) \\
&= 2^{\frac{n}{2}+1}\left(\mathbf{B}\big((2m-1)2^{-n-1}\big) - \frac{\mathbf{B}(m2^{-n}) + \mathbf{B}\big((m-1)2^{-n}\big)}{2}\right) \\
&= 2^{\frac{n}{2}}\Big[\Big(\mathbf{B}\big((2m-1)2^{-n-1}\big) - \mathbf{B}\big((m-1)2^{-n}\big)\Big) \\
&\qquad - \Big(\mathbf{B}(m2^{-n}) - \mathbf{B}\big((2m-1)2^{-n-1}\big)\Big)\Big],
\end{aligned}$$

and therefore $\{\mathbf{X}_{m,n+1} : m \geq 1\}$ is again a sequence of independent standard normal random variables. What is less obvious is that $\{\mathbf{X}_{m,n} : (m,n) \in \mathbb{Z}^+ \times \mathbb{N}\}$ is also a family of independent random variables. In fact, checking this requires us to make essential use of the fact that we are dealing with Gaussian random variables.

In preparation for proving the preceding independence assertion, say that $\mathfrak{G} \subseteq L^2(\mathbb{P};\mathbb{R})$ is a **Gaussian family** if $\mathfrak{G}$ is a linear subspace and each element of $\mathfrak{G}$ is a **centered** (i.e., mean value 0), $\mathbb{R}$-valued Gaussian random variable. My interest in Gaussian families at this point is that the linear span $\mathfrak{G}(\mathbf{B})$ of $\{(\boldsymbol{\xi}, \mathbf{B}(t))_{\mathbb{R}^N} : t \geq 0 \text{ and } \boldsymbol{\xi} \in \mathbb{R}^N\}$ is one. To see this, simply note that, for any $0 = t_0 < t_1 < \cdots t_n$ and $\boldsymbol{\xi}_1, \dots, \boldsymbol{\xi}_n \in \mathbb{R}^N$,

$$\sum_{m=1}^{n} (\boldsymbol{\xi}_m, \mathbf{B}(t_m))_{\mathbb{R}^N} = \sum_{\ell=1}^{n} \left( \sum_{m=\ell}^{n} (\boldsymbol{\xi}_m, \mathbf{B}(t_\ell) - \mathbf{B}(t_{\ell-1}))_{\mathbb{R}^N} \right)_{\mathbb{R}^N},$$

which, as a linear combination of independent centered Gaussians, is itself a centered Gaussian.

The crucial fact about Gaussian families is the content of the next lemma.

LEMMA 4.3.1. *Suppose that $\mathfrak{G} \subseteq L^2(\mathbb{P};\mathbb{R})$ is a Gaussian family. Then the closure of $\mathfrak{G}$ in $L^2(\mathbb{P};\mathbb{R})$ is again a Gaussian family. Moreover, for any $S \subseteq \mathfrak{G}$, $S$ is independent of $S^\perp \cap \mathfrak{G}$, where $S^\perp$ is the orthogonal complement of $S$ in $L^2(\mathbb{P};\mathbb{R})$.*

PROOF: The first assertion is easy since, as I noted in the introduction to Chapter 3, Gaussian random variables are closed under convergence in probability.

Turning to the second part, what I must show is that if $X_1, \dots, X_n \in S$ and $X'_1, \dots, X'_n \in S^\perp \cap \mathfrak{G}$, then (cf. part **(ii)** of Exercise 1.1.13)

$$\mathbb{E}^{\mathbb{P}}\left[\prod_{m=1}^{n} e^{\sqrt{-1}\,\xi_m X_m} \prod_{m=1}^{n} e^{\sqrt{-1}\,\xi'_m X'_m}\right] = \mathbb{E}^{\mathbb{P}}\left[\prod_{m=1}^{n} e^{\sqrt{-1}\,\xi_m X_m}\right] \mathbb{E}^{\mathbb{P}}\left[\prod_{m=1}^{n} e^{\sqrt{-1}\,\xi'_m X'_m}\right]$$

for any choice of $\{\xi_m : 1 \leq m \leq n\} \cup \{\xi'_m : 1 \leq m \leq n\} \subseteq \mathbb{R}$. But the expectation value on the left is equal to

$$\begin{aligned}
&\exp\left(-\frac{1}{2}\mathbb{E}^{\mathbb{P}}\left[\left(\sum_{m=1}^{n} (\xi_m X_m + \xi'_m X'_m)\right)^2\right]\right) \\
&= \exp\left(-\frac{1}{2}\mathbb{E}^{\mathbb{P}}\left[\left(\sum_{m=1}^{n} \xi_m X_m\right)^2\right] - \frac{1}{2}\mathbb{E}^{\mathbb{P}}\left[\left(\sum_{m=1}^{n} \xi'_m X'_m\right)^2\right]\right) \\
&= \mathbb{E}^{\mathbb{P}}\left[\prod_{m=1}^{n} e^{\sqrt{-1}\,\xi_m X_m}\right] \mathbb{E}^{\mathbb{P}}\left[\prod_{m=1}^{n} e^{\sqrt{-1}\,\xi'_m X'_m}\right],
\end{aligned}$$

since $\mathbb{E}^{\mathbb{P}}[X_m X'_{m'}] = 0$ for all $1 \le m, m' \le n$. □

Armed with Lemma 4.3.1, we can now check that $\{\mathbf{X}_{m,n} : (m,n) \in \mathbb{Z}^+ \times \mathbb{N}\}$ is independent. Indeed, since, for all $(m,n) \in \mathbb{Z}^+ \times \mathbb{N}$ and $\boldsymbol{\xi} \in \mathbb{R}^N$, $(\boldsymbol{\xi}, \mathbf{X}_{m,n})_{\mathbb{R}^N}$ a member of the Gaussian family $\mathfrak{G}(\mathbf{B})$, all that we have to do is check that, for each $(m,n) \in \mathbb{Z}^+ \times \mathbb{N}$, $\ell \in \mathbb{N}$, and $(\boldsymbol{\xi}, \boldsymbol{\eta}) \in (\mathbb{R}^N)^2$,

$$\mathbb{E}^{\mathbb{P}}\big[(\boldsymbol{\xi}, \mathbf{X}_{m,n+1})_{\mathbb{R}^N} (\boldsymbol{\eta}, \mathbf{B}(\ell 2^{-n}))_{\mathbb{R}^N}\big] = 0.$$

But, since, for $s \le t$, $\mathbf{B}(s)$ is independent of $\mathbf{B}(t) - \mathbf{B}(s)$,

$$\mathbb{E}^{\mathbb{P}}\big[(\boldsymbol{\xi}, \mathbf{B}(s))_{\mathbb{R}^N} (\boldsymbol{\eta}, \mathbf{B}(t))_{\mathbb{R}^N}\big] = \mathbb{E}^{\mathbb{P}}\big[(\boldsymbol{\xi}, \mathbf{B}(s))_{\mathbb{R}^N} (\boldsymbol{\eta}, \mathbf{B}(s))_{\mathbb{R}^N}\big] = s(\boldsymbol{\xi}, \boldsymbol{\eta})_{\mathbb{R}^N}$$

and therefore

$$\begin{aligned}
&2^{-\frac{n}{2}-1}\mathbb{E}^{\mathbb{P}}\big[(\boldsymbol{\xi}, \mathbf{X}_{m,n+1})_{\mathbb{R}^N} (\boldsymbol{\eta}, \mathbf{B}(\ell 2^{-n}))_{\mathbb{R}^N}\big] \\
&\quad = \mathbb{E}^{\mathbb{P}}\Big[\Big(\boldsymbol{\xi}, \mathbf{B}\big((2m-1)2^{-n-1}\big)\Big)_{\mathbb{R}^N} \Big(\boldsymbol{\eta}, \mathbf{B}(\ell 2^{-n})\Big)_{\mathbb{R}^N}\Big] \\
&\qquad\qquad - \frac{1}{2}\mathbb{E}^{\mathbb{P}}\Big[\Big(\boldsymbol{\xi}, \mathbf{B}(m2^{-n}) + \mathbf{B}\big((m-1)2^{-n}\big)\Big)_{\mathbb{R}^N} \Big(\boldsymbol{\eta}, \mathbf{B}(\ell 2^{-n})\Big)_{\mathbb{R}^N}\Big] \\
&\quad = 2^{-n}(\boldsymbol{\xi}, \boldsymbol{\eta})_{\mathbb{R}^N}\left[(m - \tfrac{1}{2}) \wedge \ell - \frac{m \wedge \ell + (m-1) \wedge \ell}{2}\right] = 0.
\end{aligned}$$

**§ 4.3.2. Lévy's Construction of Brownian Motion.** Lévy's idea was to invert the reasoning given in the preceding subsection. That is, start with a family $\{\mathbf{X}_{m,n} : (m,n) \in \mathbb{Z}^+ \times \mathbb{N}\}$ of independent $N(\mathbf{0}, \mathbf{I})$-random variables. Next, define $\{\mathbf{B}_n(t) : t \ge 0\}$ inductively so that $t \rightsquigarrow \mathbf{B}_n(t)$ is linear on each interval $[(m-1)2^{-n}, m2^{-n}]$, $\mathbf{B}_0(m) = \sum_{1 \le \ell \le m} \mathbf{X}_{\ell,0}$, $m \in \mathbb{N}$, $\mathbf{B}_{n+1}(m2^{-n}) = \mathbf{B}_n(m2^{-n})$ for $m \in \mathbb{N}$, and

$$\mathbf{B}_{n+1}\big((2m-1)2^{-n}\big) = \mathbf{B}_n\big((2m-1)2^{-n-1}\big) + 2^{-\frac{n}{2}-1}\mathbf{X}_{m,n+1} \quad \text{for } m \in \mathbb{Z}^+.$$

If Brownian motion exists, then the distribution of $\{\mathbf{B}_n(t) : t \ge 0\}$ is the distribution of the process obtained by polygonalizing it on each of the intervals $[(m-1)2^{-n}, m2^{-n}]$, and so the limit $\lim_{n\to\infty} \mathbf{B}_n(t)$ should exist uniformly on compacts and should be Brownian motion.

To see that this procedure works, one must first verify that the preceding definition of $\{\mathbf{B}_n(t) : t \ge 0\}$ gives a process with the correct distribution. That is, we need to show that $\{\mathbf{B}_n\big((m+1)2^{-n}\big) - \mathbf{B}_n\big(m2^{-n}\big) : m \in \mathbb{N}\}$ is a sequence of independent $N(\mathbf{0}, 2^{-n}\mathbf{I})$-random variables. But, since this sequence is contained in the Gaussian family spanned by $\{\mathbf{X}_{m,n} : (m,n) \in \mathbb{Z}^+ \times \mathbb{N}\}$, Lemma 4.3.1 says that we need only show that

$$\begin{aligned}
&\mathbb{E}^{\mathbb{P}}\Big[\Big(\boldsymbol{\xi}, \mathbf{B}_n\big((m+1)2^{-n}\big) - \mathbf{B}_n\big(m2^{-n}\big)\Big)_{\mathbb{R}^N} \\
&\qquad \times \Big(\boldsymbol{\xi}', \mathbf{B}_n\big((m'+1)2^{-n}\big) - \mathbf{B}_n\big(m'2^{-n}\big)\Big)_{\mathbb{R}^N}\Big] = 2^{-n}(\boldsymbol{\xi}, \boldsymbol{\xi}')_{\mathbb{R}^N}\delta_{m,m'}
\end{aligned}$$

for $\boldsymbol{\xi}, \boldsymbol{\xi}' \in \mathbb{R}^N$ and $m, m' \in \mathbb{N}$. When $n = 0$, this is obvious. Now assume that it is true for $n$, and observe that

$$\begin{aligned}\mathbf{B}_{n+1}(m2^{-n}) - \mathbf{B}_{n+1}\big((2m-1)2^{-n-1}\big) \\ = \frac{\mathbf{B}_n(m2^{-n}) - \mathbf{B}_n\big((m-1)2^{-n}\big)}{2} - 2^{-\frac{n}{2}-1}\mathbf{X}_{m,n+1}\end{aligned}$$

and

$$\begin{aligned}\mathbf{B}_{n+1}\big((2m-1)2^{-n-1}\big) - \mathbf{B}_{n+1}\big((m-1)2^{-n}\big) \\ = \frac{\mathbf{B}_n(m2^{-n}) - \mathbf{B}_n\big((m-1)2^{-n}\big)}{2} + 2^{-\frac{n}{2}-1}\mathbf{X}_{m,n+1}.\end{aligned}$$

Using these expressions and the induction hypothesis, it is easy to check the required equation.

Second, and more challenging, we must show that, $\mathbb{P}$-almost surely, these processes are converging uniformly on compact time intervals. For this purpose, consider the difference $t \rightsquigarrow \mathbf{B}_{n+1}(t) - \mathbf{B}_n(t)$. Since this path is linear on each interval $[m2^{-n-1}, (m+1)2^{-n-1}]$,

$$\begin{aligned}\max_{t\in[0,2^L]} \big|\mathbf{B}_{n+1}(t) - \mathbf{B}_n(t)\big| &= \max_{1\le m\le 2^{L+n+1}} \big|\mathbf{B}_{n+1}(m2^{-n-1}) - \mathbf{B}_n(m2^{-n-1})\big| \\ &= 2^{-\frac{n}{2}-1} \max_{1\le m\le 2^{L+n}} |\mathbf{X}_{m,n+1}| \le 2^{-\frac{n}{2}-1}\left(\sum_{m=1}^{2^{L+n}} |\mathbf{X}_{m,n+1}|^4\right)^{\frac{1}{4}}.\end{aligned}$$

Thus, by Jensen's Inequality,

$$\mathbb{E}^{\mathbb{P}}\big[\|\mathbf{B}_{n+1} - \mathbf{B}_n\|_{[0,2^L]}\big] \le 2^{-\frac{n}{2}-1}\left(\sum_{m=1}^{2^{L+n}} \mathbb{E}^{\mathbb{P}}\big[|\mathbf{X}_{m,n+1}|^4\big]\right)^{\frac{1}{4}} = 2^{-\frac{n-L-4}{4}} C_N,$$

where $C_N \equiv \mathbb{E}^{\mathbb{P}}\big[|\mathbf{X}_{1,0}|^4\big]^{\frac{1}{4}} < \infty$.

Starting from the preceding, it is an easy matter to show that there is a measurable $\mathbf{B} : [0,\infty)\times\Omega \longrightarrow \mathbb{R}^N$ such that $\mathbf{B}(0) = \mathbf{0}$, $\mathbf{B}(\,\cdot\,,\omega) \in C\big([0,\infty);\mathbb{R}^N\big)$ for each $\omega \in \Omega$, and $\|\mathbf{B}_n - \mathbf{B}\|_{[0,t]} \longrightarrow 0$ both $\mathbb{P}$-almost surely and in $L^1(\mathbb{P};\mathbb{R})$ for every $t \in [0,\infty)$. Furthermore, since $\mathbf{B}(m2^{-n}) = \mathbf{B}_n(m2^{-n})$ $\mathbb{P}$-almost surely for all $(m,n) \in \mathbb{N}^2$, it is clear that $\big\{\mathbf{B}\big((m+1)2^{-n}\big) - \mathbf{B}(m2^{-n}) : m \ge 0\big\}$ is a sequence of independent $N(\mathbf{0}, 2^{-n}\mathbf{I})$-random variables for all $n \in \mathbb{N}$. Hence, by continuity, it follows that $\{\mathbf{B}(t) : t \ge 0\}$ is a Brownian motion.

We have now completed the task described in the introduction to this section. However, before moving on, it is only proper to recognize that, clever as his method is, Lévy was not the first to construct a Brownian motion. Instead, it was N. Wiener who was the first. In fact, his famous[2] 1923 article "Differential Space" in *J. Math. Phys.* **#2** contains three different approaches.

§ **4.3.3. Lévy's Construction in Context.** There are elements of Lévy's construction that admit interesting generalizations, perhaps the most important of which is **Kolmogorov's Continuity Criterion**.

THEOREM 4.3.2. *Suppose that $\{X(t) : t \in [0,T]\}$ is a family of random variables taking values in a Banach space $B$, and assume that, for some $p \in [1,\infty)$, $C < \infty$, and $r \in (0,1]$,*

$$\mathbb{E}^{\mathbb{P}}\left[\|X(t) - X(s)\|_B^p\right]^{\frac{1}{p}} \le C|t-s|^{\frac{1}{p}+r} \quad \text{for all } s,\, t \in [0,T].$$

*Then there exists a family $\{\tilde{X}(t) : t \in [0,T]\}$ of random variables such that $X(t) = \tilde{X}(t)$ $\mathbb{P}$-almost surely for each $t \in [0,T]$ and $t \in [0,T] \longmapsto \tilde{X}(t,\omega) \in B$ is continuous for all $\omega \in \Omega$. In fact, for each $\alpha \in (0,r)$,*

$$\mathbb{E}^{\mathbb{P}}\left[\sup_{0\le s<t\le T}\left(\frac{\|\tilde{X}(t) - \tilde{X}(s)\|_B}{(t-s)^\alpha}\right)^p\right]^{\frac{1}{p}} \le \frac{5CT^{\frac{1}{p}+r-\alpha}}{(1-2^{-r})(1-2^{\alpha-r})}.$$

PROOF: First note that, by rescaling time, it suffices to treat the case when $T = 1$.

Given $n \ge 0$, set $M_n = \max_{1\le m\le 2^n}\left\|X(m2^{-n}) - X\big((m-1)2^{-n}\big)\right\|_B$, and observe that

$$\mathbb{E}^{\mathbb{P}}\left[M_n^p\right]^{\frac{1}{p}} \le \mathbb{E}^{\mathbb{P}}\left[\left(\sum_{m=1}^{2^n}\left\|X(m2^{-n}) - X\big((m-1)2^{-n}\big)\right\|_B^p\right)^{\frac{1}{p}}\right] \le C2^{-rn}.$$

Next, let $t \rightsquigarrow X_n(t)$ be the polygonal path obtained by linearizing $t \rightsquigarrow X(t)$ on each interval $[(m-1)2^{-n}, m2^{-n}]$, and check that

$$\begin{aligned}&\max_{t\in[0,1]}\|X_{n+1}(t) - X_n(t)\|_B \\ &\quad = \max_{1\le m\le 2^n}\left\|X\big((2m-1)2^{-n-1}\big) - \frac{X\big((m-1)2^{-n}\big) - X(m2^{-n})}{2}\right\|_B \le M_{n+1}.\end{aligned}$$

Hence, $\mathbb{E}^{\mathbb{P}}\left[\sup_{t\in[0,1]}\|X_{n+1}(t) - X_n(t)\|_B^p\right]^{\frac{1}{p}} \le C2^{-rn}$, and so there exists a measurable $\tilde{X} : [0,1]\times\Omega \longrightarrow B$ such that $t \rightsquigarrow \tilde{X}(t,\omega)$ is continuous for all $\omega \in \Omega$ and

$$\mathbb{E}^{\mathbb{P}}\left[\sup_{t\in[0,1]}\|\tilde{X}(t) - X_n(t)\|_B^p\right]^{\frac{1}{p}} \le \frac{C2^{-rn}}{1-2^{-r}}.$$

[2] Wiener's article is remarkable, but I must admit that I have never been convinced that it is complete. Undoubtedly, my skepticism are more a consequence of my own ineptitude than of his.

Moreover, because, for each $t \in [0,1]$, $\|X(\tau) - X(t)\|_B \longrightarrow 0$ in probability as $\tau \to t$, it is easy to check that, for each $t \in [0,1]$, $\tilde{X}(t) = X(t)$ $\mathbb{P}$-almost surely.

To prove the final estimate, note that for $2^{-n-1} \le t - s \le 2^{-n}$ one has that

$$\begin{aligned}\|\tilde{X}(t) - \tilde{X}(s)\|_B &\le \|\tilde{X}(t) - X_n(t)\|_B + \|X_n(t) - X_n(s)\|_B + \|X_n(s) - \tilde{X}(s)\|_B \\ &\le 2 \sup_{\tau\in[0,1]} \|\tilde{X}(\tau) - X_n(\tau)\|_B + 2^n(t-s)M_n,\end{aligned}$$

and therefore that

$$\frac{\|\tilde{X}(t) - \tilde{X}(s)\|_B}{(t-s)^\alpha} \le 2 2^{\alpha(n+1)} \sup_{\tau\in[0,1]} \|\tilde{X}(\tau) - X_n(\tau)\|_B + 2^n 2^{(\alpha-1)n} M_n.$$

But, by the estimates proved above, this means that

$$\begin{aligned}\mathbb{E}^{\mathbb{P}}\left[\sup_{0\le s<t\le 1}\left(\frac{\|\tilde{X}(t) - \tilde{X}(s)\|_B}{(t-s)^\alpha}\right)^p\right]^{\frac{1}{p}} &\le C\sum_{n=0}^{\infty}\left(2\frac{2^{\alpha(n+1)}2^{-rn}}{1-2^{-r}} + 2^{\alpha n}2^{-rn}\right) \\ &\le \frac{5C}{(1-2^{-r})(1-2^{\alpha-r})}. \quad \square\end{aligned}$$

Corollary 4.3.3. *If $\{\mathbf{B}(t) : t \ge 0\}$ is an $\mathbb{R}^N$-valued Brownian motion, then, for each $\alpha \in \left(0, \frac{1}{2}\right)$, $t \rightsquigarrow \mathbf{B}(t)$ is $\mathbb{P}$-almost surely Hölder continuous of order $\alpha$. In fact, for each $T \in (0,\infty)$,*

$$\mathbb{E}^{\mathbb{P}}\left[\sup_{0\le s<t\le T}\frac{|\mathbf{B}(t) - \mathbf{B}(s)|}{(t-s)^\alpha}\right] < \infty.$$

Proof: In view of Theorem 4.3.2, all that we have to do is note that, for each $n \in \mathbb{Z}^+$, there is a $C_n < \infty$ such that $\mathbb{E}^{\mathbb{P}}\left[|\mathbf{B}(t) - \mathbf{B}(s)|^{2n}\right] \le C_n|t-s|^n$. $\square$

§ **4.3.4. Brownian Paths Are Non-Differentiable.** Having shown that Brownian paths are Hölder continuous of every order strictly less than $\frac{1}{2}$, I will close this section by showing that they are nowhere Hölder continuous of any order strictly greater than $\frac{1}{2}$. In particular, this will prove Wiener's famous result that *Brownian paths are nowhere differentiable.* The proof that follows is due to A. Devoretzky.

Theorem 4.3.4. *Let $\{\mathbf{B}(t) : t \ge 0\}$ be an $\mathbb{R}^N$-valued Brownian motion. Then, for each $\alpha > \frac{1}{2}$,*

$$\mathbb{P}\left(\exists s \in [0,\infty)\ \varlimsup_{t\searrow s}\frac{|\mathbf{B}(t) - \mathbf{B}(s)|}{(t-s)^\alpha} < \infty\right) = 0.$$

PROOF: Because $\{\mathbf{B}(T+t)-\mathbf{B}(T) : t \geq 0\}$ is a Brownian motion for each $T \in [0,\infty)$, it suffices for us to show that

$$\mathbb{P}\left(\exists s \in [0,1)\ \overline{\lim_{t\searrow s}} \frac{|\mathbf{B}(t)-\mathbf{B}(s)|}{(t-s)^\alpha} < \infty\right) = 0.$$

To this end, note that, for every $L \in \mathbb{Z}^+$,

$$\left\{\exists s \in [0,1)\ \overline{\lim_{t\searrow s}} \frac{|\mathbf{B}(t)-\mathbf{B}(s)|}{(t-s)^\alpha} < \infty\right\}$$
$$\subseteq \bigcup_{M=1}^{\infty}\bigcup_{\nu=1}^{\infty}\bigcap_{n=\nu}^{\infty}\bigcup_{m=0}^{n}\bigcap_{\ell=0}^{L-1}\left\{\left|\mathbf{B}\left(\tfrac{m+\ell+1}{n}\right)-\mathbf{B}\left(\tfrac{m+\ell}{n}\right)\right| \leq \tfrac{M}{n^\alpha}\right\}.$$

Thus, it enough to show that there is a choice of $L$ such that

$$\lim_{n\to\infty} n\mathbb{P}\left(\left|\mathbf{B}\left(\tfrac{\ell+1}{n}\right)-\mathbf{B}\left(\tfrac{\ell}{n}\right)\right| \leq \tfrac{M}{n^\alpha},\ 0 \leq \ell < L\right) = 0.$$

But

$$\mathbb{P}\left(\left|\mathbf{B}\left(\tfrac{\ell+1}{n}\right)-\mathbf{B}\left(\tfrac{\ell}{n}\right)\right| \leq \tfrac{M}{n^\alpha},\ 0 \leq \ell < L\right)$$
$$= \gamma_{\mathbf{0},\frac{1}{n}\mathbf{I}}\left(B\left(\mathbf{0},\tfrac{M}{n^\alpha}\right)\right)^L = \left((2\pi)^{-\frac{N}{2}}\int_{B(\mathbf{0},Mn^{\frac{1}{2}-\alpha})} e^{-\frac{|\mathbf{y}|^2}{2}}\,d\mathbf{y}\right)^L \leq Cn^{(\frac{1}{2}-\alpha)NL}.$$

Hence, we need only take $L$ so that $(\alpha-\frac{1}{2})NL > 1$. □

In spite of their being non-differentiable, "differentials" of Brownian paths display remarkable regularity properties. To wit, I make the following simple observation. In its statement, $\|\cdot\|_{\mathrm{H.S.}}$ denotes the Hilbert–Schmidt norm on $\mathrm{Hom}(\mathbb{R}^N;\mathbb{R}^N)$.

THEOREM 4.3.5. *If $\{\mathbf{B}(t) : t \geq 0\}$ is an $\mathbb{R}^N$-valued Brownian motion, then, for each $T \in (0,\infty)$*

$$\lim_{n\to\infty}\sup_{t\in[0,T]}\left\|\sum_{m=1}^{[nt]}(\Delta_{m,n}\mathbf{B})\otimes(\Delta_{m,n}\mathbf{B}) - t\mathbf{I}\right\|_{\mathrm{H.S.}} = 0 \quad \mathbb{P}\text{-almost surely,}$$

*where $\Delta_{m,n}\mathbf{B} \equiv \mathbf{B}\left(\frac{m}{n}\right)-\mathbf{B}\left(\frac{m-1}{n}\right)$. In particular, $\mathbb{P}$-almost no Brownian path has locally bounded variation.*

PROOF: Let $(\mathbf{e}_1, \dots, \mathbf{e}_N)$ be an orthonormal basis for $\mathbb{R}^N$, and set $X_i(k,n) = (\mathbf{e}_i, \Delta_{k,n}\mathbf{B})_{\mathbb{R}^N}$. Then, what we have to show is that

$$(*) \qquad \lim_{n\to\infty} \sup_{1\le m\le nT} \left| \sum_{k=1}^{m} X_i(k,n)X_j(k,n) - \frac{m}{n}\delta_{i,j} \right| = 0 \quad \mathbb{P}\text{-almost surely.}$$

To this end, note that, for each $n \in \mathbb{Z}^+$ and $1 \le i \le N$, $\{X_i(k,n) : k \ge 1\}$ are mutually independent $N(0, n^{-1})$-random variables. Hence, for each $1 \le i \le N$, $\{X_i(k,n)^2 - n^{-1} : k \ge 1\}$ are independent random variables with mean value 0 and variance $3n^{-2}$, and therefore, by (1.4.22) and the second inequality in (1.3.2),

$$\mathbb{E}\left[\left| \max_{1\le m\le nT} \sum_{k=1}^{m} \left( X_i(k,n)^2 - \tfrac{1}{n} \right) \right|^4\right]$$

$$\le 4\mathbb{E}\left[\left| \sum_{1\le k\le nT} \left( X_i(k,n)^2 - \tfrac{1}{n} \right) \right|^4\right] \le \frac{12M_4T^2}{n^2},$$

where $M_4$ is the fourth moment of $X_1(1,1)^2 - 1$, and so the Borel–Cantelli Lemma can be used to check (*) when $i = j$. When $i \ne j$, the argument is essentially the same, only, because $X_i(k,n)X_j(k,n)$ has mean value 0, there is no need to subtract off its mean.

To prove the final assertion, note that if $\psi \in C([0,T];\mathbb{R})$ has bounded variation, then

$$\lim_{n\to\infty} \sum_{m=1}^{[nT]} \left( \psi\left(\tfrac{m}{n}\right) - \psi\left(\tfrac{m-1}{n}\right) \right)^2 = 0. \quad \square$$

**§ 4.3.5. General Lévy Processes.** Our original reason for constructing Brownian motion was to complete the program of constructing all the Lévy processes. In this subsection, I will do that.

Throughout this subsection, $\mu \in \mathcal{I}(\mathbb{R}^N)$ has Fourier transform

$$\begin{aligned} \exp\Bigg( &\sqrt{-1}\,(\boldsymbol{\xi}, \mathbf{m})_{\mathbb{R}^N} - \tfrac{1}{2}(\boldsymbol{\xi}, \mathbf{C}\boldsymbol{\xi})_{\mathbb{R}^N} \\ &+ \int \left[ e^{\sqrt{-1}(\boldsymbol{\xi},\mathbf{y})_{\mathbb{R}^N}} - 1 - \sqrt{-1}\mathbf{1}_{[0,1]}(|\mathbf{y}|)(\boldsymbol{\xi},\mathbf{y})_{\mathbb{R}^N} \right] M(d\mathbf{y}) \Bigg), \end{aligned} \tag{4.3.6}$$

where $\mathbf{m} \in \mathbb{R}^N$, $\mathbf{C} \in \mathrm{Hom}(\mathbb{R}^N;\mathbb{R}^N)$ is symmetric and non-negative definite, and $M \in \mathfrak{M}_2(\mathbb{R}^N)$. In addition, I will use $\mu_0$ to denote $\gamma_{\mathbf{m},\mathbf{C}}$ and $\mu_1$ to denote the element of $\mathcal{I}(\mathbb{R}^N)$ whose Fourier transform is

$$\exp\left( \int \left[ e^{\sqrt{-1}(\boldsymbol{\xi},\mathbf{y})_{\mathbb{R}^N}} - 1 - \sqrt{-1}\mathbf{1}_{[0,1]}(|\mathbf{y}|)(\boldsymbol{\xi},\mathbf{y})_{\mathbb{R}^N} \right] M(d\mathbf{y}) \right).$$

Thus, $\mu = \mu_0 \star \mu_1$.

THEOREM 4.3.7. *There is a Lévy process $\{\mathbf{Z}(t) : t \geq 0\}$ for each $\mu \in \mathcal{I}(\mathbb{R}^N)$. Furthermore, if $\mu_0$ and $\mu_1$ are as in the preceding discussion and if $\{\mathbf{Z}(t) : t \geq 0\}$ is a Lévy process for $\mu = \mu_0 \star \mu_1$, then there exist independent Lévy processes $\{\mathbf{Z}_0(t) : t \geq 0\}$ and $\{\mathbf{Z}_1(t) : t \geq 0\}$ for $\mu_0$ and $\mu_1$, respectively, such that $\mathbf{Z}(t) = \mathbf{Z}_0(t) + \mathbf{Z}_1(t)$, $t \geq 0$, $\mathbb{P}$-almost surely. In fact, if, for $r \in (0,1]$,*

$$\mathbf{Z}^{(r)}(t) = \int_{|\mathbf{y}|>r} \mathbf{y}\, j(t, d\mathbf{y}, \mathbf{Z}) - t \int_{r<|\mathbf{y}|\leq 1} \mathbf{y}\, M(d\mathbf{y}),$$

*then, for each $t \in (0, \infty)$,*

$$\mathbb{P}\big(\|\mathbf{Z}^{(r)} - \mathbf{Z}_1\|_{[0,t]} \geq \epsilon\big) \leq \frac{N^2 t}{\epsilon^2} \int_{B(\mathbf{0},r)} |\mathbf{y}|^2\, M(d\mathbf{y}).$$

PROOF: Let $\{\mathbf{B}(t) : t \geq 0\}$ be a Brownian motion and $\{\mathbf{Z}_1(t) : t \geq 0\}$ an independent Lévy process for $\mu_1$, and define $\mathbf{Z}_0(t) = t\mathbf{m} + \mathbf{C}^{\frac{1}{2}}\mathbf{B}(t)$ and $\mathbf{Z}(t) = \mathbf{Z}_0(t)+\mathbf{Z}_1(t)$. As I pointed out in the introduction to this section, $\{\mathbf{Z}_0(t) : t \geq 0\}$ is a Lévy process for $\mu_0$ and $\{\mathbf{Z}(t) : t \geq 0\}$ is a Lévy process for $\mu$. Furthermore, because $t \rightsquigarrow \mathbf{Z}_0(t)$ is continuous, $j(t, \cdot\,, \mathbf{Z}) = j(t, \cdot\,, \mathbf{Z}_1)$. Hence, by the last part of Theorem 4.2.15, we know that the last part of the present theorem holds for this choice of $\{\mathbf{Z}(t) : t \geq 0\}$. Finally, since every Lévy process for $\mu$ will have the same distribution as this one, there is nothing more to do. □

COROLLARY 4.3.8. *Let $\{\mathbf{Z}(t) : t \geq 0\}$ be a Lévy process for $\mu$. Then $t \rightsquigarrow \mathbf{Z}(t)$ is $\mathbb{P}$-almost surely continuous if and only if $M = 0$ and is $\mathbb{P}$-almost surely of locally bounded variation if and only if $\mathbf{C} = 0$ and $M \in \mathfrak{M}_1(\mathbb{R}^N)$. Finally, $t \rightsquigarrow \mathbf{Z}(t)$ is $\mathbb{P}$-almost surely an absolutely pure jump path if and only if $\mathbf{C} = 0$, $M \in \mathfrak{M}_1(\mathbb{R}^N)$, and $\mathbf{m} = \int_{B(0,1)} \mathbf{y}\, M(d\mathbf{y})$.*

PROOF: Let $\mathbf{Z}(t) = \mathbf{Z}_0(t) + \mathbf{Z}_1(t)$ be the decomposition described in Theorem 4.3.7, and let $\{j(t, \cdot\,) : t \geq 0\}$ be the jump process for $\{\mathbf{Z}(t) : t \geq 0\}$. If $M = 0$, then $\mathbf{Z}_1(t) = \mathbf{0}$, $t \geq 0$, $\mathbb{P}$-almost surely, and so $t \rightsquigarrow \mathbf{Z}(t) = \mathbf{Z}_0(t)$ is continuous $\mathbb{P}$-almost surely. Conversely, if $t \rightsquigarrow \mathbf{Z}(t)$ is continuous $\mathbb{P}$-almost surely, then $j(t, \cdot\,) = 0$, $t \geq 0$, $\mathbb{P}$-almost surely. Hence, since $\{j(t, \cdot\,) : t \geq 0\}$ is a Poisson jump process associated with $M$, we see that $M = 0$. Next, suppose that $\mathbf{C} = 0$. Then $\mathbf{Z}(t) = \mathbf{Z}_1(t) + t\mathbf{m}$, $t \geq 0$, $\mathbb{P}$-almost surely and therefore, by Corollary 4.2.16, $t \rightsquigarrow \mathbf{Z}(t)$ has locally bounded variation $\mathbb{P}$-almost surely if and only if $M \in \mathfrak{M}_1(\mathbb{R}^N)$ and is $\mathbb{P}$-almost surely an absolutely pure jump path if and only if $M \in \mathfrak{M}_1(\mathbb{R}^N)$ and $\mathbf{m} = \int_{B(\mathbf{0},1)} \mathbf{y}\, M(d\mathbf{y})$. Thus, all that remains is to show that $\mathbf{C} = 0$ if $t \rightsquigarrow \mathbf{Z}(t)$ $\mathbb{P}$-almost surely has locally bounded variation. But, if $t \rightsquigarrow \mathbf{Z}(t)$ has locally bounded variation $\mathbb{P}$-almost surely, then, by (4.1.13), $\int |\mathbf{y}| j(t, d\mathbf{y}) < \infty$, $t \geq 0$, $\mathbb{P}$-almost surely and therefore, by Lemma 4.2.13, $M \in \mathfrak{M}_1(\mathbb{R}^N)$, which, by Corollary 4.2.16, implies that $t \rightsquigarrow \mathbf{Z}_1(t)$ has locally bounded variation $\mathbb{P}$-almost surely. Since this means that $t \rightsquigarrow \mathbf{Z}_0(t)$ must also have locally bounded variation $\mathbb{P}$-almost surely, and, since $\{\mathbf{Z}_0(t) : t \geq 0\}$ has the same distribution as $\{t\mathbf{m} + \mathbf{C}^{\frac{1}{2}}\mathbf{B}(t) : t \geq 0\}$, Theorem 4.3.5 shows that this is possible only if $\mathbf{C} = 0$. □

REMARK 4.3.9. Recall the linear functional $A_\mu$ introduced in (3.2.10). As I showed in Lemma 3.2.14, the action of $A_\mu$ on $\varphi$ decomposes into a local part and a non-local part, which, with 20-20 hindsight, we can write as, respectively,

$$\left(\mathbf{m}, \nabla\varphi(\mathbf{0})\right)_{\mathbb{R}^N} + \tfrac{1}{2}\mathrm{Trace}\left(\mathbf{C}\nabla^2\varphi(\mathbf{0})\right)$$

$$\text{and} \quad \int \left[\varphi(\mathbf{y}) - \varphi(\mathbf{0}) - \mathbf{1}_{[0,1]}(|\mathbf{y}|)\left(\mathbf{y}, \nabla\varphi(\mathbf{0})\right)_{\mathbb{R}^N}\right] M(d\mathbf{y}).$$

In terms of this decomposition, Corollary 4.3.8 is saying that the local part of $A_\mu$ governs the continuous part of $\{\mathbf{Z}(t) : t \geq 0\}$ and that the non-local part governs the discontinuous part.

## Exercises for § 4.3

EXERCISE 4.3.10. This exercise deals with a few elementary facts about Brownian motion.

**(i)** Let $\{\mathbf{X}(t) : t \geq 0\}$ be an $\mathbb{R}^N$-valued stochastic process satisfying $\mathbf{X}(0,\omega) = \mathbf{0}$ and $\mathbf{X}(\,\cdot\,,\omega) \in C(\mathbb{R}^N)$ for all $\omega \in \Omega$, and show that $\{\mathbf{X}(t) : t \geq 0\}$ is an $\mathbb{R}^N$-valued Brownian motion if and only if the span of $\left\{\left(\boldsymbol{\xi}, \mathbf{X}(t)\right)_{\mathbb{R}^N} : t \geq 0 \;\&\; \boldsymbol{\xi} \in \mathbb{R}^N\right\}$ is a Gaussian family with the property that, for all $t$, $t' \in [0,\infty)$ and $\boldsymbol{\xi}$, $\boldsymbol{\xi}' \in \mathbb{R}^N$,

$$\mathbb{E}^{\mathbb{P}}\left[\left(\boldsymbol{\xi}, \mathbf{X}(t)\right)_{\mathbb{R}^N}\left(\boldsymbol{\xi}', \mathbf{X}(t')\right)_{\mathbb{R}^N}\right] = t \wedge t'(\boldsymbol{\xi}, \boldsymbol{\xi}')_{\mathbb{R}^N}.$$

**(ii)** Assuming that $\{\mathbf{B}(t) : t \geq 0\}$ is an $\mathbb{R}^N$-valued Brownian motion, show that $\{\mathcal{O}\mathbf{B}(t) : t \geq 0\}$ is also an $\mathbb{R}^N$-valued Brownian motion for any orthogonal transformation $\mathcal{O}$ on $\mathbb{R}^N$. That is, *the distribution of Brownian motion is invariant under rotation.* (See Theorem 8.3.14 for a significant generalization.)

**(iii)** Assuming that $\{\mathbf{B}(t) : t \geq 0\}$ is an $\mathbb{R}^N$-valued Brownian motion, show that $\{\lambda^{-\frac{1}{2}}\mathbf{B}(\lambda t) : t \geq 0\}$ is also an $\mathbb{R}^N$-Brownian motion for each $\lambda \in (0,\infty)$. This is called the *Brownian scaling invariance* property.

EXERCISE 4.3.11. This exercise introduces the *time inversion invariance* property of Brownian motion.

**(i)** Suppose that $\{\mathbf{B}(t) : t \geq 0\}$ is an $\mathbb{R}^N$-valued Brownian motion, and set $\mathbf{X}(t) = t\mathbf{B}\left(\frac{1}{t}\right)$ for $t > 0$. As an application of **(i)** in Exercise 4.3.10, show that $\{\mathbf{X}(t) : t > 0\}$ has the same distribution as $\{\mathbf{B}(t) : t > 0\}$, and conclude from this that $\lim_{t\searrow 0}\mathbf{X}(t) = \mathbf{0}$ $\mathbb{P}$-almost surely. In particular, if $\tilde{\mathbf{B}}(0,\omega) = \mathbf{0}$ and, for $t \in (0,\infty)$,

$$\tilde{\mathbf{B}}(t,\omega) = \begin{cases} t\mathbf{B}\left(\frac{1}{t},\omega\right) & \text{when } \lim_{\tau\to 0}\tau\mathbf{B}\left(\frac{1}{\tau},\omega\right) = \mathbf{0} \\ \mathbf{0} & \text{otherwise,} \end{cases}$$

show that $\{\tilde{\mathbf{B}}(t) : t \geq 0\}$ is an $\mathbb{R}^N$-valued Brownian motion.

**(ii)** As a consequence of part **(i)**, prove the *Brownian Strong Law of Large Numbers*: $\lim_{t\to\infty} t^{-1}\mathbf{B}(t) = \mathbf{0}$.

EXERCISE 4.3.12. Let $\{\mathbf{B}(t) : t \geq 0\}$ be an $\mathbb{R}^N$-valued Brownian motion.

**(i)** As an application of Theorem 1.4.13, show that, for any $\mathbf{e} \in \mathbb{S}^{N-1}$ and $T \in (0,\infty)$,

$$\mathbb{P}\left(\sup_{t\in[0,T]} \big|(\mathbf{e},\mathbf{B}(t))_{\mathbb{R}^N}\big| \geq R\right) \leq 2\mathbb{P}\Big(\big|(\mathbf{e},\mathbf{B}(T))_{\mathbb{R}^N}\big| \geq R\Big) \leq 2e^{-\frac{R^2}{2T}},$$

and conclude that

$$\mathbb{P}\big(\|\mathbf{B}\|_{[0,T]} \geq R\big) \leq 2Ne^{-\frac{R^2}{2NT}}. \tag{4.3.13}$$

**(ii)** Now assume that $N = 1$, and set $B^*(t) = \max_{\tau\in[0,t]} B(\tau)$. Just as in part **(i)**, use Theorem 1.4.13 to show that $\mathbb{P}\big(B^*(1) \geq a\big) \leq 2\mathbb{P}\big(B(1) \geq a\big)$ for all $a > 0$. By examining its proof, one sees that the inequality in Theorem 1.4.13 comes from not knowing how far over $a$ the partial sums jump when they first exceed level $a$. Thus, because we are now dealing with "continuous partial sums," one should suspect that the inequality can be made an equality. To verify this suspicion, let $\Gamma_n(\epsilon)$ denote the set of $\omega$ such that $|B(t,\omega) - B(s,\omega)| < \epsilon$ for all $0 \leq s < t \leq 1$ with $t - s \leq 2^{-n}$, and show that, for $0 < \epsilon < a$,

$$\{B(1) \geq a\} \cap \Gamma_n(\epsilon) \subseteq \bigcup_{m=1}^{2^n-1} \left\{\max_{0\leq \ell<m} B(\ell 2^{-n}) < a - \epsilon \leq B(m2^{-n}) \;\&\; B(1) - B(m2^{-n}) > 0\right\},$$

and conclude that $\mathbb{P}\big(\{B(1) \geq a\} \cap \Gamma_n(\epsilon)\big) \leq \frac{1}{2}\mathbb{P}\big(B^*(1) \geq a - \epsilon\big)$ for all $n \in \mathbb{N}$. Now let $n \to \infty$ and then $\epsilon \searrow 0$ to arrive at $\mathbb{P}\big(B^*(1) \geq a\big) \geq 2\mathbb{P}\big(B(1) \geq a\big)$.

**(iii)** By combining the preceding with Brownian scaling invariance, arrive at

$$\mathbb{P}\big(B^*(t) \geq a\big) = 2\mathbb{P}\big(B(t) \geq a\big) = \sqrt{\frac{2}{\pi}} \int_{at^{-\frac{1}{2}}}^{\infty} e^{-\frac{x^2}{2}}\, dx. \tag{4.3.14}$$

This beautiful result, which is sometimes called the **reflection principle for Brownian motion**, seems to have appeared first in L. Bachelier's now famous 1900 thesis, where he used what is now called "Brownian motion" to model price fluctuations on the Paris Bourse. More information about the reflection principle can be found in § 8.6.3.

EXERCISE 4.3.15. Let $\{B(t) : t \geq 0\}$ be an $\mathbb{R}$-valued Brownian motion. The goal of this exercise is to prove the **Brownian Law of the Iterated Logarithm**:

$$\varlimsup_{t\to\infty} \frac{B(t)}{\sqrt{2t\log_{(2)} t}} = 1 = \varlimsup_{t\searrow 0} \frac{B(t)}{\sqrt{2t\log_{(2)} t^{-1}}} \quad \mathbb{P}\text{-almost surely.}$$

Begin by checking that the second equality follows from the first applied to the time inverted process $\{\tilde{B}(t) : t \geq 0\}$ described in **(i)** of Exercise 4.3.11. Next, observe that

$$\varlimsup_{n\to\infty} \frac{B(n)}{\sqrt{2n\log_{(2)} n}} = 1 \quad \mathbb{P}\text{-almost surely}$$

is just the Law of the Iterated Logarithm for standard normal random variables. Thus, all that remains is to show that

$$\varlimsup_{n\to\infty} \sup_{t\in[n,n+1]} \left| \frac{B(t)}{\sqrt{2t\log_{(2)} t}} - \frac{B(n)}{\sqrt{2n\log_{(2)} n}} \right| = 0 \quad \mathbb{P}\text{-almost surely,}$$

which can be checked by a combination of the Strong Law for Brownian motion, the estimate in (4.3.13), and the easy half of the Borel–Cantelli Lemma.

EXERCISE 4.3.16. Given a stochastic process $\{X(t) : t \geq 0\}$, the stochastic process $\{\tilde{X}(t) : t \geq 0\}$ is said to be a **modification** of $\{X(t) : t \geq 0\}$ if, for each $t \in [0, \infty)$, $\tilde{X}(t) = X(t)$ $\mathbb{P}$-almost surely. Further, given a stochastic process $\{X(t) : t \geq 0\}$ with values in a metric space $(E, \rho)$, one says that $\{X(t) : t \geq 0\}$ is **stochastically continuous** if, as $t \to s$, $X(t) \longrightarrow X(s)$ in probability for each $s \in [0, \infty)$.

**(i)** Show that the simple Poisson process $\{N(t) : t \geq 0\}$ is stochastically continuous. Thus, *stochastic continuity does not imply path continuity.*

**(ii)** Let $\mathbb{Q}$ denote the set of rational real numbers. Show that an $\mathbb{R}^N$-valued, stochastically continuous stochastic process $\{X(t) : t \geq 0\}$ admits a continuous modification if and only if, for each $T > 0$, $t \in [0,T] \cap \mathbb{Q} \longmapsto X(t)$ is uniformly continuous. Conclude that a stochastically continuous process $\{X(t) : t \geq 0\}$ admits a continuous modification if and only if there exists a $\mu \in \mathbf{M}_1\big(C(\mathbb{R}^N)\big)$ such that the distribution of $\{X(t) : t \geq 0\}$ under $\mathbb{P}$ is the same as the distribution of $\{\psi(t) : t \geq 0\}$ under $\mu$. Equivalently, a stochastically continuous process $\{X(t) : t \geq 0\}$ admits a continuous modification if and only if there exists a continuous stochastic process $\{Y(t) : t \geq 0\}$, not necessarily on the same probability space, with the same distribution as $\{X(t) : t \geq 0\}$.

EXERCISE 4.3.17. It is important to realize that the insistence in Theorem 4.3.2 that the $p$th moment of $|X(t) - X(s)|$ be dominated by $|t-s|$ to a power strictly greater than $p$ is essential. To see this, recall the simple Poisson process $\{N(t) : t \geq 0\}$ in § 5.2.1, and set $X(t) = N(t) - t$. The paths of this process are right-continuous but definitely not continuous. On the other hand, show that $\mathbb{E}^{\mathbb{P}}\big[\big(N(t) - N(s) - (t-s)\big)^2\big] \leq t - s$ for $0 \leq s < t$. More generally, knowing that $\mathbb{E}\big[|X(t) - X(s)|^2\big]$ is dominated by $|t-s|$ is not enough to conclude that there is a continuous modification of $t \rightsquigarrow X(t)$.

EXERCISE 4.3.18. There is an important extension of Theorem 4.3.2 to processes that have a multidimensional parametrization. Let $B$ be a Banach space and $\{X(\mathbf{x}) : \mathbf{x} \in [0,T]^\nu\}$ a family of $B$-valued random variables with the property that

$$\mathbb{E}^{\mathbb{P}}\big[\|X(\mathbf{y}) - X(\mathbf{x})\|_B^p\big]^{\frac{1}{p}} \leq C|\mathbf{y} - \mathbf{x}|^{\frac{\nu}{p}+r}$$

for some $p \in [1,\infty)$, $r > 0$, and $C < \infty$. Show that there exists a family $\{\tilde{X}(\mathbf{x}) : \mathbf{x} \in [0,T]^\nu\}$ with the properties that $\mathbf{x} \in [0,T]^\nu \longmapsto \tilde{X}(\mathbf{x},\omega) \in B$ is continuous for all $\omega$, and, for each $\mathbf{x} \in [0,T]^\nu$, $\tilde{X}(\mathbf{x},\omega) = X(\mathbf{x},\omega)$ $\mathbb{P}$-almost surely. Further, show that, for each $\alpha \in (0,r)$, there is a universal $K(\nu,r,\alpha) < \infty$ such that

$$\mathbb{E}^{\mathbb{P}}\left[\left(\sup_{\substack{\mathbf{x},\mathbf{y}\in[0,T]^\nu \\ \mathbf{y}\neq\mathbf{x}}} \frac{\|\tilde{X}(\mathbf{y}) - \tilde{X}(\mathbf{x})\|_B}{|\mathbf{y}-\mathbf{x}|^\alpha}\right)^p\right]^{\frac{1}{p}} \leq K(\nu,r,\alpha)CT^{\frac{\nu}{p}+r-\alpha}.$$

**Hint**: First rescale time to reduce to the case when $T = 1$. Now assume that $T = 1$. Given $n \in \mathbb{N}$, take $S_n$ to be the set of pairs $(\mathbf{m},\mathbf{m}') \in \big(\{0,\dots,2^n\}^N\big)^2$ such that $m_i' \geq m_i$ for all $1 \leq i \leq \nu$ and $\sum_{i=1}^\nu (m_i' - m_i) = 1$, note that $S_n$ has no more than $\nu 2^{(n+1)\nu}$ elements, set

$$M_n = \max\big\{\|X(\mathbf{m}'2^{-n}) - X(\mathbf{m}2^{-n})\|_B : (\mathbf{m},\mathbf{m}') \in S_n\big\},$$

and show that $\mathbb{E}^{\mathbb{P}}[M_n] \leq C2^\nu \nu^{\frac{1}{p}} 2^{-rn}$. Next, let $\mathbf{x} \rightsquigarrow X_n(\mathbf{x})$ denote the $n$th dyadic multiliniarization of $x \rightsquigarrow X(\mathbf{x})$, the one that is multilinear on each dyadic cube $\prod_{i=1}^N [(m_i - 1)2^{-n}, m_i 2^{-n}]$ for $(m_1,\dots,m_N) \in \{1,\dots,2^n\}^N$. As in the proof of Theorem 4.3.2, argue that $\|X_{n+1} - X_n\|_{\mathrm{u},B} \leq M_{n+1}$, and conclude that there exists an $(\mathbf{x},\omega) \rightsquigarrow \tilde{X}(\mathbf{x},\omega)$ that is continuous in $\mathbf{x}$ for each $\omega$ and is $\mathbb{P}$-almost surely equal to $X(\mathbf{x},\,\cdot\,)$ for each $\mathbf{x}$. Finally, to derive the Hölder continuity estimate, observe that $\|X_n(\mathbf{y}) - X_n(\mathbf{x})\|_B \leq 2^n \nu^{\frac{1}{2}} |\mathbf{y}-\mathbf{x}| M_n$, and proceed as in the proof of the corresponding part of Theorem 4.3.2.

EXERCISE 4.3.19. In this exercise we will examine a couple of the implications that Theorem 4.3.5 has about any Riemann–Stieltjes type integration theory involving Brownian paths. For simplicity, I will restrict my attention to the one-dimensional case. Thus, let $\{B(t) : t \geq 0\}$ be an $\mathbb{R}$-valued Brownian motion. Because $t \rightsquigarrow B(t)$ is continuous, one knows that any function $\psi : [0,1] \longrightarrow \mathbb{R}$ of bounded variation is Riemann–Stieltjes integrable on $[0,1]$ with respect to $B \upharpoonright [0,1]$. However, as the following shows, almost no Brownian path is Riemann–Stieltjes with respect to itself. Namely, using Theorem 4.3.5, show that $\mathbb{P}$-almost surely,

$$\lim_{n\to\infty} \sum_{m=1}^{n} B\big(\tfrac{m-1}{n}\big)\Big(B\big(\tfrac{m}{n}\big) - B\big(\tfrac{m-1}{n}\big)\Big) = \frac{B(1)^2 - 1}{2},$$

$$\lim_{n\to\infty} \sum_{m=1}^{n} B\big(\tfrac{m}{n}\big)\Big(B\big(\tfrac{m}{n}\big) - B\big(\tfrac{m-1}{n}\big)\Big) = \frac{B(1)^2 + 1}{2},$$

whereas

$$\lim_{n\to\infty} \sum_{m=1}^{n} B\big(\tfrac{2m-1}{2n}\big)\Big(B\big(\tfrac{m}{n}\big) - B\big(\tfrac{m-1}{n}\big)\Big) = B(1)^2.$$

EXERCISE 4.3.20. Say that a $D(\mathbb{R}^N)$-valued process $\{\mathbf{Z}(t) : t \geq 0\}$ is a Lévy process if $\mathbf{Z}(0) = \mathbf{0}$ and it has independent, homogeneous increments. Show that every Lévy process is a Lévy process for some $\mu \in \mathcal{I}(\mathbb{R}^N)$.

EXERCISE 4.3.21. Let $\{j(t,\cdot\,) : t \geq 0\}$ be a Poisson jump process associated with some $M \in \mathfrak{M}_\infty(\mathbb{R}^N)$. In Lemma 4.2.13, we showed that when $M \in \mathfrak{M}_2(\mathbb{R}^N)$, then $\int |\mathbf{y}|\, j(t, d\mathbf{y}) < \infty$, $t \geq 0$, with positive probability only if $M \in \mathfrak{M}_1(\mathbb{R}^N)$. In this exercise, we will show that the same is true for any $M \in \mathfrak{M}_\infty(\mathbb{R}^N)$. That is, assuming that $\int |\mathbf{y}|\, j(t, d\mathbf{y}) < \infty$, $t \geq 0$, with positive probability, it is to be shown that $M \in \mathfrak{M}_1(\mathbb{R}^N)$. Here are some steps that you might want to follow.

**(i)** As an application of Kolmogorov's 0–1 Law, show that $\int |\mathbf{y}|\, j(t, d\mathbf{y}) < \infty$ with positive probability implies it is finite with probability 1.

**(ii)** Let $\mathcal{N}$ be the set of $\omega \in \Omega$ for which there is a $t > 0$ such that $\int |\mathbf{y}|\, j(t, d\mathbf{y}, \omega) = \infty$. By **(i)**, $\mathbb{P}(\mathcal{N}) = 0$. Define $\mathbf{Z}(t,\omega) = \int \mathbf{y}\, j(t, d\mathbf{y}, \omega)$ for $\omega \notin \mathcal{N}$ and $\mathbf{Z}(t,\omega) = \mathbf{0}$ for $\omega \in \mathcal{N}$, and show that $\{\mathbf{Z}(t) : t \geq 0\}$ is a Lévy process with absolutely pure jump paths.

**(iii)** Applying Theorem 4.1.8, first show that $\{\mathbf{Z}(t) : t \geq 0\}$ is a Lévy process for a $\mu$ with Lévy measure $M$, and then apply Corollary 4.3.8 to conclude that $M \in \mathfrak{M}_1(\mathbb{R}^N)$.

EXERCISE 4.3.22. Corollary 4.3.3 can be sharpened. In fact, Lévy showed that if $\{B(t) : t \geq 0\}$ is an $\mathbb{R}$-valued Brownian motion, then

$$\mathbb{P}\left(\lim_{\delta\searrow 0} \sup_{0<t-s\leq\delta} \frac{|B(t) - B(s)|}{L(\delta)} = \sqrt{2}\right) = 1,$$

where $L(\delta) \equiv \sqrt{\delta \log \delta^{-1}}$. Notice that, on the one hand, this result is in the direction that one should expect: we know (cf. Theorem 4.3.4) that Brownian paths are almost never Hölder continuous of any order greater than $\frac{1}{2}$. On the other hand, the Brownian Law of the Iterated Logarithm (cf. Exercise 4.3.15) might make one guess that their true modulus of continuity ought to be $\sqrt{\delta \log_{(2)} \delta^{-1}}$, not $L(\delta)$. However, that guess is wrong because it fails to take into account the difference between a question about what is true at a single time as opposed to what is true simultaneously for all times. The purpose of this exercise is to show how the considerations in § 4.3.3 can be used to get a statement that is related to but far less refined than Lévy's. The result to be proved here says only that

$$\mathbb{P}\left(\varlimsup_{\delta \searrow 0} \sup_{0<t-s\le\delta} \frac{|B(t)-B(s)|}{L(\delta)} \le K\right) = 1 \tag{4.3.23}$$

for some $K < \infty$.

**(i)** First show that it suffices to prove that there exists a $K < \infty$ such that

$$\mathbb{P}\left(\varlimsup_{\delta \searrow 0} \sup_{\substack{0<t-s\le\delta \\ s,t\in[0,1]}} \frac{|B(t)-B(s)|}{L(\delta)} \le K\right) = 1$$

and that this will follow from

$$\sum_{n=0}^{\infty} \mathbb{P}\left(\sup_{2^{-n-1}\le t-s\le 2^{-n}} \frac{|B(t)-B(s)|}{L(2^{-n-1})} > K\right) < \infty. \tag{*}$$

**(ii)** Define the polygonal approximation $\{B_n(t) : t \ge 0\}$ as in § 4.3.1, set $M_n = \max_{1\le m\le 2^n} |B(m2^{-n}) - B((m-1)2^{-n})|$, and show that

$$\frac{|B(t)-B(s)|}{L(2^{-n-1})} \le \frac{2\|B-B_n\|_{[0,1]}}{L(2^{-n-1})} + \frac{M_n}{L(2^{-n})} \quad \text{for } 2^{-n-1} \le t-s \le 2^{-n}.$$

**(iii)** Set $C = \sum_{n=0}^{\infty} \sqrt{(n+1)2^{-n}}$, show that $\sum_{n=m}^{\infty} L(2^{-n-1}) \le CL(2^{-m-1})$ for all $m \ge 0$, and, arguing as in the proof of Theorem 4.3.2, conclude that, for any $R > 0$,

$$\mathbb{P}\big(\|B-B_n\|_{[0,1]} \ge RL(2^{-n-1})\big) \le \sum_{m=n}^{\infty} \mathbb{P}\big(M_{m+1} \ge C^{-1}RL(2^{-m-1})\big).$$

**(iv)** Show that, for all $R > 0$ and $n \in \mathbb{N}$,

$$\mathbb{P}\big(M_n \ge RL(2^{-n})\big) \le 2^{n(1-2^{-1}R^2)+1},$$

and combine this with **(ii)** and **(iii)** to prove that (*) holds for some $K < \infty$.

# Chapter 5
## Conditioning and Martingales

Up to this point I have been dealing with random variables that are either themselves mutually independent or are built out of other random variables that are. For this reason, it has not been necessary for me to make explicit use of the concept of *conditioning*, although, as we will see shortly, this concept has been lurking silently in the background. In this chapter I will first give the modern formulation of conditional expectations and then provide an example of the way in which conditional expectations can be used.

Let $(\Omega, \mathcal{F}, \mathbb{P})$ be a probability space, and suppose that $A \in \mathcal{F}$ is a set having positive $\mathbb{P}$-measure. For reasons that are most easily understood when $\Omega$ is finite and $\mathbb{P}$ is uniform, the ratio

$$\mathbb{P}(B|A) \equiv \frac{\mathbb{P}(A \cap B)}{\mathbb{P}(A)}, \quad B \in \mathcal{F},$$

is called the **conditional probability of** $B$ **given** $A$. As one learns in an elementary course, the introduction of conditional probabilities makes many calculations much simpler; in particular, conditional probabilities help to clarify dependence relations between the events represented by $A$ and $B$. For example, $B$ is independent of $A$ precisely when $\mathbb{P}(B|A) = \mathbb{P}(B)$ or, in words, *when the condition that $A$ occurs does not change the probability that $B$ occurs.* Thus, it is unfortunate that the naïve definition of conditioning as described above does not cover many important situations. For example, suppose that $X$ and $Y$ are random variables and that one wants to talk about the conditional probability that $Y \leq b$ given that $X = a$. Unless one is very lucky and $\mathbb{P}(X = a) > 0$, dividing by $\mathbb{P}(X = a)$ is not going to do the job. As this example illustrates, it is of great importance to generalize the concept of conditional probability to include situations when the event on which one is conditioning has $\mathbb{P}$-measure 0, and the next section is devoted to Kolmogorov's elegant solution to the problem of doing so.

### § 5.1 Conditioning

In order to appreciate the idea behind Kolmogorov's solution, imagine someone told you the conditional probability that the event $B$ occurs given that the event $A$ occurs. Obviously, since you have no way of saying anything about the probability of $B$ when $A$ does not occur, she has provided you with incomplete information about $B$. Thus, before you are satisfied, you should demand to know also what is the conditional probability of $B$ given that $A$ does not occur. Of course, this second piece of information is relevant only if $A$ *is not certain*, in which case $\mathbb{P}(A) < 1$ and therefore $\mathbb{P}(B|A\complement)$ is well defined. More generally, suppose that $\mathcal{P} = \{A_1, \dots, A_N\}$ ($N$ here may be either finite or countably infinite) is a partition of $\Omega$ into elements of $\mathcal{F}$ having positive $\mathbb{P}$-measure. Then, in order to have complete information about the probability of $B \in \mathcal{F}$ relative to $\mathcal{P}$, one has to know the entire list of the numbers $\mathbb{P}(B|A_n)$, $1 \le n \le N$. Next, suppose that one attempts to describe this list in a way that does not depend explicitly on the positivity of the numbers $\mathbb{P}(A_n)$. For this purpose, consider the function

$$\omega \in \Omega \longmapsto f(\omega) \equiv \sum_{n=1}^{N} \mathbb{P}(B|A_n)\, \mathbf{1}_{A_n}(\omega).$$

Clearly, $f$ is not only $\mathcal{F}$-measurable, it is measurable with respect to the $\sigma$-algebra $\sigma(\mathcal{P})$ over $\Omega$ generated by $\mathcal{P}$. In particular (because the only $\sigma(\mathcal{P})$-measurable set of $\mathbb{P}$-measure 0 is empty), $f$ is uniquely determined by its $\mathbb{P}$-integrals $\mathbb{E}^{\mathbb{P}}[f, A]$ over sets $A \in \sigma(\mathcal{P})$. Moreover, because, for each $B \in \sigma(\mathcal{P})$ and $n$, either $A_n \subseteq B$ or $B \cap A_n = \emptyset$, we have that

$$\mathbb{E}^{\mathbb{P}}[f, A] = \sum_{n=1}^{N} \mathbb{P}(B \cap A_n) = \sum_{\{n: A_n \subseteq B\}} \mathbb{P}(A_n \cap B) = \mathbb{P}(A \cap B).$$

Hence, the function $f$ is uniquely determined by the properties that it is $\sigma(\mathcal{P})$-measurable and that

$$\mathbb{E}^{\mathbb{P}}[f, A] = \mathbb{P}(A \cap B) \quad \text{for every} \quad A \in \sigma(\mathcal{P}).$$

The beauty of this description is that it makes perfectly good sense even if some of the $A_n$'s have $\mathbb{P}$-measure 0, except in that case the description does not determine $f$ pointwise but merely up to a $\sigma(\mathcal{P})$-measurable $\mathbb{P}$-**null set** (i.e., a set of $\mathbb{P}$-measure 0), which is the very least one should expect to pay for *dividing by* 0.

§ **5.1.1. Kolmogorov's Definition.** With the preceding discussion in mind, one ought to find the following formulation reasonable. Namely, given a sub-$\sigma$-algebra $\Sigma \subseteq \mathcal{F}$ and a $(-\infty, \infty]$-valued random variable $X$ whose negative part $X^-(\equiv -(X \wedge 0))$ is $\mathbb{P}$-integrable, I will say that the random variable $X_\Sigma$ is a **conditional expectation of** $X$ **given** $\Sigma$ if $X_\Sigma$ is $(-\infty, \infty]$-valued and $\Sigma$-measurable, $(X_\Sigma)^-$ is $\mathbb{P}$-integrable, and

$$\mathbb{E}^{\mathbb{P}}[X_\Sigma, A] = \mathbb{E}^{\mathbb{P}}[X, A] \quad \text{for every } A \in \Sigma. \tag{5.1.1}$$

Obviously, having made this definition, my first duty is to show that such an $X_\Sigma$ always exists and to discover in what sense it is uniquely determined. The latter problem is dealt with in the following lemma.

LEMMA 5.1.2. *Let $\Sigma$ be a sub-$\sigma$-algebra of $\mathcal{F}$, and suppose that $X_\Sigma$ and $Y_\Sigma$ are a pair of $(-\infty,\infty]$-valued $\Sigma$-measurable random variables for which $X_\Sigma^-$ and $Y_\Sigma^-$ are both $\mathbb{P}$-integrable. Then*

$$\mathbb{E}^{\mathbb{P}}[X_\Sigma, A] \le \mathbb{E}^{\mathbb{P}}[Y_\Sigma, A] \quad \text{for every } A \in \Sigma,$$

*if and only if $X_\Sigma \le Y_\Sigma$ (a.s., $\mathbb{P}$).*

PROOF: Without loss in generality, I may and will assume that $\Sigma = \mathcal{F}$ and will therefore drop the subscript $\Sigma$; and, since the "if" implication is completely trivial, I will discuss only the minimally less trivial "only if" assertion. Thus, suppose that $\mathbb{P}$-integrals of $Y$ dominate those of $X$ and yet that $X > Y$ on a set of positive $\mathbb{P}$-measure. We could then choose an $M \in [1,\infty)$ so that $\mathbb{P}(A) \vee P(B) > 0$, where

$$A \equiv \left\{X \le M \text{ and } Y \le X - \tfrac{1}{M}\right\} \quad \text{and} \quad B \equiv \left\{X = \infty \text{ and } Y \le M\right\}.$$

But if $\mathbb{P}(A) > 0$, then

$$\mathbb{E}^{\mathbb{P}}[X, A] \le \mathbb{E}^{\mathbb{P}}[Y, A] \le \mathbb{E}^{\mathbb{P}}[X, A] - \tfrac{1}{M}P(A),$$

which, because $\mathbb{E}^{\mathbb{P}}[X, A]$ is a finite number, is impossible. At the same time, if $\mathbb{P}(B) > 0$, then

$$\infty = \mathbb{E}^{\mathbb{P}}[X, B] \le \mathbb{E}^{\mathbb{P}}[Y, B] \le M < \infty,$$

which is also impossible. □

THEOREM 5.1.3. *Let $\Sigma$ be a sub-$\sigma$-algebra of $\mathcal{F}$ and $X$ a $(-\infty,\infty]$-valued random variable for which $X^-$ is $\mathbb{P}$-integrable. Then there exists a conditional expectation value $X_\Sigma$ of $X$. Moreover, if $Y$ is a second $(-\infty,\infty]$-valued random variable and $Y \ge X$ (a.s., $\mathbb{P}$), then $Y^-$ is $\mathbb{P}$-integrable and $Y_\Sigma \ge X_\Sigma$ (a.s., $\mathbb{P}$) for any $Y_\Sigma$ that is a conditional expectation value of $Y$ given $\Sigma$. In particular, if $X = Y$ (a.s., $\mathbb{P}$), then $\{Y_\Sigma \ne X_\Sigma\}$ is a $\Sigma$-measurable, $\mathbb{P}$-null set.*[1]

PROOF: In view of Lemma 5.1.2, it suffices for me to handle the initial existence statement. To this end, let $\mathcal{G}$ denote the class of $X$ satisfying $\mathbb{E}^{\mathbb{P}}[X^-] < \infty$ for which an $X_\Sigma$ exists, and let $\mathcal{G}^+$ denote the non-negative elements of $\mathcal{G}$. If $\{X_n : n \ge 1\} \subseteq \mathcal{G}^+$ is non-decreasing and, for each $n \in \mathbb{Z}^+$, $(X_n)_\Sigma$ denotes a conditional expectation of $X_n$ given $\Sigma$, then $0 \le (X_n)_\Sigma \le (X_{n+1})_\Sigma$ (a.s., $\mathbb{P}$), and therefore I can arrange that $0 \le (X_n)_\Sigma \le (X_{n+1})_\Sigma$ everywhere. In particular, if $X$ and $X_\Sigma$ are the pointwise limits of the $X_n$'s and $(X_n)_\Sigma$'s, respectively, then the Monotone Convergence Theorem guarantees that $X_\Sigma$ is a

[1] Kolmogorov himself, and most authors ever since, have obtained the existence of conditional expectation values as a consequence of the Radon–Nikodym Theorem. Because I find projections more intuitively appealing, I prefer the approach given here.

conditional expectation of $X$ given $\Sigma$. Hence, we now know that $\mathcal{G}^+$ is closed under non-decreasing, pointwise limits, and therefore we will know that $\mathcal{G}^+$ contains all non-negative random variables $X$ as soon as we show that $\mathfrak{G}$ contains all bounded $X$'s. But if $X$ is bounded (and is therefore an element of $L^2(\mathbb{P};\mathbb{R})$) and $L_\Sigma = L^2(\Omega,\Sigma,\mathbb{P};\mathbb{R})$ is the subspace of $L^2(\mathbb{P};\mathbb{R})$ consisting of its $\Sigma$-measurable elements, then the orthogonal projection $X_\Sigma$ of $X$ onto $L_\Sigma$ is a $\Sigma$-measurable random variable that is $\mathbb{P}$-square integrable and satisfies (5.1.1).

So far I have proved that $\mathcal{G}^+$ contains all non-negative, $\mathcal{F}$-measurable $X$'s. Furthermore, if $X$ is non-negative, then (by Lemma 5.1.2) $X_\Sigma \geq 0$ (a.s., $\mathbb{P}$) and so $X_\Sigma$ is $\mathbb{P}$-integrable precisely when $X$ itself is. In particular, I can arrange to make $X_\Sigma$ take its values in $[0,\infty)$ when $X$ is non-negative and $\mathbb{P}$-integrable. Finally, to see that $X \in \mathcal{G}$ for every $X$ with $\mathbb{E}^{\mathbb{P}}[X^-] < \infty$, simply consider $X^+$ and $X^-$ separately, apply the preceding to show that $(X^\pm)_\Sigma \geq 0$ (a.s., $\mathbb{P}$) and that $(X^-)_\Sigma$ is $\mathbb{P}$-integrable, and check that the random variable

$$X_\Sigma \equiv \begin{cases} (X^+)_\Sigma - (X^-)_\Sigma & \text{when} \quad (X^\pm)_\Sigma \geq 0 \text{ and } (X^-)_\Sigma < \infty \\ 0 & \text{otherwise} \end{cases}$$

is a conditional expectation of $X$ given $\Sigma$. □

**Convention.** Because it is determined only up to a $\Sigma$-measurable $\mathbb{P}$-null set, one cannot, in general, talk about *the* conditional expectation of $X$ as a *function.* Instead, the best that one can do is say that **the conditional expectation of** $X$ is the equivalence class of $\Sigma$-measurable $X_\Sigma$'s that satisfy (5.1.1), and I will adopt the notation $\mathbb{E}^{\mathbb{P}}[X|\Sigma]$ to denote this equivalence class. On the other hand, because one is usually interested only in $\mathbb{P}$-integrals of conditional expectations, it has become common practice to ignore, for the most part, the distinction between the equivalence class $\mathbb{E}^{\mathbb{P}}[X|\Sigma]$ and the members of that equivalence class. Thus (just as one would when dealing with the Lebesgue spaces) I will abuse notation by using $\mathbb{E}^{\mathbb{P}}[X|\Sigma]$ to denote a generic element of the equivalence class $\mathbb{E}^{\mathbb{P}}[X|\Sigma]$ and will be more precise only when $\mathbb{E}^{\mathbb{P}}[X|\Sigma]$ contains some particularly distinguished member. For example, recall the random variables $T_n$ entering the definition of the simple Poisson process $\{N(t) : t \in (0,\infty)\}$ in § 4.2.1. It is then clear (cf. part **(i)** in Exercise 1.1.9) that we can take

$$\mathbb{E}^{\mathbb{P}}\Big[\mathbf{1}_{\{n\}}(N(t))\,\Big|\,\sigma(T_1,\dots,T_n)\Big] = \mathbf{1}_{[0,t]}(T_n)e^{-(t-T_n)},$$

and one would be foolish to take any other representative. More generally, I will always take non-negative representatives of $\mathbb{E}^{\mathbb{P}}[X|\Sigma]$ when $X$ itself is non-negative and $\mathbb{R}$-valued representatives when $X$ is $\mathbb{P}$-integrable. Finally, for historical reasons, it is usual to distinguish the case when $X$ is the indicator function $\mathbf{1}_B$ of a set $B \in \mathcal{F}$ and to call $\mathbb{E}^{\mathbb{P}}[\mathbf{1}_B|\Sigma]$ the **conditional probability of** $B$ **given** $\Sigma$ and to write $\mathbb{P}(B|\Sigma)$ instead of $\mathbb{E}^{\mathbb{P}}[\mathbf{1}_B|\Sigma]$. Of course, representatives of $\mathbb{P}(B|\Sigma)$ will always be assumed to take their values in $[0,1]$.

Once one has established the existence and uniqueness of conditional expectations, there is a long list of more or less obvious properties that one can easily verify. The following theorem contains some of the more important items that ought to appear on such a list.

THEOREM 5.1.4. *Let $\Sigma$ be a sub-$\sigma$-algebra of $\mathcal{F}$. If $X$ is a $\mathbb{P}$-integrable random variable and $\mathcal{C} \subseteq \Sigma$ is a $\pi$-system (cf. Exercise 1.1.12) that generates $\Sigma$, then*

$$Y = \mathbb{E}^{\mathbb{P}}[X|\Sigma] \quad (\text{a.s.}, \mathbb{P}) \iff$$
$$Y \in L^1(\Omega, \Sigma, \mathbb{P}; \mathbb{R}) \text{ and } \mathbb{E}^{\mathbb{P}}[Y, A] = \mathbb{E}^{\mathbb{P}}[X, A] \text{ for } A \in \mathcal{C} \cup \{\Omega\}.$$

*Moreover, if $X$ is any $(-\infty, \infty]$-valued random variable that satisfies $\mathbb{E}^{\mathbb{P}}[X^-] < \infty$, then each of the following relations holds $\mathbb{P}$-almost surely:*

$$\left|\mathbb{E}^{\mathbb{P}}[X|\Sigma]\right| \le \mathbb{E}^{\mathbb{P}}\left[|X|\,\middle|\Sigma\right]; \tag{5.1.5}$$

$$\mathbb{E}^{\mathbb{P}}[X|\mathcal{T}] = \mathbb{E}^{\mathbb{P}}\Big[\mathbb{E}^{\mathbb{P}}[X|\Sigma]\,\Big|\mathcal{T}\Big] \tag{5.1.6}$$

*when $\mathcal{T}$ is a sub-$\sigma$-algebra of $\Sigma$; and, when $X$ is $\mathbb{R}$-valued and $\mathbb{P}$-integrable,*

$$\mathbb{E}^{\mathbb{P}}[-X|\Sigma] = -\mathbb{E}^{\mathbb{P}}[X|\Sigma].$$

*Next, let $Y$ be a second $(-\infty, \infty]$-valued random variable with $\mathbb{E}^{\mathbb{P}}[Y^-] < \infty$. Then, $\mathbb{P}$-almost surely,*

$$\mathbb{E}^{\mathbb{P}}[\alpha X + \beta Y|\Sigma] = \alpha\mathbb{E}^{\mathbb{P}}[X|\Sigma] + \beta\mathbb{E}^{\mathbb{P}}[Y|\Sigma] \quad \text{for each } \alpha, \beta \in [0, \infty),$$

*and*

$$\mathbb{E}^{\mathbb{P}}[YX|\Sigma] = Y\,\mathbb{E}^{\mathbb{P}}[X|\Sigma] \tag{5.1.7}$$

*if $Y$ is $\Sigma$-measurable and $(XY)^-$ is $\mathbb{P}$-integrable. Finally, suppose that $\{X_n : n \ge 1\}$ is a sequence of $(-\infty, \infty]$-valued random variables. Then, $\mathbb{P}$-almost surely,*

$$\mathbb{E}^{\mathbb{P}}[X_n|\Sigma] \nearrow \mathbb{E}^{\mathbb{P}}[X|\Sigma] \quad \text{if } \mathbb{E}^{\mathbb{P}}[X_1^-] < \infty \text{ and } X_n \nearrow X \text{ (a.s., } \mathbb{P}); \tag{5.1.8}$$

*and, more generally,*

$$\mathbb{E}^{\mathbb{P}}\left[\varliminf_{n\to\infty} X_n\,\middle|\,\Sigma\right] \le \varliminf_{n\to\infty} \mathbb{E}^{\mathbb{P}}[X_n|\Sigma] \quad \text{if } X_n \ge 0 \text{ (a.s., } \mathbb{P}) \text{ for each } n \in \mathbb{Z}^+. \tag{5.1.9}$$

PROOF: To prove the first assertion, note that the set of $A \in \Sigma$ for which $\mathbb{E}^{\mathbb{P}}[X, A] = \mathbb{E}^{\mathbb{P}}[Y, A]$ is (cf. Exercise 1.1.12) a $\lambda$-system that contains $\mathcal{C}$ and therefore $\Sigma$. Next, clearly (5.1.5) is just an application of Lemma 5.1.2, while (5.1.6) and the two equations that follow it are all expressions of uniqueness. As for the next equation, one can first reduce to the case when $X$ and $Y$ are both non-negative. Then one can use uniqueness to check it when $Y$ is the indicator function of an element of $\Sigma$, use linearity to extend it to simple $\Sigma$-measurable functions, and complete the job by taking monotone limits. Finally, (5.1.8) is an immediate application of the Monotone Convergence Theorem, whereas (5.1.9) comes from the conjunction of

$$\mathbb{E}^{\mathbb{P}}\left[\inf_{n\geq m} X_n \,\middle|\, \Sigma\right] \leq \inf_{n\geq m} \mathbb{E}^{\mathbb{P}}\left[X_n \,\middle|\, \Sigma\right] \quad (\text{a.s.}, \mathbb{P}), \qquad m \in \mathbb{Z}^+,$$

with (5.1.8). □

It probably will have occurred to most readers that the properties discussed in Theorem 5.1.4 give strong evidence that, for fixed $\omega \in \Omega$, $X \longmapsto \mathbb{E}^{\mathbb{P}}[X|\Sigma](\omega)$ behaves like an integral (in the sense of Daniell) and therefore ought to be expressible in terms of integration with respect to a probability measure $\mathbb{P}_\omega$. Indeed, if one could actually talk about $X \longmapsto \mathbb{E}^{\mathbb{P}}[X|\Sigma](\omega)$ for a fixed (as opposed to $\mathbb{P}$-almost every) $\omega \in \Omega$, then there is no doubt that such a $\mathbb{P}_\omega$ would have to exist. Thus, it is reasonable to ask whether there are circumstances in which one can gain sufficient control over all the $\mathbb{P}$-null sets involved to really make sense out of $X \longmapsto \mathbb{E}^{\mathbb{P}}[X|\Sigma](\omega)$ for fixed $\omega \in \Omega$. Of course, when $\Sigma$ is generated by a countable partition $\mathcal{P}$, we already know what to do. Namely, when $\omega \in A \in \mathcal{P}$, we can take

$$\mathbb{E}^{\mathbb{P}}[X|\Sigma](\omega) = \begin{cases} 0 & \text{if } \mathbb{P}(A) = 0 \\ \frac{\mathbb{E}^{\mathbb{P}}[X, A]}{\mathbb{P}(A)} & \text{if } \mathbb{P}(A) > 0. \end{cases}$$

Even when $\Sigma$ does not arise in this way, one can often find a satisfactory representation of conditional expectations as expectations. A quite general statement of this sort is the content of Theorem 9.2.1 in Chapter 9.

§ **5.1.2. Some Extensions.** For various applications it is convenient to have two extensions of the basic theory developed in § 5.1.1. Specifically, as I will now show, the theory is not restricted to probability (or even finite) measures and can be applied to random variables that take their values in a separable Banach space. Thus, from now on, $\mu$ will be an arbitrary (non-negative) measure on $(\Omega, \mathcal{F})$ and $(E, \|\cdot\|_E)$ will be a separable Banach space; and I begin by reviewing a few elementary facts about $\mu$-integration for $E$-valued random variables.[2]

[2] The integration theory that I outline below is what functional analysts call the Bochner integral for Banach space–valued functions. There is a more subtle and intricate theory due to Pettis, but Bochner's theory seems adequate for most probabilistic considerations.

A function $X:\Omega\longrightarrow E$ is said to be **$\mu$-simple** if $X$ is $\mathcal{F}$-measurable, $X$ takes only finitely many values, and $\mu(X\neq 0)<\infty$, in which case its integral with respect to $\mu$ is the element of $E$ given by

$$\mathbb{E}^\mu[X]=\int_\Omega X(\omega)\,\mu(d\omega)\equiv\sum_{x\in E\setminus\{0\}}x\,\mu(X=x).$$

Notice that another description of $\mathbb{E}^\mu[X]$ is as the unique element of $E$ with the property that

$$\big\langle\mathbb{E}^\mu[X],x^*\big\rangle=\mathbb{E}^\mu\big[\langle X,x^*\rangle\big]\quad\text{for all } x^*\in E^*$$

(I use $E^*$ to denote the dual of $E$ and $\langle x,x^*\rangle$ to denote the action of $x^*\in E^*$ on $x\in E$), and therefore that the mapping taking $\mu$-simple $X$ to $\mathbb{E}^\mu[X]$ is linear. Next, observe that $\omega\in\Omega\longmapsto\|X(\omega)\|_E\in\mathbb{R}$ is $\mathcal{F}$-measurable if $X:\Omega\longrightarrow E$ is $\mathcal{F}$-measurable. In particular, for $\mathcal{F}$-measurable $X:\Omega\longrightarrow E$, I will set

$$\|X\|_{L^p(\mu;E)}=\begin{cases}\mathbb{E}^\mu\big[\|X\|_E^p\big]^{\frac{1}{p}} & \text{if } p\in[1,\infty)\\ \inf\big\{M:\mu\big(\|X\|_E>M\big)=0\big\} & \text{if } p=\infty\end{cases}$$

and will write $X\in L^p(\mu;E)$ when $\|X\|_{L^p(\mu;E)}<\infty$. Also, I will say the $X:\Omega\longrightarrow E$ is **$\mu$-integrable** if $X\in L^1(\mu;E)$; and I will say that $X$ is **locally $\mu$-integrable** if $\mathbf{1}_A X$ is $\mu$-integrable for every $A\in\mathcal{F}$ with $\mu(A)<\infty$.

The definition of $\mu$-integration for an $E$-valued $X$ is completed in the following lemma.

LEMMA 5.1.10. *For each $\mu$-integrable $X:\Omega\longrightarrow E$ there is a unique element $\mathbb{E}^\mu[X]\in E$ satisfying $\langle\mathbb{E}^{\mathbb{P}}[X],x^*\rangle=\mathbb{E}^{\mathbb{P}}[\langle X,x^*\rangle]$ for all $x^*\in E^*$. In particular, the mapping $X\in L^1(\mu;E)\longmapsto\mathbb{E}^\mu[X]\in E$ is linear and satisfies*

$$\big\|\mathbb{E}^\mu[X]\big\|_E\le\mathbb{E}^\mu\big[\|X\|_E\big].\tag{5.1.11}$$

*Finally, if $X\in L^p(\mu;E)$, where $p\in[1,\infty)$, then there is a sequence $\{X_n:n\ge 1\}$ of $E$-valued, $\mu$-simple functions with the property that $\|X_n-X\|_{L^p(\mu;E)}\longrightarrow 0$.*

PROOF: Clearly uniqueness, linearity, and (5.1.11) all follow immediately from the given characterization of $\mathbb{E}^\mu[X]$. Thus, all that remains is to prove existence and the final approximation assertion. In fact, once the approximation assertion is proved, then existence will follow immediately from the observation that, by (5.1.11), $\mathbb{E}^\mu[X]$ can be taken equal to $\lim_{n\to\infty}\mathbb{E}^\mu[X_n]$ if $\|X-X_n\|_{L^1(\mu;E)}\longrightarrow 0$.

To prove the approximation assertion, I begin with the case when $\mu$ is finite and $M=\sup_{\omega\in\Omega}\|X(\omega)\|_E<\infty$. Next, choose a dense sequence $\{x_\ell:\ell\ge 1\}$ in $E$, set $A_{0,n}=\emptyset$, and let

$$A_{\ell,n}=\Big\{\omega:\|X(\omega)-x_\ell\|_E<\tfrac{1}{n}\Big\}\quad\text{for } (\ell,n)\in\mathbb{Z}^+\times\mathbb{Z}^+.$$

Then, for each $n \in \mathbb{Z}^+$ there exists an $L_n \in \mathbb{Z}^+$ with the property that

$$\mu\left(\Omega \setminus \bigcup_{\ell=1}^{L_n} A_{\ell,n}\right) < \frac{1}{n^p}.$$

Hence, if $X_n : \Omega \longrightarrow E$ is defined so that

$$X_n(\omega) = x_\ell \quad \text{when } 1 \le \ell \le L_n \text{ and } \omega \in A_{\ell,n} \setminus \bigcup_{k=0}^{\ell-1} A_{k,n}$$

and $X_n(\omega) = 0$ when $\omega \notin \bigcup_1^{L_n} A_{\ell,n}$, then $X_n$ is $\mu$-simple and

$$\|X - X_n\|_{L^p(\mu;E)} \le \frac{M + \mu(E)}{n}.$$

In order to handle the general case, let $X \in L^p(\mu; E)$ and $n \in \mathbb{Z}^+$ be given. We can then find an $r_n \in (0,1]$ with the property that

$$\int_{\Omega(r_n)\complement} \|X(\omega)\|_E^p \, \mu(d\omega) \le \frac{1}{(2n)^p},$$

where

$$\Omega(r) \equiv \left\{\omega : r \le \|X(\omega)\|_E \le \tfrac{1}{r}\right\} \quad \text{for } r \in (0,1].$$

Since, for any $r \in (0,1]$, $r^p \mu(\Omega(r)) \le \|X\|^p_{L^p(\mu;E)}$, we can apply the preceding to the restrictions of $\mu$ and $X$ to $\Omega(r_n)$ and thereby find a $\mu$-simple $X_n : \Omega(r_n) \longrightarrow E$ with the property

$$\left(\int_{\Omega(r_n)} \|X(\omega) - X_n(\omega)\|_E^p \, \mu(d\omega)\right)^{\frac{1}{p}} \le \frac{1}{2n}.$$

Hence, after extending $X_n$ to $\Omega$ by taking it to be 0 off of $\Omega(r_n)$, we arrive at a $\mu$-simple $X_n$ for which $\|X - X_n\|_{L^p(\mu;E)} \le \frac{1}{n}$. □

Given an $\mathcal{F}$-measurable $X : \Omega \longrightarrow E$ and a $B \in \mathcal{F}$ for which $\mathbf{1}_B X \in L^1(\mu; E)$, I will use, depending on the context, either

$$\mathbb{E}^\mu[X, B] \quad \text{or} \quad \int_B X \, d\mu \quad \text{or} \quad \int_B X(\omega)\, \mu(d\omega)$$

to denote the quantity $\mathbb{E}^\mu[\mathbf{1}_B X]$. Also, when discussing the spaces $L^p(\mu; E)$, I will adopt the usual convention of blurring the distinction between a particular $\mathcal{F}$-measurable $X : \Omega \longrightarrow E$ belonging to $L^p(\mu; E)$ and the equivalence class of those $\mathcal{F}$-measurable $Y$'s that differ from $X$ on a $\mu$-null set. Thus, with this convention, $\|\cdot\|_{L^p(\mu;E)}$ becomes a bona fide norm (not just a seminorm) on $L^p(\mu; E)$ with respect to which $L^p(\mu; E)$ becomes a normed vector space. Finally, by the same procedure with which one proves the $L^p(\mu; \mathbb{R})$ spaces are complete, one can prove that the spaces $L^p(\mu; E)$ are complete for any separable Banach space $E$.

THEOREM 5.1.12. *Let $(\Omega, \mathcal{F}, \mu)$ be a $\sigma$-finite measure space and $X : \Omega \longrightarrow E$ a locally $\mu$-integrable function. Then*

$$\mu(X \neq 0) = 0 \iff \mathbb{E}^{\mu}[X, A] = 0 \text{ for } A \in \mathcal{F} \text{ with } \mu(A) < \infty.$$

*Next, assume that $\Sigma$ is a sub-$\sigma$-algebra for which $\mu \restriction \Sigma$ is $\sigma$-finite. Then, for each locally $\mu$-integrable $X : \Omega \longrightarrow E$, there is a $\mu$-almost everywhere unique locally $\mu$-integrable, $\Sigma$-measurable $X_\Sigma : \Omega \longrightarrow E$ such that*

$$\mathbb{E}^{\mu}[X_\Sigma, A] = \mathbb{E}^{\mu}[X, A] \quad \text{for every } A \in \Sigma \text{ with } \mu(A) < \infty. \tag{5.1.13}$$

*In particular, if $Y : \Omega \longrightarrow E$ is a second locally $\mu$-integrable function, then, for all $\alpha,\ \beta \in \mathbb{R}$,*

$$(\alpha X + \beta Y)_\Sigma = \alpha X_\Sigma + \beta Y_\Sigma \quad \text{(a.e., } \mu).$$

*Finally,*

$$\|X_\Sigma\|_E \leq \left(\|X\|_E\right)_\Sigma \quad \text{(a.e., } \mu). \tag{5.1.14}$$

*Hence, not only does (5.1.13) continue to hold for any $A \in \Sigma$ with $\mathbf{1}_A X \in L^1(\mu; E)$, but also, for each $p \in [1, \infty]$, the mapping $X \in L^p(\mu; E) \longmapsto X_\Sigma \in L^p(\mu; E)$ is a linear contraction.*

PROOF: Clearly, it is only necessary to prove the "$\Longleftarrow$" part of the first assertion. Thus, suppose that $\mu(X \neq 0) > 0$. Then, because $E$ is separable and therefore (cf. Exercise 5.1.19) $E^*$ with the weak* topology is also separable, there exists an $\epsilon > 0$ and a $x^* \in E^*$ with the property that $\mu(\langle X, x^* \rangle \geq \epsilon) > 0$, from which it follows (by $\sigma$-finiteness) that there is an $A \in \mathcal{F}$ for which $\mu(A) < \infty$ and

$$\left\langle \mathbb{E}^{\mu}[X, A], x^* \right\rangle = \mathbb{E}^{\mu}\Big[\langle X, x^* \rangle, A\Big] \neq 0.$$

I turn next to the uniqueness and other properties of $X_\Sigma$. But it is obvious that uniqueness is an immediate consequence of the first assertion and that linearity follows from uniqueness. As for (5.1.14), notice that if $x^* \in E^*$ and $\|x^*\|_{E^*} \leq 1$, then

$$\mathbb{E}^{\mu}\big[\langle X_\Sigma, x^* \rangle, A\big] = \mathbb{E}^{\mu}\big[\langle X, x^* \rangle, A\big] \leq \mathbb{E}^{\mu}\big[\|X\|_E, A\big] = \mathbb{E}^{\mu}\big[(\|X\|_E)_\Sigma, A\big]$$

for every $A \in \Sigma$ with $\mu(A) < \infty$. Hence, at least when $\mu$ is a probability measure, Theorem 5.1.3 implies that $\langle X_\Sigma, x^* \rangle \leq \left(\|X\|_E\right)_\Sigma$ (a.e., $\mu$) for each element $x^*$ from the unit ball in $E^*$; and so, because $E^*$ with the weak* topology is separable, (5.1.14) follows in this case. To handle $\mu$'s that are not probability measures, note that either $\mu(\Omega) = 0$, in which case everything is trivial, or $\mu(\Omega) \in (0, \infty)$, in which case we can renormalize $\mu$ to make it a probability measure, or $\mu(\Omega) = \infty$, in which case we can use the $\sigma$-finiteness of $\mu \restriction \Sigma$ to reduce ourselves to the countable, disjoint union of the preceding cases.

Finally, to prove the existence of $X_\Sigma$, I proceed as in the last part of the preceding paragraph to reduce myself to the case when $\mu$ is a probability measure $\mathbb{P}$. Next, suppose that $X$ is simple, let $R$ denote its range, and note that

$$X_\Sigma \equiv \sum_{x \in R} x\, \mathbb{P}(X = x \,|\, \Sigma)$$

has the required properties. In order to handle general $X \in L^1(\mathbb{P}; E)$, I use the approximation result in Lemma 5.1.10 to find a sequence $\{X_n : n \geq 1\}$ of simple functions that tend to $X$ in $L^1(\mathbb{P}; E)$. Then, since

$$(X_n)_\Sigma - (X_m)_\Sigma = \big(X_n - X_m\big)_\Sigma \quad \text{(a.s., } \mathbb{P})$$

and therefore, by (5.1.14),

$$\big\|(X_n)_\Sigma - (X_m)_\Sigma\big\|_{L^1(\mathbb{P};E)} \leq \|X_n - X_m\|_{L^1(\mathbb{P};E)},$$

we know that there exists a $\Sigma$-measurable $X_\Sigma \in L^1(\mathbb{P}; E)$ to which the sequence $\big\{(X_n)_\Sigma : n \geq 1\big\}$ converges; and clearly $X_\Sigma$ has the required properties. □

Referring to the setting in the second part of Theorem 5.1.12, I will extend the convention introduced following Theorem 5.1.3 and call the $\mu$-equivalence class of $X_\Sigma$'s satisfying (5.1.13) the **$\mu$-conditional expectation of $X$ given** $\Sigma$, will use $\mathbb{E}^\mu[X|\Sigma]$ to denote this $\mu$-equivalence class, and will, in general, ignore the distinction between the equivalence class and a generic representative of that class. In addition, if $X : \Omega \longrightarrow E$ is locally $\mu$-integrable, then, just as in Theorem 5.1.4, the following are essentially immediate consequences of uniqueness:

$$\mathbb{E}^\mu\big[YX\big|\Sigma\big] = Y\,\mathbb{E}^\mu\big[X\big|\Sigma\big] \quad \text{(a.e., } \mu) \quad \text{for } Y \in L^\infty(\Omega, \Sigma, \mu; \mathbb{R}),$$

and

$$\mathbb{E}^\mu\big[X\big|\mathcal{T}\big] = \mathbb{E}^\mu\Big[\mathbb{E}^\mu\big[X\big|\Sigma\big]\Big|\mathcal{T}\Big] \quad \text{(a.e., } \mu)$$

whenever $\mathcal{T}$ is a sub-$\sigma$-algebra of $\Sigma$ for which $\mu \restriction \mathcal{T}$ is $\sigma$-finite.

## Exercises for § 5.1

EXERCISE 5.1.15. As the proof of existence in Theorem 5.1.4 makes clear, the operation $X \in L^2(\mathbb{P}; \mathbb{R}) \longmapsto \mathbb{E}^{\mathbb{P}}[X|\Sigma]$ is just the operation of orthogonal projection from $L^2(\mathbb{P}; \mathbb{R})$ onto the space $L^2(\Omega, \Sigma, \mathbb{P}; \mathbb{R})$ of $\Sigma$-measurable elements of $L^2(\mathbb{P}; \mathbb{R})$. For this reason, one might be inclined to think that the concept of conditional expectation is basically a Hilbert space notion. However, as this exercise shows, that inclination should be resisted. The point is that, although conditional expectation is definitely an orthogonal projection, not every orthogonal projection is a conditional expectation!

(**i**) Let $L$ be a closed linear subspace of $L^2(\mathbb{P};\mathbb{R})$, and let $\Sigma_L = \sigma(\{X : X \in L\})$ be the $\sigma$-algebra over $\Omega$ generated by $X \in L$. Show that $L = L^2(\Omega, \Sigma_L, \mathbb{P};\mathbb{R})$ if and only if $\mathbf{1} \in L$ and $X^+ \in L$ whenever $X \in L$.

**Hint**: To prove the "if" assertion, let $X \in L$ be given, and show that

$$X_n \equiv \left[n(X - \alpha\mathbf{1})^+ \wedge \mathbf{1}\right] \in L \quad \text{for every } \alpha \in \mathbb{R} \text{ and } n \in \mathbb{Z}^+.$$

Conclude that $X_n \nearrow \mathbf{1}_{(\alpha,\infty)} \circ X$ must be an element of $L$.

(**ii**) Let $\Pi$ be an orthogonal projection operator on $L^2(\mathbb{P};\mathbb{R})$, set $L = \mathrm{Range}(\Pi)$, and let $\Sigma = \Sigma_L$, where $\Sigma_L$ is defined as in part (**i**). Show that $\Pi X = \mathbb{E}^{\mathbb{P}}[X|\Sigma]$ (a.s., $\mathbb{P}$) for all $X \in L^2(\mathbb{P};\mathbb{R})$ if and only if $\Pi\mathbf{1} = \mathbf{1}$ and

$$(*) \qquad \Pi(X\,\Pi Y) = (\Pi X)(\Pi Y) \quad \text{for all} \quad X, Y \in L^\infty(\mathbb{P};\mathbb{R}).$$

**Hint**: Assume that $\Pi\mathbf{1} = \mathbf{1}$ and that (*) holds. Given $X \in L^\infty(\mathbb{P};\mathbb{R})$, use induction to show that

$$\|\Pi X\|^n_{L^{2n}(\mathbb{P})} \le \|X\|^{n-1}_{L^\infty(\mathbb{P})}\|X\|_{L^2(\mathbb{P})} \quad \text{and} \quad (\Pi X)^n = \Pi\left(X(\Pi X)^{n-1}\right)$$

for all $n \in \mathbb{Z}^+$. Conclude that $\|\Pi X\|_{L^\infty(\mathbb{P})} \le \|X\|_{L^\infty(\mathbb{P})}$ and that $(\Pi X)^n \in L$, $n \in \mathbb{Z}^+$, for every $X \in L^\infty(\mathbb{P};\mathbb{R})$. Next, using the preceding together with Weierstrass's Approximation Theorem, show that $(\Pi X)^+ \in L$, first for $X \in L^\infty(\mathbb{P};\mathbb{R})$ and then for all $X \in L^2(\mathbb{P};\mathbb{R})$. Finally, apply (**i**) to arrive at $L = L^2(\Omega, \Sigma, \mathbb{P};\mathbb{R})$.

(**iii**) To emphasize the point being made here, consider once again a closed linear subspace $L$ of $L^2(\mathbb{P};\mathbb{R})$, and let $\Pi_L$ be orthogonal projection onto $L$. Given $X \in L^2(\mathbb{P};\mathbb{R})$, recall that $\Pi_L X$ is characterized as the unique element of $L$ for which $X - \Pi_L X \perp L$, and show that $\mathbb{E}^{\mathbb{P}}[X|\Sigma_L]$ is the unique element of $L^2(\Omega, \Sigma_L, \mathbb{P};\mathbb{R})$ with the property that

$$X - \mathbb{E}^{\mathbb{P}}[X|\Sigma_L] \perp f(Y_1, \dots, Y_n)$$

for all $n \in \mathbb{Z}^+$, $f \in C_{\mathrm{b}}(\mathbb{R}^n;\mathbb{R})$, and $Y_1, \dots, Y_n \in L$. In particular, $\Pi_L X = \mathbb{E}^{\mathbb{P}}[X|\Sigma_L]$ if and only if $X - \Pi_L X$ is perpendicular not only to all *linear* functions of the $Y$'s in $L$ but even to all *nonlinear* ones.

EXERCISE 5.1.16. In spite of the preceding, there is a situation in which orthogonal projection coincides with conditioning. Namely, suppose that $\mathfrak{G}$ is a closed Gaussian family in $L^2(\mathbb{P};\mathbb{R})$, and let $L$ be a closed, linear subspace of $\mathfrak{G}$. As an application of Lemma 4.3.1, show that, for any $X \in \mathfrak{G}$, the orthogonal projection $\Pi_L X$ of $X$ onto $L$ is a conditional expectation value of $X$ given the $\sigma$-algebra $\Sigma_L$ generated by the elements of $L$.

EXERCISE 5.1.17. Because most projections are not conditional expectations, it is an unfortunate fact of life that, for the most part, partial sums of Fourier series cannot be interpreted as conditional expectations. Be that as it may, there are special cases in which such an interpretation is possible. To see this, take $\Omega = [0,1)$, $\mathcal{F} = \mathcal{B}_{[0,1)}$, and $\mathbb{P}$ to be the restriction of Lebesgue measure to $[0,1)$. Next, for $n \in \mathbb{N}$, take $\mathcal{F}_n$ to be the $\sigma$-algebra generated by those $f \in C([0,1);\mathbb{C})$ that are periodic with period $2^{-n}$. Finally, set $\mathbf{e}_k(x) = \exp\big[\sqrt{-1}k2\pi x\big]$ for $k \in \mathbb{Z}$, and use elementary Fourier analysis to show that, for each $n \in \mathbb{N}$, $\{\mathbf{e}_{k2^n} : k \in \mathbb{Z}\}$ is an orthonormal basis for $L^2(\Omega, \mathcal{F}_n, \mathbb{P}; \mathbb{C})$. In particular, conclude that, for every $f \in L^2(\mathbb{P};\mathbb{C})$,

$$\mathbb{E}^{\mathbb{P}}[f \,|\, \mathcal{F}_n] = \mathbb{E}^{\mathbb{P}}[f] + \sum_{k \in \mathbb{Z}} \big(f, \mathbf{e}_{k2^n}\big)_{L^2([0,1);\mathbb{C})} \mathbf{e}_{k2^n},$$

where the convergence is in $L^2([0,1];\mathbb{C})$. (Also see Exercise 5.2.45.)

EXERCISE 5.1.18. Let $(\Omega, \mathcal{F}, \mu)$ be a measure space and $\Sigma$ a sub-$\sigma$-algebra of $\mathcal{F}$ with the property that $\mu \restriction \Sigma$ is $\sigma$-finite. Next, let $E$ be a separable Hilbert space, $p \in [1,\infty]$, $X \in L^p(\mu;E)$, and $Y$ a $\Sigma$-measurable element of $L^{p'}(\mu;E)$ ($p'$ is the Hölder conjugate of $p$). Show that

$$\mathbb{E}^{\mu}\Big[\big(Y, X\big)_E \Big| \Sigma\Big] = \Big(Y, \mathbb{E}^{\mu}[X|\Sigma]\Big)_E \quad \mu\text{-almost surely.}$$

**Hint**: First observe that it suffices to check that

$$\mathbb{E}^{\mu}\Big[\big(Y, X\big)_E\Big] = \mathbb{E}^{\mu}\Big[\Big(Y, \mathbb{E}^{\mu}[X|\Sigma]\Big)_E\Big].$$

Next, choose an orthonormal basis $\{\mathbf{e}_n : n \geq 0\}$ for $E$, and justify the steps in

$$\begin{aligned}\mathbb{E}^{\mu}\big[(Y,X)_E\big] &= \sum_1^{\infty} \mathbb{E}^{\mu}\big[(Y,\mathbf{e}_n)_E (\mathbf{e}_n, X)_E\big] \\ &= \sum_1^{\infty} \mathbb{E}^{\mu}\Big[(Y,\mathbf{e}_n)_E\, \mathbb{E}^{\mu}\big[(\mathbf{e}_n, X)_E \big| \Sigma\big]\Big] = \mathbb{E}^{\mu}\big[(Y, \mathbb{E}^{\mu}[X|\Sigma])_E\big].\end{aligned}$$

EXERCISE 5.1.19. Let $E$ be a separable Banach space, and show that, for each $R > 0$, the closed ball $\overline{B_{E^*}(0,R)}$ with the weak* topology is a compact metric space. Conclude from this that the weak* topology on $E^*$ is second countable and therefore separable.

**Hint**: Choose a countable, dense subset $\{x_n : n \geq 1\}$ in the unit ball $B_E(0,1)$, and define

$$\rho(x^*, y^*) = \sum_{n=1}^{\infty} 2^{-n} \big|\langle x_n, x^* - y^* \rangle\big| \quad \text{for } x^*, y^* \in \overline{B_{E^*}(0,R)}.$$

Show that $\rho$ is a metric for the weak* topology on $\overline{B_{E^*}(0,R)}$. Next, choose $\{x_{n_m} : m \geq 1\}$ so that $x_{n_1} = x_1$ and $x_{n_{m+1}} = x_n$ if $n$ is the first $n > n_m$ such that $x_n$ is linearly independent of $\{x_1, \dots, x_{n-1}\}$. Given a sequence $\{x^*_\ell : \ell \geq 1\}$ in $\overline{B_{E^*}(0,R)}$, use a diagonalization argument to find a subsequence $\{x^*_{\ell_k} : k \geq 1\}$ such that $a_m = \lim_{k\to\infty} \langle x_{n_m}, x^*_{\ell_k}\rangle$ exists for each $m \geq 1$. Now define $f$ on the span $S$ of $\{x_{n_m} : m \geq 1\}$ so that $f(x) = \sum_{m=1}^M \alpha_m a_m$ if $x = \sum_{m=1}^M \alpha_m x_{n_m}$, note that $f(x) = \lim_{k\to\infty} \langle x, x^*_{\ell_k}\rangle$ for $x \in S$, and conclude that $f$ is linear on $S$ and satisfies the estimate $|f(x)| \leq R\|x\|_E$ there. Since $S$ is dense in $E$, there is a unique extension of $f$ as a bounded linear functional on $E$ satisfying the same estimate, and so there exists an $x^* \in \overline{B_{E^*}(0,R)}$ such that $\langle x, x^*\rangle = \lim_{k\to\infty}\langle x, x^*_{\ell_k}\rangle$ for all $x \in S$. Finally, check that this convergence continues to hold for all $x \in E$, and conclude that $x^*_{\ell_k} \longrightarrow x^*$ in the weak* topology.

EXERCISE 5.1.20. The purpose of this exercise is to show that Bochner's theory of integration for Banach space functions relies heavily on the assumption that the Banach space be separable. In particular, the approximation procedure on which the proof of Lemma 5.1.10 fails in the absence of separability. To see this, consider the Banach space $\ell^\infty(\mu;\mathbb{R})$ of uniformly bounded sequences $\mathbf{x} = (x_0, \dots, x_n, \dots) \in \mathbb{R}^{\mathbb{N}}$ with $\|\mathbf{x}\|_{\ell^\infty(\mathbb{N};\mathbb{R})} = \sup_{n\geq 0}|x_n|$. Next, let $\{X_n : n \geq 0\}$ be a sequence of mutually independent, $\{-1,1\}$-valued, Bernoulli random with mean value 0 on some probability space $(\Omega, \mathcal{F}, \mathbb{P})$, and define $\mathbf{X} : \Omega \longrightarrow \ell^\infty(\mathbb{N};\mathbb{R})$ by $\mathbf{X}(\omega) = \big(X_0(\omega), \dots, X_n(\omega), \dots\big)$. Show that, for any simple function $\mathbf{Y} : \Omega \longrightarrow \ell^\infty(\mathbb{N};\mathbb{R})$,

$$\mathbb{P}\big(\|\mathbf{X} - \mathbf{Y}\|_{\ell^\infty(\mathbb{N};\mathbb{R})} < \tfrac{1}{4}\big) = 0.$$

**Hint**: For any $\alpha \in \mathbb{R}$, show that $\mathbb{P}\big(|X_n - \alpha| < \frac{1}{4}\big) \leq \frac{1}{2}$ and therefore that $\mathbb{P}\big(\|\mathbf{X} - \mathbf{A}\|_{\ell^\infty(\mathbb{N};\mathbb{R})} < \frac{1}{4}\big) = 0$ for any $\mathbf{A} \in \ell^\infty(\mathbb{N};\mathbb{R})$.

## § 5.2 Discrete Parameter Martingales

In this section I will introduce an interesting and useful class of stochastic processes that unifies and simplifies several branches of probability theory as well as other branches of analysis. From the analytic point of view, what I will be doing is developing an abstract version of differentiation theory (cf. Theorem 6.1.8).

Although I will want to make some extensions in § 5.3, I start in the following setting. $(\Omega, \mathcal{F}, \mathbb{P})$ is a probability space and $\{\mathcal{F}_n : n \in \mathbb{N}\}$ is a non-decreasing sequence of sub-$\sigma$-algebra's of $\mathcal{F}$. Given a measurable space $(E, \mathcal{B})$, say that the family $\{X_n : n \in \mathbb{N}\}$ of $E$-valued random variables is $\{\mathcal{F}_n : n \in \mathbb{N}\}$-**progressively measurable** if $X_n$ is $\mathcal{F}_n$-measurable for each $n \in \mathbb{N}$. Next, a family $\{X_n : n \in \mathbb{N}\}$ of $(-\infty,\infty]$-valued random variables is said to be a **$\mathbb{P}$-submartingale with respect to** $\{\mathcal{F}_n : n \in \mathbb{N}\}$ if it is $\{\mathcal{F}_n : n \in \mathbb{N}\}$-progressively measurable, and, for each $n \in \mathbb{N}$, $\mathbb{E}^{\mathbb{P}}[X_n^-] < \infty$ and $X_n \leq \mathbb{E}^{\mathbb{P}}[X_{n+1}|\mathcal{F}_n]$ (a.s., $\mathbb{P}$). It is said to be a **$\mathbb{P}$-martingale with respect to** $\{\mathcal{F}_n : n \in \mathbb{N}\}$ if $\{X_n : n \in \mathbb{N}\}$ is an $\{\mathcal{F}_n : n \in \mathbb{N}\}$-progressively measurable family of

$\mathbb{R}$-valued, $\mathbb{P}$-integrable random variables satisfying $X_n = \mathbb{E}^{\mathbb{P}}[X_{n+1}|\mathcal{F}_n]$ (a.s., $\mathbb{P}$) for each $n \in \mathbb{N}$. In the future, I will abbreviate these statements by saying that the triple $(X_n, \mathcal{F}_n, \mathbb{P})$ is a submartingale or a martingale.

**Examples.** The most trivial example of a submartingale is provided by a non-decreasing sequence $\{a_n : n \geq 0\}$. That is, if $X_n \equiv a_n$, $n \in \mathbb{N}$, then $(X_n, \mathcal{F}_n, \mathbb{P})$ is a submartingale on any probability space $(\Omega, \mathcal{F}, \mathbb{P})$ relative to any non-decreasing $\{\mathcal{F}_n : n \in \mathbb{N}\}$. More interesting examples are those which follow.[1]

**(i)** Let $\{Y_n : n \geq 1\}$ be a sequence of mutually independent $(-\infty, \infty]$-valued random variables with $\mathbb{E}^{\mathbb{P}}[Y_n^-] < \infty$, $n \in \mathbb{N}$, set $\mathcal{F}_0 = \{\emptyset, \Omega\}$, $\mathcal{F}_n = \sigma(\{Y_1, \dots, Y_n\})$ for $n \in \mathbb{Z}^+$, and define $X_n = \sum_{1 \leq m \leq n} Y_m$, where summation over the empty set is taken to be 0. Then, because $\mathbb{E}^{\mathbb{P}}[Y_{n+1}|\mathcal{F}_n] = \mathbb{E}^{\mathbb{P}}[Y_{n+1}]$ (a.s., $\mathbb{P}$) and therefore

$$\mathbb{E}^{\mathbb{P}}\big[X_{n+1}\big|\mathcal{F}_n\big] = X_n + \mathbb{E}^{\mathbb{P}}\big[Y_{n+1}\big] \quad \text{(a.s., } \mathbb{P}\text{)}$$

for every $n \in \mathbb{N}$, we see that $(X_n, \mathcal{F}_n, \mathbb{P})$ is a submartingale if and only if $\mathbb{E}^{\mathbb{P}}[Y_n] \geq 0$ for all $n \in \mathbb{Z}^+$. In fact, if the $Y_n$'s are $\mathbb{R}$-valued and $\mathbb{P}$-integrable, then the same line of reasoning shows that $(X_n, \mathcal{F}_n, \mathbb{P})$ is a martingale if and only if $\mathbb{E}^{\mathbb{P}}[Y_n] = 0$ for all $n \in \mathbb{Z}^+$. Finally, if $\{Y_n : n \geq 0\} \subseteq L^2(\mathbb{P}; \mathbb{R})$ and $\mathbb{E}^{\mathbb{P}}[Y_n] = 0$ for each $n \in \mathbb{Z}^+$, then

$$\mathbb{E}^{\mathbb{P}}\big[X_{n+1}^2 \,\big|\, \mathcal{F}_n\big] = X_n^2 + \mathbb{E}^{\mathbb{P}}\big[Y_{n+1}^2 \,\big|\, \mathcal{F}_n\big] \geq X_n^2 \quad \text{(a.s., } \mathbb{P}\text{)},$$

and so $(X_n^2, \mathcal{F}_n, \mathbb{P})$ is a submartingale.

**(ii)** If $X$ is an $\mathbb{R}$-valued, $\mathbb{P}$-integrable random variable and $\{\mathcal{F}_n : n \in \mathbb{N}\}$ is a non-decreasing sequence of sub-$\sigma$-algebras of $\mathcal{F}$, then, by (5.1.6), $(\mathbb{E}^{\mathbb{P}}[X|\mathcal{F}_n], \mathcal{F}_n, \mathbb{P})$ is a martingale.

**(iii)** If $(X_n, \mathcal{F}_n, \mathbb{P})$ is a martingale, then, by (5.1.5), $(|X_n|, \mathcal{F}_n, \mathbb{P})$ is a submartingale.

**§ 5.2.1. Doob's Inequality and Marcinkewitz's Theorem.** In view of Example (i) above, we see that partial sums of independent random variables with mean value 0 are a source of martingales and that their squares are a source of submartingales. Hence, it is reasonable to ask whether some of the important facts about such partial sums will continue to be true for all martingales, and perhaps the single most important indication that the answer may be "yes" is contained in the following generalization of Kolmogorov's Inequality (cf. Theorem 1.4.5). Like most of the foundational results in martingale theory, this one is due to J.L. Doob. It is interesting that Doob's proof is essentially the same as Kolmogorov's, only, if anything easier.

[1] For a much more interesting and complete list of examples, the reader might want to consult J. Neveu's *Discrete-parameter Martingales*, North–Holland (1975).

THEOREM 5.2.1 (**Doob's Inequality**). *Assume that $(X_n, \mathcal{F}_n, \mathbb{P})$ is a submartingale. Then, for every $N \in \mathbb{Z}^+$ and $\alpha \in (0,\infty)$,*

$$\mathbb{P}\left(\max_{0\le n\le N} X_n \ge \alpha\right) \le \frac{1}{\alpha}\mathbb{E}^{\mathbb{P}}\left[X_N,\ \max_{0\le n\le N} X_n \ge \alpha\right]. \tag{5.2.2}$$

*In particular, if the $X_n$'s are non-negative, then, for each $p \in (1,\infty)$,*

$$\mathbb{E}^{\mathbb{P}}\left[\sup_{n\in\mathbb{N}} X_n^p\right]^{\frac{1}{p}} \le \frac{p}{p-1}\sup_{n\in\mathbb{N}} \mathbb{E}^{\mathbb{P}}\left[X_n^p\right]^{\frac{1}{p}}. \tag{5.2.3}$$

PROOF: To prove (5.2.2), set $A_0 = \{X_0 \ge \alpha\}$ and

$$A_n = \left\{X_n \ge \alpha \text{ but } \max_{0\le m<n} X_m < \alpha\right\} \quad \text{for} \quad n \in \mathbb{Z}^+.$$

Then the $A_n$'s are mutually disjoint and $A_n \in \mathcal{F}_n$ for each $n \in \mathbb{N}$. Thus,

$$\begin{aligned} P\left(\max_{0\le n\le N} X_n \ge \alpha\right) &= \sum_{n=0}^N P(A_n) \le \sum_{n=0}^N \frac{\mathbb{E}^{\mathbb{P}}[X_n, A_n]}{\alpha} \\ &\le \sum_{n=0}^N \frac{\mathbb{E}^{\mathbb{P}}[X_N, A_n]}{\alpha} = \frac{1}{\alpha}\mathbb{E}^{\mathbb{P}}\left[X_N,\ \max_{0\le n\le N} X_n \ge \alpha\right]. \end{aligned}$$

Now assume that the $X_n$'s are non-negative. Given (5.2.2), (5.2.3) becomes an easy application of Exercise 1.4.18. □

Doob's inequality is an example of what analysts call a **weak-type inequality**. To be more precise, it is a *weak-type* 1–1 inequality. The terminology derives from the fact that such an inequality follows immediately from an $L^1$-norm, or *strong-type* 1–1, inequality between the objects under consideration; but, in general, it is strictly weaker. In order to demonstrate how powerful such a result can be, I will now apply Doob's Inequality to prove a theorem of Marcinkewitz. Because it is an argument to which we will return again, the reader would do well to become comfortable with the line of reasoning that allows one to pass from a *weak-type inequality*, like Doob's, to almost sure convergence results.

COROLLARY 5.2.4. *Let $X$ be an $\mathbb{R}$-valued random variable and $p \in [1,\infty)$. If $X \in L^p(\mathbb{P};\mathbb{R})$, then, for any non-decreasing sequence $\{\mathcal{F}_n : n \in \mathbb{N}\}$ of sub-$\sigma$-algebras of $\mathcal{F}$,*

$$\mathbb{E}^{\mathbb{P}}[X|\mathcal{F}_n] \longrightarrow \mathbb{E}^{\mathbb{P}}\left[X \,\middle|\, \bigvee_0^\infty \mathcal{F}_n\right] \quad \text{(a.s., } \mathbb{P}) \text{ and in } L^p(\mathbb{P};\mathbb{R}) \text{ as } n \to \infty.$$

*In particular, if $X$ is $\bigvee_0^\infty \mathcal{F}_n$-measurable, then $\mathbb{E}^{\mathbb{P}}[X|\mathcal{F}_n] \longrightarrow X$ (a.s., $\mathbb{P}$) and in $L^p(\mathbb{P};\mathbb{R})$.*

PROOF: Without loss in generality, assume that $\mathcal{F} = \bigvee_0^\infty \mathcal{F}_n$.

Given $X \in L^1(\mathbb{P};\mathbb{R})$, set $X_n = \mathbb{E}^{\mathbb{P}}[X|\mathcal{F}_n]$ for $n \in \mathbb{N}$. The key to my proof will be the inequality

$$\mathbb{P}\left(\sup_{n\in\mathbb{N}}|X_n| \geq \alpha\right) \leq \frac{1}{\alpha}\mathbb{E}^{\mathbb{P}}\left[|X|, \sup_{n\in\mathbb{N}}|X_n| \geq \alpha\right], \quad \alpha \in (0,\infty); \tag{5.2.5}$$

and, since, by (5.1.5), $|X_n| \leq \mathbb{E}^{\mathbb{P}}[|X|\,|\mathcal{F}_n]$ (a.s., $\mathbb{P}$), while proving (5.2.5) I may and will assume that $X$ and all the $X_n$'s are non-negative. But then, by (5.2.2),

$$\begin{aligned} P\left(\sup_{0\leq n\leq N} X_n > \alpha\right) &\leq \frac{1}{\alpha}\mathbb{E}^{\mathbb{P}}\left[X_N, \sup_{0\leq n\leq N} X_n > \alpha\right] \\ &= \frac{1}{\alpha}\mathbb{E}^{\mathbb{P}}\left[X, \sup_{0\leq n\leq N} X_n > \alpha\right] \end{aligned}$$

for all $N \in \mathbb{Z}^+$, and therefore (5.2.5) follows when $N \to \infty$ and one takes right limits in $\alpha$.

As my first application of (5.2.5), note that $\{X_n : n \geq 0\}$ is uniformly $\mathbb{P}$-integrable. Indeed, because $|X_n| \leq \mathbb{E}^{\mathbb{P}}[|X|\,|\mathcal{F}_n]$, we have from (5.2.5) that

$$\begin{aligned} \sup_{n\in\mathbb{N}} \mathbb{E}^{\mathbb{P}}\Big[|X_n|, |X_n| \geq \alpha\Big] &\leq \sup_{n\in\mathbb{N}} \mathbb{E}^{\mathbb{P}}\Big[|X|, |X_n| \geq \alpha\Big] \\ &\leq \mathbb{E}^{\mathbb{P}}\left[|X|, \sup_{n\in\mathbb{N}}|X_n| \geq \alpha\right] \longrightarrow 0 \end{aligned}$$

as $\alpha \to \infty$. Thus, we will know that the asserted convergence takes place in $L^1(\mathbb{P};\mathbb{R})$ as soon as we show that it happens $\mathbb{P}$-almost surely. In addition, if $X \in L^p(\mathbb{P};\mathbb{R})$ for some $p \in (1,\infty)$, then, by (5.2.5) and Exercise 1.4.18, we see that $\big\{|X_n|^p : n \in \mathbb{N}\big\}$ is uniformly $\mathbb{P}$-integrable and, therefore, that $X_n \longrightarrow X$ in $L^p(\mu;\mathbb{R})$ as soon as it does (a.s., $\mathbb{P}$). In other words, everything comes down to checking the $\mathbb{P}$-almost sure convergence for $X \in L^1(\mathbb{P};\mathbb{R})$.

To prove the $\mathbb{P}$-almost sure convergence, let $\mathcal{G}$ be the set of $X \in L^1(\mathbb{P};\mathbb{R})$ for which $X_n \longrightarrow X$ (a.s., $\mathbb{P}$). Clearly, $X \in \mathcal{G}$ if $X \in L^1(\mathbb{P};\mathbb{R})$ is $\mathcal{F}_n$-measurable for some $n \in \mathbb{N}$, and, therefore, $\mathcal{G}$ is dense in $L^1(\mathbb{P};\mathbb{R})$. Thus, all that remains is to prove that $\mathcal{G}$ is closed in $L^1(\mathbb{P};\mathbb{R})$. But if $\{X^{(k)} : k \geq 1\} \subseteq \mathcal{G}$ and $X^{(k)} \longrightarrow X$ in $L^1(\mathbb{P};\mathbb{R})$, then, by (5.2.5),

$$\begin{aligned} &\mathbb{P}\left(\sup_{n\geq N}|X_n - X| \geq 3\alpha\right) \\ \leq &\mathbb{P}\left(\sup_{n\geq N}|X_n - X_n^{(k)}| \geq \alpha\right) + \mathbb{P}\left(\sup_{n\geq N}|X_n^{(k)} - X^{(k)}| \geq \alpha\right) \\ &\qquad + \mathbb{P}\Big(|X^{(k)} - X| \geq \alpha\Big) \\ \leq &\frac{2}{\alpha}\big\|X - X^{(k)}\big\|_{L^1(\mathbb{P})} + \mathbb{P}\left(\sup_{n\geq N}|X_n^{(k)} - X^{(k)}| \geq \alpha\right) \end{aligned}$$

for every $N \in \mathbb{Z}^+$, $\alpha \in (0,\infty)$, and $k \in \mathbb{Z}^+$. Hence, by first letting $N \to \infty$ and then $k \to \infty$, we see that

$$\lim_{N\to\infty} \mathbb{P}\left(\sup_{n\geq N} |X_n - X| \geq 3\alpha\right) = 0 \quad \text{for every } \alpha \in (0,\infty);$$

and this proves that $X \in \mathcal{G}$. □

Before moving on to more sophisticated convergence results, I will spend a little time showing that Corollary 5.2.4 is already interesting. In order to introduce my main application, recall my preliminary discussion of conditioning when I was attempting to explain Kolmogorov's idea at the beginning of this chapter. As I said there, the most easily understood situation occurs when one conditions with respect to a sub-$\sigma$-algebra $\Sigma$ that is generated by a countable partition $\mathcal{P}$. Indeed, in that case one can easily verify that

$$\mathbb{E}^{\mathbb{P}}\big[X\big|\Sigma\big] = \sum_{A\in\mathcal{P}} \frac{\mathbb{E}^{\mathbb{P}}[X, A]}{\mathbb{P}(A)} \mathbf{1}_A, \tag{5.2.6}$$

where it is understood that

$$\frac{\mathbb{E}^{\mathbb{P}}[X, A]}{\mathbb{P}(A)} \equiv 0 \quad \text{when} \quad \mathbb{P}(A) = 0.$$

Unfortunately, even when $\mathcal{F}$ is countably generated, $\Sigma$ need not be (cf. Exercise 1.1.18). Furthermore, just because $\Sigma$ is countably generated, it will be seldom true that its generators can be chosen to form a countable partition. (For example, as soon as $\Sigma$ contains an uncountable number of atoms, such a partition cannot exist.) Nonetheless, if $\Sigma$ is any countably generated $\sigma$-algebra, then we can find a sequence $\{\mathcal{P}_n : n \geq 0\}$ of finite partitions with the properties that

$$\Sigma = \sigma\left(\bigcup_0^\infty \mathcal{P}_n\right) \quad \text{and} \quad \sigma(\mathcal{P}_{n-1}) \subseteq \sigma(\mathcal{P}_n), \quad n \in \mathbb{Z}^+.$$

In fact, simply choose a countable generating sequence $\{A_n : n \geq 0\}$ for $\Sigma$ and take $\mathcal{P}_n$ to be the collection of distinct sets of the form $B_0 \cap \cdots \cap B_n$, where $B_m \in \{A_m, A_m\complement\}$ for each $0 \leq m \leq n$.

THEOREM 5.2.7. *Let $\Sigma$ be a countably generated sub-$\sigma$-algebra of $\mathcal{F}$, and choose $\{\mathcal{P}_n : n \geq 0\}$ to be a sequence of finite partitions as above. Next, given $p \in [1,\infty)$ and a random variable $X \in L^p(\mathbb{P};\mathbb{R})$, define $X_n$ for $n \in \mathbb{N}$ by the right-hand side of (5.2.6) with $\mathcal{P} = \mathcal{P}_n$. Then $X_n \longrightarrow \mathbb{E}^{\mathbb{P}}[X|\Sigma]$ both $\mathbb{P}$-almost surely and in $L^p(\mathbb{P};\mathbb{R})$. Moreover, even if $\Sigma$ is not countably generated, for each separable, closed subspace $L$ of $L^p(\mathbb{P};\mathbb{R})$ there exists a sequence $\{\mathcal{P}_n : n \in \mathbb{N}\}$ of finite partitions such that*

$$\sum_{A\in\mathcal{P}_n} \frac{\mathbb{E}^{\mathbb{P}}[X, A]}{\mathbb{P}(A)} \mathbf{1}_A \longrightarrow \mathbb{E}^{\mathbb{P}}\big[X\big|\Sigma\big] \quad (\text{a.s.}, \mathbb{P}) \text{ and in } L^p(\mathbb{P};\mathbb{R})$$

*for every $X \in L$.*

PROOF: To prove the first part, simply set $\mathcal{F}_n = \sigma(\mathcal{P}_n)$, identify the $X_n$ in (5.2.6) as $\mathbb{E}^{\mathbb{P}}[X|\mathcal{F}_n]$, and finally apply Corollary 5.2.4. As for the second part, let $\Sigma(L)$ be the $\sigma$-algebra generated by $\{\mathbb{E}^{\mathbb{P}}[X|\Sigma] : X \in L\}$, note that $\Sigma(L)$ is countably generated and that

$$\mathbb{E}^{\mathbb{P}}\big[X\big|\Sigma\big] = \mathbb{E}^{\mathbb{P}}\big[X\big|\Sigma(L)\big] \quad (\text{a.s.}, \mathbb{P}) \quad \text{for each } X \in L,$$

and apply the first part with $\Sigma$ replaced by $\Sigma(L)$. □

Theorem 5.2.7 makes it easy to transfer the usual Jensen's Inequality to conditional expectations.

COROLLARY 5.2.8 (**Jensen's Inequality**). *Let $C$ be a closed, convex subset of $\mathbb{R}^N$, $\mathbf{X}$ a $C$-valued, $\mathbb{P}$-integrable random variable, and $\Sigma$ a sub-$\sigma$-algebra of $\mathcal{F}$. Then there is a $C$-valued representative $\mathbf{X}_\Sigma$ of*

$$\mathbb{E}^{\mathbb{P}}\big[\mathbf{X}\big|\Sigma\big] \equiv \begin{bmatrix} \mathbb{E}^{\mathbb{P}}\big[X_1\big|\Sigma\big] \\ \vdots \\ \mathbb{E}^{\mathbb{P}}\big[X_N\big|\Sigma\big] \end{bmatrix}.$$

*In addition, if $g : C \longrightarrow [0,\infty)$ is continuous and concave, then*

$$\mathbb{E}^{\mathbb{P}}\big[g(\mathbf{X})\big|\Sigma\big] \le g(\mathbf{X}_\Sigma) \quad (\text{a.s.}, \mathbb{P}).$$

*Finally, if $f : C \longrightarrow \mathbb{R}$ is continuous, convex, and bounded above and if $X$ is a $C$-valued, $\mathbb{P}$-integrable random variable, then $f(X)$ is $\mathbb{P}$-integrable and*

$$f\big(\mathbb{E}^{\mathbb{P}}[X|\Sigma]\big) \le \mathbb{E}^{\mathbb{P}}\big[f(X)|\Sigma\big]\big) \quad (\text{a.s.}, \mathbb{P}). \tag{5.2.9}$$

*(See Exercise 6.1.15 for Banach space–valued random variables.)*

PROOF: By the classical Jensen's Inequality, $Y \equiv g(\mathbf{X})$ is $\mathbb{P}$-integrable. Hence, by the second part of Theorem 5.2.7, we can find finite partitions $\mathcal{P}_n$, $n \in \mathbb{N}$, so that

$$\mathbf{X}_n \equiv \sum_{A \in \mathcal{P}_n} \frac{\mathbb{E}^{\mathbb{P}}[\mathbf{X}, A]}{\mathbb{P}(A)} \mathbf{1}_A \longrightarrow \mathbb{E}^{\mathbb{P}}[\mathbf{X}|\Sigma]$$

and

$$Y_n \equiv \sum_{A \in \mathcal{P}_n} \frac{\mathbb{E}^{\mathbb{P}}\big[g(\mathbf{X}), A\big]}{\mathbb{P}(A)} \mathbf{1}_A \longrightarrow \mathbb{E}^{\mathbb{P}}\big[g(\mathbf{X})\big|\Sigma\big]$$

$\mathbb{P}$-almost surely. Furthermore, again by the classical Jensen's Inequality,

$$\frac{\mathbb{E}^{\mathbb{P}}[\mathbf{X}, A]}{\mathbb{P}(A)} \in C \quad \text{and} \quad \frac{\mathbb{E}^{\mathbb{P}}\big[g(\mathbf{X}), A\big]}{\mathbb{P}(A)} \le g\left(\frac{\mathbb{E}^{\mathbb{P}}[\mathbf{X}, A]}{\mathbb{P}(A)}\right)$$

for all $A \in \mathcal{F}$ with $\mathbb{P}(A) > 0$. Hence, if $\Lambda \in \Sigma$ denotes the set of $\omega$ for which

$$\lim_{n\to\infty} \begin{bmatrix} \mathbf{X}_n(\omega) \\ Y_n(\omega) \end{bmatrix} \in \mathbb{R}^{N+1}$$

exists, $\mathbf{v}$ is a fixed element of $C$,

$$\mathbf{X}_\Sigma(\omega) \equiv \begin{cases} \lim_{n\to\infty} \mathbf{X}_n(\omega) & \text{if} \quad \omega \in \Lambda \\ \mathbf{v} & \text{if} \quad \omega \notin \Lambda, \end{cases}$$

and

$$Y(\omega) \equiv \begin{cases} \lim_{n\to\infty} Y_n(\omega) & \text{if} \quad \omega \in \Lambda \\ \mathbf{v} & \text{if} \quad \omega \notin \Lambda, \end{cases}$$

then $\mathbf{X}_\Sigma$ is a $C$-valued representative of $\mathbb{E}^{\mathbb{P}}[\mathbf{X}|\Sigma]$, $Y$ is a representative of $\mathbb{E}^{\mathbb{P}}[g(\mathbf{X})|\Sigma]$, and $Y(\omega) \le g(\mathbf{X}_\Sigma(\omega))$ for every $\omega \in \Omega$.

Turning to the final assertion, begin by observing that once one knows that $f(X) \in L^1(\mathbb{P};\mathbb{R})$, the concluding inequality follows immediately by applying the first part to the non-negative, concave function $M - f$, where $M \in \mathbb{R}$ is an upper bound of $f$. Thus, what remains to be shown is that $f^-(X) \in L^1(\mathbb{P};\mathbb{R})$. To this end, set $f_n = (-n) \vee f$ for $n \ge 1$. Then $f_n$ is bounded and convex, and so, by the preceding with $\Sigma = \{\emptyset, \Omega\}$, we know that $f_n\big(\mathbb{E}^P[X]\big) \le \mathbb{E}^{\mathbb{P}}\big[f_n(X)\big]$. Writing $f_n = f^+ - f_n^-$, this shows that $\mathbb{E}^{\mathbb{P}}\big[f_n^-(X)\big] \le M^+ - f\big(E^{\mathbb{P}}[X]\big)$ when $n \ge -f\big(E^{\mathbb{P}}[X]\big)$. Finally, note that $f_n^- = n \wedge f^-$, and conclude that $f^-(X) \in L^1(\mathbb{P};\mathbb{R})$. □

COROLLARY 5.2.10. *Let $I$ be a non-empty, closed interval in $\mathbb{R} \cup \{+\infty\}$ (i.e., either $I \subset \mathbb{R}$ is bounded on the right or $I \cap \mathbb{R}$ is unbounded on the right and $I$ includes the point $+\infty$). Then every $I$-valued random variable $X$ with $\mathbb{P}$-integrable negative part admits an $I$-valued representative of $\mathbb{E}^{\mathbb{P}}[X|\Sigma]$. Furthermore, if $f : I \longrightarrow \mathbb{R} \cup \{+\infty\}$ is a continuous, convex function and either $f$ is bounded above and $X \in L^1(\mathbb{P};\mathbb{R})$ or $f$ is bounded below and to the left (i.e., $f$ is bounded on each interval of the form $I \cap (-\infty, a]$ with $a \in I \cap \mathbb{R}$), then $f^-(X) \in L^1(\mathbb{P};\mathbb{R})$ and (5.2.9) holds. In particular, for each $p \in [1,\infty)$,*

$$\big\|\mathbb{E}^{\mathbb{P}}[X \mid \Sigma]\big\|_{L^p(\mathbb{P};\mathbb{R})} \le \|X\|_{L^p(\mathbb{P};\mathbb{R})}.$$

*Finally, if either $\big(X_n, \mathcal{F}_n, \mathbb{P}\big)$ is an $I$-valued martingale and $f$ satisfies the preceding conditions or if $\big(X_n, \mathcal{F}_n, \mathbb{P}\big)$ is an $I$-valued submartingale and $f$ is continuous, non-decreasing, convex, and bounded below, then $\big(f(X_n), \mathcal{F}_n, \mathbb{P}\big)$ is a submartingale.*

PROOF: In view of Corollary 5.2.8, we know that an $I$-valued representative of $\mathbb{E}^{\mathbb{P}}[X|\Sigma]$ exists when $X$ is $\mathbb{P}$-integrable, and the general case follows after a trivial truncation procedure.

In the case when $X$ is $\mathbb{P}$-integrable and $f$ is bounded above, $f(X) \in L^1(\mathbb{P};\mathbb{R})$ and (5.2.9) are immediate consequences of the last part of Corollary 5.2.8. To handle the case when $f$ is bounded below and to the left, first observe that either $f$ is non-increasing everywhere or there is an $a \in I \cap \mathbb{R}$ with the property that $f$ is non-increasing to the left of $a$ and non-decreasing to the right of $a$. Next, let an $I$-valued $X$ with $X^- \in L^1(\mathbb{P};\mathbb{R})$ be given, and set $X_n = X \wedge n$. Then there exists an $m \in \mathbb{Z}^+$ such that $X_n$ is $I$-valued for all $n \geq m$; and clearly, by the preceding, we know that

$$(*)\qquad f\Big(\mathbb{E}^{\mathbb{P}}\big[X_n\big|\Sigma\big]\Big) \leq \mathbb{E}^{\mathbb{P}}\big[f(X_n)\big|\Sigma\big] \quad (\text{a.s.}, \mathbb{P}) \quad \text{for all } n \geq m.$$

Moreover, in the case when $f$ is non-increasing, $\{f(X_n) : n \geq m\}$ is bounded below and non-increasing; and, in the other case, $\{f(X_n) : n \geq m \vee a\}$ is bounded below and non-decreasing. Hence, in both cases, the desired conclusion follows from (*) after an application of the version of the Monotone Convergence Theorem in (5.1.8).

To complete the proof, simply note that in either of the two cases given, the results just proved justify

$$\mathbb{E}^{\mathbb{P}}\big[f(X_n)\big|\mathcal{F}_{n-1}\big] \geq f\Big(\mathbb{E}^{\mathbb{P}}\big[X_n\big|\mathcal{F}_{n-1}\big]\Big) \geq f\big(X_{n-1}\big) \quad \mathbb{P}\text{-almost surely.} \quad \square$$

**§ 5.2.2. Doob's Stopping Time Theorem.** Perhaps the most far-reaching contribution that Doob made to martingale theory is his observation that one can "stop" a martingale without destroying the martingale property. Later, L. Snell showed that the analogous result is true for submartingales.

In order to state their results here, I need to introduce the notion of a stopping time in this setting. Namely, I will say that the function $\zeta : \Omega \longrightarrow \mathbb{N} \cup \{\infty\}$ is a **stopping time** relative to $\{\mathcal{F}_n : n \geq 0\}$ if $\{\omega : \zeta(\omega) = n\} \in \mathcal{F}_n$ for each $n \in \mathbb{N}$. In addition, given a stopping time $\zeta$, I use $\mathcal{F}_\zeta$ to denote the $\sigma$-algebra of $A \in \mathcal{F}$ such that $A \cap \{\zeta = n\} \in \mathcal{F}_n$, $n \in \mathbb{Z}^+$. Notice that $\mathcal{F}_{\zeta_1} \subseteq \mathcal{F}_{\zeta_2}$ if $\zeta_1 \leq \zeta_2$. In addition, if $\{X_n : n \in \mathbb{N}\}$ is $\{\mathcal{F}_n : n \in \mathbb{N}\}$-progressively measurable, check that the random variable $X_\zeta$ given by $X_\zeta(\omega) = X_{\zeta(\omega)}(\omega)$ is $\mathcal{F}_\zeta$-measurable on $\{\zeta < \infty\}$.

Doob used stopping times to give a mathematically rigorous formulation of the W.C. Field's assertion that "you can't cheat an honest man." That is, consider a gambler who is trying to *beat the system.* Assuming that he is playing a fair game, it is reasonable to say his gain $X_n$ after $n$ plays will evolve as a martingale. More precisely, if $\mathcal{F}_n$ contains the history of the game up to and including the $n$th play, then $(X_n, \mathcal{F}_n, \mathbb{P})$ will be a martingale. In the context of this model, a stopping time can be thought of as a feasible (i.e., one that does not require the gift of prophesy) strategy that the gambler can use to determine when he should stop playing in order to maximize his gains. When couched in these terms, the next result predicts that *there is no strategy with which the gambler can alter his expected gain.*

THEOREM 5.2.11 (**Doob's Stopping Time Theorem**). *For any submartingale (martingale) $(X_n, \mathcal{F}_n, \mathbb{P})$ that is $\mathbb{P}$-integrable and any stopping time $\zeta$, $(X_{n\wedge\zeta}, \mathcal{F}_n, P)$ is again a $\mathbb{P}$-integrable submartingale (martingale).*

PROOF: Let $A \in \mathcal{F}_{n-1}$. Then, since $A \cap \{\zeta > n-1\} \in \mathcal{F}_{n-1}$,

$$\begin{aligned}\mathbb{E}^{\mathbb{P}}\big[X_{n\wedge\zeta}, A\big] &= \mathbb{E}^{\mathbb{P}}\big[X_\zeta, A \cap \{\zeta \le n-1\}\big] + \mathbb{E}^{\mathbb{P}}\big[X_n, A \cap \{\zeta > n-1\}\big] \\ &\ge \mathbb{E}^{\mathbb{P}}\big[X_\zeta, A \cap \{\zeta \le n-1\}\big] + \mathbb{E}^{\mathbb{P}}\big[X_{n-1}, A \cap \{\zeta > n-1\}\big] = \mathbb{E}^{\mathbb{P}}\big[X_{(n-1)\wedge\zeta}, A\big];\end{aligned}$$

and, in the case of martingales, the inequality in the preceding can be replaced by an equality. □

Closely related to Doob's Stopping Time Theorem is an important variant due to G. Hunt. In order to facilitate the proof of Hunt's result, I begin with an easy but seminal observation of Doob's.

LEMMA 5.2.12 (**Doob's Decomposition**). *For each $n \in \mathbb{N}$ let $X_n$ be an $\mathcal{F}_n$-measurable, $\mathbb{P}$-integrable random variable. Then, up to a $\mathbb{P}$-null set, there is at most one sequence $\{A_n : n \ge 0\} \subseteq L^1(\mathbb{P};\mathbb{R})$ such that $A_0 = 0$, $A_n$ is $\mathcal{F}_{n-1}$-measurable for each $n \in \mathbb{Z}^+$, and $(X_n - A_n, \mathcal{F}_n, \mathbb{P})$ is a martingale. Moreover, if $(X_n, \mathcal{F}_n, \mathbb{P})$ is an integrable submartingale, then such a sequence $\{A_n : n \ge 0\}$ exists, and $A_{n-1} \le A_n$ $\mathbb{P}$-almost surely for all $n \in \mathbb{Z}^+$.*

PROOF: To prove the uniqueness assertion, suppose that $\{A_n : n \ge 0\}$ and $\{B_n : n \ge 0\}$ are two such sequences, and set $\Delta_n = B_n - A_n$. Then $\Delta_0 = 0$, $\Delta_n$ is $\mathcal{F}_{n-1}$-measurable for each $n \in \mathbb{Z}^+$, and $(\Delta_n, \mathcal{F}_n, \mathbb{P})$ is a martingale. But this means that $\Delta_n = \mathbb{E}^{\mathbb{P}}[\Delta_n \,|\, \mathcal{F}_{n-1}] = \Delta_{n-1}$ for all $n \in \mathbb{Z}^+$, and so $\Delta_n = 0$ for all $n \in \mathbb{N}$.

Now suppose that $(X_n, \mathcal{F}_n, \mathbb{P})$ is an integrable submartingale. To prove the asserted existence result, set $A_0 \equiv 0$ and

$$A_n = A_{n-1} + \mathbb{E}^{\mathbb{P}}\big[X_n - X_{n-1}\big|\mathcal{F}_{n-1}\big] \vee 0 \quad \text{for } n \in \mathbb{Z}^+. \quad \square$$

THEOREM 5.2.13 (**Hunt**). *Let $(X_n, \mathcal{F}_n, \mathbb{P})$ be a $\mathbb{P}$-integrable submartingale. Given bounded stopping times $\zeta$ and $\zeta'$ satisfying $\zeta \le \zeta'$,*

$$X_\zeta \le \mathbb{E}^{\mathbb{P}}\big[X_{\zeta'}\big|\mathcal{F}_\zeta\big] \quad \text{(a.s., } \mathbb{P}), \tag{5.2.14}$$

*and the inequality can be replaced by equality when $(X_n, \mathcal{F}_n, \mathbb{P})$ is a martingale. (Cf. Exercise 5.2.39 for unbounded stopping times.)*

PROOF: Choose $\{A_n : n \in \mathbb{N}\}$ for $(X_n, \mathcal{F}_n, \mathbb{P})$ as in Lemma 5.2.12, and set $Y_n = X_n - A_n$ for $n \in \mathbb{N}$. Then, because $A_\zeta \le A_{\zeta'}$ and $A_\zeta$ is $\mathcal{F}_\zeta$-measurable,

$$\mathbb{E}^{\mathbb{P}}\big[X_{\zeta'}\big|\mathcal{F}_\zeta\big] \ge \mathbb{E}^{\mathbb{P}}\big[Y_{\zeta'} + A_\zeta\big|\mathcal{F}_\zeta\big] = \mathbb{E}^{\mathbb{P}}\big[Y_{\zeta'}\big|\mathcal{F}_\zeta\big] + A_\zeta.$$

Hence, it suffices to prove that equality holds in (5.2.14) when $(X_n, \mathcal{F}_n, \mathbb{P})$ is a martingale. To this end, choose $N \in \mathbb{Z}^+$ to be an upper bound for $\zeta'$, let $\Gamma \in \mathcal{F}_\zeta$ be given, and note that

$$\begin{aligned}\mathbb{E}^{\mathbb{P}}\big[X_N, \Gamma\big] &= \sum_{n=0}^{N} \mathbb{E}^{\mathbb{P}}\big[X_N, \Gamma \cap \{\zeta = n\}\big] \\ &= \sum_{n=0}^{N} \mathbb{E}^{\mathbb{P}}\big[X_n, \Gamma \cap \{\zeta = n\}\big] = \mathbb{E}^{\mathbb{P}}\big[X_\zeta, \Gamma\big].\end{aligned}$$

Similarly, since $\Gamma \in \mathcal{F}_\zeta \subseteq \mathcal{F}_{\zeta'}$, $\mathbb{E}^{\mathbb{P}}[X_N, \Gamma] = \mathbb{E}^{\mathbb{P}}[X_{\zeta'}, \Gamma]$. □

§ **5.2.3. Martingale Convergence Theorem.** My next goal is to show that, even when they are not given in the form covered by Corollary 5.2.4, *martingales want to converge.* If for no other reason, such a result has got to be more difficult because one does not know ahead of time what, if it exists, the limit ought to be. Thus, the reasoning will have to be more subtle than that used in the proof of Corollary 5.2.4. I will follow Doob and base my argument on the idea that, in some sense, a martingale has got to be *nearly constant* and that a submartingale is the sum of a martingale and a non-decreasing process. In order to make mathematics out of this idea, I need to introduce a somewhat novel criterion for convergence of real numbers. Namely, given a sequence $\{x_n : n \geq 0\} \subseteq \mathbb{R}$ and a numbers $-\infty < a < b < \infty$, say that $\{x_n : n \geq 0\}$ **upcrosses the interval** $[a, b]$ **at least** $N$ **times** if there exist integers $0 \leq m_1 < n_1 < \cdots < m_N < n_N$ such that $x_{m_i} \leq a$ and $x_{n_i} \geq b$ for each $1 \leq i \leq N$ and that it **upcrosses** $[a, b]$ **precisely** $N$ **times** if it upcrosses $[a, b]$ at least $N$ but does not upcross $[a, b]$ at least $N+1$ times. Notice that $\underline{\lim}_{n\to\infty} x_n < \overline{\lim}_{n\to\infty} x_n$ if and only if there exist rational numbers $a < b$ such that $\{x_n : n \geq 0\}$ upcrosses $[a, b]$ at least $N$ times for every $N \in \mathbb{Z}^+$. Hence, $\{x_n : n \geq 0\}$ converges in $[-\infty, \infty]$ if and only if it upcrosses $[a, b]$ at most finitely often for each pair of rational numbers $a < b$.

THEOREM 5.2.15 (**Doob's Martingale Convergence Theorem**).[2] *Suppose that* $(X_n, \mathcal{F}_n, \mathbb{P})$ *is a* $\mathbb{P}$*-integrable submartingale. For* $-\infty < a < b < \infty$, *let* $U_{[a,b]}(\omega)$ *denote the precise number of times that* $\{X_n(\omega) : n \geq 0\}$ *upcrosses* $[a, b]$. *Then*

$$\mathbb{E}^{\mathbb{P}}\big[U_{[a,b]}\big] \leq \sup_{n\in\mathbb{N}} \frac{\mathbb{E}^{\mathbb{P}}\big[(X_n - a)^+\big]}{b - a}. \tag{5.2.16}$$

*In particular, if*

$$\sup_{n\in\mathbb{N}} \mathbb{E}^{\mathbb{P}}\big[X_n^+\big] < \infty, \tag{5.2.17}$$

[2] In the notes to Chapter VII of his *Stochastic Processes*, Wiley (1953), Doob gives a thorough account of the relationship between his convergence result and earlier attempts in the same direction. In particular, he points out that, in 1946, S. Anderson and B. Jessen formulated and proved a closely related convergence theorem.

*then there exists a $\mathbb{P}$-integrable random variable $X$ to which $\{X_n : n \geq 0\}$ converges $\mathbb{P}$-almost surely. (See Exercises 5.2.36 and 5.2.38 for other derivations.)*

PROOF: Set $Y_n = \frac{(X_n - a)^+}{b-a}$, and note that (by Corollary 5.2.10) $(Y_n, \mathcal{F}_n, \mathbb{P})$ is a $\mathbb{P}$-integrable submartingale. Next, let $N \in \mathbb{Z}^+$ be given, set $\zeta'_0 = 0$, and, for $k \in \mathbb{Z}^+$, define

$$\zeta_k = \inf\{n \geq \zeta'_{k-1} : X_n \leq a\} \wedge N \quad \text{and} \quad \zeta'_k = \inf\{n \geq \zeta_k : X_n \geq b\} \wedge N.$$

Proceeding by induction, it is an easy matter to check that all the $\zeta_k$'s and $\zeta'_k$'s are stopping times. Moreover, if $U^{(N)}_{[a,b]}(\omega)$ is the precise number of times $\{X_{n\wedge N}(\omega) : n \geq 0\}$ upcrosses $[a, b]$, then

$$\begin{aligned} U^{(N)}_{[a,b]} &\leq \sum_{k=1}^{N} \left(Y_{\zeta'_k} - Y_{\zeta_k}\right) = Y_N - Y_0 - \sum_{k=1}^{N} \left(Y_{\zeta_k} - Y_{\zeta'_{k-1}}\right) \\ &\leq Y_N - \sum_{k=1}^{N} \left(Y_{\zeta_k} - Y_{\zeta'_{k-1}}\right). \end{aligned}$$

Hence, since $\zeta'_{k-1} \leq \zeta_k$ and therefore, by (5.2.14), $\mathbb{E}^{\mathbb{P}}\left[Y_{\zeta_k} - Y_{\zeta'_{k-1}}\right] \geq 0$ for all $k \in \mathbb{Z}^+$, we see that $\mathbb{E}^{\mathbb{P}}[U^{(N)}_{[a,b]}] \leq \mathbb{E}^{\mathbb{P}}[Y_N]$, and clearly (5.2.16) follows from this after one lets $N \to \infty$.

Given (5.2.16), the convergence result is easy. Namely, if (5.2.17) is satisfied, then (5.2.16) implies that there is a set $\Lambda$ of full $\mathbb{P}$-measure such that $U_{[a,b]}(\omega) < \infty$ for all rational $a < b$ and $\omega \in \Lambda$; and so, by the remark preceding the statement of this theorem, for each $\omega \in \Lambda$, $\{X_n(\omega) : n \geq 0\}$ converges to some $X(\omega) \in [-\infty, \infty]$. Hence, we will be done as soon as we know that $\mathbb{E}^{\mathbb{P}}[|X|, \Lambda] < \infty$. But

$$\mathbb{E}^{\mathbb{P}}\left[|X_n|\right] = 2\mathbb{E}^{\mathbb{P}}\left[X_n^+\right] - \mathbb{E}^{\mathbb{P}}\left[X_n\right] \leq 2\mathbb{E}^{\mathbb{P}}\left[X_n^+\right] - \mathbb{E}^{\mathbb{P}}\left[X_0\right], \quad n \in \mathbb{N},$$

and therefore Fatou's Lemma plus (5.2.17) shows that $X$ is $\mathbb{P}$-integrable. □

The inequality in (5.2.16) is quite famous and is known as **Doob's Upcrossing Inequality**.

REMARK 5.2.18. The argument in the proof of Theorem 5.2.15 is so slick that it is easy to miss the point that makes it work. Namely, the whole proof turns on the inequality $\mathbb{E}^{\mathbb{P}}[Y_{\zeta_k} - Y_{\zeta'_{k-1}}] \geq 0$. At first sight, this inequality seems to be wrong, since one is inclined to think that $Y_{\zeta_k} < Y_{\zeta'_{k-1}}$. However, $Y_{\zeta_k}$ need be less than $Y_{\zeta'_{k-1}}$ only if $\zeta_k < N$, which is precisely what, with high probability, the submartingale property is preventing from happening.

Corollary 5.2.19. *Let $(X_n, \mathcal{F}_n, \mathbb{P})$ be a martingale. Then there exists an $X \in L^1(\mathbb{P};\mathbb{R})$ such that $X_n = \mathbb{E}^{\mathbb{P}}[X|\mathcal{F}_n]$ (a.s., $\mathbb{P}$) for each $n \in \mathbb{N}$ if and only if the sequence $\{X_n : n \geq 0\}$ is uniformly $\mathbb{P}$-integrable. In addition, if $p \in (1,\infty]$, then there is an $X \in L^p(\mathbb{P};\mathbb{R})$ such that $X_n = \mathbb{E}^{\mathbb{P}}[X|\mathcal{F}_n]$ (a.s., $\mathbb{P}$) for each $n \in \mathbb{N}$ if and only if $\{X_n : n \geq 0\}$ is a bounded subset of $L^p(\mathbb{P};\mathbb{R})$.*

Proof: Because of Corollary 5.2.4 and (5.2.3), I need only check the "if" statement in the first assertion. But, if $\{X_n : n \geq 0\}$ is uniformly $\mathbb{P}$-integrable, then (5.2.17) holds and therefore $X_n \longrightarrow X$ (a.s., $\mathbb{P}$) for some $\mathbb{P}$-integrable $X$. Moreover, uniform integrability together with almost sure convergence implies convergence in $L^1(\mathbb{P};\mathbb{R})$, and therefore, by (5.1.5), for each $m \in \mathbb{N}$,

$$X_m = \lim_{n\to\infty} \mathbb{E}^{\mathbb{P}}\big[X_n\big|\mathcal{F}_m\big] = \mathbb{E}^{\mathbb{P}}\big[X\big|\mathcal{F}_m\big] \quad \text{(a.s., } \mathbb{P}). \quad \square$$

Just as Corollary 5.2.4 led us to an intuitively appealing way to construct conditional expectations, so does Doob's Theorem gives us an appealing approximation procedure for Radon–Nikodym derivatives.

Theorem 5.2.20 (**Jessen**). *Let $\mathbb{P}$ and $\mathbb{Q}$ be a pair of probability measures on the measurable space $(\Omega, \mathcal{F})$ and $\{\mathcal{F}_n : n \in \mathbb{N}\}$ a non-decreasing sequence of sub-$\sigma$-algebras whose union generates $\mathcal{F}$. For each $n \in \mathbb{N}$, let $\mathbb{Q}_{n,\mathrm{a}}$ and $\mathbb{Q}_{n,\mathrm{s}}$ denote, respectively, the absolutely continuous and singular parts of $\mathbb{Q}_n \equiv \mathbb{Q} \restriction \mathcal{F}_n$ with respect to $\mathbb{P}_n \equiv \mathbb{P} \restriction \mathcal{F}_n$, and set $X_n = \frac{d\mathbb{Q}_{n,\mathrm{a}}}{d\mathbb{P}_n}$. Also, let $\mathbb{Q}_\mathrm{a}$ and $\mathbb{Q}_\mathrm{s}$ be the absolutely and singular continuous parts of $\mathbb{Q}$ with respect to $\mathbb{P}$, and set $Y = \frac{d\mathbb{Q}_\mathrm{a}}{d\mathbb{P}}$. Then $X_n \longrightarrow Y$ (a.s., $\mathbb{P}$). In particular, $\mathbb{Q} \perp \mathbb{P}$ if and only if $X_n \longrightarrow 0$ (a.s., $\mathbb{P}$). Moreover, if $\mathbb{Q}_n \ll \mathbb{P}_n$ for each $n \in \mathbb{N}$, then $\mathbb{Q} \ll \mathbb{P}$ if and only if $\{X_n : n \geq 0\}$ is uniformly $\mathbb{P}$-integrable, in which case $X_n \longrightarrow Y$ in $L^1(\mathbb{P};\mathbb{R})$ as well as $\mathbb{P}$-almost surely. Finally, if $\mathbb{Q}_n \sim \mathbb{P}_n$ (i.e., $\mathbb{P}_n \ll \mathbb{Q}_n$ as well as $\mathbb{Q}_n \ll \mathbb{P}_n$) for each $n \in \mathbb{N}$ and $G \equiv \{\lim_{n\to\infty} X_n \in (0,\infty)\}$, then $\mathbb{Q}_\mathrm{a}(A) = \mathbb{Q}(A \cap G)$ for all $A \in \mathcal{F}$, and therefore $\mathbb{Q}(G) = 1 \iff \mathbb{Q} \ll \mathbb{P}$ and $\mathbb{Q}(G) = 0 \iff \mathbb{Q} \perp \mathbb{P}$.*

Proof: Without loss in generality, I will assume throughout that all the $X_n$'s as well as $Y \equiv \frac{dQ_\mathrm{a}}{dP}$ take values in $[0,\infty)$; and clearly, $\mathbb{E}^{\mathbb{P}}[X_n]$, $n \in \mathbb{N}$, and $\mathbb{E}^{\mathbb{P}}[Y]$ are all dominated by 1.

First note that

$$\mathbb{Q}_{n,\mathrm{s}}(A) = \sup\Big\{\mathbb{Q}(A \cap B) : B \in \mathcal{F}_n \text{ and } \mathbb{P}(B) = 0\Big\} \quad \text{for} \quad A \in \mathcal{F}_n.$$

Hence, $\mathbb{Q}_{n,\mathrm{s}} \restriction \mathcal{F}_{n-1} \geq \mathbb{Q}_{n-1,\mathrm{s}}$ for each $n \in \mathbb{Z}^+$, and so

$$\mathbb{E}^{\mathbb{P}}\big[X_n,\, A\big] = \mathbb{Q}_{n,\mathrm{a}}(A) \leq \mathbb{Q}_{n-1,\mathrm{a}}(A) = \mathbb{E}^{\mathbb{P}}\big[X_{n-1},\, A\big]$$

for all $n \in \mathbb{Z}^+$ and $A \in \mathcal{F}_{n-1}$. In other words, $(-X_n, \mathcal{F}_n, \mathbb{P})$ is a non-positive submartingale. Moreover, in the case when $\mathbb{Q}_n \ll P_n$, $n \in \mathbb{N}$, the same argument shows that $(X_n, \mathcal{F}_n, \mathbb{P})$ is a non-negative martingale. Thus, in either case, there is a non-negative, $\mathbb{P}$-integrable random variable $X$ with the property that $X_n \longrightarrow X$ (a.s., $\mathbb{P}$). In order to identify $X$ as $Y$, use Fatou's Lemma to see that, for any $m \in \mathbb{N}$ and $A \in \mathcal{F}_m$,

$$\mathbb{E}^{\mathbb{P}}[X, A] \le \varliminf_{n\to\infty} \mathbb{E}^{\mathbb{P}}[X_n, A] = \varliminf_{n\to\infty} \mathbb{Q}_{n,\mathrm{a}}(A) \le \mathbb{Q}(A);$$

and therefore $\mathbb{E}^{\mathbb{P}}[X, A] \le \mathbb{Q}(A)$, first for $A \in \bigcup_0^\infty \mathcal{F}_m$ and then for every $A \in \mathcal{F}$. In particular, by choosing $B \in \mathcal{F}$ so that $\mathbb{Q}_{\mathrm{s}}(B) = 0 = \mathbb{P}(B\complement)$, we have that

$$\mathbb{E}^{\mathbb{P}}[X, A] = \mathbb{E}^{\mathbb{P}}[X, A \cap B] \le \mathbb{Q}(A \cap B) = \mathbb{Q}_{\mathrm{a}}(A) = \mathbb{E}^{\mathbb{P}}[Y, A] \quad \text{for all } A \in \mathcal{F},$$

which means that $X \le Y$ (a.s., $\mathbb{P}$). On the other hand, if $Y_n = \mathbb{E}^{\mathbb{P}}[Y|\mathcal{F}_n]$ for $n \in \mathbb{N}$, then

$$\mathbb{E}^{\mathbb{P}}[Y_n, A] = \mathbb{Q}_{\mathrm{a}}(A) \le \mathbb{Q}_{n,\mathrm{a}}(A) = \mathbb{E}^{\mathbb{P}}[X_n, A] \quad \text{for all } A \in \mathcal{F}_n,$$

and therefore $Y_n \le X_n$ (a.s., $\mathbb{P}$) for each $n \in \mathbb{N}$. Thus, since $Y_n \longrightarrow Y$ and $X_n \longrightarrow X$ $\mathbb{P}$-almost surely, this means that $Y \le X$ (a.s., $\mathbb{P}$).

Next, assume that $\mathbb{Q}_n \ll \mathbb{P}_n$ for each $n \in \mathbb{N}$ and therefore that $(X_n, \mathcal{F}_n, \mathbb{P})$ is a non-negative martingale. If $\{X_n : n \ge 0\}$ is uniformly $\mathbb{P}$-integrable, then $X_n \longrightarrow Y$ in $L^1(\mathbb{P};\mathbb{R})$ and therefore $\mathbb{Q}_s(\Omega) = 1 - \mathbb{E}^{\mathbb{P}}[Y] = 0$. Hence, $\mathbb{Q} \ll \mathbb{P}$ when $\{X_n : n \ge 0\}$ is uniformly $\mathbb{P}$-integrable. Conversely, if $\mathbb{Q} \ll \mathbb{P}$, then it is easy to see that $X_n = \mathbb{E}^{\mathbb{P}}[Y|\mathcal{F}_n]$ for each $n \in \mathbb{N}$, and therefore, by Corollary 5.2.4, that $\{X_n : n \ge 0\}$ is uniformly $\mathbb{P}$-integrable.

Finally, assume that $\mathbb{Q}_n \sim \mathbb{P}_n$ for each $n \in \mathbb{N}$. Then, the $X_n$'s can be chosen to take their values in $(0,\infty)$ and $Y_n \equiv \frac{1}{X_n} = \frac{d\mathbb{P}_n}{d\mathbb{Q}_n}$. Hence, if $\mathbb{P}_{\mathrm{a}}$ and $\mathbb{P}_{\mathrm{s}}$ are the absolutely continuous and singular parts of $\mathbb{P}$ relative to $\mathbb{Q}$ and if $Y \equiv \varliminf_{n\to\infty} Y_n$, then $Y = \frac{d\mathbb{P}_{\mathrm{a}}}{d\mathbb{Q}}$ and so $\mathbb{P}_{\mathrm{a}}(A) = \mathbb{E}^{\mathbb{Q}}[Y, A]$ for all $A \in \mathcal{F}$. Thus, when $B \in \mathcal{F}$ is chosen so that $\mathbb{P}_{\mathrm{s}}(B) = 0 = \mathbb{Q}(B\complement)$, then, since $Y = \frac{1}{X}$ on $G$ and $\mathbb{E}^{\mathbb{P}}[X, C \cap G] = \mathbb{E}^{\mathbb{P}}[X, C]$ for all $C \in \mathcal{F}$, it is becomes clear that

$$\begin{aligned}\mathbb{Q}(A \cap G) &= \mathbb{E}^{\mathbb{Q}}[XY, A \cap G] = \mathbb{E}^{\mathbb{P}_{\mathrm{a}}}[X, A \cap G] \\ &= \mathbb{E}^{\mathbb{P}}[X, A \cap G \cap B] = \mathbb{E}^{\mathbb{P}}[X, A \cap B] = \mathbb{Q}_{\mathrm{a}}(A \cap B) = \mathbb{Q}_{\mathrm{a}}(A)\end{aligned}$$

for all $A \in \mathcal{F}$. □

**§ 5.2.4. Reversed Martingales and De Finetti's Theory.** For some applications it is important to know what happens if one runs a submartingale or martingale backwards. Thus, again let $(\Omega, \mathcal{F}, \mathbb{P})$ be a probability space, only this time suppose that $\{\mathcal{F}_n : n \in \mathbb{N}\}$ is a sequence of sub-$\sigma$-algebras that is *non-increasing*. Given a sequence $\{X_n : n \ge 0\}$ of $(-\infty,\infty]$-valued random variables, I will say that the triple $(X_n, \mathcal{F}_n, \mathbb{P})$ is a **reversed submartingale** or a **reversed martingale** depending on whether, for each $n \in \mathbb{N}$, $X_n$ is $\mathcal{F}_n$-measurable and either $X_n^- \in L^1(\mathbb{P};\mathbb{R})$ and $X_{n+1} \le \mathbb{E}^{\mathbb{P}}[X_n \,|\, \mathcal{F}_{n+1}]$ or $X_n \in L^1(\mathbb{P};\mathbb{R})$ and $X_{n+1} = \mathbb{E}^{\mathbb{P}}[X_n \,|\, \mathcal{F}_{n+1}]$.

THEOREM 5.2.21. *If $(X_n, \mathcal{F}_n, \mathbb{P})$ is a reversed submartingale, then*

$$\mathbb{P}\left(\sup_{n\in\mathbb{N}} X_n \geq R\right) \leq \frac{1}{R}\mathbb{E}^{\mathbb{P}}\left[X_0,\ \sup_{n\in\mathbb{N}} X_n \geq R\right], \quad R \in (0,\infty). \tag{5.2.22}$$

*In particular, if $(X_n, \mathcal{F}_n, \mathbb{P})$ is a non-negative reversed submartingale and $X_0 \in L^1(\mathbb{P};\mathbb{R})$, then $\{X_n : n \geq 0\}$ is uniformly $\mathbb{P}$-integrable and*

$$\left\|\sup_{n\in\mathbb{N}} X_n\right\|_{L^p(\mathbb{P};\mathbb{R})} \leq \frac{p}{p-1}\|X_0\|_{L^p(\mathbb{P};\mathbb{R})} \quad \text{when } p \in (1,\infty). \tag{5.2.23}$$

*Moreover, if $(X_n, \mathcal{F}_n, \mathbb{P})$ is a reversed martingale, then $(|X_n|, \mathcal{F}_n, \mathbb{P})$ is a reversed submartingale. Finally, if $(X_n, \mathcal{F}_n, \mathbb{P})$ is a reversed submartingale and $X_0 \in L^1(\mathbb{P};\mathbb{R})$, then there is a $\mathcal{F}_\infty \equiv \bigcap_{n=0}^\infty \mathcal{F}_n$-measurable $X : \Omega \longrightarrow [-\infty,\infty]$ to which $X_n$ converges $\mathbb{P}$-almost surely. In fact, $X$ will be $\mathbb{P}$-integrable if $\sup_{n\geq 0} \mathbb{E}^{\mathbb{P}}[|X_n|] < \infty$; and if $(X_n, \mathcal{F}_n, \mathbb{P})$ is either a non-negative reversed submartingale or a reversed martingale with $X_0 \in L^p(\mathbb{P};\mathbb{R})$ for some $p \in [1,\infty)$, then $X_n \longrightarrow X$ in $L^p(\mathbb{P};\mathbb{R})$.*

PROOF: More or less everything here follows immediately from the observation that $(X_n, \mathcal{F}_n, \mathbb{P})$ is a reversed submartingale or a reversed martingale if and only if, for each $N \in \mathbb{Z}^+$, $(X_{N-n\wedge N}, \mathcal{F}_{N-n\wedge N}, \mathbb{P})$ is a submartingale or a martingale. Indeed, by this observation and (5.2.2) applied to $(X_{N-n\wedge N}, \mathcal{F}_{N-n\wedge N}, \mathbb{P})$,

$$\mathbb{P}\left(\max_{0\leq n\leq N} X_n > R\right) \leq \frac{1}{R}\mathbb{E}^{\mathbb{P}}\left[X_0,\ \max_{0\leq n\leq N} X_n > R\right]$$

for every $N \geq 1$. When $N \to \infty$, the left-hand side of the preceding tends to $\mathbb{P}\left(\sup_{n\in\mathbb{N}} X_n > R\right)$ and

$$\begin{aligned}\mathbb{E}^{\mathbb{P}}\left[X_0,\ \max_{0\leq n\leq N} X_n > R\right] &= \mathbb{E}^{\mathbb{P}}\left[X_0^+,\ \max_{0\leq n\leq N} X_n > R\right] - \mathbb{E}^{\mathbb{P}}\left[X_0^-,\ \max_{0\leq n\leq N} X_n > R\right]\\ \longrightarrow \mathbb{E}^{\mathbb{P}}\left[X_0^+,\ \sup_{n\in\mathbb{N}} X_n > R\right] - \mathbb{E}^{\mathbb{P}}\left[X_0^-,\ \sup_{n\in\mathbb{N}} X_n > R\right] &= \mathbb{E}^{\mathbb{P}}\left[X_0,\ \sup_{n\in\mathbb{N}} X_n > R\right],\end{aligned}$$

since $X_0^+$ is non-negative, and therefore the Monotone Convergence Theorems applies, and $X_0^-$ is integrable, and therefore Lebesgue's Dominated Convergence Theorem applies. Thus (5.2.22) follows after one takes right limits in $R$. Starting from (5.2.22) and applying Exercise 1.4.18, (5.2.23) follows for non-negative, reversed submartingales. Moreover, because it is obvious that $(|X_n|, \mathcal{F}_n, \mathbb{P})$ is a reversed submartingale when $(X_n, \mathcal{F}_n, \mathbb{P})$ is a reversed martingale, (5.2.23) holds for reversed martingales as well.

Next, suppose that $(X_n, \mathcal{F}_n, \mathbb{P})$ is a non-negative, reversed submartingale or a reversed martingale. Then

$$\sup_{n\in\mathbb{N}} \mathbb{E}^{\mathbb{P}}\big[|X_n|, |X_n| \geq R\big] \leq \sup_{n\in\mathbb{N}} \mathbb{E}^{\mathbb{P}}\big[|X_0|, |X_n| \geq R\big] \leq \mathbb{E}^{\mathbb{P}}\left[|X_0|, \sup_{n\in\mathbb{N}} |X_n| \geq R\right],$$

which, by (5.2.22), tends to 0 as $R \to \infty$. Thus, $\{X_n : n \geq 0\}$ is uniformly $\mathbb{P}$-integrable.

It remains to prove the convergence assertions, and again the key is the same observation about reversing time to convert reversed submartingales into submartingales. However, before seeing how it applies, first say that $\{x_n : n \geq 0\}$ downcrosses $[a, b]$ at least $N$ times if there exist $0 \leq m_1 < n_1 < \cdots < m_N < n_N$ such that $x_{m_i} \geq b$ and $x_{n_i} \leq a$ for each $1 \leq i \leq N$. Clearly, the same argument that I used for upcrossings applies to downcrossings and shows that $\{x_n : n \geq 0\}$ converges in $[-\infty, \infty]$ if and only if it downcrosses $[a, b]$ finitely often for each rational pair $a < b$. In addition, $\{x_n : 0 \leq n \leq N\}$ downcrosses $[a, b]$ the same number of times as $\{x_{N-n} : 0 \leq n \leq N\}$ upcrosses it. Hence, if $D^{(N)}_{[a,b]}(\omega)$ is the number of times $\{X_{n\wedge N} : n \geq 0\}$ downcrosses $[a, b]$, then this observation together with the estimate in the proof of Theorem 5.2.15 for $\mathbb{E}^{\mathbb{P}}[U^{(N)}_{[a,b]}]$ show that

$$\mathbb{E}^{\mathbb{P}}\big[D^{(N)}_{[a,b]}\big] \leq \frac{\mathbb{E}^{\mathbb{P}}\big[(X_0 - a)^+\big]}{b - a}.$$

Starting from here, the argument used to prove Theorem 5.2.15 shows that there exits a $\mathcal{F}_\infty$-measurable $X : \Omega \longrightarrow [-\infty, \infty]$ to which $\{X_n : n \geq 0\}$ converges $\mathbb{P}$-almost surely. Once one has this almost sure convergence result, the rest of the theorem is an easy application of standard measure theory and the uniform integrability estimates proved above. □

An important application of reversed martingales is provided by De Finetti's theory of exchangeable random variables. To describe his theory, let $\Sigma$ denote the group of all *finite permutations* of $\mathbb{Z}^+$. That is, an element $\pi$ of $\Sigma$ is an isomorphism of $\mathbb{Z}^+$ that moves only a finite number of integers. Alternatively, $\Sigma = \bigcup_{m=1}^\infty \Sigma_m$, where $\Sigma_m$ is the group of isomorphisms $\pi$ of $\mathbb{Z}^+$ with the property that $n = \pi(n)$ for all $n > m$. Next, let $(E, \mathcal{B})$ be a measurable space, and, for each $\pi \in \Sigma$, define $S_\pi : E^{\mathbb{Z}^+} \longrightarrow E^{\mathbb{Z}^+}$ so that

$$S_\pi \mathbf{x} = \big(x_{\pi(1)}, \dots, x_{\pi(n)}, \dots\big) \quad \text{if } \mathbf{x} = (x_1, \dots, x_n, \dots).$$

Obviously, each $S_\pi$ is a $\mathcal{B}^{\mathbb{Z}^+}$-measurable isomorphism from $E^{\mathbb{Z}^+}$ onto itself. Also, if

$$\mathcal{A}_m \equiv \big\{B \in \mathcal{B}^{\mathbb{Z}^+} : B = S_\pi B \text{ for all } \pi \in \Sigma_m\big\} \quad \text{for } m \in \mathbb{Z}^+,$$

then the $\mathcal{A}_m$'s form a non-increasing sequence of sub-$\sigma$-algebras of $\mathcal{B}^{\mathbb{Z}^+}$, and

$$\bigcap_{m=1}^\infty \mathcal{A}_m = \mathcal{A}_\infty \equiv \big\{B \in \mathcal{B}^{\mathbb{Z}^+} : B = S_\pi B \text{ for all } \pi \in \Sigma\big\}.$$

Now suppose that $\{X_n : n \geq 1\}$ is a sequence of $E$-valued random variables on the probability space $(\Omega, \mathcal{F}, \mathbb{P})$, and set $\mathbf{X}(\omega) = \big(X_1(\omega), \dots, X_n(\omega), \dots\big) \in E^{\mathbb{Z}^+}$. The $X_n$'s are said to be **exchangeable random variables** if $\mathbf{X}$ has the same $\mathbb{P}$-distribution as $S_\pi \mathbf{X}$ for every $\pi \in \Sigma$. The central result of De Finetti's theory is **De Finetti's Strong Law**, which states that, for any exchangable random variables and $g : E \longrightarrow \mathbb{R}$ satisfying $g \circ X_1 \in L^1(\mathbb{P}; \mathbb{R})$,

$$\mathbb{E}^{\mathbb{P}}\left[g \circ X_1 \,\middle|\, \mathbf{X}^{-1}(\mathcal{A}_\infty)\right] = \lim_{n\to\infty} \frac{1}{n} \sum_1^n g \circ X_m, \tag{5.2.24}$$

where the convergence is $\mathbb{P}$-almost sure and in $L^1(\mathbb{P}; \mathbb{R})$.

To prove (5.2.24), observe that, for any $1 \leq m \leq n$, $\mathbb{E}^{\mathbb{P}}[g \circ X_m \,|\, \mathbf{X}^{-1}(\mathcal{A}_n)] = \mathbb{E}^{\mathbb{P}}[g \circ X_1 \,|\, \mathbf{X}^{-1}(\mathcal{A}_n)]$, which immediately leads to

$$\mathbb{E}^{\mathbb{P}}\left[g \circ X_1 \,\middle|\, \mathbf{X}^{-1}(\mathcal{A}_n)\right] = \mathbb{E}^{\mathbb{P}}\left[\frac{1}{n} \sum_{m=1}^n g \circ X_m \,\middle|\, \mathbf{X}^{-1}(\mathcal{A}_n)\right] = \frac{1}{n} \sum_{m=1}^n g \circ X_m.$$

Hence, (5.2.24) follows as an application of Theorem 5.2.21.

De Finetti's Strong Law makes it important to get a handle on the $\sigma$-algebra $\mathbf{X}^{-1}(\mathcal{A}_\infty)$. In particular, one would like to know when $\mathbf{X}^{-1}(\mathcal{A}_\infty)$ is trivial in the sense that each of its elements has probability 0 or 1, in which case (5.2.24) self-improves to the statement that

$$\lim_{n\to\infty} \frac{1}{n} \sum_1^n g \circ X_m = \mathbb{E}^{\mathbb{P}}[g \circ X_1] \quad \mathbb{P}\text{-almost surely and in } L^1(\mathbb{P}; \mathbb{R}). \tag{5.2.25}$$

The following lemma is the crucial step toward gaining an understanding of $\mathbf{X}^{-1}(\mathcal{A}_\infty)$.

LEMMA 5.2.26. *Refer to the preceding, and let $\mathcal{T} = \bigcap_{m=1}^\infty \sigma\big(\{X_n : n \geq m\}\big)$ be the tail $\sigma$-algebra determined by $\{X_n : n \geq 1\}$. Then $\mathcal{T} \subseteq \mathbf{X}^{-1}(\mathcal{A}_\infty)$ and $\mathbf{X}^{-1}(\mathcal{A}_\infty)$ is contained in the completion of $\mathcal{T}$ with respect to $\mathbb{P}$. In particular, for each $F \in L^1(\mathbb{P}; \mathbb{R})$,*

$$\mathbb{E}^{\mathbb{P}}\left[F \,\middle|\, \mathbf{X}^{-1}(\mathcal{A}_\infty)\right] = \mathbb{E}^{\mathbb{P}}\left[F \,\middle|\, \mathcal{T}\right] \quad (\text{a.s.}, \mathbb{P}). \tag{5.2.27}$$

PROOF: The inclusion $\mathcal{T} \subseteq \mathbf{X}^{-1}(\mathcal{A}_\infty)$ is obvious. Thus, what remains to be proved is that, for any $F \in L^1(\mathbb{P}; \mathbb{R})$, $\mathbb{E}^{\mathbb{P}}[F \,|\, \mathbf{X}^{-1}(\mathcal{A}_\infty)]$ is, up to a $\mathbb{P}$-null set, $\mathcal{T}$-measurable. To this end, begin by observing that it suffices to check this for $F$'s that are $\sigma\big(\{X_n : 1 \leq m \leq N\}\big)$-measurable for some $N \in \mathbb{Z}^+$. Indeed, since $\mathbf{X}^{-1}(\mathcal{A}_\infty) \subseteq \sigma\big(\{X_n : n \geq 1\}\big)$, we know that

$$\begin{aligned}
\mathbb{E}^{P}\left[F \,\middle|\, \mathbf{X}^{-1}(\mathcal{A}_\infty)\right] &= \mathbb{E}^{\mathbb{P}}\Big[\mathbb{E}^{\mathbb{P}}\left[F \,\middle|\, \sigma\big(\{X_n : n \geq 1\}\big)\right] \Big|\, \mathbf{X}^{-1}(\mathcal{A}_\infty)\Big] \\
&= \lim_{N\to\infty} \mathbb{E}^{\mathbb{P}}\Big[\mathbb{E}^{\mathbb{P}}\left[F \,\middle|\, \sigma\big(\{X_m : 1 \leq m \leq N\}\big)\right] \Big|\, \mathbf{X}^{-1}(\mathcal{A}_\infty)\Big].
\end{aligned}$$

Now suppose that $F$ is $\sigma(\{X_m : 1 \le m \le N\})$-measurable. Then there exists a $g : E^N \longrightarrow \mathbb{R}$ such that $F = g(X_1, \dots, X_N)$. If $N = 1$, then, because $\lim_{n\to\infty} \frac{1}{n}\sum_{m=1}^n g \circ X_m$ is $\mathcal{T}$-measurable, (5.2.24) says that $\mathbb{E}^{\mathbb{P}}[F \,|\, \mathbf{X}^{-1}(\mathcal{A}_\infty)]$ is $\mathcal{T}$-measurable. To get the same conclusion when $N \ge 2$, I want to apply the same reasoning, only now with $E$ replaced by $E^N$. To be precise, define

$$\mathcal{A}_\infty^{(N)} = \{B \in \mathcal{B}^{\mathbb{Z}^+} : B = S_\sigma B \text{ for all } \pi \in \Sigma^{(N)}\}, \text{ where}$$
$$\Sigma^{(N)} = \{\pi \in \Sigma : \pi(\ell N + m) = \pi(\ell N + 1) + m - 1 \text{ for all } \ell \in \mathbb{N} \text{ and } 1 \le m < N\}$$

is the group of finite permutations that transform $\mathbb{Z}^+$ in blocks of length $N$. By (5.2.24) applied with $E^N$ replacing $E$, we find that $\mathbb{E}^{\mathbb{P}}[F \,|\, \mathbf{X}^{-1}(\mathcal{A}_\infty^{(N)})] = \mathbb{E}^{\mathbb{P}}[F \,|\, \mathcal{T}]$ $\mathbb{P}$-almost surely. Hence, since $\mathbf{X}^{-1}(\mathcal{A}_\infty) \subseteq \mathbf{X}^{-1}(\mathcal{A}_\infty^{(N)})$, (5.2.27) holds for every $\sigma(\{X_n : 1 \le n \le N\})$-measurable $F \in L^1(\mathbb{P};\mathbb{R})$. □

The best known consequence of Lemma 5.2.26 is the **Hewitt–Savage 0–1 Law**, which says that $\mathbf{X}^{-1}(\mathcal{A}_\infty)$ is trivial if the $X_n$'s are independent and identically distributed. Clearly, their result is an immediate consequence of Lemma 5.2.26 together with Kolmogorov's 0–1 Law.

Seeing as the Strong Law of Large Numbers follows from (5.2.24) combined with the Hewitt–Savage 0–1 Law, one might think that (5.2.24) represents an extension of the strong law. However, that is not really the case, since it can be shown that $\mathbf{X}^{-1}(\mathcal{A}_\infty)$ is trivial only if the $X_n$'s are independent. On the other hand, the derivation of the strong law via (5.2.24) extends without alteration to the Banach space setting (cf. part **(ii)** of Exercise 6.1.16).

**§ 5.2.5. An Application to a Tracking Algorithm.** In this subsection I will apply the considerations in § 5.2.1 to the analysis of a tracking algorithm. The origin of this algorithm is an idea which Jan Mycielski introduced as a model for learning. However, the treatment here derives from a variation, suggested by Roy O. Davies, of Mycielski's model. Because I do not understand learning theory, I prefer to think of Mycielski's algorithm as a tracking algorithm.

Let $(E, \mathcal{B})$ be a measurable space for which there exists a nested sequence $\{\mathcal{P}_k : k \ge 0\}$ of finite or countable partitions such that $\mathcal{P}_0 = \{E\}$ and $\mathcal{B} = \sigma\left(\bigcup_{k=0}^\infty \mathcal{P}_k\right)$. Given $k \ge 1$ and $Q \in \mathcal{P}_k$, let $\breve{Q}$ be the "parent" of $Q$ in the sense that $\breve{Q}$ is the unique element of $\mathcal{P}_{k-1}$ which contains $Q$. Also, for each $x \in E$ and $k \ge 0$, use $Q_k(x)$ to denote the unique $Q \in \mathcal{P}_k$ such that $Q \ni x$. Further, let $\mu$ be a probability measure on $(E, \mathcal{B})$ with the property that, for some $\theta \in (0,1)$, $0 < \mu(Q) \le (1-\theta)\mu(\breve{Q})$ for each $Q \in \bigcup_{k=0}^\infty \mathcal{P}_k$

Next, let $(\Omega, \mathcal{F}, \mathbb{P})$ be a probability space on which there exists a sequence $\{X_n : n \ge 1\}$ of mutually independent $E$-valued random variables with distribution $\mu$. In addition, let $\{Z_n : n \ge 1\}$ be a sequence of $E$-valued random variables with the property that, for each $n \ge 1$, $Z_n$ is independent of $\sigma(\{X_m : 1 \le m \le n\})$, let $\nu_n$ be the distribution of $Z_n$, and assume that

$\nu_n \ll \mu$ with $K_r \equiv \sup_{n\geq 1}\left\|\frac{d\nu_n}{d\mu}\right\|_{L^r(\mu;\mathbb{R})} < \infty$ for some $r \in (1,\infty)$. Finally, define $\{Y_n : n \geq 1\}$ by the prescription that $Y_n(\omega) = X_m(\omega)$ if $X_m(\omega)$ is the first element of $\{X_1(\omega), \dots, X_n(\omega)\}$ which is "closest" to $Z_n(\omega)$ in the sense that, for some $k \geq 0$, $X_j(\omega) \notin Q_k(Z_n(\omega))$ for $1 \leq j < m$, $X_m(\omega) \in Q_k(Z_n(\omega))$, and $X_j(\omega) \notin Q_{k+1}(Z_n(\omega))$ for $m < j \leq n$.

The goal here is to show that the $Y_n$'s "search out" the $Z_n$'s in the sense that, for any $\mathcal{B}$-measurable $f : E \longrightarrow \mathbb{R}$,

$$\lim_{n\to\infty} \mathbb{P}\big(|f(Y_n) - f(Z_n)| \geq \epsilon\big) = 0 \quad \text{for all } \epsilon > 0. \tag{5.2.28}$$

At least in the case when $\nu_n = \mu$, Mycielski has an alternative, in some sense simpler, derivation of (5.2.28).

The strategy which I will use is the following. For each $k \geq 1$ and $f \in L^1(\mu;\mathbb{R})$, define $f_k : E \longrightarrow \mathbb{R}$ so that

$$f_k(x) = \frac{1}{\mu(Q_k(x))}\int_{Q_k(x)} f(y)\,\mu(dy).$$

Obviously $f_k(Y_n(\omega)) = f_k(Z_n(\omega))$ if $Y_n(\omega) \in Q_k(Z_n(\omega))$. Moreover, as I will show below, $\lim_{n\to\infty} \mathbb{P}(Y_n \notin Q_k(Z_n)) = 0$ for each $k \geq 0$. Thus, the key step is to show that

$$\lim_{k\to\infty}\sup_{n\geq 1} \mathbb{P}\big(|f(Y_n) - f_k(Y_n)| \geq \epsilon\big) = 0 \quad \text{for all } \epsilon > 0.$$

Notice that, because $f_k = \mathbb{E}^\mu\big[f \,|\, \sigma(\mathcal{P}_k)\big]$, this would be obvious from Corollary 5.2.4 if the $Y_n$ were replaced by $X_n$. Thus, the problem comes down to showing that the distributions of $Y_n$'s are uniformly sufficiently close to $\mu$.

For each $n \geq 1$, define

$$\begin{aligned}\Pi_n(z,\Gamma) &= \sum_{k=0}^{\infty}\sum_{j=1}^{n}\big(1-\mu(Q_k(z))\big)^{j-1}\mu\big((Q_k(z)\setminus Q_{k+1}(z))\cap\Gamma\big)\big(1-\mu(Q_{k+1}(z))\big)^{n-j}\\ &= \sum_{k=0}^{\infty}\Delta^n(Q_{k+1}(z))\frac{\mu\big((Q_k(z)\setminus Q_{k+1}(z))\cap\Gamma\big)}{\mu\big(Q_k(z)\setminus Q_{k+1}(z)\big)},\end{aligned}$$

where $\Delta^n(Q) \equiv \big(1-\mu(Q)\big)^n - \big(1-\mu(\breve{Q})\big)^n$. Then

$$\mathbb{P}\big((Z_n, Y_n) \in B\big) = \iint \mathbf{1}_B(z,y)\,\Pi_n(z,dy)\nu_n(dz). \tag{5.2.29}$$

In particular, if $\mu_n$ is the distribution of $Y_n$, then

$$\mu_n(\Gamma) = \int \Pi_n(z,\Gamma)\,\nu_n(dz) = \sum_{k=0}^{\infty}\sum_{Q\in\mathcal{P}_{k+1}}\Delta^n(Q)\nu_n(Q)\frac{\mu\big((\breve{Q}\setminus Q)\cap\Gamma\big)}{\nu(\breve{Q}\setminus Q)}.$$

In addition, because $\big(Q_\ell(z)\setminus Q_{\ell+1}(z)\big)\cap Q_k(z)=\emptyset$ if $\ell<k$ and is equal to $Q_\ell(z)\setminus Q_{\ell+1}(z)$ when $\ell\ge k$,

$$\begin{aligned}\Pi_n\big(z,Q_k(z)\big)&=\sum_{\ell=k}^{\infty}\Delta^n\big(Q_{\ell+1}(z)\big)\\&=\lim_{L\to\infty}\Big(\big(1-\mu(Q_{L+1}(z))\big)^n-\big(1-\mu(Q_k(z)\big)^n\Big)=1-\big(1-\mu(Q_k(z))\big)^n.\end{aligned}$$

Thus, if $r'$ is the Hölder conjugate of $r$, then

$$\mathbb{P}\big(Y_n\notin Q_k(Z_n)\big)=\int\big(1-\mu(Q_k(z))\big)^n\nu_n(dz)\le K_r\left(\int\big(1-\mu(Q_k(z))\big)^{nr'}\mu(dz)\right)^{\frac{1}{r'}},$$

and so, by Lebesgue's Dominated Convergence Theorem,

$$\lim_{n\to\infty}\mathbb{P}\big(Y_n\notin Q_k(Z_n)\big)=0\quad\text{for all } k\ge 0. \tag{5.2.30}$$

Given an $f\in L^1(\mu;\mathbb{R})$ and $Q\in\bigcup_{k=0}^{\infty}\mathcal{P}_k$, set

$$Af(Q)=\frac{1}{\mu(Q)}\int_Q f\,d\mu\quad\text{and}\quad Mf(Q)=\sup\left\{A|f|(Q'):\,Q\subseteq Q'\in\bigcup_{k=0}^{\infty}\mathcal{P}_k\right\}.$$

Clearly,

$$x\in Q\implies Mf(Q)\le f^*(x)\equiv\sup_{k\ge0}A|f|\big(Q_k(x)\big),$$

and, because $Af\big(Q_k(x)\big)=\mathbb{E}^{\mu}\big[f\,\big|\,\sigma(\mathcal{P}_k)\big](x)$, Doob's Inequality (5.2.3) implies that $\|f^*\|_{L^p(\mu;\mathbb{R})}\le\frac{p}{p-1}\|f\|_{L^p(\mu;\mathbb{R})}$ for all $p\in(1,\infty]$.

**Lemma 5.2.31.** *For any $f\in L^1(\mu;\mathbb{R})$,*

$$\int|f|\,d\mu_n\le\theta^{-1}\int f^*\,d\nu_n. \tag{5.2.32}$$

*In particular, if $q\in[1,\infty)$ and $f\in L^{qr'}(\mu;\mathbb{R})$, then*

$$\|f\|_{L^q(\mu_n;\mathbb{R})}\le\left(\frac{rK_r}{\theta}\right)^{\frac{1}{q}}\|f\|_{L^{qr'}(\mu;\mathbb{R})}. \tag{5.2.33}$$

**Proof:** Without loss in generality, I will assume throughout that $f\ge0$.

To prove (5.2.32), first note that

$$\begin{aligned}\int f\,d\mu_n&=\sum_{k=0}^{\infty}\sum_{Q\in\mathcal{P}_{k+1}}\Delta^n(Q)\nu_n(Q)\frac{1}{\mu(\breve{Q}\setminus Q)}\int_{\breve{Q}\setminus Q}f\,d\mu\\&\le\theta^{-1}\sum_{k=0}\sum_{Q\in\mathcal{P}_{k+1}}\Delta^n(Q)\nu_n(Q)Mf(\breve{Q}),\end{aligned}$$

since

$$\frac{1}{\mu(\breve{Q}\setminus Q)}\int_{\breve{Q}\setminus Q} f\,d\mu \le \theta^{-1}Af(\breve{Q}) \le \theta^{-1}Mf(\breve{Q}).$$

Next, for each $k \ge 0$,

$$\begin{aligned}
\sum_{Q\in\mathcal{P}_{k+1}} \Delta^n(Q)\nu_n(Q)Mf(\breve{Q}) &= \sum_{Q\in\mathcal{P}_{k+1}} \big(1-\mu(Q)\big)^n\nu_n(Q)Mf(\breve{Q}) \\
&\qquad - \sum_{Q\in\mathcal{P}_{k+1}} \big(1-\mu(\breve{Q})\big)^n\nu_n(Q)Mf(\breve{Q}) \\
&= \sum_{Q\in\mathcal{P}_{k+1}} \big(1-\mu(Q)\big)^n\nu_n(Q)Mf(\breve{Q}) - \sum_{Q\in\mathcal{P}_{k}} \big(1-\mu(Q)\big)^n\nu_n(Q)Mf(Q) \\
&\le \sum_{Q\in\mathcal{P}_{k+1}} \big(1-\mu(Q)\big)^n\nu_n(Q)Mf(Q) - \sum_{Q\in\mathcal{P}_{k}} \big(1-\mu(Q)\big)^n\nu_n(Q)Mf(Q),
\end{aligned}$$

and therefore

$$\begin{aligned}
\theta\int f\,d\mu_n &\le \lim_{K\to\infty}\sum_{k=0}^{K}\Bigg(\sum_{Q\in\mathcal{P}_{k+1}} \big(1-\mu(Q)\big)^n\nu_n(Q)Mf(Q) \\
&\qquad\qquad - \sum_{Q\in\mathcal{P}_{k}} \big(1-\mu(Q)\big)^n\nu_n(Q)Mf(Q)\Bigg) \\
&= \lim_{K\to\infty}\sum_{Q\in\mathcal{P}_{K+1}} \big(1-\mu(Q)\big)^n\nu_n(Q)Mf(Q) \le \int f^*\,d\nu_n.
\end{aligned}$$

Given (5.2.32), (5.2.33) is an easy application of Hölder's Inequality and the estimate coming from (5.2.3) on the $L^p(\mu;\mathbb{R})$-norm of $f^*$ in terms of that of $f$. Namely,

$$\begin{aligned}
\int f^q\,d\mu_n &\le \theta^{-1}\int (f^q)^*\,d\nu_n \le K_r\left(\int \big((f^q)^*\big)^{r'}\,d\mu\right)^{\frac{1}{r'}} \\
&\le \theta^{-1}K_r\frac{r'}{r'-1}\|f^q\|_{L^{r'}(\mu;\mathbb{R})} = \frac{rK_r}{\theta}\|f\|^q_{L^{qr'}(\mu;\mathbb{R})}. \qquad \square
\end{aligned}$$

THEOREM 5.2.34. *For each $\mathcal{B}$-measurable $f: E \longrightarrow \mathbb{R}$, (5.2.28) holds. Moreover, if $q\in(1,\infty)$ and $f\in L^{qr'}(\mu;\mathbb{R})$, then*

$$\lim_{n\to\infty}\mathbb{E}^{\mathbb{P}}\big[|f(Y_n)-f(Z_n)|^q\big] = 0 \quad \text{for each } p\in[1,q). \tag{5.2.35}$$

*(See Exercise 6.1.19 for a related result.)*

PROOF: It is easy to prove (5.2.28) from (5.2.35). Indeed, given $\delta > 0$, choose $R > 0$ so that $\mu(|f| \geq R) < \delta$, and set $f^R = f\mathbf{1}_{[-R,R]}(f)$. Then, by (5.2.35), $\lim_{n\to\infty} \mathbb{P}(|f^R(Y_n) - f^R(Z_n)| \geq \epsilon) = 0$ for all $\epsilon > 0$. Hence,

$$\begin{aligned}&\varlimsup_{n\to\infty} \mathbb{P}(|f(Y_n) - f(Z_n)| \geq 3\epsilon)\\ &\qquad\leq \varlimsup_{n\to\infty} \mu_n(|f - f^R| \geq \epsilon) + \varlimsup_{n\to\infty} \nu_n(|f - f^R| \geq \epsilon).\end{aligned}$$

By Hölder's Inequality,

$$\nu_n(|f - f^R| \geq \epsilon) \leq K_r \mu(|f - f^R| \geq \epsilon)^{\frac{1}{r'}} < K_r \delta^{\frac{1}{r'}},$$

and, by (5.2.33) with $q = 1$,

$$\mu_n(|f - f^R| \geq \epsilon) \leq \frac{rK_r}{\theta}\mu(|f - f^R| \geq \epsilon)^{\frac{1}{r'}} < \frac{rK_r}{\theta}\delta^{\frac{1}{r'}}.$$

The proof of (5.2.35) follows the strategy outlined earlier. That is,

$$\begin{aligned}&\mathbb{E}^{\mathbb{P}}\big[|f(Y_n) - f(Z_n)|^p\big]^{\frac{1}{p}}\\ &\qquad\leq \|f - f_k\|_{L^p(\mu_n;\mathbb{R})} + \mathbb{E}^{\mathbb{P}}\big[|f_k(Y_n) - f_k(Z_n)|^p\big]^{\frac{1}{p}} + \|f_k - f\|_{L^p(\nu_n;\mathbb{R})}.\end{aligned}$$

By (5.2.33),

$$\|f - f_k\|_{L^p(\mu_n;\mathbb{R})} \leq \left(\frac{rK_r}{\theta}\right)^{\frac{1}{p}} \|f - f_k\|_{L^{pr'}(\mu;\mathbb{R})},$$

and, by Hölder's Inequality,

$$\|f - f_k\|_{L^p(\nu_n;\mathbb{R})} \leq K_r^{\frac{1}{p}} \|f - f_k\|_{L^{pr'}(\mu;\mathbb{R})}.$$

Since, by Corollary 5.2.4, $\|f - f_k\|_{L^{pr'}(\mu;\mathbb{R})} \longrightarrow 0$ as $k \to \infty$, all that remains is to show that, for each $k \geq 0$, $\mathbb{E}^{\mathbb{P}}\big[|f_k(Y_n) - f_k(Z_n)|^p\big]^{\frac{1}{p}} \longrightarrow 0$. But

$$\begin{aligned}\mathbb{E}^{\mathbb{P}}\big[|f_k(Y_n) - f_k(Z_n)|^p\big]^{\frac{1}{p}} &= \mathbb{E}^{\mathbb{P}}\big[|f_k(Y_n) - f_k(Z_n)|^p,\, Y_n \notin Q_k(Z_n)\big]^{\frac{1}{p}}\\ &\leq \mathbb{E}^{\mathbb{P}}\big[|f_k(Y_n) - f_k(Z_n)|^q\big]^{\frac{1}{q}} \mathbb{P}(Y_n \notin Q_k(Z_n))^{\frac{1}{p}-\frac{1}{q}}.\end{aligned}$$

By (5.2.30), the final factor tends to 0 as $n \to \infty$. Hence, since, by Hölder's Inequality and (5.2.33),

$$\begin{aligned}&\mathbb{E}^{\mathbb{P}}\big[|f_k(Y_n) - f_k(Z_n)|^q\big]^{\frac{1}{q}} \leq \|f_k\|_{L^q(\mu_n;\mathbb{R})} + \|f_k\|_{L^q(\nu_n;\mathbb{R})}\\ &\quad\leq \left[\left(\frac{r}{\theta}\right)^{\frac{1}{q}} + 1\right] K_r^{\frac{1}{q}} \|f_k\|_{L^{qr'}(\mu;\mathbb{R})} \leq \left[\left(\frac{r}{\theta}\right)^{\frac{1}{q}} + 1\right] K_r^{\frac{1}{q}} \|f\|_{L^{qr'}(\mu;\mathbb{R})},\end{aligned}$$

the proof is complete. □

## Exercises for § 5.2

EXERCISE 5.2.36. In this exercise I will outline a quite independent derivation of the convergence assertion in Doob's Martingale Convergence Theorem. The key observations here are first that, given Doob's Inequality (cf. (5.2.2)), the result is nearly trivial for martingales having uniformly bounded second moments and second that everything can be reduced to that case.

**(i)** Let $(M_n, \mathcal{F}_n, \mathbb{P})$ be a martingale which is $L^2$-bounded (i.e., $\sup_{n\in\mathbb{N}} \mathbb{E}^{\mathbb{P}}[M_n^2] < \infty$). Note that

$$\mathbb{E}^{\mathbb{P}}\big[M_n^2\big] - \mathbb{E}^{\mathbb{P}}\big[M_{m-1}^2\big] = \mathbb{E}^{\mathbb{P}}\Big[\big(M_n - M_{m-1}\big)^2\Big] \quad \text{for} \quad 1 \le m \le n;$$

and starting from this, show that there is an $M \in L^2(\mathbb{P};\mathbb{R})$ such that $M_n \longrightarrow M$ in $L^2(\mathbb{P};\mathbb{R})$. Next apply (5.2.5) to the submartingale $\Big(\big|M_{n\vee m} - M_m\big|, \mathcal{F}_n, \mathbb{P}\Big)$ to show that, for every $\epsilon > 0$,

$$\mathbb{P}\left(\sup_{n\ge m} \big|M_n - M_m\big| \ge \epsilon\right) \le \frac{1}{\epsilon}\mathbb{E}^{\mathbb{P}}\Big[\big|M - M_m\big|\Big] \longrightarrow 0 \quad \text{as} \quad m \to \infty,$$

and conclude that $M_n \longrightarrow M$ (a.s., $\mathbb{P}$).

**(ii)** Let $(X_n, \mathcal{F}_n, \mathbb{P})$ be a non-negative submartingale with the property that $\sup_{n\in\mathbb{N}} \mathbb{E}^{\mathbb{P}}[X_n^2] < \infty$, define the sequence $\{A_n : n \in \mathbb{N}\}$ accordingly, as in Lemma 5.2.12, and set $M_n = X_n - A_n$, $n \in \mathbb{N}$. Then $(M_n, \mathcal{F}_n, \mathbb{P})$ is a martingale, and clearly both $M_n$ and $A_n$ are $\mathbb{P}$-square integrable for each $n \in \mathbb{N}$. In fact, check that

$$\begin{aligned}\mathbb{E}^{\mathbb{P}}\big[M_n^2 - M_{n-1}^2\big] &= \mathbb{E}^{\mathbb{P}}\big[(M_n - M_{n-1})(X_n + X_{n-1})\big] \\ &= \mathbb{E}^{\mathbb{P}}\big[X_n^2 - X_{n-1}^2\big] - \mathbb{E}^{\mathbb{P}}\big[(A_n - A_{n-1})(X_n + X_{n-1})\big] \le \mathbb{E}^{\mathbb{P}}\big[X_n^2 - X_{n-1}^2\big],\end{aligned}$$

and therefore that

$$\mathbb{E}^{\mathbb{P}}\big[M_n^2\big] \le \mathbb{E}^{\mathbb{P}}\big[X_n^2\big] \quad \text{and} \quad \mathbb{E}^{\mathbb{P}}\big[A_n^2\big] \le 4\mathbb{E}^{\mathbb{P}}\big[X_n^2\big] \quad \text{for every } n \in \mathbb{N}.$$

Finally, show that there exist $M \in L^2(\mathbb{P};\mathbb{R})$ and $A \in L^2(\mathbb{P};[0,\infty))$ such that $M_n \longrightarrow M$, $A_n \nearrow A$, and, therefore, $X_n \longrightarrow X \equiv M + A$ both $\mathbb{P}$-almost surely and in $L^2(\mathbb{P};\mathbb{R})$.

**(iii)** Let $(X_n, \mathcal{F}_n, \mathbb{P})$ be a non-negative martingale, set $Y_n = e^{-X_n}$, $n \in \mathbb{N}$, use Corollary 5.2.10 to see that $(Y_n, \mathcal{F}_n, \mathbb{P})$ is a uniformly bounded, non-negative, submartingale, and apply part **(ii)** to conclude that $\{X_n : n \ge 0\}$ converges $\mathbb{P}$-almost surely to a non-negative $X \in L^1(\mathbb{P};\mathbb{R})$.

**(iv)** Let $(X_n, \mathcal{F}_n, \mathbb{P})$ be a martingale for which

$$\sup_{n\in\mathbb{N}} \mathbb{E}^{\mathbb{P}}\big[|X_n|\big] < \infty. \tag{5.2.37}$$

For each $m \in \mathbb{N}$, define $Y^{\pm}_{n,m} = \mathbb{E}^{\mathbb{P}}\big[X^{\pm}_{n\vee m}\big|\mathcal{F}_m\big] \vee 0$ for $n \in \mathbb{N}$. Show that $Y^{\pm}_{n+1,m} \geq Y^{\pm}_{n,m}$ (a.s., $\mathbb{P}$), define $Y^{\pm}_m = \lim_{n\to\infty} Y^{\pm}_{n,m}$, check that both $(Y^+_m, \mathcal{F}_m, \mathbb{P})$ and $(Y^-_m, \mathcal{F}_m, \mathbb{P})$ are non-negative martingales with $\mathbb{E}^{\mathbb{P}}\big[Y^+_0 + Y^-_0\big] \leq \sup_{n\in\mathbb{N}} \mathbb{E}^{\mathbb{P}}\big[|X_n|\big]$, and note that $X_m = Y^+_m - Y^-_m$ (a.s., $\mathbb{P}$) for each $m \in \mathbb{N}$. In other words, *every martingale $(X_n, \mathcal{F}_n, \mathbb{P})$ satisfying (5.2.37) admits a* **Hahn decomposition**[3] *as the difference of two non-negative martingales whose sum has expectation value dominated by the left-hand side of* (5.2.37). Finally, use this observation together with **(iii)** to see that every such martingale converges $\mathbb{P}$-almost surely to some $X \in L^1(\mathbb{P};\mathbb{R})$.

**(v)** By combining the final assertion in **(iv)** together with Doob's Decomposition in Lemma 5.2.12, give another proof of the convergence assertion in Theorem 5.2.15.

EXERCISE 5.2.38. In this exercise we will develop another way to reduce Doob's Martingale Convergence Theorem to the case of $L^2$-bounded martingales. The technique here is due to R. Gundy and derives from the ideas introduced by Calderón and Zygmund in connection with their famous work on weak-type 1–1 estimates for singular integrals.

**(i)** Let $\{Z_n : n \in \mathbb{N}\}$ be a $\{\mathcal{F}_n : n \in \mathbb{N}\}$-progressively measurable, $[0, R]$-valued sequence with the property that $(-Z_n, \mathcal{F}_n, \mathbb{P})$ is a submartingale. Next, choose $\{A_n : n \in \mathbb{N}\}$ for $(-Z_n, \mathcal{F}_n, \mathbb{P})$ as in Lemma 5.2.12, note that $A_n$'s can be chosen so that $0 \leq A_n - A_{n-1} \leq R$ for all $n \in \mathbb{Z}^+$, and set $M_n = Z_n + A_n$, $n \in \mathbb{N}$. Check that $(M_n, \mathcal{F}_n, \mathbb{P})$ is a non-negative martingale with $M_n \leq (n+1)R$ for each $n \in \mathbb{N}$. Next, show that

$$\begin{aligned}
\mathbb{E}^{\mathbb{P}}\big[M_n^2 - M_{n-1}^2\big] &= \mathbb{E}^{\mathbb{P}}\big[(M_n - M_{n-1})(Z_n + Z_{n-1})\big] \\
&= \mathbb{E}^{\mathbb{P}}\big[Z_n^2 - Z_{n-1}^2\big] + \mathbb{E}^{\mathbb{P}}\big[(A_n - A_{n-1})(Z_n + Z_{n-1})\big] \\
&\leq \mathbb{E}^{\mathbb{P}}\big[Z_n^2 - Z_{n-1}^2\big] + 2R\,\mathbb{E}^{\mathbb{P}}\big[A_n - A_{n-1}\big],
\end{aligned}$$

and conclude that $\mathbb{E}^{\mathbb{P}}[A_n^2] \leq \mathbb{E}^{\mathbb{P}}[M_n^2] \leq 3R\mathbb{E}^{\mathbb{P}}[Z_0]$ for all $n \in \mathbb{N}$.

**(ii)** Let $(X_n, \mathcal{F}_n, \mathbb{P})$ be a non-negative martingale. Show that, for each $R \in (0, \infty)$, $X_n = M_n^{(R)} - A_n^{(R)} + \Delta_n^{(R)}$, $n \in \mathbb{N}$, where $(M_n^{(R)}, \mathcal{F}_n, \mathbb{P})$ is a non-negative martingale satisfying $\sup_{n\geq 0} \mathbb{E}^{\mathbb{P}}\big[(M_n^{(R)})^2\big] \leq 3R\,\mathbb{E}^{\mathbb{P}}[X_0]$; $\{A_n^{(R)} : n \in \mathbb{N}\}$ is a non-decreasing sequence of random variables with the properties that $A_0^{(R)} \equiv 0$,

[3] This useful observation was made by Klaus Krickeberg.

$A_n^{(R)}$ is $\mathcal{F}_{n-1}$-measurable, and $\sup_{n\geq 1}\mathbb{E}^{\mathbb{P}}\big[\big(A_n^{(R)}\big)^2\big] \leq 3R\mathbb{E}^{\mathbb{P}}[X_0]$; and $\{\Delta_n^{(R)} : n \in \mathbb{N}\}$ is a $\{\mathcal{F}_n : n \in \mathbb{N}\}$-progressively measurable sequence with the property that

$$\mathbb{P}(\exists n \in \mathbb{N}\ \Delta_n^{(R)} \neq 0) \leq \frac{1}{R}\mathbb{E}^{\mathbb{P}}[X_0].$$

**Hint**: Set $Z_n^{(R)} = X_n \wedge R$ and $\Delta_n^{(R)} = X_n - Z_n^{(R)}$ for $n \in \mathbb{N}$, apply part **(i)** to $\{Z_n^{(R)} : n \in \mathbb{N}\}$, and use Doob's Inequality to estimate the probability that $\Delta_n^{(R)} \neq 0$ for some $n \in \mathbb{N}$.

**(iii)** Let $(X_n, \mathcal{F}_n, \mathbb{P})$ be any martingale. Using **(ii)** above and part **(iv)** of Exercise 5.2.36, show that, for each $R \in (0,\infty)$, $X_n = M_n^{(R)} + V_n^{(R)} + \Delta_n^{(R)}$, $n \in \mathbb{N}$, where $(M_n^{(R)}, \mathcal{F}_n, \mathbb{P})$ is a martingale satisfying $\mathbb{E}^{\mathbb{P}}\big[\big(M_n^{(R)}\big)^2\big] \leq 12\,R\mathbb{E}^{\mathbb{P}}[|X_n|]$; $\{V_n^{(R)} : n \in \mathbb{N}\}$ is a sequence of random variables satisfying $V_0^{(R)} \equiv 0$, $V_n^{(R)}$ is $\mathcal{F}_{n-1}$-measurable, and

$$\mathbb{E}^{\mathbb{P}}\left[\left(\sum_1^n \big|V_m^{(R)} - V_{m-1}^{(R)}\big|\right)^2\right] \leq 12R\mathbb{E}^{\mathbb{P}}[|X_n|]$$

for $n \in \mathbb{Z}^+$; and $\{\Delta_n : \in \mathbb{N}\}$ is an $\{\mathcal{F}_n : n \in \mathbb{N}\}$-progressively measurable sequence satisfying

$$\mathbb{P}\left(\exists 0 \leq m \leq n\ \Delta_m^{(R)} \neq 0\right) \leq \frac{2}{R}\mathbb{E}^{\mathbb{P}}[|X_n|].$$

The preceding representation is called the **Calderón–Zygmund decomposition of the martingale** $(X_n, \mathcal{F}_n, \mathbb{P})$.

**(iv)** Let $(X_n, \mathcal{F}_n, \mathbb{P})$ be a martingale that satisfies (5.2.37), and use part **(iii)** above together with part **(i)** of Exercise 5.2.36 to show that, for each $R \in (0,\infty)$, $\{X_n : n \geq 0\}$ converges off of a set whose $\mathbb{P}$-measure is no more than $\frac{2}{R}$ times the supremum over $n \in \mathbb{N}$ of $\mathbb{E}^{\mathbb{P}}[|X_n|]$. In particular, when combined with Lemma 5.2.12, the preceding line of reasoning leads to the advertised alternate proof of the convergence result in Theorem 5.2.15.

EXERCISE 5.2.39. In this exercise we will extend Hunt's Theorem (cf. Theorem 5.2.13) to allow unbounded stopping times. To this end, let $(X_n, \mathcal{F}_n, \mathbb{P})$ be a uniformly $\mathbb{P}$-integrable submartingale on the probability space $(\Omega, \mathcal{F}, \mathbb{P})$, and set $M_n = X_n - A_n$, $n \in \mathbb{N}$, where $\{A_n : n \in \mathbb{N}\}$ is the sequence produced in Lemma 5.2.12. After checking that $(M_n, \mathcal{F}_n, \mathbb{P})$ is a uniformly $\mathbb{P}$-integrable martingale, show that, for any stopping time $\zeta$: $X_\zeta = \mathbb{E}^{\mathbb{P}}[M_\infty|\mathcal{F}_\zeta] + A_\zeta$ (a.s., $\mathbb{P}$), where $X_\infty$, $M_\infty$, and $A_\infty$ are, respectively, the $\mathbb{P}$-almost sure limits of $\{X_n : n \geq 0\}$, $\{M_n : n \geq 0\}$, and $\{A_n : n \geq 0\}$. In particular, if $\zeta$ and $\zeta'$ are a pair of stopping times and $\zeta \leq \zeta'$, conclude that $X_\zeta \leq \mathbb{E}^{\mathbb{P}}[X_{\zeta'}|\mathcal{F}_\zeta]$ (a.s., $\mathbb{P}$) and that the inequality is an equality holds in the case of martingales.

EXERCISE 5.2.40. There are times when submartingales converge even though they are not bounded in $L^1(\mathbb{P};\mathbb{R})$. For example, suppose that $(X_n,\mathcal{F}_n,\mathbb{P})$ is a submartingale for which there exists a non-decreasing function $\rho:\mathbb{R}\longmapsto\mathbb{R}$ with the properties that $\rho(R)\geq R$ for all $R$ and $X_{n+1}\leq\rho(X_n)$ (a.e., $\mathbb{P}$) for each $n\in\mathbb{N}$.

**(i)** Set $\zeta_R(\omega)=\inf\{n\in\mathbb{N}: X_n(\omega)\geq R\}$ for $R\in(0,\infty)$, and note that

$$\sup_{n\in\mathbb{N}} X_{n\wedge\zeta_R}\leq X_0\vee\rho(R)\quad(\text{a.e.}, \mathbb{P}).$$

In particular, if $X_0$ is $\mathbb{P}$-integrable, show that $\{X_n(\omega): n\geq 0\}$ converges in $\mathbb{R}$ for $\mathbb{P}$-almost every $\omega$ for which the sequence $\{X_n(\omega): n\geq 0\}$ is bounded above.
**Hint**: After observing that $\sup_{n\in\mathbb{N}}\mathbb{E}^{\mathbb{P}}[X^+_{n\wedge\zeta_R}]<\infty$ for every $R\in(0,\infty)$, conclude that, for each $R\in(0,\infty)$, $\{X_n: n\geq 0\}$ converges $\mathbb{P}$-almost everywhere on $\{\zeta_R=\infty\}$.

**(ii)** Let $\{Y_n: n\geq 1\}$ be a sequence of mutually independent, $\mathbb{P}$-integrable random variables, assume that $\mathbb{E}^{\mathbb{P}}[Y_n]\geq 0$ for $n\in\mathbb{N}$ and $\sup_{n\in\mathbb{N}}\|Y_n^+\|_{L^\infty(\mathbb{P};\mathbb{R})}<\infty$, and set $S_n=\sum_1^n Y_m$. Show that $\{S_n: n\geq 0\}$ is either $\mathbb{P}$-almost surely unbounded above or $\mathbb{P}$-almost surely convergent in $\mathbb{R}$.

**(iii)** Let $\{\mathcal{F}_n: n\in\mathbb{N}\}$ be a non-decreasing sequence of sub-$\sigma$-algebras and $A_n$ an element of $\mathcal{F}_n$ for each $n\in\mathbb{N}$. Show that the set of $\omega\in\Omega$ for which either

$$\sum_{n=0}^{\infty}\mathbf{1}_{A_n}(\omega)<\infty\text{ but }\sum_{n=1}^{\infty}P(A_n\,|\,\mathcal{F}_{n-1})(\omega)=\infty$$

or

$$\sum_{n=0}^{\infty}\mathbf{1}_{A_n}(\omega)=\infty\text{ but }\sum_{n=1}^{\infty}P(A_n\,|\,\mathcal{F}_{n-1})(\omega)<\infty$$

has $\mathbb{P}$-measure 0. In particular, note that this gives another derivation of the second part of the Borel–Cantelli Lemma (cf. Lemma 1.1.3).

EXERCISE 5.2.41. For each $n\in\mathbb{N}$, let $(E_n,\mathcal{B}_n)$ be a measurable space and $\mu_n$ and $\nu_n$ a pair of probability measures on $(E_n,\mathcal{B}_n)$ with the property that $\nu_n\ll\mu_n$. Prove **Kakutani's Theorem**, which says that (cf. Exercise 1.1.14) either $\prod_{n\in\mathbb{N}}\nu_n\perp\prod_{n\in\mathbb{N}}\mu_n$ or $\prod_{n\in\mathbb{N}}\nu_n\ll\prod_{n\in\mathbb{N}}\mu_n$.
**Hint**: Set

$$\Omega=\prod_{n\in\mathbb{N}}E_n,\quad\mathcal{F}=\prod_{n\in\mathbb{N}}\mathcal{B}_n,\quad\mathbb{P}=\prod_{n\in\mathbb{N}}\mu_n,\quad\text{and }\mathbb{Q}=\prod_{n\in\mathbb{N}}\nu_n.$$

Next, take $\mathcal{F}_n=\pi_n^{-1}\left(\prod_0^n\mathcal{B}_m\right)$, where $\pi_n$ is the natural projection from $\Omega$ onto $\prod_0^n E_m$, set $\mathbb{P}_n=\mathbb{P}\restriction\mathcal{F}_n$ and $\mathbb{Q}_n=\mathbb{Q}\restriction\mathcal{F}_n$, and note that

$$X_n(x)\equiv\frac{d\mathbb{Q}_n}{d\mathbb{P}_n}(x)=\prod_0^n f_m(x_m),\quad x\in\Omega,$$

where $f_n \equiv \frac{d\nu_n}{d\mu_n}$. In particular, when $\nu_n \sim \mu_n$ for each $n \in \mathbb{N}$, use Kolmogorov's 0–1 Law (cf. Theorem 1.1.2) to see that $\mathbb{Q}(G) \in \{0,1\}$, where $G \equiv \{\lim_{n\to\infty} X_n \in (0,\infty)\}$, and combine this with the last part of Theorem 5.2.20 to conclude that $\mathbb{Q} \not\perp \mathbb{P} \implies \mathbb{Q} \ll \mathbb{P}$. Finally, to remove the assumption that $\nu_n \sim \mu_n$ for all $n$'s, define $\tilde{\nu}_n$ on $(E_n, \mathcal{B}_n)$ by $\tilde{\nu}_n = (1-2^{-n-1})\nu_n + 2^{-n-1}\mu_n$, check that $\tilde{\nu}_n \sim \mu_n$ and $\mathbb{Q} \ll \tilde{\mathbb{Q}} \equiv \prod_{n\in\mathbb{N}} \tilde{\nu}_n$, and use the preceding to complete the proof.

EXERCISE 5.2.42. Let $(\Omega, \mathcal{F})$ be a measurable space and $\Sigma$ a sub-$\sigma$-algebra of $\mathcal{F}$. Given a pair of probability measures $\mathbb{P}$ and $\mathbb{Q}$ on $(\Omega, \mathcal{F})$, let $X_\Sigma$ and $Y_\Sigma$ be non-negative Radon–Nikodym derivatives of, respectively, $\mathbb{P}_\Sigma \equiv \mathbb{P} \restriction \Sigma$ and $\mathbb{Q}_\Sigma \equiv \mathbb{Q} \restriction \Sigma$ with respect to $(\mathbb{P}_\Sigma + \mathbb{Q}_\Sigma)$, and define

$$(\mathbb{P}, \mathbb{Q})_\Sigma = \int X_\Sigma^{\frac{1}{2}} Y_\Sigma^{\frac{1}{2}} \, d(\mathbb{P} + \mathbb{Q}).$$

**(i)** Show that if $\mu$ is any $\sigma$-finite measure on $(\Omega, \Sigma)$ with the property that $\mathbb{P}_\Sigma \ll \mu$ and $\mathbb{Q}_\Sigma \ll \mu$, then the number $(\mathbb{P}, \mathbb{Q})_\Sigma$ given above is equal to

$$\int \left(\frac{d\mathbb{P}_\Sigma}{d\mu}\right)^{\frac{1}{2}} \left(\frac{d\mathbb{Q}_\Sigma}{d\mu}\right)^{\frac{1}{2}} d\mu.$$

Also, check that $\mathbb{P}_\Sigma \perp \mathbb{Q}_\Sigma$ if and only if $(\mathbb{P}, \mathbb{Q})_\Sigma = 0$.

**(ii)** Suppose that $\{\mathcal{F}_n : n \in \mathbb{N}\}$ is a non-decreasing sequence of sub-$\sigma$-algebras of $\mathcal{F}$, and show that $(\mathbb{P}, \mathbb{Q})_{\mathcal{F}_n} \longrightarrow (\mathbb{P}, \mathbb{Q})_{\bigvee_0^\infty \mathcal{F}_n}$.

**(iii)** Referring to part **(ii)**, assume that $\mathbb{Q} \restriction \mathcal{F}_n \ll P \restriction \mathcal{F}_n$ for each $n \in \mathbb{N}$, let $X_n$ be a non-negative Radon–Nikodym derivative of $\mathbb{Q} \restriction \mathcal{F}_n$ with respect to $\mathbb{P} \restriction \mathcal{F}_n$, and show that $\mathbb{Q} \restriction \bigvee_0^\infty \mathcal{F}_n$ is singular to $\mathbb{P} \restriction \bigvee_0^\infty \mathcal{F}_n$ if and only if $\mathbb{E}^{\mathbb{P}}[\sqrt{X_n}] \longrightarrow 0$ as $n \to \infty$.

**(iv)** Let $\{\sigma_n :, n \geq 0\} \subseteq (0, \infty)$, and, for each $n \in \mathbb{N}$, let $\mu_n$ and $\nu_n$ be Gaussian measures on $\mathbb{R}$ with variance $\sigma_n^2$. If $a_n$ and $b_n$ are the mean values of, respectively, $\mu_n$ and $\nu_n$, show that

$$\prod_{n\in\mathbb{N}} \nu_n \sim \prod_{n\in\mathbb{N}} \mu_n \quad \text{or} \quad \prod_{n\in\mathbb{N}} \nu_n \perp \prod_{n\in\mathbb{N}} \mu_n$$

depending on whether $\sum_0^\infty \sigma_n^{-2}(b_n - a_n)^2$ converges or diverges.

EXERCISE 5.2.43. Let $\{X_n : n \in \mathbb{Z}^+\}$ be a sequence of identically distributed, mutually independent, integrable, mean value 0, $\mathbb{R}$-valued random variables on the probability space $(\Omega, \mathcal{F}, \mathbb{P})$, and set $S_n = \sum_1^n X_m$ for $n \in \mathbb{Z}^+$. In Exercise

1.4.28 we showed that $\underline{\lim}_{n\to\infty} |S_n| < \infty$ $\mathbb{P}$-almost surely. Here we will show that

$$\underline{\lim}_{n\to\infty} |S_n| = 0 \quad \mathbb{P}\text{-almost surely.} \tag{5.2.44}$$

As was mentioned before, this result was proved first by K.L. Chung and W.H. Fuchs. The basic observation behind the present proof is due to A. Perlin, who noticed that, by the Hewitt–Savage 0–1 Law, $\underline{\lim}_{n\to\infty} |S_n| = L$ $\mathbb{P}$-almost surely for some $L \in [0,\infty)$. Thus, the problem is to show that $L = 0$, and we will do this by an simple argument invented by A. Yushkevich.

**(i)** Assuming that $L > 0$, use the Hewitt–Savage 0–1 Law to show that

$$\mathbb{P}\Big(|S_n - x| < \tfrac{L}{3} \text{ i.o.}\Big) = 0 \quad \text{for any } x \in \mathbb{R},$$

where "i.o." stands for "infinitely often" and means here "for infinitely many $n$'s."

**Hint**: Set $\rho = \frac{L}{3}$. Begin by observing that, because $\{S_{m+n} - S_m : n \in \mathbb{Z}^+\}$ has the same $\mathbb{P}$-distribution as $\{S_n : n \in \mathbb{Z}^+\}$, $\mathbb{P}(|S_{m+n} - S_m| < 2\rho \text{ i.o.}) = 0$ for any $m \in \mathbb{Z}^+$. Thus, since $|S_{m+n} - x| \geq |S_{m+n} - S_m| - |S_m - x|$, $\mathbb{P}(|S_n - x| < \rho \text{ i.o.}) \leq \mathbb{P}(|S_m - x| \geq \rho)$ for any $m \in \mathbb{Z}^+$. Moreover, by the Hewitt–Savage 0–1 Law, $\mathbb{P}(|S_n - x| < \rho \text{ i.o.}) \in \{0,1\}$. Hence, either $\mathbb{P}(|S_n - x| < \rho \text{ i.o.}) = 0$, or one has the contradiction that $\mathbb{P}(|S_m - x| < \rho) = 0$ for all $m \in \mathbb{Z}^+$ and yet $\mathbb{P}(|S_n - x| < \rho \text{ i.o.}) = 1$.

**(ii)** Still assuming that $L > 0$, argue that

$$\mathbb{P}\big(|S_n - L| < \tfrac{L}{3} \text{ i.o.}\big) \vee \mathbb{P}\big(|S_n + L| < \tfrac{L}{3} \text{ i.o.}\big) = 1,$$

which, in view of **(i)**, is a contradiction. Conclude that (5.2.44) holds.

**(iii)** Knowing (5.2.44) and the Hewitt–Savage 0–1 Law, show that, for each $x \in \mathbb{R}$ and $\epsilon > 0$, one has the dichotomy

$$P\big(|S_n - x| < \epsilon\big) = 0 \text{ for all } n \geq 1 \quad \text{or} \quad P\big(|S_n - x| < \epsilon \text{ i.o.}\big) = 1.$$

EXERCISE 5.2.45. Here is a rather frivolous application of reversed martingales. Let $(\Omega, \mathcal{F}, \mathbb{P})$, $\{\mathcal{F}_n : n \in \mathbb{N}\}$, and $\{\mathbf{e}_k : k \in \mathbb{Z}\}$ be as in part **(v)** of Exercise 5.1.17. Next, take $S_m = \{(2k+1)2^m : k \in \mathbb{Z}\}$ for each $m \in \mathbb{N}$, and, for $f \in L^2([0,1);\mathbb{C})$, set

$$\Delta_m(f) = \sum_{\ell \in S_m} \big(f, \mathbf{e}_\ell\big)_{L^2([0,1);\mathbb{C})} \mathbf{e}_\ell,$$

where the convergence is in $L^2(([0,1];\mathbb{C})$. By Exercise 5.1.17,

$$f - \mathbb{E}^{\mathbb{P}}\big[f \,\big|\, \mathcal{F}_{n+1}\big] = \sum_{m=0}^{n} \Delta_m(f).$$

After noting that $\{\mathcal{F}_n : n \in \mathbb{N}\}$ is non-increasing, use the convergence result for reversed martingales in Theorem 5.2.21 to see that the expansion

$$f = \big(f, \mathbf{1}\big)_{L^2([0,1);\mathbb{C})} + \sum_{m=0}^{\infty} \Delta_m(f)$$

converges both almost everywhere as well as in $L^2([0,1);\mathbb{C})$.[4]

[4] When $f$ is a function with the property that $(f, \mathfrak{e}_\ell)_{L^2([0,1);\mathbb{C})} = 0$ for all $\ell \in \mathbb{Z}\backslash\{2^m : m \in \mathbb{N}\}$, the preceding almost everywhere convergence result can be interpreted as saying that the Fourier series of $f$ converges almost everywhere, a result that was discovered originally by Kolmogorov. The proof suggested here is based on fading memories of a conversation with N. Varopolous. Of course, ever since L. Carleson's definitive theorem on the almost every convergence of the Fourier series of an arbitrary square integrable function, the interest in this result of Kolmogorov is mostly historical.

# Chapter 6
## Some Extensions and Applications of Martingale Theory

Many of the results obtained in § 5.2 admit easy extensions to both infinite measures and Banach space–valued random variables. Furthermore, in many applications, these extensions play a useful, and occasionally essential, role. In the first section of this chapter, I will develop some of these extensions, and in the second section I will show how these extensions can be used to derive Birkhoff's Individual Ergodic Theorem. The final section is devoted to Burkholder's Inequality for martingales, an estimate that is second in importance only to Doob's Inequality.

### § 6.1 Some Extensions

Throughout the discussion that follows, $(\Omega, \mathcal{F}, \mu)$ will be a measure space and $\{\mathcal{F}_n : n \in \mathbb{N}\}$ will be a non-decreasing sequence of sub-$\sigma$-algebras with the property that $\mu \restriction \mathcal{F}_0$ is $\sigma$-finite. In particular, this means that the conditional expectation of a locally $\mu$-integrable random variable given $\mathcal{F}_n$ is well defined (cf. Theorem 5.1.12) even if the random variable takes values in a separable Banach space $E$. Thus, I will say that the sequence $\{X_n;\, n \in \mathbb{N}\}$ of $E$-valued random variables is a **$\mu$-martingale with respect to** $\{\mathcal{F}_n : n \in \mathbb{N}\}$, or, more briefly, that the triple $(X_n, \mathcal{F}_n, \mu)$ is a **martingale**, if $\{X_n : n \in \mathbb{N}\}$ is $\{\mathcal{F}_n : n \in \mathbb{N}\}$-progressively measurable, each $X_n$ is locally $\mu$-integrable, and

$$X_{n-1} = \mathbb{E}^{\mu}\big[X_n \big| \mathcal{F}_{n-1}\big] \quad \text{(a.e., } \mu) \quad \text{for each } n \in \mathbb{Z}^+.$$

Furthermore, when $E = \mathbb{R}$, I will say that $\{X_n : n \in \mathbb{N}\}$ is a **$\mu$-submartingale with respect to** $\{\mathcal{F}_n : n \in \mathbb{N}\}$ (equivalently, the triple $(X_n, \mathcal{F}_n, \mu)$ is a **submartingale**) if $\{X_n : n \in \mathbb{N}\}$ is $\{\mathcal{F}_n : n \in \mathbb{N}\}$-progressively measurable, each $X_n$ is locally $\mu$-integrable, and

$$X_{n-1} \le \mathbb{E}^{\mu}\big[X_n \big| \mathcal{F}_{n-1}\big] \quad \text{(a.e., } \mu) \quad \text{for each } n \in \mathbb{Z}^+.$$

**§ 6.1.1. Martingale Theory for a $\sigma$-Finite Measure Space.** Without any real effort, I can now prove the following variants of each of the basic results in § 5.2.

Theorem 6.1.1. *Let $(X_n, \mathcal{F}_n, \mu)$ be an $\mathbb{R}$-valued $\mu$-submartingale. Then, for each $N \in \mathbb{N}$ and $A \in \mathcal{F}_0$ on which $X_N$ is $\mu$-integrable,*

$$\mu\left(\left\{\max_{0\le n\le N} X_n \ge \alpha\right\} \cap A\right) \le \frac{1}{\alpha}\mathbb{E}^\mu\left[X_N, \left\{\max_{0\le n\le N} X_n \ge \alpha\right\} \cap A\right] \tag{6.1.2}$$

*for all $\alpha \in (0, \infty)$; and so, when all the $X_n$'s are non-negative, for every $p \in (1, \infty)$ and $A \in \mathcal{F}_0$,*

$$\mathbb{E}^\mu\left[\sup_{n\in\mathbb{N}} |X_n|^p, A\right]^{\frac{1}{p}} \le \frac{p}{p-1} \sup_{n\in\mathbb{N}} \mathbb{E}^\mu\left[|X_n|^p, A\right]^{\frac{1}{p}}.$$

*Furthermore, for each stopping time $\zeta$, $(X_{n\wedge\zeta}, \mathcal{F}_n, \mu)$ is a submartingale or a martingale depending on whether $(X_n, \mathcal{F}_n, \mu)$ is a submartingale or a martingale. In addition, for any pair of bounded stopping times $\zeta \le \zeta'$,*

$$X_\zeta \le \mathbb{E}^\mu\left[X_{\zeta'} \middle| \mathcal{F}_\zeta\right] \quad \text{(a.e., } \mu),$$

*and the inequality is an equality in the martingale case. Finally, given $a < b$ and $A \in \mathcal{F}_0$,*

$$\mathbb{E}^\mu\left[U_{[a,b]}, A\right] \le \sup_{n\in\mathbb{N}} \frac{\mathbb{E}^\mu\left[(X_n - a)^+, A\right]}{b - a},$$

*where $U_{[a,b]}(\omega)$ denotes the precise number of times that $\{X_n(\omega) : n \ge 1\}$ upcrosses $[a, b]$ (cf. the discussion preceding Theorem 5.2.15), and therefore*

$$\sup_{n\in\mathbb{N}} \mathbb{E}^\mu\left[X_n^+, A\right] < \infty \text{ for every } A \in \mathcal{F}_0 \text{ with } \mu(A) < \infty$$
$$\implies X_n \longrightarrow X \quad \text{(a.e., } \mu),$$

*where $X$ is $\bigvee_0^\infty \mathcal{F}_n$-measurable and locally $\mu$-integrable. In fact, in the case of martingales, there is a $\bigvee_0^\infty \mathcal{F}_n$-measurable, locally $\mu$-integrable $X$ such that*

$$X_n = \mathbb{E}^\mu\left[X \middle| \mathcal{F}_n\right] \quad \text{(a.e., } \mu) \quad \text{for all } n \in \mathbb{N}$$

*if and only if $\{X_n : n \ge 0\}$ is uniformly $\mu$-integrable on each $A \in \mathcal{F}_0$ with $\mu(A) < \infty$, in which case $X$ is $\mu$-integrable if and only if $X_n \longrightarrow X$ in $L^1(\mu; \mathbb{R})$. On the other hand, when $p \in (1, \infty)$, $X \in L^p(\mu; \mathbb{R})$ if and only if $\{X_n : n \ge 0\}$ is bounded in $L^p(\mu; \mathbb{R})$, in which case $X_n \longrightarrow X$ in $L^p(\mu; \mathbb{R})$.*

Proof: Obviously, there is no problem unless $\mu(\Omega) = \infty$. However, even then, each of these results follows immediately from its counterpart in § 5.2 once one makes the following trivial observation. Namely, given $\Omega' \in \mathcal{F}_0$ with $\mu(\Omega') \in (0, \infty)$, set

$$\mathcal{F}' = \mathcal{F}[\Omega'], \quad \mathcal{F}'_n = \mathcal{F}_n[\Omega'], \quad X'_n = X_n \restriction \Omega', \quad \text{and } \mathbb{P} = \frac{\mu \restriction \mathcal{F}'}{\mu(\Omega')}.$$

Then $(X'_n, \mathcal{F}'_n, \mathbb{P}')$ is a submartingale or a martingale depending on whether the original $(X_n, \mathcal{F}_n, \mu)$ was a submartingale or a martingale. Hence, when $\mu(\Omega) = \infty$, simply choose a sequence $\{\Omega_k : k \geq 1\}$ of mutually disjoint, $\mu$-finite elements of $\mathcal{F}_0$ so that $\Omega = \bigcup_1^\infty \Omega_k$, work on each $\Omega_k$ separately, and, at the end, sum the results. □

I will now spend a little time seeing how Theorem 6.1.1 can be applied to give a simple proof of the **Hardy–Littlewood Maximal Inequality**. To state their result, define the maximal function $\mathbf{M}f$ for $f \in L^1(\mathbb{R}^N;\mathbb{R})$ by

$$\mathbf{M}f(\mathbf{x}) = \sup_{Q \ni \mathbf{x}} \frac{1}{|Q|} \int_Q |f(\mathbf{y})|\, d\mathbf{y}, \quad \mathbf{x} \in \mathbb{R}^N,$$

where $Q$ is used to denote a generic **cube**

$$Q = \prod_{j=1}^N [a_j, a_j + r) \quad \text{with } \mathbf{a} = (a_1, \dots, a_N) \in \mathbb{R}^N \text{ and } r > 0. \tag{6.1.3}$$

As is easily checked, $\mathbf{M}f : \mathbb{R}^N \longrightarrow [0,\infty]$ is lower semicontinuous and therefore certainly Borel measurable. Furthermore, if we restrict our attention to *nicely meshed* families of cubes, then it is easy to relate $\mathbf{M}f$ to martingales. More precisely, for each $n \in \mathbb{Z}$, the **$n$th standard dyadic partition of $\mathbb{R}^N$** is the partition $\mathcal{P}_n$ of $\mathbb{R}^N$ into the cubes

$$C_n(\mathbf{k}) \equiv \prod_{i=1}^N \left[\frac{k_i}{2^n}, \frac{k_i+1}{2^n}\right), \quad \mathbf{k} \in \mathbb{Z}^N. \tag{6.1.4}$$

These partitions are nicely meshed in the sense that the $(n+1)$st is a refinement of the $n$th. Equivalently, if $\mathcal{F}_n$ denotes the $\sigma$-algebra over $\mathbb{R}^N$ generated by the partition $\mathcal{P}_n$, then $\mathcal{F}_n \subseteq \mathcal{F}_{n+1}$. Moreover, if $f \in L^1(\mathbb{R}^N;\mathbb{R})$ and

$$X_n^f(\mathbf{x}) \equiv 2^{nN} \int_{C_n(\mathbf{k})} |f(\mathbf{y})|\, dy \quad \text{for } \mathbf{x} \in C_n(\mathbf{k}) \text{ and } \mathbf{k} \in \mathbb{Z}^N,$$

then, for each $n \in \mathbb{Z}$,

$$X_n^f = \mathbb{E}^{\lambda_{\mathbb{R}^N}}\big[|f|\,\big|\mathcal{F}_n\big] \quad (\text{a.e.}, \lambda_{\mathbb{R}^N}),$$

where $\lambda_{\mathbb{R}^N}$ denotes Lebesgue measure on $\mathbb{R}^N$. In particular, for each $m \in \mathbb{Z}$,

$$\big(X_{m+n}^f, \mathcal{F}_{m+n}, \lambda_{\mathbb{R}^N}\big), \quad n \in \mathbb{N},$$

is a non-negative martingale; and so, by applying (6.1.2) for each $m \in \mathbb{Z}$ and then letting $m \searrow -\infty$, we see that

$$\left|\left\{\mathbf{x} : \mathbf{M}^{(\mathbf{0})}f(\mathbf{x}) \geq \alpha\right\}\right| \leq \frac{1}{\alpha} \int_{\{\mathbf{M}^{(\mathbf{0})}f \geq \alpha\}} |f(\mathbf{y})|\, dy, \quad \alpha \in (0,\infty), \tag{6.1.5}$$

where

$$\mathbf{M}^{(\mathbf{0})}f(\mathbf{x}) = \sup\left\{\frac{1}{|Q|}\int_Q |f(\mathbf{y})|\,d\mathbf{y} : \mathbf{x} \in Q \in \bigcup_{n\in\mathbb{Z}} \mathcal{P}_n\right\}$$

and I have used $|\Gamma|$ to denote $\lambda_{\mathbb{R}^N}(\Gamma)$, the Lebesgue measure of $\Gamma$.

At first sight, one might hope that it should be possible to pass directly from (6.1.5) to analogous estimates on the level sets of $\mathbf{M}f$. However, the passage from (6.1.5) to control on $\mathbf{M}f$ is not as easy as it might appear at first: the "sup" in the definition of $\mathbf{M}f$ involves many more cubes than the one in the definition of $\mathbf{M}^{(\mathbf{0})}f$. For this reason I will have to introduce additional families of meshed partitions. Namely, for each $\boldsymbol{\eta} \in \{0,1\}^N$, set

$$\mathcal{P}_n(\boldsymbol{\eta}) = \left\{\frac{(-1)^n\boldsymbol{\eta}}{3\times 2^n} + C_n(\mathbf{k}) : \mathbf{k} \in \mathbb{Z}^N\right\},$$

where $C_n(\mathbf{k})$ is the cube described in (6.1.4). It is then an easy matter to check that, for each $\boldsymbol{\eta} \in \{0,1\}^N$, $\{\mathcal{P}_n(\boldsymbol{\eta}) : n \in \mathbb{Z}\}$ is a family of meshed partitions of $\mathbb{R}^N$. Furthermore, if

$$\big[\mathbf{M}^{(\boldsymbol{\eta})}f\big](\mathbf{x}) = \sup\left\{\frac{1}{|Q|}\int_Q |f(\mathbf{y})|\,d\mathbf{y} : \mathbf{x} \in Q \in \bigcup_{n\in\mathbb{Z}} \mathcal{P}_n(\boldsymbol{\eta})\right\}, \quad \mathbf{x} \in \mathbb{R}^N,$$

then exactly the same argument that (when $\boldsymbol{\eta} = \mathbf{0}$) led us to (6.1.5) can now be used to get

$$(*) \qquad \Big|\Big\{\mathbf{x} \in \mathbb{R}^N : \mathbf{M}^{(\boldsymbol{\eta})}f(\mathbf{x}) \geq \alpha\Big\}\Big| \leq \frac{1}{\alpha}\int_{\{\mathbf{M}^{(\boldsymbol{\eta})}f\geq\alpha\}} |f(\mathbf{y})|\,d\mathbf{y}$$

for each $\boldsymbol{\eta} \in \{0,1\}^N$ and $\alpha \in (0,\infty)$. Finally, if $Q$ is given by (6.1.3) and $r \leq \frac{1}{3\,2^n}$, then it is possible to find an $\boldsymbol{\eta} \in \{0,1\}^N$ and a $C \in \mathcal{P}_n(\boldsymbol{\eta})$ for which $Q \subseteq C$. (To see this, first reduce to the case when $N = 1$.) Hence,

$$\max_{\boldsymbol{\eta}\in\{0,1\}^N} \mathbf{M}^{(\boldsymbol{\eta})}f \leq \mathbf{M}f \leq 6^N \max_{\boldsymbol{\eta}\in\{0,1\}^N} \mathbf{M}^{(\boldsymbol{\eta})}f.$$

After combining this with the estimate in (*), we arrive at the following version of the **Hardy–Littlewood Maximal Inequality**:

$$(6.1.6) \qquad \Big|\Big\{\mathbf{x} \in \mathbb{R}^N : \mathbf{M}f(\mathbf{x}) \geq \alpha\Big\}\Big| \leq \frac{(12)^N}{\alpha}\int_{\mathbb{R}^N} |f(\mathbf{y})|\,d\mathbf{y}.$$

At the same time, (*) implies that

$$\max_{\boldsymbol{\eta}\in\{0,1\}^N} \|\mathbf{M}^{(\boldsymbol{\eta})}f\|_{L^p(\mathbb{R}^N;\mathbb{R})} \leq \frac{p}{p-1}\|f\|_{L^p(\mathbb{R}^N;\mathbb{R})}, \quad p \in (1,\infty].$$

To check this, first note that it suffices to do so when $f$ vanishes outside of the ball $B(0,R)$ for some $R>0$. Second, assuming that $f=0$ off of $B(\mathbf{0},R)$, observe that (*) implies that

$$\left|\left\{\mathbf{x}\in B(\mathbf{0},R):\mathbf{M}^{(\eta)}f(\mathbf{x})\geq\alpha\right\}\right|\leq\frac{1}{\alpha}\int_{\{\mathbf{M}^{(\eta)}f\geq\alpha\}\cap B(\mathbf{0},R)}|f(\mathbf{y})|\,d\mathbf{y}.$$

Next, even though the result in Exercise 1.4.18 was stated for probability measures, it applies equally well to any finite measure. Thus, we now know that

$$\|\mathbf{M}^{(\eta)}f\|_{L^p(\mathbb{R}^N;\mathbb{R})}=\lim_{R\to\infty}\left(\int_{B(\mathbf{0},R)}(\mathbf{M}^{(\eta)}f)^p(\mathbf{x})\,d\mathbf{x}\right)^{\frac{1}{p}}\leq\frac{p}{p-1}\|f\|_{L^p(\mathbb{R}^N;\mathbb{R})},$$

and so we can repeat the argument just made to obtain

$$\|\mathbf{M}f\|_{L^p(\mathbb{R}^N;\mathbb{R})}\leq\frac{(12)^Np}{p-1}\|f\|_{L^p(\mathbb{R}^N;\mathbb{R})}\quad\text{for } p\in(1,\infty]. \tag{6.1.7}$$

In this connection, notice that there is no hope of getting this sort of estimate when $p=1$, since it is clear that

$$\varliminf_{|\mathbf{x}|\to\infty}|\mathbf{x}|^N\,\mathbf{M}f(\mathbf{x})>0$$

whenever $f$ does not vanish $\lambda_{\mathbb{R}^N}$-almost everywhere.

The inequality in (6.1.6) plays the same role in classical analysis as Doob's Inequality plays in martingale theory. For example, by essentially the same argument as I used to pass from Doob's Inequality to Corollary 5.2.4, we obtain the following version of the famous **Lebesgue Differentiation Theorem**.

THEOREM 6.1.8. *For each $f\in L^1(\mathbb{R}^N;\mathbb{R})$,*

$$\lim_{B\searrow\{\mathbf{x}\}}\frac{1}{|B|}\int_B|f(\mathbf{y})-f(\mathbf{x})|\,d\mathbf{y}=0\quad\text{for }\lambda_{\mathbb{R}^N}\text{-almost every }\mathbf{x}\in\mathbb{R}^N, \tag{6.1.9}$$

*where, for each $\mathbf{x}\in\mathbb{R}^N$, the limit is taken over balls $B$ that contain $\mathbf{x}$ and tend to $\mathbf{x}$ in the sense that their radii shrink to 0. In particular,*

$$f(\mathbf{x})=\lim_{B\searrow\{\mathbf{x}\}}\frac{1}{|B|}\int_B f(\mathbf{y})\,d\mathbf{y}\quad\text{for }\lambda_{\mathbb{R}^N}\text{-almost every }\mathbf{x}\in\mathbb{R}^N.$$

PROOF: I begin with the observation that, for each $f\in L^1(\mathbb{R}^N;\mathbb{R})$,

$$\tilde{\mathbf{M}}f(\mathbf{x})\equiv\sup_{B\ni\mathbf{x}}\frac{1}{|B|}\int_B|f(\mathbf{y})|\,d\mathbf{y}\leq\kappa_N\mathbf{M}f(\mathbf{x}),\quad\mathbf{x}\in\mathbb{R}^N,$$

where $\kappa_n = \frac{2^N}{\boldsymbol{\Omega}_N}$ with $\boldsymbol{\Omega}_N = |B(\mathbf{0},1)|$. Second, notice that (6.1.9) for *every* $\mathbf{x} \in \mathbb{R}^N$ is trivial when $f \in C_c(\mathbb{R}^N;\mathbb{R})$. Hence, all that remains is to check that if $f_n \longrightarrow f$ in $L^1(\mathbb{R}^N;\mathbb{R})$ and if (6.1.9) holds for each $f_n$, then it holds for $f$. To this end, let $\epsilon > 0$ be given and check that, because of the preceding and (6.1.6),

$$
\begin{aligned}
&\left|\left\{\mathbf{x} : \varlimsup_{B \searrow \{\mathbf{x}\}} \frac{1}{|B|}\int_B \big|f(\mathbf{y}) - f(\mathbf{x})\big|\,d\mathbf{y} \geq \epsilon\right\}\right| \\
&\quad \leq \left|\left\{\mathbf{x} : \tilde{\mathbf{M}}(f - f_n)(\mathbf{x}) \geq \frac{\epsilon}{3}\right\}\right| \\
&\qquad + \left|\left\{\mathbf{x} : \varlimsup_{B \searrow \{\mathbf{x}\}} \frac{1}{|B|}\int_B \big|f_n(\mathbf{y}) - f_n(\mathbf{x})\big|\,d\mathbf{y} \geq \frac{\epsilon}{3}\right\}\right| \\
&\qquad + \left|\left\{\mathbf{x} : \big|f_n(\mathbf{x}) - f(\mathbf{x})\big| \geq \frac{\epsilon}{3}\right\}\right| \\
&\quad \leq \frac{3}{\epsilon}\big(1 + (12)^N \kappa_N\big)\|f - f_n\|_{L^1(\mathbb{R}^N)}
\end{aligned}
$$

for every $n \in \mathbb{Z}^+$. Hence, after letting $n \to \infty$, we get (6.1.9) $f$. □

Although applications like Lebesgue's Differentiation Theorem might make one think that (6.1.6) is most interesting because of what it says about averages over small cubes, its implications for large cubes are also significant. In fact, as I will show in § 6.2, it allows one to prove Birkhoff's Individual Ergodic Theorem (cf. Theorem 6.2.7), which may be viewed as a result about *differentiation at infinity*. The link between ergodic theory and the Hardy–Littlewood Inequality is provided by the following deterministic version of the Maximal Ergodic Lemma (cf. Lemma 6.2.1). Namely, let $\{a_{\mathbf{k}} : \mathbf{k} \in \mathbb{Z}^N\}$ be a summable subset of $[0,\infty)$, and set

$$
\overline{S}_n(\mathbf{k}) = \frac{1}{(2n)^N}\sum_{\mathbf{j} \in Q_n} a_{\mathbf{j}+\mathbf{k}}, \quad n \in \mathbb{N} \text{ and } \mathbf{k} \in \mathbb{Z}^N,
$$

where $Q_n = \{\mathbf{j} \in \mathbb{Z}^N : -n \leq j_i < n \text{ for } 1 \leq i \leq N\}$. By applying (6.1.6) and (6.1.7) to the function $f$ given by (cf. (6.1.4)) $f(\mathbf{x}) = a_{\mathbf{k}}$ when $\mathbf{x} \in C_0(\mathbf{k})$, we see that

$$
\text{(6.1.10)} \qquad \operatorname{card}\left\{\mathbf{k} \in \mathbb{Z}^N : \sup_{n \in \mathbb{Z}^+} \overline{S}_n(\mathbf{k}) \geq \alpha\right\} \leq \frac{(12)^N}{\alpha}\sum_{\mathbf{k} \in \mathbb{Z}^N} a_{\mathbf{k}}, \quad \alpha \in (0,\infty)
$$

and

$$
\text{(6.1.11)} \qquad \left(\sum_{\mathbf{k} \in \mathbb{Z}^N} \sup_{n \in \mathbb{Z}^+} |\overline{S}_n(\mathbf{k})|^p\right)^{\frac{1}{p}} \leq \frac{(12)^N p}{p-1}\left(\sum_{\mathbf{k} \in \mathbb{Z}^N} |a_{\mathbf{k}}|^p\right)^{\frac{1}{p}} \quad \text{for } p \in (1,\infty].
$$

The inequality in (6.1.10) is called **Hardy's Inequality**. Actually, Hardy worked in one dimension and was drawn to this line of research by his passion for the game of cricket. What Hardy wanted to find is the optimal order in which to arrange batters to maximize the average score per inning. Thus, he worked with a non-negative sequence $\{a_k : k \geq 0\}$ in which $a_k$ represented the expected number of runs scored by player $k$, and what he showed is that, for each $\alpha \in (0, \infty)$,

$$\left|\left\{k \in \mathbb{N} : \sup_{n \in \mathbb{Z}^+} \overline{S}_n(k) \geq \alpha\right\}\right|$$

is maximized when $\{a_k : k \geq 0\}$ is non-increasing, from which it is an easy application of Markov's Inequality to prove that

$$\left|\left\{k \in \mathbb{N} : \sup_{n \in \mathbb{Z}^+} \overline{S}_n(k) \geq \alpha\right\}\right| \leq \frac{1}{\alpha} \sum_0^\infty a_k, \quad \alpha \in (0, \infty).$$

Although this sharpened result can also be obtained as a corollary the *Sunrise Lemma*,[1] Hardy's approach remains the most appealing.

§ **6.1.2. Banach Space–Valued Martingales.** I turn next to martingales with values in a separable Banach space. Actually, everything except the easiest aspects of this topic becomes extremely complicated and technical very quickly, and, for this reason, I will restrict my attention to those results that do not involve any deep properties of the geometry of Banach spaces. In fact, the only general theory with which I will deal is contained in the following.

THEOREM 6.1.12. *Let $E$ be a separable Banach space and $(X_n, \mathcal{F}_n, \mu)$ an $E$-valued martingale. Then $(\|X_n\|_E, \mathcal{F}_n, \mu)$ is a non-negative submartingale and therefore, for each $N \in \mathbb{Z}^+$ and all $\alpha \in (0, \infty)$,*

$$\mu\left(\sup_{0 \leq n \leq N} \|X_n\|_E \geq \alpha\right) \leq \frac{1}{\alpha} \mathbb{E}^\mu\left[\|X_N\|_E, \sup_{0 \leq n \leq N} \|X_n\|_E \geq \alpha\right]. \tag{6.1.13}$$

*In particular, for each $p \in (1, \infty]$,*

$$\left\|\sup_{n \in \mathbb{N}} \|X_n\|_E\right\|_{L^p(\mu;E)} \leq \frac{p}{p-1} \sup_{n \in \mathbb{N}} \|X_n\|_{L^p(\mu;E)}. \tag{6.1.14}$$

*Finally, if $X_n = \mathbb{E}^\mu[X \mid \mathcal{F}_n]$, where $X \in L^p(\mu; E)$ for some $p \in [1, \infty)$, then*

$$X_n \longrightarrow \mathbb{E}^\mu\left[X \,\Big|\, \bigvee_0^\infty \mathcal{F}_n\right] \quad \text{both (a.e., } \mu) \text{ and in } L^p(\mu; E).$$

[1] See Theorem 3.3.1 in my *Essentials of Integration Theory for Analysis*, # 262 in the Springer-Verlag G.T.M. series (2011).

PROOF: The fact $(\|X_n\|_E, \mathcal{F}_n, \mu)$ is a submartingale is an easy application of the inequality in (5.1.14); and, given this fact, the inequalities in (6.1.13) and (6.1.14) follow from the corresponding inequalities in Theorem 6.1.1.

While proving the convergence statement, I may and will assume that $\mathcal{F} = \bigvee_0^\infty \mathcal{F}_n$. Now let $X \in L^p(\mu; E)$ be given, and set $X_n = \mathbb{E}^\mu[X|\mathcal{F}_n]$, $n \in \mathbb{N}$. Because of (6.1.13) and (6.1.14), we know (cf. the proofs of Corollary 5.2.4 and Theorem 6.1.8) that the set of $X$ for which $X_n \longrightarrow X$ (a.e., $\mu$) is a closed subset of $L^p(\mu; E)$. Moreover, if $X$ is $\mu$-simple, then the $\mu$-almost everywhere convergence of $X_n$ to $X$ follows easily from the $\mathbb{R}$-valued result. Hence, we now know that $X_n \longrightarrow X$ (a.s, $\mu$) for each $X \in L^1(\mu; E)$. In addition, because of (6.1.14), when $p \in (1, \infty)$, the convergence in $L^p(\mu; E)$ follows by Lebesgue's Dominated Convergence Theorem. Finally, to prove the convergence in $L^1(\mu; E)$ when $X \in L^1(\mu; E)$, note that, by Fatou's Lemma,

$$\|X\|_{L^1(\mu;E)} \le \varliminf_{n\to\infty} \|X_n\|_{L^1(\mu;E)},$$

whereas (5.1.14) guarantees that

$$\|X\|_{L^1(\mu;E)} \ge \varlimsup_{n\to\infty} \|X_n\|_{L^1(\mu;E)}.$$

Hence, because

$$\Big| \|X_n\|_E - \|X\|_E - \|X_n - X\|_E \Big| \le 2\|X\|_E,$$

the convergence in $L^1(\mu; E)$ is again an application of Lebesgue's Dominated Convergence Theorem. □

Going beyond the convergence result in Theorem 6.1.12 to get an analog of Doob's Martingale Convergence Theorem is hard. For one thing, a naïve analog is not even true for general separable Banach spaces, and a rather deep analysis of the geometry of Banach spaces is required in order to determine exactly when it is true. (See Exercise 6.1.18 for a case in which it is.)

## Exercises for § 6.1

EXERCISE 6.1.15. In this exercise we will develop Jensen's Inequality in the Banach space setting. Thus, $(\Omega, \mathcal{F}, \mathbb{P})$ will be a probability space, $C$ will be a closed, convex subset of the separable Banach space $E$, and $X$ will be a $C$-valued element of $L^1(\mathbb{P}; E)$.

**(i)** Show that there exists a sequence $\{X_n : n \ge 1\}$ of $C$-valued, simple functions that tend to $X$ both $\mathbb{P}$-almost surely and in $L^1(\mathbb{P}; E)$.

**(ii)** Show that $\mathbb{E}^{\mathbb{P}}[X] \in C$ and that

$$\mathbb{E}^{\mathbb{P}}[g(X)] \le g(\mathbb{E}^{\mathbb{P}}[X])$$

for every continuous, concave $g : C \longrightarrow [0, \infty)$.

**(iii)** Given a sub-$\sigma$-algebra $\Sigma$ of $\mathcal{F}$, follow the argument in Corollary 5.2.8 to show that there exists a sequence $\{\mathcal{P}_n : n \geq 0\}$ of finite, $\Sigma$-measurable partitions with the property that

$$\sum_{A \in \mathcal{P}_n} \frac{\mathbb{E}^{\mathbb{P}}[X, A]}{\mathbb{P}(A)} \mathbf{1}_A \longrightarrow \mathbb{E}^{\mathbb{P}}[X|\Sigma] \quad \text{both } \mathbb{P}\text{-almost surely and in } L^1(\mathbb{P}; E).$$

In particular, conclude that there is a representative $X_\Sigma$ of $\mathbb{E}^{\mathbb{P}}[X|\Sigma]$ that is $C$-valued and satisfies

$$\mathbb{E}^{\mathbb{P}}\big[g(X)\big|\Sigma\big] \leq g(X_\Sigma) \quad (\text{a.s.}, \mathbb{P})$$

for each continuous, convex $g : C \longrightarrow [0, \infty)$.

Exercise 6.1.16. Again let $(\Omega, \mathcal{F}, \mathbb{P})$ be a probability space and $E$ be a separable Banach space. Further, suppose that $\{\mathcal{F}_n : n \geq 0\}$ is a *non-increasing* sequence of sub-$\sigma$-algebras of $\mathcal{F}$, and set $\mathcal{F}_\infty = \bigcap_0^\infty \mathcal{F}_n$. Finally, let $X \in L^1(\mathbb{P}; E)$.

**(i)** Show that

$$\mathbb{E}^{\mathbb{P}}\big[X\big|\mathcal{F}_n\big] \longrightarrow \mathbb{E}^{\mathbb{P}}[X|\mathcal{F}_\infty] \quad \text{both } \mathbb{P}\text{-almost surely and in } L^p(\mathbb{P}; E)$$

for any $p \in [1, \infty)$ with $X \in L^p(\mathbb{P}; E)$.

**Hint**: Use (6.1.13) and the approximation result in Theorem 5.1.10 to reduce to the case when $X$ is simple. When $X$ is simple, get the result as an application of the convergence result for $\mathbb{R}$-valued, reversed martingales in Theorem 5.2.21.

**(ii)** Using part **(i)** and following the line of reasoning suggested at the end of § 5.2.4, give a proof of The Strong Law of Large Numbers for Banach space–valued random variables.[2] (See Exercises 6.2.18 and 9.1.18 for entirely different approaches.)

Exercise 6.1.17. As we saw in the proof of Theorem 6.1.8, the Hardy–Littlewood maximal function can be used to dominate other quantities of interest. As a further indication of its importance, I will use it in this exercise to prove the analog of Theorem 6.1.8 for a large class of approximate identities. That is, let $\psi \in L^1(\mathbb{R}^N; \mathbb{R})$ with $\int_{\mathbb{R}^N} \psi(\mathbf{x})\, d\mathbf{x} = 1$ be given, and set

$$\psi_t(\mathbf{x}) = t^{-N}\psi\left(\tfrac{\mathbf{x}}{t}\right), \quad t \in (0, \infty) \text{ and } \mathbf{x} \in \mathbb{R}^N.$$

Then $\{\psi_t : t > 0\}$ forms an **approximate identity** in the sense that, as tempered distributions, $\psi_t \longrightarrow \delta_{\mathbf{0}}$ as $t \searrow 0$. In fact, because

$$\|\psi_t \star f\|_{L^p(\mathbb{R}^N;\mathbb{R})} \leq \|\psi\|_{L^1(\mathbb{R}^N;\mathbb{R})} \|f\|_{L^p(\mathbb{R}^N;\mathbb{R})}, \quad t \in (0, \infty) \text{ and } p \in [1, \infty],$$

[2] This proof, which seems to have been the first, of the Strong Law for Banach spaces was given by E. Mourier in "Eléments aléatoires dans un espace de Banach," *Ann. Inst. Poincaré* **13**, pp. 166–244 (1953).

and

$$\psi_t \star f(\mathbf{x}) = \int_{\mathbb{R}^N} \psi(\mathbf{y})\, f(\mathbf{x} - t\mathbf{y})\, d\mathbf{y},$$

it is easy to see that, for each $p \in [1, \infty)$,

$$\lim_{t \searrow 0} \left\| \psi_t \star f - f \right\|_{L^p(\mathbb{R}^N;\mathbb{R})} = 0$$

first for $f \in C_c(\mathbb{R}^N;\mathbb{R})$ and then for all $f \in L^p(\mathbb{R}^N;\mathbb{R})$.

The purpose of this exercise is to sharpen the preceding under the assumption that

$$\psi(\mathbf{x}) = \alpha\big(|\mathbf{x}|\big), \quad \mathbf{x} \in \mathbb{R}^N \setminus \{\mathbf{0}\} \quad \text{for some } \alpha \in C^1\big((0,\infty);\mathbb{R}\big) \text{ with}$$
$$A \equiv \int_{(0,\infty)} r^N |\alpha'(r)|\, dr < \infty.$$

Notice that when $\alpha$ is non-negative and non-increasing, integration by parts shows that $A = N$.

(**i**) Let $f \in C_c(\mathbb{R}^N;\mathbb{R})$ be given, and set

$$\tilde{f}(r, \mathbf{x}) = \frac{1}{|B(\mathbf{x}, r)|} \int_{B(\mathbf{x},r)} f(\mathbf{y})\, d\mathbf{y} \quad \text{for } r \in (0,\infty) \text{ and } \mathbf{x} \in \mathbb{R}^N.$$

Using integration by parts and the given hypotheses, show that

$$\psi_t \star f(\mathbf{x}) = -\tfrac{1}{N} \int_{(0,\infty)} r^N \alpha'(r)\, \tilde{f}(tr, \mathbf{x})\, dr,$$

and conclude that

$$\left|\psi_t \star f(\mathbf{x})\right| \le \tfrac{A}{N} \tilde{\mathbf{M}} f(\mathbf{x}),$$

where $\tilde{\mathbf{M}} f$ is the quantity introduced at the beginning of the proof of Theorem 6.1.8. In particular, conclude that there is a constant $K_N \in (0,\infty)$, depending only on $N \in \mathbb{Z}^+$, such that

$$\mathbf{M}_\psi f(\mathbf{x}) \equiv \sup_{t \in (0,\infty)} \left|\psi_t \star f(\mathbf{x})\right| \le K_N A\, \mathbf{M} f(\mathbf{x}), \quad \mathbf{x} \in \mathbb{R}^N.$$

(**ii**) Starting from the conclusion in (**i**), show that

$$\left|\{\mathbf{x} : \mathbf{M}_\psi f(\mathbf{x}) \ge R\}\right| \le \frac{(12)^N K_N A \|f\|_{L^1(\mathbb{R}^N)}}{R}, \quad f \in L^1(\mathbb{R}^N;\mathbb{R}),$$

and that for $p \in (1,\infty]$,

$$\|\mathbf{M}_\psi f\|_{L^p(\mathbb{R}^N;\mathbb{R})} \le \frac{(12)^N K_N A p}{p-1}\|f\|_{L^p(\mathbb{R}^N;\mathbb{R})}, \quad f \in L^p(\mathbb{R}^N;\mathbb{R}).$$

Finally, proceeding as in the proof of Theorem 6.1.8, use the first of these to prove that, for $f \in L^1(\mathbb{R}^N;\mathbb{R})$ and Lebesgue almost every $\mathbf{x} \in \mathbb{R}^N$,

$$\begin{aligned}&\overline{\lim_{t\searrow 0}}\,|\psi_t \star f(\mathbf{x}) - f(\mathbf{x})|\\ &\quad \le \overline{\lim_{t\searrow 0}}\int_{\mathbb{R}^N} |\psi_t(\mathbf{y})(f(\mathbf{x}-\mathbf{y}) - f(\mathbf{x}))|\,d\mathbf{y} = 0.\end{aligned}$$

Two of the most familiar examples to which the preceding applies are the Gauss kernel $g_t(\mathbf{x}) = (2\pi t)^{-\frac{N}{2}}\exp(-\frac{|\mathbf{x}|^2}{2})$ and the Poisson kernel (cf. (3.3.19)) $\Pi_t^{\mathbb{R}^N}$. In both these cases, $A = N$.

EXERCISE 6.1.18. Let $E$ be a separable Hilbert space and $(X_n, \mathcal{F}, \mathbb{P})$ an $E$-valued martingale on some probability space $(\Omega, \mathcal{F}, \mathbb{P})$ satisfying the condition

$$\sup_{n\in\mathbb{Z}^+} \mathbb{E}^{\mathbb{P}}\left[\|X_n\|_E^2\right] < \infty.$$

Proceeding as in (**i**) of Exercise 5.2.36, first prove that there is a $\bigvee_1^\infty \mathcal{F}_n$-measurable $X \in L^2(\mathbb{P};E)$ to which $\{X_n : n \ge 1\}$ converges in $L^2(\mathbb{P};E)$, next check that

$$X_n = \mathbb{E}^{\mathbb{P}}\left[X \,\middle|\, \mathcal{F}_n\right] \quad (\text{a.s.}, \mathbb{P}) \text{ for each } n \in \mathbb{Z}^+,$$

and finally apply the last part of Theorem 6.1.12 to see that $X_n \longrightarrow X$ $\mathbb{P}$-almost surely.

EXERCISE 6.1.19. This exercise deals with a variation, proposed by Jan Mycielski, on the sort of search algorithm discussed in § 5.2.5. Let $G$ be a non-empty, bounded, open subset of $\mathbb{R}^N$ with the property that $\lambda_{\mathbb{R}^N}(B(x,r)\cap G) \ge \alpha\Omega_N r^d$ for some $\alpha > 0$ and all $x \in G$ and $0 < r \le \operatorname{diam}(G)$, and define $\mu$ on $(G, \mathcal{B}_G)$ by $\mu(\Gamma) = \frac{\lambda_{\mathbb{R}^N}(\Gamma\cap G)}{\lambda_{\mathbb{R}^N}(G)}$. Next, let $(\Omega, \mathcal{F}, \mathbb{P})$ be a probability space on which there exists sequences $\{X_n : n \ge 1\}$ and $\{Z_n : n \ge 1\}$ of $G$-valued random variables with the properties that the $X_n$'s are mutually independent and have distribution $\mu$, $Z_n$ is independent of $\{X_1, \dots, X_n\}$ and has distribution $\nu_n \ll \mu$ for each $n \ge 1$, and $K_r \equiv \sup_{n\ge 1}\left\|\frac{d\nu_n}{d\mu}\right\|_{L^r(\mu;\mathbb{R})} < \infty$ for some $r \in (1,\infty)$. Without loss in generality, assume that $n \ne n' \implies X_n(\omega) \ne X_{n'}(\omega)$ for all $\omega \in \Omega$. For each $n \ge 1$, let $Y_n(\omega)$ be the last element of $\{X_1(\omega), \dots, X_n(\omega)\}$ which is closest to $Z_n(\omega)$. That is, if $\Sigma_n$ is the permutation group on $\{1, \dots, n\}$ and, for $\pi \in \Sigma_n$,

$$A_n(\pi) = \{\omega : |X_{\pi(m)}(\omega) - Z_n(\omega)| < |X_{\pi(m-1)}(\omega) - Z_n(\omega)| : \text{ for } 2 \le m \le n\},$$

then $Y_n = Z_{\pi(n)}$ on $A_n(\pi)$. Show that for all Borel measurable $f : G \longrightarrow \mathbb{R}$, $|f(Y_n) - f(Z_n)| \longrightarrow 0$ in $\mathbb{P}$-probability. Here are some steps that you might want to follow.

**(i)** Given $f \in L^1(\mu;\mathbb{R})$, show that

$$\mathbf{M}_G f(x) \equiv \sup_{r>0} \left\{ \frac{1}{|B(x,r)\cap G|} \int_{B(x,r)\cap G} |f|\,d\mu \right\} \le \alpha^{-1}\mathbf{M}f(x)$$

and therefore that there is a $C<\infty$ such that $\|\mathbf{M}_G f\|_{L^p(\mu;\mathbb{R})} \le \frac{Cp}{p-1}\|f\|_{L^p(\mu;\mathbb{R})}$ for all $p \in (1,\infty]$.

**(ii)** Given $n \ge 1$ and $z \in G$, set

$$A_n(z) = \{\omega : |X_m(\omega) - z| < |X_{m-1}(\omega) - z| : \text{ for } 2 \le m \le n\},$$

and show that

$$\mathbb{E}^{\mathbb{P}}\big[f(Y_n)\big] = n! \int \mathbb{E}^{\mathbb{P}}\big[f(X_n),\, A_n(z)\big]\,\nu_n(dz).$$

Next, for $n \ge 2$, set $r_n(\omega) = |X_{n-1}(\omega) - z|$, and show that

$$\mathbb{E}^{\mathbb{P}}\big[f(X_n),\, A_n(z)\big] = \mathbb{E}^{\mathbb{P}}\left[\int_{B(z,r_n)} f\,d\mu,\, A_{n-1}(z)\right] \le \mathbf{M}_G f(z)\mathbb{P}\big(A_n(z)\big),$$

and conclude from this that

$$\mathbb{E}^{\mathbb{P}}\big[f(Y_n)\big] \le \int \mathbf{M}_G f\,d\nu_n.$$

**(iii)** Given the conclusion drawn at the end of **(ii)**, proceed as in the derivation of Theorem 5.2.34 from Lemma 5.2.31 to get the desired result.

## § 6.2 Elements of Ergodic Theory

Among the two or three most important general results about dynamical systems is D. Birkhoff's Individual Ergodic Theorem. In this section, I will present a generalization, due to N. Wiener, of Birkhoff's basic theorem.

The setting in which I will prove the Ergodic Theorem will be the following. $(\Omega,\mathcal{F},\mu)$ will be a $\sigma$-finite measure space on which there exits a semigroup $\{\mathbf{\Sigma}^{\mathbf{k}} : \mathbf{k} \in \mathbb{N}^N\}$ of measurable, **$\mu$-measure preserving transformations.** That is, for each $\mathbf{k} \in \mathbb{N}^N$, $\mathbf{\Sigma}^{\mathbf{k}}$ is an $\mathcal{F}$-measurable map from $\Omega$ into itself, $\mathbf{\Sigma}^{\mathbf{0}}$ is the identity map, $\mathbf{\Sigma}^{\mathbf{k}+\boldsymbol{\ell}} = \mathbf{\Sigma}^{\mathbf{k}} \circ \mathbf{\Sigma}^{\boldsymbol{\ell}}$ for all $\mathbf{k}$, $\boldsymbol{\ell} \in \mathbb{N}^N$, and

$$\mu(\Gamma) = \mu\big((\mathbf{\Sigma}^{\mathbf{k}})^{-1}(\Gamma)\big) \quad \text{for all } \mathbf{k} \in \mathbb{N} \text{ and } \Gamma \in \mathcal{F}.$$

Further, $E$ will be a separable Banach space with norm $\|\cdot\|_E$, and, given a function $F : \Omega \longrightarrow E$, I will be considering the averages

$$\mathbf{A}_n F(\omega) \equiv \frac{1}{n^N} \sum_{\mathbf{k}\in Q_n^+} F \circ \mathbf{\Sigma}^{\mathbf{k}}(\omega), \quad n \in \mathbb{Z}^+,$$

where $Q_n^+$ is the cube $\{\mathbf{k} \in \mathbb{N}^N : \|\mathbf{k}\|_\infty < n\}$ and $\|\mathbf{k}\|_\infty \equiv \max_{1\le j\le N} k_j$. My goal (cf. Theorem 6.2.7) is to show that, for each $p \in [1,\infty)$ and $F \in L^p(\mu;E)$, $\{\mathbf{A}_n F : n \ge 1\}$ converges $\mu$-almost everywhere. In fact, when either $\mu$ is finite or $p \in (1,\infty)$, I will show that the convergence is also in $L^p(\mu;E)$.

§ **6.2.1. The Maximal Ergodic Lemma.** Because he was thinking in terms of dynamical systems and therefore did not take full advantage of measure theory, Birkhoff's own proof of his theorem is rather cumbersome. Later, F. Riesz discovered a proof which has become the model for all later proofs. Specifically, he introduced what is now called the Maximal Ergodic Inequality, which is an inequality that plays the same role here that Doob's Inequality played in the derivation of Corollary 5.2.4. In order to cover Wiener's extension of Birkhoff's theorem, I will derive a multiparameter version of the Maximal Ergodic Inequality, which, as the proof shows, is really just a clever application of Hardy's Inequality.[1]

LEMMA 6.2.1 (**Maximal Ergodic Lemma**). *For each $n \in \mathbb{Z}^+$ and $p \in [1, \infty]$, $\mathbf{A}_n$ is a contraction on $L^p(\mu; E)$. Moreover, for each $F \in L^p(\mu; E)$,*

$$\mu\left(\sup_{n\geq 1} \|\mathbf{A}_n F\|_E \geq \lambda\right) \leq \frac{(24)^N}{\lambda}\|F\|_{L^1(\mu;E)}, \quad \lambda \in (0,\infty), \tag{6.2.2}$$

*or*

$$\left\|\sup_{n\geq 1} \|\mathbf{A}_n F\|_E\right\|_{L^p(\mu)} \leq \frac{(24)^N p}{p-1}\|F\|_{L^p(\mu;E)}, \tag{6.2.3}$$

*depending on whether $p = 1$ or $p \in (1, \infty)$.*

PROOF: First observe that, because $\|\mathbf{A}_n F\|_E \leq \mathbf{A}_n\|F\|_E$, it suffices to prove all of these assertions in the case when $E = \mathbb{R}$ and $F$ is non-negative. Thus, I will restrict myself to this case. Since $F \circ \boldsymbol{\Sigma}^{\mathbf{k}}$ has the same distribution as $F$ itself, the first assertion is trivial. To prove (6.2.2) and (6.2.3), let $n \in \mathbb{Z}^+$ be given, apply (6.1.10) and (6.1.11) to

$$a_{\mathbf{k}}(\omega) \equiv \begin{cases} F \circ \boldsymbol{\Sigma}^{\mathbf{k}}(\omega) & \text{if } \mathbf{k} \in Q_{2n}^+ \\ 0 & \text{if } \mathbf{k} \notin Q_{2n}^+, \end{cases}$$

and conclude that

$$\begin{aligned} C_n(\omega) &\equiv \operatorname{card}\left(\left\{\mathbf{k} \in Q_n^+ : \max_{1\leq m\leq n} \mathbf{A}_m\big(F \circ \boldsymbol{\Sigma}^{\mathbf{k}}\big)(\omega) \geq \lambda\right\}\right) \\ &\leq \frac{(12)^N}{\lambda} \sum_{\mathbf{k}\in Q_{2n}^+} F \circ \boldsymbol{\Sigma}^{\mathbf{k}}(\omega) \end{aligned}$$

[1] The idea of using Hardy's Inequality was suggested to P. Hartman by J. von Neumann and appears for the first time in Hartman's "On the ergodic theorem," *Am. J. Math.* **69**, pp. 193–199 (1947).

and

$$\sum_{\mathbf{k}\in Q_n^+} \max_{1\le m\le n} \Big(\mathbf{A}_m(F\circ\mathbf{\Sigma}^{\mathbf{k}})(\omega)\Big)^p \le \left(\frac{(12)^N p}{p-1}\right)^p \sum_{\mathbf{k}\in Q_{2n}^+} \Big(F\circ\mathbf{\Sigma}^{\mathbf{k}}(\omega)\Big)^p.$$

Hence, by Tonelli's Theorem,

$$\begin{aligned}\sum_{\mathbf{k}\in Q_n^+} \mu\left(\max_{1\le m\le n} \mathbf{A}_m(F\circ\mathbf{\Sigma}^{\mathbf{k}}) \ge \lambda\right) &= \int C_n(\omega)\,\mu(d\omega)\\ &\le \frac{(12)^N}{\lambda} \sum_{\mathbf{k}\in Q_{2n}^+} \int F\circ\mathbf{\Sigma}^{\mathbf{k}} f\,d\mu\end{aligned}$$

and, similarly,

$$\sum_{\mathbf{k}\in Q_n^+} \int \max_{1\le m\le n} \Big(\mathbf{A}_m(F\circ\mathbf{\Sigma}^{\mathbf{k}})\Big)^p d\mu \le \left(\frac{(12)^N p}{p-1}\right)^p \sum_{\mathbf{k}\in Q_{2n}^+} \int \Big(F\circ\mathbf{\Sigma}^{\mathbf{k}}\Big)^p d\mu.$$

Finally, since the distributions of $\max_{1\le m\le n} \mathbf{A}_m\big(F\circ\mathbf{\Sigma}^{\mathbf{k}}\big)$ and $F\circ\mathbf{\Sigma}^{\mathbf{k}}$ do not depend on $\mathbf{k}\in\mathbb{N}^N$, the preceding lead immediately to

$$\mu\left(\max_{1\le m\le n} \mathbf{A}_m F \ge \lambda\right) \le \frac{(24)^N}{\lambda}\|F\|_{L^1(\mu)}$$

and

$$\left\|\max_{1\le m\le n} \mathbf{A}_m F\right\|_{L^p(\mu)} \le \frac{2^{\frac{N}{p}}(12)^N p}{p-1}\|F\|_{L^p(\mu)}$$

for all $n\in\mathbb{Z}^+$. Thus, (6.2.2) and (6.2.3) follow after one lets $n\to\infty$. □

Given (6.2.2) and (6.2.3), I adopt again the strategy used in the proof of Corollary 5.2.4. That is, I must begin by finding a dense subset of each $L^p$-space on which the desired convergence results can be checked by hand, and for this purpose I will have to introduce the notion of invariance.

A set $\Gamma\in\mathcal{F}$ is said to be **invariant**, and I write $\Gamma\in\mathfrak{I}$ if $\Gamma=(\mathbf{\Sigma}^{\mathbf{k}})^{-1}(\Gamma)$ for every $\mathbf{k}\in\mathbb{N}^N$. As is easily checked, $\mathfrak{I}$ is a sub-$\sigma$-algebra of $\mathcal{F}$. In addition, it is clear that $\Gamma\in\mathcal{F}$ is invariant if $\Gamma=(\mathbf{\Sigma}^{\mathbf{e}_j})^{-1}(\Gamma)$ for each $1\le j\le N$, where $\{\mathbf{e}_i : 1\le i\le N\}$ is the standard orthonormal basis in $\mathbb{R}^N$. Finally, if $\overline{\mathfrak{I}}$ is the $\mu$-completion of $\mathfrak{I}$ relative to $\mathcal{F}$ in the sense that $\Gamma\in\overline{\mathfrak{I}}$ if and only if $\Gamma\in\mathcal{F}$ and there is $\tilde{\Gamma}\in\mathfrak{I}$ such that $\mu(\Gamma\Delta\tilde{\Gamma})=0$ ($A\Delta B\equiv(A\backslash B)\cup(B\backslash A)$ is the **symmetric difference** between the sets $A$ and $B$), then an $\mathcal{F}$-measurable $F:\Omega\longrightarrow E$ is $\overline{\mathfrak{I}}$-measurable if and only if $F=F\circ\mathbf{\Sigma}^{\mathbf{k}}$ (a.e., $\mu$) for each $\mathbf{k}\in\mathbb{N}^N$. Indeed, one need only check this equivalence for indicator functions of sets. But if $\Gamma \in \mathcal{F}$ and $\mu(\Gamma\Delta\tilde{\Gamma}) = 0$ for some $\tilde{\Gamma} \in \mathfrak{I}$, then

$$\mu\Big(\Gamma\Delta(\mathbf{\Sigma}^{\mathbf{k}})^{-1}(\Gamma)\Big) \le \mu\big((\mathbf{\Sigma}^{\mathbf{k}})^{-1}(\Gamma\Delta\tilde{\Gamma})\big) + \mu(\Gamma\Delta\tilde{\Gamma}) = 0,$$

and so $\Gamma \in \overline{\mathfrak{I}}$. Conversely, if $\Gamma \in \overline{\mathfrak{I}}$, set

$$\tilde{\Gamma} = \bigcup_{\mathbf{k}\in\mathbb{N}^N} (\mathbf{\Sigma}^{\mathbf{k}})^{-1}(\Gamma),$$

and check that $\tilde{\Gamma} \in \mathfrak{I}$ and $\mu(\Gamma\Delta\tilde{\Gamma}) = 0$.

LEMMA 6.2.4. *Let $\mathfrak{I}(E)$ be the subspace of $\overline{\mathfrak{I}}$-measurable elements of $L^2(\mu;E)$. Then, $\mathfrak{I}(E)$ is a closed linear subspace of $L^2(\mu;E)$. Moreover, if $\Pi_{\mathfrak{I}(\mathbb{R})}$ denotes orthogonal projection from $L^2(\mu;\mathbb{R})$ onto $\mathfrak{I}(\mathbb{R})$, then there exists a unique linear contraction $\Pi_{\mathfrak{I}(E)} : L^2(\mu;E) \longrightarrow \mathfrak{I}(E)$ with the property that $\Pi_{\mathfrak{I}(E)}(\mathbf{a}f) = \mathbf{a}\Pi_{\mathfrak{I}(\mathbb{R})}f$ for $\mathbf{a} \in E$ and $f \in L^2(\mu;\mathbb{R})$. Finally, for each $F \in L^2(\mu;E)$,*

$$\mathbf{A}_n F \longrightarrow \Pi_{\mathfrak{I}(E)}F \quad \text{(a.e., } \mu) \text{ and in } L^2(\mu;E). \tag{6.2.5}$$

PROOF: I begin with the case when $E = \mathbb{R}$. The first step is to identify the orthogonal complement $\mathfrak{I}(\mathbb{R})^\perp$ of $\mathfrak{I}(\mathbb{R})$. To this end, let $\mathcal{N}$ denote the subspace of $L^2(\mu;\mathbb{R})$ consisting of elements having the form $g - g\circ\mathbf{\Sigma}^{\mathbf{e}_j}$ for some $g \in L^2(\mu;\mathbb{R}) \cap L^\infty(\mu;\mathbb{R})$ and $1 \le j \le N$. Given $f \in \mathfrak{I}(\mathbb{R})$, observe that

$$\big(f, g - g\circ\mathbf{\Sigma}^{\mathbf{e}_j}\big)_{L^2(\mu;\mathbb{R})} = (f,g)_{L^2(\mu;\mathbb{R})} - \big(f\circ\mathbf{\Sigma}^{\mathbf{e}_j}, g\circ\mathbf{\Sigma}^{\mathbf{e}_j}\big)_{L^2(\mu;\mathbb{R})} = 0.$$

Hence, $\mathcal{N} \subseteq \mathfrak{I}(\mathbb{R})^\perp$. On the other hand, if $f \in L^2(\mu;\mathbb{R})$ and $f \perp \mathcal{N}$, then it is clear that $f \perp f - f\circ\mathbf{\Sigma}^{\mathbf{e}_j}$ for each $1 \le j \le N$ and therefore that

$$\begin{aligned}
&\|f - f\circ\mathbf{\Sigma}^{\mathbf{e}_j}\|^2_{L^2(\mu;\mathbb{R})} \\
&\quad = \|f\|^2_{L^2(\mu;\mathbb{R})} - 2\big(f, f\circ\mathbf{\Sigma}^{\mathbf{e}_j}\big)_{L^2(\mu;\mathbb{R})} + \|f\circ\mathbf{\Sigma}^{\mathbf{e}_j}\|^2_{L^2(\mu;\mathbb{R})} \\
&\quad = 2\Big(\|f\|^2_{L^2(\mu;\mathbb{R})} - \big(f, f\circ\mathbf{\Sigma}^{\mathbf{e}_j}\big)_{L^2(\mu;\mathbb{R})}\Big) = 2\big(f, f - f\circ\mathbf{\Sigma}^{\mathbf{e}_j}\big)_{L^2(\mu;\mathbb{R})} = 0.
\end{aligned}$$

Thus, for each $1 \le j \le N$, $f = f\circ\mathbf{\Sigma}^{\mathbf{e}_j}$ $\mu$-almost everywhere; and, by induction on $\|\mathbf{k}\|_\infty$, one concludes that $f = f\circ\mathbf{\Sigma}^{\mathbf{k}}$ $\mu$-almost everywhere for all $\mathbf{k} \in \mathbb{N}^N$. In other words, we have now shown that $\mathfrak{I}(\mathbb{R}) = \mathcal{N}^\perp$ or, equivalently, that $\overline{\mathcal{N}} = \mathfrak{I}(\mathbb{R})^\perp$.

Continuing with $E = \mathbb{R}$, next note that if $f \in \overline{\mathfrak{I}(\mathbb{R})}$, then $\mathbf{A}_n f = f$ (a.e., $\mu$) for each $n \in \mathbb{Z}^+$. Hence, (6.2.5) is completely trivial in this case. On the other hand, if $g \in L^2(\mu;\mathbb{R}) \cap L^\infty(\mu;\mathbb{R})$ and $f = g - g\circ\mathbf{\Sigma}^{\mathbf{e}_j}$, then

$$n^N\mathbf{A}_n f = \sum_{\{\mathbf{k}\in Q_n^+ : k_j = 0\}} g\circ\mathbf{\Sigma}^{\mathbf{k}} - \sum_{\{\mathbf{k}\in Q_n^+ : k_j = n-1\}} g\circ\mathbf{\Sigma}^{\mathbf{k}+\mathbf{e}_j},$$

and so, with $p \in \{2,\infty\}$,

$$\left\|\mathbf{A}_n f\right\|_{L^p(\mu;\mathbb{R})} \leq \frac{2\|g\|_{L^p(\mu;\mathbb{R})}}{n} \longrightarrow 0 \quad \text{as } n \to \infty.$$

Hence, in this case also, (6.2.5) is easy. Finally, to complete the proof for $E = \mathbb{R}$, simply note that, by (6.2.3) with $p = 2$ and $E = \mathbb{R}$, the set of $f \in L^2(\mu;\mathbb{R})$ for which (6.2.5) holds is a closed linear subspace of $L^2(\mu;\mathbb{R})$ and that we have already verified (6.2.5) for $f \in \mathfrak{I}(\mathbb{R})$ and $f$ from a dense subspace of $\mathfrak{I}(\mathbb{R})^\perp$.

Turning to general $E$'s, first note that $\Pi_{\mathfrak{I}(E)}F$ is well defined for $\mu$-simple $F$'s. Indeed, if $F = \sum_1^\ell \mathbf{a}_i \mathbf{1}_{\Gamma_i}$ for some $\{\mathbf{a}_i : 1 \leq i \leq \ell\} \subseteq E$ and $\{\Gamma_i : 1 \leq i \leq \ell\}$ of mutually disjoint elements of $\mathcal{F}$ with finite $\mu$-measure, then

$$\Pi_{\mathfrak{I}(E)}F = \sum_1^\ell \mathbf{a}_i \Pi_{\mathfrak{I}(\mathbb{R})}\mathbf{1}_{\Gamma_i}$$

and so

$$\begin{aligned}\left\|\Pi_{\mathfrak{I}(E)}F\right\|^2_{L^2(\mu;E)} &\leq \int \left(\sum_1^\ell \|\mathbf{a}_i\|_E \Pi_{\mathfrak{I}(\mathbb{R})}\mathbf{1}_{\Gamma_i}\right)^2 d\mu \\ &= \left\|\Pi_{\mathfrak{I}(\mathbb{R})}\left(\sum_1^\ell \|\mathbf{a}_i\|_E \mathbf{1}_{\Gamma_i}\right)\right\|^2_{L^2(\mu;\mathbb{R})} \leq \sum_1^\ell \|\mathbf{a}_i\|_E^2 \mu(\Gamma_i) = \|F\|^2_{L^2(\mu;E)}.\end{aligned}$$

Thus, since the space of $\mu$-simple functions is dense in $L^2(\mu;E)$, it is clear that $\Pi_{\mathfrak{I}(E)}$ not only exists but is also unique.

Finally, to check (6.2.5) for general $E$'s, note that (6.2.5) for $E$-valued, $\mu$-simple $F$'s is an immediate consequence of (6.2.5) for $E = \mathbb{R}$. Thus, we already know (6.2.5) for a dense subspace of $L^2(\mu;E)$, and so the rest is another elementary application of (6.2.3). □

**§ 6.2.2. Birkhoff's Ergodic Theorem.** For any $p \in [1,\infty)$, let $\mathfrak{I}^p(E)$ denote the subspace of $\mathfrak{I}$-measurable elements of $L^p(\mu;E)$. Clearly $\mathfrak{I}^p(E)$ is closed for every $p \in [1,\infty)$. Moreover, since

$$\mu(\Omega) < \infty \implies \Pi_{\mathfrak{I}(E)}F = \mathbb{E}^\mu\big[F\big|\mathfrak{I}\big], \tag{6.2.6}$$

when $\mu$ is finite $\Pi_{\mathfrak{I}(E)}$ extends automatically as a linear contraction from $L^p(\mu;E)$ onto $\mathfrak{I}^p(E)$ for each $p \in [1,\infty)$, the extension being given by the right-hand side of (6.2.6). However, when $\mu(E) = \infty$, there is a problem. Namely, because $\mu \restriction \mathfrak{I}$ will seldom be $\sigma$-finite, it will not be possible to condition $\mu$ with respect to $\mathfrak{I}$. Be that as it may, (6.2.5) provides an extension of $\Pi_{\mathfrak{I}(E)}$. Namely, from (6.2.5) and Fatou's Lemma, it is clear that, for each $p \in [1,\infty)$,

$$\left\|\Pi_{\mathfrak{I}(E)}F\right\|_{L^p(\mu;E)} \leq \|F\|_{L^p(\mu;E)}, \quad F \in L^p(\mu;E) \cap L^2(\mu;E),$$

and therefore the desired existence of the extension follows by continuity.

THEOREM 6.2.7 (**Birkhoff's Individual Ergodic Theorem**). *For each $p \in [1,\infty)$ and $F \in L^p(\mu;E)$,*

$$\mathbf{A}_n F \longrightarrow \Pi_{\mathfrak{I}(E)} F \quad \text{(a.e., } \mu). \tag{6.2.8}$$

*Moreover, if either $p \in (1,\infty)$ or $p=1$ and $\mu(\Omega) < \infty$, then the convergence in (6.2.8) is also in $L^p(\mu;E)$. Finally, if $\mu(\Gamma) \wedge \mu(\Gamma\complement) = 0$ for every $\Gamma \in \mathfrak{I}$, then (6.2.8) can be replaced by*

$$\lim_{n\to\infty} \mathbf{A}_n F = \begin{cases} \dfrac{\mathbb{E}^\mu[F]}{\mu(\Omega)} & \text{if } \mu(\Omega) \in (0,\infty) \\ 0 & \text{if } \mu(\Omega) = \infty \end{cases} \quad \text{(a.e., } \mu),$$

*and the convergence is in $L^p(\mu;E)$ when either $p \in (1,\infty)$ or $p=1$ and $\mu(\Omega) < \infty$.*

PROOF: As I said above, the proof is now an easy application of the strategy used to prove Corollary 5.2.4. Namely, by (6.2.2), the set of $F \in L^1(\mu;E)$ for which (6.2.8) holds is closed and, by (6.2.5), it includes $L^1(\mu;E) \cap L^\infty(\mu;E)$. Hence, (6.2.8) is proved for $p=1$. On the other hand, when $p \in (1,\infty)$, (6.2.3) applies and shows first that the set of $F \in L^p(\mu;E)$ for which (6.2.8) holds is closed in $L^p(\mu;E)$ and second that $\mu$-almost everywhere convergence already implies convergence in $L^p(\mu;E)$. Hence, we have proved that (6.2.8) holds and that the convergence is in $L^p(\mu;E)$ when $p \in (1,\infty)$. In addition, when $\mu(\Gamma) \wedge \mu(\Gamma\complement) = 0$ for all $\Gamma \in \mathfrak{I}$, it is clear that the only elements of $\mathfrak{I}^p(E)$ are $\mu$-almost everywhere constant, which, in the case when $\mu(\Omega) < \infty$, means (cf. (6.2.6)) that $\Pi_{\mathfrak{I}(E)} F = \frac{\mathbb{E}^\mu[F]}{\mu(\Omega)}$, and, when $\mu(\Omega) = \infty$, means that $\mathfrak{I}^p(E) = \{0\}$ for all $p \in [1,\infty)$.

In view of the preceding, all that remains is to discuss the $L^1(\mu;E)$ convergence in the case when $p=1$ and $\mu(\Omega) < \infty$. To this end, observe that, because the $\mathbf{A}_n$'s are all contractions in $L^1(\mu;E)$, it suffices to prove $L^1(\mu;E)$ convergence for $E$-valued, $\mu$-simple $F$'s. But $L^1(\mu;E)$ convergence for such $F$'s reduces to showing that $\mathbf{A}_n f \longrightarrow \Pi_{\mathfrak{I}(\mathbb{R})} f$ in $L^1(\mu;\mathbb{R})$ for non-negative $f \in L^\infty(\mu;\mathbb{R})$. Finally, if $f \in L^1(\mu;[0,\infty))$, then

$$\|\mathbf{A}_n f\|_{L^1(\mu;\mathbb{R})} = \|f\|_{L^1(\mu;\mathbb{R})} = \|\Pi_{\mathfrak{I}(\mathbb{R})} f\|_{L^1(\mu;\mathbb{R})}, \quad n \in \mathbb{Z}^+,$$

where, in the last equality, I used (6.2.6); and this, together with (6.2.8), implies (cf. the final step in the proof of Theorem 6.1.12) convergence in $L^1(\mu;\mathbb{R})$. □

I will say that semigroup $\{\mathbf{\Sigma}^{\mathbf{k}} : \mathbf{k} \in \mathbb{N}^N\}$ is **ergodic** on $(\Omega,\mathcal{F},\mu)$ if, in addition to being $\mu$-measure preserving, $\mu(\Gamma) \wedge \mu(\Gamma\complement) = 0$ for every invariant $\Gamma \in \mathfrak{I}$.

**Classic Example.** In order to get a feeling for what the Ergodic Theorem is saying, take $\mu$ to be Lebesgue measure on the interval $[0,1)$ and, for a given $\alpha \in (0,1)$, define $\boldsymbol{\Sigma}_\alpha : [0,1) \longrightarrow [0,1)$ so that

$$\boldsymbol{\Sigma}_\alpha(\omega) \equiv \omega + \alpha - lfloor\omega + \alpha\rfloor = \omega + \alpha \bmod 1.$$

If $\alpha$ is rational and $m$ is the smallest element of $\mathbb{Z}^+$ with the property that $m\alpha \in \mathbb{Z}^+$, then it is clear that, for any $F$ on $[0,1)$, $F \circ \boldsymbol{\Sigma}_\alpha = F$ if and only if $F$ has period $\frac{1}{m}$. Hence, if $F \in L^2([0,1);\mathbb{C})$ and

$$c_\ell(F) \equiv \int_{[0,1)} F(\omega)e^{-\sqrt{-1}\,2\pi\ell\omega}\,d\omega, \quad \ell \in \mathbb{Z},$$

then elementary Fourier analysis leads to the conclusion that, in this case,

$$\lim_{n\to\infty} \mathbf{A}_n F(\omega) = \sum_{\ell\in\mathbb{Z}} c_{m\ell}(F)e^{\sqrt{-1}\,2m\ell\pi\omega} \text{ for Lebesgue-almost every } \omega \in [0,1).$$

On the other hand, if $\alpha$ is irrational, then $\{\boldsymbol{\Sigma}_\alpha^k : k \in \mathbb{N}\}$ is $\mu$-ergodic on $[0,1)$. To see this, suppose that $F \in \mathfrak{I}(\mathbb{C})$. Then (cf. the preceding and use Parseval's Identity)

$$0 = \left\|F - F\circ\boldsymbol{\Sigma}_\alpha\right\|^2_{L^2([0,1);\mathbb{C})} = \sum_{\ell\in\mathbb{Z}} \left|c_\ell(F) - c_\ell(F\circ\boldsymbol{\Sigma}_\alpha)\right|^2.$$

But, clearly,

$$c_\ell(F\circ\boldsymbol{\Sigma}_\alpha) = e^{\sqrt{-1}\,2\pi\ell\alpha}c_\ell(F), \quad \ell \in \mathbb{Z},$$

and so (because $\alpha$ is irrational) $c_\ell(F) = 0$ for each $\ell \neq 0$. In other words, the only elements of $\mathfrak{I}(\mathbb{C})$ are $\mu$-almost everywhere constant. Thus, for each irrational $\alpha \in (0,1)$, $p \in [1,\infty)$, separable Banach space $E$, and $F \in L^p([0,1);E)$,

$$\lim_{n\to\infty} \mathbf{A}_n F = \int_{[0,1)} F(\omega)\,d\omega \text{ Lebesgue-almost everywhere and in } L^p(\mu;E).$$

Finally, notice that the situation changes radically when one moves from $[0,1)$ to $[0,\infty)$ and again takes $\mu$ to be Lebesgue measure and $\alpha \in (0,1)$ to be irrational. If I extend the definition of $\boldsymbol{\Sigma}_\alpha$ by taking $\boldsymbol{\Sigma}_\alpha(\omega) = \lfloor\omega\rfloor + \boldsymbol{\Sigma}_\alpha(\omega - \lfloor\omega\rfloor)$ for $\omega \in [0,\infty)$, then it is clear that invariant functions are those that are constant on each interval $[m, m+1)$ and that, Lebesgue-almost surely, $\mathbf{A}_n f(\omega) \longrightarrow \int_{\lfloor\omega\rfloor}^{\lfloor\omega\rfloor+1} f(\eta)\,d\eta$. On the other hand, if one defines $\boldsymbol{\Sigma}_\alpha(\omega) = \omega + \alpha$, then every invariant set that has non-zero measure will have infinite measure, and so, now, every choice of $\alpha \in (0,1)$ (not just irrational ones) will give rise to an ergodic system. In particular, one will have, for each $p \in [1,\infty)$ and $F \in L^p(\mu;E)$,

$$\lim_{n\to\infty} \mathbf{A}_n F = 0 \quad \text{Lebesgue-almost everywhere,}$$

and the convergence will be in $L^p(\mu;E)$ when $p \in (1,\infty)$.

§ **6.2.3. Stationary Sequences.** For applications to probability theory, it is useful to reformulate these considerations in terms of stationary families of random variables. Thus, let $(\Omega, \mathcal{F}, \mathbb{P})$ be a probability space and $(E, \mathcal{B})$ be a measurable space ($E$ need not be a Banach space). Given a family $\mathfrak{F} = \{X_{\mathbf{k}} : \mathbf{k} \in \mathbb{N}^N\}$ of $E$-valued random variables on $(\Omega, \mathcal{F}, \mathbb{P})$, I will say that $\mathfrak{F}$ is $\mathbb{P}$-**stationary** (or simply **stationary**) if, for each $\boldsymbol{\ell} \in \mathbb{N}^N$, the family

$$\mathfrak{F}_{\boldsymbol{\ell}} \equiv \left\{X_{\mathbf{k}+\boldsymbol{\ell}} : \mathbf{k} \in \mathbb{N}^N\right\}$$

has the same (joint) distribution under $\mathbb{P}$ as $\mathfrak{F}$ itself. Clearly, one can test for stationarity by checking that the distribution of $\mathfrak{F}_{\mathbf{e}_j}$ is the same as that of $\mathfrak{F}$ for each $1 \le j \le N$. In order to apply the considerations of § 6.2.1 to stationary families, note that all questions about the properties of $\mathfrak{F}$ can be phrased in terms of the following **canonical setting** . Namely, set $\mathbf{E} = E^{\mathbb{N}^N}$ and define $\mu$ on $\left(\mathbf{E}, \mathcal{B}^{\mathbb{N}^N}\right)$ to be the image measure $\mathfrak{F}_*\mathbb{P}$. In other words, for each $\Gamma \in \mathcal{B}^{\mathbb{N}^N}$, $\mu(\Gamma) = \mathbb{P}(\mathfrak{F} \in \Gamma)$. Next, for each $\boldsymbol{\ell} \in \mathbb{N}^N$, define $\boldsymbol{\Sigma}^{\boldsymbol{\ell}} : \mathbf{E} \longrightarrow \mathbf{E}$ to be the natural shift transformation on $\mathbf{E}$ given by $\boldsymbol{\Sigma}^{\boldsymbol{\ell}}(\mathbf{x})_{\mathbf{k}} = x_{\mathbf{k}+\boldsymbol{\ell}}$ for all $\mathbf{k} \in \mathbb{N}^N$. Obviously, *stationarity of* $\mathfrak{F}$ *is equivalent to the statement that* $\{\boldsymbol{\Sigma}^{\mathbf{k}} : \mathbf{k} \in \mathbb{N}^N\}$ *is* $\mu$*-measure preserving.* Moreover, if $\mathfrak{I}$ is the $\sigma$-algebra of **shift invariant** elements $\Gamma \in \mathcal{B}^{\mathbb{N}^N}$ (i.e., $\Gamma = \left(\boldsymbol{\Sigma}^{\mathbf{k}}\right)^{-1}(\Gamma)$ for all $\mathbf{k} \in \mathbb{N}^N$), then, by Theorem 6.2.7, for any separable Banach space $B$, any $p \in [1, \infty)$, and any $F \in L^p(\mathbb{P}; B)$,

$$\lim_{n\to\infty} \frac{1}{n^N} \sum_{\mathbf{k}\in Q_n^+} F \circ \mathfrak{F}_{\mathbf{k}} = \mathbb{E}^{\mathbb{P}}\left[F \circ \mathfrak{F} \,\middle|\, \mathfrak{F}^{-1}(\mathfrak{I})\right] \text{ (a.s., } \mathbb{P}) \text{ and in } L^p(\mathbb{P}; B).$$

In particular, when $\left\{\boldsymbol{\Sigma}^{\mathbf{k}} : \mathbf{k} \in \mathbb{N}^N\right\}$ is ergodic on $\left(\mathbf{E}, \mathcal{B}^{\mathbb{N}^N} \mu\right)$, I will say that the family $\mathfrak{F}$ is **ergodic** and conclude that the preceding can be replaced by

$$\lim_{n\to\infty} \frac{1}{n^N} \sum_{\mathbf{k}\in Q_n^+} F \circ \mathfrak{F}_{\mathbf{k}} = \mathbb{E}^{\mathbb{P}}\left[F \circ \mathfrak{F}\right] \text{ (a.s., } \mathbb{P}) \text{ and in } L^p(\mathbb{P}; B). \tag{6.2.9}$$

So far I have discussed *one-sided* stationary families, that is, families indexed by $\mathbb{N}^N$. However, for various reasons (cf. Theorem 6.2.11) it is useful to know that *one can usually embed a one-sided stationary family into a two-sided one.* In terms of the *semigroup* of shifts, this corresponds to the trivial observation that the semigroup $\left\{\boldsymbol{\Sigma}^{\mathbf{k}} : \mathbf{k} \in \mathbb{N}^N\right\}$ on $\mathbf{E} = E^{\mathbb{N}^N}$ can be viewed as a sub-semigroup of the *group* of shifts $\left\{\boldsymbol{\Sigma}^{\mathbf{k}} : \mathbf{k} \in \mathbb{Z}^N\right\}$ on $\hat{\mathbf{E}} = E^{\mathbb{Z}^N}$. With these comments in mind, I will prove the following.

LEMMA 6.2.10. *Assume that $E$ is a complete, separable, metric space and that $\mathfrak{F} = \{X_{\mathbf{k}} : \mathbf{k} \in \mathbb{N}^N\}$ is a stationary family of $E$-valued random variables on the probability space $(\Omega, \mathcal{F}, \mathbb{P})$. Then there exists a probability space $(\hat{\Omega}, \hat{\mathcal{F}}, \hat{\mathbb{P}})$ and a family $\hat{\mathfrak{F}} = \left\{\hat{X}_{\mathbf{k}} : \mathbf{k} \in \mathbb{Z}^N\right\}$ with the property that, for each $\boldsymbol{\ell} \in \mathbb{Z}^N$,*

$$\hat{\mathfrak{F}}_{\boldsymbol{\ell}} \equiv \left\{\hat{X}_{\mathbf{k}+\boldsymbol{\ell}} : \mathbf{k} \in \mathbb{N}^N\right\}$$

*has the same distribution under $\hat{\mathbb{P}}$ as $\mathfrak{F}$ has under $\mathbb{P}$.*

PROOF: When formulated correctly, this theorem is an essentially trivial application of Kolmogorov's Extension Theorem (cf. part (**iii**) of Exercise 9.1.17). Namely, for $n \in \mathbb{N}$, set

$$\Lambda_n = \{\mathbf{k} \in \mathbb{Z}^N : k_j \geq -n \text{ for } 1 \leq j \leq N\},$$

and define $\Phi_n : E^{\Lambda_0} \longrightarrow E^{\Lambda_n}$ so that

$$\Phi_n(\mathbf{x})_{\mathbf{k}} = x_{\mathbf{n}+\mathbf{k}} \quad \text{for } \mathbf{x} \in E^{\Lambda_0} \text{ and } \mathbf{k} \in \Lambda_n, \text{ where } \mathbf{n} \equiv (n, \dots, n)$$

Next, take $\mu_0$ on $E^{\Lambda_0}$ to be the $\mathbb{P}$-distribution of $\mathfrak{F}$ and, for $n \geq 1$, $\mu_n$ on $E^{\Lambda_n}$ to be $(\Phi_n)_*\mu_0$. Using stationarity, one can easily check that, for each $n \geq 0$ and $\mathbf{k} \in \mathbb{N}^N$, $\mu_n$ is invariant under the obvious extension of $\mathbf{\Sigma}^{\mathbf{k}}$ to $E^{\Lambda_n}$. In particular, if one identifies $E^{\Lambda_{n+1}}$ with $E^{\Lambda_{n+1}\setminus\Lambda_n} \times E^{\Lambda_n}$, then

$$\mu_{n+1}\big(E^{\Lambda_{n+1}\setminus\Lambda_n} \times \Gamma\big) = \mu_n(\Gamma) \quad \text{for all } \Gamma \in \mathcal{B}_{E^{\Lambda_n}}.$$

Hence the $\mu_n$'s are consistently defined on the spaces $E^{\Lambda_n}$, and therefore Kolmogorov's Extension Theorem applies and guarantees the existence of a unique Borel probability measure $\mu$ on $E^{\mathbb{Z}^N}$ with the property that

$$\mu\big(E^{\mathbb{Z}^N\setminus\Lambda_n} \times \Gamma\big) = \mu_n(\Gamma) \quad \text{for all } n \geq 0 \text{ and } \Gamma \in \mathcal{B}_{E^{\Lambda_n}}.$$

Moreover, since each $\mu_n$ is $\mathbf{\Sigma}^{\mathbf{k}}$-invariant for all $\mathbf{k} \in \mathbb{N}^N$, it is clear that $\mu$ is also. Thus, because $\mathbf{\Sigma}^{\mathbf{k}}$ is invertible on $E^{\mathbb{Z}^N}$ and $\mathbf{\Sigma}^{-\mathbf{k}}$ is its inverse, it follows that $\mu$ is invariant under $\mathbf{\Sigma}^{\mathbf{k}}$ for all $\mathbf{k} \in \mathbb{Z}^N$.

To complete the proof at this point, simply take $\hat{\Omega} = E^{\mathbb{Z}^N}$, $\hat{\mathcal{F}} = \mathcal{B}_{\hat{\Omega}}$, $\hat{\mathbb{P}} = \mu$, and $\hat{X}_{\mathbf{k}}(\hat{\omega}) = \hat{\omega}_{\mathbf{k}}$ for $\mathbf{k} \in \mathbb{Z}^N$. □

As an example of the advantage that Lemma 6.2.10 affords, I present the following beautiful observation made originally by M. Kac.

THEOREM 6.2.11. *Let $(E, \mathcal{B})$ be a measurable space and $\{X_k : k \in \mathbb{N}\}$ a stationary sequence of $E$-valued random variables on the probability space $(\Omega, \mathcal{F}, \mathbb{P})$. Given $\Gamma \in \mathcal{B}$, define the return time $\rho_\Gamma(\omega) = \inf\{k \geq 1 : X_k(\omega) \in \Gamma\}$. Then, $\mathbb{E}^{\mathbb{P}}[\rho_\Gamma, X_0 \in \Gamma] = \mathbb{P}(X_k \in \Gamma$ for some $k \in \mathbb{N})$. In particular, if $\{X_k : k \in \mathbb{N}\}$ is ergodic, then*

$$\mathbb{P}(X_0 \in \Gamma) > 0 \implies \mathbb{E}^{\mathbb{P}}[\rho_\Gamma, X_0 \in \Gamma] = 1.$$

PROOF: Set $U_k = \mathbf{1}_\Gamma \circ X_k$ for $k \in \mathbb{N}$. Then $\{U_k : k \in \mathbb{N}\}$ is a stationary sequence of $\{0, 1\}$-valued random variables. Hence, by Lemma 6.2.10, we can find a probability space $(\hat{\Omega}, \hat{\mathcal{F}}, \hat{\mathbb{P}})$ on which there is a family $\{\hat{U}_k : k \in \mathbb{Z}\}$ of $\{0, 1\}$-valued random variables with the property that, for every $n \in \mathbb{Z}$, $(\hat{U}_n, \dots, \hat{U}_{n+k}, \dots)$ has the same distribution under $\hat{\mathbb{P}}$ as $(U_0, \dots, U_k, \dots)$ has under $\mathbb{P}$. In particular,

$$P(\rho_\Gamma \geq 1,\, X_0 \in \Gamma) = \hat{\mathbb{P}}(\hat{U}_0 = 1) \text{ and}$$
$$P(\rho_\Gamma \geq n+1,\, X_0 \in \Gamma) = \hat{\mathbb{P}}(\hat{U}_{-n} = 1, \hat{U}_{-n+1} = 0, \dots, \hat{U}_0 = 0), \quad n \in \mathbb{Z}^+.$$

Thus, if

$$\lambda_\Gamma(\hat{\omega}) \equiv \inf\{k \in \mathbb{N} : U_{-k}(\hat{\omega}) = 1\},$$

then

$$\mathbb{P}(\rho_\Gamma \geq n,\, X_0 \in \Gamma) = \hat{\mathbb{P}}(\lambda_\Gamma = n-1), \quad n \in \mathbb{Z}^+,$$

and so

$$\mathbb{E}^{\mathbb{P}}[\rho_\Gamma,\, X_0 \in \Gamma] = \hat{\mathbb{P}}(\lambda_\Gamma < \infty).$$

Now observe that

$$\hat{\mathbb{P}}(\lambda_\Gamma > n) = \hat{\mathbb{P}}(\hat{U}_{-n} = 0, \dots, \hat{U}_0 = 0) = \mathbb{P}(X_0 \notin \Gamma, \dots, X_n \notin \Gamma),$$

from which it is clear that

$$\hat{\mathbb{P}}(\lambda_\Gamma < \infty) = \mathbb{P}(\exists k \in \mathbb{N}\; X_k \in \Gamma).$$

Finally, assume that $\{X_k : k \in \mathbb{N}\}$ is ergodic and that $\mathbb{P}(X_0 \in \Gamma) > 0$. Because, by (6.2.9), $\sum_0^\infty \mathbf{1}_\Gamma(X_k) = \infty$ $\mathbb{P}$-almost surely, it follows that, $\mathbb{P}$-almost surely, $X_k \in \Gamma$ for some $k \in \mathbb{N}$. □

It should be noticed that, although there are far more elementary proofs, when $\{X_n : n \geq 0\}$ is an irreducible, ergodic Markov chain on a countable state space $E$, then Kac's theorem proves that the stationary measure at the state $x \in E$ is the reciprocal of the expected time that the chain takes to return to $x$ when it starts at $x$.

**§ 6.2.4. Continuous Parameter Ergodic Theory.** I turn now to the setting of continuously parametrized semigroups of transformations. Thus, again $(\Omega, \mathcal{F}, \mu)$ is a $\sigma$-finite measure space and $\{\mathbf{\Sigma}^{\mathbf{t}} : \mathbf{t} \in [0,\infty)^N\}$ is a measurable semigroup of $\mu$-measure preserving transformations on $\Omega$. That is, $\mathbf{\Sigma}^{\mathbf{0}}$ is the identity, $\mathbf{\Sigma}^{\mathbf{s}+\mathbf{t}} = \mathbf{\Sigma}^{\mathbf{s}} \circ \mathbf{\Sigma}^{\mathbf{t}}$,

$$(\mathbf{t}, \omega) \in [0,\infty)^N \times \Omega \longmapsto \mathbf{\Sigma}^{\mathbf{t}}(\omega) \in \Omega \quad \text{is } \mathcal{B}_{[0,\infty)^N} \times \mathcal{F}\text{-measurable},$$

and $(\mathbf{\Sigma}^{\mathbf{t}})_* \mu = \mu$ for every $\mathbf{t} \in [0,\infty)^N$. Next, given an $\mathcal{F}$-measurable $F$ with values in some separable Banach space $E$, let $\mathfrak{G}(F)$ be the set of $\omega \in \Omega$ with the property that

$$\int_{[0,T)^N} \left\| F \circ \mathbf{\Sigma}^{\mathbf{t}}(\omega) \right\|_E d\mathbf{t} < \infty \quad \text{for all } T \in (0,\infty).$$

Clearly,

$$\omega \in \mathfrak{G}(F) \implies \boldsymbol{\Sigma}^{\mathbf{t}}(\omega) \in \mathfrak{G}(F) \quad \text{for every } \mathbf{t} \in [0,\infty)^N.$$

In addition, if $F \in L^p(\mu; E)$ for some $p \in [1,\infty)$, then

$$\int_\Omega \left( \int_{[0,T)^N} \|F \circ \boldsymbol{\Sigma}^{\mathbf{t}}(\omega)\|_E^p \, d\mathbf{t} \right) \mu(d\omega) = T^N \|F\|_{L^p(\mu;E)}^p < \infty,$$

and so

$$F \in \bigcup_{p\in[1,\infty)} L^p(\mu; E) \implies \mu\Big(\mathfrak{G}(F)\complement\Big) = 0.$$

Next, for each $T \in (0,\infty)$, define

$$\mathcal{A}_T F(\omega) = \begin{cases} T^{-N} \int_{[0,T)^N} F \circ \boldsymbol{\Sigma}^{\mathbf{t}}(\omega)\, d\mathbf{t} & \text{if } \omega \in \mathfrak{G}(F) \\ 0 & \text{if } \omega \notin \mathfrak{G}(F), \end{cases}$$

and note that, as a consequence of the invariance of $\mathfrak{G}(F)$,

$$\big(\mathcal{A}_T F\big) \circ \boldsymbol{\Sigma}^{\mathbf{t}} = \mathcal{A}_T\big(F \circ \boldsymbol{\Sigma}^{\mathbf{t}}\big) \quad \text{for all } \mathbf{t} \in [0,\infty)^N.$$

Finally, use $\hat{\mathfrak{I}}$ to denote the $\sigma$-algebra of $\Gamma \in \mathcal{F}$ with the property that $\Gamma = (\boldsymbol{\Sigma}^{\mathbf{t}})^{-1}(\Gamma)$ for each $\mathbf{t} \in [0,\infty)^N$, and say that $\{\boldsymbol{\Sigma}^{\mathbf{t}} : \mathbf{t} \in [0,\infty)^N\}$ is **ergodic** if $\mu(\Gamma) \wedge \mu(\Gamma\complement) = 0$ for every $\Gamma \in \hat{\mathfrak{I}}$.

THEOREM 6.2.12. *Let $(\Omega, \mathcal{F}, \mu)$ be a $\sigma$-finite measure space and $\{\boldsymbol{\Sigma}^{\mathbf{t}} : \mathbf{t} \in [0,\infty)^N\}$ be a measurable semigroup of $\mu$-measure preserving transformations on $\Omega$. Then, for each separable Banach space $E$, $p \in [1,\infty)$, and $T \in (0,\infty)$, $\mathcal{A}_T$ is a contraction on $L^p(\mu; E)$. Next, set $\Pi_{\hat{\mathfrak{I}}(E)} = \Pi_{\mathfrak{I}(E)} \circ \mathcal{A}_1$, where $\Pi_{\mathfrak{I}(E)}$ is defined in terms of $\{\boldsymbol{\Sigma}^{\mathbf{k}} : \mathbf{k} \in \mathbb{N}^N\}$ as in Theorem 6.2.7. Then, for each $p \in [1,\infty)$ and $F \in L^p(\mu; E)$,*

$$\text{(6.2.13)} \qquad \lim_{T\to\infty} \mathcal{A}_T F = \Pi_{\hat{\mathfrak{I}}(E)} F \quad \text{(a.e., } \mu).$$

*Moreover, if $p \in (1,\infty)$ or $p = 1$ and $\mu(\Omega) < \infty$, then the convergence is also in $L^p(\mu; E)$. In fact, if $\mu(\Omega) < \infty$, then*

$$\lim_{T\to\infty} \mathcal{A}_T F = \mathbb{E}^\mu\big[F \,\big|\, \hat{\mathfrak{I}}\big] \quad \text{(a.e., } \mu) \text{ and in } L^p(\mu : E).$$

*Finally, if $\{\boldsymbol{\Sigma}^{\mathbf{t}} : \mathbf{t} \in [0,\infty)^N\}$ is ergodic, then (6.2.13) can be replaced by*

$$\lim_{T\to\infty} \mathcal{A}_T F = \frac{\mathbb{E}^\mu[F]}{\mu(\Omega)} \quad \text{(a.e., } \mu),$$

*where it is understood that the ratio is 0 when the denominator is infinite.*

PROOF: The first step is the observation that

$$\mu\left(\sup_{T>0}\left\|\mathcal{A}_T F\right\|_E \geq \lambda\right) \leq \frac{(24)^N}{\lambda}\|F\|_{L^1(\mu;E)}, \quad \lambda \in (0,\infty) \tag{6.2.14}$$

and

$$\left\|\sup_{T>0}\left\|\mathcal{A}_T F\right\|_E\right\|_{L^p(\mu;E)} \leq \frac{(24)^N p}{p-1}\|F\|_{L^p(\mu;E)} \quad \text{for } p \in (1,\infty). \tag{6.2.15}$$

Indeed, because of $(\mathcal{A}_T F)\circ \boldsymbol{\Sigma}^{\mathbf{t}} = \mathcal{A}_T(F\circ\boldsymbol{\Sigma}^{\mathbf{t}})$, (6.2.14) is derived from (6.1.6) in precisely the same way as I derived (6.2.2) from (6.1.10), and (6.2.15) comes from (6.1.7) just as (6.2.3) came from (6.1.7).

Given (6.2.14) and (6.2.15), we know that it suffices to prove (6.2.13) for a dense subset of $L^1(\mu;E)$. Thus, let $F$ be a uniformly bounded element of $L^1(\mu;E)$ and set $\hat{F} = \mathcal{A}_1 F$. Because

$$\left\|T^N \mathcal{A}_T F(\omega) - n^N \mathbf{A}_n \hat{F}(\omega)\right\|_E \leq \int_{[0,n+1)^N\setminus[0,n)^N} \|F\circ\boldsymbol{\Sigma}^{\mathbf{t}}(\omega)\|_E\, d\mathbf{t}$$

for $n \leq T \leq n+1$,

$$\lim_{n\to\infty}\left\|\sup_{n\leq T\leq n+1}\left\|\mathcal{A}_T F - \mathbf{A}_n\hat{F}\right\|_E\right\|_{L^p(\mu;\mathbb{R})} = 0 \quad \text{for every } p\in[1,\infty].$$

Hence, for $F \in L^1(\mu;E)\cap L^\infty(\mu;E)$, (6.2.13) follows from (6.2.8). As for the case when $\mu(\Omega)<\infty$, all that we have to do is check that $\Pi_{\hat{\mathfrak{J}}(E)}F = \mathbb{E}^\mu\left[F\middle|\hat{\mathfrak{J}}\right]$ (a.e., $\mu$). However, from (6.2.13), it is easy to see that $\Pi_{\hat{\mathfrak{J}}(E)}F$ is measurable with respect to the $\mu$-completion of $\hat{\mathfrak{J}}$, and so it suffices to show that

$$\mathbb{E}^\mu[F,\Gamma] = \mathbb{E}^\mu[\mathcal{A}_1 F,\Gamma] \quad \text{for all } \Gamma\in\hat{\mathfrak{J}}.$$

But, if $\Gamma \in \hat{\mathfrak{J}}$, then

$$\begin{aligned}\mathbb{E}^\mu[\mathcal{A}_1 F,\Gamma] &= \int_{[0,1)^N}\mathbb{E}^\mu\left[F\circ\boldsymbol{\Sigma}^{\mathbf{t}},\Gamma\right]d\mathbf{t}\\ &= \int_{[0,1)^N}\mathbb{E}^\mu\left[F\circ\boldsymbol{\Sigma}^{\mathbf{t}},\left(\boldsymbol{\Sigma}^{\mathbf{t}}\right)^{-1}(\Gamma)\right]d\mathbf{t} = \mathbb{E}^\mu[F,\Gamma].\end{aligned}$$

Finally, assume that $\left\{\boldsymbol{\Sigma}^{\mathbf{t}} : \mathbf{t}\in[0,\infty)^N\right\}$ is $\mu$-ergodic. When $\mu(\Omega)<\infty$, the asserted result follows immediately from the preceding; and when $\mu(\Omega)=\infty$, it follows from the fact that $\Pi_{\hat{\mathfrak{J}}(E)}F$ is measurable with respect to the $\mu$-completion of $\hat{\mathfrak{J}}$. □

## Exercises for § 6.2

EXERCISE 6.2.16. Given an irrational $\alpha \in (0,1)$ and an $\epsilon \in (0,1)$, let $N_n(\alpha,\epsilon)$ be the number of $1 \le m \le n$ with the property that

$$\left|\alpha - \frac{\ell}{m}\right| \le \frac{\epsilon}{2m} \quad \text{for some } \ell \in \mathbb{Z}.$$

As an application of the considerations in the Classic Example given at the end of § 6.1, show that

$$\varliminf_{n\to\infty} \frac{N_n(\alpha,\epsilon)}{n} \ge \epsilon.$$

**Hint**: Let $\delta \in \left(0,\frac{\epsilon}{2}\right)$ be given, take $f$ equal to the indicator function of $[0,\delta) \cup (1-\delta,1)$, and observe that $N_n(\alpha,\epsilon) \ge \sum_{k=1}^{n} f \circ \mathbf{\Sigma}_\alpha^k(\omega)$ so long as $0 \le \omega \le \frac{\epsilon}{2} - \delta$.

EXERCISE 6.2.17. Assume that $\mu(\Omega) < \infty$ and that $\left\{\mathbf{\Sigma}^{\mathbf{k}} : \mathbf{k} \in \mathbb{N}^N\right\}$ is ergodic. Given a non-negative $\mathcal{F}$-measurable function $f$, show that

$$\begin{aligned} \varlimsup_{n\to\infty} \mathbf{A}_n f < \infty \text{ on a set of positive } \mu\text{-measure} &\implies f \in L^1(\mu;\mathbb{R}) \\ &\implies \lim_{n\to\infty} \mathbf{A}_n f = \frac{\mathbb{E}^\mu[f]}{\mu(\Omega)} \quad (\text{a.e.}, \mu). \end{aligned}$$

EXERCISE 6.2.18. Let $\mathfrak{F} = \left\{X_{\mathbf{k}} : \mathbf{k} \in \mathbb{N}^N\right\}$ be a stationary family of random variables on the probability space $(\Omega,\mathcal{F},\mathbb{P})$ with values in the measurable space $(E,\mathcal{B})$, and let $\mathfrak{I}$ denote the $\sigma$-algebra of shift invariant $\Gamma \in \mathcal{B}_E^{\mathbb{N}^N}$.

**(i)** Take

$$\mathcal{T} \equiv \bigcap_{n\ge 0} \sigma\bigl(X_{\mathbf{k}} : k_j \ge n \text{ for all } 1 \le j \le N\bigr),$$

the tail $\sigma$-algebra determined by $\left\{X_{\mathbf{k}} : \mathbf{k} \in \mathbb{N}^N\right\}$. Show that $\mathfrak{F}^{-1}(\mathfrak{I}) \subseteq \mathcal{T}$, and conclude that $\left\{X_{\mathbf{k}} : \mathbf{k} \in \mathbb{N}^N\right\}$ is ergodic if $\mathcal{T}$ is $\mathbb{P}$-trivial (i.e., $\mathbb{P}(\Gamma) \in \{0,1\}$ for all $\Gamma \in \mathcal{T}$).

**(ii)** By combining **(i)**, Kolmogorov's 0–1 Law, and the Individual Ergodic Theorem, give another derivation of The Strong Law of Large Numbers for independent, identically distributed, integrable random variables with values in a separable Banach space.

EXERCISE 6.2.19. Let $\{X_k : k \in \mathbb{N}\}$ be a stationary, ergodic sequence of $\mathbb{R}$-valued, integrable random variables on $(\Omega,\mathcal{F},\mathbb{P})$. Using the reasoning suggested in Exercise 1.4.28, prove Guivarc'h's lemma:

$$\mathbb{E}^{\mathbb{P}}[X_1] = 0 \implies \varliminf_{n\to\infty} \left|\sum_{k=0}^{n-1} X_k\right| < \infty.$$

## § 6.3 Burkholder's Inequality

Given a martingale $\left(X_n, \mathcal{F}_n, \mathbb{P}\right)$ with $X_0 = 0$ and a sequence $\{\sigma_n : n \geq 0\}$ of bounded functions with the property that $\sigma_n$ is $\mathcal{F}_n$-measurable for $n \geq 0$, determine $\{Y_n : n \geq 0\}$ by $Y_0 = 0$ and $Y_n - Y_{n-1} = \sigma_{n-1}(X_n - X_{n-1})$ for $n \geq 1$. It is clear that $\left(Y_n, \mathcal{F}_n, \mathbb{P}\right)$ is again a martingale. In addition, if the absolute values of all the $\sigma_n$'s are bounded by some constant $\sigma < \infty$ and $X_n$ is square $\mathbb{P}$-integrable, then one can easily check that

$$\mathbb{E}^{\mathbb{P}}\left[Y_n^2\right] = \sum_{m=1}^{n} \mathbb{E}^{\mathbb{P}}\left[\sigma_n^2 (X_n - X_{n-1})^2\right] \leq \sigma^2 \sum_{m=1}^{n} \mathbb{E}^{\mathbb{P}}\left[(X_n - X_{n-1})^2\right] = \sigma^2 \mathbb{E}^{\mathbb{P}}\left[X_n^2\right].$$

On the other hand, it is not at all clear how to compare the size of $Y_n$ to that of $X_n$ in any of the $L^p$ spaces other than $p = 2$.

The problem of finding such a comparison was given a definitive solution by D. Burkholder, and I will present his solution in this section. Actually, Burkholder solved the problem twice. His first solution was a beautiful adaptation of general ideas and results that had been developed over the years to solve related problems in probability theory and analysis and, as such, did not yield the optimal solution. His second approach is designed specifically to address the problem at hand and bears little or no resemblance to familiar techniques. It is entirely original, remarkably elementary and effective, but somewhat opaque. The approach is the outgrowth of many years of deep thinking that Burkholder devoted to the topic, and the reader who wants to understand the path that led him to it should consult the explanation that he wrote.[1]

**§ 6.3.1. Burkholder's Comparison Theorem.** Burkholder's basic result is the following comparison theorem.

THEOREM 6.3.1 (**Burkholder**). *Let $(\Omega, \mathcal{F}, \mathbb{P})$ be a probability space, $\{\mathcal{F}_n : n \in \mathbb{N}\}$ a non-decreasing sequence of sub-$\sigma$-algebras of $\mathcal{F}$, and $E$ and $F$ a pair of (real or complex) separable Hilbert spaces. Next, suppose that $\left(X_n, \mathcal{F}_n, \mathbb{P}\right)$ and $\left(Y_n, \mathcal{F}_n, \mathbb{P}\right)$ are, respectively, $E$- and $F$-valued martingales. If*

$$\|Y_0\|_F \leq \|X_0\|_E \text{ and } \|Y_n - Y_{n-1}\|_F \leq \|X_n - X_{n-1}\|_E,\ n \in \mathbb{Z}^+,$$

*$\mathbb{P}$-almost surely, then, for each $p \in (1, \infty)$ and $n \in \mathbb{N}$,*

$$\left\|Y_n\right\|_{L^p(\mathbb{P};F)} \leq B_p \left\|X_n\right\|_{L^p(\mathbb{P};E)}, \quad \text{where } B_p \equiv (p-1) \vee \frac{1}{p-1}. \tag{6.3.2}$$

As I said before, the derivation of Theorem 6.3.1 is both elementary and mysterious. I begin with the trivial observation that, without loss in generality,

---

[1] For those who want to know the secret behind this proof, Burkholder has revealed it in his article "Explorations in martingale theory and its applications" for the 1989 Saint-Flour Ecole d'Eté lectures published by Springer-Verlag, LNM **1464** (1991).

I may assume that both $E$ and $F$ are complex Hilbert spaces, since we can always complexify them, and, in addition, that $E = F$, since, if that is not already the case, I can embed them in $E \oplus F$. Thus, I will be making these assumptions throughout.

The heart of the proof lies in the computations contained in the following two lemmas.

LEMMA 6.3.3. *Let $p \in (1, \infty)$ be given, set*

$$\alpha_p = \begin{cases} p^{2-p}\,(p-1)^{p-1} & \text{if } p \in [2,\infty) \\ p^{2-p} & \text{if } p \in (1,2], \end{cases}$$

*and define $u : E^2 \longrightarrow \mathbb{R}$ by (cf. (6.3.2))*

$$u(x,y) = \big(\|y\|_E - B_p\|x\|_E\big)\big(\|y\|_E + \|x\|_E\big)^{p-1}.$$

*Then*

$$\|y\|_E^p - \big(B_p\|x\|_E\big)^p \le \alpha_p\, u(x,y), \quad (x,y) \in E^2.$$

PROOF: When $p = 2$, there is nothing to do. Thus, I will assume that $p \in (1,\infty) \setminus \{2\}$.

Observe that it suffices to show that, for all $(x,y) \in E^2$ satisfying $\|x\|_E + \|y\|_E = 1$, depending on whether $p \in (2,\infty)$ or $p \in (1,2)$,

$$(*)\qquad \|y\|_E^p - \big((p-1)\|x\|_E\big)^p \begin{cases} \le p^{2-p}\,(p-1)^{p-1}\big(\|y\|_E - (p-1)\|x\|_E\big) \\ \ge p^{2-p}\,(p-1)^{p-1}\big(\|y\|_E - (p-1)\|x\|_E\big). \end{cases}$$

Indeed, when $p \in (2,\infty)$, (*) is precisely the result desired, and, when $p \in (1,2)$, (*) gives the desired result after one divides through by $(p-1)^p$ and reverses the roles of $x$ and $y$.

I begin the verification of (*) by checking that

$$(**)\qquad p^{2-p}\,(p-1)^{p-1} \begin{cases} > 1 & \text{if } p \in (2,\infty) \\ < 1 & \text{if } p \in (1,2). \end{cases}$$

To this end, set $f(p) = (p-1)\log(p-1) - (p-2)\log p$ for $p \in (1,\infty)$. Then $f$ is strictly convex on $(1,2)$ and strictly concave on $(2,\infty)$. Thus, $f \restriction (1,2)$ cannot achieve a maximum and, therefore, since $\lim_{p\searrow 1} f(p) = 0 = f(2)$, $f < 0$ on $(1,2)$. Similarly, $f \restriction (2,\infty)$ cannot achieve a minimum and, therefore, since $f(2) = 0$ while $\lim_{p\nearrow\infty} f(p) = \infty$, we have that $f > 0$ on $(2,\infty)$.

Next, observe that proving (*) comes down to checking that, for $s \in [0,1]$,

$$\Phi(s) \equiv p^{2-p}\,(p-1)^{p-1}\,(1-ps) - (1-s)^p + (p-1)^p\, s^p \begin{cases} \ge 0 & \text{if } p \in (2,\infty) \\ \le 0 & \text{if } p \in (1,2). \end{cases}$$

To this end, note that, by (**), $\Phi(0) > 0$ when $p \in (2,\infty)$ and $\Phi(0) < 0$ when $p \in (1,2)$. Also, for $s \in (0,1)$,

$$\Phi'(s) = p\Big[(p-1)^p s^{p-1} + (1-s)^{p-1} - p^{2-p}(p-1)^{p-1}\Big]$$

and

$$\Phi''(s) = p(p-1)\Big[(p-1)^p s^{p-2} - (1-s)^{p-2}\Big].$$

In particular, we see that $\Phi\big(\frac{1}{p}\big) = \Phi'\big(\frac{1}{p}\big) = 0$. In addition, depending on whether $p \in (2,\infty)$ or $p \in (1,2)$, $\lim_{s\searrow 0}\Phi''(s)$ is negative or positive, $\Phi''$ is strictly increasing or decreasing on $(0,1)$, and $\lim_{s\nearrow 1}\Phi''(1)$ is positive or negative. Hence, there exists a unique $t = t_p \in (0,1)$ with the property that

$$\Phi'' \restriction (0,t)\begin{cases} <0 & \text{if } p \in (2,\infty) \\ >0 & \text{if } p \in (1,2) \end{cases} \quad\text{and}\quad \Phi'' \restriction (t,1)\begin{cases} >0 & \text{if } p \in (2,\infty) \\ <0 & \text{if } p \in (1,2). \end{cases}$$

Moreover, because $\Phi''(t) = 0$, it is easy to see that $t \in \Big(0, \frac{1}{p}\Big)$.

Now suppose that $p \in (2,\infty)$ and consider $\Phi$ on each of the intervals $\big[\frac{1}{p}, 1\big]$, $\big[t, \frac{1}{p}\big]$, and $\big[0, t\big]$ separately. Because both $\Phi$ and $\Phi'$ vanish at $\frac{1}{p}$ while $\Phi'' > 0$ on $\big(\frac{1}{p}, 1\big)$, it is clear that $\Phi > 0$ on $\big(\frac{1}{p}, 1\big]$. Next, because $\Phi'\big(\frac{1}{p}\big) = 0$ and $\Phi'' \restriction \big(t, \frac{1}{p}\big) > 0$, we know that $\Phi$ is strictly decreasing on $\big(t, \frac{1}{p}\big)$ and therefore that $\Phi \restriction \big[t, \frac{1}{p}\big) > \Phi\big(\frac{1}{p}\big) = 0$. Finally, because $\Phi'' \restriction (0,t) < 0$ while $\Phi(0) \wedge \Phi(t) \geq 0$, we also know that $\Phi \restriction (0,t) > 0$. The argument when $p \in (1,2)$ is similar, only this time all the signs are reversed. □

Lemma 6.3.4. *Again let $p \in (1,\infty)$ be given, and define $u : E^2 \longrightarrow \mathbb{R}$ as in Lemma 6.3.3. In addition, define the functions $v$ and $w$ on $E^2 \setminus \{0,0\}$ by*

$$v(x,y) = p\big(\|y\|_E + \|x\|_E\big)^{p-2}\big(\|y\|_E + (2-p)\|x\|_E\big)$$

*and*

$$w(x,y) = p(1-p)\big(\|y\|_E + \|x\|_E\big)^{p-2}\|x\|_E.$$

*Then, for $(x,y) \in E^2$ and $(k,h) \in E^2$ satisfying*

$$\min_{t\in[0,1]}\big(\|y+th\|_E \wedge \|x+tk\|_E\big) > 0 \quad\text{and}\quad \|h\|_E \leq \|k\|_E,$$

*one has*

$$u(x+k, y+h) - u(x,y) \leq v(x,y)\,\Re\Big(\tfrac{y}{\|y\|_E}, h\Big)_E + w(x,y)\,\Re\Big(\tfrac{x}{\|x\|_E}, k\Big)_E$$

*when $p \in [2,\infty)$ and*

$$(p-1)\big[u(x+k,y+h) - u(x,y)\big] \leq -w(y,x)\,\Re\Big(\tfrac{y}{\|y\|_E}, h\Big)_E - v(y,x)\,\Re\Big(\tfrac{x}{\|x\|_E}, k\Big)_E$$

*when $p \in (1,2]$.*

PROOF: Set

$$\Phi(t) = \Phi\big(t;(x,k),(y,h)\big)$$
$$\equiv \big(\|y+th\|_E - (p-1)\|x+tk\|_E\big)\big(\|x+tk\|_E + \|y+th\|_E\big)^{p-1},$$

and observe that

$$u(x+tk, y+th) = \begin{cases} \Phi\big(t;(x,k),(y,h)\big) & \text{if} \quad p \in [2,\infty) \\ -(p-1)^{-1}\,\Phi\big(t;(y,h),(x,k)\big) & \text{if} \quad p \in (1,2). \end{cases}$$

Hence, it suffices for us to check that

$$\Phi'(t) = v(x+tk, y+th)\mathfrak{Re}\left(\tfrac{y+th}{\|y+th\|_E}, h\right)_E + w(x+tk, y+th)\mathfrak{Re}\left(\tfrac{x+tk}{\|x+tk\|_E}, k\right)_E$$

and prove that

$$\Phi''\big(t;(x,k),(y,h)\big)\begin{cases} \le 0 & \text{if } p \in [2,\infty) \text{ and } \|h\|_E \le \|k\|_E \\ \ge 0 & \text{if } p \in (1,2] \text{ and } \|h\|_E \ge \|k\|_E. \end{cases}$$

To prove the preceding, set $x(t) = x + tk$, $y(t) = y + th$, $\Psi(t) \doteq \|x(t)\|_E + \|y(t)\|_E$, $a(t) = \frac{\mathfrak{Re}\left(x(t),k\right)_E}{\|x(t)\|_E}$, and $b(t) = \frac{\mathfrak{Re}\left(y(t),h\right)_E}{\|y(t)\|_E}$. One then has that

$$\Phi'(t) = p\Psi(t)^{p-2}\left[(1-p)\|x(t)\|_E\, a(t) + \Big(\|y(t)\|_E + (2-p)\|x(t)\|_E\Big)\, b(t)\right]$$
$$= p\Big[(1-p)\Psi(t)^{p-2}\,\|x(t)\|_E\,\big(a(t)+b(t)\big) + \Psi(t)^{p-1}\, b(t)\Big].$$

In particular, the first expression establishes the required form for $\Phi'(t)$. In addition, from the second expression, we see that

$$-\frac{\Phi''(t)}{p} = (p-1)(p-2)\,\Psi(t)^{p-3}\,\|x(t)\|_E\,\big(a(t)+b(t)\big)^2$$
$$+ (p-1)\Psi(t)^{p-2}\left[a(t)\big(a(t)+b(t)\big) + \tfrac{\|x(t)\|_E}{\|y(t)\|_E}\, b_\perp(t)^2 + a_\perp(t)^2\right]$$
$$- \Psi(t)^{p-2}\left[(p-1)\big(a(t)+b(t)\big)\, b(t) + \Psi(t)\, \tfrac{b_\perp(t)^2}{\|y(t)\|_E}\right]$$
$$= (p-1)(p-2)\,\Psi(t)^{p-3}\,\|x(t)\|_E\,\big(a(t)+b(t)\big)^2$$
$$+ (p-1)\Psi(t)^{p-2}\big(\|k\|_E^2 - \|h\|_E^2\big) + (p-2)\Psi(t)^{p-1}\,\tfrac{b_\perp(t)^2}{\|y(t)\|_E},$$

where $a_\perp(t) = \sqrt{\|k\|_E^2 - a(t)^2}$ and $b_\perp(t) = \sqrt{\|h\|_E^2 - b(t)^2}$. Hence the required properties of $\Phi''(t)$ have also been established. □

PROOF OF THEOREM 6.3.1: Set $K_n = X_n - X_{n-1}$ and $H_n = Y_n - Y_{n-1}$ for $n \in \mathbb{Z}^+$. I will assume that there is an $\epsilon > 0$ with the property that

$$\left\|X_0(\omega) - \operatorname{span}\{K_n(\omega) : n \in \mathbb{Z}^+\}\right\|_E \geq \epsilon$$

and

$$\left\|Y_0(\omega) - \operatorname{span}\{H_n(\omega) : n \in \mathbb{Z}^+\}\right\|_E \geq \epsilon$$

for all $\omega \in \Omega$. Indeed, if this is not already the case, then I can replace $E$ by $\mathbb{R} \times E$ (or, when $E$ is complex, $\mathbb{C} \times E$) and $X_n(\omega)$ and $Y_n(\omega)$, respectively, by

$$X_n^{(\epsilon)}(\omega) \equiv \big(\epsilon, X_n(\omega)\big) \quad \text{and} \quad Y_n^{(\epsilon)}(\omega) \equiv \big(\epsilon, Y_n(\omega)\big),$$

for each $n \in \mathbb{N}$. Clearly, (6.3.2) for each $X_n^{(\epsilon)}$ and $Y_n^{(\epsilon)}$ implies (6.3.2) for $X_n$ and $Y_n$ after one lets $\epsilon \searrow 0$. Finally, because there is nothing to do when the right-hand side of (6.3.2) is infinite, let $p \in (1, \infty)$ be given, and assume that $X_n \in L^p(\mathbb{P}; E)$ for each $n \in \mathbb{N}$. In particular, if $u$ is the function defined in Lemma 6.3.3 and $v$ and $w$ are those defined in Lemma 6.3.4, then

$$u(X_n, Y_n) \in L^1(\mathbb{P}; \mathbb{R}) \quad \text{and} \quad v(X_n, Y_n),\, w(X_n, Y_n) \in L^{p'}(\mathbb{P}; \mathbb{R})$$

for all $n \in \mathbb{N}$, where $p' = \frac{p}{p-1}$ is the Hölder conjugate of $p$.

Note that, by Lemma 6.3.3, it suffices for us to show that $A_n \equiv \mathbb{E}^{\mathbb{P}}\big[u(X_n, Y_n)\big] \leq 0$, $n \in \mathbb{N}$. Since $u(X_0, Y_0) \leq 0$ $\mathbb{P}$-almost surely, there is no question that $A_0 \leq 0$. Next, assume that $A_n \leq 0$, and, depending on whether $p \in [2, \infty)$ or $p \in (1, 2]$, use the appropriate part of Lemma 6.3.4 to see that

$$\begin{aligned} A_{n+1} \leq & \mathbb{E}^{\mathbb{P}}\left[v(X_n, Y_n)\mathfrak{Re}\left(\tfrac{Y_n}{\|Y_n\|_E}, H_{n+1}\right)_E\right] \\ & + \mathbb{E}^{\mathbb{P}}\left[w(X_n, Y_n)\mathfrak{Re}\left(\tfrac{X_n}{\|X_n\|_E}, K_{n+1}\right)_E\right] \end{aligned}$$

or

$$\begin{aligned} A_{n+1} \leq & - \mathbb{E}^{\mathbb{P}}\left[w(Y_n, X_n)\mathfrak{Re}\left(\tfrac{Y_n}{\|Y_n\|_E}, H_{n+1}\right)_E\right] \\ & - \mathbb{E}^{\mathbb{P}}\left[v(Y_n, X_n)\mathfrak{Re}\left(\tfrac{X_n}{\|X_n\|_E}, K_{n+1}\right)_E\right]. \end{aligned}$$

But, since $v(X_n, Y_n)\frac{Y_n}{\|Y_n\|_E}$ is $\mathcal{F}_n$-measurable, $\mathbb{E}^{\mathbb{P}}[H_{n+1}|\mathcal{F}_n] = 0$, and therefore (cf. Exercise 5.1.18)

$$\mathbb{E}^{\mathbb{P}}\left[v(X_n, Y_n)\mathfrak{Re}\left(\tfrac{Y_n}{\|Y_n\|_E}, H_{n+1}\right)_E\right] = 0.$$

Since the same reasoning shows that each of the other terms on the right-hand side vanishes, we have now proved that $A_{n+1} \leq 0$. □

As an immediate consequence of Theorem (6.3.2), we have the following answer to the question raised at the beginning of this section.

COROLLARY 6.3.5. *Suppose that $(X_n, \mathcal{F}_n, \mathbb{P})$ is a martingale with values in a separable (real or complex) Hilbert space $E$. Further, let $F$ be a second separable, Hilbert space, and suppose that $\{\boldsymbol{\sigma}_n : n \geq 0\}$ is a sequence of $\mathrm{Hom}(E; F)$-valued random variables with the properties that $\boldsymbol{\sigma}_0$ is constant, $\boldsymbol{\sigma}_n$ is $\mathcal{F}_n$-measurable for $n \geq 1$, and $\|\boldsymbol{\sigma}_n\|_{\mathrm{op}} \leq \sigma < \infty$ (a.s., $\mathbb{P}$) for some constant $\sigma < \infty$ and all $n \in \mathbb{N}$. If $\|Y_0\|_F \leq \sigma \|X_0\|_E$ and $Y_n - Y_{n-1} = \boldsymbol{\sigma}_{n-1}(X_n - X_{n-1})$ for $n \geq 1$, then $(Y_n, \mathcal{F}_n, \mathbb{P})$ is an $F$-valued martingale and, for each $p \in (1, \infty)$, (cf. (6.3.2))*

$$\|Y_n\|_{L^p(\mathbb{P};F)} \leq \sigma B_p \|X_n\|_{L^p(\mathbb{P};E)}, \quad n \in \mathbb{N}.$$

§ **6.3.2. Burkholder's Inequality.** In many applications, the most useful form of Burkholder's result is as a generalization to $p \neq 2$ of the obvious equality

$$\mathbb{E}^{\mathbb{P}}\left[|X_n - X_0|^2\right] = \mathbb{E}^{\mathbb{P}}\left[\sum_{m=1}^{n} |X_m - X_{m-1}|^2\right].$$

This is the form of his inequality which is best known and, as such, is called **Burkholder's Inequality**. Notice that his inequality can be viewed as a vast generalization of Khinchine's Inequality (2.3.27), although it applies only when $p \in (1, \infty)$.

THEOREM 6.3.6 (**Burkholder's Inequality**). *Let $(\Omega, \mathcal{F}, \mathbb{P})$ and $\{\mathcal{F}_n : n \in \mathbb{N}\}$ be as in Theorem 6.3.1, and let $(X_n, \mathcal{F}_n, \mathbb{P})$ be a martingale with values in the separable Hilbert space $E$. Then, for each $p \in (1, \infty)$,*

$$\begin{aligned}
&\frac{1}{B_p} \sup_{n \in \mathbb{N}} \left\|X_n - X_0\right\|_{L^p(\mathbb{P};E)} \\
&\qquad \leq \mathbb{E}^{\mathbb{P}}\left[\left(\sum_1^\infty \left\|X_n - X_{n-1}\right\|_E^2\right)^{\frac{p}{2}}\right]^{\frac{1}{p}} \qquad (6.3.7)\\
&\qquad \leq B_p \sup_{n \in \mathbb{N}} \left\|X_n - X_0\right\|_{L^p(\mathbb{P};E)},
\end{aligned}$$

*with $B_p$ as in (6.3.2).*

PROOF: Let $F = \ell^2(\mathbb{N}; E)$ be the separable Hilbert space of sequences

$$y = (x_0, \ldots, x_n, \ldots) \in E^{\mathbb{N}}$$

satisfying

$$\|y\|_F \equiv \left(\sum_0^\infty \|x_n\|_E^2\right)^{\frac{1}{2}} < \infty,$$

and define

$$Y_n(\omega) = (X_0(\omega), X_1(\omega) - X_0(\omega), \dots, X_n(\omega) - X_{n-1}(\omega), 0, 0, \dots) \in F$$

for $\omega \in \Omega$ and $n \in \mathbb{N}$. Obviously, $(Y_n, \mathcal{F}_n, \mathbb{P})$ is an $F$-valued martingale. Moreover,

$$\|X_0\|_E = \|Y_0\|_F \quad \text{and} \quad \|X_n - X_{n-1}\|_E = \|Y_n - Y_{n-1}\|_F, \quad n \in \mathbb{N},$$

and therefore the right-hand side of (6.3.7) is implied by (6.3.2) while the left-hand side also follows from (6.3.2) when the roles of the $X_n$'s and $Y_n$'s are reversed. □

## Exercises for § 6.3

EXERCISE 6.3.8. Because it arises repeatedly in the theory of stochastic integration, one of the most frequent applications of Burkholder's Inequality is to situations in which $E$ is a separable Hilbert space and $(X_n, \mathcal{F}_n, \mathbb{P})$ is an $E$-valued martingale for which one has an estimate of the form

$$K_p \equiv \sup_{n \in \mathbb{Z}^+} \left\| \mathbb{E}^{\mathbb{P}}\left[ \|X_n - X_{n-1}\|_E^{2p} \,\middle|\, \mathcal{F}_{n-1} \right]^{\frac{1}{2p}} \right\|_{L^\infty(\mathbb{P};\mathbb{R})} < \infty$$

for some $p \in [1, \infty)$. To see how such an estimate gets used, let $F$ be a second separable Hilbert space and suppose that $\{\boldsymbol{\sigma}_n : n \in \mathbb{N}\}$ is a sequence of $\mathrm{Hom}(E; F)$-valued random variables with the properties that, for each $n \in \mathbb{N}$, $\boldsymbol{\sigma}_n$ is $\mathcal{F}_n$-measurable and $a_n \equiv \mathbb{E}^{\mathbb{P}}\left[\|\boldsymbol{\sigma}_n\|_{\mathrm{op}}^{2p}\right]^{\frac{1}{2p}} < \infty$. Set $Y_0 = 0$ and

$$Y_n = \sum_{m=1}^{n} \boldsymbol{\sigma}_{m-1}(X_m - X_{m-1}) \quad \text{for } n \in \mathbb{Z}^+,$$

and show that

$$\|Y_n\|_{L^{2p}(\mathbb{P};F)} \le (2p-1) n^{\frac{1}{2}} K_p \left( \frac{1}{n} \sum_{m=0}^{n-1} a_m^{2p} \right)^{\frac{1}{2p}}.$$

EXERCISE 6.3.9. Return to the setting in Exercise 5.2.45, and let $\lambda_{[0,1)}$ denote Lebesgue measure on $[0, 1)$. Given $f \in L^2(\lambda_{[0,1)}; \mathbb{C})$, show that, for each $p \in (1, \infty)$,

$$\begin{aligned}
(p-1) \wedge \frac{1}{p-1} & \left\| f - (f, \mathbf{1})_{L^2(\lambda_{[0,1)};\mathbb{C})} \right\|_{L^p([0,1);\mathbb{C})} \\
& \le \left( \int_{[0,1)} \left( \sum_{m=0}^{\infty} |\Delta_m(f)|^2 \right)^{\frac{p}{2}} dt \right)^{\frac{1}{p}} \\
& \le (p-1) \vee \frac{1}{p-1} \left\| f - (f, \mathbf{1})_{L^2(\lambda_{[0,1)};\mathbb{C})} \right\|_{L^p(\lambda_{[0,1)};\mathbb{C})}.
\end{aligned}$$

For functions $f$ with $(f, e_\ell)_{L^2(\lambda_{[0,1)};\mathbb{C})} = 0$ unless $\ell = \pm 2^m$ for some $m \in \mathbb{N}$, this estimate is a case of a famous theorem proved by Littlewood and Paley in order to generalize Parseval's Identity to cover $p \neq 2$. Unfortunately, the argument here is far too weak to give their inequality for general $f$'s.

EXERCISE 6.3.10. In connection with the preceding exercise, it is interesting to note that there is an orthonormal basis for $L^2(\lambda_{[0,1)};\mathbb{R})$ that, as distinguished from the trigonometric functions, can be nearly completely understood in terms of martingale analysis. Namely, recall the Rademacher functions $\{R_n : n \in \mathbb{Z}^+\}$ introduced in § 1.1.2. Next, use $\mathfrak{F}$ to denote the set of all finite subsets $F$ of $\mathbb{Z}^+$, and define the **Walsh function** $W_F$ for $F \in \mathfrak{F}$ by

$$W_F = \begin{cases} \mathbf{1} & \text{if} \quad F = \emptyset \\ \prod_{m\in F} R_m & \text{if} \quad F \neq \emptyset. \end{cases}$$

Finally, set $A_0 = \emptyset$ and $A_n = \{1, \dots, n\}$ for $n \in \mathbb{Z}^+$.

**(i)** For each $n \in \mathbb{N}$, let $\mathcal{F}_n$ be the $\sigma$-algebra generated by the partition

$$\left\{\left[\tfrac{k}{2^n}, \tfrac{k+1}{2^n}\right) : 0 \le k < 2^n\right\}.$$

Show that, for each $n \in \mathbb{Z}^+$, $\{W_F : F \subseteq A_n\}$ is an orthonormal basis for the subspace $L^2([0,1), \mathcal{F}_n, \lambda_{[0,1)};\mathbb{R})$, and conclude from this that $\{W_F : F \in \mathfrak{F}\}$ forms an orthonormal basis for $L^2(\lambda_{[0,1)};\mathbb{R})$.

**(ii)** Let $f \in L^1([0,1);\mathbb{R})$ be given, and set

$$X_n^f = \sum_{F\subseteq A_n} \left( \int_{[0,1)} f(t)\, W_F(t)\, dt \right) W_F \quad \text{for } n \in \mathbb{N}.$$

Using the result in **(i)**, show that $X_n^f = \mathbb{E}^{\lambda_{[0,1)}}[f|\mathcal{F}_n]$ and therefore that $(X_n^f, \mathcal{F}_n, \lambda_{[0,1)})$ is a martingale. In particular, $X_n^f \longrightarrow f$ both (a.e., $\lambda_{[0,1)}$) as well as in $L^1(\lambda_{[0,1)};\mathbb{R})$.

**(iii)** Show that for each $p \in (1,\infty)$ and $f \in L^1(\lambda_{[0,1)};\mathbb{R})$ with mean value 0,

$$\begin{aligned}
&(p-1)\wedge(p-1)^{-1}\|f\|_{L^p([0,1);\mathbb{R})} \\
&\le \left[ \int_{[0,1)} \left( \sum_{n=1}^{\infty} \left[ \sum_{F\subseteq A_n\setminus A_{n-1}} \left( \int_{[0,1)} f(s)\, W_F(s)\, ds \right) W_F(t) \right]^2 \right)^{\frac{p}{2}} dt \right]^{\frac{1}{p}} \\
&\le (p-1)\vee(p-1)^{-1}\|f\|_{L^p([0,1);\mathbb{R})}.
\end{aligned}$$

EXERCISE 6.3.11. Although Burkholder's Inequality is extremely useful, it does not give particularly good estimates in the case of martingales with bounded increments. For such martingales, the following line of reasoning, which was introduced by K. Azuma and W. Hoeffding, is useful.

**(i)** For any $a \in \mathbb{R}$ and $x \in [-1,1]$, show that

$$e^{ax} \le \frac{1+x}{2}e^{a} + \frac{1-x}{2}e^{-a} = \cosh a + x \sinh a.$$

**(ii)** Suppose that $\{Y_1, \dots, Y_n\}$ are $[-1,1]$-valued random variables on the probability space $(\Omega, \mathcal{F}, \mathbb{P})$ with the property that, for each $1 \le m \le n$,

$$\mathbb{E}^{\mathbb{P}}\big[Y_{j_1}\cdots Y_{j_m}\big] = 0 \quad \text{for all } 1 \le j_1 < \cdots < j_m \le n.$$

Show that, for any $\{a_j\}_1^n \subseteq \mathbb{R}$,

$$\mathbb{E}^{\mathbb{P}}\left[\exp\left(\sum_{j=1}^{n} a_j Y_j\right)\right] \le \prod_{j=1}^{n} \cosh a_j \le \exp\left(\frac{1}{2}\sum_{j=1}^{n} a_j^2\right),$$

and conclude that

$$P\left(\sum_{j=1}^{n} a_j Y_j \ge R\right) \le \exp\left(-\frac{R^2}{2\sum_{j=1}^{n} a_j^2}\right), \quad R \in [0,\infty).$$

**(iii)** Suppose that $(X_n, \mathcal{F}_n, \mathbb{P})$ is a bounded martingale with $X_0 \equiv 0$, and set $D_n \equiv \|X_n - X_{n-1}\|_{L^\infty(\mathbb{P})}$. Show that

$$P(X_n \ge R) \le \exp\left(-\frac{R^2}{2\sum_{j=1}^{n} D_j^2}\right), \quad R \in [0,\infty).$$

# Chapter 7
## Continuous Parameter Martingales

It turns out that many of the ideas and results introduced in § 5.2 can be easily transferred to the setting of processes depending on a *continuous parameter*. In addition, the resulting theory is intimately connected with Lévy processes, and particularly Brownian. In this chapter, I will give a brief introduction to this topic and some of the techniques to which it leads.[1]

### § 7.1 Continuous Parameter Martingales

There is a huge number of annoying technicalities which have to be addressed in order to give a mathematically correct description of the continuous time theory of martingales. Fortunately, for the applications which I will give here, I can keep them to a minimum.

**§ 7.1.1. Progressively Measurable Functions.** Let $(\Omega, \mathcal{F})$ be a measurable space and $\{\mathcal{F}_t : t \in [0,\infty)\}$ a non-decreasing family of sub-$\sigma$-algebras. I will say that a function $X$ on $[0,\infty)\times\Omega$ into a measurable space $(E, \mathcal{B})$ is **progressively measurable** with respect to $\{\mathcal{F}_t : t \in [0,\infty)\}$ if $X \upharpoonright [0,T]\times\Omega$ is $\mathcal{B}_{[0,T]} \times \mathcal{F}_T$-measurable for every $T \in [0,\infty)$. When $E$ is a metric space, I will say that $X : [0,\infty)\times\Omega \longrightarrow E$ is **right-continuous** if $X(s,\omega) = \lim_{t\searrow s} X(t,\omega)$ for every $(s,\omega) \in [0,\infty)\times\Omega$ and will say that it is **continuous** if $X(\,\cdot\,,\omega)$ is continuous for all $\omega \in \Omega$.

REMARK 7.1.1. The reader might have been expecting a slightly different definition of progressive measurability here. Namely, he might have thought that one would say that $X$ is $\{\mathcal{F}_t : t \in [0,\infty)\}$-progressively measurable if it is $\mathcal{B}_{[0,\infty)} \times \mathcal{F}$-measurable and $\omega \in \Omega \longmapsto X(t,\omega) \in E$ is $\mathcal{F}_t$-measurable for each $t \in [0,\infty)$. Indeed, in extrapolating from the discrete parameter setting, this would be the first definition at which one would arrive. In fact, it was the notion with which Doob and Itô originally worked; and such functions were said by them to be **adapted** to $\{\mathcal{F}_t : t \in [0,\infty)\}$. However, it came to be realized that there are various problems with the notion of adaptedness. For example, even if $X$ is adapted and $f : E \longrightarrow \mathbb{R}$ is a bounded, $\mathcal{B}$-measurable function, the

[1] A far more thorough treatment can be found in D. Revuz and M. Yor's treatise *Continuous Martingales and Brownian Motion*, Springer-Verlag, Grundlehren der Mathematishen **#293** (1999).

function $(t,\omega) \rightsquigarrow Y(t,\omega) \equiv \int_0^t f\big(X(s,\omega)\big)\,ds \in \mathbb{R}$ need not be adapted. On the other hand, if $X$ is progressively measurable, then $Y$ will be also.

The following simple lemma should help to explain the virtue of progressive measurability and its relationship to adaptedness.

LEMMA 7.1.2. *Let $\mathcal{PM}$ denote the set of $A \subseteq [0,\infty) \times \Omega$ with the property that $\big([0,t]\times\Omega\big) \cap A \in \mathcal{B}_{[0,t]} \times \mathcal{F}_t$ for every $t \geq 0$. Then $\mathcal{PM}$ is a sub-$\sigma$-algebra of $\mathcal{B}_{[0,\infty)} \times \mathcal{F}$ and $X$ is progressively measurable if and only if it is $\mathcal{PM}$-measurable. Furthermore, if $E$ is a separable metric space and $X : [0,\infty) \times \Omega \longrightarrow E$ is a right-continuous function, then $X$ is progressively measurable if it is adapted.*

PROOF: Checking that $\mathcal{PM}$ is a $\sigma$-algebra is easy. Furthermore, for any $X : [0,\infty) \times \Omega \longrightarrow E$, $T \in [0,\infty)$, and $\Gamma \in \mathcal{B}$,

$$\begin{aligned}&\big\{(t,\omega) \in [0,T]\times\Omega : X(t,\omega) \in \Gamma\big\}\\ &\qquad = \big([0,T]\times\Omega\big) \cap \big\{(t,\omega) \in [0,\infty)\times\Omega : X(t,\omega) \in \Gamma\big\},\end{aligned}$$

and so $X$ is $\big\{\mathcal{F}_t : t \in [0,\infty)\big\}$-progressively measurable if and only if it is $\mathcal{PM}$-measurable. Hence, the first assertion has been proved.

Next, suppose that $X$ is a right-continuous, adapted function. To see that $X$ is progressively measurable, let $t \in [0,\infty)$ be given, and define

$$X_n^t(\tau,\omega) = X\left(\tfrac{[2^n\tau]+1}{2^n} \wedge t, \omega\right), \quad \text{for } (\tau,\omega) \in [0,\infty)\times\Omega \text{ and } n \in \mathbb{N}.$$

Obviously, $X_n^t$ is $\mathcal{B}_{[0,t]} \times \mathcal{F}_t$-measurable for every $n \in \mathbb{N}$ and $X_n^t(\tau,\omega) \longrightarrow X(\tau,\omega)$ as $n \to \infty$ for every $(\tau,\omega) \in [0,t]\times\Omega$. Hence, $X \upharpoonright [0,t]\times\Omega$ is $\mathcal{B}_{[0,t]} \times \mathcal{F}_t$-measurable, and so $X$ is progressively measurable. □

§ **7.1.2. Martingales: Definition and Examples.** Given a probability space $(\Omega,\mathcal{F},P)$ and a non-decreasing family of sub-$\sigma$-algebras $\big\{\mathcal{F}_t : t \in [0,\infty)\big\}$, I will say that $X : [0,\infty)\times\Omega \longrightarrow (-\infty,\infty]$ is a **submartingale with respect to** $\big\{\mathcal{F}_t : t \in [0,\infty)\big\}$ or, equivalently, that $\big(X(t),\mathcal{F}_t,\mathbb{P}\big)$ is a **submartingale** if $X$ is a right-continuous, progressively measurable function with the properties that $X(t)^-$ is $\mathbb{P}$-integrable for every $t \in [0,\infty)$ and

$$X(s) \leq \mathbb{E}^{\mathbb{P}}\big[X(t)\big|\mathcal{F}_s\big] \quad (\text{a.s.}, \mathbb{P}) \quad \text{for all} \quad 0 \leq s \leq t < \infty.$$

When both $\big(X(t),\mathcal{F}_t,\mathbb{P}\big)$ and $\big(-X(t),\mathcal{F}_t,\mathbb{P}\big)$ are submartingales, I will say either that $X$ is a **martingale with respect to** $\big\{\mathcal{F}_t : t \in [0,\infty)\big\}$ or simply that $\big(X(t),\mathcal{F}_t,\mathbb{P}\big)$ is a **martingale**. Finally, if $Z : [0,\infty)\times\Omega \longrightarrow \mathbb{C}$ is a right-continuous, progressively measurable function, then $\big(Z(t),\mathcal{F}_t,\mathbb{P}\big)$ is said to be a **(complex) martingale** if both $\big(\mathfrak{Re}\,Z(t),\mathcal{F}_t,\mathbb{P}\big)$ and $\big(\mathfrak{Im}\,Z(t),\mathcal{F}_t,\mathbb{P}\big)$ are.

The next two results show that Lévy processes provide a rich source of continuous parameter martingales.

THEOREM 7.1.3. *Let $\mu \in \mathcal{I}(\mathbb{R}^N)$ with $\hat{\mu}(\boldsymbol{\xi}) = e^{\ell_\mu(\boldsymbol{\xi})}$, where $\ell_\mu(\boldsymbol{\xi})$ equals*

$$\sqrt{-1}\,(\boldsymbol{\xi},\mathbf{m})_{\mathbb{R}^N} - \frac{1}{2}(\boldsymbol{\xi},\mathbf{C}\boldsymbol{\xi})_{\mathbb{R}^N} + \int_{\mathbb{R}^N}\Big(e^{\sqrt{-1}(\boldsymbol{\xi},\mathbf{y})_{\mathbb{R}^N}} - 1 - \mathbf{1}_{[0,1]}(|\mathbf{y}|)(\boldsymbol{\xi},\mathbf{y})_{\mathbb{R}^N}\Big)\,M(d\mathbf{y}).$$

*If $(\Omega,\mathcal{F},\mathbb{P})$ is a probability space and $\mathbf{Z} : [0,\infty)\times\Omega \longrightarrow \mathbb{R}^N$ is a $\mathcal{B}_{[0,\infty)}\times\mathcal{F}$-measurable map with the properties that $\mathbf{Z}(0,\omega) = \mathbf{0}$ and $\mathbf{Z}(\,\cdot\,,\omega) \in D(\mathbb{R}^N)$ for every $\omega \in \Omega$, then $\{\mathbf{Z}(t) : t \geq 0\}$ is a Lévy process for $\mu$ if and only if, for each $\boldsymbol{\xi} \in \mathbb{R}^N$,*

$$\Big(\exp\big(\sqrt{-1}(\boldsymbol{\xi},\mathbf{Z}(t))_{\mathbb{R}^N} - t\ell_\mu(\boldsymbol{\xi})\big), \mathcal{F}_t, \mathbb{P}\Big) \text{ is a martingale,} \tag{7.1.4}$$

*where $\mathcal{F}_t = \sigma\big(\{\mathbf{Z}(\tau) : \tau \in [0,t]\}\big)$.*

PROOF: If $\{\mathbf{Z}(t) : t \geq 0\}$ is a Lévy process for $\mu$, then, because $\mathbf{Z}(t) - \mathbf{Z}(s)$ is independent of $\mathcal{F}_s$ and has characteristic function $e^{(t-s)\ell_\mu(\boldsymbol{\xi})}$,

$$\begin{aligned}
&\mathbb{E}^{\mathbb{P}}\Big[\exp\big[\sqrt{-1}(\boldsymbol{\xi},\mathbf{Z}(t))_{\mathbb{R}^N} - t\ell_\mu(\boldsymbol{\xi})\big]\,\Big|\,\mathcal{F}_s\Big] \\
&\quad = \exp\big[\sqrt{-1}(\boldsymbol{\xi},\mathbf{Z}(s))_{\mathbb{R}^N} - s\ell_\mu(\boldsymbol{\xi})\big]\, e^{(s-t)\ell_\mu(\boldsymbol{\xi})}\mathbb{E}^{\mathbb{P}}\Big[e^{\sqrt{-1}(\boldsymbol{\xi},\mathbf{Z}(t)-\mathbf{Z}(s))_{\mathbb{R}^N}}\Big] \\
&\quad = \exp\big[\sqrt{-1}(\boldsymbol{\xi},\mathbf{Z}(s))_{\mathbb{R}^N} - s\ell_\mu(\boldsymbol{\xi})\big].
\end{aligned}$$

To prove the converse assertion, observe that the defining distributional property of a Lévy process for $\mu$ can be summarized as the statement that $\mathbf{Z}(0,\omega) = 0$ and, for each $0 \leq s < t$, $\mathbf{Z}(t) - \mathbf{Z}(s)$ is independent of $\sigma\big(\{\mathbf{Z}(\tau) : \tau \in [0,t]\}\big)$ and has distribution $\mu_{t-s}$, where $\widehat{\mu_\tau} = e^{\tau\ell_\mu}$. Hence, since (7.1.4) implies that

$$\mathbb{E}^{\mathbb{P}}\Big[\exp\Big(\sqrt{-1}(\boldsymbol{\xi},\mathbf{Z}(t)-\mathbf{Z}(s))_{\mathbb{R}^N}\Big)\,\Big|\,\mathcal{F}_s\Big] = e^{(t-s)\ell_\mu(\boldsymbol{\xi})}, \quad \boldsymbol{\xi} \in \mathbb{R}^N,$$

there is nothing more to do. □

Another, and often more useful, way to capture the same result is to introduce the **Lévy operator**

$$\begin{aligned}
\mathcal{L}^\mu\varphi(x) &= \tfrac{1}{2}\mathrm{Trace}\big(\mathbf{C}\nabla^2\varphi(\mathbf{x})\big) + \big(\mathbf{m},\nabla\varphi(\mathbf{x})\big)_{\mathbb{R}^N} \\
&\quad + \int_{\mathbb{R}^N}\Big[\varphi(\mathbf{x}+\mathbf{y}) - \varphi(\mathbf{x}) - \mathbf{1}_{[0,1]}(|\mathbf{y}|)\big(\mathbf{y},\nabla\varphi(\mathbf{x})\big)_{\mathbb{R}^N}\Big]\,M(d\mathbf{y})
\end{aligned} \tag{7.1.5}$$

for $\varphi \in C^2_{\mathrm{b}}(\mathbb{R}^N;\mathbb{C})$.

THEOREM 7.1.6. *Assume that $\mu \in \mathcal{I}(\mathbb{R}^N)$ and that $\{\mathbf{Z}(t) : t \geq 0\}$ is a Lévy process for $\mu$. Then, for every $F \in C^{1,2}_{\mathrm{b}}([0,\infty)\times\mathbb{R}^N;\mathbb{C})$,*

$$\left(F(t,\mathbf{Z}(t)) - \int_0^t (\partial_\tau + \mathcal{L}_\mu)F(\tau,\mathbf{Z}(\tau))\,d\tau, \mathcal{F}_t, \mathbb{P}\right)$$

*is a martingale, where* $\mathcal{F}_t = \sigma(\{\mathbf{Z}(\tau) : \tau \in [0,t]\})$ *and* $\mathcal{L}^\mu$ *is the operator described in (7.1.5). Conversely, if* $\mathbf{Z}$ *is a progressively measurable function satisfying* $\mathbf{Z}(0,\omega) = \mathbf{0}$ *and* $\mathbf{Z}(\cdot,\omega) \in D(\mathbb{R}^N)$ *for each* $\omega \in \Omega$*, and if*

$$\left(\varphi(\mathbf{Z}(t)) - \int_0^t \mathcal{L}^\mu\varphi(\mathbf{Z}(\tau))\,d\tau, \mathcal{F}_t, \mathbb{P}\right)$$

*is a martingale for each* $\varphi \in C_c^\infty(\mathbb{R}^N;\mathbb{R})$*, then* $\{\mathbf{Z}(t) : t \geq 0\}$ *is a Lévy process for* $\mu$*.*

PROOF: Begin by noting that it suffices to handle the case when $F$ is the restriction to $[0,\infty) \times \mathbb{R}^N$ of an element of the Schwartz test function space $\mathscr{S}(\mathbb{R} \times \mathbb{R}^N;\mathbb{C})$. Indeed, because $\|\mathcal{L}^\mu\varphi\|_u \leq C\|\varphi\|_{C_b^2(\mathbb{R}^N;\mathbb{C})}$ for some $C < \infty$, the result for $F \in C_b^{1,2}([0,\infty) \times \mathbb{R}^N;\mathbb{C})$ follows, via an obvious approximation procedure, from the result for $F \in \mathscr{S}(\mathbb{R} \times \mathbb{R}^N;\mathbb{C})$. Next observe that it suffices to treat $F \in \mathscr{S}(\mathbb{R}^N;\mathbb{C})$. To see this, simply interpret the process $t \in [0,\infty) \longmapsto (t, \mathbf{Z}_\mu(t)) \in \mathbb{R}^{N+1}$ as a Lévy process for $\delta_1 \times \mu$.

Now let $\varphi \in \mathscr{S}(\mathbb{R}^N;\mathbb{C})$ be given. The key to proving the required result is the identity

$$(*) \qquad \frac{d}{dt}\varphi \star \check{\mu}_t = (\mathcal{L}^\mu\varphi) \star \check{\mu}_t,$$

where $\check{\mu}_t$ is the distribution of $-\mathbf{x}$ under $\mu_t$, the measure determined by $\widehat{\mu_t} = e^{t\ell_\mu}$. The easiest way to check (*) is to work via Fourier transform and to use (3.2.10) to verify that

$$\frac{d}{dt}\widehat{\varphi \star \check{\mu}_t}(\boldsymbol{\xi}) = \ell_\mu(-\boldsymbol{\xi})\hat{\varphi}(\boldsymbol{\xi})e^{t\ell_\mu(-\boldsymbol{\xi})} = \widehat{\mathcal{L}^\mu\varphi}(\boldsymbol{\xi})e^{t\ell_\mu(-\boldsymbol{\xi})},$$

which is equivalent to (*). To see how (*) applies, observe that

$$\mathbb{E}^{\mathbb{P}}\left[\varphi(\mathbf{Z}(t)) \,\middle|\, \mathcal{F}_s\right] = \varphi \star \check{\mu}_{t-s}(\mathbf{Z}(s)),$$

and therefore that, for any $A \in \mathcal{F}_s$,

$$\begin{aligned}\mathbb{E}^{\mathbb{P}}\left[\varphi(\mathbf{Z}(t)), A\right] - \mathbb{E}^{\mathbb{P}}\left[\varphi(\mathbf{Z}(s)), A\right] &= \int_s^t \mathbb{E}^{\mathbb{P}}\left[(\mathcal{L}^\mu\varphi) \star \check{\mu}_{\tau-s}(\mathbf{Z}(s)), A\right] d\tau \\ = \int_s^t \mathbb{E}^{\mathbb{P}}\left[\mathcal{L}^\mu\varphi(\mathbf{Z}(\tau)), A\right] d\tau &= \mathbb{E}^{\mathbb{P}}\left[\int_s^t \mathcal{L}^\mu\varphi(\mathbf{Z}(\tau))\,d\tau, A\right],\end{aligned}$$

which, after rearrangement, is the asserted martingale property.

To prove the converse assertion, again begin with the observation that, by an easy approximation procedure, one can prove the martingale property for all

$\varphi \in C^2_{\mathrm{b}}(\mathbb{R}^N;\mathbb{C})$ as soon as one knows it for $\varphi \in C^\infty_{\mathrm{c}}(\mathbb{R}^N;\mathbb{R})$. In particular, one can take $\varphi(\mathbf{x}) = e^{\sqrt{-1}(\boldsymbol{\xi},\mathbf{x})_{\mathbb{R}^N}}$; in which case $\mathcal{L}^\mu\varphi = \ell_\mu(\boldsymbol{\xi})\varphi$, and therefore, for any $A \in \mathcal{F}_s$, one gets that

$$u(t) \equiv \mathbb{E}^{\mathbb{P}}\Big[\exp\Big(\sqrt{-1}(\boldsymbol{\xi},\mathbf{Z}(t))_{\mathbb{R}^N}\Big)_{\mathbb{R}^N}, A\Big] = u(s) + \ell_\mu(\boldsymbol{\xi})\int_s^t u(\tau)\,d\tau.$$

Since this means that $u(t) = e^{(t-s)\ell_\mu(\boldsymbol{\xi})}u(s)$, it follows that $\{\mathbf{Z}(t) : t \geq 0\}$ satisfies (7.1.4) and is therefore a Lévy process for $\mu$. $\square$

As an immediate consequence of the preceding we have the following characterizations of the distribution of a Lévy process. In the statement that follows, $\mathcal{F}_t$ is the $\sigma$-algebra over $D(\mathbb{R}^N)$ generated by $\{\boldsymbol{\psi}(\tau) : \tau \in [0,t]\}$.

THEOREM 7.1.7. *Given $\mu \in \mathcal{I}(\mathbb{R}^N)$, let $\mathbb{Q}^\mu \in \mathbf{M}_1\big(D(\mathbb{R}^N)\big)$ be the distribution of a Lévy process for $\mu$. Then $\mathbb{Q}^\mu$ is the unique $\mathbb{P} \in \mathbf{M}_1\big(D(\mathbb{R}^N)\big)$ that satisfies either one of the properties:*

$$\Big(\exp\Big[\sqrt{-1}\,(\boldsymbol{\xi},\boldsymbol{\psi}(t))_{\mathbb{R}^N} + t\ell_\mu(\boldsymbol{\xi})\Big], \mathcal{F}_t, \mathbb{P}\Big)$$

*is a martingale with mean value 1 for each $\boldsymbol{\xi} \in \mathbb{R}^N$,*

*or*

$$\Big(\varphi(\boldsymbol{\psi}(t)) - \varphi(\mathbf{0}) - \int_0^t \mathcal{L}^\mu\varphi(\boldsymbol{\psi}(\tau))\,d\tau, \mathcal{F}_t, \mathbb{P}\Big)$$

*is a martingale with mean value 0 for each $\varphi \in C^\infty_{\mathrm{c}}(\mathbb{R}^N;\mathbb{R})$.*

§ **7.1.3. Basic Results.** In this subsection I run through some of the results from § 5.2 that transfer immediately to the continuous parameter setting.

LEMMA 7.1.8. *Let the interval $I$ and the function $f : I \longrightarrow \mathbb{R} \cup \{\infty\}$ be as in Corollary 5.2.10. If either $\big(X(t),\mathcal{F}_t,\mathbb{P}\big)$ is an $I$-valued martingale or $\big(X(t),\mathcal{F}_t,\mathbb{P}\big)$ is an $I$-valued submartingale and $f$ is non-decreasing and bounded below, then $\big(f\circ X(t),\mathcal{F}_t,\mathbb{P}\big)$ is a submartingale.*

PROOF: The fact that the parameter is continuous plays no role here, and so this result is already covered by the argument in Corollary 5.2.10. $\square$

THEOREM 7.1.9 (**Doob's Inequality**). *Let $\big(X(t),\mathcal{F}_t,\mathbb{P}\big)$ be a submartingale. Then, for every $\alpha \in (0,\infty)$ and $T \in [0,\infty)$,*

$$P\left(\sup_{t\in[0,T]} X(t) \geq \alpha\right) \leq \frac{1}{\alpha}\mathbb{E}^P\left[X(T),\ \sup_{t\in[0,T]} X(t) \geq \alpha\right].$$

*In particular, for non-negative submartingales and $T \in [0,\infty)$,*

$$\mathbb{E}^{\mathbb{P}}\left[\sup_{t\in[0,T]} X(t)^p\right]^{\frac{1}{p}} \le \frac{p}{p-1}\mathbb{E}^{\mathbb{P}}\big[X(T)^p\big]^{\frac{1}{p}}, \quad p \in (0,\infty).$$

PROOF: Because of Exercise 1.4.18, I need only prove the first assertion. To this end, let $T \in (0,\infty)$ and $n \in \mathbb{N}$ be given, apply Theorem 5.2.1 to the discrete parameter submartingale $\left(X\left(\frac{mT}{2^n}\right), \mathcal{F}_{\frac{mT}{2^n}}, \mathbb{P}\right)$, and observe that

$$\sup\left\{X\left(\tfrac{mT}{2^n}\right) : 0 \le m \le 2^n\right\} \nearrow \sup_{t\in[0,T]} X(t) \quad \text{as } n \to \infty. \quad \square$$

THEOREM 7.1.10 (**Doob's Martingale Convergence Theorem**). *Assume that $(X(t), \mathcal{F}_t, \mathbb{P})$ is a $\mathbb{P}$-integrable submartingale. If*

$$\sup_{t\in[0,\infty)} \mathbb{E}^{\mathbb{P}}\big[X(t)^+\big] < \infty,$$

*then there exists an $\mathcal{F}_\infty \equiv \bigvee_{t\ge 0}\mathcal{F}_t$-measurable $X = X(\infty) \in L^1(\mathbb{P};\mathbb{R})$ to which $X(t)$ converges $\mathbb{P}$-almost surely as $t \to \infty$. Moreover, when $(X(t), \mathcal{F}_t, \mathbb{P})$ is either a non-negative submartingale or a martingale, the convergence takes place in $L^1(\mathbb{P};\mathbb{R})$ if and only if the family $\{X(t) : t \in [0,\infty)\}$ is uniformly $\mathbb{P}$-integrable, in which case $X(t) \le \mathbb{E}^{\mathbb{P}}[X \mid \mathcal{F}_t]$ or $X(t) = \mathbb{E}^{\mathbb{P}}[X \mid \mathcal{F}_t]$ (a.s., $\mathbb{P}$) for all $t \in [0,\infty)$, and*

$$\text{(7.1.11)} \qquad \mathbb{P}\left(\sup_{t\ge 0}|X(t)| \ge \alpha\right) \le \frac{1}{\alpha}\mathbb{E}^{\mathbb{P}}\left[|X|, \sup_{t\ge 0}|X(t)| \ge \alpha\right].$$

*Finally, again when $(X(t), \mathcal{F}_t, \mathbb{P})$ is either a non-negative submartingale or a martingale, for each $p \in (1,\infty)$ the family $\{|X(t)|^p : t \in [0,\infty)\}$ is uniformly $\mathbb{P}$-integrable if and only if $\sup_{t\in[0,\infty)} \|X(t)\|_{L^p(\mathbb{P};\mathbb{R})} < \infty$, in which case $X(t) \longrightarrow X$ in $L^p(\mathbb{P};\mathbb{R})$.*

PROOF: To prove the initial convergence assertion, note that, by Theorem 5.2.15 applied to the discrete parameter process $(X(n), \mathcal{F}_n, \mathbb{P})$, there is an $\bigvee_{n\in\mathbb{N}}\mathcal{F}_n$-measurable $X \in L^1(\mathbb{P};\mathbb{R})$ to which $X(n)$ converges $\mathbb{P}$-almost surely. Hence, we need only check that $\lim_{t\to\infty} X(t)$ exists in $[-\infty,\infty]$ $\mathbb{P}$-almost surely. To this end, define $U^{(n)}_{[a,b]}(\omega)$ for $n \in \mathbb{N}$ and $a < b$ to be the precise number of times that the sequence $\left\{X\left(\frac{m}{2^n},\omega\right) : m \in \mathbb{N}\right\}$ upcrosses the interval $[a,b]$ (cf. the paragraph preceding Theorem 5.2.15), observe that $U^{(n)}_{[a,b]}(\omega)$ is non-decreasing as $n$ increases, and set $U_{[a,b]}(\omega) = \lim_{n\to\infty} U^{(n)}_{[a,b]}(\omega)$. Note that if $U_{[a,b]}(\omega) < \infty$, then (by right-continuity), there is an $s \in [0,\infty)$ such that either $X(t,\omega) \le b$ for all $t \ge s$ or $X(t,\omega) \ge a$ for all $t \ge s$. Hence, we will know that $X(t,\omega)$ converges in $[-\infty,\infty]$ for $\mathbb{P}$-almost every $\omega \in \Omega$ as soon as we show that $\mathbb{E}^{\mathbb{P}}\big[U_{[a,b]}\big] < \infty$ for every pair $a < b$. In addition, by (5.2.16), we know that

$$\sup_{n\in\mathbb{N}} \mathbb{E}^{\mathbb{P}}\Big[U_{[a,b]}^{(n)}\Big] \le \sup_{t\in[0,\infty)} \frac{\mathbb{E}^{\mathbb{P}}\big[(X(t)-a)^+\big]}{b-a} < \infty,$$

and so the required estimate follows from the Monotone Convergence Theorem.

Now assume that $\big(X(t),\mathcal{F}_t,\mathbb{P}\big)$ is either a non-negative submartingale or a martingale. Given the preceding, it is clear that $X(t) \longrightarrow X$ in $L^1(\mathbb{P};\mathbb{P})$ if $\big\{X(t) : t \in [0,\infty)\big\}$ is uniformly $\mathbb{P}$-integrable. Conversely, suppose that $X(t) \longrightarrow X$ in $L^1(\mathbb{P};\mathbb{R})$. Then, for any $T \in [0,\infty)$,

$$(*) \qquad |X(T)| \le \lim_{t\to\infty} \mathbb{E}^{\mathbb{P}}\big[|X(t)| \,\big|\, \mathcal{F}_T\big] = \mathbb{E}^{\mathbb{P}}\big[|X| \,\big|\, \mathcal{F}_T\big].$$

In particular, from Theorem 7.1.9,

$$\mathbb{P}\left(\sup_{t\in[0,T]} |X(t)| \ge \alpha\right) \le \frac{1}{\alpha}\mathbb{E}^{\mathbb{P}}\left[|X|,\ \sup_{t\in[0,T]} |X(t)| \ge \alpha\right]$$

for every $T \in (0,\infty)$. Hence, (7.1.11) follows when one lets $T \to \infty$. But, again from (*),

$$\mathbb{E}^{\mathbb{P}}\big[|X(T)|,\, |X(T)| \ge \alpha\big] \le \mathbb{E}^{\mathbb{P}}\big[|X|,\, |X(T)| \ge \alpha\big] \le \mathbb{E}^{\mathbb{P}}\big[|X|,\, \sup_{t\ge 0}|X(t)| \ge \alpha\big],$$

and therefore, since, by (7.1.11), $\mathbb{P}\big(\sup_{t\ge 0}|X(t)| \ge \alpha\big) \longrightarrow 0$ as $\alpha \to \infty$, we can conclude that $\{X(t) : t \ge 0\}$ is uniformly $\mathbb{P}$-integrable.

Finally, if $\{X(T) : T \ge 0\}$ is bounded in $L^p(\mathbb{P};\mathbb{R})$ for some $p \in (1,\infty)$, then, by the last part of Theorem 7.1.9, $\sup_{t\ge 0}|X(t)|^p$ is $\mathbb{P}$-integrable and therefore $X(t) \longrightarrow X$ in $L^p(\mathbb{P};\mathbb{R})$. □

**§ 7.1.4. Stopping Times and Stopping Theorems.** A **stopping time** relative to a non-decreasing family $\{\mathcal{F}_t : t \ge 0\}$ of $\sigma$-algebras is a map $\zeta : \Omega \longrightarrow [0,\infty]$ with the property that $\{\zeta \le t\} \in \mathcal{F}_t$ for every $t \ge 0$. Given a stopping time $\zeta$, I will associate with it the $\sigma$-algebra $\mathcal{F}_\zeta$ consisting of those $A \subseteq \Omega$ such that $A \cap \{\zeta \le t\} \in \mathcal{F}_t$ for every $t \ge 0$. Note that, because $\{\zeta < t\} = \bigcup_{n=0}^{\infty}\{\zeta \le (1-2^{-n})t\}$, $\{\zeta < t\} \in \mathcal{F}_t$ for all $t \ge 0$.

Here are a few useful facts about stopping times.

LEMMA 7.1.12. *Let $\zeta$ be a stopping time. Then $\zeta$ is $\mathcal{F}_\zeta$-measurable, and, for any progressively measurable function $X$ with values in a measurable space $(E,\mathcal{B})$, the function $\omega \rightsquigarrow X(\zeta,\omega) \equiv X\big(\zeta(\omega),\omega\big)$ is $\mathcal{F}_\zeta$-measurable on $\{\zeta < \infty\}$ in* *the sense that* $\{\omega : \zeta(\omega) < \infty \ \&\ X(\zeta,\omega) \in \Gamma\} \in \mathcal{F}_\zeta$ *for all* $\Gamma \in \mathcal{B}$. *In addition,* $f \circ \zeta$ *is again a stopping time if* $f : [0,\infty] \longrightarrow [0,\infty]$ *is a non-decreasing, right-continuous function satisfying* $f(\tau) \geq \tau$ *for all* $\tau \in [0,\infty]$. *Next, suppose that* $\zeta_1$ *and* $\zeta_2$ *are a pair of stopping times. Then* $\zeta_1 + \zeta_2$, $\zeta_1 \wedge \zeta_2$, *and* $\zeta_1 \vee \zeta_2$ *are all stopping times, and* $\mathcal{F}_{\zeta_1 \wedge \zeta_2} \subseteq \mathcal{F}_{\zeta_1} \cap \mathcal{F}_{\zeta_2}$. *Finally, for any* $A \in \mathcal{F}_{\zeta_1}$, $A \cap \{\zeta_1 \leq \zeta_2\} \in \mathcal{F}_{\zeta_1 \wedge \zeta_2}$.

PROOF: Since $\{\zeta \leq s\} \cap \{\zeta \leq t\} = \{\zeta \leq s \wedge t\} \in \mathcal{F}_t$, it is clear that $\zeta$ is $\mathcal{F}_\zeta$-measurable. Next, suppose that $X$ is a progressively measurable function. To prove that $X(\zeta)$ is $\mathcal{F}_\zeta$-measurable, begin by checking that $\{\omega : (\zeta(\omega),\omega) \in A\} \in \mathcal{F}_t$ for any $A \in \mathcal{B}_t \times \mathcal{F}_t$. Indeed, this is obvious when $A = [0,s] \times B$ for $s \in [0,t]$ and $B \in \mathcal{F}_t$ and, since these generate $\mathcal{B}_{[0,t]} \times \mathcal{F}_t$, it follows in general. Now, for any $t \geq 0$ and $\Gamma \in \mathcal{B}$,

$$A(t,\Gamma) \equiv \{(\tau,\omega) \in [0,\infty) \times \Omega : (\tau, X(\tau,\omega)) \in [0,t] \times \Gamma\} \in \mathcal{B}_{[0,t]} \times \mathcal{F}_t,$$

and therefore

$$\{X(\zeta) \in \Gamma\} \cap \{\zeta \leq t\} = \{\omega : (\zeta(\omega),\omega) \in A(t,\Gamma)\} \in \mathcal{F}_t.$$

As for $f \circ \zeta$ when $f$ satisfies the stated conditions, simply note that

$$\{f \circ \zeta \leq t\} = \{\zeta \leq f^{-1}(t)\} \in \mathcal{F}_t,$$

where $f^{-1}(t) \equiv \inf\{\tau : f(\tau) \geq t\} \leq t$.

Next suppose that $\zeta_1$ and $\zeta_2$ are two stopping times. It is trivial to see that $\zeta_1 \wedge \zeta_2$ and $\zeta_1 \vee \zeta_2$ are again stopping times. In addition, if $\mathbb{Q}$ denotes the set of rational numbers, then

$$\{\zeta_1 + \zeta_2 > t\} = \{\zeta_1 > t\} \cup \bigcup_{q \in \mathbb{Q} \cap [0,1]} \{\zeta_1 \geq qt \ \&\ \zeta_2 > (1-q)t\} \in \mathcal{F}_t.$$

Thus, $\zeta_1 + \zeta_2$ is a stopping time. To prove the final assertions, begin with the observation that if $\zeta_1 \leq \zeta_2$, then $A \cap \{\zeta_2 \leq t\} = (A \cap \{\zeta_1 \leq t\}) \cap \{\zeta_2 \leq t\} \in \mathcal{F}_t$ for all $A \in \mathcal{F}_{\zeta_1}$ and $t \geq 0$, and therefore $\mathcal{F}_{\zeta_1} \subseteq \mathcal{F}_{\zeta_2}$. Next, for any $\zeta_1$ and $\zeta_2$, $\{\zeta_1 \leq \zeta_2\} \in \mathcal{F}_{\zeta_2}$ since

$$\{\zeta_1 > \zeta_2\} \cap \{\zeta_2 \leq t\} = \bigcup_{q \in \mathbb{Q} \cap [0,1]} \{\zeta_1 > qt\} \cap \{\zeta_2 \leq qt\} \in \mathcal{F}_t.$$

Finally, if $A \in \mathcal{F}_{\zeta_1}$, then

$$(A \cap \{\zeta_1 \leq \zeta_2\}) \cap \{\zeta_1 \wedge \zeta_2 \leq t\} = (A \cap \{\zeta_1 \leq t\}) \cap \{\zeta_1 \leq t \wedge \zeta_2\},$$

and therefore, since $A \cap \{\zeta_1 \leq t\} \in \mathcal{F}_t$ and $\{\zeta_1 \leq t \wedge \zeta_2\} \in \mathcal{F}_{t \wedge \zeta_2} \subseteq \mathcal{F}_t$, we have that $A \cap \{\zeta_1 \leq \zeta_2\} \in \mathcal{F}_{\zeta_1 \wedge \zeta_2}$. □

In order to prove the continuous parameter analog of Theorems 5.2.13 and 5.2.11, I will need the following uniform integrability result.

LEMMA 7.1.13. *If $\big(X(t),\mathcal{F}_t,\mathbb{P}\big)$ is either a martingale or a non-negative, integrable submartingale, then, for each $T>0$, the set*

$$\{X(\zeta) : \zeta \text{ is a stopping time dominated by } T\}$$

*is uniformly $\mathbb{P}$-integrable. Furthermore, if, in addition, $\{X(t) : t\geq 0\}$ is uniformly $\mathbb{P}$-integrable and (cf. Theorem 7.1.10) $X(\infty)=\lim_{t\to\infty}X(t)$ (a.s., $\mathbb{P}$), then $\{X(\zeta) : \zeta \text{ is a stopping time}\}$ is uniformly $\mathbb{P}$-integrable.*

PROOF: Throughout, without loss in generality, I will assume that $\big(X(t),\mathcal{F}_t,\mathbb{P}\big)$ is a non-negative, integrable submartingale.

Given a stopping time $\zeta\leq T$, define $\zeta_n=\frac{\lfloor 2^n\zeta\rfloor+1}{2^n}$ for $n\geq 0$. By Lemma 7.1.12, $\zeta_n$ is again a stopping time. Thus, by Theorem 5.2.13 applied to the discrete parameter submartingale $\big(X(m2^{-n}),\mathcal{F}_{m2^{-n}},\mathbb{P}\big)$,

$$X(\zeta_n)\leq\mathbb{E}^{\mathbb{P}}\big[X\big(2^{-n}(\lfloor 2^nT\rfloor+1)\big)\,\big|\,\mathcal{F}_{\zeta_n}\big]\leq\mathbb{E}^{\mathbb{P}}\big[X(T+1)\,\big|\,\mathcal{F}_{\zeta_n}\big],$$

and so

$$\begin{aligned}\mathbb{E}^{\mathbb{P}}\big[X(\zeta_n),\,X(\zeta_n)\geq\alpha\big]&\leq\mathbb{E}^{\mathbb{P}}\big[X(T+1),\,X(\zeta_n)\geq\alpha\big]\\&\leq\mathbb{E}^{\mathbb{P}}\left[X(T+1),\ \sup_{t\in[0,T+1]}X(t)\geq\alpha\right].\end{aligned}$$

Starting from here, noting that $\zeta_n\searrow\zeta$ as $n\to\infty$, and applying Fatou's Lemma, we arrive at

$$(*)\qquad \mathbb{E}^{\mathbb{P}}\big[X(\zeta),\,X(\zeta)>\alpha\big]\leq\mathbb{E}^{\mathbb{P}}\left[X(T+1),\ \sup_{t\in[0,T+1]}X(t)\geq\alpha\right].$$

Hence, since, by Theorem 7.1.9, $\mathbb{P}\left(\sup_{t\in[0,T+1]}X(t)\geq\alpha\right)$ tends to 0 as $\alpha\to\infty$, this proves the first assertion. When $\{X(t) : t\geq 0\}$ is uniformly integrable, we can replace (*) by

$$\mathbb{E}^{\mathbb{P}}\big[X(\zeta\wedge T),\,X(\zeta\wedge T)>\alpha\big]\leq\mathbb{E}^{\mathbb{P}}\left[X(\infty),\,\sup_{t\geq 0}X(t)\geq\alpha\right]$$

for any stopping time $\zeta$ and $T>0$. Hence, after another application of Fatou's Lemma, we get

$$\mathbb{E}^{\mathbb{P}}\big[X(\zeta),\,X(\zeta)>\alpha\big]\leq\mathbb{E}^{\mathbb{P}}\left[X(\infty),\,\sup_{t\geq 0}X(t)\geq\alpha\right].$$

At the same time, the first inequality in Theorem 7.1.9 can be replaced by

$$\mathbb{P}\left(\sup_{t\geq 0}X(t)\geq\alpha\right)\leq\frac{1}{\alpha}\mathbb{E}^{\mathbb{P}}\left[X(\infty),\,\sup_{t\geq 0}X(t)\geq\alpha\right]\leq\frac{1}{\alpha}\mathbb{E}^{\mathbb{P}}[X(\infty)],$$

and so the asserted uniform integrability follows. □

It turns out that in the continuous time context, Doob's Stopping Time Theorem is most easily seen as a corollary of Hunt's. Thus, I will begin with Hunt's.

THEOREM 7.1.14 (**Hunt**). *Let $(X(t), \mathcal{F}_t, \mathbb{P})$ be either a non-negative, integrable submartingale or a martingale. If $\zeta_1$ and $\zeta_2$ are bounded stopping times and $\zeta_1 \leq \zeta_2$, then $X(\zeta_1) \leq \mathbb{E}^{\mathbb{P}}\big[X(\zeta_2)\,\big|\,\mathcal{F}_{\zeta_1}\big]$, and equality holds in the martingale case. Moreover, when $\{X(t) : t \geq 0\}$ is uniformly $\mathbb{P}$-integrable and $X(\infty) \equiv \lim_{t\to\infty} X(t)$, then the same result holds for arbitrary stopping times $\zeta_1 \leq \zeta_2$.*

PROOF: Given $\zeta_1 \leq \zeta_2 \leq T$, define $(\zeta_i)_n = 2^{-n}\big(\lfloor 2^n\zeta_i\rfloor + 1\big)$ for $n \geq 0$, note that $(\zeta_i)_n$ is a $\{\mathcal{F}_{m2^{-n}} : m \geq 0\}$-stopping time and that $\mathcal{F}_{\zeta_1} \subseteq \mathcal{F}_{(\zeta_1)_n}$, and apply Theorem 5.2.13 to the discrete parameter submartingale $\big(X(m2^{-n}), \mathcal{F}_{m2^{-n}}, \mathbb{P}\big)$ in order to see that

$$\mathbb{E}^{\mathbb{P}}\Big[X\big((\zeta_1)_n\big), A\Big] \leq \mathbb{E}^{\mathbb{P}}\Big[X\big((\zeta_2)_n\big), A\Big], \quad A \in \mathcal{F}_{\zeta_1},$$

with equality in the martingale case. Because of right-continuity and Lemma 7.1.13, $X\big((\zeta_i)_n\big) \longrightarrow X(\zeta_i)$ in $L^1(\mathbb{P};\mathbb{R})$, and so we have now shown that $X(\zeta_1) \leq \mathbb{E}^{\mathbb{P}}\big[X(\zeta_2)\,\big|\,\mathcal{F}_{\zeta_1}\big]$, with equality in the martingale case.

When $\{X(t) : t \geq 0\}$ is uniformly $\mathbb{P}$-integrable and $\zeta_1 \leq \zeta_2$ are unbounded, $\{X(\zeta_i \wedge T) : T \geq 0\}$ is uniformly $\mathbb{P}$-integrable for $i \in \{1,2\}$. Hence, for any $A \in \mathcal{F}_{\zeta_1}$ and $0 \leq t \leq T$,

$$\mathbb{E}^{\mathbb{P}}\big[X(T\wedge\zeta_1), A\cap\{\zeta_1 \leq t\}\big] \leq \mathbb{E}^{\mathbb{P}}\big[X(T\wedge\zeta_2), A\cap\{\zeta_1 \leq t\}\big],$$

with equality in the martingale case. Letting first $T$ and then $t$ tend to infinity, one gets the same relationship for $X(\zeta_1)$ and $X(\zeta_2)$, initially with $A\cap\{\zeta_1 < \infty\}$ and then, trivially, with $A$ alone. □

THEOREM 7.1.15 (**Doob's Stopping Time Theorem**). *If $(X(t), \mathcal{F}_t, \mathbb{P})$ is either a non-negative, integrable submartingale or a martingale, then, for every stopping time $\zeta$, $\big(X(t\wedge\zeta), \mathcal{F}_t, \mathbb{P}\big)$ is either an integrable submartingale or a martingale.*

PROOF: Given $0 \leq s < t$ and $A \in \mathcal{F}_s$, note that $A\cap\{\zeta > s\} \in \mathcal{F}_{s\wedge\zeta}$ and therefore, by Hunt's Theorem applied to the stopping times $s\wedge\zeta$ and $t\wedge\zeta$, that

$$\begin{aligned}\mathbb{E}^{\mathbb{P}}\big[X(t\wedge\zeta), A\big] &= \mathbb{E}^{\mathbb{P}}\big[X(\zeta), A\cap\{\zeta\leq s\}\big] + \mathbb{E}^{\mathbb{P}}\big[X(t\wedge\zeta), A\cap\{\zeta > s\}\big]\\ &\geq \mathbb{E}^{\mathbb{P}}\big[X(\zeta), A\cap\{\zeta\leq s\}\big] + \mathbb{E}^{\mathbb{P}}\big[X(s\wedge\zeta), A\cap\{\zeta > s\}\big] = \mathbb{E}^{\mathbb{P}}\big[X(s\wedge\zeta), A\big],\end{aligned}$$

where the inequality is an equality in the martingale case. □

To demonstrate just how powerful these results are, I give the following extension of the independent increment property of Lévy processes. In its statement, the maps $\delta_t : D(\mathbb{R}^N) \longrightarrow D(\mathbb{R}^N)$ for $t \in [0,\infty)$ are defined so that $\delta_t\psi(\tau) = \psi(\tau + t) - \psi(t)$, $\tau \in [0,\infty)$. Also, $\mathcal{F}_t = \sigma\big(\{\psi(\tau) : \tau \in [0,t]\}\big)$, $\zeta$ is a stopping time relative to $\{\mathcal{F}_t : t \in [0,\infty)\}$, and $\delta_\zeta$ is the map on $\{\psi : \zeta(\psi) < \infty\}$ into $D(\mathbb{R}^N)$ given by $\delta_\zeta\psi = \delta_{\zeta(\psi)}\psi$.

THEOREM 7.1.16. *Given $\mu \in \mathcal{I}(\mathbb{R}^N)$, let $\mathbb{Q}^\mu \in \mathbf{M}_1\big(D(\mathbb{R}^N)\big)$ be the distribution of the Lévy process for $\mu$. Then, for each stopping time $\zeta$ and $\mathcal{F}_{D(\mathbb{R}^N)} \times \mathcal{F}_\zeta$-measurable functions $F : D(\mathbb{R}^N) \times D(\mathbb{R}^N) \longrightarrow [0,\infty)$,*

$$\int_{\{\zeta<\infty\}} F(\delta_\zeta \boldsymbol{\psi}, \boldsymbol{\psi})\, \mathbb{Q}^\mu(d\boldsymbol{\psi}) = \iint \mathbf{1}_{[0,\infty)}\big(\zeta(\boldsymbol{\psi}')\big) F(\boldsymbol{\psi}, \boldsymbol{\psi}')\, Q^\mu(d\boldsymbol{\psi}) Q^\mu(d\boldsymbol{\psi}').$$

PROOF: By elementary measure theory, all that we have to show is that, for each $B \in \mathcal{F}_\zeta$ contained in $\{\zeta < \infty\}$, $\mathbb{Q}^\mu\big((\delta_\zeta^{-1}\Gamma) \cap B\big) = \mathbb{Q}^\mu(\Gamma)\mathbb{Q}^\mu(B)$.

Given $B \in \mathcal{F}_\zeta$ contained in $\{\zeta < \infty\}$ with $\mathbb{Q}^\mu(B) > 0$, choose $T > 0$ so that $\mathbb{Q}^\mu(B_T) > 0$ when $B_T = B \cap \{\zeta \le T\}$, and define $\mathbb{Q}_T \in \mathbf{M}_1\big(D(\mathbb{R}^N)\big)$ so that

$$\mathbb{Q}_T(\Gamma) = \frac{\mathbb{Q}^\mu\big((\delta_\zeta^{-1}\Gamma) \cap B_T\big)}{\mathbb{Q}^\mu(B_T)}.$$

If we show that $\mathbb{Q}_T = \mathbb{Q}^\mu$, then we will know that

$$\begin{aligned}\mathbb{Q}^\mu\big((\delta_\zeta^{-1}\Gamma) \cap B\big) &= \lim_{T\to\infty} \mathbb{Q}^\mu\big((\delta_\zeta^{-1}\Gamma) \cap B_T\big)\\ &= \mathbb{Q}^\mu(\Gamma) \lim_{T\to\infty} \mathbb{Q}^\mu(B_T) = \mathbb{Q}^\mu(\Gamma)\mathbb{Q}^\mu(B)\end{aligned}$$

and therefore will be done.

By Theorem 7.1.6, checking that $\mathbb{Q}_T = \mathbb{Q}^\mu$ comes down to showing that, for any $0 \le s < t$, $\boldsymbol{\xi} \in \mathbb{R}^N$, and $A \in \mathcal{F}_s$,

$$\mathbb{E}^{\mathbb{Q}_T}\Big[e^{\sqrt{-1}(\mathbf{x},\boldsymbol{\psi}(t))_{\mathbb{R}^N} - t\ell_\mu(\boldsymbol{\xi})}, A\Big] = \mathbb{E}^{\mathbb{Q}_T}\Big[e^{\sqrt{-1}(\mathbf{x},\boldsymbol{\psi}(s))_{\mathbb{R}^N} - s\ell_\mu(\boldsymbol{\xi})}, A\Big].$$

To this end, note that, by Theorem 7.1.14 applied to $s + \zeta \wedge T$ and $t + \zeta \wedge T$,

$$\begin{aligned}&\mathbb{Q}^\mu(B_T)\mathbb{E}^{\mathbb{Q}_T}\Big[e^{\sqrt{-1}(\mathbf{x},\boldsymbol{\psi}(t))_{\mathbb{R}^N} - t\ell_\mu(\boldsymbol{\xi})}, A\Big]\\ &= \mathbb{E}^{\mathbb{Q}^\mu}\Big[e^{-\sqrt{-1}(\boldsymbol{\xi},\boldsymbol{\psi}(\zeta))_{\mathbb{R}^N} + \zeta\ell_\mu(\boldsymbol{\xi})} e^{\sqrt{-1}(\boldsymbol{\xi},\boldsymbol{\psi}(t+\zeta))_{\mathbb{R}^N} - (t+\zeta)\ell_\mu(\boldsymbol{\xi})}, (\delta_\zeta^{-1}A) \cap B_T\Big]\\ &= \mathbb{E}^{\mathbb{Q}^\mu}\Big[e^{-\sqrt{-1}(\boldsymbol{\xi},\boldsymbol{\psi}(\zeta))_{\mathbb{R}^N} + \zeta\ell_\mu(\boldsymbol{\xi})} e^{\sqrt{-1}(\boldsymbol{\xi},\boldsymbol{\psi}(s+\zeta))_{\mathbb{R}^N} - (s+\zeta)\ell_\mu(\boldsymbol{\xi})}, (\delta_\zeta^{-1}A) \cap B_T\Big]\\ &= \mathbb{Q}^\mu(B_T)\mathbb{E}^{\mathbb{Q}_T}\Big[e^{\sqrt{-1}(\mathbf{x},\boldsymbol{\psi}(s))_{\mathbb{R}^N} - s\ell_\mu(\boldsymbol{\xi})}, A\Big],\end{aligned}$$

since $\boldsymbol{\psi} \rightsquigarrow e^{-\sqrt{-1}(\boldsymbol{\xi},\boldsymbol{\psi}(\zeta))_{\mathbb{R}^N} + \zeta\ell_\mu(\boldsymbol{\xi})} \mathbf{1}_A(\delta_\zeta \boldsymbol{\psi}) \mathbf{1}_{B_T}(\boldsymbol{\psi})$ is $\mathcal{F}_{s+\zeta\wedge T}$-measurable. □

**§ 7.1.5. An Integration by Parts Formula.** In this subsection I will derive a simple result that has many interesting applications.

THEOREM 7.1.17. *Suppose $V : [0,\infty) \times \Omega \longrightarrow \mathbb{C}$ is a right-continuous, progressively measurable function, and let $|V|(t,\omega) \in [0,\infty]$ denote the total variation $\mathrm{var}_{[0,t]}\big(V(\,\cdot\,,\omega)\big)$ of $V(\,\cdot\,,\omega)$ on the interval $[0,t]$. Then $|V| : [0,\infty) \times \Omega \longrightarrow [0,\infty]$ is a non-decreasing, progressively measurable function that is right-continuous* *on each interval $[0,t)$ for which $|V|(t,\omega)<\infty$. Next, suppose that $(X(t),\mathcal{F}_t,\mathbb{P})$ is a $\mathbb{C}$-valued martingale with the property that, for each $(t,\omega)\in(0,\infty)\times\Omega$, the product $\|X(\,\cdot\,,\omega)\|_{[0,t]}|V|(t,\omega)<\infty$, and define*

$$Y(t,\omega)=\begin{cases}\int_{(0,t]}X(s,\omega)\,V(ds,\omega) & \text{if } |V|(t,\omega)<\infty\\ 0 & \text{otherwise,}\end{cases}$$

*where, in the case when $|V|(t,\omega)<\infty$, the integral is the Lebesgue integral of $X(\,\cdot\,,\omega)$ on $[0,t]$ with respect to the $\mathbb{C}$-valued measure determined by $V(\,\cdot\,,\omega)$. If*

$$\mathbb{E}^{\mathbb{P}}\Big[\|X\|_{[0,T]}\Big(|V|(T)+\big|V(0)\big|\Big)\Big]<\infty \quad \text{for all } T\in(0,\infty),$$

*then $\big(X(t)V(t)-Y(t),\mathcal{F}_t,\mathbb{P}\big)$ is a martingale.*

PROOF: Without loss in generality, I will assume that both $X$ and $V$ are $\mathbb{R}$-valued. To see that $|V|$ is $\{\mathcal{F}_t:t\in[0,\infty)\}$-progressively measurable, simply observe that, by right-continuity,

$$|V|(t,\omega)=\sup_{n\in\mathbb{N}}\sum_{k=0}^{[2^n t]}\Big|V\big(\tfrac{k+1}{2^n}\wedge t,\omega\big)-V\big(\tfrac{k}{2^n},\omega\big)\Big|;$$

and to see that $|V|(\,\cdot\,,\omega)$ is right-continuous on $[0,t)$ whenever $|V|(t,\omega)<\infty$, recall that the magnitude of the jumps (from the right and left) of the variation of a function coincide with those of the function itself.

I turn now to the second part. Certainly $Y$ is $\{\mathcal{F}_t:t\in[0,\infty)\}$-progressively measurable. In addition, because $\|X(\,\cdot\,,\omega)\|_{[0,t]}|V|(t,\omega)<\infty$ for all $(t,\omega)\in[0,\infty)\times\Omega$, for any $\omega\in\Omega$ one has that

$$Y(t,\omega)=0 \quad\text{or}\quad Y(t,\omega)=\int_{(0,t]}X(s,\omega)\,V(ds,\omega)\quad \text{for all } t\in[0,\infty);$$

and so, in either case, $Y(\,\cdot\,,\omega)$ is right-continuous and $Y(t,\omega)-Y(s,\omega)$ can be computed as

$$\lim_{n\to\infty}\sum_{k=[2^n s]}^{[2^n t]}X\big(\tfrac{k+1}{2^n}\wedge t,\omega\big)\Big(V\big(\tfrac{k+1}{2^n}\wedge t,\omega\big)-V\big(\tfrac{k}{2^n}\vee s,\omega\big)\Big).$$

In fact, under the stated integrability condition, the convergence in the preceding takes place in $L^1(\mathbb{P};\mathbb{R})$ for every $t\in[0,\infty)$; and therefore, for any $0\le s\le t<\infty$

and $A \in \mathcal{F}_s$,

$$\begin{aligned}
&\mathbb{E}^{\mathbb{P}}\big[Y(t) - Y(s), A\big] \\
&= \lim_{n\to\infty} \sum_{k=\lfloor 2^n s\rfloor}^{\lfloor 2^n \rfloor} \mathbb{E}^{\mathbb{P}}\left[X\left(\tfrac{k+1}{2^n}\wedge t,\omega\right)\left(V\left(\tfrac{k+1}{2^n}\wedge t,\omega\right) - V\left(\tfrac{k}{2^n}\vee s,\omega\right)\right), A\right] \\
&= \lim_{n\to\infty} \sum_{k=\lfloor 2^n s\rfloor}^{\lfloor 2^n t\rfloor} \mathbb{E}^{\mathbb{P}}\left[X(t)\left(V\left(\tfrac{k+1}{2^n}\wedge t,\omega\right) - V\left(\tfrac{k}{2^n}\vee s,\omega\right)\right), A\right] \\
&= \mathbb{E}^{\mathbb{P}}\Big[X(t)\big(V(t) - V(s)\big), A\Big] = \mathbb{E}^{\mathbb{P}}\Big[X(t)V(t) - X(s)V(s), A\Big],
\end{aligned}$$

and clearly this is equivalent to the asserted martingale property. □

We will make frequent practical applications of Theorem 7.1.17 later, but here I will show that it enables us to prove that there is an important dichotomy between continuous martingales and functions of bounded variation. However, before doing so, I need to make a small, technical digression.

A function $\zeta : \Omega \longrightarrow [0,\infty]$ is an **extended stopping time** relative to $\{\mathcal{F}_t : t \in [0,\infty)\}$ if $\{\zeta < t\} \in \mathcal{F}_t$ for every $t \in (0,\infty)$. Since $\{\zeta < t\} \in \mathcal{F}_t$ for any stopping time $\zeta$, it is clear that every stopping time is an extended stopping time. On the other hand, not every extended stopping time is a stopping time. To wit, if $X : [0,\infty) \times \Omega \longrightarrow \mathbb{R}$ is a right-continuous, progressively measurable function relative to $\big\{\{\sigma\big(X(\tau) : \tau \in [0,t]\big\}\big) : t \ge 0\big\}$, then $\zeta = \inf\{t \ge 0 : X(t) > 1\}$ will always be an extended stopping time but will seldom be a stopping time.

LEMMA 7.1.18. *For each $t \ge 0$, set $\mathcal{F}_{t+} = \bigcap_{\tau > t} \mathcal{F}_\tau$. Then $\zeta : \Omega \longrightarrow [0,\infty]$ is an extended stopping time if and only if it is a stopping time relative to $\{\mathcal{F}_{t+} : t \ge 0\}$. Moreover, if $\big(X(t), \mathcal{F}_t, \mathbb{P}\big)$ is either a non-negative, integrable submartingale or a martingale, then so is $\big(X(t), \mathcal{F}_{t+}, \mathbb{P}\big)$. In particular, if $\zeta$ is an extended stopping time, then $\big(X(t\wedge\zeta), \mathcal{F}_{t+}, \mathbb{P}\big)$ is a non-negative, integrable submartingale or a martingale.*

PROOF: The first assertion is immediate from $\{\zeta \le t\} = \bigcap_{\tau > t}\{\zeta < \tau\}$. To prove the second assertion, apply right-continuity and the first uniform integrability result in Lemma 7.1.13 to see that if $0 \le s < t$ and $A \in \mathcal{F}_{s+}$, then

$$\mathbb{E}^{\mathbb{P}}\big[X(s), A\big] = \lim_{\tau \searrow s} \mathbb{E}^{\mathbb{P}}\big[X(\tau), A\big] \le \mathbb{E}^{\mathbb{P}}\big[X(t), A\big],$$

where the inequality is an equality in the martingale case. □

THEOREM 7.1.19. *Suppose that $\big(X(t), \mathcal{F}_t, \mathbb{P}\big)$ is a continuous martingale, and let $|X|(t,\omega) = \mathrm{var}_{[0,t]}\big(X(\,\cdot\,,\omega)\big)$ denote the variation of $X(\,\cdot\,,\omega) \upharpoonright [0,t]$. Then*

$$\mathbb{P}\big(\exists t > 0\ 0 < |X|(t,\omega) < \infty\big) = 0.$$

*Equivalently, for $\mathbb{P}$-almost every $\omega$ and all $t > 0$, either $X(\tau,\omega) = X(0,\omega)$ for $\tau \in [0,t]$ or $|X|(t,\omega) = \infty$.*

PROOF: Without loss in generality, I will assume that $X(0,\omega) \equiv 0$. Given $R > 0$, let $\zeta_R(\omega) = \sup\{t \geq 0 : |X|(t,\omega) \leq R\}$, and set $X_R(t) = X(t \wedge \zeta_R)$. Then $\zeta_R$ is an extended stopping time, and so, by Lemma 7.1.18, $(X_R(t), \mathcal{F}_{t+}, \mathbb{P})$ is a bounded martingale. Hence, by Theorem 7.1.17,

$$\left(X_R(t)^2 - \int_0^t X_R(\tau)\, X_R(d\tau), \mathcal{F}_{t+}, \mathbb{P}\right)$$

is also a martingale, and so

$$\mathbb{E}^{\mathbb{P}}\left[X_R(t)^2\right] = \mathbb{E}^{\mathbb{P}}\left[\int_0^t X_R(\tau)\, X_R(d\tau)\right].$$

On the other hand, since $X_R(\,\cdot\,)$ is continuous, and therefore, by Fubini's Theorem,

$$X_R(t)^2 = \iint\limits_{[0,t]^2} X_R(d\tau_1) X_R(d\tau_2) = 2\int_0^t X_R(\tau)\, X_R(d\tau),$$

we also know that

$$\mathbb{E}^{\mathbb{P}}\left[X_R(t)^2\right] = 2\mathbb{E}^{\mathbb{P}}\left[\int_0^t X_R(\tau)\, X_R(d\tau)\right].$$

Hence, $\mathbb{E}^{\mathbb{P}}\left[X_R(t)^2\right] = 0$ for all $t > 0$, which means that $X_R(\,\cdot\,) \equiv 0$ $\mathbb{P}$-almost surely. □

The preceding result leads immediately to the following analog of the uniqueness statement in Lemma 5.2.12.

COROLLARY 7.1.20. *Let $X : \Omega \longrightarrow \mathbb{R}$ be a right-continuous, progressively measurable function. Then, up to a $\mathbb{P}$-null set, there is at most one continuous, progressively measurable $A : \Omega \longrightarrow R$ such that $A(0,\omega) = 0$, $A(\,\cdot\,,\omega)$ is of locally bounded variation for $\mathbb{P}$-almost every $\omega \in \Omega$, and $\big(X(t) - A(t), \mathcal{F}_t, \mathbb{P}\big)$ is a martingale.*

The role of continuity here seems minor, but it is crucial. Namely, continuity was used in Theorem 7.1.19 only when I wanted to know that $X_R(t)^2 = 2\int_0^t X_R(\tau)\, X_R(d\tau)$. On the other hand, it is critical. For example, if $\{N(t) : t \geq 0\}$ is the simple Poisson process in § 4.2 and $\mathcal{F}_t = \sigma(\{N(\tau) : \tau \in [0,t]\})$, then it is easy to check that $\big(N(t) - t, \mathcal{F}_t, \mathbb{P}\big)$ is a martingale, all of whose paths are of locally bounded variation.

## Exercises for § 7.1

EXERCISE 7.1.21. The definition of stopping times and their associated $\sigma$-algebras that I have adopted is due to E.B. Dynkin. Earlier, less ubiquitous but more transparent, definitions appear in the work of Doob and Hunt under the name of *optional stopping times.* To explain these earlier definitions, let $E$ be a complete, separable metric space and $\Psi$ a non-empty collection of right-continuous paths $\psi : [0,\infty) \longrightarrow E$ with the property that for all $\psi \in \Psi$ and $t \in [0,\infty)$, the stopped path $\psi^t$ given by $\psi^t(\tau) = \psi(t\wedge\tau)$ is again in $\Psi$. Similarly, given a function $\zeta : \Psi \longrightarrow [0,\infty]$, define $\psi^\zeta$ so that $\psi^\zeta(t) = \psi\big(t\wedge\zeta(\psi)\big)$. Finally, for each $t \in [0,\infty)$, define the $\sigma$-algebras $\mathcal{F}_t$ over $\Psi$ to be the one generated by $\{\psi(\tau) : \tau \in [0,t]\}$, and take $\mathcal{F} = \bigvee_{t\geq 0}\mathcal{F}_t$. In terms of these quantities, an **optional stopping time** is an $\mathcal{F}$-measurable map $\zeta : \Psi \longrightarrow [0,\infty]$ such that $\zeta(\psi) \leq t \implies \zeta(\psi) = \zeta(\psi^t)$, and the associated $\sigma$-algebra is $\sigma\big(\{\psi^\zeta(t) : t \geq 0\}\big)$. The goal of this exercise is to show that $\zeta$ is an optional stopping time if and only if it is a stopping time and that its associated $\sigma$-algebra is $\mathcal{F}_\zeta$.

**(i)** It is an easy matter (cf. Exercise 4.1.9) to check that $f : \Omega \longrightarrow \mathbb{R}$ is $\mathcal{F}$-measurable if and only if there exists a $\mathcal{B}^{\mathbb{Z}^+}$-measurable $F : E^{\mathbb{Z}^+} \longrightarrow \mathbb{R}$ and a sequence $\{t_m : m \in \mathbb{Z}^+\}$ such that $f(\psi) = F\big(\psi(t_1), \dots, \psi(t_m), \dots\big)$, from which it is clear that an $\mathcal{F}$-measurable $f$ will be $\mathcal{F}_t$-measurable for some $t \in [0,\infty)$ if and only if $f(\psi) = f(\psi^t)$. Use this to show that every optional stopping time is a stopping time.

**(ii)** Show that $\zeta : \Psi \longrightarrow [0,\infty]$ is a stopping time relative to $\{\mathcal{F}_t : t \in [0,\infty)\}$ if and only if it is $\mathcal{F}$-measurable and, for each $t \in [0,\infty)$, $\{\psi : \zeta(\psi) \leq t\} = \{\psi : \zeta(\psi^t) \leq t\}$. In addition, if $\zeta$ is a stopping time, show that $\zeta(\psi) < \infty \implies \zeta(\psi) = \zeta(\psi^\zeta)$, and therefore that $\zeta(\psi) \leq t \implies \zeta(\psi) = \zeta(\psi^t)$ for all $t \in [0,\infty)$. Thus, $\zeta$ is an optional stopping time if and only if it is a stopping time.

**Hint**: In proving the second part, check that $\{\zeta = t\} \in \mathcal{F}_t$, and conclude that $\mathbf{1}_{\{t\}}\big(\zeta(\psi)\big) = \mathbf{1}_{\{t\}}\big(\zeta(\psi^t)\big)$ for all $(t,\psi) \in [0,\infty) \times \Psi$.

**(iii)** If $\zeta$ is a stopping time, show that $\mathcal{F}_\zeta = \sigma\big(\{\psi^\zeta(t) : t \geq 0\}\big)$. Besides having intuitive value, this shows that, at least in the situation here, $\mathcal{F}_\zeta$ is countably generated.

**Hint**: Using right-continuity, first show that $\psi \rightsquigarrow \psi^\zeta$ is $\mathcal{F}$-measurable. Next, given a $\mathcal{B}$-measurable $f : E \longrightarrow \mathbb{R}$ and $t \in [0,\infty)$, use **(ii)** to show that

$$\mathbf{1}_{[0,t]}\big(\zeta(\psi)\big)f\big(\psi^\zeta(\tau)\big) = \mathbf{1}_{[0,t]}\big(\zeta(\psi^t)\big)f\Big(\psi\big(\tau \wedge \zeta(\psi^t)\big)\Big), \quad \tau \in [0,\infty),$$

and conclude that $\sigma\big(\{\psi^\zeta(t) : t \geq 0\}\big) \subseteq \mathcal{F}_\zeta$. To prove the opposite inclusion, show that if $f : \Psi \longrightarrow \mathbb{R}$ is $\mathcal{F}_\zeta$-measurable, then, for each $t \in [0,\infty)$, $\mathbf{1}_{\{t\}}\big(\zeta(\psi)\big)f(\psi) = \mathbf{1}_{\{t\}}\big(\zeta(\psi^t)\big)f(\psi^t)$, and thereby arrive at $f(\psi) = f(\psi^\zeta)$. Finally, use this together with Exercise 4.1.9 to show that $f$ is $\sigma\big(\{\psi^\zeta(t) : t \geq 0\}\big)$-measurable.

EXERCISE 7.1.22. Let $(\Omega, \mathcal{F}, \mathbb{P})$ be a probability space and $\{\mathcal{F}_t : t \in [0,\infty)\}$ is non-decreasing family of sub-$\sigma$-algebras of $\mathcal{F}$. Denote by $\overline{\mathcal{F}}$ and $\overline{\mathcal{F}_t}$ the completions of $\mathcal{F}$ and $\mathcal{F}_t$ with respect to $\mathbb{P}$. If $\big(X(t), \mathcal{F}_t, \mathbb{P}\big)$ is a submartingale or martingale, show that $\big(X(t), \overline{\mathcal{F}_t}, \mathbb{P}\big)$ is also.

EXERCISE 7.1.23. Let $\mu \in \mathcal{I}(\mathbb{R}^N)$ be given as in Exercise 3.2.23, and extend $\ell_\mu$ to $\mathbb{C}^N$ accordingly. If $\{\mathbf{Z}(t) : t \geq 0\}$ is a Lévy process for $\mu$, show that (7.1.4) continues to hold for all $\boldsymbol{\xi} \in \mathbb{C}^N$.

EXERCISE 7.1.24. In Exercise 3.3.12, we discussed one-sided stable laws, and in Exercise 4.3.12 we showed that $\mathbb{P}\big(\max_{\tau\in[0,t]} B(\tau) \geq a\big) = 2\mathbb{P}\big(B(t) \geq a\big)$, where $\{B(t) : t \geq 0\}$ is an $\mathbb{R}$-valued Brownian motion. In this exercise, we will examine the relationship between these two.

**(i)** Set $\zeta^a(\psi) = \inf\{t \geq 0 : \psi(t) \geq a\}$, and show that the result in Exercise 4.3.12 can be rewritten as

$$\mathcal{W}^{(1)}(\zeta^a \leq t) = \sqrt{\frac{2}{\pi}} \int_{at^{-\frac{1}{2}}}^{\infty} e^{-\frac{y^2}{2}}\, dy.$$

Now use the results in Exercise 3.3.14 (especially (3.3.16)) to conclude that the $\mathcal{W}^{(1)}$-distribution of $\zeta^a$ is $\nu^{\frac{1}{2}}_{2^{\frac{1}{2}}a}$, the one-sided $\frac{1}{2}$-stable law "at time $2^{\frac{1}{2}}a$."

**(ii)** Here is another, more conceptual way to understand the conclusion drawn in (i) that the $\mathcal{W}^{(1)}$-distribution is a one-sided $\frac{1}{2}$-stable law. Namely, begin by showing that if $\psi(0) = 0$ and $\zeta^a(\psi) < \infty$, then $\zeta^{a+b}(\psi) = \zeta^a(\psi) + \zeta^b(\delta_{\zeta^a}\psi)$. As an application of Theorem 7.1.16, conclude from this that if $\beta_a$ denotes the $\mathcal{W}^{(1)}$-distribution of $\zeta^a$, then $\beta_{a+b} = \beta_a \star \beta_b$. In particular, this means that $\beta \equiv \beta_1$ is infinitely divisible and that $\widehat{\beta_a} = e^{a\ell_\beta}$, where $\ell_\beta$ is the exponent appearing in the Lévy–Khinchine formula for $\hat{\beta}$.

**(iii)** Next, use Brownian scaling to see that, for all $\lambda > 0$, $\zeta^{\lambda a}$ has the same $\mathcal{W}^{(1)}$-distribution as $\lambda^2\zeta^a$, and use this together with part **(iii)** of Exercise 3.3.12 to see that the distribution of $\zeta^1$ is $\nu_c^{\frac{1}{2}}$ for some $c > 0$.

**(iv)** Although we know from (i) that the constant $c$ must be $2^{\frac{1}{2}}$, here is an easier way to find it. Use Exercise 7.1.23 to see that $\big(e^{\lambda\psi(t)-\frac{1}{2}\lambda^2 t}, \mathcal{F}_t, \mathcal{W}^{(1)}\big)$ for every $\lambda \in \mathbb{R}$, and apply Doob's Stopping Time Theorem and the fact that $\mathcal{W}^{(1)}(\zeta^a < \infty) = 1$ to verify the identity $\mathbb{E}^{\mathcal{W}^{(1)}}\big[e^{-\frac{1}{2}\lambda^2\zeta^a}\big] = e^{-\lambda a}$ for $\lambda > 0$. Hence, the Laplace transform of $\nu_c^{\frac{1}{2}}$ is $e^{-\sqrt{2\lambda}}$, which, by the calculation in part **(iii)** of Exercise 3.3.12, means that $c = 2^{\frac{1}{2}}$. Of course, this calculation makes the preceding parts of this exercise unnecessary. Nonetheless, it is interesting to see the Brownian explanation for the properties of the one-sided, $\frac{1}{2}$-stable laws.

EXERCISE 7.1.25. An important corollary of Theorem 7.1.16 is the following formula. Working in the setting of that theorem, show that, for any stopping time $\zeta$ and $t \in (0,\infty)$ and $\Gamma \in \mathcal{B}_{\mathbb{R}^N}$,

$$\mathbb{Q}^\mu\big(\{\psi : \psi(t) \in \Gamma \;\&\; \zeta(\psi) \le t\}\big) = \mathbb{E}^{\mathbb{Q}^\mu}\Big[\mu_{t-\zeta}\big(\Gamma - \psi(\zeta)\big),\, \zeta \le t\Big],$$

where, as usual, $\mu_\tau$ is determined by $\widehat{\mu_\tau} = e^{\tau \ell_\mu}$. As a consequence,

$$\mathbb{Q}^\mu\big(\{\psi : \psi(t) \in \Gamma \;\&\; \zeta(\psi) > t\}\big) = \mu_t(\Gamma) - \mathbb{E}^{\mathbb{Q}^\mu}\Big[\mu_{t-\zeta}\big(\Gamma - \psi(\zeta)\big),\, \zeta \le t\Big],$$

which is a quite general, generic statement of what is called **Duhamel's Formula**.

## § 7.2 Brownian Motion and Martingales

In this section we will see that continuous martingales and Brownian motion are intimately related concepts. In addition, we will find that martingale theory, and especially Doob's and Hunt's Stopping Time Theorems, provides a powerful tool with which to study Brownian paths.

**§ 7.2.1. Lévy's Characterization of Brownian Motion.** When applied to $\mu = \gamma_{0,\mathbf{I}}$, Theorem 7.1.6 says that a progressively measurable function $\mathbf{B} : [0,\infty) \times \Omega \longrightarrow \mathbb{R}^N$ with $\mathbf{B}(0,\omega) = \mathbf{0}$ and $\mathbf{B}(\,\cdot\,,\omega) \in D(\mathbb{R}^N)$ is a Brownian motion if and only if

$$\left(\varphi(\mathbf{B}(t)) - \int_0^t \tfrac{1}{2}\Delta\varphi(\mathbf{B}(\tau))\, d\tau, \mathcal{F}_t, \mathbb{P}\right)$$

is a martingale for all $\varphi \in C_c^\infty(\mathbb{R}^N;\mathbb{R})$. In this subsection, I, following Lévy,[1] will give another martingale characterization of Brownian motion, this time involving many fewer test functions. On the other hand, we will have to assume ahead of time that $\mathbf{B}(\,\cdot\,,\omega) \in C(\mathbb{R}^N)$ for every $\omega \in \Omega$.

THEOREM 7.2.1 (**Lévy**). *Let $\mathbf{B} : [0,\infty) \times \Omega \longrightarrow \mathbb{R}^N$ be a progressively measurable function satisfying $\mathbf{B}(0,\omega) = \mathbf{0}$ and $\mathbf{B}(\,\cdot\,,\omega) \in C(\mathbb{R}^N)$ for every $\omega \in \Omega$. Then $\big(\mathbf{B}(t), \mathcal{F}_t, \mathbb{P}\big)$ is a Brownian motion if and only if*

$$\left(\big(\boldsymbol{\xi}, \mathbf{B}(t)\big)_{\mathbb{R}^N} + \big(\boldsymbol{\eta}, \mathbf{B}(t)\big)^2_{\mathbb{R}^N} - \frac{t|\boldsymbol{\eta}|^2}{2}, \mathcal{F}_t, \mathbb{P}\right)$$

*is a martingale for every $\boldsymbol{\xi}, \boldsymbol{\eta} \in \mathbb{R}^N$.*

---

[1] Lévy's Theorem is Theorem 11.9 in Chapter VII of Doob's *Stochastic Processes*, Wiley (1953). Doob uses a clever but somewhat opaque Central Limit argument. The argument given here is far simpler and is adapted from the one introduced by H. Kunita and S. Watanabe in their article "On square integrable martingales," *Nagoya Math. J.* **30** (1967).

PROOF: First suppose that $\big(\mathbf{B}(t),\mathcal{F}_t,\mathbb{P}\big)$ is a Brownian motion. Then, because $\mathbf{B}(t)-\mathbf{B}(s)$ is independent of $\mathcal{F}_s$ and has distribution $\gamma_{\mathbf{0},\mathbf{I}}$,

$$\mathbb{E}^{\mathbb{P}}\big[\mathbf{B}(t)-\mathbf{B}(s)\,\big|\,\mathcal{F}_s\big]=\mathbf{0}\quad\text{and}\quad\mathbb{E}^{\mathbb{P}}\big[\mathbf{B}(t)\otimes\mathbf{B}(t)-\mathbf{B}(s)\otimes\mathbf{B}(s)\,\big|\,\mathcal{F}_s\big]=(t-s)\mathbf{I}.$$

Hence, the necessity is obvious.

To prove the sufficiency, Theorem 7.1.3 says that it is enough to prove that

$$(*)\qquad \begin{aligned}&\mathbb{E}^{\mathbb{P}}\left[\exp\left[\sqrt{-1}\,\big(\boldsymbol{\xi},\mathbf{B}(t)\big)_{\mathbb{R}^N}+\tfrac{t|\boldsymbol{\xi}|^2}{2}\right],\,A\right]\\ &\qquad=\mathbb{E}^{\mathbb{P}}\left[\exp\left[\sqrt{-1}\,\big(\boldsymbol{\xi},\mathbf{B}(s)\big)_{\mathbb{R}^N}+\tfrac{s|\boldsymbol{\xi}|^2}{2}\right],\,A\right]\end{aligned}$$

for $0\le s<t$ and $A\in\mathcal{F}_s$. The challenge is to learn how to do this by taking full advantage of the assumed continuity. To this end, let $\epsilon\in(0,1]$ be given, set $\zeta_0\equiv s$, and use induction to define

$$\zeta_n=\Big(\inf\Big\{t\ge\zeta_{n-1}:\,\big|\mathbf{B}(t)-\mathbf{B}(\zeta_{n-1})\big|\ge\epsilon\Big\}\Big)\wedge\big(\zeta_{n-1}+\epsilon\big)\wedge t$$

for $n\in\mathbb{Z}^+$. Proceeding by induction, one can easily check that $\{\zeta_n:\,n\ge 0\}$ is a non-decreasing sequence of $[s,t]$-valued stopping times. Hence, by Theorem 7.1.14 and our assumption,

$$(**)\qquad \mathbb{E}^{\mathbb{P}}\Big[\Delta_n\,\Big|\,\mathcal{F}_{\zeta_{n-1}}\Big]=0=\mathbb{E}^{\mathbb{P}}\Big[\Delta_n^2-\delta_n\,\Big|\,\mathcal{F}_{\zeta_{n-1}}\Big],$$

where

$$\begin{aligned}\Delta_n(\omega)&\equiv\Big(\boldsymbol{\xi},\mathbf{B}\big(\zeta_n(\omega),\omega\big)-\mathbf{B}\big(\zeta_{n-1}(\omega),\omega\big)\Big)_{\mathbb{R}^N}\\ \delta_n(\omega)&\equiv|\boldsymbol{\xi}|^2\big(\zeta_n(\omega)-\zeta_{n-1}(\omega)\big).\end{aligned}$$

Moreover, because $\mathbf{B}(\,\cdot\,,\omega)$ is continuous, we know that, for each $\omega\in\Omega$, $|\Delta_n(\omega)|\le\epsilon|\boldsymbol{\xi}|$, $\delta_n(\omega)\le\epsilon|\boldsymbol{\xi}|^2$, and $\zeta_n(\omega)=t$ for all but a finite number of $n$'s. In particular, we can write the difference between the left and the right sides of (*) as the sum over $n\in\mathbb{Z}^+$ of $\mathbb{E}^{\mathbb{P}}\big[D_nM_n,\,A\big]$, where

$$\begin{aligned}D_n&\equiv\exp\left[\sqrt{-1}\,\Delta_n+\tfrac{\delta_n}{2}\right]-1\\ M_n&\equiv\exp\left[\sqrt{-1}\,\big(\boldsymbol{\xi},\mathbf{B}(\zeta_{n-1})\big)_{\mathbb{R}^N}+\tfrac{|\boldsymbol{\xi}|^2}{2}\zeta_{n-1}\right].\end{aligned}$$

By Taylor's Theorem,

$$\left|D_n-\Big(\sqrt{-1}\,\Delta_n+\tfrac{\delta_n}{2}\Big)-\tfrac12\Big(\sqrt{-1}\,\Delta_n+\tfrac{\delta_n}{2}\Big)^2\right|\le\tfrac16 e^{\frac{1+|\boldsymbol{\xi}|^2}{2}}\left|\sqrt{-1}\,\Delta_n+\tfrac{\delta_n}{2}\right|^3.$$

Hence, after rearranging terms, we see that $D_n = \sqrt{-1}\,\Delta_n - \frac{1}{2}(\Delta_n^2 - \delta_n) + E_n$, where, by our estimates on $\Delta_n$ and $\delta_n$,

$$|E_n| \le \tfrac{1}{2}|\Delta_n\delta_n| + \tfrac{\delta_n^2}{8} + \tfrac{2}{3}e^{\frac{1+|\boldsymbol{\xi}|^2}{2}}\left(|\Delta_n|^3 + \tfrac{\delta_n^3}{8}\right) \le \epsilon(1+|\boldsymbol{\xi}|^2)e^{\frac{1+|\boldsymbol{\xi}|^2}{2}}(\Delta_n^2 + \delta_n);$$

and so, after taking (**) into account,we arrive at

$$\left|\sum_1^\infty \mathbb{E}^{\mathbb{P}}[D_n M_m,\, A]\right| = \left|\sum_1^\infty \mathbb{E}^{\mathbb{P}}[E_n M_n,\, A]\right|$$
$$\le 2\epsilon(1+|\boldsymbol{\xi}|^2)e^{\frac{1+|\boldsymbol{\xi}|^2}{2}}\sum_1^\infty \mathbb{E}^{\mathbb{P}}[\delta_n|M_n|,\, A] \le 2\epsilon(1+|\boldsymbol{\xi}|^2)(t-s)e^{\frac{1+|\boldsymbol{\xi}|^2}{2}(1+t)}.$$

In other words, we have now proved that, for every $\epsilon \in (0,1]$, the difference between the two sides of (*) is dominated by $2\epsilon(1+|\boldsymbol{\xi}|^2)(t-s)e^{\frac{1+|\boldsymbol{\xi}|^2}{2}(1+t)}$, and so the equality in (*) has been established. □

As in Theorem 7.1.19, the subtlety here is in the use of the continuity assumption. Indeed, the same example that demonstrated its importance there does so again here. Namely, if $\{N(t) : t \ge 0\}$ is a simple Poisson process and $X(t) = N(t) - t$, then both $\big(X(t), \mathcal{F}_t, \mathbb{P}\big)$ and $\big(X(t)^2 - t, \mathcal{F}_t, \mathbb{P}\big)$ are martingales, but $\big(X(t), \mathcal{F}_t, \mathbb{P}\big)$ is certainly not a Brownian motion.

**§ 7.2.2. Doob–Meyer Decomposition, an Easy Case.** The continuous parameter analog of Lemma 5.2.12 is a highly non-trivial result, one that was proved by P.A. Meyer and led him to his profound analysis of stochastic processes. Nonetheless, there is an important case in which Meyer's result is relatively easy to prove, and that is the case proved in this subsection. However, before getting to that result, there is a rather fussy matter to be dealt with.

LEMMA 7.2.2. *For each $n \in \mathbb{N}$, let $X_n : [0,\infty) \longrightarrow \mathbb{R}$ be a right-continuous, progressively measurable function with the property that $X_n(\,\cdot\,,\omega)$ is continuous for $\mathbb{P}$-almost every $\omega \in \Omega$. If*

$$\lim_{m\to\infty}\sup_{n>m}\|X_n(\,\cdot\,,\omega) - X_m(\,\cdot\,,\omega)\|_{[0,t]} = 0 \text{ (a.s., } \mathbb{P}) \text{ for each } t \in (0,\infty),$$

*then there is a right-continuous, progressively measurable $X : [0,\infty) \longrightarrow \mathbb{R}$ such that $X(\,\cdot\,,\omega)$ is continuous and $X_n(\,\cdot\,,\omega) \longrightarrow X(\,\cdot\,,\omega)$ uniformly on compacts for $\mathbb{P}$-almost every $\omega \in \Omega$.*

PROOF: Set $A = \{(t,\omega) : \lim_{m\to\infty}\sup_{n>m}\|X_n(\,\cdot\,,\omega) - X_m(\,\cdot\,,\omega)\|_{[0,t]} = 0\}$. Then $A$ is progressively measurable. Next, define $\zeta(\omega) = \sup\{t \ge 0 : (t,\omega) \in A\}$, note that $\{\zeta < t\} \in \mathcal{F}_t$ for each $t \in (0,\infty)$, and set $B = \{(t,\omega) : \zeta(\omega) \ge t\}$. Then, $B$ is again progressively measurable. To see this, first note that

$$\{(\tau,\omega) \in [0,t]\times\Omega : \tau \wedge \zeta(\omega) < s\} = \begin{cases} \Omega \in \mathcal{F}_t & \text{if } t \le s \\ \{\zeta < s\} \in \mathcal{F}_t & \text{if } t > s, \end{cases}$$

and so $(\tau,\omega) \rightsquigarrow \tau\wedge\zeta(\omega)$ and therefore also $(\tau,\omega) \rightsquigarrow \tau\wedge\zeta(\omega)-\tau$ are progressively measurable functions. Hence, since $B = \{(\tau,\omega) : \tau \wedge \zeta(\omega) - \tau \geq 0\}$, $B$ is progressively measurable.

Now define

$$X(t,\omega) = \begin{cases} \lim_{n\to\infty} X_n(t,\omega) & \text{if } (t,\omega) \in A \\ 0 & \text{if } (t,\omega) \in B \setminus A \\ X(\zeta(\omega),\omega) & \text{if } (t,\omega) \notin B. \end{cases}$$

Clearly $X(\,\cdot\,,\omega)$ is right-continuous. Moreover, because $\zeta = \infty$ (a.s., $\mathbb{P}$), $X(\,\cdot\,,\omega)$ is continuous and $X_n(\,\cdot\,,\omega) \longrightarrow X(\,\cdot\,,\omega)$ uniformly on compacts for $\mathbb{P}$-almost every $\omega \in \Omega$. Thus, it only remains to check that $X$ is progressively measurable. For this purpose, let $\Gamma \in \mathcal{B}_{\mathbb{R}}$ be given, and set $C = \{(t,\omega) : X(t,\omega) \in \Gamma\}$. Because $A$ and the $X_n$'s are progressively measurable, it is clear that $C \cap A$ is progressively measurable. Similarly, because $B \setminus A$ is progressively measurable and $C \cap (B \setminus A)$ equals $B \setminus A$ or $\emptyset$ depending on whether $0 \in \Gamma$ or $0 \notin \Gamma$, $C \cap (B \setminus A)$, and therefore $C \cap B$, are progressively measurable. Hence, we now know that $X \restriction B$ is progressively measurable. Finally, we showed earlier that $(t,\omega) \rightsquigarrow t \wedge \zeta(\omega)$ is progressively measurable, and therefore so is $(t,\omega) \in [0,\infty) \times \Omega \longmapsto (t \wedge \zeta(\omega),\omega) \in B$. Thus, because $X(t,\omega) = X(t \wedge \zeta(\omega),\omega)$, we are done. $\square$

THEOREM 7.2.3. *Let $\big(X(t),\mathcal{F}_t,\mathbb{P}\big)$ be an $\mathbb{R}$-valued, square integrable martingale with the property that $X(\,\cdot\,,\omega)$ is continuous for $\mathbb{P}$-almost every $\omega \in \Omega$. Then there is a $\mathbb{P}$-almost surely unique progressively measurable function $\langle X\rangle : [0,\infty) \times \Omega \longrightarrow [0,\infty)$ such that $\langle X\rangle(0,\omega) = 0$ and $\langle X\rangle(\,\cdot\,,\omega)$ is continuous and non-decreasing for $\mathbb{P}$-almost every $\omega \in \Omega$, and $\big(X(t)^2 - \langle X\rangle(t),\mathcal{F}_t,\mathbb{P}\big)$ is a martingale.*

PROOF: The uniqueness is an immediate consequence of Corollary 7.1.20.

The proof of existence, which is based on a suggestion I got from K. Itô, is very much like that of Theorem 7.2.1. Without loss in generality, I will assume that $X(0) \equiv 0$.

I begin by reducing to the case when $X$ is $\mathbb{P}$-almost surely bounded. To this end, suppose that we know the result in this case. Given a general $X$ and $n \in \mathbb{N}$, define $\zeta_n = \inf\{t \geq 0 : |X(t)| \geq n\}$ and $X_n(t) = X(t\wedge\zeta_n)$. Then, $|X_n(\,\cdot\,,\omega)| \leq n$ and, by Doob's Inequality, $\zeta_n(\omega) \nearrow \infty$ for $\mathbb{P}$-almost every $\omega \in \Omega$. Moreover, by Corollary 7.1.15, $\big(X_n(t),\mathcal{F}_t,\mathbb{P}\big)$ is a martingale. Thus, by our assumption, for each $n$, we know $\langle X_n\rangle$ exists. In addition, by Corollary 7.1.15 and uniqueness, we know (cf. Exercise 7.2.10) that, $\mathbb{P}$-almost surely, $\langle X_m\rangle(t) = \langle X_n\rangle(t \wedge \zeta_m)$ for all $m \leq n$ and $t \geq 0$. Now define $\langle X\rangle$ so that $\langle X\rangle(t) = \langle X_n\rangle(t)$ for $\zeta_n \leq t < \zeta_{n+1}$. Then $\langle X\rangle$ is progressively measurable and right-continuous, $\langle X\rangle(0) = 0$, and, $\mathbb{P}$-almost surely, $\langle X\rangle$ is continuous and non-decreasing. Furthermore, $\big(X(t \wedge$

$\zeta_n)^2 - \langle X\rangle(t\wedge\zeta_n), \mathcal{F}_t, \mathbb{P})$ is a martingale for each $n \in \mathbb{N}$. Finally, note that, by Doob's Inequality,

$$\mathbb{E}^{\mathbb{P}}\big[\|\langle X\rangle\|_{[0,t]}\big] \le \mathbb{E}^{\mathbb{P}}\big[\|X\|^2_{[0,t]}\big] \le 4\mathbb{E}^{\mathbb{P}}\big[|X(t)|^2\big],$$

and so, as $n \to \infty$, $X(t\wedge\zeta_n)^2 - \langle X\rangle(t\wedge\zeta_n) \longrightarrow X(t)^2 - \langle X\rangle(t)$ in $L^1(\mathbb{P};\mathbb{R})$. Hence, $\big(X(t)^2 - \langle X\rangle(t), \mathcal{F}_t, \mathbb{P}\big)$ is a martingale.

I now assume that $|X(\,\cdot\,,\omega)| \le C < \infty$ for $\mathbb{P}$-almost every $\omega \in \Omega$. For each $n \in \mathbb{N}$, use induction to define $\{\zeta_{k,n} : k \ge 0\}$ so that $\zeta_{0,n} = 0$, $\zeta_{k,0} = k$, and, for $(k,n) \in (\mathbb{Z}^+)^2$, $\zeta_{k,n}$ is equal to

$$\Big(\inf\{\zeta_{\ell,n-1} : \zeta_{\ell,n-1} > \zeta_{k-1,n}\}\Big) \\ \wedge \Big(\inf\{t \ge \zeta_{k-1,n} : (t-\zeta_{k-1,n}) \vee |X(t) - X(\zeta_{k-1,n})| \ge \tfrac{1}{n}\}\Big).$$

Working by induction, one sees that, for each $n \in \mathbb{N}$, $\{\zeta_{k,n} : k \ge 0\}$ is a non-decreasing sequence of bounded stopping times. Moreover, because $X(\,\cdot\,,\omega)$ is $\mathbb{P}$-almost surely continuous, we know that, for each $n \in \mathbb{N}$, $\lim_{k\to\infty}\zeta_{k,n}(\omega) = \infty$ $\mathbb{P}$-almost every $\omega \in \Omega$. Finally, the sequences $\{\zeta_{k,n} : k \ge 0\}$ are nested in the sense that $\{\zeta_{k,n-1} : k \ge 0\} \subseteq \{\zeta_{k,n} : k \ge 0\}$ for each $n \in \mathbb{Z}^+$.

Set $X_{k,n} = X(\zeta_{k,n})$ and, for $k \ge 1$, $\Delta_{k,n}(t) = X(t\wedge\zeta_{k,n}) - X(t\wedge\zeta_{k-1,n})$. Then $X(t)^2 = 2M_n(t) + \langle X\rangle_n(t)$, where

$$M_n(t) = \sum_{k=1}^{\infty} X_{k-1,n}\Delta_{k,n}(t) \quad\text{and}\quad \langle X\rangle_n(t) = \sum_{k=1}^{\infty}\Delta_{k,n}(t)^2.$$

Of course, for $\mathbb{P}$-almost every $\omega \in \Omega$, all but a finite number of terms in each of these sums vanish. In addition, one should observe that $\langle X\rangle_n(s) \le \langle X\rangle_n(t)$ if $s \ge 0$ and $t - s > \frac{1}{n}$.

I now want to show that $\big(M_n(t), \mathcal{F}_t, \mathbb{P}\big)$ is a $\mathbb{P}$-almost surely continuous martingale for all $n \in \mathbb{N}$, and the first step is to show for each $(k,n) \in \mathbb{Z}^+ \times \mathbb{N}$, $\big(X_{k-1,n}\Delta_{k,n}(t), \mathcal{F}_t, \mathbb{P}\big)$ is a $\mathbb{P}$-almost surely continuous martingale. Indeed, if $0 \le s < t$ and $A \in \mathcal{F}_s$, then

$$\mathbb{E}^{\mathbb{P}}\big[X_{k-1,n}\Delta_{k,n}(t),\, A\big] = \mathbb{E}^{\mathbb{P}}\big[X_{k-1,n}\Delta_{k,n}(t),\, A \cap \{\zeta_{k-1,n} \le s\}\big] \\ + \mathbb{E}^{\mathbb{P}}\big[X_{k-1,n}\Delta_{k,n}(t),\, A \cap \{\zeta_{k-1,n} > s\}\big].$$

Next, check that

$$\begin{aligned} &\mathbb{E}^{\mathbb{P}}\big[X_{k-1,n}\Delta_{k,n}(t),\, A \cap \{\zeta_{k-1,n} \le s\}\big] \\ &= \mathbb{E}^{\mathbb{P}}\big[X_{k-1,n}\big(X(\zeta_{k,n}) - X(\zeta_{k-1,n})\big),\, A \cap \{\zeta_{k,n} \le s\}\big] \\ &\qquad + \mathbb{E}^{\mathbb{P}}\Big[X_{k-1,n}\Big(X\big((t\wedge\zeta_{k,n})\vee s\big) - X(\zeta_{k-1,n})\Big),\, A \cap \{\zeta_{k-1,n} \le s < \zeta_{k,n}\}\Big] \\ &= \mathbb{E}^{\mathbb{P}}\big[X_{k-1,n}\Delta_{k,n}(s),\, A \cap \{\zeta_{k,n} \le s\}\big] \\ &\qquad + \mathbb{E}^{\mathbb{P}}\big[X_{k-1,n}\big(X(s) - X(\zeta_{k-1,n})\big),\, A \cap \{\zeta_{k-1,n} \le s < \zeta_{k,n}\}\big] \\ &= \mathbb{E}^{\mathbb{P}}\big[X_{k-1,n}\Delta_{k,n}(s),\, A \cap \{\zeta_{k-1,n} \le s\}\big], \end{aligned}$$

where, in the passage to the second to last equality, I have used the fact that $X_{k-1,n}\mathbf{1}_A\mathbf{1}_{[\zeta_{k-1,n},\zeta_{k,n})}(s)$ is $\mathcal{F}_s$-measurable and applied Theorem 7.1.14. At the same time

$$\begin{aligned}
&\mathbb{E}^{\mathbb{P}}\big[X_{k-1,n}\Delta_{k,n}(t),\, A\cap\{\zeta_{k-1,n}>s\}\big]\\
&\quad=\mathbb{E}^{\mathbb{P}}\big[X_{k-1,n}\big(X(t\wedge\zeta_{k,n})-X(t\wedge\zeta_{k-1,n})\big),\, A\cap\{s<\zeta_{k-1,n}\le t\}\big]\\
&\quad=\mathbb{E}^{\mathbb{P}}\big[X_{k-1,n}\big(X(t)-X(t)\big),\, A\cap\{s<\zeta_{k-1,n}\le t\}\big]\\
&\quad=0=\mathbb{E}^{\mathbb{P}}\big[X_{k-1,n}\Delta_{k,n}(s),\, A\cap\{\zeta_{k-1,n}>s\}\big],
\end{aligned}$$

where I have used the fact that $X_{k-1,n}\mathbf{1}_A\mathbf{1}_{(s,t]}(\zeta_{k-1,n})$ is $\mathcal{F}_{t\wedge\zeta_{\zeta_{k-1,n}}}$-measurable and again applied Theorem 7.1.14 in getting the second to last line. After combining these, one sees that $\mathbb{E}^{\mathbb{P}}\big[X_{k-1,n}\Delta_{k,n}(t),\, A\big]=\mathbb{E}^{\mathbb{P}}\big[X_{k-1,n}\Delta_{k,n}(s),\, A\big]$, which means that $\big(X_{k-1,n}\Delta_{k,n}(t),\mathcal{F}_t,\mathbb{P}\big)$ is a $\mathbb{P}$-almost surely continuous martingale.

Given the preceding, it is clear that, for each $n$ and $\ell$, $\big(M_n(t\wedge\zeta_{\ell,n}),\mathcal{F}_t,\mathbb{P}\big)$ is a $\mathbb{P}$-almost surely continuous, square integrable martingale. In addition, for $k\ne k'$, $X_{k-1}\Delta_{k,n}(t\wedge\zeta_{\ell,n})$ is orthogonal to $X_{k'-1}\Delta_{k',n}(t\wedge\zeta_{\ell,n})$ in $L^2(\mathbb{P};\mathbb{R})$. Thus

$$\begin{aligned}
&\mathbb{E}^{\mathbb{P}}\left[\sup_{0\le\tau\le t\wedge\zeta_{\ell,n}} M_n(\tau)^2\right]\le 4\mathbb{E}^{\mathbb{P}}\big[M_n(t\wedge\zeta_{\ell,n})^2\big]\\
&\quad=4\sum_{k=1}^{\ell}\mathbb{E}^{\mathbb{P}}\big[X_{k-1,n}^2\Delta_{k,n}(t\wedge\zeta_{\ell,n})^2\big]\le 4C^2\sum_{k=1}^{\ell}\mathbb{E}^{\mathbb{P}}\big[\Delta_{k,n}(t\wedge\zeta_{\ell,n})^2\big]\\
&\quad=4C^2\mathbb{E}^{\mathbb{P}}\big[X(t\wedge\zeta_{\ell,n})^2\big]\le 4C^2\mathbb{E}^{\mathbb{P}}\big[X(t)^2\big],
\end{aligned}$$

from which it is easy to see that $\big(M_n(t),\mathcal{F}_t,\mathbb{P}\big)$ is a $\mathbb{P}$-almost surely continuous, square integrable martingale.

I will now show that $\lim_{m\to\infty}\sup_{n>m}\|M_n-M_m\|_{[0,t]}=0$ $\mathbb{P}$-almost surely and in $L^2(\mathbb{P};\mathbb{R})$ for each $t\in[0,\infty)$. To this end, define $Y_{k-1,n}^{(m)}$ so that $Y_{k-1,n}^{(m)}(\omega)$ $=X_{k-1,n}(\omega)-X_{\ell-1,m}(\omega)$ when $\zeta_{\ell-1,m}(\omega)\le\zeta_{k-1,n}(\omega)<\zeta_{\ell,m}(\omega)$. Then $Y_{k-1,n}^{(m)}$ is $\mathcal{F}_{k-1,n}$-measurable, $|Y_{k-1,n}^{(m)}|\le\frac{1}{m}$ (a.s., $\mathbb{P}$), and $M_n-M_m=\sum_{k=1}^{\infty}Y_{k-1,n}^{(m)}\Delta_{k,n}$. Hence, by the same reasoning as above,

$$\mathbb{E}^{\mathbb{P}}\big[\|M_n-M_m\|_{[0,t]}^2\big]\le 4\sum_{k=1}^{\infty}\mathbb{E}^{\mathbb{P}}\big[(Y_{k-1,n}^{(m)})^2\Delta_{k,n}(t)^2\big]\le\frac{4}{m^2}\mathbb{E}^{\mathbb{P}}\big[X(t)^2\big],$$

which is more than enough to get the asserted convergence result.

We can now apply Lemma 7.2.2 to produce a right-continuous, progressively measurable $M:[0,\infty)\times\Omega\longrightarrow\mathbb{R}$ which is $\mathbb{P}$-almost surely continuous and to which $\{M_n : n \geq 1\}$ converges uniformly on compacts, both $\mathbb{P}$-almost surely and in $L^2(\mathbb{P};\mathbb{R})$. In particular, $\big(M(t), \mathcal{F}_t, \mathbb{P}\big)$ is a square integrable martingale. Finally, set $\langle X\rangle = (X^2 - 2M)^+$. Obviously, $\langle X\rangle = X^2 - 2M$ (a.s., $\mathbb{P}$), and $\langle X\rangle$ is right-continuous, progressively measurable, and $\mathbb{P}$-almost surely continuous. In addition, because, $\mathbb{P}$-almost surely, $\langle X\rangle_n \longrightarrow \langle X\rangle$ uniformly on compacts and $\langle X\rangle_n(s) \leq \langle X\rangle_n(t)$ when $t - s > \frac{1}{n}$, it follows that $\langle X\rangle(\,\cdot\,,\omega)$ is non-decreasing for $\mathbb{P}$-almost every $\omega \in \Omega$. □

REMARK 7.2.4. The reader may be wondering why I chose to complicate the preceding statement and proof by insisting that $\langle X\rangle$ be progressively measurable with respect to the original family of $\sigma$-algebras $\{\mathcal{F}_t : t \in [0,\infty)\}$. Indeed, Exercise 7.1.22 shows that I could have replaced all the $\sigma$-algebras with their completions, and, if I had done so, there would have been no reason not to have taken $X(\,\cdot\,,\omega)$ to be continuous and $\langle X\rangle(\,\cdot\,,\omega)$ to be continuous and non-decreasing for every $\omega \in \Omega$. However, there is a price to be paid for completing $\sigma$-algebras. In the first place, when one does, all statements become dependent on the particular $\mathbb{P}$ with which one is dealing. Secondly, because completed $\sigma$-algebras are nearly never countably generated, certain desirable properties can be lost by introducing them. See, for example, Theorem 9.2.1.

By combining Theorem 7.2.3 with Theorem 7.2.1, one can show that, up to time re-parametrization, all continuous martingales are Brownian motions. In order to avoid technical difficulties, I will prove this result only in the simplest case.

COROLLARY 7.2.5. *Let $\big(X(t), \mathcal{F}_t, \mathbb{P}\big)$ be a continuous, square integrable martingale with the properties that, for $\mathbb{P}$-almost every $\omega \in \Omega$, $\langle X\rangle(\,\cdot\,,\omega)$ is strictly increasing and $\lim_{t\to\infty}\langle X\rangle(t,\omega) = \infty$. Then there exists a Brownian motion $\big(B(t), \mathcal{F}'_t, \mathbb{P}\big)$ such that $X(t) = X(0) + B\big(\langle X\rangle(t)\big)$, $t \in [0,\infty)$ $\mathbb{P}$-almost surely. In particular,*

$$\varlimsup_{t\to\infty} \frac{X(t)}{\sqrt{2\langle X\rangle(t)\log_{(2)}\langle X\rangle(t)}} = 1 = -\varliminf_{t\to\infty} \frac{X(t)}{\sqrt{2\langle X\rangle(t)\log_{(2)}\langle X\rangle(t)}}$$

*$\mathbb{P}$-almost surely.*

PROOF: Clearly, given the first part, the last assertion is a trivial application of Exercise 4.3.15.

After replacing $\mathcal{F}$ and the $\mathcal{F}_t$'s by their completions and applying Exercise 7.1.22, I may and will assume that $X(0,\omega) = 0$, $X(\,\cdot\,,\omega)$ is continuous, $\langle X\rangle(\,\cdot\,,\omega)$ is continuous and strictly increasing, and $\lim_{t\to\infty}\langle X\rangle(t,\omega) = \infty$ for every $\omega \in \Omega$. Next, for each $(t,\omega) \in [0,\infty)$, set $\zeta_t(\omega) = \langle X\rangle^{-1}(t,\omega)$, where $\langle X\rangle^{-1}(\cdot,\omega)$ is the inverse of $\langle X\rangle(\,\cdot\,,\omega)$. Clearly, for each $\omega \in \Omega$, $t \rightsquigarrow \zeta_t(\omega)$ is a continuous, strictly increasing function that tends to infinity as $t \to \infty$. Moreover, because $\langle X\rangle$ is progressively measurable, $\zeta_t$ is a stopping time for each $t \in [0,\infty)$. Now set

$B(t) = X(\zeta_t)$. Since it is obvious that $X(t) = B(\langle X\rangle(t))$, all that I have to show is that $(B(t), \mathcal{F}'_t, \mathbb{P})$ is a Brownian motion for some non-decreasing family $\{\mathcal{F}'_t : t \geq 0\}$ of sub-$\sigma$-algebras.

Trivially, $B(0,\omega) = 0$ and $B(\,\cdot\,,\omega)$ is continuous for all $\omega \in \Omega$. In addition, $B(t)$ is $\mathcal{F}_{\zeta_t}$-measurable, and so $B$ is progressively measurable with respect to $\{\mathcal{F}_{\zeta_t} : t \geq 0\}$. Thus, by Theorem 7.2.1, I will be done once I show that $(B(t), \mathcal{F}_{\zeta_t}, \mathbb{P})$ and $(B(t)^2 - t, \mathcal{F}_{\zeta_t}, \mathbb{P})$ are martingales. To this end, first observe that

$$\begin{aligned}\mathbb{E}^{\mathbb{P}}\left[\sup_{\tau\in[0,\zeta_t]} X(\tau)^2\right] &= \lim_{T\to\infty} \mathbb{E}^{\mathbb{P}}\left[\sup_{\tau\in[0,T\wedge\zeta_t]} X(\tau)^2\right] \\ &\leq 4 \lim_{T\to\infty} \mathbb{E}^{\mathbb{P}}\big[X(T\wedge\zeta_t)^2\big] \leq 4 \lim_{T\to\infty} \mathbb{E}^{\mathbb{P}}\big[\langle X\rangle(T\wedge\zeta_t)\big] \leq 4t.\end{aligned}$$

Thus, $\lim_{T\to\infty} X(T\wedge\zeta_t) \longrightarrow B(t)$ in $L^2(\mathbb{P};\mathbb{R})$. Now let $0 \leq s < t$ and $A \in \mathcal{F}_{\zeta_s}$ be given. Then, for each $T > 0$, $A_T \equiv A \cap \{\zeta_s \leq T\} \in \mathcal{F}_{T\wedge\zeta_s}$, and so, by Theorem 7.1.14,

$$\mathbb{E}^{\mathbb{P}}\big[X(T\wedge\zeta_t),\, A_T\big] = \mathbb{E}^{\mathbb{P}}\big[X(T\wedge\zeta_s),\, A_T\big]$$

and

$$\mathbb{E}^{\mathbb{P}}\big[X(T\wedge\zeta_t)^2 - \langle X\rangle(T\wedge\zeta_t),\, A_T\big] = \mathbb{E}^{\mathbb{P}}\big[X(T\wedge\zeta_s)^2 - \langle X\rangle(T\wedge\zeta_s),\, A_T\big].$$

Now let $T \to \infty$, and apply the preceding convergence assertion to get the desired conclusion. □

§ **7.2.3. Burkholder's Inequality Again.** In this subsection we will see what Burkholder's Inequality looks like in the continuous parameter setting, a result whose importance for the theory of stochastic integration is hard to overstate.

THEOREM 7.2.6 (**Burkholder**). *Let $(X(t), \mathcal{F}_t, \mathbb{P})$ be a $\mathbb{P}$-almost surely continuous, square integrable martingale. Then, for each $p \in (1,\infty)$ and $t \in [0,\infty)$ (cf. (6.3.2)),*

$$(7.2.7)\quad B_p^{-1}\|X(t) - X(0)\|_{L^p(\mathbb{P};\mathbb{R})} \leq \mathbb{E}^{\mathbb{P}}\big[\langle X(t)\rangle^{\frac{p}{2}}\big]^{\frac{1}{p}} \leq B_p\|X(t) - X(0)\|_{L^p(\mathbb{P};\mathbb{R})}.$$

PROOF: After completing the $\sigma$-algebras if necessary, I may (cf. Exercise 7.1.22) and will assume that $X(\,\cdot\,,\omega)$ is continuous and that $\langle X\rangle(\,\cdot\,,\omega)$ is continuous and non-decreasing for every $\omega \in \Omega$. In addition, I may and will assume that $X(0) = 0$. Finally, I will assume that $X$ is bounded. To justify this last assumption, let $\zeta_n = \inf\{t \geq 0 : |X(t)| \geq n\}$, set $X_n(t) = X(t\wedge\zeta_n)$, and use Exercise 7.2.10 to see that one can take $\langle X_n\rangle = \langle X\rangle(t\wedge\zeta_n)$. Hence, if we know (7.2.7) for bounded martingales, then

$$B_p^{-1}\|X(t\wedge\zeta_n)\|_{L^p(\mathbb{P};\mathbb{R})} \leq \mathbb{E}^{\mathbb{P}}\big[\langle X\rangle(t\wedge\zeta_n)^{\frac{p}{2}}\big]^{\frac{1}{p}} \leq B_p\|X(t\wedge\zeta_n)\|_{L^p(\mathbb{P};\mathbb{R})}$$

for all $n \geq 1$. Since $\langle X \rangle$ is non-decreasing, we can apply Fatou's Lemma to the preceding and thereby get

$$\|X(t)\|_{L^p(\mathbb{P};\mathbb{R})} \leq \varliminf_{n\to\infty} \|X(t \wedge \zeta_n)\|_{L^p(\mathbb{P};\mathbb{R})} \leq B_p \mathbb{E}^{\mathbb{P}}\left[\langle X \rangle(t)^{\frac{p}{2}}\right]^{\frac{1}{p}},$$

which is the left-hand side of (7.2.7). To get the right-hand side, note that either $\|X(t)\|_{L^p(\mathbb{P};\mathbb{R})} = \infty$, in which case there is nothing to do, or $\|X(t)\|_{L^p(\mathbb{P};\mathbb{R})} < \infty$, in which case, by the second half of Theorem 7.1.9, $X(t \wedge \zeta_n) \longrightarrow X(t)$ in $L^p(\mathbb{P};\mathbb{R})$ and therefore

$$\begin{aligned}\mathbb{E}^{\mathbb{P}}\left[\langle X \rangle(t)^{\frac{p}{2}}\right]^{\frac{1}{p}} &= \lim_{n\to\infty} \mathbb{E}^{\mathbb{P}}\left[\langle X \rangle(t \wedge \zeta_n)^{\frac{p}{2}}\right]^{\frac{1}{p}} \\ &\leq B_p \lim_{n\to\infty} \|X(t \wedge \zeta_n)\|_{L^p(\mathbb{P};\mathbb{R})} = B_p \|X(t)\|_{L^p(\mathbb{P};\mathbb{R})}.\end{aligned}$$

Proceeding under the above assumptions and referring to the notation in the proof of Theorem 7.2.3, begin by observing that, for any $t \in [0,\infty)$ and $n \in \mathbb{N}$, Theorem 7.1.14 shows that $\left(X(t \wedge \zeta_{k,n}), \mathcal{F}_{t\wedge\zeta_{k,n}}, \mathbb{P}\right)$ is a discrete parameter martingale indexed by $k \in \mathbb{N}$. In addition, $\zeta_{k,n} = t$ for all but a finite number of $k$'s. Hence, by (6.3.7) applied to $\left(X(t \wedge \zeta_{k,n}), \mathcal{F}_{t\wedge\zeta_{k,n}}, \mathbb{P}\right)$,

$$B_p^{-1}\|X(t)\|_{L^p(\mathbb{P};\mathbb{R})} \leq \mathbb{E}^{\mathbb{P}}\left[\langle X \rangle_n(t)^{\frac{p}{2}}\right]^{\frac{1}{p}} \leq B_p\|X(t)\|_{L^p(\mathbb{P};\mathbb{R})} \quad \text{for all } n \in \mathbb{N}.$$

In particular, this shows that $\sup_{n\geq 0} \|\langle X \rangle_n(t)\|_{L^p(\mathbb{P};\mathbb{R})} < \infty$ for every $p \in (1,\infty)$, and therefore, since $\langle X \rangle_n(t) \longrightarrow \langle X \rangle(t)$ (a.s., $\mathbb{P}$), this is more than enough to verify that $\mathbb{E}^{\mathbb{P}}\left[\langle X \rangle_n(t)^{\frac{p}{2}}\right] \longrightarrow \mathbb{E}^{\mathbb{P}}\left[\langle X \rangle(t)^{\frac{p}{2}}\right]$ for every $p \in (1,\infty)$. □

## Exercises for § 7.2

EXERCISE 7.2.8. Let $\left(X(t), \mathcal{F}_t, \mathbb{P}\right)$ be a square integrable, continuous martingale. Following the strategy used to prove Theorem 7.2.1, show that

$$\left(F(X(t)) - \int_0^t \tfrac{1}{2}\partial_x^2 F(X(\tau))\langle X \rangle(d\tau), \mathcal{F}_t, \mathbb{P}\right)$$

is a martingale for every $F \in C_b^2(\mathbb{R};\mathbb{C})$.

**Hint**: Begin by using cutoffs and mollification to reduce to the case when $F \in C_c^\infty(\mathbb{R};\mathbb{R})$. Next, given $s < t$ and $\epsilon > 0$, introduce the stopping times $\zeta_0 = s$ and

$$\zeta_n = \inf\{t \geq \zeta_{n-1} : |X(t) - X(\zeta_{n-1})| \geq \epsilon\} \wedge (\zeta_{n-1} + \epsilon) \wedge (\langle X \rangle(\zeta_{n-1}) + \epsilon) \wedge t$$

for $n \geq 1$. Now proceed as in the proof of Theorem 7.2.1.

EXERCISE 7.2.9. Let $(X(t), \mathcal{F}_t, \mathbb{P})$ be a continuous, square integrable martingale with $X(0) = 0$, and assume that there exists a non-decreasing function $A : [0, \infty) \longrightarrow [0, \infty)$ such that $\langle X \rangle(t) \le A(t)$ (a.s., $\mathbb{P}$) for each $t \in [0, \infty)$. The goal of this exercise is to show that $(E(t), \mathcal{F}_t, \mathbb{P})$ is a martingale when

$$E(t) = \exp\left[X(t) - \tfrac{1}{2}\langle X \rangle(t)\right].$$

(i) Given $R \in (0, \infty)$, set $\zeta_R = \inf\{t \ge 0 : |X(t)| \ge R\}$, and show that

$$\left( e^{X(t\wedge\zeta_R)} - \frac{1}{2}\int_0^{t\wedge\zeta_R} e^{X(\tau)}\, d\langle X \rangle, \mathcal{F}_t, \mathbb{P} \right)$$

is a martingale.

**Hint**: Choose $F \in C_c^\infty(\mathbb{R};\mathbb{R})$ so that $F(x) = e^x$ for $x \in [-2R, 2R]$, apply Exercise 7.2.8 to this $F$, and then use Doob's Stopping Time Theorem.

(ii) Apply Theorem 7.1.17 to the martingale in (i) and $e^{-\frac{1}{2}\langle X \rangle(t\wedge\zeta_R)}$ to show that $(E(t\wedge\zeta_R), \mathcal{F}_t, \mathbb{P})$ is a martingale.

(iii) By replacing $X$ and $R$ with $2X$ and $2R$ in (ii), show that

$$\mathbb{E}^{\mathbb{P}}\left[E(t\wedge\zeta_R)^2\right] \le e^{A(t)}\mathbb{E}^{\mathbb{P}}\left[e^{2X(t\wedge\zeta_R) - 2\langle X \rangle(t\wedge\zeta_R)}\right] = e^{A(t)}.$$

Conclude that $\{E(t\wedge\zeta_R) : R \in (0,\infty)\}$ is uniformly $\mathbb{P}$-integrable and therefore that $(E(t), \mathcal{F}_t, \mathbb{P})$ is a martingale.

EXERCISE 7.2.10. If $(X(t), \mathcal{F}_t, \mathbb{P})$ is a $\mathbb{P}$-almost surely continuous, square integrable martingale, $\zeta$ is a stopping time, and $Y(t) = X(t\wedge\zeta)$, show that $\langle Y \rangle(t) = \langle X \rangle(t\wedge\zeta)$, $t \ge 0$, $\mathbb{P}$-almost surely.

EXERCISE 7.2.11. Continuing in the setting of Exercise 7.2.9, first show that, for every $\lambda \in \mathbb{R}$, $(E_\lambda(t), \mathcal{F}_t, \mathbb{P})$ is a martingale, where

$$E_\lambda(t) = \exp\left[\lambda X(t) - \tfrac{\lambda^2}{2}\langle X \rangle(t)\right].$$

Next, use Doob's Inequality to see that, for each $\lambda \ge 0$,

$$\mathbb{P}\left( \sup_{\tau\in[0,t]} X(\tau) \ge R \right) \le \mathbb{P}\left( \sup_{\tau\in[0,t]} E_\lambda(\tau) \ge e^{\lambda R - \frac{\lambda^2}{2}A(t)} \right) \le e^{-\lambda R + \frac{\lambda^2}{2}A(t)}.$$

Starting from this, conclude that

$$\mathbb{P}\left(\|X\|_{[0,t]} \ge R\right) \le 2e^{-\frac{R^2}{2A(t)}}. \tag{7.2.12}$$

Finally, given this estimate, show that the conclusion in Exercise 7.2.8 continues to hold for any $F \in C^2(\mathbb{R};\mathbb{C})$ whose second derivative has at most exponential growth.

EXERCISE 7.2.13. Given a pair of square integrable, continuous martingales $\big(X(t), \mathcal{F}_t, \mathbb{P}\big)$ and $\big(Y(t), \mathcal{F}_t, \mathbb{P}\big)$, set $\langle X, Y\rangle = \frac{\langle X+Y\rangle - \langle X-Y\rangle}{4}$, and show that $\big(X(t)Y(t) - \langle X, Y\rangle(t), \mathcal{F}_t, \mathbb{P}\big)$ is a martingale. Further, show that $\langle X, Y\rangle$ is uniquely determined up to a $\mathbb{P}$-null set by this property together with the facts that $\langle X, Y\rangle(0,\omega) = 0$ and $\langle X, Y\rangle(\,\cdot\,,\omega)$ is continuous and has locally bounded variation for $\mathbb{P}$-almost every $\omega \in \Omega$.

EXERCISE 7.2.14. Let $\big(\mathbf{B}(t), \mathcal{F}_t, \mathbb{P}\big)$ be an $\mathbb{R}^N$-valued Brownian motion. Given $f, g \in C^{1,2}_{\mathrm{b}}\big([0,\infty)\times\mathbb{R}^N;\mathbb{R}\big)$, set

$$X(t) = f\big(t, \mathbf{B}(t)\big) - \int_0^t \big(\partial_\tau + \tfrac{1}{2}\Delta\big) f\big(\tau, \mathbf{B}(\tau)\big)\, d\tau,$$

$$Y(t) = g\big(t, \mathbf{B}(t)\big) - \int_0^t \big(\partial_\tau + \tfrac{1}{2}\Delta\big) g\big(\tau, \mathbf{B}(\tau)\big)\, d\tau,$$

and show that

$$\langle X, Y\rangle(t) = \int_0^t \nabla f \cdot \nabla g\big(\tau, \mathbf{B}(\tau)\big)\, d\tau.$$

**Hint**: First reduce to the case when $f = g$. Second, write $X(t)^2$ as

$$f\big(t, \mathbf{B}(t)\big)^2 - 2X(t)\int_0^t \big(\partial_\tau + \tfrac{1}{2}\Delta\big) f\big(\tau, \mathbf{B}(\tau)\big)\, d\tau - \left(\int_0^t \big(\partial_\tau + \tfrac{1}{2}\Delta\big) f\big(\tau, \mathbf{B}(\tau)\big)\, d\tau\right)^2,$$

and apply Theorem 7.1.17 to the second term.

## § 7.3 The Reflection Principle Revisited

In Exercise 4.3.12 we saw that Lévy's Reflection Principle (Theorem 1.4.13) has a sharpened version when applied to Brownian motion. In this section I will give another, more powerful way of discussing the reflection principle for Brownian motion.

**§ 7.3.1. Reflecting Symmetric Lévy Processes.** In this subsection, $\mu$ will be used to denote a symmetric, infinitely divisible law. Equivalently (cf. Exercise 3.3.11), $\hat{\mu} = e^{\ell_\mu(\boldsymbol{\xi})}$, where

$$\ell_\mu(\boldsymbol{\xi}) = -\frac{1}{2}\big(\boldsymbol{\xi}, \mathbf{C}\boldsymbol{\xi}\big)_{\mathbb{R}^N} + \int_{\mathbb{R}^N}\Big(\cos(\boldsymbol{\xi}, \mathbf{y})_{\mathbb{R}^N} - 1\Big)\, M(d\mathbf{y})$$

for some non-negative definite, symmetric $\mathbf{C}$ and symmetric Lévy measure $M$.

LEMMA 7.3.1. *Let $\{\mathbf{Z}(t) : t \geq 0\}$ be a Lévy process for $\mu$, and set $\mathcal{F}_t = \sigma(\{\mathbf{Z}(\tau) : \tau \in [0,t]\})$. If $\zeta$ is a stopping time relative to $\{\mathcal{F}_t : t \in [0,\infty)\}$ and*

$$\tilde{\mathbf{Z}}(t) \equiv 2\mathbf{Z}(t\wedge\zeta) - \mathbf{Z}(t) = \begin{cases} \mathbf{Z}(t) & \text{if } \zeta > t \\ 2\mathbf{Z}(\zeta) - \mathbf{Z}(t) & \text{if } \zeta \leq t, \end{cases}$$

*then $\{\tilde{\mathbf{Z}}(t) : t \geq 0\}$ is again a Lévy process for $\mu$.*

PROOF: According to Theorem 7.1.3, all that I have to show is that

$$\Big(\exp\big[\sqrt{-1}\,(\boldsymbol{\xi}, \tilde{\mathbf{Z}}(t))_{\mathbb{R}^N} - t\ell_\mu(\boldsymbol{\xi})\big], \mathcal{F}_t, \mathbb{P}\Big)$$

is a martingale for all $\boldsymbol{\xi} \in \mathbb{R}^N$. Thus, let $0 \leq s < t$ and $A \in \mathcal{F}_s$ be given. Then, by Theorem 7.1.14 and the fact that $\ell_\mu(-\boldsymbol{\xi}) = \ell_\mu(\boldsymbol{\xi})$,

$$\begin{aligned}
&\mathbb{E}^{\mathbb{P}}\Big[\exp\big[\sqrt{-1}\,(\boldsymbol{\xi}, \tilde{\mathbf{Z}}(t))_{\mathbb{R}^N} - t\ell_\mu(\boldsymbol{\xi})\big],\, A\cap\{\zeta\leq s\}\Big] \\
&= \mathbb{E}^{\mathbb{P}}\Big[e^{2\sqrt{-1}(\boldsymbol{\xi},\mathbf{Z}(s\wedge\zeta))_{\mathbb{R}^N}}\exp\big[-\sqrt{-1}\,(\boldsymbol{\xi}, \mathbf{Z}(t))_{\mathbb{R}^N} - t\ell_\mu(\boldsymbol{\xi})\big],\, A\cap\{\zeta\leq s\}\Big] \\
&= \mathbb{E}^{\mathbb{P}}\Big[e^{2\sqrt{-1}(\boldsymbol{\xi},\mathbf{Z}(s\wedge\zeta))_{\mathbb{R}^N}}\exp\big[-\sqrt{-1}\,(\boldsymbol{\xi}, \mathbf{Z}(s))_{\mathbb{R}^N} - s\ell_\mu(\boldsymbol{\xi})\big],\, A\cap\{\zeta\leq s\}\Big] \\
&= \mathbb{E}^{\mathbb{P}}\Big[\exp\big[\sqrt{-1}\,(\boldsymbol{\xi}, \tilde{\mathbf{Z}}(s))_{\mathbb{R}^N} - t\ell_\mu(\boldsymbol{\xi})\big],\, A\cap\{\zeta\leq s\}\Big].
\end{aligned}$$

Similarly,

$$\begin{aligned}
&\mathbb{E}^{\mathbb{P}}\Big[\exp\big[\sqrt{-1}\,(\boldsymbol{\xi}, \tilde{\mathbf{Z}}(t))_{\mathbb{R}^N} - t\ell_\mu(\boldsymbol{\xi})\big],\, A\cap\{\zeta> s\}\Big] \\
&= \mathbb{E}^{\mathbb{P}}\Big[e^{2\sqrt{-1}(\boldsymbol{\xi},\mathbf{Z}(t\wedge\zeta))_{\mathbb{R}^N}}\exp\big[-\sqrt{-1}\,(\boldsymbol{\xi}, \mathbf{Z}(t))_{\mathbb{R}^N} - t\ell_\mu(\boldsymbol{\xi})\big],\, A\cap\{\zeta> s\}\Big] \\
&= \mathbb{E}^{\mathbb{P}}\Big[\exp\big[\sqrt{-1}(\boldsymbol{\xi}, \mathbf{Z}(t\wedge\zeta))_{\mathbb{R}^N} - (t\wedge\zeta)\ell_\mu(\boldsymbol{\xi})\big],\, A\cap\{\zeta> s\}\Big] \\
&= \mathbb{E}^{\mathbb{P}}\Big[\exp\big[\sqrt{-1}(\boldsymbol{\xi}, \mathbf{Z}(s\wedge\zeta))_{\mathbb{R}^N} - (s\wedge\zeta)\ell_\mu(\boldsymbol{\xi})\big],\, A\cap\{\zeta> s\}\Big] \\
&= \mathbb{E}^{\mathbb{P}}\Big[\exp\big[\sqrt{-1}(\boldsymbol{\xi}, \tilde{\mathbf{Z}}(s))_{\mathbb{R}^N} - s\ell_\mu(\boldsymbol{\xi})\big],\, A\cap\{\zeta> s\}\Big]. \quad \square
\end{aligned}$$

Obviously, the process $\{\tilde{\mathbf{Z}}(t) : t \geq 0\}$ in Lemma 7.3.1 is the one obtained by reflecting (i.e., reversing the direction of $\{\mathbf{Z}(t) : t \geq 0\}$) at time $\zeta$, and the lemma says that the distribution of the resulting process is the same as that of the original one. Most applications of this result are to situations when one knows more or less precisely where the process is at the time when it is reflected. For example, suppose $N = 1$, $a \in (0,\infty)$, and $\zeta_a = \inf\{t \geq 0 : Z(t) \geq a\}$. Noting that, because $\tilde{Z}(t) = Z(t)$ for $t \leq \zeta_a$ and therefore that $\zeta_a = \inf\{t \geq 0 : \tilde{Z}(t) \geq a\}$, we have that

$$\begin{aligned}
\mathbb{P}\big(Z(t) \leq x \;\&\; \zeta_a \leq t\big) &= \mathbb{P}\big(2Z(\zeta_a) - Z(t) \leq x \;\&\; \zeta_a \leq t\big) \\
&= \mathbb{P}\big(Z(t) \geq 2Z(\zeta_a) - x \;\&\; \zeta_a \leq t\big).
\end{aligned}$$

Hence, if $x \le a$, and therefore $Z(t) \ge 2Z(\zeta_a) - x \implies \zeta_a \le t$ when $\zeta_a < \infty$, then

$$\mathbb{P}\big(Z(t) \le x \;\&\; \zeta_a \le t\big) = \mathbb{P}\big(Z(t) \ge 2Z(\zeta_a) - x \;\&\; \zeta_a < \infty\big) \quad \text{for } x \le a.$$

Applying this when $x = a$ and using $\mathbb{P}(\zeta_a \le t) = \mathbb{P}\big(Z(t) \le a \;\&\; \zeta_a \le t\big) + \mathbb{P}\big(Z(t) > a\big)$, one gets $\mathbb{P}\big(\zeta_a \le t\big) \le 2\mathbb{P}\big(Z(t) \ge a\big)$, a conclusion that also could have been reached via Theorem 1.4.13.

§ **7.3.2. Reflected Brownian Motion.** The considerations in the preceding subsection are most interesting when applied to $\mathbb{R}$-valued Brownian motion. Thus, let $\big(B(t), \mathcal{F}_t, \mathbb{P}\big)$ be an $\mathbb{R}$-valued Brownian motion. To appreciate the improvements that can be made in the calculations just made, again take $\zeta_a = \inf\{t \ge 0 : B(t) \ge a\}$ for some $a > 0$. Then, because Brownian paths are continuous, $\zeta_a < \infty \implies B(\zeta_a) = a$ and so, since $\mathbb{P}(\zeta_a < \infty) = 1$, we can say that

$$\text{(7.3.2)} \quad \mathbb{P}\big(B(t) \le x \;\&\; \zeta_a \le t\big) = \mathbb{P}\big(B(t) \ge 2a - x\big) \quad \text{for } (t, x) \in [0, \infty) \times (-\infty, a].$$

In particular, by taking $x = a$ and using $\mathbb{P}\big(B(t) \ge a\big) = \mathbb{P}\big(B(t) \ge a \;\&\; \zeta_a \le t\big)$, we recover the result in Exercise 4.3.12 that

$$\mathbb{P}\big(\zeta_a \le t\big) = 2\mathbb{P}\big(B(t) \ge a\big).$$

A more interesting application of Lemma 7.3.1 to Brownian motion is to the case when $\zeta$ is the exit time from an interval other than a half-line.

THEOREM 7.3.3. *Let $a_1 < 0 < a_2$ be given, define $\zeta^{(a_1,a_2)} = \inf\{t \ge 0 : B(t) \notin (a_1, a_2)\}$, and set $A_i(t) = \{\zeta^{(a_1,a_2)} \le t \;\&\; B(\zeta^{(a_1,a_2)}) = a_i\}$ for $i \in \{1, 2\}$. Then, for $\Gamma \in \mathcal{B}_{[a_1,\infty)}$,*

$$\begin{aligned} 0 &\le \mathbb{P}\big(\{B(t) \in \Gamma\} \cap A_1(t)\big) - \mathbb{P}\big(\{B(t) \in 2(a_2 - a_1) + \Gamma\} \cap A_1(t)\big) \\ &= \mathbb{P}\big(B(t) \in 2a_1 - \Gamma\big) - \mathbb{P}\big(B(t) \in 2(a_2 - a_1) + \Gamma\big) \end{aligned}$$

*and, for $\Gamma \in \mathcal{B}_{(-\infty,a_2]}$,*

$$\begin{aligned} 0 &\le \mathbb{P}\big(\{B(t) \in \Gamma\} \cap A_2(t)\big) - \mathbb{P}\big(\{B(t) \in -2(a_2 - a_1) + \Gamma\} \cap A_2(t)\big) \\ &= \mathbb{P}\big(B(t) \in 2a_2 - \Gamma\big) - \mathbb{P}\big(B(t) \in -2(a_2 - a_1) + \Gamma\big). \end{aligned}$$

*Hence, for $\Gamma \in \mathcal{B}_{[a_1,\infty)}$, $\mathbb{P}\big(\{B(t) \in \Gamma\} \cap A_1(t)\big)$ equals*

$$\sum_{m=1}^{\infty} \Big[\gamma_{0,t}\big(\Gamma - 2a_1 + 2(m-1)(a_2 - a_1)\big) - \gamma_{0,t}\big(\Gamma + 2m(a_2 - a_1)\big)\Big]$$

*and, for $\Gamma \in \mathcal{B}_{(-\infty,a_2]}$, $\mathbb{P}\big(\{B(t) \in \Gamma\} \cap A_2(t)\big)$ equals*

$$\sum_{m=1}^{\infty} \Big[\gamma_{0,t}\big(\Gamma - 2a_2 - 2(m-1)(a_2 - a_1)\big) - \gamma_{0,t}\big(\Gamma - 2m(a_2 - a_1)\big)\Big],$$

*where in both cases the convergence is uniform with respect $t$ in compacts and $\Gamma \in \mathcal{B}_{(a_1,a_2)}$.*

PROOF: Suppose $\Gamma \in \mathcal{B}_{[a_1,\infty)}$. Then, by Lemma 7.3.1,

$$\begin{aligned}\mathbb{P}\big(\{B(t)\in\Gamma\}\cap A_1(t)\big) &= \mathbb{P}\big(\{2a_1 - B(t)\in\Gamma\}\cap A_1(t)\big)\\ &= \mathbb{P}\big(B(t)\in 2a_1-\Gamma\big) - \mathbb{P}\big(\{B(t)\in 2a_1-\Gamma\}\cap A_2(t)\big),\end{aligned}$$

since $B(t)\in 2a_1-\Gamma \implies B(t)\le a_1 \implies \zeta^{(a_1,a_2)}\le t$. Similarly,

$$\begin{aligned}\mathbb{P}\big(\{B(t)\in\Gamma\}\cap A_2(t)\big) &= \mathbb{P}\big(\{2a_2 - B(t)\in\Gamma\}\cap A_1(t)\big)\\ &= \mathbb{P}\big(B(t)\in 2a_2-\Gamma\big) - \mathbb{P}\big(\{B(t)\in 2a_2-\Gamma\}\cap A_1(t)\big)\end{aligned}$$

when $\Gamma\in\mathcal{B}_{(-\infty,a_2]}$. Hence, since $2a_1-\Gamma\subseteq(-\infty,a_1]\subseteq(-\infty,a_2]$ if $\Gamma\in\mathcal{B}_{[a_1,\infty)}$,

$$\begin{aligned}\mathbb{P}\big(\{B(t)\in\Gamma\}\cap A_1(t)\big) &= \mathbb{P}\big(B(t)\in 2a_1-\Gamma\big) - \mathbb{P}\big(B(t)\in 2(a_2-a_1)+\Gamma\big)\\ &\quad + \mathbb{P}\big(\{B(t)\in 2(a_2-a_1)+\Gamma\}\cap A_1(t)\big)\end{aligned}$$

when $\Gamma\in\mathcal{B}_{[a_1,\infty)}$. Similarly, when $\Gamma\in\mathcal{B}_{(-\infty,a_2]}$,

$$\begin{aligned}\mathbb{P}\big(\{B(t)\in\Gamma\}\cap A_2(t)\big) &= \mathbb{P}\big(B(t)\in 2a_2-\Gamma\big) - \mathbb{P}\big(B(t)\in -2(a_2-a_1)+\Gamma\big)\\ &\quad + \mathbb{P}\big(\{B(t)\in -2(a_2-a_1)+\Gamma\}\cap A_2(t)\big).\end{aligned}$$

To check that

$$\mathbb{P}\big(\{B(t)\in\Gamma\}\cap A_1(t)\big) - \mathbb{P}\big(\{B(t)\in 2(a_2-a_1)+\Gamma\}\cap A_1(t)\big)\ge 0 \text{ when } \Gamma\in\mathcal{B}_{[a_1,\infty)},$$

first use Theorem 7.1.16 to see that

$$\mathbb{P}\big(\{B(t)\in\Gamma\}\cap A_1(t)\big) = \mathbb{E}^{\mathbb{P}}\big[\gamma_{0,t-\zeta^{(a_1,a_2)}}(\Gamma-a_1),\,A_1(t)\big].$$

Second, observe that, because $\Gamma\subseteq[a_1,\infty)$, $\gamma_{0,\tau}\big(2(a_2-a_1)+\Gamma\big)\le\gamma_{0,\tau}(\Gamma)$ for all $\tau\ge 0$. The case when $\Gamma\in\mathcal{B}_{(-\infty,a_2]}$ and $A_1(t)$ is replaced by $A_2(t)$ is handled in the same way.

Given the preceding, one can use induction to check that $\mathbb{P}\big(\{B(t)\in\Gamma\}\cap A_1(t)\big)$ equals

$$\begin{aligned}&\sum_{m=1}^{M}\Big[\mathbb{P}\big(B(t)\in 2a_1-2(m-1)(a_2-a_1)-\Gamma\big) - \mathbb{P}\big(B(t)\in 2m(a_2-a_1)+\Gamma\big)\Big]\\ &\qquad + \mathbb{P}\big(\{B(t)\in 2M(a_2-a_1)+\Gamma\}\cap A_1(t)\big)\end{aligned}$$

for all $\Gamma\in\mathcal{B}_{[a_1,\infty)}$. The same line of reasoning applies when $\Gamma\in\mathcal{B}_{(-\infty,a_2]}$ and $A_1(t)$ is replaced by $A_2(t)$. □

Perhaps the most useful consequence of the preceding is the following corollary.

**Corollary 7.3.4.** *Given a $c \in \mathbb{R}$ and an $r \in (0,\infty)$, set $I = (c-r, c+r)$ and*

$$P^I(t,x,\Gamma) = \mathbb{P}\big(\{x+B(t)\in\Gamma\}\cap\{\zeta^I > t\}\big), \quad x \in I \text{ and } \Gamma \in \mathcal{B}_I.$$

*Then*

$$P^I(s+t,x,\Gamma) = \int_I P^I(t,z,\Gamma)\,P^I(s,x,dz). \tag{7.3.5}$$

*Next, set*

$$\tilde g(t,x) = \sum_{m\in\mathbb{Z}} g(t,x+4m), \quad \text{where } g(t,x) = (2\pi t)^{-\frac12} e^{-\frac{x^2}{2t}}$$

*and*

$$p^{(-1,1)}(t,x,y) = \tilde g(t,y-x) - \tilde g(t,y+x+2) \quad \text{for } (t,x,y) \in (0,\infty)\times[-1,1]^2.$$

*Then $p^{(-1,1)}$ is a smooth function that is symmetric in $(x,y)$, strictly positive on $(0,\infty)\times(0,1)^2$, and vanishes when $x \in \{-1,1\}$. Finally, if*

$$p^I(t,x,y) = r^{-1}p^{(-1,1)}\big(r^{-2}t, r^{-1}(x-c), r^{-1}(y-c)\big), \quad (t,x,y)\in(0,\infty)\times I^2,$$

*then*

$$p^I(s+t,x,y) = \int_I p^I(s,x,z)p^I(t,z,y)\,dz \tag{7.3.6}$$

*and, for $(t,x) \in (0,\infty)\times I$, $P^I(t,x,dy) = p^I(t,x,y)\,dy$.*

**Proof:** Begin by applying Theorem 7.1.16 to check that $P^I(s+t,x,\Gamma)$ equals

$$\begin{aligned}
&\mathcal{W}^{(1)}\big(\{x+\psi(s)+\delta_s\psi(t)\in\Gamma\}\cap\{x+\psi(s)+\delta_s\psi(\tau)\in I,\ \tau\in[0,t-s]\}\\
&\qquad\qquad \cap\{x+\psi(\sigma)\in I,\ \sigma\in[0,s]\}\big)\\
&= \mathbb{E}^{\mathcal{W}^{(1)}}\big[P^I(t,x+\psi(s),\Gamma),\ \{x+\psi(\sigma)\in I,\ \sigma\in[0,s]\}\big]\\
&= \int_I P^I(t,z,\Gamma)\,P^I(s,x,dz).
\end{aligned}$$

Next, set $a_1 = r^{-1}(c-x)-1$ and $a_2 = r^{-1}(c-x)+1$. Then

$$\begin{aligned}
P^I(t,x,\Gamma) &= \mathbb{P}\big(\{B(t)\in\Gamma-x\}\cap\{B(\tau)\in(ra_1,ra_2),\ \tau\in[0,t]\}\big)\\
&= \mathbb{P}\big(\{B(r^{-2}t)\in r^{-1}(\Gamma-x)\}\cap\{B(r^{-2}\tau)\in(a_1,a_2),\ \tau\in[0,t]\}\big)\\
&= \mathbb{P}\big(B(r^{-2}t)\in r^{-1}(\Gamma-x)\ \&\ \zeta^{(a_1,a_2)} > r^{-2}t\big)\\
&= \mathbb{P}\big(B(r^{-2}t)\in r^{-1}(\Gamma-x)\big) - \mathbb{P}\big(B(r^{-2}t)\in r^{-1}(\Gamma-x)\ \&\ \zeta^{(a_1,a_2)} \le r^{-2}t\big),
\end{aligned}$$

where, in the passage to the second line, I have used Brownian scaling. Now, use the last part of Theorem 7.3.3, the symmetry of $\gamma_{0,r^{-2}t}$, and elementary rearrangement of terms to arrive first at

$$P^I(t,x,\Gamma) = \sum_{m\in\mathbb{Z}} \Big[\gamma_{r^{-2}t}\big(4m + r^{-1}(\Gamma - x)\big) - \gamma_{r^{-2}t}\big(4m + 2 + r^{-1}(\Gamma + x - 2c)\big)\Big],$$

and then at $P^I(t,x,dy) = p^I(t,x,y)\,dy$. Given this and (7.3.5), (7.3.6) is obvious.

Turning to the properties of $p^{(-1,1)}(t,x,y)$, both its symmetry and smoothness are clear. In addition, as the density for $P^{(-1,1)}(t,x,\cdot\,)$, it is non-negative, and, because $x \rightsquigarrow \tilde{g}(t,x)$ is periodic with period 4, it is easy to see that $p^{(-1,1)}(t,\pm 1,y) = 0$. Thus, everything comes down to proving that $p^{(-1,1)}(t,x,y) > 0$ for $(t,x,y) \in (0,\infty)\times(-1,1)^2$. To this end, first observe that, after rearranging terms, one can write $p^{(-1,1)}(t,x,y)$ as

$$\begin{aligned} &g(t,y-x) - g(t,y+x) + g(t,2-x-y) \\ &\quad + \sum_{m=1}^{\infty}\Big[\big(g(t,y-x+4m) - g(t,y+x+2+4m)\big) \\ &\qquad\qquad + \big(g(t,y-x-4m) - g(t,y+x-2-4m)\big)\Big]. \end{aligned}$$

Since each of the terms in the sum over $m \in \mathbb{Z}^+$ is positive, we have that

$$p^{(-1,1)}(t,x,y) > g(t,y-x)\big[1 - 2e^{-\frac{2(1-|x|)(1-|y|)}{t}}\big] \geq \big(1 - \tfrac{2}{e}\big) g(t,y-x)$$

if $t \leq 2(1-|x|)(1-|y|)$. Hence, for each $\theta \in (0,1)$, $p^{(-1,1)}(t,x,y) > 0$ for all $(t,x,y) \in [0,2\theta^2]\times[-1+\theta,1-\theta]^2$. Finally, to handle $x,y \in [-1+\theta,1-\theta]$ and $t > 2\theta^2$, apply (7.3.6) with $I = (-1,1)$ to see that

$$p^{(-1,1)}\big((m+1)\theta^2,x,y\big) \geq \int_{|z|\leq(1-\theta)} p^{(-1,1)}(\theta^2,x,z)p^{(-1,1)}(m\theta^2,z,y)\,dz,$$

and use this and induction to see that $p^{(-1,1)}(m\theta^2,x,y) > 0$ for all $m \geq 1$. Thus, if $n \in \mathbb{Z}^+$ is chosen so that $n\theta^2 < t \leq (n+1)\theta^2$, then another application of (7.3.6) shows that

$$p^{(-1,1)}(t,x,y) \geq \int_{|z|\leq(1-\theta)} p^{(-1,1)}(t-n\theta^2,x,z)p^{(-1,1)}(n\theta^2,z,y)\,dz > 0. \quad \square$$

## Exercises for § 7.3

EXERCISE 7.3.7. Suppose that $G$ is a non-empty, open subset of $\mathbb{R}^N$, define $\zeta_x^G : C(\mathbb{R}^N) \longrightarrow [0,\infty]$ by

$$\zeta_{\mathbf{x}}^G(\boldsymbol{\psi}) = \inf\{t \geq 0 : \mathbf{x} + \boldsymbol{\psi}(t) \notin G\},$$

and set

$$P^G(t,\mathbf{x},\Gamma) = \mathcal{W}^{(N)}(\{\boldsymbol{\psi} : \mathbf{x} + \boldsymbol{\psi}(t) \in \Gamma \;\&\; \zeta_{\mathbf{x}}^G(\boldsymbol{\psi}) > t\})$$

for $(t,\mathbf{x}) \in (0,\infty) \times G$ and $\Gamma \in \mathcal{B}_G$.

**(i)** Show that

$$P^G(s+t,\mathbf{x},\Gamma) = \int_G P^G(t,\mathbf{z},\Gamma)\, P^G(s,\mathbf{x},d\mathbf{y}).$$

**(ii)** As an application of Exercise 7.1.25, show that

$$P^G(t,\mathbf{x},\Gamma) = \gamma_{\mathbf{0},t\mathbf{I}}(\Gamma - \mathbf{x}) - \mathbb{E}^{\mathcal{W}^{(N)}}\left[\gamma_{\mathbf{0},(t-\zeta_{\mathbf{x}}^G)\mathbf{I}}(\Gamma - \mathbf{x} - \boldsymbol{\psi}(\zeta_{\mathbf{x}}^G)),\, \zeta_{\mathbf{x}}^G \leq t\right].$$

This is the probabilistic version of Duhamel's Formula, which we will see again in § 10.3.1.

**(iii)** As a consequence of **(ii)**, show that there is a Borel measurable function $p^G : (0,\infty) \times G^2 \longrightarrow [0,\infty)$ such that $(t,\mathbf{y}) \rightsquigarrow p^G(t,\mathbf{x},\mathbf{y})$ is continuous for each $\mathbf{x} \in G$ and $P^G(t,\mathbf{x},d\mathbf{y}) = p^G(t,\mathbf{x},\mathbf{y})\,d\mathbf{y}$ for each $(t,\mathbf{x}) \in (0,\infty) \times G$. In particular, use this in conjunction with **(i)** to conclude that

$$p^G(s+t,\mathbf{x},\mathbf{y}) = \int_G p^G(t,\mathbf{z},\mathbf{y})p^G(s,\mathbf{x},\mathbf{z})\,d\mathbf{z}.$$

**Hint**: Keep in mind that $(\tau,\boldsymbol{\xi}) \rightsquigarrow (2\pi\tau)^{-\frac{N}{2}}e^{-\frac{|\boldsymbol{\xi}|^2}{2\tau}}$ is smooth and bounded as long as $\boldsymbol{\xi}$ stays away from the origin.

**(iv)** Given $\mathbf{c} = (c_1,\dots,c_N) \in \mathbb{R}^N$ and $r > 0$, let $Q(\mathbf{c},r)$ denote the open cube $\prod_{i=1}^N (c_i - r, c_i + r)$, and show that (cf. Corollary 7.3.4)

$$p^{Q(\mathbf{c},r)}(t,\mathbf{x},\mathbf{y}) = \prod_{i=1}^N p^{(c_i-r,c_i+r)}(t,x_i,y_i)$$

for $\mathbf{x} = (x_1,\dots,x_N)$, $\mathbf{y} = (y_1,\dots,y_N) \in Q(\mathbf{c},r)$. In particular, conclude that $p^{Q(\mathbf{c},r)}(t,\mathbf{x},\mathbf{y})$ is uniformly positive on compact subsets of $(0,\infty) \times Q(\mathbf{c},r)^2$.

**(v)** Assume that $G$ is connected, and show that $p^G(t,\mathbf{x},\mathbf{y})$ is uniformly positive on compact subsets of $(0,\infty) \times G^2$.

**Hint**: If $Q(\mathbf{c},r) \subseteq G$, show that $p^G(t,\mathbf{x},\mathbf{y}) \geq p^{Q(\mathbf{c},r)}(t,\mathbf{x},\mathbf{y})$ on $(0,\infty)\times Q(\mathbf{c},r)^2$.

# Chapter 8
## Gaussian Measures on a Banach Space

As I said at the end of § 4.3.2, the distribution of Brownian motion is called Wiener measure because Wiener was the first to construct it. Wiener's own thinking about his measure had little or nothing in common with the Lévy–Khinchine program. Instead, he looked upon his measure as a Gaussian measure on an infinite dimensional space, and most of what he did with his measure is best understood from that perspective. Thus, in this chapter, we will look at Wiener measure from a strictly Gaussian point of view. More generally, we will be dealing here with measures on a real Banach space $E$ that are centered Gaussian in the sense that, for each $x^*$ in the dual space $E^*$, $x \in E \longmapsto \langle x, x^* \rangle \in \mathbb{R}$ is a centered Gaussian random variable. Not surprisingly, such a measure will be said to be a **centered Gaussian measure** on $E$.

Although the ideas that I will use are already implicit in Wiener's work, it was I. Segal and his school, especially L. Gross,[1] who gave them the form presented here.

### § 8.1 The Classical Wiener Space

In order to motivate what follows, it is helpful to first understand Wiener measure from the point of view which I will be adopting here.

**§ 8.1.1. Classical Wiener Measure.** Up until now I have been rather casual about the space from which Brownian paths come. Namely, because Brownian paths are continuous, I have thought of their distribution as being a probability on the space $C(\mathbb{R}^N) = C\big([0,\infty);\mathbb{R}^N\big)$. In general, there is no harm done by choosing $C(\mathbb{R}^N)$ as the sample space for Brownian paths. However, for my purposes here, I need my sample spaces to be separable Banach spaces, and, although it is a complete, separable metric space, $C(\mathbb{R}^N)$ is not a Banach space. With this in mind, define $\Theta(\mathbb{R}^N)$ to be the space of continuous paths $\boldsymbol{\theta} : [0,\infty) \longrightarrow \mathbb{R}^N$ with the properties that $\boldsymbol{\theta}(0) = 0$ and $\lim_{t\to\infty} t^{-1}|\boldsymbol{\theta}(t)| = 0$.

[1] See I.E. Segal's "Distributions in Hilbert space and canonical systems of operators," *T.A.M.S.*, **88** (1958) and L. Gross's "Abstract Wiener spaces," *Proc. 5th Berkeley Symp. on Prob. & Stat.*, **2** (1965), Univ. of California Press. A good exposition of this topic can be found in H.-H. Kuo's *Gaussian Measures in Banach Spaces*, Springer-Verlag, Math. Lec. Notes., # **463** (1975).

LEMMA 8.1.1. *The map*

$$\psi \in C(\mathbb{R}^N) \longmapsto \|\psi\|_{\Theta(\mathbb{R}^N)} \equiv \sup_{t\geq 0} \frac{|\psi(t)|}{1+t} \in [0,\infty]$$

*is lower semicontinuous, and the pair* $\left(\Theta(\mathbb{R}^N), \|\cdot\|_{\Theta(\mathbb{R}^N)}\right)$ *is a separable Banach space that is continuously embedded as a Borel measurable subset of* $C(\mathbb{R}^N)$. *In particular,* $\mathcal{B}_{\Theta(\mathbb{R}^N)}$ *coincides with* $\mathcal{B}_{C(\mathbb{R}^N)}[\Theta(\mathbb{R}^N)] = \left\{A \cap \Theta(\mathbb{R}^N) : A \in \mathcal{B}_{C(\mathbb{R}^N)}\right\}$. *Moreover, the dual space* $\Theta(\mathbb{R}^N)^*$ *of* $\Theta(\mathbb{R}^N)$ *can be identified with the space of* $\mathbb{R}^N$*-valued, Borel measures* $\boldsymbol{\lambda}$ *on* $[0,\infty)$ *with the properties that* $\boldsymbol{\lambda}(\{0\}) = 0$ *and*[1]

$$\|\boldsymbol{\lambda}\|_{\Theta(\mathbb{R}^N)^*} \equiv \int_{[0,\infty)} (1+t)\,|\boldsymbol{\lambda}|(dt) < \infty,$$

*when the duality relation is given by*

$$\langle \boldsymbol{\theta}, \boldsymbol{\lambda}\rangle = \int_{[0,\infty)} \boldsymbol{\theta}(t)\cdot\boldsymbol{\lambda}(dt).$$

*Finally, if* $(\mathbf{B}(t), \mathcal{F}_t, \mathbb{P})$ *is an* $\mathbb{R}^N$*-valued Brownian motion, then* $\mathbf{B} \in \Theta(\mathbb{R}^N)$ $\mathbb{P}$*-almost surely and*

$$\mathbb{E}^{\mathbb{P}}\left[\|\mathbf{B}\|^2_{\Theta(\mathbb{R}^N)}\right] \leq 32N.$$

PROOF: It is obvious that the inclusion map taking $\Theta(\mathbb{R}^N)$ into $C(\mathbb{R}^N)$ is continuous. To see that $\|\cdot\|_{\Theta(\mathbb{R}^N)}$ is lower semicontinuous on $C(\mathbb{R}^N)$ and that $\Theta(\mathbb{R}^N) \in \mathcal{B}_{C(\mathbb{R}^N)}$, note that, for any $s \in [0,\infty)$ and $R \in (0,\infty)$,

$$A(s,R) \equiv \left\{\psi \in C(\mathbb{R}^N) : \big|\psi(t)\big| \leq R(1+t) \text{ for } t \geq s\right\}$$

is closed in $C(\mathbb{R}^N)$. Hence, since $\|\psi\|_{\Theta(\mathbb{R}^N)} \leq R \iff \psi \in A(0,R)$, $\|\cdot\|_{\Theta(\mathbb{R}^N)}$ is lower semicontinuous. In addition, since $\{\psi \in C(\mathbb{R}^N) : \psi(0) = \mathbf{0}\}$ is also closed,

$$\Theta(\mathbb{R}^N) = \bigcap_{n=1}^{\infty}\bigcup_{m=1}^{\infty}\left\{\psi \in A\left(m, \tfrac{1}{n}\right) : \psi(0) = \mathbf{0}\right\} \in \mathcal{B}_{C(\mathbb{R}^N)}.$$

In order to analyze the space $\left(\Theta(\mathbb{R}^N), \|\cdot\|_{\Theta(\mathbb{R}^N)}\right)$, define

$$F : \Theta(\mathbb{R}^N) \longrightarrow C_0(\mathbb{R};\mathbb{R}^N) \equiv \left\{\psi \in C(\mathbb{R};\mathbb{R}^N) : \lim_{|s|\to\infty} |\psi(s)| = 0\right\}$$

by

$$[F(\boldsymbol{\theta})](s) = \frac{\boldsymbol{\theta}(e^s)}{1+e^s}, \quad s \in \mathbb{R}.$$

[1] I use $|\boldsymbol{\lambda}|$ to denote the variation measure determined by $\boldsymbol{\lambda}$.

As is well known, $C_0(\mathbb{R};\mathbb{R}^N)$ with the uniform norm is a separable Banach space, and it is obvious that $F$ is an isometry from $\Theta(\mathbb{R}^N)$ onto $C_0(\mathbb{R};\mathbb{R}^N)$. Moreover, by the Riesz Representation Theorem for $C_0(\mathbb{R};\mathbb{R}^N)$, one knows that the dual of $C_0(\mathbb{R};\mathbb{R}^N)$ is isometric to the space of totally finite, $\mathbb{R}^N$-valued measures on $(\mathbb{R};\mathcal{B}_{\mathbb{R}})$ with the norm given by total variation. Hence, the identification of $\Theta(\mathbb{R}^N)^*$ reduces to the obvious interpretation of the adjoint map $F^*$ as a mapping from totally finite $\mathbb{R}^N$-valued measures onto the space of $\mathbb{R}^N$-valued measures that do not charge $\mathbf{0}$ and whose variation measure integrates $(1+t)$.

Because of the Strong Law in part (**ii**) of Exercise 4.3.11, it is clear that almost every Brownian path is in $\Theta(\mathbb{R}^N)$. In addition, by the Brownian scaling property and Doob's Inequality (cf. Theorem 7.1.9),

$$\begin{aligned}\mathbb{E}^{\mathbb{P}}\left[\|\mathbf{B}\|^2_{\Theta(\mathbb{R}^N)}\right] &\le \sum_{n=0}^{\infty} 4^{-n+1}\mathbb{E}^{\mathbb{P}}\left[\sup_{0\le t\le 2^n}|\mathbf{B}(t)|^2\right]\\ &= \sum_{n=0}^{\infty} 2^{-n+2}\mathbb{E}^{\mathbb{P}}\left[\sup_{0\le t\le 1}|\mathbf{B}(t)|^2\right] \le 32\mathbb{E}^{\mathbb{P}}\left[|\mathbf{B}(1)|^2\right] = 32N. \quad \square\end{aligned}$$

In view of Lemma 8.1.1, we now know that the distribution of $\mathbb{R}^N$-valued Brownian motion induces a Borel measure $\mathcal{W}^{(N)}$ on the separable Banach space $\Theta(\mathbb{R}^N)$, and throughout this chapter I will refer to this measure as the **classical Wiener measure**.

My next goal is to characterize, in terms of $\Theta(\mathbb{R}^N)$, exactly which measure on $\Theta(\mathbb{R}^N)$ Wiener's is, and for this purpose I will use the following simple fact about Borel probability measures on a separable Banach space.

LEMMA 8.1.2. *Let $E$ with norm $\|\cdot\|_E$ be a separable, real Banach space, and use*

$$(x,x^*)\in E\times E^* \longmapsto \langle x,x^*\rangle\in\mathbb{R}$$

*to denote the duality relation between $E$ and its dual space $E^*$. Then the Borel field $\mathcal{B}_E$ coincides with the $\sigma$-algebra generated by the maps $x\in E\longmapsto\langle x,x^*\rangle$ as $x^*$ runs over $E^*$. In particular, if, for $\mu\in\mathbf{M}_1(E)$, one defines its* **Fourier transform** $\hat{\mu}: E^*\longrightarrow\mathbb{C}$ *by*

$$\hat{\mu}(x^*) = \int_E \exp\left[\sqrt{-1}\,\langle x,x^*\rangle\right]\mu(dx), \quad x^*\in E^*,$$

*then $\hat{\mu}$ is a continuous function of weak* convergence on bounded subsets of $\Theta^*$, and $\hat{\mu}$ uniquely determines $\mu$ in the sense that if $\nu$ is a second element of $\mathbf{M}_1(\Theta)$ and $\hat{\mu}=\hat{\nu}$, then $\mu=\nu$.*

PROOF: Since it is clear that each of the maps $x\in E\longmapsto\langle x,x^*\rangle\in\mathbb{R}$ is continuous and therefore $\mathcal{B}_E$-measurable, the first assertion will follow as soon as we show that the norm $x \rightsquigarrow \|x\|_E$ can be expressed as a measurable function of these maps. But, because $E$ is separable, we know (cf. Exercise 5.1.19) that the closed unit ball $\overline{B_{E^*}(0,1)}$ in $E^*$ is separable with respect to the weak* topology and therefore that we can find a sequence $\{x_n^* :, n \geq 1\} \subseteq \overline{B_{E^*}(0,1)}$ so that

$$\|x\|_\Theta = \sup_{n \in \mathbb{Z}^+} \langle x, x_n^* \rangle, \quad x \in E.$$

Turning to the properties of $\hat{\mu}$, note that its continuity on bounded subsets with respect to weak* convergence is an immediate consequence of Lebesgue's Dominated Convergence Theorem and the fact that the weak* topology on bounded subsets admits a metric. Furthermore, in view of the preceding, we will know that $\hat{\mu}$ completely determines $\mu$ as soon as we show that, for each $n \in \mathbb{Z}^+$ and $X^* = (x_1^*, \dots, x_n^*) \in (E^*)^n$, $\hat{\mu}$ determines the marginal distribution $\mu_{X^*} \in \mathbf{M}_1(\mathbb{R}^N)$ of

$$x \in E \longmapsto \Big( \langle x, x_1^* \rangle, \dots, \langle x, x_n^* \rangle \Big) \in \mathbb{R}^n$$

under $\mu$. But this is clear (cf. Lemma 2.3.3), since

$$\widehat{\mu_{X^*}}(\boldsymbol{\xi}) = \hat{\mu}\left( \sum_{m=1}^n \xi_m x_m^* \right) \quad \text{for } \boldsymbol{\xi} = (\xi_1, \dots, \xi_n) \in \mathbb{R}^n. \quad \square$$

I will now compute the Fourier transform of $\mathcal{W}^{(N)}$. To this end, first recall that, for an $\mathbb{R}^N$-valued Brownian motion, $\{(\boldsymbol{\xi}, \mathbf{B}(t))_{\mathbb{R}^N} : t \geq 0 \text{ and } \boldsymbol{\xi} \in \mathbb{R}^N\}$ spans a Gaussian family $\mathfrak{G}(\mathbf{B})$ in $L^2(\mathbb{P};\mathbb{R})$. Hence, $\operatorname{span}(\{(\boldsymbol{\xi}, \boldsymbol{\theta}(t)) : t \geq 0 \text{ and } \boldsymbol{\xi} \in \mathbb{R}^N\})$ is a Gaussian family in $L^2(\mathcal{W}^{(N)};\mathbb{R})$. From this, combined with an easy limit argument using Riemann sum approximations, one sees that, for any $\boldsymbol{\lambda} \in \Theta(\mathbb{R}^N)^*$, $\boldsymbol{\theta} \rightsquigarrow \langle \boldsymbol{\theta}, \boldsymbol{\lambda} \rangle$ is a centered Gaussian random variable under $\mathcal{W}^{(N)}$. Furthermore, because, for $0 \leq s \leq t$,

$$\mathbb{E}^{\mathcal{W}^{(N)}}\left[ (\boldsymbol{\xi}, \boldsymbol{\theta}(s))_{\mathbb{R}^N} (\boldsymbol{\eta}, \boldsymbol{\theta}(t))_{\mathbb{R}^N} \right] = \mathbb{E}^{\mathcal{W}^{(N)}}\left[ (\boldsymbol{\xi}, \boldsymbol{\theta}(s))_{\mathbb{R}^N} (\boldsymbol{\eta}, \boldsymbol{\theta}(s))_{\mathbb{R}^N} \right] = s(\boldsymbol{\xi}, \boldsymbol{\eta})_{\mathbb{R}^N},$$

we can apply Fubini's Theorem to see that

$$\mathbb{E}^{\mathcal{W}^{(N)}}[\langle \boldsymbol{\theta}, \boldsymbol{\lambda} \rangle^2] = \iint_{[0,\infty)^2} s \wedge t \, \boldsymbol{\lambda}(ds) \cdot \boldsymbol{\lambda}(dt).$$

Therefore, we now know that $\mathcal{W}^{(N)}$ is characterized by its Fourier transform

$$\widehat{\mathcal{W}^{(N)}}(\boldsymbol{\lambda}) = \exp\left[ -\frac{1}{2} \iint_{[0,\infty)^2} s \wedge t \, \boldsymbol{\lambda}(ds) \cdot \boldsymbol{\lambda}(dt) \right], \quad \boldsymbol{\lambda} \in \Theta(\mathbb{R}^N)^*. \tag{8.1.3}$$

Equivalently, we have shown that $\mathcal{W}^{(N)}$ is the centered Gaussian measure on $\Theta(\mathbb{R}^N)$ with the property that, for each $\boldsymbol{\lambda} \in \Theta(\mathbb{R}^N)^*$, $\boldsymbol{\theta} \rightsquigarrow \langle \boldsymbol{\theta}, \boldsymbol{\lambda} \rangle$ is a centered Gaussian random variable with variance equal to $\iint_{[0,\infty)^2} s \wedge t \, \boldsymbol{\lambda}(ds) \cdot \boldsymbol{\lambda}(dt)$.

§ **8.1.2. The Classical Cameron–Martin Space.** From the Gaussian standpoint, it is extremely unfortunate that the natural home for Wiener measure is a Banach space rather than a Hilbert space. Indeed, in finite dimensions, every centered, Gaussian measure with non-degenerate covariance can be thought of as the canonical, or standard, Gaussian measure on a Hilbert space. Namely, if $\gamma_{\mathbf{0},\mathbf{C}}$ is the Gaussian measure on $\mathbb{R}^N$ with mean $\mathbf{0}$ and non-degenerate covariance $\mathbf{C}$, consider $\mathbb{R}^N$ as a Hilbert space $H$ with inner product $(\mathbf{g},\mathbf{h})_H = (\mathbf{g},\mathbf{Ch})_{\mathbb{R}^N}$, and take $\lambda_H$ to be the natural Lebesgue measure there: the one that assigns measure 1 to a unit cube in $H$ or, equivalently, the one obtained by pushing the usual Lebesgue measure $\lambda_{\mathbb{R}^N}$ forward under the linear transformation $\mathbf{C}^{\frac{1}{2}}$. Then we can write

$$\gamma_{\mathbf{0},\mathbf{C}}(d\mathbf{h}) = \frac{1}{(2\pi)^{\frac{N}{2}}} e^{-\frac{\|\mathbf{h}\|_H^2}{2}} \lambda_H(d\mathbf{h})$$

and

$$\widehat{\gamma_{\mathbf{0},\mathbf{C}}}(\mathbf{h}) = e^{-\frac{\|\mathbf{h}\|_H^2}{2}}.$$

As was already pointed out in Exercise 3.1.11, in infinite dimensions there is no precise analog of the preceding canonical representation (cf. Exercise 8.1.7 for further corroboration of this point). Nonetheless, a good deal of insight can be gained by seeing how close one can come. In order to guess on which Hilbert space it is that $\mathcal{W}^{(N)}$ would like to live, I will describe R. Feynman's highly questionable but remarkably powerful way of thinking about such matters. Namely, given $n \in \mathbb{Z}^+$, $0 = t_0 < t_1 < \cdots < t_n$, and a set $A \in \left(\mathcal{B}_{\mathbb{R}^N}\right)^n$, we know that $\mathcal{W}^{(N)}$ assigns $\{\boldsymbol{\theta} : (\boldsymbol{\theta}(t_1), \dots, \boldsymbol{\theta}(t_n)) \in A\}$ probability

$$\frac{1}{Z(t_1,\dots,t_n)} \int_A \exp\left[-\sum_{m=1}^{n} \frac{|\mathbf{y}_m - \mathbf{y}_{m-1}|^2}{t_m - t_{m-1}}\right] d\mathbf{y}_1 \cdots d\mathbf{y}_n,$$

where $\mathbf{y}_0 \equiv \mathbf{0}$ and $Z(t_1,\dots,t_n) = \prod_{m=1}^n \left(2\pi(t_m - t_{m_1})\right)^{\frac{N}{2}}$. Now rename the variable $\mathbf{y}_m$ as $\boldsymbol{\theta}(t_m)$, and rewrite the preceding as $Z(t_1,\dots,t_n)^{-1}$ times

$$\int_A \exp\left[-\sum_{m=1}^{n} \frac{t_m - t_{m-1}}{2}\left(\frac{|\boldsymbol{\theta}(t_m) - \boldsymbol{\theta}(t_{m-1})|}{t_m - t_{m-1}}\right)^2\right] d\boldsymbol{\theta}(t_1)\cdots d\boldsymbol{\theta}(t_n).$$

Obviously, nothing very significant has happened yet, since nothing very exciting has been done yet. However, if we now close our eyes, suspend our disbelief, and *pass to the limit* as $n$ tends to infinity and the $t_k$'s become dense, we arrive at *Feynman's representation*[2] of Wiener's measure:

$$\mathcal{W}^{(N)}(d\boldsymbol{\theta}) = \frac{1}{Z} \exp\left[-\frac{1}{2}\int_{[0,\infty)} \left|\dot{\boldsymbol{\theta}}(t)\right|^2 dt\right] d\boldsymbol{\theta}, \tag{8.1.4}$$

[2] In truth, Feynman himself never dabbled in considerations so mundane as the ones that follow. He was interested in the Schödinger equation, and so he had a factor $\sqrt{-1}$ multiplying the exponent.

where $\dot{\boldsymbol{\theta}}$ denotes the velocity (i.e., derivative) of $\boldsymbol{\theta}$. Of course, when we reopen our eyes and take a look at (8.1.4), we see that it is riddled with flaws. Not even one of the ingredients on the right-hand side of (8.1.4) makes sense! In the first place, the constant $Z$ must be 0 (or maybe $\infty$). Secondly, since the image of the "measure $d\boldsymbol{\theta}$" under

$$\boldsymbol{\theta} \in \Theta(\mathbb{R}^N) \longmapsto \big(\boldsymbol{\theta}(t_1)\dots,\boldsymbol{\theta}(t_n)\big) \in \big(\mathbb{R}^N\big)^n$$

is Lebesgue measure for every $n \in \mathbb{Z}^+$ and $0 < t_1 \cdots < t_n$, $d\boldsymbol{\theta}$ must be the nonexistent *translation invariant measure* on the infinite dimensional space $\Theta(\mathbb{R}^N)$. Finally, the integral in the exponent only makes sense if $\boldsymbol{\theta}$ is differentiable in some sense, but almost no Brownian path is. Nonetheless, ridiculous as it is, (8.1.4) is exactly the expression at which one would arrive if one were to make a sufficiently naïve interpretation of the notion that Wiener measure is the standard Gauss measure on the Hilbert space $\mathbf{H}(\mathbb{R}^N)$ consisting of absolutely continuous $\mathbf{h} : [0,\infty) \longrightarrow \mathbb{R}^N$ with $\mathbf{h}(0) = \mathbf{0}$ and

$$\|\mathbf{h}\|_{\mathbf{H}^1(\mathbb{R}^N)} = \|\dot{\mathbf{h}}\|_{L^2([0,\infty);\mathbb{R}^N)} < \infty.$$

Of course, the preceding discussion is entirely heuristic. However, now that we know that $\mathbf{H}^1(\mathbb{R}^N)$ is the Hilbert space at which to look, it is easy to provide a mathematically rigorous statement of the connection between $\Theta(\mathbb{R}^N)$, $\mathcal{W}^{(N)}$, and $\mathbf{H}^1(\mathbb{R}^N)$. To this end, observe that $\mathbf{H}(\mathbb{R}^N)$ is continuously embedded in $\Theta(\mathbb{R}^N)$ as a dense subspace. Indeed, if $\mathbf{h} \in \mathbf{H}^1(\mathbb{R}^N)$, then $|\mathbf{h}(t)| \le t^{\frac{1}{2}}\|\mathbf{h}\|_{\mathbf{H}^1(\mathbb{R}^N)}$, and so not only is $\mathbf{h} \in \Theta(\mathbb{R}^N)$ but also $\|\mathbf{h}\|_{\Theta(\mathbb{R}^N)} \le \frac{1}{2}\|\mathbf{h}\|_{\mathbf{H}^1(\mathbb{R}^N)}$. In addition, since $C_c^\infty\big((0,\infty);\mathbb{R}^N\big)$ is already dense in $\Theta(\mathbb{R}^N)$, the density of $\mathbf{H}^1(\mathbb{R}^N)$ in $\Theta(\mathbb{R}^N)$ is clear. Knowing this, abstract reasoning (cf. Lemma 8.2.3) guarantees that $\Theta(\mathbb{R}^N)^*$ can be identified as a subspace of $\mathbf{H}^1(\mathbb{R}^N)$. That is, for each $\boldsymbol{\lambda} \in \Theta(\mathbb{R}^N)^*$, there is a $\mathbf{h}_{\boldsymbol{\lambda}} \in \mathbf{H}^1(\mathbb{R}^N)$ with the property that $\big(\mathbf{h},\mathbf{h}_{\boldsymbol{\lambda}}\big)_{\mathbf{H}^1(\mathbb{R}^N)} = \langle \mathbf{h},\boldsymbol{\lambda}\rangle$ for all $\mathbf{h} \in \mathbf{H}^1(\mathbb{R}^N)$, and in the present setting it is easy to give a concrete representation of $\mathbf{h}_{\boldsymbol{\lambda}}$. In fact, if $\boldsymbol{\lambda} \in \Theta(\mathbb{R}^N)^*$, then, for any $\mathbf{h} \in \mathbf{H}^1(\mathbb{R}^N)$,

$$\begin{aligned}\langle \mathbf{h},\boldsymbol{\lambda}\rangle &= \int_{(0,\infty)} \mathbf{h}(t)\cdot\boldsymbol{\lambda}(dt) = \int_{(0,\infty)}\left(\int_{(0,t)} \dot{\mathbf{h}}(\tau)\,d\tau\right)\cdot\boldsymbol{\lambda}(dt)\\ &= \int_{(0,\infty)} \dot{\mathbf{h}}(\tau)\cdot\boldsymbol{\lambda}\big((\tau,\infty)\big)\,d\tau = \big(\mathbf{h},\mathbf{h}_\lambda\big)_{\mathbf{H}^1(\mathbb{R}^N)},\end{aligned}$$

where

$$\mathbf{h}_{\boldsymbol{\lambda}}(t) = \int_{(0,t]} \boldsymbol{\lambda}\big((\tau,\infty)\big)\,d\tau.$$

Moreover,

$$\|\mathbf{h}_{\boldsymbol{\lambda}}\|^2_{\mathbf{H}^1(\mathbb{R}^N)} = \int_{(0,\infty)} |\boldsymbol{\lambda}((\tau,\infty))|^2\, d\tau = \int_{(0,\infty)} \left( \iint\limits_{(\tau,\infty)^2} \boldsymbol{\lambda}(ds)\cdot\boldsymbol{\lambda}(dt) \right) d\tau$$

$$= \iint\limits_{(0,\infty)^2} s\wedge t\, \boldsymbol{\lambda}(ds)\cdot\boldsymbol{\lambda}(dt).$$

Hence, by (8.1.3),

$$\widehat{\mathcal{W}^{(N)}}(\boldsymbol{\lambda}) = \exp\left( -\frac{\|\mathbf{h}_{\boldsymbol{\lambda}}\|^2_{\mathbf{H}(\mathbb{R}^N)}}{2} \right), \quad \boldsymbol{\lambda}\in\Theta(\mathbb{R}^N)^*. \tag{8.1.5}$$

Although (8.1.5) is far less intuitively appealing than (8.1.4), it provides a mathematically rigorous way in which to think of $\mathcal{W}^{(N)}$ as the standard Gaussian measure on $\mathbf{H}^1(\mathbb{R}^N)$. Furthermore, there is another way to understand why one should accept (8.1.5) as evidence for this way of thinking about $\mathcal{W}^{(N)}$. Indeed, given $\boldsymbol{\lambda}\in\Theta(\mathbb{R}^N)^*$, write

$$\langle\boldsymbol{\theta},\boldsymbol{\lambda}\rangle = \lim_{T\to\infty}\int_{[0,T]} \boldsymbol{\theta}(t)\cdot\boldsymbol{\lambda}(dt) = -\lim_{T\to\infty}\int_0^T \boldsymbol{\theta}(t)\cdot d\boldsymbol{\lambda}\big((t,\infty)\big),$$

where the integral in the last expression is taken in the sense of Riemann–Stieltjes. Next, apply the integration by part formula[3] to conclude that $t\rightsquigarrow \boldsymbol{\lambda}\big((t,\infty)\big)$ is Riemann–Stieltjes integrable with respect to $t\rightsquigarrow\boldsymbol{\theta}(t)$ and that

$$-\int_0^T \boldsymbol{\theta}(t)\cdot d\boldsymbol{\lambda}\big((t,\infty)\big) = -\boldsymbol{\theta}(T)\cdot\boldsymbol{\lambda}\big((T,\infty)\big) + \int_0^T \boldsymbol{\lambda}\big((t,\infty)\big)\cdot d\boldsymbol{\theta}(t).$$

Hence, since

$$\lim_{T\to\infty} |\boldsymbol{\theta}(T)||\boldsymbol{\lambda}|(T,\infty) \le \lim_{T\to\infty}\frac{|\boldsymbol{\theta}(T)|}{1+T}\int_{(0,\infty)} (1+t)\,|\boldsymbol{\lambda}|(dt) = 0,$$

$$\langle\boldsymbol{\theta},\boldsymbol{\lambda}\rangle = \lim_{T\to\infty}\int_0^T \dot{\mathbf{h}}_{\boldsymbol{\lambda}}(t)\cdot d\boldsymbol{\theta}(t), \tag{8.1.6}$$

where again the integral is in the sense of Riemann–Stieltjes. Thus, if one somewhat casually writes $d\boldsymbol{\theta}(t) = \dot{\boldsymbol{\theta}}(t)\,dt$, one can believe that $\langle\boldsymbol{\theta},\boldsymbol{\lambda}\rangle$ provides a reasonable interpretation of $(\boldsymbol{\theta},\mathbf{h}_{\boldsymbol{\lambda}})_{\mathbf{H}(\mathbb{R}^N)}$ for all $\boldsymbol{\theta}\in\Theta(\mathbb{R}^N)$, not just those that are in $\mathbf{H}^1(\mathbb{R}^N)$.

Because R. Cameron and T. Martin were the first mathematicians to systematically exploit the consequences of this line of reasoning, I will call $\mathbf{H}^1(\mathbb{R}^N)$ the **Cameron–Martin space** for classical Wiener measure.

[3] See, for example, Theorem 1.2.3 in my *Essentials of Integration Theory for Analysis*, # 262 in the Springer-Verlag G.T.M. series (2011).

## Exercises for § 8.1

EXERCISE 8.1.7. Let $H$ be a separable Hilbert space, and, for each $n \in \mathbb{Z}^+$ and subset $\{g_1, \dots, g_n\} \subseteq H$, let $\mathcal{A}(g_1, \dots, g_n)$ denote the $\sigma$-algebra over $H$ generated by the mapping

$$h \in H \longmapsto \big((h, g_1)_H, \dots, (h, g_n)_H\big) \in \mathbb{R}^n,$$

and check that

$$\mathcal{A} = \bigcup \{\mathcal{A}(g_1, \dots, g_n) : n \in \mathbb{Z}^+ \text{ and } g_1, \dots, g_n \in H\}$$

is an algebra that generates $\mathcal{B}_H$. Show that there *always* exists a *finitely additive* $\mathcal{W}_H$ on $\mathcal{A}$ that is uniquely determined by the properties that it is $\sigma$-additive on $\mathcal{A}(g_1, \dots, g_n)$ for every $n \in \mathbb{Z}^+$ and $\{g_1, \dots, g_n\} \subseteq H$ and that

$$\int_H \exp\Big[\sqrt{-1}\,(h, g)_H\Big]\, \mathcal{W}_H(dh) = \exp\left[-\frac{\|g\|_H^2}{2}\right], \quad g \in H.$$

On the other hand, as we already know, this finitely additive measure admits a countably additive extension to $\mathcal{B}_H$ if and only if $H$ is finite dimensional.

## § 8.2 A Structure Theorem for Gaussian Measures

Say that a centered Gaussian measure $\mathcal{W}$ on a separable Banach space $E$ is **non-degenerate** if $\mathbb{E}^{\mathcal{W}}\big[\langle x, x^*\rangle^2\big] > 0$ unless $x^* = 0$. (See Exercise 8.2.11.) In this section I will show that any non-degenerate, centered Gaussian measure $\mathcal{W}$ on a separable Banach space $E$ shares the same basic structure that $\mathcal{W}^{(N)}$ has on $\Theta(\mathbb{R}^N)$. In particular, I will show that there is always a Hilbert space $H \subseteq E$ for which $\mathcal{W}$ is the standard Gauss measure in the same sense that $\mathcal{W}^{(N)}$ was shown in § 8.1.2 to be the standard Gauss measure for $\mathbf{H}^1(\mathbb{R}^N)$.

**§ 8.2.1. Fernique's Theorem.** In order to carry out my program, I need a basic integrability result about Banach space–valued, Gaussian random variables. The one that I will use is due to X. Fernique, and his is arguably the most singularly beautiful result in the theory of Gaussian measures on a Banach space.

THEOREM 8.2.1 (**Fernique's Theorem**). *Let $E$ be a real, separable Banach space, and suppose that $X$ is an $E$-valued random variable that is centered and Gaussian in the sense that, for each $x^* \in E^*$, $\langle X, x^*\rangle$ is a centered, $\mathbb{R}$-valued Gaussian random variable. If $R = \inf\{r : \mathbb{P}(\|X\|_E \le r) \ge \frac{9}{10})\}$, then*

$$\mathbb{E}\left[e^{\frac{\|X\|_E^2}{18R^2}}\right] \le K \equiv e^{\frac{1}{2}} + \sum_{n=0}^{\infty} \left(\frac{e}{3}\right)^{2^n}. \tag{8.2.2}$$

*(See Corollary 8.4.3 for a sharpened statement.)*

PROOF: After enlarging the sample space if necessary, I may and will assume that there is an $E$-valued random variable $X'$ that is independent of $X$ and has the same distribution as $X$. Set $Y = 2^{-\frac{1}{2}}(X + X')$ and $Y' = 2^{-\frac{1}{2}}(X - X')$. Then the pair $(Y, Y')$ has the same distribution as the pair $(X, X')$. Indeed, by Lemma 8.1.2, this comes down to showing that the $\mathbb{R}^2$-valued random variable $\big(\langle Y, x^*\rangle, \langle Y', x^*\rangle\big)$ has the same distribution as $\big(\langle X, x^*\rangle, \langle X', x^*\rangle\big)$, and that is an elementary application of the additivity property of independent Gaussians.

Turning to the main assertion, let $0 < s \le t$ be given, and use the preceding to justify

$$\begin{aligned}
&\mathbb{P}(\|X\|_E \le s)\mathbb{P}(\|X\|_E \ge t) = \mathbb{P}(\|X\|_E \le s \;\&\; \|X'\|_E \ge t)\\
&\quad= \mathbb{P}(\|X - X'\|_E \le 2^{\frac{1}{2}} s \;\&\; \|X + X'\|_E \ge 2^{\frac{1}{2}} t)\\
&\quad\le \mathbb{P}\big(\big|\|X\|_E - \|X'\|_E\big| \le 2^{\frac{1}{2}} s \;\&\; \|X\|_E + \|X'\|_E \ge 2^{\frac{1}{2}} t\big)\\
&\quad\le \mathbb{P}\big(\|X\|_E \wedge \|X'\|_E \ge 2^{-\frac{1}{2}}(t-s)\big) = \mathbb{P}\big(\|X\|_E \ge 2^{-\frac{1}{2}}(t-s)\big)^2.
\end{aligned}$$

Now suppose that $\mathbb{P}(\|X\| \le R) \ge \frac{9}{10}$, and define $\{t_n : n \ge 0\}$ by $t_0 = R$ and $t_n = R + 2^{\frac{1}{2}} t_{n-1}$ for $n \ge 1$. Then

$$\mathbb{P}(\|X\|_E \le R)\mathbb{P}(\|X\|_E \ge t_n) \le \mathbb{P}\big(\|X\|_E \ge t_{n-1}\big)^2$$

and therefore

$$\frac{\mathbb{P}(\|X\|_E \ge t_n)}{\mathbb{P}(\|X\|_E \le R)} \le \left(\frac{\mathbb{P}(\|X\|_E \ge t_{n-1})}{\mathbb{P}(\|X\|_E \le R)}\right)^2$$

for $n \ge 1$. Working by induction, one gets from this that

$$\frac{\mathbb{P}(\|X\|_E \ge t_n)}{\mathbb{P}(\|X\|_E \le R)} \le \left(\frac{\mathbb{P}(\|X\|_E \ge R)}{\mathbb{P}(\|X\|_E \le R)}\right)^{2^n}$$

and therefore, since $t_n = R\frac{2^{\frac{n+1}{2}}-1}{2^{\frac{1}{2}}-1} \le 32^{\frac{n+1}{2}} R$, that $\mathbb{P}(\|X\|_E \ge 32^{\frac{n}{2}} R) \le 3^{-2^n}$. Hence,

$$\begin{aligned}
\mathbb{E}^{\mathbb{P}}\left[e^{\frac{\|X\|_E^2}{18R^2}}\right] &\le e^{\frac{1}{2}}\mathbb{P}(\|X\|_E \le 3R) + \sum_{n=0}^{\infty} e^{2^n}\mathbb{P}(32^{\frac{n}{2}} R \le \|X\|_E \le 32^{\frac{n+1}{2}} R)\\
&\le e^{\frac{1}{2}} + \sum_{n=0}^{\infty}\left(\frac{e}{3}\right)^{2^n} = K. \quad \square
\end{aligned}$$

§ **8.2.2. The Basic Structure Theorem.** I will now abstract the relationship, proved in § 8.1.2, between $\Theta(\mathbb{R}^N)$, $\mathbf{H}^1(\mathbb{R}^N)$, and $\mathcal{W}^{(N)}$, and for this purpose I will need the following simple lemma.

LEMMA 8.2.3. *Let $E$ be a separable, real Banach space, and suppose that $H \subseteq E$ is a real Hilbert space that is continuously embedded as a dense subspace of $E$.*

**(i)** *For each $x^* \in E^*$ there is a unique $h_{x^*} \in H$ with the property that $(h, h_{x^*})_H = \langle h, x^* \rangle$ for all $h \in H$, and the map $x^* \in E^* \longmapsto h_{x^*} \in H$ is linear, continuous, one-to-one, and onto a dense subspace of $H$.*

**(ii)** *If $x \in E$, then $x \in H$ if and only if there is a $K < \infty$ such that $|\langle x, x^* \rangle| \leq K\|h_{x^*}\|_H$ for all $x^* \in E^*$. Moreover, for each $h \in H$, $\|h\|_H = \sup\{\langle h, x^* \rangle : x^* \in E^* \ \& \ \|h_{x^*}\|_H \leq 1\}$.*

**(iii)** *If $L^*$ is a weak* dense subspace of $E^*$, then there exists a sequence $\{x_n^* : n \geq 0\} \subseteq L^*$ such that $\{h_{x_n^*} : n \geq 0\}$ is an orthonormal basis for $H$. Moreover, if $x \in E$, then $x \in H$ if and only if $\sum_{n=0}^{\infty} \langle x, x_n^* \rangle^2 < \infty$. Finally,*

$$(h, h')_H = \sum_{n=0}^{\infty} \langle h, x_n^* \rangle \langle h', x_n^* \rangle \quad \text{for all } h, h' \in H.$$

PROOF: Because $H$ is continuously embedded in $E$, there exists a $C < \infty$ such that $\|h\|_E \leq C\|h\|_H$. Thus, if $x^* \in E^*$ and $f(h) = \langle h, x^* \rangle$, then $f$ is linear and $|f(h)| \leq \|h\|_E \|x^*\|_{E^*} \leq C\|x^*\|_{E^*}\|h\|_H$, and so, by the Riesz Representation Theorem for Hilbert spaces, there exists a unique $h_{x^*} \in H$ such that $f(h) = (h, h_{x^*})_H$. In fact, $\|h_{x^*}\|_H \leq C\|x^*\|_{E^*}$, and uniqueness can be used to check that $x^* \rightsquigarrow h_{x^*}$ is linear. To see that $x^* \rightsquigarrow h_{x^*}$ is one-to-one, it suffices to show that $x^* = 0$ if $h_{x^*} = 0$. But if $h_{x^*} = 0$, then $\langle h, x^* \rangle = 0$ for all $h \in H$, and therefore, because $H$ is dense in $E$, $x^* = 0$. Because I will use it later, I will prove slightly more than the density of just $\{h_{x^*} : x^* \in E^*\}$ in $H$. Namely, for any weak* dense subset $S^*$ of $E^*$, $\{h_{x^*} : x* \in S^*\}$ is dense in $H$. Indeed, if this were not the case, then there would exist an $h \in H \setminus \{0\}$ with the property that $\langle h, x^* \rangle = (h, h_{x^*})_H = 0$ for all $x^* \in S$. But, since $S^*$ is weak* dense in $E^*$, this would lead to the contradiction that $h = 0$. Thus, **(i)** is now proved.

Obviously, if $h \in H$, then $|\langle h, x^* \rangle| = |(h, h_{x^*})_H| \leq \|h_{x^*}\|_H \|h\|_H$ for $x^* \in E^*$. Conversely, if $x \in E$ and $|\langle x, x^* \rangle| \leq K\|h_{x^*}\|_H$ for some $K < \infty$ and all $x^* \in E^*$, set $f(h_{x^*}) = \langle x, x^* \rangle$ for $x^* \in E^*$. Then, because $x^* \rightsquigarrow h_{x^*}$ is one-to-one, $f$ is a well-defined, linear functional on $\{h_{x^*} : x^* \in E^*\}$. Moreover, $|f(x^*)| \leq K\|h_{x^*}\|_H$, and therefore, since $\{h_{x^*} : x^* \in E^*\}$ is dense, $f$ admits a unique extension as a continuous, linear functional on $H$. Hence, by Riesz's theorem, there is an $h \in H$ such that

$$\langle x, x^* \rangle = f(h_{x^*}) = (h, h_{x^*})_H = \langle h, x^* \rangle, \quad x^* \in E^*,$$

which means that $x = h \in H$. In addition, if $h \in H$, then $\|h\|_H = \sup\{\langle h, x^* \rangle : \|h_{x^*}\|_H \leq 1\}$ follows from the density of $\{h_{x^*} : x^* \in E^*\}$, and this completes the proof of **(ii)**.

Turning to (**iii**), remember that, by Exercise 5.1.19, the weak* topology on bounded subsets of $E^*$ is second countable. Hence, the weak* topology on bounded subsets of $L^*$ is also second countable, and therefore $L^*$ itself is separable. Thus, we can find a sequence in $L^*$ that is weak* dense in $E^*$, and then, proceeding as in the hint given for Exercise 5.1.19, extract a subsequence of linearly independent elements whose span $S^*$ is weak* dense in $E^*$. Starting with this subsequence, apply the Grahm–Schmidt orthogonalization procedure to produce a sequence $\{x_n^* : n \geq 0\}$ whose span is $S^*$ and for which $\{h_{x_n^*} : n \geq 0\}$ is orthonormal in $H$. Moreover, because the span of $\{h_{x_n^*} : n \geq 0\}$ is equal to $\{h_{x^*} : x^* \in S^*\}$, which, by what we proved earlier, is dense in $H$, $\{h_{x_n^*} : n \geq 0\}$ is an orthonormal basis in $H$. Knowing this, it is immediate that

$$(h, h')_H = \sum_{n=0}^{\infty} (h, h_{x_n})_H (h', h_{x_n})_H = \sum_{n=0}^{\infty} \langle h, x_n^* \rangle \langle h', x_n^* \rangle.$$

In particular, $\|h\|_H^2 = \sum_{n=0}^{\infty} \langle h, x_n^* \rangle^2$. Finally, if $x \in E$ and $\sum_{n=0}^{\infty} \langle x, x_n^* \rangle^2 < \infty$, set $g = \sum_{m=0}^{\infty} \langle x, x_n^* \rangle h_{x_n^*}$. Then $g \in H$ and $\langle x - g, x^* \rangle = 0$ for all $x^* \in S^*$. Hence, since $S^*$ is weak* dense in $E^*$, $x = g \in H$. $\square$

Given a separable real Hilbert space $H$, a separable real Banach space $E$, and a $\mathcal{W} \in \mathbf{M}_1(E)$, I will say that the triple $(H, E, \mathcal{W})$ is an **abstract Wiener space** if $H$ is continuously embedded as a dense subspace of $E$ and $\mathcal{W} \in \mathbf{M}_1(E)$ has Fourier transform

$$\widehat{\mathcal{W}}(x^*) = e^{-\frac{\|h_{x^*}\|_H^2}{2}} \quad \text{for all } x^* \in E^*. \tag{8.2.4}$$

The terminology is justified by the fact, demonstrated at the end of § 8.1.2, that $\left(\mathbf{H}^1(\mathbb{R}^N), \Theta(\mathbb{R}^N), \mathcal{W}^{(N)}\right)$ is an abstract Wiener space. The concept of an abstract Wiener space was introduced by Gross, although his description was somewhat different from the one just given (cf. Theorem 8.3.9 for a reconciliation of mine with his definition).

THEOREM 8.2.5. *Suppose that $E$ is a separable, real Banach space and that $\mathcal{W} \in \mathbf{M}_1(E)$ is a centered Gaussian measure that is non-degenerate. Then there exists a unique Hilbert space $H$ such that $(H, E, \mathcal{W})$ is an abstract Wiener space.*

PROOF: By Fernique's Theorem, we know that $C \equiv \sqrt{\mathbb{E}^{\mathcal{W}}\left[\|x\|_E^2\right]} < \infty$.

To understand the proof of existence, it is best to start with the proof of uniqueness. Thus, suppose that $H$ is a Hilbert space for which $(E, H, \mathcal{W})$ is an abstract Wiener space. Then, for all $x^*$, $y^* \in E^*$, $\langle h_{x^*}, y^* \rangle = (h_{x^*}, h_{y^*})_H = \langle h_{y^*}, x^* \rangle$. In addition,

$$\langle h_{x^*}, x^* \rangle = \|h_{x^*}\|_H^2 = \int \langle x, x^* \rangle^2 \, \mathcal{W}(dx),$$

and so, by the symmetry just established,

$$(*)\qquad \langle h_{x^*}, y^*\rangle = \big(h_{x^*}, h_{y^*}\big)_H = \int \langle x, x^*\rangle\langle x, y^*\rangle\,\mathcal{W}(dx),$$

for all $x^*$, $y^* \in E^*$. Next observe that

$$(**)\qquad \int \big\|\langle x, x^*\rangle\, x\big\|_E\,\mathcal{W}(dx) \le C\|h_{x^*}\|_H,$$

and therefore that the integral $\int x\langle x, x^*\rangle\,\mathcal{W}(dx)$ is a well-defined element of $E$. Moreover, by (*),

$$\langle h_{x^*}, y^*\rangle = \left\langle \int x\langle x, x^*\rangle\,\mathcal{W}(dx), y^*\right\rangle \quad \text{for all } y^* \in E^*,$$

and so

$$(***)\qquad h_{x^*} = \int x\langle x, x^*\rangle\,\mathcal{W}(dx).$$

Finally, given $h \in H$, choose $\{x_n^* : n \ge 1\} \subseteq E^*$ so that $h_{x_n^*} \longrightarrow h$ in $H$. Then

$$\lim_{m\to\infty}\sup_{n>m}\big\|\langle\,\cdot\,, x_n^*\rangle - \langle\,\cdot\,, x_m^*\rangle\big\|_{L^2(\mathcal{W};\mathbb{R})} = \lim_{m\to\infty}\sup_{n>m}\|h_{x_n^*} - h_{x_m^*}\|_H = 0,$$

and so, if $\Psi$ denotes the closure of $\{\langle\,\cdot\,, x^*\rangle : x^* \in E^*\}$ in $L^2(\mathcal{W};\mathbb{R})$ and $F : \Psi \longrightarrow E$ is given by

$$F(\psi) = \int x\psi(x)\,\mathcal{W}(dx), \quad \psi \in \Psi,$$

then $h = F(\psi)$ for some $\psi \in \Psi$. Conversely, if $\psi \in \Psi$ and $\{x_n^* : n \ge 1\}$ is chosen so that $\langle\,\cdot\,, x_n^*\rangle \longrightarrow \psi$ in $L^2(\mathcal{W};\mathbb{R})$, then $\{h_{x_n^*} : n \ge 1\}$ converges in $H$ to some $h \in H$ and it converges in $E$ to $F(\psi)$. Hence, $F(\psi) = h \in H$. In other words, $H = F(\Psi)$.

The proof of existence is now a matter of checking that if $\Psi$ and $F$ are defined as above and if $H = F(\Psi)$ with $\|F(\psi)\|_H = \|\psi\|_{L^2(\mathcal{W};\mathbb{R})}$, then $(H, E, \mathcal{W})$ is an abstract Wiener space. To this end, observe that

$$\langle F(\psi), x^*\rangle = \int \langle x, x^*\rangle\psi(x)\,\mathcal{W}(dx) = \big(F(\psi), h_{x^*}\big)_H,$$

and therefore both (*) and (***) hold for this choice of $H$. Further, given (*), it is clear that $\|h_{x^*}\|_H^2$ is the variance of $\langle\,\cdot\,, x^*\rangle$ and therefore that (8.2.4) holds. At the same time, just as in the derivation of (**), $\|F(\psi)\|_E \le C\|\psi\|_{L^2(\mathcal{W};\mathbb{R})} = C\|F(\psi)\|_H$, and so $H$ is continuously embedded inside $E$. Finally, by the Hahn–Banach Theorem, to show that $H$ is dense in $E$ it suffices to check that the only $x^* \in E^*$ such that $\langle F(\psi), x^*\rangle = 0$ for all $\psi \in \Psi$ is $x^* = 0$. But when $\psi = \langle\,\cdot\,, x^*\rangle$, $\langle F(\psi), x^*\rangle = \int \langle x, x^*\rangle^2\,\mathcal{W}(dx)$, and therefore, because $\mathcal{W}$ is non-degenerate, such an $x^*$ would have to be 0. □

§ **8.2.3. The Cameron–Marin Space.** Given a centered, non-degenerate Gaussian measure $\mathcal{W}$ on $E$, the Hilbert space $H$ for which $(H, E, \mathcal{W})$ is an abstract Wiener space is called its **Cameron–Martin space**. Here are a couple of important properties of the Cameron–Martin subspace.

THEOREM 8.2.6. *If $(H, E, \mathcal{W})$ is an abstract Wiener space, then the map $x^* \in E^* \longmapsto h_{x^*} \in H$ is continuous on bounded subsets from the weak* topology on $E^*$ into the strong topology on $H$. In particular, for each $R > 0$, $\{h_{x^*} : x^* \in \overline{B_{E^*}(0,R)}\}$ is a compact subset of $H$, $\overline{B_H(0,R)}$ is a compact subset of $E$, and so $H \in \mathcal{B}_E$. Moreover, when $E$ is infinite dimensional, $\mathcal{W}(H) = 0$. Finally, there is a unique linear, isometric map $\mathcal{I} : H \longrightarrow L^2(\mathcal{W};\mathbb{R})$ such that $\mathcal{I}(h_{x^*}) = \langle \,\cdot\,, x^*\rangle$ for all $x^* \in E^*$, and $\{\mathcal{I}(h) : h \in H\}$ is a Gaussian family in $L^2(\mathcal{W};\mathbb{R})$.*

PROOF: To prove the initial assertion, remember that $x^* \rightsquigarrow \widehat{\mathcal{W}}(x^*)$ is continuous with respect to the weak* topology on bounded subsets. Hence, if $\{x_k^* : k \geq 1\}$ is a bounded sequence which is weak* convergent to $x^*$, then

$$\exp\left(-\frac{\|h_{x_k^*} - h_{x^*}\|_H^2}{2}\right) = \widehat{\mathcal{W}}(x_k^* - x^*) \longrightarrow 1,$$

and so $h_{x_k^*} \longrightarrow h_{x^*}$ in $H$.

Given the first assertion, the compactness of $\{h_{x^*} : x^* \in \overline{B_{E^*}(0,R)}\}$ in $H$ follows from the compactness (cf. Exercise 5.1.19) of $\overline{B_{E^*}(0,R)}$ in the weak* topology. To see that $\overline{B_H(0,R)}$ is compact in $E$, again apply Exercise 5.1.19 to check that $\overline{B_H(0,R)}$ is compact in the weak topology on $H$. Therefore, all that we have to show is that the embedding map $h \in \overline{B_H(0,R)} \longmapsto h \in E$ is continuous from the weak topology on $H$ into the strong topology on $E$. Thus, suppose that $h_k \longrightarrow h$ weakly in $H$. Because $\{h_{x^*} : x^* \in \overline{B_{E^*}(0,1)}\}$ is compact in $H$, for each $\epsilon > 0$ there exist an $n \in \mathbb{Z}^+$ and a $\{x_1^*, \dots, x_n^*\} \subseteq \overline{B_{E^*}(0,1)}$ such that

$$\{h_{x^*} : x^* \in \overline{B_{E^*}(0,1)}\} \subseteq \bigcup_1^n B_H(h_{x_m^*}, \epsilon).$$

Now choose $\ell$ so that $\max_{1\leq m\leq n} |\langle h_k - h, x_m^*\rangle| < \epsilon$ for all $k \geq \ell$. Then, for any $x^* \in \overline{B_{E^*}(0,1)}$ and all $k \geq \ell$,

$$|\langle h_k - h, x^*\rangle| \leq \epsilon + \min_{1\leq m\leq n} \left|\big(h_k - h, h_{x^*} - h_{x_m^*}\big)_H\right| + 2R\epsilon \leq (2R+1)\epsilon,$$

which proves that $\|h_k - h\|_H \leq (2R+1)\epsilon$ for all $k \geq \ell$.

Because $H = \bigcup_1^\infty \overline{B_H(0,n)}$ and $\overline{B_H(0,n)}$ is a compact subset of $E$ for each $n \in \mathbb{Z}^+$, it is clear that $H \in \mathcal{B}_E$. To see that $\mathcal{W}(H) = 0$ when $E$ is infinite dimensional, choose $\{x_n^* : n \geq 0\}$ as in the final part of Lemma 8.2.3, and set $X_n(x) = \langle x, x_n^*\rangle$. Then the $X_n$'s are an infinite sequence of independent, centered, Gaussians with mean value 1, and so $\sum_{n=0}^\infty X_n^2 = \infty$ $\mathcal{W}$-almost surely. Hence, by part **(iii)** of Lemma 8.2.3, $\mathcal{W}$-almost no $x$ is in $H$.

Turning to the map $\mathcal{I}$, define $\mathcal{I}(h_{x^*}) = \langle \,\cdot\,, x^*\rangle$. Then, for each $x^*$, $\mathcal{I}(h_{x^*})$ is a centered Gaussian with variance $\|h_{x^*}\|_H^2$, and so $\mathcal{I}$ is a linear isometry from

$\{h_{x^*} : x^* \in E^*\}$ into $L^2(\mathcal{W};\mathbb{R})$. Hence, since $\{h_{x^*} : x^* \in E^*\}$ is dense in $H$, $\mathcal{I}$ admits a unique extension as a linear isometry from $H$ into $L^2(\mathcal{W};\mathbb{R})$. Moreover, as the $L^2(\mathcal{W};\mathbb{R})$-limit of centered Gaussians, $\mathcal{I}(h)$ is a centered Gaussian for each $h \in H$. □

The map $\mathcal{I}$ in Theorem 8.2.6 was introduced for the classical Wiener space by Paley and Wiener, and so I will call it the **Paley–Wiener map**. To appreciate its importance here, observe that $\{h_{x^*} : x^* \in E^*\}$ is the subspace of $g \in H$ with the property that $h \in H \longmapsto (h,g)_H \in \mathbb{R}$ admits a continuous extension to $E$. Even though, when $\dim(H) = \infty$, no such continuous extension exists for general $g \in H$, $\mathcal{I}(g)$ can be thought of as an extension of $h \rightsquigarrow (h,g)_H$, albeit one that is defined only up to a $\mathcal{W}$-null set. Of course, one has to be careful when using this interpretation, since, when $H$ is infinite dimensional, $\mathcal{I}(g)(x)$ for a given $x \in E$ is not well-defined simultaneously of all $g \in H$. Nonetheless, by adopting it, one gets further evidence for the idea that $\mathcal{W}$ wants to be the standard Gauss measure on $H$. Namely, because

$$\text{(8.2.7)} \qquad \mathbb{E}^{\mathcal{W}}\left[e^{\sqrt{-1}\,\mathcal{I}(h)}\right] = e^{-\frac{\|h\|_H^2}{2}}, \quad h \in H,$$

if $\mathcal{W}$ lived on $H$, then it would certainly be the standard Gauss measure there.

Perhaps the most important application of the Paley–Wiener map is the following theorem about the behavior of Gaussian measures under translation. That is, if $y \in E$ and $\tau_y : E \longrightarrow E$ is given by $\tau_y(x) = x + y$, we will be looking at the measure $(\tau_y)_*\mathcal{W}$ and its relationship to $\mathcal{W}$. Using the reasoning suggested above, the result is easy to guess. Namely, if $\mathcal{W}$ really lived on $H$ and were given by a Feynman-type representation

$$\mathcal{W}(dh) = \frac{1}{Z} e^{-\frac{\|h\|_H^2}{2}} \lambda_H(dh),$$

then $(\tau_g)_*\mathcal{W}$ should have the Feynman representation

$$\frac{1}{Z} e^{-\frac{\|h-g\|_H^2}{2}} \lambda_H(dh),$$

which could be rewritten as

$$\big[(\tau_g)_*\mathcal{W}\big](dh) = \exp\left[(h,g)_H - \tfrac{1}{2}\|g\|_H^2\right] \mathcal{W}(dh).$$

Hence, if we assume that $\mathcal{I}(g)$ gives us the correct interpretation of $(\,\cdot\,,g)_H$, we are led to guess that, at least for $g \in H$,

$$\text{(8.2.8)} \quad \big[(\tau_g)_*\mathcal{W}(dx)\big](dh) = R_g(x)\,\mathcal{W}(dx), \quad \text{where } R_g = \exp\big[\mathcal{I}(g) - \tfrac{1}{2}\|g\|_H^2\big].$$

That (8.2.8) is correct was proved for the classical Wiener space by Cameron and Martin, and for this reason it is called the **Cameron–Martin formula**. In fact, one has the following result, the second half of which is due to Segal.

THEOREM 8.2.9. *If $(H, E, \mathcal{W})$ is an abstract Wiener space, then, for each $g \in H$, $(\tau_g)_*\mathcal{W} \ll \mathcal{W}$ and the $R_g$ in (8.2.8) is the corresponding Radon–Nikodym derivative. Conversely, if $(\tau_y)_*\mathcal{W}$ is not singular with respect to $\mathcal{W}$, then $y \in H$.*

PROOF: Let $g \in H$, and set $\mu = (\tau_g)_*\mathcal{W}$. Then

$$(*) \qquad \hat{\mu}(x^*) = \mathbb{E}^{\mathcal{W}}\big[e^{\sqrt{-1}\langle x+g,x^*\rangle}\big] = \exp\big[\sqrt{-1}\langle g, x^*\rangle - \tfrac{1}{2}\|h_{x^*}\|_H^2\big].$$

Now define $\nu$ by the right-hand side of (8.2.8). Clearly $\nu \in \mathbf{M}_1(E)$. Thus, we will have proved the first part once we show that $\hat{\nu}$ is given by the right-hand side of (*). To this end, observe that, for any $h_1, h_2 \in H$,

$$\mathbb{E}^{\mathcal{W}}\big[e^{\xi_1\mathcal{I}(h_1)+\xi_2\mathcal{I}(h_2)}\big] = \exp\left[\frac{\xi_1^2}{2}\|h_1\|_H^2 + \xi_1\xi_2\big(h_1,h_2\big)_H + \frac{\xi_2^2}{2}\|h_2\|_H^2\right]$$

for all $\xi_1, \xi_2 \in \mathbb{C}$. Indeed, this is obvious when $\xi_1$ and $\xi_2$ are pure imaginary, and, since both sides are entire functions of $(\xi_1, \xi_2) \in \mathbb{C}^2$, it follows in general by analytic continuation. In particular, by taking $h_1 = g$, $\xi_1 = 1$, $h_2 = h_{x^*}$, and $\xi_2 = \sqrt{-1}$, it is easy to check that the right-hand side of (*) is equal to $\hat{\nu}(x^*)$.

To prove the second assertion, begin by recalling from Lemma 8.2.3 that if $y \in E$, then $y \in H$ if and only if there is a $K < \infty$ with the property that $|\langle y, x^*\rangle| \le K$ for all $x^* \in E^*$ with $\|h_{x^*}\|_H = 1$. Now suppose that $(\tau_{x^*})_*\mathcal{W} \not\perp \mathcal{W}$, and let $R$ be the Radon–Nikodym derivative of its absolutely continuous part. Given $x^* \in E^*$ with $\|h_{x^*}\|_H = 1$, let $\mathcal{F}_{x^*}$ be the $\sigma$-algebra generated by $x \rightsquigarrow \langle x, x^*\rangle$, and check that $(\tau_y)_*\mathcal{W} \restriction \mathcal{F}_{x^*} \ll \mathcal{W} \restriction \mathcal{F}_{x^*}$ with Radon–Nikodym derivative

$$Y(x) = \exp\left(\langle y, x^*\rangle\langle x, x^*\rangle - \frac{\langle y, x^*\rangle^2}{2}\right).$$

Hence,

$$Y \ge \mathbb{E}^{\mathcal{W}}\big[R\,\big|\,\mathcal{F}_{x^*}\big] \ge \mathbb{E}^{\mathcal{W}}\big[R^{\frac{1}{2}}\,\big|\,\mathcal{F}_{x^*}\big]^2,$$

and so (cf. Exercise 8.2.19)

$$\exp\left(-\frac{\langle y, x^*\rangle^2}{8}\right) = \mathbb{E}^{\mathcal{W}}\big[Y^{\frac{1}{2}}\big] \ge \alpha \equiv \mathbb{E}^{\mathcal{W}}\big[R^{\frac{1}{2}}\big] \in (0,1].$$

Since this means that $\langle y, x^*\rangle^2 \le 8\log\frac{1}{\alpha}$, the proof is complete. □

## Exercises for § 8.2

EXERCISE 8.2.10. Let $\mathbf{C} \in \mathrm{Hom}(\mathbb{R}^N; \mathbb{R}^N$ be a positive definite and symmetric, take $E = \mathbb{R}^N$ to be the standard Euclidean metric, and let $H = \mathbb{R}^N$ with the Hilbert inner product $(\mathbf{x}, \mathbf{y})_H = (\mathbf{x}, \mathbf{C}^{-1}\mathbf{y})_{\mathbb{R}^N}$. Show that $\big(H, E, \gamma_{\mathbf{0},\mathbf{C}}\big)$ is an abstract Wiener space.

EXERCISE 8.2.11. Let $E$ be a separable Banach space and $\mathcal{W}$ a centered Gaussian measure on $E$, but do not assume that $\mathcal{W}$ is non-degenerate. Denote by $\mathcal{N}$ the set of $x^* \in E^*$ for which $\mathbb{E}^{\mathcal{W}}\big[\langle x, x^*\rangle^2\big] = 0$, and set

$$\hat{E} = \big\{x \in E : \langle x, x^*\rangle = 0 \text{ for all } x^* \in \mathcal{N}\big\}.$$

Show that $\hat{E}$ is closed, that $\mathcal{W}(\hat{E}) = 1$, and that $\mathcal{W} \restriction \hat{E}$ is a non-degenerate, centered Gaussian measure on $\hat{E}$.

**Hint**: Since $\mathcal{W}\big(\{x \in E : \langle x, x^*\rangle \neq 0\}\big) = 0$ for each $x^* \in \mathcal{N}$, the only question is whether one can choose a countable subset $C \subseteq \mathcal{N}$ such that $x \in \hat{E}$ if and only if $\langle x, x^*\rangle = 0$ for all $x^* \in C$. For this purpose, recall that, by Exercise 5.1.19, bounded subsets of $E^*$ with the weak* topology are second countable and therefore that $\mathcal{N}$ is separable with respect to the weak* topology.

EXERCISE 8.2.12. Let $\{x_n : n \geq 0\}$ be a sequence in the separable Banach space $E$ with the property that $\sum_{n=0}^{\infty} \|x_n\|_E < \infty$. Show that $\sum_{n=0}^{\infty} |\xi_n| \|x_n\| < \infty$ for $\gamma_{0,1}^{\mathbb{N}}$-almost every $\xi \in \mathbb{R}^{\mathbb{N}}$, and define $X : \mathbb{R}^{\mathbb{N}} \longrightarrow E$ so that $X(\xi) = \sum_{n=0}^{\infty} \xi_n x_n$ if $\sum_{n=0}^{\infty} |\xi_n| \|x_n\|_E < \infty$ and $X(\xi) = 0$ otherwise. Show that the distribution $\mu$ of $X$ is a centered, Gaussian measure on $E$. In addition, show that $\mu$ is non-degenerate if and only if the span of $\{x_n : n \geq 0\}$ is dense in $E$.

EXERCISE 8.2.13. Here an application of Fernique's Theorem to functional analysis. Let $E$ and $F$ be a pair of separable Banach spaces and $\psi$ a Borel measurable, linear map from $E$ to $F$. Given a centered, Gaussian $E$-valued random variable $X$, use Exercise 2.3.21 see that $\psi \circ X$ is an $F$-valued, centered Gaussian random variable, and apply Fernique's Theorem to conclude that $\psi \circ X$ is a square integrable and has mean value 0. Next, suppose that $\psi$ is not continuous, and choose $\{x_n : n \geq 0\} \subseteq E$ and $\{y_n : n \geq 0\} \subseteq F^*$ so that $\|x_n\|_E = 1 = \|y_n *\|_{F^*}$ and $\langle \psi(x_n), y_n^*\rangle \geq (n+1)^3$. Using Exercise 8.2.12, show that there exist centered, Gaussian $F$-valued random variables $\{X_n : n \geq 0\}$, and $X$ under $\gamma_{0,1}^{\mathbb{N}}$ such that $X_n(\xi) = (n+1)^{-2}\xi_n x_n$, $X(\xi) = \sum_{n=0}^{\infty} X_n(\xi)$, and $X^n(\xi) = X(\xi) - X_n(\xi)$ for $\gamma_{0,1}^{\mathbb{N}}$-almost every $\xi \in \mathbb{R}^{\mathbb{N}}$. Show that

$$\begin{aligned}\int \|\psi \circ X(\xi)\|_F^2 \, \gamma_{0,1}^{\mathbb{N}}(d\xi) &\geq \int \langle \psi \circ X(\xi), y_n^*\rangle \, \gamma_{0,1}^{\mathbb{N}}(d\xi) \\ &\geq \int \langle \psi \circ X_n(\xi), y_n^*\rangle \, \gamma_{0,1}^{\mathbb{N}}(d\xi) \geq (n+1),\end{aligned}$$

and thereby arrive at the contradiction that $\psi \circ X \notin L^2(\gamma_{0,1}^{\mathbb{N}}; F)$. Conclude that *every Borel measurable, linear map from E to F is continuous.* Notice that, as a consequence, we know that the Paley–Wiener integral $\mathcal{I}(h)$ of an $h$ in the Cameron–Martin space is equal $\mathcal{W}$-almost everywhere to a Borel measurable, linear function if and only if $h = h_{x^*}$ for some $x^* \in E^*$.

EXERCISE 8.2.14. Let $\mathcal{W}$ be a centered, Gaussian measure on a separable Banach space $E$, and set $\sigma = \sqrt{\sum_{m=1}^n a_m^2}$, where $a_1, \dots, a_n \in \mathbb{R}$. If $X_1, \dots, X_n$ are mutually independent, $E$-valued random variables with distribution $\mathcal{W}$ on some probability space $(\Omega, \mathcal{F}, \mathbb{P})$, show that the $\mathbb{P}$-distribution of $S \equiv \sum_{m=1}^n a_m X_m$ is the same as the $\mathcal{W}$-distribution of $x \rightsquigarrow \sigma x$. In particular,

$$\mathbb{E}^{\mathbb{P}}\left[\|S\|_E^p\right] = \sigma^p \mathbb{E}^{\mathcal{W}}\left[\|x\|_E^p\right] \quad \text{for all } p \in [0,\infty).$$

**Hint**: Using Exercise 8.2.11, reduce to the case when $\mathcal{W}$ is non-degenerate. For this case, let $H$ be the Cameron–Martin space for $\mathcal{W}$ on $E$, and show that

$$\mathbb{E}^{\mathbb{P}}\left[e^{\sqrt{-1}\langle S,x^*\rangle}\right] = e^{-\frac{\sigma^2}{2}\|h_{x^*}\|_H^2} \quad \text{for all } x^* \in E^*.$$

EXERCISE 8.2.15. Referring to the setting in Lemma 8.2.3, show that there is a sequence $\{\|\cdot\|_E^{(n)} : n \geq 0\}$ of norms on $E$ each of which is commensurate with $\|\cdot\|_E$ (i.e., $C_n^{-1}\|\cdot\| \leq \|\cdot\|_E^{(N)} \leq C_n\|\cdot\|$ for some $C_n \in [1,\infty)$) such that, for each $R > 0$,

$$\overline{B_H(0,R)} = \{x \in E : \|x\|_E^{(n)} \leq R \text{ for all } n \geq 0\}.$$

**Hint**: Choose $\{x_m^* : m \geq 0\} \subseteq E^*$ so that $\{h_{x_m^*} : m \geq 0\}$ is an orthonormal basis for $H$, define $P_n : E \longrightarrow H$ by $P_n x = \sum_{m=0}^n \langle x, x_m^*\rangle h_{x_m^*}$, and set

$$\|x\|_E^{(n)} = \sqrt{\|P_n x\|_H^2 + \|x - P_n x\|_E^2}.$$

EXERCISE 8.2.16. Referring to the setting in Fernique's Theorem, observe that all powers of $\|X\|_E$ are integrable, and set $\sigma^2 = \mathbb{E}\left[\|X\|_E^2\right]$. Show that

$$\mathbb{E}\left[e^{\frac{\|X\|_E^2}{180\sigma^2}}\right] \leq K.$$

In particular, for any $n \geq 1$, conclude that

$$\mathbb{E}\left[\|X\|_E^{2n}\right] \leq (180)^n n! K \sigma^{2n},$$

which is remarkably close to the equality that holds when $E = \mathbb{R}$. See Corollary 8.4.3 for a sharper statement.

EXERCISE 8.2.17. Again let $E$ be a separable, real Banach space. Suppose that $\{X_n : n \geq 1\}$ is a sequence for centered, Gaussian $E$-valued random variables on some probability space $(\Omega, \mathcal{F}, \mathbb{P})$ and that $X_n \longrightarrow X$ in $\mathbb{P}$-probability. Show that $X$ is again a centered, Gaussian random variable and that there exists a $\lambda > 0$ for which $\sup_{n\geq 1} \mathbb{E}^{\mathbb{P}}\left[e^{\lambda\|X_n\|_E^2}\right] < \infty$. Conclude, in particular, that $X_n \longrightarrow X$ in $L^p(\mathbb{P}; E)$ for every $p \in [1,\infty)$.

EXERCISE 8.2.18. Given $\boldsymbol{\lambda} \in \Theta(\mathbb{R}^N)^*$, I pointed out at the end of § 8.1.2 that the Paley–Wiener integral $[\mathcal{I}(\mathbf{h}_{\boldsymbol{\lambda}})](\boldsymbol{\theta})$ can be interpreted as the Riemann–Stieltjes integral of $\boldsymbol{\lambda}\big((s,\infty)\big)$ with respect to $\boldsymbol{\theta}(s)$. In this exercise, I will use this observation as the starting point for what is called **stochastic integration.**

**(i)** Given $\boldsymbol{\lambda} \in \Theta(\mathbb{R}^N)^*$ and $t > 0$, set $\boldsymbol{\lambda}^t(d\tau) = \mathbf{1}_{[0,t)}(\tau)\boldsymbol{\lambda}(d\tau) + \delta_t\boldsymbol{\lambda}\big([t,\infty)\big)$, and show that for all $\boldsymbol{\theta} \in \Theta(\mathbb{R}^N)$

$$\langle \boldsymbol{\theta}, \boldsymbol{\lambda}^t \rangle = \int_0^t \boldsymbol{\lambda}\big((\tau,\infty)\big) \cdot d\boldsymbol{\theta}(\tau),$$

where the integral on the right is taken in the sense of Riemann–Stieltjes. In particular, conclude that $t \rightsquigarrow \langle \boldsymbol{\theta}, \boldsymbol{\lambda}^t \rangle$ is continuous for each $\boldsymbol{\theta}$.

**(ii)** Given $\mathbf{f} \in C^1_c\big([0,\infty);\mathbb{R}^N\big)$, set $\boldsymbol{\lambda}_{\mathbf{f}}(d\tau) = -\dot{\mathbf{f}}(\tau)\,d\tau$, and show that

$$\langle \boldsymbol{\theta}, \boldsymbol{\lambda}^t_{\mathbf{f}} \rangle = \int_0^t \mathbf{f}(\tau) \cdot d\boldsymbol{\theta}(\tau),$$

where again the integral on the right is Riemann–Stieltjes. Use this to see that the process

$$\left\{ \int_0^t \mathbf{f}(\tau) \cdot d\boldsymbol{\theta}(\tau) : t \geq 0 \right\}$$

has the same distribution under $\mathcal{W}^{(N)}$ as

$$(*) \qquad \left\{ B\left( \int_0^t |\mathbf{f}(\tau)|^2\,d\tau \right) : t \geq 0 \right\},$$

where $\{B(t) : t \geq 0\}$ is an $\mathbb{R}$-valued Brownian motion.

**(iii)** Given $\mathbf{f} \in L^2_{\text{loc}}\big([0,\infty);\mathbb{R}^N\big)$ and $t > 0$, set $\mathbf{h}^t_{\mathbf{f}}(\tau) = \int_0^{t\wedge\tau} \mathbf{f}(s)\,ds$. Show that the $\mathcal{W}^{(N)}$-distribution of the process $\{\mathcal{I}(\mathbf{h}^t_{\mathbf{f}}) : t \geq 0\}$ is the same as that of the process in (*). In particular, conclude (cf. part **(ii)** of Exercise 4.3.16) that there is a continuous modification of the process $\{\mathcal{I}(\mathbf{h}^t_{\mathbf{f}}) : t \geq 0\}$. For reasons made clear in **(ii)**, such a continuous modification is denoted by

$$\left\{ \int_0^t \mathbf{f}(\tau) \cdot d\boldsymbol{\theta}(\tau) : t \geq 0 \right\}.$$

Of course, unless $\mathbf{f}$ has bounded variation, the integrals in the preceding are no longer interpretable as Riemann–Stieltjes integrals. In fact, they not even defined $\boldsymbol{\theta}$ by $\boldsymbol{\theta}$ but only as a stochastic process. For this reason, they are called **stochastic integrals**.

EXERCISE 8.2.19. Define $R_g$ as in (8.2.8), and show that

$$\mathbb{E}^{\mathcal{W}}\big[R_g^p\big]^{\frac{1}{p}} = \exp\left[\frac{(p-1)\|g\|_H^2}{2}\right] \quad \text{for all } p \in (0,\infty).$$

EXERCISE 8.2.20. Here is another way to think about Segal's half of Theorem 8.2.9. Using Lemma 8.2.3, choose $\{x_n^* : n \geq 0\} \subseteq E^*$ so that $\{h_{x_n^*} : n \geq 0\}$ is an orthonormal basis for $H$. Next, define $F : E \longrightarrow \mathbb{R}^{\mathbb{N}}$ so that $F(x)_n = \langle x, x_n^* \rangle$ for each $n \in \mathbb{N}$, and show that $F_*\mathcal{W} = \gamma_{0,1}^{\mathbb{N}}$ and $(F \circ \tau_y)_*\mathcal{W} = \prod_0^\infty \gamma_{a_n,1}$, where $a_n = \langle y, x_n^* \rangle$. Conclude from this that $(\tau_y)_*\mathcal{W} \perp \mathcal{W}$ if $\gamma_{0,1}^{\mathbb{N}} \perp \prod_0^\infty \gamma_{a_n,1}$. Finally, use this together with Exercise 5.2.42 to see that $(\tau_y)_*\mathcal{W} \perp \mathcal{W}$ if $\sum_0^\infty a_m^2 = \infty$, which, by Lemma 8.2.3, will be the case if $y \notin H$.

## § 8.3 From Hilbert to Abstract Wiener Space

Up to this point I have been assuming that we already have at hand a non-degenerate, centered Gaussian measure $\mathcal{W}$ on a Banach space $E$, and, on the basis of this assumption, I produced the associated Cameron–Martin space $H$. In this section, I will show how one can go in the opposite direction. That is, I will start with a separable, real Hilbert space $H$ and show how to go about finding a separable, real Banach space $E$ for which there exists a $\mathcal{W} \in \mathbf{M}_1(E)$ such that $(H, E, \mathcal{W})$ is an abstract Wiener space. Although I will not adopt his approach, the idea of carrying out such a program is Gross's.

**Warning**: From now on, unless the contrary is explicitly stated, I will be assuming that the spaces with which I am dealing are all infinite dimensional, separable, and real.

**§ 8.3.1. An Isomorphism Theorem.** Because, at an abstract level, all infinite dimensional, separable Hilbert spaces are the same, one should expect that, in a related sense, the set of all abstract Wiener spaces for which one Hilbert space is the Cameron–Martin space is the same as the set of all abstract Wiener spaces for which any other Hilbert space is the Cameron–Martin space. The following simple result verifies this conjecture.

THEOREM 8.3.1. *Let $H$ and $H'$ be a pair of Hilbert spaces, and suppose that $F$ is a linear isometry from $H$ onto $H'$. Further, suppose that $(H, E, \mathcal{W})$ is an abstract Wiener space. Then there exists a separable, real Banach space $E' \supseteq H'$ and a linear isometry $\tilde{F}$ from $E$ onto $E'$ such that $\tilde{F} \restriction H = F$ and $(H', E', \tilde{F}_*\mathcal{W})$ is an abstract Wiener space.*

PROOF: Define $\|h'\|_{E'} = \|F^{-1}h'\|_E$ for $h' \in H'$, and let $E'$ be the Banach space obtained by completing $H'$ with respect to $\|\cdot\|_{E'}$. Trivially, $H'$ is continuously embedded in $E'$ as a dense subspace, and $F$ admits a unique extension $\tilde{F}$ as an isometry from $E$ onto $E'$. Moreover, if $(x')^* \in (E')^*$ and $\tilde{F}^\top$ is the adjoint map from $(E')^*$ onto $E^*$, then

$$\begin{aligned}\big(h', h'_{(x')^*}\big)_{H'} &= \langle h', (x')^* \rangle = \langle F^{-1}h', \tilde{F}^\top (x')^* \rangle \\ &= \big(F^{-1}h', h_{\tilde{F}^\top (x')^*}\big)_H = \big(h', F h_{\tilde{F}^\top (x')^*}\big)_{H'},\end{aligned}$$

and so $h'_{(x')^*} = F h_{\tilde{F}^\top(x')^*}$. Hence,

$$\mathbb{E}^{\tilde{F}_*\mathcal{W}}\left[e^{\sqrt{-1}\,\langle x',(x')^*\rangle}\right] = \mathbb{E}^{\mathcal{W}}\left[e^{\sqrt{-1}\,\langle \tilde{F}x,(x')^*\rangle}\right] = \mathbb{E}^{\mathcal{W}}\left[e^{\sqrt{-1}\,\langle x,\tilde{F}^\top(x')^*\rangle}\right]$$
$$= e^{-\frac{1}{2}\|h_{\tilde{F}^\top(x')^*}\|_H^2} = e^{-\frac{1}{2}\|F^{-1}h'_{(x')^*}\|_H^2} = e^{-\frac{1}{2}\|h'_{(x')^*}\|_{H'}^2},$$

which completes the proof that $(H', E', \tilde{F}_*\mathcal{W})$ is an abstract Wiener space. □

Theorem 8.3.1 says that there is a one-to-one correspondence between the abstract Wiener spaces associated with one Hilbert space and the abstract Wiener spaces associated with any other. In particular, it allows us to prove the theorem of Gross which states that every Hilbert space is the Cameron–Martin space for some abstract Wiener space.

COROLLARY 8.3.2. *Given a separable, real Hilbert space $H$, there exists a separable Banach space $E$ and a $\mathcal{W} \in \mathbf{M}_1(E)$ such that $(H, E, \mathcal{W})$ is an abstract Wiener space.*

PROOF: Let $F : H^1(\mathbb{R}) \longrightarrow H$ be an isometric isomorphism, and use Theorem 8.3.1 to construct a separable Banach space $E$ and an isometric, isomorphism $\tilde{F} : \Theta(\mathbb{R}) \longrightarrow E$ so that $(H, E, \mathcal{W})$ is an abstract Wiener space when $\mathcal{W} = \tilde{F}_*\mathcal{W}^{(1)}$. □

It is important to recognize that although a non-degenerate, centered Gaussian measure on a Banach space $E$ determines a unique Cameron–Martin space $H$, a given $H$ will be the Cameron–Martin space for an uncountable number of abstract Wiener spaces. For example, in the classical case when $H = \mathbf{H}^1(\mathbb{R}^N)$, we could have replaced $\Theta(\mathbb{R}^N)$ by a subspace which reflected the fact that almost every Brownian path is locally Hölder continuous of any order less than a half. We will see a definitive, general formulation of this point in Corollary 8.3.10.

**§ 8.3.2. Wiener Series.** The proof that I gave of Corollary 8.3.2 is too nonconstructive to reveal much about the relationship between $H$ and the abstract Wiener spaces for which it is the Cameron–Martin space. Thus, in this subsection I will develop another, entirely different way of constructing abstract Wiener spaces for a Hilbert space.

The approach here has its origins in one of Wiener's own constructions of Brownian motion and is based on the following line of reasoning. Given $H$, choose an orthonormal basis $\{h_n : n \geq 0\}$. If there were a standard Gauss measure $\mathcal{W}$ on $H$, then the random variables $\{X_n : n \geq 0\}$ given by $X_n(h) = (h, h_n)_H$ would be independent, standard normal, $\mathbb{R}$-valued random variables, and, for each $h \in H$, $\sum_0^\infty X_n(h)h_n$ would converge in $H$ to $h$. Even though $\mathcal{W}$ cannot live on $H$, this line of reasoning suggests that a way to construct an abstract Wiener space is to start with a sequence $\{X_n : n \geq 0\}$ of $\mathbb{R}$-valued, independent standard normal random variables on some probability space, find a Banach space $E$ in which $\sum_0^\infty X_n h_n$ converges with probability 1, and take $\mathcal{W}$ on $E$ to the distribution of this series.

To convince oneself that this line of reasoning has a chance of leading somewhere, one should observe that Lévy's construction corresponds to a particular choice of the orthonormal basis $\{\mathbf{h}_m : m \geq 0\}$.[1] To see this, determine $\{\dot{h}_{k,n} : (k,n) \in \mathbb{N}^2\}$ by

$$\dot{h}_{k,0} = \mathbf{1}_{[k,k+1)} \text{ and } \dot{h}_{k,n} = 2^{\frac{n-1}{2}} \begin{cases} 1 & \text{on } [k2^{1-n}, (2k+1)2^{-n}) \\ -1 & \text{on } [(2k+1)2^{-n}, (k+1)2^{1-n}) \\ 0 & \text{elsewhere} \end{cases}$$

for $n \geq 1$. Clearly, the $\dot{h}_{k,n}$'s are orthonormal in $L^2([0,\infty);\mathbb{R})$. In addition, for each $n \in \mathbb{N}$, the span of $\{\dot{h}_{k,n} : k \in \mathbb{N}\}$ equals that of $\{\mathbf{1}_{[k2^{-n},(k+1)2^{-n})} : k \in \mathbb{N}\}$. Perhaps the easiest way to check this is to do so by dimension counting. That is, for a given $(\ell, n) \in \mathbb{N}^2$, note that

$$\{\dot{h}_{\ell,0}\} \cup \{\dot{h}_{k,m} : \ell 2^{m-1} \leq k < (\ell+1)2^{m-1} \text{ and } 1 \leq m \leq n\}$$

has the same number of elements as $\{\mathbf{1}_{[k2^{-n},(k+1)2^{-n})} : \ell 2^n \leq k < (\ell+1)2^n\}$ and that the first set is contained in the span of the second. As a consequence, we know that $\{\dot{h}_{k,n} : (k,n) \in \mathbb{N}^2\}$ is an orthonormal basis in $L^2([0,\infty);\mathbb{R})$, and so, if $h_{k,n}(t) = \int_0^t \dot{h}_{k,n}(\tau)\,d\tau$ and $(\mathbf{e}_1, \ldots, \mathbf{e}_N)$ is an orthonormal basis in $\mathbb{R}^N$, then

$$\{\mathbf{h}_{k,n,i} \equiv h_{k,n}\mathbf{e}_i : (k,n,i) \in \mathbb{N}^2 \times \{1,\ldots,N\}\}$$

is an orthonormal basis, known as the **Haar basis**, in $\mathbf{H}^1(\mathbb{R}^N)$. Finally, if $\{X_{k,n,i} : (k,n,i) \in \mathbb{N}^2 \times \{1,\ldots,N\}\}$ is a family of independent, $N(0,1)$-random variables and $\mathbf{X}_{k,n} = \sum_{i=1}^N X_{k,n,i}\mathbf{e}_i$, then

$$\sum_{m=0}^{n}\sum_{k=0}^{\infty}\sum_{i=1}^{N} X_{k,m,i}\mathbf{h}_{k,m,i}(t) = \sum_{m=0}^{n}\sum_{k=0}^{\infty} h_{k,m}(t)\mathbf{X}_{k,m}$$

is precisely the polygonalization that I denoted by $\mathbf{B}_n(t)$ in Lévy's construction (cf. § 4.3.2).

The construction by Wiener, alluded to above, was essentially the same, only he chose a different basis for $\mathbf{H}^1(\mathbb{R}^N)$. Wiener took $\dot{h}_{k,0}(t) = \mathbf{1}_{[k,k+1)}(t)$ for $k \in \mathbb{N}$ and $\dot{h}_{k,n}(t) = 2^{\frac{1}{2}}\mathbf{1}_{[k,k+1)}(t)\cos(\pi n(t-k))$ for $(k,n) \in \mathbb{N} \times \mathbb{Z}^+$, which means that he was looking at the series

$$\sum_{k=0}^{\infty}((t-k)^+ \wedge 1)\mathbf{1}_{[k,k+1)}(t)\mathbf{X}_{k,0} + \sum_{(k,n)\in\mathbb{N}\times\mathbb{Z}^+} \mathbf{1}_{[k,k+1)}(t)\frac{2^{\frac{1}{2}}\sin(\pi n(t-k))}{\pi n}\mathbf{X}_{k,n},$$

---

[1] The observation that Lévy's construction (cf. § 4.3.2) can be interpreted in terms of a Wiener series is due to Z. Ciesielski. To be more precise, initially Ciesielski himself was thinking entirely in terms of orthogonal series and did not realize that he was giving a re-interpretation of Lévy's construction. Only later did the connection become clear.

where again $\{\mathbf{X}_{k,n} : (k,n) \in \mathbb{N}^2\}$ is a family of independent, $\mathbb{R}^N$-valued, $N(\mathbf{0},\mathbf{I})$-random variables. The reason why Lévy's choice is easier to handle than Wiener's is that, in Lévy's case, for each $n \in \mathbb{Z}^+$ and $t \in [0,\infty)$, $h_{k,n}(t) \neq 0$ for precisely one $k \in \mathbb{N}$. Wiener's choice has no such property.

With these preliminaries, the following theorem should come as no surprise.

THEOREM 8.3.3. *Let $H$ be an infinite dimensional, separable, real Hilbert space and $E$ a Banach space into which $H$ is continuously embedded as a dense subspace. If for some orthonormal basis $\{h_m : m \geq 0\}$ in $H$ the series*

$$\sum_{m=0}^{\infty} \xi_m h_m \text{ converges in } E \tag{8.3.4}$$

*for $\gamma_{0,1}^{\mathbb{N}}$-almost every $\boldsymbol{\xi} = (\xi_0, \dots, \xi_m, \dots) \in \mathbb{R}^{\mathbb{N}}$*

*and if $S : \mathbb{R}^{\mathbb{N}} \longrightarrow E$ is given by*

$$S(\boldsymbol{\xi}) = \begin{cases} \sum_{m=0}^{\infty} \xi_m h_m & \text{when the series converges in } E \\ 0 & \text{otherwise,} \end{cases}$$

*then $(H, E, \mathcal{W})$ with $\mathcal{W} = S_* \gamma_{0,1}^{\mathbb{N}}$ is an abstract Wiener space. Conversely, if $(H, E, \mathcal{W})$ is an abstract Wiener space and $\{h_m : m \geq 0\}$ is an orthogonal sequence in $H$ such that, for each $m \in \mathbb{N}$, either $h_m = 0$ or $\|h_m\|_H = 1$, then*

$$\mathbb{E}^{\mathcal{W}}\left[\sup_{n \geq 0} \left\| \sum_{m=0}^{n} \mathcal{I}(h_m) h_m \right\|_E^p \right] < \infty \quad \text{for all } p \in [1,\infty), \tag{8.3.5}$$

*and, for $\mathcal{W}$-almost every $x \in E$, $\sum_{m=0}^{\infty} [\mathcal{I}(h_m)](x) h_m$ converges in $E$ to the $\mathcal{W}$-conditional expectation value of $x$ given $\sigma(\{\mathcal{I}(h_m) : m \geq 0\})$. Moreover,*

$$\sum_{m=0}^{\infty} [\mathcal{I}(h_m)](x) h_m \text{ is } \mathcal{W}\text{-independent of } x - \sum_{m=0}^{\infty} [\mathcal{I}(h_m)](x) h_m.$$

*Finally, if $\{h_m : m \geq 0\}$ is an orthonormal basis in $H$, then, for $\mathcal{W}$-almost every $x \in E$, $\sum_{m=0}^{\infty} [\mathcal{I}(h_m)](x) h_m$ converges in $E$ to $x$, and the convergence is also in $L^p(\mathcal{W}; E)$ for every $p \in [1,\infty)$.*

PROOF: First assume that (8.3.4) holds for some orthonormal basis, and set $S_n(\boldsymbol{\xi}) = \sum_{m=0}^{n} \xi_m h_m$ and $\mathcal{W} = S_* \gamma_{0,1}^{\mathbb{N}}$. Then, because $S_n(\boldsymbol{\xi}) \longrightarrow S(\boldsymbol{\xi})$ in $E$ for $\gamma_{0,1}^{\mathbb{N}}$-almost every $\boldsymbol{\xi} \in \mathbb{R}^{\mathbb{N}}$,

$$\widehat{\mathcal{W}}(x^*) = \lim_{n\to\infty} \mathbb{E}^{\gamma_{0,1}^{\mathbb{N}}}\left[e^{\sqrt{-1}\langle S_n, \lambda\rangle}\right] = \lim_{n\to\infty} \prod_{m=0}^{n} e^{-\frac{1}{2}(h_{x^*}, h_m)_H^2} = e^{-\frac{1}{2}\|h_{x^*}\|_H^2},$$

which proves that $(H, E, \mathcal{W})$ is an abstract Wiener space.

Next suppose that $(H, E, \mathcal{W})$ is an abstract Wiener space and that $\{h_m : m \geq 0\}$ is an orthogonal sequence with $\|h_m\|_H \in \{0,1\}$ for each $m \geq 0$. By Theorem 8.2.1, $x \in L^p(\mathcal{W}; E)$ for every $p \in [1,\infty)$. Next, for each $n \in \mathbb{N}$, set $\mathcal{F}_n = \sigma(\{\mathcal{I}(h_m) : 0 \leq m \leq n\})$. Clearly, $\mathcal{F}_n \subseteq \mathcal{F}_{n+1}$ and $\mathcal{F} \equiv \bigvee_{n=0}^{\infty} \mathcal{F}_n$ is the $\sigma$-algebra generated by $\{\mathcal{I}(h_m) : m \geq 0\}$. Moreover, if $S_n = \sum_{m=0}^{n} \mathcal{I}(h_m)h_m$, then, since $\{\mathcal{I}(h_m) : m \geq 0\}$ is a Gaussian family and $\langle x - S_n(x), x^* \rangle$ is perpendicular in $L^2(\mathcal{W}; \mathbb{R})$ to $\mathcal{I}(h_m)$ for all $x^* \in E^*$ and $0 \leq m \leq n$, $x - S_n(x)$ is $\mathcal{W}$-independent of $\mathcal{F}_n$. Thus $S_n = \mathbb{E}^{\mathcal{W}}[x \,|\, \mathcal{F}_n]$, and so, by Theorem 6.1.12, we know both that (8.3.5) holds and that $S_n \longrightarrow \mathbb{E}^{\mathcal{W}}[x \,|\, \mathcal{F}]$ $\mathcal{W}$-almost surely. In addition, the $\mathcal{W}$-independence of $S_n(x)$ from $x - S_n(x)$ implies that the limit quantities possess the same independence property.

In order to complete the proof at this point, all that I have to do is show that $x = \mathbb{E}^{\mathcal{W}}[x \,|\, \mathcal{F}]$ $\mathcal{W}$-almost surely when $\{h_m : m \geq 0\}$ is an orthonormal basis. Equivalently, I must check that $\mathcal{B}_E$ is contained in the $\mathcal{W}$-completion $\overline{\mathcal{F}}^{\mathcal{W}}$ of $\mathcal{F}$. To this end, note that, for each $h \in H$, because $\sum_{m=0}^{n}(h, h_m)_H h_m$ converges in $H$ to $h$,

$$\sum_{m=0}^{n} (h, h_m)_H \mathcal{I}(h_m) = \mathcal{I}\left(\sum_{m=0}^{n} (h, h_m)_H h_m\right) \longrightarrow \mathcal{I}(h) \quad \text{in } L^2(\mathcal{W}; \mathbb{R}).$$

Hence, $\mathcal{I}(h)$ is $\overline{\mathcal{F}}^{\mathcal{W}}$-measurable for every $h \in H$. In particular, this means that $x \rightsquigarrow \langle x, x^* \rangle$ is $\overline{\mathcal{F}}^{\mathcal{W}}$-measurable for every $x^* \in E^*$, and so, since $\mathcal{B}_E$ is generated by $\{\langle \cdot\,, x^* \rangle : x^* \in E^*\}$, $\mathcal{B}_E \subseteq \overline{\mathcal{F}}^{\mathcal{W}}$. □

It is important to acknowledge that the preceding theorem does not give another proof of Wiener's theorem that Brownian motion exists. Instead, it simply says that, knowing it exists, there are lots of ways in which to construct it. See Exercise 8.3.21 for a more satisfactory proof of the same conclusion in the classical case, one that does not require the a priori existence of $\mathcal{W}^{(N)}$.

The following result shows that, in some sense, a non-degenerate, centered, Gaussian measure $\mathcal{W}$ on a Banach space does not fit on a smaller space.

Corollary 8.3.6. *If $\mathcal{W}$ is a non-degenerate, centered Gaussian measure on a separable Banach space $E$, then $E$ is the support of $\mathcal{W}$ in the sense that $\mathcal{W}$ assigns positive probability to every non-empty open subset of $E$.*

Proof: Let $H$ be the Cameron–Martin space for $\mathcal{W}$. Since $H$ is dense in $E$, it suffices to show that $\mathcal{W}\big(B_E(g,r)\big) > 0$ for every $g \in H$ and $r > 0$. Moreover, since, by the Cameron–Martin formula (8.2.8) (cf. Exercise 8.2.19)

$$\begin{aligned}\mathcal{W}\big(B_E(0,r)\big) &= (\tau_{-g})_*\mathcal{W}\big(B_E(g,r)\big) = \mathbb{E}^{\mathcal{W}}\big[R_{-g},\, B_E(g,r)\big] \\ &\leq e^{\frac{\|g\|_H^2}{2}} \sqrt{\mathcal{W}\big(B_E(g,r)\big)},\end{aligned}$$

I need only show that $\mathcal{W}(B_E(0,r)) > 0$ for all $r > 0$. To this end, choose an orthonormal basis $\{h_m : m \geq 0\}$ in $H$, and set $S_n = \sum_{m=0}^{n} \mathcal{I}(h_m)h_m$. Then, by Theorem 8.3.3, $x \rightsquigarrow S_n(x)$ is $\mathcal{W}$-independent of $x \rightsquigarrow x - S_n(x)$ and $S_n(x) \longrightarrow x$ in $E$ for $\mathcal{W}$-almost every $x \in E$. Hence, $\mathcal{W}(\{\|x - S_n(x)\|_E < \frac{r}{2}\}) \geq \frac{1}{2}$ for some $n \in \mathbb{N}$, and therefore

$$\mathcal{W}(B_E(0,r)) \geq \tfrac{1}{2}\mathcal{W}(\|S_n\|_E < \tfrac{r}{2}).$$

But $\|S_n\|_E^2 \leq C\|S_n\|_H^2 = \sum_{m=0}^{n} \mathcal{I}(h_m)^2$ for some $C < \infty$, and so

$$\mathcal{W}(\|S_n\|_E < \tfrac{r}{2}) \geq \gamma_{0,1}^{n+1}\Big(B_{\mathbb{R}^{n+1}}\big(\mathbf{0}, \tfrac{r}{2C}\big)\Big) > 0 \quad \text{for any } r > 0. \quad \square$$

**§ 8.3.3. Orthogonal Projections.** Associated with any closed, linear subspace $L$ of a Hilbert space $H$, there is an orthogonal projection map $\Pi_L : H \longrightarrow L$ determined by the property that, for each $h \in H$, $h - \Pi_L h \perp L$. Equivalently, $\Pi_L h$ is the element of $L$ that is closest to $h$. In this subsection I will show that if $(H, E, \mathcal{W})$ is an abstract Wiener space and $L$ is a finite dimensional subspace of $H$, then $\Pi_L$ admits a $\mathcal{W}$-almost surely unique extension $P_L$ to $E$. In addition, I will show that $P_L x \longrightarrow x$ in $L^2(\mathcal{W}; E)$ as $L \nearrow H$.

LEMMA 8.3.7. *Let $(H, E, \mathcal{W})$ be an abstract Wiener space and $\{h_m : m \geq 0\}$ an orthonormal basis in $H$. Then, for each $h \in H$, $\sum_{m=0}^{\infty}(h, h_m)_H \mathcal{I}(h_m)$ converges to $\mathcal{I}(h)$ $\mathcal{W}$-almost surely and in $L^p(\mathcal{W}; \mathbb{R})$ for every $p \in [1, \infty)$.*

PROOF: Define the $\sigma$-algebras $\mathcal{F}_n$ and $\mathcal{F}$ as in the proof of Theorem 8.3.3. Then, by the same argument as I used there, one can identify $\sum_{m=0}^{n}(h, h_m)_H \mathcal{I}(h_m)$ as $\mathbb{E}^{\mathcal{W}}[\mathcal{I}(h) \,|\, \mathcal{F}_n]$. Thus, since $\overline{\mathcal{F}}^{\mathcal{W}} \supseteq \mathcal{B}_E$, the required convergence statement is an immediate consequence of Corollary 5.2.4. $\square$

THEOREM 8.3.8. *Let $(H, E, \mathcal{W})$ be an abstract Wiener space. For each finite dimensional subspace $L$ of $H$ there is a $\mathcal{W}$-almost surely unique map $P_L : E \longrightarrow H$ such that, for every $h \in H$ and $\mathcal{W}$-almost every $x \in E$, $(h, P_L x)_H = \mathcal{I}(\Pi_L h)(x)$, where $\Pi_L$ denotes orthogonal projection from $H$ onto $L$. In fact, if $\{g_1, \dots, g_{\dim(L)}\}$ is an orthonormal basis for $L$, then $P_L x = \sum_1^{\dim(L)}[\mathcal{I}(g_i)](x)g_i$, and so $P_L x \in L$ for $\mathcal{W}$-almost every $x \in E$. In particular, the distribution of $x \in E \longmapsto P_L x \in L$ under $\mathcal{W}$ is the same as that of $(\xi_1, \dots, \xi_{\dim(L)}) \in \mathbb{R}^{\dim(L)} \longmapsto \sum_1^{\dim(L)} \xi_i g_i \in L$ under $\gamma_{0,1}^{\dim(L)}$. Finally, $x \rightsquigarrow P_L x$ is $\mathcal{W}$-independent of $x \rightsquigarrow x - P_L x$.*

PROOF: Set $\ell = \dim(L)$. It suffices to note that

$$\mathcal{I}(\Pi_L h) = \mathcal{I}\left(\sum_{k=1}^{\ell}(h, g_k)_H g_k\right) = \sum_{k=1}^{\ell}(h, g_k)_H \mathcal{I}(g_k) = \left(\sum_{k=1}^{\ell}\mathcal{I}(g_k)g_k, h\right)_H$$

for all $h \in H$ $\square$

We now have the preparations needed to prove a result which shows that my definition of an abstract Wiener space is the same as Gross's. Specifically, Gross's own definition was based on the property proved in the following.

THEOREM 8.3.9. *Let $(H, E, \mathcal{W})$ be an abstract Wiener space and $\{h_n : n \geq 0\}$ an orthonormal basis for $H$, and set $L_n = \operatorname{span}(\{h_0, \dots, h_n\})$. Then, for all $\epsilon > 0$ there exists an $n \in \mathbb{N}$ such that $\mathbb{E}^{\mathcal{W}}\big[\|P_L x\|_E^2\big] \leq \epsilon^2$ whenever $L$ is a finite dimensional subspace that is perpendicular to $L_n$.*

PROOF: Without loss in generality, I will assume that $\|\cdot\|_E \leq \|\cdot\|_H$.

Arguing by contradiction, I will show that if the asserted property does not hold, then there would exist an orthonormal basis $\{f_n : n \geq 0\}$ for $H$ such that $\sum_0^\infty \mathcal{I}(f_n) f_n$ fails to converge in $L^2(\mathcal{W}; E)$. Thus, suppose that there exists an $\epsilon > 0$ such that for all $n \in \mathbb{N}$ there exists a finite dimensional $L \perp L_n$ with $\mathbb{E}^{\mathcal{W}}\big[\|P_L x\|_E^2\big] \geq \epsilon^2$. Under this assumption, define $\{n_m : m \geq 0\} \subseteq \mathbb{N}$, $\{\ell_m : m \geq 0\} \subseteq \mathbb{N}$, and $\{\{f_0, \dots, f_{n_m}\} : m \geq 0\} \subseteq L_{n_m}$ inductively by the following prescription. First, take $n_0 = 0 = \ell_0$ and $f_0 = h_0$. Next, knowing $n_m$ and $\{f_0, \dots, f_{n_m}\}$, choose a finite dimensional subspace $L \perp L_{n_m}$ so that $\mathbb{E}^{\mathcal{W}}\big[\|P_L x\|_E^2\big] \geq \epsilon^2$, set $\ell_m = \dim(L)$, and let $\{g_{m,1}, \dots, g_{m,\ell_m}\}$ be an orthonormal basis for $L$. For any $\delta > 0$ there exists an $n \geq n_m + \ell_m$ such that

$$\sum_{i,j=1}^{\ell_m} \Big| \big(\Pi_{L_n} g_{m,i}, \Pi_{L_n} g_{m,j}\big)_H - \delta_{i,j} \Big| \leq \delta.$$

In particular, if $\delta \in (0,1)$, then the elements of $\{\Pi_{L_n} g_{m,i} : 1 \leq i \leq \ell_m\}$ are linearly independent and the orthonormal set $\{\tilde{g}_{m,i} : 1 \leq i \leq \ell_m\}$ obtained from them via the Gram–Schmidt orthogonalization procedure satisfies (cf. Exercise 8.3.16)

$$\sum_{i=1}^{\ell_m} \|\tilde{g}_{m,i} - \Pi_{L_n} g_{m,i}\|_H \leq K_{\ell_m} \sum_{i,j=1}^{\ell_m} \Big| \big(\Pi_{L_n} g_{m,i}, \Pi_{L_n} g_{m,j}\big) - \delta_{i,j} \Big|$$

for some $K_m < \infty$ which depends only on $\ell_m$. Moreover, and because $L \perp L_{n_m}$, $\tilde{g}_{m,i} \perp L_{n_m}$ for all $1 \leq i \leq \ell_m$. Hence, we can find an $n_{m+1} \geq n_m + \ell_m$ so that $\operatorname{span}(\{h_n : n_m < n \leq n_{m+1}\})$ admits an orthonormal basis $\{f_{n_m+1}, \dots, f_{n_{m+1}}\}$ with the property that $\sum_1^{\ell_m} \|g_{m,i} - f_{n_m+i}\|_H \leq \frac{\epsilon}{4}$.

Clearly $\{f_n : n \geq 0\}$ is an orthonormal basis for $H$. On the other hand,

$$\begin{aligned} \mathbb{E}^{\mathcal{W}}\left[\left\| \sum_{n=n_m+1}^{n_m+\ell_m} \mathcal{I}(f_n) f_n \right\|_E^2\right]^{\frac{1}{2}} &\geq \epsilon - \mathbb{E}^{\mathcal{W}}\left[\left\| \sum_1^{\ell_m} \big(\mathcal{I}(g_{m,i}) g_{m,i} - \mathcal{I}(f_{n_m+i}) f_{n_m+i}\big) \right\|_E^2\right]^{\frac{1}{2}} \\ &\geq \epsilon - \sum_1^{\ell_m} \mathbb{E}^{\mathcal{W}}\Big[\big\|\mathcal{I}(g_{m,i}) g_{m,i} - \mathcal{I}(f_{n_m+i}) f_{n_m+i}\big\|_H^2\Big]^{\frac{1}{2}}, \end{aligned}$$

and so, since $\mathbb{E}^{\mathcal{W}}\Big[\big\|\mathcal{I}(g_{i,m}) g_{m,i} - \mathcal{I}(f_{n_m+i}) f_{n_m+i}\big\|_H^2\Big]^{\frac{1}{2}}$ is dominated by

$$\begin{aligned} &\mathbb{E}^{\mathcal{W}}\Big[\big\|\big(\mathcal{I}(g_{m,i}) - \mathcal{I}(f_{n_m+i})\big) g_{m,i}\big\|_H^2\Big]^{\frac{1}{2}} + \mathbb{E}^{\mathcal{W}}\big[\mathcal{I}(f_{n_m+i})^2\big]^{\frac{1}{2}} \|g_{m,i} - f_{n_m+i}\|_H \\ &\quad \leq 2\|g_{m,i} - f_{n_m+i}\|_H, \end{aligned}$$

we have that

$$\mathbb{E}^{\mathcal{W}}\left[\left\|\sum_{n_m+1}^{n_m+\ell_m} \mathcal{I}(f_n)f_n\right\|_E^2\right]^{\frac{1}{2}} \geq \frac{\epsilon}{2} \quad \text{for all } m \geq 0,$$

and this means that $\sum_0^\infty \mathcal{I}(f_n)f_n$ cannot be converging in $L^2(\mathcal{W};E)$. □

Besides showing that my definition of an abstract Wiener space is the same as Gross's, Theorem 8.3.9 allows us to prove a very convincing statement, again due to Gross, of just how non-unique is the Banach space for which a given Hilbert space is the Cameron–Martin space.

Corollary 8.3.10. *If $(H, E, \mathcal{W})$ is an abstract Wiener space, then there exists a separable Banach space $E_0$ that is continuously embedded in $E$ as a measurable subset and has the properties that $\mathcal{W}(E_0) = 1$, bounded subsets of $E_0$ are relatively compact in $E$, and $(H, E_0, \mathcal{W} \restriction E_0)$ is again an abstract Wiener space.*

Proof: Again I will assume that $\|\cdot\|_E \leq \|\cdot\|_H$.

Choose $\{x_n^* : n \geq 0\} \subseteq E^*$ so that $\{h_n : n \geq 0\}$ is an orthonormal basis in $H$ when $h_n = h_{x_n^*}$, and set $L_n = \operatorname{span}(\{h_0, \dots, h_n\})$. Next, using Theorem 8.3.9, choose an increasing sequence $\{n_m : m \geq 0\}$ so that $n_0 = 0$ and $\mathbb{E}^{\mathcal{W}}\left[\|P_L x\|_E^2\right]^{\frac{1}{2}} \leq 2^{-m}$ for $m \geq 1$ and finite dimensional $L \perp L_{n_m}$, and define $Q_\ell$ for $\ell \geq 0$ on $E$ into $H$ so that

$$Q_0 x = \langle x, x_0^* \rangle h_0 \quad \text{and} \quad Q_\ell x = \sum_{n=n_{\ell-1}+1}^{n_\ell} \langle x, x_n^* \rangle h_n \quad \text{when } \ell \geq 1.$$

Finally, set $S_m = P_{L_{n_m}} = \sum_{\ell=0}^m Q_\ell$, and define $E_0$ to be the set of $x \in E$ such that

$$\|x\|_{E_0} \equiv \|Q_0 x\|_E + \sum_{\ell=1}^\infty \ell^2 \|Q_\ell x\|_E < \infty \quad \text{and} \quad \|S_m x - x\|_E \longrightarrow 0.$$

To show that $\|\cdot\|_{E_0}$ is a norm on $E_0$ and that $E_0$ with norm $\|\cdot\|_{E_0}$ is a Banach space, first note that if $x \in E_0$, then

$$\|x\|_E = \lim_{m\to\infty} \|S_m x\|_E \leq \|Q_0 x\|_E + \lim_{m\to\infty} \sum_{\ell=1}^m \|Q_\ell x\|_E \leq \|x\|_{E_0},$$

and therefore $\|\cdot\|_{E_0}$ is certainly a norm on $E_0$. Next, suppose that the sequence $\{x_k : k \geq 1\} \subseteq E_0$ is a Cauchy sequence with respect to $\|\cdot\|_{E_0}$. By the preceding, we know that $\{x_k : k \geq 1\}$ is also Cauchy convergent with respect to

$\|\cdot\|_E$, and so there exists an $x \in E$ such that $x_k \longrightarrow x$ in $E$. We need to show that $x \in E_0$ and that $\|x_k - x\|_{E_0} \longrightarrow 0$. Because $\{x_k : k \geq 1\}$ is bounded in $E_0$, it is clear that $\|x\|_{E_0} < \infty$. In addition, for any $m \geq 0$ and $k \geq 1$,

$$\begin{aligned}\|x - S_m x\|_E &= \lim_{\ell\to\infty} \|x_\ell - S_m x_\ell\|_E \leq \varliminf_{\ell\to\infty} \|x_\ell - S_m x_\ell\|_{E_0} \\ &= \varliminf_{\ell\to\infty} \sum_{n>m} n^2 \|Q_n x_\ell\|_E \leq \sum_{n>m} n^2 \|Q_n x_k\|_E + \sup_{\ell>k} \|x_\ell - x_k\|_{E_0}.\end{aligned}$$

Thus, by choosing $k$ for a given $\epsilon > 0$ so that $\sup_{\ell>k} \|x_\ell - x_k\|_{E_0} < \epsilon$, we conclude that $\varlimsup_{m\to\infty} \|x - S_m x\|_E < \epsilon$ and therefore that $S_m x \longrightarrow x$ in $E$. Hence, $x \in E_0$. Finally, to see that $x_k \longrightarrow x$ in $E_0$, simply note that

$$\begin{aligned}\|x - x_k\|_{E_0} &= \|Q_0(x - x_k)\|_E + \sum_{m=1}^{\infty} m^2 \|Q_m(x - x_k)\|_E \\ &\leq \varliminf_{\ell\to\infty} \left( \|Q_0(x_\ell - x_k)\|_E + \sum_{m=1}^{\infty} m^2 \|Q_m(x_\ell - x_k)\|_E \right) \leq \sup_{\ell>k} \|x_\ell - x_k\|_{E_0},\end{aligned}$$

which tends to 0 as $k \to \infty$.

To show that bounded subsets of $E_0$ are relatively compact in $E$, it suffices to show that if $\{x_\ell : \ell \geq 1\} \subseteq \overline{B_{E_0}(0, R)}$, then there is an $x \in E$ to which a subsequence converges in $E$. For this purpose, observe that, for each $m \geq 0$, there is a subsequence $\{x_{\ell_k} : k \geq 1\}$ along which $S_m x_{\ell_k}$ converges in $L_{n_m}$. Hence, by a diagonalization argument, $\{x_{\ell_k} : k \geq 1\}$ can be chosen so that $\{S_m x_{\ell_k} : k \geq 1\}$ converges in $L_{n_m}$ for all $m \geq 0$. Since, for $1 \leq j < k$,

$$\begin{aligned}\|x_{\ell_k} - x_{\ell_j}\|_E &\leq \|S_m x_{\ell_k} - S_m x_{\ell_j}\|_E + \sum_{n>m} \|Q_n(x_{\ell_k} - x_{\ell_j})\|_E \\ &\leq \|S_m x_{\ell_k} - S_m x_{\ell_j}\|_E + 2R \sum_{n>m} \frac{1}{n^2},\end{aligned}$$

it follows that $\{x_{\ell_k} : k \geq 1\}$ is Cauchy convergent in $E$ and therefore that it converges in $E$.

I must still show that $E_0 \in \mathcal{B}_E$ and that $(H, E_0, \mathcal{W}_0)$ is an abstract Wiener space when $\mathcal{W}_0 = \mathcal{W} \upharpoonright E_0$. To see the first of these, observe that $x \in E \longmapsto \|x\|_{E_0} \in [0, \infty]$ is lower semicontinuous and that $\{x : \|S_m x - x\|_E \longrightarrow 0\} \in \mathcal{B}_E$. In addition, because, by Theorem 8.3.3, $\|S_m x - x\|_E \longrightarrow 0$ for $\mathcal{W}$-almost every $x \in E$, we will know that $\mathcal{W}(E_0) = 1$ once I show that $\mathcal{W}(\|x\|_{E_0} < \infty) = 1$, which follows immediately from

$$\begin{aligned}\mathbb{E}^{\mathcal{W}}\big[\|x\|_{E_0}\big] &= \mathbb{E}^{\mathcal{W}}\big[\|Q_0 x\|_E\big] + \sum_1^{\infty} m^2 \mathbb{E}^{\mathcal{W}}\big[\|Q_m x\|_E\big] \\ &\leq \mathbb{E}^{\mathcal{W}}\big[\|Q_0 x\|_E\big] + \sum_1^{\infty} m^2 \mathbb{E}^{\mathcal{W}}\big[\|Q_m x\|_E^2\big]^{\frac{1}{2}} < \infty.\end{aligned}$$

The next step is to check that $H$ is continuously embedded in $E_0$. Certainly $h \in H \implies \|S_m h - h\|_E \le \|S_m h - h\|_H \longrightarrow 0$. Next suppose that $h \in H \setminus \{0\}$ and that $h \perp L_{n_m}$, and let $L$ be the line spanned by $h$. Then $P_L x = \|h\|_H^{-2}[\mathcal{I}(h)](x)h$, and so, because $L \perp L_{n_m}$,

$$\frac{1}{2^m} \ge \mathbb{E}^{\mathcal{W}}\big[\mathcal{I}(h)^2\big]^{\frac{1}{2}} \frac{\|h\|_E}{\|h\|_H^2} = \frac{\|h\|_E}{\|h\|_H}.$$

Hence, we now know that $h \perp L_{n_m} \implies \|h\|_E \le 2^{-m}\|h\|_H$. In particular, $\|Q_{m+1}h\|_E \le 2^{-m}\|Q_{m+1}h\|_H \le 2^{-m}\|h\|_H$ for all $m \ge 0$ and $h \in H$, and so

$$\|h\|_{E_0} = \|Q_0 h\|_E + \sum_{m=1}^{\infty} m^2 \|Q_m h\|_E \le \left(1 + 2\sum_{m=1}^{\infty} \frac{m^2}{2^m}\right)\|h\|_H = 25\|h\|_H.$$

To complete the proof, I must show that $H$ is dense in $E_0$ and that, for each $y^* \in E_0^*$, $\widehat{\mathcal{W}_0}(y^*) = e^{-\frac{1}{2}\|h_{y^*}\|_H^2}$, where $\mathcal{W}_0 = \mathcal{W} \restriction E_0$ and $h_{y^*} \in H$ is determined by $\left(h, h_{y^*}\right)_H = \langle h, y^* \rangle$ for $h \in H$. Both these facts rely on the observation that

$$\|x - S_m x\|_{E_0} = \sum_{n>m} n^2 \|Q_n x\|_E \longrightarrow 0 \quad \text{for all } x \in E_0.$$

Knowing this, the density of $H$ in $E_0$ is obvious. Finally, if $y^* \in E_0^*$, then, by the preceding and Lemma 8.3.7,

$$\begin{aligned}\langle x, y^* \rangle &= \lim_{m\to\infty} \langle S_m x, y^* \rangle = \lim_{m\to\infty} \sum_{n=0}^{n_m} \langle x, x_n^* \rangle \langle h_n, y^* \rangle \\ &= \lim_{m\to\infty} \sum_{n=0}^{n_m} \left(h_{y^*}, h_n\right)_H \big[\mathcal{I}(h_n)\big](x) = \big[\mathcal{I}(h_{y^*})\big](x)\end{aligned}$$

for $\mathcal{W}_0$-almost every $x \in E_0$. Hence $\langle \cdot, y^* \rangle$ under $\mathcal{W}_0$ is a centered Gaussian with variance $\|h_{y^*}\|_H^2$. □

§ **8.3.4. Pinned Brownian Motion.** Theorem 8.3.8 has a particularly interesting application to the classical abstract Wiener space $\big(\mathbf{H}^1(\mathbb{R}^N), \Theta(\mathbb{R}^N), \mathcal{W}^{(N)}\big)$. Namely, suppose that $0 = t_0 < t_1 < \cdots < t_n$, and let $L$ be the span of $\big\{h_{t_m}\mathbf{e} : 1 \le m \le n \text{ and } \mathbf{e} \in \mathbb{S}^{N-1}\big\}$, where $h_t(\tau) \equiv t \wedge \tau$. In this case,

$$P_L \boldsymbol{\theta} = \sum_{m=1}^{n} \frac{h_{t_m} - h_{t_{m-1}}}{t_m - t_{m-1}} \big(\boldsymbol{\theta}(t_m) - \boldsymbol{\theta}(t_{m-1})\big),$$

and so

$$\begin{aligned}\boldsymbol{\theta}_{(t_1,\dots,t_n)}(t) &\equiv [\boldsymbol{\theta} - P_L\boldsymbol{\theta}](t) \\ &= \begin{cases} \boldsymbol{\theta}(t) - \boldsymbol{\theta}(t_{m-1}) - \frac{t - t_{m-1}}{t_m - t_{m-1}}\big(\boldsymbol{\theta}(t_m) - \boldsymbol{\theta}(t_{m-1})\big) & \text{if } t \in [t_{m-1}, t_m] \\ \boldsymbol{\theta}(t) - \boldsymbol{\theta}(t_n) & \text{if } t \in [t_n, \infty).\end{cases}\end{aligned} \tag{8.3.11}$$

Thus, if $(\boldsymbol{\theta},\vec{\mathbf{y}}) \in \Theta(\mathbb{R}^N)\times(\mathbb{R}^N)^n \longmapsto \boldsymbol{\theta}_{(t_1,\dots,t_n),\vec{\mathbf{y}}} \in \Theta(\mathbb{R}^N)$ is given by

$$\boldsymbol{\theta}_{(t_1,\dots,t_n),\vec{\mathbf{y}}} = \boldsymbol{\theta}_{(t_1,\dots,t_n)} + \sum_{m=1}^{n} \frac{h_{t_m}-h_{t_{m-1}}}{t_m-t_{m-1}}(\mathbf{y}_m-\mathbf{y}_{m-1}),$$

where $\vec{\mathbf{y}}=(\mathbf{y}_1,\dots,\mathbf{y}_n)$ and $\mathbf{y}_0\equiv\mathbf{0}$, then, for any Borel measurable $F:\Theta(\mathbb{R}^N)\times(\mathbb{R}^N)^n\longrightarrow[0,\infty)$,

$$\begin{aligned}&\int_{\Theta(\mathbb{R}^N)} F\Big(\boldsymbol{\theta},(\boldsymbol{\theta}(t_1),\dots,\boldsymbol{\theta}(t_n))\Big)\,\mathcal{W}^{(N)}(d\boldsymbol{\theta})\\ (8.3.12)\qquad &= \int_{(\mathbb{R}^N)^n}\bigg(\int_{\Theta(\mathbb{R}^N)} F\big(\boldsymbol{\theta}_{(t_1,\dots,t_n),\vec{\mathbf{y}}},\vec{\mathbf{y}}\big)\,\mathcal{W}^{(N)}(d\boldsymbol{\theta})\bigg)\,\gamma_{\mathbf{0},\mathbf{C}(t_1,\dots,t_n)}(d\vec{\mathbf{y}}),\end{aligned}$$

where $\mathbf{C}(t_1,\dots,t_n)_{(m,i),(m'i')} = t_m\wedge t_{m'}\delta_{i,i'}$ for $1\le m,m'\le n$ and $1\le i,i'\le N$ is the covariance of $\boldsymbol{\theta}\rightsquigarrow(\boldsymbol{\theta}(t_1),\dots,\boldsymbol{\theta}(t_n))$ under $\mathcal{W}^{(N)}$. Equivalently, if

$$\check{\boldsymbol{\theta}}_{(t_1,\dots,t_n),\vec{\mathbf{y}}} = \boldsymbol{\theta}_{(t_1,\dots,t_n)} + \sum_{m=1}^{n} \frac{h_{t_m}-h_{t_{m-1}}}{t_m-t_{m-1}}\mathbf{y}_m,$$

then

$$\begin{aligned}&\int_{\Theta(\mathbb{R}^N)} F\Big(\boldsymbol{\theta},(\boldsymbol{\theta}(t_1)-\boldsymbol{\theta}(t_0),\dots,\boldsymbol{\theta}(t_n)-\boldsymbol{\theta}(t_{n-1}))\Big)\,\mathcal{W}^{(N)}(d\boldsymbol{\theta})\\ (8.3.13)\qquad &= \int_{(\mathbb{R}^N)^n}\bigg(\int_{\Theta(\mathbb{R}^N)} F\big(\check{\boldsymbol{\theta}}_{(t_1,\dots,t_n),\vec{\mathbf{y}}},\vec{\mathbf{y}}\big)\,\mathcal{W}^{(N)}(d\boldsymbol{\theta})\bigg)\,\gamma_{\mathbf{0},\mathbf{D}(t_1,\dots,t_n)}(d\vec{\mathbf{y}}),\end{aligned}$$

where $\mathbf{D}(t_1,\dots,t_n)_{(m,i),(m',i')} = (t_m-t_{m-1})\delta_{m,m'}\delta_{i,i'}$ for $1\le m,m'\le n$ and $1\le i,i'\le N$ is the covariance matrix for $(\boldsymbol{\theta}(t_1)-\boldsymbol{\theta}(t_0),\dots,\boldsymbol{\theta}(t_n)-\boldsymbol{\theta}(t_{n-1}))$ under $\mathcal{W}^{(N)}$.

There are several comments that should be made about these conclusions. In the first place, it is clear from (8.3.11) that $t\rightsquigarrow\boldsymbol{\theta}_{(t_1,\dots,t_n)}(t)$ returns to the origin at each of the times $\{t_m: 1\le m\le n\}$. In addition, the excursions $\boldsymbol{\theta}_{(t_1,\dots,t_n)}\restriction[t_{m-1},t_m]$, $1\le m\le n$, are independent of each other and of $\boldsymbol{\theta}_{(t_1,\dots,t_n)}\restriction[t_n,\infty)$. Secondly, if $\mathcal{W}^{(N)}_{(t_1,\dots,t_n),\vec{\mathbf{y}}}$ denotes the $\mathcal{W}^{(N)}$-distribution of $\boldsymbol{\theta}\rightsquigarrow\boldsymbol{\theta}_{(t_1,\dots,t_n),\vec{\mathbf{y}}}$, then (8.3.12) says that

$$\boldsymbol{\theta}\rightsquigarrow\mathcal{W}^{(N)}_{(t_1,\dots,t_n),(\boldsymbol{\theta}(t_1),\dots,\boldsymbol{\theta}(t_n))}$$

is a regular conditional probability distribution (cf. § 9.2) of $\mathcal{W}^{(N)}$ given the $\sigma$-algebra generated by $\{\boldsymbol{\theta}(t_1),\dots,\boldsymbol{\theta}(t_n)\}$. Expressed in more colloquial terms, the process $\{\boldsymbol{\theta}_{(t_1,\dots,t_n),\vec{\mathbf{y}}}(t): t\ge 0\}$ is **Brownian motion pinned to the points** $\{\mathbf{y}_m: 1\le m\le n\}$ **at times** $\{t_m: 1\le m\le n\}$.

**§ 8.3.5. Orthogonal Invariance.** Consider the standard Gauss distribution $\gamma_{\mathbf{0},\mathbf{I}}$ on $\mathbb{R}^N$. Obviously, $\gamma_{\mathbf{0},\mathbf{I}}$ is orthogonal invariant. That is, if $\mathcal{O}$ is an orthogonal transformation on $\mathbb{R}^N$, then $\gamma_{\mathbf{0},\mathbf{I}}$ is invariant under the transformation $T_{\mathcal{O}} : \mathbb{R}^N \longrightarrow \mathbb{R}^N$ given by $T_{\mathcal{O}}\mathbf{x} = \mathcal{O}\mathbf{x}$. On the other hand, none of these transformations can be ergodic, since any radial function on $\mathbb{R}^N$ is invariant under $T_{\mathcal{O}}$ for every $\mathcal{O}$.

Now think about the analogous situation when $\mathbb{R}^N$ is replaced by an infinite dimensional Hilbert space $H$ and $(H, E, \mathcal{W})$ is an associated abstract Wiener space. As I am about to show, $\mathcal{W}$ still invariant with respect to orthogonal transformations on $H$. On the other hand, because $\|x\|_H = \infty$ for $\mathcal{W}$-almost every $x \in E$, there are no non-trivial radial functions now, a fact that leaves open the possibility that some orthogonal transformation of $H$ give rise to ergodic transformations for $\mathcal{W}$. The purpose of this subsection is to investigate these matters, and I begin with the following formulation of the orthogonal invariance of $\mathcal{W}$.

THEOREM 8.3.14. *Let $(H, E, \mathcal{W})$ be an abstract Wiener space and $\mathcal{O}$ an orthogonal transformation on $H$. Then there is a $\mathcal{W}$-almost surely unique, Borel measurable map $T_{\mathcal{O}} : E \longrightarrow E$ such that $\mathcal{I}(h) \circ T_{\mathcal{O}} = \mathcal{I}(\mathcal{O}^\top h)$ $\mathcal{W}$-almost surely for each $h \in H$. Moreover, $\mathcal{W} = (T_{\mathcal{O}})_*\mathcal{W}$.*

PROOF: To prove uniqueness, note that if $T$ and $T'$ both satisfy the defining property for $T_{\mathcal{O}}$, then, for each $x^* \in E^*$,

$$\langle Tx, x^*\rangle = \mathcal{I}(h_{x^*})(Tx) = \mathcal{I}(\mathcal{O}^\top h_{x^*}) = \mathcal{I}(h_{x^*})(T'x) = \langle T'x, x^*\rangle$$

for $\mathcal{W}$-almost every $x \in E$. Hence, since $E^*$ is separable in the weak* topology, $Tx = T'x$ for $\mathcal{W}$-almost every $x \in E$.

To prove existence, choose an orthonormal basis $\{h_m : m \geq 0\}$ for $H$, and let $C$ be the set of $x \in E$ for which both $\sum_{m=0}^\infty [\mathcal{I}(h_m)](x) h_m$ and $\sum_{m=0}^\infty [\mathcal{I}(h_m)](x)\mathcal{O}h_m$ converge in $E$. By Theorem 8.3.3, we know that $\mathcal{W}(C) = 1$ and that

$$x \rightsquigarrow T_{\mathcal{O}}x \equiv \begin{cases} \sum_{m=0}^\infty [\mathcal{I}(h_m)](x)\mathcal{O}h_m & \text{if } x \in C \\ 0 & \text{if } x \notin C \end{cases}$$

has distribution $\mathcal{W}$. Hence, all that remains is to check that $\mathcal{I}(h) \circ T_{\mathcal{O}} = \mathcal{I}(\mathcal{O}^\top h)$ $\mathcal{W}$-almost surely for each $h \in H$. To this end, let $x^* \in E^*$, and observe that

$$\begin{aligned} [\mathcal{I}(h_{x^*})](T_{\mathcal{O}}x) = \langle T_{\mathcal{O}}x, x^*\rangle &= \sum_{m=0}^\infty \left(h_{x^*}, \mathcal{O}h_m\right)_H [\mathcal{I}(h_m)](x) \\ &= \sum_{m=0}^\infty \left(\mathcal{O}^\top h_{x^*}, h_m\right)_H [\mathcal{I}(h_m)](x) \end{aligned}$$

for $\mathcal{W}$-almost every $x \in E$. Thus, since, by Lemma 8.3.7, the last of these series convergences $\mathcal{W}$-almost surely to $\mathcal{I}(\mathcal{O}^\top h_{x^*})$, we have that $\mathcal{I}(h_{x^*}) \circ T_{\mathcal{O}} =$

$\mathcal{I}(\mathcal{O}^\top h_{x^*})$ $\mathcal{W}$-almost surely. To handle general $h \in H$, simply note that both $h \in H \longmapsto \mathcal{I}(h) \circ T_{\mathcal{O}} \in L^2(\mathcal{W};\mathbb{R})$ and $h \in H \longmapsto \mathcal{I}(\mathcal{O}^\top h) \in L^2(\mathcal{W};\mathbb{R})$ are isometric, and remember that $\{h_{x^*} : x^* \in E^*\}$ is dense in $H$. □

I next want to discuss the possibility of $T_{\mathcal{O}}$ being ergodic for some orthogonal transformations $\mathcal{O}$. First notice that $T_{\mathcal{O}}$ cannot be ergodic if $\mathcal{O}$ has a non-trivial, finite dimensional invariant subspace $L$, since if $\{h_1, \dots, h_n\}$ were an orthonormal basis for $L$, then $\sum_{m=1}^n \mathcal{I}(h_m)^2$ would be a non-constant, $T_{\mathcal{O}}$-invariant function. Thus, the only candidates for ergodicity are $\mathcal{O}$'s that have no non-trivial, finite dimensional, invariant subspaces. In a more general and highly abstract context, I. Segal[2] showed that the existence of a non-trivial, finite dimensional subspace for $\mathcal{O}$ is the only obstruction to $T_{\mathcal{O}}$ being ergodic. Here I will show less.

THEOREM 8.3.15. *Let $(H, E, \mathcal{W})$ be an abstract Wiener space. If $\mathcal{O}$ is an orthogonal transformation on $H$ with the property that, for every $g, h \in H$, $\underline{\lim}_{n\to\infty} \big(\mathcal{O}^n g, h\big)_H = 0$, then $T_{\mathcal{O}}$ is ergodic.*

PROOF: What I have to show is that any $T_{\mathcal{O}}$-invariant element $\Phi \in L^2(\mathcal{W};\mathbb{R})$ is $\mathcal{W}$-almost surely constant, and for this purpose it suffices to check that

$$(*) \qquad \underline{\lim}_{n\to\infty} \big|\mathbb{E}^{\mathcal{W}}\big[(\Phi \circ T_{\mathcal{O}}^n)\Phi\big]\big| = 0$$

for all $\Phi \in L^2(\mathcal{W};\mathbb{R})$ with mean value 0. In fact, if $\{h_m : m \geq 1\}$ is an orthonormal basis for $H$, then it suffices to check (*) when

$$\Phi(x) = F\big([\mathcal{I}(h_1)](x), \dots, [\mathcal{I}(h_N)](x)\big)$$

for some $N \in \mathbb{Z}^+$ and bounded, Borel measurable $F : \mathbb{R}^N \longrightarrow \mathbb{R}$. The reason why it is sufficient to check it for such $\Phi$'s is that, because $T_{\mathcal{O}}$ is $\mathcal{W}$-measure preserving, the set of $\Phi$'s for which (*) holds is closed in $L^2(\mathcal{W};\mathbb{R})$. Hence, if we start with any $\Phi \in L^2(\mathcal{W};\mathbb{R})$ with mean value 0, we can first approximate it in $L^2(\mathcal{W};\mathbb{R})$ by bounded functions with mean value 0 and then condition these bounded approximates with respect to $\sigma\big(\{\mathcal{I}(h_1), \dots, \mathcal{I}(h_N)\}\big)$ to give them the required form.

Now suppose that $\Phi = F\big(\mathcal{I}(h_1), \dots, \mathcal{I}(h_N)\big)$ for some $N$ and bounded, measurable $F$. Then

$$\mathbb{E}^{\mathcal{W}}\big[\Phi \circ T_{\mathcal{O}}^n \Phi\big] = \iint_{\mathbb{R}^N \times \mathbb{R}^N} F(\boldsymbol{\xi}) F(\boldsymbol{\eta})\, \gamma_{\mathbf{0},\mathbf{C}_n}(d\boldsymbol{\xi} \times d\boldsymbol{\eta}),$$

[2] See I.E. Segal's "Ergodic subsgroups of the orthogonal group on a real Hilbert Space," *Annals of Math.* **66** #2, pp. 297–303 (1957). For a treatment in the setting here, see my article "Some thoughts about Segals ergodic theorem," *Colloq. Math.* **118** #1, pp. 89-105 (2010).

where

$$\mathbf{C}_n = \begin{pmatrix} \mathbf{I} & \mathbf{B}_n \\ \mathbf{B}_n^\top & \mathbf{I} \end{pmatrix} \quad \text{with } \mathbf{B}_n = \Big(\big((h_k, \mathcal{O}^n h_\ell)_H\big)\Big)_{1\le k,\ell\le N}$$

and the block structure corresponds to $\mathbb{R}^N \times \mathbb{R}^N$. Finally, by our hypothesis about $\mathcal{O}$, we can find a subsequence $\{n_m : m \ge 0\}$ such that $\lim_{m\to\infty} \mathbf{B}_{n_m} = \mathbf{0}$, from which it is clear that $\gamma_{\mathbf{0},\mathbf{C}_{n_m}}$ tends to $\gamma_{\mathbf{0},\mathbf{I}} \times \gamma_{\mathbf{0},\mathbf{I}}$ in variation and therefore

$$\lim_{m\to\infty} \mathbb{E}^{\mathcal{W}}\big[(\Phi \circ T_{\mathcal{O}}^{n_m})\Phi\big] = \mathbb{E}^{\mathcal{W}}[\Phi]^2 = 0. \quad \square$$

Perhaps the best tests for whether an orthogonal transformation satisfies the hypothesis in Theorem 8.3.15 come from spectral theory. To be more precise, if $H_c$ and $\mathcal{O}_c$ are the space and operator obtained by complexifying $H$ and $\mathcal{O}$, the Spectral Theorem for normal operators allows one to write

$$\mathcal{O}_c = \int_0^{2\pi} e^{\sqrt{-1}\alpha}\, dE_\alpha,$$

where $\{E_\alpha : \alpha \in [0, 2\pi)\}$ is a resolution of the identity in $H_c$ by orthogonal projection operators. The spectrum of $\mathcal{O}_c$ is said to be **absolutely continuous** if, for each $h \in H_c$, the non-decreasing function $\alpha \rightsquigarrow (E_\alpha h, h)_{H_c}$ is absolutely continuous, which, by polarization, means that $\alpha \rightsquigarrow (E_\alpha h, h')_{H_c}$ is absolutely continuous for all $h, h' \in H_c$. The reason for introducing this concept here is that, by combining the Riemann–Lebesgue Lemma with Theorem 8.3.15, one can prove that *$T_{\mathcal{O}}$ is ergodic if the spectrum of $\mathcal{O}_c$ is absolutely continuous.*[3] Indeed, given $h, h' \in H$, let $f$ be the Radon–Nikodym derivative of $\alpha \rightsquigarrow (E_\alpha h, h')_{H_c}$, and apply the Riemann–Lebesgue Lemma to see that

$$\big(\mathcal{O}^n h, h'\big)_H = \int_0^{2\pi} e^{\sqrt{-1}n\alpha} f(\alpha)\, d\alpha \longrightarrow 0 \quad \text{as } n \to \infty.$$

See Exercises 8.3.24, 8.3.25, and 8.5.15 for a more concrete examples.

## Exercises for § 8.3

EXERCISE 8.3.16. The purpose of this exercise is to provide the linear algebraic facts that I used in the proof of Theorem 8.3.9. Namely, I want to show that if a set $\{h_1, \dots, h_n\} \subseteq H$ is approximately orthonormal, then the vectors $h_i$ differ by very little from their Gram–Schmidt orthogonalization.

[3] This conclusion highlights the poverty of the result here in comparison to Segal's result, which says that $T_{\mathcal{O}}$ is ergodic as soon as, except for the eigenvalue 1, the spectum of $\mathcal{O}_c$ is continuous.

**(i)** Suppose that $A = \big((a_{ij})\big)_{1\le i,j\le n} \in \mathbb{R}^n \otimes \mathbb{R}^n$ is a lower triangular matrix whose diagonal entries are non-negative. Show that there is a $C_n < \infty$, depending only on $n$, such that $\|\mathbf{I}_{\mathbb{R}^n} - A\|_{\text{op}} \le C_n \|\mathbf{I}_{\mathbb{R}^n} - AA^\top\|_{\text{op}}$.

**Hint**: Begin by showing that the inequality is trivial when either $n = 1$ or $\|AA^\top\|_{\text{op}} \ge 2$. To prove it when $n \ge 2$ and $\|AA^\top\|_{\text{op}} \le 2$, set $\Delta = \mathbf{I}_{\mathbb{R}^n} - AA^\top$, show that

$$|a_{\ell\ell} a_{n\ell}| \le |\Delta_{n\ell}| + (AA^\top)_{nn}^{\frac{1}{2}} \left(\sum_{j=1}^{\ell-1} a_{jj}^2\right)^{\frac{1}{2}} \quad \text{for } 1 \le \ell < n$$

$$|1 - a_{nn}| \le |1 - a_{nn}^2| \le \Delta_{nn} + \sum_{\ell=1}^{n-1} a_{n\ell}^2,$$

and proceed by induction on $n$.

**(ii)** Let $\{h_1, \dots, h_n\} \subseteq H$, set $B = \big(((h_i, h_j)_H)\big)_{1\le i,j\le n}$, and assume that $\|\mathbf{I}_{\mathbb{R}^n} - B\|_{\text{op}} < 1$. Show that the $h_i$'s are linearly independent.

**(iii)** Continuing part **(ii)**, let $\{f_1, \dots, f_n\}$ be the orthonormal set obtained from the $h_i$'s by the Gram–Schmidt orthogonalization procedure, and let $A$ be the matrix whose $(i,j)$th entry is $(h_i, f_j)_H$. Show that $A$ is lower triangular and that its diagonal entries are non-negative. In addition, show that $AA^\top = B$.

**(iv)** By combining **(i)** and **(iii)**, show that there is a $K_n < \infty$, depending only on $n$, such that

$$\sum_{i=1}^{n} \|h_i - f_i\|_H \le K_n \sum_{i,j=1}^{n} \big|\delta_{i,j} - (h_i, h_j)_H\big|.$$

**Hint**: Note that $h_i = \sum_{j=1}^n a_{ij} f_j$ and therefore that

$$\|h_i - f_i\|_H^2 = \sum_{j=1}^{n} \big(\mathbf{I}_{\mathbb{R}^n} - A\big)_{ij}^2 \le n \|\mathbf{I}_{\mathbb{R}^n} - A\|_{\text{op}}^2.$$

EXERCISE 8.3.17. Given a Hilbert space $H$, the problem of determining for which Banach spaces $H$ arises as the Cameron–Martin space is an extremely delicate one. For example, one might hope that $H$ will be the Cameron–Martin space for $E$ if $H$ is dense in $E$ and its closed unit ball $\overline{B_H(0,1)}$ is compact in $E$. However, this is not the case. For example, take $H = \ell^2(\mathbb{N};\mathbb{R})$ and let $E$ be the completion of $H$ with respect to $\|\boldsymbol{\xi}\|_E \equiv \sqrt{\sum_{n=0}^\infty \frac{\xi_n^2}{n+1}}$. Show that $\overline{B_H(0,1)}$ is compact as a subset of $E$ but that there is no $\mathcal{W} \in \mathbf{M}_1(E)$ for which $(H, E, \mathcal{W})$ is an abstract Wiener space.

**Hint**: The first part is an easy application of the standard diagonalization argument combined with the obvious fact that $\sum_{n\geq m}\frac{\xi_n^2}{n+1}\leq\frac{1}{m+1}\|\xi\|_{\ell^2(\mathbb{N};\mathbb{R})}$. To prove the second part, note that in order for $\mathcal{W}$ to exist it would be necessary for $\sum_{n=0}^{\infty}\frac{\xi_n^2}{n+1}$ to be $\gamma_{0,1}^{\mathbb{N}}$-almost surely convergent.

EXERCISE 8.3.18. Let $(H,E,\mathcal{W})$ be an abstract Wiener space, and assume that $H$ is infinite dimensional. As was pointed out, $\{h_{x^*}: x^*\in E^*\}$ is the subspace of $g\in H$ for which there exists a $C<\infty$ with the property that $|(h,g)_H|\leq C\|h\|_E$ for all $h\in H$. Show that for each $g\in H$ there is separable Banach space $E_g$ that is continuously embedded as a Borel subset of $E$ such that $\mathcal{W}(E_g)=1$, $(H,E_g,\mathcal{W}\restriction E_g)$ is an abstract Wiener space, and $|(h,g)_H|\leq\|h\|_{E_g}$ for all $h\in H$.

**Hint**: Refer to the notation used in the proof of Corollary 8.3.10. Choose $n_m\nearrow\infty$ so that $n_0=0$ and, for $m\geq 1$, $\|\Pi_{L_{n_m}^{\perp}}g\|_H\leq 2^{-m}$ and $\mathbb{E}^{\mathcal{W}}\big[\|P_Lx\|_E^2\big]^{\frac12}\leq 2^{-m}$ for finite dimensional $L\perp L_{n_m}$. Next, define $E_g$ to be the space of $x\in E$ with the properties that $P_{L_{n_m}}x\longrightarrow x$ in $E$ and

$$\|x\|_{E_g}\equiv\sum_{\ell=0}\Big(\|Q_\ell x\|_E+\big|(Q_\ell x,g)_H\big|\Big)<\infty,$$

where $Q_0x=\langle x,x_0^*\rangle h_{x_0^*}$ and $Q_\ell x=\sum_{n=n_{\ell-1}+1}^{n_\ell}\langle x,x_n^*\rangle h_{x_n^*}$ for $\ell\geq 1$. Using the reasoning in the proof of Corollary 8.3.10, show that $E_g$ has the required properties.

EXERCISE 8.3.19. Let $N=1$. Using Theorem 8.3.3, take Wiener's choice of orthonormal basis and check that there are independent, standard normal random variables $\{X_m: m\geq 1\}$ under $\mathcal{W}^{(1)}$ such that, for $\mathcal{W}^{(1)}$-almost almost every $\theta$,

$$\theta(t)=tX_0(\theta)+2^{\frac12}\sum_{m=1}^{\infty}X_m(\theta)\frac{\sin(\pi mt)}{m\pi},\quad t\in[0,1],$$

where the convergence is uniform. From this, show that, $\mathcal{W}^{(1)}$-almost surely,

$$\int_0^1\theta(t)^2\,dt=\frac{X_0(\theta)^2}{3}+\frac{1}{\pi^2}\sum_{m=1}^{\infty}\frac{X_m(\theta)^2+(-1)^{m+1}\sqrt{8}X_0(\theta)X_m(\theta)}{m^2},$$

where the convergence of the series is absolute. Using the preceding, conclude that, for any $\alpha\in(0,\infty)$,

$$\mathbb{E}^{\mathcal{W}^{(1)}}\left[-\alpha\int_0^1\theta(t)^2\,dt\right]=\left[\prod_{m=1}^{\infty}\left(1+\frac{2\alpha}{m^2\pi^2}\right)\right]^{-\frac12}\left[1+4\alpha\sum_{m=1}^{\infty}\frac{1}{m^2\pi^2+2\alpha}\right]^{-\frac12}.$$

Finally, recall Euler's product formula

$$\sinh z = z\prod_{m=1}^{\infty}\left(1+\frac{z^2}{m^2\pi^2}\right), \quad z\in\mathbb{C},$$

and arrive first at

$$\mathbb{E}^{\mathcal{W}^{(1)}}\left[\exp\left(-\alpha\int_0^1 \theta(t)^2\,dt\right)\right] = \left[\cosh\sqrt{2\alpha}\right]^{-\frac{1}{2}},$$

and then, after an application of Brownian rescaling, at

$$\mathbb{E}^{\mathcal{W}^{(1)}}\left[\exp\left(-\alpha\int_0^T \theta(t)^2\,dt\right)\right] = \left[\cosh\sqrt{2\alpha}\,T\right]^{-\frac{1}{2}}.$$

This is a famous calculation that can be made using many different methods. We will return to it in § 10.1.3. See, in addition, Exercise 8.4.7.

**Hint**: Use Euler's product formula to see that

$$\frac{d}{dt}\log\frac{\sinh t}{t} = 2t\sum_{n=1}^{\infty}\frac{1}{n^2\pi^2+t^2} \quad\text{for}\quad t\in\mathbb{R}.$$

EXERCISE 8.3.20. Related to the preceding exercise, but easier, is finding the Laplace transform of the variance

$$V_T(\theta)\equiv\frac{1}{T}\int_0^T \theta(t)^2\,dt-\left(\frac{1}{T}\int_0^T\theta(t)\,dt\right)^2$$

of a Brownian path over the interval $[0,T]$. To do this calculation, first use Brownian scaling to show that

$$\mathbb{E}^{\mathcal{W}^{(1)}}\left[e^{-\alpha V_T}\right] = \mathbb{E}^{\mathcal{W}^{(1)}}\left[e^{-\alpha T V_1}\right].$$

Next, use elementary Fourier series to show that (cf. part (**iii**) of Exercise 8.2.18)

$$V_1(\theta) = 2\sum_{k=1}^{\infty}\left(\int_0^1\theta(t)\cos(k\pi t)\,dt\right)^2 = \sum_{k=1}^{\infty}\frac{\left(\int_0^1 f_k(t)\,d\theta(t)\right)^2}{k^2\pi^2},$$

where $f_k(t) = 2^{\frac{1}{2}}\sin(k\pi t)$ for $k\ge 1$. Since the $f_k$'s are orthonormal as elements of $L^2([0,\infty);\mathbb{R})$, this leads to

$$\mathbb{E}^{\mathcal{W}^{(1)}}\left[e^{-\alpha V_1}\right] = \prod_{k=1}^{\infty}\left(1+\frac{2\alpha}{k^2\pi^2}\right)^{-\frac{1}{2}}.$$

Now apply Euler's formula to arrive at

$$\mathbb{E}^{\mathcal{W}}\left[e^{-\alpha V_T}\right] = \sqrt{\frac{\sqrt{2\alpha T}}{\sinh(\sqrt{2\alpha T})}}.$$

Finally, using Wiener's choice of basis, show that $\theta \rightsquigarrow V_1(\theta)$ has the same distribution as $\theta \rightsquigarrow \int_0^1\left(\theta(t)-t\theta(1)\right)^2 dt$ under $\mathcal{W}^{(1)}$, a fact for which I would like but do not have any conceptual explanation.

EXERCISE 8.3.21. The purpose of this exercise is to show that, without knowing ahead of time that $\mathcal{W}^{(N)}$ lives on $\Theta(\mathbb{R}^N)$, for the Hilbert space $\mathbf{H}^1(\mathbb{R}^N)$ one can give a proof that any Wiener series converges $\gamma_{0,1}^{\mathbb{N}}$-almost surely in $\Theta(\mathbb{R}^N)$. Thus, let $\{\mathbf{h}_m : m \geq 0\}$ be an orthonormal basis in $\mathbf{H}(\mathbb{R}^N)$ and, for $n \in \mathbb{N}$ and $\boldsymbol{\omega} = (\omega_0, \dots, \omega_m, \dots) \in \mathbb{R}^{\mathbb{N}}$, set $\mathbf{S}_n(t, \boldsymbol{\omega}) = \sum_{m=0}^{n} \omega_m \mathbf{h}_m(t)$. The goal is to show that $\{\mathbf{S}_n(\,\cdot\,, \boldsymbol{\omega}) : n \geq 0\}$ converges in $\Theta(\mathbb{R}^N)$ for $\gamma_{0,1}^{\mathbb{N}}$-almost every $\boldsymbol{\omega} \in \mathbb{R}^{\mathbb{N}}$.

**(i)** For $\boldsymbol{\xi} \in \mathbb{R}^N$, set $\mathbf{h}_{t,\boldsymbol{\xi}}(\tau) = t \wedge \tau \boldsymbol{\xi}$, check that $\big(\boldsymbol{\xi}, \mathbf{S}_n(t)\big)_{\mathbb{R}^N} = \big(\mathbf{h}_{t,\boldsymbol{\xi}}, \mathbf{S}_n(t)\big)_{\mathbf{H}^1(\mathbb{R}^N)}$, and apply Theorem 1.4.2 to show that $\lim_{n\to\infty} \big(\boldsymbol{\xi}, \mathbf{S}_n(t)\big)_{\mathbb{R}^N}$ exists both $\gamma_{0,1}^{\mathbb{N}}$-almost surely and in $L^2(\gamma_{0,1}^{\mathbb{N}}; \mathbb{R})$ for each $(t, \boldsymbol{\xi}) \in [0,\infty) \times \mathbb{R}^N$. Conclude from this that, for each $t \in [0,\infty)$, $\lim_{n\to\infty} \mathbf{S}_n(t)$ exists both $\gamma_{0,1}^{\mathbb{N}}$-almost surely and in $L^2(\gamma_{0,1}^{\mathbb{N}}; \mathbb{R}^N)$.

**(ii)** On the basis of part **(i)**, show that we will be done once we know that, for $\gamma_{0,1}^{\mathbb{N}}$-almost every $\mathbf{x} \in \mathbb{R}^N$, $\{\mathbf{S}_n(\,\cdot\,, \mathbf{x}) : n \geq 0\}$ is equicontinuous on finite intervals and that $\sup_{n\geq 0} t^{-1}|\mathbf{S}_n(t, \mathbf{x})| \longrightarrow 0$ as $t \to \infty$. Show that both these will follow from the existence of a $C < \infty$ such that

$$(*) \qquad \mathbb{E}^{\gamma_{0,1}^{\mathbb{N}}}\left[\sup_{0\leq s<t\leq T}\ \sup_{n\geq 0} \frac{\big|\mathbf{S}_n(t) - \mathbf{S}_n(s)\big|}{(t-s)^{\frac{1}{8}}}\right] \leq CT^{\frac{3}{8}} \quad \text{for all } T \in (0,\infty).$$

**(iii)** As an application of Theorem 4.3.2, show that $(*)$ will follow once one checks that

$$\mathbb{E}^{\gamma_{0,1}^{\mathbb{N}}}\left[\sup_{n\geq 0} |\mathbf{S}_n(t) - \mathbf{S}_n(s)|^4\right] \leq B(t-s)^2, \quad 0 \leq s < t,$$

for some $B < \infty$. Next, apply (6.1.14) to see that

$$\mathbb{E}^{\gamma_{0,1}^{\mathbb{N}}}\left[\sup_{n\geq 0} |\mathbf{S}_n(t) - \mathbf{S}_n(s)|^4\right] \leq \left(\frac{4}{3}\right)^4 \sup_{n\geq 0} \mathbb{E}^{\gamma_{0,1}^{\mathbb{N}}}\left[|\mathbf{S}_n(t) - \mathbf{S}_n(s)|^4\right].$$

In addition, because $\mathbf{S}_n(t) - \mathbf{S}_n(s)$ is a centered Gaussian, argue that

$$\mathbb{E}^{\gamma_{0,1}^{\mathbb{N}}}\left[|\mathbf{S}_n(t) - \mathbf{S}_n(s)|^4\right] \leq 3\mathbb{E}^{\gamma_{0,1}^{\mathbb{N}}}\left[|\mathbf{S}_n(t) - \mathbf{S}_n(s)|^2\right]^2.$$

Finally, repeat the sort of reasoning used in **(i)** to check that

$$\mathbb{E}^{\gamma_{0,1}^{\mathbb{N}}}\left[|\mathbf{S}_n(t) - \mathbf{S}_n(s)|^2\right] \leq N(t-s) \quad \text{for } 0 \leq s < t.$$

EXERCISE 8.3.22. In this exercise we discuss some properties of pinned Brownian motion. Given $T > 0$, set $\boldsymbol{\theta}_T(t) = \boldsymbol{\theta}(t) - \frac{t\wedge T}{T}\boldsymbol{\theta}(T)$. As I pointed out at the end of § 8.3.2, the $\mathcal{W}^{(N)}$-distribution of $\boldsymbol{\theta}_T$ is that of a Brownian motion conditioned to be back at $\mathbf{0}$ at time $T$. Next take $\Theta_T(\mathbb{R}^N)$ to be the space of continuous paths $\boldsymbol{\theta} : [0,T] \longrightarrow \mathbb{R}^N$ satisfying $\boldsymbol{\theta}(0) = \mathbf{0} = \boldsymbol{\theta}(T)$, and let $\mathcal{W}_T^{(N)}$ denote the $\mathcal{W}^{(N)}$-distribution of $\boldsymbol{\theta} \in \Theta(\mathbb{R}^N) \longmapsto \boldsymbol{\theta}_T \restriction [0,T] \in \Theta_T(\mathbb{R}^N)$.

**(i)** Show that the $\mathcal{W}^{(N)}$-distribution of $\{\boldsymbol{\theta}_T(t) : t \geq 0\}$ is the same as that of $\{T^{\frac{1}{2}}\boldsymbol{\theta}_1(T^{-1}t) : t \geq 0\}$.

**(ii)** Set $\mathbf{H}^1_T(\mathbb{R}^N) = \{\mathbf{h} \restriction [0,T] : \mathbf{h} \in \mathbf{H}^1(\mathbb{R}^N) \ \&\ \mathbf{h}(T) = 0\}$, and define $\|\mathbf{h}\|_{\mathbf{H}^1_T(\mathbb{R}^N)} = \|\dot{\mathbf{h}}\|_{L^2([0,T];\mathbb{R}^N)}$. Show that the triple $\left(\mathbf{H}^1_T(\mathbb{R}^N), \Theta_T(\mathbb{R}^N), \mathcal{W}^{(N)}_T\right)$ is an abstract Wiener space. In addition, show that $\mathcal{W}^{(N)}_T$ is invariant under **time reversal**. That is, $\{\boldsymbol{\theta}(t) : t \in [0,T]\}$ and $\{\boldsymbol{\theta}(T-t) : t \in [0,T]\}$ have the same distribution under $\mathcal{W}^{(N)}_T$.

**Hint**: Begin by identifying $\Theta_T(\mathbb{R}^N)^*$ as the space of finite, $\mathbb{R}^N$-valued Borel measures $\boldsymbol{\lambda}$ on $[0,T]$ such that $\boldsymbol{\lambda}(\{0\}) = \mathbf{0} = \boldsymbol{\lambda}(\{T\})$.

EXERCISE 8.3.23. Say that $D \subseteq E^*$ is determining if $x = y$ whenever $\langle x, x^* \rangle = \langle y, x^* \rangle$ for all $x^* \in D$. Next, referring to Theorem 8.3.14, suppose that $\mathcal{O}$ is an orthogonal transformation on $H$ and that $F : E \longmapsto E$ has the properties that $F \restriction H = \mathcal{O}$ and that $x \rightsquigarrow \langle F(x), x^* \rangle$ is continuous for all $x^*$'s from a determining set $D$. Show that $T_{\mathcal{O}}x = F(x)$ for $\mathcal{W}$-almost every $x \in E$.

EXERCISE 8.3.24. Consider $\left(\mathbf{H}^1(\mathbb{R}^N), \Theta(\mathbb{R}^N), \mathcal{W}^{(N)}\right)$, the classical Wiener space. Given $\alpha \in (0,\infty)$, define $\mathcal{O}_\alpha : \mathbf{H}^1(\mathbb{R}^N) \longrightarrow \mathbf{H}^1(\mathbb{R}^N)$ by $[\mathcal{O}_\alpha \mathbf{h}](t) = \alpha^{-\frac{1}{2}}\mathbf{h}(\alpha t)$, show that $\mathcal{O}_\alpha$ is an orthogonal transformation, and apply Exercise 8.3.23 to see that $T_{\mathcal{O}_\alpha}$ is the Brownian scaling map $S_\alpha$ given by $S_\alpha\boldsymbol{\theta}(t) = \alpha^{-\frac{1}{2}}\boldsymbol{\theta}(\alpha t)$ discussed in part **(iii)** of Exercise 4.3.10. The main goal of this exercise is to apply Theorem 8.3.15 to show that $T_{\mathcal{O}_\alpha}$ is ergodic for every $\alpha \in (0,\infty)\setminus\{1\}$.

**(i)** Given an orthogonal transformation $\mathcal{O}$ on $\mathbf{H}^1(\mathbb{R}^N)$, show that $\left(\mathcal{O}^n\mathbf{h}, \mathbf{h}'\right)_{\mathbf{H}^1(\mathbb{R}^N)}$ tends to 0 for all $\mathbf{h}, \mathbf{h}' \in \mathbf{H}^1(\mathbb{R}^N)$ if $\lim_{n\to\infty}\left(\mathcal{O}^n\mathbf{h}, \mathbf{h}'\right)_{\mathbf{H}^1(\mathbb{R}^N)} = 0$ for all $\mathbf{h}, \mathbf{h}' \in \mathbf{H}^1(\mathbb{R}^N)$ with $\dot{\mathbf{h}}, \dot{\mathbf{h}}' \in C^\infty_c\left((0,\infty);\mathbb{R}^N\right)$.

**(ii)** Complete the program by showing that $\left(\mathcal{O}^n_\alpha\mathbf{h}, \mathbf{h}'\right)_{\mathbf{H}^1(\mathbb{R}^N)}$ tends to 0 for all $\alpha \in (0,\infty)\setminus\{1\}$ and $\mathbf{h}, \mathbf{h}' \in \mathbf{H}^1(\mathbb{R}^N)$ with $\dot{\mathbf{h}}, \dot{\mathbf{h}}' \in C^\infty_c\left((0,\infty);\mathbb{R}^N\right)$.

**(iii)** There is another way to think about the operator $\mathcal{O}_\alpha$. Namely, let $\lambda_{\mathbb{R}^N}$ be Lebesgue measure on $\mathbb{R}$, define $U : \mathbf{H}^1(\mathbb{R}^N) \longrightarrow L^2(\lambda_{\mathbb{R}^N};\mathbb{R}^N)$ by $Uh(x) = e^{\frac{x}{2}}\dot{h}(e^x)$, and show that $U$ is an isometry from $\mathbf{H}^1(\mathbb{R}^N)$ onto $L^2(\lambda_{\mathbb{R}^N};\mathbb{R}^N)$. Further, show that $U \circ \mathcal{O}_\alpha = \tau_{\log\alpha} \circ U$, where $\tau_\alpha : L^2(\lambda_{\mathbb{R}^N};\mathbb{R}^N) \longrightarrow L^2(\lambda_{\mathbb{R}^N};\mathbb{R}^N)$ is the translation map $\tau_\alpha f(x) = f(x+\alpha)$. Conclude from this that

$$\left(\mathcal{O}^n_\alpha\mathbf{h}, \mathbf{h}'\right)_{\mathbf{H}^1(\mathbb{R}^N)} = (2\pi)^{-1}\int_{\mathbb{R}} e^{-\sqrt{-1}n\xi\log\alpha}\left(\widehat{U\mathbf{h}}(\xi), \widehat{U\mathbf{h}'}\right)_{\mathbb{C}^N} d\xi,$$

and use this, together with the Riemann–Lebesgue Lemma, to give a second proof that $\left(\mathcal{O}^n_\alpha\mathbf{h}, \mathbf{h}'\right)_{\mathbf{H}^1(\mathbb{R}^N)}$ tends to 0 as $n \to \infty$ when $\alpha \neq 1$.

**(iv)** As a consequence of the above and Theorem 6.2.7, show that for each $\alpha \in (0,\infty) \setminus \{1\}$, $q \in [1,\infty)$, and $F \in L^q(\mathcal{W}^{(N)};\mathbb{C})$,

$$\lim_{n\to\infty} \frac{1}{n}\sum_{m=0}^{n-1} F(S_{\alpha^n}\boldsymbol{\theta}) = \mathbb{E}^{\mathcal{W}^{(N)}}[F] \ \mathcal{W}^{(N)}\text{-almost surely and in } L^q(\mathcal{W}^{(N)};\mathbb{C}).$$

Next, replace Theorem 6.2.7 by Theorem 6.2.12 to show that

$$\lim_{t\to\infty} \frac{1}{\log t}\int_1^t \tau^{-1} F(S_\tau \boldsymbol{\theta})\, d\tau = \mathbb{E}^{\mathcal{W}^{(N)}}[F]$$

$\mathcal{W}^{(N)}$-almost surely and in $L^q(\mathcal{W}^{(N)};\mathbb{C})$. In particular, use this to show that, for $n \in \mathbb{N}$,

$$\lim_{t\to\infty} \frac{1}{\log t}\int_1^t \tau^{-\frac{n}{2}-1}\theta(\tau)^n\, d\tau = \begin{cases} \prod_{m=1}^{\frac{n}{2}}(2m-1) & \text{if } n \text{ is even} \\ 0 & \text{if } n \text{ is odd.}\end{cases}$$

EXERCISE 8.3.25. Here is a second reasonably explicit example to which Theorem 8.3.15 applies. Again consider the classical case when $H = \mathbf{H}^1(\mathbb{R}^N)$, and assume that $N \in \mathbb{Z}^+$ is even. Choose a skew-symmetric $A \in \mathrm{Hom}(\mathbb{R}^N;\mathbb{R}^N)$ whose kernel is $\{\mathbf{0}\}$. That is, $A^\top = -A$ and $A\mathbf{x} = \mathbf{0} \implies \mathbf{x} = \mathbf{0}$.

**(i)** Define $\mathcal{O}_A$ on $\mathbf{H}^1(\mathbb{R}^N)$ by

$$\mathcal{O}_A\mathbf{h}(t) = \int_0^t e^{\tau A}\dot{\mathbf{h}}(\tau)\, d\tau,$$

and show that $\mathcal{O}_A$ is an orthogonal transformation that satisfies the hypotheses in Theorem 8.3.15.

**Hint**: Using elementary spectral theory, show that there exist non-zero, real numbers $\alpha_1, \dots, \alpha_{\frac{N}{2}}$ and an orthonormal basis $(\mathbf{e}_1, \dots, \mathbf{e}_N)$ in $\mathbb{R}^N$ such that $A\mathbf{e}_{2m-1} = \alpha_m \mathbf{e}_{2m}$ and $A\mathbf{e}_{2m} = -\alpha_m\mathbf{e}_{2m-1}$ for $1 \le m \le \frac{N}{2}$. Thus, if $L_m$ is the space spanned by $\mathbf{e}_{2m-1}$ and $\mathbf{e}_{2m}$, then $L_m$ is invariant under $A$ and the action of $e^{\tau A}$ on $L_m$ in terms of this basis is given by

$$\begin{pmatrix} \cos(\alpha_m\tau) & -\sin(\alpha_m\tau) \\ \sin(\alpha_m\tau) & \cos(\alpha_m\tau)\end{pmatrix}.$$

Finally, observe that $\mathcal{O}_A^n = \mathcal{O}_{nA}$, and apply the Riemann–Lebesgue Lemma.

**(ii)** With the help of Exercise 8.3.23, show that

$$T_{\mathcal{O}_A}\boldsymbol{\theta}(t) = \int_0^t e^{\tau A}\, d\boldsymbol{\theta}(\tau),$$

where the integral is taken in the sense of Riemann–Stieltjes.

## § 8.4 A Large Deviations Result and Strassen's Theorem

In this section I will prove the analog of Corollary 1.3.13 for non-degenerate, centered Gaussian measures on a Banach space. Once we have that result, I will apply it to prove Strassen's Theorem, which is the law of the iterated logarithm for such measures.

**§ 8.4.1. Large Deviations for Abstract Wiener Space.** The goal of this subsection is to derive the following result.

THEOREM 8.4.1. *Let $(H, E, \mathcal{W})$ be an abstract Wiener space, and, for $\epsilon > 0$, denote by $\mathcal{W}_\epsilon$ the $\mathcal{W}$-distribution of $x \rightsquigarrow \epsilon^{\frac{1}{2}} x$. Then, for each $\Gamma \in \mathcal{B}_E$,*

$$(8.4.2)\qquad \begin{aligned} - \inf_{h \in \Gamma^\circ} \frac{\|h\|_H^2}{2} &\le \varliminf_{\epsilon \searrow 0} \epsilon \log \mathcal{W}_\epsilon(\Gamma) \\ &\le \varlimsup_{\epsilon \searrow 0} \epsilon \log \mathcal{W}_\epsilon(\Gamma) \le - \inf_{h \in \overline{\Gamma}} \frac{\|h\|_H^2}{2}. \end{aligned}$$

The original version of Theorem 8.4.1 was proved by M. Schilder for the classical Wiener measure using a method that does not extend easily to the general case. The statement that I have given is due to Donsker and S.R.S. Varadhan, and my proof derives from an approach (which very much resembles the arguments given in § 1.3 to prove Cramér's Theorem) that was introduced into this context by Varadhan.

The lower bound is an easy application of the Cameron–Martin formula. Indeed, all that I have to do is show that if $h \in H$ and $r > 0$, then

$$(*)\qquad \varliminf_{\epsilon \searrow 0} \epsilon \log \mathcal{W}_\epsilon\big(B_E(h, r)\big) \ge - \frac{\|h\|_H^2}{2}.$$

To this end, note that, for any $x^* \in E^*$ and $\delta > 0$,

$$\begin{aligned} \mathcal{W}_\epsilon\big(B_E(h_{x^*}, \delta)\big) &= \mathcal{W}\big(B_E(\epsilon^{-\frac{1}{2}} h_{x^*}, \epsilon^{-\frac{1}{2}} \delta)\big) \\ &= \mathbb{E}^{\mathcal{W}}\Big[e^{-\epsilon^{-\frac{1}{2}} \langle x, x^* \rangle - \frac{1}{2\epsilon}\|h_{x^*}\|_H^2}, B_E(0, \epsilon^{-\frac{1}{2}} \delta)\Big] \\ &\ge e^{-\delta \epsilon^{-1} \|x^*\|_{E^*} - \frac{1}{2\epsilon}\|h_{x^*}\|_H^2} \mathcal{W}\big(B_E(0, \epsilon^{-\frac{1}{2}} \delta)\big), \end{aligned}$$

which means that

$$B_E(h_{x^*}, \delta) \subseteq B_E(h, r) \implies \varliminf_{\epsilon \searrow 0} \epsilon \log \mathcal{W}_\epsilon\big(B_E(h, r)\big) \ge -\delta \|x^*\|_{E^*} - \frac{\|h_{x^*}\|_H^2}{2},$$

and therefore, after letting $\delta \searrow 0$ and remembering that $\{h_{x^*} : x \in E^*\}$ is dense in $H$, that (*) holds.

The proof of the upper bound in (8.4.2) is a little more involved. The first step is to show that it suffices to treat the case when $\Gamma$ is relatively compact. To this end, refer to Corollary 8.3.10, and set $C_R$ equal to the closure in $E$ of $B_{E_0}(0,R)$. By Fernique's Theorem applied to $\mathcal{W}$ on $E_0$, we know that $\mathbb{E}^{\mathcal{W}}\left[e^{\alpha\|x\|_{E_0}^2}\right] \le K < \infty$ for some $\alpha > 0$. Hence

$$\mathcal{W}_\epsilon(E \setminus C_R) = \mathcal{W}(E \setminus C_{\epsilon^{-\frac{1}{2}}R}) \le K e^{-\alpha\frac{R^2}{\epsilon}},$$

and so, for any $\Gamma \in \mathcal{B}_E$ and $R > 0$,

$$\mathcal{W}_\epsilon(\Gamma) \le 2\mathcal{W}_\epsilon(\Gamma \cap C_R) \vee \left(K e^{-\alpha\frac{R^2}{\epsilon}}\right).$$

Thus, if we can prove the upper bound for relatively compact $\Gamma$'s, then, because $\Gamma \cap C_R$ is relatively compact, we will know that, for all $R > 0$,

$$\varlimsup_{\epsilon\searrow 0} \epsilon \log \mathcal{W}_\epsilon(\Gamma) \le -\left[\left(\inf_{h\in\overline{\Gamma}} \frac{\|h\|_H^2}{2}\right) \wedge (\alpha R^2)\right],$$

from which the general result is immediate.

To prove the upper bound when $\Gamma$ is relatively compact, I will show that, for any $y \in E$,

$$(**) \qquad \varlimsup_{r\searrow 0} \varlimsup_{\epsilon\searrow 0} \epsilon \log \mathcal{W}_\epsilon\big(B_E(y,r)\big) \le \begin{cases} -\frac{\|y\|_H^2}{2} & \text{if } y \in H \\ -\infty & \text{if } y \notin H. \end{cases}$$

To see that (**) is enough, assume that it is true and let $\Gamma \in \mathcal{B}_E \setminus \{\emptyset\}$ be relatively compact. Given $\beta \in (0,1)$, for each $y \in \overline{\Gamma}$ choose $r(y) > 0$ and $\epsilon(y) > 0$ so that

$$\mathcal{W}_\epsilon\big(B_E(y, r(y))\big) \le \begin{cases} e^{-\frac{(1-\beta)}{2\epsilon}\|y\|_H^2} & \text{if } y \in H \\ e^{-\frac{1}{\beta\epsilon}} & \text{if } y \notin H \end{cases}$$

for all $0 < \epsilon \le \epsilon(y)$. Because $\Gamma$ is relatively compact, we can find $N \in \mathbb{Z}^+$ and $\{y_1, \dots, y_N\} \subseteq \overline{\Gamma}$ such that $\Gamma \subseteq \bigcup_1^N B_E(y_n, r_n)$, where $r_n = r(y_n)$. Then, for sufficiently small $\epsilon > 0$,

$$\mathcal{W}_\epsilon(\Gamma) \le N \exp\left(-\left[\left(\frac{1-\beta}{2\epsilon} \inf_{h\in\overline{\Gamma}} \|h\|_H^2\right) \wedge \frac{1}{\epsilon\beta}\right]\right),$$

and so

$$\varlimsup_{\epsilon\searrow 0} \epsilon \log \mathcal{W}_\epsilon(\Gamma) \le -\left[\left(\frac{1-\beta}{2} \inf_{h\in\overline{\Gamma}} \|h\|_H^2\right) \wedge \frac{1}{\beta}\right].$$

Now let $\beta \searrow 0$.

Finally, to prove (**), observe that

$$\mathcal{W}_\epsilon\big(B_E(y,r)\big) = \mathcal{W}\Big(B_E\big(\tfrac{y}{\sqrt{\epsilon}}, \tfrac{r}{\sqrt{\epsilon}}\big)\Big) = \mathbb{E}^{\mathcal{W}}\Big[e^{-\epsilon^{-\frac{1}{2}}\langle x,x^*\rangle} e^{\epsilon^{-\frac{1}{2}}\langle x,x^*\rangle}, B_E\big(\tfrac{y}{\sqrt{\epsilon}}, \tfrac{r}{\sqrt{\epsilon}}\big)\Big]$$

$$\le e^{-\epsilon^{-1}(\langle y,x^*\rangle - r\|x^*\|_{E^*})}\mathbb{E}^{\mathcal{W}}\big[e^{\epsilon^{-\frac{1}{2}}\langle x,x^*\rangle}\big] = e^{-\epsilon^{-1}\left(\langle y,x^*\rangle - \frac{\|h_{x^*}\|_H^2}{2} - r\|x^*\|_{E^*}\right)},$$

for all $x^* \in E$. Hence,

$$\overline{\lim_{r\searrow 0}}\,\overline{\lim_{\epsilon\searrow 0}}\,\epsilon \log \mathcal{W}_\epsilon\big(B_E(y,r)\big) \le -\sup_{x^*\in E^*}\Big(\langle y,x^*\rangle - \tfrac{1}{2}\|h_{x^*}\|_H^2\Big).$$

Finally, note that the preceding supremum is the same as half the supremum of $\langle y,x^*\rangle$ over $x^*$ with $\|h_{x^*}\|_H = 1$, which, by Lemma 8.2.3, is equal to $\frac{\|y\|_H^2}{2}$ if $y \in H$ and to $\infty$ if $y \notin H$.

An interesting corollary of Theorem 8.4.1 is the following sharpening, due to Donsker and Varadhan, of Fernique's Theorem.

COROLLARY 8.4.3. *Let $\mathcal{W}$ be a non-degenerate, centered, Gaussian measure on the separable Banach space $E$, let $H$ be the associated Cameron–Martin space, and determine $\Sigma > 0$ by $\Sigma^{-1} = \inf\{\|h\|_H : \|h\|_E = 1\}$. Then*

$$\lim_{R\to\infty} R^{-2}\log \mathcal{W}\big(\|x\|_E \ge R\big) = -\frac{1}{2\Sigma^2}.$$

*In particular, $\mathbb{E}^{\mathcal{W}}\big[e^{\frac{\alpha^2}{2}\|x\|_E^2}\big]$ is finite if $\alpha < \Sigma^{-1}$ and infinite if $\alpha \ge \Sigma^{-1}$.*

PROOF: Set $f(r) = \inf\{\|h\|_H : \|h\|_E \ge r\}$. Clearly $f(r) = rf(1)$ and $f(1) = \Sigma^{-1}$. Thus, by the upper bound in (8.4.2), we know that

$$\overline{\lim_{R\to\infty}}\, R^{-2}\log \mathcal{W}\big(\|x\|_E \ge R\big) = \overline{\lim_{R\to\infty}}\, R^{-2}\log \mathcal{W}_{R^{-2}}\big(\|x\|_E \ge 1\big) \le -\frac{f(1)^2}{2} = \frac{\Sigma^{-2}}{2}.$$

Similarly, by the lower bound in (8.4.2), for any $\delta \in (0,1)$,

$$\underline{\lim_{R\to\infty}}\, R^{-2}\log \mathcal{W}\big(\|x\|_E \ge R\big) \ge \underline{\lim_{R\to\infty}}\, R^{-2}\log \mathcal{W}\big(\|x\|_E > R\big)$$

$$\ge -\inf\left\{\frac{\|h\|_H^2}{2} : \|h\|_E > R\right\} \ge -\frac{f(1+\delta)^2}{2} = -(1+\delta)^2\frac{1}{2\Sigma^2},$$

and so we have now proved the first assertion.

Given the first assertion, it is obvious that $\mathbb{E}^{\mathcal{W}}\big[e^{\frac{\alpha^2\|x\|_E^2}{2}}\big]$ is finite when $\alpha < \Sigma^{-1}$ and infinite when $\alpha > \Sigma^{-1}$. The case when $\alpha = \Sigma^{-1}$ is more delicate. To handle it, I first show that $\Sigma = \sup\{\|h_{x^*}\|_H : \|x^*\|_{E^*} = 1\}$. Indeed, if $x^* \in E^*$ and $\|x^*\|_{E^*} = 1$, set $g = \frac{h_{x^*}}{\|h_{x^*}\|_E}$, note that $\|g\|_E = 1$, and check that

$1 \geq \langle g, x^* \rangle = \big(g, h_{x^*}\big)_H = \|g\|_H \|h_{x^*}\|_H$. Hence $\|h_{x^*}\|_H \leq \|g\|_H^{-1} \leq \Sigma$. Next, suppose that $h \in H$ with $\|h\|_E = 1$. Then, by the Hahn–Banach Theorem, there exists a $x^* \in E^*$ with $\|x^*\|_{E^*} = 1$ and $\langle h, x^* \rangle = 1$. In particular, $\|h\|_H \|h_{x^*}\|_H \geq \big(h, h_{x^*}\big)_H = \langle h, x^* \rangle = 1$, and therefore $\|h\|_H^{-1} \leq \|h_{x^*}\|_H$, which, together with the preceding, completes the verification.

The next step is to show that there exists an $x^* \in E^*$ with $\|x^*\|_{E^*} = 1$ such that $\|h_{x^*}\|_H = \Sigma$. To this end, choose $\{x_k^* : k \geq 1\} \subseteq E^*$ with $\|x_k^*\|_{E^*} = 1$ so that $\|h_{x_k^*}\|_H \longrightarrow \Sigma$. Because $\overline{B_{E^*}(0,1)}$ is compact in the weak* topology and, by Theorem 8.2.6, $x^* \in \overline{B_{E^*}(0,1)} \longmapsto h_{x^*} \in H$ is continuous from the weak* topology into the strong topology, we can assume that $\{x_k^* : k \geq 1\}$ is weak* convergent to some $x^* \in \overline{B_{E^*}(0,1)}$ and that $\|h_{x^*}\|_H = \Sigma$, which is possible only if $\|x^*\|_{E^*} = 1$. Finally, knowing that this $x^*$ exists, note that $\langle \cdot\,, x^* \rangle$ is a centered Gaussian under $\mathcal{W}$ with variance $\Sigma^2$. Hence, since $\|x\|_E \geq |\langle x, x^* \rangle|$,

$$\mathbb{E}^{\mathcal{W}}\left[e^{\frac{\|x\|_E^2}{2\Sigma^2}}\right] \geq \int_{\mathbb{R}} e^{\frac{\xi^2}{2\Sigma^2}}\, \gamma_{0,\Sigma^2}(d\xi) = \infty. \qquad \square$$

**§ 8.4.2. Strassen's Law of the Iterated Logarithm.** Just as in § 1.5 we were able to prove a law of the iterated logarithm on the basis of the large deviation estimates in § 1.3, so here the estimates in the preceding subsection will allow us to prove a law of the iterated for centered Gaussian random variables on a Banach space. Specifically, I will prove the following theorem, whose statement is modeled on V. Strassen's famous law of the iterated for Brownian motion (cf. § 8.6.3).

Recall from § 1.5 the notation $\Lambda_n = \sqrt{2n \log_{(2)}(n \vee 3)}$ and $\tilde{S}_n = \frac{S_n}{\Lambda_n}$, where $S_n = \sum_1^n X_m$.

THEOREM 8.4.4. *Suppose that $\mathcal{W}$ is a non-degenerate, centered, Gaussian measure on the Banach space $E$, and let $H$ be its Cameron–Martin space. If $\{X_n : n \geq 1\}$ is a sequence of independent, $E$-valued, $\mathcal{W}$-distributed random variables on some probability space $(\Omega, \mathcal{F}, \mathbb{P})$, then, $\mathbb{P}$-almost surely, the sequence $\{\tilde{S}_n : n \geq 1\}$ is relatively compact in $E$ and the closed unit ball $\overline{B_H(0,1)}$ in $H$ coincides with its set of limit points. Equivalently, $\mathbb{P}$-almost surely, $\lim_{n\to\infty} \|\tilde{S}_n - \overline{B_H(0,1)}\|_E = 0$ and, for each $h \in \overline{B_H(0,1)}$, $\underline{\lim}_{n\to\infty} \|\tilde{S}_n - h\|_E = 0$.*

Because, by Theorem 8.2.6, $\overline{B_H(0,1)}$ is compact in $E$, the equivalence of the two formulations is obvious, and so I will concentrate on the second formulation.

I begin by showing that $\lim_{n\to\infty} \|\tilde{S}_n - \overline{B_H(0,1)}\|_E = 0$ $\mathbb{P}$-almost surely, and the fact that underlies my proof is the estimate that, for each open subset $G$ of $E$ and $\alpha < \inf\{\|h\|_H : h \notin G\}$, there is an $M \in (0,\infty)$ with the property that

$$(*) \qquad \mathbb{P}\left(\frac{S_n}{\Lambda} \notin G\right) \leq \exp\left[-\frac{\alpha^2 \Lambda^2}{2n}\right] \quad \text{for all } n \in \mathbb{Z}^+ \text{ and } \Lambda \geq M\sqrt{n}.$$

To check (*), first note (cf. Exercise 8.2.14) that the distribution of $S_n$ under $\mathbb{P}$ is the same as that of $x \rightsquigarrow n^{\frac{1}{2}}x$ under $\mathcal{W}$ and therefore that $\mathbb{P}\left(\frac{\tilde{S}_n}{\Lambda} \notin G\right) = \mathcal{W}_{\frac{n}{\Lambda^2}}(G\complement)$. Hence, (*) is really just an application of the upper bound in (8.4.2). Given (*), I proceed in very much the same way as I did at the analogous place in § 1.5. Namely, for any $\beta \in (1,2)$,

$$\begin{aligned}
\overline{\lim_{n\to\infty}} \|\tilde{S}_n - \overline{B_H(0,1)}\|_E &\le \overline{\lim_{m\to\infty}} \max_{\beta^{m-1}\le n\le\beta^m} \|\tilde{S}_n - \overline{B_H(0,1)}\|_E \\
&\le \overline{\lim_{m\to\infty}} \max_{\beta^{m-1}\le n\le\beta^m} \frac{\|S_n - \overline{B_H(0,\Lambda_{[\beta^{m-1}]})}\|_E}{\Lambda_n} \\
&\le \overline{\lim_{m\to\infty}} \max_{1\le n\le\beta^m} \left\| \frac{S_n}{\Lambda_{[\beta^{m-1}]}} - \overline{B_H(0,1)} \right\|_E .
\end{aligned}$$

At this point in § 1.5 (cf. the proof of Lemma 1.5.3), I applied Lévy's reflection principle to get rid of the "max." However, Lévy's argument works only for $\mathbb{R}$-valued random variables, and so here I will replace his estimate by one based on the idea in Exercise 1.4.25.

LEMMA 8.4.5. *Let $\{Y_m : m \ge 1\}$ be mutually independent, $E$-valued random variables, and set $S_n = \sum_{m=1}^n Y_m$ for $n \ge 1$. Then, for any closed $F \subseteq E$ and $\delta > 0$,*

$$\mathbb{P}\left(\max_{1\le m\le n} \|S_m - F\|_E \ge 2\delta\right) \le \frac{\mathbb{P}(\|S_n - F\|_E \ge \delta)}{1 - \max_{1\le m\le n}\mathbb{P}(\|S_n - S_m\|_E \ge \delta)}.$$

PROOF: Set

$$A_m = \{\|S_m - F\|_E \ge 2\delta \text{ and } \|S_k - F\|_E < 2\delta \text{ for } 1 \le k < m\}.$$

Following the hint for Exercise 1.4.25, observe that

$$\begin{aligned}
&\mathbb{P}\left(\max_{1\le m\le n} \|S_m - F\|_E \ge 2\delta\right) \min_{1\le m\le n} \mathbb{P}(\|S_n - S_m\|_E < \delta) \\
&\le \sum_{m=1}^n \mathbb{P}(A_m \cap \{\|S_n - S_m\|_E < \delta\}) \le \sum_{m=1}^n \mathbb{P}(A_m \cap \{\|S_n - F\|_E \ge \delta\}),
\end{aligned}$$

which, because the $A_m$'s are disjoint, is dominated by $\mathbb{P}(\|S_n - F\|_E \ge \delta)$. □

Applying the preceding to the situation at hand, we see that

$$\begin{aligned}
&\mathbb{P}\left(\max_{1\le n\le\beta^m} \left\|\frac{S_n}{\Lambda_{[\beta^{m-1}]}} - \overline{B_H(0,1)}\right\|_E \ge 2\delta\right) \\
&\le \frac{\mathbb{P}\left(\left\|\frac{S_{[\beta^m]}}{\Lambda_{[\beta^{m-1}]}} - \overline{B_H(0,1)}\right\|_E \ge \delta\right)}{1 - \max_{1\le n\le\beta^m}\mathbb{P}(\|S_n\|_E \ge \delta\Lambda_{[\beta^{m-1}]})}.
\end{aligned}$$

After combining this with the estimate in (*), it is an easy matter to show that, for each $\delta > 0$, there is a $\beta \in (1,2)$ such that

$$\sum_{m=1}^{\infty} \mathbb{P}\left( \max_{\beta^{m-1} \le n \le \beta^m} \left\| \frac{S_n}{\Lambda_{[\beta^{m-1}]}} - \overline{B_H(0,1)} \right\|_E \ge 2\delta \right) < \infty,$$

from which it should be clear why $\overline{\lim}_{n\to\infty} \|\tilde{S}_n - \overline{B_H(0,1)}\|_E = 0$ $\mathbb{P}$-almost surely.

The proof that, $\mathbb{P}$-almost surely, $\underline{\lim}_{n\to\infty} \|\tilde{S}_n - h\|_E = 0$ for all $h \in \overline{B_H(0,1)}$ differs in no substantive way from the proof of the analogous assertion in the second part of Theorem 1.5.9. Namely, because $\overline{B_H(0,1)}$ is separable, it suffices to work with one $h \in \overline{B_H(0,1)}$ at a time. Furthermore, just as I did there, I can reduce the problem to showing that, for each $k \ge 2$, $\epsilon > 0$, and $h$ with $\|h\|_H < 1$,

$$\sum_{m=1}^{\infty} \mathbb{P}\left( \left\| \tilde{S}_{k^m - k^{m-1}} - h \right\|_E < \epsilon \right) = \infty.$$

But, if $\|h\|_H < \alpha < 1$, then (8.4.2) says that

$$\mathbb{P}\left( \left\| \tilde{S}_{k^m - k^{m-1}} - h \right\|_E < \epsilon \right) = \mathcal{W}_{\frac{k^m - k^{m-1}}{\Lambda^2_{k^m - k^{m-1}}}} \big( B_E(h,\epsilon) \big) \ge e^{-\alpha^2 \log_{(2)}(k^m - k^{m-1})}$$

for all large enough $m$'s.

## Exercises for § 8.4

EXERCISE 8.4.6. Let $(H, E, \mathcal{W})$ be an abstract Wiener space, and assume that $\dim(H) = \infty$. If $\mathcal{W}_\epsilon$ is defined for $\epsilon > 0$ as in Theorem 8.4.1, show that $\mathcal{W}_{\epsilon_1} \perp \mathcal{W}_{\epsilon_2}$ if $\epsilon_2 \ne \epsilon_1$.

**Hint**: Choose $\{x_m^* : m \ge 0\} \subseteq E^*$ so that $\{h_{x_m^*} : m \ge 0\}$ is an orthonormal basis in $H$, and show that

$$\lim_{n\to\infty} \frac{1}{n} \sum_{m=0}^{n-1} \langle x, x_m^* \rangle^2 = \epsilon \quad \mathcal{W}_\epsilon\text{-almost surely.}$$

EXERCISE 8.4.7. Show that the $\Sigma$ in Corollary 8.4.3 is $\frac{1}{2}$ in the case of the classical abstract Wiener space $\big(\mathbf{H}^1(\mathbb{R}^N), \Theta(\mathbb{R}^N), \mathcal{W}^{(N)}\big)$ and therefore that

$$\lim_{R\to\infty} R^{-2} \log \mathcal{W}^{(N)}\big( \|\boldsymbol{\theta}\|_{\Theta(\mathbb{R}^N)} \ge R \big) = -2.$$

Next, show that

$$\lim_{R\to\infty} R^{-2} \log \mathcal{W}^{(N)}\left( \sup_{\tau\in[0,t]} |\boldsymbol{\theta}(\tau)| \ge R \right) = -\frac{1}{2t}$$

and that

$$\lim_{R\to\infty} R^{-2}\log \mathcal{W}^{(N)}\left(\sup_{\tau\in[0,t]}|\boldsymbol{\theta}(\tau)|\ge R\,\middle|\,\boldsymbol{\theta}(t)=0\right)=-\frac{2}{t}.$$

Finally, show that

$$\lim_{R\to\infty} R^{-1}\log \mathcal{W}^{(N)}\left(\int_0^t |\boldsymbol{\theta}(\tau)|^2\,d\tau\ge R\right)=-\frac{\pi^2}{8t^2}$$

and that

$$\lim_{R\to\infty} R^{-1}\log \mathcal{W}^{(N)}\left(\int_0^t |\boldsymbol{\theta}(\tau)|^2\,d\tau\ge R\,\middle|\,\boldsymbol{\theta}(t)=0\right)=-\frac{\pi^2}{2t^2}.$$

**Hint**: In each case after the first, Brownian scaling can be used to reduce the problem to the case when $t=1$, and the challenge is to find the optimal constant $C$ for which $\|h\|_E\le C\|h\|_H$, $h\in H$ for the appropriate abstract Wiener space $(E,H,\mathcal{W})$. In the second case $E=C_0([0,1]:\mathbb{R}^N)\equiv\{\boldsymbol{\theta}\restriction[0,1]:\boldsymbol{\theta}\in\Theta(\mathbb{R}^N)\}$ and $H=\{\boldsymbol{\eta}\restriction[0,1]:\boldsymbol{\eta}\in\mathbf{H}^1(\mathbb{R}^N)\}$, in the third (cf. part (**ii**) of Exercise 8.3.22) $E=\Theta_1(\mathbb{R}^N)$ and $H=\mathbf{H}^1_1(\mathbb{R}^N)$ , in the fourth $E=L^2([0,1];\mathbb{R}^N)$ and $H=\{\boldsymbol{\eta}\restriction[0,1]:\boldsymbol{\eta}\in\mathbf{H}^1(\mathbb{R}^N)\}$, and in the fifth $E=L^2([0,1];\mathbb{R}^N)$ and $H=\mathbf{H}^1_1(\mathbb{R}^N)$. The optimization problems when $E=\Theta(\mathbb{R}^N)$ or $C_0([0,1];\mathbb{R}^N)$ are rather easy consequences of $|\boldsymbol{\eta}(t)|\le t^{\frac12}\|\boldsymbol{\eta}\|_{\mathbf{H}^1(\mathbb{R}^N)}$. When $E=\Theta_1(\mathbb{R}^N)$, one should start with the observation that if $\boldsymbol{\eta}\in\mathbf{H}^1_1(\mathbb{R}^N)$, then $2\|\boldsymbol{\eta}\|_{\mathrm{u}}\le\|\dot{\boldsymbol{\eta}}\|_{L^1([0,1];\mathbb{R}^N)}\le\|\boldsymbol{\eta}\|_{\mathbf{H}^1_1(\mathbb{R}^N)}$. In the final two cases, one can either use elementary variational calculus or one can make use of, respectively, the orthonormal bases

$$\{2^{\frac12}\sin(n+\tfrac12)\pi\tau:\,n\ge0\}\text{ and }\{2^{\frac12}\sin n\pi\tau:\,n\ge1\}\text{ in }L^2([0,1];\mathbb{R}).$$

EXERCISE 8.4.8. Suppose that $f\in C(E;\mathbb{R})$, and show, as a consequence of Theorem 8.4.4, that

$$\varliminf_{n\to\infty} f(\tilde S_n)=\min\{f(h):\|h\|_H\le1\}\text{ and }\varlimsup_{n\to\infty} f(\tilde S_n)=\max\{f(h):\|h\|_H\le1\}$$

$\mathcal{W}^{\mathbb{N}}$-almost surely.

## § 8.5 Euclidean Free Fields

In this section I will give a very cursory introduction to a family of abstract Wiener spaces that played an important role in the attempt to give a mathematically rigorous construction of quantum fields. From the physical standpoint, the fields treated here are "trivial" in the sense that they model "free" (i.e., non-interacting) fields. Nonetheless, they are interesting from a mathematical standpoint and, if nothing else, show how profoundly properties of a process are effected by the dimension of its parameter set.

I begin with the case when the parameter set is one dimensional and the resulting process can be seen as a minor variant of Brownian motion. As we will see, the intractability of the higher dimensional analogs increases with the number of dimensions.

§ **8.5.1. The Ornstein–Uhlenbeck Process.** Given $\mathbf{x} \in \mathbb{R}^N$ and $\boldsymbol{\theta} \in \Theta(\mathbb{R}^N)$, consider the integral equation

$$\text{(8.5.1)} \qquad \mathbf{U}(t,\mathbf{x},\boldsymbol{\theta}) = \mathbf{x} + \boldsymbol{\theta}(t) - \frac{1}{2}\int_0^t \mathbf{U}(\tau,\mathbf{x},\boldsymbol{\theta})\,d\tau, \quad t \geq 0.$$

A completely elementary argument (e.g., via Gronwall's Inequality) shows that, for each $\mathbf{x}$ and $\boldsymbol{\theta}$, there is at most one solution. Furthermore, integration by parts allows one to check that if

$$\mathbf{U}(t,\mathbf{0},\boldsymbol{\theta}) = e^{-\frac{t}{2}}\int_0^t e^{\frac{\tau}{2}}\,d\boldsymbol{\theta}(\tau),$$

where the integral is taken in the sense of Riemann-Stieltjes, then

$$\mathbf{U}(t,\mathbf{x},\boldsymbol{\theta}) = e^{-\frac{t}{2}}\mathbf{x} + \mathbf{U}(t,\mathbf{0},\boldsymbol{\theta})$$

is one, and therefore the one and only, solution.

The stochastic process $\{\mathbf{U}(t,\mathbf{x}) : t \geq 0\}$ under $\mathcal{W}^{(N)}$ was introduced by L. Ornstein and G. Uhlenbeck[1] and is known as the **Ornstein–Uhlenbeck process** starting from $\mathbf{x}$. From our immediate point of view, its importance is that it leads to a completely tractable example of a free field.

Intuitively, $\mathbf{U}(t,\mathbf{0},\boldsymbol{\theta})$ is a Brownian motion that has been subjected to a linear restoring force. Thus, locally it should behave very much like a Brownian motion. However, over long time intervals it should feel the effect of the restoring force, which is always pushing it back toward the origin. To see how these intuitive ideas are reflected in the distribution of $\{\mathbf{U}(t,\mathbf{0},\boldsymbol{\theta}) : t \geq 0\}$, I begin by using Exercise 8.2.18 to identify $\big(\mathbf{e},\mathbf{U}(t,\mathbf{0})\big)_{\mathbb{R}^N}$ as $e^{-\frac{t}{2}}\mathcal{I}(\mathbf{h}^t_{\mathbf{e}})$ for each $\mathbf{e} \in \mathbb{S}^{N-1}$, where $\mathbf{h}^t_{\mathbf{e}}(\tau) = 2(e^{\frac{t\wedge\tau}{2}} - 1)\mathbf{e}$. Hence, the span of $\big\{\big(\boldsymbol{\xi},\mathbf{U}(t,\mathbf{0})\big)_{\mathbb{R}^N} : t \geq 0 \;\&\; \boldsymbol{\xi} \in \mathbb{R}^N\big\}$ is a Gaussian family in $L^2(\mathcal{W}^{(N)};\mathbb{R})$, and

$$\mathbb{E}^{\mathcal{W}^{(N)}}\big[\mathbf{U}(s,\mathbf{0}) \otimes \mathbf{U}(t,\mathbf{0})\big] = \big(e^{-\frac{|t-s|}{2}} - e^{-\frac{s+t}{2}}\big)\mathbf{I}.$$

The key to understanding the process $\{\mathbf{U}(t,\mathbf{0}) : t \geq 0\}$ is the observation that it has the same distribution as the process $\big\{e^{-\frac{t}{2}}\mathbf{B}(e^t - 1) : t \geq 0\big\}$, where

[1] In their article "On the theory of Brownian motion," *Phys. Reviews* **36** **#3**, pp. 823-841 (1930), L. Ornstein and G. Uhlenbeck introduced this process in an attempt to reconcile some of the more disturbing properties of Wiener paths with physical reality.

$\{\mathbf{B}(t) : t \geq 0\}$ is a Brownian motion, a fact that follows immediately from the observation that they are Gaussian families with the same covariance structure. In particular, by combining this with the Law of the Iterated Logarithm proved in Exercise 4.3.15, we see that, for each $\mathbf{e} \in \mathbb{S}^{N-1}$,

$$\overline{\lim_{t\to\infty}} \frac{(\mathbf{e}, \mathbf{U}(t,\mathbf{x}))_{\mathbb{R}^N}}{\sqrt{2\log t}} = 1 = -\varliminf_{t\to\infty} \frac{(\mathbf{e}, \mathbf{U}(t,\mathbf{x}))_{\mathbb{R}^N}}{\sqrt{2\log t}} \tag{8.5.2}$$

$\mathcal{W}^{(N)}$-almost surely, which confirms the suspicion that the restoring force dampens the Brownian excursions out toward infinity.

A second indication that $\mathbf{U}(\,\cdot\,,\mathbf{x})$ tends to spend more time than Brownian paths do near the origin is that its distribution at time $t$ will be $\gamma_{e^{-\frac{t}{2}}\mathbf{x},(1-e^{-t})\mathbf{I}}$, and so, as distinguished from Brownian motion itself, its distribution as time $t$ tends to a limit, namely $\gamma_{\mathbf{0},\mathbf{I}}$. This observation suggests that it might be interesting to look at an *ancient* Ornstein–Uhlenbeck process, one that already has been running for an infinite amount of time. To be more precise, since the distribution of an ancient Ornstein–Uhlenbeck at time 0 would be $\gamma_{\mathbf{0},\mathbf{I}}$, what we should look at is the process that we get by making the $\mathbf{x}$ in $\mathbf{U}(\,\cdot\,,\mathbf{x},\boldsymbol{\theta})$ a standard normal random variable. Thus, I will say that a stochastic process $\{\mathbf{U}_A(t) : t \geq 0\}$ is an **ancient Ornstein–Uhlenbeck process** if its distribution is that of $\{\mathbf{U}(t,\mathbf{x},\boldsymbol{\theta}) : t \geq 0\}$ under $\gamma_{\mathbf{0},\mathbf{I}} \times \mathcal{W}^{(N)}$.

If $\{\mathbf{U}_A(t) : t \geq 0\}$ is an ancient Ornstein–Uhlenbeck process, then it is clear that $\{(\boldsymbol{\xi}, \mathbf{U}_A(t))_{\mathbb{R}^N} : t \geq 0 \;\&\; \boldsymbol{\xi} \in \mathbb{R}^N\}$ spans a Gaussian family with covariance

$$\mathbb{E}^{\mathbb{P}}\big[\mathbf{U}_A(s) \otimes \mathbf{U}_A(t)\big] = e^{-\frac{|t-s|}{2}}\mathbf{I}.$$

As a consequence, we see that if $\{\mathbf{B}(t) : t \geq 0\}$ is a Brownian motion, then $\{e^{-\frac{t}{2}}\mathbf{B}(e^t) : t \geq 0\}$ is an ancient Ornstein–Uhlenbeck process. In addition, as we suspected, the ancient Ornstein–Uhlenbeck process is a **stationary process** in the sense that, for each $T > 0$, the distribution of $\{\mathbf{U}_A(t+T) : t \geq 0\}$ is the same as that of $\{\mathbf{U}_A(t) : t \geq 0\}$, which can be checked either by using the preceding representation in terms of Brownian motion or by observing that its covariance is a function of $t - s$.

In fact, even more is true: it is time reversible in the sense that, for each $T > 0$, $\{\mathbf{U}_A(t) : t \in [0,T]\}$ has the same distribution as $\{\mathbf{U}_A(T-t) : t \in [0,T]\}$. This observation suggests that we can give the ancient Ornstein–Uhlenbeck its past by running it backwards. That is, define $\mathbf{U}_R : [0,\infty) \times \mathbb{R}^N \times \Theta(\mathbb{R}^N)^2 \longrightarrow \mathbb{R}^N$ by

$$\mathbf{U}_R(t,\mathbf{x},\boldsymbol{\theta}_+,\boldsymbol{\theta}_-) = \begin{cases} \mathbf{U}(t,\mathbf{x},\boldsymbol{\theta}_+) & \text{if } t \geq 0 \\ \mathbf{U}(-t,\mathbf{x},\boldsymbol{\theta}_-) & \text{if } t < 0, \end{cases}$$

and consider the process $\{\mathbf{U}_R(t,\mathbf{x},\boldsymbol{\theta}_+,\boldsymbol{\theta}_-) : t \in \mathbb{R}\}$ under $\gamma_{\mathbf{0},\mathbf{I}} \times \mathcal{W}^{(N)} \times \mathcal{W}^{(N)}$. This process also spans a Gaussian family, and it is still true that

$$\mathbb{E}^{\gamma_{\mathbf{0},\mathbf{I}}\times\mathcal{W}^{(N)}\times\mathcal{W}^{(N)}}\big[\mathbf{U}_R(s) \otimes \mathbf{U}_R(t)\big] = u(s,t)\mathbf{I}, \text{ where } u(s,t) \equiv e^{-\frac{|t-s|}{2}}, \tag{8.5.3}$$

only now for all $s, t \in \mathbb{R}$. One advantage of having added the past is that the statement of reversibility takes a more appealing form. Namely, $\{\mathbf{U}_R(t) : t \in \mathbb{R}\}$ is **reversible** in the sense that its distribution is the same whether one runs it forward or backward in time. That is, $\{\mathbf{U}_R(-t) : t \in \mathbb{R}\}$ has the same distribution as $\{\mathbf{U}_R(t) : t \in \mathbb{R}\}$. For this reason, I will say that $\{\mathbf{U}_R(t) : t \geq 0\}$ is a **reversible Ornstein–Uhlenbeck process** if its distribution is the same as that of $\{\mathbf{U}_R(t, \mathbf{x}, \boldsymbol{\theta}_+, \boldsymbol{\theta}_-) : t \geq 0\}$ under $\gamma_{\mathbf{0},\mathbf{I}} \times \mathcal{W}^{(N)} \times \mathcal{W}^{(N)}$.

An alternative way to realize a reversible Ornstein–Uhlenbeck process is to start with an $\mathbb{R}^N$-valued Brownian motion $\{\mathbf{B}(t) : t \geq 0\}$ and consider the process $\{e^{-\frac{t}{2}}\mathbf{B}(e^t) : t \in \mathbb{R}\}$. Clearly $\left\{\left(\boldsymbol{\xi}, e^{-\frac{t}{2}}\mathbf{B}(e^t)\right)_{\mathbb{R}^N} : (t, \boldsymbol{\xi}) \in \mathbb{R} \times \mathbb{R}^N\right\}$ is a Gaussian family with covariance given by (8.5.3). It is amusing to observe that, when one uses this realization, the reversibility of the Ornstein–Uhlenbeck process is equivalent to the time inversion invariance (cf. Exercise 4.3.11) of the original Brownian motion.

### § 8.5.2. Ornstein–Uhlenbeck as an Abstract Wiener Space.

So far, my treatment of the Ornstein–Uhlenbeck process has been based on its relationship to Brownian motion. Here I will look at it as an abstract Wiener space.

Begin with the one-sided process $\{\mathbf{U}(t, \mathbf{0}) : t \geq 0\}$. Seeing as this process has the same distribution as $\left\{e^{-\frac{t}{2}}\mathbf{B}(e^t - 1) : t \geq 0\right\}$, it is reasonably clear that the Hilbert space associated with this process should be the space $\mathbf{H}^U(\mathbb{R}^N)$ of functions $\mathbf{h}^U(t) = e^{-\frac{t}{2}}\mathbf{h}(e^t - 1)$, $\mathbf{h} \in \mathbf{H}^1(\mathbb{R}^N)$. Thus, define the map $F^U : \mathbf{H}^1(\mathbb{R}^N) \longrightarrow \mathbf{H}^U(\mathbb{R}^N)$ accordingly, and introduce the Hilbert norm $\|\cdot\|_{\mathbf{H}^U(\mathbb{R}^N)}$ on $\mathbf{H}^U(\mathbb{R}^N)$ that makes $F^U$ into an isometry. Equivalently,

$$\begin{aligned}\|\mathbf{h}^U\|^2_{\mathbf{H}^U(\mathbb{R}^N)} &= \int_{[0,\infty)} \left[\frac{d}{ds}\Big((1+s)^{\frac{1}{2}}\mathbf{h}^U\big(\log(1+s)\big)\Big)\right]^2 ds \\ &= \|\dot{\mathbf{h}}^U\|^2_{L^2([0,\infty);\mathbb{R}^N)} + \left(\dot{\mathbf{h}}^U, \mathbf{h}^U\right)_{L^2([0,\infty);\mathbb{R}^N)} + \tfrac{1}{4}\|\mathbf{h}^U\|^2_{L^2([0,\infty);\mathbb{R}^N)}.\end{aligned}$$

Note that

$$\left(\dot{\mathbf{h}}^U, \mathbf{h}^U\right)_{L^2([0,\infty);\mathbb{R}^N)} = \tfrac{1}{2}\int_{[0,\infty)} \frac{d}{dt}|\mathbf{h}^U(t)|^2\,dt = \tfrac{1}{2}\lim_{t\to\infty}|\mathbf{h}^U(t)|^2 = 0.$$

To check the final equality, observe that it is equivalent to $\lim_{t\to\infty} t^{-\frac{1}{2}}|\mathbf{h}(t)| = 0$ for $\mathbf{h} \in \mathbf{H}^1(\mathbb{R}^N)$. Hence, since $\sup_{t>0} t^{-\frac{1}{2}}|\mathbf{h}(t)| \leq \|\mathbf{h}\|_{\mathbf{H}^1(\mathbb{R}^N)}$ and $\lim_{t\to\infty} t^{-\frac{1}{2}}|\mathbf{h}(t)| = 0$ if $\dot{\mathbf{h}}$ has compact support, the same result is true for all $\mathbf{h} \in \mathbf{H}^1(\mathbb{R}^N)$. In particular,

$$\|\mathbf{h}^U\|_{\mathbf{H}^U(\mathbb{R}^N)} = \sqrt{\|\dot{\mathbf{h}}^U\|^2_{L^2([0,\infty);\mathbb{R}^N)} + \tfrac{1}{4}\|\mathbf{h}^U\|^2_{L^2([0,\infty);\mathbb{R}^N)}}.$$

If we were to follow the prescription in Theorem 8.3.1, we would next complete $\mathbf{H}^U(\mathbb{R}^N)$ with respect to the norm $\sup_{t\geq 0} e^{-\frac{t}{2}}|\mathbf{h}^U(t)|$. However, we already know from (8.5.2) that $\{\mathbf{U}(t,0) : t \geq 0\}$ lives on $\Theta^U(\mathbb{R}^N)$, the space of $\boldsymbol{\theta} \in \Theta(\mathbb{R}^N)$ such that $\lim_{t\to\infty}(\log t)^{-1}|\boldsymbol{\theta}(t)| = 0$ with Banach norm

$$\|\boldsymbol{\theta}\|_{\Theta^U(\mathbb{R}^N)} \equiv \sup_{t\geq 0}\big(\log(e+t)\big)^{-1}|\boldsymbol{\theta}(t)|,$$

and so we will adopt $\Theta^U(\mathbb{R}^N)$ as the Banach space for $\mathbf{H}^U(\mathbb{R}^N)$. Clearly, the dual space $\Theta^U(\mathbb{R}^N)^*$ of $\Theta^U(\mathbb{R}^N)$ can be identified with the space of $\mathbb{R}^N$-valued Borel measures $\boldsymbol{\lambda}$ on $[0,\infty)$ that give 0 mass to $\{\mathbf{0}\}$ and satisfy $\|\boldsymbol{\lambda}\|_{\Lambda^U(\mathbb{R}^N)} \equiv \int_{[0,\infty)} \log(e+t)\,|\boldsymbol{\lambda}|(dt) < \infty$.

THEOREM 8.5.4. *Let $\mathcal{U}_0^{(N)} \in \mathbf{M}_1\big(\Theta^U(\mathbb{R}^N)\big)$ be the distribution of $\{\mathbf{U}(t,\mathbf{0}) : t \geq 0\}$ under $\mathcal{W}^{(N)}$. Then $\big(\mathbf{H}^U(\mathbb{R}^N), \Theta^U(\mathbb{R}^N), \mathcal{U}_0^{(N)}\big)$ is an abstract Wiener space.*

PROOF: Since $C_c^\infty\big((0,\infty);\mathbb{R}^N\big)$ is contained in $\mathbf{H}^U(\mathbb{R}^N)$ and is dense in $\Theta^U(\mathbb{R}^N)$, we know that $\mathbf{H}^U(\mathbb{R}^N)$ is dense in $\Theta^U(\mathbb{R}^N)$. In addition, because $\boldsymbol{\eta}^U(t) = e^{-\frac{t}{2}}\boldsymbol{\eta}(e^t-1)$, where $\boldsymbol{\eta} \in \mathbf{H}^1(\mathbb{R}^N)$, and $\|\boldsymbol{\eta}^U\|_{\mathbf{H}^U(\mathbb{R}^N)} = \|\boldsymbol{\eta}\|_{\mathbf{H}^1(\mathbb{R}^N)}$, $\|\boldsymbol{\eta}^U\|_{\Theta^U(\mathbb{R}^N)} \leq \frac{1}{2}\|\boldsymbol{\eta}^U\|_{\mathbf{H}^U(\mathbb{R}^N)}$ follows from $|\boldsymbol{\eta}(t)| \leq t^{\frac{1}{2}}\|\boldsymbol{\eta}\|_{\mathbf{H}^1(\mathbb{R}^N)}$. Hence, $\mathbf{H}^U(\mathbb{R}^N)$ is continuously embedded in $\Theta^U(\mathbb{R}^N)$.

To complete the proof, remember our earlier calculation of the covariance of $\{\mathbf{U}(t;\mathbf{0}) : t \geq 0\}$, and use it to check that

$$\mathbb{E}^{\mathcal{U}_0^{(N)}}\big[\langle\boldsymbol{\theta},\boldsymbol{\lambda}\rangle^2\big] = \iint\limits_{[0,\infty)^2} u_0(s,t)\,\boldsymbol{\lambda}(ds)\cdot\boldsymbol{\lambda}(dt), \quad \text{where } u_0(s,t) \equiv e^{-\frac{|s-t|}{2}} - e^{-\frac{s+t}{2}}.$$

Hence, what I need to show is that if $\boldsymbol{\lambda} \in \Theta^U(\mathbb{R}^N)^* \longrightarrow \mathbf{h}_{\boldsymbol{\lambda}}^U \in \mathbf{H}^U(\mathbb{R}^N)$ is the map determined by $\langle\mathbf{h}^U,\boldsymbol{\lambda}\rangle = \big(\mathbf{h}^U,\mathbf{h}_{\boldsymbol{\lambda}}^U\big)_{\mathbf{H}^U(\mathbb{R}^N)}$, then

$$\|\mathbf{h}_{\boldsymbol{\lambda}}^U\|_{\mathbf{H}^U(\mathbb{R}^N)}^2 = \iint\limits_{[0,\infty)^2} u_0(s,t)\,\boldsymbol{\lambda}(ds)\cdot\boldsymbol{\lambda}(dt). \tag{8.5.5}$$

In order to do this, we must first know how $\mathbf{h}_{\boldsymbol{\lambda}}^U$ is constructed from $\boldsymbol{\lambda}$. But if (8.5.5) is going to hold, then, by polarization,

$$\begin{aligned}\big(\mathbf{e},\mathbf{h}_{\boldsymbol{\lambda}}^U(\tau)\big)_{\mathbb{R}^N} = \langle\mathbf{h}_{\boldsymbol{\lambda}}^U,\delta_\tau\mathbf{e}\rangle &= \iint\limits_{[0,\infty)^2} u_0(s,t)\,\delta_\tau(ds)\big(\mathbf{e},\boldsymbol{\lambda}(dt)\big)_{\mathbb{R}^N}\\ &= \left(\mathbf{e},\int_{[0,\infty)} u_0(\tau,t)\,\boldsymbol{\lambda}(dt)\right)_{\mathbb{R}^N}.\end{aligned}$$

Thus, one should guess that $\mathbf{h}^U_{\boldsymbol{\lambda}}(\tau) = \int_{[0,\infty)} u_0(\tau,t)\,\boldsymbol{\lambda}(dt)$ and must check that, with this choice, $\mathbf{h}^U_{\boldsymbol{\lambda}} \in \mathbf{H}^U(\mathbb{R}^N)$, (8.5.5) holds, and, for all $\mathbf{h}^U \in \mathbf{H}^U(\mathbb{R}^N)$, $\langle \mathbf{h}^U, \boldsymbol{\lambda}\rangle = \big(\mathbf{h}^U, \mathbf{h}^U_{\boldsymbol{\lambda}}\big)_{\mathbf{H}^U(\mathbb{R}^N)}$.

The key to proving all these is the equality

$$(*) \qquad \int_{[0,\infty)} \dot{\mathbf{h}}^U(\tau)\partial_\tau u_0(\tau,t)\,d\tau + \tfrac{1}{4}\int_{[0,\infty)} \mathbf{h}^U(\tau) u_0(\tau,t)\,d\tau = \mathbf{h}^U(t),$$

which is an elementary application of integration by parts. Applying (*) with $N=1$ to $h^U = u_0(\,\cdot\,,s)$, we see that

$$\int_{[0,\infty)} \Big(\partial_\tau u_0(s,\tau)\partial_\tau u_0(t,\tau) + \tfrac{1}{4}u_0(s,\tau)u_0(t,\tau)\Big)\,d\tau = u_0(s,t),$$

from which it follows easily both that $\mathbf{h}^U_{\boldsymbol{\lambda}} \in \mathbf{H}^U(\mathbb{R}^N)$ and that (8.5.5) holds. In addition, if $\mathbf{h}^U \in \mathbf{H}^U(\mathbb{R}^N)$, then $\langle \mathbf{h}^U, \boldsymbol{\lambda}\rangle = \big(\mathbf{h}^U, \mathbf{h}^U_{\boldsymbol{\lambda}}\big)_{\mathbf{H}^U(\mathbb{R}^N)}$ follows from (*) after one integrates both sides of the preceding with respect to $\boldsymbol{\lambda}(dt)$. □

I turn next to the reversible case. By the considerations in § 8.4.1, we know that the distribution $\mathcal{U}_R^{(N)}$ of $\{\mathbf{U}_R(t) : t \ge 0\}$ under $\gamma_{0,1} \times \mathcal{W}^{(N)} \times \mathcal{W}^{(N)}$ is a Borel measure on the space Banach space $\Theta^U(\mathbb{R};\mathbb{R}^N)$ of continuous $\boldsymbol{\theta} : \mathbb{R} \longrightarrow \mathbb{R}^N$ such that $\lim_{|t|\to\infty} (\log t)^{-1}|\boldsymbol{\theta}(t)| = 0$ with norm

$$\|\boldsymbol{\theta}\|_{\Theta^U(\mathbb{R};\mathbb{R}^N)} \equiv \sup_{t\in\mathbb{R}} \big(\log(e+|t|)\big)^{-1}|\boldsymbol{\theta}(t)| < \infty.$$

Furthermore, it should be clear that one can identify $\Theta^U(\mathbb{R};\mathbb{R}^N)^*$ with the space of $\mathbb{R}^N$-valued Borel measures $\boldsymbol{\lambda}$ on $\mathbb{R}$ satisfying

$$\|\boldsymbol{\lambda}\|_{\Lambda^U(\mathbb{R};\mathbb{R}^N)} \equiv \int_{\mathbb{R}} \log(e+|t|)\,|\boldsymbol{\lambda}|(dt) < \infty.$$

THEOREM 8.5.6. *Take $\mathbf{H}^1(\mathbb{R};\mathbb{R}^N)$ to be the separable Hilbert space of absolutely continuous $\mathbf{h} : \mathbb{R} \longrightarrow \mathbb{R}^N$ satisfying*

$$\|\mathbf{h}\|_{\mathbf{H}^1(\mathbb{R};\mathbb{R}^N)} \equiv \sqrt{\|\dot{\mathbf{h}}\|^2_{L^2(\mathbb{R}:\mathbb{R}^N)} + \tfrac{1}{4}\|\mathbf{h}\|^2_{L^2(\mathbb{R}:\mathbb{R}^N)}} < \infty.$$

*Then $\big(\mathbf{H}^1(\mathbb{R};\mathbb{R}^N), \Theta^U(\mathbb{R};\mathbb{R}^N), \mathcal{U}_R^{(N)}\big)$ is an abstract Wiener space.*

PROOF: Set $u(s,t) \equiv e^{-\frac{|s-t|}{2}}$, and let $\boldsymbol{\lambda} \in \Lambda^U(\mathbb{R};\mathbb{R}^N)$. By the same reasoning as I used in the preceding proof,

$$\langle \mathbf{h}, \boldsymbol{\lambda}\rangle = \big(\mathbf{h}, \mathbf{h}_{\boldsymbol{\lambda}}\big)_{\mathbf{H}^1(\mathbb{R};\mathbb{R}^N)}$$

and

$$\|\mathbf{h}_{\boldsymbol{\lambda}}\|^2_{\mathbf{H}^1(\mathbb{R};\mathbb{R}^N)} = \iint\limits_{\mathbb{R}\times\mathbb{R}} u(s,t)\,\boldsymbol{\lambda}(ds)\cdot\boldsymbol{\lambda}(dt)$$

when $\mathbf{h}_{\boldsymbol{\lambda}}(\tau) = \int_{\mathbb{R}} u(\tau,t)\,\boldsymbol{\lambda}(dt)$. Hence, since $\{(\boldsymbol{\xi},\boldsymbol{\theta}(t))_{\mathbb{R}^N} : t \geq 0 \;\&\; \boldsymbol{\xi} \in \mathbb{R}^N\}$ spans a Gaussian family in $L^2(\mathcal{U}_R^{(N)};\mathbb{R})$ and $u(s,t)\mathbf{I} = \mathbb{E}^{\mathcal{U}_R^{(N)}}[\boldsymbol{\theta}(s)\otimes\boldsymbol{\theta}(t)]$, the proof is complete. □

**§ 8.5.3. Higher Dimensional Free Fields.** Thinking a la Feynman, Theorem 8.5.6 is saying that $\mathcal{U}_R^{(N)}$ wants to be the measure on $H^1(\mathbb{R};\mathbb{R})$ given by

$$\frac{1}{(\sqrt{2\pi})^{\dim(\mathbf{H}^1(\mathbb{R};\mathbb{R}^N))}}\exp\left[-\frac{1}{2}\int_{\mathbb{R}}\left(|\dot{\mathbf{h}}(t)|^2 + \tfrac{1}{4}|\mathbf{h}(t)|^2\right)dt\right]\lambda_{\mathbf{H}^1(\mathbb{R};\mathbb{R}^N)}(d\mathbf{h}),$$

where $\lambda_{\mathbf{H}^1(\mathbb{R};\mathbb{R}^N)}$ is the Lebesgue measure on $\mathbf{H}^1(\mathbb{R};\mathbb{R}^N)$.

I am now going to look at the analogous situation when $N = 1$ but the parameter set $\mathbb{R}$ is replaced by $\mathbb{R}^\nu$ for some $\nu \geq 2$. That is, I want to look at the measure that Feynman would have written as

$$\frac{1}{(\sqrt{2\pi})^{\dim(H^1(\mathbb{R}^\nu;\mathbb{R}))}}\exp\left[-\frac{1}{2}\int_{\mathbb{R}^\nu}\left(|\nabla h(\mathbf{x}))|^2 + \tfrac{1}{4}|h(\mathbf{x})|^2\right)d\mathbf{x}\right]\lambda_{H^1(\mathbb{R}^\nu;\mathbb{R})}(dh),$$

where $H^1(\mathbb{R}^\nu;\mathbb{R})$ is the separable Hilbert space obtained by completing the Schwartz test function space $\mathscr{S}(\mathbb{R}^\nu;\mathbb{R})$ with respect to the Hilbert norm

$$\|h\|_{H^1(\mathbb{R}^\nu;\mathbb{R})} \equiv \sqrt{\|\nabla h\|^2_{L^2(\mathbb{R}^\nu;\mathbb{R})} + \tfrac{1}{4}\|h\|^2_{L^2(\mathbb{R}^\nu;\mathbb{R})}}.$$

When $\nu = 1$ this is exactly the Hilbert space $H^1(\mathbb{R};\mathbb{R})$ described in Theorem 8.5.6 for $N = 1$. When $\nu \geq 2$, generic elements of $H^1(\mathbb{R}^\nu;\mathbb{R})$ are better than generic elements of $L^2(\mathbb{R}^\nu;\mathbb{R})$ but are not enough better to be continuous. In fact, they are not even well-defined pointwise, and matters get worse as $\nu$ gets larger. Thus, although Feynman's representation is already questionable when $\nu = 1$, its interpretation when $\nu \geq 2$ is even more fraught with difficulties. As we will see, these difficulties are reflected mathematically by the fact that, in order to construct an abstract Wiener space for $H^1(\mathbb{R}^\nu;\mathbb{R})$ when $\nu \geq 2$, we will have to resort to Banach spaces whose elements are generalized functions (i.e., distributions in the sense of L. Schwartz).[2]

[2] The need to deal with generalized functions is the primary source of the difficulties that mathematicians have when they attempt to construct non-trivial quantum fields. Without going into any details, suffice it to say that in order to construct interacting (i.e., non-Gaussian) fields, one has to take non-linear functions of a Gaussian field. However, if the Gaussian field is distribution valued, it is not at all clear how to apply a non-linear function to it.

The approach that I will adopt is based on the following subterfuge. The space $H^1(\mathbb{R}^\nu;\mathbb{R})$ is one of a continuously graded family of spaces known as **Sobolev spaces**. Sobolev spaces are graded according to the number of derivatives "better or worse" than $L^2(\mathbb{R}^\nu;\mathbb{R})$ their elements are. To be more precise, for each $s \in \mathbb{R}$, define the **Bessel operator** $B^s$ on $\mathscr{S}(\mathbb{R}^\nu;\mathbb{C})$ so that

$$\widehat{B^s\varphi}(\boldsymbol{\xi}) = \left(\tfrac{1}{4} + |\boldsymbol{\xi}|^2\right)^{-\frac{s}{2}}\hat{\varphi}(\boldsymbol{\xi}).$$

When $s = -2m$, it is clear that $B^s = \left(\frac{1}{4} - \Delta\right)^m$, and so, in general, it is reasonable to think of $B^s$ as an operator that, depending on whether $s \le 0$ or $s \ge 0$, involves taking or restoring derivatives of order $|s|$. In particular, $\|\varphi\|_{H^1(\mathbb{R}^\nu;\mathbb{R})} = \|B^{-1}\varphi\|_{L^2(\mathbb{R}^\nu;\mathbb{R})}$ for $\varphi \in \mathscr{S}(\mathbb{R}^\nu;\mathbb{R})$. More generally, define the Sobolev space $H^s(\mathbb{R}^\nu;\mathbb{R})$ to be the separable Hilbert space obtained by completing $\mathscr{S}(\mathbb{R}^\nu;\mathbb{R})$ with respect to

$$\|h\|_{H^s(\mathbb{R}^\nu;\mathbb{R})} \equiv \|B^{-s}h\|_{L^2(\mathbb{R}^\nu;\mathbb{R})} = \sqrt{\frac{1}{(2\pi)^\nu}\int_{\mathbb{R}^\nu}\left(\tfrac{1}{4} + |\boldsymbol{\xi}|^2\right)^s|\hat{h}(\boldsymbol{\xi})|^2\,d\boldsymbol{\xi}}.$$

Obviously, $H^0(\mathbb{R}^\nu;\mathbb{R})$ is just $L^2(\mathbb{R}^\nu;\mathbb{R})$. When $s > 0$, $H^s(\mathbb{R}^\nu;\mathbb{R})$ is a subspace of $L^2(\mathbb{R}^\nu;\mathbb{R})$, and the quality of its elements will improve as $s$ gets larger. However, when $s < 0$, some elements of $H^s(\mathbb{R}^\nu;\mathbb{R})$ will be strictly worse than elements of $L^2(\mathbb{R}^\nu;\mathbb{R})$, and their quality will deteriorate as $s$ becomes more negative. Nonetheless, for every $s \in \mathbb{R}$, $H^s(\mathbb{R}^\nu;\mathbb{R}) \subseteq \mathscr{S}'(\mathbb{R}^\nu;\mathbb{R})$, where $\mathscr{S}'(\mathbb{R}^\nu;\mathbb{R})$, whose elements are called real-valued **tempered distributions**, is the dual space of $\mathscr{S}(\mathbb{R}^\nu;\mathbb{R})$. In fact, with a little effort, one can check that an alternative description of $H^s(\mathbb{R}^\nu;\mathbb{R})$ is as the subspace of $u \in \mathscr{S}'(\mathbb{R}^\nu;\mathbb{R})$ with the property that $B^{-s}u \in L^2(\mathbb{R}^\nu;\mathbb{R})$. Equivalently, $H^s(\mathbb{R}^\nu;\mathbb{R})$ is the isometric image in $\mathscr{S}'(\mathbb{R}^\nu;\mathbb{R})$ of $L^2(\mathbb{R}^\nu;\mathbb{R})$ under the map $B^s$, and, more generally, $H^{s_2}(\mathbb{R}^\nu;\mathbb{R})$ is the isometric image of $H^{s_1}(\mathbb{R}^\nu;\mathbb{R})$ under $B^{s_2-s_1}$. Thus, by Theorem 8.3.1, once we understand the abstract Wiener spaces for any one of the spaces $H^s(\mathbb{R}^\nu;\mathbb{R})$, understanding the abstract Wiener spaces for any of the others comes down to understanding the action of the Bessel operators, a task that, depending on what one wants to know, can be highly non-trivial.

LEMMA 8.5.7. *The space $H^{\frac{\nu+1}{2}}(\mathbb{R}^\nu;\mathbb{R})$ is continuously embedded as a dense subspace of the separable Banach space $C_0(\mathbb{R}^\nu;\mathbb{R})$ whose elements are continuous functions that tend to 0 at infinity and whose norm is the uniform norm. Moreover, given a totally finite, signed Borel measure $\lambda$ on $\mathbb{R}^\nu$, the function*

$$h_\lambda(\mathbf{x}) \equiv K_\nu\int_{\mathbb{R}^\nu} e^{-\frac{|\mathbf{x}-\mathbf{y}|}{2}}\,\lambda(d\mathbf{y}), \quad \text{with } K_\nu \equiv \frac{\pi^{\frac{1-\nu}{2}}}{\Gamma\left(\frac{\nu+1}{2}\right)},$$

*is an element of $H^{\frac{\nu+1}{2}}(\mathbb{R}^\nu;\mathbb{R})$,*

$$\|h_\lambda\|^2_{H^{\frac{\nu+1}{2}}(\mathbb{R}^\nu;\mathbb{R})} = K_\nu\iint_{\mathbb{R}^\nu\times\mathbb{R}^\nu} e^{-\frac{|\mathbf{x}-\mathbf{y}|}{2}}\,\lambda(d\mathbf{x})\lambda(d\mathbf{y}),$$

*and*

$$\langle h, \lambda\rangle = \left(h, h_\lambda\right)_{H^{\frac{\nu+1}{2}}(\mathbb{R}^\nu;\mathbb{R})} \qquad \text{for each } h \in H^{\frac{\nu+1}{2}}(\mathbb{R}^\nu;\mathbb{R}).$$

PROOF: To prove the initial assertion, use the Fourier inversion formula to write

$$h(\mathbf{x}) = (2\pi)^{-\nu} \int_{\mathbb{R}^\nu} e^{-\sqrt{-1}(\mathbf{x},\boldsymbol{\xi})_{\mathbb{R}^\nu}} \hat{h}(\boldsymbol{\xi})\, d\boldsymbol{\xi}$$

for $h \in \mathscr{S}(\mathbb{R}^\nu;\mathbb{R})$, and derive from this the estimate

$$\|h\|_{\mathrm{u}} \le (2\pi)^{-\frac{\nu}{2}} \left(\int_{\mathbb{R}^\nu} \left(\tfrac{1}{4} + |\boldsymbol{\xi}|^2\right)^{-\frac{\nu+1}{2}} d\boldsymbol{\xi}\right)^{\frac{1}{2}} \|h\|_{H^{\frac{\nu+1}{2}}(\mathbb{R}^\nu;\mathbb{R})}.$$

Hence, since $H^{\frac{\nu+1}{2}}(\mathbb{R}^\nu;\mathbb{R})$ is the completion of $\mathscr{S}(\mathbb{R}^\nu;\mathbb{R})$ with respect to the norm $\|\cdot\|_{H^{\frac{\nu+1}{2}}(\mathbb{R}^\nu;\mathbb{R})}$, it is clear that $H^{\frac{\nu+1}{2}}(\mathbb{R}^\nu;\mathbb{R})$ is continuously embedded in $C_0(\mathbb{R}^\nu;\mathbb{R})$. In addition, since $\mathscr{S}(\mathbb{R}^\nu;\mathbb{R})$ is dense in $C_0(\mathbb{R}^\nu;\mathbb{R})$, $H^{\frac{\nu+1}{2}}(\mathbb{R}^\nu;\mathbb{R})$ is also.

To carry out the next step, let $\lambda$ be given, and observe that the Fourier transform of $B^{\nu+1}\lambda$ is $\left(\frac{1}{4} + |\boldsymbol{\xi}|^2\right)^{-\frac{\nu+1}{2}} \hat{\lambda}(\boldsymbol{\xi})$ and therefore that

$$\begin{aligned} B^{\nu+1}\lambda(\mathbf{x}) &= \frac{1}{(2\pi)^\nu} \int_{\mathbb{R}^\nu} \frac{e^{-\sqrt{-1}(\mathbf{x},\boldsymbol{\xi})_{\mathbb{R}^\nu}} \hat{\lambda}(\boldsymbol{\xi})}{\left(\frac{1}{4} + |\boldsymbol{\xi}|^2\right)^{\frac{\nu+1}{2}}}\, d\boldsymbol{\xi} \\ &= \frac{1}{(2\pi)^\nu} \int_{\mathbb{R}^\nu} \left( \int_{\mathbb{R}^\nu} \frac{e^{\sqrt{-1}(\mathbf{y}-\mathbf{x},\boldsymbol{\xi})_{\mathbb{R}^\nu}}}{\left(\frac{1}{4} + |\boldsymbol{\xi}|^2\right)^{\frac{\nu+1}{2}}}\, d\boldsymbol{\xi} \right) \lambda(d\mathbf{y}). \end{aligned}$$

Now use (3.3.19) (with $N = \nu$ and $t = \frac{1}{2}$) to see that

$$\frac{1}{(2\pi)^\nu} \int_{\mathbb{R}^\nu} \frac{e^{\sqrt{-1}(\mathbf{y}-\mathbf{x},\boldsymbol{\xi})_{\mathbb{R}^\nu}}}{\left(\frac{1}{4} + |\boldsymbol{\xi}|^2\right)^{\frac{\nu+1}{2}}}\, d\boldsymbol{\xi} = K_\nu e^{-\frac{|\mathbf{y}-\mathbf{x}|}{2}},$$

and thereby arrive at $h_\lambda = B^{\nu+1}\lambda$. In particular, this shows that

$$\|h_\lambda\|^2_{H^{\frac{\nu+1}{2}}(\mathbb{R}^\nu;\mathbb{R})} = \frac{1}{(2\pi)^\nu} \int_{\mathbb{R}^\nu} \frac{|\hat{\lambda}(\boldsymbol{\xi})|^2}{\left(\frac{1}{4} + |\boldsymbol{\xi}|^2\right)^{\frac{\nu+1}{2}}}\, d\boldsymbol{\xi} < \infty.$$

Now let $h \in \mathscr{S}(\mathbb{R}^\nu;\mathbb{R})$, and use the preceding to justify

$$\langle h, \lambda\rangle = \langle B^{-\frac{\nu+1}{2}} h, B^{-\frac{\nu+1}{2}} B^{\nu+1}\lambda\rangle = \left(h, h_\lambda\right)_{H^{\frac{\nu+1}{2}}(\mathbb{R}^\nu;\mathbb{R})}.$$

Since both sides are continuous with respect to convergence in $H^{\frac{\nu+1}{2}}(\mathbb{R}^\nu;\mathbb{R})$, we have now proved that $\langle h, \lambda\rangle = \left(h, h_\lambda\right)_{H^{\frac{\nu+1}{2}}(\mathbb{R}^\nu;\mathbb{R})}$ for all $h \in H^{\frac{\nu+1}{2}}(\mathbb{R}^\nu;\mathbb{R})$. In particular,

$$\|h_\lambda\|^2_{H^{\frac{\nu+1}{2}}(\mathbb{R}^\nu;\mathbb{R})} = \langle h_\lambda, \lambda\rangle = K_\nu \iint_{\mathbb{R}^\nu\times\mathbb{R}^\nu} e^{-\frac{|\mathbf{y}-\mathbf{x}|}{2}} \lambda(d\mathbf{x})\lambda(d\mathbf{y}). \qquad \square$$

THEOREM 8.5.8. *Let $\Theta^{\frac{\nu+1}{2}}(\mathbb{R}^\nu;\mathbb{R})$ be the space of continuous $\theta:\mathbb{R}^\nu \longrightarrow \mathbb{R}$ satisfying $\lim_{|\mathbf{x}|\to\infty}\big(\log(e+|\mathbf{x}|)\big)^{-1}|\theta(\mathbf{x})| = 0$, and turn $\Theta^{\frac{\nu+1}{2}}(\mathbb{R}^\nu;\mathbb{R})$ into a separable Banach space with norm $\|\theta\|_{\Theta^{\frac{\nu+1}{2}}(\mathbb{R}^\nu;\mathbb{R})} = \sup_{\mathbf{x}\in\mathbb{R}^N}\big(\log(e+|\mathbf{x}|)\big)^{-1}|\theta(\mathbf{x})|$. Then $H^{\frac{\nu+1}{2}}(\mathbb{R}^\nu;\mathbb{R})$ is continuously embedded as a dense subspace of $\Theta^{\frac{\nu+1}{2}}(\mathbb{R}^\nu;\mathbb{R})$, and there is a $\mathcal{W}_{H^{\frac{\nu+1}{2}}(\mathbb{R}^\nu;\mathbb{R})} \in \mathbf{M}_1\big(\Theta^{\frac{\nu+1}{2}}(\mathbb{R}^\nu;\mathbb{R})\big)$ such that*

$$\Big(H^{\frac{\nu+1}{2}}(\mathbb{R}^\nu;\mathbb{R}),\Theta^{\frac{\nu+1}{2}}(\mathbb{R}^\nu;\mathbb{R}),\mathcal{W}_{H^{\frac{\nu+1}{2}}(\mathbb{R}^\nu;\mathbb{R})}\Big)$$

*is an abstract Wiener space. Moreover, for each $\alpha\in\big(0,\frac12\big)$, $\mathcal{W}_{H^{\frac{\nu+1}{2}}(\mathbb{R}^\nu;\mathbb{R})}$-almost every $\theta$ is Hölder continuous of order $\alpha$ and, for each $\alpha>\frac12$, $\mathcal{W}_{H^{\frac{\nu+1}{2}}(\mathbb{R}^\nu;\mathbb{R})}$-almost no $\theta$ is anywhere Hölder continuous of order $\alpha$.*

PROOF: The initial part of the first assertion follows from the first part of Lemma 8.5.7 plus the essentially trivial fact that $C_0(\mathbb{R}^\nu;\mathbb{R})$ is continuously embedded as a dense subspace of $\Theta^{\frac{\nu+1}{2}}(\mathbb{R}^\nu;\mathbb{R})$. Further, by the second part of that same lemma combined with Theorem 8.3.3, we will have proved the second part of the first assertion here once we show that, when $\{h_m : m\ge 0\}$ is an orthonormal basis in $H^{\frac{\nu+1}{2}}(\mathbb{R}^\nu;\mathbb{R})$, the Wiener series $\sum_{m=0}^\infty \omega_m h_m$ converges in $\Theta^{\frac{\nu+1}{2}}(\mathbb{R}^\nu;\mathbb{R})$ for $\gamma_{0,1}^{\mathbb{N}}$-almost every $\boldsymbol{\omega}=(\omega_0,\dots,\omega_m,\dots)\in\mathbb{R}^{\mathbb{N}}$. Thus, set $S_n(\boldsymbol{\omega})=\sum_{m=0}^n \omega_m h_m$ for $n\ge 1$. More or less mimicking the steps outlined in Exercise 8.3.21, I will begin by showing that, for each $\alpha\in\big(0,\frac12\big)$ and $R\in[1,\infty)$,

$$(*)\qquad \sup_{\mathbf{z}\in\mathbb{R}^\nu}\mathbb{E}^{\gamma_{0,1}^{\mathbb{N}}}\left[\sup_{n\ge 0}\ \sup_{\substack{\mathbf{x},\mathbf{y}\in Q(\mathbf{z},R)\\ \mathbf{x}\neq\mathbf{y}}}\frac{|S_n(\mathbf{y})-S_n(\mathbf{x})|}{|\mathbf{y}-\mathbf{x}|^\alpha}\right]<\infty,$$

where $Q(\mathbf{z},R)=\mathbf{z}+[-R,R)^\nu$. Indeed, by the argument given in that exercise combined with the higher dimensional analog of Kolmogorov's continuity criterion in Exercise 4.3.18, (*) will follow once we show that

$$\mathbb{E}^{\gamma_{0,1}^{\mathbb{N}}}\big[|S_n(\mathbf{y})-S_n(\mathbf{x})|^2\big]\le C|\mathbf{y}-\mathbf{x}|,\quad \mathbf{x},\mathbf{y}\in\mathbb{R}^\nu,$$

for some $C<\infty$. To this end, set $\lambda=\delta_{\mathbf{y}}-\delta_{\mathbf{x}}$, and apply Lemma 8.5.7 to check

$$\begin{aligned}\mathbb{E}^{\gamma_{0,1}^{\mathbb{N}}}\big[|S_n(\mathbf{y})-S_n(\mathbf{x})|^2\big] &= \sum_{m=0}^n\big(h_m,h_\lambda\big)^2_{H^{\frac{\nu+1}{2}}(\mathbb{R}^\nu;\mathbb{R})}\\ &\le \|h_\lambda\|^2_{H^{\frac{\nu+1}{2}}(\mathbb{R}^\nu;\mathbb{R})} = 2K_\nu\big(1-e^{-\frac{|\mathbf{y}-\mathbf{x}|}{2}}\big).\end{aligned}$$

Knowing (*), it becomes an easy matter to see that there exists a measurable $S:\mathbb{R}^\nu\times\mathbb{R}^{\mathbb{N}}\longrightarrow\mathbb{R}$ such that $\mathbf{x}\rightsquigarrow S(\mathbf{x},\boldsymbol{\omega})$ is continuous of each $\boldsymbol{\omega}$ and

$S_n(\cdot,\boldsymbol{\omega}) \longrightarrow S(\cdot,\boldsymbol{\omega})$ uniformly on compacts for $\gamma_{0,1}^{\mathbb{N}}$-almost every $\boldsymbol{\omega} \in \mathbb{R}^{\mathbb{N}}$. In fact, because of (*), it suffices to check that $\lim_{n\to\infty} S_n(\mathbf{x})$ exists $\gamma_{0,1}^{\mathbb{N}}$-almost surely for each $\mathbf{x} \in \mathbb{R}^\nu$, and this follows immediately from Theorem 1.4.2 plus

$$\sum_{m=0}^{\infty} \operatorname{Var}\big(\omega_m h_m(\mathbf{x})\big) = \sum_{m=0}^{\infty} \big(h_m, h_{\delta_{\mathbf{x}}}\big)^2_{H^{\frac{\nu+1}{2}}(\mathbb{R}^\nu;\mathbb{R})} = \|h_{\delta_{\mathbf{x}}}\|^2_{H^{\frac{\nu+1}{2}}(\mathbb{R}^\nu;\mathbb{R})} = K_\nu.$$

Furthermore, again from (*), we know that, $\gamma_{0,1}^{\mathbb{N}}$-almost every $\boldsymbol{\omega}$, $\mathbf{x} \rightsquigarrow S(\mathbf{x},\boldsymbol{\omega})$ is $\alpha$-Hölder continuous so long as $\alpha \in \left(0,\frac{1}{2}\right)$.

I must still check that, $\gamma_{0,1}^{\mathbb{N}}$-almost surely, the convergence of $S_n(\cdot,\boldsymbol{\omega})$ to $S(\cdot,\boldsymbol{\omega})$ is taking place in $\Theta^{\frac{\nu+1}{2}}(\mathbb{R}^\nu;\mathbb{R})$, and, in view of the fact that we already know that, $\gamma_{0,1}^{\mathbb{N}}$-almost surely, it is taking place uniformly on compacts, this reduces to showing that

$$\lim_{|\mathbf{x}|\to\infty} \big(\log(e+|\mathbf{x}|)\big)^{-1} \sup_{n\ge 0} |S_n(\mathbf{x})| \longrightarrow 0 \quad \gamma_{0,1}^{\mathbb{N}}\text{-almost surely.}$$

For this purpose, observe that (*) says that

$$\sup_{\mathbf{z}\in\mathbb{R}^\nu} \mathbb{E}^{\gamma_{0,1}^{\mathbb{N}}}\left[\sup_{n\ge 0} \|S_n\|_{\mathrm{u},Q(\mathbf{z},1)}\right] < \infty,$$

where $\|\cdot\|_{\mathrm{u},C}$ denotes the uniform norm over a set $C \subseteq \mathbb{R}^\nu$. At this point, I would like to apply Fernique's Theorem (Theorem 8.2.1) to the Banach space $\ell^\infty\big(\mathbb{N}; C_{\mathrm{b}}(Q(\mathbf{z},1);\mathbb{R})\big)$ and thereby conclude that there exists an $\alpha > 0$ such that

$$(**) \qquad B \equiv \sup_{\mathbf{z}\in\mathbb{R}^\nu} \mathbb{E}^{\gamma_{0,1}^{\mathbb{N}}}\left[\exp\left(\alpha \sup_{n\ge 0} \|S_n\|^2_{\mathrm{u},Q(\mathbf{z},1)}\right)\right] < \infty.$$

However, $\ell^\infty\big(\mathbb{N}; C_{\mathrm{b}}(Q(\mathbf{z},1);\mathbb{R})\big)$ is not separable. Nonetheless, there are two ways to get around this technicality. The first is to observe that the only place separability was used in the proof of Fernique's Theorem was at the beginning, where I used it to guarantee that $\mathcal{B}_E$ is generated by the maps $x \rightsquigarrow \langle x, x^*\rangle$ as $x^*$ runs over $E^*$ and therefore that the distribution of $X$ is determined by the distribution of $\{\langle X, x^*\rangle : x^* \in E^*\}$. But, even though $\ell^\infty\big(\mathbb{N}; C_{\mathrm{b}}(Q(\mathbf{z},1);\mathbb{R})\big)$ is not separable, one can easily check that it nevertheless possesses this property. The second way to deal with the problem is to apply his theorem to $\ell^\infty\big(\{0,\dots,N\}; C_{\mathrm{b}}(Q(\mathbf{z},1);\mathbb{R})\big)$, which is separable, and to note that the resulting estimate can be made uniform in $N \in \mathbb{N}$. Either way, one arrives at (**).

Now set $\psi(t) = e^{\alpha t^2} - 1$ for $t \ge 0$. Then $\psi^{-1}(s) = \sqrt{\alpha^{-1}\log(1+s)}$, and

$$\begin{aligned}
\sup_{n\ge 0} \|S_n\|_{\mathrm{u},Q(\mathbf{0},M)} &= \max\left\{\sup_{n\ge 0} \|S_n\|_{\mathrm{u},Q(\mathbf{m},1)} : \mathbf{m} \in Q(\mathbf{0},M)\cap\mathbb{Z}^\nu\right\} \\
&\le \psi^{-1}\left(\sum_{\mathbf{m}\in Q(\mathbf{0},M)\cap\mathbb{Z}^\nu} \psi\left(\sup_{n\ge 0} \|S_n\|_{\mathrm{u},Q(\mathbf{m},1)}\right)\right).
\end{aligned}$$

Thus, because $\psi^{-1}$ is concave, Jensen's Inequality applies and yields

$$\mathbb{E}^{\gamma_{0,1}^{\mathbb{N}}}\left[\sup_{n\geq 0}\|S_n\|_{\mathrm{u},Q(\mathbf{0},M)}\right] \leq \psi^{-1}\big((2M)^\nu B\big),$$

and therefore

$$\mathbb{E}^{\gamma_{0,1}^{\mathbb{N}}}\left[\sup_{|\mathbf{x}|\geq R}\sup_{n\geq 0}\frac{S_n(\mathbf{x})}{\log(e+|\mathbf{x}|)}\right] \leq \sum_{m\geq(\log R)^{\frac{1}{4}}} \frac{\mathbb{E}^{\gamma_{0,1}^{\mathbb{N}}}\left[\sup_{n\geq 0}\|S_n\|_{\mathrm{u},Q(\mathbf{0},e^{m^4})}\right]}{\log(e+e^{(m-1)^4})}$$

$$\leq \sum_{m\geq(\log R)^{\frac{1}{4}}} \frac{\sqrt{\log(1+2^\nu e^{\nu(m+1)^4}B)}}{\sqrt{\alpha}\log(e+e^{(m-1)^4})} \longrightarrow 0 \quad \text{as } R\to\infty.$$

To complete the proof, I must show that, for any $\alpha>\frac{1}{2}$, $\mathcal{W}_{H^{\frac{\nu+1}{2}}(\mathbb{R}^\nu;\mathbb{R})}$-almost no $\theta$ is anywhere Hölder continuous of order $\alpha$, and for this purpose I will proceed as in the proof of Theorem 4.3.4. Because the $\{\theta(\mathbf{x}+\mathbf{y}) : \mathbf{x}\in\mathbb{R}^\nu\}$ has the same $\mathcal{W}_{H^{\frac{\nu+1}{2}}(\mathbb{R}^\nu;\mathbb{R})}$-distribution for all $\mathbf{y}$, it suffices for me to show that, $\mathcal{W}_{H^{\frac{\nu+1}{2}}(\mathbb{R}^\nu;\mathbb{R})}$-almost surely, there is no $\mathbf{x}\in Q(\mathbf{0},1)$ at which $\theta$ is Hölder continuous of order $\alpha>\frac{1}{2}$. Now suppose that $\alpha\in\left(\frac{1}{2},1\right)$, and observe that, for any $L\in\mathbb{Z}^+$ and $\mathbf{e}\in\mathbb{S}^{\nu-1}$, the set $H(\alpha)$ of $\theta$'s that are $\alpha$-Hölder continuous at some $\mathbf{x}\in Q(\mathbf{0},1)$ is contained in

$$\bigcup_{M=1}^\infty\bigcap_{n=1}^\infty\bigcup_{\mathbf{m}\in Q(\mathbf{0},n)\cap\mathbb{Z}^\nu}\bigcap_{\ell=1}^L\left\{\theta : \left|\theta\left(\tfrac{\mathbf{m}+\ell\mathbf{e}}{n}\right)-\theta\left(\tfrac{\mathbf{m}+(\ell-1)\mathbf{e}}{n}\right)\right|\leq\tfrac{M}{n^\alpha}\right\}.$$

Hence, again using translation invariance, we see that we need only show that there is an $L\in\mathbb{Z}^+$ such that, for each $M\in\mathbb{Z}^+$,

$$n^\nu\mathcal{W}_{H^{\frac{\nu+1}{2}}(\mathbb{R}^\nu;\mathbb{R})}\left(\left\{\theta : \left|\theta\left(\tfrac{\ell\mathbf{e}}{n}\right)-\theta\left(\tfrac{(\ell-1)\mathbf{e}}{n}\right)\right|\leq\tfrac{M}{n^\alpha},\ 1\leq\ell\leq L\right\}\right)$$

tends to 0 as $n\to\infty$. To this end, set $U(t,\theta)=K_\nu^{-\frac{1}{2}}\theta(t\mathbf{e})$, and observe that the $\mathcal{W}_{H^{\frac{\nu+1}{2}}(\mathbb{R}^\nu;\mathbb{R})}$-distribution of $\{U(t) : t\geq 0\}$ is that of an $\mathbb{R}$-valued ancient Ornstein–Uhlenbeck process. Thus, what I have to estimate is

$$\mathbb{P}\left(\left|e^{-\frac{\ell}{2n}}B\left(e^{\frac{\ell}{n}}\right)-e^{-\frac{\ell-1}{2n}}B\left(e^{\frac{\ell-1}{n}}\right)\right|\leq\tfrac{M}{n^\alpha},\ 1\leq\ell\leq L\right),$$

where $\big(B(t),\mathcal{F}_t,\mathbb{P}\big)$ is an $\mathbb{R}$-valued Brownian motion. But clearly this probability is dominated by the sum of

$$\mathbb{P}\left(\left|B\left(e^{\frac{\ell}{n}}\right)-B\left(e^{\frac{\ell-1}{n}}\right)\right|\leq\tfrac{Me^{\frac{\ell}{2n}}}{2n^\alpha},\ 1\leq\ell\leq L\right)$$

and

$$\mathbb{P}\Big(\exists 1 \le \ell \le L \ \ (1-e^{-\frac{1}{2n}})\big|B(e^{\frac{\ell-1}{n}})\big| \ge \frac{Me^{\frac{\ell}{2n}}}{2n^\alpha}\Big).$$

The second of these is easily dominated by $2Le^{-\frac{M^2n^{2(1-\alpha)}}{8}}$, which, since $\alpha < 1$, means that it causes no problems. As for the first, one can use the independence of Brownian increments and Brownian scaling to dominate it by the $L$th power of $\mathbb{P}\big(\big|B(1)-B(e^{-\frac{1}{n}})\big| \le M(2n^\alpha)^{-1}\big)$. Hence, I can take any $L$ such that $(\alpha-\frac{1}{2})L > \nu$. □

As a consequence of the preceding and Theorem 8.3.1, we have the following corollary.

COROLLARY 8.5.9. *Given $s \in \mathbb{R}$, set*

$$\Theta^s(\mathbb{R}^\nu;\mathbb{R}) = \big\{B^{s-\frac{\nu+1}{2}}\theta : \theta \in \Theta^{\frac{\nu+1}{2}}(\mathbb{R}^\nu;\mathbb{R})\big\},$$
$$\|\theta\|_{\Theta^s(\mathbb{R}^\nu;\mathbb{R})} = \|B^{\frac{\nu+1}{2}-s}\theta\|_{\Theta^{\frac{\nu+1}{2}}(\mathbb{R}^\nu;\mathbb{R})},$$

*and*

$$\mathcal{W}_{H^s(\mathbb{R}^\nu;\mathbb{R})} = (B^{s-\frac{\nu+1}{2}})_*\mathcal{W}_{H^{\frac{\nu+1}{2}}(\mathbb{R}^\nu;\mathbb{R})}.$$

*Then $\Theta^s(\mathbb{R}^\nu;\mathbb{R})$ is a separable Banach space in which $H^s(\mathbb{R}^\nu;\mathbb{R})$ is continuously embedded as a dense subspace, and $\big(H^s(\mathbb{R}^\nu;\mathbb{R}), \Theta^s(\mathbb{R}^\nu;\mathbb{R}), \mathcal{W}_{H^s(\mathbb{R}^\nu;\mathbb{R})}\big)$ is an abstract Wiener space.*

## Exercises for § 8.5

EXERCISE 8.5.10. In this exercise we will show how to use the Ornstein–Uhlenbeck process to prove **Poincaré's Inequality**

$$\text{(8.5.11)} \qquad \operatorname{Var}_{\gamma_{0,1}}(\varphi) = \|\varphi - \langle\varphi,\gamma_{0,1}\rangle\|^2_{L^2(\gamma_{0,1};\mathbb{R})} \le \|\varphi'\|^2_{L^2(\gamma_{0,1};\mathbb{R})}$$

for the standard Gaussian distribution on $\mathbb{R}$. I will outline the proof of (8.5.11) for $\varphi \in \mathscr{S}(\mathbb{R};\mathbb{R})$, but the estimate immediately extends to any $\varphi \in L^2(\gamma_{0,1};\mathbb{R})$ whose (distributional) first derivative is again in $L^2(\gamma_{0,1};\mathbb{R})$.

**(i)** For $\varphi \in \mathscr{S}(\mathbb{R};\mathbb{R})$, set

$$u_\varphi(t,x) = \mathbb{E}^{\mathcal{W}^{(1)}}\big[\varphi(U(t,x))\big],$$

where $\{U(t,x) : t \ge 0\}$ is the one-sided, $\mathbb{R}$-valued Ornstein–Uhlenbeck process starting at $x$. Show that $u'_\varphi(t,x) = e^{-\frac{t}{2}}u_{\varphi'}(t,x)$ and that

$$\lim_{t\searrow 0} u_\varphi(t,\cdot\,) = \varphi \quad\text{and}\quad \lim_{t\to\infty} u_\varphi(t,\cdot\,) = \langle\varphi,\gamma_{0,1}\rangle \quad\text{in } L^2(\gamma_{0,1};\mathbb{R}).$$

Show that another expression for $u_\varphi$ is

$$u_\varphi(t,x) = \big(2\pi(1-e^{-t})\big)^{-\frac{1}{2}}\int_{\mathbb{R}} \varphi(y)\exp\left(-\frac{(y-e^{-\frac{t}{2}}x)^2}{2(1-e^{-t})}\right)dy.$$

Using this second expression, show that $u_\varphi(t,\cdot\,) \in \mathscr{S}(\mathbb{R};\mathbb{R})$ and that $t \in [0,\infty) \longmapsto u_\varphi(t,\cdot\,) \in \mathscr{S}(\mathbb{R};\mathbb{R})$ is continuous. In addition, show that $\dot{u}_\varphi(t,x) = \frac{1}{2}\big(u''_\varphi(t,x) - xu'_\varphi(t,x)\big)$.

**(ii)** For $\varphi_1, \varphi_2 \in C^2(\mathbb{R};\mathbb{R})$ whose second derivative are tempered, show that

$$\left(\varphi_1, \varphi_2'' - x\varphi_2\right)_{L^2(\gamma_{0,1};\mathbb{R})} = -\left(\varphi_1', \varphi_2'\right)_{L^2(\gamma_{0,1};\mathbb{R})},$$

and use this together with **(i)** to show that, for any $\varphi \in \mathscr{S}(\mathbb{R};\mathbb{R})$,

$$\langle u_\varphi(t,\,\cdot\,), \gamma_{0,1}\rangle = \langle \varphi, \gamma_{0,1}\rangle \text{ and } \frac{d}{dt}\|u_\varphi(t,\,\cdot\,)\|^2_{L^2(\gamma_{0,1};\mathbb{R})} = -e^{-t}\|u_{\varphi'}(t,\,\cdot\,)\|^2_{L^2(\gamma_{0,1};\mathbb{R})}.$$

Conclude that $\|u_\varphi(t,\,\cdot\,)\|_{L^2(\gamma_{0,1};\mathbb{R})} \le \|\varphi\|_{L^2(\gamma_{0,1};\mathbb{R})}$ and

$$\frac{d}{dt}\|u_\varphi(t,\,\cdot\,)\|^2_{L^2(\gamma_{0,1};\mathbb{R})} \ge -e^{-t}\|\varphi'\|^2_{L^2(\gamma_{0,1};\mathbb{R})}.$$

Finally, integrate the preceding inequality to arrive at (8.5.11).

EXERCISE 8.5.12. In this exercise I will outline how the ideas in Exercise 8.5.10 can be used to give another derivation of the logarithmic Sobolev Inequality (2.4.42). Again, I restrict my attention to $\varphi \in \mathscr{S}(\mathbb{R};\mathbb{R})$, since the general case can be easily obtained from this by taking limits.

**(i)** Begin by showing that (2.4.42) for $\varphi \in \mathscr{S}(\mathbb{R};\mathbb{R})$ once one knows that

$$(*) \qquad \langle \varphi \log \varphi \rangle_{\gamma_{0,1}} \le \frac{1}{2}\left\langle \frac{(\varphi')^2}{\varphi} \right\rangle_{\gamma_{0,1}}$$

for uniformly positive $\varphi \in \mathbb{R} \oplus \mathscr{S}(\mathbb{R};\mathbb{R})$.

**(ii)** Given a uniformly positive $\varphi \in \mathbb{R} \oplus \mathscr{S}(\mathbb{R};\mathbb{R})$, use the results in Exercise 8.5.10 to show that

$$\frac{d}{dt}\langle u_\varphi(t,\,\cdot\,)\log u_\varphi(t,\,\cdot\,)\rangle_{\gamma_{0,1}} = -\frac{e^{-t}}{2}\left\langle \frac{u_{\varphi'}(t,\,\cdot\,)^2}{u_\varphi(t,\,\cdot\,)} \right\rangle_{\gamma_{0,1}}.$$

**(iii)** Continuing **(ii)**, apply Schwarz's inequality to check that

$$\frac{u_{\varphi'}(t,x)^2}{u_\varphi(t,x)} \le u_{\frac{(\varphi')^2}{\varphi}}(t,x),$$

and combine this with **(ii)** to get

$$\frac{d}{dt}\langle u_\varphi(t,\,\cdot\,)\log u_\varphi(t,\,\cdot\,)\rangle_{\gamma_{0,1}} \ge -\frac{e^{-t}}{2}\left\langle \frac{(\varphi')^2}{\varphi} \right\rangle_{\gamma_{0,1}}.$$

Finally, integrate this to arrive at (*).

EXERCISE 8.5.13. Although it should be clear that the arguments given in Exercises 8.5.10 and 8.5.12 work equally well in $\mathbb{R}^N$ and yield (8.5.11) and (2.4.42) with $\gamma_{0,1}$ replaced by $\gamma_{\mathbf{0},\mathbf{I}}$ and $(\varphi')^2$ replaced by $|\nabla\varphi|^2$, it is significant that each of these inequalities for $\mathbb{R}$ implies its $\mathbb{R}^N$ analog. Indeed, show that Fubini's Theorem is all that one needs to pass to the higher dimensional results. The reason why this remark is significant is that it allows one to prove infinite dimensional versions of both Poincaré's Inequality and the logarithmic Sobolev Inequality, and both of these play a crucial role in infinite dimensional analysis. In fact, Nelson's interest in hypercontractive estimates sprung from his brilliant insight that hypercontractive estimates would allow him to construct a non-trivial (i.e., non-Gaussian), Euclidean invariant quantum field for $\mathbb{R}^2$.

EXERCISE 8.5.14. It is interesting to see what happens if one changes the sign of the second term on the right-hand side of (8.5.1), thereby converting the centripetal force into a centrifugal one.

(**i**) Show that, for each $\boldsymbol{\theta} \in \Theta(\mathbb{R}^N)$, the unique solution to

$$\mathbf{V}(t,\boldsymbol{\theta}) = \boldsymbol{\theta}(t) + \tfrac{1}{2}\int_0^t \mathbf{V}(\tau,\boldsymbol{\theta})\,d\tau, \quad t \geq 0,$$

is

$$\mathbf{V}(t,\boldsymbol{\theta}) = e^{\frac{t}{2}}\int_0^t e^{-\frac{\tau}{2}}\,d\boldsymbol{\theta}(\tau),$$

where the integral is taken in the sense of Riemann–Stieltjes.

(**ii**) Show that $\left\{(\boldsymbol{\xi},\mathbf{V}(t,\,\cdot\,))_{\mathbb{R}^N} : (t,\boldsymbol{\xi}) \in [0,\infty)\times\mathbb{R}^N\right\}$ under $\mathcal{W}^{(N)}$ is a Gaussian family with covariance

$$v(s,t) = e^{\frac{s+t}{2}} - e^{\frac{|t-s|}{2}}.$$

(**iii**) Let $\{\mathbf{B}(t) : t \geq 0\}$ be an $\mathbb{R}^N$-valued Brownian motion, and show that the distribution of

$$\left\{e^{\frac{t}{2}}\mathbf{B}\left(1-e^{-t}\right) : t \geq 0\right\}$$

is the $\mathcal{W}^{(N)}$-distribution of $\{\mathbf{V}(t) : t \geq 0\}$. Next, let $\Theta^V(\mathbb{R}^N)$ be the space of continuous $\boldsymbol{\theta} : [0,\infty) \longrightarrow \mathbb{R}^N$ with the properties that

$$\boldsymbol{\theta}(0) = \mathbf{0} = \lim_{t\to\infty} e^{-t}|\boldsymbol{\theta}(t)|,$$

and set $\|\boldsymbol{\theta}\|_{\Theta^V(\mathbb{R}^N)} \equiv \sup_{t\geq 0} e^{-t}|\boldsymbol{\theta}(t)|$. Show that $\left(\Theta^V(\mathbb{R}^N); \|\cdot\|_{\Theta^V(\mathbb{R}^N)}\right)$ is a separable Banach space and that there exists a unique $\mathcal{V}^{(N)} \in \mathbf{M}_1\left(\Theta^V(\mathbb{R}^N)\right)$ such that the distribution of $\{\boldsymbol{\theta}(t) : t \geq 0\}$ under $\mathcal{V}^{(N)}$ is the same as the distribution of $\{\mathbf{V}(t) : t \geq 0\}$ under $\mathcal{W}^{(N)}$.

**(iv)** Let $\mathbf{H}^V(\mathbb{R}^N)$ be the space of absolutely continuous $\mathbf{h} : [0,\infty) \longrightarrow \mathbb{R}^N$ with the properties that $\mathbf{h}(0) = \mathbf{0}$ and $\dot{\mathbf{h}} - \frac{1}{2}\mathbf{h} \in L^2([0,\infty);\mathbb{R}^N)$. Show that $\mathbf{H}^V(\mathbb{R}^N)$ with norm

$$\|\mathbf{h}\|_{\mathbf{H}^V(\mathbb{R}^N)} \equiv \left\|\dot{\mathbf{h}} - \tfrac{1}{2}\mathbf{h}\right\|_{L^2([0,\infty);\mathbb{R}^N)}$$

is a separable Hilbert space that is continuously embedded in $\Theta^V(\mathbb{R}^N)$ as a dense subspace. Finally, show that $\left(\mathbf{H}^V(\mathbb{R}^N), \Theta^V(\mathbb{R}^N), \mathcal{V}^{(N)}\right)$ is an abstract Wiener space.

**(v)** There is a subtlety here that is worth mentioning. Namely, show that $\mathbf{H}^U(\mathbb{R}^N)$ is isometrically embedded in $\mathbf{H}^V(\mathbb{R}^N)$. On the other hand, as distinguished from elements of $\mathbf{H}^U(\mathbb{R}^N)$, it is not true that $\|\dot{\boldsymbol{\eta}} - \frac{1}{2}\boldsymbol{\eta}\|^2_{L^2(R;\mathbb{R}^N)} = \|\dot{\boldsymbol{\eta}}\|^2_{L^2(\mathbb{R};\mathbb{R}^N)} + \frac{1}{4}\|\boldsymbol{\eta}\|^2_{L^2(\mathbb{R};\mathbb{R}^N)}$, the point being that whereas the elements $\mathbf{h}$ of $\mathbf{H}^V(\mathbb{R}^N)$ with $\dot{\mathbf{h}} \in C_c((0,\infty);\mathbb{R}^N)$ are dense in $\mathbf{H}^U(\mathbb{R}^N)$, they are not dense in $\mathbf{H}^V(\mathbb{R}^N)$.

EXERCISE 8.5.15. Given $\mathbf{x} \in \mathbb{R}^\nu$ and a slowly increasing $\varphi \in C(\mathbb{R}^\nu;\mathbb{R})$, define $\tau_{\mathbf{x}}\varphi \in C(\mathbb{R}^\nu;\mathbb{R})$ so that $\tau_{\mathbf{x}}\varphi(\mathbf{y}) = \varphi(\mathbf{x}+\mathbf{y})$ for $\mathbf{y} \in \mathbb{R}^\nu$. Next, extend $\tau_{\mathbf{x}}$ to $\mathscr{S}'(\mathbb{R}^\nu;\mathbb{R})$ so that $\langle \varphi, \tau_{\mathbf{x}}u\rangle = \langle \tau_{-\mathbf{x}}\varphi, u\rangle$ for $\varphi \in \mathscr{S}(\mathbb{R}^\nu;\mathbb{R})$, and check that this is a legitimate extension in the sense that it is consistent with the original definition when applied to $u$'s that are slowly increasing, continuous functions. Finally, given $s \in \mathbb{R}$, define $\mathcal{O}_{\mathbf{x}} : H^s(\mathbb{R}^\nu;\mathbb{R}) \longrightarrow H^s(\mathbb{R}^\nu;\mathbb{R})$ by $\mathcal{O}_{\mathbf{x}}h = \tau_{\mathbf{x}}h$.

**(i)** Show that $B^s \circ \tau_{\mathbf{x}} = \tau_{\mathbf{x}} \circ B^s$ for all $s \in \mathbb{R}$ and $\mathbf{x} \in \mathbb{R}^\nu$.

**(ii)** Given $s \in \mathbb{R}$, define $\mathcal{O}_{\mathbf{x}} = \tau_{\mathbf{x}} \restriction H^s(\mathbb{R}^\nu;\mathbb{R})$, and show that $\mathcal{O}_{\mathbf{x}}$ is an orthogonal transformation.

**(iii)** Referring to Theorem 8.3.14 and Corollary 8.5.9, show that the measure preserving transformation $T_{\mathcal{O}_{\mathbf{x}}}$ that $\mathcal{O}_{\mathbf{x}}$ determines on $\left(\Theta^s(\mathbb{R}^\nu;\mathbb{R}), \mathcal{W}_{H^s(\mathbb{R}^\nu;\mathbb{R})}\right)$ is the restriction of $\tau_{\mathbf{x}}$ to $\Theta^s(\mathbb{R}^\nu;\mathbb{R})$.

**(iv)** If $\mathbf{x} \neq \mathbf{0}$, show that $T_{\mathcal{O}_{\mathbf{x}}}$ is ergodic on $\left(\Theta^s(\mathbb{R}^\nu;\mathbb{R}), \mathcal{W}_{H^s(\mathbb{R}^\nu;\mathbb{R})}\right)$.

## § 8.6 Brownian Motion on a Banach Space

In this concluding section I will discuss Brownian motion on a Banach space. More precisely, given a non-degenerate, centered, Gaussian measure $\mathcal{W}$ on a separable Banach space $E$, we will see that there exists an $E$-valued stochastic process $\{B(t) : t \geq 0\}$ with the properties that $B(0) = 0$, $t \rightsquigarrow B(t)$ is continuous, and, for all $0 \leq s < t$, $B(t) - B(s)$ is independent of $\sigma(\{B(\tau) : \tau \in [0,s]\})$ and has distribution (cf. the notation in § 8.4) $\mathcal{W}_{t-s}$.

**§ 8.6.1. Abstract Wiener Formulation.** Let $\mathcal{W}$ on $E$ be as above, use $H$ to denote its Cameron–Martin space, and take $H^1(H)$ to be the Hilbert space of absolutely continuous $h : [0,\infty) \longrightarrow H$ such that $h(0) = 0$ and $\|h\|_{H^1(H)} = \|\dot{h}\|_{L^2([0,\infty);H)} < \infty$. Finally, let $\Theta(E)$ be the space of continuous $\theta : [0,\infty) \longrightarrow$

$E$ satisfying $\lim_{t\to\infty} \frac{\|\theta(t)\|_E}{t} = 0$, and turn $\Theta(E)$ into a Banach space with norm $\|\theta\|_{\Theta(E)} = \sup_{t\ge 0}(1+t)^{-1}\|\theta(t)\|_E$. By exactly the same line of reasoning as I used when $E = \mathbb{R}^N$, one can show that $\Theta(E)$ is a separable Banach space in which $H^1(E)$ is continuously embedded as a dense subspace. My goal is to prove the following statement.

THEOREM 8.6.1. *With $H^1(H)$ and $\Theta(E)$ as above, there is a unique $\mathcal{W}^{(E)} \in \mathbf{M}_1(\Theta(E))$ such that $(H^1(H), \Theta(E), \mathcal{W}^{(E)})$ is an abstract Wiener space.*

Choose an orthonormal basis $\{h^1_m : m \ge 0\}$ in $H^1(\mathbb{R})$, and, for $n \ge 0$, $t \ge 0$, and $\mathbf{x} = (x_0, \dots, x_m, \dots) \in E^{\mathbb{N}}$, set $S_n(t,\mathbf{x}) = \sum_{m=0}^n h^1_m(t)x_m$. I will show that, $\mathcal{W}^{\mathbb{N}}$-almost surely, $\{S_n(\,\cdot\,,\mathbf{x}) : n \ge 0\}$ converges in $\Theta(E)$, and, for the most part, the proof follows the same basic line of reasoning as that suggested in Exercise 8.3.21 when $E = \mathbb{R}^N$. However, there is a problem here that we did not encounter there. Namely, unless $E$ is finite dimensional, bounded subsets will not necessarily be relatively compact in $E$. Hence, local uniform equicontinuity plus local boundedness is not sufficient to guarantee that a collection of $E$-valued paths is relatively compact in $C([0,\infty);E)$, and that is the reason why we have to work a little harder here.

LEMMA 8.6.2. *For $\mathcal{W}^{\mathbb{N}}$-almost every $\mathbf{x} \in E^{\mathbb{N}}$, $\{S_n(\,\cdot\,,\mathbf{x}) : n \ge 0\}$ is relatively compact in $\Theta(E)$.*

PROOF: Choose $E_0 \subseteq E$, as in Corollary 8.3.10, so that bounded subsets of $E_0$ are relatively compact in $E$ and $(H, E_0, \mathcal{W} \restriction E_0)$ is again an abstract Wiener space. Without loss in generality, I will assume that $\|\cdot\|_E \le \|\cdot\|_{E_0}$, and, by Fernique's Theorem, we know that $C \equiv \mathbb{E}^{\mathcal{W}_0}\big[\|x\|^4_{E_0}\big] < \infty$.

Since (cf. Exercise 8.2.14) $S_n(t,\mathbf{x}) - S_n(s,\mathbf{x}) = \sum_{m=0}^n \big(h^1_t - h^1_s, h^1_m\big)_{H^1(\mathbb{R})} x_m$, where $h^1_\tau = \,\cdot\, \wedge \tau$, the $\mathcal{W}_0^{\mathbb{N}}$-distribution of $S_n(t) - S_n(s)$ is $\mathcal{W}_{\epsilon_n}$, where $\epsilon_n = \sum_0^n \big(h^1_t - h^1_s, h^1_m\big)^2_{H^1(\mathbb{R})} \le t - s$. Hence, $\mathbb{E}^{\mathcal{W}^{\mathbb{N}}}\big[\|S_n(t) - S_n(s)\|^4_{E_0}\big] \le C(t-s)^2$. In addition, $\{\|S_n(t) - S_n(s)\|_{E_0} : n \ge 1\}$ is a submartingale, and so, by Doob's Inequality plus Kolmogorov's Continuity Criterion, there exists a $K < \infty$ such that, for each $T > 0$,

$$(*) \qquad \mathbb{E}^{\mathcal{W}^{\mathbb{N}}}\left[\sup_{n\ge 0}\ \sup_{0\le s<t\le T} \frac{\|S_n(t) - S_n(s)\|_{E_0}}{(t-s)^{\frac{1}{8}}}\right] \le KT^{\frac{3}{4}}.$$

From (*) and $S_n(0) = 0$, we know that, $\mathcal{W}^{\mathbb{N}}$-almost surely, $\{S_n(\,\cdot\,,\mathbf{x}) : n \ge 0\}$ is uniformly $\|\cdot\|_{E_0}$-bounded and uniformly $\|\cdot\|_{E_0}$-equicontinuous on each interval $[0,T]$. Since this means that, for every $T > 0$, $\{S_n(t,\mathbf{x}) : n \ge 0\ \&\ t \in [0,T]\}$ is relatively compact in $E$ and $\{S_n(\,\cdot\,,\mathbf{x}) \restriction [0,T] : n \ge 0\}$ is uniformly $\|\cdot\|_E$-equicontinuous $\mathcal{W}^{\mathbb{N}}$-almost surely, the Ascoli–Arzela Theorem guarantees that, $\mathcal{W}^{\mathbb{N}}$-almost surely, $\{S_n(\,\cdot\,,\mathbf{x}) : n \ge 0\}$ is relatively compact in $C([0,\infty);E)$ with the topology of uniform convergence on compacts. Thus, in order to complete the proof, all that I have to show is that, $\mathcal{W}^{\mathbb{N}}$-almost surely,

$$\lim_{T\to\infty} \sup_{n\geq 0} \sup_{t\geq T} \frac{\|S_n(t,\mathbf{x})\|_E}{t} = 0.$$

But,

$$\sup_{t\geq 2^k} \frac{\|S_n(t,\mathbf{x})\|_E}{t} \leq \sum_{\ell\geq k} \sup_{2^\ell\leq t\leq 2^{\ell+1}} \frac{\|S_n(t,\mathbf{x})\|_E}{t} \leq \sum_{\ell\geq k} 2^{-\frac{7\ell}{8}} \sup_{0\leq t\leq 2^{\ell+1}} \frac{\|S_n(t,\mathbf{x})\|_E}{t^{\frac{1}{8}}},$$

and therefore, by (*),

$$\mathbb{E}^{\mathcal{W}^{\mathbb{N}}}\left[\sup_{n\geq 0} \sup_{t\geq 2^k} \frac{\|S_n(t,\mathbf{x})\|_E}{t}\right] \leq \frac{2^{\frac{3}{4}}K}{2^{\frac{1}{8}}-1} 2^{-\frac{k}{8}}. \quad \square$$

Now that we have the requisite compactness of $\{S_n : n \geq 0\}$, convergence comes to checking a criterion of the sort given in the following simple lemma.

LEMMA 8.6.3. *Suppose that $\{\theta_n : n \geq 0\}$ is a relatively compact sequence in $\Theta(E)$. If $\lim_{n\to\infty}\langle \theta_n(t), x^*\rangle$ exists for each $t$ in a dense subset of $[0,\infty)$ and $x^*$ in a weak* dense subset of $E^*$, then $\{\theta_n : n \geq 0\}$ converges in $\Theta(E)$.*

PROOF: For a relatively compact sequence to be convergent, it is necessary and sufficient that every convergent subsequence have the same limit. Thus, suppose that $\theta$ and $\theta'$ are limit points of $\{\theta_n : n \geq 0\}$. Then, by hypothesis, $\langle \theta(t), x^*\rangle = \langle \theta'(t), x^*\rangle$ for $t$ in a dense subset of $[0,\infty)$ and $x^*$ in a weak* dense subset of $E^*$. But this means that the same equality holds for all $(t, x^*) \in [0,\infty) \times E^*$ and therefore that $\theta = \theta'$. $\square$

PROOF OF THEOREM 8.6.1: In view of Lemmas 8.6.2 and 8.6.3 and the separability of $E^*$ in the weak* topology, we will know that $\{S_n(\,\cdot\,,\mathbf{x}) : n \geq 0\}$ converges in $\Theta(E)$ for $\mathcal{W}^{\mathbb{N}}$-almost every $\mathbf{x} \in E^{\mathbb{N}}$ once we show that, for each $(t, x^*) \in [0,\infty) \times E^*$, $\{\langle S_n(t,\mathbf{x}), x^*\rangle : n \geq 0\}$ converges in $\mathbb{R}$ for $\mathcal{W}^{\mathbb{N}}$-almost every $\mathbf{x} \in E^{\mathbb{N}}$. But if $x^* \in E^*$, then $\langle S_n(t,\mathbf{x}), x^*\rangle = \sum_0^n \langle x_m, x^*\rangle h_m^1(t)$, the random variables $\mathbf{x} \rightsquigarrow \langle x_m, x^*\rangle h_m^1(t)$ are independent, centered Gaussians under $\mathcal{W}^{\mathbb{N}}$ with variance $\|h_{x^*}\|_H^2 h_m^1(t)^2$, and $\sum_0^\infty h_m^1(t)^2 = \|h_t\|_{H^1(\mathbb{R})}^2 = t$. Thus, by Theorem 1.4.2, we have the required convergence.

Next, define $B : [0,\infty) \times E^{\mathbb{N}} \longrightarrow E$ so that

$$B(t,\mathbf{x}) = \begin{cases} \lim_{n\to\infty} S_n(t,\mathbf{x}) & \text{if } \{S_n(\,\cdot\,,\mathbf{x}) : n \geq 0\} \text{ converges in } \Theta(E) \\ 0 & \text{otherwise.} \end{cases}$$

Given $\lambda \in \Theta(E)^*$, determine $h_\lambda \in H^1(H)$ by $\left(h, h_\lambda\right)_{H^1(H)} = \langle h, \lambda\rangle$ for all $h \in H^1(H)$. I want to show that, under $\mathcal{W}^{\mathbb{N}}$, $\mathbf{x} \rightsquigarrow \langle B(\,\cdot\,,\mathbf{x}), \lambda\rangle$ is a centered Gaussian with variance $\|h_\lambda\|^2_{H^1(H)}$. To this end, define $x^*_m \in E^*$ so that[1] $\langle x, x^*_m\rangle = \langle h^1_m x, \lambda\rangle$ for $x \in E$. Then,

$$\langle B(\cdot,\mathbf{x}),\lambda\rangle = \lim_{n\to\infty}\langle S_n(\cdot,\mathbf{x}),\lambda\rangle = \lim_{n\to\infty}\sum_0^n \langle x_m, x^*_m\rangle \quad \mathcal{W}^{\mathbb{N}}\text{-almost surely.}$$

Hence, $\langle B(\cdot,\mathbf{x}),\lambda\rangle$ is certainly a centered Gaussian under $\mathcal{W}^{\mathbb{N}}$, and, because we are dealing with Gaussian random variables, almost sure convergence implies $L^2$-convergence. To compute its variance, choose an orthonormal basis $\{h_k : k \ge 0\}$ for $H$, and note that, for each $m \ge 0$,

$$\mathbb{E}^{\mathcal{W}^{\mathbb{N}}}\big[\langle x_m, x^*_m\rangle^2\big] = \|h_{x^*_m}\|^2_H = \sum_{k=0}^\infty \langle h^1_m h_k, \lambda\rangle^2.$$

Thus, since $\{h^1_m h_k : (m,k) \in \mathbb{N}^2\}$ is an orthonormal basis in $H^1(H)$,

$$\mathbb{E}^{\mathcal{W}^{\mathbb{N}}}\big[\langle B(\cdot),\lambda\rangle^2\big] = \sum_{m,k=0}^\infty \langle h^1_m h_k,\lambda\rangle^2 = \sum_{m,k=0}^\infty \big(h^1_m h_k, h_\lambda\big)^2_{H^1(H)} = \|h_\lambda\|^2_{H^1(H)}.$$

Finally, to complete the proof, all that remains is to take $\mathcal{W}^{(E)}$ to be the $\mathcal{W}^{\mathbb{N}}$-distribution of $\mathbf{x} \rightsquigarrow B(\cdot,\mathbf{x})$. □

§ **8.6.2. Brownian Formulation.** Let $(H, E, \mathcal{W})$ be an abstract Wiener space. Given a probability space $(\Omega, \mathcal{F}, \mathbb{P})$, a non-decreasing family of sub-$\sigma$-algebras $\{\mathcal{F}_t : t \ge 0\}$, and a measurable map $B : [0,\infty) \times \Omega \longrightarrow E$, say that the triple $\big(B(t), \mathcal{F}_t, \mathbb{P}\big)$ is a **$\mathcal{W}$-Brownian motion** if

(1) $\{B(t) : t \ge 0\}$ is $\{\mathcal{F}_t : t \ge 0\}$-progressively measurable,
(2) $B(0,\omega) = 0$ and $B(\cdot,\omega) \in C\big([0,\infty); E\big)$ for $\mathbb{P}$-almost every $\omega$,
(3) $B(1)$ has distribution $\mathcal{W}$, and, for all $0 \le s < t$, $B(t) - B(s)$ is independent of $\mathcal{F}_s$ and has the same distribution as $(t-s)^{\frac{1}{2}} B(1)$.

LEMMA 8.6.4. *Suppose that $\{B(t) : t \ge 0\}$ satisfies conditions (1) and (2). Then $\big(B(t), \mathcal{F}_t, \mathbb{P}\big)$ is a $\mathcal{W}$-Brownian motion if and only if $\big(\langle B(t), x^*\rangle, \mathcal{F}_t, \mathbb{P}\big)$ is an $\mathbb{R}$-valued Brownian motion for each $x^* \in E^*$ with $\|h_{x^*}\|_H = 1$. In addition, if $\big(B(t), \mathcal{F}_t, \mathbb{P}\big)$ is a $\mathcal{W}$-Brownian motion, then the span $\mathfrak{G}(B)$ of $\{\langle B(t), x^*\rangle : (t, x^*) \in [0,\infty) \times E^*\}$ is a Gaussian family in $L^2(\mathbb{P};\mathbb{R})$ and*

$$\mathbb{E}^{\mathbb{P}}\big[\langle B(t_1), x^*_1\rangle\langle B(t_2), x^*_2\rangle\big] = (t_1 \wedge t_2)\big(h_{x^*_1}, h_{x^*_2}\big)_H. \tag{8.6.5}$$

*Conversely, if $\mathfrak{G}(B)$ is a Gaussian family in $L^2(\mathbb{P};\mathbb{R})$ and (8.6.5) holds, then $\big(B(t), \mathcal{F}_t, \mathbb{P}\big)$ is a $\mathcal{W}$-Brownian motion when $\mathcal{F}_t = \sigma\big(\{B(\tau) : \tau \in [0,t]\}\big)$.*

---

[1] Given $h^1 \in H^1(\mathbb{R})$ and $x \in E$, I use $h^1 x$ to denote the element $\theta$ of $\Theta(E)$ determined by $\theta(t) = h^1(t)x$.

PROOF: If $(B(t), \mathcal{F}_t, \mathbb{P})$ is a $\mathcal{W}$-Brownian motion and $x^* \in E^*$ with $\|h_{x^*}\|_H = 1$, then $\langle B(t), x^*\rangle - \langle B(s), x^*\rangle = \langle B(t) - B(s), x^*\rangle$ is independent of $\mathcal{F}_s$ and is a centered Gaussian with variance $(t-s)$. Thus, $(\langle B(t), x^*\rangle, \mathcal{F}_t, \mathbb{P})$ is an $\mathbb{R}$-valued Brownian motion.

Next assume that $(\langle B(t), x^*\rangle, \mathcal{F}_t, \mathbb{P})$ is an $\mathbb{R}$-valued Brownian motion for every $x^*$ with $\|h_{x^*}\|_H = 1$. Then $\langle B(t) - B(s), x^*\rangle$ is independent of $\mathcal{F}_s$ for every $x^* \in E^*$, and so, since $\mathcal{B}_E$ is generated by $\{\langle \cdot, x^*\rangle : x^* \in E^*\}$, $B(t) - B(s)$ is independent of $\mathcal{F}_s$. In addition, $\langle B(t) - B(s), x^*\rangle$ is a centered Gaussian with variance $(t-s)\|h_{x^*}\|_H^2$, and therefore $B(1)$ has distribution $\mathcal{W}$ and $B(t) - B(s)$ has the same distribution as $(t-s)^{\frac{1}{2}}B(1)$. Thus, $(B(t), \mathcal{F}_t, \mathbb{P})$ is a $\mathcal{W}$-Brownian motion.

Again assume that $(B(t), \mathcal{F}_t, \mathbb{P})$ is a $\mathcal{W}$-Brownian motion. To prove that $\mathfrak{G}(B)$ is a Gaussian family for which (8.6.5) holds, it suffices to show that, for all $0 \le t_1 \le t_2$ and $x_1^*, x_2^* \in E^*$, $\langle B(t_1), x_1^*\rangle + \langle B(t_2), x_2^*\rangle$ is a centered Gaussian with covariance $t_1\|h_{x_1*} + h_{x_2^*}\|_H^2 + (t_2 - t_1)\|h_{x_2^*}\|_H^2$. Indeed, we would then know not only that $\mathfrak{G}(B)$ is a Gaussian family but also that the variance of $\langle B(t_1), x_1^*\rangle \pm \langle B(t_2), x_2^*\rangle$ is $t_1\|h_{x_1*} \pm h_{x_2^*}\|_H^2 + (t_2 - t_1)\|h_{x_2^*}\|_H^2$, from which (8.6.5) is immediate. But

$$\langle B(t_1), x_1^*\rangle + \langle B(t_2), x_2^*\rangle = \langle B(t_1), x_1^* + x_2^*\rangle + \langle B(t_2) - B(t_1), x_2^*\rangle,$$

and the terms on the right are independent, centered Gaussians, the first with variance $t_1\|h_{x_1^*} + h_{x_2^*}\|_H^2$ and the second with variance $(t_2 - t_1)\|h_{x_2^*}\|_H^2$.

Finally, take $\mathcal{F}_t = \sigma(\{B(\tau) : \tau \in [0,t]\})$, and assume that $\mathfrak{G}(B)$ is a Gaussian family satisfying (8.6.5). Given $x^*$ with $\|h_{x^*}\|_H = 1$ and $0 \le s < t$, we know that $\langle B(t) - B(s), x^*\rangle = \langle B(t), x^*\rangle - \langle B(s), x^*\rangle$ is orthogonal in $L^2(\mathbb{P};\mathbb{R})$ to $\langle B(\tau), y^*\rangle$ for every $\tau \in [0,s]$ and $y^* \in E^*$. Hence, since $\mathcal{F}_s$ is generated by $\{\langle B(\tau), y^*\rangle : (\tau, y^*) \in [0,s] \times E^*\}$, we know that $\langle B(t) - B(s), x^*\rangle$ is independent of $\mathcal{F}_s$. In addition, $\langle B(t) - B(s), x^*\rangle$ is a centered Gaussian with variance $t-s$, and so we have proved that $(\langle B(t), x^*\rangle, \mathcal{F}_t, \mathbb{P})$ is an $\mathbb{R}$-valued Brownian motion. Now apply the first part of the lemma to conclude that $(B(t), \mathcal{F}_t, \mathbb{P})$ is a $\mathcal{W}$-Brownian motion. □

THEOREM 8.6.6. *Refer to the notation in Theorem 8.6.1. When $\Omega = \Theta(E)$, $\mathcal{F} = \mathcal{B}_E$, and $\mathcal{F}_t = \sigma(\{\theta(\tau) : \tau \in [0,t]\})$, $(\theta(t), \mathcal{F}_t, \mathcal{W}^{(E)})$ is a $\mathcal{W}$-Brownian motion. Conversely, if $(B(t), \mathcal{F}_t, \mathbb{P})$ is any $\mathcal{W}$-Brownian motion, then $B(\cdot, \omega) \in \Theta(E)$ $\mathbb{P}$-almost surely and $\mathcal{W}^{(E)}$ is the $\mathbb{P}$-distribution of $\omega \rightsquigarrow B(\cdot, \omega)$.*

PROOF: To prove the first assertion, let $t_1, t_2 \in [0,\infty)$ and $x_1^*, x_2^* \in E^*$ be given, and define $\lambda_i \in \Theta(E)^*$ so that $\langle \theta, \lambda_i\rangle = \langle \theta(t_i), x_i^*\rangle$ for $i \in \{1,2\}$. Then (cf. the notation in the proof of Theorem 8.6.1) $h_{\lambda_i} = h_{t_i}^1 h_{x_i^*}$, and so

$$\mathbb{E}^{\mathcal{W}^{(E)}}\left[\langle \theta(t_1), x_1^*\rangle\langle \theta(t_2), x_2^*\rangle\right] = \left(h_{\lambda_1} h_{\lambda_2}\right)_{H^1(H)} = (t_1 \wedge t_2)\left(h_{x_1^*}, h_{x_2^*}\right)_H.$$

Starting from this, it is an easy matter to check that the span of $\{\langle\theta(t), x^*\rangle : (t, x^*) \in [0,\infty) \times E^*\}$ is a Gaussian family in $L^2(\mathcal{W}^{(E)};\mathbb{R})$ that satisfies (8.6.5).

To prove the converse, begin by observing that, because $\mathfrak{G}(B)$ is a Gaussian family satisfying (8.6.5), the distribution of $\omega \in \Omega \longmapsto B(\,\cdot\,,\omega) \in C\big([0,\infty);E\big)$ under $\mathbb{P}$ is the same as that of $\theta \in \Theta(E) \longmapsto \theta(\,\cdot\,) \in C\big([0,\infty);E\big)$ under $\mathcal{W}^{(E)}$. Hence

$$\mathbb{P}\left(\varlimsup_{t\to\infty} \frac{\|B(t)\|_E}{t} = 0\right) = \mathcal{W}^{(E)}\left(\varlimsup_{t\to\infty} \frac{\|\theta(t)\|_E}{t} = 0\right) = 1,$$

and so $B(\,\cdot\,,\omega) \in \Theta(E)$ $\mathbb{P}$-almost surely and the distribution of $\omega \rightsquigarrow B(\,\cdot\,,\omega)$ on $\Theta(E)$ is $\mathcal{W}^{(E)}$. $\square$

§ **8.6.3. Strassen's Theorem Revisited.** What I called Strassen's Theorem in § 8.4.2 is not the form in which Strassen himself presented it. Instead, his formulation was in terms of rescaled $\mathbb{R}$-valued Brownian motion, not partial sums of independent random variables. The true statement in the present setting of Strassen's Theorem is the following.

THEOREM 8.6.7 (**Strassen**). *Given $\theta \in \Theta(E)$, define $\tilde{\theta}_n(t) = \frac{\theta(nt)}{\Lambda_n}$ for $n \geq 1$ and $t \in [0,\infty)$, where $\Lambda_n = \sqrt{2n\log_{(2)}(n \vee 3)}$. Then, for $\mathcal{W}^{(E)}$-almost every $\theta$, the sequence $\{\tilde{\theta}_n : n \geq 0\}$ is relatively compact in $\Theta(E)$ and $\overline{B_{H^1(H)}(0,1)}$ is its set of limit points. Equivalently, for $\mathcal{W}^{(E)}$-almost every $\theta$,*

$$\varlimsup_{n\to\infty} \|\tilde{\theta}_n - \overline{B_{H^1(H)}(0,1)}\|_{\Theta(E)} = 0$$

*and, for each $h \in \overline{B_{H^1(H)}(0,1)}$, $\varliminf_{n\to\infty} \|\tilde{\theta}_n - h\|_{\Theta(E)} = 0$.*

Not surprisingly, the proof differs only slightly from that of Theorem 8.4.4. In proving the $\mathcal{W}^{(E)}$-almost sure convergence of $\{\tilde{\theta}_n : n \geq 1\}$ to $\overline{B_{H^1(H)}(0,1)}$, there are two new ingredients here. The first is the use of the Brownian scaling invariance property (cf. Exercise 8.6.8), which says that the $\mathcal{W}^{(E)}$ is invariant under the scaling maps $S_\alpha : \Theta(E) \longrightarrow \Theta(E)$ given by $S_\alpha\theta = \alpha^{-\frac{1}{2}}\theta(\alpha\,\cdot\,)$ for $\alpha > 0$ and is easily proved as a consequence of the fact that these maps are isometric from $H^1(H)$ onto itself. The second new ingredient is the observation that, for any $R > 0$, $r \in (0,1]$, and $\theta \in \Theta(E)$, $\|\theta(r\,\cdot\,) - B_{H^1(H)}(0,R)\|_{\Theta(E)} \leq \|\theta - B_{H^1(H)}(0,R)\|_{\Theta(E)}$. To see this, let $h \in B_{H^1(H)}(0,R)$ be given, and check that $h(r\,\cdot\,)$ is again in $B_H(0,R)$ and that $\|\theta(r\,\cdot\,) - h(r\,\cdot\,)\|_{\Theta(E)} \leq \|\theta - h\|_{\Theta(E)}$.

Taking these into account and applying (8.4.2), one can now justify

$$\begin{aligned}
&\mathcal{W}^{(E)}\left(\max_{\beta^{m-1}\le n\le\beta^m}\left\|\tilde\theta_n-\overline{B_{H^1(H)}(0,1)}\right\|_{\Theta(E)}\ge\delta\right)\\
&=\mathcal{W}^{(E)}\left(\max_{\beta^{m-1}\le n\le\beta^m}\left\|\frac{\beta^{\frac{m}{2}}\theta(n\beta^{-m}\cdot)}{\Lambda_n}-\overline{B_{H^1(H)}(0,1)}\right\|_{\Theta(E)}\ge\delta\right)\\
&\le\mathcal{W}^{(E)}\left(\max_{\beta^{m-1}\le n\le\beta^m}\left\|\theta(\beta^{-m}n\,\cdot)-\overline{B_{H^1(H)}\left(0,\frac{\Lambda_{\lfloor\beta^{m-1}\rfloor}}{\beta^{\frac{m}{2}}}\right)}\right\|_{\Theta(E)}\ge\frac{\delta\Lambda_{\lfloor\beta^{m-1}\rfloor}}{\beta^{\frac{m}{2}}}\right)\\
&\le\mathcal{W}^{(E)}\left(\left\|\theta-\overline{B_{H^1(H)}\left(0,\frac{\Lambda_{\lfloor\beta^{m-1}\rfloor}}{\beta^{\frac{m}{2}}}\right)}\right\|_{\Theta(E)}\ge\frac{\delta\Lambda_{\lfloor\beta^{m-1}\rfloor}}{\beta^{\frac{m}{2}}}\right)\\
&=\mathcal{W}^{(E)}\left(\left\|\beta^{\frac{m}{2}}\Lambda^{-1}_{\lfloor\beta^{m-1}\rfloor}\theta-\overline{B_{H^1(H)}(0,1)}\right\|_{\Theta(E)}\ge\delta\right)\\
&=\mathcal{W}^{(E)}_{\beta^m\Lambda^{-2}_{\lfloor\beta^{m-1}\rfloor}}\left(\|\theta-\overline{B_{H^1(H)}(0,1)}\|_{\Theta(E)}\ge\delta\right)\le\exp\left(-\frac{R^2\lfloor\beta^{m-1}\rfloor}{\beta^m}\log_{(2)}\lfloor\beta^{m-1}\rfloor\right)
\end{aligned}$$

for all $\beta\in(1,2)$, $R<\inf\{\|h\|_{H^1(H)}:\|h\|_{\Theta(E)}\ge\delta\}$, and sufficiently large $m\ge 1$. Armed with this information, one can simply repeat the argument given at the analogous place in the proof of Theorem 8.4.4.

The proof that, $\mathcal{W}^{(E)}$-almost surely, $\tilde\theta_n$ approaches every $h\in C$ infinitely often also requires only minor modification. To begin, one remarks that if $A\subseteq\Theta(E)$ is relatively compact, then

$$\lim_{T\to\infty}\sup_{\theta\in A}\sup_{t\notin[T^{-1},T]}\frac{\|\theta(t)\|_E}{1+t}=0.$$

Thus, since, by the preceding, for $\mathcal{W}^{(E)}$-almost every $\theta$, the union of $\{\theta_n : n\ge 1\}$ and $\overline{B_{H^1(H)}(0,1)}$ is relatively compact in $\Theta(E)$, it suffices to prove that

$$\lim_{n\to\infty}\sup_{t\in[k^{-1},k]}\frac{\|(\tilde\theta_n(t)-\tilde\theta_n(k^{-1}))-(h(t)-h(k^{-1}))\|_E}{1+t}=0\ \mathcal{W}^{(E)}\text{-almost surely}$$

for each $h\in B_{H^1(H)}(0,1)$ and $k\ge 2$. Because, for a fixed $k\ge 2$, the random variables $(\tilde\theta_{k^{2m}}-\tilde\theta_{k^{2m}}(k^{-1}))\restriction[k^{-1},k]$, $m\ge 1$, are $\mathcal{W}^{(E)}$-independent random variables, we can use the Borel–Cantelli Lemma as in § 8.4.2 and thereby reduce the problem to showing that, if $\check\theta_{k^m}(t)=\tilde\theta_{k^m}(t+k^{-1})-\tilde\theta_{k^m}(k^{-1})$, then

$$\sum_{m=1}^{\infty}\mathcal{W}^{(E)}\left(\|\check\theta_{k^{2m}}-h\|_{\Theta(E)}\le\delta\right)=\infty$$

for each $\delta>0$, $k\ge 2$, and $h\in B_{H^1(H)}(0,1)$. Finally, since $\mathcal{W}^{(E)}{}_{(k^m\Lambda^{-1}_{k^{2m}})^2}$ is the $\mathcal{W}^{(E)}$ distribution of $\theta\rightsquigarrow\check\theta_{k^{2m}}$, the rest of the argument is the same as the one given in § 8.4.2.

### Exercises for § 8.6

EXERCISE 8.6.8. Let $\big(H^1(H), \Theta(E), \mathcal{W}^{(E)}\big)$ be as in Theorem 8.6.1.

(i) Given $\alpha > 0$, define $S_\alpha : \Theta(E) \longrightarrow \Theta(E)$ so that $S_\alpha\theta(t) = \alpha^{-\frac{1}{2}}\theta(\alpha t)$, $t \in [0,\infty)$, and show that $(S_\alpha)_*\mathcal{W}^{(E)} = \mathcal{W}^{(E)}$. Again, this property is called **Brownian scaling invariance**.

(ii) Define $I : \Theta(E) \longrightarrow C\big([0,\infty); E\big)$ so the $I\theta(0) = 0$ and $I\theta(t) = t\theta(t^{-1})$ for $t > 0$. Show that $I$ is an isometry from $\Theta(E)$ onto itself and that $I \restriction H^1(H)$ is an isometry on $H$ onto itself. Finally, use this to prove the **Brownian time inversion invariance** property: $I_*\mathcal{W}^{(E)} = \mathcal{W}^{(E)}$.

EXERCISE 8.6.9. Let $H^U(H)$ be the Hilbert space of absolutely continuous $h^U : \mathbb{R} \longrightarrow H$ with the property that

$$\|h\|_{H^U(H)} = \sqrt{\|\dot{h}^U\|^2_{L^2(\mathbb{R};H)} + \tfrac{1}{4}\|h^U\|^2_{L^2(\mathbb{R};H)}} < \infty,$$

and take $\Theta^U(E)$ to be the Banach space of continuous $\theta^U : \mathbb{R} \longrightarrow E$ satisfying $\lim_{|t|\to\infty} \frac{\|\theta^U(t)\|}{\log t} = 0$ with norm $\|\theta^U\|_{\Theta^U(E)} = \sup_{t\in\mathbb{R}} \frac{\|\theta^U(t)}{\log(e+|t|)}$. If $F : \Theta(E) \longrightarrow C(\mathbb{R}; E)$ is given by $[F(\theta)](t) = e^{-\frac{t}{2}}\theta(e^t)$, show that $F$ takes $\Theta(E)$ continuously into $\Theta^U(E)$ and that $\big(H^U(H), \Theta^U(E), \mathcal{U}^{(E)}\big)$ is an abstract Wiener space when $\mathcal{U}^{(E)}_R = F_*\mathcal{W}^{(E)}$. Of course, one should recognize the measure $\mathcal{U}^{(E)}_R$ as the distribution of an $E$-valued, reversible, Ornstein–Uhlenbeck process.

EXERCISE 8.6.10. A particularly interesting case of the construction in Exercise 8.6.9 is when $H = H^1(\mathbb{R}^N)$ and $E = \Theta(\mathbb{R}^N)$. Working in that setting, define $\mathbf{B} : \mathbb{R} \times [0,\infty) \times \Theta^U\big(\Theta(E)\big) \longrightarrow \mathbb{R}^N$ by $\mathbf{B}\big((s,t),\theta\big) = [\theta(s)](t)$, and show that, for each $s \in \mathbb{R}$, $\big(\mathbf{B}(s,t), \mathcal{F}_{(s,t)}, \mathcal{U}^{\Theta(\mathbb{R}^N)}_R\big)$ is an $\mathbb{R}^N$-valued Brownian motion when $\mathcal{F}_{(s,t)} = \sigma\big(\{\mathbf{B}(s,\tau) : \tau \in [0,t]\}\big)$. Next, for each $t \in [0,\infty)$, show that the $\mathcal{U}^{\Theta(E)}_R$-distribution of $\theta \rightsquigarrow \mathbf{B}(\,\cdot\,,t)$ is that of $\sqrt{t}$ times a reversible, $\mathbb{R}^N$-valued Ornstein–Uhlenbeck process.

EXERCISE 8.6.11. Continuing in the same setting as in the preceding, set $\sigma^2 = \mathbb{E}^{\mathcal{W}^{(E)}}\big[\|\theta\|^2_{\Theta(E)}\big]$, and combine the result in Exercise 8.2.16 with Brownian scaling invariance to show that

$$\mathcal{W}^{(E)}\left(\sup_{\tau\in[0,t]} \|\theta(t)\|_E \geq R\right) \leq K \exp\left[-\frac{R^2}{180\sigma^2 t}\right],$$

where $K$ is the constant in Fernique's Theorem. Next, use this together with Theorem 8.4.4 and the reasoning in Exercise 4.3.16 to show that

$$\varlimsup_{t\to\infty} \frac{\|\theta(t)\|_E}{\sqrt{2t\log_{(2)} t}} = L = \varlimsup_{t\searrow 0} \frac{\|\theta(t)\|_E}{\sqrt{2t\log_{(2)} \frac{1}{t}}} \quad \mathcal{W}^{(E)}\text{-almost surely},$$

where $L = \sup\big\{\|h\|_E : h \in \overline{B_H(0,1)}\big\}$.

EXERCISE 8.6.12. It should be recognized that Theorem 8.4.4 is an immediate corollary of Theorem 8.6.7. To see this, check that $\{\theta(n) : n \geq 1\}$ has the same distribution under $\mathcal{W}^{(E)}$ as $\{S_n : n \geq 1\}$ has under $\mathcal{W}^{\mathbb{N}}$ and that $\overline{B_H(0,1)} = \{h(1) : h \in \overline{B_{H^1(H)}}\}$, and use these to show that Theorem 8.4.4 follows from Theorem 8.6.7.

EXERCISE 8.6.13. For $\theta \in \Theta(E)$ and $n \in \mathbb{Z}^+$, define $\breve{\theta}_n \in \Theta(E)$ so that

$$\breve{\theta}_n(t) = \sqrt{\frac{n}{\log_{(2)}(n \vee 3)}}\,\theta\left(\frac{t}{n}\right), \quad t \in [0,\infty),$$

and show that, $\mathcal{W}^{(E)}$-almost surely, $\{\breve{\theta}_n : n \geq 1\}$ is relatively compact in $\Theta(E)$ and that $\overline{B_{H^1(H)}(0,1)}$ is the set of its limit points.

**Hint**: Referring to **(ii)** in Exercise 8.6.8, show that it suffices to prove these properties for the sequence $\{(I\theta)\breve{}_n : n \geq 1\}$. Next check that

$$\left\|(I\theta)\breve{}_n - Ih\right\|_{\Theta(E)} = \left\|\tilde{\theta}_n - h\right\|_{\Theta(E)} \quad \text{for } h \in H^1(H),$$

and use Theorem 8.6.7 and the fact that $I$ is an isometry of $H^1(H)$ onto itself.

# Chapter 9

## Convergence of Measures on a Polish Space

In Chapters 2 and 3, I introduced a notion of convergence on $\mathbf{M}_1(\mathbb{R}^N)$ that is appropriate when discussing either Central Limit phenomena or the sort of limits that arose in connection with infinitely divisible laws. In this chapter, I will give a systematic treatment of this sort of convergence and show how it extends to probability measures on any **Polish space**, that is, any complete, separable, metric space. Unfortunately, this extension will entail an excursion into territory that borders on *abstract nonsense*, although I hope to avoid crossing that border. In any case, just as Banach's great achievement was the ingenious use for infinite dimensional vector spaces of completeness to replace local compactness, so here we will have to learn how to substitute compactness by completeness in measure theoretic arguments.

### § 9.1 Prohorov–Varadarajan Theory

The goal in this section is to generalize results like Lemma 2.1.7 and Theorem 3.1.1 to a very abstract setting.

**§ 9.1.1. Some Background.** When discussing the convergence of probability measures on a measurable space $(E, \mathcal{B})$, one always has at least two senses in which the convergence may take place, and (depending on additional structure that the space may possess) one may have more. To be more precise, let $B(E;\mathbb{R}) \equiv B\big((E,\mathcal{B});\mathbb{R}\big)$ be the space of bounded, $\mathbb{R}$-valued, $\mathcal{B}$-measurable functions on $E$, use $\mathbf{M}_1(E) \equiv \mathbf{M}_1(E,\mathcal{B})$ to denote the space of all probability measures on $(E,\mathcal{B})$, and define the **duality relation**

$$\langle \varphi, \mu \rangle = \int_E \varphi \, d\mu \quad \text{for } \varphi \in B(E;\mathbb{R}) \text{ and } \mu \in \mathbf{M}_1(E).$$

Next, again use $\|\varphi\|_{\mathrm{u}} \equiv \sup_{x\in E} |\varphi(x)|$ to denote the uniform norm of $\varphi \in B(E;\mathbb{R})$, and consider the neighborhood basis at $\mu \in \mathbf{M}_1(E)$ determined by the sets

$$U(\mu, r) = \big\{\nu \in \mathbf{M}_1(E) : \big|\langle \varphi, \nu\rangle - \langle \varphi, \mu \rangle\big| < r \text{ for } \varphi \in B(E,\mathbb{R}) \text{ with } \|\varphi\|_{\mathrm{u}} \le 1\big\}$$

as $r$ runs over $(0,\infty)$. For obvious reasons, the topology defined by these neighborhoods $U$ is called the **uniform topology** on $\mathbf{M}_1(E)$. In order to develop some feeling for the uniform topology, I will begin by examining a few of its elementary properties.

LEMMA 9.1.1. *Define the variation distance between elements $\mu$ and $\nu$ of $\mathbf{M}_1(E)$ by*

$$\|\nu-\mu\|_{\mathrm{var}} = \sup\Big\{\big|\langle\varphi,\mu\rangle - \langle\varphi,\nu\rangle\big| : \varphi\in B(E;\mathbb{R}) \text{ with } \|\varphi\|_{\mathrm{u}}\le 1\Big\}.$$

*Then $(\mu,\nu)\in\mathbf{M}_1(E)^2 \longmapsto \|\mu-\nu\|_{\mathrm{var}}$ is a metric on $\mathbf{M}_1(E)$ that is compatible with the uniform topology. Moreover, if $\mu,\nu\in\mathbf{M}_1(E)$ are two elements of $\mathbf{M}_1(E)$ and $\lambda$ is any element of $\mathbf{M}_1(E)$ with respect to which both $\mu$ and $\nu$ are absolutely continuous (e.g., $\frac{\mu+\nu}{2}$), then*

$$\|\mu-\nu\|_{\mathrm{var}} = \|g-f\|_{L^1(\lambda;\mathbb{R})}, \quad \text{where } f=\frac{d\mu}{d\lambda} \text{ and } g=\frac{\partial\nu}{\partial\lambda}. \tag{9.1.2}$$

*In particular, $\|\mu-\nu\|_{\mathrm{var}}\le 2$, and equality holds precisely when $\nu\perp\mu$ (i.e., they are singular). Finally, the metric $(\mu,\nu)\in\mathbf{M}_1(E)^2\longmapsto\|\mu-\nu\|_{\mathrm{var}}$ is complete.*

PROOF: The first assertion needing comment is the one in (9.1.2). But, for every $\varphi\in B(E;\mathbb{R})$ with $\|\varphi\|_{\mathrm{u}}\le 1$,

$$\big|\langle\varphi,\nu\rangle - \langle\varphi,\mu\rangle\big| = \left|\int_E \varphi(g-f)\,d\lambda\right| \le \|g-f\|_{L^1(\lambda;\mathbb{R})},$$

and equality holds when $\varphi = \mathrm{sgn}\circ(g-f)$. To prove the assertion that follows (9.1.2), note that

$$\|g-f\|_{L^1(\lambda;\mathbb{R})} \le \|f\|_{L^1(\lambda;\mathbb{R})} + \|g\|_{L^1(\lambda;\mathbb{R})} = 2$$

and that the inequality is strict if and only if $fg>0$ on a set of strictly positive $\lambda$-measure or, equivalently, if and only if $\mu\not\perp\nu$. Thus, all that remains is to check the completeness assertion. To this end, let $\{\mu_n : n\ge 1\}\subseteq\mathbf{M}_1(E)$ satisfying

$$\lim_{m\to\infty}\sup_{n\ge m}\|\mu_n-\mu_m\|_{\mathrm{var}} = 0$$

be given, and set $\lambda=\sum_{n=1}^\infty 2^{-n}\mu_n$. Clearly, $\lambda$ is an element of $\mathbf{M}_1(E)$ with respect to which each $\mu_n$ is absolutely continuous. Moreover, if $f_n=\frac{d\mu_n}{d\lambda}$, then, by (9.1.2), $\{f_n : n\ge 1\}$ is a Cauchy convergent sequence in $L^1(\lambda;\mathbb{R})$. Hence, since $L^1(\lambda;\mathbb{R})$ is complete, there is an $f\in L^1(\lambda;\mathbb{R})$ to which the $f_n$'s converge in $L^1(\lambda;\mathbb{R})$. Obviously, we may choose $f$ to be non-negative, and certainly it has $\lambda$-integral 1. Thus, the measure $\mu$ given by $d\mu = f\,d\lambda$ is an element of $\mathbf{M}_1(E)$, and, by (9.1.2), $\|\mu_n-\mu\|_{\mathrm{var}}\longrightarrow 0$. □

As a consequence of Lemma 9.1.1, we see that the uniform topology on $\mathbf{M}_1(E)$ admits a complete metric and that convergence in this topology is intimately related to $L^1$-convergence in the $L^1$-space of an appropriate element of $\mathbf{M}_1(E)$.

In fact, $\mathbf{M}_1(E)$ looks in the uniform topology like a *galaxy* that is broken into many *constellations*, each constellation consisting of measures that are all absolutely continuous with respect to some fixed measure. In particular, there will usually be too many *constellations* for $\mathbf{M}_1(E)$ in the uniform topology to be separable. To wit, if $E$ is uncountable and $\{x\} \in \mathcal{B}$ for every $x \in E$, then the **point masses** $\delta_x$, $x \in E$, (i.e., $\delta_x(\Gamma) = \mathbf{1}_\Gamma(x)$) form an uncountable subset of $\mathbf{M}_1(E)$ and $\|\delta_y - \delta_x\|_{\text{var}} = 2$ for $y \neq x$. Hence, in this case, $\mathbf{M}_1(E)$ cannot be covered by a countable collection of open $\|\cdot\|_{\text{var}}$-balls of radius 1.

As I said at the beginning of this section, the uniform topology is not the only one available. Indeed, for many purposes and, in particular, for probability theory, it is *too rigid* a topology to be useful. For this reason, it is often convenient to consider a more lenient topology on $\mathbf{M}_1(E)$. The first one that comes to mind is the one that results from eliminating the *uniformity* in the uniform topology. That is, given a $\mu \in \mathbf{M}_1(E)$, define

$$(9.1.3) \quad S(\mu, \delta; \varphi_1, \dots, \varphi_n) \equiv \left\{\nu \in \mathbf{M}_1(E) : \max_{1 \le k \le n} \left|\langle \varphi_k, \nu\rangle - \langle \varphi_k, \mu\rangle\right| < \delta\right\}$$

for $n \in \mathbb{Z}^+$, $\varphi_1, \dots, \varphi_n \in B(E;\mathbb{R})$, and $\delta > 0$. Clearly these sets $S$ determine a Hausdorff topology on $\mathbf{M}_1(E)$ in which the net $\{\mu_\alpha : \alpha \in A\}$ converges to $\mu$ if and only if $\lim_\alpha \langle \varphi, \mu_\alpha\rangle = \langle \varphi, \mu\rangle$ for every $\varphi \in B(E;\mathbb{R})$. For historical reasons, in spite of the fact that it is obviously *weaker* than the uniform topology, this topology on $\mathbf{M}_1(E)$ is sometimes called the **strong topology**, although, in some of the statistics literature, it is also known as the **$\boldsymbol{\tau}$-topology**.

A good understanding of the relationship between the strong and uniform topologies is most easily gained through functional analytic considerations that will not be particularly important for what follows. Nonetheless, it will be useful to recognize that, except in very special circumstances, the strong topology is *strictly weaker* than the uniform topology. For example, take $E = [0, 1]$ with its Borel field, and consider the probability measures $\mu_n(dt) = (1 + \sin(2n\pi t))\, dt$ for $n \in \mathbb{Z}^+$. Noting that, since $|\sin(2n\pi t) - \sin(2m\pi t)| \le 2$ and therefore

$$\begin{aligned} \tfrac{1}{2}\|\mu_n - \mu_m\|_{\text{var}} &= \int_0^1 \frac{|\sin(2n\pi t) - \sin(2m\pi t)|}{2}\, dt \\ &\ge \frac{1}{4}\int_0^1 \left|\sin(2n\pi t) - \sin(2m\pi t)\right|^2 dt = \frac{1}{4} \end{aligned}$$

for $m \neq n$, one sees that $\{\mu_n : n \ge 1\}$ not only fails to converge in the uniform topology, it does not even have any limit points as $n \to \infty$. On the other hand, because $\left\{2^{\frac{1}{2}} \sin(2n\pi t) : n \ge 1\right\}$ is orthonormal in $L^2(\lambda_{[0,1]};\mathbb{R})$, Bessel's Inequality says that

$$2\sum_{n=1}^{\infty}\left(\int_{[0,1]} \varphi(t)\, \sin(2n\pi t)\, dt\right)^2 \le \|\varphi\|^2_{L^2(\lambda_{[0,1]})} \le \|\varphi\|_{\text{u}}^2 < \infty$$

and therefore $\langle\varphi,\mu_n\rangle \longrightarrow \langle\varphi,\lambda_{[0,1]}\rangle$ for every $\varphi \in B([0,1];\mathbb{R})$. In other words, $\{\mu_n : n \geq 1\}$ *converges to* $\lambda_{[0,1]}$ *in the strong topology, but it converges to nothing at all in the uniform topology.*

§ **9.1.2. The Weak Topology.** Although the strong topology is weaker than the uniform and can be effectively used in various applications, it is still not weak enough for most probabilistic applications. Indeed, even when $E$ possesses a good topological structure and $\mathcal{B} = \mathcal{B}_E$ is the Borel field over $E$, the strong topology on $\mathbf{M}_1(E)$ shows no respect for the topology on $E$. For example, suppose that $E$ is a metric space and, for each $x \in E$, consider the point mass $\delta_x$ on $\mathcal{B}_E$. Then, no matter how *close* $x \in E \setminus \{x\}$ gets to $y$ in the sense of the topology on $E$, $\delta_x$ is not getting *close* to $\delta_y$ in the strong topology on $\mathbf{M}_1(E)$. More generally (cf. Exercise 9.1.15), measures cannot be close in the strong topology unless their *sets of small measure* are essentially the same. Thus, for example, the convergence that is occurring in The Central Limit Theorem (cf. Theorem 2.1.8) cannot, in general, be taking place in the strong topology; and since The Central Limit Theorem is an archetypal example of the sort of convergence result at which probabilists look, it is only sensible for us to take a hint from the result that we got there.

Thus, let $E$ be a metric space, set $\mathcal{B} = \mathcal{B}_E$, and consider the neighborhood basis at $\mu \in \mathbf{M}_1(E)$ given by the sets $S(\mu,\delta;\varphi_1,\dots,\varphi_n)$ in (9.1.3) when the $\varphi_k$'s are *restricted to be elements of* $C_{\mathrm{b}}(E;\mathbb{R})$. The topology that results is *much weaker* than the strong topology, and is therefore justifiably called the **weak topology** on $\mathbf{M}_1(E)$. (The reader who is familiar with the language of functional analysis will, with considerable justice, complain about this terminology. Indeed, if one thinks of $C_{\mathrm{b}}(E;\mathbb{R})$ as a Banach space and of $\mathbf{M}_1(E)$ as a subspace of its dual space $C_{\mathrm{b}}(E;\mathbb{R})^*$, then the topology that I am calling the weak topology is what a functional analyst would call the weak* topology. However, because it is the most commonly accepted choice of probabilists, I will continue to use the term *weak* instead of the more correct term *weak**.) In particular, the weak topology respects the topology on $E$: $\delta_y$ tends to $\delta_x$ in the weak topology on $\mathbf{M}_1(E)$ if and only if $y \longrightarrow x$ in $E$. Lemma 2.3.3 provides further evidence that the weak topology is well adapted to the sort of analysis encountered in probability theory, since, by that lemma, weak convergence of $\{\mu_n : n \geq 1\} \subseteq \mathbf{M}_1(\mathbb{R}^N)$ to $\mu$ is equivalent to pointwise convergence of $\widehat{\mu_n}(\boldsymbol{\xi})$ to $\hat{\mu}(\boldsymbol{\xi})$.

Besides being well adapted to probabilistic analysis, the weak topology turns out to have many intrinsic virtues that are not shared by either the uniform or strong topologies. In particular, as we will see shortly, when $E$ is a separable metric space, the weak topology on $\mathbf{M}_1(E)$ is not only a metric topology, which (cf. Exercise 9.1.15) the strong topology seldom is, but it is even separable, which, as we have seen, the uniform topology seldom is. In order to check these properties, we will first have to review some elementary facts about separable metric spaces.

Given a metric $\rho$ for a topological space $E$, I will use $U^{\rho}_{\mathrm{b}}(E;\mathbb{R})$ to denote the space of bounded, $\rho$-uniformly continuous $\mathbb{R}$-valued functions on $E$ and will endow $U_b^\rho(E;\mathbb{R})$ with the topology determined by the uniform norm. Thus, $U_b^\rho(E;\mathbb{R})$ becomes in this way a closed subspace of $C_b(E;\mathbb{R})$.

LEMMA 9.1.4. *Let $E$ be a separable metric space. Then $E$ is homeomorphic to a subset of $[0,1]^{\mathbb{Z}^+}$. In particular:*

**(i)** *If $E$ is compact, then the space $C(E;\mathbb{R})$ is separable with respect to the uniform metric.*

**(ii)** *Even when $E$ is not compact, it nonetheless admits a metric $\hat{\rho}$ with respect to which it becomes a totally bounded metric space.*

**(iii)** *If $\hat{\rho}$ is a totally bounded metric on $E$, then $U_b^{\hat{\rho}}(E;\mathbb{R})$ is separable.*

PROOF: Let $\rho$ be any metric on $E$, and choose $\{p_n : n \geq 1\}$ to be a countable, dense subset of $E$. Next, define $\mathbf{h} : E \longrightarrow [0,1]^{\mathbb{Z}^+}$ to be the mapping whose $n$th coordinate is given by

$$h_n(x) = \frac{\rho(x,p_n)}{1+\rho(x,p_n)}, \quad x \in E.$$

It is then an easy matter to check that $\mathbf{h}$ is homeomorphic onto a subset of $[0,1]^{\mathbb{Z}^+}$.

To prove (**i**), I will first check it for compact subsets $K$ of $E = [0,1]^{\mathbb{Z}^+}$. To this end, denote by $\mathcal{P}$ the space of polynomials $p : [0,1]^{\mathbb{Z}^+} \longrightarrow \mathbb{R}$. That is, $\mathcal{P}$ consists of finite, $\mathbb{R}$-linear combinations of the monomials $\boldsymbol{\xi} \in [0,1]^{\mathbb{Z}^+} \longmapsto \xi_{k_1}^{n_1} \cdots \xi_{k_\ell}^{n_\ell}$, where $\ell \geq 1$, $1 \leq k_1 < \cdots < k_\ell$, and $\{n_1, \dots, n_\ell\} \subseteq \mathbb{N}$. Clearly, if $\mathcal{P}_0$ is the subset of $\mathcal{P}$ consisting of those $p$'s with rational coefficients, then $\mathcal{P}_0$ is countable, and $\mathcal{P}_0$ is dense in $\mathcal{P}$. Thus, it suffices to show that $\{p \restriction K : p \in \mathcal{P}\}$ is dense in $C(K;\mathbb{R})$. But $\mathcal{P}$ is obviously an algebra. In addition, if $\boldsymbol{\xi}$ and $\boldsymbol{\eta}$ are distinct points in $[0,1]^{\mathbb{Z}^+}$, it is an easy (in fact, a one dimensional) matter to see that there is a $p \in \mathcal{P}$ for which $p(\boldsymbol{\xi}) \neq p(\boldsymbol{\eta})$. Hence, the desired density follows from the Stone–Weierstrass Approximation Theorem. Finally, for an arbitrary compact metric space $E$, define $\mathbf{h} : E \longrightarrow [0,1]^{\mathbb{Z}^+}$ as above, note that $K \equiv \mathbf{h}(E)$ is compact, and conclude that the map $\varphi \in C(K;\mathbb{R}) \longmapsto \varphi \circ \mathbf{h} \in C(E;\mathbb{R})$ is a homeomorphism between the uniform topologies on these spaces. Since we already know that $C(K;\mathbb{R})$ is separable, this completes (**i**).

The proof of (**ii**) is easy. Namely, define

$$D(\mathbf{x},\boldsymbol{\eta}) = \sum_{n=1}^{\infty} \frac{|\xi_n - \eta_n|}{2^n} \quad \text{for } \mathbf{x}, \boldsymbol{\eta} \in [0,1]^{\mathbb{Z}^+}.$$

Clearly, $D$ is a metric for $[0,1]^{\mathbb{Z}^+}$, and therefore

$$(x,y) \in E^2 \longmapsto \hat{\rho}(x,y) \equiv D(\mathbf{h}(x), \mathbf{h}(y))$$

is a metric for $E$. At the same time, since $[0,1]^{\mathbb{Z}^+}$ is compact, and therefore the restriction of $D$ to any subset is totally bounded, it is clear that $\hat{\rho}$ is totally bounded on $E$.

To prove **(iii)**, let $\hat{E}$ denote the completion of $E$ with respect to the totally bounded metric $\hat{\rho}$. Then, because $E$ is dense in $\hat{E}$, $\hat{E}$ is both complete and totally bounded and therefore compact. In addition, $\hat{\varphi} \in C(\hat{E};\mathbb{R}) \longmapsto \hat{\varphi} \restriction E \in U_{\mathrm{b}}^{\hat{\rho}}(E;\mathbb{R})$ is a surjective homeomorphism; and so **(iii)** now follows from **(i)**. □

One of the main reasons why Lemma 9.1.4 will be important to us is that it will enable us to show that, for separable metric spaces $E$, the weak topology on $\mathbf{M}_1(E)$ is also a separable metric topology. However, thus far we do not even know that the neighborhood bases are countably generated, and so, for a moment longer, I must continue to consider nets when discussing convergence. In order to indicate that a net $\{\mu_\sigma : \alpha \in A\} \subseteq \mathbf{M}_1(E)$ is converging weakly (i.e., in the weak topology) to $\mu$, I will write $\mu_\alpha \Longrightarrow \mu$.

THEOREM 9.1.5. *Let $E$ be any metric space and $\{\mu_\alpha : \alpha \in A\}$ a net in $\mathbf{M}_1(E)$. Given any $\mu \in \mathbf{M}_1(E)$, the following statements are equivalent:*

**(i)** $\mu_\alpha \Longrightarrow \mu$.

**(ii)** *If $\rho$ is any metric for $E$, then $\langle\varphi,\mu_\alpha\rangle \longrightarrow \langle\varphi,\mu\rangle$ for every $\varphi \in U_{\mathrm{b}}^{\rho}(E;\mathbb{R})$.*

**(iii)** *For every closed set $F \subseteq E$, $\overline{\lim_\alpha}\, \mu_\alpha(F) \le \mu(F)$.*

**(iv)** *For every open set $G \subseteq E$, $\underline{\lim_\alpha}\, \mu_\alpha(G) \ge \mu(G)$.*

**(v)** *For every upper semicontinuous function $f : E \longrightarrow \mathbb{R}$ that is bounded above, $\overline{\lim_\alpha}\langle f,\mu_\alpha\rangle \le \langle f,\mu\rangle$.*

**(vi)** *For every lower semicontinuous function $f : E \longrightarrow \mathbb{R}$ that is bounded below, $\underline{\lim_\alpha}\langle f,\mu_\alpha\rangle \ge \langle f,\mu\rangle$.*

**(vii)** *For every $f \in B(E;\mathbb{R})$ that is continuous at $\mu$-almost every $x \in E$, $\langle f,\mu_\alpha\rangle \longrightarrow \langle f,\mu\rangle$.*

*Finally, assume that $E$ is separable, and let $\hat{\rho}$ be a totally bounded metric for $E$. Then there exists a countable subset $\{\varphi_n : n \ge 1\} \subseteq U_{\mathrm{b}}^{\hat{\rho}}(E;[0,1])$ that is dense in $U_{\mathrm{b}}^{\hat{\rho}}(E;\mathbb{R})$, and therefore the mapping $\mathbf{H} : \mathbf{M}_1(E) \longrightarrow [0,1]^{\mathbb{Z}^+}$ given by $\mathbf{H}(\mu) = \big(\langle\varphi_1,\mu\rangle,\dots,\langle\varphi_n,\mu\rangle,\dots\big)$ is a homeomorphism from the weak topology on $\mathbf{M}_1(E)$ into $[0,1]^{\mathbb{Z}^+}$. In particular, when $E$ is separable, $\mathbf{M}_1(E)$ with the weak topology is itself a separable metric space and, in fact, one can take*

$$(\mu,\nu) \in \mathbf{M}_1(E)^2 \longmapsto R(\mu,\nu) \equiv \sum_{n=1}^{\infty} \frac{\big|\langle\varphi_n,\mu\rangle - \langle\varphi_n,\nu\rangle\big|}{2^n}$$

*to be a metric for $\mathbf{M}_1(E)$.*

PROOF: The implications

$$\textbf{(vii)} \implies \textbf{(i)} \implies \textbf{(ii)}, \qquad \textbf{(iii)} \iff \textbf{(iv)}, \quad \text{and } \textbf{(v)} \iff \textbf{(vi)}$$

are all trivial. Thus, the first part will be complete once I check that **(ii)** $\implies$ **(iii)**, **(iv)** $\implies$ **(vi)**, and that **(v)** together with **(vi)** imply **(vii)**. To see the first of these, let $F$ be a closed subset of $E$, and set

$$\psi_n(x) = 1 - \left(\frac{\rho(x,F)}{1+\rho(x,F)}\right)^{\frac{1}{n}} \quad \text{for } n \in \mathbb{Z}^+ \text{ and } x \in E.$$

It is then clear that $\psi_n \in U_b^\rho(E;\mathbb{R})$ for each $n \in \mathbb{Z}^+$ and that $1 \geq \psi_n(x) \searrow \mathbf{1}_F(x)$ as $n \to \infty$ for each $x \in E$. Thus, The Monotone Convergence Theorem followed by **(ii)** imply that

$$\mu(F) = \lim_{n\to\infty} \langle \psi_n, \mu\rangle = \lim_{n\to\infty} \lim_\alpha \langle \psi_n, \mu_\alpha\rangle \geq \overline{\lim_\alpha}\, \mu_\alpha(F).$$

In proving that **(iv)** $\implies$ **(vi)**, I may and will assume that $f$ is a non-negative, lower semicontinuous function. For $n \in \mathbb{N}$, define

$$f_n = \sum_{\ell=0}^{\infty} \frac{\ell \wedge 4^n}{2^n} \mathbf{1}_{I_{\ell,n}} \circ f = \frac{1}{2^n} \sum_{\ell=0}^{4^n} \mathbf{1}_{J_{\ell,n}} \circ f,$$

where

$$I_{\ell,n} = \left(\frac{\ell}{2^n}, \frac{\ell+1}{2^n}\right] \quad \text{and} \quad J_{\ell,n} = \left(\frac{\ell}{2^n}, \infty\right).$$

It is then clear that $0 \leq f_n \nearrow f$ and therefore that $\langle f_n, \mu\rangle \longrightarrow \langle f, \mu\rangle$ as $n \to \infty$. At the same time, by lower semicontinuity, the sets $\{f \in J_{\ell,n}\}$ are open, and so **(iv)** implies

$$\langle f_n, \mu\rangle \leq \underline{\lim_\alpha} \langle f_n, \mu_\alpha\rangle \leq \underline{\lim_\alpha} \langle f, \mu_\alpha\rangle$$

for each $n \in \mathbb{Z}^+$. After letting $n \to \infty$, one sees that **(iv)** $\implies$ **(vi)**.

Turning to the proof that **(v)** & **(vi)** $\implies$ **(vii)**, suppose that $f \in B(E;\mathbb{R})$ is continuous at $\mu$-almost every $x \in E$, and define

$$\underline{f}(x) = \underline{\lim_{y\to x}}\, f(y) \quad \text{and} \quad \overline{f}(x) = \overline{\lim_{y\to x}}\, f(y) \quad \text{for } x \in E.$$

It is then an easy matter to check that $\underline{f} \leq f \leq \overline{f}$ everywhere and that equality holds $\mu$-almost surely. Furthermore, $\underline{f}$ is lower semicontinuous, $\overline{f}$ is upper semicontinuous, and both are bounded. Hence, by **(v)** and **(vi)**,

$$\overline{\lim_\alpha} \langle f, \mu_\alpha\rangle \leq \overline{\lim_\alpha} \langle \overline{f}, \mu_\alpha\rangle \leq \langle \overline{f}, \mu\rangle = \langle \underline{f}, \mu\rangle \leq \underline{\lim_\alpha} \langle \underline{f}, \mu_\alpha\rangle \leq \underline{\lim_\alpha} \langle f, \mu_\alpha\rangle;$$

and so I have now completed the proof that conditions (**i**) through (**vii**) are equivalent.

Now assume that $E$ is separable, and let $\hat{\rho}$ be a totally bounded metric for $E$. By (**iii**) of Lemma 9.1.4, $U_b^{\hat{\rho}}(E;\mathbb{R})$ is separable. Hence, we can find a countable set $\{\varphi_n : n \geq 1\}$ that is dense in $U_b^{\hat{\rho}}(E;\mathbb{R})$. In particular, by the equivalence of (**i**) and (**ii**) above, we see that $\langle \varphi_n, \mu_\alpha\rangle \longrightarrow \langle \varphi_n, \mu\rangle$ for all $n \in \mathbb{Z}^+$ if and only if $\mu_\alpha \Longrightarrow \mu$, which is to say that the corresponding map $\mathbf{H} : \mathbf{M}_1(E) \longrightarrow [0,1]^{\mathbb{Z}^+}$ is a homeomorphism. Since $[0,1]^{\mathbb{Z}^+}$ is a compact metric space and $D$ (cf. the proof of (**ii**) in Lemma 9.1.4) is a metric for it, we also see that the $R$ described is a totally bounded metric for $\mathbf{M}_1(E)$. In particular, $\mathbf{M}_1(E)$ is separable. Finally, since, by (**ii**) in Lemma 9.1.4, it is always possible to find a totally bounded metric for $E$, the last assertion needs no further comment. □

The reader would do well to pay close attention to what (**iii**) and (**iv**) say about the nature of weak convergence. Namely, even though $\mu_\alpha \Longrightarrow \mu$, it is possible that some or all of the mass that the $\mu_\alpha$'s assign to the interior of a set may gravitate to the boundary in the limit. This phenomenon is most easily understood by taking $E = \mathbb{R}$, $\mu_\alpha$ to be the unit point mass $\delta_\alpha$ at $\alpha \in [0,1)$, checking that $\delta_\alpha \Longrightarrow \delta_1$, and noting that $\delta_1\big((0,1)\big) = 0 < 1 = \delta_\alpha\big((0,1)\big)$ for each $\alpha \in [0,1)$.

REMARK 9.1.6. Those who find nets distasteful will be pleased to learn that, from now on, I will be restricting my attention to separable metric spaces $E$ and therefore need only discuss sequential convergence when working with the weak topology on $\mathbf{M}_1(E)$. Furthermore, unless the contrary is explicitly stated, *I will always be thinking of the weak topology when working with* $\mathbf{M}_1(E)$.

Given a separable metric space $E$, I next want to find conditions that guarantee that a subset of $\mathbf{M}_1(E)$ is compact; and at this point it will be convenient to have introduced the notation $K \subset\subset E$ to indicate that $K$ is a compact subset of $E$. The key to my analysis is the following extension of the sort of Riesz Representation result in Theorem 3.1.1 combined with a crucial observation made by S. Ulam.[1]

LEMMA 9.1.7. *Let $E$ be a separable metric space, $\rho$ a metric for $E$, and $\Lambda$ a non-negative linear functional on $U_b^{\rho}(E;\mathbb{R})$ (i.e., $\Lambda$ is a linear map that assigns a non-negative value to a non-negative $\varphi \in U_b^{\rho}(E;\mathbb{R})$) with $\Lambda(\mathbf{1}) = 1$. Then, in order for there to be a (necessarily unique) $\mu \in \mathbf{M}_1(E)$ satisfying $\Lambda(\varphi) = \langle \varphi, \mu\rangle$ for all $\varphi \in U_b^{\rho}(E;\mathbb{R})$, it is sufficient that, for every $\epsilon > 0$, there exist a $K \subset\subset E$*

[1] It is no accident that Ulam was the first to make this observation. Indeed, the term *Polish space* was coined by Bourbaki in recognition of the contribution made to this subject by the Polish school in general and C. Kuratowski in particular (cf. Kuratowski's *Topologie, Vol. I*, Warszawa–Lwow (1933)). Ulam had studied with Kuratowski.

*such that*

$$|\Lambda(\varphi)| \le \sup_{x\in K} |\varphi(x)| + \epsilon\|\varphi\|_{\mathrm{u}}, \quad \varphi \in U^{\rho}_{\mathrm{b}}(E;\mathbb{R}). \tag{9.1.8}$$

*Conversely, if $E$ is a Polish space and $\mu \in \mathbf{M}_1(E)$, then for every $\epsilon > 0$ there is a $K \subset\subset E$ such that $\mu(K) \ge 1-\epsilon$. In particular, if $\mu \in \mathbf{M}_1(E)$ and $\Lambda(\varphi) = \langle \varphi, \mu\rangle$ for $\varphi \in C_{\mathrm{b}}(E;\mathbb{R})$, then, for each $\epsilon > 0$, (9.1.8) holds for some $K \subset\subset E$.*

PROOF: I begin with the trivial observation that, because $\Lambda$ is non-negative and $\Lambda(\mathbf{1}) = 1$, $|\Lambda(\varphi)| \le \|\varphi\|_{\mathrm{u}}$. Next, according to the Daniell theory of integration, the first statement will be proved as soon as we know that $\Lambda(\varphi_n) \searrow 0$ whenever $\{\varphi_n : n \ge 1\}$ is a non-increasing sequence of functions from $U^{\rho}_{\mathrm{b}}(E;[0,\infty))$ that tend pointwise to 0 as $n \to \infty$. To this end, let $\epsilon > 0$ be given, and choose $K \subset\subset E$ so that (9.1.8) holds. One then has that

$$\varlimsup_{n\to\infty} |\Lambda(\varphi_n)| \le \lim_{n\to\infty} \sup_{x\in K} |\varphi_n(x)| + \epsilon\|\varphi_1\|_{\mathrm{u}} = \epsilon\|\varphi_1\|_{\mathrm{u}},$$

since, by Dini's Lemma, $\varphi_n \searrow 0$ uniformly on compact subsets of $E$.

Turning to the second part, assume that $E$ is Polish, and use $B(x,r)$ to denote the open ball of radius $r > 0$ around $x \in E$, computed with respect to a complete metric $\rho$ for $E$. Next, let $\{p_k : k \ge 1\}$ be a countable dense subset of $E$, and set $B_{k,n} = B\left(p_k, \frac{1}{n}\right)$ for $k,\, n \in \mathbb{Z}^+$. Given $\mu \in \mathbf{M}_1(E)$ and $\epsilon > 0$, we can choose, for each $n \in \mathbb{Z}^+$, an $\ell_n \in \mathbb{Z}^+$ so that

$$\mu\left(\bigcup_{k=1}^{\ell_n} B_{k,n}\right) \ge 1 - \frac{\epsilon}{2^n}.$$

Hence, if

$$C_n \equiv \bigcup_{k=1}^{\ell_n} \overline{B}_{k,n} \quad \text{and} \quad K = \bigcap_{n=1}^{\infty} C_n,$$

then $\mu(K) \ge 1-\epsilon$. At the same time, it is obvious that, on the one hand, $K$ is closed (and therefore $\rho$-complete) and that, on the other hand, $K \subseteq \bigcup_{k=1}^{\ell_n} B\left(p_k, \frac{2}{n}\right)$ for every $n \in \mathbb{Z}^+$. Hence, $K$ is both complete and totally bounded with respect to $\rho$ and, as such, is compact. □

As Lemma 9.1.7 makes clear, probability measures on a Polish space like to be *nearly concentrated on a compact set*. Following Prohorov and Varadarajan,[2]

[2] See Yu. V. Prohorov's article "Convergence of random processes and limit theorems in probability theory," *Theory of Prob. & Appl.*, which appeared in 1956. Independently, V.S. Varadarajan developed essentially the same theory in "Weak convergence of measures on a separable metric spaces," *Sankhyă*, which was published in 1958. Although Prohorov got into print first, subsequent expositions, including this one, rely heavily on Varadarajan.

what we are about to see is that, for a Polish space $E$, relatively compact subsets of $\mathbf{M}_1(E)$ are those whose elements are *nearly concentrated on the same compact set of* $E$. More precisely, given a separable metric space $E$, say that $M \subseteq \mathbf{M}_1(E)$ is **tight** if, for every $\epsilon > 0$, there exists a $K \subset\subset E$ such that $\mu(K) \geq 1 - \epsilon$ for all $\mu \in M$.

THEOREM 9.1.9. *Let $E$ be a separable metric space and $M \subseteq \mathbf{M}_1(E)$. Then $\overline{M}$ is compact if $M$ is tight. Conversely, when $E$ is Polish, $M$ is tight if $\overline{M}$ is compact.*[3]

PROOF: Since it is clear, from (**iii**) in Theorem 9.1.5, that $\overline{M}$ is tight if and only if $M$ is, I will assume throughout that $M$ is closed in $\mathbf{M}_1(E)$.

To prove the first statement, take $\hat{\rho}$ to be a totally bounded metric on $E$, choose $\{\varphi_n : n \geq 1\} \subseteq U_{\mathrm{b}}^{\hat{\rho}}(E;[0,1])$ accordingly, as in the last part of Theorem 9.1.5, and let $\varphi_0 = \mathbf{1}$. Given a sequence $\{\mu_\ell : \ell \geq 1\} \subseteq \mathbf{M}_1(E)$, we can use a standard diagonalization procedure to extract a subsequence $\{\mu_{\ell_k} : k \geq 1\}$ such that

$$\Lambda(\varphi_n) \equiv \lim_{k\to\infty} \langle \varphi_n, \mu_{\ell_k} \rangle$$

exists for each $n \in \mathbb{N}$. Since $\Lambda(\varphi) \equiv \lim_{k\to\infty}\langle \varphi, \mu_{\ell_k}\rangle$ continues to exist for every $\varphi$ in the uniform closure of the span of $\{\varphi_n : n \geq 1\}$, we now see that $\Lambda$ determines a non-negative linear functional on $U_{\mathrm{b}}^{\hat{\rho}}(E;\mathbb{R})$ and that $\Lambda(\mathbf{1}) = 1$. Moreover, because $M$ is tight, we can find, for any $\epsilon > 0$, a $K \subset\subset E$ such that $\mu(K) \geq 1 - \epsilon$ for every $\mu \in M$, and therefore (9.1.8) holds with this choice of $K$. Hence, by Lemma 9.1.7, we know that there is a $\mu \in \mathbf{M}_1(E)$ for which $\Lambda(\varphi) = \langle \varphi, \mu \rangle$, $\varphi \in U_{\mathrm{b}}^{\hat{\rho}}(E;\mathbb{R})$. Because this means that $\langle \varphi, \mu_{\ell_k}\rangle \longrightarrow \langle \varphi, \mu \rangle$ for every $\varphi \in U_{\mathrm{b}}^{\hat{\rho}}(E;\mathbb{R})$, the equivalence of (**i**) and (**ii**) in Theorem 9.1.5 allows us to conclude that $\mu_{\ell_k} \Longrightarrow \mu$.

Finally, suppose that $E$ is Polish and that $M$ is compact in $\mathbf{M}_1(E)$. To see that $M$ must be tight, repeat the argument used to prove the second part of Lemma 9.1.7. Thus, choose $B_{k,n}$ for $k,\, n \in \mathbb{Z}^+$ as in the proof there, and set

$$f_{\ell,n}(\mu) = \mu\left(\bigcup_{k=1}^{\ell} B_{k,n}\right) \quad \text{for } \ell,\, n \in \mathbb{Z}^+.$$

By (**iv**) in Theorem 9.1.5, $\mu \in \mathbf{M}_1(E) \longmapsto f_{\ell,n}(\mu) \in [0,1]$ is lower semicontinuous. Moreover, for each $n \in \mathbb{Z}^+$, $f_{\ell,n} \nearrow 1$ as $\ell \nearrow \infty$. Thus, by Dini's Lemma, we can choose, for each $n \in \mathbb{Z}^+$, one $\ell_n \in \mathbb{Z}^+$ so that $f_{\ell_n,n}(\mu) \geq 1 - \frac{\epsilon}{2^n}$ for all

[3] For the reader who wishes to investigate just how far these results can be pushed before they start of break down, a good place to start is Appendix III in P. Billingsley's *Convergence of Probability Measures*, Wiley (1968). In particular, although it is reasonably clear that completeness is more or less essential for the necessity, the havoc that results from dropping separability may come as a surprise.

$\mu \in M$; and at this point the rest of the argument is precisely the same as the one given at the end of the proof of Lemma 9.1.7. □

**§ 9.1.3. The Lévy Metric and Completeness of $\mathbf{M}_1(E)$.** We have now seen that $\mathbf{M}_1(E)$ inherits properties from $E$. To be more specific, if $E$ is a metric space, then $\mathbf{M}_1(E)$ is separable or compact if $E$ itself is. What I want to show next is that completeness also gets transferred. That is, I will show that $\mathbf{M}_1(E)$ is Polish if $E$ is. In order to do this, I will need a lemma that is of considerable importance in its own right.

LEMMA 9.1.10. *Let $E$ be a Polish space and $\Phi$ a bounded subset of $C_b(E;\mathbb{R})$ that is equicontinuous at each $x \in E$. (That is, for each $x \in E$, $\sup_{\varphi\in\Phi} |\varphi(y) - \varphi(x)| = 0$ as $y \to x$.) If $\{\mu_n : n \geq 1\} \cup \{\mu\} \subseteq \mathbf{M}_1(E)$ and $\mu_n \Longrightarrow \mu$, then*

$$\lim_{n\to\infty} \sup_{\varphi\in\Phi} \Big| \langle \varphi, \mu_n\rangle - \langle \varphi, \mu\rangle \Big| = 0.$$

PROOF: Let $\epsilon > 0$ be given, and use the second part of Theorem 9.1.9 to choose $K \subset\subset E$ so that

$$\left(\sup_{\varphi\in\Phi} \|\varphi\|_{\mathrm{u}}\right)\left(\sup_{n\in\mathbb{Z}^+} \mu_n(K\complement)\right) < \frac{\epsilon}{4}.$$

By (**iv**) of Theorem 9.1.5, $\mu(K\complement)$ satisfies the same estimate. Next, choose a metric $\rho$ for $E$ and a countable dense set $\{p_k : k \geq 1\}$ in $K$. Using equicontinuity together with compactness, find $\ell \in \mathbb{Z}^+$ and $\delta_1, \dots, \delta_\ell > 0$ so that $K \subseteq \{x : \rho(x, p_k) < \delta_k \text{ for some } 1 \leq k \leq \ell\}$ and

$$\sup_{\varphi\in\Phi} \big|\varphi(x) - \varphi(p_k)\big| < \frac{\epsilon}{4} \quad \text{for } 1 \leq k \leq \ell \text{ and } x \in K \text{ with } \rho(x, p_k) < 2\delta_k.$$

Because $r \in (0,\infty) \longmapsto \mu(\{y \in K : \rho(y,x) \leq r\}) \in [0,1]$ is non-decreasing for each $x \in K$, we can find, for each $1 \leq k \leq \ell$, an $r_k \in (\delta_k, 2\delta_k)$ such that $\mu(\partial B_k) = 0$ when $B_k \equiv \{x \in K : \rho(x, p_k) < r_k\}$. Finally, set $A_1 = B_1$ and $A_{k+1} = B_{k+1} \setminus \bigcup_{j=1}^k B_j$ for $1 \leq k < \ell$. Then, $K \subseteq \bigcup_{k=1}^\ell A_k$, the $A_k$'s are disjoint, and, for each $1 \leq k \leq \ell$,

$$\sup_{\varphi\in\Phi} \sup_{x\in A_k} \big|\varphi(x) - \varphi(p_k)\big| < \frac{\epsilon}{4} \quad \text{and} \quad \mu(\partial A_k) = 0.$$

Hence, by (**vii**) in Theorem 9.1.5 applied to the $\mathbf{1}_{A_k}$'s,

$$\varlimsup_{n\to\infty} \sup_{\varphi\in\Phi} \Big|\langle\varphi,\mu_n\rangle - \langle\varphi,\mu\rangle\Big| < \epsilon + \varlimsup_{n\to\infty} \sum_{k=1}^{\ell} \sup_{\varphi\in\Phi} |\varphi(p_k)|\, \big|\mu_n(A_k) - \mu(A_k)\big| = \epsilon. \quad \square$$

THEOREM 9.1.11. *Let $E$ be a Polish space and $\rho$ a complete metric for $E$. Given $(\mu,\nu) \in \mathbf{M}_1(E)^2$, define*

$$L(\mu,\nu) = \inf\Big\{\delta : \mu(F) \le \nu\big(F^{(\delta)}\big) + \delta \\ \text{and } \nu(F) \le \mu\big(F^{(\delta)}\big) + \delta \text{ for all closed } F \subseteq E\Big\},$$

*where $F^{(\delta)}$ denotes the set of $x \in E$ that lie a $\rho$-distance less than $\delta$ from $F$. Then $L$ is a complete metric for $\mathbf{M}_1(E)$, and therefore $\mathbf{M}_1(E)$ is Polish.*

PROOF: It is clear that $L$ is symmetric and that it satisfies the triangle inequality. Thus, we will know that it is a metric for $\mathbf{M}_1(E)$ as soon as we show that $L(\mu_n,\mu) \longrightarrow 0$ if and only if $\mu_n \Longrightarrow \mu$. To this end, first suppose that $L(\mu_n,\mu) \longrightarrow 0$. Then, for every closed $F$, $\mu(F^{(\delta)}) + \delta \ge \overline{\lim}_{n\to\infty} \mu_n(F)$ for all $\delta > 0$; and therefore, by countable additivity, $\mu(F) \ge \overline{\lim}_{n\to\infty}\mu_n(F)$ for every closed $F$. Hence, by the equivalence of (**i**) and (**iii**) in Theorem 9.1.5, $\mu_n \Longrightarrow \mu$. Now suppose that $\mu_n \Longrightarrow \mu$, and let $\delta > 0$ be given. Given a closed $F$ in $E$, define

$$\psi_F(x) = \frac{\rho\big(x, F^{(\delta)}\complement\big)}{\rho\big(x, F^{(\delta)}\complement\big) + \rho(x,F)} \quad \text{for} \quad x \in E.$$

It is then an easy matter to check that both

$$\mathbf{1}_F \le \psi_F \le \mathbf{1}_{F^{(\delta)}} \quad \text{and} \quad \big|\psi_F(x) - \psi_F(y)\big| \le \frac{\rho(x,y)}{\delta}.$$

In particular, by Lemma 9.1.10, we can choose $m \in \mathbb{Z}^+$ so that

$$\sup_{n\ge m}\, \sup\Big\{\Big|\langle\psi_F,\mu_n\rangle - \langle\psi_F,\mu\rangle\Big| : F \text{ closed in } E\Big\} < \delta,$$

from which it is an easy matter to see that, for all $n \ge m$,

$$\mu(F) \le \mu_n\big(F^{(\delta)}\big) + \delta \quad \text{and} \quad \mu_n(F) \le \mu\big(F^{(\delta)}\big) + \delta.$$

In other words, $\sup_{n\ge m} L(\mu_n,\mu) \le \delta$, and, since $\delta > 0$ was arbitrary, we have shown that $L(\mu_n,\mu) \longrightarrow 0$.

In order to finish the proof, I must show that if $\{\mu_n : n \ge 1\} \subseteq \mathbf{M}_1(E)$ is $L$-Cauchy convergent, then it is tight. Thus, let $\epsilon > 0$ be given, and choose, for each $\ell \in \mathbb{Z}^+$, an $m_\ell \in \mathbb{Z}^+$ and a $K_\ell \subset\subset E$ so that

$$\sup_{n\ge m_\ell} L\big(\mu_n,\mu_{m_\ell}\big) \le \frac{\epsilon}{2^{\ell+1}} \quad \text{and} \quad \max_{1\le n\le m_\ell} \mu_n\big(K_\ell\complement\big) \le \frac{\epsilon}{2^{\ell+1}}.$$

Setting $\epsilon_\ell = \frac{\epsilon}{2^\ell}$, one then has that $\sup_{n\in\mathbb{Z}^+} \mu_n\big(K_\ell^{(\epsilon_\ell)}\complement\big) \le \epsilon_\ell$ for each $\ell \in \mathbb{Z}^+$. In particular, if $K \equiv \bigcap_{\ell=1}^\infty \overline{K_\ell^{(\epsilon_\ell)}}$, then $\mu_n(K) \ge 1 - \epsilon$ for all $n \in \mathbb{Z}^+$. Finally, because each $K_\ell$ is compact, it is easy to see that $K$ is both $\rho$-complete and totally bounded and therefore also compact. □

When $E = \mathbb{R}$, P. Lévy was the first to construct a complete metric on $\mathbf{M}_1(E)$, and it is for this reason that I will call the metric $L$ described in Theorem 9.1.11 the **Lévy metric** determined by $\rho$. Using an abstract argument, Varadarajan showed that $\mathbf{M}_1(E)$ must be Polish whenever $E$ is, and the explicit construction that I have used is essentially the one first produced by Prohorov.

Before closing this subsection, it seems appropriate to introduce and explain some of the more classical terminology connected with applications of weak convergence to probability theory. For this purpose, let $(\Omega, \mathcal{F}, \mathbb{P})$ be a probability space and $E$ a metric space. Given a sequence $\{X_n : n \geq 1\}$ of $E$-valued random variables on $(\Omega, \mathcal{F}, \mathbb{P})$, one says that the $\{X_n : n \geq 1\}$ **tends in law** (or **in distribution**) to the $E$-valued random variable $X$ and writes $X_n \xrightarrow{\mathcal{L}} X$ if (cf. Exercise 1.1.16) $(X_n)_*\mathbb{P} \Longrightarrow X_*\mathbb{P}$. The idea here is that, when the measures under consideration are the distributions of random variables, one wants to think of weak convergence of the distributions as determining a kind of convergence of the corresponding random variables. Thus, one can add *convergence in law* to the list of possible ways in which random variables might converge. In order to elucidate the relationship between convergence in law, $\mathbb{P}$-almost sure convergence, and convergence in $\mathbb{P}$-measure, it will be convenient to have the following lemma.

LEMMA 9.1.12. *Let $(\Omega, \mathcal{F}, \mathbb{P})$ be a probability space and $E$ a metric space. Given any $E$-valued random variables $\{X_n : n \geq 1\} \cup \{X\}$ on $(\Omega, \mathcal{F}, \mathbb{P})$ and any pair of topologically equivalent metrics $\rho$ and $\sigma$ for $E$, $\rho(X_n, X) \longrightarrow 0$ in $\mathbb{P}$-measure if and only if $\sigma(X_n, X) \longrightarrow 0$ in $\mathbb{P}$-measure. In particular, convergence in $\mathbb{P}$-measure does not depend on the choice of metric, and so one can write $X_n \longrightarrow X$ in $\mathbb{P}$-measure without specifying a metric. Moreover, if $X_n \longrightarrow X$ in $\mathbb{P}$-measure, then $X_n \xrightarrow{\mathcal{L}} X$. In fact, if $E$ is a Polish space and $L$ is the Lévy metric on $\mathbf{M}_1(E)$ associated with a complete metric $\rho$ for $E$, then*

$$L(X_*\mathbb{P}, Y_*\mathbb{P}) \leq \delta \vee \mathbb{P}\big(\rho(X,Y) \geq \delta\big)$$

*for all $\delta > 0$ and $E$-valued random variables $X$ and $Y$.*

PROOF: To prove the first assertion, suppose that

$$\rho(X_n, X) \longrightarrow 0 \text{ in } \mathbb{P}\text{-measure but that } \sigma(X_n, X) \not\longrightarrow 0 \text{ in } \mathbb{P}\text{-measure.}$$

After passing to a subsequence if necessary, we could then arrange that $\rho(X_n, X) \longrightarrow 0$ (a.s., $\mathbb{P}$) but $\mathbb{P}\big(\sigma(X_n, X) \geq \epsilon\big) \geq \epsilon$ for all $n \in \mathbb{Z}^+$ and some $\epsilon > 0$. But this is impossible, since then we would have that $\sigma(X_n, X) \longrightarrow 0$ $\mathbb{P}$-almost surely but not in $\mathbb{P}$-measure. Hence, we now know that convergence in $\mathbb{P}$-measure does not depend on the choice of metric. To complete the first part, suppose that $\rho(X_n, X) \longrightarrow 0$ in $\mathbb{P}$-measure. Then, for every $\varphi \in U^\rho_{\mathrm{b}}(E;\mathbb{R})$ and $\delta > 0$,

$$\begin{aligned}\varlimsup_{n\to\infty} \left|\mathbb{E}^{\mathbb{P}}[\varphi(X_n)] - \mathbb{E}^{\mathbb{P}}[\varphi(X)]\right| &\le \varlimsup_{n\to\infty} \mathbb{E}^{\mathbb{P}}\left[|\varphi(X_n) - \varphi(X)|\right] \\ &\le \epsilon(\delta) + \|\varphi\|_{\mathrm{u}} \varlimsup_{n\to\infty} \mathbb{P}\Big(\rho(X_n, X) \ge \delta\Big) = \epsilon(\delta),\end{aligned}$$

where

$$\epsilon(\delta) \equiv \sup\left\{|\varphi(y) - \varphi(x)| : \rho(x,y) \le \delta\right\} \longrightarrow 0 \quad \text{as} \quad \delta \searrow 0.$$

Thus, by **(ii)** in Theorem 9.1.5, $(X_n)_*\mathbb{P} \Longrightarrow X_*\mathbb{P}$.

Now assume that $E$ is Polish, and take $\rho$ and $L$ accordingly. Then, for any closed set $F$ and $\delta > 0$,

$$\begin{aligned}X_*\mathbb{P}(F) = \mathbb{P}(X \in F) &\le \mathbb{P}\big(\rho(Y, F) < \delta\big) + \mathbb{P}\big(\rho(X,Y) \ge \delta\big) \\ &= Y_*\mathbb{P}\big(F^{(\delta)}\big) + \mathbb{P}\big(\rho(X,Y) \ge \delta\big).\end{aligned}$$

Hence, since the same is true when the roles of $X$ and $Y$ are reversed, the asserted estimate for $L(X_*\mathbb{P}, Y_*\mathbb{P})$ holds. □

As a demonstration of the sort of use to which one can put these ideas, I present the following version of the **Principle of Accompanying Laws**.

THEOREM 9.1.13. *Let $E$ be a Polish space and, for each $k \in \mathbb{Z}^+$, let $\{Y_{k,n} : n \ge 1\}$ be a sequence of $E$-valued random variables on the probability space $(\Omega, \mathcal{F}, \mathbb{P})$. Further, assume that, for each $k \in \mathbb{Z}^+$, there is a $\mu_k \in \mathbf{M}_1(E)$ such that $Y^*_{k,n}\mathbb{P} \Longrightarrow \mu_k$ as $n \to \infty$. Finally, let $\rho$ be a complete metric for $E$, and suppose that $\{X_n : n \ge 1\}$ is a sequence of $E$-valued random variables on $(\Omega, \mathcal{F}, \mathbb{P})$ with the property that*

$$\lim_{k\to\infty} \varlimsup_{n\to\infty} \mathbb{P}\Big(\rho(X_n, Y_{k,n}) \ge \delta\Big) = 0 \quad \text{for every } \delta > 0. \tag{9.1.14}$$

*Then there is a $\mu \in \mathbf{M}_1(E)$ such that $\mu_k \Longrightarrow \mu$ as $k \to \infty$ and $(X_n)_*\mathbb{P} \Longrightarrow \mu$ as $n \to \infty$. In particular, if, as $n \to \infty$, $Y_n \xrightarrow{\mathcal{L}} X$ and $\mathbb{P}\big(\rho(X_n, Y_n) \ge \delta\big) \longrightarrow 0$ for each $\delta > 0$, then $X_n \xrightarrow{\mathcal{L}} X$.*

PROOF: Let $L$ be the Lévy metric associated with a complete metric $\rho$ for $E$. By the second part of Lemma 9.1.12,

$$\sup_{\ell \ge k} L\big((Y_{\ell,n})_*\mathbb{P}, (X_n)_*\mathbb{P}\big) \le \delta \vee \left(\sup_{\ell\ge k} \varlimsup_{n\to\infty} \mathbb{P}\big(\rho(Y_{\ell,n}, X_n) \ge \delta\big)\right),$$

and therefore, by (9.1.14),

$$\varlimsup_{k\to\infty} \varlimsup_{n\to\infty} L\big((Y_{\ell,n})_*\mathbb{P}, (X_n)_*\mathbb{P}\big) = 0. \tag{*}$$

Thus, since for any $k \in \mathbb{Z}^+$,

$$\sup_{\ell \geq k} L(\mu_\ell, \mu_k) = \sup_{\ell \geq k} \lim_{n\to\infty} L\big((Y_{\ell,n})_*\mathbb{P}, (Y_{k,n})_*\mathbb{P}\big),$$

$\{\mu_k : k \geq 1\}$ is an $L$-Cauchy sequence and, as such, converges to some $\mu$. Finally, for every $k \in \mathbb{Z}^+$,

$$L\big(\mu, (X_n)_*\mathbb{P}\big) \leq L(\mu, \mu_k) + L\big(\mu_k, (Y_{k,n})_*\mathbb{P}\big) + L\big((Y_{k,n})_*\mathbb{P}, (X_n)_*\mathbb{P}\big),$$

and so

$$\varlimsup_{n\to\infty} L\big(\mu, (X_n)_*\mathbb{P}\big) \leq L(\mu, \mu_k) + \varlimsup_{n\to\infty} L\big((Y_{k,n})_*\mathbb{P}, (X_n)_*\mathbb{P}\big).$$

Thus, after letting $k \to \infty$ and applying (*), one concludes that $(X_n)_*\mathbb{P} \Longrightarrow \mu$. □

## Exercises for § 9.1

EXERCISE 9.1.15. Let $(E, \mathcal{B})$ be a measurable space with the property that $\{x\} \in \mathcal{B}$ for all $x \in E$. In this exercise, we will investigate the strong topology in a little more detail. In particular, in part **(iv)**, we will show that when $\mu \in \mathbf{M}_1(E)$ is **non-atomic** (i.e., $\mu(\{x\}) = 0$ for every $x \in E$), then there is no countable neighborhood basis of $\mu$ in the strong topology. Obviously, this means that *the strong topology for* $\mathbf{M}_1(E)$ *admits no metric whenever* $\mathbf{M}_1(E)$ *contains a non-atomic element.*

**(i)** Show that, in general,

$$\|\nu - \mu\|_{\text{var}} = 2\max\{\nu(A) - \mu(A) : A \in \mathcal{B}\}$$

and that in the case when $E$ is a metric space, $\mathcal{B}$ its Borel field, and $\rho$ a metric for $E$,

$$\|\nu - \mu\|_{\text{var}} = \sup\{\langle \varphi, \nu\rangle - \langle \varphi, \mu\rangle : \varphi \in U_{\text{b}}^{\rho}(E;\mathbb{R}) \text{ and } \|\varphi\|_{\text{u}} \leq 1\}.$$

**(ii)** Show that if $\{\mu_n : n \geq 1\}$ is a sequence in $\mathbf{M}_1(E)$ that tends in the strong topology to $\mu \in \mathbf{M}_1(E)$, then $\mu \ll \sum_{n=1}^{\infty} 2^{-n}\mu_n$.

**(iii)** Given $\mu \in \mathbf{M}_1(E)$, show that $\mu$ admits a countable neighborhood basis in the strong topology if and only if there exists a countable $\{\varphi_k : k \geq 1\} \subseteq B(E;\mathbb{R})$ such that, for any net $\{\mu_\alpha : \alpha \in A\} \subseteq \mathbf{M}_1(E)$, $\mu_\alpha \longrightarrow \mu$ in the strong topology as soon as $\lim_\alpha \langle \varphi_k, \mu_\alpha\rangle = \langle \varphi_k, \mu\rangle$ for every $k \in \mathbb{Z}^+$.

**(iv)** Referring to Exercises 1.1.14 and 1.1.16, set $\Omega = E^{\mathbb{Z}^+}$ and $\mathcal{F} = \mathcal{B}^{\mathbb{Z}^+}$. Next, let $\mu \in \mathbf{M}_1(E)$ be given, and define $\mathbb{P} = \mu^{\mathbb{Z}^+}$ on $(\Omega, \mathcal{F})$. Show that, for any $\varphi \in B(E;\mathbb{R})$, the random variables $\mathbf{x} \in \Omega \longmapsto X_n^\varphi(\mathbf{x}) \equiv \varphi(x_n)$, $n \in \mathbb{Z}^+$, are mutually $\mathbb{P}$-independent and all have distribution $\varphi_*\mu$. In particular, use the Strong Law of Large Numbers to conclude that

$$\lim_{n\to\infty} \frac{1}{n} \sum_{m=1}^{n} X_m^\varphi(\mathbf{x}) = \langle \varphi, \mu \rangle$$

for each $\mathbf{x}$ outside of a $\mathbb{P}$-null set.

Now assume that $\mu$ is non-atomic, and suppose that $\mu$ admitted a countable neighborhood basis in the strong topology. Choose $\{\varphi_k : k \geq 1\} \subseteq B(E;\mathbb{R})$ accordingly, as in **(iii)**, and (using the preceding) conclude that there exists at least one $\mathbf{x} \in \Omega$ for which the measures $\mu_n$ given by $\mu_n \equiv \frac{1}{n}\sum_{m=1}^{n} \delta_{x_m}$, $n \in \mathbb{Z}^+$, converge in the strong topology to $\mu$. Finally, apply **(ii)** to see that this is impossible.

EXERCISE 9.1.16. Throughout this exercise, $E$ is a separable metric space.

**(i)** We already know that $\mathbf{M}_1(E)$ is separable; however, our proof was non-constructive. Show that if $\{p_k : k \geq 1\}$ is a dense subset of $E$, then the set of all convex combinations $\sum_{k=1}^{n} \alpha_k \delta_{p_k}$, where $n \in \mathbb{Z}^+$ and $\{\alpha_k : 1 \leq k \leq n\} \subset [0,1] \cap \mathbb{Q}$ with $\sum_1^n \alpha_k = 1$, is a countable dense set in $\mathbf{M}_1(E)$.

**(ii)** For $\ell \in \mathbb{Z}^+$, let $\pi_\ell$ be the natural projection map from $\mathbf{E}$ onto $E_\ell$, and show that $\mathbf{K} \subset\subset \mathbf{E}$ if

$$\mathbf{K} = \bigcap_{\ell \in \mathbb{Z}^+} \pi_\ell^{-1}(K_\ell), \quad \text{where} \quad K_\ell \subset\subset E_\ell \text{ for each } \ell \in \mathbb{Z}^+.$$

Conclude from this that $\mathbf{A} \subseteq \mathbf{M}_1(\mathbf{E})$ is tight if and only if $\{(\pi_\ell)_*\mu : \mu \in \mathbf{A}\} \subseteq \mathbf{M}_1(E_\ell)$ is tight for every $\ell \in \mathbb{Z}^+$. Next, set $\mathbf{E}_\ell = \prod_{k=1}^{\ell} E_k$, and let $\boldsymbol{\pi}_\ell$ denote the natural projection map from $\mathbf{E}$ onto $\mathbf{E}_\ell$. Show that for each $\varphi \in U_\mathrm{b}^{\mathbf{R}}(\mathbf{E};\mathbb{R})$ and $\epsilon > 0$ there is an $\ell \geq 1$ such that $|\varphi(\mathbf{y}) - \varphi(\mathbf{x})| < \epsilon$ for all $\mathbf{x}, \mathbf{y} \in \mathbf{E}$ with $\boldsymbol{\pi}_\ell(\mathbf{x}) = \boldsymbol{\pi}_\ell(\mathbf{y})$. Use this to show that $\mu_n \Longrightarrow \mu$ in $\mathbf{M}_1(\mathbf{E})$ if and only if $\langle \varphi \circ \boldsymbol{\pi}_\ell, \mu_n \rangle \longrightarrow \langle \varphi \circ \boldsymbol{\pi}_\ell, \mu \rangle$ for every $\ell \geq 1$ and $\varphi \in C_\mathrm{b}(\mathbf{E}_\ell;\mathbb{R})$.

**(iii)** Let $\mu_{[1,\ell]}$ be an element of $\mathbf{M}_1(\mathbf{E}_\ell)$, and assume that the $\mu_{[1,\ell]}$'s are **consistent** in the sense that, for every $\ell \in \mathbb{Z}^+$,

$$\mu_{[1,\ell+1]}(\Gamma \times E_{\ell+1}) = \mu_{[1,\ell]}(\Gamma) \quad \text{for all } \Gamma \in \mathcal{B}_{\mathbf{E}_\ell}.$$

Show that there is a unique $\mu \in \mathbf{M}_1(\mathbf{E})$ such that $\mu_{[1,\ell]} = (\boldsymbol{\pi}_\ell)_*\mu$ for every $\ell \in \mathbb{Z}^+$.

**Hint**: Let $\hat{\rho}$ be a totally bounded metric on $E$, and use $\hat{E}$ to denote the $\hat{\rho}$-completion of $E$. Show that if $\{x_n : n \geq 1\} \subseteq E$ has the properties that $x_n \longrightarrow \hat{x} \in \hat{E}$ and $\lim_{n\to\infty} \varphi(x_n)$ exists for every $\varphi \in C_b(E;\mathbb{R})$, then $\hat{x} \in E$. (Suppose not, set $\psi(x) = \frac{1}{\hat{\rho}(x,\hat{x})}$, and consider functions of the form $f \circ \psi$ for $f \in C_b(\mathbb{R};\mathbb{R})$.) Finally, assuming that $C_b(E;\mathbb{R})$ is separable, and, using a diagonalization procedure, show that every sequence $\{x_n : n \geq 1\} \subseteq E$ admits a subsequence $\{x_{n_m} : m \geq 1\}$ that converges to some $\hat{x} \in \hat{E}$ and $\lim_{m\to\infty} \varphi(x_{n_m})$ exists for every $\varphi \in C_b(E;\mathbb{R})$.

(**iv**) Let $\{M_n : n \geq 1\}$ be a sequence of finite, non-negative measures on $(E, \mathcal{B})$. Assuming that $\{M_n : n \geq 1\}$ is **tight** in the sense that $\{M_n(E) : n \geq 1\}$ is bounded and that, for each $\epsilon > 0$, there is a $K \subset\subset E$ such that $\sup_n M_n(K\complement) \leq \epsilon$, show that there is a subsequence $\{M_{n_k} : k \geq 1\}$ and a finite measure $M$ such that

$$\int_E \varphi\, dM = \lim_{k\to\infty} \int_E \varphi\, dM_{n_k}, \quad \text{for all } \varphi \in C_b(E;\mathbb{R}).$$

Conversely, if $E$ is Polish and there is a finite measure $M$ such that $\int_E \varphi\, dM_n \longrightarrow \int_E \varphi\, dM$ for every $\varphi \in C_b(E;\mathbb{R})$, show that $\{M_n : n \geq 1\}$ is tight.

EXERCISE 9.1.17. Let $\{E_\ell : \ell \geq 1\}$ be a sequence of Polish spaces, set $\mathbf{E} = \prod_1^\infty E_\ell$, and give $\mathbf{E}$ the product topology.

(**i**) For each $\ell \in \mathbb{Z}^+$, let $\rho_\ell$ be a complete metric for $E_\ell$, and define

$$\mathbf{R}(\mathbf{x},\mathbf{y}) = \sum_{\ell=1}^\infty \frac{1}{2^\ell} \frac{\rho_\ell(x_\ell, y_\ell)}{1 + \rho_\ell(x_\ell, y_\ell)} \quad \text{for } \mathbf{x}, \mathbf{y} \in \mathbf{E}.$$

Show that $\mathbf{R}$ is a complete metric for $\mathbf{E}$, and conclude that $\mathbf{E}$ is a Polish space. In addition, check that $\mathcal{B}_{\mathbf{E}} = \prod_1^\infty \mathcal{B}_{E_\ell}$.

(**ii**) For $\ell \in \mathbb{Z}^+$, let $\pi_\ell$ be the natural projection map from $\mathbf{E}$ onto $E_\ell$, and show that $\mathbf{K} \subset\subset \mathbf{E}$ if and only if

$$\mathbf{K} = \bigcap_{\ell\in\mathbb{Z}^+} \pi_\ell^{-1}(K_\ell), \quad \text{where} \quad K_\ell \subset\subset E_\ell \text{ for each } \ell \in \mathbb{Z}^+.$$

Also, show that the span of the functions

$$\prod_{k=1}^\ell \varphi_k \circ \pi_k, \quad \text{where } \ell \in \mathbb{Z}^+ \text{ and } \varphi_k \in U_b^{\rho_k}(E_k;\mathbb{R}),\ 1 \leq k \leq \ell,$$

is dense in $U_b^{\mathbf{R}}(\mathbf{E};\mathbb{R})$. In particular, conclude from these that $\mathbf{A} \subseteq \mathbf{M}_1(\mathbf{E})$ is tight if and only if $\{(\pi_\ell)_*\mu : \mu \in \mathbf{A}\} \subseteq \mathbf{M}_1(E_\ell)$ is tight for every $\ell \in \mathbb{Z}^+$ and that $\mu_n \Longrightarrow \mu$ in $\mathbf{M}_1(\mathbf{E})$ if and only if

$$\left\langle \prod_{k=1}^\ell \varphi_k \circ \pi_k, \mu_n \right\rangle \longrightarrow \left\langle \prod_{k=1}^\ell \varphi_k \circ \pi_k, \mu \right\rangle$$

for every $\ell \in \mathbb{Z}^+$ and choice of $\varphi_k \in U_b^{\rho_k}(E_k;\mathbb{R})$, $1 \leq k \leq \ell$.

**(iii)** For each $\ell \in \mathbb{Z}^+$, set $\mathbf{E}_\ell = \prod_{k=1}^{\ell} E_k$, and let $\boldsymbol{\pi}_\ell$ denote the natural projection map from $\mathbf{E}$ onto $\mathbf{E}_\ell$. Next, let $\mu_{[1,\ell]}$ be an element of $\mathbf{M}_1(\mathbf{E}_\ell)$, and assume that the $\mu_{[1,\ell]}$'s are **consistent** in the sense that, for every $\ell \in \mathbb{Z}^+$,

$$\mu_{[1,\ell+1]}(\Gamma \times E_{\ell+1}) = \mu_{[1,\ell]}(\Gamma) \quad \text{for all } \Gamma \in \mathcal{B}_{\mathbf{E}_\ell}.$$

Show that there is a unique $\mu \in \mathbf{M}_1(\mathbf{E})$ such that $\mu_{[1,\ell]} = (\boldsymbol{\pi}_\ell)_*\mu$ for every $\ell \in \mathbb{Z}^+$.

**Hint**: Choose and fix an $\mathbf{e} \in \mathbf{E}$, and define $\Phi_\ell : \mathbf{E}_\ell \longrightarrow \mathbf{E}$ so that

$$\Big(\Phi_\ell(x_1, \dots, x_\ell)\Big)_n = \begin{cases} x_n & \text{if } \ n \le \ell \\ e_n & \text{otherwise.} \end{cases}$$

Show that $\{(\Phi_\ell)_*\mu_{[1,\ell]} : \ell \in \mathbb{Z}^+\} \in \mathbf{M}_1(\mathbf{E})$ is tight and that any limit must be the desired $\mu$.

The conclusion drawn in **(iii)** is the renowned **Kolmogorov Extension** (or **Consistency) Theorem**. Notice that, at least for Polish spaces, it represents a vast generalization of the result obtained in Exercise 1.1.14.

EXERCISE 9.1.18. In this exercise we will use the theory of weak convergence to develop variations on The Strong Law of Large Numbers (cf. Theorem 1.4.9). Thus, let $E$ be a Polish space, $(\Omega, \mathcal{F}, P)$ a probability space, and $\{X_n : n \ge 1\}$ a sequence of mutually independent $E$-valued random variables on $(\Omega, \mathcal{F}, P)$ with common distribution $\mu \in \mathbf{M}_1(E)$. Next, define the *empirical distribution function*

$$\omega \in \Omega \longmapsto \mathbf{L}_n(\omega) \equiv \frac{1}{n}\sum_{m=1}^{n} \delta_{X_m(\omega)} \in \mathbf{M}_1(E),$$

and observe that, for any $\varphi \in B(E;\mathbb{R})$,

$$\langle \varphi, \mathbf{L}_n(\omega)\rangle = \frac{1}{n}\sum_{m=1}^{n} \varphi(X_m(\omega)), \quad n \in \mathbb{Z}^+ \text{ and } \omega \in \Omega.$$

As a consequence of the Strong Law, show that

$$\mathbf{L}_n(\omega) \Longrightarrow \mu \quad \text{for} \quad P\text{-almost every } \omega \in \Omega, \tag{9.1.19}$$

which is **The Strong Law of Large Numbers for the empirical distribution.**

Now show that (9.1.19) provides another (cf. Exercises 6.1.16 and 6.2.18) proof of the Strong Law of Large Numbers for Banach space–valued random variables. Thus, let $E$ be a real, separable, Banach space with dual space $E^*$, and set $\overline{S}_n(\omega) = \frac{1}{n}\sum_1^n X_m(\omega)$ for $n \in \mathbb{Z}^+$ and $\omega \in \Omega$.

(**i**) As a preliminary step, begin with the case when

$$(*)\qquad \mu\Big(B_E(0,R)\complement\Big) = 0 \quad \text{for some} \quad R \in (0,\infty).$$

Choose $\eta \in C_b(\mathbb{R};\mathbb{R})$ so that $\eta(t) = t$ for $t \in [-R,R]$ and $\eta(t) = 0$ when $|t| \geq R+1$, and define $\psi_{x^*} \in C_b(E;\mathbb{R})$ for $x^* \in E^*$ by $\psi_{x^*}(x) = \eta(\langle x, x^*\rangle)$, $x \in E$, where $\langle x, x^*\rangle$ is used here to denote the action of $x^* \in E^*$ on $x \in E$. Taking (*) into account and applying (9.1.19) and Lemma 9.1.10, show that

$$\lim_{n\to\infty} \sup_{\|x^*\|_{E^*}\leq 1} \left| \langle \psi_{x^*}, \mathbf{L}_n(\omega)\rangle - \int_E \langle x, x^*\rangle\, \mu(dx)\right| = 0$$

for $\mathbb{P}$-almost every $\omega \in \Omega$, and conclude from this that

$$\lim_{n\to\infty} \big\|\overline{S}_n(\omega) - \mathbf{m}\big\|_E = 0 \quad \text{for } \mathbb{P}\text{-almost every } \omega \in \Omega,$$

where (cf. Lemma 5.1.10) $\mathbf{m} = \mathbb{E}^{\mu}[x]$.

(**ii**) The next step is to replace the boundedness assumption in (*) by the hypothesis that $x \rightsquigarrow \|x\|_E$ is $\mu$-integrable. Assuming that it is, define, for $R \in (0,\infty)$, $n \in \mathbb{Z}^+$, and $\omega \in \Omega$,

$$X_n^{(R)}(\omega) = \begin{cases} X_n(\omega) & \text{if } \big\|X_n(\omega)\big\|_E < R \\ 0 & \text{otherwise} \end{cases}$$

and $Y_n^{(R)}(\omega) = X_n(\omega) - X_n^{(R)}(\omega)$. Next, set $\overline{S}_n^{(R)} = \frac{1}{n}\sum_1^n X_m^{(R)}$, $n \in \mathbb{Z}^+$, and, from (**i**), note that $\{\overline{S}_n^{(R)}(\omega) : n \geq 1\}$ converges in $E$ for $\mathbb{P}$-almost every $\omega \in \Omega$. In particular, if $\epsilon > 0$ is given and $R \in (0,\infty)$ is chosen so that

$$\int_{\{\|x\|_E \geq R\}} \|x\|_E\, \mu(dx) < \frac{\epsilon}{8},$$

use the preceding and Theorem 1.4.9 to verify the computation

$$\begin{aligned}
&\varlimsup_{m\to\infty} P\Big(\sup_{n\geq m} \big\|\overline{S}_n - \overline{S}_m\big\|_E \geq \epsilon\Big) \\
&\leq \lim_{m\to\infty} P\Big(\sup_{n\geq m} \big\|\overline{S}_n^{(R)} - \overline{S}_m^{(R)}\big\| \geq \frac{\epsilon}{2}\Big) \\
&\qquad + 2\varlimsup_{m\to\infty} P\left(\sup_{n\geq m}\left\|\frac{1}{n}\sum_1^n Y_k^{(R)}\right\|_E \geq \frac{\epsilon}{4}\right) \\
&\leq 2\varlimsup_{m\to\infty} P\left(\sup_{n\geq m} \frac{1}{n}\sum_1^n \|Y_k^{(R)}\|_E \geq \frac{\epsilon}{4}\right) = 0,
\end{aligned}$$

and from this conclude that $\overline{S}_n \longrightarrow \mathbb{E}^{\mu}[x]$ $\mathbb{P}$-almost surely.

**(iii)** Finally, repeat the argument given in the proof of Theorem 1.4.9 to show that $\|\mathbf{x}\|$ is $\mu$-integrable if $\{\overline{S}_n : n \geq 1\}$ converges in $E$ on a set of positive $\mathbb{P}$-measure.[4]

## § 9.2 Regular Conditional Probability Distributions

As I mentioned in the discussion following Theorem 5.1.4, there are quite general situations in which conditional expectation values can be computed as expectation values. The following is a basic result in that direction.

THEOREM 9.2.1. *Suppose that $\Omega$ is a Polish space and that $\mathcal{F} = \mathcal{B}_\Omega$. Then, for every sub-$\sigma$-algebra $\Sigma$ of $\mathcal{F}$, there is a $\mathbb{P}$-almost surely unique $\Sigma$-measurable map $\omega \in \Omega \longmapsto \mathbb{P}^\Sigma_\omega \in \mathbf{M}_1(\Omega)$ with the property that*

$$\mathbb{P}(A \cap B) = \int_A \mathbb{P}^\Sigma_\omega(B)\,\mathbb{P}(d\omega) \quad \text{for all } A \in \Sigma \text{ and } B \in \mathcal{F}.$$

*In particular, for each $(-\infty, \infty]$-valued random variable $X$ that is bounded below, $\omega \in \Omega \longmapsto \mathbb{E}^{\mathbb{P}^\Sigma_\omega}[X]$ is a conditional expectation value of $X$ given $\Sigma$. Finally, if $\Sigma$ is countably generated, then there is a $\mathbb{P}$-null set $\mathcal{N} \in \Sigma$ with the property that $\mathbb{P}^\Sigma_\omega(A) = \mathbf{1}_A(\omega)$ for all $\omega \notin \mathcal{N}$ and $A \in \Sigma$.*

PROOF: To prove the uniqueness, suppose $\omega \in \Omega \longmapsto \mathbb{Q}^\Sigma_\omega \in \mathbf{M}_1(\Omega)$ were a second such mapping. We would then know that, for each $B \in \mathcal{F}$, $\mathbb{Q}^\Sigma_\omega(B) = \mathbb{P}^\Sigma_\omega(B)$ for $\mathbb{P}$-almost every $\omega \in \Omega$. Hence, since $\mathcal{F}$ (as the Borel field over a second countable topological space) is countably generated, we could find one $\Sigma$-measurable $\mathbb{P}$-null set off of which $\mathbb{Q}^\Sigma_\omega = \mathbb{P}^\Sigma_\omega$. Similarly, to prove the final assertion when $\Sigma$ is countably generated, note (cf. (5.1.7)) that, for each $A \in \Sigma$, $\mathbb{P}^\Sigma_\omega(A) = \mathbf{1}_A(\omega) = \delta_\omega(A)$ for $\mathbb{P}$-almost every $\omega \in \Omega$. Thus, once again countability allows us to choose one $\Sigma$-measurable $\mathbb{P}$-null set $\mathcal{N}$ such that $\mathbb{P}^\Sigma_\omega \restriction \Sigma = \delta_\omega \restriction \Sigma$ if $\omega \notin \mathcal{N}$.

I turn next to the question of existence. For this purpose, first choose (cf. **(ii)** of Lemma 9.1.4) $\rho$ to be a totally bounded metric for $\Omega$, and let $\mathcal{U} = U^\rho_b(\Omega; \mathbb{R})$ be the space of bounded, $\rho$-uniformly continuous, $\mathbb{R}$-valued functions on $\Omega$. Then (cf. **(iii)** of Lemma 9.1.4) $\mathcal{U}$ is a separable Banach space with respect to the uniform norm. In particular, we can choose a sequence $\{f_n : n \geq 0\} \subseteq \mathcal{U}$ so that $f_0 = \mathbf{1}$, the functions $f_0, \dots, f_n$ are linearly independent for each $n \in \mathbb{Z}^+$, and the linear span $\mathcal{S}$ of $\{f_n : n \geq 0\}$ is dense in $\mathcal{U}$. Set $g_0 = \mathbf{1}$, and, for each $n \in \mathbb{Z}^+$, let $g_n$ be some fixed representative of $\mathbb{E}^{\mathbb{P}}[f_n \,|\, \Sigma]$. Next, set

$$\mathfrak{R} = \left\{\boldsymbol{\alpha} \in \mathbb{R}^{\mathbb{N}} : \exists m \in \mathbb{N}\ \alpha_n = 0 \text{ for all } n \geq m\right\}$$

[4] The beautiful argument that I have just outlined is due to Ranga Rao. See his 1963 article "The law of large numbers for $D[0,1]$-valued random variables," *Theory of Prob. & Appl.* **VIII #1**, where he shows that this method applies even outside the separable context.

and define

$$f_{\boldsymbol{\alpha}} = \sum_{n=0}^{\infty} \alpha_n f_n \quad \text{and} \quad g_{\boldsymbol{\alpha}} = \sum_{n=0}^{\infty} \alpha_n g_n$$

for $\boldsymbol{\alpha} \in \mathfrak{R}$. Because of the linear independence of the $f_n$'s, we know that $f_{\boldsymbol{\alpha}} = f_{\boldsymbol{\beta}}$ if and only if $\boldsymbol{\alpha} = \boldsymbol{\beta}$. Hence, for each $\omega \in \Omega$, we can define the (not necessarily continuous) linear functional $\Lambda_\omega : \mathcal{S} \longrightarrow \mathbb{R}$ so that

$$\Lambda_\omega(f_{\boldsymbol{\alpha}}) = g_{\boldsymbol{\alpha}}(\omega), \qquad \boldsymbol{\alpha} \in \mathfrak{R}.$$

Clearly, $\Lambda_\omega(\mathbf{1}) = \mathbf{1}$ for all $\omega \in \Omega$. On the other hand, we cannot say that $\Lambda_\omega$ is always non-negative as a linear functional on $\mathcal{S}$. In fact, the best we can do is extract a $\Sigma$-measurable $\mathbb{P}$-null set $\mathcal{N}$ so that $\Lambda_\omega$ is a non-negative linear functional on $\mathcal{S}$ whenever $\omega \notin \mathcal{N}$. To this end, let $\mathbb{Q}$ denote the rational reals and set

$$\mathfrak{Q}^+ = \left\{\boldsymbol{\alpha} \in \mathfrak{R} \cap \mathbb{Q}^{\mathbb{N}} : f_{\boldsymbol{\alpha}} \geq 0\right\}.$$

Since $g_{\boldsymbol{\alpha}} \geq 0$ (a.s., $\mathbb{P}$) for every $\boldsymbol{\alpha} \in \mathfrak{Q}^+$ and $\mathfrak{Q}^+$ is countable,

$$\mathcal{N} \equiv \left\{\omega \in \Omega : \exists \boldsymbol{\alpha} \in \mathfrak{Q}^+ \;\; g_{\boldsymbol{\alpha}}(\omega) < 0\right\}$$

is a $\Sigma$-measurable, $\mathbb{P}$-null set. In addition, it is obvious that, for every $\omega \notin \mathcal{N}$, $\Lambda_\omega(f) \geq 0$ whenever $f$ is a non-negative element of $\mathcal{S}$. In particular, for $\omega \notin \mathcal{N}$,

$$\|f\|_{\mathrm{u}} \pm \Lambda_\omega(f) = \Lambda_\omega\big(\|f\|_{\mathrm{u}}\mathbf{1} \pm f\big) \geq 0, \quad f \in \mathcal{S},$$

and therefore $\Lambda_\omega$ admits a unique extension as a non-negative, continuous linear functional on $\mathcal{U}$ that takes $\mathbf{1}$ to 1. Furthermore, it is an easy matter to check that, for every $f \in \mathcal{U}$, the function

$$g(\omega) = \begin{cases} \Lambda_\omega(f) & \text{for } \omega \notin \mathcal{N} \\ \mathbb{E}^{\mathbb{P}}[f] & \text{for } \omega \in \mathcal{N} \end{cases}$$

is a conditional expectation value of $f$ given $\Sigma$.

At this point, all that remains is to show that, for $\mathbb{P}$-almost every $\omega \notin \mathcal{N}$, $\Lambda_\omega$ is given by integration with respect to a $\mathbb{P}_\omega \in \mathbf{M}_1(\Omega)$. In particular, by the Riesz Representation Theorem, there is nothing more to do in the case when $\Omega$ is compact. To treat the case when $\Omega$ is not compact, I will use Lemma 9.1.7. Namely, first choose (cf. the last part of Lemma 9.1.7) a non-decreasing sequence of sets $K_n \subset\subset \Omega$, $n \in \mathbb{Z}^+$, with the property that $\mathbb{P}(K_n\complement) \leq \frac{1}{2^n}$. Next, define

$$\eta_{m,n}(\omega) = \frac{m\,\rho(\omega, K_n)}{1 + m\,\rho(\omega, K_n)} \quad \text{for } m,\, n \in \mathbb{Z}^+.$$

Clearly, $\eta_{m,n} \in \mathcal{U}$ for each pair $(m,n)$ and $0 \le \eta_{m,n} \nearrow \mathbf{1}_{K_n\complement}$ as $m \to \infty$ for each $n \in \mathbb{Z}^+$. Thus, by The Monotone Convergence Theorem, for each $n \in \mathbb{Z}^+$,

$$\begin{aligned}\int_{\mathcal{N}\complement} \sup_{m\in\mathbb{Z}^+} \Lambda_\omega(\eta_{m,n})\,\mathbb{P}(d\omega) &= \lim_{m\to\infty}\int_{\mathcal{N}\complement} \Lambda_\omega(\eta_{m,n})\,\mathbb{P}(d\omega)\\ &= \lim_{m\to\infty} \mathbb{E}^{\mathbb{P}}\big[\eta_{m,n}\big] \le \frac{1}{2^n};\end{aligned}$$

and so, by the Borel–Cantelli Lemma, we can find a $\Sigma$-measurable $\mathbb{P}$-null set $\mathcal{N}' \supseteq \mathcal{N}$ such that

$$M(\omega) \equiv \sup_{n\in\mathbb{Z}^+} n\left(\sup_{m\in\mathbb{Z}^+}\Lambda_\omega(\eta_{m,n})\right) < \infty \quad \text{for every } \omega \notin \mathcal{N}'.$$

Hence, if $\omega \notin \mathcal{N}'$, then, for every $f \in \mathcal{U}$ and $n \in \mathbb{Z}^+$,

$$\begin{aligned}\big|\Lambda_\omega(f)\big| &\le \big|\Lambda_\omega\big((1-\eta_{m,n})\,f\big)\big| + \big|\Lambda_\omega(\eta_{m,n}\,f)\big|\\ &\le \big\|(1-\eta_{m,n})\,f\big\|_{\mathrm{u}} + \frac{M(\omega)}{n}\,\|f\|_{\mathrm{u}}\end{aligned}$$

for all $m \in \mathbb{Z}^+$. But $\big\|(1-\eta_{m,n})\,f\big\|_{\mathrm{u}} \longrightarrow \|f\|_{\mathrm{u},K_n}$ as $m \to \infty$, and so we now see that the condition in (9.1.8) is satisfied by $\Lambda_\omega$ for every $\omega \notin \mathcal{N}'$. In other words, I have shown that, for each $\omega \notin \mathcal{N}'$, there is a unique $\mathbb{P}^\Sigma_\omega \in \mathbf{M}_1(\Omega)$ such that $\Lambda_\omega(f) = \mathbb{E}^{\mathbb{P}^\Sigma_\omega}[f]$ for all $f \in \mathcal{U}$. Finally, if we complete the definition of the map $\omega \in \Omega \longmapsto \mathbb{P}^\Sigma_\omega$ by taking $\mathbb{P}^\Sigma_\omega = \mathbb{P}$ for $\omega \in \mathcal{N}'$, then this map is $\Sigma$-measurable and

$$\mathbb{E}^{\mathbb{P}}[f, A] = \int_\Omega \mathbb{E}^{\mathbb{P}^\Sigma_\omega}[f]\,\mathbb{P}(d\omega), \quad A \in \Sigma,$$

first for all $f \in \mathcal{U}$ and thence for all $\mathcal{F}$-measurable $f$'s that are bounded below. □

If $\mathbb{P}$ is a probability measure on $(\Omega, \mathcal{F})$ and $\Sigma$ is a sub-$\sigma$-algebra of $\mathcal{F}$, then a **conditional probability distribution of $\mathbb{P}$ given** $\Sigma$ is a map $(\omega, B) \longmapsto \mathbb{P}^\Sigma_\omega(B)$ such that $\mathbb{P}^\Sigma_\omega$ is a probability measure on $(\Omega, \mathcal{F})$ for each $\omega \in \Omega$ and $\omega \rightsquigarrow \mathbb{P}^\Sigma_\omega(B)$ a conditional probability of $B$ given $\Sigma$ for all $B \in \mathcal{F}$. If, in addition, for $\omega$ outside a $\Sigma$-measurable, $\mathbb{P}$-null set and all $A \in \Sigma$, $\mathbb{P}^\Sigma(A) = \mathbf{1}_A(\omega)$, then the conditional probability distribution is said to be **regular**. Notice that, although they may not always exist, conditional probability distributions are always unique up to a $\Sigma$-measurable, $\mathbb{P}$-null set so long as $\mathcal{F}$ is countably generated. Moreover, Theorem 9.2.1 says that they will always exist if $\Omega$ is Polish and $\mathcal{F} = \mathcal{B}_\Omega$. Finally, whenever a conditional probability distribution of $\mathbb{P}$ given $\Sigma$ exists, the argument leading to the last part of Theorem 9.2.1 when $\Sigma$ is countably generated is completely general and shows that a regular version can be found.

§ **9.2.1. Fibering a Measure.** When $\Omega$ is a product space $E_1 \times E_2$ of two Polish spaces and $\Sigma$ is the $\sigma$-algebra generated by the second coordinate, then the conclusion of Theorem 9.2.1 takes a particularly pleasing form.

THEOREM 9.2.2. *Let $E_1$ and $E_2$ be a pair of Polish spaces, and take $\Omega$ to be the Polish space $E_1 \times E_2$. Given $\mu \in \mathbf{M}_1(\Omega)$, use $\mu_2$ to denote the marginal distribution of $\mu$ on $E_2$: $\mu_2(\Gamma) = \mu(E_1 \times \Gamma)$ for $\Gamma \in \mathcal{B}_{E_2}$. Then there is a Borel measurable map $x_2 \in E_2 \longmapsto \mu(x_2, \cdot) \in \mathbf{M}_1(E_1)$ such that $\mu(dx_1 \times dx_2) = \mu(x_2, dx_1)\,\mu_2(dx_2)$.*

PROOF: Referring to Theorem 9.2.1, take $\mathbb{P} = \mu$, $\Sigma = \{E_1 \times \Gamma : \Gamma \in \mathcal{B}_{E_2}\}$, and let $\omega \in \Omega \longmapsto \mathbb{P}^\Sigma_\omega \in \mathbf{M}_1(\Omega)$ be the map guaranteed by the result there. Next, choose and fix a point $x_1^0 \in E_1$. Then, because $\omega \rightsquigarrow \mathbb{P}^\Sigma_\omega$ is $\Sigma$-measurable, we know that $\mathbb{P}^\Sigma_{(x_1,x_2)} = \mathbb{P}^\Sigma_{(x_1^0,x_2)}$. In addition, because $\Sigma$ is countably generated, the final part of Theorem 9.2.1 guarantees that there exists a $\mu_2$-null set $B \in \mathcal{B}_{E_2}$ such that $\mathbb{P}^\Sigma_{(x_1^0,x_2)}(E_1 \times \{x_2\}) = 1$ for all $x_2 \notin B$. Hence, if we define $x_2 \rightsquigarrow \mu(x_2, \cdot)$ by $\mu(x_2, \Gamma) = \mathbb{P}^\Sigma_{(x_1^0,x_2)}(\Gamma \times E_2)$, then, for any Borel measurable $\varphi : E_1 \times E_2 \longrightarrow [0,\infty)$, $\langle \varphi, \mu \rangle$ equals

$$\int \left( \int \varphi(\omega')\mathbb{P}^\Sigma_\omega(d\omega') \right) \mathbb{P}(d\omega) = \int_{E_2} \left( \int_{E_1} \varphi(x_1, x_2)\,\mu(x_2, dx_1) \right) \mu_2(dx_2). \quad \square$$

In the older literature, the result in Theorem 9.2.2 would be called a **fibering** of $\mu$. The name derives from the idea that $\mu$ on $E_1 \times E_2$ can be decomposed into its "vertical component" $\mu_2$ and its "restrictions" $\mu(x_2, \cdot)$ to "horizontal fibers" $E_1 \times \{x_2\}$. Alternatively, Theorem 9.2.2 can be interpreted as saying that any $\mu \in \mathbf{M}_1(E_1 \times E_2)$ can be decomposed into its marginal distribution on $E_2$ and a transition probability $x_2 \in E_2 \longmapsto \mu(x_2, \cdot) \in \mathbf{M}_1(E_1)$. The two extreme cases are when the coordinates are independent, in which case $\mu(x_2, \cdot)$ is independent of $x_2$, and the case when the coordinates are equal, in which case $\mu(x_2, \cdot) = \delta_{x_2}$.

As an application of Theorem 9.2.2, I present the following important special case of a more general result that indicates just how remarkably fungible non-atomic measures are.

COROLLARY 9.2.3. *Let $\lambda_{[0,1)}$ denote Lebesgue measure on $[0,1)$. For each $N \in \mathbb{Z}^+$ and $\mu \in \mathbf{M}_1(\mathbb{R}^N)$, there is a Borel measurable map $f : [0,1) \longrightarrow \mathbb{R}^N$ such that $\mu = f_*\lambda_{[0,1)}$.*

PROOF: I will work by induction on $N \in \mathbb{Z}^+$. When $N = 1$, take

$$f(u) = \inf\{t \in \mathbb{R} : \mu((-\infty, t]) \ge u\}, \quad u \in [0,1).$$

Next, assume the result is true for $N$, take $E_1 = \mathbb{R}$ and $E_2 = \mathbb{R}^N$ in Theorem 9.2.2, and, given $\mu \in \mathbf{M}_1(\mathbb{R}^N)$, define $\mu_2 \in \mathbf{M}_1(\mathbb{R}^N)$ and $\mathbf{y} \in \mathbb{R}^N \longmapsto \mu(\mathbf{y}, \cdot) \in \mathbf{M}_1(\mathbb{R})$ accordingly. By the induction hypothesis, $\mu_2 = f_2(\cdot)_*\lambda_{[0,1)}$ for some $f_2 : [0,1) \longrightarrow \mathbb{R}^N$. Thus, if $g : [0,1)^2 \longrightarrow \mathbb{R} \times \mathbb{R}^N$ is given by

$$g(u_1, u_2) = \Big( \inf\{t \in \mathbb{R} : \mu\big(f_2(u_2), (-\infty, t]\big) \ge u_1\}, f_2(u_2) \Big)$$

for $(u_1, u_2) \in [0,1)^2$, then $g$ is Borel measurable on $[0,1)^2$ and $\mu = g_*\lambda^2_{[0,1)}$. Finally, by Lemma 1.1.6 or part **(ii)** of Exercise 1.1.11, we know that there is a Borel measurable map $u \in [0,1) \longmapsto \mathbf{U}(u) = \big(U_1(u), U_2(u)\big) \in [0,1)^2$ such that $\mathbf{U}_*\lambda_{[0,1)} = \lambda^2_{[0,1)}$, and so we can take $f(u) = g \circ \mathbf{U}$. □

**§ 9.2.2. Representing Lévy Measures via the Itô Map.** There is another way of thinking about the construction of the Poisson jump processes, one that is based on Corollary 9.2.3 and the transformation property described in Lemma 4.2.12. The advantage of this approach is that it provides a method of coupling Lévy processes corresponding to different Lévy measures. Indeed, it is this coupling procedure that underlies K. Itô's construction of Markov processes modeled on Lévy processes.[1]

Let $M_0(d\mathbf{y}) = |\mathbf{y}|^{-N-1}\,d\mathbf{y}$, which is the Lévy measure for a (cf. Corollary 3.3.9) symmetric 1-stable law. My first goal is to show that every $M \in \mathfrak{M}_\infty(\mathbb{R}^N)$ can be realized as (cf. the notation in Lemma 4.2.6) $M_0^F$ for some Borel measurable $F : \mathbb{R}^N \longrightarrow \mathbb{R}^N$ satisfying $F(\mathbf{0}) = \mathbf{0}$.[2]

THEOREM 9.2.4. *For each $M \in \mathfrak{M}_\infty(\mathbb{R}^N)$ there exists a Borel measurable map $F : \mathbb{R}^N \longrightarrow \mathbb{R}^N$ such that $F(\mathbf{0}) = \mathbf{0}$ and*

$$M(\Gamma) = M_0^F \equiv M_0\big(F^{-1}(\Gamma \setminus \{\mathbf{0}\})\big), \quad \Gamma \in \mathcal{B}_{\mathbb{R}^N}.$$

PROOF: I begin with the case when $N = 1$. Given $M \in \mathfrak{M}_\infty(\mathbb{R})$, define $\rho(r, \pm 1)$ for $r > 0$ by

$$\begin{aligned}\rho(r,1) &= \sup\{\rho \in [0,\infty) : M\big([\rho,\infty)\big) \ge r^{-1}\}\\ \rho(r,-1) &= \sup\{\rho \in [0,\infty) : M\big((-\infty,-\rho]\big) \ge r^{-1}\},\end{aligned}$$

where I have taken the supremum over the empty set to be 0. Applying Exercise 9.2.6 with $\nu(dr) = r^{-2}\lambda_{(0,\infty)}(dr)$, one sees that $M = M_0^F$ when $F(0) = 0$ and $F(y) = \rho\big(|y|, \frac{y}{|y|}\big)$ for $y \in \mathbb{R} \setminus \{0\}$.

Now assume that $N \ge 2$, and let $M \in \mathfrak{M}_\infty(\mathbb{R}^N)$. If $M = 0$, simply take $F \equiv 0$. If $M \ne 0$, choose a non-decreasing function $h : (0,\infty) \longrightarrow (0,\infty)$ so that

$$\int h\big(|\mathbf{y}|\big)\, M(d\mathbf{y}) = 1,$$

and define $\mu \in \mathbf{M}_1\big((0,\infty) \times \mathbb{S}^{N-1}\big)$ so that

$$\langle \varphi, \mu \rangle = \int_{\mathbb{R}^N} h\big(|\mathbf{y}|\big)\varphi(\mathbf{y}) M(d\mathbf{y}).$$

[1] See K. Itô's *On stochastic differential equations*, Memoirs of the A.M.S. **4** (1951) or my *Markov Processes from K. Itô's Perspective*, Princeton Univ. Press, Annals of Math. Studies **155** (2003).

[2] There is nothing sacrosanct about the choice of $M_0$ as my reference measure. For instance, it should be obvious that one can choose any Lévy measure $M$ with the property that $M_0 = M^F$ for some Borel measurable $F : \mathbb{R}^N \longrightarrow \mathbb{R}^N$ that takes $\mathbf{0}$ to $\mathbf{0}$.

Using $\mu_2$ to denote the marginal distribution of $\mu$ on $\mathbb{S}^{N-1}$, apply Corollary 9.2.3 to find a Borel measurable $\mathbf{f} : [0,1) \longrightarrow \mathbb{R}^N$ so that $\mu_2 = \mathbf{f}_*\lambda_{[0,1)}$. Since $\mu_2$ lives on $\mathbb{S}^{N-1}$, I may and will assume that $\mathbf{f}(u) \in \mathbb{S}^{N-1}$ for all $u \in [0,1)$. Next, use Theorem 9.2.2 to find a measurable map $\boldsymbol{\eta} \in \mathbb{S}^{N-1} \longmapsto \mu(\boldsymbol{\eta}, \cdot) \in \mathbf{M}_1((0,\infty))$ so that $\mu(dr \times d\boldsymbol{\eta}) = \mu(\boldsymbol{\eta}, dr)\,\mu_2(d\boldsymbol{\eta})$, and define $\rho : (0,\infty) \times \mathbb{S}^{N-1} \longrightarrow [0,\infty)$ by

$$\rho(r, \boldsymbol{\eta}) = \sup\left\{\rho \in [0,\infty) : \int_{[\rho,\infty)} \frac{1}{h(r)}\,\mu(\boldsymbol{\eta}, dr) \geq \frac{\omega_{N-1}}{r}\right\}.$$

Then, again by Exercise 9.2.6, but this time with $\nu(dr) = \omega_{N-1}r^{-2}\lambda_{(0,\infty)}(dr)$, for any continuous $\varphi : \mathbb{R}^N \longrightarrow [0,\infty)$ that vanishes in a neighborhood of $\mathbf{0}$,

$$\int_{(0,\infty)} \frac{\varphi(r\boldsymbol{\eta})}{h(r)}\,\mu(\boldsymbol{\eta}, dr) = \omega_{N-1}\int_{(0,\infty)} \varphi\big(\rho(r,\boldsymbol{\eta})\boldsymbol{\eta}\big)r^{-2}\,dr, \quad \boldsymbol{\eta} \in \mathbb{S}^{N-1},$$

and so

$$\begin{aligned}\int_{\mathbb{R}^N} \varphi(\mathbf{y})\,M(dy) &= \omega_{N-1}\int_{\mathbb{S}^{N-1}}\left(\int_{(0,\infty)} \varphi\big(\rho(r,\boldsymbol{\eta})\boldsymbol{\eta}\big)r^{-2}\,dr\right)\mu_2(d\boldsymbol{\eta})\\ &= \omega_{N-1}\int_{[0,1)}\left(\int_{(0,\infty)} \varphi\big(\rho(r,\boldsymbol{\eta})\mathbf{f}(t)\big)r^{-2}\,dr\right)\lambda_{[0,1)}(dt).\end{aligned}$$

Finally, define $g : \mathbb{S}^{N-1} \longrightarrow [0, \omega_{N-1})$ by $g(\boldsymbol{\eta}) = \lambda_{\mathbb{S}^{N-1}}(\{\boldsymbol{\eta}' \in \mathbb{S}^{N-1} : \eta_1' \leq \eta_1\})$, note that $\omega_{N-1}\lambda_{[0,1)} = g_*\lambda_{\mathbb{S}^{N-1}}$, and conclude that $M = M_0^F$ when

$$F(\mathbf{0}) = 0 \quad\text{and}\quad F(\mathbf{y}) = \rho\big(|\mathbf{y}|, \tfrac{\mathbf{y}}{|\mathbf{y}|}\big)\mathbf{f}\circ g\big(\tfrac{\mathbf{y}}{|\mathbf{y}|}\big) \text{ for } \mathbf{y} \in \mathbb{R}^N \setminus \{\mathbf{0}\}. \quad \square$$

We can now prove the following theorem, which is the simplest example of Itô's procedure.

THEOREM 9.2.5. *Let $\{j_0(t,\cdot) : t \geq 0\}$ be a Poisson jump process associated with $M_0$. Then, for each $M \in \mathfrak{M}_\infty(\mathbb{R}^N)$, there is a Borel measurable map $F : \mathbb{R}^N \longrightarrow \mathbb{R}^N$ with $F(\mathbf{0}) = \mathbf{0}$ and a Poisson jump process $\{j(t,\cdot) : t \geq 0\}$ associated with $M$ such that $j(t,\cdot) = j_0^F(t,\cdot)$, $t \geq 0$, $\mathbb{P}$-almost surely.*

PROOF: Choose $F$ as in Theorem 9.2.4 so that $M = M_0^F$. For $R > 0$, set $F_R(\mathbf{y}) = \mathbf{1}_{[R,\infty)}(\mathbf{y})F(\mathbf{y})$. By Lemma 4.2.12, we know that $\{j_0^{F_R}(t,\cdot) : t \geq 0\}$ is a Poisson jump process associated with $M^{F_R}$. In particular, for each $r > 0$,

$$\mathbb{E}^{\mathbb{P}}\big[j_0^F(t, \mathbb{R}^N \setminus B(\mathbf{0}, r))\big] = \lim_{R\searrow 0}\mathbb{E}^{\mathbb{P}}\big[j_0^{F_R}(t, \mathbb{R}^N \setminus B(\mathbf{0}, r))\big] = M\big(\mathbb{R}^N \setminus B(0,r)\big) < \infty.$$

Hence, there exists a $\mathbb{P}$-null set $\mathcal{N}$ such that $t \rightsquigarrow j_0^F(t,\cdot,\omega)$ is a jump function for all $\omega \notin \mathcal{N}$. Finally, if $j(t,\cdot,\omega) = j_0^F(t,\cdot,\omega)$ when $\omega \notin \mathcal{N}$ and $j(t,\cdot,\omega) = 0$ for $\omega \in \mathcal{N}$, then $\{j(t,\cdot) : t \geq 0\}$ is a jump process associated with $M$ and $j(t,\cdot) = j_0^F(t,\cdot)$, $t \geq 0$, for $\mathbb{P}$-almost every $\omega \in \Omega$. $\square$

### Exercises for § 9.2

EXERCISE 9.2.6. Let $\nu$ be an infinite non-negative, non-atomic, Borel measure on $[0,\infty)$ with the property that $\nu([r_2,\infty)) < \nu([r_1,\infty)) < \infty$ for all $0 < r_1 < r_2 < \infty$. Given any other non-negative Borel measure on $[0,\infty)$ with the properties that $\mu(\{0\}) = 0$ and $\mu([r,\infty)) < \infty$ for all $r > 0$, define

$$\rho(r) = \sup\{\rho \in (0,\infty) : \mu([\rho,\infty)) \geq \nu([r,\infty))\}, \quad r \geq 0,$$

where the supremum over the empty set is taken to be 0. Show that $\mu([t,\infty)) = \nu(\{r : \rho(r) \geq t\})$ for all $t > 0$, and therefore that $\langle \varphi, \mu \rangle = \langle \varphi \circ \rho, \nu \rangle$ for all Borel measurable $\varphi : [0,\infty) \longrightarrow [0,\infty)$ that vanish at 0.

**Hint**: Determine $g : (0,\infty) \longrightarrow (0,\infty)$ so that $\nu([g(r),\infty)) = r$, and check that $\{r : \rho(r) \geq t\} = [g(\mu([t,\infty))),\infty)$ for all $t > 0$.

## § 9.3 Donsker's Invariance Principle

The content of this section is my main justification for presenting the material in § 9.1. Namely, as we saw in Chapter 8, there is good reason to think that Wiener measure is the infinite dimensional version of the standard Gauss measure in $\mathbb{R}^N$, and as such one might suspect that there is a version of The Central Limit Theorem that applies to it. In this section I will prove such a Central Limit Theorem for Wiener measure. The result is due to M. Donsker and is known as **Donsker's Invariance Principle** (cf. Theorem 9.3.1).

Before getting started, I need to make a couple of simple preparatory remarks. In the first place, I will be thinking of Wiener measure $\mathcal{W}^{(N)}$ as a Borel probability measure on $C(\mathbb{R}^N) = C([0,\infty);\mathbb{R}^N)$ with the topology of uniform convergence on compact intervals. Equivalently, $C(\mathbb{R}^N)$ is given the topology for which

$$\rho(\psi,\psi') = \sum_{n=1}^{\infty} \frac{1}{2^n} \frac{\|\psi - \psi'\|_{[0,n]}}{1 + \|\psi - \psi'\|_{[0,n]}}$$

is a metric, which, just as in the case of $D(\mathbb{R}^N)$ (cf. 4.1.1), is complete on $C(\mathbb{R}^N)$ and, as distinguished from $D(\mathbb{R}^N)$, is separable there. One way to check separability is to note that the set of paths $\psi$ that, for some $n \in \mathbb{N}$, are linear on $[(m-1)2^{-n}, m2^{-n}]$ and satisfy $\psi(m2^{-n}) \in \mathbb{Q}^N$ for all $m \in \mathbb{Z}^+$ is a countable, dense subset. In particular, this means that $C(\mathbb{R}^N)$ is a Polish space, and so the theory developed in § 9.1 applies to it. In addition, the Borel field $\mathcal{B}_{C(\mathbb{R}^N)}$ coincides with $\sigma(\{\psi(t) : t \geq 0\})$, the $\sigma$-algebra that $C(\mathbb{R}^N)$ inherits as a subset of $(\mathbb{R}^N)^{[0,\infty)}$ (cf. § 4.1). Indeed, since $\psi \rightsquigarrow \psi(t)$ is continuous for every $t \geq 0$, it is obvious that $\sigma(\{\psi(t) : t \geq 0\}) \subseteq \mathcal{B}_{C(\mathbb{R}^N)}$. At the same time, since $\|\psi\|_{[0,t]} = \sup\{|\psi(\tau) : \tau \in [0,t] \cap \mathbb{Q}\}$, it is easy to check that open balls are $\sigma(\{\psi(t) : t \geq 0\})$-measurable. Hence, since every open set is the countable union of open balls, $\mathcal{B}_{C(\mathbb{R}^N)} \subseteq \sigma(\{\psi(t) : t \geq 0\})$. Knowing that these $\sigma$-algebras coincide, we know that two probability measures $\mu, \nu \in \mathbf{M}_1(C(\mathbb{R}^N))$ are equal if they determine the same distribution on $(\mathbb{R}^N)^{[0,\infty)}$, that is, if, for each $n \in \mathbb{Z}^+$ and $0 = t_0 < t_1 < t_n$, the distribution of $\psi \in C(\mathbb{R}^N) \longmapsto (\psi(t_0, \dots, \psi(t_n)) \in (\mathbb{R}^N)^n$ is the same under $\mu$ and $\nu$.

§ **9.3.1. Donsker's Theorem.** Let $(\Omega, \mathcal{F}, \mathbb{P})$ be a probability space, and suppose that $\{\mathbf{X}_n : n \geq 1\}$ is a sequence of independent, $\mathbb{P}$-uniformly square integrable random variables (i.e., as $R \to \infty$, $\mathbb{E}^{\mathbb{P}}[|\mathbf{X}_n|^2, |\mathbf{X}_n| \geq R] \longrightarrow 0$ uniformly in $n$) with mean value $\mathbf{0}$ and covariance $\mathbf{I}$. Given $n \geq 1$, define $\omega \in \Omega \longmapsto \mathbf{S}_n(\,\cdot\,,\omega) \in C(\mathbb{R}^N)$ so that $\mathbf{S}_n(0) = \mathbf{0}$, $\mathbf{S}_n(\frac{m}{n}) = n^{-\frac{1}{2}} \sum_{k=1}^m \mathbf{X}_k$, and $\mathbf{S}_n(\,\cdot\,,\omega)$ is linear on each interval $[\frac{m-1}{n}, \frac{m}{n}]$ for all $m \in \mathbb{Z}^+$. Donsker's theorem is the following.

THEOREM 9.3.1 (**Donsker's Invariance Principle**). *If $\mu_n = (\mathbf{S}_n)_*\mathbb{P} \in \mathbf{M}_1(C(\mathbb{R}^N))$ is the distribution of $\omega \in \Omega \longmapsto \mathbf{S}_n(\,\cdot\,,\omega) \in C(\mathbb{R}^N)$ under $\mathbb{P}$, then $\mu_n \Longrightarrow \mathcal{W}^{(N)}$. Equivalently, for any bounded, continuous $\Phi : C(\mathbb{R}^N) \longrightarrow \mathbb{C}$,*

$$\lim_{n\to\infty} \mathbb{E}^{\mathbb{P}}[\Phi \circ \mathbf{S}_n] = \langle \Phi, \mathcal{W}^{(N)} \rangle.$$

Proving this result comes down to showing that $\{\mu_n : n \geq 1\}$ is tight and that every limit point is $\mathcal{W}^{(N)}$. The second of these is a rather elementary application of the Central Limit Theorem, and, at least when the $\mathbf{X}_n$'s have uniformly bounded fourth moments, the first is an application of Kolmogorov's Continuity Criterion. Finally, to remove the fourth moment assumption, I will use the Principle of Accompanying Laws. It should be noticed that, at no point in the proof, do I make use of the a priori existence of Wiener measure. Thus, Theorem 9.3.1 provides another derivation of its existence, a derivation that includes an an extremely ubiquitous approximation procedure.

LEMMA 9.3.2. *Any limit point of $\{\mu_n : n \geq 1\}$ is $\mathcal{W}^{(N)}$.*

PROOF: Since a probability on $C(\mathbb{R}^N)$ is uniquely determined by its finite dimensional time marginals, and because $\psi(0) = \mathbf{0}$ with probability 1 under all the $\mu_n$'s as well as $\mathcal{W}^{(N)}$, it suffices to show that, for each $\ell \in \mathbb{Z}^+$ and $0 = t_0 < t_1 < \cdots < t_\ell$,

$$\big(\mathbf{S}_n(t_1), \mathbf{S}_n(t_2) - \mathbf{S}_n(t_1), \dots, \mathbf{S}_n(t_\ell) - \mathbf{S}_n(t_{\ell-1})\big)_*\mathbb{P} \Longrightarrow \gamma_{\mathbf{0},\tau_1\mathbf{I}} \times \cdots \times \gamma_{\mathbf{0},\tau_\ell\mathbf{I}},$$

where $\tau_k = t_k - t_{k-1}$, $1 \leq k \leq \ell$. To this end, for $1 \leq k \leq \ell$ and $n > \frac{1}{\tau_k}$, set

$$\mathbf{\Delta}_n(k) = n^{-\frac{1}{2}} \sum_{j=\lfloor nt_{k-1}\rfloor+1}^{\lfloor nt_k \rfloor} \mathbf{X}_j,$$

where, as usual, I use the notation $\lfloor t \rfloor$ to denote the integer part of $t$. Noting that

$$\begin{aligned}
&\left|\mathbf{S}_n(t_k) - \mathbf{S}_n(t_{k-1}) - \boldsymbol{\Delta}_n(k)\right| \\
&\quad\le \left|\mathbf{S}_n(t_k) - \mathbf{S}_n\left(\frac{\lfloor nt_k \rfloor}{n}\right)\right| + \left|\mathbf{S}_n(t_{k-1}) - \mathbf{S}_n\left(\frac{\lfloor nt_{k-1} \rfloor}{n}\right)\right| \\
&\quad\le \frac{\left|\mathbf{X}_{\lfloor nt_k \rfloor+1}\right| + \left|\mathbf{X}_{\lfloor nt_{k-1} \rfloor+1}\right|}{n^{\frac{1}{2}}},
\end{aligned}$$

one sees that, for any $\epsilon > 0$,

$$\begin{aligned}
\mathbb{P}\left(\sum_{k=1}^{\ell}\left|\mathbf{S}_n(t_k) - \mathbf{S}_n(t_{k-1}) - \boldsymbol{\Delta}_n(k)\right|^2 \ge \epsilon^2\right) &\le \mathbb{P}\left(\sum_{k=0}^{\ell}\left|\mathbf{X}_{\lfloor nt_k \rfloor+1}\right|^2 \ge \frac{n\epsilon^2}{4}\right) \\
\le \frac{4}{n\epsilon^2}\sum_{k=0}^{\ell}\mathbb{E}^{\mathbb{P}}\left[\left|\mathbf{X}_{\lfloor nt_k \rfloor+1}\right|^2\right] = \frac{4(\ell+1)N}{n\epsilon^2} \longrightarrow 0
\end{aligned}$$

as $n \to \infty$. Hence, by the Principle of Accompanying Laws (cf. Theorem 9.1.13), we need only check that

$$\left(\boldsymbol{\Delta}_n(1), \dots, \boldsymbol{\Delta}_n(\ell)\right)_*\mathbb{P} \Longrightarrow \gamma_{\tau_1}^N \times \cdots \times \gamma_{\tau_\ell}^N.$$

Moreover, since

$$\left(\boldsymbol{\Delta}_n(1), \dots, \boldsymbol{\Delta}_n(\ell)\right)_*\mathbb{P} = \left(\boldsymbol{\Delta}_n(1)\right)_*\mathbb{P} \times \cdots \times \left(\boldsymbol{\Delta}_n(\ell)\right)_*\mathbb{P}$$

for all sufficiently large $n$'s, this reduces to checking $\left(\boldsymbol{\Delta}_n(k)\right)_*\mathbb{P} \Longrightarrow \gamma_{\mathbf{0},\tau_k\mathbf{I}}$ for each $1 \le k \le \ell$. Finally, given $1 \le k \le \ell$, set $M_n(k) = \lfloor nt_k \rfloor - \lfloor nt_{k-1} \rfloor$, and use Theorem 2.3.8 to see that, as $n \to \infty$,

$$\mathbb{E}^{\mathbb{P}}\left[\exp\left(\frac{\sqrt{-1}}{M_n(k)^{\frac{1}{2}}}\sum_{j=1}^{M_n(k)}\left(\boldsymbol{\xi}, \mathbf{X}_{\lfloor nt_k \rfloor+j}\right)_{\mathbb{R}^N}\right)\right] \longrightarrow \exp\left[-\frac{|\boldsymbol{\xi}|^2}{2}\right]$$

uniformly for $\boldsymbol{\xi}$ in compact subsets of $\mathbb{R}^N$. Hence, since $\frac{M_n(k)}{n} \longrightarrow \tau_k$, we now see that, for any fixed $\boldsymbol{\xi} \in \mathbb{R}^N$,

$$\mathbb{E}^{\mathbb{P}}\left[\exp\left(\sqrt{-1}\left(\boldsymbol{\xi}, \boldsymbol{\Delta}_n(k)\right)_{\mathbb{R}^N}\right)\right] \longrightarrow \exp\left[-\frac{\tau_k|\boldsymbol{\xi}|^2}{2}\right] = \widehat{\gamma_{\mathbf{0},\tau_k\mathbf{I}}}(\boldsymbol{\xi}),$$

and therefore $\left(\boldsymbol{\Delta}_n(k)\right)_*\mathbb{P} \Longrightarrow \gamma_{\mathbf{0},\tau_k\mathbf{I}}$. □

I turn next to the problem of showing that $\{\mu_n : n \geq 1\}$ is tight. By the Ascoli–Arzelaá Theorem, any subset $K \subseteq C(\mathbb{R}^N)$ of the form

$$\bigcap_{\ell=1}^{\infty} \left\{ \psi : |\psi(0)| \vee \sup_{0 \leq s < t \leq \ell} \frac{|\psi(t) - \psi(s)|}{(t-s)^{\alpha}} \leq R_\ell \right\}$$

is compact for any $\alpha > 0$ and $\{R_\ell : \ell \geq 1\} \subseteq [0, \infty)$. Thus, since $\mu_n(\psi(0) = \mathbf{0}) = 1$, all that we have to do is show that, for each $T > 0$,

$$\sup_{n \geq 1} \mathbb{E}^{\mathbb{P}} \left[ \sup_{1 \leq s < t \leq T} \frac{|\mathbf{S}_n(t) - \mathbf{S}_n(s)|}{(t-s)^{\frac{1}{8}}} \right] < \infty,$$

and, by Theorem 4.3.2, this would follow if we knew that

$$(*) \qquad \sup_{n \geq 1} \mathbb{E}^{\mathbb{P}} \left[ |\mathbf{S}_n(t) - \mathbf{S}_n(s)|^4 \right] \leq C(t-s)^2, \quad s, t \in [0, \infty),$$

for some $C < \infty$.

I will prove (*) under the assumption that, for some $M < \infty$ and all $n \geq 1$, $\mathbb{E}^{\mathbb{P}}\left[|\mathbf{X}_n|^4\right] \leq M$. To do this, note that when $k - 1 \leq ns < nt \leq k$,

$$\mathbb{E}^{\mathbb{P}} \left[ \left| \mathbf{S}_n(t) - \mathbf{S}_n(s) \right|^4 \right] = n^2 (t-s)^4 \mathbb{E}^{\mathbb{P}} \left[ \left| \mathbf{X}_k \right|^4 \right] \leq M(t-s)^2.$$

On the other hand, when $k - 1 \leq ns \leq k \leq \ell \leq nt \leq \ell + 1$,

$$\begin{aligned}
&\mathbb{E}^{\mathbb{P}} \left[ \left| \mathbf{S}_n(t) - \mathbf{S}_n(s) \right|^4 \right] \\
&\leq 27 \mathbb{E}^{\mathbb{P}} \left[ \left| \mathbf{S}_n(t) - \mathbf{S}_n\left(\tfrac{\ell}{n}\right) \right|^4 \right] + 27 \mathbb{E}^{\mathbb{P}} \left[ \left| \mathbf{S}_n\left(\tfrac{\ell}{n}\right) - \mathbf{S}_n\left(\tfrac{k}{n}\right) \right|^4 \right] \\
&\qquad + 27 \mathbb{E}^{\mathbb{P}} \left[ \left| \mathbf{S}_n\left(\tfrac{k}{n}\right) - \mathbf{S}_n(s) \right|^4 \right] \\
&\leq 27 M n^2 \left( t - \frac{\ell}{n} \right)^4 + \frac{27}{n^2} \mathbb{E}^{\mathbb{P}} \left[ \left| \sum_{j=1}^{\ell-k} \mathbf{X}_{k+j} \right|^4 \right] + 27 M n^2 \left( \frac{k}{n} - s \right)^4 \\
&\leq 54 M (t-s)^2 + \frac{81 N^2 M (\ell - k)^2}{n^2} \leq 135 N^2 M (t-s)^2,
\end{aligned}$$

where, in the passage to the final line, I have taken $\{\mathbf{e}_i : 1 \leq i \leq N\}$ to be an orthonormal basis for $\mathbb{R}^N$ and used the estimate

$$\begin{aligned}
\mathbb{E}^{\mathbb{P}} \left[ \left| \sum_{j=1}^{\ell-k} \mathbf{X}_{k+j} \right|^4 \right] &= \mathbb{E}^{\mathbb{P}} \left[ \left( \sum_{i=1}^{N} \left( \sum_{j=1}^{\ell-k} (\mathbf{e}_i, \mathbf{X}_{k+j})_{\mathbb{R}^N} \right)^2 \right)^2 \right] \\
&\leq N \sum_{i=1}^{N} \mathbb{E}^{\mathbb{P}} \left[ \left( \sum_{j=1}^{\ell-k} (\mathbf{e}_i, \mathbf{X}_{k+j})_{\mathbb{R}^N} \right)^4 \right] \leq 3 N^2 M (\ell - k)^2
\end{aligned}$$

coming from the second inequality in (1.3.2).

In order to remove the assumption on the fourth moment, I will apply the Principle of Accompanying Laws. Namely, because the $\mathbf{X}_n$'s are uniformly square $\mathbb{P}$-integrable, one can use a truncation procedure to find functions $\{f_{n,\delta} : n \in \mathbb{Z}^+ \text{ and } \delta > 0\} \subseteq C_{\mathrm{b}}(\mathbb{R}^N, \mathbb{R}^N)$ with the properties that, for each $\delta > 0$, $\sup_{n\in\mathbb{Z}^+} \|f_{n,\delta}\|_{\mathrm{u}} < \infty$,

$$\sup_{n\in\mathbb{Z}^+} \mathbb{E}^{\mathbb{P}}\Big[\big|\mathbf{X}_n - f_{n,\delta}\circ \mathbf{X}_n\big|^2\Big] < \delta,$$

and, for every $n \in \mathbb{Z}^+$, the random variable $\mathbf{X}_{n,\delta} \equiv f_{n,\delta} \circ \mathbf{X}_n$ has mean value $\mathbf{0}$ and covariance $\mathbf{I}$. Next, for each $\delta > 0$, define the maps $\omega \in \Omega \longmapsto \mathbf{S}_{n,\delta}(\,\cdot\,,\omega) \in C(\mathbb{R}^N)$ relative to $\{\mathbf{X}_{n,\delta} : n \geq 1\}$, and set $\mu_{n,\delta} = (\mathbf{S}_{n,\delta})_*\mathbb{P}$. Then, by the preceding, we know that $\mu_{n,\delta} \Longrightarrow \mathcal{W}^{(N)}$ for each $\delta > 0$. Hence, by Theorem 9.1.13, we will have proved that $\mu_n \Longrightarrow \mathcal{W}^{(N)}$ as soon as we show that

$$\varlimsup_{\delta\searrow 0} \sup_{n\in\mathbb{Z}^+} \mathbb{P}\left(\sup_{0\leq t\leq T} \big|\mathbf{S}_n(t) - \mathbf{S}_{n,\delta}(t)\big| \geq \epsilon\right) = 0$$

for every $T \in \mathbb{Z}^+$ and $\epsilon > 0$. To this end, first observe that, because $\mathbf{S}_n(\,\cdot\,)$ and $\mathbf{S}_{n,\delta}(\,\cdot\,)$ are linear on each interval $[(m-1)2^{-n}, m2^{-n}]$,

$$\sup_{t\in[0,T]} \big|\mathbf{S}_n(t) - \mathbf{S}_{n,\delta}(t)\big| = \max_{1\leq m\leq nT} \frac{1}{n^{\frac{1}{2}}} \left|\sum_{k=1}^m \mathbf{Y}_{k,\delta}\right|,$$

where $\mathbf{Y}_{k,\delta} \equiv \mathbf{X}_k - \mathbf{X}_{k,\delta}$. Next, note that

$$\begin{aligned}
&\mathbb{P}\left(\max_{1\leq m\leq nT} \frac{1}{n^{\frac{1}{2}}} \left|\sum_{k=1}^m \mathbf{Y}_{k,\delta}\right| \geq \epsilon\right) \\
&\qquad \leq N \max_{\mathbf{e}\in\mathbb{S}^{N-1}} \mathbb{P}\left(\max_{1\leq m\leq nT} \left|\sum_{k=1}^m \big(\mathbf{e}, \mathbf{Y}_{k,\delta}\big)_{\mathbb{R}^N}\right| \geq \frac{n^{\frac{1}{2}}\epsilon}{N^{\frac{1}{2}}}\right).
\end{aligned}$$

Finally, by Kolmogorov's Inequality,

$$\mathbb{P}\left(\max_{1\leq m\leq nT} \left|\sum_{k=1}^m \big(\mathbf{e}, \mathbf{Y}_{k,\delta}\big)_{\mathbb{R}^N}\right| \geq \frac{n^{\frac{1}{2}}\epsilon}{N^{\frac{1}{2}}}\right) \leq \frac{NT\delta}{\epsilon^2}$$

for every $\mathbf{e} \in \mathbb{S}^{N-1}$.

**§ 9.3.2. Rayleigh's Random Flights Model.** Here is a more picturesque scheme for approximating Brownian motion. Imagine the path $t \rightsquigarrow \mathbf{R}(t)$ of a bird that starts at the origin, flies in a randomly chosen direction at unit speed for a unit exponential random time, then switches to a new randomly chosen direction for a second unit exponential time, etc. Next, given $\epsilon > 0$, rescale time and space so that the path becomes $t \rightsquigarrow \mathbf{R}_\epsilon(t)$, where $\mathbf{R}_\epsilon(t) \equiv \epsilon^{\frac{1}{2}}\mathbf{R}(\epsilon^{-1}t)$. I will show that, as $\epsilon \searrow 0$, the distribution of $\{\mathbf{R}_\epsilon(t) : t \geq 0\}$ becomes Brownian motion. This model was introduced by Rayleigh and is called his random flights model.

In the following, $\{\tau_m : m \geq 1\}$ is a sequence of mutually independent, unit exponential random variables from which their partial sums $\{T_n : n \geq 0\}$ and the associated simple Poisson process $\{N(t) : t \geq 0\}$ are defined as in § 4.2.1. Finally, given $\epsilon > 0$, set $N_\epsilon(t) = N(\epsilon^{-1}t)$.

LEMMA 9.3.3. *Let $\{\mathbf{X}_n : n \geq 1\}$ a sequence of mutually independent $\mathbb{R}^N$-valued, uniformly square $\mathbb{P}$-integrable random variables with mean value $\mathbf{0}$ and covariance $\mathbf{I}$, and define $\{\mathbf{S}_n(t) : t \geq 0\}$ accordingly, as in Theorem 9.3.1. (Note that the $\mathbf{X}_n$'s are not assumed to be independent of the $\tau_n$'s.) Next, define*

$$\mathbf{X}_\epsilon(t,\omega) = \sqrt{\epsilon}\sum_{m=1}^{N_\epsilon(t,\omega)} \mathbf{X}_m, \quad (t,\omega) \in [0,\infty) \times \Omega.$$

*Then, for all $r \in (0,\infty)$ and $T \in [0,\infty)$,*

$$\lim_{\epsilon \searrow 0} \mathbb{P}\left( \sup_{t\in[0,T]} \left|\mathbf{X}_\epsilon(t) - \mathbf{S}_{n_\epsilon}(t)\right| \geq r \right) = 0, \quad \text{where } n_\epsilon \equiv \lfloor \epsilon^{-1} \rfloor.$$

PROOF: Note that

$$\begin{aligned}\mathbf{X}_\epsilon(t,\omega) - \mathbf{S}_{n_\epsilon}(t,\omega) &= (\sqrt{\epsilon n_\epsilon} - 1)\,\mathbf{S}_{n_\epsilon}\left(\frac{N_\epsilon(t,\omega)}{n_\epsilon},\omega\right) \\ &\quad + \left(\mathbf{S}_{n_\epsilon}\left(\frac{N_\epsilon(t,\omega)}{n_\epsilon},\omega\right) - \mathbf{S}_{n_\epsilon}(t,\omega)\right).\end{aligned}$$

Hence, for every $\delta \in (0,1]$,

$$\begin{aligned}&\mathbb{P}\left( \sup_{t\in[0,T]} \left|\mathbf{X}_\epsilon(t) - \mathbf{S}_{n_\epsilon}(t)\right| \geq r \right) \\ &\quad \leq \mathbb{P}\left( \sup_{t\in[0,T+\delta]} \left|\mathbf{S}_{n_\epsilon}(t)\right| \geq \frac{r}{2\epsilon} \right) + \mathbb{P}\left( \sup_{t\in[0,T]} \left|\frac{N_\epsilon(t)}{n_\epsilon} - t\right| \geq \delta \right) \\ &\quad + \mathbb{P}\left( \sup_{s\in[0,T]} \sup_{|t-s|\leq\delta} \left|\mathbf{S}_{n_\epsilon}(t) - \mathbf{S}_{n_\epsilon}(s)\right| \geq \frac{r}{2} \right).\end{aligned}$$

But, by Theorem 9.3.1 and the converse statement in Theorem 9.1.9, we know that the first term tends to 0 as $\epsilon \searrow 0$ uniformly in $\delta \in (0,1]$ and that the third term tends to 0 as $\delta \searrow 0$ uniformly in $\epsilon \in (0,1]$. Thus, all that remains is to note that, by Exercise 4.2.20,

$$\lim_{\epsilon \searrow 0} \mathbb{P}\left( \sup_{t\in[0,T]} \left|\epsilon N_\epsilon(t) - t\right| \geq \delta \right) = 0. \qquad \square \tag{9.3.4}$$

Now suppose that $\{\boldsymbol{\theta}_n : n \geq 1\}$ is a sequence of mutually independent $\mathbb{R}^N$-valued random variables that satisfy the conditions that

$$M \equiv \sup_{n\in\mathbb{Z}^+} \mathbb{E}^{\mathbb{P}}\left[|\tau_n \boldsymbol{\theta}_n|^4\right] < \infty,$$

$$\mathbb{E}^{\mathbb{P}}\left[\tau_n\boldsymbol{\theta}_n\right] = \mathbf{0}, \quad \text{and} \quad \mathbb{E}^{\mathbb{P}}\left[(\tau_n\boldsymbol{\theta}_n)\otimes(\tau_n\boldsymbol{\theta}_n)\right] = \mathbf{I}, \quad n \in \mathbb{Z}^+.$$

Finally, define $\omega \in \Omega \longmapsto \mathbf{R}(\,\cdot\,,\omega) \in C(\mathbb{R}^N)$ by

$$\mathbf{R}(t,\omega) = \big(t - T_{N(t,\omega)}(\omega)\big)\boldsymbol{\theta}_{N(t,\omega)+1}(\omega) + \sum_{m=1}^{N(t,\omega)} \tau_m(\omega)\boldsymbol{\theta}_m(\omega).$$

The process $\{\mathbf{R}(t) : t \geq 0\}$ is my interpretation of Rayleigh's random flights model. A typical choice of the $\boldsymbol{\theta}_n$'s would be to make them independent of the holding times (i.e., the $\tau_n$'s) and to choose them to be uniformly distributed over the sphere $\mathbb{S}^{N-1}\left(\sqrt{N}\right)$.

Theorem 9.3.5. *Referring to the preceding, set*

$$\mathbf{R}_\epsilon(t,\omega) = \sqrt{\epsilon}\,\mathbf{R}\left(\tfrac{t}{\epsilon},\omega\right), \quad (t,\omega) \in [0,\infty)\times\Omega.$$

*Then* $\left(\mathbf{R}_\epsilon\right)_*\mathbb{P} \Longrightarrow \mathcal{W}^{(N)}$ *as* $\epsilon \searrow 0$.

Proof: Set $\mathbf{X}_n = \tau_n\boldsymbol{\theta}_n$, and, using the same notation as in Lemma 9.3.3, observe that

$$\left|\mathbf{R}_\epsilon(t) - \mathbf{X}_\epsilon(t)\right| \leq \sqrt{\epsilon}\left|\mathbf{X}_{N_\epsilon(t)+1}\right|.$$

Hence, by Lemma 9.3.3 and Theorems 9.3.1 and 9.1.13, all that we have to do is check that

$$\lim_{\epsilon\searrow 0} \mathbb{P}\left( \sup_{t\in[0,T]} \left|\sqrt{\epsilon}\,\mathbf{X}_{N_\epsilon(t)+1}\right| \geq r \right) = 0$$

for every $r \in (0,\infty)$ and $T \in [0,\infty)$. To this end, set $T_\epsilon = \frac{1+T}{\epsilon}$. Then, by (9.3.4), we have that

$$\lim_{\epsilon\searrow 0}\mathbb{P}\left( \sup_{t\in[0,T]} \left|\sqrt{\epsilon}\,\mathbf{X}_{N_\epsilon(t)+1}\right| \geq r \right) = \lim_{\epsilon\searrow 0} \mathbb{P}\left( \max_{0\leq n\leq T_\epsilon} |\mathbf{X}_{n+1}| \geq \frac{r}{\sqrt{\epsilon}} \right)$$

$$\leq \lim_{\epsilon\searrow 0} \frac{\sqrt{\epsilon}}{r}\mathbb{E}^{\mathbb{P}}\left[\left(\sum_{0\leq n\leq T_\epsilon} |\mathbf{X}_{n+1}|^4\right)^{\frac{1}{4}}\right] \leq \lim_{\epsilon\searrow 0} \frac{\big(M\epsilon(2+T)\big)^{\frac{1}{4}}}{r} = 0. \qquad \square$$

## Exercise for § 9.3

EXERCISE 9.3.6. Let $\{\mu_n : n \geq 1\} \subseteq \mathbf{M}_1\big(C(\mathbb{R}^N)\big)$, and, for each $T \in (0,\infty)$, let $\mu_n^T \in \mathbf{M}_1\big(C([0,T];E)\big)$ denote the distribution of

$$\psi \in C(\mathbb{R}^N) \longmapsto \psi \restriction [0,T] \in C\big([0,T];\mathbb{R}^N\big) \text{ under } \mu_n.$$

Show that there is a $\mu \in \mathbf{M}_1\big(C(\mathbb{R}^N)\big)$ to which $\{\mu_n : n \geq 1\}$ converges in $\mathbf{M}_1\big(C(\mathbb{R}^N)\big)$ if and only if, for each $T \in (0,\infty)$, there is a $\mu^T \in \mathbf{M}_1\big(C([0,T];\mathbb{R}^N)\big)$ with the property that

$$\mu_n^T \Longrightarrow \mu^T \quad \text{in } \mathbf{M}_1\big(C([0,T];\mathbb{R}^N)\big),$$

in which case $\mu^T$ is the distribution of

$$\psi \in C(\mathbb{R}^N) \longmapsto \psi \restriction [0,T] \in C\big([0,T];\mathbb{R}^N\big) \text{ under } \mu.$$

In particular, weak convergence of measures on $C(\mathbb{R}^N)$ is really a *local* property.

EXERCISE 9.3.7. Donsker's own proof of Theorem 9.3.1 was entirely different from the one given here. Instead it was based on a special case of his result, a case that had been proved already (with a very difficult argument) by P. Erdös and M. Kac. The result of Erdös and Kac was that if $\{X_n : n \geq 1\}$ is a sequence of independent, uniformly square integrable random variables with mean value 0 and variance 1, then, for all $a \geq 0$,

$$\lim_{n\to\infty} \mathbb{P}\left( \max_{1\leq m\leq n} n^{-\frac{1}{2}} \sum_{k=1}^{m} X_k \geq a \right) = \sqrt{\frac{2}{\pi}} \int_a^\infty e^{-\frac{x^2}{2}}\, dx.$$

Prove their result as an application of Donsker's Theorem and part **(iii)** of Exercise 4.3.11. According to Kac, it was G. Uhlenbeck who first suggested that their result might be a consequence of a more general "invariance" principle.

EXERCISE 9.3.8. Here is another version of Rayleigh's random flights model. Again let $\{\tau_k : k \geq 1\}$, $\{T_m : m \geq 0\}$, and $\{N(t) : t \geq 0\}$ be as in § 4.2.2, and set

$$R(t) = \int_0^t (-1)^{N(s)}\, ds \quad \text{and} \quad R_\epsilon(t) = \sqrt{\epsilon}\, R\left(\tfrac{t}{\epsilon}\right).$$

Show that $\left(R_\epsilon\right)_* \mathbb{P} \Longrightarrow \mathcal{W}^{(1)}$ as $\epsilon \searrow 0$.

**Hint**: Set $\beta_k = 0$ or $1$ according to whether $k \in \mathbb{N}$ is even or odd, and note that

$$\sum_{k=1}^{n} (-1)^k \tau_k = \sum_{k=1}^{n} \beta_k\big(\tau_{k+1} - \tau_k\big) - \beta_n \tau_n = \sum_{1\leq k\leq \frac{n}{2}} \big(\tau_{2k} - \tau_{2k-1}\big) - \beta_n \tau_{n+1}.$$

Now proceed as in the derivations of Lemma 9.3.3 and Theorem 9.3.5.

# Chapter 10
## Wiener Measure and Partial Differential Equations

In this chapter I will give a somewhat sketchy survey of the bridge between Brownian motion and partial differential equations. Like all good bridges, it is valuable when crossed starting at either end. For those starting from the probability side, it provides a computational tool with which the evaluation of many otherwise intractable Wiener integrals is reduced to finding the solution to a partial differential equation. For aficionados of partial differential equations, it provides a representation of solutions that often reveals properties that are not at all apparent in more conventional, purely analytic, representations.

### § 10.1 Martingales and Partial Differential Equations

The origin of all the connections between Brownian motion and partial differential equations is the observation that the Gauss kernel

$$g^{(N)}(t,\mathbf{x}) = (2\pi t)^{-\frac{N}{2}} e^{-\frac{|\mathbf{x}|^2}{2t}}, \quad (t,\mathbf{x}) \in (0,\infty) \times \mathbb{R}^N, \tag{10.1.1}$$

is simultaneously the density for the Gaussian distribution $\gamma_{\mathbf{0},t\mathbf{I}}$ and the solution to the **heat equation** $\partial_t u = \frac{1}{2}\Delta u$ in $(0,\infty)\times\mathbb{R}$ with initial condition $\delta_{\mathbf{0}}$. More precisely, if $\varphi \in C_{\mathrm{b}}(\mathbb{R}^N;\mathbb{R})$, then

$$u_\varphi(t,\mathbf{x}) = \int_{\mathbb{R}^N} g^{(N)}(t,\mathbf{y}-\mathbf{x})\varphi(\mathbf{y})\,d\mathbf{y}$$

is the one and only bounded $u \in C^{1,2}\big((0,\infty)\times\mathbb{R}^N;\mathbb{R}\big)$ that solves the **Cauchy initial value problem**

$$\partial_t u = \tfrac{1}{2}\Delta u \text{ in } (0,\infty)\times\mathbb{R}^N \text{ with } \lim_{t\searrow 0} u(t,\,\cdot\,) = \varphi \text{ uniformly on compacts.}$$

Checking that $u_\varphi$ solves this problem is an elementary computation. Showing that it is the only solution is less straightforward. Purely analytic proofs can be based on the weak minimum principle. If one assumes more about $u$, then a probabilistic proof can be based on Theorem 7.1.6. Indeed, if one assumes that

$u \in C_{\mathrm{b}}^{1,2}([0,\infty)\times\mathbb{R}^N;\mathbb{C})$, then that theorem shows that, when $(\mathbf{B}(t),\mathcal{F}_t,\mathbb{P})$ is a Brownian motion, for each $T>0$, $(u(T-t\wedge T,\mathbf{x}+\mathbf{B}(t\wedge T)\mathcal{F}_t,\mathbb{P})$ is a martingale. Thus,

$$u(T,\mathbf{x}) = \mathbb{E}^{\mathbb{P}}[\varphi(\mathbf{B}(T))] = \int_{\mathbb{R}^N} \varphi(\mathbf{x}+\mathbf{y})\,\gamma_{\mathbf{0},t\mathbf{I}}(d\mathbf{y}) = u_\varphi(T,\mathbf{x}).$$

In Theorem 10.1.2, I will prove a refinement of Theorem 7.1.6 that will enable me (cf. the discussion following Corollary 10.1.3) to remove the assumption that the derivatives of $u$ are bounded.

As the preceding line of reasoning indicates, the advantage that probability theory provides comes from lifting questions about a partial differential equation to a pathspace setting, and martingales provide one of the most powerful machines with which to do the requisite lifting. In this section I will refine and exploit that machine.

**§ 10.1.1. Localizing and Extending Martingale Representations.** The purpose of this subsection is to combine Theorems 7.1.6 and 7.1.17 with Corollary 7.1.15 to obtain a quite general method for representing solutions to partial differential equations as Wiener integrals.

For the purposes of this chapter, it is best to think of Wiener measure $\mathcal{W}^{(N)}$ as a Borel measure on the Polish space $C(\mathbb{R}^N)\equiv C([0,\infty);\mathbb{R}^N)$ and to take $\{\mathcal{F}_t : t\ge 0\}$ with $\mathcal{F}_t=\sigma(\{\psi(\tau):\tau\in[0,t]\})$ as the standard choice of a non-decreasing family of $\sigma$-algebras. The reason for using $C(\mathbb{R}^N)$ instead of (cf. § 8.1.3) $\Theta(\mathbb{R}^N)$ is that we will want to consider the translates $\mathcal{W}_{\mathbf{x}}^{(N)}$ of $\mathcal{W}^{(N)}$ by $\mathbf{x}\in\mathbb{R}^N$. That is, $\mathcal{W}_{\mathbf{x}}^{(N)}$ is the distribution of $\psi\rightsquigarrow\mathbf{x}+\psi$ under $\mathcal{W}^{(N)}$. Since it is clear that the map $\mathbf{x}\in\mathbb{R}^N\longmapsto\mathcal{W}_{\mathbf{x}}^{(N)}\in\mathbf{M}_1(C(\mathbb{R}^N))$ is continuous, there is no doubt that it is Borel measurable.

THEOREM 10.1.2. *Let $\mathfrak{G}$ be a non-empty, open subset of $\mathbb{R}\times\mathbb{R}^N$, and, for $s\in\mathbb{R}$, define $\zeta_s^{\mathfrak{G}} : C(\mathbb{R}^N)\longrightarrow[0,\infty]$ by*

$$\zeta_s^{\mathfrak{G}}(\psi)=\inf\{t\ge 0:\ (s+t,\psi(t))\notin\mathfrak{G}\}.$$

*Further, suppose that $V:\mathfrak{G}\longrightarrow\mathbb{R}$ is a Borel measurable function that is bounded above on the whole of $\mathfrak{G}$ and bounded below on each compact subset of $\mathfrak{G}$, and set*

$$E_s^V(t,\psi)=\exp\left(\int_0^{t\wedge\zeta_s^{\mathfrak{G}}} V(s+\tau,\psi(\tau))\,d\tau\right).$$

*If $w\in C^{1,2}(\mathfrak{G};\mathbb{R})\cap C_{\mathrm{b}}(\overline{\mathfrak{G}};\mathbb{R})$ satisfies $(\partial_t+\frac{1}{2}\Delta+V)w\ge f$ on $\mathfrak{G}$, where $f:\mathfrak{G}\longrightarrow\mathbb{R}$ is a bounded, Borel measurable function, then*

$$\Bigg(E_s^V(t,\psi)w(s+t\wedge\zeta_s^{\mathfrak{G}}(\psi),\psi(t\wedge\zeta_s^{\mathfrak{G}})) - \int_0^{t\wedge\zeta_s^{\mathfrak{G}}(\psi)} E^V(\tau,\psi)f(s+\tau,\psi(\tau)),\mathcal{F}_t,\mathcal{W}_{\mathbf{x}}^{(N)}\Bigg)$$

*is a submartingale for every $(s, \mathbf{x}) \in \mathfrak{G}$. In particular, if $\left(\partial_t + \frac{1}{2}\Delta + V\right)w = f$ on $\mathfrak{G}$, then the preceding triple is a martingale.*

PROOF: Without loss in generality, I may and will assume that $s = 0$.

Choose a sequence $\{\mathfrak{G}_n : n \geq 0\}$ of open sets such that $(0, \mathbf{x}) \in \mathfrak{G}_0$, $\mathfrak{G}_n \subseteq \mathfrak{G}_{n+1}$, $\overline{\mathfrak{G}_n}$ is a compact subset of $\mathfrak{G}$ for each $n \in \mathbb{N}$, and $\mathfrak{G} = \bigcup_{n=0}^{\infty} \mathfrak{G}_n$. At the same time, for each $n \in \mathbb{N}$, choose $\eta_n \in C^\infty(\mathbb{R} \times \mathbb{R}^N; [0, 1])$ so that $\eta_n = 1$ on $\overline{\mathfrak{G}_n}$ and $\eta_n$ vanishes off a compact subset of $\mathfrak{G}$, and define $w_n$ and $V_n$ so that $w_n = \eta_n w$ and $V_n = \eta_n V$ on $\mathfrak{G}$ and $w_n$ and $V_n$ vanish off of $\mathfrak{G}$. Clearly, $w_n \in C^{1,2}_{\mathrm{b}}(\mathbb{R} \times \mathbb{R}^N; \mathbb{R})$ and $V_n$ is bounded and measurable.

By Theorem 7.1.6, we know that $\left(M_n(t), \mathcal{F}_t, \mathcal{W}^{(N)}_{\mathbf{x}}\right)$ is a martingale, where

$$M_n(t, \psi) = w_n(t, \psi(t)) - \int_0^t g_n(\tau, \psi(\tau))\, d\tau \quad \text{with } g_n = \partial_t w_n + \tfrac{1}{2}\Delta w_n.$$

Thus, if

$$E_n(t, \psi) = \exp\left(\int_0^t V_n(\tau, \psi(\tau))\, d\tau\right),$$

then, by Theorem 7.1.17,

$$\left(E_n(t, \psi) M_n(t, \psi) - \int_0^t E_n(\tau, \psi) M_n(\tau, \psi) V_n(\tau, \psi)\, d\tau, \mathcal{F}_t, \mathcal{W}^{(N)}_{\mathbf{x}}\right)$$

is also a martingale. In addition,

$$\begin{aligned}
&\int_0^t E_n(\tau, \psi) V_n(\tau, \psi) \left(\int_0^\tau g_n(\sigma, \psi(\sigma))\, d\sigma\right) d\tau \\
&\quad = \int_0^t g_n(\sigma, \psi(\sigma)) \left(\int_\sigma^t E_n(\tau, \psi) V_n(\tau, \psi(\tau))\, d\tau\right) d\sigma \\
&\quad = E_n(t, \psi) \int_0^t g_n(\sigma, \psi(\sigma))\, d\sigma - \int_0^t E_n(\sigma, \psi) g_n(\sigma, \psi(\sigma))\, d\sigma,
\end{aligned}$$

and therefore

$$\begin{aligned}
E_n(t, \psi) M_n(t, \psi) &- \int_0^t E_n(\tau, \psi) M_n(\tau, \psi) V_n(\tau, \psi)\, d\tau \\
&= E_n(t, \psi) w_n(t, \psi(t)) - \int_0^t E_n(\tau, \psi) f_n(\tau, \psi(\tau))\, d\tau,
\end{aligned}$$

where $f_n = g_n + V_n w_n$. Hence, we now know that

$$\left(E_n(t, \psi) w_n(t, \psi(t)) - \int_0^t E_n(\tau, \psi) f_n(\tau, \psi(\tau))\, d\tau, \mathcal{F}_t, \mathcal{W}^{(N)}_{\mathbf{x}}\right)$$

is a martingale.

Finally, define $\zeta_0^{\mathfrak{G}_n}$ for $\mathfrak{G}_n$ in the same way as $\zeta_0^{\mathfrak{G}}$ was defined for $\mathfrak{G}$. Since $f_n \geq f$ on $\mathfrak{G}_n$, an application of Theorem 7.1.15 gives the desired result with $\zeta_0^{\mathfrak{G}_n}$ in place of $\zeta_0^{\mathfrak{G}}$, and, because $\zeta_0^{\mathfrak{G}_n} \nearrow \zeta_0^{\mathfrak{G}}$, this completes the proof. □

Perhaps the most famous application of Theorem 10.1.2 is the **Feynman–Kac formula**,[1] a version of which is the content of the following corollary.

COROLLARY 10.1.3. *Let $V : [0,T] \times \mathbb{R}^N \longrightarrow \mathbb{R}$ be a Borel measurable function that is uniformly bounded above everywhere and bounded below uniformly on compacts. If $u \in C^{1,2}\big((0,T)\times\mathbb{R}^N;\mathbb{R}\big)$ is bounded and satisfies the Cauchy initial value problem*

$$\partial_t u = \tfrac{1}{2}\Delta u + Vu + f \text{ in } (0,T)\times\mathbb{R}^N \quad \text{with } \lim_{t\searrow 0} u(t,\cdot\,) = \varphi \text{ uniformly on compacts}$$

*for some bounded, Borel measurable $f : [0,T]\times\mathbb{R}^N \longrightarrow \mathbb{R}$ and $\varphi \in C_{\mathrm{b}}(\mathbb{R}^N;\mathbb{R})$, then*

$$u(T,\mathbf{x}) = \mathbb{E}^{\mathcal{W}_{\mathbf{x}}^{(N)}}\left[e^{\int_0^T V(\tau,\psi(\tau))\,d\tau}\varphi\big(\psi(T)\big)\right] + \mathbb{E}^{\mathcal{W}_{\mathbf{x}}^{(N)}}\left[\int_0^T e^{\int_0^t V(\tau,\psi(\tau))\,d\tau} f\big(t,\omega(t)\big)\,dt\right].$$

PROOF: Given Theorem 10.1.2, there is hardly anything to do. Indeed, here $\mathfrak{G} = (0,T)\times\mathbb{R}^N$ and so $\zeta_0^{\mathfrak{G}} = T$. Thus, by Theorem 10.1.2 applied to $w(t,\cdot\,) = u(T-t,\cdot\,)$, we know that

$$\left(e^{\int_0^{t\wedge T} V(\tau,\psi(\tau))\,d\tau}u\big(T - t\wedge T,\psi(t)\big) - \int_0^{t\wedge T} e^{\int_0^\tau V(\sigma,\psi(\sigma))\,d\sigma} f\big(\tau,\psi(\tau)\big)\,d\tau, \mathcal{F}_t, \mathcal{W}_{\mathbf{x}}^{(N)}\right)$$

is a martingale. Hence,

$$u(T,\mathbf{x}) = \lim_{t\nearrow T}\mathbb{E}^{\mathcal{W}^{(N)}}\left[e^{\int_0^t V(\tau,\psi(\tau))\,d\tau}u\big(T-t,\psi(t)\big) + \int_0^t e^{\int_0^\tau V(\sigma,\psi(\sigma))\,d\sigma} f\big(\tau,\psi(\tau)\big)\,d\tau\right],$$

[1] In the same spirit as he wrote down (8.1.4), Feynman expressed solutions to Schrödinger's equation in terms of path-integrals. After hearing Feynman lecture on his method, Kac realized that one could transfer Feynman's ideas from the Schrödinger to the heat context and thereby arrive at a mathematically rigorous but far less exciting theory.

from which the asserted equality follows immediately. □

As a special case of the preceding, we obtain the missing uniqueness statement in the introduction to this section. Namely, if $u \in C^{1,2}\big((0,\infty)\times\mathbb{R}^N;\mathbb{C}\big)$ is a bounded solution to the heat equation with initial value $\varphi$, then, by considering the real and imaginary parts of $u$ separately, Corollary 10.1.3 implies that

$$u(t,\mathbf{x}) = \mathbb{E}^{\mathcal{W}^{(N)}_{\mathbf{x}}}\big[\varphi(\boldsymbol{\psi}(t))\big] = \int_{\mathbb{R}^N} \varphi(\mathbf{y})g(t,\mathbf{y}-\mathbf{x})\,d\mathbf{y}.$$

**§ 10.1.2. Minimum Principles.** In this subsection I will show how Theorem 10.1.2 leads to an elegant derivation of the basic minimum principle for solutions to equations like the heat equation. Actually, there are two such minimum principles, one of which says that solutions achieve their minimum value at the boundary of the region in which they are defined and the other of which says that only solutions that are constant can achieve a minimum value on the interior. The first of these principles is called the **weak minimum principle**, and the second is called the **strong minimum principle**.

THEOREM 10.1.4. *Let $\mathfrak{G}$ be a non-empty open subset of $\mathbb{R}\times\mathbb{R}^N$, and let $V$ be a function of the sort described in Theorem 10.1.2. Further, suppose that $(s,\mathbf{x})\in\mathfrak{G}$ is a point at which*

$$\mathcal{W}^{(N)}_{\mathbf{x}}\Big(\exists\, t\in(0,\infty)\ \big(s-t,\boldsymbol{\psi}(t)\big)\notin\mathfrak{G}\Big) = 1. \tag{10.1.5}$$

*If $u\in C^{1,2}(\mathfrak{G};\mathbb{R})$ is bounded below and satisfies $\partial_t u - \frac{1}{2}\Delta u - Vu \ge 0$ in $\mathfrak{G}$ and if $\varliminf_{(t,\mathbf{y})\to(t_0,\mathbf{y}_0)} u(t,\mathbf{y}) \ge 0$ for every $(t_0,\mathbf{y}_0)\in\partial\mathfrak{G}$ with $t_0 < s$, then $u(s,\mathbf{x})\ge 0$.*

PROOF: Without loss in generality, I will assume that $s=0$.

Set $\tilde{\mathfrak{G}} = \{(t,\mathbf{y}) : (-t,\mathbf{y})\in\mathfrak{G}\}$ and define $w$ on $\tilde{\mathfrak{G}}$ by $w(t,\mathbf{y}) = u(-t,\mathbf{y})$. Next, choose an exhaustion $\{\mathfrak{G}_n : n\ge 0\}$ of $\mathfrak{G}$ as in the proof of Theorem 10.1.2, and set $\widetilde{\mathfrak{G}_n} = \{(t,\mathbf{y}) : (-t,\mathbf{y})\in\mathfrak{G}_n\}$. By Theorem 10.1.2, we know that

$$w(0,\mathbf{x}) \ge \mathbb{E}^{\mathcal{W}^{(N)}_{\mathbf{x}}}\Big[e^{\int_0^{\zeta_n(\boldsymbol{\psi})} V(-\tau,\boldsymbol{\psi}(\tau))\,d\tau}\, w\big(\zeta_n(\boldsymbol{\psi}),\boldsymbol{\psi}(\zeta_n)\big)\Big],$$

where $\zeta_n(\boldsymbol{\psi}) = \inf\{t\ge 0 : \ (t,\boldsymbol{\psi}(t))\notin\widetilde{\mathfrak{G}_n}\}\wedge n$. Moreover, by (10.1.5), for $\mathcal{W}^{(N)}_{\mathbf{x}}$-almost every $\boldsymbol{\psi}$, $\big(-\zeta_n(\boldsymbol{\psi}),\boldsymbol{\psi}(\zeta_n)\big)$ tends to a point in $\{(t,x)\in\partial\mathfrak{G} : t<0\}$ as $n\to\infty$, and therefore

$$\varliminf_{n\to\infty} w\big(\zeta_n(\boldsymbol{\psi}),\boldsymbol{\psi}(\zeta_n)\big) = \varliminf_{n\to\infty} u\big(-\zeta_n(\boldsymbol{\psi}),\boldsymbol{\psi}(\zeta_n)\big) \ge 0 \quad \mathcal{W}^{(N)}_{\mathbf{x}}\text{-almost surely.}$$

Hence, by Fatou's Lemma, we see that $u(0,\mathbf{x}) = w(0,\mathbf{x}) \ge 0$. □

THEOREM 10.1.6. *In the same setting as the preceding, suppose that* $u \in C^{1,2}(\mathfrak{G};\mathbb{R})$ *satisfies* $\partial_t u - \frac{1}{2}\Delta u - Vu \geq 0$ *in* $\mathfrak{G}$. *If* $(s,\mathbf{x}) \in \mathfrak{G}$ *and* $0 = u(s,\mathbf{x}) \leq u(t,\mathbf{y})$ *for all* $(t,\mathbf{y}) \in \mathfrak{G}$ *with* $t \leq s$, *then* $u(s-t,\boldsymbol{\psi}(t)) = 0$ *for all* $(t,\boldsymbol{\psi}) \in [0,\infty) \times C(\mathbb{R}^N)$ *such that* $\boldsymbol{\psi}(0) = \mathbf{x}$ *and* $\big(s-\tau,\boldsymbol{\psi}(\tau)\big) \in \mathfrak{G}$ *for all* $\tau \in [0,t]$. *In particular, if* $G$ *is a connected, open subset of* $\mathbb{R}^N$, $V$ *is independent of time, and* $u \in C^2\big(G;[0,\infty)\big)$ *satisfies* $\frac{1}{2}\Delta u + Vu \leq 0$, *then either* $u \equiv 0$ *or* $u > 0$ *everywhere on* $G$.

PROOF: Again, without loss in generality, I assume that $s = 0$. In addition, I may and will assume that $\mathbf{x} = \mathbf{0}$, $V$ is uniformly bounded, and $u \in C_b\big(\overline{\mathfrak{G}};[0,\infty)\big)$. To see that these latter assumptions cause no loss in generality, one can use an exhaustion argument of the same sort as was used in the proof of Theorem 10.1.2.

Given $(t,\boldsymbol{\psi}) \in (0,\infty) \times C(\mathbb{R}^N)$ with $\boldsymbol{\psi}(0) = \mathbf{0}$ and $\big(-\tau,\boldsymbol{\psi}(\tau)\big) \in \mathfrak{G}$ for $\tau \in [0,t]$, suppose that $u\big(-t,\boldsymbol{\psi}(t)\big) > 0$. In order to get a contradiction, choose $r > 0$ so that $u(-t,\mathbf{y}) \geq r$ if $|\mathbf{y} - \boldsymbol{\psi}(t)| \leq r$ and so that $\big(-\tau,\boldsymbol{\psi}'(\tau)\big) \in \mathfrak{G}$ if $\tau \in [0,t]$ and $\|\boldsymbol{\psi}' - \boldsymbol{\psi}\|_{[0,t]} \leq r$. If $\tilde{\mathfrak{G}} = \{(t,\mathbf{y}) : (-t,\mathbf{y}) \in \mathfrak{G}\}$, then, just as in the proof of Theorem 10.1.2,

$$
\begin{aligned}
0 = u(0,\mathbf{0}) &\geq \int e^{\int_0^{t\wedge\zeta_0^{\tilde{\mathfrak{G}}}(\boldsymbol{\psi}')} V(-\tau,\boldsymbol{\psi}'(\tau))\,d\tau}\, u\big(-t\wedge\zeta_0^{\tilde{\mathfrak{G}}}(\boldsymbol{\psi}'),\boldsymbol{\psi}'(t\wedge\zeta_0^{\tilde{\mathfrak{G}}})\big)\,\mathcal{W}^{(N)}(d\boldsymbol{\psi}') \\
&\geq re^{-t\|V\|_{\mathrm{u}}}\mathcal{W}^{(N)}\big(\{\boldsymbol{\psi}' : \|\boldsymbol{\psi}' - \boldsymbol{\psi}\|_{[0,t]} \leq r\}\big).
\end{aligned}
$$

Since, by Corollary 8.3.6, $\mathcal{W}^{(N)}\big(\{\boldsymbol{\psi}' : \|\boldsymbol{\psi}' - \boldsymbol{\psi}\|_{[0,t]} \leq r\}\big) > 0$, we have the required contradiction.

Turning to the final assertion, take $\mathfrak{G} = \mathbb{R} \times G$, and observe that for all $(\mathbf{x},\mathbf{y}) \in G^2$ there is a $\boldsymbol{\psi}$ such that $\boldsymbol{\psi}(0) = \mathbf{x}$, $\boldsymbol{\psi}(1) = \mathbf{y}$, and $\boldsymbol{\psi}(\tau) \in G$ for all $\tau \in [0,1]$. □

At first glance, one might think that the strong minimum principle overshadows the weak minimum principle and makes it obsolete. However, that is not entirely true. Specifically, before one can apply the strong minimum principle, one has to know that a minimum is actually achieved. In many situations, continuity plus compactness provide the necessary existence. However, when compactness is absent, special considerations have to be brought to bear. The weak minimum principle does not suffer from this problem. On the other hand, it suffers from a related problem. Namely, one has to know ahead of time that (10.1.5) holds. As we will see below, this is usually not too serious a problem, but it should be kept in mind.

**§ 10.1.3. The Hermite Heat Equation.** In the preceding subsection I gave an example of how probability theory can give information about solutions to partial differential equations. In this subsection, it will be a differential equation that gives us information about probability theory. To be precise, I, following M. Kac, will give in this subsection his derivation of the formulas that we derived by purely Gaussian techniques in Exercise 8.3.19, and in the next section I will give his treatment of a closely related problem.[2]

Closed form solutions to the Cauchy initial value problem are available for very few $V$'s, but there is a famous one for which they are. Namely, when $V = -\frac{1}{2}|\mathbf{x}|^2$, a great deal is known. Indeed, already in the nineteenth century, Hermite knew how to analyze the operator $\frac{1}{2}\Delta - \frac{1}{2}|\mathbf{x}|^2$. As a result, this operator is often called the **Hermite operator** by mathematicians, although physicists call it the **harmonic oscillator** because it arises in quantum mechanics as minus the Hamiltonian for an oscillator that satisfies Hook's law. Be that as it may, set (cf. (10.1.1))

$$h(t,\mathbf{x},\mathbf{y}) = e^{-\frac{Nt+|\mathbf{x}|^2}{2}} g^{(N)}\left(\frac{1-e^{-2t}}{2}, \mathbf{y} - e^{-t}\mathbf{x}\right) e^{\frac{|\mathbf{y}|^2}{2}} \tag{10.1.7}$$

for $(t,\mathbf{x},\mathbf{y}) \in (0,\infty) \times \mathbb{R}^N \times \mathbb{R}^N$. By using the fact that $g^{(N)}$ solves the heat equation and tends to $\delta_{\mathbf{0}}$ as $t \searrow 0$, one can apply elementary calculus to check that

$$\begin{aligned} &\partial_t h(t,\cdot,\mathbf{y}) = \left(\tfrac{1}{2}\Delta - \tfrac{1}{2}|\mathbf{x}|^2\right) h(t,\cdot,\mathbf{y}) \quad \text{in } (0,\infty)\times\mathbb{R}^N \\ &\qquad\text{and } \lim_{t\searrow 0} h(t,\mathbf{x},\mathbf{y}) = \delta_{\mathbf{y}-\mathbf{x}} \end{aligned} \qquad \text{for each } \mathbf{y}\in\mathbb{R}^N.$$

Now let $\varphi \in C_{\mathrm{b}}(\mathbb{R}^N;\mathbb{R})$ be given, and set

$$u_\varphi(t,\mathbf{x}) = \int_{\mathbb{R}^N} \varphi(\mathbf{y}) h(t,\mathbf{x},\mathbf{y})\, d\mathbf{y}.$$

Then, $u_\varphi$ is a bounded solution to $\partial_t u = \frac{1}{2}\Delta u - \frac{1}{2}|\mathbf{x}|^2 u$ that tends to $\varphi$ as $t \searrow 0$. Hence, as an immediate consequence of Corollary 10.1.3, we see that

$$u_\varphi(t,\mathbf{x}) = \mathbb{E}^{\mathcal{W}_{\mathbf{x}}^{(N)}}\left[e^{-\frac{1}{2}\int_0^t |\psi(\tau)|^2\,d\tau} \varphi\big(\psi(t)\big)\right].$$

By taking $\varphi = \mathbf{1}$ and performing a tedious, but completely standard, Gaussian computation, one can use this to derive

$$\mathbb{E}^{\mathcal{W}_{\mathbf{x}}^{(N)}}\left[e^{-\frac{1}{2}\int_0^t |\psi(\tau)|^2\,d\tau}\right] = (\cosh t)^{-\frac{N}{2}} \exp\left(-\frac{|\mathbf{x}|^2}{2}\tanh t\right),$$

which, together with Brownian scaling, vastly generalizes the result in Exercise 8.3.19.

---

[2] See Kac's "On some connections between probability theory and differential and integral equations," *Proc. 2nd Berkeley Symp. on Prob. & Stat.* Univ. of California Press (1951), where he gives several additional, intriguing applications of Corollary 10.1.3.

§ **10.1.4. The Arcsine Law.** As I said at the beginning of the last subsection, there are very few $V$'s for which one can write down explicit solutions to equations of the form $\partial_t u = \frac{1}{2}\Delta u + Vu$. On the other hand, when $V$ is independent of time one can often, particularly when $N = 1$, write down a closed form expression for the Laplace transform $U_\lambda = \int_0^\infty e^{-\lambda t} u(t, \cdot\,)\, dt$ of $u$. Indeed, if $u$ is a bounded solution to $\partial_t u = \frac{1}{2}\Delta u + Vu$, then it is an elementary exercise to check that

$$(\lambda - \tfrac{1}{2}\Delta - V)U_\lambda = f,$$

and when $N = 1$ this is an ordinary differential equation. Moreover, when $U_\lambda \in C^2(\mathbb{R}^N;\mathbb{R})$ is bounded, one can apply Corollary 10.1.3 to see that

$$\begin{aligned} U_\lambda(\mathbf{x}) =& \mathbb{E}^{\mathcal{W}_{\mathbf{x}}^{(N)}}\left[e^{\int_0^T V_\lambda(\psi(\tau))\,d\tau} U_\lambda(\psi(T))\right] \\ &+ \mathbb{E}^{\mathcal{W}_{\mathbf{x}}^{(N)}}\left[\int_0^T e^{\int_0^t V_\lambda(\psi(\tau))\,d\tau} f(\psi(t))\,dt\right] \quad \text{for } T > 0, \end{aligned}$$

where $V_\lambda = V - \lambda$. Hence, if $V_\lambda$ is uniformly negative and one lets $T \to \infty$, one gets

$$U_\lambda(\mathbf{x}) = \mathbb{E}^{\mathcal{W}_{\mathbf{x}}^{(N)}}\left[\int_0^\infty e^{\int_0^t V_\lambda(\psi(\tau))\,d\tau} f(\psi(t))\,dt\right].$$

The preceding remark is the origin of Kac's derivation of Lévy's **Arcsine Law** for Wiener measure.

THEOREM 10.1.8. *For every $T \in (0,\infty)$ and $\alpha \in [0,1]$,*

$$\mathcal{W}^{(1)}\left(\left\{\psi \in C(\mathbb{R}) : \frac{1}{T}\int_0^T \mathbf{1}_{[0,\infty)}(\psi(t))\,dt \le \alpha\right\}\right) = \frac{2}{\pi}\arcsin(\sqrt{\alpha}).$$

PROOF: First note that, by Brownian scaling, it suffices to prove the result when $T = 1$. Next, set

$$F(\alpha) = \mathcal{W}\left(\left\{\psi \in C(\mathbb{R}) : \int_0^1 \mathbf{1}_{[0,\infty)}(\psi(s))\,ds \le \alpha\right\}\right), \quad \alpha \in [0,\infty),$$

and let $\mu$ denote the element of $\mathbf{M}_1([0,\infty))$ for which $F$ is the distribution function. We are going to compute $F(\alpha)$ by looking at the double Laplace transform

$$G(\lambda) \equiv \int_{(0,\infty)} e^{-\lambda t} g(t)\,dt, \quad \lambda \in (0,\infty),$$

where

$$g(t) \equiv \int_{[0,\infty)} e^{-t\alpha}\,\mu(d\alpha), \quad t \in (0,\infty);$$

and, by another application of the Brownian scaling property, we see that

$$G(\lambda) = \int_0^\infty \left( \int \exp\left[ -\int_0^t \left(\lambda + \mathbf{1}_{[0,\infty)}(\psi(s))\right) ds \right] \mathcal{W}^{(1)}(d\psi) \right) dt$$
$$= \mathbb{E}^{\mathcal{W}^{(1)}}\left[ \int_0^\infty e^{\int_0^t V_\lambda(\psi(\tau))\, d\tau}\, dt \right] \quad \text{where } V_\lambda \equiv -\lambda - \mathbf{1}_{[0,\infty)}.$$

At this point, the strategy is to calculate $G(\lambda)$ with the help of the idea explained above. For this purpose, I begin by seeking as good a solution $x \in \mathbb{R} \longmapsto u_\lambda(x) \in \mathbb{R}$ as I can find to the equation $\frac{1}{2}u'' + V_\lambda u = -1$. By considering this equation separately on the left and right half-lines and then matching, in so far as possible, at 0, one finds that the best choice of bounded $u_\lambda$ will be to take

$$u_\lambda(x) = \begin{cases} A_\lambda \exp\left[ -\sqrt{2(1+\lambda)}\, x \right] + \frac{1}{1+\lambda} & \text{if } x \in [0,\infty) \\ B_\lambda \exp\left[ \sqrt{2\lambda}\, x \right] + \frac{1}{\lambda} & \text{if } x \in (-\infty, 0), \end{cases}$$

where

$$A_\lambda = \left( \frac{1}{\lambda(1+\lambda)} \right)^{\frac{1}{2}} - \frac{1}{1+\lambda} \quad \text{and} \quad B_\lambda = \left( \frac{1}{\lambda(1+\lambda)} \right)^{\frac{1}{2}} - \frac{1}{\lambda}.$$

(The choice of sign in the exponent is dictated by my desire to have $u_\lambda$ bounded.) If $u_\lambda$ were twice continuously differentiable, I could apply the reasoning above directly and thereby arrive at $G(\lambda) = u_\lambda(0)$. However, because the second derivative of $u_\lambda$ is discontinuous at 0, I have to work a little harder.

Notice that, although the second derivative of $u_\lambda$ has a discontinuity at 0, $u'_\lambda$ is nonetheless uniformly Lipschitz continuous everywhere. Hence, by taking $\rho \in C^\infty_{\mathrm{c}}(\mathbb{R};[0,\infty))$ with Lebesgue integral 1 and setting

$$u_{\lambda,n}(x) = n \int_{\mathbb{R}} u_\lambda(x-y)\rho(ny)\, dy, \quad n \in \mathbb{Z}^+,$$

we see that $u_{\lambda,n} \in C^\infty_{\mathrm{b}}(\mathbb{R};\mathbb{R})$ for each $n \in \mathbb{Z}^+$, $u_{\lambda,n} \longrightarrow u_\lambda$ uniformly on $\mathbb{R}$ as $n \to \infty$, $\sup_{n\in\mathbb{Z}^+} \|u_{\lambda,n}\|_{C^2_{\mathrm{b}}(\mathbb{R};\mathbb{R})} < \infty$, and, as $n \to \infty$,

$$f_n \equiv \frac{1}{2}u''_{\lambda,n} - \left(\lambda + \mathbf{1}_{[0,\infty)}\right)u_{\lambda,n} \longrightarrow -1 \quad \text{on } \mathbb{R}\setminus\{0\}.$$

Thus, since the argument that I attempted to apply to $u_\lambda$ works for $u_{\lambda,n}$, we know that

$$u_{\lambda,n}(0) = \mathbb{E}^{\mathcal{W}^{(1)}}\left[ \int_0^\infty e^{\int_0^t V_\lambda(\psi(\tau))\, d\tau} f_n(\psi(t))\, d\tau\, dt \right].$$

In addition, because

$$\mathbb{E}^{\mathcal{W}^{(1)}}\left[\int_0^\infty \mathbf{1}_{\{0\}}\big(\psi(t)\big)\,dt\right] = \int_0^\infty \gamma_{0,t}\big(\{0\}\big)\,dt = 0,$$

$$\mathbb{E}^{\mathcal{W}^{(1)}}\left[\int_0^\infty e^{\int_0^t V_\lambda(\psi(\tau))\,d\tau} f_n\big(\psi(t)\big)\,dt\right] \longrightarrow G(\lambda).$$

Hence, the conclusion $u_\lambda(0) = G(\lambda)$ has now been rigorously verified.

Knowing that $G(\lambda) = \big(\lambda(1-\lambda)\big)^{-\frac{1}{2}}$, the rest of the calculation is easy. Indeed, since

$$\int_0^\infty t^{-\frac{1}{2}} e^{-\lambda t}\,dt = \sqrt{\frac{\pi}{\lambda}},$$

the multiplication rule for Laplace transforms tells us that

$$g(t) = \frac{1}{\pi}\int_0^t \frac{e^{-s}}{\sqrt{s(t-s)}}\,ds = \frac{1}{\pi}\int_0^1 \frac{e^{-t\alpha}}{\sqrt{\alpha(1-\alpha)}}\,d\alpha;$$

and so we now find that

$$F(\alpha) = \frac{1}{\pi}\int_0^{\alpha\wedge 1} \frac{1}{\sqrt{\beta(1-\beta)}}\,d\beta = \frac{2}{\pi}\arcsin\big(\sqrt{\alpha\wedge 1}\big). \quad \square$$

Just as Donsker's Invariance Principle enabled us in Exercise 9.3.7 to derive the Erdös–Kac Theorem from the reflection principle for Brownian motion, it now allows us to transfer the Arcsine Law for Wiener measure to the Arcsine Law for sums of independent random variables.

COROLLARY 10.1.9. *If $\{X_n : n \geq 1\}$ is a sequence of independent, uniformly square $\mathbb{P}$-integrable random variables with mean value* 0 *and variance* 1 *on some probability space $(\Omega, \mathcal{F}, \mathbb{P})$, then, for every $\alpha \in [0,1]$,*

$$\lim_{n\to\infty} \mathbb{P}\left(\left\{\omega : \frac{N_n(\omega)}{n} \leq \alpha\right\}\right) = \frac{2}{\pi}\arcsin\big(\sqrt{\alpha}\big),$$

*where $N_n(\omega)$ is the number of $m \in \mathbb{Z}^+ \cap [0,n]$ for which $S_m(\omega) \equiv \sum_{\ell=1}^m X_\ell(\omega)$ is non-negative.*

PROOF: Thinking of $\frac{N_n(\omega)}{n}$ as a Riemann approximation to (cf. the notation in § 9.2.1)

$$\int_0^1 \mathbf{1}_{[0,\infty)}\big(S_n(t,\omega)\big)\,dt,$$

one should guess that, in view of Theorem 9.3.1 and Theorem 9.1.13, there should be very little left to be done. However, once again there are continuity issues that have to be dealt with. Thus, for each $f \in C\big(\mathbb{R};[0,1]\big)$ and $n \in \mathbb{Z}^+$, introduce the functions $F^f$ and $F^f_n$ on $C(\mathbb{R})$ given by

$$F^f(\psi) = \int_0^1 f\big(\psi(t)\big)\,dt \quad \text{and} \quad F^f_n(\psi) = \frac{1}{n}\sum_{m=1}^n f\big(\psi(\tfrac{m}{n})\big)$$

for any $f \in C\big(\mathbb{R};[0,1]\big)$. Since $F^f_n \longrightarrow F^f$ uniformly on compacts, Theorem 9.3.1 plus Lemma 9.1.10 show that the distribution of

$$\omega \in \Omega \longmapsto A^f_n(\omega) \equiv \frac{1}{n}\sum_{m=1}^n f\left(\frac{S_m(\omega)}{n^{\frac{1}{2}}}\right)$$

under $\mathbb{P}$ tends weakly to that of $\psi \in C(\mathbb{R}) \longmapsto F^f(\psi)$ under $\mathcal{W}^{(1)}$. Next, for each $\delta \in (0,\infty)$, choose continuous functions $f^\pm_\delta$ so that $\mathbf{1}_{(\delta,\infty)} \le f^+_\delta \le \mathbf{1}_{[0,\infty)}$ and $\mathbf{1}_{[0,\infty)} \le f^-_\delta \le \mathbf{1}_{[-\delta,\infty)}$, and conclude that

$$\varlimsup_{n\to\infty} \mathbb{P}\left(\frac{N_n}{n} \le \alpha\right) \le \mathcal{W}^{(1)}\left(F^{f^+_\delta} \le \alpha\right)$$

and

$$\varliminf_{n\to\infty} \mathbb{P}\left(\frac{N_n}{n} < \alpha\right) \ge \mathcal{W}^{(1)}\left(F^{f^-_\delta} < \alpha\right)$$

for every $\delta > 0$. Passing to the limit as $\delta \searrow 0$, we arrive at

$$\varlimsup_{n\to\infty} \mathbb{P}\left(\frac{N_n}{n} \le \alpha\right) \le \mathcal{W}^{(1)}\left(\left\{\psi : \int_0^1 \mathbf{1}_{(0,\infty)}\big(\psi(t)\big)\,dt \le \alpha\right\}\right)$$

and

$$\varliminf_{n\to\infty} \mathbb{P}\left(\frac{N_n}{n} < \alpha\right) \ge \mathcal{W}^{(1)}\left(\left\{\psi : \int_0^1 \mathbf{1}_{[0,\infty)}\big(\psi(t)\big)\,dt < \alpha\right\}\right).$$

Finally, since

$$\int\left(\int_0^1 \mathbf{1}_{\{0\}}\big(\psi(t)\big)\,dt\right)\mathcal{W}^{(1)}(d\psi) = \int_0^1 \mathcal{W}^{(1)}\big(\psi(t) = 0\big)\,dt = 0,$$

and $\alpha \in [0,1] \longmapsto \arcsin\big(\sqrt{\alpha}\big)$ is continuous, the asserted result follows. □

REMARK 10.1.10. The renown of the Arcsine Law stems, in large part, from the following counterintuitive deduction that can be drawn from it. Namely, given $\delta \in \left(0, \frac{1}{2}\right)$, guess which $\alpha$ maximizes $\lim_{n\to\infty} \mathbb{P}\left(\frac{N_n}{n} \in (\alpha-\delta, \alpha+\delta) \bmod 1\right)$ for a fixed $\delta$. Because of The Law of Large Numbers (in more common parlance, "The Law of Averages"), most people are inclined to guess that the maximum should occur at $\alpha = \frac{1}{2}$. Thus, it is surprising that, since

$$\alpha \in [0,1] \longmapsto \frac{1}{\sqrt{\alpha(1-\alpha)}} \in [0,\infty]$$

is convex and has its minimum at $\frac{1}{2}$, the Arcsine Law makes the exact opposite prediction! The point is, of course, that the sequence of partial sums $\{S_n(\omega) : n \geq 1\}$ is most likely to make long excursions above and below 0 but tends to spend relatively little time in a neighborhood of 0. In other words, although one may be correct to feel that "my luck has got to change," one had better be prepared to wait a long time.

A more technical point is one raised by S. Sternberg. The arcsine distribution is familiar to people who study iterated maps and is important to them because (cf. Exercise 10.1.15) it is the one and only absolutely continuous probability distribution on $[0,1]$ that is invariant under $x \in [0,1] \longmapsto 4x(1-x) \in [0,1]$. Sternberg asked whether a derivation of Theorem 10.1.8 can be based on this invariance property. Taking $T_+ = \int_0^1 \mathbf{1}_{[0,\infty)}\big(\psi(s)\big)\,ds$ and $S = \int_0^1 \operatorname{sgn}\big(\psi(s)\big)\,ds$, and noting that $4T_+(1-T_+) = 1 - S^2$, one way to phrase Sternberg's question is to ask is whether there is a *pure thought* way to check that $T_+$ and $1-S^2$ have the same distribution under $\mathcal{W}^{(1)}$ and that this distribution is absolutely continuous. I have posed this problem to several experts but, as yet, none of them has come up with a satisfactory solution.

§ **10.1.5. Recurrence and Transience of Brownian Motion.** In this subsection I will use solutions to partial differential equations to examine the long time behavior of Brownian motion.

THEOREM 10.1.11. *For $r \in [0,\infty)$, define*

$$\zeta_r(\psi) = \inf\{t \in [0,\infty) : |\psi(t)| = r\}, \quad \psi \in C(\mathbb{R}^N).$$

*Then*

$$\begin{aligned}\mathbb{E}^{\mathcal{W}^{(N)}_{\mathbf{x}}}[\zeta_r] &= \frac{r^2 - |\mathbf{x}|^2}{N} \\ \mathbb{E}^{\mathcal{W}^{(N)}_{\mathbf{x}}}[\zeta_r^2] &= \frac{(N+4)r^2 - N|\mathbf{x}|^2}{N^2(N+2)}(r^2 - |\mathbf{x}|^2)\end{aligned} \qquad \text{for } |\mathbf{x}| < r.$$

*In addition, if* $0 < r < |\mathbf{x}| < R < \infty$, *then*

$$\mathcal{W}_{\mathbf{x}}^{(N)}(\zeta_r < \zeta_R) = \begin{cases} \dfrac{R - |x|}{R - r} & \text{if } N = 1 \\ \dfrac{\log R - \log|\mathbf{x}|}{\log R - \log r} & \text{if } N = 2 \\ \left(\dfrac{r}{|\mathbf{x}|}\right)^{N-2} \dfrac{R^{N-2} - |\mathbf{x}|^{N-2}}{R^{N-2} - r^{N-2}} & \text{if } N \geq 3. \end{cases}$$

*In particular,*

$$\mathcal{W}_{x}^{(1)}(\zeta_0 < \infty) = 1 \quad \text{for all } x \in \mathbb{R},$$

$$\mathcal{W}_{\mathbf{x}}^{(2)}(\zeta_0 < \infty) = 0, \ \mathbf{x} \neq \mathbf{0}, \quad \text{but} \quad \mathcal{W}_{\mathbf{x}}^{(2)}(\zeta_r < \infty) = 1, \ \mathbf{x} \in \mathbb{R}^2 \text{ and } r > 0,$$

*and*

$$\mathcal{W}_{\mathbf{x}}^{(N)}(\zeta_r < \infty) = \left(\frac{r}{|\mathbf{x}|}\right)^{N-2}, \ 0 < r < |\mathbf{x}|, \quad \text{when } N \geq 3.$$

PROOF: To prove the first two equalities, set $f(t, \mathbf{x}) = |\mathbf{x}|^2 - Nt$, use Theorem 10.1.2 to show that

$$\left(f\big(t \wedge \zeta_r, \boldsymbol{\psi}(t \wedge \zeta_r)\big), \mathcal{F}_t, \mathcal{W}_{\mathbf{x}}^{(N)}\right)$$

and

$$\left(f\big(t \wedge \zeta_r, \boldsymbol{\psi}(t \wedge \zeta_r)\big)^2 - 4\int_0^{t\wedge\zeta_r} |\boldsymbol{\psi}(s)|^2\, ds, \mathcal{F}_t, \mathcal{W}_{\mathbf{x}}^{(N)}\right)$$

are continuous martingales, and conclude that

$$N\mathbb{E}^{\mathcal{W}_{\mathbf{x}}^{(N)}}\big[t \wedge \zeta_r\big] = \mathbb{E}^{\mathcal{W}_{\mathbf{x}}^{(N)}}\Big[\big|\boldsymbol{\psi}(t \wedge \zeta_r)\big|^2\Big] - |\mathbf{x}|^2, \quad t \in [0, \infty),$$

and

$$\begin{aligned} N^2\mathbb{E}^{\mathcal{W}_{\mathbf{x}}^{(N)}}\big[(t \wedge \zeta_r)^2\big] =& |\mathbf{x}|^4 + 4\mathbb{E}^{\mathcal{W}_{\mathbf{x}}^{(N)}}\left[\int_0^{t\wedge\zeta_r} |\boldsymbol{\psi}(s)|^2\, ds\right] \\ &+ 2N\mathbb{E}^{\mathcal{W}_{\mathbf{x}}^{(N)}}\Big[(t \wedge \zeta_r)\big|\boldsymbol{\psi}(t \wedge \zeta_r)\big|^2\Big] - \mathbb{E}^{\mathcal{W}_{\mathbf{x}}^{(N)}}\Big[\big|\boldsymbol{\psi}(t \wedge \zeta_r)\big|^4\Big] \end{aligned}$$

for all $t \in [0, \infty)$. Now assume that $|\mathbf{x}| \leq r$, and use the first of these to show that $N\mathbb{E}^{\mathcal{W}_{\mathbf{x}}^{(N)}}[\zeta_r] \leq r^2$. Thus $\mathcal{W}_{\mathbf{x}}^{(N)}(\zeta_r < \infty) = 1$, and so $N\mathbb{E}^{\mathcal{W}_{\mathbf{x}}^{(N)}}[\zeta_r] = r^2 - |\mathbf{x}|^2$ follows when $t \to \infty$. To get the second equality, use Theorem 10.1.2 to show that

$$\left( |\psi(t \wedge \zeta_r)|^4 - (4 + 2N) \int_0^{t \wedge \zeta_r} |\psi(s)|^2 \, ds, \mathcal{F}_t, \mathcal{W}_{\mathbf{x}}^{(N)} \right)$$

is a continuous martingale and therefore that, for $\mathbf{x} \in B(\mathbf{0}, r)$,

$$(4 + 2N)\mathbb{E}^{\mathcal{W}_{\mathbf{x}}^{(N)}} \left[ \int_0^{t \wedge \zeta_r} |\psi(s)|^2 \, ds \right] = \mathbb{E}^{\mathcal{W}_{\mathbf{x}}^{(N)}} \left[ |\psi(t \wedge \zeta_r)|^4 \right] - |\mathbf{x}|^4,$$

plug this into the above, and pass to the limit as $t \longrightarrow \infty$.

Turning to the second part of the theorem, for each $N \in \mathbb{Z}^+$ choose $f_N \in C_{\mathrm{b}}^\infty(\mathbb{R}^N; \mathbb{R})$ so that, on an open neighborhood $G$ of the annulus $\overline{B(\mathbf{0}, R)} \setminus B(\mathbf{0}, r)$, $f_N$ is equal to the corresponding expression on the right-hand side of the equality under consideration, note that $\Delta f_N = 0$ on $G$, and conclude (via Theorem 10.1.2) that

$$\left( f_N(t \wedge \zeta_r \wedge \zeta_R), \mathcal{F}_t, \mathcal{W}_{\mathbf{x}}^{(N)} \right)$$

is a bounded, continuous martingale. In particular, after one lets $t \to \infty$ and notes that, by the first part of this theorem, $\mathcal{W}_{\mathbf{x}}^{(N)}(\zeta_R < \infty) = 1$ for $|\mathbf{x}| < R$, this leads to

$$\mathcal{W}_{\mathbf{x}}^{(N)}(\zeta_r < \zeta_R) = \mathbb{E}^{\mathcal{W}_{\mathbf{x}}^{(N)}} \left[ f_N(\zeta_r \wedge \zeta_R) \right] = f_N(\mathbf{x}), \quad 0 < r < |\mathbf{x}| < R,$$

as required. Given this, the rest of the theorem follows easily when one lets $R \nearrow \infty$ and, in the case when $N = \{1, 2\}$, $r \searrow 0$. □

The second part of Theorem 10.1.11 says something significant about the global behavior of Brownian paths and the dependence of that behavior on dimension. Namely, when $N \in \{1, 2\}$, it says that, no matter where it is started, a Brownian path will hit any non-empty open set with probability 1. As will be shown in Theorem 10.2.3, this property implies the seemly stronger statement that, with probability 1, a Brownian path will visit every non-empty open set infinitely often and will spend infinite time in each. For this reason, Brownian motion in one and two dimensions is said to be **recurrent**. By contrast, when $N \geq 3$, Theorem 10.1.11 says that, with positive probability, a Brownian path will never visit a closed ball in which it was not started. Moreover, if it is started outside of a ball, then the probability of its ever hitting that ball goes to 0 as the diameter of the set goes to 0. As I am about to show, this latter property leads to the conclusion that, with probability 1, a Brownian path in three or more dimensions tends to infinity.

Corollary 10.1.12. *If $N \geq 3$, then*

$$\mathcal{W}_{\mathbf{x}}^{(N)}\left(\lim_{t\to\infty} |\psi(t)| = \infty\right) = 1, \quad \mathbf{x} \in \mathbb{R}^N.$$

Proof: Given $r > 0$, apply Theorem 10.1.2 to see that (cf. the notation in Theorem 10.1.11)

$$\left(\big|\psi(t \wedge \zeta_r)\big|^{-N+2}, \mathcal{F}_t, \mathcal{W}_{\mathbf{x}}^{(N)}\right)$$

is a bounded, non-negative martingale for every $|\mathbf{x}| > r > 0$. Hence, by Theorem 7.1.14, for any $0 \leq s \leq t < \infty$ and $A \in \mathcal{F}_s$,

$$\begin{aligned}|\mathbf{x}|^{-N+2} &\geq \mathbb{E}^{\mathcal{W}_{\mathbf{x}}^{(N)}}\left[\big|\psi(s)\big|^{-N+2}, A \cap \{\zeta_r(\psi) > s\}\right] \\ &= \mathbb{E}^{\mathcal{W}_{\mathbf{x}}^{(N)}}\left[\big|\psi(t \wedge \zeta_r)\big|^{-N+2}, A \cap \{\zeta_r(\psi) > s\}\right];\end{aligned}$$

and, because $N \geq 3$ and therefore $\zeta_r \nearrow \infty$ $\left(\text{a.s.}, \mathcal{W}_{\mathbf{x}}^{(N)}\right)$ as $r \searrow 0$, an application of the Monotone Convergence Theorem and Fatou's Lemma leads to

$$|\mathbf{x}|^{-N+2} \geq \mathbb{E}^{\mathcal{W}_{\mathbf{x}}^{(N)}}\left[\big|\psi(s)\big|^{-N+2}, A\right] \geq \mathbb{E}^{\mathcal{W}_{\mathbf{x}}^{(N)}}\left[\big|\psi(t)\big|^{-N+2}, A\right]$$

for all $0 \leq s \leq t < \infty$, $A \in \mathcal{F}_s$, and $\mathbf{x} \neq \mathbf{0}$. In particular, this proves that

$$\left(-\big|\psi(t)\big|^{-N+2}, \mathcal{F}_t, \mathcal{W}_{\mathbf{x}}^{(N)}\right)$$

is a non-positive submartingale for every $\mathbf{x} \neq \mathbf{0}$ and therefore, by Theorem 7.1.10, that $\lim_{t\to\infty} \big|\psi(t)\big|$ exists in $[0, \infty]$ for $\mathcal{W}_{\mathbf{x}}^{(N)}$-almost every $\psi \in C(\mathbb{R}^N)$. At the same time,

$$\mathcal{W}_{\mathbf{x}}^{(N)}\left(\big|\psi(t)\big| \leq R\right) = \gamma_{\mathbf{0},t\mathbf{I}}\left(\{\mathbf{y} : |\mathbf{y} - \mathbf{x}| \leq R\}\right) \longrightarrow 0$$

as $t \to \infty$ for every $R \in (0, \infty)$ and $\mathbf{x} \in \mathbb{R}^N$; and so we now know that, at least when $\mathbf{x} \neq \mathbf{0}$, $|\psi(t)| \longrightarrow \infty$ for $\mathcal{W}_{\mathbf{x}}^{(N)}$-almost every $\psi \in C(\mathbb{R}^N)$. Finally, since

$$\mathcal{W}_{\mathbf{0}}^{(N)}\left(\inf_{t \geq T+1} |\psi(t)| \leq R\right) = \int_{\mathbb{R}^N \setminus \{\mathbf{0}\}} \mathcal{W}_{\mathbf{x}}^{(N)}\left(\inf_{t \geq T} |\psi(t)| \leq R\right) \gamma_{\mathbf{0},\mathbf{I}}(d\mathbf{x}),$$

the same result also holds when $\mathbf{x} = \mathbf{0}$. □

The conclusion drawn in the preceding is sometimes summarized as the statement that Brownian motion in three or more dimensions is **transient**.

## Exercises for § 10.1

EXERCISE 10.1.13. Referring to § 8.4.1, define $\mathbf{U}(t,\mathbf{x},\boldsymbol{\theta})$ by (8.5.1), and let $\mathcal{U}_{\mathbf{x}}^{(N)} \in \mathbf{M}_1(C(\mathbb{R}^N))$ denote the $\mathcal{W}^{(N)}$-distribution of $\boldsymbol{\theta} \rightsquigarrow \mathcal{U}^{(N)}(\,\cdot\,,\mathbf{x},\boldsymbol{\theta})$. Given a non-empty open set $\mathfrak{G} \subseteq \mathbb{R}\times\mathbb{R}^N$, define $\zeta_s^{\mathfrak{G}}(\boldsymbol{\psi})$ as in Theorem 10.1.2, and show that for each $w \in C^{1,2}(\mathfrak{G};\mathbb{R})\cap C_{\mathrm{b}}(\overline{\mathfrak{G}};\mathbb{R})$ and $f\in C_{\mathrm{b}}(\overline{\mathfrak{G}};\mathbb{R})$ satisfying

$$\partial_t w(t,\mathbf{y}) - \frac{1}{2}\Delta w(t,\mathbf{y}) + \frac{1}{2}\big(\mathbf{y},\nabla w(t,\mathbf{y})\big)_{\mathbb{R}^N} \ge f \quad \text{in } \mathfrak{G},$$

$$\left( w\big(s+t\wedge\zeta_s^{\mathfrak{G}}(\boldsymbol{\psi}),\boldsymbol{\psi}(t\wedge\zeta_s^{\mathfrak{G}})\big) - \int_0^{t\wedge\zeta_s^{\mathfrak{G}}(\boldsymbol{\psi})} f\big(s+\tau,\boldsymbol{\psi}(\tau)\big)\,d\tau, \mathcal{F}_t, \mathcal{U}_{\mathbf{x}}^{(N)}\right)$$

is a submartingale for all $(s,\mathbf{x})\in\mathfrak{G}$.

EXERCISE 10.1.14. Let $h$ be the function described in (10.1.7), and show that

$$\mathbb{E}^{\mathcal{W}_{\mathbf{x}}^{(N)}}\left[e^{-\frac{1}{2}\int_0^T |\boldsymbol{\psi}(\tau)|^2\,d\tau}\,\Big|\,\sigma(\{\boldsymbol{\psi}(T)\})\right] = \frac{h(T,\mathbf{x},\boldsymbol{\psi}(T))}{g^{(N)}(T,\boldsymbol{\psi}(T)-\mathbf{x})}.$$

Next, referring to Exercise 8.3.21, set $\ell_{T,\mathbf{x},\mathbf{y}}(t) = \frac{T-t}{T}\mathbf{x} + \frac{t}{T}\mathbf{y}$ for $t\in[0,T]$, let $\mathcal{W}_{T,\mathbf{x},\mathbf{y}}^{(N)} \in \mathbf{M}_1(C([0,T];\mathbb{R}^N))$ denote the $\mathcal{W}^{(N)}$-distribution of $\boldsymbol{\theta}\rightsquigarrow \ell_{T,\mathbf{x},\mathbf{y}} + \boldsymbol{\theta}_T \restriction [0,T]$, and show that

$$\mathbb{E}^{\mathcal{W}_{T,\mathbf{x},\mathbf{y}}^{(N)}}\left[e^{-\frac{1}{2}\int_0^T |\boldsymbol{\psi}(\tau)|^2\,d\tau}\right] = \frac{h(t,\mathbf{x},\mathbf{y})}{g^{(N)}(T,\mathbf{y}-\mathbf{x})}.$$

EXERCISE 10.1.15. The purpose of this exercise is to examine the assertion made in Remark 10.1.10 about the characterization of the arcsine distribution (i.e., the Borel probability measure on $[0,1]$ with distribution function $x\in[0,1]\longmapsto F(x) = \frac{2}{\pi}\arcsin\sqrt{x}\in[0,1]$). Specifically, the goal is to show that the arcsine distribution is the one and only Borel probability measure on $[0,1]$ that is absolutely continuous with respect to Lebesgue measure and invariant under $x\in[0,1]\longmapsto 4x(1-x)\in[0,1]$.

**(i)** Define $x\in[0,1]\longmapsto \Phi(x) = \left(\sin\frac{\pi x}{2}\right)^2 \in[0,1]$, and show that a Borel probability measure $\mu$ on $[0,1]$ is invariant under $x\rightsquigarrow 4x(1-x)$ if and only if $\Phi_*\mu$ is invariant under $x\rightsquigarrow 2x \bmod 1$. Conclude that the desired characterization of the arcsine distribution is equivalent to showing that Lebesgue measure $\lambda_{[0,1]}$ on $[0,1]$ is the one and only Borel probability measure on $[0,1]$ that is absolutely continuous with respect to Lebesgue measure and invariant under $x\rightsquigarrow 2x \bmod 1$.

**(ii)** Suppose that $\mu$ is a Borel probability measure on $[0,1]$ that is invariant under $x\rightsquigarrow 2x \bmod 1$ and assigns probability 0 to $\{0\}$. Set $F(x)=\mu([0,x])$, the distribution function for $\mu$, and use induction on $n\ge 0$ to show that

$$F(x) = \sum_{m=0}^{2^n-1}\Big(F(m2^{-n}+x2^{-n}) - F(m2^{-n})\Big)$$

for $x\in[0,1]$.

**(iii)** Now add the assumption that $\mu \ll \lambda_{[0,1]}$, let $f$ be the corresponding Radon–Nikodym derivative, and extend $f$ to $\mathbb{R}$ by taking $f = 0$ off of $[0,1]$. Given $0 \le x < x+y \le 1$, conclude that

$$\big|F(x+y) - F(x) - F(y)\big| \le \int_{\mathbb{R}} \big|f(t + x2^{-n}) - f(t)\big|\,dt \longrightarrow 0$$

as $n \to \infty$. In other words, $F(x+y) = F(x) + F(y)$ whenever $0 \le x < x+y \le 1$. Finally, after combining this with the facts that $F(0) = 0$, $F(1) = 1$, and $F$ is continuous, conclude that $F(x) = x$ for $x \in [0,1]$. In view of part **(i)**, this completes the proof that the arcsine distribution admits the asserted characterization.

**(vi)** To see that absolute continuity is absolutely essential in the preceding considerations, consider any Borel probability measure $M$ on $\{0,1\}^{\mathbb{Z}^+}$ that is stationary in the sense that the $M$-distribution of

$$\boldsymbol{\omega} \in \{0,1\}^{\mathbb{Z}^+} \longmapsto (\omega_2, \dots, \omega_{n+1}, \dots) \in \{0,1\}^{\mathbb{Z}^+}$$

is again $M$. Show that the $M$-distribution $\mu$ of

$$\boldsymbol{\omega} \in \{0,1\}^{\mathbb{Z}^+} \longmapsto \sum_{n=1}^{\infty} 2^{-n}\omega_n \in [0,1]$$

is invariant under $x \rightsquigarrow 2x \bmod 1$. In particular, this means that, for each $p \in (0,1) \setminus \{\frac{1}{2}\}$, the $\mu_p$ described in Exercise 1.4.29 is a non-atomic, Borel probability measure on $[0,1]$ that is invariant under $x \rightsquigarrow 2x \bmod 1$ but singular to Lebesgue measure.

## § 10.2 The Markov Property and Potential Theory

In this section I will discuss the Markov property for Wiener measure and show how it can be used as a tool for connecting Brownian motion to partial differential equations.

**§ 10.2.1. The Markov Property for Wiener Measure.** The introduction of the translates $\mathcal{W}_{\mathbf{x}}^{(N)}$'s facilitates the statement of the following important interpretation of Theorem 7.1.16. In its statement, and elsewhere, $\Sigma_t : C(\mathbb{R}^N) \longrightarrow C(\mathbb{R}^N)$ is the **time-shift map** determined by $\Sigma_t\psi(\tau) = \psi(t+\tau)$, $\tau \in [0,\infty)$, and when $\zeta$ is a stopping time, $\Sigma_\zeta$ is the map on $\{\psi : \zeta(\psi) < \infty\} \longrightarrow C(\mathbb{R}^N)$ given by $\Sigma_\zeta\psi(\tau) = \psi\big(\zeta(\psi) + \tau\big)$.

THEOREM 10.2.1. *If $\zeta$ is a stopping time and $F : C(\mathbb{R}^N) \times C(\mathbb{R}^N) \longrightarrow [0,\infty)$ is a $\mathcal{F}_\zeta \times \mathcal{F}_{C(\mathbb{R}^N)}$-measurable function, then*

$$\begin{aligned}&\int_{\{\psi:\zeta(\psi)<\infty\}} F(\psi, \Sigma_\zeta\psi)\,\mathcal{W}_{\mathbf{x}}^{(N)}(d\psi)\\ &\qquad = \int_{\{\psi:\zeta(\psi)<\infty\}} \left(\int_{C(\mathbb{R}^N)} F(\psi,\psi')\,\mathcal{W}_{\psi(\zeta)}^{(N)}(d\psi')\right)\mathcal{W}_{\mathbf{x}}^{(N)}(d\psi).\end{aligned} \tag{10.2.2}$$

PROOF: Given Theorem 7.1.16, the proof is mostly a matter of notation. In the first place, by replacing $F(\psi,\psi')$ with $F(\mathbf{x}+\psi,\psi')$, one can reduce to the case when $\mathbf{x}=\mathbf{0}$. Thus, I will assume that $\mathbf{x}=\mathbf{0}$. Secondly, $\Sigma_\zeta\psi=\psi(\zeta)+\delta_\zeta\psi$ if $\zeta(\psi)<\infty$. Hence,

$$\int_{\{\psi:\zeta(\psi)<\infty\}} F(\psi,\Sigma_\zeta\psi)\,\mathcal{W}^{(N)}(d\psi)=\int_{\{\psi:\zeta(\psi)<\infty\}} F(\psi,\psi(\zeta)+\delta_\zeta\psi)\,\mathcal{W}^{(N)}(d\psi).$$

Now define $\tilde F(\psi,\psi')=\mathbf{1}_{[0,\infty)}(\zeta(\psi))F(\psi,\psi(\zeta)+\psi')$, note that $\tilde F$ is again $\mathcal{F}_\zeta\times\mathcal{B}_{C(\mathbb{R}^N)}$-measurable, and apply Theorem 7.1.16 to reach the desired conclusion. □

Theorem 10.2.1 is a statement of the **Markov property** for Wiener measure. More precisely, because it involves stopping times, and not just fixed times, it is often called the **strong Markov property**.

**§ 10.2.2. Recurrence in One and Two Dimensions.** As my first application of the Markov property, I will prove the statement made following Theorem 10.1.11 about the recurrence of Brownian motion when $N\in\{1,2\}$.

THEOREM 10.2.3. *If $N\in\{1,2\}$, then, for all $\mathbf{x}\in\mathbb{R}^N$,*

$$\mathcal{W}^{(N)}_{\mathbf{x}}\left(\int_0^\infty \mathbf{1}_{B(\mathbf{c},r)}(\psi(t))\,dt=\infty \text{ for all } \mathbf{c}\in\mathbb{R}^N \text{ and } r\in(0,\infty)\right)=1.$$

PROOF: Because $\mathbb{R}^N$ is separable, it is easy to use countable additivity to see that the asserted result will be proved once we show that

$$(*)\quad \mathcal{W}^{(N)}_{\mathbf{x}}\left(\int_0^\infty \mathbf{1}_{B(\mathbf{0},r)}(\psi(t))\,dt=\infty\right)=1 \quad\text{for all } \mathbf{x}\in\mathbb{R}^N \text{ and } r\in(0,\infty).$$

Define $\{\zeta^n: n\ge 0\}$ so that $\zeta^0(\psi)=\inf\{t\ge 0: |\psi(t)|\le\frac{r}{2}\}$ and

$$\zeta^n(\psi)=\begin{cases}\inf\{t\ge\zeta^{n-1}(\psi): |\psi(t)|\ge r\} & \text{for odd } n\ge 1\\ \inf\{t\ge\zeta^{n-1}(\psi): |\psi(t)|\le\frac{r}{2}\} & \text{for even } n\ge 2,\end{cases}$$

with the understanding that $\zeta^{n-1}(\psi)=\infty \implies \zeta^n(\psi)=\infty$. Clearly, all the $\zeta^n$'s are stopping times. In addition, $\zeta^{2n}(\psi)<\infty \implies \zeta^{2n+1}(\psi)=\zeta^{2n}(\psi)+\zeta^{B(\mathbf{0},r)}\circ\Sigma_{\zeta^{2n}}(\psi)$ for all $n\ge 0$, and $\zeta^{2n-1}(\psi)<\infty\implies\zeta^{2n}(\psi)=\zeta^{2n-1}(\psi)+\zeta_{\frac{r}{2}}\circ\Sigma_{\zeta^{2n-1}}(\psi)$ for $n\ge 1$, where $\zeta^{B(\mathbf{0},r)}(\psi)=\inf\{t\ge 0: |\psi(t)|\ge r\}$ and, as in Theorem 10.1.11, $\zeta_{\frac{r}{2}}(\psi)=\inf\{t\ge 0: |\psi(t)|=\frac{r}{2}\}$. Hence, by Theorem 10.2.1,

$$(**)\quad\begin{aligned}\mathcal{W}^{(N)}_{\mathbf{x}}(\zeta^{2n+1}\le t\,|\,\mathcal{F}_{\zeta^{2n}})&=\mathcal{W}^{(N)}_{\psi(\zeta^{2n})}(\zeta^{B(\mathbf{0},r)}\le t) && \text{if } \zeta_{2n}(\psi)<\infty,\\ \mathcal{W}^{(N)}_{\mathbf{x}}(\zeta^{2n}\le t\,|\,\mathcal{F}_{\zeta^{2n-1}})&=\mathcal{W}^{(N)}_{\psi(\zeta^{2n-1})}(\zeta_{\frac{r}{2}}\le t) && \text{if } \zeta_{2n-1}(\psi)<\infty.\end{aligned}$$

Note that, because $N \in \{1,2\}$, Theorem 10.1.11 says that both $\zeta^{B(\mathbf{0},r)}$ and $\zeta_{\frac{r}{2}}$ are $\mathcal{W}^{(N)}_{\mathbf{y}}$-almost surely finite for all $\mathbf{y} \in \mathbb{R}^N$. Thus, by induction, $\zeta^n < \infty$ $\mathcal{W}^{(N)}_{\mathbf{x}}$-almost surely for all $n \ge 0$.

Next set

$$X_n(\psi) \equiv \begin{cases} \zeta^{2n+1}(\psi) - \zeta^{2n}(\psi) & \text{if } \zeta^{2n}(\psi) < \infty \\ 0 & \text{if } \zeta^{2n}(\psi) = \infty. \end{cases}$$

By the preceding, we know that, for each $n \ge 0$, $X_n > 0$ $\mathcal{W}^{(N)}_{\mathbf{x}}$-almost surely. In addition, it is obvious that

$$\int_0^\infty \mathbf{1}_{B(\mathbf{0},r)}\big(\psi(t)\big)\,dt \ge \sum_{n=0}^\infty X_n(\psi).$$

Hence, if we show that the $X_n$'s are mutually independent and identically distributed under $\mathcal{W}^{(N)}_{\mathbf{x}}$, then (*) will follow from The Strong Law of Large Numbers. But, by (**), we will know that the $X_n$'s have both these properties once we show that $\mathcal{W}^{(N)}_{\mathbf{y}}(\zeta^{B(\mathbf{0},r)} \le t)$ is the same for all $\mathbf{y} \in \mathbb{R}^N$ with $|\mathbf{y}| = \frac{r}{2}$. To this end, let $\mathbf{y}_i$, $i \in \{1,2\}$ with $|\mathbf{y}_i| = \frac{r}{2}$ be given, and choose an orthogonal transformation $\mathcal{O}$ of $\mathbb{R}^N$ so that $\mathbf{y}_2 = \mathcal{O}\mathbf{y}_1$. Then, $\mathcal{W}^{(N)}_{\mathbf{y}_2}$ is the distribution of $\psi \rightsquigarrow \mathcal{O}\psi$ under $\mathcal{W}^{(N)}_{\mathbf{y}_1}$, and so $\mathcal{W}^{(N)}_{\mathbf{y}_2}(\zeta^{B(\mathbf{0},r)} \le t) = \mathcal{W}^{(N)}_{\mathbf{y}_1}(\zeta^{B(\mathbf{0},r)} \le t)$. □

**§ 10.2.3. The Dirichlet Problem.** There are many ways in which the Markov property can be used to relate Brownian motion to partial differential equations, but among the most compelling is the one that was discovered by S. Kakutani and developed by Doob.[1] What Kakutani discovered is that the capacitory potential (cf. § 11.4.1) of a set $K \subseteq \mathbb{R}^2$ at a point $\mathbf{x} \in \mathbb{R}^2 \setminus K$ is equal to the probability that a Brownian motion started at $\mathbf{x}$ ever hits $K$. What Doob did is extend Kakutani's result to $\mathbb{R}^N$ and show that it is a very special case of a result that identifies the distribution of the place where a Brownian motion hits the boundary of a set as the harmonic measure (cf. § 11.1.4) for that set. In this subsection, I will give a brief introduction to these ideas. A much more thorough account is given in Chapter 11.

Let $G$ be a non-empty, connected open subset of $\mathbb{R}^N$. Given an $f \in C_b(G;\mathbb{R})$, one says that $u \in C^2(G;\mathbb{R})$ solves the **Dirichlet problem** for $f$ in $G$ if $u$ is

[1] Kakutani's 1944 article, "Two dimensional Brownian motion and harmonic functions," *Proc. Imp. Acad. Tokyo*, **20**, together with his 1949 article, "Markoff process and the Dirichlet problem," *Proc. Imp. Acad. Tokyo*, **21**, are generally accepted as the first place in which a definitive connection between harmonic functions and Brownian motion was established. However, it was not until with Doob's "Semimartingales and subharmonic functions," *T.A.M.S.*, **77**, in 1954 that the connection was completed. It is ironic that this connection was not made by Wiener himself. Indeed, Wiener's early fame as an analyst was based on his contributions to potential theory. However, in spite of his claims to the contrary, I know of no evidence that he discovered the connection between his measure and potential theory.

harmonic (i.e., $\Delta u = 0$) in $G$ and, for each $\mathbf{a} \in \partial G$, $u(\mathbf{x}) \longrightarrow f(\mathbf{a})$ if as $\mathbf{x} \in G$ tends to $\mathbf{a}$. Assuming that (10.1.5) holds with $\mathfrak{G} = \mathbb{R} \times G$, the weak minimum principle shows that there is at most one solution to the Dirichlet problem for each $f \in C_b(G;\mathbb{R})$. However, the corresponding question about the existence of solutions is not so easily resolved. The following preliminary result will get us started. In its statement, and elsewhere, a **harmonic function** on a non-empty, open subset $G$ of $\mathbb{R}$ is a $u \in C^2(G;\mathbb{R})$ that satisfies $\Delta u = 0$. Also, if $\mu$ is a non-zero, finite measure on $E$ and $f : E \longrightarrow \mathbb{R}$ is a $\mu$-integrable function, I will write

$$⨍ f\,d\mu \equiv \frac{1}{\mu(E)} \int f\,d\mu$$

to denote the average value of $f$ with respect to $\mu$. Finally, $\zeta^G : C(\mathbb{R}^N) \longrightarrow [0,\infty]$ given by $\zeta^G(\psi) = \inf\{t \ge 0 : \psi(t) \notin G\}$ is the **first exit time** from $G$.

THEOREM 10.2.4. *Let $G$ be a non-empty, open subset of $\mathbb{R}^N$. If $u \in C_b(\bar{G};\mathbb{R})$, $u \restriction G$ is harmonic, and $\mathbf{x}$ is an element of $G$ for which*

$$\mathcal{W}^{(N)}_{\mathbf{x}}(\zeta^G < \infty) = 1, \tag{10.2.5}$$

*then*

$$u(\mathbf{x}) = \mathbb{E}^{\mathcal{W}^{(N)}_{\mathbf{x}}}\big[u(\psi(\zeta^G)),\, \zeta^G(\psi) < \infty\big]. \tag{10.2.6}$$

*In particular, if $u$ is harmonic on $G$, then*[2]

$$\overline{B(\mathbf{x},r)} \subset\subset G \implies u(\mathbf{x}) = ⨍_{\mathbb{S}^{N-1}} u(\mathbf{x}+r\boldsymbol{\omega})\,\lambda_{\mathbb{S}^{N-1}}(d\boldsymbol{\omega}). \tag{10.2.7}$$

*Conversely, if $u : G \longrightarrow \mathbb{R}$ is a locally bounded (i.e., bounded on compacts), Borel measurable function that satisfies (10.2.7), then $u \in C^\infty(G;\mathbb{R})$ and $u$ is harmonic. Finally, if $f : \partial G \longrightarrow \mathbb{R}$ is a bounded, Borel measurable function, then the function $u : G \longrightarrow \mathbb{R}$ given by*

$$u(\mathbf{x}) = \mathbb{E}^{\mathcal{W}^{(N)}_{\mathbf{x}}}\Big[f(\psi(\zeta^G)),\, \zeta^G(\psi) < \infty\Big] \quad \text{for } \mathbf{x} \in G$$

*$u$ is a bounded, harmonic function on $G$.*

PROOF: Suppose that $u \in C_b(\bar{G};\mathbb{R})$ is harmonic on $G$. By Theorem 10.1.2,

$$\Big(u(\psi(t\wedge\zeta^G)), \mathcal{F}_t, \mathcal{W}^{(N)}_{\mathbf{x}}\Big) \quad \text{is a martingale.}$$

Hence, $u(\mathbf{x}) = \mathbb{E}^{\mathcal{W}^{(N)}_{\mathbf{x}}}\big[u(\psi(t\wedge\zeta^G))\big]$, and so, after letting $t \to \infty$ and taking (10.2.5) into account, one gets (10.2.6).

[2] Remember that I use the notation $\Gamma \subset\subset G$ to mean that $\overline{\Gamma}$ is a compact subset of $G$.

Now assume that $u$ is harmonic in $G$ and that $\overline{B(\mathbf{x},r)} \subset\subset G$. By applying (10.2.6) to $u \restriction \overline{B(\mathbf{x},r)}$ and noting that (cf. the first part of Theorem 10.1.11) $\mathcal{W}^{(N)}_{\mathbf{x}}(\zeta^{B(\mathbf{x},r)} < \infty) = 1$, one has

$$u(\mathbf{x}) = \mathbb{E}^{\mathcal{W}^{(N)}_{\mathbf{x}}}\left[u(\psi(\zeta^{B(\mathbf{x},r)})),\, \zeta^{B(\mathbf{x},r)}(\psi) < \infty\right].$$

Hence, the proof of (10.2.7) reduces to the observation that the distribution of $\psi \in \{\zeta^{B(\mathbf{x},r)} < \infty\} \longmapsto \psi(\zeta^{B(\mathbf{x},r)}) \in \partial B(\mathbf{x},r)$ under $\mathcal{W}^{(N)}_{\mathbf{x}}$ is same as that of $\psi \in \{\zeta^{B(\mathbf{0},r)} < \infty\} \longmapsto \mathbf{x}+\psi(\zeta^{B(\mathbf{0},r)})$ under $\mathcal{W}^{(N)}$ and that (cf. Exercise 4.3.10) the distribution of $\psi \in \{\zeta^{B(\mathbf{0},r)} < \infty\} \longmapsto \psi(\zeta^{B(\mathbf{0},r)})$ under $\mathcal{W}^{(N)}$ is rotation invariant.

Turning to the converse assertion, suppose that $u : G \longrightarrow \mathbb{R}$ is a locally bounded, Borel measurable function for which (10.2.7) holds. To see that $u \in C^\infty(G;\mathbb{R})$, extend $u$ to $\mathbb{R}^N$ so that it is 0 off of $G$, and choose a $\rho \in C^\infty_{\mathrm{c}}(\mathbb{R};[0,\infty))$ with support in $(0,1)$ and total integral 1. Using (10.2.7) together with Fubini's Theorem, one sees that, as long as $\overline{B(\mathbf{x},r)} \subset\subset G$,

$$\begin{aligned} u(\mathbf{x}) &= \int_0^1 \rho(t)\left(\fint_{\mathbb{S}^{N-1}} u(\mathbf{x}+tr\boldsymbol{\omega})\,\lambda_{\mathbb{S}^{N-1}}(d\boldsymbol{\omega})\right)dt \\ &= \frac{1}{\omega_{N-1}r}\int_{\mathbb{R}^N} |\mathbf{y}-\mathbf{x}|^{1-N}\rho(r^{-1}|\mathbf{y}-\mathbf{x}|)u(\mathbf{y})\,d\mathbf{y}, \end{aligned}$$

from which it is clear that $u \in C^\infty(G;\mathbb{R})$. Further, knowing that $u$ is smooth and satisfies (10.2.7), it is easy to see that it is harmonic. Indeed, by Taylor's Theorem, we know that

$$\begin{aligned} &\fint_{\mathbb{S}^{N-1}} u(\mathbf{x}+r\boldsymbol{\omega})\,\lambda_{\mathbb{S}^{N-1}}(d\boldsymbol{\omega}) - u(\mathbf{x}) = \fint_{\mathbb{S}^{N-1}} (u(\mathbf{x}+r\boldsymbol{\omega}) - u(\mathbf{x})\,\lambda_{\mathbb{S}^{N-1}}(d\boldsymbol{\omega}) \\ &\quad = \tfrac{r^2}{2}\Delta u(\mathbf{x}) + o(r^2), \end{aligned}$$

since, for any orthonormal basis $(\mathbf{e}_1,\dots,\mathbf{e}_N)$ in $\mathbb{R}^N$ and $1 \le i \le N$,

$$\fint_{\mathbb{S}^{N-1}} (\mathbf{e}_i,\boldsymbol{\omega})^2_{\mathbb{R}^N}\,\lambda_{\mathbb{S}^{N-1}}(d\boldsymbol{\omega}) = \frac{1}{N}\fint_{\mathbb{S}^{N-1}} |\boldsymbol{\omega}|^2\,\lambda_{\mathbb{S}^{N-1}}(d\boldsymbol{\omega})$$

and, when $1 \le i \ne j \le N$,

$$\int_{\mathbb{S}^{N-1}} (\mathbf{e}_i,\boldsymbol{\omega})_{\mathbb{R}^N}\,\lambda_{\mathbb{S}^{N-1}}(d\boldsymbol{\omega}) = \int_{\mathbb{S}^{N-1}} (\mathbf{e}_i,\boldsymbol{\omega})_{\mathbb{R}^N}(\mathbf{e}_j,\boldsymbol{\omega})_{\mathbb{R}^N}\,\lambda_{\mathbb{S}^{N-1}}(d\boldsymbol{\omega}) = 0.$$

Hence, after dividing through by $r^2$ and letting $r \searrow 0$, we see that (10.2.7) implies $\Delta u(\mathbf{x}) = 0$.

To complete the proof, let $u$ be as in the final assertion. Because $\zeta^G = \zeta^{B(\mathbf{x},r)} + \zeta^G \circ \Sigma_{\zeta^{B(\mathbf{x},r)}}$ if $\overline{B(\mathbf{x},r)} \subset\subset G$ and $\zeta^{B(\mathbf{x},r)} < \infty$, Theorem 10.2.1 implies that

$$\overline{B(\mathbf{x},r)} \subset\subset G \implies u(\mathbf{x}) = \mathbb{E}^{\mathcal{W}_{\mathbf{x}}^{(N)}}\left[u\big(\psi(\zeta^{B(\mathbf{x},r)})\big),\, \zeta^{B(\mathbf{x},r)}(\psi) < \infty\right];$$

and, as we have seen earlier, this implies that (10.2.7) holds. □

An easy, but important, corollary of the preceding is that *if $G$ is connected, then (10.2.5) for one $\mathbf{x} \in G$ implies (10.2.5) for all $\mathbf{x} \in G$.* Indeed, take $u(\mathbf{x}) = \mathbf{1} - \mathcal{W}_{\mathbf{x}}^{(N)}(\zeta^G < \infty)$, apply Theorem 10.2.4 to see that $u$ is a $[0,1]$-valued harmonic function in $G$, and apply the strong minimum principle to see that $u(\mathbf{x}) = 0$ for some $\mathbf{x} \in G$ implies $u = 0$ throughout $G$. A second easy corollary is that if (10.2.5) holds and if $u$ solves the Dirichlet problem in $G$ for some $f \in C_b(\partial G; \mathbb{R})$, then

$$u(\mathbf{x}) = \mathbb{E}^{\mathcal{W}^{(N)}}\left[f\big(\psi(t)\big),\, \zeta^G(\psi) < \infty\right]. \tag{10.2.8}$$

Thus, if (10.2.5) holds for all $\mathbf{x} \in G$ and we are going to solve the Dirichlet problem for $f$, then we have no choice but to show that the $u$ given by (10.2.8) is a solution. Furthermore, because of the last part of Theorem 10.2.4, we already know that this $u$ is harmonic in $G$. Thus, all that remains is to find conditions under which the $u$ in (10.2.8) will take the correct boundary values.

It should be reasonably clear, and will be verified shortly (cf. Theorem 10.2.14), that if $f$ is continuous at $\mathbf{a} \in \partial G$ and if

$$\lim_{\substack{\mathbf{x}\to\mathbf{a}\\ \mathbf{x}\in G}} \mathcal{W}_{\mathbf{x}}^{(N)}(\zeta^G \geq \delta) = 0 \quad \text{for all } \delta > 0, \tag{10.2.9}$$

then the function $u$ in (10.2.8) tends to $f(\mathbf{a})$ as $\mathbf{x} \to \mathbf{a}$ through $G$. For this reason, I will say that $\mathbf{a} \in \partial G$ is a **regular point** if (10.2.9) holds, in which case I write $\mathbf{a} \in \partial_{\text{reg}}\partial G$.

In order to give a probabilistic criterion with which to test the regularity of a point, I need to introduce some notation. Given $s \in [0,\infty)$, set $\zeta_s^G(\psi) = \inf\{t \geq s : \psi(t) \notin G\}$, the first time $\psi$ exits from $G$ after time $s$, and define $\zeta_{0+}^G(\psi) = \lim_{s\searrow 0} \zeta_s^G(\psi)$, the first positive time $\psi$ exits from $G$. Notice that, for $s \in (0,\infty)$,

$$\zeta_s^G = s + \zeta^G \circ \Sigma_s \quad \text{and} \quad \zeta_{0+}^G \geq s \implies \zeta_{0+}^G = \zeta_s^G. \tag{10.2.10}$$

Lemma 10.2.11. *Regularity is a local property in the sense that, for each $r \in (0,\infty)$, $\mathbf{a} \in \partial_{\text{reg}}\mathfrak{G}$ if and only if $\mathbf{a} \in \partial_{\text{reg}}\big(G \cap B(\mathbf{a},r)\big)$. Furthermore,*

$$\mathbf{a} \in \partial_{\text{reg}}G \iff \mathbf{a} \in \partial G \text{ and } \mathcal{W}_{\mathbf{a}}^{(N)}(\zeta_{0+}^{\mathfrak{G}} > 0) = 0, \tag{10.2.12}$$

*and so $\partial_{\text{reg}}G$ is Borel measurable. Finally, if $\mathbf{a} \in \partial_{\text{reg}}G$, then, for each $\delta > 0$,*

$$\lim_{\substack{\mathbf{x}\to\mathbf{a}\\ \mathbf{x}\in G}} \mathcal{W}_{\mathbf{x}}^{(N)}\Big(\big(\zeta^G, \boldsymbol{\psi}(\zeta^G)\big) \in (0,\delta) \times B(\mathbf{a},\delta)\Big) = 1. \tag{10.2.13}$$

PROOF: Set $G(\mathbf{a},r) = G \cap B_{\mathbb{R}^N}(\mathbf{a},r)$. Since it is obvious that $\zeta^{G(\mathbf{a},r)}$ is dominated by $\zeta^G$, there is no question that $\mathbf{a} \in \partial_{\text{reg}}G \implies \mathbf{a} \in \partial_{\text{reg}}G(\mathbf{a},r)$. On the other hand, if $\mathbf{a} \in \partial_{\text{reg}}G(\mathbf{a},r)$ and $\epsilon > 0$, then, for all $0 < \delta < \epsilon$,

$$\begin{aligned}
\varlimsup_{\substack{\mathbf{x}\to\mathbf{a}\\ \mathbf{x}\in G}} \mathcal{W}_{\mathbf{x}}^{(N)}(\zeta^G \ge \epsilon) &\le \varlimsup_{\substack{\mathbf{x}\to\mathbf{a}\\ \mathbf{x}\in G}} \mathcal{W}_{\mathbf{x}}^{(N)}(\zeta^G \ge \delta)\\
&\le \varlimsup_{\substack{\mathbf{x}\to\mathbf{a}\\ \mathbf{x}\in G(\mathbf{a},r)}} \mathcal{W}_{\mathbf{x}}^{(N)}(\zeta^{G(\mathbf{a},r)} \ge \delta) + \varlimsup_{\substack{\mathbf{x}\to\mathbf{a}\\ \mathbf{x}\in G}} \mathcal{W}_{\mathbf{x}}^{(N)}(\zeta^{B_{\mathbb{R}^N}(\mathbf{a},r)} \le \delta)\\
&\le \mathcal{W}^{(N)}\left(\sup_{t\in[0,\delta]} |\boldsymbol{\psi}(t)| \ge \frac{r}{2}\right) \longrightarrow 0 \quad \text{as } \delta \searrow 0.
\end{aligned}$$

Hence, we have now also proved that $\mathbf{a} \in \partial_{\text{reg}}G(\mathbf{a},r) \implies \mathbf{a} \in \partial_{\text{reg}}G$.

Next, let $\mathbf{a} \in \partial G$. To check the equivalence in (10.2.12), use the first part of (10.2.10) and the Markov property to see that

$$\mathbf{x} \in \mathbb{R}^N \longmapsto \mathcal{W}_{\mathbf{x}}^{(N)}(\zeta_s^G \ge \delta) = \int_{\mathbb{R}^N} \mathcal{W}_{\mathbf{y}}^{(N)}(\zeta^G \ge \delta - s)\, g^{(N)}(s, \mathbf{y}-\mathbf{x})\, d\mathbf{y} \in [0,1]$$

is a continuous function for every $s \in (0,\infty)$, and therefore that

$$\mathbf{x} \in \mathbb{R}^N \longmapsto \mathcal{W}_{\mathbf{x}}^{(N)}(\zeta_{0+}^G \ge \delta) = \lim_{s\searrow 0} \mathcal{W}_{\mathbf{x}}^{(N)}(\zeta_s^G \ge \delta)$$

is upper semicontinuous for all $\delta \ge 0$. In particular, if $\mathcal{W}_{\mathbf{a}}^{(N)}(\zeta_{0+}^G > 0) = 0$, then, because $\zeta^G(\boldsymbol{\psi}) = \zeta_{0+}^G(\boldsymbol{\psi})$ when $\boldsymbol{\psi}(0) \in G$, it follows that

$$\varlimsup_{\substack{\mathbf{x}\to\mathbf{a}\\ \mathbf{x}\in G}} \mathcal{W}_{\mathbf{x}}^{(N)}(\zeta^G \ge \delta) = \varlimsup_{\substack{\mathbf{x}\to\mathbf{a}\\ \mathbf{x}\in G}} \mathcal{W}_{\mathbf{x}}^{(N)}(\zeta_{0+}^G \ge \delta) = 0$$

for every $\delta > 0$. To prove the converse, suppose that $\mathbf{a} \in \partial_{\text{reg}}G$, let positive $\epsilon$ and $\delta$ be given, and choose $r > 0$ so that

$$\mathcal{W}_{\mathbf{x}}^{(N)}(\zeta^G \ge \delta) \le \epsilon \quad \text{for } \mathbf{x} \in G \cap B(\mathbf{a},r).$$

Then, by the second part of (10.2.10), the Markov property, and (4.3.13), for each $s \in (0,\delta)$ one has

$$\begin{aligned}
\mathcal{W}_{\mathbf{a}}^{(N)}(\zeta_{0+}^G \ge 2\delta) &\le \mathbb{E}^{\mathcal{W}_{\mathbf{a}}^{(N)}}\Big[\mathcal{W}_{\boldsymbol{\psi}(s)}^{(N)}(\zeta^G \ge \delta),\, \boldsymbol{\psi}(s) \in G\Big]\\
&\le \epsilon + \mathcal{W}_{\mathbf{a}}^{(N)}\big(\boldsymbol{\psi}(s) \notin B(\mathbf{a},r)\big) \le \epsilon + 2Ne^{-\frac{r^2}{2Ns}},
\end{aligned}$$

from which $\mathcal{W}_{\mathbf{a}}^{(N)}(\zeta_{0+}^G > 0) = 0$ follows when first $s \searrow 0$ and then $\epsilon \searrow 0$.

Now, assume that $\mathbf{a} \in \partial_{\text{reg}} G$, and observe that, for each $0 < \epsilon < \delta$,

$$\begin{aligned}
&\mathcal{W}_{\mathbf{x}}^{(N)}\Big(\psi(\zeta^G) \notin B(\mathbf{a},\delta) \text{ or } \zeta^G \geq \delta\Big) \\
&\qquad \leq \mathcal{W}_{\mathbf{x}}^{(N)}(\zeta^G \geq \epsilon) + \mathcal{W}_{\mathbf{x}}^{(N)}\left(\sup_{t\in[0,\epsilon]} |\psi(t) - \mathbf{a}| \geq \delta\right).
\end{aligned}$$

Hence, (10.2.9) and (4.3.13) together imply that

$$\varlimsup_{\substack{\mathbf{x}\to\mathbf{a}\\ \mathbf{x}\in G}} \mathcal{W}_{\mathbf{x}}^{(N)}\Big(\psi(\zeta^G) \notin B(\mathbf{x},\delta) \text{ or } \zeta^G \geq \delta\Big) \leq 2N \exp\left[-\frac{\delta^2}{2N\epsilon}\right],$$

from which (10.2.13) follows after one lets $\epsilon \searrow 0$. □

In view of the last part of Theorem 10.2.4 and (10.2.13), the following statement is obvious.

THEOREM 10.2.14. *Let $G$ be a non-empty open subset of $\mathbb{R}^N$ and $f : \partial G \longrightarrow \mathbb{R}$ a bound, Borel measurable function. If $u$ is given by (10.2.8), then $u$ is a bounded harmonic function in $G$, and, for every $\mathbf{a} \in \partial_{\text{reg}} G$ at which $f$ is continuous, $u(\mathbf{x}) \longrightarrow f(\mathbf{a})$ as $\mathbf{x} \to \mathbf{a}$ through $G$.*

Before closing this brief introduction to one of the most successful applications of probability theory to partial differential equations, it seems only appropriate to check that the conclusion in Theorem 10.2.14 is equivalent to the classical one at which analysts arrived. To be precise, recall the famous program, initiated by O. Perron and completed by Wiener, M. Brélot, and others, for solving the Dirichlet problem. Namely, given a bounded, non-empty open set $G$ in $\mathbb{R}^N$ and an $f \in C(\partial G; \mathbb{R})$, consider the set $\mathcal{U}(f)$ of lower semicontinuous functions $w : G \longrightarrow \mathbb{R}$ that are bounded below and satisfy the super-mean value property

$$\overline{B(\mathbf{x},r)} \subset\subset G \implies w(\mathbf{x}) \geq \fint_{\mathbb{S}^{N-1}} w(\mathbf{x} + r\omega)\, \lambda_{\mathbb{S}^{N-1}}(d\omega),$$

and the boundary condition

$$\varliminf_{\substack{\mathbf{x}\to\mathbf{a}\\ \mathbf{x}\in G}} w(\mathbf{x}) \geq f(\mathbf{a}) \quad \text{for all } \mathbf{a} \in \partial G.$$

At the same time, define $\mathcal{L}(f)$ to be the set of $v : G \longrightarrow \mathbb{R}$ such that $-v \in \mathcal{U}(-f)$. Finally, given $\mathbf{a} \in \partial G$, say that $\mathbf{a}$ admits a **barrier** if, for some $r > 0$, there exists an $\eta \in C^2\big(G \cap B(\mathbf{a}, r); (0,\infty)\big)$ such that

$$\lim_{\substack{x\to\mathbf{a}\\ x\in G\cap B(\mathbf{a},r)}} \eta(\mathbf{x}) = 0 \quad \text{and} \quad \Delta\eta \leq -\epsilon \text{ for some } \epsilon > 0.$$

A famous theorem[3] proved by Wiener states that

$$\inf\{w(\mathbf{x}) : w \in \mathcal{U}(f)\} = \sup\{v(\mathbf{x}) : v \in \mathcal{L}(f)\} \quad \text{for all } \mathbf{x} \in G$$

and that if $H_f(\mathbf{x})$ denotes this common value, then $\mathbf{x} \rightsquigarrow H_f(\mathbf{x})$ is a bounded harmonic function on $G$ with the property that

$$\lim_{\substack{\mathbf{x}\to\mathbf{a}\\ \mathbf{x}\in G}} H_f(\mathbf{x}) = f(\mathbf{a}) \quad \text{for } \mathbf{a} \in \partial G \text{ that admit a barrier.}$$

THEOREM 10.2.15. *Referring to the preceding paragraph, the function $H_f$ described there coincides with the function $u$ in (10.2.8). In addition, a boundary point $\mathbf{a} \in \partial G$ is regular (i.e., (10.2.9) holds) if and only if it admits a barrier.*

PROOF: To prove the first part, all that I have to do is check that $v \le u \le w$ for all $v \in \mathcal{L}(f)$ and $w \in \mathcal{U}(f)$. For this purpose, set $r(\mathbf{x}) = \frac{1}{2}|\mathbf{x} - G\complement|$, and define $\{\zeta^n : n \ge 0\}$ so that $\zeta^0 = 0$ and

$$\zeta^{n+1}(\boldsymbol{\psi}) = \inf\{t \ge \zeta^n(\boldsymbol{\psi}) : |\boldsymbol{\psi}(t) - \boldsymbol{\psi}(\zeta^n)| \ge r(\boldsymbol{\psi}(\zeta^n))\} \quad \text{for } n \ge 0,$$

with the usual understanding that $\zeta^n(\boldsymbol{\psi}) = \infty \implies \zeta^{n+1}(\boldsymbol{\psi}) = \infty$. An easy inductive argument shows that all the $\zeta^n$'s are stopping times. In addition, it is clear that $\zeta^n \le \zeta^{n+1} \le \zeta^G$. I now want to show that $\zeta^G(\boldsymbol{\psi}) < \infty \implies \zeta^n(\boldsymbol{\psi}) \nearrow \zeta^G(\boldsymbol{\psi})$. To this end, suppose that $\sup_{n\ge 0} \zeta^n(\boldsymbol{\psi}) < \zeta^G(\boldsymbol{\psi}) < \infty$, in which case there exists an $\epsilon > 0$ such that $r(\boldsymbol{\psi}(\zeta^n)) \ge \epsilon$ for all $n \ge 0$. But this would mean that $\{\zeta^n(\boldsymbol{\psi}) : n \ge 0\}$ is a bounded sequence for which $\inf_{n\ge 0} |\boldsymbol{\psi}(\zeta^{n+1}) - \boldsymbol{\psi}(\zeta^n)| \ge \epsilon$, which contradicts the continuity of $\boldsymbol{\psi}$. Finally, choose a reference point $\mathbf{y} \in G$, and set $\mathbf{X}_n(\boldsymbol{\psi})$ equal to $\boldsymbol{\psi}(\zeta^n)$ or $\mathbf{y}$ according to whether $\zeta^n(\boldsymbol{\psi}) < \infty$ or not, $R_n(\boldsymbol{\psi}) = r(\mathbf{X}_n(\boldsymbol{\psi}))$, and $B_n(\boldsymbol{\psi}) = B(\mathbf{X}_n(\boldsymbol{\psi}), R_n(\boldsymbol{\psi}))$, the ball around $\mathbf{X}_n(\boldsymbol{\psi})$ of radius $R_n(\boldsymbol{\psi})$, and observe that

$$\zeta^n(\boldsymbol{\psi}) < \infty \implies \zeta^{n+1}(\boldsymbol{\psi}) = \zeta^n(\boldsymbol{\psi}) + \zeta^{B_n(\boldsymbol{\psi})} \circ \Sigma_{\zeta^n}(\boldsymbol{\psi}).$$

With these preparations at hand, let $w \in \mathcal{U}(f)$ and $\mathbf{x} \in G$ be given. By Theorem 10.2.1 and the preceding,

$$\begin{aligned}
&\mathbb{E}^{\mathcal{W}^{(N)}_{\mathbf{x}}}\left[w(\boldsymbol{\psi}(\zeta^{n+1})),\, \zeta^{n+1}(\boldsymbol{\psi}) < \infty\right] \\
&= \int_{\{\boldsymbol{\psi}:\zeta^n(\boldsymbol{\psi})<\infty\}} \left( \int_{\{\boldsymbol{\psi}':\zeta^{B_n(\boldsymbol{\psi})}(\boldsymbol{\psi}')<\infty\}} w(\boldsymbol{\psi}'(\zeta^{B_n(\boldsymbol{\psi})}))\, \mathcal{W}^{(N)}_{X_n(\boldsymbol{\psi})}(d\boldsymbol{\psi}') \right) \mathcal{W}^{(N)}_{\mathbf{x}}(\boldsymbol{\psi}) \\
&= \mathbb{E}^{\mathcal{W}^{(N)}_{\mathbf{x}}}\left[ \fint_{\mathbb{S}^{N-1}} w(X_n(\boldsymbol{\psi}) + R_n(\boldsymbol{\psi})\boldsymbol{\omega})\, \lambda_{\mathbb{S}^{N-1}}(d\boldsymbol{\omega}),\, \zeta^n(\boldsymbol{\psi}) < \infty \right] \\
&\le \mathbb{E}^{\mathcal{W}^{(N)}_{\mathbf{x}}}\left[w(\boldsymbol{\psi}(\zeta^n)),\, \zeta^n(\boldsymbol{\psi}) < \infty\right],
\end{aligned}$$

[3] See O.D. Kellogg's *Foundations of Potential Theory*, Dover Publ. (1963).

where, in the passage to the second to last line, I have used the fact, established earlier, that the exit place from a ball of a Brownian path started at its center is uniformly distributed. Hence, by Fatou's Lemma and the boundary condition satisfied by $w$,

$$\begin{aligned} w(\mathbf{x}) &\geq \varliminf_{n\to\infty} \mathbb{E}^{\mathcal{W}_{\mathbf{x}}^{(N)}}\big[w(\psi(\zeta^n)),\, \zeta^n(\psi)<\infty\big] \\ &\geq \mathbb{E}^{\mathcal{W}_{\mathbf{x}}^{(N)}}\big[f(\psi(\zeta^G)),\, \zeta^G(\psi)<\infty\big] = u(\mathbf{x}). \end{aligned}$$

Thus, we have now shown that $w \geq u$ for all $w \in \mathcal{U}(f)$. Of course, if $v \in \mathcal{L}(f)$, then, because $-v \in \mathcal{U}(-f)$, we also know that $-v \geq -u$ and therefore that $v \leq u$.

I turn next to the second part of the theorem. Set $m(\mathbf{x}) = \mathbb{E}^{\mathcal{W}_{\mathbf{x}}^{(N)}}[\zeta^G]$, $\mathbf{x} \in G$. Clearly $m$ is positive. Moreover, if $m(\mathbf{x}) \longrightarrow 0$ as $\mathbf{x} \to \mathbf{a}$ through $G$, then $\mathbf{a}$ is regular. Conversely, suppose $\mathbf{a}$ is regular. Since

$$m(\mathbf{x}) \leq \delta + \mathbb{E}^{\mathcal{W}_{\mathbf{x}}^{(N)}}\big[\zeta^G, \zeta^G \geq \delta\big] \leq \delta + \mathcal{W}_{\mathbf{x}}^{(N)}(\zeta^G \geq \delta)^{\frac{1}{2}}\mathbb{E}^{\mathcal{W}_{\mathbf{x}}^{(N)}}\big[(\zeta^G)^2\big]^{\frac{1}{2}},$$

it will follow that $m(\mathbf{x}) \longrightarrow 0$ as $\mathbf{x} \to \mathbf{a}$ through $G$ once we check that $\mathbf{x} \rightsquigarrow \mathbb{E}^{\mathcal{W}_{\mathbf{x}}^{(N)}}\big[(\zeta^G)^2\big]$ is bounded on $G$. But, if $R$ is the diameter of $G$ and $\mathbf{c} \in \mathbb{R}^N$ is chosen so that $G \subseteq B(\mathbf{c}, R)$, then $\zeta^G \leq \zeta^{B(\mathbf{c},R)}$, and, by the first part of Theorem 10.1.11 and translation invariance, $\mathbf{x} \rightsquigarrow \mathbb{E}^{\mathcal{W}_{\mathbf{x}}^{(N)}}\big[(\zeta^{B(\mathbf{c},R)})^2\big]$ is bounded on $B(\mathbf{c}, R)$. Hence, we now know that $\mathbf{a} \in \partial G$ is regular if and only if $m(\mathbf{x}) \longrightarrow 0$ as $\mathbf{x} \to \mathbf{a}$ through $G$. To complete the proof at this point, set $\tilde{m}(\mathbf{x}) = \mathbb{E}^{\mathcal{W}_{\mathbf{x}}^{(N)}}\big[\zeta^{B(\mathbf{c},R)}\big]$, and observe that, since $\zeta^{B(\mathbf{c},R)} = \zeta^G + \zeta^{B(\mathbf{c},R)} \circ \Sigma_{\zeta^G}$ when $\zeta^G < \infty$,

$$\tilde{m}(\mathbf{x}) - m(\mathbf{x}) = \mathbb{E}^{\mathcal{W}_{\mathbf{x}}^{(N)}}\big[\tilde{m}(\psi(\zeta^G)),\, \zeta^G(\psi) < \infty\big].$$

Thus $\tilde{m} - m$ is harmonic on $G$ and so, by the first part of Theorem 10.1.11, $\Delta m = \Delta \tilde{m} = -2$ on $G$. Hence, if $\mathbf{a}$ is regular, then $m$ is a barrier at $\mathbf{a}$. Conversely, suppose that $\mathbf{a}$ admits a barrier $\eta \in C^2\big(G \cap B(\mathbf{a}, r); (0, \infty)\big)$. Because of the locality property proved in Lemma 10.2.11, I will, without loss in generality, assume that $B(\mathbf{a}, r) \supseteq G$. Choose a sequence $\{G_n : n \geq 1\}$ of open sets so that $\bar{G}_n \subset\subset G_{n+1}$ for each $n$ and $G_n \nearrow G$. Then, by Theorem 10.1.2, for $\mathbf{x} \in G_n$ and $t \geq 0$,

$$\begin{aligned} \eta(\mathbf{x}) &\geq \eta(\mathbf{x}) - \mathbb{E}^{\mathcal{W}_{\mathbf{x}}^{(N)}}\big[\eta(\psi(t \wedge \zeta^{G_n})\big] \\ &= -\mathbb{E}^{\mathcal{W}_{\mathbf{x}}^{(N)}}\left[\int_0^{t\wedge\zeta^{G_n}(\psi)} \tfrac{1}{2}\Delta\eta(\psi(\tau))\,d\tau\right] \geq \frac{\epsilon}{2}\mathbb{E}^{\mathcal{W}_{\mathbf{x}}^{(N)}}\big[t \wedge \zeta^{G_n}\big]. \end{aligned}$$

Hence, after letting first $t$ and then $n$ tend to infinity, we see that $m(\mathbf{x}) \leq \frac{2}{\epsilon}\eta(\mathbf{x})$ for all $\mathbf{x} \in G$; and, since $\eta(\mathbf{x}) \longrightarrow 0$ as $\mathbf{x}$ tends to $\mathbf{a}$ through $G$, it follows that $\mathbf{a} \in \partial_{\text{reg}} G$. □

The argument used to prove the first part of Theorem 10.2.15 is a probabilistic implementation of what analysts call the "balayage" procedure for solving the Dirichlet problem.

## Exercises for § 10.2

EXERCISE 10.2.16. Suppose that $G$ is a non-empty, open subset of $\mathbb{R}^M \times \mathbb{R}^N$ and that $(\mathbf{x},\mathbf{y}) \in G \longmapsto u(\mathbf{x},\mathbf{y}) \in \mathbb{R}$ is a Borel measurable function that is harmonic with respect to $\mathbf{x}$ and $\mathbf{y}$ separately (i.e., $u(\,\cdot\,,\mathbf{y})$ is harmonic on $\{\mathbf{x} : (\mathbf{x},\mathbf{y}) \in G\}$ for each $\mathbf{y} \in G$ and $u(\mathbf{x},\,\cdot\,)$ is harmonic on $\{\mathbf{y} : (\mathbf{x},\mathbf{y}) \in G\}$ for each $\mathbf{x} \in G$). Assuming that $u$ is bounded below on compact subsets of $G$, show that $u$ is harmonic on $G$.

**Hint**: Clearly, all that one has to show is that $u$ is smooth on $G$. In addition, without loss in generality, one can assume that $u$ can be extended to $\mathbb{R}^M \times \mathbb{R}^N$ as a non-negative, Borel measurable function. Making this assumption, proceed as in the proof of Theorem 10.2.4 to show that if $\rho \in C_c^\infty\big((0,1);\mathbb{R}\big)$ has total integral 1 and $\overline{B_{\mathbb{R}^M}(\mathbf{x},r) \times B_{\mathbb{R}^N}(\mathbf{y},r)} \subset\subset G$, then $u(\mathbf{x},\mathbf{y})$ equals

$$\frac{1}{\omega_{M-1}\omega_{N-1}r^2} \iint\limits_{\mathbb{R}^M \times \mathbb{R}^N} |\mathbf{x}-\boldsymbol{\xi}|^{1-M}|\mathbf{y}-\boldsymbol{\eta}|^{1-N}\rho\big(r^{-1}|\mathbf{x}-\boldsymbol{\xi}|\big)\rho\big(r^{-1}|\mathbf{y}-\boldsymbol{\eta}|\big)u(\boldsymbol{\xi},\boldsymbol{\eta})\,d\mathbf{x}d\boldsymbol{\eta}.$$

Thus, all that remains is to justify differentiating under the integral.

EXERCISE 10.2.17. Show that the only functions $u \in C^2\big(\mathbb{R}^2;[0,\infty)\big)$ satisfying $\Delta u \le 0$ are constant. This result is a manifestation of recurrence. Indeed, show that it is completely false when $\mathbb{R}^2$ is replaced by either the half-space $\mathbb{R} \times (0,\infty)$ or $R^3$.

**Hint**: Using the sort of reasoning in the proof of Corollary 10.1.12, show that $\big(-u(\psi(t)),\mathcal{F}_t,\mathcal{W}^{(2)}\big)$ is a non-positive submartingale and, as a consequence, that $\lim_{t\to\infty} u\big(\psi(t)\big)$ exists $\mathcal{W}^{(2)}$-almost surely. Now, using Theorem 10.2.3, show that this is possible only if $u$ is constant. To handle the last part, let $f \in C_c^\infty\big(\mathbb{R};[0,\infty)\big)$, and consider the function on $\mathbb{R}^3$ given by

$$u(\mathbf{x}) = |\mathbf{x}|^{-1}\int_0^{|\mathbf{x}|}\left(\int_\sigma^\infty \rho f(\rho)\,d\rho\right)d\sigma.$$

EXERCISE 10.2.18. The goal of this exercise is to prove **Blumenthal's 0–1 Law**, which states that if $A \in \mathcal{F}_{0+} \equiv \bigcap_{t>0}\mathcal{F}_t$, then $\mathcal{W}_{\mathbf{x}}^{(N)}(A) \in \{0,1\}$ for each $\mathbf{x} \in \mathbb{R}^N$.

**(i)** If $F : C(\mathbb{R}^N) \longrightarrow \mathbb{R}$ is a bounded, continuous function, show that, for any $A \in \mathcal{F}_{0+}$ and $\mathbf{x} \in \mathbb{R}^N$,

$$\begin{aligned}\mathbb{E}^{\mathcal{W}_{\mathbf{x}}^{(N)}}[F,A] &= \lim_{t\searrow 0}\mathbb{E}^{\mathcal{W}_{\mathbf{x}}^{(N)}}[F\circ\Sigma_t,A]\\ &= \lim_{t\searrow 0}\int_A \mathbb{E}^{\mathcal{W}_{\psi(t)}^{(N)}}[F]\,W_{\mathbf{x}}^{(N)}(d\psi) = \mathbb{E}^{\mathcal{W}_{\mathbf{x}}^{(N)}}[F]\mathcal{W}_{\mathbf{x}}^{(N)}(A).\end{aligned}$$

(**ii**) For any $A \in \mathcal{B}_{C(\mathbb{R}^N)}$, show that $\mathbb{E}^{\mathcal{W}_{\mathbf{x}}^{(N)}}[F, A] = \mathbb{E}^{\mathcal{W}_{\mathbf{x}}^{(N)}}[F]\mathcal{W}_{\mathbf{x}}^{(N)}(A)$ for all bounded, Borel measurable $F : C(\mathbb{R}^N) \longrightarrow \mathbb{R}$ if it holds for all bounded, continuous ones.

(**iii**) By combining (**i**) and (**ii**), show that $\mathcal{W}_{\mathbf{x}}^{(N)}(A) = \mathcal{W}_{\mathbf{x}}^{(N)}(A)^2$ for all $A \in \mathcal{F}_{0+}$ and $\mathbf{x} \in \mathbb{R}^N$.

EXERCISE 10.2.19. Let $G$ be a non-empty, open subset of $\mathbb{R}^N$. In this exercise, we will develop a criterion for checking the regularity of boundary points.

(**i**) As an application of Exercise 4.3.15, show that, for $\mathcal{W}_x^{(1)}$-almost every $\psi \in C(\mathbb{R})$, $\forall t > 0\ \exists \tau \in (0, t)\ \psi(\tau) = x$. Next use this to see that when $N = 1$ every boundary point of every open set is regular.

(**ii**) As an application of Blumenthal's 0–1 Law, show that $\mathcal{W}_{\mathbf{x}}^{(N)}(\zeta_{0+}^G > 0) \in \{0, 1\}$ for all $\mathbf{x} \in \mathbb{R}^N$. Next, using this together with (10.2.12), show that $\mathbf{a}$ is regular if and only if $\mathcal{W}_{\mathbf{a}}^{(N)}(\zeta_{0+} = 0) > 0$.

(**iii**) Assume that $\mathbf{a} \in \partial G$ has positive, upper Lebesgue density in $G\complement$. That is,

$$\varlimsup_{r \searrow 0} \frac{|B(\mathbf{a}, r) \cap G\complement|}{|B(\mathbf{a}, r)|} > 0,$$

where $|\Gamma|$ denotes the Lebesgue measure of $\Gamma \in \mathcal{B}_{\mathbb{R}^N}$. Show that $\mathbf{a}$ is regular. In particular, because, for any Borel set $\Gamma$, the set of $\mathbf{x} \in \Gamma$ with upper Lebesgue density less than 1 has Lebesgue measure 0, this proves that $\partial G \setminus \partial_{\text{reg}} G$ has Lebesgue measure 0. (See the Lemma 11.1.9 for another proof of this fact.)

**Hint**: Show that, for all $t > 0$,

$$\mathcal{W}_{\mathbf{a}}^{(N)}(\zeta_{0+} \le t) \ge \mathcal{W}_{\mathbf{a}}^{(N)}(\psi(t) \notin G) \ge \frac{\Omega_N e^{-\frac{1}{2}}}{(2\pi)^{\frac{N}{2}}} \frac{|B(\mathbf{a}, t^{\frac{1}{2}}) \cap G\complement|}{|B(\mathbf{a}, t^{\frac{1}{2}})|},$$

where $\Omega_N \equiv |B(\mathbf{0}, 1)|$.

(**iv**) Use (**ii**) to prove the **exterior cone condition** for regularity. That is, show that $\mathbf{a} \in \partial G$ is regular if there is an $\omega \in \mathbb{S}^{N-1}$ and an $\alpha \in (0, 1]$ such that the cone

$$\left\{ \mathbf{y} \in \mathbb{R}^N : 0 < |\mathbf{y} - \mathbf{a}| < \alpha\ \&\ \frac{(\mathbf{y} - \mathbf{a}, \omega)_{\mathbb{R}^N}}{|\mathbf{y} - \mathbf{a}|} < \alpha \right\}$$

is contained in $G\complement$

(**v**) If $F$ is a closed subset of $\mathbb{R}^N$, $r > 0$, and $G = \{\mathbf{x} \in \mathbb{R}^N : |\mathbf{x} - F| > r\}$, show that every boundary point of $G$ satisfies the exterior cone condition and is therefore regular.

EXERCISE 10.2.20. Let $G$ be a non-empty, open subset of $\mathbb{R}^N$. In this exercise we will give a probabilistic justification of the famous Courant–Friedrichs–Lewy[4] finite difference scheme for solving the Dirichlet problem. To this end, let $\{\mathbf{X}_n : n \geq 1\}$ be a sequence of independent, identically distributed $\mathbb{R}^N$-valued random variables with mean $\mathbf{0}$ and covariance $\mathbf{I}$, define $\{\{\mathbf{S}_n(t) : t \geq 0\} : n \geq 0\}$ as in § 9.2.1, and let $\mathbb{P}_{n,\mathbf{x}} \in \mathbf{M}_1(C(\mathbb{R}^N))$ be the distribution of $\mathbf{x} + \mathbf{S}_n(\cdot)$. By Donsker's Invariance Principle, we know that, as $n \to \infty$, $\{\mathbb{P}_{n,\mathbf{x}_n} : n \geq 0\}$ tends weakly to $\mathcal{W}_{\mathbf{x}}^{(N)}$ if $\mathbf{x}_n \to \mathbf{x}$. Thus, one might hope that, for $f \in C_{\mathrm{b}}(G;\mathbb{R})$,

$$(10.2.21) \quad \mathbb{E}^{\mathbb{P}_{n,\mathbf{x}}}\left[f(\psi(\zeta^G)),\, \zeta^G(\psi) < \infty\right] \longrightarrow \mathbb{E}^{\mathcal{W}_{\mathbf{x}}^{(N)}}\left[f(\psi(\zeta^G)),\, \zeta^G(\psi) < \infty\right]$$

uniformly on compacts. On the other hand, in order to justify (10.2.21), one has to confront the problem that $\psi \rightsquigarrow \zeta^G(\psi)$ is, in general, only a lower semicontinuous function, not a continuous one. Thus, we must find conditions under which $\zeta^G$ is $\mathcal{W}_{\mathbf{x}}^{(N)}$-almost surely continuous.

**(i)** Let $\zeta^{\bar{G}}(\psi) = \inf\{t \geq 0 : \psi(t) \notin \bar{G}\}$. Obviously, $\zeta^G \leq \zeta^{\bar{G}}$. Show that $\psi \rightsquigarrow \zeta^{\bar{G}}(\psi)$ is upper semicontinuous and that $\zeta^G$ is continuous at any $\psi$ for which $\zeta^G(\psi) = \zeta^{\bar{G}}(\psi) < \infty$.

**(ii)** Say that $\mathbf{a} \in \partial G$ is strongly regular if $\mathcal{W}_{\mathbf{a}}^{(N)}(\zeta^{\bar{G}} = 0) = 1$. If every $\mathbf{a} \in \partial G$ is strongly regular and if $\mathbf{x} \in G$ is a point at which (10.2.5) holds, show that $\mathcal{W}_{\mathbf{x}}^{(N)}(\zeta^G = \zeta^{\bar{G}} < \infty) = 1$. Thus, (10.2.21) holds in this situation.

**Hint**: Use $\zeta^G < \infty \implies \zeta^{\bar{G}} = \zeta^G + \zeta^{\bar{G}} \circ \Sigma_{\zeta^G}$.

**(iii)** Using Blumenthal's 0–1 Law and the technique described in the Hint for part **(iii)** of Exercise 10.2.19, show that $\mathcal{W}_{\mathbf{a}}^{(N)}(\zeta^{\bar{G}} = 0) = 1$ if $\mathbf{a} \in \partial G$ has positive, upper Lebesgue density in $\mathbb{R}^N \setminus \bar{G}$, that is, if

$$\varlimsup_{r \searrow 0} \frac{|B(\mathbf{a}, r) \cap (\mathbb{R}^N \setminus \bar{G})|}{|B(\mathbf{a}, r)|} > 0.$$

Thus, if (10.2.5) holds for all $\mathbf{x} \in G$ and every $\mathbf{a} \in \partial G$ has positive, upper Lebesgue density in $\mathbb{R}^N \setminus \bar{G}$, then (10.2.21) holds uniformly for $\mathbf{x}$ in compact subsets of $G$.

---

[4] This type of approximation was carried out originally by H. Phillips and N. Wiener in "Nets and Dirichlet problem," *J. Math. Phys.* **2** in 1923. Ironically, the authors do not appear to have made the connection between their procedure and probability theory. In 1928, a more complete analysis was carried out in the famous article "Über die partiellen Differenzengleichungen der Phsik," *Ann. Math.* **5** #2, of R. Courant, K. Friedrichs, and H. Lewy. Interestingly, these authors do allude to a possible probabilistic interpretation, although their method (based on energy considerations) makes nô use of probability theory.

EXERCISE 10.2.22. Although, as the preceding exercise shows, probability theory provides approximation schemes with which to solve the Dirichlet problem, it is less successful when it comes to writing down explicit expressions for solutions. Nonetheless, there is a situation in which probability theory does lead to an explicit answer. To wit, consider the "upper half-space" $H = \mathbb{R}^N \times (0,\infty)$ in $\mathbb{R}^{N+1}$. Given $y \in (0,\infty)$, show that, for $\mathbf{x} \in \mathbb{R}^N$ and $\Gamma \in \mathcal{B}_{\mathbb{R}^N}$,

$$\mathcal{W}^{(N+1)}_{(\mathbf{x},y)}\big(\boldsymbol{\psi}(\zeta^H) \in \Gamma \times \{0\}\big) = \mathbb{E}^{\mathcal{W}^{(1)}}\big[\gamma^{(N)}_{\mathbf{x},\zeta^y\mathbf{I}}(\Gamma)\big],$$

where $\zeta^y(\psi) = \inf\{t \geq 0 : \psi(t) \geq y\}$. Next, recall from Exercise 7.1.24 that the $\mathcal{W}^{(1)}$-distribution of $\zeta^y$ is the one-sided, $\frac{1}{2}$-stable law $\nu^{\frac{1}{2}}_{2^{\frac{1}{2}}y}$. Thus

$$\mathcal{W}^{(N+1)}_{(\mathbf{x},y)}\big(\boldsymbol{\psi}(\zeta^H) \in \Gamma \times \{0\}\big) = \int_\Gamma \pi^{\mathbb{R}^N}_y(\mathbf{y}-\mathbf{x})\,d\mathbf{y},$$

where

$$\pi^{\mathbb{R}^N}_y(\mathbf{y}) = \int_{(0,\infty)} \gamma_{\mathbf{0},t\mathbf{I}}(\mathbf{y})\,\nu^{\frac{1}{2}}_{2^{\frac{1}{2}}y}(dt).$$

Finally, referring to Exercise 3.3.17, conclude that $\pi^{\mathbb{R}^N}_y$ is the Cauchy distribution in (3.3.19). This, of course, explains the convention, alluded to in (**ii**) of Exercise 3.3.17, of analysts calling $\pi^{\mathbb{R}^N}_y$ the Poisson kernel for the upper half-space.

## § 10.3 Other Heat Kernels

As we saw in § 10.1, from the perspective of someone studying partial differential equations, the function $(t,\mathbf{x},\mathbf{y}) \in (0,\infty) \times \mathbb{R}^N \times \mathbb{R}^N \longmapsto g^{(N)}(t,\mathbf{y}-\mathbf{x}) \in (0,\infty)$ is the **heat kernel**, or, equivalently, the fundamental solution, to the classical heat equation $\partial_t u = \frac{1}{2}\Delta u$ in $(0,\infty) \times \mathbb{R}^N$. That is, if $\varphi \in C_b(\mathbb{R}^N;\mathbb{R})$, then

$$u(t,\mathbf{x}) = \int_{\mathbb{R}^N} \varphi(\mathbf{y}) g^{(N)}(t,\mathbf{y}-\mathbf{x})\,d\mathbf{y}$$

is the unique bounded solution to the classical heat equation that tends to $\varphi$ as $t \searrow 0$. Of course, from a probabilistic perspective, $g^{(N)}(t,\mathbf{y}-\mathbf{x})$ is the probability (in the sense of densities) of a Brownian path going from $\mathbf{x}$ to $\mathbf{y}$ during a time interval of length $t$.

In this section I will construct other functions that, on the one hand, are the fundamental solution to a heat equation and, at the same time, the density for the probability of a Brownian motion making transitions under various conditions.

### § 10.3.1. A General Construction.

For each $t > 0$, let $E_t : C(\mathbb{R}^N) \longrightarrow [0,\infty)$ be a $\mathcal{F}_t$-measurable function with the property that

$$E_{s+t}(\boldsymbol{\psi}) = E_s(\boldsymbol{\psi})E_t(\Sigma_s\boldsymbol{\psi}) \quad \text{for } s,\, t \in (0,\infty) \text{ and } \boldsymbol{\psi} \in C(\mathbb{R}^N), \tag{10.3.1}$$

and define

$$q(t,\mathbf{x},\mathbf{y}) = \mathbb{E}^{\mathcal{W}^{(N)}}\Big[E_t\big(\mathbf{x}(1-\ell_t)+\boldsymbol{\theta}_t+\mathbf{y}\ell_t\big)\Big]g^{(N)}(t,\mathbf{y}-\mathbf{x}),\tag{10.3.2}$$
$$\text{for } (t,\mathbf{x},\mathbf{y}) \in (0,\infty)\times\mathbb{R}^N\times\mathbb{R}^N,$$

where $\ell_t(\tau) = \frac{\tau\wedge t}{t}$, $\tau\in[0,\infty)$, and $\boldsymbol{\theta}_t = \boldsymbol{\theta}-\boldsymbol{\theta}(t)\ell_t$, $\boldsymbol{\theta}\in\Theta(\mathbb{R}^N)$. Clearly $(\mathbf{x},\mathbf{y})\in(\mathbb{R}^N)^2\longmapsto q(t,\mathbf{x},\mathbf{y})\in[0,\infty)$ is Borel measurable for each $t>0$.

My goal in this subsection is to prove the following theorem.

THEOREM 10.3.3. *For each $t\in(0,\infty)$ and Borel measurable $\varphi:\mathbb{R}^N\longrightarrow\mathbb{R}$ that is bounded below,*

$$\int_{\mathbb{R}^N}\varphi(\mathbf{y})q(t,\mathbf{x},\mathbf{y})\,d\mathbf{y} = \mathbb{E}^{\mathcal{W}^{(N)}_{\mathbf{x}}}\Big[E_t(\boldsymbol{\psi})\varphi(\boldsymbol{\psi}(t))\Big].$$

*Moreover, for all $s,t\in(0,\infty)$ and $\mathbf{x},\mathbf{y}\in\mathbb{R}^N$, $q$ satisfies the Chapman–Kolmogorov equation*

$$q(s+t,x,y) = \int_{\mathbb{R}^N} q(s,\mathbf{x},\mathbf{z})q(t,\mathbf{z},\mathbf{y})\,d\mathbf{z}.$$

*Finally, if, for each $t>0$, $E_t$ is reversible in the sense that*

$$E_t(\boldsymbol{\psi}) = E_t(\breve{\boldsymbol{\psi}}^t),\quad \boldsymbol{\psi}\in C(\mathbb{R}^N),$$

*where $\breve{\boldsymbol{\psi}}^t(\tau)=\boldsymbol{\psi}(t-\tau\wedge t)$, $\tau\in[0,\infty)$, then $q(t,\mathbf{x},\mathbf{y})=q(t,\mathbf{y},\mathbf{x})$ for all $(t,\mathbf{x},\mathbf{y})\in(0,\infty)\times(\mathbb{R}^N)^2$.*

PROOF: The first assertion is an easy application of (8.3.12) with $n=1$. Namely, by that result,

$$\mathbb{E}^{\mathcal{W}^{(N)}_{\mathbf{x}}}\big[E_t(\boldsymbol{\psi})\varphi(\boldsymbol{\psi}(t))\big] = \mathbb{E}^{\mathcal{W}^{(N)}}\big[E_t(\mathbf{x}+\boldsymbol{\theta})\varphi(\mathbf{x}+\boldsymbol{\theta}(t))\big]$$
$$= \int_{\mathbb{R}^N}\mathbb{E}^{\mathcal{W}^{(N)}}\big[E_t(\mathbf{x}+\boldsymbol{\theta}_t+\mathbf{y}\ell_t)\varphi(\mathbf{x}+\mathbf{y})\big]g^{(N)}(t,\mathbf{y})\,d\mathbf{y} = \int_{\mathbb{R}^N}\varphi(\mathbf{y})q(t,\mathbf{x},\mathbf{y})\,d\mathbf{y}.$$

To prove the Chapman–Kolmogorov equation, set $\bar q(t,\mathbf{x},\mathbf{y}) = \frac{q(t,\mathbf{x},\mathbf{y})}{g^{(N)}(t,\mathbf{y}-\mathbf{x})}$. Then, another application of (8.3.12), this time with $n=2$, $t_1=s$, and $t_2=s+t$, shows that $\bar q(s+t,\mathbf{x},\mathbf{y})$ equals

$$\iint_{\mathbb{R}^N\times\mathbb{R}^N}\mathbb{E}^{\mathcal{W}^{(N)}}\Big[E_{s+t}\big(x+\boldsymbol{\theta}_{(s,s+t,(\boldsymbol{\xi},\boldsymbol{\eta}))}+(\mathbf{y}-\mathbf{x}-\boldsymbol{\eta})\ell_{s+t}\big)\Big]g^{(N)}(s,\boldsymbol{\xi})g^{(N)}(t,\boldsymbol{\eta}-\boldsymbol{\xi})\,d\boldsymbol{\xi}d\boldsymbol{\eta}.$$

Next note that, by (10.3.1), $E_{s+t}\big(x+\boldsymbol{\theta}_{(s,s+t,(\boldsymbol{\xi},\boldsymbol{\eta}))}+(\mathbf{y}-\mathbf{x}-\boldsymbol{\eta})\ell_{s+t}\big)$ equals

$$E_s\Big(\mathbf{x}+\boldsymbol{\theta}_s+\big(\boldsymbol{\xi}+\tfrac{s}{s+t}(\mathbf{y}-\mathbf{x}-\boldsymbol{\eta})\big)\ell_s\Big)$$
$$\times E_t\Big(x+\boldsymbol{\xi}+\tfrac{s}{s+t}(\mathbf{y}-\mathbf{x}-\boldsymbol{\eta})+(\delta_s\boldsymbol{\theta})_t$$
$$+\big(\mathbf{y}-\big(\mathbf{x}+\boldsymbol{\xi}+\tfrac{s}{s+t}(\mathbf{y}-\mathbf{x}-\boldsymbol{\eta})\big)\big)\ell_t(\,\cdot\,-s)\Big).$$

Therefore, since $E_s$ is $\mathcal{F}_s$-measurable and $\boldsymbol{\theta}_s \restriction [0,s]$ is $\mathcal{W}^{(N)}$-independent of $(\delta_s\boldsymbol{\theta})_t$,

$$\begin{aligned}&\mathbb{E}^{\mathcal{W}^{(N)}}\Big[E_{s+t}\big(x+\boldsymbol{\theta}_{(s,s+t,(\boldsymbol{\xi},\boldsymbol{\eta}))}+(\mathbf{y}-\mathbf{x}-\boldsymbol{\eta})\ell_{s+t}\big)\Big]\\&\qquad=\bar{q}\big(s,\mathbf{x},\mathbf{x}+\boldsymbol{\xi}+\tfrac{s}{s+t}(\mathbf{y}-\mathbf{x}-\boldsymbol{\eta})\big)\bar{q}\big(t,\mathbf{x}+\boldsymbol{\xi}+\tfrac{s}{s+t}(\mathbf{y}-\mathbf{x}-\boldsymbol{\eta}),\mathbf{y}\big).\end{aligned}$$

Plugging this into the expression for $\bar{q}(s+t,\mathbf{x},\mathbf{y})$ and making the change of variable $\boldsymbol{\xi}\to\mathbf{x}+\boldsymbol{\xi}+\frac{s}{s+t}(\mathbf{y}-\mathbf{x}-\boldsymbol{\eta})$, one finds that $\bar{q}(s+t,\mathbf{x},\mathbf{y})$ equals

$$\iint\limits_{\mathbb{R}^N\times\mathbb{R}^N}\bar{q}(s,\mathbf{x},\boldsymbol{\xi})\bar{q}(t,\boldsymbol{\xi},\mathbf{y})g^{(N)}(s,\alpha\boldsymbol{\eta}+\boldsymbol{\xi}-\mathbf{c})g^{(N)}(t,\beta\boldsymbol{\eta}-\boldsymbol{\xi}+\mathbf{c})\,d\boldsymbol{\xi}d\boldsymbol{\eta},$$

where $\alpha=\frac{s}{s+t}$, $\beta=\frac{t}{s+t}$, and $\mathbf{c}=\frac{t\mathbf{x}+s\mathbf{y}}{s+t}$. At the same time, by Exercise 10.3.34,

$$\begin{aligned}\int_{\mathbb{R}^N}g^{(N)}(s,\alpha\boldsymbol{\eta}+\boldsymbol{\xi}-\mathbf{c})g^{(N)}(t,\beta\boldsymbol{\eta}-\boldsymbol{\xi}+\mathbf{c})\,d\boldsymbol{\eta}&=g^{(N)}\big(\tfrac{st}{s+t},\boldsymbol{\xi}-\mathbf{c}\big)\\&=\frac{g^{(N)}(s,\boldsymbol{\xi}-\mathbf{x})g^{(N)}(t,\boldsymbol{\eta}-\boldsymbol{\xi})}{g^{(N)}(s+t,\mathbf{y}-\mathbf{x})},\end{aligned}$$

and so we are done.

To prove the last assertion, simply note that when $E_t$ is reversible, $\bar{q}(t,\mathbf{x},\mathbf{y})$ equals

$$\mathbb{E}^{\mathcal{W}^{(N)}}\Big[E_t\big(\mathbf{x}(1-\ell_t)+\boldsymbol{\theta}_t+\mathbf{y}\ell_t\big)\Big]=\mathbb{E}^{\mathcal{W}^{(N)}}\Big[E_t\big(\mathbf{y}(1-\ell_t)+(\breve{\boldsymbol{\theta}}_t)^t+\mathbf{y}\ell_t\big)\Big]=\bar{q}(t,\mathbf{y},\mathbf{x}),$$

since, by part **(ii)** of Exercise 8.3.22, $\boldsymbol{\theta}\rightsquigarrow(\breve{\boldsymbol{\theta}}_t)^t\restriction[0,t]$ has the same distribution under $\mathcal{W}^{(N)}$ as $\boldsymbol{\theta}\rightsquigarrow\boldsymbol{\theta}_t\restriction[0,t]$. □

**§ 10.3.2. The Dirichlet Heat Kernel.** Let $G$ be a non-empty, open subset of $\mathbb{R}^N$, and set $E_t^G(\boldsymbol{\psi})=\mathbf{1}_{(t,\infty)}\big(\zeta^G(\boldsymbol{\psi})\big)$. Obviously $E_t$ is $\mathcal{F}_t$-measurable and (10.3.1) holds. In addition, if $p^G(t,\mathbf{x},\mathbf{y})$ is used to denote the associated $q$ given in (10.3.2), then, $p^G(t,\mathbf{x},\mathbf{y})=0$ unless $\mathbf{x},\,\mathbf{y}\in G$, and, by Theorem 10.3.3,

$$\text{(10.3.4)}\qquad \int_G\varphi(\mathbf{y})p^G(t,\mathbf{x},\mathbf{y})\,d\mathbf{y}=\mathbb{E}^{\mathcal{W}_{\mathbf{x}}^{(N)}}\Big[\varphi(\boldsymbol{\psi}(t)),\,\zeta^G(\boldsymbol{\psi})>t\Big],\quad\text{for }(t,\mathbf{x})\in(0,\infty)\times G,$$

$$\text{(10.3.5)}\quad p^G(s+t,\mathbf{x},\mathbf{y})=\int_G p^G(s,\mathbf{x},\mathbf{z})p^G(t,\mathbf{z},\mathbf{y})\,d\mathbf{z},\quad(s,\mathbf{x}),\,(t,\mathbf{y})\in(0,\infty)\times G,$$

and

$$\text{(10.3.6).}\qquad p^G(t,\mathbf{x},\mathbf{y})=p^G(t,\mathbf{y},\mathbf{x})\quad\text{for }(t,\mathbf{x},\mathbf{y})\in(0,\infty)\times G^2.$$

In order to show that $p^G$ is smooth on $(0,\infty)\times G^2$, I will use the Duhamel formula contained in the following.

THEOREM 10.3.7. *For all $(t,\mathbf{x},\mathbf{y}) \in (0,\infty)\times G^2$,*

$$(10.3.8)\quad p^G(t,\mathbf{x},\mathbf{y}) = g^{(N)}(t,\mathbf{y}-\mathbf{x}) - \mathbb{E}^{\mathcal{W}^{(N)}_{\mathbf{x}}}\Big[g^{(N)}(t,\mathbf{y}-\boldsymbol{\psi}(\zeta^G)),\ \zeta^G(\boldsymbol{\psi})<t\Big].$$

PROOF: Given $\alpha\in(0,1)$, set

$$q_\alpha(t,\mathbf{x},\mathbf{y}) = \mathcal{W}^{(N)}\Big(\mathbf{x}+\boldsymbol{\theta}_t(\tau)+(\mathbf{y}-\mathbf{x})\ell_t(\tau)\in G,\ \tau\in[0,\alpha t]\Big)g^{(N)}(t,\mathbf{y}-\mathbf{x}).$$

Clearly $q_\alpha(t,\mathbf{x},\mathbf{y}) \searrow p^G(t,\mathbf{x},\mathbf{y})$ as $\alpha\nearrow 1$ for each $(t,\mathbf{x},\mathbf{y})\in(0,\infty)\times G^2$. Thus, it suffices for us to know that, for $\alpha\in(0,1)$ and $(t,\mathbf{x},\mathbf{y})\in(0,\infty)\times G^2$,

$$(*)\quad q_\alpha(t,\mathbf{x},\mathbf{y}) = g^{(N)}(t,\mathbf{y}-\mathbf{x}) - \mathbb{E}^{\mathcal{W}^{(N)}_{\mathbf{x}}}\Big[g^{(N)}(t,\mathbf{y}-\boldsymbol{\psi}(\zeta^G)),\ \zeta^G(\boldsymbol{\psi})\le\alpha t\Big].$$

Further, by the same argument as was used to prove the first assertion in Theorem 10.3.3, for any $\varphi\in C_c(G;\mathbb{R})$,

$$\begin{aligned}\int_G \varphi(\mathbf{y})q_\alpha(t,\mathbf{x},\mathbf{y})\,d\mathbf{y} &= \mathbb{E}^{\mathcal{W}^{(N)}_{\mathbf{x}}}\Big[\varphi(\boldsymbol{\psi}(\zeta^G)),\ \zeta^G(\boldsymbol{\psi})>\alpha t\Big]\\ &= \int_G \varphi(\mathbf{y})g^{(N)}(\mathbf{y}-\mathbf{x})\,d\mathbf{y}\\ &\qquad - \mathbb{E}^{\mathcal{W}^{(N)}_{\mathbf{x}}}\Big[\int_G \varphi(\mathbf{y})g^{(N)}(t-\zeta^G(\boldsymbol{\psi}),\mathbf{y}-\boldsymbol{\psi}(\zeta^G)),\ \zeta^G(\boldsymbol{\psi})\le\alpha t\Big],\end{aligned}$$

where, in the passage to the second line, I have applied the same reasoning as was suggested in part (**i**) of Exercise 7.3.7. Hence, (*) will follow once $\mathbf{y}\rightsquigarrow\bar q_\alpha(t,\mathbf{x},\mathbf{y}) \equiv \frac{q_\alpha(t,\mathbf{x},\mathbf{y})}{g^{(N)}(t,\mathbf{y}-\mathbf{x})}$ is shown to be continuous on $G$. To this end, argue as in the last part of Theorem (10.3.1) and apply the Markov property to show that $\bar q_\alpha(t,\mathbf{x},\mathbf{y})$ equals

$$\begin{aligned}&\mathcal{W}^{(N)}\big(\mathbf{x}+\boldsymbol{\theta}_t(t-\tau)+(\mathbf{y}-\mathbf{x})\ell_t(t-\tau)\in G,\ \tau\in[(1-\alpha)t,t]\big)\\ &= \mathcal{W}^{(N)}\big(\mathbf{y}+\boldsymbol{\theta}_t(\tau)+(\mathbf{x}-\mathbf{y})\ell_t(\tau)\in G,\ \tau\in[(1-\alpha)t,t]\big)\\ &= \int_{\mathbb{R}^N} g^{(N)}((1-\alpha)t,\mathbf{z}-\mathbf{y})\mathcal{W}^{(N)}_{\mathbf{z}}\Big(\boldsymbol{\psi}(\tau)+(\mathbf{y}-\boldsymbol{\psi}(t))\ell_t(\tau)\in G\Big)\,d\mathbf{z},\end{aligned}$$

which is certainly continuous with respect to $\mathbf{y}$. □

The importance of (10.3.8) is that it provides vital information used in the proof of the next theorem.

THEOREM 10.3.9. *For each $(t,\mathbf{x},\mathbf{y})\in(0,\infty)\times G^2$, $p^G(t,\mathbf{x},\mathbf{y})>0$ or $p^G(t,\mathbf{x},\mathbf{y}) = 0$ according to whether $\mathbf{y}$ is or is not in the same connected component of $G$ as $\mathbf{x}$. Furthermore, for each $K\subset\subset G$,*

$$\lim_{t\searrow 0}\sup_{\mathbf{x}\in K}\left|1-\int_{G\cap B(\mathbf{x},r)} p^G(t,\mathbf{x},\mathbf{y})\,d\mathbf{y}\right| = 0\quad\text{for all } r>0,$$

and

$$\lim_{\substack{(t,\mathbf{x})\to(s,\mathbf{a})\\ \mathbf{x}\in G}} \sup_{\mathbf{y}\in K} p^G(t,\mathbf{x},\mathbf{y}) = 0 \quad \text{for } (s,\mathbf{a}) \in (0,\infty)\times\partial_{\text{reg}}G.$$

*Finally, $p^G$ is smooth on $(0,\infty)\times G^2$, and, for each $m \geq 1$, $\partial_t^m p^G(t,\mathbf{x},\mathbf{y}) = 2^{-m}\Delta_{\mathbf{x}}^m p^G(t,\mathbf{x},\mathbf{y}) = 2^{-m}\Delta_{\mathbf{y}}^m p^G(t,\mathbf{x},\mathbf{y})$ on $(0,\infty)\times G\times G$.*

PROOF: Obviously, $p^G(t,\mathbf{x},\mathbf{y}) = 0$ unless $\mathbf{x}$ and $\mathbf{y}$ lie in the same connected component of $G$. On the other hand, if $\mathbf{x}$ and $\mathbf{y}$ lie in the same connected component of $G$, then there is a smooth $\mathbf{f} : [0,t] \longrightarrow G$ such that $\mathbf{f}(0) = \mathbf{x}$ and $\mathbf{f}(t) = \mathbf{y}$. Thus, if $\mathbf{h}(\tau) = \mathbf{f}(\tau\wedge t) - \mathbf{x}\big(1-\ell_t(\tau\wedge t)\big) - \mathbf{y}\ell_t(\tau\wedge t)$ for $\tau\in[0,\infty)$, then $\mathbf{h}\in\Theta(\mathbb{R}^N)$ and there is an $r>0$ such that $\mathbf{x}\big(1-\ell_t(\tau)\big)+\boldsymbol{\theta}_t(\tau)+\mathbf{y}\ell_t(\tau)\in G$, $\tau\in[0,t]$, for $\boldsymbol{\theta}\in B_{\Theta(\mathbb{R}^N)}(\mathbf{h},r)$. Hence, by Corollary 8.3.6,

$$p^G(t,\mathbf{x},\mathbf{y}) \geq \mathcal{W}^{(N)}\big(B_{\Theta(\mathbb{R}^N)}(\mathbf{h},r)\big) > 0.$$

Next, let $K \subset\subset G$ be given. Because, by (10.3.8), $p^G(t,\mathbf{x},\mathbf{y}) \leq g^{(N)}(\mathbf{y}-\mathbf{x})$,

$$\begin{aligned}\int_{G\cap B(\mathbf{x},r)} p^G(t,\mathbf{x},\mathbf{y})\,d\mathbf{y} &\geq \int_G p^G(t,\mathbf{x},\mathbf{y})\,dy - \int_{\mathbb{R}^N\setminus B(\mathbf{x},r)} g^{(N)}(\mathbf{y}-\mathbf{x})\,d\mathbf{y}\\ &= \mathcal{W}_{\mathbf{x}}^{(N)}(\zeta^G > t) - \int_{\mathbb{R}^N\setminus B(\mathbf{0},r)} g^{(N)}(\mathbf{y})\,d\mathbf{y},\end{aligned}$$

and therefore

$$\varlimsup_{t\searrow 0}\sup_{\mathbf{x}\in K}\left|1-\int_{G\cap B(\mathbf{x},r)} p^G(t,\mathbf{x},\mathbf{y})\,d\mathbf{y}\right| \leq \varlimsup_{t\searrow 0}\sup_{\mathbf{x}\in K}\mathcal{W}_{\mathbf{x}}^{(N)}(\zeta^G\leq t),$$

which, by (4.3.13), is 0. Also, again by (10.3.8),

$$\begin{aligned}p^G(t,\mathbf{x},\mathbf{y}) =& \mathbb{E}^{\mathcal{W}_{\mathbf{x}}^{(N)}}\Big[g^{(N)}(t,\mathbf{y}-\mathbf{x}) - g^{(N)}\big(t-\zeta^G(\boldsymbol{\psi}),\mathbf{y}-\boldsymbol{\psi}(\zeta^G)\big),\,\zeta^G<t\Big]\\ &+ g^{(N)}(t,\mathbf{y}-\mathbf{x})\mathcal{W}_{\mathbf{x}}^{(N)}(\zeta^G\geq t).\end{aligned}$$

Thus, as an application of (10.2.13), it is an easy matter to see that $p^G(t,\mathbf{x},\mathbf{y})$ tends to 0 uniformly in $\mathbf{y}\in K$ as $(t,\mathbf{x})\to(s,\mathbf{a})\in(0,\infty)\times\partial_{\text{reg}}G$.

To prove the asserted smoothness properties, begin with the observation that, for any multi-index $\alpha\in\mathbb{N}^N$[1]

$$\partial_{\mathbf{x}}^\alpha g^{(N)}(t,\mathbf{x}) = t^{-\frac{N+\|\alpha\|}{2}}P_\alpha\big(t^{-\frac{1}{2}}\mathbf{x}\big)e^{-\frac{|\mathbf{x}|^2}{2t}},$$

[1] I use the conventional multi-index notation for partial derivatives. Thus, if $\alpha = (\alpha_1,\dots,\alpha_N)\in\mathbb{N}^N$, then $\partial_{\mathbf{y}}^\alpha = \partial_{y_1}^{\alpha_1}\cdots\partial_{y_N}^{\alpha_N}$ and $\|\alpha\| = \sum_{i=1}^N\alpha_i$.

where $P_\alpha$ is an $\|\alpha\|$th order polynomial. Hence, by considering the cases $t \le |\mathbf{x}|^2$ and $t \ge |\mathbf{x}|^2$ separately, one finds that

$$\max_{\|\alpha\|=n} \left|\partial^\alpha_{\mathbf{x}} g^{(N)}(t,\mathbf{x})\right| \le \frac{C_n}{(t+|\mathbf{x}|^2)^{\frac{N+n}{2}}} e^{-\frac{|\mathbf{x}|^2}{4t}} \tag{10.3.10}$$

for some $C_n < \infty$. Therefore, if $\|\alpha\| = n$, then, by (10.3.8),

$$\begin{aligned}&\mathbb{E}^{\mathcal{W}^{(N)}_{\mathbf{x}}}\left[\left|\partial^\alpha_{\mathbf{y}} g^{(N)}\big(t-\zeta^G(\boldsymbol{\psi}), \mathbf{y}-\boldsymbol{\psi}(\zeta^G)\big)\right|, \zeta^G(\boldsymbol{\psi}) < t\right] \\ &\qquad \le \frac{C_n}{|\mathbf{y}-\partial G|^{N+n}} \mathbb{E}^{\mathcal{W}^{(N)}_{\mathbf{x}}}\left[e^{-\frac{|\mathbf{y}-\boldsymbol{\psi}(\zeta^G)|^2}{4t}}, \zeta^G(\boldsymbol{\psi}) < t\right].\end{aligned}$$

At the same time,

$$\mathbb{E}^{\mathcal{W}^{(N)}_{\mathbf{x}}}\left[e^{-\frac{|\mathbf{y}-\boldsymbol{\psi}(\zeta^G)|^2}{4t}}, \zeta^G(\boldsymbol{\psi}) < t\right] \le e^{-\frac{|\mathbf{y}-\mathbf{x}|^2}{16t}} + \mathcal{W}^{(N)}\left(\|\boldsymbol{\psi}\|_{[0,t]} \ge \frac{|\mathbf{y}-\mathbf{x}|}{2}\right),$$

and so we now see (cf. (4.3.13)) that, for some other choice of $C_n < \infty$,

$$\left|\partial^\alpha_{\mathbf{y}} p^G(t,\mathbf{x},\mathbf{y})\right| \le C_n\left(\frac{1}{(t+|\mathbf{y}-\mathbf{x}|^2)^{\frac{N+n}{2}}} + \frac{1}{|\mathbf{y}-\partial G|^{(N+n)}}\right) e^{-\frac{|\mathbf{y}-\mathbf{x}|^2}{16Nt}} \tag{10.3.11}$$

when $\|\alpha\| = n$.

Combining (10.3.11) with the symmetry of $p^G$, we have

$$\left|\partial^\alpha_{\mathbf{x}} p^G(t,\mathbf{x},\mathbf{y})\right| \le C_n\left(\frac{1}{(t+|\mathbf{y}-\mathbf{x}|^2)^{\frac{N+n}{2}}} + \frac{1}{|\mathbf{x}-\partial G|^{(N+n)}}\right) e^{-\frac{|\mathbf{y}-\mathbf{x}|^2}{16Nt}}. \tag{10.3.12}$$

In addition, from (10.3.5),

$$p^G(t,\mathbf{x},\mathbf{y}) = \int_G p^G\left(\tfrac{t}{2},\mathbf{x},\mathbf{z}\right) p^G\left(\tfrac{t}{2},\mathbf{z},\mathbf{y}\right) d\mathbf{z},$$

and so, by (10.3.12) and (10.3.11), we see that $(\mathbf{x},\mathbf{y}) \rightsquigarrow p^G(t,\mathbf{x},\mathbf{y})$ is smooth for each $t \in (0,\infty)$.

To check the assertions about the time derivatives, first observe that for any bounded $\varphi \in C^2(G;\mathbb{R})$ and $(\mathbf{x},\mathbf{y}) \in G^2$,

$$\begin{aligned}\lim_{h\searrow 0}\frac{1}{h}\left[\int_G p^G(h,\mathbf{x},\mathbf{y})\varphi(\mathbf{y})\,d\mathbf{y} - \varphi(\mathbf{x})\right] &= \tfrac{1}{2}\Delta\varphi(\mathbf{x}) \\ \lim_{h\searrow 0}\frac{1}{h}\left[\int_G p^G(h,\mathbf{x},\mathbf{y})\varphi(\mathbf{x})\,d\mathbf{x} - \varphi(\mathbf{y})\right] &= \tfrac{1}{2}\Delta\varphi(\mathbf{y}).\end{aligned}$$

To see this, use the symmetry of $p^G$ to show that the second of these follows from the first one. To prove the first one, use $p^G(h,\mathbf{x},\mathbf{y}) \le g^{(N)}(h,\mathbf{y}-\mathbf{x})$ and (10.3.8) to show that, for any $\tilde{\varphi} \in C^2_c(\mathbb{R}^N;\mathbb{R})$ that equals $\varphi$ in a neighborhood of $\mathbf{x}$,

$$\left| \int_G p^G(h,\mathbf{x},\mathbf{y})\varphi(\mathbf{y})\,d\mathbf{y} - \varphi(\mathbf{x}) - \int_G g^{(N)}(h,\mathbf{y}-\mathbf{x})\tilde{\varphi}(\mathbf{y})\,d\mathbf{y} \right|$$

tends to 0 faster than any power of $h$. Thus, since

$$\frac{1}{h}\left( \int_G g^{(N)}(h,\mathbf{y}-\mathbf{x})\tilde{\varphi}(\mathbf{y})\,d\mathbf{y} - \varphi(\mathbf{x}) \right) \longrightarrow \tfrac{1}{2}\Delta\varphi(\mathbf{x}),$$

the assertion is proved. Given the preceding, we know that

$$\frac{1}{h}\Big[p^G(t+h,\mathbf{x},\mathbf{y}) - p^G(t,\mathbf{x},\mathbf{y})\Big] = \frac{1}{h}\left[\int_G p^G(h,\mathbf{x},\mathbf{z})p^G(t,\mathbf{z},\mathbf{y})\,d\mathbf{z} - p^G(t,\mathbf{x},\mathbf{y})\right]$$

tends to $\frac{1}{2}\Delta_{\mathbf{x}}p^G(t,\mathbf{x},\mathbf{y})$. Thus, $\partial_t p^G(t,\mathbf{x},\mathbf{y}) = \frac{1}{2}\Delta_{\mathbf{x}}p^G(t,\mathbf{x},\mathbf{y})$. Similarly, using

$$p^G(t+h,\mathbf{x},\mathbf{y}) = \int_G p^G(t,\mathbf{x},\mathbf{z})p^G(h,\mathbf{z},\mathbf{y})\,d\mathbf{z} = \int_G p^G(h,\mathbf{y},\mathbf{z})p^G(t,\mathbf{x},\mathbf{z})\,d\mathbf{z},$$

one gets $\partial_t p^G(t,\mathbf{x},\mathbf{y}) = \frac{1}{2}\Delta_{\mathbf{y}}p^G(t,\mathbf{x},\mathbf{y})$. Finally, assume the result for $m$, use (10.3.11) to justify

$$\partial_t^m p^G(t+h,\mathbf{x},\mathbf{y}) = 2^{-m}\int_G p^G(h,\mathbf{x},\mathbf{z})\Delta_{\mathbf{y}}^m p^G(t,\mathbf{z},\mathbf{y})\,d\mathbf{z},$$

differentiate this with respect to $h$, and let $h \searrow 0$ to arrive at

$$\begin{aligned}\partial_t^{m+1}p^G(t,\mathbf{x},\mathbf{y}) &= 2^{-m-1}\Delta_{\mathbf{x}}\Delta_{\mathbf{y}}^m p^G(t,\mathbf{x},\mathbf{y}) \\ &= 2^{-m-1}\begin{cases} \Delta_{\mathbf{x}}^{m+1}p^G(t,\mathbf{x},\mathbf{y}) \\ \Delta_{\mathbf{y}}^m\Delta_{\mathbf{x}}p^G(t,\mathbf{x},\mathbf{y}) = \Delta_{\mathbf{y}}^{m+1}p^G(t,\mathbf{x},\mathbf{y}). \end{cases}\end{aligned}$$ □

The following result provides the justification for my calling $p^G$ the **Dirichlet heat kernel** on $G$.

COROLLARY 10.3.13. *For each $\varphi \in C_b(G;\mathbb{R})$, the function*

$$u(t,\mathbf{x}) = \mathbb{E}^{\mathcal{W}_{\mathbf{x}}^{(N)}}\left[\varphi(\boldsymbol{\psi}(t)),\, \zeta^G(\boldsymbol{\psi}) > t\right] = \int_G \varphi(\mathbf{y})p^G(t,\mathbf{x},\mathbf{y})\,d\mathbf{y}$$

*is a smooth solution to the boundary value problem*

$$\begin{aligned} &\partial_t u(t,\mathbf{x}) = \tfrac{1}{2}\Delta u(t,\mathbf{x}) && \text{in } (0,\infty)\times G, \\ &\lim_{t\searrow 0} u(t,\,\cdot\,) = \varphi && \text{uniformly on compacts,} \\ &\lim_{\substack{(t,\mathbf{x})\to(s,\mathbf{a}) \\ \mathbf{x}\in G}} u(t,\mathbf{x}) = 0 && \text{for } (s,\mathbf{a}) \in (0,\infty)\times\partial_{\text{reg}}G. \end{aligned}$$

*Moreover, if $\partial G = \partial_{\text{reg}}G$, then $u$ is the only bounded solution to this boundary value problem.*

PROOF: That the $u$ in the first part is a bounded, smooth solution follows easily from (10.3.12) and the last part of Theorem 10.3.9. To prove the uniqueness assertion when $\partial G = \partial_{\text{reg}}G$, choose $\{G_n : n \geq 1\}$ to be a non-decreasing sequence of open sets so that $\overline{G_n} \subseteq G$ and $G = \bigcup_{n\geq 1} G_n$. Given a bounded solution $u$, apply Theorem 10.1.2 to see that, for each $n \geq 1$, $u(t, \mathbf{x})$ equals

$$\begin{aligned} &\mathbb{E}^{\mathcal{W}_{\mathbf{x}}^{(N)}}\left[\varphi(\boldsymbol{\psi}(t)),\, \zeta^{G_n}(\boldsymbol{\psi}) > t\right] + \mathbb{E}^{\mathcal{W}_{\mathbf{x}}^{(N)}}\left[u\big(t - \zeta^{G_n}(\boldsymbol{\psi}), \boldsymbol{\psi}(\zeta^{G_n})\big),\, \zeta^{G_n}(\boldsymbol{\psi}) \leq t\right] \\ &= \mathbb{E}^{\mathcal{W}_{\mathbf{x}}^{(N)}}\left[\varphi(\boldsymbol{\psi}(t)),\, \zeta^{G}(\boldsymbol{\psi}) > t\right] + \mathbb{E}^{\mathcal{W}_{\mathbf{x}}^{(N)}}\left[u\big(t - \zeta^{G_n}(\boldsymbol{\psi}), \boldsymbol{\psi}(\zeta^{G_n})\big),\, \zeta^{G}(\boldsymbol{\psi}) < t\right] \\ &\quad + \mathbb{E}^{\mathcal{W}_{\mathbf{x}}^{(N)}}\left[u\big(t - \zeta^{G_n}(\boldsymbol{\psi}), \boldsymbol{\psi}(\zeta^{G_n})\big) - \varphi(\boldsymbol{\psi}(t)),\, \zeta^{G_n}(\boldsymbol{\psi}) \leq t < \zeta^{G}(\boldsymbol{\psi})\right]. \end{aligned}$$

Since $\zeta^{G_n} \nearrow \zeta^G$, the second and third terms on the right tend to 0 as $n \to \infty$. □

REMARK 10.3.14. The uniqueness part of Corollary 10.3.13 continues to hold even if $\partial G \neq \partial_{\text{reg}}G$. Indeed, the only place at which I used the assumption that $\partial G = \partial_{\text{reg}}G$ was where I wanted to know that $u\big(t - \zeta^{G_n}(\boldsymbol{\psi}), \boldsymbol{\psi}(\zeta^{G_n}(\boldsymbol{\psi})\big) \longrightarrow 0$ on $\{\zeta^G < t\}$, and for this I needed to be sure that $\zeta^G(\boldsymbol{\psi}) < \infty \implies \boldsymbol{\psi}(\zeta^G) \in \partial_{\text{reg}}G$. However, it would have been enough to know that $\boldsymbol{\psi}(\zeta^G) \in \partial_{\text{reg}}G$ for $\mathcal{W}_{\mathbf{x}}^{(N)}$-almost every $\boldsymbol{\psi} \in \{\zeta^G < \infty\}$, and this is always the case. Because the proof that, for $\mathbf{x} \in G$, $\mathcal{W}_{\mathbf{x}}^{(N)}\big(\boldsymbol{\psi}(\zeta^G) \notin \partial_{\text{reg}}G\big) = 0$ is not simple, and since Corollary 10.3.13 covers most applications, I have chosen to settle for the weaker statement here and postpone the proof of the general case until the next chapter (cf. § 11.1).

**§ 10.3.3. Feynman–Kac Heat Kernels.** In this subsection I will put some of the considerations in §§ 10.1.3 and 10.1.4 into a more general framework.

Let $V : \mathbb{R}^N \longrightarrow \mathbb{R}$ be a Borel measurable function that is bounded above, and define

$$\begin{aligned} &q^V(t, \mathbf{x}, \mathbf{y}) \\ &= \mathbb{E}^{\mathcal{W}^{(N)}}\left[\exp\left(\int_0^t V\big(\mathbf{x} + \boldsymbol{\theta}_t + (\mathbf{y} - \mathbf{x})\ell_t(\tau)\, d\tau\right)\right] g^{(N)}(t, \mathbf{y} - \mathbf{x}). \end{aligned} \tag{10.3.15}$$

By applying Theorem 10.3.3 with

$$E_t(\boldsymbol{\psi}) \equiv \exp\left(\int_0^t V\big(\boldsymbol{\psi}(\tau)\big)\, d\tau\right),$$

we see that

$$\int_{\mathbb{R}^N} \varphi(\mathbf{y}) q^V(t, \mathbf{x}, \mathbf{y})\, d\mathbf{y} = \mathbb{E}^{\mathcal{W}_{\mathbf{x}}^{(N)}}\left[\exp\left(\int_0^t V\big(\boldsymbol{\psi}(\tau)\big)\, d\tau\right) \varphi(\boldsymbol{\psi}(t))\right] \tag{10.3.16}$$

for $(t, \mathbf{x}) \in (0, \infty) \times \mathbb{R}^N$ and Borel measurable $\varphi$'s that are bounded below,

$$q^V(t, \mathbf{x}, \mathbf{y}) = \int_{\mathbb{R}^N} q^V(s, \mathbf{x}, \mathbf{z}) q^V(t, \mathbf{z}, \mathbf{y})\, d\mathbf{z} \tag{10.3.17}$$

for $(s,\mathbf{x}),\ (t,\mathbf{y}) \in (0,\infty)\times\mathbb{R}^N$, and

$$q^V(t,\mathbf{x},\mathbf{y}) = q^V(t,\mathbf{y},\mathbf{x}) \quad \text{for } (t,\mathbf{x},\mathbf{y}) \in (0,\infty)\times(\mathbb{R}^N)^2. \tag{10.3.18}$$

As a consequence of (10.3.16) and Theorem 10.1.2, we know that $q^V(t,\mathbf{x},\cdot\,)$ is intimately related to the operator $\frac{1}{2}\Delta + V$. Indeed, by that theorem, we know that if $u \in C_{\mathrm{b}}^{1,2}\big((0,\infty)\times\mathbb{R}^N;\mathbb{R}\big)$ satisfies the Cauchy initial value problem

$$\partial_t u = \tfrac{1}{2}\Delta u + V u \quad \text{with } \lim_{t\searrow 0} u(t,\,\cdot\,) = \varphi \quad \text{uniformly on compacts} \tag{10.3.19}$$

for some $\varphi \in C_{\mathrm{b}}(\mathbb{R}^N;\mathbb{R})$, then

$$u(t,\mathbf{x}) = \int_{\mathbb{R}^N} \varphi(\mathbf{y})\, q^V(t,\mathbf{x},\mathbf{y})\, d\mathbf{y} \quad (t,\mathbf{x}) \in (0,\infty)\times\mathbb{R}^N. \tag{10.3.20}$$

I now want to make an analysis of $q^V(t,\mathbf{x},\,\cdot\,)$ which, among other things, will enable me to show (cf. Corollary 10.3.22) that, under suitable conditions on $V$, the right-hand side of (10.3.20) is necessarily a solution to (10.3.19). For this reason, I will call $q^V$ the **Feynman–Kac heat kernel with potential** $V$.

THEOREM 10.3.21. *Assume that $V \in C^n(\mathbb{R}^N;\mathbb{R})$ is bounded above and that, for some $C_n < \infty$,*

$$\max_{\|\alpha\|\le n} \big|\partial_{\mathbf{x}}^\alpha V(\mathbf{x})\big| \le C_n\big(1 + V^-(\mathbf{x})\big), \quad \mathbf{x} \in \mathbb{R}^N.$$

*Then $q^V(t,\,\cdot\,,\mathbf{y}) \in C^n(\mathbb{R}^N;\mathbb{R})$ for every $(t,\mathbf{y}) \in (0,\infty)\times\mathbb{R}$, $\partial_{\mathbf{x}}^\alpha q^V(t,\mathbf{x},\,\cdot\,) \in C^n(\mathbb{R}^N;\mathbb{R})$ for each $\alpha \in \mathbb{N}^N$ with $\|\alpha\| \le n$, and there exists a $C < \infty$ such that, when $\|\alpha\| \vee \|\beta\| \le n$,*

$$\big|\partial_{\mathbf{x}}^\alpha \partial_{\mathbf{y}}^\beta q^V(t,\mathbf{x},\mathbf{y})\big| \le C\big(1 + (t + |\mathbf{y}-\mathbf{x}|^2)^{-\frac{N+\|\alpha\|+\|\beta\|}{2}}\big) e^{t\|V^+\|_{\mathrm{u}} - \frac{|\mathbf{y}-\mathbf{x}|^2}{4t}}.$$

*Finally, if $n \ge 2$ and $m \le \frac{n}{2}$, then*

$$\partial_t^m q^V(t,\mathbf{x},\mathbf{y}) = \big(\tfrac{1}{2}\Delta_{\mathbf{x}} + V(\mathbf{x})\big)^m q^V(t,\mathbf{x},\mathbf{y}) = \big(\tfrac{1}{2}\Delta_{\mathbf{y}} + V(\mathbf{y})\big)^m q^V(t,\mathbf{x},\mathbf{y}).$$

PROOF: To prove the differentiability properties of $q^V(t,\mathbf{x},\mathbf{y})$ with respect to $\mathbf{x}$, let $\|\alpha\| \le n$ be given, use (10.3.15) to see that $\partial_{\mathbf{x}}^\alpha q^V(t,\mathbf{x},\mathbf{y})$ is a finite linear combination of terms of the form

$$\mathbb{E}^{\mathcal{W}^{(N)}}\left[\prod_{k=1}^{\ell} \int_0^t \big(1 - \tfrac{\tau}{t}\big)^{\|\alpha^{(k)}\|} \big(\partial^{\alpha^{(k)}} V\big)\big(\psi_{t,\mathbf{x},\mathbf{y}}(\tau)\big) E^V(t,\mathbf{x},\mathbf{y},\psi)\, d\tau\right] \times \partial_{\mathbf{x}}^{\alpha^{(0)}} g^{(N)}(t,\mathbf{y}-\mathbf{x}),$$

where $\psi_{t,\mathbf{x},\mathbf{y}}(\tau) = \mathbf{x} + \psi_t(\tau) + \frac{\tau}{t}(\mathbf{y}-\mathbf{x})$, $E^V(t,\mathbf{x},\mathbf{y},\psi) = e^{\int_0^t V(\psi_{t,\mathbf{x},\mathbf{y}}(\tau))\,d\tau}$, and $\sum_{k=0}^{\ell} \alpha^{(k)} = \alpha$. Since, by our hypotheses, each of the integrands in these terms is bounded by a constant times $e^{t\|V^+\|_{\mathrm{u}}}$, the asserted estimate for $\partial_{\mathbf{x}}^{\alpha} q^V(t,\mathbf{x},\mathbf{y})$ follows from this and (10.3.10).

The rest of the proof is similar to, but easier than, that of Theorem 10.3.9. Specifically, one uses $q^V(t,\mathbf{x},\mathbf{y}) = q^V(t,\mathbf{y},\mathbf{x})$ and

$$q^V(t,\mathbf{x},\mathbf{y}) = \int_{\mathbb{R}^N} q^V\left(\tfrac{t}{2},\mathbf{x},\mathbf{z}\right) q^V\left(\tfrac{t}{2},\mathbf{z},\mathbf{y}\right) d\mathbf{z}$$

to prove the existence of and estimate for $\partial_{\mathbf{x}}^{\alpha}\partial_{\mathbf{y}}^{\beta} q^V(t,\mathbf{x},\mathbf{y})$. Also, knowing these results about the spacial derivatives, one deals with the time derivatives in the same way as I did at the end of that theorem. The details are left to the reader. □

COROLLARY 10.3.22. *Let $V$ be as in Theorem 10.3.21, and assume that $n \geq 2$. Then, for each $\varphi \in C_{\mathrm{b}}(\mathbb{R}^N;\mathbb{R})$, the function*

$$u(t,\mathbf{x}) = \mathbb{E}^{\mathcal{W}_{\mathbf{x}}^{(N)}}\left[e^{\int_0^t V(\psi(\tau))\,d\tau} \varphi(t)\right] = \int_{\mathbb{R}^N} \varphi(\mathbf{y}) q^V(t,\mathbf{x},\mathbf{y})\,d\mathbf{y}$$

*is the unique $u \in C^{1,2}\big((0,\infty)\times\mathbb{R}^N;\mathbb{R}\big)$ that is bounded on $(0,T)\times\mathbb{R}^N$ for each $T > 0$ and satisfies (10.3.19).*

PROOF: The only assertion that has not already been proved is that the $u$ described takes on the correct initial value. However, because $q^V(t,\mathbf{x},\mathbf{y}) \leq e^{\|V^+\|_{\mathrm{u}}} g^{(N)}(t,\mathbf{y}-\mathbf{x})$, it is clear that, for each $r > 0$,

$$\lim_{t\searrow 0} \sup_{\mathbf{x}\in\mathbb{R}^N} \int_{B(\mathbf{x},r)\complement} q^V(t,\mathbf{x},\mathbf{y})\,d\mathbf{y} = 0.$$

Hence, all that remains is to check that, for each $R > 0$,

$$\lim_{t\searrow 0} \sup_{|\mathbf{x}|\leq R} \left|1 - \int_{\mathbb{R}^N} q^V(t,\mathbf{x},\mathbf{y})\,dy\right| = 0.$$

But if $K(R) = \sup_{|\mathbf{y}|\leq 2R} |V(\mathbf{y})|$, then

$$\begin{aligned}\sup_{|\mathbf{x}|\leq R} \left|1 - \int_{\mathbb{R}^N} q^V(t,\mathbf{x},\mathbf{y})\,dy\right| &\leq \mathbb{E}^{\mathcal{W}^{(N)}\mathbf{x}}\left[\left|1 - e^{\int_0^t V(\psi(\tau))\,d\tau}\right|\right] \\ &\leq tK(R)e^{tK(R)} + \left(1 + e^{t\|V^+\|_{\mathrm{u}}}\right)\mathcal{W}^{(N)}\left(\|\psi\|_{[0,t]} \geq R\right),\end{aligned}$$

which, by (4.3.13), gives the desired conclusion. □

**§ 10.3.4. Ground States and Associated Measures on Pathspace.** From a probabilistic standpoint, the heat kernel $q^V(t,\mathbf{x},\mathbf{y})$ is flawed by the fact that it is not a probability density. However, in many cases this flaw can be removed by what physicists call switching to the **ground state representation**.

This terminology and the ideas underlying it are best understood when expressed in terms of operators. Thus, let $V \in C(\mathbb{R}^N;\mathbb{R})$ be bounded above, refer to the preceding subsection, and define the operator

$$\mathbf{Q}_t^V\varphi(\mathbf{x}) = \int_{\mathbb{R}^N} \varphi(\mathbf{y})q^V(t,\mathbf{x},\mathbf{y})\,d\mathbf{y} \quad \text{for } t \geq 0 \text{ and } \varphi \in C_b(\mathbb{R}^N;\mathbb{R}).$$

We know that $\mathbf{Q}_t^V$ is a bounded map from $C_b(\mathbb{R}^N;\mathbb{R})$ into itself. In addition, by (10.3.17), $\{\mathbf{Q}_t^V : t \geq 0\}$ is a semigroup. That is, $\mathbf{Q}_{s+t}^V = \mathbf{Q}_t^V \circ \mathbf{Q}_s^V$. Also, by Corollary 10.3.22, we know that if

$$V \in C^2(\mathbb{R}^N;\mathbb{R}) \text{ and } \max_{\|\alpha\|\leq 2} |\partial^\alpha V| \leq C\big(1+V^-\big), \tag{10.3.23}$$

then $(t,\mathbf{x}) \rightsquigarrow \mathbf{Q}_t^V\varphi(\mathbf{x})$ is a solution to (10.3.19).

I will say that $\rho : \mathbb{R}^N \longrightarrow \mathbb{R}$ is a **ground state** for $V$ if $\rho$ is a (strictly) positive, continuous function that satisfies the equation $e^{t\lambda}\rho = \mathbf{Q}_t^V\rho$ for some $\lambda \in \mathbb{R}$ and all $t \geq 0$, in which case $\lambda$ will be called the **eigenvalue** associated with $\rho$.

LEMMA 10.3.24. *Let $V$ be as above, and assume that $\rho \in C\big(\mathbb{R}^N;[0,\infty)\big)$ does not vanish identically. If $e^{t\lambda}\rho = \mathbf{Q}_t^V\rho$ for all $t \geq 0$, then $\rho$ is a ground state with associated eigenvalue $\lambda$. In fact, $\rho \in C_b^2\big(\mathbb{R}^N;(0,\infty)\big)$ if $\rho$ is bounded and $V \in C^2(\mathbb{R}^N;\mathbb{R})$ satisfies (10.3.23). Next, if $\rho$ is a twice continuously differentiable ground state with associated eigenvalue $\lambda$, then $\frac{1}{2}\Delta\rho + V\rho = \lambda\rho$. Conversely, if $\rho$ is a twice continuously differentiable, bounded solution to $\frac{1}{2}\Delta\rho + V\rho = \lambda\rho$, then $\rho$ is a ground state with associated eigenvalue $\lambda$.*

PROOF: Since I can always replace $V$ by $V-\lambda$, I may and will assume that $\lambda = 0$ throughout. Also, observe that if $\rho \in C\big(\mathbb{R}^N;[0,\infty)\big)$ satisfies $\rho = \mathbf{Q}_1^V\rho$, then, because $q^V(1,\mathbf{x},\mathbf{y}) > 0$ everywhere, $\rho > 0$ everywhere unless $\rho \equiv 0$. Hence, the first assertion is proved.

Next suppose that $\rho$ is a twice continuously differentiable ground state with eigenvalue 0. To see that $\frac{1}{2}\Delta\rho + V\rho = 0$, it suffices to show that

$$\big(\tfrac{1}{2}\Delta\varphi + V\varphi, \rho\big)_{L^2(\mathbb{R}^N;\mathbb{R})} = 0 \quad \text{for all } \varphi \in C_c^\infty(\mathbb{R}^N;\mathbb{R}).$$

To this end, let $\varphi \in C_c^\infty(\mathbb{R}^N;\mathbb{R})$ be given, and apply symmetry, Theorem 10.1.2, and Fubini's Theorem to justify

$$\begin{aligned}
0 &= \big(\varphi, \mathbf{Q}_1^V\rho - \rho\big)_{L^2(\mathbb{R}^N;\mathbb{R})} = \big(\mathbf{Q}_1^V\varphi - \varphi, \rho\big)_{L^2(\mathbb{R}^N;\mathbb{R})} \\
&= \int_0^1 \Big(\mathbf{Q}_\tau^V\big(\tfrac{1}{2}\Delta\varphi + V\varphi\big), \rho\Big)_{L^2(\mathbb{R}^N;\mathbb{R})} d\tau \\
&= \int_0^1 \Big(\tfrac{1}{2}\Delta\varphi + V\varphi, \mathbf{Q}_\tau^V\rho\Big)_{L^2(\mathbb{R}^N;\mathbb{R})} d\tau = \big(\tfrac{1}{2}\Delta\varphi + V\varphi, \rho\big)_{L^2(\mathbb{R}^N;\mathbb{R})}.
\end{aligned}$$

Finally, suppose that $\rho$ is a bounded, twice continuously differentiable solution to $\frac{1}{2}\Delta\rho + V\rho = 0$. Then, by Corollary 10.1.3 applied to the time-independent function $u(t,\,\cdot\,) = \rho$, we know that $\rho = \mathbf{Q}_t^V \rho$ for all $t \geq 0$. Thus, by the initial observation, $\rho$ is a ground state with associated eigenvalue 0. □

THEOREM 10.3.25. *Let $V \in C(\mathbb{R}^N;\mathbb{R})$ be bounded above, assume that $\rho$ is a ground state for $V$ with associated eigenvalue $\lambda$, and set*

$$p^\rho(t,\mathbf{x},\mathbf{y}) = e^{-t\lambda}\rho(\mathbf{x})^{-1} q^V(t,\mathbf{x},\mathbf{y})\rho(\mathbf{y}) \quad \text{for } (t,\mathbf{x},\mathbf{y}).$$

*Then $p^\rho$ is a strictly positive, continuous function, $p^\rho(t,\mathbf{x},\,\cdot\,)$ has total integral 1 for all $(t,\mathbf{x}) \in (0,\infty)\times\mathbb{R}^N$,*

$$\lim_{t\searrow 0}\sup_{|\mathbf{x}|\leq R}\int_{B(\mathbf{0},r)} p^\rho(t,\mathbf{x},\mathbf{y})\,d\mathbf{y} = 0 \quad \text{for all } r,\ R \in (0,\infty),$$

*and*

$$p^\rho(s+t,\mathbf{x},\mathbf{y}) = \int_{\mathbb{R}^N} p^\rho(t,\mathbf{z},\mathbf{y})p^\rho(t,\mathbf{x},\mathbf{z})\,d\mathbf{z}.$$

*Finally, if $V \in C^2(\mathbb{R}^N;\mathbb{R})$ satisfies (10.3.23), then $\mathbf{x} \rightsquigarrow p^\rho(t,\mathbf{x},\mathbf{y})$ is twice continuously differentiable for each $(t,\mathbf{y}) \in (0,\infty)\times\mathbb{R}^N$, $\mathbf{y} \rightsquigarrow \partial_{\mathbf{x}}^\alpha p^\rho(t,\mathbf{x},\mathbf{y})$ is twice continuously differentiable for each $\alpha$ with $\|\alpha\| \leq 2$ and $(t,\mathbf{x}) \in (0,\infty)\times\mathbb{R}^N$, and*

$$\begin{aligned}\partial_t p^\rho(t,\mathbf{x},\mathbf{y}) &= \tfrac{1}{2}\Delta_{\mathbf{x}} p^\rho(t,\mathbf{x},\mathbf{y}) + \big(\nabla_{\mathbf{x}}(\log\rho), \nabla_{\mathbf{x}} p^\rho(t,\mathbf{x},\mathbf{y})\big)_{\mathbb{R}^N}\\ &= \tfrac{1}{2}\Delta_{\mathbf{y}} p^\rho(t,\mathbf{x},\mathbf{y}) - \operatorname{div}_{\mathbf{y}}\big(p^\rho(t,\mathbf{x},\mathbf{y})\nabla\log\rho(\mathbf{y})\big)\end{aligned}$$

*for all $(t,\mathbf{x},\mathbf{y}) \in (0,\infty)\times\mathbb{R}^N\times\mathbb{R}^N$. In particular, for each $\varphi \in C_{\mathrm{b}}(\mathbb{R}^N;\mathbb{R})$, the function*

$$u(t,\mathbf{x}) = \int_{\mathbb{R}^N}\varphi(\mathbf{y})p^\rho(t,\mathbf{x},\mathbf{y})\,d\mathbf{y}$$

*is the one and only bounded $u \in C^{1,2}\big((0,\infty)\times\mathbb{R}^N;\mathbb{R}\big)$ that satisfies*

$$\begin{aligned}&\partial_t u(t,\mathbf{x}) = \tfrac{1}{2}\Delta u(t,\mathbf{x}) + \big(\nabla\log\rho(\mathbf{x}), \nabla u(t,\mathbf{x})\big)_{\mathbb{R}^N} \quad \text{in } (0,\infty)\times\mathbb{R}^N\\ &\lim_{t\searrow 0} u(t,\mathbf{x}) = \varphi(\mathbf{x}) \quad \text{uniformly on compacts.}\end{aligned}$$

PROOF: The only assertion that is not an immediate consequence of Theorem 10.3.21, Corollary 10.3.22, and the preceding lemma is the uniqueness in the final part, which is an easy consequence of the corresponding uniqueness statement in Corollary 10.3.22. Indeed, if $u$ is a bounded solution to the given Cauchy initial value problem and $w(t,\,\cdot\,) = \rho u(t,\,\cdot\,)$, then $w$ is a bounded solution to $\partial_t w = \frac{1}{2}\Delta w + (V-\lambda)w$ with initial condition $\rho\varphi$. Hence, by the uniqueness result in Corollary 10.3.22, $w(t,\,\cdot\,) = \mathbf{Q}_t^V(\rho\varphi)$, and so $u(t,\,\cdot\,) = \int_{\mathbb{R}^N}\varphi(\mathbf{y})p^\rho(t,\mathbf{x},\mathbf{y})\,d\mathbf{y}$. □

The advantage that $p^\rho(t,\mathbf{x},\mathbf{y})$ has over $q^V(t,\mathbf{x},\mathbf{y})$ is that we can construct measures on $C(\mathbb{R}^N)$ that bear the same relationship to it as the Wiener measures $\mathcal{W}_{\mathbf{x}}^{(N)}$ bear to the classical heat kernel $g^{(N)}(t,\mathbf{y}-\mathbf{x})$.

THEOREM 10.3.26. *Let $V \in C(\mathbb{R}^N;\mathbb{R})$ be bounded above, and assume that $\rho$ is a ground state for $V$ with associated eigenvalue $\lambda$. Then, for each $\mathbf{x} \in \mathbb{R}^N$, there is a unique $\mathbb{P}^\rho_{\mathbf{x}} \in \mathbf{M}_1\big(C(\mathbb{R}^N)\big)$ such that, for each $n \geq 1$, $0 = t_0 \leq t_1 < \cdots < t_m$, and $\Gamma, \cdots, \Gamma_n \in \mathcal{B}_{\mathbb{R}^N}$,*

$$\mathbb{P}^\rho_{\mathbf{x}}\big(\psi(t_m) \in \Gamma_m,\ 1 \leq m \leq n\big) = \int_{\Gamma_1\times\cdots\times\Gamma_n} \cdots \int \prod_{m=1}^n p^\rho(t_m - t_{m-1}, \mathbf{y}_{m-1}, \mathbf{y}_m)\, d\mathbf{y}_1 \cdots d\mathbf{y}_n,$$

*where $\mathbf{y}_0 = \mathbf{x}$. In fact, if*

$$R^\rho(t,\psi) = e^{-t\lambda}\rho\big(\psi(0)\big)^{-1}E^V(t,\psi)\rho\big(\psi(t)\big) \quad \text{where } E^V(t,\psi) = e^{\int_0^t V((\psi(\tau))\,d\tau},$$

*then*

$$\mathbb{P}^\rho_{\mathbf{x}}(A) = \mathbb{E}^{\mathcal{W}^{(N)}_{\mathbf{x}}}\big[R^\rho(t),\, A\big] \quad \text{for all } t \geq 0 \text{ and } A \in \mathcal{F}_t.$$

*Finally, $\mathbf{x} \rightsquigarrow \mathbb{P}^\rho_{\mathbf{x}}$ is continuous, and, for any stopping time $\zeta$,*

$$\int_{\{\zeta(\psi)<\infty\}} F(\psi, \Sigma_\zeta\psi)\, \mathbb{P}^\rho_{\mathbf{x}}(d\psi) = \int_{\{\zeta(\psi)<\infty\}} \left(\int F(\psi,\psi')\, \mathbb{P}^\rho_{\psi(\zeta)}(d\psi')\right) \mathbb{P}^\rho_{\mathbf{x}}(d\psi)$$

*whenever $F : C(\mathbb{R}^N) \times C(\mathbb{R}^N) \longrightarrow \mathbb{R}$ is a $\mathcal{F}_\zeta \times \mathcal{B}_{C(\mathbb{R}^N)}$-measurable function that is bounded below.*

PROOF: I begin by showing the $\big(R^\rho(t), \mathcal{F}_t, \mathcal{W}^{(N)}_{\mathbf{x}}\big)$ is a martingale. Indeed, $\mathbb{E}^{\mathcal{W}^{(N)}_{\mathbf{x}}}\big[R^\rho(t)\big] = 1$ for all $(t,\mathbf{x}) \in [0,\infty) \times \mathbb{R}^N$. In addition, $R^\rho(s+t,\psi) = R^\rho(s,\psi)R^\rho(t,\Sigma_s\psi)$, and so, by (10.2.2),

$$\mathbb{E}^{\mathcal{W}^{(N)}_{\mathbf{x}}}\big[R^\rho(s+t),\, A\big] = \int_A R^\rho(s,\psi)\mathbb{E}^{\mathcal{W}^{(N)}_{\psi(s)}}\big[R^\rho(t)\big]\, \mathcal{W}^{(N)}_{\mathbf{x}}(d\psi) = \mathbb{E}^{\mathcal{W}^{(N)}_{\mathbf{x}}}\big[R^\rho(s),\, A\big]$$

for $A \in \mathcal{F}_s$.

Determine $\mu_{t,\mathbf{x}} \in \mathbf{M}_1\big(C(\mathbb{R}^N)\big)$ by $\mu_{t,\mathbf{x}}(d\psi) = R(t,\psi)\mathcal{W}^{(N)}_{\mathbf{x}}(d\psi)$. By the preceding, $\mu_{t_1,\mathbf{x}} \restriction \mathcal{F}_{t_1} = \mu_{t_2,\mathbf{x}} \restriction \mathcal{F}_{t_1}$ for all $0 \leq t_1 \leq t_2$, and so (cf. Exercise 9.3.6) there is a unique $\mathbb{P}^\rho_{\mathbf{x}} \in \mathbf{M}_1\big(C(\mathbb{R}^N)\big)$ whose restriction to $\mathcal{F}_t$ is the same as that of $\mu_{t,\mathbf{x}}$ for all $t \geq 0$.

To see that $\mathbf{x} \rightsquigarrow \mathbb{P}^\rho_{\mathbf{x}}$ is continuous, it suffices to check that

$$\lim_{\mathbf{y}\to\mathbf{x}} R^\rho(t,\mathbf{y}+\psi) = R^\rho(t,\mathbf{x}+\psi) \quad \text{in } L^1(\mathcal{W}^{(N)};\mathbb{R}).$$

But clearly this convergence is taking place pointwise for each $\psi \in C(\mathbb{R}^N)$. In addition, $R^\rho(t,\cdot) \geq 0$ and, for each $\mathbf{z} \in \mathbb{R}^N$, $R^\rho(t,\mathbf{z}+\psi)$ has $\mathcal{W}^{(N)}$-integral 1. Hence, the convergence is also taking place in $L^1(\mathcal{W}^{(N)};\mathbb{R})$.

Now suppose that $\zeta$ is a stopping time and that $\zeta \le T$ for some $T \in (0,\infty)$. Then, for any $\mathcal{F}_\zeta \times \mathcal{F}_T$-measurable $F : C(\mathbb{R}^N)^2 \longrightarrow \mathbb{R}$ that is bounded below,

$$\begin{aligned}
&\int F(\boldsymbol{\psi}, \Sigma_\zeta \boldsymbol{\psi})\, \mathbb{P}^\rho_{\mathbf{x}}(d\boldsymbol{\psi}) \\
&= \int R^\rho\big(\zeta(\boldsymbol{\psi}), \boldsymbol{\psi}\big) R^\rho\big(2T - \zeta(\boldsymbol{\psi}), \Sigma_\zeta \boldsymbol{\psi}\big) F(\boldsymbol{\psi}, \Sigma_\zeta \boldsymbol{\psi})\, \mathcal{W}^{(N)}_{\mathbf{x}}(d\boldsymbol{\psi}) \\
&= \int R^\rho\big(\zeta(\boldsymbol{\psi}), \boldsymbol{\psi}\big) \left( \int R^\rho\big(2T - \zeta(\boldsymbol{\psi}), \boldsymbol{\psi}'\big) F(\boldsymbol{\psi}, \boldsymbol{\psi}') \mathcal{W}^{(N)}_{\boldsymbol{\psi}(\zeta)}(d\boldsymbol{\psi}') \right) \mathcal{W}^{(N)}_{\mathbf{x}}(d\boldsymbol{\psi}) \\
&= \int R^\rho\big(\zeta(\boldsymbol{\psi}), \boldsymbol{\psi}\big) \left( \int F(\boldsymbol{\psi}, \boldsymbol{\psi}')\, \mathbb{P}^\rho_{\boldsymbol{\psi}(\zeta)}(d\boldsymbol{\psi}') \right) \mathcal{W}^{(N)}_{\mathbf{x}}(d\boldsymbol{\psi}) \\
&= \int \left( \int F(\boldsymbol{\psi}, \boldsymbol{\psi}')\, \mathbb{P}^\rho_{\boldsymbol{\psi}(\zeta)}(d\boldsymbol{\psi}') \right) \mathbb{P}^\rho_{\mathbf{x}}(d\boldsymbol{\psi}),
\end{aligned}$$

where I have again used (10.2.2) and, in the final step, Hunt's Theorem (cf. Theorem 7.1.14) to replace $R^\rho\big(\zeta(\boldsymbol{\psi}), \boldsymbol{\psi}\big)$ by $R^\rho(T, \boldsymbol{\psi})$. Starting from this, one can easily remove the condition that $\zeta$ is bounded and extend the result to all $F$'s that are $\mathcal{F}_\zeta \times \mathcal{B}_{C(\mathbb{R}^N)}$-measurable and bounded below.

To complete the proof, observe that, as a special case of the preceding,

$$\mathbb{E}^{\mathbb{P}^\rho_{\mathbf{x}}}\big[\varphi\big(\boldsymbol{\psi}(s+t)\big), A\big] = \mathbb{E}^{\mathbb{P}^\rho_{\mathbf{x}}}\left[ \int_{\mathbb{R}^N} p^\rho\big(t, \boldsymbol{\psi}(s), \mathbf{y}\big)\, d\mathbf{y}, A \right]$$

for all $s, t \in [0,\infty)$, $A \in \mathcal{F}_s$, and bounded Borel measurable $\varphi : \mathbb{R}^N \longrightarrow \mathbb{R}$. Hence, proceeding by induction on $n$ and applying the preceding at each stage, one can readily show that $\mathbb{P}^\rho_{\mathbf{x}}$ is related to $p^\rho(t, \mathbf{x}, \mathbf{y})$ in the way described in the initial assertion. □

Corollary 10.3.27. *Let everything be as in Theorem 10.3.26, only this time assume that $\rho$ is twice continuously differentiable. Then, for any bounded $\varphi \in C^{1,2}\big([0,\infty) \times \mathbb{R}^N; \mathbb{R}\big)$ such that $f \equiv \partial_t \varphi + \frac{1}{2}\Delta + \big(\nabla \log \rho, \nabla \varphi\big)_{\mathbb{R}^N}$ is bounded,*

$$\left( \varphi\big(t, \boldsymbol{\psi}(t)\big) - \int_0^t f\big(\tau, \boldsymbol{\psi}(\tau)\big)\, d\tau, \mathcal{F}_t, \mathbb{P}^\rho_{\mathbf{x}} \right)$$

*is a martingale for all $\mathbf{x} \in \mathbb{R}^N$.*

Proof: By replacing $V$ with $V - \lambda$, I can reduce to the case when $\lambda = 0$. Hence, I will assume that $\lambda = 0$.

To prove the asserted martingale property, set $E^V(t, \boldsymbol{\psi}) = e^{\int_0^t V((\boldsymbol{\psi}(\tau))\, d\tau}$, and remember that, by Lemma 10.3.24, $\frac{1}{2}\Delta\rho + V\rho = 0$. Thus, $\frac{1}{2}\Delta(\rho\varphi) + V(\rho\varphi) = f\rho$, and so, by Theorem 10.1.2,

$$\begin{aligned}
\Big( E^V&\big(t \wedge \zeta^{B(\mathbf{x},R)}(\boldsymbol{\psi}), \boldsymbol{\psi}\big)(\rho\varphi)\big(t \wedge \zeta^{B(\mathbf{x},R)}(\boldsymbol{\psi}), \boldsymbol{\psi}(t \wedge \zeta^{B(\mathbf{x},R)})\big) \\
&- \int_0^{t \wedge \zeta^{B(\mathbf{x},R)}(\boldsymbol{\psi})} E^V(\tau, \boldsymbol{\psi})(\rho f)\big(\tau, \boldsymbol{\psi}(\tau)\big)\, d\tau, \mathcal{F}_t, \mathcal{W}^{(N)}_{\mathbf{x}} \Big)
\end{aligned}$$

is a martingale for every $R > 0$. Equivalently,

$$\left(R^\rho\big(t\wedge\zeta^{B(\mathbf{x},R)}(\boldsymbol{\psi}),\boldsymbol{\psi}\big)\varphi\big(\boldsymbol{\psi}(t\wedge\zeta^{B(\mathbf{x},R)})\big) - \int_0^{t\wedge\zeta^{B(\mathbf{x},R)}(\boldsymbol{\psi})} R^\rho(\tau,\boldsymbol{\psi})f\big(\boldsymbol{\psi}(\tau)\,d\tau, \mathcal{F}_t, \mathcal{W}_{\mathbf{x}}^{(N)}\right)$$

is a martingale. Hence, since $\big(R^\rho(t),\mathcal{F}_t,\mathcal{W}_{\mathbf{x}}^{(N)}\big)$ is a martingale, one can apply Theorem 7.1.14 to see that

$$\left(\varphi\big(\boldsymbol{\psi}(t\wedge\zeta^{B(\mathbf{x},R)})\big) - \int_0^{t\wedge\zeta^{B(\mathbf{x},R)}(\boldsymbol{\psi})} f\big(\boldsymbol{\psi}(\tau)\big)\,d\tau, \mathcal{F}_t, \mathbb{P}_{\mathbf{x}}^\rho\right)$$

is a martingale. Finally, since $\varphi$ and $f$ are bounded, we can let $R\to\infty$ and thereby get the required conclusion. □

In order to understand the relationship between Brownian paths and the paths as seen by the measure $\mathbb{P}_{\mathbf{x}}^\rho$, I will need the following general lemma.

LEMMA 10.3.28. *Let $\mathbf{b}:\mathbb{R}^N\longrightarrow\mathbb{R}^N$ be a continuous function, and set*

$$\mathbf{B}(t,\boldsymbol{\psi}) = \boldsymbol{\psi}(t) - \mathbf{x} - \int_0^t \mathbf{b}\big(\boldsymbol{\psi}(\tau)\big)\,d\tau.$$

*If $\mathbb{P}\in\mathbf{M}_1\big(C(\mathbb{R}^N)\big)$ has the properties that $\mathbb{P}(\boldsymbol{\psi}(0)=\mathbf{x}) = 1$ and*

$$\left(\varphi\big(\boldsymbol{\psi}(t)\big) - \int_0^t \left(\tfrac{1}{2}\Delta\varphi + \big(\mathbf{b},\nabla\varphi\big)_{\mathbb{R}^N}\right)\big(\boldsymbol{\psi}(\tau)\big)\,d\tau, \mathcal{F}_t, \mathbb{P}\right)$$

*is a martingale for all $\varphi\in C_c^\infty(\mathbb{R}^N;\mathbb{R})$, then $\big(\mathbf{B}(t),\mathcal{F}_t,\mathbb{P}\big)$ is a Brownian motion.*

PROOF: Without loss in generality, I will assume that $\mathbf{x}=\mathbf{0}$.

Given $\boldsymbol{\xi}\in\mathbb{R}^N$ and $R>0$, set $e_{\boldsymbol{\xi}}(\mathbf{y}) = e^{\sqrt{-1}(\boldsymbol{\xi},\mathbf{y})_{\mathbb{R}^N}}$,

$$f(\mathbf{y}) = \frac{|\boldsymbol{\xi}|^2}{2} - \sqrt{-1}\big(\boldsymbol{\xi},\mathbf{b}(\mathbf{y})\big)_{\mathbb{R}^N}, \text{ and } E_{\boldsymbol{\xi}}^R(t,\boldsymbol{\psi}) = \exp\left(\int_0^{t\wedge\zeta^{B(\mathbf{0},R)}(\boldsymbol{\psi})} f\big(\boldsymbol{\psi}(\tau)\big)\,d\tau\right).$$

By choosing $\varphi\in C_c^\infty(\mathbb{R}^N;\mathbb{C})$ so that $\varphi = e_{\boldsymbol{\xi}}$ on $B(\mathbf{0},2R)$ and applying Doob's Stopping Time Theorem, we know that $\big(M_{\boldsymbol{\xi}}^R(t),\mathcal{F}_t,\mathbb{P}\big)$ is a martingale, where

$$M_{\boldsymbol{\xi}}^R(t,\boldsymbol{\psi}) = e_{\boldsymbol{\xi}}\big(\boldsymbol{\psi}(t\wedge\zeta^{B(\mathbf{0},R)})\big) + \int_0^{t\wedge\zeta^{B(\mathbf{0},R)}(\boldsymbol{\psi})} f\big(\boldsymbol{\psi}(\tau)\big)e_{\boldsymbol{\xi}}\big(\boldsymbol{\psi}(\tau)\big)\,d\tau.$$

Thus, by Theorem 7.1.17,

$$\left(E^R_{\boldsymbol{\xi}}(t)M^R_{\boldsymbol{\xi}}(t) - \int_0^{t\wedge\zeta^{B(\mathbf{0},R)}(\boldsymbol{\psi})} M^R_{\boldsymbol{\xi}}(\tau,\boldsymbol{\psi})f(\boldsymbol{\psi}(\tau))E^R_{\boldsymbol{\xi}}(\tau,\boldsymbol{\psi})\,d\tau, \mathcal{F}_t, \mathbb{P}\right)$$

is also a martingale. At the same time, after performing elementary calculus operations, one sees that

$$\begin{aligned}\exp\Big(\sqrt{-1}\big(\boldsymbol{\xi},\mathbf{B}(t\wedge\zeta^{B(\mathbf{0},R)}(\boldsymbol{\psi}))_{\mathbb{R}^N} + \frac{|\boldsymbol{\xi}|^2}{2}\big(t\wedge\zeta^{B(\mathbf{0},R)}(\boldsymbol{\psi})\big)\Big) \\ = E^R_{\boldsymbol{\xi}}(t)M^R_{\boldsymbol{\xi}}(t) - \int_0^{t\wedge\zeta^{B(\mathbf{0},R)}(\boldsymbol{\psi})} M^R_{\boldsymbol{\xi}}(\tau,\boldsymbol{\psi})f(\boldsymbol{\psi}(\tau))E^R_{\boldsymbol{\xi}}(\tau,\boldsymbol{\psi})\,d\tau.\end{aligned}$$

Hence

$$\left(\exp\Big(\sqrt{-1}\big(\boldsymbol{\xi},\mathbf{B}(t\wedge\zeta^{B(\mathbf{0},R)}(\boldsymbol{\psi}))\big)_{\mathbb{R}^N} + \frac{|\boldsymbol{\xi}|^2}{2}\big(t\wedge\zeta^{B(\mathbf{0},R)}(\boldsymbol{\psi})\big)\Big), \mathcal{F}_t, \mathbb{P}\right)$$

is a martingale for every $R > 0$, and so, after letting $R \to \infty$, we know, by Theorem 7.1.7, that $\big(B(t), \mathcal{F}_t, \mathbb{P}\big)$ is a Brownian motion. □

It is important to be clear about what Lemma 10.3.28 says and what it does not say. It says that there is a progressively measurable $\mathbf{B} : [0,\infty) \times C(\mathbb{R}^N) \longrightarrow \mathbb{R}^N$ such that $\big(\mathbf{B}(t), \mathcal{F}_t, \mathbb{P}\big)$ is a Brownian motion and

$$(*) \qquad \boldsymbol{\psi}(t) = \mathbf{x} + \mathbf{B}(t,\boldsymbol{\psi}) + \int_0^t \mathbf{b}\big(\boldsymbol{\psi}(\tau)\big)\,d\tau, \quad (t,\boldsymbol{\psi}) \in [0,\infty) \times C(\mathbb{R}^N).$$

In the probabilistic literature, this would be summarized by saying that $\mathbb{P}$ is the distribution of a Brownian motion with **drift** $\mathbf{b}$. What Lemma 10.3.28 does *not* say is that one can always use (*) to reconstruct $\boldsymbol{\psi}$ from $\mathbf{B}(\,\cdot\,,\boldsymbol{\psi})$. More precisely, $\boldsymbol{\psi}$ is not necessarily a measurable function of $\mathbf{B}(\,\cdot\,,\boldsymbol{\psi})$. Indeed, without additional assumptions on $\mathbf{b}$, $\boldsymbol{\psi}$ will *not* be a measurable function of $\mathbf{B}$. Nonetheless, if $\mathbf{b}$ is locally Lipschitz continuous, then it will be. To see this, take $\eta \in C^\infty_c\big(\mathbb{R}^N;[0,1]\big)$ so that $\eta = 1$ on $B(\mathbf{0},2)$ and 0 off of $B(\mathbf{0},3)$, and set $\mathbf{b}^R(\mathbf{y}) = \eta\big(\frac{\mathbf{y}}{R}\big)\mathbf{b}(\mathbf{y})$. Then $\mathbf{b}^R$ is uniformly Lipschitz continuous, and so, by completely standard methods (e.g., Picard iteration), one can show that there is a continuous map $\boldsymbol{\varphi} \in C(\mathbb{R}^N) \longmapsto \mathbf{X}^R(\,\cdot\,,\boldsymbol{\varphi}) \in C(\mathbb{R}^N)$ such that, for each $\boldsymbol{\varphi} \in C(\mathbb{R}^N)$,

$$\mathbf{X}^R(t,\boldsymbol{\varphi}) = \boldsymbol{\varphi}(t) + \int_0^t \mathbf{b}^R\big(\mathbf{X}^R(\tau,\boldsymbol{\varphi})\big)\,d\tau, \quad t \ge 0.$$

Moreover, if $\boldsymbol{\psi} \in C(\mathbb{R}^N)$ and

$$\boldsymbol{\psi}(t) = \boldsymbol{\varphi}(t) + \int_0^t \mathbf{b}^R\big(\boldsymbol{\psi}(\tau)\big)\,d\tau, \quad t \in [0,T],$$

then $\psi \restriction [0,T] = \mathbf{X}^R(\,\cdot\,,\varphi) \restriction [0,T]$. Hence, if

$$A(\mathbf{b}) = \{\varphi \in C(\mathbb{R}^N) : \forall t \geq 0\; \exists R > 0\; \|\mathbf{X}^R(\,\cdot\,,\varphi)\|_{[0,t]} \leq R\},$$

then $A(\mathbf{b}) \in \mathcal{B}_{C(\mathbb{R}^N)}$, and I can define the Borel measurable map $\varphi \in C(\mathbb{R}^N) \longmapsto \mathbf{X}_{\mathbf{b}}(\,\cdot\,,\varphi) \in C(\mathbb{R}^N)$ given by

$$\mathbf{X}_{\mathbf{b}}(t,\varphi) = \begin{cases} \mathbf{X}^R(t,\varphi) & \text{if } \varphi \in A(\mathbf{b}) \text{ and } \|\mathbf{X}^R(\,\cdot\,,\varphi)\|_{[0,t]} \leq R \\ \varphi(t) & \text{if } \varphi \notin A(\mathbf{b}). \end{cases}$$

In particular, when $\mathbf{b}$ is locally Lipschitz continuous, Lemma 10.3.28 says that $\mathbf{x}+\mathbf{B}(\,\cdot\,,\psi) \in A(\mathbf{b})$ and $\psi(t) = \mathbf{X}_{\mathbf{b}}\big(t, \mathbf{x}+\mathbf{B}(\,\cdot\,,\psi)\big)$ for all $(t,\psi) \in [0,\infty)\times C(\mathbb{R}^N)$.

COROLLARY 10.3.29. *Let everything be as in Corollary 10.3.27, $\mathbf{b}^\rho = \nabla \log \rho$, and define the set $A(\mathbf{b}^\rho)$ and the map $\mathbf{X}_{\mathbf{b}^\rho}$ accordingly, as in the preceding discussion. Then $\mathcal{W}^{(N)}_{\mathbf{x}}\big(A(\mathbf{b}^\rho)\big) = 1$ and $\mathbb{P}^\rho_{\mathbf{x}} = (\mathbf{X}_{\mathbf{b}^\rho})_*\mathcal{W}^{(N)}_{\mathbf{x}}$ for all $\mathbf{x} \in \mathbb{R}^N$.*

PROOF: Define $\psi \rightsquigarrow \mathbf{B}(\,\cdot\,,\psi)$ in terms of $\mathbf{b}^\rho$ as in Corollary 10.3.27. Then, by that corollary, we know that $\mathcal{W}^{(N)}_{\mathbf{x}}$ is the distribution of $\psi \rightsquigarrow \mathbf{x}+\mathbf{B}(\,\cdot\,,\psi)$ under $\mathbb{P}^\rho_{\mathbf{x}}$. Therefore, since $\mathbf{x}+\mathbf{B}(\,\cdot\,,\psi) \in A(\mathbf{b}^\rho)$ and $\psi(t) = \mathbf{X}_{\mathbf{b}^\rho}\big(t,\mathbf{x}+\mathbf{B}(\,\cdot\,,\psi)\big)$ for all $(t,\psi) \in [0,\infty) \times C(\mathbb{R}^N)$, the desired conclusions follow immediately. □

**§ 10.3.5. Producing Ground States.** As yet I have not addressed the problem of producing ground states. In this subsection I will provide two approaches. The first of these gives a criterion that guarantees the existence of a ground state for a given $V$. The second goes in the opposite direction. It is the essentially trivial remark that there are many $\rho \in C^2\big(\mathbb{R}^N;(0,\infty)\big)$ such that $\rho$ is the ground state of some $V$.

The first approach is an application of elementary spectral theory and is based on the observation that, because $q^V(t,\mathbf{x},\mathbf{y}) = q^V(t,\mathbf{y},\mathbf{x})$, $\mathbf{Q}^V_t$ is symmetric on $L^2(\mathbb{R}^N;\mathbb{R})$ in the sense that

$$\text{(10.3.30)} \qquad \big(\varphi_1, \mathbf{Q}^V_t\varphi_2\big)_{L^2(\mathbb{R}^N;\mathbb{R})} = \big(\varphi_2, \mathbf{Q}^V_t\varphi_1\big)_{L^2(\mathbb{R}^N;\mathbb{R})}$$

for all $\varphi_1,\ \varphi_2 \in C_{\mathrm{c}}(\mathbb{R}^N;\mathbb{R})$.

The fact that $\mathbf{Q}^V_t$ is symmetric on $L^2(\mathbb{R}^N;\mathbb{R})$ has profound implications, a few of which are contained in the following lemma.

LEMMA 10.3.31. *For each $q \in [1,\infty)$ and $t \in (0,\infty)$, $\mathbf{Q}^V_t \restriction C^\infty_{\mathrm{c}}(\mathbb{R}^N;\mathbb{R})$ admits a unique extension (which I again denote by $\mathbf{Q}^V_t$) as a bounded linear operator on $L^q(\mathbb{R}^N;\mathbb{R})$ into itself with norm at most $e^{t\|V^+\|_{\mathrm{u}}}$. Moreover, for each $t > 0$, $\mathbf{Q}^V_t$ is non-negative definite and self-adjoint on $L^2(\mathbb{R}^N;\mathbb{R})$, and, for each $q \in [1,\infty)$, $\mathbf{Q}^V_t$ takes $L^q(\mathbb{R}^N;\mathbb{R})$ into $C_{\mathrm{b}}(\mathbb{R}^N;\mathbb{R})$ and*

$$\|\mathbf{Q}^V_t\varphi(\mathbf{x})\|_{\mathrm{u}} \leq (2\pi t)^{-\frac{N}{2q}} e^{t\|V^+\|_{\mathrm{u}}} \|\varphi\|_{L^q(\mathbb{R}^N;\mathbb{R})}.$$

*Finally,*

$$\iint_{\mathbb{R}^N\times\mathbb{R}^N} q^V(t,\mathbf{x},\mathbf{y})^2\,d\mathbf{x}\,d\mathbf{y} = \int_{\mathbb{R}^N} q^V(2t,\mathbf{x},\mathbf{x})\,d\mathbf{x} \le (4\pi t)^{-\frac{N}{2}}\int_{\mathbb{R}^N} e^{2tV(\mathbf{x})}\,d\mathbf{x}.$$

PROOF: Given $q\in[1,\infty)$ and a Borel measurable $\varphi:\mathbb{R}^N\longrightarrow[0,\infty)$, we have, by Jensen's Inequality, that

$$\begin{aligned}(\mathbf{Q}_t^V\varphi(\mathbf{x}))^q &= \mathbb{E}^{\mathcal{W}_{\mathbf{x}}^{(N)}}\left[e^{\int_0^t V(\boldsymbol{\psi}(\tau))\,d\tau}\varphi(\boldsymbol{\psi}(t))\right]^q\\ &\le \mathbb{E}^{\mathcal{W}_{\mathbf{x}}^{(N)}}\left[e^{q\int_0^t V(\boldsymbol{\psi}(\tau))\,d\tau}\varphi(\boldsymbol{\psi}(t))^q\right] \le e^{qt\|V^+\|_{\mathrm{u}}}\int_{\mathbb{R}^N}\varphi(\mathbf{y})^q g^{(N)}(t,\mathbf{y}-\mathbf{x})\,d\mathbf{y}.\end{aligned}$$

Hence, since $g^{(N)}(t,\,\cdot\,)$ has $L^1(\mathbb{R}^N;\mathbb{R})$ norm 1,

$$\|\mathbf{Q}_t^V\varphi\|_{L^q(\mathbb{R}^N;\mathbb{R})} \le e^{t\|V^+\|_{\mathrm{u}}}\|\varphi\|_{L^q(\mathbb{R}^N;\mathbb{R})},$$

and so we have proved the first assertion. In addition, if $q'$ is the Hölder conjugate of $q$, then

$$\|\mathbf{Q}_t^V\varphi\|_{\mathrm{u}} \le e^{t\|V^+\|_{\mathrm{u}}}\|g^{(N)}(t,\,\cdot\,)\|_{L^{q'}(\mathbb{R}^N;\mathbb{R})}\|\varphi\|_{L^q(\mathbb{R}^N;\mathbb{R})} \le \frac{e^{t\|V^+\|_{\mathrm{u}}}}{(2\pi t)^{-\frac{N}{2q}}}\|\varphi\|_{L^q(\mathbb{R}^N;\mathbb{R})}.$$

Thus, since $\mathbf{Q}_t^V$ maps $C_{\mathrm{c}}^\infty(\mathbb{R}^N;\mathbb{R})$ into $C_{\mathrm{b}}(\mathbb{R}^N;\mathbb{R})$, it also takes $L^q(\mathbb{R}^N;\mathbb{R})$ there.

Because (10.3.30) holds for elements of $C_{\mathrm{c}}(\mathbb{R}^N;\mathbb{R})$, the preceding estimates make it clear that it continues to hold for elements of $L^2(\mathbb{R}^N;\mathbb{R})$. That is, $\mathbf{Q}_t^V$ is self-adjoint on $L^2(\mathbb{R}^N;\mathbb{R})$. To see that it is non-negative definite, simply observe that

$$\left(\varphi,\mathbf{Q}_t^V\varphi\right)_{L^2(\mathbb{R}^N;\mathbb{R})} = \left(\mathbf{Q}_{\frac{t}{2}}^V\varphi,\mathbf{Q}_{\frac{t}{2}}^V\varphi\right)_{L^2(\mathbb{R}^N;\mathbb{R})} \ge 0.$$

Turning to the final estimate, note that (cf. (10.3.17))

$$\int_{\mathbb{R}^N} q^V(t,\mathbf{x},\mathbf{y})^2\,d\mathbf{y} = \int_{\mathbb{R}^N} q^V(t,\mathbf{x},\mathbf{y})q^V(t,\mathbf{y},\mathbf{x})\,d\mathbf{y} = q^V(2t,\mathbf{x},\mathbf{x}).$$

At the same time, by Jensen's Inequality,

$$\begin{aligned}q^V(2t,\mathbf{x},\mathbf{x}) &= \mathbb{E}^{\mathcal{W}^{(N)}}\left[e^{\int_0^{2t} V(\mathbf{x}+\boldsymbol{\psi}_{2t}(\tau))\,d\tau}\right]g^{(N)}(2t,\mathbf{0})\\ &\le (4\pi t)^{-\frac{N}{2}}\frac{1}{2t}\int_0^{2t}\mathbb{E}^{\mathcal{W}^{(N)}}\left[e^{2tV(\mathbf{x}+\boldsymbol{\psi}_{2t}(\tau))}\right]d\tau,\end{aligned}$$

and, by Tonelli's Theorem,

$$\int_{\mathbb{R}^N}\mathbb{E}^{\mathcal{W}^{(N)}}\left[e^{2tV(\mathbf{x}+\boldsymbol{\psi}(\tau))}\right]d\mathbf{x} = \int_{\mathbb{R}^N} e^{2tV(\mathbf{x})}\,d\mathbf{x}. \quad \square$$

In the language of functional analysis, the last part of Lemma 10.3.31 says that $\mathbf{Q}_T^V$ is **Hilbert–Schmidt** and therefore compact if $e^{2TV}\in L^1(\mathbb{R}^N;\mathbb{R})$. As a consequence, the elementary theory of compact, self-adjoint operators allows us to make the conclusions drawn in the following theorem.

THEOREM 10.3.32. *Assume that $e^{TV} \in L^2(\mathbb{R}^N;\mathbb{R})$ for some $T \in (0,\infty)$. Then there is a unique $\rho \in C_b\big(\mathbb{R}^N;(0,\infty)\big) \cap L^2(\mathbb{R}^N;\mathbb{R})$ such that*

$$\|\rho\|_{L^2(\mathbb{R}^N;\mathbb{R})} = 1 \quad \text{and} \quad e^{t\lambda}\rho = \mathbf{Q}_t^V \rho \text{ for some } \lambda \in \mathbb{R} \text{ and all } t \in (0,\infty).$$

*Moreover, if $V \in C^2(\mathbb{R}^N;\mathbb{R})$ satisfies (10.3.23), then $p^\rho(t,\,\cdot\,,\mathbf{y}) \in C^2(\mathbb{R}^N;\mathbb{R})$ and*

$$\partial_t p^\rho(t,\mathbf{x},\mathbf{y}) = \tfrac{1}{2}\Delta_{\mathbf{x}} p^\rho(t,\mathbf{x},\mathbf{y}) + \big(\nabla \log\rho(\mathbf{x}), \nabla_{\mathbf{x}} p^\rho(t,\mathbf{x},\mathbf{y})\big)_{\mathbb{R}^N} \text{ in } (0,\infty)\times\mathbb{R}^N\times\mathbb{R}^N.$$

PROOF: The spectral theory of compact, self-adjoint operators guarantees that the operator $\mathbf{Q}_T^V$ has a completely discrete spectrum and that its largest eigenvalue is

$$\alpha(T) = \sup\{(\varphi, \mathbf{Q}_T^V\varphi)_{L^2(\mathbb{R}^N;\mathbb{R})} : \|\varphi\|_{L^2(\mathbb{R}^N;\mathbb{R})} = 1\}.$$

Now let $\rho$ be an $L^2(\mathbb{R}^N;\mathbb{R})$-normalized eigenvector for $\mathbf{Q}_T^V$ with eigenvalue $\alpha(T)$. Because $\alpha(T)\rho = \mathbf{Q}_T^V\rho$, we know that $\rho$ can be taken to be continuous. In addition, by the preceding paragraph,

$$\iint_{\mathbb{R}^N\times\mathbb{R}^N} \rho(\mathbf{x})q^V(T,\mathbf{x},\mathbf{y})\rho(\mathbf{y})\,d\mathbf{x}\,d\mathbf{y} = \alpha(T) \geq \iint_{\mathbb{R}^N\times\mathbb{R}^N} |\rho(\mathbf{x})|q^V(T,\mathbf{x},\mathbf{y})|\rho(\mathbf{y})|\,d\mathbf{x}\,d\mathbf{y},$$

which, because $q^V(T,\mathbf{x},\mathbf{y}) > 0$ for all $(\mathbf{x},\mathbf{y})$, is possible only if $\alpha(T) > 0$ and $\rho$ never changes sign. Therefore we can be take $\rho$ to be non-negative. But, if $\rho \geq 0$, then, since $p^\rho(T,\mathbf{x},\mathbf{y}) > 0$ everywhere and $\alpha(T)\rho = \mathbf{Q}_T^V\rho$, $\rho > 0$ everywhere. Thus, we have now shown that every normalized eigenvector for $\mathbf{Q}_T^V$ with eigenvalue $\alpha(T)$ is a bounded, continuous function that, after a change of sign, can be taken to be strictly positive. In particular, if $\rho_1$ and $\rho_2$ were linearly independent, normalized eigenvectors of $\mathbf{Q}_T^V$ with eigenvalue $\alpha(T)$, then

$$g = \frac{\rho_2 - (\rho_1,\rho_2)_{L^2(\mathbb{R}^N;\mathbb{R})}\rho_1}{\|\rho_2 - (\rho_1,\rho_2)_{L^2(\mathbb{R}^N;\mathbb{R})}\rho_1\|_{L^2(\mathbb{R}^N;\mathbb{R})}}$$

would also be such an eigenvector, and this one would be orthogonal to $\rho_1$. On the other hand, since neither $\rho_1$ nor $g$ changes sign, $(\rho_1, g)_{L^2(\mathbb{R}^N;\mathbb{R})} \neq 0$. In summary, we now know that there is, up to sign, a unique $L^2(\mathbb{R}^N;\mathbb{R})$-normalized eigenvector $\rho$ for $\mathbf{Q}_T^V$ with eigenvalue $\alpha(T)$ and that $\rho$ can be taken to be strictly positive, bounded, and continuous.

To complete the proof, I must show that $\mathbf{Q}_t^V\rho = e^{t\lambda}\rho$, where $\lambda = \frac{1}{T}\log\alpha(T)$. To this end, set $\rho_t = \mathbf{Q}_t^V\rho$ for $t > 0$. Then $\rho_t \in C_b\big(\mathbb{R}^N;(0,\infty)\big)$ for each $t > 0$ and $t \rightsquigarrow \rho_t(\mathbf{x})$ is continuous for each $\mathbf{x} \in \mathbb{R}^N$. Moreover, $\mathbf{Q}_T^V\rho_t = \mathbf{Q}_t^V \circ \mathbf{Q}_T^V\rho = \alpha(T)\rho_t$. Hence, by the uniqueness proved above, $\rho_t = \alpha(t)\rho$ for some $\alpha(t) \in \mathbb{R}$. In addition, because $t \rightsquigarrow \rho_t(\mathbf{x})$ is continuous and strictly positive, so is $t \rightsquigarrow \alpha(t)$. Finally,

$$\alpha(s+t) = \big(\rho, \mathbf{Q}^V_{s+t}\rho\big)_{L^2(\mathbb{R}^N;\mathbb{R})} = \alpha(s)\big(\rho, \mathbf{Q}^V_t \rho\big)_{L^2(\mathbb{R}^N;\mathbb{R})} = \alpha(s)\alpha(t),$$

which means that $\alpha(t) = e^{t\beta}$ for some $\beta \in \mathbb{R}$, and, because $\alpha(T) = e^{T\lambda}$, this completes the proof of everything except the final statement, which is an immediate consequence of Theorem 10.3.21. □

If nothing else, Theorem 10.3.32 helps to explain the terminology that I have been using. In Schrödinger mechanics, the function $\rho$ in Theorem 10.3.32 is called the **ground state** because it is the wave function corresponding to the lowest energy level of the quantum mechanical Hamiltonian $-\frac{1}{2}\Delta - V$. From our standpoint, its importance is that it shows that lots of $V$'s admit a ground state.

I turn now to the second method for producing ground states. Namely, suppose that $\rho \in C^2\big(\mathbb{R}^N; (0,\infty)\big)$. Then, it is obvious that $\frac{1}{2}\Delta\rho + V\rho = 0$, where

$$V = -\frac{\Delta\rho}{2\rho} = -\frac{\Delta \log\rho + |\nabla \log\rho|^2}{2}.$$

THEOREM 10.3.33. *Let $U \in C^2(\mathbb{R}^N;\mathbb{R})$, and assume that both $U$ and $V^U \equiv -\frac{1}{2}\big(\Delta U + |\nabla U|^2\big)$ are bounded above. Then, for each $\mathbf{x} \in \mathbb{R}^N$, there is a unique $\mathbb{P}^U_{\mathbf{x}} \in \mathbf{M}_1\big(C(\mathbb{R}^N)\big)$ such that $\mathbb{P}^U_{\mathbf{x}}\big(\psi(0) = \mathbf{x}\big) = 1$ and*

$$\left(\varphi(\psi(t)) - \frac{1}{2}\int_0^t \Big(\Delta\varphi + (\nabla U, \nabla\varphi)_{\mathbb{R}^N}\Big)(\psi(\tau))\,d\tau, \mathcal{F}_t, \mathbb{P}^U_{\mathbf{x}}\right)$$

*is a martingale for all $\varphi \in C^\infty_{\mathrm{c}}(\mathbb{R}^N;\mathbb{R})$. Moreover, for each $\mathbf{x} \in \mathbb{R}^N$,*

$$\left(\psi(t) - \mathbf{x} - \int_0^t \nabla U\big(\psi(\tau)\big)\,d\tau, \mathcal{F}_t, \mathbb{P}^U_{\mathbf{x}}\right)$$

*is a Brownian motion and*

$$\mathbb{P}^U_{\mathbf{x}}(A) = e^{-U(\mathbf{x})}\mathbb{E}^{\mathcal{W}^{(N)}_{\mathbf{x}}}\Big[e^{U((\psi(t)) + \int_0^t V^U(\psi(\tau))\,d\tau}, A\Big] \quad \text{for all } t \ge 0 \text{ and } A \in \mathcal{F}_t.$$

*Finally, $\mathbf{x} \rightsquigarrow \mathbb{P}^U_{\mathbf{x}}$ is continuous and, for any stopping time $\zeta$ and any $\mathcal{F}_\zeta \times \mathcal{B}_{C(\mathbb{R}^N)}$-measurable $F : C(\mathbb{R}^N) \times C(\mathbb{R}^N)$ that is bounded below,*

$$\int_{\{\zeta(\psi)<\infty\}} F(\psi, \Sigma_\zeta\psi)\,\mathbb{P}^U_{\mathbf{x}}(d\psi) = \int_{\{\zeta(\psi)<\infty\}} \left(\int F(\psi,\psi')\,\mathbb{P}^U_{\psi(\zeta)}(d\psi')\right)\mathbb{P}^U_{\mathbf{x}}(d\psi).$$

PROOF: By Lemma 10.3.24, $\rho \equiv e^U$ is a ground state for $V^U$ with associated eigenvalue 0. Thus, the existence of $\mathbb{P}^U_{\mathbf{x}}$ follows immediately from Theorem 10.3.26 with $\rho = e^U$, as does the relation between this choice of $\mathbb{P}^U_{\mathbf{x}}$ and $\mathcal{W}^{(N)}_{\mathbf{x}}$ as well as the Markov property in the final statement. Moreover, by Lemma 10.3.28, we know that any $\mathbb{P}$ satisfying the initial condition and the stated martingale property is related to Brownian motion in the stated way. Therefore, all that remains is to show that this relationship to Brownian motion determines $\mathbb{P}$. But, by Corollary 10.3.29 with $\rho = e^U$, we know that such a $\mathbb{P}$ equals $(\mathbf{X}_{\nabla U})_*\mathcal{W}^{(N)}_{\mathbf{x}}$, where $\mathbf{X}_{\nabla U}$ is the mapping described in the paragraph preceding that corollary. □

## Exercises for § 10.3

EXERCISE 10.3.34. Given $\alpha$, $\beta \in (0,\infty)$ and $\mathbf{a}$, $\mathbf{b} \in \mathbb{R}^N$, show that

$$\int_{\mathbb{R}^N} g^{(N)}(s, \alpha\boldsymbol{\eta} + \mathbf{a}) g^{(N)}(t, \beta\boldsymbol{\eta} + \mathbf{b})\, d\boldsymbol{\eta} = g^{(N)}(\beta^2 s + \alpha^2 t, \alpha\mathbf{b} - \beta\mathbf{a}).$$

**Hint**: Note that

$$g^{(N)}(s, \alpha\boldsymbol{\eta} + \mathbf{a}) g^{(N)}(t, \beta\boldsymbol{\eta} + \mathbf{b}) = \frac{1}{\alpha\beta} g^{(N)}\big(\tfrac{s}{\alpha^2}, \boldsymbol{\eta} + \tfrac{\mathbf{a}}{\alpha}\big) g^{(N)}\big(\tfrac{t}{\beta^2}, \boldsymbol{\eta} + \tfrac{\mathbf{b}}{\beta}\big).$$

EXERCISE 10.3.35. When $N = 1$, the considerations in § 7.2.2 can be used to give a reasonably explicit formula for $p^G(t, x, y)$. Namely, show that

$$p^{(0,\infty)}(t, x, y) = g^{(1)}(t, y - x) - g^{(1)}(t, x + y) \quad \text{for } (t, x, y) \in (0,\infty) \times (0,\infty)^2,$$

where $g^{(1)}(\tau, \xi) = (2\pi\tau)^{-\frac{1}{2}} e^{-\frac{\xi^2}{2\tau}}$. In addition, referring to Corollary 7.3.4, show that, for $c \in \mathbb{R}$, $r > 0$, and $(x, y) \in (c - r, c + r)$,

$$p^{(c-r,c+r)}(t, x, y) = r^{-1}\tilde{g}^{(1)}\big(r^{-2}t, r^{-1}(y - x)\big) - r^{-1}\tilde{g}^{(1)}\big(r^{-2}t, r^{-1}(x + y + 2 - 2c)\big)\big),$$

where $\tilde{g}(\tau, \xi) = \sum_{m\in\mathbb{Z}} g^{(1)}(t, x + 4m)$.

EXERCISE 10.3.36. Set $Q(\mathbf{a}, R) = \prod_{i=1}^N [a_i - R, a_i + R]$ for $\mathbf{a} \in \mathbb{R}^N$ and $R > 0$. Show that

$$\int_{Q(\mathbf{a},R)} p^{Q(\mathbf{a},R)}(t, \mathbf{x}, \mathbf{y}) \prod_{i=1}^N \sin \frac{\pi(y_i - a_i + R)}{2R}\, d\mathbf{y} = e^{-\frac{N\pi^2}{8R^2}t} \prod_{i=1}^N \sin \frac{\pi(x_i - a_i + R)}{2R}$$

for $(t, \mathbf{x}, \mathbf{y}) \in (0,\infty) \times Q(\mathbf{a}, R)^2$. Conclude that

$$\lim_{t\to\infty} \frac{1}{t} \log \mathcal{W}^{(N)}_{\mathbf{x}}(\zeta^{Q(\mathbf{a},R)} > t) = -\frac{N\pi^2}{8R^2} \quad \text{for } \mathbf{x} \in Q(\mathbf{a}, R).$$

**Hint**: First observe that it suffices to handle $\mathbf{a} = \mathbf{0}$, $R = 1$, and $N = 1$. To prove the first part, set $u(t,x) = e^{\frac{\pi^2 t}{8}} \sin \frac{\pi(x+1)}{2}$, and show that $\left(u(t,\psi(t)), \mathcal{F}_t, \mathcal{W}_x^{(1)}\right)$ is a martingale. Given the first part, $\underline{\lim}_{t\to\infty} \frac{1}{t} \log \mathcal{W}_x^{(1)}(\zeta^{(-1,1)} > t) \geq -\frac{\pi^2}{8}$ is clear. To get the inequality in the opposite direction, note that $p^{(-1,1)}(t,x,y) \leq p^{(-R,R)}(t,x,y)$ if $R > 1$, and use this to see that, for $R > 1$ and $(t,x) \in (0,\infty) \times (-1,1)$,

$$\int_{(-1,1)} p^{(-1,1)}(t,x,y) \sin \frac{\pi(y+R)}{2R}\, dy \leq e^{-\frac{\pi^2}{8R^2}t} \sin \frac{\pi(x+R)}{2R}.$$

EXERCISE 10.3.37. Let $G$ be a non-empty, bounded, connected, open subset of $\mathbb{R}^N$, and set $w(t) = \sup_{\mathbf{x}\in G} \mathcal{W}_{\mathbf{x}}^{(N)}(\zeta^G > t)$ for $t > 0$. The purpose of this exercise is to show that $\lambda^G \equiv -\lim_{t\to\infty} \frac{1}{t} \log w(t)$ exists and is an element of $(0,\infty)$.

**(i)** Show that $w$ is sub-multiplicative in the sense that $w(s+t) \leq w(s)w(t)$, and conclude from this that $\lim_{t\to\infty} \frac{1}{t} \log w(t) = \sup_{T>0} \frac{1}{T} f(T) \in [-\infty, 0]$.
**Hint**: Set $f(t) = \log w(t)$. Because $w$ takes values in $(0,1]$ and is non-increasing, $f$ is non-positive and bounded on compacts. Further, $f$ is sub-additive: $f(s+t) \leq f(s)+f(t)$. Thus, given, $T > 0$, $f(t) \leq \left\lfloor \frac{t}{T} \right\rfloor f(T)$, and so $\overline{\lim}_{t\to\infty} \frac{1}{t} f(t) \leq \frac{1}{T} f(T)$ for every $T > 0$. Conclude from this that $\lim_{t\to\infty} \frac{1}{t} f(t) = \sup_{T>0} \frac{1}{T} f(T) \in [-\infty, 0]$.

**(ii)** Refer to the notation in Exercise 10.3.36, set $R_1 = \sup\{r \geq 0 : Q(\mathbf{a}, r) \subseteq G \text{ for some } \mathbf{a} \in G\}$, and show that $\lambda^G \equiv -\lim_{t\to\infty} \frac{1}{t} \log w(t) \leq \frac{N\pi^2}{8R_1^2}$. In particular, $\lambda^G < \infty$.

**(iii)** Let $R_2$ be the diameter of $G$, choose $\mathbf{a} \in \mathbb{R}^N$ so that $G \subseteq B(\mathbf{a}, R_2)$, and use the first part of Theorem 10.1.11 to show that $\mathbb{E}^{\mathcal{W}_{\mathbf{x}}^{(N)}}[\zeta^G] \leq \frac{R_2^2}{N}$ for all $\mathbf{x} \in G$. In particular, conclude that $w(2N^{-1}R_2^2) \leq \frac{1}{2}$ and therefore that $\lambda^G \geq \frac{N \log 2}{2R_2^2} > 0$.

EXERCISE 10.3.38. Again let $G$ be a bounded, connected, open subset of $\mathbb{R}^N$. Using spectral theory, the conclusions drawn in Exercise 10.3.37 can be sharpened. Namely, this exercise outlines a proof that

$$p^G(t,\mathbf{x},\mathbf{y}) = \sum_{n=0}^{\infty} e^{-t\lambda_n} \varphi_n(\mathbf{x})\varphi_n(\mathbf{y}), \tag{10.3.39}$$

where $\{\lambda_n : n \geq 0\} \subseteq (0,\infty)$ is a non-decreasing sequence that tends to $\infty$, $\{\varphi_n : n \geq 0\} \subseteq C_b(G;\mathbb{R})$ is an orthonormal basis in $L^2(G;\mathbb{R})$ of smooth functions, $\lambda_0 < \lambda_1$, $\varphi_0 > 0$, and the convergence is uniform on $[\epsilon,\infty) \times G^2$ for each $\epsilon > 0$. Finally, from (10.3.39), it will follow that

$$\left|e^{t\lambda_0} p(t,\mathbf{x},\mathbf{y}) - \varphi_0(\mathbf{x})\varphi_0(\mathbf{y})\right| \leq \delta^{-1} e^{-t\delta}, \quad (t,\mathbf{x},\mathbf{y}) \in [1,\infty) \times G^2,$$

for some $\delta > 0$. In particular, this means that $\lambda_0$ here is equal to $\lambda^G$ in Exercise 10.3.37.

**(i)** Let $\mathbf{P}_t^G$ be the operator on $C_b(G;\mathbb{R})$ whose kernel is $p^G(t,\mathbf{x},\mathbf{y})$, and show that $\mathbf{P}_t^G$ admits a unique extension to $L^2(G;\mathbb{R})$ as a self-adjoint contraction. Further, show that $\{\mathbf{P}_t^G : t > 0\}$ is a continuous semigroup of non-negative definite, self-adjoint contractions on $L^2(G;\mathbb{R})$. Finally, show that

$$\iint_{G\times G} p^G(t,\mathbf{x},\mathbf{y})^2\,d\mathbf{x}d\mathbf{y} = \int_G p^G(2t,\mathbf{x},\mathbf{x})\,d\mathbf{x} \le \frac{|G|}{(4\pi t)^{\frac{N}{2}}},$$

and therefore that each $\mathbf{P}_t^G$ is Hilbert–Schmidt.

**(ii)** Knowing that the operators $\mathbf{P}_t^G$ form a continuous semigroup of self-adjoint, Hilbert–Schmidt (and therefore compact), non-negative definite contractions, standard spectral theory[2] guarantees that there exists a non-decreasing sequence $\{\lambda_n : n \ge 0\} \subseteq [0,\infty)$ tending to $\infty$ and an orthonormal basis $\{\varphi_n : n \ge 0\}$ in $L^2(G;\mathbb{R})$ such that $e^{-t\lambda_n}\varphi_n = \mathbf{P}_t^G\varphi_n$ for all $t \in (0,\infty)$ and $n \ge 0$. Conclude from this that $\varphi_n$ can be taken to be smooth and bounded. In addition, show that $\mathbf{P}_t^G\varphi_0 \longrightarrow 0$ uniformly, and therefore that $\lambda_0 > 0$.

**(iii)** Show that

$$(*) \qquad \big(\varphi, \mathbf{P}_t^G\varphi'\big)_{L^2(G;\mathbb{R})} = \sum_{n=0}^{\infty} e^{-t\lambda_n}\big(\varphi,\varphi_n\big)_{L^2(G;\mathbb{R})}\big(\varphi',\varphi_n\big)_{L^2(G;\mathbb{R})}$$

for $\varphi, \varphi' \in L^2(G;\mathbb{R})$, and conclude that

$$e^{-\lambda_0} = \sup\big\{\big(\varphi, \mathbf{P}_1^G\varphi\big)_{L^2(G;\mathbb{R})} : \|\varphi\|_{L^2(G;\mathbb{R})} = 1\big\}.$$

Use (cf. the proof of Theorem 10.3.32) this to show that if $\lambda_n = \lambda_0$, then $\varphi_n$ never changes sign and can therefore be taken to be non-negative. In particular, show that this means that $\lambda_1 > \lambda_0$ and that $\varphi_0 > 0$.

**(iv)** Starting from (*), show that

$$\sum_{n=0}^{\infty} e^{-t\lambda_n}\big(\varphi,\varphi_n\big)^2_{L^2(G;\mathbb{R})} = \int_{G\times G} \varphi(\mathbf{x})p^G(t,\mathbf{x},\mathbf{y})\varphi(\mathbf{y})\,d\mathbf{x}d\mathbf{y} \le (2\pi t)^{-\frac{N}{2}}\|\varphi\|^2_{L^1(G;\mathbb{R})}$$

[2] What is needed here is the variant of Stone's Theorem that applies to semigroups. The technical question which his theorem addresses is that of finding a simultaneous diagonalization of the operators $\mathbf{P}_t^G$. Because we are dealing here with compact operators, this question can be reduced to one about operators in finite dimensions, where it is quite easy to handle. For a general statement, see, for example, K. Yoshida's *Functional Analysis and its Applications*, Springer-Verlag (1971).

for any $\varphi \in L^2(G;\mathbb{R})$, and use this to show that, for any $M \in \mathbb{N}$ and $\varphi, \varphi' \in L^2(G;\mathbb{R})$,

$$\sum_{n=M}^{\infty} e^{-t\lambda_n}\left|\left(\varphi,\varphi_n\right)_{L^2(G;\mathbb{R})}\right|\left|\left(\varphi',\varphi_n\right)_{L^2(G;\mathbb{R})}\right| \le \frac{e^{-\frac{t\lambda_M}{2}}}{(\pi t)^{\frac{N}{2}}}\|\varphi\|_{L^1(G;\mathbb{R})}\|\varphi'\|_{L^1(G;\mathbb{R})}.$$

Next, given $\mathbf{x}, \mathbf{y} \in G$, set $R = |\mathbf{x} - \partial G| \wedge |\mathbf{y} - \partial G|$, and, for $0 < r \le R$, apply the preceding to see that

$$\sum_{n=M}^{\infty} e^{-t\lambda_n}\left|\fint_{B(\mathbf{x},r)} \varphi_n(\mathbf{z})\,d\mathbf{z}\right|\left|\fint_{B(\mathbf{y},r)} \varphi_n(\mathbf{z})\,d\mathbf{z}\right| \le \frac{e^{-\frac{t\lambda_M}{2}}}{(\pi t)^{\frac{N}{2}}}.$$

Finally, by combining this with (*), reach the conclusion that

$$\left|p^G(t,\mathbf{x},\mathbf{y}) - \sum_{n=0}^{M-1} e^{-t\lambda_n}\varphi_n(\mathbf{x})\varphi_n(\mathbf{y})\right| \le \frac{e^{-\frac{t\lambda_M}{2}}}{(\pi t)^{\frac{N}{2}}},$$

which, because $\lambda_M \longrightarrow \infty$, certainly implies the asserted convergence result.

**(v)** To complete program, set $\theta = 1 - \frac{\lambda_0}{\lambda_1} \in (0,1)$. Show that

$$\left|e^{t\lambda_0}p(t,\mathbf{x},\mathbf{y}) - \varphi_0(\mathbf{x})\varphi_0(\mathbf{y})\right| \le \left(\sum_{n=1}^{\infty} e^{-\theta t\lambda_n}\varphi_n(\mathbf{x})^2\right)^{\frac{1}{2}}\left(\sum_{n=1}^{\infty} e^{-\theta t\lambda_n}\varphi_n(\mathbf{y})^2\right)^{\frac{1}{2}}$$

$$\le e^{-\frac{\theta t\lambda_1}{2}} p^G\left(\tfrac{\theta t}{2},\mathbf{x},\mathbf{x}\right)^{\frac{1}{2}} p^G\left(\tfrac{\theta t}{2},\mathbf{y},\mathbf{y}\right)^{\frac{1}{2}} \le \frac{e^{-\frac{\theta t\lambda_1}{2}}}{(\pi\theta t)^{\frac{N}{2}}}.$$

EXERCISE 10.3.40. M. Kac[3] made an interesting application of (10.3.39) to a problem raised originally by the physicist H. Lorentz and solved, remarkably quickly, by H. Weyl. What Lorentz noticed is that, if one takes Planck's theory of black body radiation seriously, then the distribution of high frequencies emitted should depend only on the volume of the radiator. In order to state Lorentz's question in mathematical terms, let $G$ be a non-empty, bounded, connected, open subset of $\mathbb{R}^N$, let $\{\lambda_n : n \ge 0\}$ be the eigenvalues, arranged in non-decreasing order, of $-\frac{1}{2}\Delta$ with zero boundary conditions, and use $\mathbf{N}(\lambda)$ to denote the number of $n \ge 0$ such that $\lambda_n \le \lambda$. What Lorentz predicted was that the rate at which $\mathbf{N}(\lambda)$ grows as $\lambda \to \infty$ depends only on the volume $|G|$ of $G$ and on nothing else about $G$. Thus, the original interest in the result was that the asymptotic distribution of high frequencies is so insensitive to the shape of the

[3] See Kac's wonderful article "Can one hear the shape of drum?," *Am. Math. Monthly* **73 # 4**, pp. 1–23 (1966), or, better yet, borrow the movie from the A.M.S.

radiator. When Kac took up the problem, he turned it around. Namely, he asked what geometric information, besides the volume, is encoded in the eigenvalues. When he explained his program to L. Bers, Bers rephrased the problem in the terms that Kac adopted for his title. Audiophiles will be disappointed to learn that, according to C. Gordon, D. Webb, and S. Wolpert's,[4] one cannot hear the shape of a drum, even a two dimensional one.

This exercise outlines Kac's argument for proving **Weyl's asymptotic formula**

$$N(\lambda) \sim \frac{|G|\lambda^{\frac{N}{2}}}{(2\pi)^{\frac{N}{2}}\Gamma(\frac{N+1}{2})},$$

in the sense that the ratio of the two sides tends to 1 as $\lambda \to \infty$.

**(i)** Refer to Exercise 10.3.38, and show that, for each $n \geq 0$,

$$\tfrac{1}{2}\Delta\varphi_n = -\lambda_n\varphi_n \text{ and } \lim_{\substack{\mathbf{x}\in G\\ \mathbf{x}\to\mathbf{a}}} \varphi_n(\mathbf{x}) = 0 \text{ for } \mathbf{a} \in \partial_{\text{reg}}G.$$

Thus, I will interpret the $\lambda_n$'s in Exercise 10.3.38 as the frequencies referred to in Lorentz's problem.

**(ii)** Using (10.3.39), show that

$$\int_{(0,\infty)} e^{-t\lambda}\, N(d\lambda) = \sum_{n=0}^{\infty} e^{-t\lambda_n} = \int_G p^G(t,\mathbf{x},\mathbf{x})\, d\mathbf{x},$$

where $N(d\lambda)$ denotes integration with respect the purely atomic measure on $(0,\infty)$ determined by the non-decreasing function $\lambda \rightsquigarrow N(\lambda)$.

**(iii)** Using (10.3.8), show that

$$1 \geq (2\pi t)^{\frac{N}{2}} p^G(t,\mathbf{x},\mathbf{x}) \geq 1 - E(t,\mathbf{x}),$$

where $E(t,\mathbf{x}) \geq 0$ and, as $t \searrow 0$, $E(t,\mathbf{x}) \longrightarrow 0$ uniformly on compact subsets of $G$. Conclude that

$$\lim_{t\searrow 0} (2\pi t)^{\frac{N}{2}} \int_{(0,\infty)} e^{-t\lambda}\, N(d\lambda) = |G|.$$

At this point, Kac invoked Karamata's Tauberian Theorem,[5] which relates the asymptotics at infinity of an increasing function to the asymptotics at zero of

[4] See their 1992 announcement in *B.A.M.S.*, new series **27** (**2**), "One cannot hear the shape of a drum."

[5] See, for example, Theorem 1.7.6 in N. Bingham, C. Goldie, and J. Teugel's *Regularly Varying Functions*, Cambridge U. Press (1987).

its Laplace transform. Given the preceding, Karamata's theorem yields Weyl's asymptotic formula. It should be pointed out that the weakness of Kac's method is its reliance on the Laplace transform and Tauberian theory, which gives only the principal term in the asymptotics. Further information can be obtained using Fourier methods, which, in terms of partial differential equations, means that one is replacing the heat equation by the wave equation, an equation about which probability theory has embarrassingly little to say.

EXERCISE 10.3.41. It will have occurred to most readers that the relation between the Hermite heat kernel in (10.1.7) and the Ornstein–Uhlenbeck process in § 8.4.1 is the archetypal example of what we have been doing in this section. This exercise gives substance to this remark.

**(i)** Set $\rho_\pm(\mathbf{x}) = e^{\pm\frac{|\mathbf{x}|^2}{2}}$, and show that $\frac{1}{2}\Delta\rho_\pm - \frac{1}{2}|\mathbf{x}|^2\rho_\pm = \pm\frac{N}{2}\rho_\pm$. By Lemma 10.3.24, $\rho_-$ is a ground state for $-\frac{|\mathbf{x}|^2}{2}$ with associated eigenvalue $-\frac{N}{2}$, a fact that also can be verified by direct computation using (10.1.7). Show that the measure $\mathbb{P}^{\rho_-}_{\mathbf{x}}$ is the distribution under $\mathcal{W}^{(N)}$ of $\{2^{-\frac{1}{2}}\mathbf{U}(2t, 2^{\frac{1}{2}}\mathbf{x}, \boldsymbol{\theta}) : t \geq 0\}$, where $\mathbf{U}(t, \mathbf{x}, \boldsymbol{\theta})$ is the Ornstein–Uhlenbeck process described in (8.5.1).

**(ii)** Although it does not follow from Lemma 10.3.24, use (10.1.7) to show that $\rho_+$ is also a ground state for $-\frac{|\mathbf{x}|^2}{2}$ with associated $\frac{N}{2}$. (See Exercise 10.3.43.) Also, show that $\mathbb{P}^{\rho_+}_{\mathbf{x}}$ is the $\mathcal{W}^{(N)}$-distribution of $\{\boldsymbol{\theta} \in e^t\mathbf{x}+2^{-\frac{1}{2}}\mathbf{V}(2t, \boldsymbol{\theta}) : t \geq 0\}$, where $\{\mathbf{V}(t, \boldsymbol{\theta}) : t \geq 0\}$ is the process discussed in Exercise 8.5.14.

EXERCISE 10.3.42. Recall the Hermite polynomials $H_n(x) = (-1)^n e^{\frac{x^2}{2}} \frac{d^n}{dx^n} e^{-\frac{x^2}{2}}$ in § 2.4.1. Show that the Hermite functions (although these are not precisely the ones introduced in § 2.4, they are obtained from those by rescaling)

$$\tilde{h}_n(x) = \frac{2^{\frac{1}{4}}}{(n!)^{\frac{1}{2}}} e^{-\frac{x^2}{2}} H_n(2^{\frac{1}{2}}x), \quad n \geq 0,$$

form an orthonormal basis in $L^2(\mathbb{R};\mathbb{R})$ and that

$$\int_{\mathbb{R}} \tilde{h}_n(y) h(t,x,y)\,dy = e^{-(n+\frac{1}{2})t}\tilde{h}_n(x), \quad n \geq 0 \text{ and } (t,x) \in (0,\infty)\times\mathbb{R},$$

where $h(t,x,y)$ is the function in (10.1.7) when $N = 1$. As a consequence, if

$$\tilde{h}_{\mathbf{n}}(\mathbf{x}) = \prod_{i=1}^{N} h_{n_i}(x_i) \quad \text{for } \mathbf{n} \in \mathbb{N}^N \text{ and } \mathbf{x} \in \mathbb{R}^N,$$

show that $\{\tilde{h}_{\mathbf{n}} : \mathbf{n} \in \mathbb{N}^N\}$ is an orthonormal basis in $L^2(\mathbb{R}^N;\mathbb{R})$ and

$$\int_{\mathbb{R}} \tilde{h}_{\mathbf{n}}(\mathbf{y}) h(t,\mathbf{x},\mathbf{y})\,d\mathbf{y} = e^{-(\|\mathbf{n}\|+\frac{N}{2})t}\tilde{h}_{\mathbf{n}}(\mathbf{x}), \quad \mathbf{n} \in \mathbb{N}^N \text{ and } (t,\mathbf{x}) \in (0,\infty)\times\mathbb{R}^N.$$

**Hint**: Remember that

$$\sum_{n=0}^{\infty} \frac{\lambda^n}{n!} H_n(x) = e^{\lambda x - \frac{\lambda^2}{2}}.$$

EXERCISE 10.3.43. Part (**ii**) of Exercise 10.3.41 might lead one to question the necessity of the boundedness assumption made in Lemma 10.3.24. However, that would be a mistake because, in general, a positive solution to $\frac{1}{2}\Delta\rho + V\rho = \lambda\rho$ need not be a ground state. For example, in this exercise we will show that although $\rho(x) = e^{\frac{x^4}{4}}$ satisfies $\frac{1}{2}\partial_x^2\rho + V\rho = 0$ when $V = -\frac{1}{2}(x^6 + 3x^2)$, this $\rho$ is *not* a ground state for $V$. The proof is based on the following idea. If $\rho$ were a ground state, then Theorems 10.3.26 and its corollaries would apply, and so we would know that the equation

$$(*) \qquad X(t,\psi) = \psi(t) + \int_0^t X(\tau,\psi)^3\,d\tau$$

would have a solution on $[0,\infty)$ for $\mathcal{W}_{\mathbf{x}}^{(1)}$-almost every $\psi \in C(\mathbb{R})$ for every $x \in \mathbb{R}$. The following steps show that this is impossible.

(**i**) Suppose that $\psi_1, \psi_2 \in C(\mathbb{R})$ and that $0 \le \psi_1(t) \le \psi_2(t)$ for $t \in [0,1]$. If $X(\,\cdot\,,\psi_2)$ exists on $[0,1]$, show that $X(\,\cdot\,,\psi_1)$ exists on $[0,1]$.

**Hint**: Define $X_0(t,\psi) = \psi(t)$ and $X_{n+1}(t,\psi) = \psi(t) + \int_0^t X_n(\tau,\psi)^3\,d\tau$. First show that if $0 \le \psi_1(t) \le \psi_2(t)$, then $0 \le X_n(\,\cdot\,,\psi_1) \le X_n(\,\cdot\,,\psi_2)$. Second, if $\sup_{n\ge 0} \|X_n(\,\cdot\,,\psi)\|_{[0,T]} < \infty$, show that $X_n(\,\cdot\,,\psi)$ converges uniformly on $[0,T]$ to the unique solution to (*) on $[0,T]$.

(**ii**) Show that if $\psi(t) \ge 1$ for $t \in [0,1]$, then $X(t,\psi) \ge (1-2t)^{-\frac{1}{2}}$ for $t \in [0,\frac{1}{2}]$ and therefore $X(\,\cdot\,,\psi)$ fails to exist after time $\frac{1}{2}$.

(**iii**) Show that $\mathcal{W}_2^{(1)}\big(\psi(t) \ge 1 \text{ for } t \in [0,1]\big) > 0$, and conclude from this that $\rho$ cannot be a ground state for $V$.

# Chapter 11
## Some Classical Potential Theory

In this concluding chapter I will discuss a few refinements and extensions of the material in §§ 10.2 and 10.3. Even so, I will be barely scratching the surface. The interested reader should consult J.L. Doob's thorough account in *Classical Potential Theory and Its Probabilistic Counterpart*, published by Springer–Verlag in 1984, or S. Port and C. Stones's *Brownian Motion and Classical Potential Theory*, published by Academic Press in 1978.

### § 11.1 Uniqueness Refined

In this section I will refine some of the uniqueness statements made in § 10.2. The improved statements result from the removal of the defect mentioned in Remark 10.3.14. To be precise, recall that if $G$ is an open subset of $\mathbb{R}^N$, then $\zeta_s^G(\psi) = \inf\{t \geq s : \psi(t) \notin G\}$, $\zeta_{0+}^G = \lim_{s \searrow 0} \zeta_s^G$, and (cf. Lemma 10.2.11) $\partial_{\text{reg}} G$ is the set of $\mathbf{x} \in \partial G$ such that $\mathcal{W}_{\mathbf{x}}^{(N)}(\zeta_{0+}^G = 0) = 1$. The main result proved in this section is Theorem 11.1.15, which states that, for any $\mathbf{x} \in G$ and $\mathcal{W}_{\mathbf{x}}^{(N)}$-almost all $\psi \in C(\mathbb{R}^N)$, $\zeta^G(\psi) < \infty \implies \psi(\zeta^G) \in \partial_{\text{reg}} G$. However, I will begin by amending the treatment that I gave in § 10.3 of the Dirichlet heat kernel $p^G(t, \mathbf{x}, \mathbf{y})$.

**§ 11.1.1. The Dirichlet Heat Kernel Again.** In § 10.3, I introduced the Dirichlet heat kernel $p^G(t, \mathbf{x}, \mathbf{y})$. At the time, I was concerned with it only when $(\mathbf{x}, \mathbf{y}) \in G \times G$, and so I defined it in such a way that it was 0 outside $G \times G$. When $G$ is regular in the sense that $\partial G = \partial_{\text{reg}} G$, this choice is the obvious one, since (cf. Theorem 10.3.9) it is the one that makes $p^G(t, \cdot, \mathbf{y})$ continuous on $\mathbb{R}$ for each $(t, \mathbf{y}) \in (0, \infty) \times \mathbb{R}^N$. However, when $G$ is not regular, it is too crude for the analysis here. Instead, from now on I will take

$$p^G(t, \mathbf{x}, \mathbf{y}) = \tag{11.1.1}$$
$$\mathcal{W}^{(N)}\Big(\mathbf{x}(1 - \ell_t(\tau)) + \boldsymbol{\theta}_t(\tau) + \mathbf{y}\ell_t(\tau) \in G,\ \tau \in (0, t)\Big) g^{(N)}(t, \mathbf{y} - \mathbf{x}),$$

where $\ell_t(\tau) = \frac{\tau \wedge t}{t}$ and $\boldsymbol{\theta}_t(\tau) = \boldsymbol{\theta}(\tau) - \boldsymbol{\theta}(t)\ell_t(\tau)$. Notice that the difference between this definition and the one in § 10.3.2 results from the replacement of the closed interval $[0, t]$ there by the open interval $(0, t)$ here. That is, in § 10.3.2, $p^G(t, \mathbf{x}, \mathbf{y})$ was given by

$$\mathcal{W}^{(N)}\Big(\mathbf{x}(1 - \ell_t(\tau)) + \boldsymbol{\theta}_t(\tau) + \mathbf{y}\ell_t(\tau) \in G,\ \tau \in [0, t]\Big) g^{(N)}(t, \mathbf{y} - \mathbf{x}).$$

Of course, unless $\mathbf{x}, \mathbf{y} \in \partial G$, the difference between these two disappears. On the other hand, when either $\mathbf{x}$ or $\mathbf{y}$ is an element of $\partial G$, there is a subtle, but crucial, difference.

To relate the preceding definition to the considerations in § 10.3.1, set $E_t^\circ(\boldsymbol{\psi}) = \mathbf{1}_{[t,\infty)}\big(\zeta_{0+}^G(\boldsymbol{\psi})\big)$. Then (11.1.1) is equivalent to saying that $p^G(t,\mathbf{x},\boldsymbol{\psi}) = q^\circ(t,\mathbf{x},\mathbf{y})$ when $q^\circ(t,\mathbf{x},\mathbf{y})$ is defined in terms of $E_t^\circ$ via (10.3.2). Hence, just as in the proof of Theorem 10.3.3, one can use the results in § 8.3.3 to check that $p^G(t,\mathbf{x},\mathbf{y}) = p^G(t,\mathbf{y},\mathbf{x})$ is again true but that (10.3.4) has to be replaced by

$$\int_{\mathbb{R}^N} \varphi(\mathbf{y}) p^G(t,\mathbf{x},\mathbf{y})\,d\mathbf{y} = \mathbb{E}^{\mathcal{W}_{\mathbf{x}}^{(N)}}\big[\varphi(\boldsymbol{\psi}(t)),\, \zeta_{0+}^G(\boldsymbol{\psi}) \geq t\big]. \tag{11.1.2}$$

However, the analog here of the Chapman–Kolmogorov equation (10.3.5) presents something of challenge. To understand this challenge, note that $t \rightsquigarrow E_t^\circ$ fails to satisfy (10.3.1). Indeed,

$$E_{s+t}^\circ(\boldsymbol{\psi}) = \mathbf{1}_G\big(\boldsymbol{\psi}(s)\big) E_s^\circ(\boldsymbol{\psi}) E_t^\circ(\boldsymbol{\psi}). \tag{11.1.3}$$

Thus, repeating the argument used in the proof of Theorem 10.3.3 to derive (10.3.5), one finds that

$$p^G(s+t,\mathbf{x},\mathbf{y}) = \int_G p^G(s,\mathbf{x},\mathbf{z}) p^G(t,\mathbf{z},\mathbf{y})\,d\mathbf{z}, \tag{11.1.4}$$

which, because the integral is over $G$ and not $\mathbb{R}^N$, is a flawed version of the Chapman–Kolmogorov equation. In order to remove this flaw, I will need the following lemma.

Lemma 11.1.5. *For each $(t,\mathbf{x}) \in (0,\infty) \times \mathbb{R}^N$,*

$$\mathcal{W}_{\mathbf{x}}^{(N)}(\zeta^G = t) = 0 = \mathcal{W}_{\mathbf{x}}^{(N)}(\zeta_{0+}^G = t),$$

*and therefore*

$$\int_{\mathbb{R}^N} \varphi(\mathbf{y}) p^G(t,\mathbf{x},\mathbf{y}) = \mathcal{W}_{\mathbf{x}}^{(N)}\Big[\varphi(\boldsymbol{\psi}(t)),\, \zeta_{0+}^G(\boldsymbol{\psi}) > t\Big] \tag{11.1.6}$$

*for all Borel measurable $\varphi : \mathbb{R}^N \longrightarrow \mathbb{R}$ that are bounded below. In particular, $p^G(t,\mathbf{x},\mathbf{y}) = 0$ for Lebesgue-almost every $\mathbf{y} \notin G$.*

Proof: Set

$$\rho(\xi) = \int_{\mathbb{R}^N} \mathcal{W}_{\mathbf{y}}^{(N)}(\zeta^G > \xi)\,\gamma_{\mathbf{0},\mathbf{I}}(d\mathbf{y}), \quad \xi \in (0,\infty).$$

Obviously, $\rho$ is a right-continuous, non-increasing, $[0,1]$-valued function, and, as such, it has only countably many discontinuities. Hence, there is a countable set $\Lambda \subseteq (0,\infty)$ such that

$$\xi \notin \Lambda \implies \mathcal{W}^{(N)}_{\mathbf{y}}(\zeta^G = \xi) = 0 \quad \text{for Lebesgue-almost every } \mathbf{y} \in \mathbb{R}^N.$$

Now let $(t,\mathbf{x}) \in (0,\infty) \times \mathbb{R}^N$ be given, and choose $s \in (0,t)$ so that $t-s \notin \Lambda$. Then, by the Markov property and (10.2.10),

$$\begin{aligned}\mathcal{W}^{(N)}_{\mathbf{x}}(\zeta^G_{0+} = t) &= \mathcal{W}^{(N)}_{\mathbf{x}}(\zeta^G_{0+} > s \;\&\; \zeta^G \circ \Sigma_s = t-s) \leq \mathcal{W}^{(N)}_{\mathbf{x}}(\zeta^G \circ \Sigma_s = t-s)\\ &= \int_{\mathbb{R}^N} \mathcal{W}^{(N)}_{\mathbf{x}+\mathbf{y}}(\zeta^G = t-s)\,\gamma_{\mathbf{0},s\mathbf{I}}(d\mathbf{y}) = 0.\end{aligned}$$

In addition, because $\zeta^G(\boldsymbol{\psi}) = t \implies \zeta^G_{0+}(\boldsymbol{\psi}) = t$ when $t > 0$, it follows that $\mathcal{W}^{(N)}_{\mathbf{x}}(\zeta^G = t) = 0$ also.

Given the preceding, it is clear how to pass from (11.1.2) to (11.1.6). Finally, by applying (11.1.6) with $\varphi = \mathbf{1}_{G\complement}$, we see that

$$\int_{G\complement} p^G(t,\mathbf{x},\mathbf{y})\,d\mathbf{y} = \mathcal{W}^{(N)}_{\mathbf{x}}\big(\boldsymbol{\psi}(t) \notin G \;\&\; \zeta^G_{0+}(\boldsymbol{\psi}) > t\big) = 0,$$

which says that $p^G(t,\mathbf{x},\,\cdot\,)$ vanishes Lebesgue-almost everywhere on $G\complement$. □

Because of the final part of Lemma 11.1.5, we can now replace the preceding flawed version of the Chapman–Kolmogorov equation by

$$\text{(11.1.7)}\quad p^G(s+t,\mathbf{x},\mathbf{y}) = \int_{\mathbb{R}^N} p^G(s,\mathbf{x},\mathbf{z})p^G(t,\mathbf{z},\mathbf{y})\,d\mathbf{z}, \quad (t,\mathbf{x},\mathbf{y}) \in (0,\infty)\times(\mathbb{R}^N)^2.$$

Before completing this discussion, I want to develop a Duhamel formula for $p^G$. That is, I want to show that

$$\text{(11.1.8)}\quad \begin{aligned}p^G(t,\mathbf{x},\mathbf{y}) =& g^{(N)}(t,\mathbf{y}-\mathbf{x})\\ &- \mathbb{E}^{\mathcal{W}^{(N)}_{\mathbf{x}}}\Big[g^{(N)}\big(t-\zeta^G_{0+}(\boldsymbol{\psi}),\mathbf{y}-\boldsymbol{\psi}(\zeta^G_{0+})\big),\,\zeta^G_{0+}(\boldsymbol{\psi}) < t\Big]\end{aligned}$$

for all $(t,\mathbf{x},\mathbf{y}) \in (0,\infty)\times(\mathbb{R}^N)^2$, and the idea is very much the same as the one used to prove (10.3.8). Thus, for $\alpha \in (0,1)$, set

$$q^\circ_\alpha(t,\mathbf{x},\mathbf{y}) = \mathcal{W}^{(N)}\Big(\mathbf{x}\big(1-\ell_t(\tau)\big)+\boldsymbol{\theta}_t(\tau)+\mathbf{y}\ell_t(\tau) \in G,\ \tau \in (0,\alpha t)\Big)g^{(N)}(t,\mathbf{y}-\mathbf{x}).$$

Obviously, $q^\circ_\alpha(t,\mathbf{x},\mathbf{y}) \searrow p^G(t,\mathbf{x},\mathbf{y})$ as $\alpha \nearrow 1$. In addition, proceeding as in the proof of Theorem 10.3.3, one finds that $q^\circ_\alpha(t,\mathbf{x},\,\cdot\,)$ is continuous and that

$$\text{(*)}\quad \int_{\mathbb{R}^N} \varphi(\mathbf{y})q^\circ_\alpha(t,\mathbf{x},\mathbf{y})\,d\mathbf{y} = \mathbb{E}^{\mathcal{W}^{(N)}_{\mathbf{x}}}\Big[\varphi(\boldsymbol{\psi}(t)),\ \zeta^G_{0+}(\boldsymbol{\psi}) \geq \alpha t\Big].$$

Now use the Markov property to justify

$$\begin{aligned}
&\int_{\mathbb{R}^N} \varphi(\mathbf{y}) g^{(N)}(t, \mathbf{y}-\mathbf{x})\, d\mathbf{y} \\
&\quad = \mathbb{E}^{\mathcal{W}_{\mathbf{x}}^{(N)}}\Big[\varphi(\boldsymbol{\psi}(t)),\, \zeta_s^G(\boldsymbol{\psi}) \geq \alpha t\Big] + \mathbb{E}^{\mathcal{W}_{\mathbf{x}}^{(N)}}\Big[\varphi(\boldsymbol{\psi}(t)),\, \zeta_s^G(\boldsymbol{\psi}) < \alpha t\Big] \\
&\quad = \mathbb{E}^{\mathcal{W}_{\mathbf{x}}^{(N)}}\Big[\varphi(\boldsymbol{\psi}(t)),\, \zeta_s^G(\boldsymbol{\psi}) \geq \alpha t\Big] \\
&\qquad + \mathbb{E}^{\mathcal{W}_{\mathbf{x}}^{(N)}}\left[\int_{\mathbb{R}^N} \varphi(\mathbf{y}) g^{(N)}\big(t-\zeta_s^G(\boldsymbol{\psi}), \mathbf{y}-\boldsymbol{\psi}(\zeta_s^G)\big)\, d\mathbf{y},\, \zeta_s^G(\boldsymbol{\psi}) < \alpha t\right].
\end{aligned}$$

for all $\alpha \in (0,1)$, $t \in (0,\infty)$ and $s \in (0,\alpha t)$. Thus, by (*), after letting $s \searrow 0$, we see that

$$\begin{aligned}
&\int_{\mathbb{R}^N} \varphi(\mathbf{y}) q_\alpha(t,\mathbf{x},\mathbf{y})\, d\mathbf{y} \\
&\quad = \int_{\mathbb{R}^N} \varphi(\mathbf{y}) g^{(N)}(t,\mathbf{y}-\mathbf{x})\, d\mathbf{y} \\
&\qquad - \mathbb{E}^{\mathcal{W}_{\mathbf{x}}^{(N)}}\left[\int_{\mathbb{R}^N} \varphi(\mathbf{y}) g\big(t-\zeta_{0+}^G(\boldsymbol{\psi}), \mathbf{y}-\boldsymbol{\psi}(\zeta_{0+}^G)\big)\, d\mathbf{y},\, \zeta_{0+}^G(\boldsymbol{\psi}) < \alpha t\right].
\end{aligned}$$

Because $q_\alpha^\circ(t,\mathbf{x},\cdot)$ is continuous, this means that

$$q_\alpha^\circ(t,\mathbf{x},\mathbf{y}) = g^{(N)}(t,\mathbf{y}-\mathbf{x}) - \mathbb{E}^{\mathcal{W}_{\mathbf{x}}^{(N)}}\left[g^{(N)}\big(t-\zeta_{0+}^G(\boldsymbol{\psi}), \mathbf{y}-\boldsymbol{\psi}(\zeta_{0+}^G)\big),\, \zeta_{0+}^G(\boldsymbol{\psi}) < \alpha t\right],$$

and so (11.1.8) follows when one lets $\alpha \nearrow 1$.

**§ 11.1.2. Exiting Through $\partial_{\text{reg}}G$.** The purpose of this subsection is to prove that when Brownian motion exits from a region, it does so through regular points. My proof of this fact follows the reasoning in the book, cited above, by Port and Stone.

LEMMA 11.1.9. *Let $G$ be a non-empty, connected open subset of $\mathbb{R}^N$, and define $p^G$ by (11.1.1). Then, for each $(t,\mathbf{x},\mathbf{a}) \in (0,\infty)\times\mathbb{R}^N\times\text{reg}(G) \equiv \partial_{\text{reg}}G \cup (\mathbb{R}^N\setminus\overline{G})$, $p^G(t,\mathbf{x},\mathbf{a}) = 0$. On the other hand, if $(t,\mathbf{x}) \in (0,\infty)\times G$, then $p^G(t,\mathbf{x},\mathbf{a}) > 0$ for all $\mathbf{a} \in \partial G \setminus \partial_{\text{reg}}G$. In particular, $\partial G\setminus\partial_{\text{reg}}G$ has Lebesgue measure 0.*

PROOF: Obviously, $p^G(t,\mathbf{x},\mathbf{a}) = 0$ if $a \in \mathbb{R}^N\setminus\bar{G}$. Next, suppose that $\mathbf{a} \in \partial_{\text{reg}}G$. Then, by (11.1.8), $p^G(t,\mathbf{a},\mathbf{x}) = 0$ for all $(t,\mathbf{x}) \in (0,\infty)\times\mathbb{R}^N$, and so, by symmetry, the same is true of $p^G(t,\mathbf{x},\mathbf{a})$.

To go in the other direction when $(t,\mathbf{x}) \in (0,\infty)\times G$, let $\mathbf{a} \in \partial G$ be given, and begin with the observation that $(t,\mathbf{x}) \in (0,\infty)\times G \longmapsto p^G(t,\mathbf{x},\mathbf{a})$ is in $C^{1,2}\big((0,\infty)\times G;[0,\infty)\big)$ and satisfies $\partial_t p^G(t,\mathbf{x},\mathbf{a}) = \frac{1}{2}\Delta_{\mathbf{x}} p^G(t,\mathbf{x},\mathbf{a})$. To check this, use (11.1.4) to write

$$p^G(t,\mathbf{x},\mathbf{a}) = \int_G p^G(t-s,\mathbf{x},\mathbf{z}) p^G(s,\mathbf{z},\mathbf{a})\, d\mathbf{z}$$

for any $0 < s < t$, and note that $p^G(s, \cdot, \mathbf{a})$ is bounded. Hence, the desired conclusions follow from (10.3.12) and the argument used to prove the last part of Theorem 10.3.9. Next, suppose that $p^G(t_0, \mathbf{x}_0, \mathbf{a}) = 0$ for some $(t_0, \mathbf{x}_0) \in (0, \infty) \times G$. Then, by the strong minimum principle (cf. Theorem 10.1.6), $p^G(t, \mathbf{x}, \mathbf{a}) = 0$ for all $(t, \mathbf{x}) \in (0, t_0) \times G$. But this, by (11.1.2) and symmetry, means that, for $t \in (0, t_0)$,

$$\mathcal{W}_{\mathbf{a}}^{(N)}(\zeta_{0+}^G \geq t) = \int_{\mathbb{R}^N} p^G(t, \mathbf{a}, \mathbf{y})\, d\mathbf{y} = \int_G p^G(t, \mathbf{x}, \mathbf{a})\, d\mathbf{x} = 0,$$

where I have used the final part of Lemma 11.1.5 to get the second equality. Hence, $p^G(t_0, \mathbf{x}_0, \mathbf{a}) = 0 \implies \mathbf{a} \in \partial_{\mathrm{reg}} G$.

Finally, because, by the preceding and symmetry, for any $\mathbf{x} \in G$, $\partial G \setminus \partial_{\mathrm{reg}} G$ is contained in $\{\mathbf{y} \notin G : p(1, \mathbf{x}, \mathbf{y}) > 0\}$, and, by Lemma 11.1.5, the latter set has Lebesgue measure 0, it is clear the $\partial G \setminus \partial_{\mathrm{reg}} G$ has Lebesgue measure 0. □

I next introduce the function

$$v^G(\mathbf{x}) \equiv \mathbb{E}^{\mathcal{W}_{\mathbf{x}}^{(N)}}\left[e^{-\zeta_{0+}^G}\right], \quad \mathbf{x} \in \mathbb{R}^N. \tag{11.1.10}$$

Since, by the Markov property,

$$e^{-s} \int_{\mathbb{R}^N} g^{(N)}(s, \mathbf{y} - \mathbf{x}) \mathbb{E}^{\mathcal{W}_{\mathbf{y}}^{(N)}}\left[e^{-\zeta^G}\right] d\mathbf{y} = \mathbb{E}^{\mathcal{W}_{\mathbf{x}}^{(N)}}\left[e^{-\zeta_s^G}\right] \nearrow v^G(\mathbf{x})$$

as $s \searrow 0$, it is clear that $v^G$ is lower semicontinuous. In addition, it is obvious that $v^G \leq 1$ everywhere and that

$$\{\mathbf{x} \in \mathbb{R}^N : v^G(\mathbf{x}) = 1\} = \mathrm{reg}(G) = \partial_{\mathrm{reg}} G \cup (\mathbb{R}^N \setminus \overline{G}).$$

LEMMA 11.1.11. *Define the Borel measure $\nu^G$ on $\mathbb{R}^N$ by*[1]

$$\nu^G(\Gamma) = \int_{\mathbb{R}^N} \mathbb{E}^{\mathcal{W}_{\mathbf{x}}^{(N)}}\left[e^{-\zeta_{0+}^G(\boldsymbol{\psi})}, \boldsymbol{\psi}(\zeta_{0+}^G) \in \Gamma\right] d\mathbf{x}.$$

*Then $\nu^G$ is supported on $G\complement$, and if*

$$r(\mathbf{x}) = \int_{(0,\infty)} e^{-t} g^{(N)}(t, \mathbf{x})\, dt, \quad \mathbf{x} \in \mathbb{R}^N,$$

*then*

$$v^G(\mathbf{x}) = \int_{\mathbb{R}^N} r(\mathbf{y} - \mathbf{x})\, \nu^G(d\mathbf{y}), \quad \mathbf{x} \in \mathbb{R}^N. \tag{11.1.12}$$

*In particular, $\nu^G$ is always locally finite and is therefore finite in the case when $G\complement$ is compact. Finally, for any non-empty, open set $H \subset \mathbb{R}^N$,*

$$G\complement \subseteq \mathrm{reg}(H) \implies v^G(\mathbf{x}) = \mathbb{E}^{\mathcal{W}_{\mathbf{x}}^{(N)}}\left[e^{-\zeta_{0+}^H} v^G\left(\boldsymbol{\psi}(\zeta_{0+}^H)\right)\right], \quad \mathbf{x} \in \mathbb{R}^N, \tag{11.1.13}$$

*where $\mathrm{reg}(H) = \partial_{\mathrm{reg}} H \cup (\mathbb{R}^N \setminus \bar{H})$.*

[1] Below I use the convention that $e^{-\zeta_{0+}^G(\boldsymbol{\psi})} = 0$ when $\zeta_{0+}^G(\boldsymbol{\psi}) = \infty$. Thus, the problem of giving $\boldsymbol{\psi}(\zeta_{0+}^G)$ meaning when $\zeta_{0+}^G(\boldsymbol{\psi}) = \infty$ does not arise in integrals having $e^{-\zeta_{0+}^G(\boldsymbol{\psi})}$ as a factor in their integrands.

PROOF: Clearly $\nu^G$ is supported on $G\complement$. To prove (11.1.12), note that the symmetry of $p^G(t,\mathbf{x},\mathbf{y})$ together with (11.1.8) imply that

$$\begin{aligned}\mathbb{E}^{\mathcal{W}^{(N)}_{\mathbf{x}}}&\Big[g^{(N)}\big(t-\zeta^G_{0+}(\boldsymbol{\psi}),\mathbf{y}-\boldsymbol{\psi}(\zeta^G_{0+})\big),\,\zeta^G_{0+}(\boldsymbol{\psi})<t\Big]\\&=\mathbb{E}^{\mathcal{W}^{(N)}_{\mathbf{y}}}\Big[g^{(N)}\big(t-\zeta^G_{0+}(\boldsymbol{\psi}),\mathbf{x}-\boldsymbol{\psi}(\zeta^G_{0+})\big),\,\zeta^G_{0+}(\boldsymbol{\psi})<t\Big]\end{aligned}$$

for all $(t,\mathbf{x},\mathbf{y})\in(0,\infty)\times\mathbb{R}^N\times\mathbb{R}^N$. Hence, after multiplying by $e^{-t}$ and integrating with respect to $t\in(0,\infty)$, one arrives at

$$\mathbb{E}^{\mathcal{W}^{(N)}_{\mathbf{x}}}\Big[e^{-\zeta^G_{0+}(\boldsymbol{\psi})}\,r\big(\boldsymbol{\psi}(\zeta^G_{0+})-\mathbf{y}\big)\Big]=\mathbb{E}^{\mathcal{W}^{(N)}_{\mathbf{y}}}\Big[e^{-\zeta^G_{0+}(\boldsymbol{\psi})}\,r\big(\boldsymbol{\psi}(\zeta^G_{0+})-\mathbf{x}\big)\Big].$$

But

$$\int_{\mathbb{R}^N} r(\mathbf{x}-\mathbf{y})\,d\mathbf{y}=1,\quad \mathbf{x}\in\mathbb{R}^N,$$

and so (11.1.12) follows after one integrates the preceding over $\mathbf{y}\in\mathbb{R}^N$ and applies Tonelli's Theorem.

Given (11.1.12) and the fact that $r$ is uniformly positive on compacts, it becomes obvious that $\nu^G$ must be always locally finite and finite when $G\complement$ is compact. Thus, all that remains is to check (11.1.13). But clearly, after multiplying (11.1.8) with $G=H$ throughout by $e^{-t}$ and integrating with respect to $t\in(0,\infty)$, one gets

$$r(\mathbf{x}-\mathbf{y})=\int_{(0,\infty)}e^{-t}p^H(t,\mathbf{x},\mathbf{y})\,dt+\mathbb{E}^{\mathcal{W}^{(N)}_{\mathbf{x}}}\Big[e^{-\zeta^G_{0+}}\,r\big(\boldsymbol{\psi}(\zeta^G_{0+})-\mathbf{y}\big)\Big].$$

Hence, since, by the first part of Lemma 11.1.9 with $G=H$, $p^H(t,\mathbf{x},\,\cdot\,)$ vanishes on $\mathrm{reg}(H)$, (11.1.13) follows after one integrates the preceding with respect to $\nu^G(d\mathbf{y})$ and uses (11.1.12). □

LEMMA 11.1.14. *If $G\complement$ is compact and, for some $\theta\in[0,1)$, $v^G\restriction G\complement\le\theta$, then $\mathcal{W}^{(N)}_{\mathbf{x}}(\zeta^G_{0+}<\infty)=0$ for every $\mathbf{x}\in\mathbb{R}^N$.*

PROOF: I begin by checking that $v^G\le\theta$ everywhere. Thus, suppose that $H=\{\mathbf{x}\in\mathbb{R}^N:\,v^G(\mathbf{x})>\theta+\epsilon\}\neq\emptyset$ for some $\epsilon>0$. Because $v^G$ is lower semicontinuous, $H$ is open. I will derive a contradiction by first showing that $G\complement\subseteq\mathrm{reg}(H)$ and then applying (11.1.13). To carry out the first step, use (11.1.12) to see that, for any $s\in(0,\infty)$,

$$\begin{aligned}v^G(\mathbf{x})&\ge\int_s^\infty e^{-t}\left(\int_{\mathbb{R}^N}g^{(N)}(t,\mathbf{y}-\mathbf{x})\,\nu^G(d\mathbf{y})\right)dt\\&=e^{-s}\int_{(0,\infty)}e^{-t}\left(\int_{\mathbb{R}^N}g^{(N)}(s,\mathbf{y}-\mathbf{x})v^G(\mathbf{y})\,d\mathbf{y}\right)dt\\&\ge e^{-s}(\theta+\epsilon)\mathcal{W}^{(N)}_{\mathbf{x}}\big(\boldsymbol{\psi}(s)\in H\big)\ge e^{-s}(\theta+\epsilon)\mathcal{W}^{(N)}_{\mathbf{x}}\big(\zeta^H_{0+}>s\big),\end{aligned}$$

and so, after letting $s \searrow 0$, we have that $v^G(\mathbf{x}) \geq (\theta+\epsilon)\mathcal{W}^{(N)}_{\mathbf{x}}(\zeta^H_{0+} > 0)$. In particular, if $\mathbf{x} \notin G$, then $\theta \geq (\theta+\epsilon)\mathcal{W}^{(N)}_{\mathbf{x}}(\zeta^H_{0+} > 0)$, which means that $\mathbf{x} \notin G \implies \mathcal{W}^{(N)}_{\mathbf{x}}(\zeta^H_{0+} > 0) < 1$. Hence, because (cf. part **(ii)** of Exercise 10.2.19) $\mathcal{W}^{(N)}_{\mathbf{x}}(\zeta^H_{0+} > 0) \in \{0,1\}$, this means that $x \notin G \implies \mathbf{x} \in \operatorname{reg}(H)$ and therefore that (11.1.13) applies. But if $\mathbf{x} \in H$, (11.1.13) yields the contradiction

$$\theta + \epsilon < v^G(\mathbf{x}) = \mathbb{E}^{\mathcal{W}^{(N)}_{\mathbf{x}}}\left[e^{-\zeta^H_{0+}}v^G\big(\boldsymbol{\psi}(\zeta^H_{0+})\big)\right] < \theta + \epsilon,$$

since $\zeta^H_{0+}(\boldsymbol{\psi}) < \infty \implies \boldsymbol{\psi}(\zeta^H_{0+}) \notin H$. That is, I have shown that $H$ must be empty.

Knowing that $v^G \leq \theta$ everywhere, I now want to argue that $\nu^G(\mathbb{R}^N) \leq \theta\nu^G(\mathbb{R}^N)$. Since $\nu^G(\mathbb{R}^N) < \infty$, this will show that $\nu^G = 0$ and therefore, by (11.1.12), that $v^G \equiv 0$, which is the same as saying that $\mathcal{W}^{(N)}_{\mathbf{x}}(\zeta^G_{0+} < \infty) = 0$ everywhere. Thus, let $K = G\complement$, and set $K_n = \{\mathbf{x} : \operatorname{dist}(\mathbf{x}, K) \leq n^{-1}\}$ and $G_n = K_n\complement$ for $n \geq 1$. Clearly, $K \subseteq \mathbb{R}^N \setminus \overline{G_n} \subseteq \operatorname{reg}(G_n)$, and so, by (11.1.12) and Tonelli's Theorem,

$$\nu^G(\mathbb{R}^N) = \int_{\mathbb{R}^N} v^{G_n}(\mathbf{x})\,\nu^G(d\mathbf{x}) = \int_{\mathbb{R}^N} v^G(\mathbf{y})\,\nu^{G_n}(d\mathbf{y}) \leq \theta\nu^{G_n}(\mathbb{R}^N).$$

Thus, all that we have to do is check that $\nu^{G_n}(\mathbb{R}^N) \searrow \nu^G(\mathbb{R}^N)$ when $n \to \infty$. But

$$\nu^{G_n}(\mathbb{R}^N) = \int_{\mathbb{R}^N} v^{G_n}(\mathbf{x})\,d\mathbf{x}$$

and $\nu^{G_1}(\mathbb{R}^N) < \infty$. Hence, by the Monotone Convergence Theorem, it is enough for us to know that $v^{G_n}(\mathbf{x}) \searrow v^G(\mathbf{x})$ for Lebesgue-almost every $\mathbf{x} \in \mathbb{R}^N$. Because $\mathbf{x} \in G_n$ implies $\zeta^{G_n}_{0+} = \zeta^{G_n} \nearrow \zeta^G = \zeta^G_{0+}$ $\mathcal{W}^{(N)}_{\mathbf{x}}$-almost surely, $1 \geq v^{G_n} \searrow v^G$ on $G$. At the same time, $1 \geq v^{G_n} \geq v^G = 1$ on $\operatorname{reg}(G)$, and, by the last part of Lemma 11.1.9, $G\complement \setminus \operatorname{reg}(G) = \partial G \setminus \partial_{\text{reg}}G$ has Lebesgue measure 0. □

THEOREM 11.1.15. *For every open $G \subset \mathbb{R}^N$,*

$$\mathcal{W}^{(N)}_{\mathbf{x}}\Big(\zeta^G_{0+}(\boldsymbol{\psi}) < \infty \;\&\; \boldsymbol{\psi}(\zeta^G_{0+}) \notin \partial_{\text{reg}}G\Big) = 0 \quad \textit{for all } \mathbf{x} \in G.$$

PROOF: Suppose not. Because $\mathcal{W}^{(N)}_{\mathbf{y}}(\zeta^G_{0+} > 0) \in \{0,1\}$ for all $\mathbf{y} \in \mathbb{R}^N$, we could then find an $\mathbf{x} \in G$ and a $\delta > 0$ for which

$$\mathcal{W}^{(N)}_{\mathbf{x}}\Big(\zeta^G_{0+}(\boldsymbol{\psi}) < \infty \;\&\; \boldsymbol{\psi}(\zeta^G_{0+}) \in \Gamma_\delta\Big) > 0,$$

where

$$\Gamma_\delta = \left\{\mathbf{y} \in \partial G : \mathcal{W}^{(N)}_{\mathbf{y}}(\zeta^G_{0+} \geq \delta) \geq \tfrac{1}{2}\right\}.$$

But then there would exist a compact $K \subseteq \Gamma_\delta$ for which

$$\mathcal{W}^{(N)}_{\mathbf{x}}\big(\zeta^{K\complement}_{0+} < \infty\big) \geq \mathcal{W}^{(N)}_{\mathbf{x}}\Big(\zeta^{G}_{0+}(\boldsymbol{\psi}) < \infty \;\&\; \boldsymbol{\psi}(\zeta^{G}_{0+}) \in K\Big) > 0.$$

On the other hand, because $K\complement \supseteq G$, $v^{K\complement} \leq v^{G}$ everywhere, and therefore, because $v^G(\mathbf{y}) \leq \frac{1}{2}\big(1+e^{-\delta}\big) < 1$ for $\mathbf{y} \in K$, Lemma 11.1.14 would say that $\mathcal{W}^{(N)}_{\mathbf{x}}\big(\zeta^{K\complement}_{0+} < \infty\big) = 0$, which is obviously a contradiction. $\square$

**§ 11.1.3. Applications to Questions of Uniqueness.** My main reason for wanting the result in Theorem 11.1.15 is that it allows me to improve on the uniqueness results that were proved in §§ 10.2.3 and 10.3.1. For example, by the comment in Remark 10.3.14, we can now remove the assumption that $\partial G = \partial_{\mathrm{reg}} G$ from the uniqueness assertion in Corollary 10.3.13.

THEOREM 11.1.16. *Let $G$ be an open subset of $\mathbb{R}^N$ and $\varphi \in C_{\mathrm{b}}(G;\mathbb{R})$. Then*

$$(t,\mathbf{x}) \in (0,\infty)\times G \longmapsto \mathbb{E}^{\mathcal{W}^{(N)}_{\mathbf{x}}}\big[\varphi(\boldsymbol{\psi}(t)),\, \zeta^G(\boldsymbol{\psi}) > t\big] = \int_G \varphi(y)p^G(t,\mathbf{x},\mathbf{y})\,d\mathbf{y} \in \mathbb{R}$$

*is the one and only bounded, smooth solution to the boundary value problem described in Corollary 10.3.13.*

More interesting are the improvements that Theorem 11.1.15 allows me to make to the results in § 10.2.3.

THEOREM 11.1.17. *Given an open $G \subseteq \mathbb{R}^N$ and a bounded Borel measurable $f : \partial G \longrightarrow \mathbb{R}$, set*

$$(11.1.18) \qquad u_f(\mathbf{x}) = \mathbb{E}^{\mathcal{W}^{(N)}_{\mathbf{x}}}\big[f(\boldsymbol{\psi}(\zeta^G)),\, \zeta^G(\boldsymbol{\psi}) < \infty\big], \quad \text{for } \mathbf{x} \in G.$$

*Then $u_f$ is a bounded harmonic function on $G$ and $\lim_{\substack{\mathbf{x}\to\mathbf{a}\\ \mathbf{x}\in G}} u_f(\mathbf{x}) = f(\mathbf{a})$ whenever $\mathbf{a} \in \partial_{\mathrm{reg}} G$ is a point at which $f$ is continuous. Furthermore, if $f \in C_{\mathrm{b}}\big(\partial G;[0,\infty)\big)$ and $u$ is an element of $C^2\big(G;[0,\infty)\big)$ that satisfies*

$$\Delta u \leq 0 \;\text{ in } G \quad\text{and}\quad \varliminf_{\substack{\mathbf{x}\to\mathbf{a}\\ \mathbf{x}\in G}} u(\mathbf{x}) \geq f(\mathbf{a}) \quad \text{for } \mathbf{a} \in \partial_{\mathrm{reg}} G,$$

*then $u_f \leq u$. In particular, if $f \in C_{\mathrm{b}}(\partial G;\mathbb{R})$, then $u_f$ is the one and only harmonic function $u$ on $G$ with the properties that*

$$\big|u(\mathbf{x})\big| \leq C\mathcal{W}^{(N)}_{\mathbf{x}}\big(\zeta^G < \infty\big) \quad \text{for all } \mathbf{x} \in G,$$

*for some $C < \infty$ and*

$$\lim_{\substack{\mathbf{x}\to\mathbf{a}\\ \mathbf{x}\in G}} u(\mathbf{x}) = f(\mathbf{a}) \quad \text{for each } \mathbf{a} \in \partial_{\mathrm{reg}} G.$$

PROOF: The initial assertions are covered already by Theorem 10.2.14. Next, let $f \in C_b(\partial G;\mathbb{R})$ be given, and suppose that $u$ is an element of $C^2(G;[0,\infty))$ which satisfies the conditions in the second assertion. To prove that $u_f \le u$, set $\mathcal{F}_t = \sigma\big(\{\psi(\tau) : \tau \in [0,t]\}\big)$, and choose a sequence of bounded, open subsets $G_n$ so that $\overline{G_n} \subseteq G$ and $G_n \nearrow G$. Then, for each $n \ge 1$, $\big(-u(\psi(t\wedge\zeta^{G_n}), \mathcal{F}_t, \mathcal{W}^{(N)}_{\mathbf{x}}\big)$ is a submartingale, and so we know that, for each $\mathbf{x} \in G$, $u(\mathbf{x})$ dominates

$$\varliminf_{T\nearrow\infty}\ \lim_{n\to\infty} \mathbb{E}^{\mathcal{W}^{(N)}_{\mathbf{x}}}\big[u(\psi(T\wedge\zeta^{G_n}))\big] \ge \varliminf_{T\nearrow\infty}\ \lim_{n\to\infty} \mathbb{E}^{\mathcal{W}^{(N)}_{\mathbf{x}}}\big[u(\psi(\zeta^{G_n})),\ \zeta^G \le T\big]$$
$$\ge \mathbb{E}^{\mathcal{W}^{(N)}_{\mathbf{x}}}\Big[f(\psi(\zeta^G)),\ \zeta^G < \infty\Big] = u_f(\mathbf{x}),$$

where, in the passage to the last line, I have used Fatou's Lemma and Theorem 11.1.15.

Finally, let $f \in C_b(\partial G;\mathbb{R})$ be given. What I still have to show is that if $u$ is a harmonic function on $G$ which tends to $f$ at points in $\partial_{\text{reg}}G$ and satisfies $|u(\mathbf{x})| \le C\mathcal{W}^{(N)}_{\mathbf{x}}(\zeta^G < \infty)$ for some $C < \infty$, then $u = u_f$. Thus, suppose $u$ is such a function, and set $M = C + \|f\|_u$. Then, by the preceding, we have both that

$$M\mathcal{W}^{(N)}_{\mathbf{x}}(\zeta^G < \infty) + u(\mathbf{x}) \ge u_{M\mathbf{1}+f}(\mathbf{x}) = M\mathcal{W}^{(N)}_{\mathbf{x}}(\zeta^G < \infty) + u_f(\mathbf{x})$$

and that

$$M\mathcal{W}^{(N)}_{\mathbf{x}}(\zeta^G < \infty) - u(\mathbf{x}) \ge u_{M\mathbf{1}-f}(\mathbf{x}) = M\mathcal{W}^{(N)}_{\mathbf{x}}(\zeta^G < \infty) - u_f(\mathbf{x}),$$

which means, of course, that $u = u_f$. □

As an immediate consequence of Theorem 11.1.17, we have the following.

COROLLARY 11.1.19. *Assume that*

$$\mathcal{W}^{(N)}_{\mathbf{x}}(\zeta^G < \infty) = 1 \quad \textit{for all } \mathbf{x} \in G. \tag{11.1.20}$$

*Then, for each $f \in C_b(\partial G;\mathbb{R})$ the function $u_f$ in (11.1.18) is the one and only bounded, harmonic function $u$ on $G$ which satisfies $\lim_{\substack{\mathbf{x}\to\mathbf{a}\\ \mathbf{x}\in G}} u(\mathbf{x}) = f(\mathbf{a})$ for every $\mathbf{a} \in \partial_{\text{reg}}G$. In particular, this will be the case if $G$ is contained in a half-space.*

In order to go further, it will be helpful to have the following lemma.

LEMMA 11.1.21. *Let $G$ be a non-empty, connected, open set in $\mathbb{R}^N$. Then*

$$\partial_{\text{reg}}G = \emptyset \iff \mathcal{W}^{(N)}_{\mathbf{x}}(\zeta^G < \infty) = 0 \quad \textit{for all } \mathbf{x} \in G.$$

*On the other hand, if $\partial_{\text{reg}}G \ne \emptyset$ and $\mathbf{b} \in \partial G$, then*

$$\mathbf{b} \notin \partial_{\text{reg}}G \iff \lim_{r\searrow 0}\ \varlimsup_{\substack{\mathbf{x}\to\mathbf{b}\\ \mathbf{x}\in G}} \mathcal{W}^{(N)}_{\mathbf{x}}\Big(\psi(\zeta^G) \notin B_{\mathbb{R}^N}(\mathbf{b},r)\ \&\ \zeta^G < \infty\Big) > 0.$$

PROOF: The equivalence

$$\partial_{\text{reg}}G = \emptyset \iff \mathcal{W}^{(N)}_{\mathbf{x}}(\zeta^G < \infty) = 0, \quad \mathbf{x} \in G,$$

follows immediately from Theorems 11.1.15 and 11.1.17.

Now assume that $\partial_{\text{reg}}G \neq \emptyset$, and let $\mathbf{b} \in \partial G$. If $\mathbf{b} \in \partial_{\text{reg}}G$, then

$$\lim_{r\searrow 0} \lim_{\substack{\mathbf{x}\to\mathbf{b}\\ \mathbf{x}\in G}} \mathcal{W}^{(N)}_{\mathbf{x}}\Big(\psi(\zeta^G) \notin B_{\mathbb{R}^N}(\mathbf{b}, r)\ \&\ \zeta^G < \infty\Big) = 0$$

follows from (10.2.13). Thus, suppose that $\mathbf{b} \notin \partial_{\text{reg}}G$. Choose $\mathbf{a} \in \partial_{\text{reg}}G$, and set $B = B_{\mathbb{R}^N}(\mathbf{b}, r)$, where $0 < r \leq \frac{1}{2}|\mathbf{a} - \mathbf{b}|$. One can then construct an $f \in C(\partial G; [0,1])$ with the properties that $f = 0$ on $B \cap \partial G$ and $f(\mathbf{a}) = 1$. In particular,

$$0 \leq u_f(\mathbf{x}) \leq \mathcal{W}^{(N)}_{\mathbf{x}}\Big(\psi(\zeta^G) \notin B\ \&\ \zeta^G < \infty\Big) \leq 1 \quad \text{for all } \mathbf{x} \in G,$$

and so we need only check that $\overline{\lim}_{\substack{\mathbf{x}\to\mathbf{b}\\ \mathbf{x}\in G}} u_f(\mathbf{x}) > 0$. To this end, first note that, since

$$\lim_{\substack{\mathbf{x}\to\mathbf{a}\\ \mathbf{x}\in G}} u_f(\mathbf{x}) = f(\mathbf{a}) = 1,$$

the Strong Minimum Principle (cf. Theorem 10.1.6) says that $u_f > 0$ everywhere in $G$. Next, because $\mathbf{b}$ is not regular, we can find a $\delta > 0$ and a sequence $\{\mathbf{x}_n : n \geq 1\} \subseteq G$ such that $\mathbf{x}_n \to \mathbf{b}$ and

$$\epsilon \equiv \inf_{n\in\mathbb{Z}^+} \mathcal{W}^{(N)}_{\mathbf{x}_n}(\zeta^G > \delta) > 0.$$

Moreover, by the Markov property, we know that

$$u_f(\mathbf{x}_n) \geq \mathbb{E}^{\mathcal{W}^{(N)}_{\mathbf{x}_n}}\Big[f\big(\psi(\zeta^G)\big),\, \delta < \zeta^G < \infty\Big] = \int_G u_f(\mathbf{y})\, p^G(\delta, \mathbf{x}_n, \mathbf{y})\, d\mathbf{y}.$$

At the same time, we know that $p^G(\delta, \mathbf{x}_n, \mathbf{y}) \leq g^{(N)}(\delta, \mathbf{y} - \mathbf{x}_n)$, and therefore that

$$\sup_{n\in\mathbb{Z}^+} \int_{G\setminus K} p^G(\delta, \mathbf{x}_n, \mathbf{y})\, d\mathbf{y} \leq \frac{\epsilon}{2}$$

for some compact subset $K$ of $G$. Hence,

$$\overline{\lim_{\substack{\mathbf{x}\to\mathbf{b}\\ \mathbf{x}\in G}}}\, u_f(\mathbf{x}) \geq \overline{\lim_{n\to\infty}}\, u_f(\mathbf{x}_n) \geq \frac{\epsilon}{2} \inf_{\mathbf{y}\in K} u_f(\mathbf{y}) > 0. \quad \square$$

As a consequence of Lemma 11.1.21, I will now show that solutions to the Dirichlet problem will not, in general, approach the correct value at points outside of $\partial_{\text{reg}}G$.

THEOREM 11.1.22. *Let $G$ be a connected open set in $\mathbb{R}^N$, and assume that $\partial_{\text{reg}}G \neq \emptyset$. If $\mathbf{b} \in \partial G \setminus \partial_{\text{reg}}G$, then there exists an $f \in C\big(\partial G; [0,1]\big)$ which has the property that*

$$\lim_{\substack{\mathbf{x}\to\mathbf{b}\\ \mathbf{x}\in G}} u_f(\mathbf{x}) \neq f(\mathbf{b}).$$

PROOF: Given $\mathbf{b}$, use Lemma 11.1.21 to find an $r \in (0,\infty)$ so that $\partial G \not\subseteq B(\mathbf{b},r)$ and

$$\varlimsup_{\substack{\mathbf{x}\to\mathbf{b}\\ \mathbf{x}\in G}} \mathcal{W}_{\mathbf{x}}^{(N)}\Big(\psi(\zeta^G) \notin B(\mathbf{b},r)\ \&\ \zeta^G < \infty\Big) > 0,$$

and construct $f$ so that $f \equiv 1$ on $\partial G \cap B(\mathbf{b},r)\complement$ and $f(\mathbf{b}) = 0$. Then $f(\mathbf{b}) < \varlimsup_{\substack{\mathbf{x}\to\mathbf{b}\\ \mathbf{x}\in G}} u_f(\mathbf{x})$. □

I next take a closer look at the conditions under which we can assert the uniqueness of solutions to the Dirichlet problem. To begin, observe that, by Corollary 11.1.19, the situation is quite satisfactory when (11.1.20) holds. In fact, the same line of reasoning which I used there shows that the same conclusion holds as soon as one knows that $\mathcal{W}_{\mathbf{x}}^{(N)}(\zeta^G < \infty)$ is bounded below by a positive constant; and therefore, because $\mathbf{x} \in G \longmapsto \mathcal{W}_{\mathbf{x}}^{(N)}(\zeta^G < \infty)$ is a bounded harmonic function which tends to 1 at $\partial_{\text{reg}}G$, Theorem 11.1.17 tells us that

$$\inf_{\mathbf{x}\in G} \mathcal{W}_{\mathbf{x}}^{(N)}(\zeta^G < \infty) > 0 \implies \inf_{\mathbf{x}\in G} \mathcal{W}_{\mathbf{x}}^{(N)}(\zeta^G < \infty) = 1. \tag{11.1.23}$$

I will close this discussion of the Dirichlet problem with two results which reflect the transience of Brownian paths in three and higher dimensions and their recurrence in one and two dimensions.

THEOREM 11.1.24. *Assume that $N \geq 3$, and let $G$ be a nonempty, connected, open subset of $\mathbb{R}^N$. If $f \in C_c(\partial G;\mathbb{R})$, then $u_f$ is the one and only bounded harmonic function $u$ on $G$ which tends to $f$ at $\partial_{\text{reg}}G$ and satisfies*

$$\lim_{\substack{|\mathbf{x}|\to\infty\\ \mathbf{x}\in G}} u(\mathbf{x}) = 0. \tag{11.1.25}$$

PROOF: We already know that $u_f$ is a bounded harmonic function which tends to $f$ at $\partial_{\text{reg}}G$, but we must still show that it satisfies (11.1.25). For this purpose, choose $r \in (0,\infty)$ so that $f$ is supported in $B(\mathbf{0},r)$. Then (cf. the last part of Theorem 10.1.11), because $N \geq 3$,

$$\big|u_f(\mathbf{x})\big| \leq \|f\|_{\text{u}} \mathcal{W}_{\mathbf{x}}^{(N)}(\zeta_r < \infty) \longrightarrow 0 \quad \text{as } |\mathbf{x}| \to \infty.$$

To prove that $u_f$ is the only such function $u$, select bounded open sets $G_n \nearrow G$ with $\overline{G_n} \subset\subset G$, and note that, for each $T \in (0,\infty)$,

$$\begin{aligned} u(\mathbf{x}) &= \lim_{n\to\infty} \mathbb{E}^{\mathcal{W}_{\mathbf{x}}^{(N)}}\Big[u\big(\psi(T\wedge\zeta^{G_n})\big)\Big] \\ &= \mathbb{E}^{\mathcal{W}_{\mathbf{x}}^{(N)}}\Big[f\big(\psi(\zeta^G)\big),\, \zeta^G \leq T\Big] + \mathbb{E}^{\mathcal{W}_{\mathbf{x}}^{(N)}}\Big[u\big(\psi(T)\big),\, T < \zeta^G < \infty\Big] \\ &\qquad + \mathbb{E}^{\mathcal{W}_{\mathbf{x}}^{(N)}}\Big[u\big(\psi(T)\big),\, \zeta^G = \infty\Big]. \end{aligned}$$

Clearly,

$$u_f(\mathbf{x}) = \lim_{T\nearrow\infty} \mathbb{E}^{\mathcal{W}_{\mathbf{x}}^{(N)}}\Big[f(\psi(\zeta^G)),\, \zeta^G \le T\Big]$$

and

$$\lim_{T\nearrow\infty} \mathbb{E}^{\mathcal{W}_{\mathbf{x}}^{(N)}}\Big[u(\psi(T)),\, T < \zeta^G < \infty\Big] = 0.$$

Finally, because $N \ge 3$ and, therefore, by Corollary 10.1.12, $|\psi(T)| \longrightarrow \infty$ as $T \nearrow \infty$ for $\mathcal{W}_{\mathbf{x}}^{(N)}$-almost every $\psi \in C(\mathbb{R}^N)$, (11.1.25) guarantees that

$$\lim_{T\nearrow\infty} \mathbb{E}^{\mathcal{W}_{\mathbf{x}}^{(N)}}\Big[u(\psi(T)),\, \zeta^G = \infty\Big] = 0,$$

which completes the proof that $u = u_f$. □

The situation when $N \in \{1,2\}$ is more complicated.

THEOREM 11.1.26. *If $N \in \{1,2\}$, then for every non-empty, open set $G$ in $\mathbb{R}^N$,*

$$\mathcal{W}_{\mathbf{x}}^{(N)}\big(\zeta^G < \infty\big) = 1 \text{ for all } \mathbf{x} \in G \quad \text{or} \quad \mathcal{W}_{\mathbf{x}}^{(N)}\big(\zeta^G < \infty\big) = 0 \text{ for all } \mathbf{x} \in G,$$

*depending on whether $\partial_{\text{reg}} G \ne \emptyset$ or $\partial_{\text{reg}} G = \emptyset$. Moreover, if $\partial_{\text{reg}} G = \emptyset$, then the only functions $u \in C^2\big(G; [0,\infty)\big)$ satisfying $\Delta u \le 0$ are constant. In particular, either $\partial_{\text{reg}} G = \emptyset$, and there are no non-constant, nonnegative harmonic functions on $G$, or $\partial_{\text{reg}} G \ne \emptyset$, and, for each $f \in C_{\text{b}}(\partial G; \mathbb{R})$, $u_f$ is the unique bounded harmonic function on $G$ which tends to $f$ at $\partial_{\text{reg}} G$.*

PROOF: Suppose that $\mathcal{W}_{\mathbf{x}_0}^{(N)}\big(\zeta^G < \infty\big) < 1$ for some $\mathbf{x}_0 \in G$, and choose open sets $G_n \nearrow G$ so that $\mathbf{x}_0 \in G_1$ and $\overline{G_n} \subset\subset G$ for all $n \in \mathbb{Z}^+$. Given $u \in C^2\big(G; [0,\infty)\big)$ with $\Delta u \le 0$, set

$$X_n(t,\psi) = \mathbf{1}_{(t,\infty]}\big(\zeta^{G_n}(\psi)\big)\, u\big(\psi(t)\big) \quad \text{for } (t,\psi) \in [0,\infty) \times C(\mathbb{R}^N).$$

Then $\big(-X_n(t), \mathcal{F}_t, \mathcal{W}_{\mathbf{x}_0}^{(N)}\big)$ is a non-positive, right-continuous, $\mathcal{W}_{\mathbf{x}_0}^{(N)}$-submartingale when $\mathcal{F}_t = \sigma\big(\{\psi(\tau) : \tau \in [0,t]\}\big)$. Hence, since

$$X_n(t,\psi) \nearrow X(t,\psi) \equiv \mathbf{1}_{(t,\infty]}\big(\zeta^G\big)\, u\big(\psi(t)\big) \quad \text{pointwise as } n \to \infty,$$

an application of The Monotone Convergence Theorem allows us to conclude that $\big(-X(t), \mathcal{F}_t, \mathcal{W}_{\mathbf{x}_0}^{(N)}\big)$ is also a non-positive, continuous, submartingale. In particular, by Theorem 7.1.10, this means that

$$\lim_{t\to\infty} u\big(\psi(t)\big) \text{ exists for } \mathcal{W}_{\mathbf{x}_0}^{(N)}\text{-almost every } \psi \in \{\zeta^G = \infty\}.$$

At the same time, by Theorem 10.2.3, we know that, for $\mathcal{W}^{(N)}_{\mathbf{x}_0}$-almost every $\psi \in C(\mathbb{R}^N)$,

$$\int_0^\infty \mathbf{1}_U(\psi(t))\,dt = \infty \quad \text{for all open } U \neq \emptyset.$$

Hence, since $\mathcal{W}^{(N)}_{\mathbf{x}_0}(\zeta^G = \infty) > 0$, there exists a $\psi_0 \in C(\mathbb{R}^N)$ with the properties that $\psi(0) = \mathbf{x}_0$, $\zeta^G(\psi_0) = \infty$,

$$\int_0^\infty \mathbf{1}_U(\psi_0(t))\,dt = \infty \quad \text{for all open } U \neq \emptyset, \quad \text{and } \lim_{t\to\infty} u(\psi_0(t)) \text{ exists,}$$

which is possible only if $u$ is constant. In other words, we have now proved that when $\mathcal{W}^{(N)}_{\mathbf{x}_0}(\zeta^G < \infty) < 1$ for some $\mathbf{x}_0 \in G$, then the only $u \in C^2(G;[0,\infty))$ with $\Delta u \leq 0$ are constant.

Given the preceding paragraph, the rest is easy. Indeed, if $\partial_{\text{reg}}G = \emptyset$, then Theorem 11.1.15 already implies that $\mathcal{W}^{(N)}_{\mathbf{x}}(\zeta^G < \infty) = 0$ for all $\mathbf{x} \in G$. On the other hand, if $\mathbf{a} \in \partial_{\text{reg}}G$ but $\mathcal{W}^{(N)}_{\mathbf{x}_0}(\zeta^G < \infty) < 1$ for some $\mathbf{x}_0 \in G$, then the preceding paragraph applied to $\mathbf{x} \rightsquigarrow \mathcal{W}^{(N)}_{\mathbf{x}}(\zeta^G < \infty)$ says that $\mathcal{W}^{(N)}_{\mathbf{x}}(\zeta^G < \infty)$ is constant, which leads to the contradiction

$$1 > \mathcal{W}^{(N)}_{\mathbf{x}_0}(\zeta^G < \infty) = \lim_{\substack{\mathbf{x}\to\mathbf{a}\\ \mathbf{x}\in G}} \mathcal{W}^{(N)}_{\mathbf{x}}(\zeta^G < \infty) = 1. \quad \square$$

§ **11.1.4. Harmonic Measure.** We now have a rather complete abstract analysis of when the Dirichlet problem can be solved. Indeed, we know that, at least when $f \in C_c(\partial G;\mathbb{R})$, one cannot do better than take one's solution to be the function $u_f$ given by (11.1.18). For this reason, I will call

$$\Pi^G(\mathbf{x},\Gamma) \equiv \mathcal{W}^{(N)}_{\mathbf{x}}\big(\psi(\zeta^G) \in \Gamma,\ \zeta^G(\psi) < \infty\big) \tag{11.1.27}$$

the **harmonic measure for $G$ based at $\mathbf{x} \in G$ of the set $\Gamma \in \mathcal{B}_{\partial G}$**. Obviously, Theorem 11.1.15 says that $\Pi^G(\mathbf{x}, \partial G \setminus \partial_{\text{reg}}G) = 0$, and

$$u_f(\mathbf{x}) = \int_{\partial G} f(\boldsymbol{\eta})\,\Pi^G(\mathbf{x}, d\boldsymbol{\eta}).$$

This connection between harmonic measure and Wiener's measure is due to Doob,[2], and Doob's work was the starting point for what, in the hands of G. Hunt,[3] became an isomorphism between potential theory and the theory of Markov processes.

[2] As I said in the footnote in § 10.2.3, it was S. Kakutani who took the initial step in establishing this connection, and it was Doob who made it definitive.

[3] In 1957, Hunt published a series of three articles: "Markov processes and potentials, parts I, II, & III," *Ill. J. Math.* **1** & **2**. In these articles, he literally created the modern theory of Markov processes and established their relationship to potential theory. To see just how far Hunt's ideas can be elaborated, see M. Sharpe's *General Theory of Markov Processes*, Acad. Press Series in Pure & Appl. Math. **133** (1988).

Although (11.1.27) provides an intuitively appealing formula for the harmonic measure $\Pi^G(\mathbf{x}, \cdot\,)$, it hardly can be considered *explicit*. Thus, in this subsection I will write down two important examples in which explicit formulas for the harmonic measure are readily available. The first example is the one discussed in Exercise 10.2.22, namely, when $G$ is a half-space. To be precise, if $N = 1$ and $G = (0,\infty)$, then, because one-dimensional Wiener paths hit points, it is clear that $\Pi^{(0,\infty)}(\mathbf{x}, \cdot\,)$ is nothing but the point mass $\delta_0$ for all $\mathbf{x} \in (0,\infty)$. On the other hand, if $N \geq 2$ and $G = \mathbb{R}^N_+ \equiv \mathbb{R}^{N-1} \times (0,\infty)$, then we know from Exercise 10.2.22 and (3.3.19) that, for $y \in (0,\infty)$,

$$\Pi^{\mathbb{R}^N_+}\big((\mathbf{0}, y), d\boldsymbol{\omega}\big) = \frac{2}{\omega_{N-1}} \frac{y}{\big(y^2 + |\boldsymbol{\omega}|^2\big)^{\frac{N}{2}}} \lambda_{\mathbb{R}^{N-1}}(d\boldsymbol{\omega}), \quad y \in (0,\infty),$$

where $\omega_{N-1}$ is the surface area of $\mathbb{S}^{N-1}$ and I have identified $\partial\mathbb{R}^N_+$ with $\mathbb{R}^{N-1}$ and used $\lambda_{\mathbb{R}^{N-1}}$ to denote Lebesgue measure on $\mathbb{R}^{N-1}$. Hence, after a trivial translation,

$$\Pi^{\mathbb{R}^N_+}\big((\mathbf{x}, y), d\boldsymbol{\omega}\big) = \frac{2}{\omega_{N-1}} \frac{y}{\big(y^2 + |\mathbf{x} - \boldsymbol{\omega}|^2\big)^{\frac{N}{2}}} \lambda_{\mathbb{R}^{N-1}}(d\boldsymbol{\omega})$$
$$\text{for} \quad (\mathbf{x}, y) \in \mathbb{R}^{N-1} \times (0,\infty).$$

Moreover, by using further translation plus Wiener rotation invariance (cf. (**ii**) in Exercise 4.3.10), one can pass easily from the preceding to an explicit expression of the harmonic measure for an arbitrary half-space.

In the preceding, we were able to derive an expression giving the harmonic measure for half-spaces directly from probabilistic considerations. Unfortunately, half-spaces are essentially the only regions for which probabilistic reasoning yields such explicit expressions. Indeed, embarrassing as it is to admit, it must recognized that, when it comes to explicit expressions, the time-honored techniques of clever changes of variables followed by separation of variables are more powerful than anything which comes out of (11.1.27). To wit, I have been unable to give a truly *probabilistic derivation* of the classical formula given in the following.

THEOREM 11.1.28 (**Poisson Formula**). *Use $\lambda_{\mathbb{S}^{N-1}}$ to denote the surface measure on the unit sphere $\mathbf{S}^{N-1}$ in $\mathbb{R}^N$, and define*

$$\pi^{(N)}(\mathbf{x}, \boldsymbol{\omega}) = \frac{1}{\omega_{N-1}} \frac{1 - |\mathbf{x}|^2}{|\mathbf{x} - \boldsymbol{\omega}|^N} \quad \text{for } (\mathbf{x}, \boldsymbol{\omega}) \in B(\mathbf{0}, 1) \times \mathbb{S}^{N-1}.$$

*Then:*

$$\Pi^{B(\mathbf{0},1)}(\mathbf{x}, d\boldsymbol{\omega}) = \pi^{(N)}(\mathbf{x}, \boldsymbol{\omega})\, \lambda_{\mathbb{S}^{N-1}}(d\boldsymbol{\omega}), \quad \text{for } \mathbf{x} \in B(\mathbf{0}, 1).$$

*More generally, if* $\mathbf{c} \in \mathbb{R}^N$, $r \in (0,\infty)$, *and* $\lambda_{\mathbb{S}^{N-1}(\mathbf{c},r)}$ *denotes the surface measure on the sphere* $\mathbb{S}^{N-1}(\mathbf{c},r) \equiv \partial B(\mathbf{c},r)$, *then*

$$\Pi^{B(\mathbf{c},r)}(\mathbf{x},d\boldsymbol{\omega}) = \frac{1}{\omega_{N-1}r}\frac{r^2-|\mathbf{x}-\mathbf{c}|^2}{|\mathbf{x}-\boldsymbol{\omega}|^N}\lambda_{\mathbb{S}^{N-1}(\mathbf{c},r)}(d\boldsymbol{\omega}), \quad \mathbf{x} \in B(\mathbf{c},r).$$

*Equivalently, for each open* $G$ *in* $\mathbb{R}^N$, *harmonic function* $u$ *on* $G$, $\overline{B(\mathbf{c},r)} \subset\subset G$, *and* $\mathbf{x} \in B(\mathbf{c},r)$,

$$u(\mathbf{x}) = \int_{\mathbb{S}^{N-1}} u(\mathbf{c}+r\boldsymbol{\omega})\,\pi^{(N)}\left(\tfrac{\mathbf{x}-\mathbf{c}}{r},\boldsymbol{\omega}\right)\lambda_{\mathbb{S}^{N-1}}(d\boldsymbol{\omega}).$$

*In particular, if* $\{u_n : n \geq 1\}$ *is a sequence of harmonic functions on the open set* $G$ *and if* $u_n \longrightarrow u$ *boundedly and pointwise on compact subsets of* $G$, *then* $u$ *is harmonic on* $G$ *and* $u_n \longrightarrow u$ *uniformly on compact subsets. (See Exercise 11.2.22 for another approach.)*

PROOF: Set $B = B(\mathbf{0},1)$. Clearly, everything except the final assertion follows by scaling and translation once we identify $\pi^{(N)}$ as the density for $\Pi^B$. To make this identification, first check, by direct calculation, that $\pi^{(N)}(\,\cdot\,,\boldsymbol{\omega})$ is harmonic in $B$ for each $\boldsymbol{\omega} \in \mathbb{S}^{N-1}$. Hence, in order to complete the proof, all that we have to do is check that

$$\lim_{\substack{\mathbf{x}\to\mathbf{a}\\ \mathbf{x}\in B}} \int_{\mathbb{S}^{N-1}} f(\boldsymbol{\omega})\,\pi^{(N)}(\mathbf{x},\boldsymbol{\omega})\,\lambda_{\mathbb{S}^{N-1}}(d\boldsymbol{\omega}) = f(\mathbf{a})$$

for every $f \in C(\mathbb{S}^{N-1};\mathbb{R})$ and $\mathbf{a} \in \mathbb{S}^{N-1}$. Since, for each $\delta > 0$, it is clear that

$$\lim_{\substack{\mathbf{x}\to\mathbf{a}\\ \mathbf{x}\in B}} \int_{\mathbb{S}^{N-1}\cap B(\mathbf{a},\delta)\complement} \pi^{(N)}(\mathbf{x},\boldsymbol{\omega})\,\lambda_{\mathbb{S}^{N-1}}(d\boldsymbol{\omega}) = 0,$$

we will be done as soon as we show that

$$\int_{\mathbb{S}^{N-1}} \pi^{(N)}(\mathbf{x},\boldsymbol{\omega})\,\lambda_{\mathbb{S}^{N-1}}(d\boldsymbol{\omega}) = 1 \quad \text{for all } \mathbf{x} \in B.$$

But, because, for each $\boldsymbol{\xi} \in \mathbb{S}^{N-1}$, $\pi^{(N)}(\,\cdot\,,\boldsymbol{\xi})$ is harmonic in $B$ and, by (10.2.7),

$$\Pi^{B(\mathbf{0},r)}(\mathbf{0},\,\cdot\,) = \frac{\lambda_{\mathbb{S}^{N-1}(\mathbf{0},r)}}{\omega_{N-1}r^{N-1}} \quad \text{for each } r \in (0,\infty),$$

we have that, for $r \in [0,1)$ and $\boldsymbol{\xi} \in \mathbb{S}^{N-1}$,

$$1 = \omega_{N-1}\pi^{(N)}(\mathbf{0},\boldsymbol{\xi}) = \int_{\mathbb{S}^{N-1}} \pi^{(N)}(r\boldsymbol{\omega},\boldsymbol{\xi})\,\lambda_{\mathbb{S}^{N-1}}(d\boldsymbol{\omega})$$

$$= \int_{\mathbb{S}^{N-1}} \pi^{(N)}(r\boldsymbol{\xi},\boldsymbol{\omega})\,\lambda_{\mathbb{S}^{N-1}}(d\boldsymbol{\omega}),$$

where, in the final step, I have used the easily verified identity

$$\pi^{(N)}(r\boldsymbol{\xi},\boldsymbol{\omega}) = \pi^{(N)}(r\boldsymbol{\omega},\boldsymbol{\xi}) \quad \text{for all } r \in [0,1) \text{ and } (\boldsymbol{\xi},\boldsymbol{\omega}) \in \left(\mathbb{S}^{N-1}\right)^2.$$

Thus, by writing $\mathbf{x} = r\boldsymbol{\xi}$, we obtain the desired identity. □

When $N = 2$, one gets the following dividend from Theorem 11.1.28.

COROLLARY 11.1.29. *Set* $\mathbf{D}(r) = B(\mathbf{0}, r)$ *in* $\mathbb{R}^2$ *for* $r \in (0, \infty)$. *Then*

$$\Pi^{\mathbb{R}^2 \setminus \overline{\mathbf{D}(r)}}(\mathbf{x}, d\boldsymbol{\omega}) = \frac{r|\mathbf{x}|^2}{2\pi} \frac{|\mathbf{x}|^2 - r^2}{\big||\mathbf{x}|^2\boldsymbol{\omega} - r^2\mathbf{x}\big|^2} \lambda_{\mathbb{S}^1(\mathbf{0},r)}(d\boldsymbol{\omega}) \tag{11.1.30}$$

*for each* $\mathbf{x} \notin \overline{\mathbf{D}(r)}$. *In particular, if* $u \in C_{\mathrm{b}}(\mathbb{R}^2 \setminus \mathbf{D}(r); \mathbb{R})$ *is harmonic on* $\mathbb{R}^2 \setminus \overline{\mathbf{D}(r)}$, *then*

$$u(\mathbf{x}) = \frac{|\mathbf{x}|^2}{2\pi} \int_{\mathbb{S}^1} \frac{|\mathbf{x}|^2 - r^2}{\big||\mathbf{x}|^2\boldsymbol{\omega} - r\mathbf{x}\big|^2} u(r\boldsymbol{\omega}) \lambda_{\mathbb{S}^1}(d\boldsymbol{\omega}),$$

*and so*

$$\lim_{|\mathbf{x}| \to \infty} u(\mathbf{x}) = \frac{1}{2\pi} \int_{\mathbb{S}^1} u(r\boldsymbol{\omega}) \, \lambda_{\mathbb{S}^1}(d\boldsymbol{\omega}). \tag{11.1.31}$$

PROOF: After an easy scaling argument, I may and will assume that $r = 1$. Thus, set $\mathbf{D} = \mathbf{D}(1)$, and assume that $u \in C_{\mathrm{b}}(\mathbb{R}^2 \setminus \mathbf{D}; \mathbb{R})$ is harmonic in $\mathbb{R}^2 \setminus \overline{\mathbf{D}}$. Next, set $v(\mathbf{x}) = u\left(\frac{\mathbf{x}}{|\mathbf{x}|^2}\right)$ for $\mathbf{x} \in \mathbf{D} \setminus \{\mathbf{0}\}$. Obviously, $v$ is bounded and continuous. In addition, by using polar coordinates, one can easily check that $v$ is harmonic in $\mathbf{D} \setminus \{\mathbf{0}\}$. In particular, if $\rho \in (0,1)$ and $G(\rho) \equiv B \setminus \overline{B(\mathbf{0}, \rho)}$, then

$$v(\mathbf{x}) = \mathbb{E}^{\mathcal{W}_{\mathbf{x}}^{(N)}}\Big[v(\boldsymbol{\psi}(\zeta_1)),\, \zeta_1 < \zeta_\rho\Big] + \mathbb{E}^{\mathcal{W}_{\mathbf{x}}^{(N)}}\Big[v(\boldsymbol{\psi}(\zeta_\rho)),\, \zeta_\rho < \zeta_1\Big], \quad \mathbf{x} \in G(\rho),$$

where the notation is that in Theorem 10.1.11. Hence, because, by that theorem, $\zeta_\rho \nearrow \infty$ (a.s., $\mathcal{W}_{\mathbf{x}}^{(N)}$) as $\rho \searrow 0$, this leads to

$$v(\mathbf{x}) = \mathbb{E}^{\mathcal{W}_{\mathbf{x}}^{(N)}}\Big[v(\boldsymbol{\psi}(\zeta_1)),\, \zeta_1 < \infty\Big] = \frac{1}{2\pi} \int_{\mathbb{S}^1} \frac{1 - |\mathbf{x}|^2}{|\boldsymbol{\omega} - \mathbf{x}|^2} u(\boldsymbol{\omega}) \lambda_{\mathbb{S}^1}(d\boldsymbol{\omega})$$

for all $\mathbf{x} \in \mathbf{D} \setminus \{\mathbf{0}\}$. Finally, given the preceding, the rest comes down to a simple matter of bookkeeping. □

As a second application of Poisson's formula, I make the following famous observation, which can be viewed as a quantitative version of the Strong Minimum Principle (cf. Theorem 10.1.6) for harmonic functions.

COROLLARY 11.1.32 (**Harnack's Principle**). *For any* $\mathbf{c} \in \mathbb{R}^N$ *and* $r \in (0, \infty)$,

$$\begin{aligned} &\frac{r^{N-2}(r - |\mathbf{x} - \mathbf{c}|)}{(r + |\mathbf{x} - \mathbf{c}|)^{N-1}} \Pi^{B(\mathbf{c},r)}(\mathbf{c}, \cdot\,) \\ &\qquad \le \Pi^{B(\mathbf{c},r)}(\mathbf{x}, \cdot\,) \le \frac{r^{N-2}(r + |\mathbf{x} - \mathbf{c}|)}{(r - |\mathbf{x} - \mathbf{c}|)^{N-1}} \Pi^{B(\mathbf{c},r)}(\mathbf{c}, \cdot\,). \end{aligned} \tag{11.1.33}$$

*for all* $\mathbf{x} \in B(\mathbf{c}, r)$. *Hence, if* $u$ *is a non-negative, harmonic function on* $B(\mathbf{c}, r)$, *then*

$$\frac{r^{N-2}(r - |\mathbf{x}-\mathbf{c}|)}{(r + |\mathbf{x}-\mathbf{c}|)^{N-1}} u(\mathbf{c}) \le u(\mathbf{x}) \le \frac{r^{N-2}(r + |\mathbf{x}-\mathbf{c}|)}{(r - |\mathbf{x}-\mathbf{c}|)^{N-1}} u(\mathbf{c}). \tag{11.1.34}$$

*In particular, if* $G$ *is a connected region in* $\mathbb{R}^N$ *and* $\{u_n : n \ge 1\}$ *is a non-decreasing sequence of harmonic functions on* $G$, *then either* $\lim_{n\to\infty} u(\mathbf{x}) = \infty$ *for every* $\mathbf{x} \in G$ *or there is a harmonic function* $u$ *on* $G$ *to which* $\{u_n : n \ge 1\}$ *converges uniformly on compact subsets of* $G$.

PROOF: The inequalities in (11.1.33) are immediate consequences of Poisson's formula and the triangle inequality; and, given (11.1.33), the inequalities in (11.1.34) comes from integrating the inequalities in (11.1.33). Finally, let a connected, open set $G$ and a nondecreasing sequence $\{u_n : n \ge 1\}$ of harmonic functions be given. By replacing $u_n$ with $u_n - u_0$ if necessary, I may and will assume that all the $u_n$'s are nonnegative. Next, for each $\mathbf{x} \in G$, set $u(\mathbf{x}) = \lim_{n\to\infty} u_n(\mathbf{x}) \in [0, \infty]$. Because (11.1.34) holds for each of the $u_n$'s and $B(\mathbf{c}, r) \subset\subset G$, the Monotone Convergence Theorem allows us to conclude that it also holds for $u$ itself. Hence, we know that both $\{\mathbf{x} \in G : u(\mathbf{x}) = \infty\}$ and $\{\mathbf{x} \in G : u(\mathbf{x}) < \infty\}$ are open subsets of $G$, and so one of them must be empty. Finally, assume that $u < \infty$ everywhere on $G$, and suppose that $\overline{B(\mathbf{c}, 2r)} \subset\subset G$. Then, by the right-hand side of (11.1.34), the $u_n$'s are uniformly bounded on $B\left(\mathbf{c}, \frac{3r}{2}\right)$, and so, by the last part of Theorem 11.1.28, we know that $u$ is harmonic and that $u_n \longrightarrow u$ uniformly on $B(\mathbf{c}, r)$. □

Notice that, by taking $\mathbf{c} = \mathbf{0}$ and letting $r \nearrow \infty$ in (11.1.34), one gets an easy derivation of the following general statement, of which we already know a sharper version (cf. Theorem 11.1.26) when $N \in \{1, 2\}$.

COROLLARY 11.1.35 (**Liouville Theorem**). *The only nonnegative harmonic functions on* $\mathbb{R}^N$ *are constant.*

## Exercises for § 11.1

EXERCISE 11.1.36. Once I reduced the problem to that of studying $v$ on $\overline{\mathbf{D}}\setminus\{\mathbf{0}\}$, the rest of the argument which I used in the proof of (11.1.31) was based on a general principle. Namely, given an open $G$, a $K \subset\subset G$, and a harmonic function on $G \setminus K$, one says that $K$ is a **removable singularity** for $u$ in $G$ if $u$ admits a unique harmonic extension to the whole of $G$.

(**i**) Let $K \subset\subset \mathbb{R}^N$, and take $\sigma_K(\psi) = \inf\{t > 0 : \psi(t) \in K\}$ to be the first positive entrance time of $\psi \in C(\mathbb{R}^N)$ into $K$. Given an open $G \supset\supset K$, show that

$$\mathcal{W}_{\mathbf{x}}^{(N)}(\sigma_K < \zeta^G) = 0 \quad \text{for all } \mathbf{x} \in G \setminus K \tag{11.1.37}$$

if and only if $K \cap \partial_{\text{reg}}(G \setminus K) = \emptyset$, and use the locality proved in Lemma 10.2.11 to conclude that (11.1.37) for some $G \supset\supset K$ is equivalent to $K \cap \partial_{\text{reg}}(G \setminus K) = \emptyset$ for all $G \supset\supset K$. In particular, conclude that (11.1.37) holds for some $G \supset\supset K$ if and only if

$$\mathcal{W}_{\mathbf{x}}^{(N)}\Big(\exists t \in [0,\infty) \ \psi(t) \in K\Big) = 0 \quad \text{for all } \mathbf{x} \notin K. \tag{11.1.38}$$

**(ii)** Let $K \subset\subset \mathbb{R}^N$ be given, and assume that (11.1.38) holds. Given $G \supset\supset K$ and a $u \in C(G;\mathbb{R})$ which is harmonic on $G \setminus K$, show that $K$ is a removable singularity for $u$ in $G$.

**Hint**: Begin by choosing a bounded open set $H \supset\supset K$ so that $\overline{H} \subset\subset G$. Next, set

$$\sigma_n(\psi) = \inf\Big\{t > 0 : \operatorname{dist}(\psi(t), K) \le \tfrac{1}{2n}\operatorname{dist}(K, H\complement)\Big\},$$

and define $u_n$ on $H$ by

$$u_n(\mathbf{x}) = \mathbb{E}^{\mathcal{W}_{\mathbf{x}}^{(N)}}\Big[u(\psi(\zeta^H)), \zeta^H < \sigma_n\Big].$$

Show that, on the one hand, $u_n \longrightarrow u$ on $H \setminus K$, while, on the other hand,

$$\lim_{n\to\infty} u_n(\mathbf{x}) = \mathbb{E}^{\mathcal{W}_{\mathbf{x}}^{(N)}}\Big[u(\psi(\zeta^H)), \zeta^H < \infty\Big]$$

for all $\mathbf{x} \in H$.

**(iii)** Let $K$ be a compact subset of $\mathbb{R}^N$ and a connected $G \supset\supset K$ be given. Assuming either that $N \ge 3$ or that $\partial_{\text{reg}} G \ne \emptyset$, show that (11.1.38) holds if $K$ is a removable singularity in $G$ for every bounded, harmonic function on $G \setminus K$.

**Hint**: Consider the function $\mathbf{x} \in G \setminus K \longmapsto \mathcal{W}_{\mathbf{x}}^{(N)}(\sigma_K < \zeta^G) \in [0,1]$, and use the Strong Minimum Principle.

**(iv)** Let $G$ be a non-empty, open subset of $\mathbb{R}^N$, where $N \ge 2$, and set $D = \{(\mathbf{x},\mathbf{x}) : \mathbf{x} \in G\}$, the diagonal in $G^2$. Given a $u \in C(G^2;\mathbb{R})$ which is harmonic on $G \setminus D$, show that $u$ is harmonic on $G^2$.

**Hint**: Show that

$$\begin{aligned}
&\mathcal{W}_{\mathbf{x},\mathbf{y}}^{(2N)}\big(\exists t \in [0,\infty) \ \psi(t) \in D\big) \\
&\quad \le \int_{C(\mathbb{R}^N)} \mathcal{W}_{\mathbf{y}}^{(N)}\big(\exists t \in (0,\infty) \ \psi(t) = \varphi(t)\big)\, \mathcal{W}_{\mathbf{x}}^{(N)}(d\varphi) = 0
\end{aligned}$$

for $(\mathbf{x},\mathbf{y}) \in G^2 \setminus D$.

EXERCISE 11.1.39. For each $r \in (0,\infty)$, let $S(r)$ denote the open vertical strip $(-r,r)\times\mathbb{R}$ in $\mathbb{R}^2$. Clearly,

$$\zeta^{S(r)}(\psi) = \zeta_r^{(1)}(\psi) \equiv \inf\{t \geq 0 : |\psi_1(t)| \geq r\},$$

and so the harmonic measure for $S(r)$, based at any point in $S(r)$, will be supported on $\{(x,y) : x = \pm r \text{ and } y \in \mathbb{R}\}$. In particular, if $u \in C_{\mathrm{b}}(\overline{S(r)};\mathbb{R})$ is bounded and harmonic on $S(r)$, then

$$\|u\|_{\mathrm{u}} \leq \sup_{y\in\mathbb{R}} |u(r,y)| \vee |u(-r,y)|. \tag{11.1.40}$$

The estimate in (11.1.40) is a primitive version of the Phragmén–Lindelöf maximum principle. To get a sharper version, one has to relax the global boundedness condition on $u \restriction S(r)$. To see what can be expected, consider the function

$$u_r(\mathbf{z}) \equiv \left(\sin\frac{\pi(x+r)}{2r}\right)\left(\cosh\frac{\pi y}{2r}\right) \quad \text{for } \mathbf{z} = (x,y) \in \mathbb{R}^2.$$

Obviously, $u_r$ is harmonic everywhere but (11.1.40) fails dramatically. Hence, even if boundedness is not necessary for (11.1.40), something is: the function cannot be allowed to grow, as $|y|\to\infty$, as fast as $u_r$ does. What follows is the outline of a proof that those harmonic functions which grow strictly slower than $u_r$ do satisfy (11.1.40). More precisely, it will be shown that, for $u \in C(\overline{S(r)};\mathbb{R})$ which are harmonic on $S(r)$,

$$\sup_{(x,y)\in S(r)} \exp\left[-\frac{\theta\pi|y|}{2r}\right]\big|u(x,y)\big| < \infty \text{ for some } \theta \in [0,1)$$
$$\implies u \text{ satisfies (11.1.40)},$$

which is the true **Phragmén–Lindelöf principle**

**(i)** Given $R \in (0,\infty)$, set $\zeta_R^{(i)}(\psi) = \inf\{t \geq 0 : |\psi_i(t)| \geq R\}$, and show that, for any $u \in C(\overline{S(r)};\mathbb{R})$ which is harmonic on $S(r)$,

$$u(\mathbf{z}) = \mathbb{E}^{\mathcal{W}_{\mathbf{z}}^{(2)}}\left[u\left(\psi(\zeta_r^{(1)})\right), \zeta_r^{(1)} \leq \zeta_R^{(2)}\right] + \mathbb{E}^{\mathcal{W}_{\mathbf{z}}^{(2)}}\left[u\left(\psi(\zeta_R^{(2)})\right), \zeta_R^{(2)} < \zeta_r^{(1)}\right]$$

for $\mathbf{z} \in S(r,R) \equiv (-r,r)\times(-R,R)$. Conclude that (11.1.40) holds as long as

$$\varliminf_{R\to\infty} \sup_{|x|\leq 1} \big|u(x,R)\big| \vee \big|u(x,-R)\big| \mathcal{W}_{\mathbf{z}}^{(2)}\left(\zeta_R^{(2)} < \zeta_r^{(1)}\right) = 0, \quad z \in S(r).$$

Thus, the desired conclusion comes down to showing that, for each $\rho \in (r,\infty)$,

$$\lim_{R\to\infty} \exp\left[\frac{\pi R}{2\rho}\right] \mathcal{W}_{\mathbf{z}}^{(2)}\left(\zeta_R^{(2)} < \zeta_r^{(1)}\right) = 0, \quad \mathbf{z} \in S(r). \tag{*}$$

**(ii)** To prove (*), let $\rho \in (r,\infty)$ be given. Show that, for $R \in (0,\infty)$ and $\mathbf{z} \in S(r,R)$,

$$\begin{aligned} u_\rho(\mathbf{z}) &\geq \left(\cosh \tfrac{\pi R}{2\rho}\right) \mathbb{E}^{\mathcal{W}_{\mathbf{z}}^{(2)}}\left[\sin \frac{\pi\left(\psi_1\left(\zeta_R^{(2)}\right)+\rho\right)}{2\rho},\, \zeta_R^{(2)} < \zeta_r^{(1)}\right] \\ &\geq \left(\cosh \tfrac{\pi R}{2\rho}\right)\left(\cos \tfrac{\pi r}{2\rho}\right) \mathcal{W}_{\mathbf{z}}^{(2)}\left(\zeta_R^{(2)} < \zeta_r^{(1)}\right), \end{aligned}$$

and from this get (*).

## § 11.2 The Poisson Problem and Green Functions

Let $G$ be an open subset of $\mathbb{R}^N$ and $f$ a smooth function on $G$. The basic problem which motivates the contents of this section is that of analyzing solutions $u$ to the **Poisson problem**

$$\tfrac{1}{2}\Delta u = -f \text{ in } G \quad \text{and} \quad \lim_{\mathbf{x}\to\mathbf{a}} u(\mathbf{x}) = 0 \text{ for } \mathbf{a} \in \partial_{\mathrm{reg}} G. \tag{11.2.1}$$

Notice that, at least when $G$ is bounded, or, more generally, whenever (11.1.20) holds, there is at most one bounded $u \in C^2(G;\mathbb{R})$ which satisfies (11.2.1). Indeed, if there were two, then their difference would be a bounded harmonic function on $G$ satisfying boundary condition 0 at $\partial_{\mathrm{reg}} G$, which, because of (11.1.20) and Corollary 11.1.19, means that this difference vanishes. Moreover, when $N \geq 3$, even if (11.1.20) fails, one can (cf. Theorem 11.1.24) recover uniqueness by adding to (11.2.1) the condition that

$$\lim_{\substack{|\mathbf{x}|\to\infty \\ \mathbf{x}\in G}} u(\mathbf{x}) = 0. \tag{11.2.2}$$

In view of the preceding discussion, the *problem* in Poisson's problem is that of proving that solutions exist. In order to get a feeling for what is involved, given $f \in C_{\mathrm{c}}(G;\mathbb{R})$, define

$$u_T(\mathbf{x}) = \mathbb{E}^{\mathcal{W}_{\mathbf{x}}^{(N)}}\left[\int_{\frac{1}{T}}^{T} \mathbf{1}_{[0,\zeta^G)}(t) f(\boldsymbol{\psi}(t))\,dt\right] = \int_{\frac{1}{T}}^{T} \left(\int_G f(\mathbf{y}) p^G(t,\mathbf{x},\mathbf{y})\,\mathbf{y}\right) dt$$

for $T \in (1,\infty)$ and $\mathbf{x} \in G$. Then, by Corollary 10.3.13,

$$\tfrac{1}{2}\Delta u_T = \int_G f(\mathbf{y})\left(p^G(T,\mathbf{x},\mathbf{y}) - p^G(T^{-1},\mathbf{x},\mathbf{y})\right) d\mathbf{y}$$
$$\text{and } \lim_{\substack{\mathbf{x}\to\mathbf{a} \\ \mathbf{x}\in G}} u_T(\mathbf{x}) = 0 \text{ for } \mathbf{a} \in \partial_{\mathrm{reg}} G.$$

Hence, at least when (11.1.20) holds and therefore $\int_G p^G(T,\mathbf{x},\mathbf{y}) f(\mathbf{y})\,d\mathbf{y} \longrightarrow 0$ as $T \nearrow \infty$, it is reasonable to hope that $u = \lim_{T\to\infty} u_T$ exists and will be the desired solution to (11.2.1). On the other hand, it is neither obvious that the limit will exist nor, even if it does exist, in what sense either the smoothness properties or (11.2.2) will survive the limit procedure.

Motivated by these considerations, I now define the **Green function** to be the function $g^G$ given by

$$g^G(\mathbf{x},\mathbf{y}) = \int_{(0,\infty)} p^G(t,\mathbf{x},\mathbf{y})\,dt, \quad (\mathbf{x},\mathbf{y}) \in G^2. \tag{11.2.3}$$

My goal in this section is to show that, in great generality, $g^G$ is the fundamental solution to (11.2.1) in the sense that $\mathbf{x} \rightsquigarrow \int_G f(\mathbf{y})g^G(\mathbf{x},\mathbf{y})\,d\mathbf{y}$ solves (11.2.1).

**§ 11.2.1. Green Functions when $N \geq 3$.** The transience of Brownian motion in $\mathbb{R}^N$ for $N \geq 3$ greatly simplifies the analysis of $g^G$ there. The basic reason why is that

$$\begin{aligned} g^{\mathbb{R}^N}(\mathbf{x},\mathbf{y}) &\equiv \int_0^\infty g^{(N)}(t,\mathbf{y}-\mathbf{x})\,dt = (2\pi)^{-\frac{N}{2}} \int_0^\infty t^{-\frac{N}{2}} e^{-\frac{|\mathbf{y}-\mathbf{x}|^2}{2t}}\,dt \\ &= \frac{\Gamma(\frac{N}{2}-1)}{2\pi^{\frac{N}{2}}|\mathbf{y}-\mathbf{x}|^{N-2}}, \end{aligned}$$

and therefore (cf. part (**i**) in Exercise 2.1.13)

$$g^{\mathbb{R}^N}(\mathbf{x},\mathbf{y}) = \frac{2|\mathbf{y}-\mathbf{x}|^{2-N}}{(N-2)\omega_{N-1}}, \tag{11.2.4}$$

where $\omega_{N-1}$ is the area of $\mathbb{S}^{N-1}$. In particular, when $N \geq 3$, $g^{\mathbb{R}^N}(\mathbf{x},\,\cdot\,)$ is smooth and has bounded derivatives of all orders in $\mathbb{R}^N \setminus B(\mathbf{x},r)$ for each $r > 0$. Next, by integrating both sides of (10.3.8) with respect to $t \in (0,\infty)$, we obtain, for any $G$, the Duhamel formula

$$g^G(\mathbf{x},\mathbf{y}) = g^{\mathbb{R}^N}(\mathbf{x},\mathbf{y}) - \mathbb{E}^{\mathcal{W}_{\mathbf{x}}^{(N)}}\left[g^{\mathbb{R}^N}\left(\psi(\zeta_{0+}^G),\mathbf{y}\right),\, \zeta_{0+}^G < \infty\right], \tag{11.2.5}$$

which means that $g^G(\mathbf{x},\,\cdot\,)$ is bounded on compact subsets of $G\setminus\{\mathbf{x}\}$. In addition, for each $\mathbf{y} \in G$, the second term on the right of (11.2.5) is a harmonic function of $\mathbf{x} \in G$, and so, for any $f \in C_c(G;\mathbb{R})$, $\int_G f(\mathbf{y})g^G(\,\cdot\,,\mathbf{y})\,d\mathbf{y}$ makes sense and differs from $\int_{\mathbb{R}^N} f(\mathbf{y})g^{\mathbb{R}^N}(\,\cdot\,,\mathbf{y})\,d\mathbf{y}$ by a function that is harmonic on $G$.

Now define

$$\mathbf{G}^G f(\mathbf{x}) = \int_G f(\mathbf{y})g^G(\mathbf{x},\mathbf{y})\,d\mathbf{y} \quad \text{for } f \in C_c(G;\mathbb{R}) \text{ and } \mathbf{x} \in G. \tag{11.2.6}$$

What we still have to find are conditions under which $\mathbf{G}^G f$ solves (11.2.1) and satisfies (11.2.2). From (11.2.5) and Theorem 10.2.14, it is clear that $\mathbf{G}^G f(\mathbf{x})$

tends to 0 as $\mathbf{x}$ tends to $\partial_{\text{reg}}G$. In addition, since $|\mathbf{G}^G f| \le \mathbf{G}^{\mathbb{R}^N}|f|$ and $\mathbf{G}^{\mathbb{R}^N}|f|$ tends to 0 at infinity, $\mathbf{G}^G f$ satisfies (11.2.2). Hence, the remaining question is whether $\frac{1}{2}\Delta\mathbf{G}^G f = -f$ on $G$. As an initial step, suppose that $\mathbf{G}^G f \in C^2_{\text{b}}(G;\mathbb{R})$, and note that, for each $\mathbf{x} \in G$,

$$\begin{aligned}\tfrac{1}{2}\Delta\mathbf{G}^G f(\mathbf{x}) &= \lim_{s\searrow 0}\frac{1}{s}\left[\int_G p^G(s,\mathbf{x},\mathbf{y})\mathbf{G}^G f(\mathbf{y})\,d\mathbf{y} - \mathbf{G}^G f(\mathbf{x})\right]\\ &= -\lim_{s\searrow 0}\frac{1}{s}\int_0^s\left(\int_G f(\mathbf{y})p^G(t,\mathbf{x},\mathbf{y})\,d\mathbf{y}\right)dt = -f(\mathbf{x}).\end{aligned}$$

Thus, what we need to know is whether $\mathbf{G}^G f \in C^2_{\text{b}}(G;\mathbb{R})$. By the considerations above, we already know that $\mathbf{G}^G f \in C^2_{\text{b}}(G;\mathbb{R})$ if and only if $\mathbf{G}^{\mathbb{R}^N} f$ is. Moreover, if $f \in C^2_{\text{c}}(G;\mathbb{R})$, then $\partial^\alpha\mathbf{G}^{\mathbb{R}^N} f = \mathbf{G}^{\mathbb{R}^N}\partial^\alpha f$ for any $\alpha$ with $\|\alpha\| \le 2$. In addition,

$$\mathbf{G}^{\mathbb{R}^N}\partial_{x_i}\partial_{x_j} f(\mathbf{x}) = \frac{2}{\omega_{N-1}}\int_{\mathbb{R}^N}\frac{(y_i - x_i)\partial_{y_j} f(\mathbf{y})}{|\mathbf{x}-\mathbf{y}|^N}\,d\mathbf{y}.$$

Hence, by starting with $f$'s that are in $C^2_{\text{c}}(G;\mathbb{R})$ and applying an obvious approximation argument, we see that $\mathbf{G}^G f \in C^2_{\text{b}}(G;\mathbb{R})$ whenever $f \in C^1_{\text{c}}(G;\mathbb{R})$.[1]

THEOREM 11.2.7. *Assume that $N \ge 3$ and that $G$ is a non-empty, open subset of $\mathbb{R}^N$. Then, for each $f \in C^1_{\text{c}}(G;\mathbb{R})$, the function $\mathbf{G}^G f$ in (11.2.6) is the unique bounded, twice differentiable solution to (11.2.1) which satisfies (11.2.2).*

REMARK 11.2.8. Notice that the Duhamel formula in (11.2.5) could have been guessed. To be precise, $g^{\mathbb{R}^N}$ is a *fundamental solution* for $-\frac{1}{2}\Delta$ in $\mathbb{R}^N$ in the sense that $\frac{1}{2}\Delta\mathbf{G}^{\mathbb{R}^N} f = -f$ all test functions $f \in C^1_{\text{c}}(\mathbb{R}^N;\mathbb{R})$, and $g^G$ is to be a fundamental solution for $-\frac{1}{2}\Delta$ in $G$ with 0 boundary data in the sense that it should be the kernel for the solution operator which solves the Poisson problem in (11.2.1). Based on these remarks, one should guess that a reasonable approach to the construction of $g^G$ would be to *correct* $g^{\mathbb{R}^N}(\,\cdot\,,\mathbf{y})$ for each $\mathbf{y} \in G$ by subtracting off a harmonic function which has $g^{\mathbb{R}^N}(\,\cdot\,,\mathbf{y})$ as its boundary value, and this is, of course, precisely what is being done in (11.2.5).

**§ 11.2.2. Green Functions when $N \in \{1,2\}$.** Because (cf. Theorem 10.2.3) Brownian paths in one and two dimensions spend infinite time in every non-empty open set, the reasoning in § 11.2.1 is too crude to handle the Poisson problem in these dimensions. In particular, when $N \in \{1,2\}$, $g^{\mathbb{R}^N}$ will be identically infinite, and so (11.2.5) does us no good. To overcome this difficulty, I will

[1] It turns out that if $f$ is Hölder continuous of some order, then $\mathbf{G}^G f$ will be twice continuously differentiable and its second derivatives will be Hölder continuous of the same order as $f$. Such results are called Schauder estimates. See, for example, N.V. Krylov's *Lectures on Elliptic and Parabolic Equations in Hölder Spaces*, A.M.S. Graduate Studies in Math. **12** (1996).

use a generalization of (11.2.5). Namely, let $H$ be an open set that contains $G$. Then, by the Markov property, it is easy to check that

$$P^H(t,\mathbf{x},\Gamma) = P^G(t,\mathbf{x},\Gamma) + \mathbb{E}^{\mathcal{W}_{\mathbf{x}}^{(N)}}\Big[P^H\big(t-\zeta^G(\psi),\psi(\zeta^G),\Gamma\big),\,\zeta^G(\psi)<\infty\Big]$$

for $(t,\mathbf{x}) \in (0,\infty) \times G$ and $\Gamma \in \mathcal{B}_G$. As a consequence, we have that

$$p^H(t,\mathbf{x},\mathbf{y}) = p^G(t,\mathbf{x},\mathbf{y}) + \mathbb{E}^{\mathcal{W}_{\mathbf{x}}^{(N)}}\Big[p^H\big(t-\zeta^H(\psi),\psi(\zeta^G),\mathbf{y}\big),\,\zeta^G(\psi)<\infty\Big]$$

for $(t,\mathbf{x},\mathbf{y}) \in (0,\infty)\times G^2$. Hence, after integrating with respect to $t \in (0,\infty)$, one obtains

$$g^H(\mathbf{x},\mathbf{y}) = g^G(\mathbf{x},\mathbf{y}) + \mathbb{E}^{\mathcal{W}_{\mathbf{x}}^{(N)}}\Big[g^H\big(t-\zeta^G(\psi),\mathbf{y}\big),\,\zeta^G(\psi)<\infty\Big] \tag{11.2.9}$$

for all $(\mathbf{x},\mathbf{y}) \in G^2$.

Of course, (11.2.9) is useful only when $g^H$ is finite and calculable, and so we have to find a suitable class of $H$'s for which it is. Thus, I will start by calculating $g^{(0,\infty)}(x,y)$ for $x,y \in (0,\infty)$. To this end, recall from Exercise 10.3.35 that $p^{(0,\infty)}(t,x,y) = g^{(1)}(t,y-x) - g^{(1)}(t,x+y)$. Next, check that

$$\int_0^\infty t^{-\frac12}\big(e^{-\frac{a^2}{2t}}-1\big)\,dt = -\frac{a^2}{2}\int_0^\infty t^{-\frac32}e^{-\frac{a^2}{2t}}\,dt = -2^{\frac12}|a|\Gamma\big(\tfrac12\big) = -(2\pi)^{\frac12}|a|$$

for any $a \in \mathbb{R}$. Thus,

$$\begin{aligned} g^{(0,\infty)}(x,y) &= \int_0^\infty \big(g(t,y-x)-g(t,y+x)\big)\,dt \\ &= (2\pi)^{-\frac12}\int_0^\infty t^{-\frac12}\Big(e^{-\frac{(y-x)^2}{2t}}-1\Big)\,dt - (2\pi)^{-\frac12}\int_0^\infty t^{-\frac12}\Big(e^{-\frac{(y+x)^2}{2t}}-1\Big)\,dt \\ &= |x+y| - |x-y| = 2(x\wedge y). \end{aligned}$$

More generally, by translation and reflection, we see that, for any $c \in \mathbb{R}$,

$$g^{(c,\infty)}(x,y) = 2(x-c)\wedge(y-c) \quad \text{for } x,\, y \in (c,\infty) \tag{11.2.10}$$

and that $g^{(-\infty,c)}(x,y) = g^{(-c,\infty)}(-x,-y)$.

Now suppose that $G \neq \mathbb{R}$ is a non-empty, connected, open subset of $\mathbb{R}$. Then, either $\pm G = (c,\infty)$ for some $c \in \mathbb{R}$ or $G = (a,b)$ for some $-\infty < a < b < \infty$. Since we already know an exact expression in the first of these cases, assume that $G = (a,b)$. By taking $H = (a,\infty)$ in (11.2.9), we know that

$$g^{(a,b)}(x,y) = g^{(a,\infty)}(x,y) - \mathbb{E}^{\mathcal{W}_x^{(1)}}\Big[g^{(a,\infty)}\big(\psi(\zeta^{(a,b)}),y\big),\,\zeta^{(a,b)}(\psi)<\infty\Big].$$

Since $\mathcal{W}_x^{(1)}(\zeta^{(a,b)} < \infty) = 1$ for all $x \in \mathbb{R}$ and the boundary of $(a,b)$ is regular, Corollary 11.1.19 together with (11.2.10) say that, as a function of $x \in (a,b)$, the second term on the right equals $-u$, where $u'' = 0$, $\lim_{x\searrow 0} u(x) = 0$, and $\lim_{x\nearrow b} u(x) = 2(y-a)$. Hence, $u(x) = \frac{2(x-a)(y-a)}{b-a}$, and so

$$g^{(a,b)}(x,y) = \frac{2}{b-a}(x \wedge y - a)(b - x \vee y). \tag{11.2.11}$$

Starting from these, it is an easy matter to check by hand that if $G \neq \mathbb{R}$ is any open interval and $f \in C_c(G;\mathbb{R})$, $\mathbf{G}^G f$ is bounded and solves (11.2.1). Moreover, because (11.1.20) holds, $\mathbf{G}^G f$ is the only such solution.

When $N = 2$, matters are significantly more complicated but much more interesting. I will begin by considering the $\mathbb{R}^2$ analog of $(0,\infty)$, which is the upper half-space $\mathbb{R}^2_+ = \{(x_1,x_2) : x_2 > 0\}$. It should be clear that, for $\mathbf{x} = (x_1,x_2)$ and $\mathbf{y} = (y_1,y_2)$,

$$p^{\mathbb{R}^2_+}(t,\mathbf{x},\mathbf{y}) = g^{(1)}(t, y_1 - x_1)p^{(0,\infty)}(t,y_1,y_2) = \frac{1}{2\pi t}\left[e^{-\frac{|\mathbf{y}-\mathbf{x}|^2}{2t}} - e^{-\frac{|\check{\mathbf{y}}-\mathbf{x}|^2}{2t}}\right],$$

where $\check{\mathbf{y}} = (y_1,-y_2)$. Therefore,

$$\begin{aligned} 2\pi \int_{(0,\infty)} & p^{\mathbb{R}^2_+}(t,\mathbf{x},\mathbf{y})\,dt \\ &= \lim_{T\nearrow\infty} \int_0^T \frac{1}{t}\left(\exp\left[-\frac{|\mathbf{y}-\mathbf{x}|^2}{2t}\right] - \exp\left[-\frac{|\check{\mathbf{y}}-\mathbf{x}|^2}{2t}\right]\right) dt \\ &= \lim_{T\nearrow\infty} \int_{|\check{\mathbf{y}}-\mathbf{x}|^{-2}}^{|\mathbf{y}-\mathbf{x}|^{-2}} \frac{1}{t} e^{-\frac{1}{2tT}}\,dt, \end{aligned}$$

which means that $g^{\mathbb{R}^2_+}(\mathbf{x},\mathbf{y}) = -\frac{1}{\pi}\log\frac{|\mathbf{y}-\mathbf{x}|}{|\check{\mathbf{y}}-\mathbf{x}|}$. Hence, by (11.2.9), we know that

$$g^G(\mathbf{x},\mathbf{y}) = g^{\mathbb{R}^2_+}(\mathbf{x},\mathbf{y}) - \mathbb{E}^{\mathcal{W}_{\mathbf{x}}^{(2)}}\left[g^{\mathbb{R}^2_+}(\psi(\zeta^G_{0+}),\mathbf{y}),\ \zeta^G_{0+} < \infty\right]$$

if $G \subseteq \mathbb{R}^2_+$. Furthermore, because $\mathbf{x} \rightsquigarrow \log|\check{\mathbf{y}}-\mathbf{x}|$ is harmonic in $G$, one can pass from the preceding to

$$g^G(\mathbf{x},\mathbf{y}) = -\frac{1}{\pi}\log|\mathbf{y}-\mathbf{x}| + \frac{1}{\pi}\mathbb{E}^{\mathcal{W}_{\mathbf{x}}^{(2)}}\left[\log|\mathbf{y} - \psi(\zeta^G_{0+})|,\ \zeta^G_{0+}(\psi) < \infty\right], \tag{11.2.12}$$

first for $G \subseteq \mathbb{R}^2_+$ and then, after translation and rotation, for $G$ contained in any half-space. In addition, by the same argument as I used in § 11.2.1, one can use (11.2.12) to check that if $G$ is contained in a half-space, then $\mathbf{G}^G f$ solves Poisson's problem for every $f \in C^1_c(G;\mathbb{R})$.

To handle regions that are not contained in a half-space, one needs to work harder.

LEMMA 11.2.13. *If $G$ is an open subset of $\mathbb{R}^2$ for which $\partial_{\text{reg}}G \neq \emptyset$, then, for each $K \subset\subset G$,*

$$\sup_{\mathbf{y}\in G}\int_K g^G(\mathbf{x},\mathbf{y})\,d\mathbf{x} = \sup_{\mathbf{x}\in G}\int_K g^G(\mathbf{x},\mathbf{y})\,d\mathbf{y} < \infty$$

*and*

$$\sup_{(\mathbf{x},\mathbf{y})\in K^2} \mathbb{E}^{\mathcal{W}^{(2)}_{\mathbf{x}}}\Big[\big|\log|\mathbf{y}-\boldsymbol{\psi}(\zeta^G)|\big|,\, \zeta^G(\boldsymbol{\psi})<\infty\Big] < \infty.$$

*In addition, for each $\mathbf{c} \in G$ and $r > 0$,*

$$(\mathbf{x},\mathbf{y}) \rightsquigarrow u_r(\mathbf{x},\mathbf{y}) \equiv \mathbb{E}^{\mathcal{W}^{(2)}_{\mathbf{x}}}\Big[\log|y-\boldsymbol{\psi}(\zeta^G)|,\, \zeta^G(\boldsymbol{\psi}) < \zeta^{B(\mathbf{c},r)}(\boldsymbol{\psi})\Big]$$

*is harmonic on $\big(G\cap B(\mathbf{c},r)\big)^2$, and, as $r\to\infty$, $\{u_r : r>0\}$ tends uniformly on compact subsets of $G^2$ to the function*

$$(\mathbf{x},\mathbf{y}) \in G^2 \longmapsto u(\mathbf{x},\mathbf{y}) \equiv \mathbb{E}^{\mathcal{W}^{(2)}_{\mathbf{x}}}\Big[\log|y-\boldsymbol{\psi}(\zeta^G)|,\, \zeta^G(\boldsymbol{\psi})<\infty\Big] \in \mathbb{R}.$$

*In particular, $u$ is harmonic on $G^2$.*

PROOF: Since $g^G$ is symmetric, the first equality is obvious. While proving the associated finiteness assertion, I may and will assume that $G$ is connected. In addition, it suffices for me to prove

$$\sup_{\mathbf{x}\in G}\mathbb{E}^{\mathcal{W}^{(2)}_{\mathbf{x}}}\left[\int_0^{\zeta^G(\boldsymbol{\psi})} \mathbf{1}_{B(\mathbf{c},r)}\big(\boldsymbol{\psi}(t)\big)\,dt\right] < \infty$$

for all $\mathbf{c} \in G$ and $r>0$ with $\overline{B(\mathbf{c},2r)} \subset\subset G$. Given such a ball, set $B = B(\mathbf{c},r)$ and $2B = B(\mathbf{c},2r)$, and define $\{\zeta_n : n \geq 0\}$ inductively by $\zeta_0 = 0$ and, for $n\geq 1$, $\zeta_{2n-1} = \inf\{t \geq \zeta_{2(n-1)} : \boldsymbol{\psi}(t) \in \overline{B}\}$ and $\zeta_{2n} = \inf\{t\geq \zeta_{2n-1} : \boldsymbol{\psi}(t) \notin 2B\}$. If $u(\mathbf{x}) = \mathcal{W}^{(2)}_{\mathbf{x}}(\zeta_1 < \zeta^G)$, then $u$ is a $[0,1]$-valued, harmonic function on $G\setminus\overline{B}$ that tends to $0$ as $\mathbf{x}$ tends to $\partial_{\text{reg}}G$ and to $1$ as $\mathbf{x}$ tends to $\partial B$. Thus, since $\partial_{\text{reg}}G\neq\emptyset$, the Minimum Principle says that $u(\mathbf{x}) \in (0,1)$ for all $\mathbf{x}\in G\setminus\overline{B}$. In particular, this means that $\alpha \equiv \max\{u(\mathbf{x}) : |\mathbf{x}-\mathbf{c}| = 2r\} \in (0,1)$. At the same time, by the Markov property,

$$\mathcal{W}^{(2)}_{\mathbf{x}}(\zeta_{2n+1}<\zeta^G) = \mathbb{E}^{\mathcal{W}^{(2)}_{\mathbf{x}}}\big[u(\boldsymbol{\psi}(\zeta_{2n})),\, \zeta_{2n}(\boldsymbol{\psi}) < \zeta^G(\boldsymbol{\psi})\big] \leq \alpha\mathcal{W}^{(2)}_{\mathbf{x}}(\zeta_{2n-1}<\zeta^G),$$

and so $\mathcal{W}^{(2)}_{\mathbf{x}}(\zeta_{2n-1}<\zeta^G) \leq \alpha^{n-1}$ for $n\in\mathbb{Z}^+$. Hence, if $f(\mathbf{y}) = \mathbb{E}^{\mathcal{W}^{(2)}_{\mathbf{y}}}[\zeta^{2B}]$, then

$$\begin{aligned}\mathbb{E}^{\mathcal{W}^{(2)}_{\mathbf{x}}}\left[\int_0^{\zeta^G}\mathbf{1}_B\big(\boldsymbol{\psi}(t)\big)\,dt\right] &= \sum_{n=1}^\infty \mathbb{E}^{\mathcal{W}^{(2)}_{\mathbf{x}}}\left[\int_{\zeta_{2n-1}}^{\zeta_{2n}}\mathbf{1}_B\big(\boldsymbol{\psi}(t)\big)\,dt,\, \zeta_{2n-1}(\boldsymbol{\psi})<\zeta^G\right]\\ &\leq \sum_{n=1}^\infty \mathbb{E}^{\mathcal{W}^{(2)}_{\mathbf{x}}}\big[f(\boldsymbol{\psi}(\zeta_{2n-1})),\, \zeta_{2n-1}(\boldsymbol{\psi})<\zeta^G(\boldsymbol{\psi})\big] \leq \frac{\|f\|_{\text{u}}}{1-\alpha}.\end{aligned}$$

Since, by Theorem 10.1.11, $f$ is bounded, this completes the proof.

Turning to the second part, begin by observing that, for each $r > 0$ and $\mathbf{x} \in G(r) \equiv G \cap B(\mathbf{c}, r)$, $u_r(\mathbf{x}, \cdot\,)$ is a harmonic function on $G(r)$. Next, given $\mathbf{y} \in G(r)$, define $f$ on $\partial G(r)$ so that $f(\boldsymbol{\xi}) = \log|\mathbf{y} - \boldsymbol{\xi}|$ or $0$ according to whether $\boldsymbol{\xi}$ is or is not an element of $\partial G(r) \setminus \partial B(\mathbf{c}, r)$. Then

$$u_r(\mathbf{x}, \mathbf{y}) = \mathbb{E}^{\mathcal{W}_{\mathbf{x}}^{(2)}}\left[f(\boldsymbol{\psi}(\zeta^{G(r)}), \zeta^{G(r)}(\boldsymbol{\psi}) < \infty\right],$$

and so $u_r(\,\cdot\,, \mathbf{y})$ is also harmonic on $G(r)$. Hence, since $u_r$ is locally bounded on $G(r)^2$, Exercise 10.2.16 applies and says that $u_r$ is harmonic on $G(r)^2$. To complete the proof, let $B$ be an open ball whose closure is contained in $G$, set $D = \operatorname{dist}(\bar{B}, G\complement)$, and choose $R > 0$ so that $\bar{B} \subseteq G(R)$. Then, for each $r > R$,

$$\begin{aligned} v_r(\mathbf{x}, \mathbf{y}) &\equiv u_r(\mathbf{x}, \mathbf{y}) - \log D\mathcal{W}_{\mathbf{x}}^{(2)}\left(\zeta^G < \zeta^{B(\mathbf{c},r)}\right) \\ &= \mathbb{E}^{\mathcal{W}_{\mathbf{x}}^{(2)}}\left[\log\frac{|\mathbf{y} - \boldsymbol{\psi}(\zeta^G)|}{D}, \zeta^G(\boldsymbol{\psi}) < \zeta^{B(\mathbf{c},r)}(\boldsymbol{\psi})\right] \end{aligned}$$

is a non-negative, harmonic function on $B^2$, and, for each $(\mathbf{x}, \mathbf{y}) \in B^2$, $v_r(\mathbf{x}, \mathbf{y})$ is non-decreasing as a function of $r > R$. Thus, by Harnack's Principle (cf. Corollary 11.1.32), either $\lim_{r\to\infty} v_r = \infty$ on $B^2$ or $v_r$ tends uniformly on compact subsets of $B^2$ to a harmonic function $v$. Since

$$\lim_{r\to\infty} \sup_{\mathbf{x}\in B}\left|\mathcal{W}_{\mathbf{x}}^{(2)}\left(\zeta^G < \zeta^{B(\mathbf{c},r)}\right) - \mathcal{W}_{\mathbf{x}}^{(2)}\left(\zeta^G < \infty\right)\right| = 0,$$

it is clear that the latter case implies that

$$\begin{aligned} &\sup_{(\mathbf{x},\mathbf{y})\in K^2} \mathbb{E}^{\mathcal{W}_{\mathbf{x}}^{(2)}}\left[\left|\log|\mathbf{y} - \zeta^G(\boldsymbol{\psi})|\right|, \zeta^G(\boldsymbol{\psi}) < \infty\right] \\ &\qquad \le \varlimsup_{r\to\infty} \sup_{(\mathbf{x},\mathbf{y})\in K^2} v_r(\mathbf{x}, \mathbf{y}) + |\log D| < \infty \end{aligned}$$

for $K \subset\subset B$ and that $u_r$ tends to $u$ uniformly on compact subsets of $B$. Hence, all that remains is for me to rule out the possibility that $\lim_{r\to\infty} v_r = \infty$ on $B^2$.

Equivalently, I must show that $\varlimsup_{r\to\infty} u_r(\mathbf{x}, \mathbf{y}) < \infty$ for some $(\mathbf{x}, \mathbf{y}) \in B^2$. For this purpose, note that, because $G(r)$ is bounded, and therefore contained in some half-space, (11.2.12) applies and says that

$$\begin{aligned} \pi g^{G(r)}(\mathbf{x}, \mathbf{y}) + \log|\mathbf{y} - \mathbf{x}| &= \mathbb{E}^{\mathcal{W}_{\mathbf{x}}^{(2)}}\left[\log\left|\mathbf{y} - \boldsymbol{\psi}(\zeta^{G(r)})\right|, \zeta^{G(r)}(\boldsymbol{\psi}) < \infty\right] \\ &= u_r(\mathbf{x}, \mathbf{y}) + \mathbb{E}^{\mathcal{W}_{\mathbf{x}}^{(2)}}\left[\log\left|\mathbf{y} - \boldsymbol{\psi}(\zeta^{B(\mathbf{c},r)})\right|, \zeta^{B(\mathbf{c},r)}(\boldsymbol{\psi}) < \zeta^G(\boldsymbol{\psi}) < \infty\right]. \end{aligned}$$

Hence, for sufficiently large $r$'s and all $(\mathbf{x}, \mathbf{y}) \in B^2$,

$$u_r(\mathbf{x}, \mathbf{y}) \le \pi g^{G(r)}(\mathbf{x}, \mathbf{y}) + \frac{1}{\pi}\log|\mathbf{y} - \mathbf{x}| \le \pi g^G(\mathbf{x}, \mathbf{y}) + \frac{1}{\pi}\log|\mathbf{y} - \mathbf{x}|,$$

which, by the first part of this lemma, means that $\varlimsup_{r\to\infty} u_r$ cannot be infinite everywhere on $B^2$. □

THEOREM 11.2.14. *Let $G$ be a non-empty, open subset of $\mathbb{R}^2$ for which $\partial_{\text{reg}}G \neq \emptyset$. Then, (11.1.20) holds,*

$$\sup_{\mathbf{x},\mathbf{y}\in K} \mathbb{E}^{\mathcal{W}_{\mathbf{x}}^{(2)}}\left[\left|\log|\mathbf{y}-\boldsymbol{\psi}(\zeta^G)|\right|, \zeta^G<\infty\right]<\infty \quad \text{for } K \subset\subset G,$$

*and*

$$(\mathbf{x},\mathbf{y}) \in G^2 \longmapsto \mathbb{E}^{\mathcal{W}_{\mathbf{x}}^{(2)}}\left[\log|\mathbf{y}-\boldsymbol{\psi}(\zeta^G)|, \zeta^G<\infty\right] \in \mathbb{R}$$

*is a harmonic function. In addition, for each $\mathbf{c} \in G$, the limit*

$$h^G(\mathbf{x}) \equiv \lim_{r\to\infty} \frac{\log r}{\pi} \mathcal{W}_{\mathbf{x}}^{(2)}\left(\zeta^{B(\mathbf{c},r)} \leq \zeta^G\right), \quad \mathbf{x} \in G, \tag{11.2.15}$$

*exists, is uniform with respect to $\mathbf{x}$ in compact subsets of $G$ and independent of $\mathbf{c} \in G$, and determines a harmonic function of $\mathbf{x} \in G$. Finally,*

$$g^G(\mathbf{x},\mathbf{y}) = -\frac{1}{\pi}\log|\mathbf{y}-\mathbf{x}| + \frac{1}{\pi}\mathbb{E}^{\mathcal{W}_{\mathbf{x}}^{(2)}}\left[\log|\mathbf{y}-\boldsymbol{\psi}(\zeta^G)|, \zeta^G<\infty\right] + h^G(\mathbf{x}) \tag{11.2.16}$$

*for all distinct $\mathbf{x}$'s and $\mathbf{y}$'s from $G$, and so either $h^G \equiv 0$ or $G$ is unbounded and*

$$g^G(\,\cdot\,,\mathbf{y}) \longrightarrow h^G \quad \text{uniformly on compacts as } |\mathbf{y}| \to \infty \text{ through } G. \tag{11.2.17}$$

PROOF: Note that, because $N = 2$, Theorem 11.1.26 guarantees that (11.1.20) follows from $\partial_{\text{reg}}G \neq \emptyset$, and the rest of the initial assertion is covered by Lemma 11.2.13.

To prove the remaining assertions, let $\mathbf{c} \in G$ be given, set $G(r) = G \cap B(\mathbf{c}, r)$, and set $g_r(\mathbf{x},\mathbf{y}) = g^{G(r)}(\mathbf{x},\mathbf{y})$ for $(\mathbf{x},\mathbf{y}) \in G(r)^2$. By (11.2.12),

$$g_r(\mathbf{x},\mathbf{y}) = -\frac{1}{\pi}\log|\mathbf{y}-\mathbf{x}| + \frac{1}{\pi}\mathbb{E}^{\mathcal{W}_{\mathbf{x}}^{(2)}}\left[\log|\mathbf{y}-\boldsymbol{\psi}(\zeta^{G(r)})|, \zeta^{G(r)}(\boldsymbol{\psi})<\infty\right].$$

In particular, for each $(\mathbf{x},\mathbf{y}) \in G(r)^2$, $g_r(\,\cdot\,,\mathbf{y})$ is harmonic on $G(r)\setminus\{\mathbf{y}\}$ and $g_r(\mathbf{x},\,\cdot\,)$ is harmonic on $G(r) \setminus \{\mathbf{x}\}$. Hence, by Exercise 10.2.16, $g_r$ is a non-negative, harmonic function on $\widehat{G(r)^2} \equiv \{(\mathbf{x},\mathbf{y}) \in G(r)^2 : \mathbf{x}\neq\mathbf{y}\}$. At the same time, because $p^{G(r)}(t,\mathbf{x},\mathbf{y})$ is non-decreasing in $r$ for each $(t,\mathbf{x},\mathbf{y}) \in (0,\infty)\times G(r)^2$, we know that $g_r$ is non-decreasing in $r$. Hence, by Harnack's Principle (cf. Corollary 11.1.32), either $\lim_{r\nearrow\infty} g_r$ is everywhere infinite on $\widehat{G^2} \equiv \{(\mathbf{x},\mathbf{y}) \in G^2 : \mathbf{x}\neq\mathbf{y}\}$ or $g_r$ converges uniformly on compact subsets of $\widehat{G^2}$ to a harmonic function. Because

$$g^G(\mathbf{x},\mathbf{y}) = \int_{(0,\infty)} p^G(t,\mathbf{x},\mathbf{y})\,dt = \lim_{r\nearrow\infty}\int_{(0,\infty)} p^{G(r)}(t,\mathbf{x},\mathbf{y})\,dt = \lim_{r\nearrow\infty} g_r(\mathbf{x},\mathbf{y}),$$

we conclude from the first part of Lemma 11.2.13 that only the second alternative is possible. Thus, we now know that $g^G$ is harmonic on $\widehat{G^2}$ and that

$$(*)\qquad g_r(\mathbf{x},\mathbf{y}) \nearrow g^G(\mathbf{x},\mathbf{y}) \quad \text{uniformly on compact subsets of } \widehat{G^2}.$$

To go further, first notice that the expression in (11.2.12) for $g_r$ can be rewritten as

$$(**)\qquad \begin{aligned} \pi g_r(\mathbf{x},\mathbf{y}) &= -\log|\mathbf{y}-\mathbf{x}| + u_r(\mathbf{x},\mathbf{y}) \\ &\quad + \mathbb{E}^{\mathcal{W}_{\mathbf{x}}^{(2)}}\Big[\log\big|\mathbf{y}-\psi(\zeta^{B(\mathbf{c},r)})\big|,\ \zeta^{B(\mathbf{c},r)}(\psi) \le \zeta^G(\psi) < \infty\Big], \end{aligned}$$

where

$$u_r(\mathbf{x},\mathbf{y}) = \mathbb{E}^{\mathcal{W}_{\mathbf{x}}^{(2)}}\Big[\log\big|\mathbf{y}-\psi(\zeta^G)\big|,\ \zeta^G(\psi) < \zeta^{B(\mathbf{c},r)}(\psi)\Big] \quad \text{for } (\mathbf{x},\mathbf{y}) \in G(r)^2.$$

By the second part of Lemma 11.2.13, we know that each $u_r$ is harmonic on $G(r)^2$ and that, as $r\to\infty$, $\{u_r : r > 0\}$ tends uniformly on compact subsets of $G^2$ to the harmonic function

$$(\mathbf{x},\mathbf{y}) \rightsquigarrow u(\mathbf{x},\mathbf{y}) \equiv \mathbb{E}^{\mathcal{W}_{\mathbf{x}}^{(2)}}\Big[\log\big|\mathbf{y}-\psi(\zeta^G)\big|,\ \zeta^G(\psi) < \infty\Big].$$

Moreover, by combining this with (*) and (**), we also know that the third term on the right of (**) converges uniformly on compact subsets of $\widehat{G^2}$ to a harmonic function on $\widehat{G^2}$. At the same time, as $r\to\infty$,

$$\begin{aligned} &\mathbb{E}^{\mathcal{W}_{\mathbf{x}}^{(2)}}\Big[\log\big|\mathbf{y}-\psi(\zeta^{B(\mathbf{c},r)})\big|,\ \zeta^{B(\mathbf{c},r)}(\psi) \le \zeta^G(\psi) < \infty\Big] \\ &\qquad - \log r\,\mathcal{W}_{\mathbf{x}}^{(2)}\Big(\zeta^{B(\mathbf{c},r)} \le \zeta^G(\psi) < \infty\Big) \\ &\qquad = \mathbb{E}^{\mathcal{W}_{\mathbf{x}}^{(2)}}\left[\log\left(\frac{|\mathbf{y}-\psi(\zeta^{B(\mathbf{c},r)})|}{r}\right),\ \zeta^{B(\mathbf{c},r)} \le \zeta^G(\psi) < \infty\right] \longrightarrow 0 \end{aligned}$$

uniformly for $(\mathbf{x},\mathbf{y})$ in compact subsets of $G^2$. Thus, the asserted limit in (11.2.15) exists, the function $h^G$ is harmonic on $G$, and (11.2.16) holds.

Finally, to complete the proof, note that if $G$ is bounded, then (11.2.12) holds and therefore $h^G$ must be identically 0. Now, assume that $G$ is unbounded. To prove (11.2.17), use (11.2.16) to write

$$h^G(\mathbf{x}) = g^G(\mathbf{x},\mathbf{y}) + \frac{1}{\pi}\mathbb{E}^{\mathcal{W}_{\mathbf{x}}^{(2)}}\left[\log\left(\frac{|\mathbf{y}-\psi(\zeta^G)|}{|\mathbf{y}-\mathbf{x}|}\right),\ \zeta^G(\psi) < \infty\right],$$

and apply Lebesgue's Dominated Convergence Theorem together with the integrability estimate in the second part of Lemma 11.2.13 to see that, as $|\mathbf{y}|\to\infty$ through $G$, the second term tends to 0 uniformly for $\mathbf{x}$ in compact subsets of $G$. □

REMARK 11.2.18. The appearance of the extra term $h^G$ in (11.2.16) is, of course, a reflection of the fact that, for unbounded regions in $\mathbb{R}^2$, we do not know a priori which harmonic function (cf. Remark 11.2.8) should be used to correct $-\frac{1}{\pi}\log|\mathbf{y}-\mathbf{x}|$. When $N \geq 3$, the obvious choice was the one that behaved the same way at $\infty$ as $g^{\mathbb{R}^N}$ itself (i.e., the one that tends to 0 at $\infty$). Actually, as (11.2.17) makes explicit, the same principle applies to the case when $N = 2$, although now 0 may not be that limiting behavior. To see that, in general, $h^G$ is not identically 0, consider the open disk $\mathbf{D}(R) = \{\mathbf{x} : |\mathbf{x}| < R\}$, and take $G = \mathbb{R}^2 \setminus \overline{\mathbf{D}(R)}$. Then it is an easy matter to check that, for $R < |\mathbf{x}| < r$,

$$\mathcal{W}^{(2)}_{\mathbf{x}}\big(\zeta^{\mathbf{D}(r)} < \zeta^G\big) = \frac{\log \frac{|\mathbf{x}|}{R}}{\log \frac{r}{R}}.$$

Hence, by (11.2.15), we see that

$$h^{\mathbb{R}^2\setminus\overline{\mathbf{D}(R)}}(\mathbf{x}) = \frac{1}{\pi}\log\frac{|\mathbf{x}|}{R}, \quad \mathbf{x} \notin \overline{\mathbf{D}(R)}.$$

As we are about to see, for $G$'s whose complements are compact, the conclusion drawn about $h^G$ at the end of Remark 11.2.18 is typical, at least as $|\mathbf{x}| \to \infty$.

COROLLARY 11.2.19. *Let everything be as in Theorem 11.2.14, and assume that $K \equiv \mathbb{R}^2 \setminus G$ is compact. Then, for each $R \in (0, \infty)$ with the property that $K \subset\subset \mathbf{D}(R)$, one has that*

$$\begin{aligned} h^G(\mathbf{x}) - \frac{1}{\pi}\log\frac{|\mathbf{x}|}{R} &= \frac{|\mathbf{x}|^2}{2\pi}\int_{\mathbb{S}^1} \frac{|\mathbf{x}|^2 - R^2}{\big||\mathbf{x}|^2\boldsymbol{\omega} - R\mathbf{x}\big|^2} h^G(R\boldsymbol{\omega})\,\lambda_{\mathbb{S}^1}(d\boldsymbol{\omega}) \\ &\longrightarrow \frac{1}{2\pi}\int_{\mathbb{S}^1} h^G(R\boldsymbol{\omega})\,\lambda_{\mathbb{S}^1}(d\boldsymbol{\omega}) \end{aligned}$$

*as $|\mathbf{x}| \to \infty$.*

PROOF: Define $\sigma : C(\mathbb{R}^N) \longrightarrow [0, \infty]$ to be the first entrance time into $\overline{\mathbf{D}(R)}$, and note (cf. the preceding discussion) that, for each $r > R$ and $R < |\mathbf{x}| < r$,

$$\begin{aligned} &\mathcal{W}^{(2)}_{\mathbf{x}}\big(\zeta^{\mathbf{D}(r)} < \zeta^G\big) \\ &= \mathcal{W}^{(2)}_{\mathbf{x}}\big(\zeta^{\mathbf{D}(r)} < \sigma\big) + \mathbb{E}^{\mathcal{W}^{(2)}_{\mathbf{x}}}\Big[\mathcal{W}^{(2)}_{\psi(\sigma)}\big(\zeta^{\mathbf{D}(r)} < \zeta^G\big),\ \sigma < \zeta^{\mathbf{D}(r)}\Big] \\ &= \frac{\log\frac{|\mathbf{x}|}{R}}{\log\frac{r}{R}} + \mathbb{E}^{\mathcal{W}^{(2)}_{\mathbf{x}}}\Big[\mathcal{W}^{(2)}_{\psi(\sigma)}\big(\zeta^{\mathbf{D}(r)} < \zeta^G\big),\ \sigma < \zeta^{\mathbf{D}(r)}\Big]. \end{aligned}$$

Hence, after multiplying the preceding through by $\frac{\log r}{\pi}$, using (11.2.15), and letting $r \to \infty$, we arrive at

$$h^G(\mathbf{x}) = \frac{1}{\pi}\log\frac{|\mathbf{x}|}{R} + \frac{1}{\pi}\mathbb{E}^{\mathcal{W}^{(2)}_{\mathbf{x}}}\Big[h^G\big(\psi(\sigma)\big),\ \sigma < \infty\Big], \quad \mathbf{x} \in \mathbb{R}^2 \setminus \overline{\mathbf{D}(R)},$$

which certainly implies that

$$\mathbf{x} \in \mathbb{R}^2 \longmapsto h^G(\mathbf{x}) - \frac{1}{\pi} \log \frac{|\mathbf{x}|}{R}$$

is a bounded function that is harmonic off of $\overline{D(R)}$. Thus, the desired result now follows from the first part of Theorem 11.1.29. □

Notice that, as a by-product, one knows that the number

$$\frac{1}{2\pi} \int_{\mathbb{S}^1} h^G(R\boldsymbol{\omega})\, \lambda_{\mathbb{S}^1}(\boldsymbol{\omega}) - \frac{1}{\pi} \log R$$

does not depend on $R$ as long as $G\complement \subset\subset B(\mathbf{0}, R)$. This number plays an important role in classical two-dimensional potential theory, where it is known as **Robin's constant** for $G$.

COROLLARY 11.2.20. *Again let everything be as in Theorem 11.2.14. Then, for each $K \subset\subset G$ and $r > 0$,*

$$\sup\left\{g^G(\mathbf{x}, \mathbf{y}) : |\mathbf{x} - \mathbf{y}| \geq r \text{ and } \mathbf{y} \in K\right\} < \infty$$

*and*

$$\lim_{\substack{\mathbf{x} \to \mathbf{a} \\ \mathbf{x} \in G}} \sup_{\mathbf{y} \in K} g^G(\mathbf{x}, \mathbf{y}) = 0 \quad \text{for each } \mathbf{a} \in \partial_{\text{reg}} G.$$

*Moreover, for each $f \in C^1_c(G; \mathbb{R})$, $\mathbf{G}^G f$ is the unique bounded solution to (11.2.1).*

PROOF: To prove the initial statements, let $\mathbf{c} \in G$ and $r > 0$ satisfying $\overline{B(\mathbf{c}, 2r)} \subset\subset G$ be given, set $B = B(\mathbf{c}, r)$, and define the first entrance time $\sigma(\boldsymbol{\psi})$ of $\boldsymbol{\psi}$ into $\overline{B}$ by $\sigma(\boldsymbol{\psi}) = \inf\{t \geq 0 : \boldsymbol{\psi}(t) \in \overline{B}\}$. By the Markov property, we see that, for any $f \in C_c(B; [0, \infty))$,

$$\begin{aligned}\int_G g^G(\mathbf{x}, \mathbf{y}) f(\mathbf{y})\, d\mathbf{y} &= \mathbb{E}^{\mathcal{W}^{(2)}_{\mathbf{x}}}\left[\int_\sigma^{\zeta^G} f(\boldsymbol{\psi}(t))\, dt,\ \sigma < \zeta^G\right] \\ &= \mathbb{E}^{\mathcal{W}^{(2)}_{\mathbf{x}}}\left[\int_G g^G(\boldsymbol{\psi}(\sigma), \mathbf{y}) f(\mathbf{y})\, d\mathbf{y},\ \sigma < \zeta^G\right].\end{aligned}$$

Hence, if $\mathbf{x} \notin 2B \equiv B(\mathbf{c}, 2r)$ and therefore $g^G(\mathbf{x}, \cdot\,) \restriction B$ is continuous, we find that

$$g^G(\mathbf{x}, \mathbf{y}) = \mathbb{E}^{\mathcal{W}^{(2)}_{\mathbf{x}}}\left[g^G(\boldsymbol{\psi}(\sigma), \mathbf{y}),\ \sigma < \zeta^G\right] \quad \text{for all } y \in B.$$

But, because $g^G \restriction (\partial(2B)) \times B$ is bounded, we now see that

$$(*) \qquad \sup_{\mathbf{y} \in B} g^G(\mathbf{x}, \mathbf{y}) \leq C \mathcal{W}^{(2)}_{\mathbf{x}}(\sigma < \zeta^G), \quad \mathbf{x} \notin 2B,$$

for some $C \in (0, \infty)$. In particular, this, combined with the obvious Heine–Borel argument, proves the first estimate. In addition, if $\mathbf{a} \in \partial_{\text{reg}} G$, then, for each $\delta > 0$,

$$\begin{aligned}\varlimsup_{\substack{\mathbf{x}\to\mathbf{a}\\ \mathbf{x}\in G}} \mathcal{W}_{\mathbf{x}}^{(2)}(\sigma < \zeta^G) &\le \varlimsup_{\substack{\mathbf{x}\to\mathbf{a}\\ \mathbf{x}\in G}} \mathcal{W}_{\mathbf{x}}^{(2)}(\sigma \le \delta) + \varlimsup_{\substack{\mathbf{x}\to\mathbf{a}\\ \mathbf{x}\in G}} \mathcal{W}_{\mathbf{x}}^{(2)}(\zeta^G > \delta)\\ &= \varlimsup_{\substack{\mathbf{x}\to\mathbf{a}\\ \mathbf{x}\in G}} \mathcal{W}_{\mathbf{x}}^{(2)}(\sigma \le \delta).\end{aligned}$$

Thus, since the last expression obviously tends to 0 as $\delta \searrow 0$, this, together with (*), implies that

$$\varlimsup_{\substack{\mathbf{x}\to\mathbf{a}\\ \mathbf{x}\in G}} \sup_{\mathbf{y}\in B} g^G(\mathbf{x}, \mathbf{y}) = 0,$$

which (again after the obvious Heine–Borel argument) means that we have also proved the second assertion.

Turning to the last part of the statement, let $f \in C^1_c(G, \mathbb{R})$ be given. By the preceding, we know that $\mathbf{G}^G f$ is bounded and tends to 0 at $\partial_{\text{reg}} G$. In addition, using Theorem 11.2.14, especially (11.2.16), and arguing as I did in the case when $N \ge 3$, it is easy to check that $\mathbf{G}^G f \in C^2(G; \mathbb{R})$ and $\frac{1}{2}\Delta \mathbf{G}^G = -f$. Thus, $\mathbf{G}^G f$ is a bounded solution to (11.2.1), and, because (11.1.20) holds, it can be the only such solution. □

## Exercises for § 11.2

EXERCISES 11.2.21. Give an explicit expression for the Green function $g^{B(\mathbf{c},R)}$ when $N \ge 2$. To this end, first use translation and scaling to see that

$$g^{B(\mathbf{c},R)}(\mathbf{x}, \mathbf{y}) = R^{2-N} g^{B(\mathbf{0},1)}\left(\frac{\mathbf{x}-\mathbf{c}}{R}, \frac{\mathbf{y}-\mathbf{c}}{R}\right)$$

for distinct $\mathbf{x}, \mathbf{y}$ from $B(\mathbf{c}, R)$. Thus, assume that $\mathbf{c} = \mathbf{0}$ and $R = 1$. Next, observe that

$$|\mathbf{x} - \mathbf{y}| = \left| |\mathbf{y}|\mathbf{x} - \tfrac{\mathbf{y}}{|\mathbf{y}|} \right| \quad \text{for } \mathbf{x} \in \mathbb{S}^{N-1} \text{ and } \mathbf{y} \in B_{\mathbb{R}^N}(\mathbf{0}, 1) \setminus \{\mathbf{0}\},$$

and use this observation together with (11.2.12) and (11.2.5) to conclude that

$$g^{B(\mathbf{0},1)}(\mathbf{x}, \mathbf{y}) = -\frac{1}{\pi} \log|\mathbf{y} - \mathbf{x}| + \frac{1}{\pi} \begin{cases} \log\left|\frac{\mathbf{y}}{|\mathbf{y}|} - |\mathbf{y}|\mathbf{x}\right| & \text{if } \mathbf{y} \ne \mathbf{0} \\ 0 & \text{if } \mathbf{y} = \mathbf{0} \end{cases}$$

when $N = 2$ and

$$g^{B(\mathbf{0},1)}(\mathbf{x}, \mathbf{y}) = g^{\mathbb{R}^N}(\mathbf{x}, \mathbf{y}) - \begin{cases} g^{\mathbb{R}^N}\left(\frac{\mathbf{y}}{|\mathbf{y}|} - |\mathbf{y}|\mathbf{x}\right) & \text{if } \mathbf{y} \ne \mathbf{0} \\ \frac{2}{(N-2)\omega_{N-1}} & \text{if } \mathbf{y} = \mathbf{0} \end{cases}$$

when $N \ge 3$.

EXERCISE 11.2.22. The derivation that I gave of Poisson's formula (cf. Theorem 11.1.28) required me to already know the answer and simply verify that it is correct. Here I outline another approach, which is the basis for a quite general procedure. To begin with, recall the classical **Green's Identity**

$$\int_G \Bigl(u\Delta v - v\Delta u\Bigr)\,d\mathbf{x} = \int_{\partial G} \bigl(u\tfrac{\partial v}{\partial \mathbf{n}} - v\tfrac{\partial u}{\partial \mathbf{n}}\bigr)\,d\lambda_{\partial G}$$

for bounded, smooth regions $G$ in $\mathbb{R}^N$ and functions $u$ and $v$ that are smooth in a neighborhood of $\overline{G}$. (In the preceding, $\frac{\partial w}{\partial \mathbf{n}}(\mathbf{x})$ is used to denote the normal derivative $\bigl(\nabla w(\mathbf{x}), \mathbf{n}(\mathbf{x})\bigr)_{\mathbb{R}^N}$, where $\mathbf{n}(\mathbf{x})$ is the outer unit normal at $\mathbf{x} \in \partial G$ and $\lambda_{\partial G}$ is the standard surface measure for $\partial G$.) Next, let $\mathbf{c}$ be an element of $B(\mathbf{0},1)$, suppose $r > 0$ satisfies $\overline{B(\mathbf{c},r)} \subset\subset B(\mathbf{0},1)$, and let $u$ be a function that is harmonic in a neighborhood of $\overline{B_{\mathbb{R}^N}(\mathbf{0},1)}$. By applying Green's Identity with $G = B_{\mathbb{R}^N}(\mathbf{0},1) \setminus \overline{B(\mathbf{c},r)}$ and $v = \frac{1}{2} g^{B(\mathbf{0},1)}(\mathbf{c},\,\cdot\,)$, use Exercise 11.2.21 to verify

$$\begin{aligned} u(\mathbf{c}) &= \lim_{r \searrow 0} r^{N-1} \int_{\mathbb{S}^{N-1}} \bigl(\boldsymbol{\omega}, \nabla v(\mathbf{c} + r\boldsymbol{\omega})\bigr)_{\mathbb{R}^N} u(\mathbf{c} + r\boldsymbol{\omega})\,\lambda_{\mathbb{S}^{N-1}}(d\boldsymbol{\omega}) \\ &= \int_{\mathbb{S}^{N-1}} \bigl(\boldsymbol{\omega}, \nabla v(\boldsymbol{\omega})\bigr)_{\mathbb{R}^N} u(\boldsymbol{\omega})\,\lambda_{\mathbb{S}^{N-1}}(d\boldsymbol{\omega}) = \int_{\mathbb{S}^{N-1}} u(\boldsymbol{\omega})\pi^{(N)}(\mathbf{c},\boldsymbol{\omega})\,\lambda_{\mathbb{S}^{N-1}}(d\boldsymbol{\omega}), \end{aligned}$$

where $\pi^{(N)}$ is the Poisson kernel given in Theorem 11.1.28. Finally, given $f \in C(\partial G;\mathbb{R})$, extend $f$ to $B_{\mathbb{R}^N}(\mathbf{0},1)\complement$ so that it is constant along rays, take

$$u_R(\mathbf{x}) = \mathbb{E}^{\mathcal{W}^{(N)}_{\mathbf{x}}}\Bigl[f\bigl(\psi(\zeta^{B(\mathbf{0},R)})\bigr),\ \zeta^{B(\mathbf{0},R)} < \infty\Bigr] \quad \text{for } R \ge 1 \text{ and } \mathbf{x} \in B(\mathbf{0},R),$$

check that, as $R \searrow 1$, $u_R \longrightarrow u_1$ uniformly on $\overline{B(\mathbf{0},1)}$, and use the preceding to conclude that

$$u_1(\mathbf{c}) = \int_{\mathbb{S}^{N-1}} f(\boldsymbol{\omega})\,\pi^{(N)}(\mathbf{c},\boldsymbol{\omega})\,\lambda_{\mathbb{S}^{N-1}}(d\boldsymbol{\omega}),$$

which is, of course, the result that was proved in Theorem 11.1.28.

## § 11.3 Excessive Functions, Potentials, and Riesz Decompositions

The origin of the Green function lies in the theory of electricity and magnetism. Namely, if $G$ is a region in $\mathbb{R}^N$ whose boundary is *grounded* and $\mathbf{y} \in G$, then $g^G(\,\cdot\,,\mathbf{y})$ should be the *electrical potential* in $G$ that results from placing a unit point charge at $\mathbf{y}$. More generally, if $\mu$ is any *distribution of charge in* $G$ (i.e., a non-negative, locally finite, Borel measure on $G$), then one can consider the **potential** $\mathbf{G}^G\mu$ given by

$$\mathbf{G}^G\mu(\mathbf{x}) = \int_G g^G(\mathbf{x},\mathbf{y})\,\mu(d\mathbf{y}), \quad \mathbf{x} \in G, \tag{11.3.1}$$

where I have implicitly assumed that either $N \ge 3$ or (11.1.20) holds. In this section I will characterize functions that arise in this way (i.e., are potentials).

**§ 11.3.1. Excessive Functions.** Throughout this subsection, $G$ will be a non-empty, connected, open region in $\mathbb{R}^N$, and I will be assuming either that $N \geq 3$ or that (11.1.20) holds. Thus, by the results obtained in §§ 8.2.1 and 8.2.2, the Green function (cf. (11.2.3)) $g^G$ satisfies (depending on whether $N = 1$, $N = 2$, or $N \geq 3$) either (11.2.10), (11.2.11), (11.2.16), or (11.2.5), and, in order to have $g^G$ defined everywhere on $G^2$, I will take $g^G(\mathbf{x}, \mathbf{x}) = \infty$, $\mathbf{x} \in G$, when $N \geq 2$.

I will say that $u$ is an **excessive function** on $G$ and will write $u \in \mathcal{E}(G)$ if $u$ is a lower semicontinuous, $[0, \infty]$-valued function that satisfies the *super mean value property*:

$$u(\mathbf{x}) \geq \frac{1}{\omega_{N-1}} \int_{\mathbb{S}^{N-1}} u(\mathbf{x} + r\boldsymbol{\omega})\, \lambda_{\mathbb{S}^{N-1}}(d\boldsymbol{\omega}) \quad \text{whenever } \overline{B_{\mathbb{R}^N}(\mathbf{x}, r)} \subseteq G.$$

As the next lemma shows, there are lots of excessive functions.

LEMMA 11.3.2. *$\mathcal{E}(G)$ is closed under non-negative linear combinations and non-decreasing limits, and $u, v \in \mathcal{E}(G) \implies u \wedge v \in \mathcal{E}(G)$. Moreover, if $u \in C^2(G; [0, \infty))$, then $u \in \mathcal{E}(G) \iff \Delta u \leq 0$. Finally, for each non-negative, locally finite, Borel measure $\mu$ on $G$ and each non-negative harmonic function $h$ on $G$, $\mathbf{G}^G \mu + h$ is an excessive function on $G$.*

PROOF: The initial assertions are obvious. To prove the next part, suppose that $u \in C^2(G; [0, \infty))$ is given. If $u \in \mathcal{E}(G)$, then

$$\tfrac{1}{2}\Delta u(\mathbf{x}) = \lim_{r \searrow 0} \frac{1}{\omega_{N-1}} \int_{\mathbb{S}^{N-1}} \big(u(\mathbf{x} + r\boldsymbol{\omega}) - u(\mathbf{x})\big)\, \lambda_{\mathbb{S}^{N-1}}(d\boldsymbol{\omega}) \leq 0$$

for each $\mathbf{x} \in G$. Conversely, if $\Delta u \leq 0$ and $\overline{B(\mathbf{x}, r)} \subset\subset G$, then

$$\begin{aligned} u(\mathbf{x}) &= \mathbb{E}^{\mathcal{W}_{\mathbf{x}}^{(N)}}\big[u(\boldsymbol{\psi}(\zeta^{B(\mathbf{x},r)})),\, \zeta^{B(\mathbf{x},r)} < \infty\big] - \mathbb{E}^{\mathcal{W}_{\mathbf{x}}^{(N)}}\left[\int_0^{\zeta^{B(\mathbf{x},r)}} \tfrac{1}{2}\Delta u(\boldsymbol{\psi}(\tau))\, d\tau\right] \\ &\geq \frac{1}{\omega_{N-1}} \int_{\mathbb{S}^{N-1}} u(\mathbf{x} + r\boldsymbol{\omega})\, \lambda_{\mathbb{S}^{N-1}}(d\boldsymbol{\omega}). \end{aligned}$$

Clearly the third assertion comes down to showing that $\mathbf{G}^G\mu$ is excessive. Moreover, by Fatou's Lemma and Tonelli's Theorem, we will know that $\mathbf{G}^G\mu$ is excessive as soon as we show that, for each $\mathbf{y} \in G$, $g^G(\,\cdot\,, \mathbf{y})$ is excessive. To this end, set $f_n = p^G\big(\frac{1}{n}, \cdot\,, \mathbf{y}\big)$ and (cf. (11.2.6)) $u_n = \mathbf{G}^G f_n$. Because

$$\int_{\frac{1}{n}}^T p^G(t, \cdot\,, \mathbf{y})\, dt \nearrow u_n \quad \text{as } T \to \infty,$$

$u_n$ is lower semicontinuous. In addition, by the Markov property and rotation invariance, $\overline{B(\mathbf{x},r)} \subset\subset G$ implies

$$u_n(\mathbf{x}) \geq \mathbb{E}^{\mathcal{W}_{\mathbf{x}}^{(N)}}\left[\int_{\zeta_r}^{\zeta^G} f_n(\boldsymbol{\psi}(t))\,dt\right] = \mathbb{E}^{\mathcal{W}_{\mathbf{x}}^{(N)}}\Big[u_n(\boldsymbol{\psi}(\zeta_r)),\,\zeta_r < \infty\Big]$$
$$= \frac{1}{\omega_{N-1}}\int_{\mathbb{S}^{N-1}} u_n(\mathbf{x}+r\mathbf{x})\,\lambda_{\mathbb{S}^{N-1}}(d\mathbf{x}),$$

where I have introduced the notation

$$\zeta_r(\boldsymbol{\psi}) = \inf\Big\{t:\,\big|\boldsymbol{\psi}(t)-\boldsymbol{\psi}(0)\big| \geq r\Big\} \tag{11.3.3}$$

and used the rotation invariance of Brownian motion. Hence, each $u_n$ is excessive, and therefore, since

$$u_n(\mathbf{x}) = \int_{\frac{1}{n}}^{\infty} p^G(t,\mathbf{x},\mathbf{y})\,dt \nearrow g^G(\mathbf{x},\mathbf{y}) \quad \text{as } n\to\infty,$$

we are done. □

**§ 11.3.2. Potentials and Riesz Decomposition.** My next goal is to prove that, apart from the trivial case when $u \equiv \infty$, every excessive function on $G$ admits a unique representation in the form $\mathbf{G}^G\mu + h$ for an appropriate choice of $\mu$ and $h$. The proof requires me to make some preparations.

LEMMA 11.3.4. *If $u \in \mathcal{E}(G)$, then either $u \equiv \infty$ or $u$ is locally integrable on $G$. Next, given a $u \in \mathcal{E}(G)$ that is not identically infinite, there exists a sequence $\{u_n : n \geq 1\} \subseteq C_c^\infty(G;\mathbb{R})$ and a non-decreasing sequence $\{G_n : n \geq 1\}$ of open subsets of $G$ with the properties that $\overline{G_n} \subset\subset G$, $G_n \nearrow G$, $u_n \leq u$, $\Delta u_n \leq 0$ on $G_n$ for each $n \geq 1$, and $u_n \longrightarrow u$ pointwise as $n\to\infty$. Moreover, if $\mu_n(d\mathbf{y}) = -\frac{1}{2}\mathbf{1}_{G_n}(\mathbf{y})\Delta u_n(\mathbf{y})\,d\mathbf{y}$, then there is a non-negative, locally finite, Borel measure $\mu$ on $G$ such that*

$$\lim_{n\to\infty}\int_G \varphi\,d\mu_n = \int_G \varphi\,d\mu \text{ for all } \varphi \in C_c(G;\mathbb{R}). \tag{11.3.5}$$

*In fact, $\mu$ is uniquely determined by the fact that $\mu = -\frac{1}{2}\Delta u$ in the sense that*

$$\int_G \tfrac{1}{2}\Delta\varphi(\mathbf{y})u(\mathbf{y})\,d\mathbf{y} = -\int_G \varphi\,d\mu \quad \text{for all } \varphi \in C_c^\infty(G;\mathbb{R}). \tag{11.3.6}$$

PROOF: To prove the first assertion, let $U$ denote the set of all $\mathbf{x} \in G$ with the property that

$$\int_{B(\mathbf{x},r)} u(\mathbf{y})\,d\mathbf{y} < \infty \quad \text{for some } r > 0 \text{ with } \overline{B(\mathbf{x},r)} \subset\subset G.$$

Obviously, $U$ is an open subset of $G$. At the same time, if $\mathbf{x} \in G \setminus U$ and $r > 0$ is chosen so that $\overline{B_{\mathbb{R}^N}(\mathbf{x}, 2r)} \subset\subset G$, then, for each $\mathbf{y} \in B(\mathbf{x}, r)$ and $s \in (0, r)$,

$$u(\mathbf{y}) \geq \frac{1}{\omega_{N-1}} \int_{\mathbb{S}^{N-1}} u(\mathbf{y} + s\boldsymbol{\omega})\, \lambda_{\mathbb{S}^{N-1}}(d\boldsymbol{\omega}),$$

and so, after integrating this with respect to $Ns^{N-1}\, ds$ over $(0, r)$, we get

$$u(\mathbf{y}) \geq \frac{1}{\Omega_{N-1} r^N} \int_{B(\mathbf{y}, r)} u(\mathbf{z})\, d\mathbf{z} \geq \frac{1}{\Omega_{N-1} r^N} \int_{B(\mathbf{x}, \delta)} u(\mathbf{z})\, d\mathbf{z} = \infty,$$

where $\delta \equiv r - |\mathbf{y} - \mathbf{x}|$. Hence, we now see that $G \setminus U$ is also open, and therefore that either $U = G$ or $U = \emptyset$ and $u \equiv \infty$.

Now assume that $u \in \mathcal{E}(G)$ is not identically infinite. To construct the required $G_n$'s and $u_n$'s, choose a reference point $\mathbf{c} \in G$, set $R = \frac{1}{2}|\mathbf{c} - G\complement|$, and take $\rho \in C^\infty_{\mathrm{c}}\big(B(\mathbf{0}, \frac{R}{4}); [0, \infty)\big)$ to be a rotationally invariant function with total integral 1. Next, for each $n \in \mathbb{Z}^+$, set

$$\begin{aligned} &G_n = \left\{\mathbf{x} \in G \cap B(\mathbf{c}, n) : |\mathbf{x} - G\complement| > \tfrac{R}{n}\right\} \text{ and} \\ &u_n(\mathbf{x}) = \int_{G_{4n}} \rho_n(\mathbf{x} - \mathbf{y}) u(\mathbf{y})\, d\mathbf{y}, \quad \mathbf{x} \in \mathbb{R}^N, \end{aligned} \tag{11.3.7}$$

where $\rho_n(\boldsymbol{\xi}) = n^N \rho(n\boldsymbol{\xi})$. Clearly, $\{u_n : n \geq 1\} \subseteq C^\infty_{\mathrm{c}}\big(G; [0, \infty)\big)$. In addition, if $\mathbf{x} \in G_n$, then, by taking advantage of the rotation invariance of $\rho$, one can check that

$$\begin{aligned} u_n(\mathbf{x}) &= \int_{(0, \frac{R}{4})} t^{N-1} \tilde{\rho}(t) \left( \int_{\mathbb{S}^{N-1}} u\left(\mathbf{x} + \tfrac{t}{n}\boldsymbol{\omega}\right) \lambda_{\mathbb{S}^{N-1}}(d\boldsymbol{\omega}) \right) dt \\ &\leq u(\mathbf{x})\, \omega_{N-1} \int_{(0, \frac{R}{4})} t^{N-1} \tilde{\rho}(t)\, dt = u(\mathbf{x}), \end{aligned}$$

where $\tilde{\rho} : \mathbb{R} \longrightarrow [0, \infty)$ is taken so that $\rho(\mathbf{x}) = \tilde{\rho}(|\mathbf{x}|)$. Similarly, if $\overline{B(\mathbf{x}, r)} \subset\subset G_n$, then

$$\begin{aligned} &\int_{\mathbb{S}^{N-1}} u_n(\mathbf{x} + r\boldsymbol{\omega})\, \lambda_{\mathbb{S}^{N-1}}(d\boldsymbol{\omega}) \\ &\quad = \int_{B(\mathbf{0}, \frac{R}{4})} \rho(\mathbf{z}) \left( \int_{\mathbb{S}^{N-1}} u\left(\mathbf{x} + \tfrac{1}{n}\mathbf{z} + r\boldsymbol{\omega}\right) \lambda_{\mathbb{S}^{N-1}}(d\boldsymbol{\omega}) \right) d\mathbf{z} \\ &\quad \leq \omega_{N-1} \int_{B(\mathbf{0}, \frac{R}{4})} \rho(\mathbf{z}) u\left(\mathbf{x} + \tfrac{1}{n}\mathbf{z}\right) d\mathbf{z} = \omega_{N-1} u_n(\mathbf{x}). \end{aligned}$$

Hence, $u_n \restriction G_n$ is a smooth element of $\mathcal{E}(G_n)$, and therefore, by the second part of Lemma 11.3.2, we know that $\Delta u_n \leq 0$ on $G_n$. To see that $u_n \longrightarrow u$ pointwise, observe that we already know that $u(\mathbf{x}) \geq \overline{\lim}_{n\to\infty} u_n(\mathbf{x})$. On the other hand, because $u$ is lower semicontinuous, an application of Fatou's Lemma yields

$$u(\mathbf{x}) \leq \lim_{n\to\infty} \int_G \rho(\mathbf{y})\, u\left(\mathbf{x} + \tfrac{1}{n}\mathbf{y}\right) d\mathbf{y} = \lim_{n\to\infty} u_n(\mathbf{x}).$$

To complete the proof, let $\mu_n$ be the measure described, and note that

$$\begin{aligned} u_n(\mathbf{x}) &= \mathbb{E}^{\mathcal{W}_{\mathbf{x}}^{(N)}}\left[u_n\left(\boldsymbol{\psi}(t \wedge \zeta^{G_n})\right)\right] - \mathbb{E}^{\mathcal{W}_{\mathbf{x}}^{(N)}}\left[\int_0^{t\wedge\zeta^{G_n}} \tfrac{1}{2}\Delta u_n(\boldsymbol{\psi}(s))\, ds\right] \\ &\geq -\mathbb{E}^{\mathcal{W}_{\mathbf{x}}^{(N)}}\left[\int_0^{t\wedge\zeta^{G_n}} \tfrac{1}{2}\Delta u_n(\boldsymbol{\psi}(s))\, ds\right] = \int_0^t \left(\int_{G_n} p^{G_n}(s,\mathbf{x},\mathbf{y})\, \mu_n(d\mathbf{y})\right) ds \end{aligned}$$

for all $n \in \mathbb{Z}^+$ and $(t,\mathbf{x}) \in (0,\infty) \times G_n$. Hence, after letting $t \nearrow \infty$, we see that

$$u(\mathbf{x}) \geq u_n(\mathbf{x}) \geq \int_{G_n} g^{G_n}(\mathbf{x},\mathbf{y})\, \mu_n(d\mathbf{y}), \quad n \in \mathbb{Z}^+ \text{ and } \mathbf{x} \in G_n.$$

In particular, because $u(\mathbf{x}) < \infty$ for Lebesgue-almost every $\mathbf{x} \in G$, this proves that, for each $K \subset\subset G$, $\sup_{n\in\mathbb{Z}^+} \mu_n(K) < \infty$, and therefore (cf. part (**iv**) of Exercise 9.1.16 and apply a diagonalization procedure) $\{\mu_n : n \geq 1\}$ is relatively compact in the sense that every subsequence $\{\mu_{n_m} : m \geq 1\}$ admits a subsequence $\{\mu_{n_{m_k}} : k \geq 1\}$ and a locally finite, non-negative, Borel measure $\mu$ on $G$ with the property that

$$\lim_{k\to\infty} \int_G \varphi\, d\mu_{n_{m_k}} = \int_G \varphi\, d\mu \quad \text{for all } \varphi \in C_c(G;\mathbb{R}).$$

At the same time, using integration by parts followed by Lebesgue's Dominated Theorem, we see that

$$\lim_{n\to\infty} \int_G \varphi\, d\mu_n = -\lim_{n\to\infty} \int_G \tfrac{1}{2}\Delta\varphi\, u_n\, d\mathbf{x} = -\int_G \tfrac{1}{2}\Delta\varphi\, u\, d\mathbf{x}, \quad \varphi \in C_c^2(G;\mathbb{R}),$$

and therefore any limit $\mu$ of $\{\mu_n : n \geq 1\}$ must satisfy (11.3.6), which proves not only that there is such a $\mu$ but also that (11.3.5) is satisfied. □

LEMMA 11.3.8. *For any lower semicontinuous $u : G \longrightarrow [0,\infty]$, $u \in \mathcal{E}(G)$ if and only if*

$$\text{(11.3.9)} \quad \mathbb{E}^{\mathcal{W}_{\mathbf{x}}^{(N)}}\left[u(\boldsymbol{\psi}(\tau)),\, \tau(\boldsymbol{\psi}) < \zeta^G(\boldsymbol{\psi})\right] \leq \mathbb{E}^{\mathcal{W}_{\mathbf{x}}^{(N)}}\left[u(\boldsymbol{\psi}(\sigma)),\, \sigma(\boldsymbol{\psi}) < \zeta^G(\boldsymbol{\psi})\right]$$

*for every pair $\sigma$ and $\tau$ of $\{\mathcal{F}_t : t \geq 0\}$-stopping times with $\sigma \leq \tau$. In particular, if $u \in \mathcal{E}(G)$ and $\overline{B(\mathbf{x},r)} \subset\subset G$, then, for any rotationally symmetric $\rho \in C_c(B(\mathbf{0},r);[0,\infty))$ with total integral 1,*

$$t \in (0,1) \longmapsto \int_{B(\mathbf{0},r)} \rho(\mathbf{y})\, u(\mathbf{x} + t\mathbf{y})\, d\mathbf{y} \in [0,\infty]$$

*is a non-increasing function.*

PROOF: Let $u \in \mathcal{E}(G)$ be given. Clearly (11.3.9) is trivial in the case when $u \equiv \infty$. Thus, assume that $u \not\equiv \infty$, and define $G_n$ and $u_n$ for $n \in \mathbb{Z}^+$ as in (11.3.7). Because $\Delta u_n \restriction G_n \leq 0$, we know that

$$\begin{aligned}\mathbb{E}^{\mathcal{W}_{\mathbf{x}}^{(N)}}&\Big[u_n\big(\boldsymbol{\psi}(\tau \wedge \zeta^{G_m} \wedge T)\big),\, \sigma(\boldsymbol{\psi}) \wedge T < \zeta^{G_m}(\boldsymbol{\psi})\Big]\\ &\leq \mathbb{E}^{\mathcal{W}_{\mathbf{x}}^{(N)}}\Big[u_n\big(\boldsymbol{\psi}(\sigma \wedge T)\big),\, \sigma(\boldsymbol{\psi}) \wedge T < \zeta^{G_m}(\boldsymbol{\psi})\Big]\end{aligned}$$

for all $1 \leq m \leq n$, $\mathbf{x} \in G_m$, and $T \in [0, \infty)$. Next, after noting that $\zeta^{G_m} < \infty$ $\mathcal{W}_{\mathbf{x}}^{(N)}$-almost surely, let $T \nearrow \infty$ in the preceding, and arrive at

$$\mathbb{E}^{\mathcal{W}_{\mathbf{x}}^{(N)}}\Big[u_n\big(\boldsymbol{\psi}(\tau \wedge \zeta^{G_m})\big),\, \sigma(\boldsymbol{\psi}) < \zeta^{G_m}(\boldsymbol{\psi})\Big] \leq \mathbb{E}^{\mathcal{W}_{\mathbf{x}}^{(N)}}\Big[u_n\big(\boldsymbol{\psi}(\sigma)\big),\, \sigma(\boldsymbol{\psi}) < \zeta^{G_m}(\boldsymbol{\psi})\Big].$$

But, because $\sigma \leq \tau$ and $u \geq u_n \geq 0$, this means that

$$\mathbb{E}^{\mathcal{W}_{\mathbf{x}}^{(N)}}\Big[u_n\big(\boldsymbol{\psi}(\tau)\big),\, \tau(\boldsymbol{\psi}) < \zeta^{G_m}(\boldsymbol{\psi})\Big] \leq \mathbb{E}^{\mathcal{W}_{\mathbf{x}}^{(N)}}\Big[u\big(\boldsymbol{\psi}(\sigma)\big),\, \sigma(\boldsymbol{\psi}) < \zeta^{G_m}(\boldsymbol{\psi})\Big],$$

which, because $0 \leq u_n \longrightarrow u$ pointwise, leads, via Fatou's Lemma, first to

$$\mathbb{E}^{\mathcal{W}_{\mathbf{x}}^{(N)}}\Big[u\big(\boldsymbol{\psi}(\tau)\big),\, \tau(\boldsymbol{\psi}) < \zeta^{G_m}(\boldsymbol{\psi})\Big] \leq \mathbb{E}^{\mathcal{W}_{\mathbf{x}}^{(N)}}\Big[u\big(\boldsymbol{\psi}(\sigma)\big),\, \sigma(\boldsymbol{\psi}) < \zeta^{G_m}(\boldsymbol{\psi})\Big]$$

and thence, by the Monotone Convergence Theorem, to (11.3.9) when $m \to \infty$.

From here, the rest is easy. Given a lower semicontinuous $u : G \longrightarrow [0, \infty]$ and $\overline{B(\mathbf{x}, r)} \subset\subset G$, we have (cf. (11.3.3))

$$\frac{1}{\omega_{N-1}} \int_{\mathbb{S}^{N-1}} u(\mathbf{x} + r\boldsymbol{\omega})\, \lambda_{\mathbb{S}^{N-1}}(d\boldsymbol{\omega}) = \mathbb{E}^{\mathcal{W}_{\mathbf{x}}^{(N)}}\Big[u\big(\boldsymbol{\psi}(\zeta_r)\big),\, \zeta_r(\boldsymbol{\psi}) < \zeta^G(\boldsymbol{\psi})\Big].$$

Thus, if, in addition, (11.3.9) holds, then

$$t \in [0, 1] \longmapsto \frac{1}{\omega_{N-1}} \int_{\mathbf{S}^{N-1}} u(\mathbf{x} + tr\boldsymbol{\omega})\, \lambda_{\mathbb{S}^{N-1}}(d\boldsymbol{\omega}) \in [0, \infty]$$

is non-increasing; and, therefore, not only is $u$ excessive but also (after passing to polar coordinates and integrating) one finds that the monotonicity described in the final assertion is true. □

THEOREM 11.3.10 (**Riesz Decomposition**). *Let $G$ be a non-empty, connected open subset of $\mathbb{R}^N$, and assume either that $N \geq 3$ or that (11.1.20) holds. If $u \in \mathcal{E}(G)$ is not identically infinite, then there exists a unique locally finite, non-negative Borel measure $\mu$ and a unique non-negative harmonic function $h$ on $G$ with the property that*

$$u(\mathbf{x}) = \mathbf{G}^G\mu(\mathbf{x}) + h(\mathbf{x}) \quad \text{for all } \mathbf{x} \in G. \tag{11.3.11}$$

*In fact, $\mu$ is uniquely determined by (11.3.6), and $h$ is the unique harmonic function on $G$ that is dominated by $u$ and has the property that $h \geq w$ for every non-negative harmonic $w$ that is dominated by $u$. (Cf. Exercise 11.3.14 as well.)*

PROOF: Take $G_n$ and $u_n$ as in (11.3.7), and define $\mu_n$ accordingly, as in Lemma 11.3.4. Then, for each $1 \le m \le n$, Lemma 11.3.4 and the final part of Lemma 11.3.8 say that $u_m \le u_n \nearrow u$ pointwise on $G_m$. In addition, for $m \le n$ and $\mathbf{x} \in G_m$,

$$u_n(\mathbf{x}) = \int_{G_m} g^{G_m}(\mathbf{x}, \mathbf{y})\, \mu_n(d\mathbf{y}) + w_{m,n}(\mathbf{x}),$$
$$\text{where } w_{m,n} = \mathbb{E}^{\mathcal{W}_{\mathbf{x}}^{(N)}}\left[u_n(\boldsymbol{\psi}(\zeta^{G_m})),\, \zeta^{G_m} < \infty\right].$$

Hence, by the Monotone Convergence Theorem, for any locally finite, non-negative, Borel measure $\nu$ on $G$,

$$(*) \qquad \int_{G_m} u(\mathbf{x})\, \nu(d\mathbf{x}) = \lim_{n\to\infty} \iint_{G_m^2} g^{G_m}(\mathbf{x}, \mathbf{y})\, \nu(d\mathbf{x}) d\mu_n(\mathbf{y}) + \int_{G_m} w_m(\mathbf{x})\, \nu(d\mathbf{x}),$$

where $w_m(\mathbf{x}) = \mathbb{E}^{\mathcal{W}_{\mathbf{x}}^{(N)}}\left[u(\boldsymbol{\psi}(\zeta^{G_m})),\, \zeta^{G_m} < \infty\right]$.

Notice (cf. Harnack's Principle) that, as the non-decreasing limit of non-negative harmonic functions $\{w_{m,n} : n \ge m\}$, $w_m$ is either identically infinite or is itself a non-negative harmonic function on $G$; and so, since $u(\mathbf{x}) < \infty$ Lebesgue-almost everywhere, (*) shows that the latter must be the case. Now let $\mathbf{a}$ be a fixed element of $G_m$, take $\rho_n$ as in (11.3.7), and, for $n \ge m$, define

$$\varphi_n(\mathbf{y}) = \begin{cases} \int_{G_m} \rho_n(\mathbf{x} - \mathbf{a}) g^{G_m}(\mathbf{x}, \mathbf{y})\, d\mathbf{x} & \text{if } \mathbf{y} \in G_m \\ 0 & \text{otherwise.} \end{cases}$$

By taking $\nu(d\mathbf{x}) = \mathbf{1}_{G^m}(\mathbf{x}) \rho_n(\mathbf{x} - \mathbf{a})\, d\mathbf{x}$ in (*), we see that, for $n \ge m$,

$$\int_{G_m} \rho_n(\mathbf{x} - \mathbf{a})\, u(\mathbf{x})\, d\mathbf{x} = \lim_{k\to\infty} \int_G \varphi_n(\mathbf{y})\, \mu_k(d\mathbf{y}) + \int_{G_m} \rho_n(\mathbf{x} - \mathbf{a})\, w_m(\mathbf{x})\, d\mathbf{x}.$$

But, since $G_m$ is the intersection of two sets, both of which (cf. part **(iv)** in Exercise 10.2.19) are regular, and is therefore regular as well, there is an $n(\mathbf{a}) \ge m$ for which $\varphi_n$ is continuous whenever $n \ge n(\mathbf{a})$. In particular, by (11.3.5), we can now say that

$$\int_{G_m} \rho_n(\mathbf{x} - \mathbf{a})\, u(\mathbf{x})\, d\mathbf{x} = \int_G \varphi_n(\mathbf{x})\, \mu(d\mathbf{x}) + \int_{G_m} \rho_n(\mathbf{x} - \mathbf{a})\, w_m(\mathbf{x})\, d\mathbf{x}$$

for all $n \ge n(\mathbf{a})$. In addition, as $n \to \infty$, the reasoning with which we showed that $u_n \longrightarrow u$ in Lemma 11.3.4 shows that the term on the left tends to $u(\mathbf{a})$.

At the same time, it is clear that the second term on the right goes to $w_m(\mathbf{a})$ and that $\{\varphi_n(\mathbf{y}) : n \geq n(\mathbf{a})\}$ tends non-decreasingly to $g^{G_m}(\mathbf{a}, \mathbf{y})$. Thus, we have now proved that

$$(**) \qquad u = G^{G_m}\mu + w_m \quad \text{on } G_m \text{ for every } m \in \mathbb{Z}^+.$$

Starting from (**), the rest of the proof is quite easy. Namely, fix $\mathbf{x} \in G$, choose $m$ so that $\mathbf{x} \in G_m$, note that, $g^{G_n}(\mathbf{x}, \cdot)$ is non-decreasing as $n \geq m$ increases, and conclude that $\mathbf{G}^{G_{n\vee m}}\mu(\mathbf{x}) \nearrow \mathbf{G}^G\mu(\mathbf{x})$. Hence, by (**) (alternatively, by (11.3.9)), we know that $w_{m\vee n}(\mathbf{x})$ tends non-increasingly to a limit $h(\mathbf{x})$, which Harnack's Principle guarantees to be harmonic as a function of $\mathbf{x} \in G$. Thus, after passing to the limit as $m \to \infty$ in (**), we conclude that (11.3.11) holds with the $\mu$ satisfying (11.3.6) and $h = \lim_{m\to\infty} H^{G_m}u$.

To prove that these quantities are unique, note that if $\nu$ is any locally finite, non-negative, Borel measure on $G$ for which $u - G^G\nu$ is a non-negative harmonic function, then, for every $\varphi \in C_c^\infty(G;\mathbb{R})$, simple integration by parts plus the symmetry of $g^G$ shows that

$$-\tfrac{1}{2}\int_G \Delta\varphi u \, d\mathbf{x} = -\tfrac{1}{2}\int_G \Delta\mathbf{G}^G\varphi \, d\nu = \int_G \varphi \, d\nu.$$

That is, $\nu$ must satisfy (11.3.6); and so we have now derived the required uniqueness result.

Finally, to check the asserted characterization of $h$, suppose that $v$ is a non-negative harmonic function that is dominated by $u$ on $G$. We then have

$$v(\mathbf{x}) = \mathbb{E}^{\mathcal{W}_\mathbf{x}^{(N)}}\left[v(\boldsymbol{\psi}(\zeta^{G_m})),\, \zeta^{G_m}(\boldsymbol{\psi}) < \infty\right] \leq w_m(\mathbf{x}) \text{ for } m \in \mathbb{Z}^+ \text{ and } \mathbf{x} \in G_m,$$

and therefore the desired conclusion follows from the fact that $w_m$ tends to $h$. □

By combining Lemma 11.3.2 with Theorem 11.3.10, we arrive at the following characterization of potentials.

COROLLARY 11.3.12. *Let everything be as in Theorem 11.3.10, and suppose that $u : G \longrightarrow [0,\infty]$ is not identically infinite. Then a necessary and sufficient condition for $u$ to be the potential $\mathbf{G}^G\mu$ of some locally finite, non-negative, Borel measure $\mu$ on $G$ is that $u$ be excessive on $G$ and have the property that the constant function 0 is the only non-negative harmonic function on $G$ that is dominated by $u$.*

Let $u$ be an excessive function on $G$ that is not identically infinite. In keeping with the electrostatic metaphor, I will call the measure $\mu$ entering the Riesz decomposition (11.3.11) of $u$ the **charge determined by** $u$. A more mathematical interpretation is provided by Schwartz's theory of distributions. Namely, when $u \in \mathcal{E}(G)$ is not identically infinite, it is (cf. Lemma 11.3.4) locally integrable on $G$, and, as such, it determines a distribution there. Moreover, in the language of distribution theory, (11.3.6) says that $\mu = -\frac{1}{2}\Delta u$. However, the following theorem provides a better way of thinking about $\mu$.

THEOREM 11.3.13. *Let $G$ be as in Theorem 11.3.10 and $u : G \longrightarrow [0,\infty]$ a lower semicontinuous function. Then $u \in \mathcal{E}(G)$ if and only if*

$$u(\mathbf{x}) \geq u_s(\mathbf{x}) \equiv \int_G u(\mathbf{y}) p^G(s,\mathbf{x},\mathbf{y})\, d\mathbf{y} \quad \text{for all } (s,\mathbf{x}) \in (0,\infty) \times G.$$

*Moreover, if $u \in \mathcal{E}(G)$ is not identically infinite and, for $s \in (0,\infty)$, $\mu_s(d\mathbf{x}) = f_s(\mathbf{x})\, d\mathbf{x}$, where $f_s(\mathbf{x}) = \frac{u(\mathbf{x}) - u_s(\mathbf{x})}{s}$, then, as $s \searrow 0$, $\{\mu_s : s > 0\}$ tends to the charge $\mu$ of $u$ in the sense that*

$$\int_G \varphi(\mathbf{x})\, \mu(d\mathbf{x}) = \lim_{s \searrow 0} \int_G \varphi(\mathbf{x})\, \mu_s(d\mathbf{x}) \quad \text{for all } \varphi \in C_c(G;\mathbb{R}).$$

PROOF: If $u \in \mathcal{E}(G)$, then, by the first part of Lemma 11.3.8 with $\tau = s$ and $\sigma = 0$, one sees that $u \geq u_s$. Conversely, suppose that $u : G \longrightarrow [0,\infty]$ is lower semicontinuous, not identically infinite, and satisfies $u \geq u_s$ for all $s > 0$. Then, since $p^G(s,\mathbf{x},\cdot\,) > 0$, $u$ is locally integrable on $G$. Thus, if $\overline{B(\mathbf{c},r)} \subset\subset G$ and

$$w_s(\mathbf{x}) = \int_{B(\mathbf{c},r)} u(\mathbf{y}) p^{B(\mathbf{c},r)}(s,\mathbf{x},\mathbf{y})\, d\mathbf{y},$$

then $w_s$ is bounded on $B(\mathbf{c},r)$ and therefore, because $p^{B(\mathbf{c},r)}$ is smooth on $(0,\infty) \times B(\mathbf{c},r)^2$ and satisfies the Chapman–Kolmogorov equation, it follows that $w_s$ is smooth on $B(\mathbf{c},r)$. In addition, because $p^{B(\mathbf{c},r)} \leq p^G$ and $u_t \leq u$, another application of the Chapman–Kolmogorov equation leads to

$$\begin{aligned} w_{s+t}(\mathbf{x}) &= \int_{B(\mathbf{c},r)} u(\mathbf{y}) p^{B(\mathbf{c},r)}(s+t,\mathbf{x},\mathbf{y})\, d\mathbf{y} \\ &\leq \int_{B(\mathbf{c},r)} p^{B(\mathbf{c},r)}(s,\mathbf{x},\mathbf{y}) u_t(\mathbf{y})\, d\mathbf{y} \leq w_s(\mathbf{x}) \end{aligned}$$

for $(s,t) \in (0,\infty)^2$ and $\mathbf{x} \in B(\mathbf{c},r)$. Hence, if $\varphi \in C^2_c\big(B(\mathbf{c},r);[0,\infty)\big)$, then

$$\int_{B(\mathbf{c},r)} -\tfrac{1}{2}\Delta w_s(\mathbf{x})\varphi(\mathbf{x})\, d\mathbf{x} = \lim_{t \searrow 0} \frac{1}{s} \int_{B(\mathbf{c},r)} \big(w_s(\mathbf{x}) - w_{s+t}(\mathbf{x})\big)\varphi(\mathbf{x})\, d\mathbf{x} \geq 0,$$

which proves that $\Delta w_s \leq 0$ on $B(\mathbf{c},r)$. Since this means that $w_s \in \mathcal{E}\big(B(\mathbf{c},r)\big)$ for each $s > 0$ and because $w_s$ is non-increasing as a function of $s$, we will know that $u \in \mathcal{E}\big(B(\mathbf{c},r)\big)$ once we show that $w_s \longrightarrow u$ pointwise on $B(\mathbf{c},r)$. But, since $w_s \leq u$, this comes down to checking $u(\mathbf{x}) \leq \underline{\lim}_{s \searrow 0} w_s(\mathbf{x})$, which follows from lower semicontinuity.

Turning to the second assertion, begin with the observation that, because $u \ge u_s$ and $u$ is lower semicontinuous, $u_s \longrightarrow u$ pointwise as $s \searrow 0$. Next, note that for $(s, \mathbf{x}) \in (0, \infty) \times G$,

$$\begin{aligned}\int_G g^G(\mathbf{x}, \mathbf{y}) f_s(\mathbf{y})\, d\mathbf{y} &= \lim_{T \to \infty} \frac{1}{s} \left[ \int_0^s u_t(\mathbf{x})\, dt - \int_T^{T+s} u_t(\mathbf{x})\, dt \right] \\ &\le \frac{1}{s} \int_0^s u_t(\mathbf{x})\, dt \le u(\mathbf{x}).\end{aligned}$$

Hence, since $u < \infty$ Lebesgue-almost everywhere on $G$, $\sup_{s>o} \mu_s(K) < \infty$ for all $K \subset\subset G$, and so $\{\mu_s : s > 0\}$ is (cf. part (**iv**) of Exercise 9.1.16) relatively sequentially compact in the sense that every subsequence admits a subsequence that converges when tested against $\varphi \in C_c(G; \mathbb{R})$. At the same time, if $\varphi \in C_c^2(G; \mathbb{R})$ and $\varphi_s(\mathbf{x}) = \int_G \varphi(\mathbf{y}) p^G(s, \mathbf{x}, \mathbf{y})\, d\mathbf{y}$, then

$$\varphi_s - \varphi = \int_0^s \left( \int_G \tfrac{1}{2} \Delta\varphi(\mathbf{y}) p^G(\tau, \cdot, \mathbf{y})\, d\mathbf{y} \right) d\tau,$$

and so, by Fubini's Theorem and the symmetry of $p^G(\tau, \mathbf{x}, \mathbf{y})$, one can justify

$$\begin{aligned}\int_G \varphi\, d\mu_s &= -\frac{1}{2s} \int_G \left( \int_0^s u_\tau(\mathbf{y})\, d\tau \right) \Delta\varphi(\mathbf{y})\, d\mathbf{y} \\ &\longrightarrow \int_G -\tfrac{1}{2} \Delta\varphi(\mathbf{y}) u(\mathbf{y})\, d\mathbf{y} = \int_G \varphi\, d\mu.\end{aligned}$$

Hence, every limit of $\{\mu_s : s > 0\}$ is $\mu$. □

## Exercises for § 11.3

EXERCISE 11.3.14. Let $G$ be a connected open set in $\mathbb{R}^N$, and assume that $N \in \{1, 2\}$. If (11.1.20) fails, show that every excessive function on $G$ is constant. Hence, the only cases not already covered by Riesz's Decomposition Theorem are trivial anyhow.

**Hint**: Using the reasoning employed to prove the first part of Lemma 11.3.4, reduce to the case when $u$ is smooth and satisfies $\Delta u \le 0$, and in this case apply the result in Theorem 11.1.26.

EXERCISE 11.3.15. Let $G$ be an open subset of $\mathbb{R}$, and assume that either $N \ge 3$ or (11.1.20) holds. If $u$ is an excessive function on $G$ that is not identically infinite and has charge $\mu$, show that $u$ is harmonic on any open $H \subseteq G$ for which $\mu(H) = 0$. In addition, show that $u$ is a potential if it is bounded and $u(\mathbf{x}) \longrightarrow 0$ as $\mathbf{x} \in G$ tends to $\partial_{\text{reg}} G \cup \{\infty\}$.

EXERCISE 11.3.16. Let $G$ be a connected, open subset of $\mathbb{R}^N$, and again assume that either $N \geq 3$ or (11.1.20) holds. If $u \in \mathcal{E}(G)$ is not identically infinite but $u$ is infinite on the compact set $K$, show that $\mathcal{W}^{(N)}_{\mathbf{x}}\big(\exists t \in (0, \zeta^G(\psi))\ \psi(t) \in K\big) = 0$ for all $\mathbf{x} \in G \setminus K$. Finally, apply part (**ii**) of Exercise 11.1.36 to conclude that $\mathcal{W}^{(N)}_{\mathbf{x}}(\exists t > 0\ \psi(t) \in K) = 0$ for all $\mathbf{x} \notin K$.

## § 11.4 Capacity

In the classical theory of electricity, a question of interest is that of determining the largest charge that can be placed on a body so that the resulting electric field nowhere exceeds 1. From a mathematical standpoint this question is the following. Let $\mathbf{M}(G)$ denote the space of non-negative, finite Borel measures on an open set $G$. Then, given $\emptyset \neq K \subset\subset \mathbb{R}^3$, what we want to know is the total mass of the $\mu_K \in \mathbf{M}(\mathbb{R}^3)$ that is supported on $K$ and solves the extremal problem

$$\mathbf{G}^{\mathbb{R}^3}\mu_K(\mathbf{x}) = \max\{\mathbf{G}^{\mathbb{R}^3}\mu(\mathbf{x}) : \mu \in \mathbf{M}(\mathbb{R}^3) \text{ with } \mu(\mathbb{R}^3 \setminus K) = 0 \text{ and } \mathbf{G}^{\mathbb{R}^3}\mu \leq 1\}$$

for all $\mathbf{x} \in \mathbb{R}^3$. Of course, it is not at all obvious that such a $\mu_K$ exists. Indeed, the proof that it always does was one of Wiener's significant contributions to classical potential theory. As we are about to see, probability provides a simple proof of Wiener's result.[1]

**§ 11.4.1. The Capacitory Potential.** Here I will show that the extremal problem described above has a solution.

THEOREM 11.4.1. *Assume that $G$ is a connected, open subset of $\mathbb{R}^N$ and that either $N \geq 3$ or (11.1.20) holds. Given $K \subset\subset G$, set*

$$p^G_K(\mathbf{x}) = \mathcal{W}^{(N)}_{\mathbf{x}}\Big(\exists t \in (0, \zeta^G(\psi))\ \psi(t) \in K\Big), \quad \mathbf{x} \in G. \tag{11.4.2}$$

*Then $p^G_K$ is a potential whose charge $\mu^G_K$ is supported on $K$. Moreover, if $\mu \in \mathbf{M}(G)$ is supported on $K$ and $\mathbf{G}^G\mu \leq 1$, then $\mathbf{G}^G\mu \leq p^G_K$.*

PROOF: I begin by checking that $p^G_K$ is excessive. For this purpose, note that, for any $s > 0$, the Markov property says that

$$\int_G p^G_K(\mathbf{y})p^G(s,\mathbf{x},\mathbf{y})\,d\mathbf{y} = \mathcal{W}^{(N)}_{\mathbf{x}}\Big(\exists t \in (s, \zeta^G(\psi))\ \psi(t) \in K\Big) \leq p^G_K(\mathbf{x}).$$

In addition, because $p^G_K$ is bounded, the left-hand side is continuous with respect to $\mathbf{x} \in G$, and clearly the middle expression tends non-decreasingly to $p^G_K(\mathbf{x})$ as $s \searrow 0$. Thus, by the first part of Theorem 11.3.13, we now know that $p^G_K \in \mathcal{E}(G)$.

[1] It is interesting to note that, although Wiener's 1924 article, "Certain notions in potential theory," *J. Math. Phys. M.I.T.* **4**, contains the first proof that an arbitrary compact set is *capacitable*, it contains no reference to his own measure.

The next step is to prove that $p_K^G$ is a potential whose charge is supported on $K$. But, because $N \geq 3$ or (11.1.20) holds, it is clear that $p_K^G(\mathbf{x})$ tends to 0 as $\mathbf{x} \in G$ tends to either $\partial_{\text{reg}} G$ or $\infty$. Hence, if $u$ is a non-negative harmonic function on $G$ that is dominated by $p_K^G$, then $u$ must be a bounded harmonic function that tends to 0 at $\partial_{\text{reg}} G \cup \{\infty\}$, and so, because $N \geq 3$ or (11.1.20) holds, $u \equiv 0$. Therefore, $p_K^G$ is a potential. By Exercise 11.3.15, to check that $\mu_K^G(G \setminus K) = 0$, it suffices to show that $p_K^G$ is harmonic on $G \setminus K$. For this purpose, assume that $\overline{B(\mathbf{x}, r)} \subset\subset (G \setminus K)$, and use the Markov property to justify

$$\frac{1}{\omega_{N-1}} \int_{\mathbb{S}^{N-1}} p_K^G(\boldsymbol{\omega})\, \lambda_{\mathbb{S}^{N-1}}(d\boldsymbol{\omega}) = \mathbb{E}^{\mathcal{W}_{\mathbf{x}}^{(N)}}\left[p_K^G\big(\boldsymbol{\psi}(\zeta^{B(\mathbf{x},r)})\big),\ \zeta^{B(\mathbf{x},r)}(\boldsymbol{\psi}) < \infty\right]$$
$$= \mathcal{W}_{\mathbf{x}}^{(N)}\Big(\exists t \in \big(\zeta^{B(\mathbf{x},r)}(\boldsymbol{\psi}), \zeta^G(\boldsymbol{\psi})\big)\ \boldsymbol{\psi}(t) \in K\Big) = p_K^G(\mathbf{x}).$$

That is, $p_K^G$ satisfies the mean value property in $G \setminus K$ and is therefore harmonic there.

To complete the proof I must still show that if $\mu \in \mathbf{M}(G)$ is supported on $K$ and $u \equiv \mathbf{G}^G \mu \leq 1$, then $u \leq p_K^G$, and I will start by showing that $u \leq p_K^G$ on $G \setminus K$. To this end, observe that $u$ is harmonic on $G \setminus K$ and that it tends to 0 at $\partial_{\text{reg}} G \cup \{\infty\}$. Thus, if $\zeta_\delta(\boldsymbol{\psi}) = \inf\{t \geq 0 : \boldsymbol{\psi}(t) \in K(\delta)\}$, where $K(\delta) = \{\mathbf{x} : |\mathbf{x} - K| \leq \delta\}$, then, for $\delta \in \big(0, \operatorname{dist}(K, G\complement)\big)$ and $\mathbf{x} \in G \setminus K(\delta)$, $u(\mathbf{x})$ is dominated by

$$\mathbb{E}^{\mathcal{W}_{\mathbf{x}}^{(N)}}\left[u\big(\boldsymbol{\psi}(\zeta_\delta)\big),\ \zeta_\delta(\boldsymbol{\psi}) < \zeta^G(\boldsymbol{\psi})\right] \leq \mathcal{W}_{\mathbf{x}}^{(N)}\Big(\exists t \in \big(0, \zeta^G(\boldsymbol{\psi})\big)\ \boldsymbol{\psi}(t) \in K(\delta)\Big).$$

But, as $\delta \searrow 0$, the last expression tends to $p_K^G(\mathbf{x})$ plus

$$\mathcal{W}_{\mathbf{x}}^{(N)}\big(\forall \delta > 0\ \zeta_\delta < \zeta^G \text{ and } \lim_{\delta \searrow 0} \zeta_\delta = \infty = \zeta^G\big),$$

and, because $N \geq 3$ or (11.1.20) holds, this additional term is 0.

We now know that $u \leq p_K^G$ on $G \setminus K$. To prove that the same inequality holds on $K$, first observe that, by part (**i**) of Exercise 10.2.19, $p_K^G \restriction K = 1 \geq u \restriction K$ when $N = 1$. Thus, assume that $N \geq 2$. In this case, $g^G(\mathbf{x}, \mathbf{x}) = \infty$ for $\mathbf{x} \in G$, and so, since $u \leq 1$, $\mu$ must be non-atomic. In particular, this means that

$$u(\mathbf{x}) = \lim_{r \searrow 0} u_r(\mathbf{x}), \quad \text{where } u_r \equiv \int_{G \setminus B(\mathbf{x},r)} g^G(\,\cdot\,, \mathbf{y})\, \mu(d\mathbf{y}).$$

But, by the preceding applied with $K \setminus B(\mathbf{x}, r)$ replacing $K$, $u_r(x) \leq p^G_{K \setminus B(\mathbf{x},r)}$, and obviously $p^G_{K \setminus B(\mathbf{x},r)} \leq p_K^G$. $\square$

The function $p_K^G$ and the measure $\mu_K^G$ are, for the reasons explained above, known as, respectively, the **capacitory potential** and the **capacitory distribution** for $K$ in $G$, and the total mass

$$\text{Cap}(K;G) \equiv \mu_K^G(K) \tag{11.4.3}$$

is called the **capacity of $K$ in $G$**. As a dividend from Theorem 11.4.1, we get the following important connection between properties of Brownian paths and classical potential theory.

COROLLARY 11.4.4. *Let everything be as in the statement of Theorem 11.4.1. Then the following are equivalent:*

**(i)** *For every* $\mathbf{x} \in G$,

$$\mathcal{W}_{\mathbf{x}}^{(N)}\Big(\exists t \in \big(0, \zeta^G(\psi)\big)\ \psi(t) \in K\Big) > 0.$$

**(ii)** *There is an* $\mathbf{x} \in G$ *for which*

$$\mathcal{W}_{\mathbf{x}}^{(N)}\Big(\exists t \in \big(0, \zeta^G(\psi)\big)\ \psi(t) \in K\Big) > 0.$$

**(iii)** *There exists a non-zero, bounded potential on $G$ whose charge is supported in $K$.*

**(iv)** $\text{Cap}(K;G) > 0$.

*Moreover,* $\text{Cap}(K;G) = 0$ *for, when* $N \geq 3$, *some* $G \supset\supset K$ *or, when* $N \in \{1,2\}$, *some* $G \supset\supset K$ *satisfying (11.1.20), if and only if* $\mathcal{W}_{\mathbf{x}}^{(N)}\big(\exists t \in (0,\infty)\ \psi(t) \in K\big) = 0$ *for all* $\mathbf{x} \notin K$.

PROOF: The only implications in the equivalence assertion that are not completely trivial are **(iii)** $\implies$ **(iv)** and **(iv)** $\implies$ **(i)**. But, by Theorem 11.4.1, **(iii)** implies that $p_K^G \not\equiv 0$ and therefore that $\mu_K^G \neq 0$. Similarly, **(iv)** implies that $\mu_K^G \neq 0$, and therefore, since $g^G > 0$ throughout $G^2$, that $p_K^G > 0$ throughout $G$.

To prove the final assertion, first suppose that $\mathcal{W}_{\mathbf{x}_0}^{(N)}\big(\exists t \in (0,\infty)\ \psi(t) \in K\big) > 0$ for some $\mathbf{x}_0 \notin K$. Then we can choose $R \in (0,\infty)$ so that $K \subset\subset B(\mathbf{0},R)$ and $p_K^{B(\mathbf{0},R)}(\mathbf{x}_0) > 0$. In particular, $\mu_K^{B(\mathbf{0},R)} \neq 0$ and

$$\mathbf{G}^{G\cap B(\mathbf{0},R)}\mu_K^{B(\mathbf{0},R)} \leq \mathbf{G}^G\mu_K^{B(\mathbf{0},R)} \leq 1.$$

At the same time, because

$$g^G(\mathbf{x},\mathbf{y}) \leq g^{G\cap B(\mathbf{0},R)}(\mathbf{x},\mathbf{y}) + \mathbb{E}^{\mathcal{W}_{\mathbf{x}}^{(N)}}\Big[g^G\big(\psi(\zeta^{G\cap B(\mathbf{0},R)})\big),\ \zeta^{G\cap B(\mathbf{0},R)}(\psi) < \infty\Big],$$

there exists (cf. Corollary 11.2.20 when $N = 2$) a $C < \infty$ such that $g^G(\mathbf{x},\mathbf{y}) \leq g^{G\cap B(\mathbf{0},R)}(\mathbf{x},\mathbf{y}) + C$ for all $\mathbf{x} \notin B(\mathbf{0},R)$ and $\mathbf{y} \in K$. Hence, $\mathbf{G}^G\mu_K^{B(\mathbf{0},R)} \leq$

$1 + C\mathrm{Cap}\big(K, B(0,R)\big)$, and so we have shown that $\mathbf{G}^G\mu_K^{B(\mathbf{0},R)}$ is a non-zero, bounded potential on $G$ whose charge is supported in $K$, which, by the preceding equivalences, means that $\mathrm{Cap}(K;G) > 0$. Conversely, if $\mathrm{Cap}(K;G) > 0$, then, again by the preceding equivalences, we know that $p_K^G > 0$ everywhere on $G$, which, of course, means that $\mathcal{W}_{\mathbf{x}}^{(N)}\big(\exists t \in (0,\infty)\ \psi(t) \in K\big) > 0$, first for all $\mathbf{x} \in G$ and then for all $\mathbf{x} \in \mathbb{R}^N$. □

The last part of the preceding allows us to use capacity to determine whether Brownian paths will hit a $K \subset\subset \mathbb{R}^N$. Indeed, we now know that they will if and only if $\mathrm{Cap}(K;G) > 0$ for some $G \supset\supset K$ satisfying our hypotheses. Thus, the ability of Brownian paths in $\mathbb{R}^N$ to hit a set is completely determined by the singularity in the Green function. Namely, they will hit $K$ with positive probability if and only if there is a non-zero $\mu$ supported on $K$ for which $\mathbf{G}^G\mu$ is bounded. When $N = 1$, there is no singularity, and so even points can be hit. When $N \geq 2$, there is a singularity, and so, in order to be hit, $K$ has to be large enough to support a measure that is sufficiently smooth to mollify the singularity in the Green function. Non-trivial (i.e., $K$'s for which $K\complement$ is the interior of its closure) examples of $K$'s that cannot be hit are hard to come by. "Lebesgue's spine" provides one in $\mathbb{R}^3$ and can be adapted to $\mathbb{R}^N$ for $N \geq 3$. When $N = 2$ one has too work much harder. The most famous example is a devilishly clever construction, known as "Littlewood's crocodile," due to J.E. Littlewood. See M. Brelot's lecture notes *Éléments de la Théorie Classique du Potenial* published in 1965 by Centre de Documentation Universitaire, Sorbonne, Paris V.

§ **11.4.2. The Capacitory Distribution.** In this subsection I will give a probabilistic representation, discovered by K.L. Chung, of the capacitory distribution $\mu_K^G$. Again I assume that $G$ is a connected open subset of $\mathbb{R}^N$ and that either $N \geq 3$ or (11.1.20) holds.

The function $\ell_K^G : C(\mathbb{R}^N) \longrightarrow [0,\infty]$ given by

$$\begin{aligned} \ell_K^G(\psi) &= \sup\big\{t \in \big(0,\zeta^G(\psi)\big) : \psi(t) \in K\big\} \\ &\qquad \big(\equiv 0 \text{ if } \big\{t \in \big(0,\zeta^G(\psi)\big) : \psi(t) \in K\big\} = \emptyset\big). \end{aligned} \tag{11.4.5}$$

is called a **quitting time**. Clearly, *$\ell_K^G$ is not a stopping time.* On the other hand, it transforms nicely under the time-shift maps $\Sigma_t$. Specifically,

$$\ell_K^G \circ \Sigma_t = \big(\ell_K^G - t\big)^+ \quad \text{for } t \in [0,\zeta^G).$$

THEOREM 11.4.6 (**Chung**).[2] *Let $G$ be a connected open subset of $\mathbb{R}^N$, assume that either $N \geq 3$ or that (11.1.20) holds, and suppose that $K \subset\subset G$ with*

[2] This result appeared originally in K.L. Chung's "Probabilistic approach in potential theory to the equilibrium problem," *Ann. Inst. Fourier Gren.* **23 # 3**, pp. 313–322 (1973). It gives the first direct probabilistic interpretation of the capacitory measure.

$\mathrm{Cap}(K;G) > 0$. *Then, for all Borel measurable* $\varphi : G \longrightarrow \mathbb{R}$ *that are bounded below and every* $\mathbf{c} \in G$,

$$\int_G \varphi \, d\mu_K^G = \mathbb{E}^{\mathcal{W}_{\mathbf{c}}^{(N)}}\left[\frac{\varphi(\boldsymbol{\psi}(\ell_K^G))}{g^G(\mathbf{c}, \boldsymbol{\psi}(\ell_K^G))}, \; \ell_K^G \in (0,\infty)\right]. \tag{11.4.7}$$

PROOF: Take $u = p_K^G$, and define $f_s$ and $\mu_s$ for $s > 0$ as in Theorem 11.3.13. Then $sf_s(\mathbf{x}) = \mathcal{W}_{\mathbf{x}}^{(N)}(0 < \ell_K^G \le s)$, and so, for any $\varphi \in C_b(G;\mathbb{R})$,

$$\begin{aligned}
\int_G g^G(\mathbf{c},\mathbf{y})\varphi(\mathbf{y})\,\mu_s(d\mathbf{y}) &= \mathbb{E}^{\mathcal{W}_{\mathbf{c}}^{(N)}}\left[\int_0^{\zeta^G} \varphi(\boldsymbol{\psi}(t))\, f_s(\boldsymbol{\psi}(t))\,dt\right] \\
&= \frac{1}{s}\int_0^\infty \mathbb{E}^{\mathcal{W}_{\mathbf{c}}^{(N)}}\left[\varphi(\boldsymbol{\psi}(t))\mathcal{W}_{\boldsymbol{\psi}(t)}^{(N)}(0 < \ell_K^G \le s), \; \zeta^G > t\right] dt \\
&= \frac{1}{s}\int_0^\infty \mathbb{E}^{\mathcal{W}_{\mathbf{c}}^{(N)}}\left[\varphi(\boldsymbol{\psi}(t)), \; t < \ell_K^G \le s + t\right] dt \\
&= \mathbb{E}^{\mathcal{W}_{\mathbf{c}}^{(N)}}\left[\frac{1}{s}\int_{(\ell_K^G - s)^+}^{\ell_K^G} \varphi(\boldsymbol{\psi}(t))\,dt, \; \ell_K^G \in (0,\infty)\right] \\
&\longrightarrow \mathbb{E}^{\mathcal{W}_{\mathbf{c}}^{(N)}}\left[\varphi(\boldsymbol{\psi}(\ell_K^G)), \; \ell_K^G \in (0,\infty)\right] \quad \text{as } s \searrow 0,
\end{aligned}$$

where, in the passage to the third line, I have applied the Markov property and used the time-shift property of $\ell_K^G$. Next, let $\eta \in C_c(G;\mathbb{R})$ be given, note that $\varphi = \frac{\eta}{g^G(\mathbf{c},\,\cdot\,)}$ is again an element of $C_c(G;\mathbb{R})$, and conclude from Theorem 11.3.13 and the preceding that (11.4.7) holds first for $\varphi$'s in $C_c(G;\mathbb{R})$ and then for all bounded, measurable $\varphi$'s on $G$. □

Aside from its intrinsic beauty, (11.4.7) has the virtue that it simplifies the proofs of various important facts about capacity. For instance, it allows one to prove a basic monotone convergence result for capacity. However, before doing so, I will need to introduce the the **energy** $\mathcal{E}^G(\mu,\nu)$, which is defined for locally finite, non-negative Borel measures $\mu$ and $\nu$ on $G$ by

$$\mathcal{E}^G(\mu,\nu) = \iint_{G^2} g^G(\mathbf{x},\mathbf{y})\,\mu(d\mathbf{x})\nu(d\mathbf{y}).$$

Clearly $\mathcal{E}^G(\mu,\nu)$ is some sort of inner product, and so it is not surprising that there is a Schwarz inequality for it.

LEMMA 11.4.8. *For any pair of locally finite, non-negative, Borel measures* $\mu$ *and* $\nu$ *on* $G$,

$$\mathcal{E}^G(\mu,\nu) \le \sqrt{\mathcal{E}^G(\mu,\mu)}\,\sqrt{\mathcal{E}^G(\nu,\nu)};$$

*and, when the factors on the right are both finite, equality holds if and only if* $a\mu - b\nu = 0$ *for some pair* $(a,b) \in [0,\infty)^2 \setminus (0,0)$.

PROOF: For each $(t, \mathbf{x}) \in (0, \infty) \times G$, set

$$f(t, \mathbf{x}) = \int_G p^G(t, \mathbf{x}, \mathbf{y})\, \mu(d\mathbf{y}) \quad \text{and} \quad g(t, \mathbf{x}) = \int_G g^G(t, \mathbf{x}, \mathbf{y})\, \nu(d\mathbf{y}),$$

and note that, by the Chapman–Kolmogorov equation, Tonelli's Theorem, and Schwarz's Inequality:

$$\begin{aligned}
\mathcal{E}^G(\mu, \nu) &= \int_{(0,\infty)} \left( \iint_{G^2} p^G(t, \mathbf{x}, \mathbf{y})\, \mu(d\mathbf{x})\nu(d\mathbf{y}) \right) dt \\
&= \iint_{(0,\infty)\times G} f\left(\tfrac{t}{2}, \mathbf{x}\right) g\left(\tfrac{t}{2}, \mathbf{x}\right) dt d\mathbf{x} \\
&\le \left( \iint_{(0,\infty)\times G} f\left(\tfrac{t}{2}, \mathbf{x}\right)^2 dt d\mathbf{x} \right)^{\frac{1}{2}} \left( \iint_{(0,\infty)\times G} g\left(\tfrac{t}{2}, \mathbf{x}\right)^2 dt d\mathbf{x} \right)^{\frac{1}{2}} \\
&= \left( \iint_{(0,\infty)\times G} f(t, \mathbf{x})\, dt d\mathbf{x} \right)^{\frac{1}{2}} \left( \iint_{(0,\infty)\times G} g(t, \mathbf{x})\, dt d\mathbf{x} \right)^{\frac{1}{2}} \\
&= \sqrt{\mathcal{E}^G(\mu, \mu)}\, \sqrt{\mathcal{E}^G(\nu, \nu)}.
\end{aligned}$$

Furthermore, when $f$ and $g$ are square integrable, then equality holds if and only if they are linearly dependent in the sense that $af - bg = 0$ Lebesgue-almost everywhere for some non-trivial choice of $a, b \in [0, \infty)$. But this means that

$$\begin{aligned}
a \int_G \varphi\, d\mu &= \lim_{T \searrow 0} \frac{a}{T} \int_0^T \left( \int_G \varphi(\mathbf{x}) p^G(t, \mathbf{x}, \mathbf{y})\, \mu(d\mathbf{x}) \right) dt \\
&= \lim_{T \searrow 0} \frac{a}{T} \iint_{(0,T]\times G} \varphi(\mathbf{x})\, f(t, \mathbf{x})\, dt d\mathbf{x} = \lim_{T \searrow 0} \frac{b}{T} \iint_{(0,T]\times G} \varphi(\mathbf{x})\, g(t, \mathbf{x})\, dt d\mathbf{x} \\
&= \lim_{T \searrow 0} \frac{b}{T} \int_0^T \left( \int_G \varphi(\mathbf{x}) p^G(t, \mathbf{x}, \mathbf{y})\, \nu(d\mathbf{x}) \right) dt = b \int_G \varphi\, d\nu
\end{aligned}$$

for every $\varphi \in C_c(G; \mathbb{R})$, and so $a\mu - b\nu = 0$. □

With this lemma, I can now give the application of Theorem 11.4.6 mentioned above.

THEOREM 11.4.9. *Let $G$ be as in Theorem 11.4.6 and $\{K_n : n \geq 1\}$ a non-increasing sequence of compact subsets of $G$. If $K = \bigcap_1^\infty K_n$, then, for every Borel measurable $\varphi : G \longrightarrow \mathbb{R}$ that is continuous in a neighborhood of $K_1$,*

$$\lim_{n\to\infty} \int_G \varphi \, d\mu_{K_n}^G = \int_G \varphi \, d\mu_K^G,$$

*and so*

$$\operatorname{Cap}(K;G) = \lim_{n\to\infty} \operatorname{Cap}(K_n;G).$$

*Finally, if $\mu$ is any non-negative Borel measure on $G$ satisfying $\mu(G \setminus K) = 0$ and $\mathbf{G}^G\mu \leq 1$, then*

$$\mathcal{E}^G(\mu,\mu) \leq \operatorname{Cap}(K;G) \quad \text{and equality holds} \iff \mu = \mu_K^G.$$

PROOF: Let $\mathbf{c} \in G\setminus K_1$ be given. In view of (11.4.7), checking the first assertion comes down to showing that, for $\mathcal{W}_{\mathbf{c}}^{(N)}$-almost every $\boldsymbol{\psi} \in C(\mathbb{R}^N)$,

$$\ell_{K_n}^G(\boldsymbol{\psi}) \longrightarrow \ell_K^G(\boldsymbol{\psi}) \in \big(0, \zeta^G(\boldsymbol{\psi})\big) \quad \text{if either}$$
$$\{\ell_{K_n}^G(\boldsymbol{\psi}) : n \geq 1\} \subseteq \big(0, \zeta^G(\boldsymbol{\psi})\big) \quad \text{or} \quad \ell_K^G(\boldsymbol{\psi}) \in \big(0, \zeta^G(\boldsymbol{\psi})\big).$$

To this end, let $\boldsymbol{\psi} \in C(\mathbb{R}^N)$ with $\boldsymbol{\psi}(0) = \mathbf{c}$ be given. If $\{\ell_{K_n}^G(\boldsymbol{\psi}) : n \geq 1\} \subseteq \big(0, \zeta^G(\boldsymbol{\psi})\big)$, then it is clear that

$$\ell_{K_n}^G(\boldsymbol{\psi}) \searrow T \geq \ell_K^G(\boldsymbol{\psi}), \quad \text{where } T \in \big(0, \zeta^G(\boldsymbol{\psi})\big).$$

In addition, by continuity, $\boldsymbol{\psi}(T) \in K$, which means first that $T \leq \ell_K^G(\boldsymbol{\psi})$ and then that $\ell_{K_n}^G(\boldsymbol{\psi}) \longrightarrow \ell_K^G(\boldsymbol{\psi}) \in \big(0, \zeta^G(\boldsymbol{\psi})\big)$. Next, observe that

$$0 < \ell_K^G(\boldsymbol{\psi}) < \zeta^G(\boldsymbol{\psi}) < \infty \implies \ell_{K_n}^G(\boldsymbol{\psi}) \in \big[\ell_K^G(\boldsymbol{\psi}), \zeta^G(\boldsymbol{\psi})\big) \quad \text{for all } n \in \mathbb{Z}^+.$$

Hence, we are done if (11.1.20) holds. On the other hand, if $N \geq 3$, then, because $\lim_{t\to\infty} |\boldsymbol{\psi}(t)| = \infty$ for $\mathcal{W}_{\mathbf{c}}^{(N)}$-almost all $\boldsymbol{\psi} \in C(\mathbb{R}^N)$, we know that, for $\mathcal{W}_{\mathbf{c}}^{(N)}$-almost every $\boldsymbol{\psi} \in C(\mathbb{R}^N)$,

$$\zeta^G(\boldsymbol{\psi}) = \infty \text{ and } \ell_K^G(\boldsymbol{\psi}) \in (0,\infty) \implies \{\ell_{K_n}^G(\boldsymbol{\psi}) : n \geq 1\} \subseteq \big(0, \zeta^G(\boldsymbol{\psi})\big);$$

and so we have now completed the proof of the first part.

To prove the final assertion, first choose compact $K_n$'s in $G$ so that $K \subset\subset (K_n)^\circ$ for each $n \in \mathbb{Z}^+$ and $K_n \searrow K$ as $n \to \infty$. Because $p_{K_n}^G \restriction K \equiv 1$ and $p_{K_n}^G \leq 1$, we have that

$$\begin{aligned}
\operatorname{Cap}(K;G) &= \int_G p_{K_n}^G(\mathbf{x})\, \mu_K^G(d\mathbf{x}) = \mathcal{E}^G(\mu_K^G, \mu_{K_n}^G) \\
&\leq \mathcal{E}^G(\mu_K^G, \mu_K^G)^{\frac{1}{2}} \mathcal{E}^G(\mu_{K_n}^G, \mu_{K_n}^G)^{\frac{1}{2}} \\
&= \mathcal{E}^G(\mu_K^G, \mu_K^G)^{\frac{1}{2}} \left(\int_G p_{K_n}^G(\mathbf{x})\, \mu_{K_n}^G(d\mathbf{x})\right)^{\frac{1}{2}} \\
&\leq \mathcal{E}^G(\mu_K^G, \mu_K^G)^{\frac{1}{2}} \operatorname{Cap}(K_n;G)^{\frac{1}{2}} \longrightarrow \mathcal{E}^G(\mu_K^G, \mu_K^G)^{\frac{1}{2}} \operatorname{Cap}(K;G)^{\frac{1}{2}}
\end{aligned}$$

as $n \to \infty$. Hence, $\mathrm{Cap}(K;G) \le \mathcal{E}^G(\mu_K^G, \mu_K^G)$. On the other hand, if $\mu(G \setminus K) = 0$ and $\mathbf{G}^G\mu \le 1$, then, by Theorem 11.4.1, $\mathbf{G}^G\mu \le p_K^G \le 1$,

$$\begin{aligned}\mathcal{E}^G(\mu,\mu) &= \int_G \mathbf{G}^G\mu\, d\mu \le \int_G p_K^G\, d\mu = \mathcal{E}^G(\mu_K^G, \mu)\\ &\le \mathcal{E}^G(\mu_K^G,\mu_K^G)^{\frac{1}{2}}\mathcal{E}^G(\mu,\mu)^{\frac{1}{2}}\\ &= \left(\int_G p_K^G\, d\mu_K^G\right)^{\frac{1}{2}} \mathcal{E}^G(\mu,\mu)^{\frac{1}{2}} \le \sqrt{\mathrm{Cap}(K;G)}\sqrt{\mathcal{E}^G(\mu,\mu)},\end{aligned}$$

and equality can hold only if $a\mu_K^G - b\mu = 0$ for some non-trivial pair $(a,b) \in [0,\infty)^2$. When one takes $\mu = \mu_K^G$, this, in conjunction with the preceding, proves that $\mathrm{Cap}(K;G) = \mathcal{E}^G(\mu_K^G, \mu_K^G)$. In addition, for any $\mu$ with $\mu(G \setminus K) = 0$ and $\mathbf{G}^G\mu \le 1$, it shows that $\mathcal{E}^G(\mu,\mu) \le \mathrm{Cap}(K;G)$ and that equality can hold only if $\mu$ and $\mu_K^G$ are related by a non-trivial linear equation, in which case $\mu = \mu_K^G$ follows immediately from the equality $\mathcal{E}^G(\mu_K^G,\mu_K^G) = \mathcal{E}^G(\mu,\mu)$. □

The result in Theorem 11.4.9, which was known to Wiener, played an important role in his analysis of classical potential theory. To be more precise, when $N = 3$ and $K\complement$ is regular, $p_K^{\mathbb{R}^3}$ is the continuous function on $\mathbb{R}^3$ that is harmonic off $K$, is 1 on $K$, and tends to 0 at infinity. Thus, it is a relatively simple problem to define the capacitory distribution for such $K$'s in $\mathbb{R}^3$. The importance to Wiener of results like that in Theorem 11.4.9 is that they enabled him (cf. Exercise 11.4.20) to make a consistent assignment of capacity to $K$'s for which $K\complement$ is not necessarily regular.

§ **11.4.3. Wiener's Test.** This subsection is devoted to another of Wiener's famous contributions to classical potential theory.

As was pointed out following Corollary 11.4.4, capacity can be used to test whether Brownian paths will hit a compact set $K$. By Lemma 11.1.21, an equivalent statement is that capacity can be used to test whether $\partial_{\mathrm{reg}}(K\complement)$ is empty or not. The result of Wiener that will be proved here can be viewed as a sharpening of this remark.

Assume that $N \ge 2$, and let an open subset $G$ of $\mathbb{R}^N$ and an $\mathbf{a} \in \partial G$ be given. For $n \in \mathbb{Z}^+$, set

$$K_n = \left\{\mathbf{y} \notin G : 2^{-n-1} \le |\mathbf{y} - \mathbf{a}| \le 2^{-n}\right\},$$

and define

$$W_n(\mathbf{a}, G) = \begin{cases} n\mathrm{Cap}\big(K_n; B(\mathbf{a},1)\big) & \text{if } N = 2\\ 2^{n(N-2)}\mathrm{Cap}\big(K_n; B(\mathbf{a},1)\big) & \text{if } N \ge 3.\end{cases} \tag{11.4.10}$$

Then **Wiener's test** says that

$$\mathbf{a} \in \partial_{\mathrm{reg}} G \iff \sum_{n=1}^{\infty} W_n(\mathbf{a}, G) = \infty. \tag{11.4.11}$$

Notice that, at least qualitatively, (11.4.11) is what one should expect in that the divergence of the series is some sort of statement that $G\complement$ *is robust at* $\mathbf{a}$.

The key to my proof of Wiener's test is the trivial observation that because

$$p_n(\mathbf{x}) \equiv p_{K_n}^{B(\mathbf{a},1)}(\mathbf{x}) = \int_{K_n} g^{B(\mathbf{a},1)}(\mathbf{x},\mathbf{y})\,\mu_{K_n}^{B(\mathbf{a},1)}(d\mathbf{y}),$$

and, depending on whether $N = 2$ or $N \geq 3$, there exists (cf. Exercise 11.2.21) an $\alpha_N \in (0,1)$ such that $\alpha_N n \leq g^{B(\mathbf{a},1)}(\mathbf{a},\mathbf{y}) \leq \alpha_N^{-1} n$ or $\alpha_N 2^{n(N-2)} \leq g^{B(\mathbf{a},1)}(\mathbf{a},\mathbf{y}) \leq \alpha_N^{-1} 2^{n(N-2)}$ for $\mathbf{y} \in B(\mathbf{a},2^{-n}) \setminus B(\mathbf{a},2^{-n-1})$, we know that

$$\alpha_N W_n(\mathbf{a},G) \leq p_n(\mathbf{a}) \leq W_n(\mathbf{a},G), \quad n \in \mathbb{Z}^+.$$

Hence, in probabilistic terms, Wiener's test comes down to the assertion that

$$\mathcal{W}_{\mathbf{a}}^{(N)}\left(\zeta_{0+}^{G} = 0\right) = 1 \iff \sum_{1}^{\infty} \mathcal{W}_{\mathbf{a}}^{(N)}(A_n) = \infty,$$

where $A_n$ is the set of $\psi \in C(\mathbb{R}^N)$ that visit $K_n$ before leaving $B(\mathbf{a},1)$. Actually, although the preceding equivalence is not obvious, the closely related statement

$$\mathcal{W}_{\mathbf{a}}^{(N)}\left(\zeta_{0+}^{G} = 0\right) = 1 \iff \mathcal{W}_{\mathbf{a}}^{(N)}\left(\overline{\lim_{n\to\infty}} A_n\right) > 0 \tag{11.4.12}$$

is essentially immediate. Indeed, if $\psi(0) = \mathbf{a}$ and $\zeta_{0+}^{G}(\psi) = 0$, then there exists a sequence of times $t_m \searrow 0$ with the property that $\psi(t_m) \in B(\mathbf{a},1) \cap G\complement$ for all $m$, from which it is clear that $\psi$ visits infinitely many $K_n$'s before leaving $B(\mathbf{a},1)$. Hence, the " $\Longrightarrow$ " in (11.4.12) is trivial. As for the opposite implication, suppose that $\psi \in C(\mathbb{R}^N)$ has the properties that $\zeta^{B(\mathbf{a},1)}(\psi) < \infty$, $\{t \in [0,\zeta^{B(\mathbf{a},1)}(\psi)) : \psi(t) = \mathbf{a}\} = \{0\}$, and that $\psi$ visits infinitely many $K_n$'s before leaving $B(\mathbf{a},1)$. We can then find a subsequence $\{n_m : m \geq 1\}$ and a convergent sequence of times $t_m > 0$ such that $\psi(t_m) \in K_{n_m}$ for each $m$. Clearly, $\lim_{m\to\infty} \psi(t_m) = \mathbf{a}$, and therefore $\lim_{m\to\infty} t_m = 0$. In other words, if $\zeta^{B(\mathbf{a},1)}(\psi) < \infty$, $\{t \in [0,\zeta^{B(\mathbf{a},1)}(\psi)) : \psi(t) = \mathbf{a}\} = \{0\}$, and $\psi \in \overline{\lim}_{n\to\infty} A_n$, then $\zeta_{0+}^{G}(\psi) = 0$. Hence, since $N \geq 2$ and therefore

$$\mathcal{W}_{\mathbf{a}}^{(N)}\left(\{\psi : \zeta^{B(\mathbf{a},1)}(\psi) < \infty \text{ and } \forall t > 0\ \psi(t) \neq \mathbf{a}\}\right) = 1,$$

we have shown that

$$\mathcal{W}_{\mathbf{a}}^{(N)}\left(\zeta_{0+}^{G} = 0\right) \geq \mathcal{W}_{\mathbf{a}}^{(N)}\left(\overline{\lim_{n\to\infty}} A_n\right);$$

and therefore, because $\mathcal{W}_{\mathbf{a}}^{(N)}\left(\zeta_{0+}^{G} = 0\right) \in \{0,1\}$, we have proved the equivalence in (11.4.12).

In view of the preceding paragraph, the proof of Wiener's test reduces to the problem of showing that

$$\text{(11.4.13)} \qquad \mathcal{W}_{\mathbf{a}}^{(N)}\left(\varlimsup_{n\to\infty} A_n\right) > 0 \iff \sum_1^\infty \mathcal{W}_{\mathbf{a}}^{(N)}(A_n) = \infty.$$

By the trivial part of the Borel–Cantelli Lemma, the "$\Longrightarrow$" implication in (11.4.13) is easy. On the other hand, because the events $\{A_n : n \ge 1\}$ are not mutually independent, the non-trivial part of that lemma does not apply and therefore cannot be used to go in the opposite direction. Nonetheless, as we will see, the following interesting variation on the Borel–Cantelli theme does apply and gives us the "$\Longleftarrow$" implication in (11.4.13).

LEMMA 11.4.14. *Let $(\Omega, \mathcal{F}, \mathbb{P})$ be a probability space and $\{A_n : n \ge 1\}$ a sequence of $\mathcal{F}$-measurable sets with the property that*

$$\mathbb{P}(A_m \cap A_n) \le C\mathbb{P}(A_m)\mathbb{P}(A_n), \quad m \in \mathbb{Z}^+ \text{ and } n \ge m + d,$$

*for some $C \in [1,\infty)$ and $d \in \mathbb{Z}^+$. Then*

$$\sum_1^\infty \mathbb{P}(A_n) = \infty \implies \mathbb{P}\left(\varlimsup_{n\to\infty} A_n\right) \ge \frac{1}{4C}.$$

PROOF: Because

$$\sum_{n=1}^\infty \mathbb{P}(A_n) = \infty \implies \sum_{n=1}^\infty \mathbb{P}(A_{nd+k}) = \infty \quad \text{for some } 0 \le k < d,$$

whereas

$$\mathbb{P}\left(\varlimsup_{n\to\infty} A_n\right) \ge \mathbb{P}\left(\varlimsup_{n\to\infty} A_{nd+k}\right) \quad \text{for each } 0 \le k < d,$$

I may and will assume that $d = 1$. Further, since

$$\mathbb{P}\left(\varlimsup_{n\to\infty} A_n\right) \ge \varlimsup_{n\to\infty} \mathbb{P}(A_n),$$

I will assume that $\mathbb{P}(A_n) \le \frac{1}{4C}$ for all $n \in \mathbb{Z}^+$. In particular, these assumptions mean that, for each $m \in \mathbb{Z}^+$, we can find an $n_m > m$ such that

$$s_m \equiv \sum_{\ell=m}^{n_m} \mathbb{P}(A_\ell) \in \left[\tfrac{3}{4C}, \tfrac{1}{C}\right].$$

Indeed, simply take $n_m$ to be the largest $n > m$ for which $\sum_{\ell=m}^n \mathbb{P}(A_\ell) \le \frac{1}{C}$. At the same time, by an easy induction argument on $n > m$, one has that

$$\mathbb{P}\left(\bigcup_{\ell=m}^n A_\ell\right) \ge \sum_{\ell=m}^n \mathbb{P}(A_\ell) - \frac{1}{2} \sum_{m \le k \ne \ell \le n} \mathbb{P}(A_k \cap A_\ell)$$

for all $n > m \geq 1$, and therefore

$$\mathbb{P}\left(\bigcup_{\ell=m}^{\infty} A_\ell\right) \geq \mathbb{P}\left(\bigcup_{\ell=m}^{n_m} A_\ell\right) \geq s_m - \frac{Cs_m^2}{2} \geq \frac{1}{4C}$$

for all $m \in \mathbb{Z}^+$. □

Proof of Wiener's Test: All that remains is to check that the sets $A_n$ appearing in (11.4.13) satisfy the hypothesis in Lemma 11.4.14 when $\mathbb{P} = \mathcal{W}_{\mathbf{a}}^{(N)}$. To this end, set

$$\sigma_n(\boldsymbol{\psi}) = \inf\left\{t \in (0,\infty) : \boldsymbol{\psi}(t) \in K_n\right\}.$$

Clearly, $A_n = \{\sigma_n < \zeta^{B(\mathbf{a},1)}\}$, and so

$$\mathcal{W}_{\mathbf{a}}^{(N)}(A_m \cap A_n) \leq \mathcal{W}_{\mathbf{a}}^{(N)}\left(\sigma_m < \sigma_n < \zeta^{B(\mathbf{a},1)}\right) + \mathcal{W}_{\mathbf{a}}^{(N)}\left(\sigma_n < \sigma_m < \zeta^{B(\mathbf{a},1)}\right)$$

for all $m \in \mathbb{Z}^+$ and $n \neq m$. But, by the Markov property,

$$\begin{aligned}\mathcal{W}_{\mathbf{a}}^{(N)}\left(\sigma_m < \sigma_n < \zeta^{B(\mathbf{a},1)}\right) &\leq \mathbb{E}^{\mathcal{W}_{\mathbf{a}}^{(N)}}\left[p_n(\boldsymbol{\psi}(\sigma_m)),\ \sigma_m(\boldsymbol{\psi}) < \zeta^{B(\mathbf{a},1)}(\boldsymbol{\psi})\right] \\ &\leq \beta(m,n)p_m(\mathbf{a}),\end{aligned}$$

where I have introduced the notation $\beta(m,n) \equiv \max_{\mathbf{x}\in K_m} p_n(\mathbf{x})$. Finally, because $p_n(\mathbf{x}) = \mathbf{G}^{B(\mathbf{a},1)}\mu_{K_n}^{B(\mathbf{a},1)}(\mathbf{x})$ and there is a $C_N < \infty$ such that $\frac{g^{B(\mathbf{a},1)}(\mathbf{x},\mathbf{y})}{g^{B(\mathbf{a},1)}(\mathbf{a},\mathbf{y})} \leq C_N$ for $\mathbf{x} \in \bigcup_{|m-n|\geq 2} K_m$ and $\mathbf{y} \in K_n$,

$$\beta(m,n) \leq C_N p_n(\mathbf{a}) \quad \text{for all } |m-n| \geq 2.$$

Hence, since $p_n(\mathbf{a}) = \mathcal{W}_{\mathbf{a}}^{(N)}(A_n)$, we have now shown that

$$\mathcal{W}_{\mathbf{a}}^{(N)}(A_m \cap A_n) \leq 2C_N\mathcal{W}_{\mathbf{a}}^{(N)}(A_m)\mathcal{W}_{\mathbf{a}}^{(N)}(A_n) \quad \text{for all } |m-n| \geq 2,$$

which means that Lemma 11.4.14 applies with $C = 2C_N$ and $d = 2$. □

**§ 11.4.4. Some Asymptotic Expressions Involving Capacity.** Assume that $K \subset\subset \mathbb{R}^N$ and that $N \geq 2$. Given $K \subset\subset \mathbb{R}^N$, define $\sigma_K(\boldsymbol{\psi}) = \zeta_{0+}^{K\complement}(\boldsymbol{\psi}) = \inf\{t > 0 : \boldsymbol{\psi}(t) \in K\}$ to be the first positive entrance time into $K$. In this subsection I will make some computations in which $\sigma_K$ and capacity play a critical role.

I begin with a result of F. Spitzer's[3] about the rate of heat transfers from the outside to the inside of a compact set. To be precise, let $K \subset\subset \mathbb{R}^N$, where $N \geq 3$, and think of

(11.4.15) $$E_K(t) \equiv \int_{K\complement} \mathcal{W}_{\mathbf{x}}^{(N)}(\sigma_K \leq t)\, d\mathbf{x}$$

as the amount of heat that flows into $K$ during $[0,t]$ from outside.

[3] See *Electrostatic capacity, heat flow, and Brownian motion*, in Z. Wahrsh. Gebiete. **3**. Recently, M. Van den Burg has written several papers in which he greatly refines Spitzer's result.

THEOREM 11.4.16 (**Spitzer**). *Assume that $N \geq 3$, and, for $K \subset\subset \mathbb{R}^N$, define $t \rightsquigarrow E_K(t)$ as in (11.4.15). Then*

$$\lim_{t\to\infty} \frac{E_K(t)}{t} = \mathrm{Cap}(K;\mathbb{R}^N).$$

PROOF: Because, by the second part of Lemma 11.1.5,

$$\mathcal{W}^{(N)}_{\mathbf{x}}(\sigma_K = t) = 0 \quad \text{for all } (t,\mathbf{x}) \in (0,\infty)\times\mathbb{R}^N,$$

we know that $t \rightsquigarrow E_K(t)$ is a bounded, non-negative, continuous, non-decreasing function.

I next observe that, for any $0 \leq h < t$,

$$E_K(t) - E_K(t-h) = \int_{\mathbb{R}^N} \mathcal{W}^{(N)}_{\mathbf{x}}(t-h < \sigma_K \leq t)\, d\mathbf{x}.$$

To see this, notice that there would be nothing to do if the integral were over $K\complement$. On the other hand, by part (**iii**) of Exercise 10.2.19, $\mathcal{W}^{(N)}_{\mathbf{x}}(\sigma_K > 0) = 0$ Lebesgue-almost everywhere on $K$, and so the integral over $K$ does not contribute anything.

I now want to replace the preceding by

$$(*) \qquad E_K(t) - E_K(t-h) = \int_{\mathbb{R}^N} \mathcal{W}^{(N)}_{\mathbf{y}}(\sigma_K \leq h \text{ and } \sigma^h_K > t)\, d\mathbf{y},$$

where

$$\sigma^h_K(\psi) \equiv \inf\{s \in (h,\infty) : \psi(s) \in K\}$$

is the first entrance time into $K$ after time $h$. To prove (*), set

$$\boldsymbol{\theta}^{(\mathbf{x},\mathbf{y})}_t(s) = \frac{t-s}{t}\mathbf{x} + \boldsymbol{\theta}_t(s) + \frac{s}{t}\mathbf{y}, \quad s \in [0,t],$$

where $\boldsymbol{\theta}_t(s) = \boldsymbol{\theta}(s) - \frac{s\wedge t}{t}\boldsymbol{\theta}(t)$. Then, by (8.3.12) and the reversibility property discussed in Exercise 8.3.22,

$$\begin{aligned}
&\mathcal{W}^{(N)}_{\mathbf{x}}(t-h < \sigma_K \leq t) \\
&\quad = \int_{\mathbb{R}^N} \mathcal{W}^{(N)}\Big(t-h < \sigma_K\big(\boldsymbol{\theta}^{(\mathbf{x},\mathbf{y})}_t\big) \leq t\Big)\, g^{(N)}(t,\mathbf{y}-\mathbf{x})\, d\mathbf{y} \\
&\quad = \int_{\mathbb{R}^N} \mathcal{W}^{(N)}\Big(\sigma_K\big(\boldsymbol{\theta}^{(\mathbf{y},\mathbf{x})}_t\big) \leq h \text{ and } \sigma^h_K\big(\boldsymbol{\theta}^{(\mathbf{y},\mathbf{x})}_t\big) > t\Big)\, g^{(N)}(t,\mathbf{y}-\mathbf{x})\, d\mathbf{y},
\end{aligned}$$

and now integrate with respect to $\mathbf{x}$ to arrive at (*) after an application of Tonelli's Theorem and another application of (8.3.12).

Starting from (*), one has that, for each $h \in [0,\infty)$,

$$\Delta_K(h) \equiv \lim_{t\to\infty} \big(E_K(t+h) - E_K(t)\big) = \int_{\mathbb{R}^N} \mathcal{W}_{\mathbf{y}}^{(N)}\Big(\sigma_K \le h \text{ and } \sigma_K^h = \infty\Big)\, d\mathbf{y},$$

the convergence being uniform for $h$ in compacts. Thus, $\Delta_K$ is non-negative and continuous, and, from its definition, it is clear that it is additive in the sense that $\Delta_K(h_1+h_2) = \Delta_K(h_1) + \Delta_K(h_2)$. Therefore, by standard results about additive functions, we now know that $\Delta_K(h) = h\Delta_K(1)$.

The problem which remains is that of evaluating $\Delta_K(1)$. First observe that, by (4.3.13),

$$\mathcal{W}_{\mathbf{y}}^{(N)}\Big(\sigma_K \le h \text{ and } \sigma_K^h = \infty\Big) \le \mathcal{W}_{\mathbf{y}}^{(N)}(\sigma_K \le h) \le 2N \exp\left[-\frac{|\mathbf{y}-K|^2}{2Nh}\right],$$

and therefore that

$$\Delta_K(1) = \lim_{h\searrow 0} \frac{\Delta_K(h)}{h} = \lim_{h\searrow 0} \frac{1}{h}\int_{B(\mathbf{0},R)} \mathcal{W}_{\mathbf{y}}^{(N)}(\sigma_K \le h \;\&\; \sigma_K^h = \infty)\, d\mathbf{y}$$

for any $R>0$ satisfying $K \subset\subset B(\mathbf{0},R)$. Second, note that

$$\begin{aligned}\mathcal{W}_{\mathbf{y}}^{(N)}(\sigma_K \le h \text{ and } \sigma_K^h = \infty) &= \mathcal{W}_{\mathbf{y}}^{(N)}(\sigma_K^h = \infty) - \mathcal{W}_{\mathbf{y}}^{(N)}(\sigma_K = \infty)\\ &= \mathcal{W}_{\mathbf{y}}^{(N)}(\sigma_K < \infty) - \mathcal{W}_{\mathbf{y}}^{(N)}(\sigma_K^h < \infty) = p_K^{\mathbb{R}^N}(\mathbf{y}) - \int_{\mathbb{R}^N} g^{(N)}(h, \mathbf{y}-\boldsymbol{\xi}) p_K^{\mathbb{R}^N}(\boldsymbol{\xi})\, d\boldsymbol{\xi}.\end{aligned}$$

Finally, combine these with Theorem 11.3.13 to arrive at $\Delta_K(1) = \operatorname{Cap}(K;\mathbb{R}^N)$.

To complete the proof, set $\rfloor t \lfloor = t - \lfloor t \rfloor$ and write

$$E_K(t) = E_K(\rfloor t\lfloor) + \sum_{n=1}^{\lfloor t \rfloor}\Big(E_K(\rfloor t\lfloor + n) - E_K(\rfloor t\lfloor + n - 1)\Big).$$

Using this together with $\Delta_K(h) = h\operatorname{Cap}(K;G)$, one obtains the desired result. □

The next two computations provide asymptotic formulas as $t \nearrow \infty$ for the quantity $\mathcal{W}_{\mathbf{x}}^{(N)}(\sigma_K \in (t,\infty))$.

**Theorem 11.4.17.**[4] *If $N \ge 3$ and $K \subset\subset \mathbb{R}^N$, then, as $t \nearrow \infty$,*

$$p_K(t,\mathbf{x}) \equiv \mathcal{W}_{\mathbf{x}}^{(N)}(\sigma_K \in (t,\infty)) \sim \frac{2\operatorname{Cap}(K;\mathbb{R}^N)(1 - p_K^{\mathbb{R}^N}(\mathbf{x}))}{(2\pi)^{\frac{N}{2}}(N-2)} t^{1-\frac{N}{2}}$$

*uniformly for $\mathbf{x}$ in compacts.*

[4] This result was conjectured by Kac and first proved by his student A. Joffe. However, I will follow the argument given by F. Spitzer in the article cited above.

PROOF: Without loss in generality (cf. Corollary 11.4.4), I will assume that $\mathrm{Cap}(K;\mathbb{R}^N) > 0$. Next, set $p_K(\mathbf{x}) = p_K^{\mathbb{R}^N}(\mathbf{x})$ and $p_K(t,\mathbf{x},\mathbf{y}) = p^{K\complement}(t,\mathbf{x},\mathbf{y})$, and note that, by the Markov property,

$$p_K(t,\mathbf{x}) = \int_{K\complement} p_K(\mathbf{y})\, p_K(t,\mathbf{x},\mathbf{y})\, d\mathbf{y}.$$

Thus, since $p_K(t,\mathbf{x},\mathbf{y}) \le (2\pi t)^{-\frac{N}{2}}$, we know that

$$\overline{\lim_{t\to\infty}} \sup_{\mathbf{x}\in\mathbb{R}^N} t^{\frac{N}{2}-1} \left| p_K(t,\mathbf{x}) - \int_{|\mathbf{y}|\ge R} p_K(\mathbf{y})\, p_K(t,\mathbf{x},\mathbf{y})\, d\mathbf{y} \right| = 0$$

for every $R > 0$ with $K \subset\subset B(\mathbf{0}, R)$. At the same time, because

$$p_K(\mathbf{y}) = \int_K g^{\mathbb{R}^3}(\mathbf{x},\mathbf{y})\, \mu_K^{\mathbb{R}^N}(d\mathbf{x}),$$

it is clear that

$$\lim_{|\mathbf{y}|\to\infty} |\mathbf{y}|^{N-2} p_K(\mathbf{y}) = \frac{2\mathrm{Cap}(K;\mathbb{R}^N)}{(N-2)\omega_{N-1}}.$$

Hence, we have now shown that

$$\overline{\lim_{t\to\infty}} \sup_{\mathbf{x}\in\mathbb{R}^N} t^{\frac{N}{2}-1} \left| p_K(t,\mathbf{x}) - \frac{2\mathrm{Cap}(K;\mathbb{R}^N)}{(N-2)\omega_{N-1}} \int_{|\mathbf{y}|\ge R} \frac{p_K(t,\mathbf{x},\mathbf{y})}{|\mathbf{y}|^{N-2}}\, d\mathbf{y} \right| = 0$$

for each $R \in (0,\infty)$ with $K \subset\subset B(\mathbf{0}, R)$, and what we must still prove is that

$$(*) \qquad \lim_{t\to\infty} \sup_{|\mathbf{x}|\le r} \left| t^{\frac{N}{2}-1} \int_{|y|\ge R} \frac{p_K(t,\mathbf{x},\mathbf{y})}{|\mathbf{y}|^{N-2}}\, d\mathbf{y} - \frac{\omega_{N-1}}{(2\pi)^{\frac{N}{2}}} \mathcal{W}_{\mathbf{x}}^{(N)}(\sigma_K = \infty) \right| = 0$$

for all positive $r$ and $R$ with $K \subset\subset B(\mathbf{0}, R)$.

To prove (*), let $r$ and $R$ be given, and use (10.3.8) to see that

$$\int_{|y|\ge R} \frac{p_K(t,\mathbf{x},\mathbf{y})}{|\mathbf{y}|^{N-2}}\, d\mathbf{y} = q(t,\mathbf{x}) - \mathbb{E}^{\mathcal{W}_{\mathbf{x}}^{(N)}}\Big[q(t-\sigma_K, \boldsymbol{\psi}(\sigma_K)),\, \sigma_K < t\Big],$$

where

$$q(t,\mathbf{x}) \equiv \int_{|\mathbf{y}|\ge R} \frac{g^{(N)}(t,\mathbf{y}-\mathbf{x})}{|\mathbf{y}|^{N-2}}\, d\mathbf{y} \quad \text{for } (t,\mathbf{x}) \in (0,\infty)\times\mathbb{R}^N.$$

After changing to polar coordinates and making a change of variables, one can easily check that, for each $T \in [0, \infty)$,

$$\lim_{t\to\infty} \sup_{\substack{0<s\leq T\\|\mathbf{x}|\leq r}} \left| t^{\frac{N}{2}-1} q(t-s,\mathbf{x}) - \frac{\omega_{N-1}}{(2\pi)^{\frac{N}{2}}} \right| = 0.$$

Thus, if, for $T \in (0, t)$, we write

$$\begin{aligned}
&t^{\frac{N}{2}-1} \int_{|y|\geq R} \frac{p_K(t,\mathbf{x},\mathbf{y})}{|\mathbf{y}|^{N-2}}\, d\mathbf{y} - \frac{\omega_{N-1}}{(2\pi)^{\frac{N}{2}}} \mathcal{W}^{(N)}_{\mathbf{x}}(\sigma_K = \infty)\\
&= \left( t^{\frac{N}{2}-1} q(t,x) - \frac{\omega_{N-1}}{(2\pi)^{\frac{N}{2}}} \right) - \mathbb{E}^{\mathcal{W}^{(N)}_{\mathbf{x}}} \left[ t^{\frac{N}{2}-1} q(t-\sigma_K, \psi(\sigma_K)) - \tfrac{\omega_{N-1}}{(2\pi)^{\frac{N}{2}}},\, \sigma_K \leq T \right]\\
&\quad - \mathbb{E}^{\mathcal{W}^{(N)}_{\mathbf{x}}} \left[ t^{\frac{N}{2}-1} q(t-\sigma_K, \psi(\sigma_K)),\, \sigma_K \in (T,t) \right] + \frac{\omega_{N-1}}{(2\pi)^{\frac{N}{2}}} \mathcal{W}^{(N)}_{\mathbf{x}}(\sigma_K \in (T,\infty)),
\end{aligned}$$

then it becomes clear that (*) will follow once we check that

(**)
$$\begin{aligned}
&\lim_{T\to\infty} \sup_{\mathbf{x}\in\mathbb{R}^N} \mathcal{W}^{(N)}_{\mathbf{x}}\Big(\sigma_K \in (T,\infty)\Big) = 0 \text{ and}\\
&\lim_{T\to\infty} \sup_{\substack{t>T\\ \mathbf{x}\in\mathbb{R}^N}} t^{\frac{N}{2}-1} \mathbb{E}^{\mathcal{W}^{(N)}_{\mathbf{x}}} \Big[ q(t-\sigma_K, \psi(\sigma_K)),\, \sigma_K \in (T,t) \Big] = 0.
\end{aligned}$$

To check the first part of (**), note that, by the Markov property,

$$\begin{aligned}
\mathcal{W}^{(N)}_{\mathbf{x}}(\sigma_K \in (T, T+1]) &= \int_{K\complement} p_K(T,\mathbf{x},\mathbf{y}) \mathcal{W}^{(N)}_{\mathbf{y}}(\sigma_K \leq 1)\, d\mathbf{y}\\
&\leq (2\pi T)^{-\frac{N}{2}} \int_{\mathbb{R}^N} \mathcal{W}^{(N)}_{\mathbf{y}}(\sigma_K \leq 1)\, d\mathbf{y} \leq CT^{-\frac{N}{2}},
\end{aligned}$$

where $C = C(N, R) \in (0, \infty)$. Hence, after writing

$$\mathcal{W}^{(N)}_{\mathbf{x}}\Big(\sigma_K \in (T,\infty)\Big) \leq \sum_{n=0}^{\infty} \mathcal{W}^{(N)}_{\mathbf{x}}\Big(\sigma_K \in (T+n, T+n+1]\Big),$$

we see that, as $T \to \infty$, $\mathcal{W}^{(N)}_{\mathbf{x}}(\sigma_K \in (T,\infty)) \longrightarrow 0$ uniformly with respect to $\mathbf{x} \in \mathbb{R}^N$.

To handle the second part of (**), note that there is a constant $A \in (0, \infty)$ for which

$$q(t,\mathbf{x}) \leq A\,(t \vee 1)^{1-\frac{N}{2}}, \quad (t,\mathbf{x}) \in (0,\infty) \times K,$$

and therefore

$$\begin{aligned} t^{\frac{N}{2}-1}\mathbb{E}^{\mathcal{W}^{(N)}_{\mathbf{x}}}&\big[q\big(t-\sigma_K,\psi(\sigma_K)\big),\ \sigma_K\in(T,t)\big]\\ &\le At^{\frac{N}{2}-1}\Bigg(\mathcal{W}^{(N)}_{\mathbf{x}}\big(\sigma_K\in(\lfloor t\rfloor-1,t]\big)\\ &\qquad\qquad+\sum_{\ell=\lfloor T\rfloor}^{\lfloor t\rfloor-1}(t-\ell)^{1-\frac{N}{2}}\mathcal{W}^{(N)}_{\mathbf{x}}\big(\sigma_K\in(\ell-1,\ell]\big)\Bigg)\\ &\le ACt^{\frac{N}{2}-1}(\lfloor t\rfloor-1)^{-\frac{N}{2}}+ACt^{\frac{N}{2}-1}\sum_{\ell=\lfloor T\rfloor}^{\lfloor t\rfloor-1}(t-\ell)^{1-\frac{N}{2}}(\ell-1)^{-\frac{N}{2}}, \end{aligned}$$

where the $C$ is the same as the one that appeared in the derivation of the first part of (**). Thus, everything comes down to verifying that

$$\lim_{m\to\infty}\sup_{n>m} n^{\frac{N}{2}-1}\sum_{\ell=m}^{n-1}(n-\ell)^{1-\frac{N}{2}}\ell^{-\frac{N}{2}}=0.$$

But, by taking $\epsilon_m=m^{\frac{2}{N}-1}$ and considering

$$\sum_{m\le\ell\le(1-\epsilon_m)n}(n-\ell)^{1-\frac{N}{2}}\ell^{-\frac{N}{2}}\quad\text{and}\quad\sum_{(1-\epsilon_m)n\le\ell\le n}(n-\ell)^{1-\frac{N}{2}}\ell^{-\frac{N}{2}}$$

separately, one finds that there is a $B\in(0,\infty)$ such that

$$n^{\frac{N}{2}-1}\sum_{\ell=m}^{n-1}(n-\ell)^{1-\frac{N}{2}}\ell^{-\frac{N}{2}}\le B\epsilon_m.\quad\square$$

As one might guess, on the basis of (11.2.15), the analogous situation in $\mathbb{R}^2$ is somewhat more delicate in that it involves logarithms.

THEOREM 11.4.18 (**Hunt**).[5] *Let $K$ be a compact subset of $\mathbb{R}^2$, define $\sigma_K$ as above, assume that $\mathcal{W}^{(2)}_{\mathbf{x}}(\sigma_K<\infty)=1$ for all $\mathbf{x}\in\mathbb{R}^2$, and use $h_K$ to denote the function $h^G$ given in (11.2.15) when $G=\mathbb{R}^2\setminus K$. Then, as $t\nearrow\infty$,*

$$\mathcal{W}^{(2)}_{\mathbf{x}}(\sigma_K>t)\sim\frac{2\pi h_K(\mathbf{x})}{\log t}\quad\textit{for each } \mathbf{x}\in\mathbb{R}^2\setminus K.$$

[5] This theorem is taken from G. Hunt's article *Some theorems concerning Brownian motion*, *T.A.M.S.* **81**, pp. 294–319 (1956). With breathtaking rapidity, it was followed by the articles referred to in § 11.1.4.

PROOF: The strategy of Hunt's proof is to deal with the Laplace transform

$$\int_0^\infty e^{-\alpha t}\,\mathcal{W}^{(2)}(\sigma_K > t)\,dt = \alpha^{-1}\Big(1 - \mathbb{E}^{\mathcal{W}^{(2)}_{\mathbf{x}}}\big[e^{-\alpha\sigma_K}\big]\Big),$$

show that

$$(*)\qquad \lim_{\alpha\searrow 0} \frac{\log\frac{1}{\alpha}}{2\pi}\Big(1 - \mathbb{E}^{\mathcal{W}^{(2)}_{\mathbf{x}}}\big[e^{-\alpha\sigma_K}\big]\Big) = h_K(\mathbf{x}),$$

and apply Karamata's Tauberian Theorem to conclude first that

$$\lim_{t\to\infty} \frac{\log t}{2\pi t}\int_0^t \mathcal{W}^{(2)}_{\mathbf{x}}(\sigma_K > \tau)\,d\tau = h_K(\mathbf{x})$$

and then, because $t \rightsquigarrow \mathcal{W}^{(2)}(\sigma_K > t)$ is non-increasing, that the asserted result holds. Thus, everything comes down to proving (*).

Set $G = \mathbb{R}^2 \setminus K$. By assumption, $G$ satisfies the hypotheses of Theorem 11.2.14. Now let $\mathbf{x} \in G$ be given, and choose $\mathbf{y} \in G\setminus\{\mathbf{x}\}$ from the same connected component of $G$ as $\mathbf{x}$. Then $p^G(t,\mathbf{x},\mathbf{y}) > 0$ for all $t \in (0,\infty)$. In addition, by (10.3.8), for each $\alpha \in (0,\infty)$,

$$\int_0^\infty e^{-\alpha t} p^G(t,\mathbf{x},\mathbf{y})\,dt$$
$$= \int_0^\infty e^{-\alpha t} g^{(2)}(t, \mathbf{y} - \psi(\sigma_K))\,dt - \mathbb{E}^{\mathcal{W}^{(N)}_{\mathbf{x}}}\left[e^{-\alpha\sigma_K}\int_0^\infty e^{-\alpha t} g^{(2)}(t,\mathbf{y}-\mathbf{x})\,dt\right].$$

Next observe that

$$\int_0^\infty e^{-\alpha t} g^{(2)}(t,\mathbf{z})\,dt = f\left(\frac{\alpha|\mathbf{z}|^2}{2}\right)$$
$$\text{where } f(\beta) \equiv \frac{1}{2\pi}\int_0^\infty t^{-1}\exp\big[-\beta t - t^{-1}\big]\,dt \text{ for } \beta > 0.$$

Writing

$$2\pi f(\beta) = \int_0^1 t^{-1}\exp\big[-\beta t - t^{-1}\big]\,dt + \int_1^\infty t^{-1}e^{-\beta t}\Big(\exp\big[-t^{-1}\big] - 1\Big)\,dt$$
$$+ \int_\beta^\infty t^{-1}e^{-t}\,dt,$$

integrating by parts, and performing elementary manipulations, we find that

$$f(\beta) = \frac{\log\frac{1}{\beta}}{2\pi} + \kappa + o(1) \quad \text{as } \beta \searrow 0,$$

where

$$\kappa = \frac{1}{\pi}\int_0^\infty e^{-t}\log t\,dt.$$

At the same time, we have that

$$\int_0^\infty e^{-\alpha t} p^G(t,\mathbf{x},\mathbf{y})\,dt \longrightarrow g^G(\mathbf{x},\mathbf{y}) \quad \text{as } \alpha \searrow 0.$$

Hence, when we plug these into the preceding, we get

$$\begin{aligned} g^G(\mathbf{x},\mathbf{y}) = -\frac{1}{\pi}\log|\mathbf{y}-\mathbf{x}| + \frac{1}{\pi}\mathbb{E}^{\mathcal{W}_{\mathbf{x}}^{(2)}}\big[\log|\mathbf{y}-\psi(\sigma_K)|,\ \sigma_K<\infty\big] \\ + \frac{\log\frac{1}{\alpha}}{2\pi}\Big(1-\mathbb{E}^{\mathcal{W}_{\mathbf{x}}^{(2)}}\big[e^{-\alpha\sigma_K}\big]\Big) + o(1) \end{aligned}$$

as $\alpha \searrow 0$. Finally, after comparing this to (11.2.16), we arrive at (*). □

Let $K \subset\subset \mathbb{R}^N$ be as in the preceding theorem, and choose some $\mathbf{c} \in K\complement$. By comparing the result just obtained to (11.2.15), we see that

$$\lim_{t\to\infty}\frac{\mathcal{W}_{\mathbf{x}}^{(2)}(\sigma_K>t)}{\mathcal{W}_{\mathbf{x}}^{(2)}\big(\sigma_K>\zeta^{B_{\mathbb{R}^2}(\mathbf{c},t)}\big)} = 2 \quad \text{for each } \mathbf{x}\in K\complement.$$

It would be interesting to know if there is a more direct route to this conclusion, in particular, one that avoids a Tauberian argument.

## Exercises for § 11.4

Exercise 11.4.19. Assume that $N \geq 2$. Given a $\mu \in \mathbf{M}(\mathbb{R}^N)$, say that $\mu$ is **tame** if

$$\sup_{\mathbf{x}\in\mathbb{R}^2} -\int_{\mathbb{R}^2}\log\big(|\mathbf{y}-\mathbf{x}|\wedge 1\big)\,\mu(d\mathbf{y}) < \infty \quad \text{when } N=2$$

$$\sup_{\mathbf{x}\in\mathbb{R}^N}\int_{\mathbb{R}^N}|\mathbf{y}-\mathbf{x}|^{2-N}\,\mu(d\mathbf{y}) < \infty \quad \text{when } N\geq 3.$$

Further, say that $\Gamma \in \mathcal{B}_{\mathbb{R}^N}$ has **capacity zero** if there is no tame $\mu \in \mathbf{M}(\mathbb{R}^N)$ for which $\mu(\Gamma) > 0$.

**(i)** If $K \subset\subset \mathbb{R}^N$, show that $K$ has capacity 0 if and only if $\mathrm{Cap}\big(K; B(\mathbf{0},R)\big) = 0$ for some $R > 0$ with $K \subset\subset B(\mathbf{0},R)$. Further, show that if $K$ has capacity 0, $G$ is open with $K \subset\subset G$, and either $N \geq 3$ or (11.1.20) holds, then $\mathrm{Cap}(K;G) = 0$.

**(ii)** If $\Gamma \in \mathcal{B}_{\mathbb{R}^N}$, show that $\Gamma$ has capacity 0 if and only if every compact $K \subseteq \Gamma$ has capacity 0.

**(iii)** For any open $G \subseteq \mathbb{R}^N$, show that $\partial G \setminus \partial_{\mathrm{reg}} G$ has capacity 0.

**(iv)** Let $G$ be a connected, open subset of $\mathbb{R}^N$, and assume that either $N \geq 3$ or (11.1.20) holds. If $u \in \mathcal{E}(G)$ is not identically infinite, show that $\{\mathbf{x} \in G : u(\mathbf{x}) = \infty\}$ has capacity 0.
**(v)** Suppose that $G$ is an open subset of $\mathbb{R}^N$ and that either $N \geq 3$ or (11.1.20) holds. If $K \subset\subset G$, show that $\{\mathbf{x} \in K : p_K^G(\mathbf{x}) < 1\}$ has capacity 0. Conclude that if $\mu \in \mathbf{M}(\mathbb{R}^N)$ is tame and $\mu(\mathbb{R}^N \setminus K) = 0$, then

$$\mu(K) = \int_G p_K^G \, d\mu = \mathcal{E}^G(\mu, \mu_K^G) = \int_G \mathbf{G}^G \mu \, d\mu_K^G.$$

EXERCISE 11.4.20. Let $G$ be an open subset of $\mathbb{R}^N$ for some $N \geq 2$, and assume that either $N \geq 3$ or that (11.1.20) holds. We know how to define $\mathrm{Cap}(K; G)$ for $K \subset\subset G$. However, the map $K \rightsquigarrow \mathrm{Cap}(K;G)$ is somewhat mysterious. In this exercise we will discuss a few of its important properties, properties that enabled G. Choquet[1] to prove that $\mathrm{Cap}(\cdot, G)$ admits a well-defined extension to all of $\mathcal{B}_G$.
**(i)** If $\mu, \nu \in \mathbf{M}(G)$ and $\mathbf{G}^G\mu \leq \mathbf{G}^G\nu$, show that $\mathcal{E}^G(\mu,\mu) \leq \mathcal{E}^G(\mu,\nu)$. In particular, conclude that $\mathrm{Cap}(K_1, G) \leq \mathrm{Cap}(K_2, G)$ for all compacts $K_1 \subseteq K_2 \subset G$. Thus the convergence in Theorem 11.4.9 is non-increasing convergence.
**(ii)** If $K_1$, $K_2 \subset\subset G$, show that

$$\begin{aligned} p_{K_1 \cup K_2}^G(\mathbf{x}) - p_{K_1}^G(\mathbf{x}) &= \mathcal{W}_{\mathbf{x}}^{(N)}\big(\sigma_{K_2} < \zeta^G \leq \sigma_{K_1}\big) \\ &\leq \mathcal{W}_{\mathbf{x}}^{(N)}\big(\sigma_{K_2} < \zeta^G \leq \sigma_{K_1 \cap K_2}\big) \leq p_{K_2}^G(\mathbf{x}) - p_{K_1 \cap K_2}^G(\mathbf{x}), \end{aligned}$$

and therefore that $p_{K_1 \cup K_2}^G + p_{K_1 \cap K_2}^G \leq p_{K_1}^G + p_{K_2}^G$.
**(iii)** By combining **(i)** and **(ii)**, arrive at

$$\mathcal{E}^G(\mu_{K_1 \cup K_2} + \mu_{K_1 \cap K_2}, \mu_{K_1 \cup K_2} + \mu_{K_1 \cap K_2}) \leq \mathcal{E}^G(\mu_{K_1 \cup K_2} + \mu_{K_1 \cap K_2}, \mu_{K_1} + \mu_{K_2}).$$

Next, apply **(v)** of the preceding exercise to see that

$$\mathcal{E}^G(\mu_{K_1 \cup K_2} + \mu_{K_1 \cap K_2}, \mu_{K_1 \cup K_2} + \mu_{K_1 \cap K_2}) = \mathrm{Cap}(K_1 \cup K_2; G) + 3\mathrm{Cap}(K_1 \cap K_2; G)$$

and

$$\mathcal{E}^G(\mu_{K_1 \cup K_2} + \mu_{K_1 \cap K_2}, \mu_{K_1} + \mu_{K_2}) = \mathrm{Cap}(K_1; G) + \mathrm{Cap}(K_2; G) + 2\mathrm{Cap}(K_1 \cap K_2; G),$$

and conclude that $\mathrm{Cap}(\cdot\,; G)$ satisfies the strong sub-additivity property

$$\mathrm{Cap}(K_1 \cup K_2; G) + \mathrm{Cap}(K_1 \cap K_2; G) \leq \mathrm{Cap}(K_1; G) + \mathrm{Cap}(K_2; G).$$

What Choquet showed is that a non-negative set function defined for compact subsets of $G$ and satisfying the monotonicity property in **(i)**, the monotone convergence property in **(ii)**, and the strong subadditivity property in **(iii)** admits a unique extension to $\mathcal{B}_G$ in such a way that these properties persist. In the articles alluded to earlier, Hunt used Choquet's result to show that the first positive entrance into a Borel set is measurable.

[1] See Choquet's *Lectures on Analysis, Vol. I*, W.A. Benjamin (1965).

# Notation

*General*

| Notation | Description | See |
|---|---|---|
| $a \wedge b$ & $a \vee b$ | The minimum and the maximum of $a$ and $b$ | |
| $a^+$ & $a^-$ | The non-negative part, $a \vee 0$, and non-positive part, $-(a \wedge 0)$, of $a \in \mathbb{R}$ | |
| $f \restriction S$ | The restriction of the function $f$ to the set $S$ | |
| $\\|\cdot\\|_{\mathrm{u}}$ | The uniform (supremum) norm | |
| $\\|\psi\\|_{[a,b]}$ | The uniform norm of the path $\psi$ restricted to the interval $[a,b]$ | (4.1.1) |
| $\mathrm{var}_{[a,b]}(\psi)$ | Variation norm of the path $\psi \restriction [a,b]$ | (4.1.2) |
| $\Gamma(t)$ | Euler Gamma function | (1.3.20) |
| $\omega_{N-1}$ | The surface area of the sphere $\mathbb{S}^{N-1}$ in $\mathbb{R}^N$ | (2.1.13) |
| $\Omega_{N-1}$ | The volume, $N^{-1}\omega_{N-1}$, of the unit ball $B(\mathbf{0};1)$ in $\mathbb{R}^N$ | |
| $\lfloor t \rfloor$ | The integer part of $t \in \mathbb{R}$ | |

*Sets and Spaces*

| | | |
|---|---|---|
| $A\complement$ | The complement of the set $A$ | |
| $A^{(\delta)}$ | The $\delta$-hull around the set $A$ | §3.1 |
| $\mathbf{1}_A$ | The indicator function of the set $A$. | §1.1 |
| $B_E(a,r)$ | The ball of radius $r$ around $a$ in $E$. When $E$ is omitted, it is assumed to be the $\mathbb{R}^N$ for some $N \in \mathbb{Z}^+$ | |
| $B(E;\mathbb{R})$ | Space of bounded, Borel measurable functions from $E$ into $\mathbb{R}$ | |
| $K \subset\subset E$ | To be read: *$K$ is a compact subset of $E$.* | |
| $\mathbb{C}$ | The complex numbers | |
| $\mathbb{N}$ | The non-negative integers: $\mathbb{N} = \{0\} \cup \mathbb{Z}^+$ | |

| | | |
|---|---|---|
| $\mathbb{S}^{N-1}$ | The unit sphere in $\mathbb{R}^N$ | |
| $\mathbb{Q}$ | The set of rational numbers | |
| $\mathbb{Z}$ & $\mathbb{Z}^+$ | Set of all integers and the subset of positive integers | |
| $C(\mathbb{R}^N)$ | The space $C([0,\infty);\mathbb{R}^N)$ of continuous paths $\psi : [0,\infty) \longrightarrow \mathbb{R}^N$ | §9.3 |
| $C_b(E;\mathbb{R})$ | The space of bounded continuous functions from $E$ into $\mathbb{R}$. | |
| $C_c(G;\mathbb{R})$ | The space of continuous, $\mathbb{R}$-valued functions having compact support in the open set $G$ | |
| $C^{1,2}(\mathbb{R}\times\mathbb{R}^N;\mathbb{R})$ | The space of functions $(t,\mathbf{x}) \in \mathbb{R}\times\mathbb{R}^N \longrightarrow \mathbb{R}$ which are continuously differentiable once in $t$ and twice in $\mathbf{x}$. | |
| $D(\mathbb{R}^N)$ | The space of right-continuous paths $\psi : [0,\infty) \longrightarrow \mathbb{R}^N$ with left-limits on $(0,\infty)$ | §4.1.1 |
| $\mathbf{H}(\mathbb{R}^N)$ | The Cameron–Martin subspace for Wiener measure on $\Theta(\mathbb{R}^N)$ | §8.1.2 |
| $L^p(\mu;E)$ | The Lebesgue space of $E$-valued functions $f$ for which $\|f\|_E^p$ is $\mu$-integrable | |
| $\mathbf{M}_1(E)$ | The space of Borel probability measures on $E$ | §9.1.2 |
| $\mathbf{M}(E)$ | The space of non-negative, finite, Borel probability measures on $E$ | |
| $\mathscr{S}(\mathbb{R}^N;\mathbb{R})$ or $\mathscr{S}(\mathbb{R}^N;\mathbb{C})$ | Real- or complex-valued Schwartz test function space on $\mathbb{R}^N$ | §3.2.3 |

## *Measure Theoretic*

| | |
|---|---|
| $\mathcal{B}_E$ | The Borel $\sigma$-algebra over $E$ |
| $B(E;\mathbb{R})$ | The space of bounded, measurable functions on $E$ |
| $\mathbb{E}^\mu[X, A]$ | To be read *the expectation value of* $X$ *with respect to* $\mu$ *on* $A$. Equivalent to $\int_A X\,d\mu$. When $A$ is unspecified, it is assumed to be the whole space |
| $\delta_a$ | The unit point mass at $a$ |

| | | |
|---|---|---|
| $\lambda_A$ | Lebesgue measure on the set $A$. Usually $A = \mathbb{R}^N$ or some interval | |
| $\mathbb{E}^{\mu}[X \mid \mathcal{F}]$ | To be read: *the conditional expectation value of* $X$ *given the* $\sigma$*-algebra* $\mathcal{F}$ | §5.1.1 |
| $\hat{f}$ | The Fourier transform of the function $f$ | §2.3.1 |
| $f \star g$ | The convolution of $f$ with $g$ | |
| $\langle \varphi, \mu \rangle$ | An alternative notation for $\mathbb{E}^{\mu}[\varphi]$ | §2.1 |
| $g^{(N)}(t, \mathbf{x})$ | The density of the Gauss distribution in $\mathbb{R}^N$ | §10.1 |

## *Wiener Measure*

| | | |
|---|---|---|
| $\gamma_{\mathbf{m},\mathbf{C}}$ | Gaussian or normal distribution with mean $\mathbf{m}$ and covariance $\mathbf{C}$ | §2.3.1 |
| $\hat{\mu}$ | The Fourier of the measure $\mu$ | §2.3.1 |
| $\mu \star \nu$ | The convolution of measures $\mu$ with $\nu$ | Chap. III |
| $\mu \ll \nu$ | The measure $\mu$ is absolutely continuous with respect to $\nu$ | |
| $\mu_n \Longrightarrow \mu$ | The sequence $\{\mu_n : n \geq 1\}$ tends weakly to $\mu$ | §9.1.2 |
| $\mu \perp \nu$ | The measure $\mu$ is singular to $\nu$ | |
| $\mathrm{med}(Y)$ | The set of medians of the random variable $Y$ | §1.4 |
| $N(\mathbf{m}, \mathbf{C})$ | Normal distributions with mean $\mathbf{m}$ and covariance $\mathbf{C}$ | §2.3.1 |
| $\Phi_* \mu$ | The pushforward (image) of $\mu$ under $\Phi$ | (1.1.16) |
| $\sigma(\{X_i : i \in \mathcal{I}\})$ | The $\sigma$-algebra generated by the set of random variables $\{X_i : i \in \mathcal{I}\}$ | |
| $\bigvee_{i \in \mathcal{I}} \mathcal{F}_i$ | The $\sigma$-algebra generated by $\bigcup_{i \in \mathcal{I}} \mathcal{F}_i$ | |
| $\delta_s$ | The differential time-shift map on $C(\mathbb{R}^N)$ | §7.1.4 |
| $\Sigma_s$ | The time-shift map on $C(\mathbb{R}^N)$ | §10.2.1 |
| $\mathcal{W}^{(N)}$ | Wiener measure on $\Theta(\mathbb{R}^N)$ or $C(\mathbb{R}^N)$ | §8.1.1 |
| $\mathcal{W}^{(N)}_{\mathbf{x}}$ | The distribution of $\mathbf{x} + \boldsymbol{\psi}$ under $\mathcal{W}^{(N)}$ | §10.1.1 |
| $(H, E, \mathcal{W}_H)$ | The abstract Wiener space with Cameron–Martin space $H$ | §8.2.2 |

*Potential Theoretic*

| | | |
|---|---|---|
| $\mathcal{E}(G)$ | The set of excessive functions on $G$ | §11.3.1 |
| $g^G(\mathbf{x}, \mathbf{y})$ | Dirichlet Green function for $G$ | §11.2 |
| $\mathbf{G}^G\mu$ | Green potential with charge $\mu$ in $G$ | (11.3.1) |
| $p^G(t, \mathbf{x}, \mathbf{y})$ | Dirichlet heat kernel for $G$ | §10.3.1 |

# Index

**M**



**N**



**O**



**P**

**Q**



**R**



**S**